Brief Contents

CHAPTER **R** REVIEW OF PREALGEBRA 1

CHAPTER **1** BUILDING BLOCKS OF ALGEBRA 71

CHAPTER **2** LINEAR EQUATIONS AND INEQUALITIES WITH ONE VARIABLE 155

CHAPTER **3** LINEAR EQUATIONS WITH TWO VARIABLES 231

CHAPTER **4** SYSTEMS OF LINEAR EQUATIONS 359

CHAPTER **5** EXPONENTS AND POLYNOMIALS 463

CHAPTER **6** FACTORING 531

CHAPTER **7** RATIONAL EXPRESSIONS AND EQUATIONS 589

CHAPTER **8** RADICAL EXPRESSIONS AND EQUATIONS 679

CHAPTER **9** QUADRATIC EQUATIONS 745

APPENDIX **A** ANSWERS TO PRACTICE PROBLEMS A-1

APPENDIX **B** ANSWERS TO SELECTED EXERCISES B-1

Index I-1
Unit Conversions REF-2
Geometric Formulas REF-3
Equation Solving Toolbox REF-4
Expression Simplifying Toolbox REF-5
Factoring Toolbox REF-6

Contents

R — Review of Prealgebra 1

R.1 Operations with Integers 2
Natural Numbers, Whole Numbers, and Integers • Number Lines • Relations between Numbers • Absolute Value • Opposite of a Number • Operations with Integers • Order of Operations

R.2 Operations with Fractions 20
Prime Numbers and Prime Factorization • Simplifying Fractions and Equivalent Fractions • Fractions on Number Lines • Addition and Subtraction of Fractions • Multiplication and Division of Fractions • Order of Operations

R.3 Operations with Decimals 34
Place Value • Relationships between Fractions and Decimals • Graphing Decimals on a Number Line • Rounding Decimals • Addition and Subtraction of Decimals • Multiplication and Division of Decimals • Order-of-Operations with Decimals

R.4 Operations with Percents 44
What Is a Percent? • Converting between Percents, Fractions, and Decimals • Problem Solving with Percents • Finding Percent Increase and Decrease

R.5 The Real Number System 51
Rational Numbers • Irrational Numbers • The Real Number System • Exact and Approximate Answers

Chapter R   Summary   60
Chapter R   Review Exercises   65
Chapter R   Test   68
Chapter R   Projects   69

1 — Building Blocks of Algebra 71

1.1 Exponents, Order of Operations, and Properties of Real Numbers 72
Exponents • Scientific Notation • Order of Operations • Properties of Real Numbers

1.2 Algebra and Working with Variables 88
Constants and Variables • Evaluating Expressions • Unit Conversions • Defining Variables • Translating Sentences into Expressions • Generating Expressions from Input–Output Tables

iv

1.3 Simplifying Expressions 109
Like Terms • Addition and Subtraction Properties • Multiplication and Distributive Properties • Simplifying Expressions

1.4 Graphs and the Rectangular Coordinate System 122
Data Tables • Bar Graphs • Scatterplots • Rectangular Coordinate System

Chapter 1   Summary   143
Chapter 1   Review Exercises   147
Chapter 1   Test   152
Chapter 1   Projects   153

2 Linear Equations and Inequalities with One Variable 155

2.1 Addition and Subtraction Properties of Equality 156
Recognizing Equations and Their Solutions • Addition and Subtraction Properties of Equality • Solving Literal Equations

2.2 Multiplication and Division Properties of Equality 171
Multiplication and Division Properties of Equality • Solving Multiple-Step Equations • Generating Equations from Applications • More on Solving Literal Equations

2.3 Solving Equations with Variables on Both Sides 189
Solving Equations with Variables on Both Sides • Solving Equations That Contain Fractions • Equations That Are Identities or Have No Solution • Translating Sentences into Equations and Solving

2.4 Solving and Graphing Linear Inequalities on a Number Line 202
Introduction to Inequalities • Solving Inequalities • Interval Notation and Number Lines • Compound Inequalities

Chapter 2   Summary   218
Chapter 2   Review Exercises   224
Chapter 2   Test   226
Chapter 2   Projects   227
Cumulative Review Chapters 1–2   229

3 Linear Equations with Two Variables 231

3.1 Graphing Equations with Two Variables 232
Using Tables to Represent Ordered Pairs and Data • Graphing Equations by Plotting Points • Graphing Nonlinear Equations by Plotting Points • Vertical and Horizontal Lines

3.2 Finding and Interpreting Slope 245
Interpreting Graphs • Determining a Rate of Change • Calculating Slope • Interpreting Slope

3.3 Slope-Intercept Form of Lines 266
Finding and Interpreting Intercepts from Graphs • Finding and Interpreting Intercepts from Equations • Slope-Intercept Form of a Line

vi   CONTENTS

3.4 Linear Equations and Their Graphs 282
Graphing from Slope-Intercept Form • Graphing from General Form • Recognizing a Linear Equation • Parallel and Perpendicular Lines

3.5 Finding Equations of Lines 299
Finding Equations of Lines Using Slope-Intercept Form • Finding Equations of Lines from Applications • Finding Equations of Lines Using Point-Slope Form • Finding Equations of Parallel and Perpendicular Lines

3.6 Modeling Linear Data 316
Finding a Graphical Model for Linear Data • Finding a Linear Model • Using a Linear Model to Make Estimates • Determining Model Breakdown

Chapter 3   Summary   340
Chapter 3   Review Exercises   349
Chapter 3   Test   355
Chapter 3   Projects   356

4 Systems of Linear Equations 359

4.1 Identifying Systems of Linear Equations 360
Introduction to Systems of Equations • Solving Systems Graphically • Types of Systems

4.2 Solving Systems Using the Substitution Method 382
Substitution Method • Inconsistent and Consistent Systems • Practical Applications of Systems of Linear Equations

4.3 Solving Systems Using the Elimination Method 400
Using the Elimination Method • More Practical Applications of Systems of Linear Equations • Substitution or Elimination?

4.4 Solving Linear Inequalities in Two Variables Graphically 416
Linear Inequalities in Two Variables • Graphing Vertical and Horizontal Inequalities

4.5 Systems of Linear Inequalities 433

Chapter 4   Summary   445
Chapter 4   Review Exercises   452
Chapter 4   Test   455
Chapter 4   Projects   456
Cumulative Review Chapters 1–4   458

5 Exponents and Polynomials 463

5.1 Rules for Exponents 464
Product Rule for Exponents • Quotient Rule for Exponents • Power Rule for Exponents • Powers of Products and Quotients

5.2 Negative Exponents and Scientific Notation 476
Negative Exponents • Using Scientific Notation in Calculations

5.3 Adding and Subtracting Polynomials   488
The Terminology of Polynomials • Adding and Subtracting Polynomials

5.4 Multiplying Polynomials   498
Multiplying Polynomials • FOIL: A Handy Acronym • Special Products

5.5 Dividing Polynomials   509
Dividing a Polynomial by a Monomial • Dividing a Polynomial by a Polynomial Using Long Division

Chapter 5   Summary   521
Chapter 5   Review Exercises   526
Chapter 5   Test   528
Chapter 5   Projects   528

6 Factoring   531

6.1 What It Means to Factor   532
Factoring Out the Greatest Common Factor • Factoring by Grouping • Factoring Completely

6.2 Factoring Trinomials   546
Factoring Trinomials of the Form $x^2 + bx + c$ by Inspection • Factoring Trinomials of the Form $ax^2 + bx + c$ • More Techniques to Factor Completely

6.3 Factoring Special Forms   556
Difference of Squares • Perfect Square Trinomials • Summary of Factoring Tools

6.4 Solving Quadratic Equations by Factoring   564
Recognizing a Quadratic Equation • Zero-Product Property • Solving Quadratic Equations by Factoring

Chapter 6   Summary   575
Chapter 6   Review Exercises   580
Chapter 6   Test   581
Chapter 6   Projects   581
Cumulative Review Chapters 1–6   584

7 Rational Expressions and Equations   589

7.1 The Basics of Rational Expressions and Equations   590
Evaluating Rational Expressions and Equations • Excluded Values • Simplifying Rational Expressions

7.2 Multiplication and Division of Rational Expressions   601
Multiplying Rational Expressions • Expanding Unit Conversions • Dividing Rational Expressions • Basics of Complex Fractions

viii CONTENTS

- **7.3** Addition and Subtraction of Rational Expressions 612
  Adding and Subtracting Rational Expressions with Common Denominators • Finding the Least Common Denominator (LCD) • Adding and Subtracting Rational Expressions with Unlike Denominators • Simplifying Complex Fractions

- **7.4** Solving Rational Equations 631
  Solving Rational Equations • Setting Up and Solving Shared Work Problems

- **7.5** Proportions, Similar Triangles, and Variation 646
  Ratios, Rates, and Proportions • Similar Triangles • Variation

  Chapter 7 Summary 665
  Chapter 7 Review Exercises 672
  Chapter 7 Test 675
  Chapter 7 Projects 676

8 Radical Expressions and Equations 679

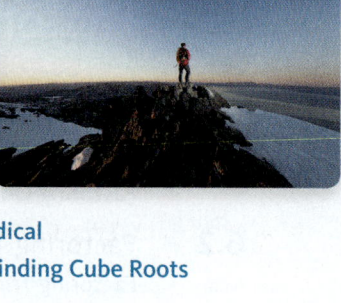

- **8.1** From Squaring a Number to Roots and Radicals 680
  Finding Square Roots • Evaluating Radical Expressions • Evaluating Radical Equations • Simplifying Radical Expressions That Contain Variables • Finding Cube Roots

- **8.2** Basic Operations with Radical Expressions 692
  Simplifying More Complicated Radical Expressions • Adding and Subtracting Radical Expressions

- **8.3** Multiplying and Dividing Radical Expressions 704
  Multiplying Radical Expressions • Dividing Radical Expressions • Rationalizing the Denominator

- **8.4** Solving Radical Equations 719
  Checking Solutions to Radical Equations • Solving Radical Equations • Solving Applications Involving Radical Equations • Solving More Radical Equations

  Chapter 8 Summary 731
  Chapter 8 Review Exercises 735
  Chapter 8 Test 737
  Chapter 8 Projects 737
  Cumulative Review Chapters 1–8 739

9 Quadratic Equations 745

- **9.1** Graphing Quadratic Equations 746
  Graphing Quadratic Equations by Plotting Points • The Relationship between the Leading Coefficient a and the Graph • Locating the Vertex • Using the Axis of Symmetry

- **9.2** Solving Quadratic Equations by Using the Square Root Property 764
  Solving Quadratic Equations by Using the Square Root Property • Using the Pythagorean Theorem

9.3 Solving Quadratic Equations by Completing the Square and by the Quadratic Formula 777
Solving Quadratic Equations by Completing the Square • Solving Quadratic Equations by Using the Quadratic Formula

9.4 Graphing Quadratic Equations Including Intercepts 791
Finding Intercepts • Putting It All Together to Sketch a Graph

9.5 Working with Quadratic Models 800
Determining Whether or Not the Graph of a Data Set Is Shaped Like a Parabola • Using Quadratic Models to Make Estimates • Determining when Model Breakdown Happens

9.6 The Basics of Functions 817
Relations • Functions • Vertical Line Test • Function Notation • Evaluating Functions

Chapter 9 Summary 830
Chapter 9 Review Exercises 838
Chapter 9 Test 842
Chapter 9 Projects 844

A Answers to Practice Problems A-1

B Answers to Selected Exercises B-1

Index I-1
Unit Conversions REF-2
Geometric Formulas REF-3
Equation Solving Toolbox REF-4
Expression Simplifying Toolbox REF-5
Factoring Toolbox REF-6

Preface

Our goal in writing this book is to help college students learn how to solve problems generated from realistic applications, even as they build a strong mathematical basis in beginning algebra. We think that focusing our book on concepts and applications makes the mathematics more useful and vivid to college students. One tried and true application that we will make more vivid is problem solving. In our textbook, problem solving will be introduced as an organic, integrated part of the course, rather than separated off from the traditional skill set.

We include applications throughout the text, rather than in a stand-alone chapter or section. Our goal is to tie applications to realistic situations, providing straightforward background explanations of the mathematics so that students have a down-to earth rationale in their heads for how to set up a problem. At the same time, traditional algebra concepts and skills are presented throughout the book and the problem sets. The applications give students a chance to practice their skills as well as to see the importance of mathematics in the real world. We think that the combination of basic mathematical skills, concepts, and applications will allow students to think critically about mathematics and communicate their results better, too.

About The Authors

MARK CLARK

Mark Clark graduated from California State University, Long Beach, with a Bachelor's and Master's in Mathematics. He is a full-time Associate Professor at Palomar College and has taught there since 1996. He is committed to teaching his students through applications and using technology to help them both understand the mathematics in context and communicate their results clearly. Intermediate algebra is one of his favorite courses to teach, and he continues to teach several sections of this course each year. Mark also loves to share his passion for teaching concepts of developmental math through applications by giving workshops and talks to other instructors at local and national conferences.

CINDY ANFINSON

Cindy Anfinson graduated from the University of California, San Diego, with a Bachelor of Arts Degree in Mathematics and is a member of Phi Beta Kappa. Under the Army Science and Technology Graduate Fellowship, she earned a Master of Science Degree in Applied Mathematics from Cornell University. She is currently an Associate Professor of Mathematics at Palomar College. At Palomar College, she has worked with the First-Year Experience and the Summer Bridge programs, she was the Mathematics Learning Center Director for a 3-year term, and has served on multiple committees, including the Basic Skills Committee and the Student Success and Equity Council.

New to This Edition

Three Toolboxes are included throughout the new edition: the Equations Solving Toolbox, the Factoring Toolbox, and the Expression Simplifying Toolbox. These Toolboxes are integrated throughout the text with visual icons so that students receive just-in-time help connecting them to the solving techniques and tools used for different types of problems. Each Toolbox emphasizes how these fundamental tools are used throughout the course. A quick reference of all three Toolboxes as well as the Geometric Formulas and Unit Conversions appears at the back of the text. The text highlights the importance of using these from the start of the course.

There is an increased emphasis on students identifying expression and equation types within simplifying, solving and graphing problems. This helps students to review previous material and connect it to the current topics.

Students are asked to provide reasons for each step they took in selected Solving exercises. This helps students to think critically about and explain the Solving process, and it connects with the Toolboxes that have been integrated throughout the text.

Vocabulary short-answer exercises have been added at the start of each exercise set. These help students learn the vocabulary of algebra and improve their communication skills.

The Exercise Sets have been updated with new data and current applications to help students see connections between mathematics and the world in which they live.

Within WebAssign, there is expanded problem coverage with an emphasis on conceptual problems, full "WatchIt" coverage with closed-captioning, and expanded "MasterIt" and "Expanded Problem" coverage with emphasis on conceptual problems.

The Annotated Instructor's Edition has been replaced with a comprehensive Instructor's Manual. Practical tips and classroom examples are provided on how to approach and pace chapters as well as integrate features such as Concept Investigations into the classroom. For every student example in the student text, there is a different instructor classroom example with accompanying answers that can be used for additional in-class practice and/or homework.

Activities that are hand-selected by the authors provide additional opportunities for Instructors to get their students involved using active learning in the classroom. WebAssign suggestions integrated throughout the Instructor Manual also tie in the digital aspects of the course so that no matter how instructors approach their class, they feel supported. Pair the Instructor's Manual with the loose-leaf version of the text to create the perfect Instructor's Edition mockup.

Ancillaries

For the Student

Online Student Solutions Manual
(ISBN: 978-1-337-61609-6)
Author: Scott Barnett
The Student Solutions Manual provides worked-out solutions to all of the odd-numbered exercises in the text.

For the Instructor

Online Complete Solutions Manual
(ISBN: 978-1-337-61608-9)
Author: Scott Barnett
The Complete Solutions Manual provides worked-out solutions to all of the problems in the text.

Instructor's Companion Website
Everything you need for your course in one place! Access and download a comprehensive Instructor's Manual that paces the chapters, includes WebAssign suggestions, and provides additional opportunities for in-class practice, homework, and activities. In addition, you can find the online chapter, PowerPoint presentations, and more on the companion site. This collection of book-specific lecture and class tools is available online via www.cengage.com/login.

STUDENT

   www.webassign.com

(Printed Access Card ISBN: 978-1-337-61614-0,
Online Access Code ISBN: 978-1-337-61613-3)

Prepare for class with confidence using WebAssign from Cengage for *Intermediate Algebra: Connecting Concepts through Applications*, 2e. This online learning platform, which includes an interactive ebook, fuels practice, so you truly absorb what you learn—and are better prepared come test time. Videos and tutorials walk you through concepts and deliver instant feedback and grading, so you always know where you stand in class. Focus your study time and get extra practice where you need it most. Study smarter with WebAssign!

Ask your instructor today how you can get access to WebAssign, or learn about self-study options at www.webassign.com.

INSTRUCTOR

   www.webassign.com/cengage

(Printed Access Card ISBN: 978-1-337-61614-0,
Online Access Code ISBN: 978-1-337-61613-3)

WebAssign from Cengage for *Intermediate Algebra: Connecting Concepts through Applications*, 2e is a fully customizable online solution, including an interactive ebook, for STEM disciplines that empowers you to help your students learn, not just do homework. Insightful tools save you time and highlight exactly where your students are struggling. Decide when and what type of help students can access while working on assignments—and incentivize independent work so help features aren't abused. Meanwhile, your students get an engaging experience, instant feedback and better outcomes. A total win-win!

To try a sample assignment, learn about LMS integration or connect with our digital course support visit www.webassign.com/cengage.

Acknowledgments

We would like to thank our reviewers and users for their many helpful suggestions for improving the text. In particular, we thank Karen Mifflin and Gina Hayes for their suggestions and work on the solutions. We are extremely grateful to Scott Barnett and James Brust for helping with the accuracy checking and solutions for this text. We also thank the editorial, production, and marketing staffs of Cengage, Frank Snyder, Michael Lepera, Samantha Gomez, Alison Duncan, Pamela Polk, and Jaime Manz; for all of their help and support during the development and production of this edition. Thanks also to Vernon Boes, Diane Beasley, Irene Morris, Leslie Lahr, and Lisa Torri for their work on the design and art program, and to the Lumina Datamatics staff for their copyediting and proofreading expertise. We especially want to thank Danielle Derbenti for believing in us and mentoring our development as authors. Our gratitude also goes to Katy Gabel who had an amazing amount of patience with us throughout production. We truly appreciate all the hard work and efforts of the entire team.

Mark Clark
Cindy Anfinson

Review of Prealgebra

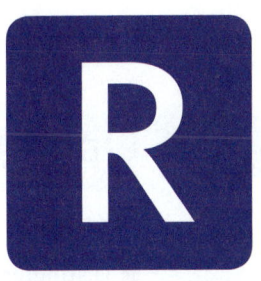

R.1	Operations with Integers
R.2	Operations with Fractions
R.3	Operations with Decimals
R.4	Operations with Percents
R.5	The Real Number System

Mathematics has been used by humans throughout recorded history and even in prehistoric times. The Ishango bone was found in 1960 in Africa, in an area near the border between Uganda and the Democratic Republic of Congo. The bone is that of a baboon, and it is currently estimated to be more than 20,000 years old. The tally marks on the bone were originally thought to be counting marks, but some scientists now believe that the marks indicate knowledge of multiplication and division by two. This bone shows that prehistoric peoples had a firm understanding of counting and perhaps multiplication.

R.1 Operations with Integers

LEARNING OBJECTIVES
- Distinguish natural numbers, whole numbers, and integers.
- Use number lines.
- Write inequalities.
- Find the absolute value of an integer.
- Find the opposite.
- Use the order-of-operations agreement for integers.

■ Natural Numbers, Whole Numbers, and Integers

People have been doing mathematics continuously throughout recorded history. There is strong evidence that in prehistory (before recorded history), people were using mathematics. First they invented numbers and counting. Then people invented ways of combining numbers. Arithmetic is the study of combining numbers through the operations of addition, subtraction, multiplication, and division.

Humans like to classify objects into various sets. For example, in the animal kingdom, animals have been classified into sets such as the reptiles, amphibians, birds, mammals, and fish. In the plant world, some classifications are ferns, conifers, and flowering plants. Mathematicians also like to classify numbers into sets.

The first set of numbers that people used are the **natural (or counting) numbers**. Ancient peoples did not see numbers as ideas, as we do today. They used numbers to count things up. For example, "two sheep" is how ancient peoples would have used the value 2. It is a big step to make the mental shift from "two sheep" to "two things" and then to "2" as an idea with no "things" attached to it. Zero (0) is not in the set of counting numbers, as the ancients would have had no reason to count a set of no sheep.

Zero was a difficult concept for people to grasp, even though zero is taken for granted today. Zero seems to have developed from two different needs. One is the use of zero as a placeholder, and the other is a symbol to represent that no objects are present. As an example of zero as a placeholder, in our number system, 102 is a different number from 12. The value of zero holds the place of tens, and the zero is necessary to distinguish between the numbers 102 and 12. It appears that zero as a placeholder developed in both India and Mexico (in the Mayan civilization) independently.

Zero also represents the number 0 or, as the ancients would have seen it, no objects are present. This idea of zero as a symbol was published in the Bakhshali manuscript, dating to the 3rd or 4th century C.E. When we add zero to the set of natural numbers, we call this new set the **whole numbers**.

> ### ■ What's That Mean?
> **Sets**
> Mathematicians call a collection of objects a *set*.
>
> The curly brace symbols, { }, in mathematics indicate a set.
>
> The symbols, . . . , denote a pattern. These symbols mean that the set continues in this way indefinitely.
>
> A set with no elements is called the empty set. It is written { } or ∅.

> **DEFINITIONS**
> **Natural Numbers** The natural numbers consist of the set {1, 2, 3, . . .}.
> **Whole Numbers** The whole numbers consist of the set {0, 1, 2, 3, . . .}.

When we count the number of wheels on a wheeled vehicle (such as a bike or a car), it is clear that the vehicle can have 1, 2, 3, 4, or more wheels. Counting the number of wheels is an example of using values from the set of *natural* numbers.

Suppose we want to count up the number of people in a room. The room can contain 0, 1, 2, . . . people. The value of 0 needs to be included, since the room could be empty. The number of people in the room takes on values from the set of *whole* numbers.

The use of negative numbers first appeared in ancient China (200 B.C.E. to 220 C.E.) as answers to certain equations. Negative numbers also appeared when traders balanced their accounts and had to show a loss. As with the number 0, it took a while for the concept of negative numbers to be widely accepted. Today we use negative numbers in different fields, such as the measurement of extremely cold temperatures, double-entry bookkeeping, and describing profit–loss situations. The set of positive and negative whole numbers is called the set of **integers**. In weather forecasts, the temperature of the day is often measured in integers. The temperature on a cold winter's day in Fargo, North Dakota, could be $-30°F$. The summer temperature in San Diego, California, could be $75°F$. Pure oxygen freezes at about $-361°F$.

> **DEFINITION**
>
> **Integers** The integers consist of the set $\{\ldots, -3, -2, -1, 0, 1, 2, 3, \ldots\}$.

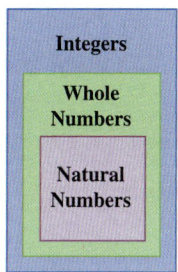

The diagram at the right helps us to visualize the number systems. The natural numbers are contained in the set of whole numbers, and the set of whole numbers is contained in the set of integers. This diagram is a visual way to see that if a number is a natural number, then it is also a whole number and an integer.

CONCEPT INVESTIGATION
What kind of number is that?

1. Place each number in the *smallest* possible set it fits into.
   - **a.** -16
   - **b.** 29
   - **c.** 0
   - **d.** $5°F$
   - **e.** 65 mph
   - **f.** $-12°C$

2. List all the natural numbers between -5 and 4, including the end points -5 and 4.

3. List all the whole numbers between -7 and 3, including the end points -7 and 3.

4. List all the integers between -3 and 5, including the end points -3 and 5.

Example 1   Classifying numbers

Determine the set(s), natural numbers, whole numbers, or integers, to which all possible numbers described by the statement could belong.

a. The number of dogs boarding at a kennel during a 2-week period

b. The average daily low temperature, in degrees Fahrenheit, in Fairbanks, Alaska, over the course of a year

c. The number of tires on a truck used for long-haul trucking (long-distance transportation of goods) in the United States

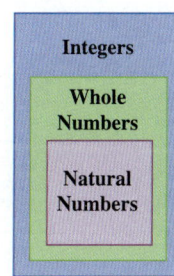

SOLUTION

a. This value (the number of dogs) is a whole number. The number of dogs boarding at a kennel could be 0, 1, 2, ... Since the kennel could be empty on any given day (0 dogs), the value 0 is included as a possibility. These values are also integers.

b. This is an example from the set of integers. The average daily low in Fairbanks ranged from −13°F in January to 50°F in July. The temperatures range from negative numbers to positive numbers.

c. This is an example from the natural numbers, whole numbers, or integers. Trucks can have four or more tires, so the number of tires could be 4, 5, 6, ….

PRACTICE PROBLEM FOR EXAMPLE 1

Determine the set(s), natural numbers, whole numbers, or integers, to which all possible numbers described by the statement could belong.

a. The number of children attending kindergarten at Park View Elementary School

b. The average daily low temperature, in degrees Fahrenheit, in Minneapolis, Minnesota, over the course of a year

c. The number of people in the quiet study room in the library at Irvine Valley Community College

■ Number Lines

> **What's That Mean?**
>
> **Scale**
>
> The *scale* of a number line is the distance between the consistent and evenly spaced tick marks on the number line.

Number lines are a useful way to visualize numbers. A number line is like a ruler. It is straight with a *consistent scale*. This means the tick marks on the ruler are the same distance apart, and the distance between the marks is called the **scale**. Numbers get larger (increase) as we move to the right on the number line. Numbers get smaller (decrease) as we move to the left on the number line. The positive numbers are to the right of 0, and the negative numbers are to the left of 0.

The number line below has scale of 1. This is because the tick marks on the line are 1 unit apart.

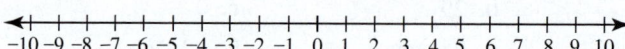

The number line below has scale of 10. The tick marks on the line are 10 units apart.

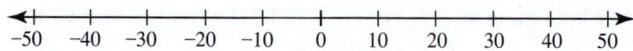

The number line below has scale of 0.5. The tick marks on the line are a half unit apart.

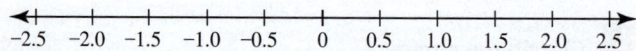

Example 2 Finding the scale

Determine the scale of the following number line.

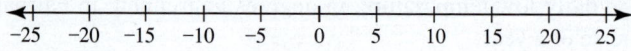

SOLUTION
The tick marks on the number line are 5 units apart. The scale of this number line is 5.

PRACTICE PROBLEM FOR EXAMPLE 2

Determine the scale of the following number line.

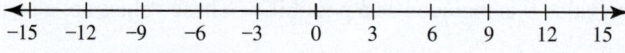

Each number corresponds to one point on the number line. To graph a number on the number line, we locate its position and draw in a darkened circle.

Example 3 Graphing points

Graph the numbers $-3, 0, 1$, and 4 on the following number line.

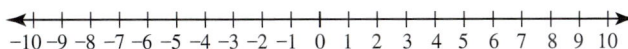

SOLUTION
The numbers are graphed as darkened circles on the number line.

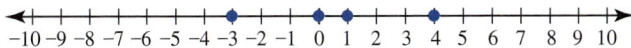

PRACTICE PROBLEM FOR EXAMPLE 3
Graph the numbers $-6, 8, -3$, and 7 on the following number line.

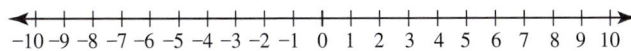

Relations Between Numbers

When we look at a number line, we can see that some numbers are larger than others. Mathematicians express this idea by saying that the integers are a number system that has *order*. Order means that we can organize a set of numbers from smallest to largest. The symbols given in the following table are mathematical notation used to express order in numbers.

Symbols	In Words	Example
$a < b$	a is less than b	$5 < 10$
$a > b$	a is greater than b	$2 > -2$
$a \leq b$	a is less than or equal to b	$-1 \leq 0$
$a \geq b$	a is greater than or equal to b	$-2 \geq -4$
$a < b < c$	b is between a and c, not including a or c	$3 < 5 < 7$
$a \leq b \leq c$	b is between a and c, including a and c	$1 \leq 1 \leq 3$
$a = b$	a is equal to b	$2 = 2$
$a \neq b$	a is not equal to b	$3 \neq 5$
$a \approx b$	a is approximately equal to b	$0.249 \approx 0.25$

> **What's That Mean?**
>
> **Notation**
> A set of symbols that represents something can also be called *notation*. For example, musical notation is formed by musical notes and other musical symbols.

> ## Connecting the Concepts
>
> **Can something be approximately equal to?**
> In the last row of the table, we wrote $a \approx b$. The $\approx$ symbol means "approximately equal to." It lets the reader know that this solution is not exact. The solution has been rounded or chopped off in some way. When we find a decimal approximation to an exact answer, we will use the approximately equal to symbol $\approx$.

The symbols $\leq, <, \geq, >$ and $\neq$ are called *inequality symbols*, and the first six rows in the table are called *inequalities*.

Example 4 Classifying inequalities

Fill in the blank column with $<$ or $>$. Draw a number line to visualize the correct relationship between the values. This is especially helpful in comparing two negative numbers.

a.	-2		-5
b.	-6		-1
c.	0		7

6 CHAPTER R Review of Prealgebra

SOLUTION
Use the following number line to determine which number is larger. Remember that the numbers to the right are larger.

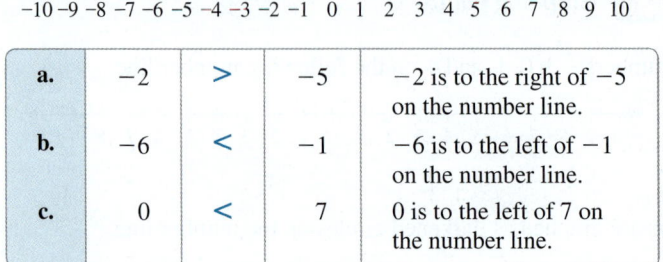

a.	−2	>	−5	−2 is to the right of −5 on the number line.
b.	−6	<	−1	−6 is to the left of −1 on the number line.
c.	0	<	7	0 is to the left of 7 on the number line.

PRACTICE PROBLEM FOR EXAMPLE 4
Fill in the blank column with < or >. If needed, draw a number line to visualize the correct relationship between the values.

a.	−5		−8
b.	3		−3
c.	−2		0

■ Absolute Value

■ CONCEPT INVESTIGATION
What is the distance between a number and zero?

Use the following number line to answer the questions.

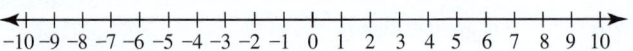

1. What is the distance between −6 and 0?
2. What is the distance between 6 and 0?
3. What is the distance between 0 and 0?
4. Can a distance be positive?
5. Can a distance be negative?
6. Can a distance be zero?

■ **What's That Mean?**

Nonnegative
When a quantity takes on only values that are 0 or positive (that is, greater than or equal to 0), mathematicians will say the quantity is *nonnegative*.

To discuss distance on a number line, we use the idea of **absolute value**. Absolute value is defined to be a distance measurement, so it must be a positive number or 0 (nonnegative). Distances cannot be negative. Therefore, absolute value never results in a negative number.

DEFINITION

Absolute Value   The absolute value of a number is the distance between that number and 0.
Note   The | | symbols are the notation for absolute value.

For example, $|-4| = 4$, since the distance between -4 and 0 is 4. See the number line below.

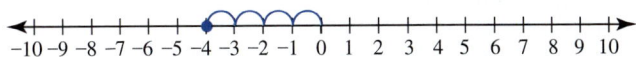

Example 5 Finding absolute value

Find the value of the absolute value expressions.

a. $|5|$
b. $|-7|$

SOLUTION

a. $|5| = 5$, since the distance between 5 and 0 is 5 units.

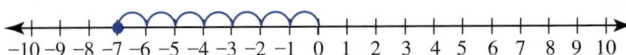

b. $|-7| = 7$, since the distance between -7 and 0 is 7 units.

PRACTICE PROBLEM FOR EXAMPLE 5

Find the value of the absolute value expressions.

a. $|10|$
b. $|-16|$

Example 6 Ordering numbers

Place the following numbers in increasing order (from smallest to largest):

$$|-2|, |0|, |3|, -5, |12|$$

SOLUTION
First, find any absolute values.

$$|-2| = 2$$
$$|0| = 0$$
$$|3| = 3$$
$$-5 = -5$$
$$|12| = 12$$

Order the numbers from smallest to largest.

$$-5, 0, 2, 3, 12$$

Convert back to the original notation.

$$-5, |0|, |-2|, |3|, |12|$$

PRACTICE PROBLEM FOR EXAMPLE 6

Place the following numbers in increasing order (from smallest to largest):

$$|6|, |-4|, -2, |0|, 8$$

8 CHAPTER R Review of Prealgebra

■ Opposite of a Number

Look at the following number line. The numbers 7 and −7 are the same distance from 0, so they have the same absolute value.

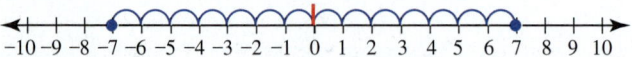

Numbers that have the same absolute value but different signs are called **opposites**. Another way to read the number −7 (negative 7) is the opposite of 7.

> **DEFINITION**
>
> **Opposite of a Number** The opposite of any number has the same absolute value but differs in sign.
> **Note** Read −a as the "opposite of a," not as "negative a."

Example 7 Rewriting an expression in sentence form

Rewrite the math expression as a sentence. Then find the value of the expression.

a. −(−7)   **b.** −(6)

SOLUTION

a. The expression translates as "the opposite of negative seven."

$-(-7) = 7$

b. The expression translates as "the opposite of six."

$-(6) = -6$

PRACTICE PROBLEM FOR EXAMPLE 7

Rewrite the math expression as a sentence. Then find the value of the expression.

a. −(8)   **b.** −(−1)   **c.** −(0)

What's That Mean?

Expression

In English, an *expression* can mean a word or a phrase. In mathematics, an *expression* is a mathematical phrase. It is a combination of numbers and symbols. An expression does not contain an equal sign.

What's That Mean?

Signed Numbers

In algebra, a *signed number* refers to a number with either a positive or negative sign. Positive numbers are written with no sign. Thus, 5 = +5. Negative numbers are always written with the negative sign, such as −5.

■ Operations with Integers

We can use the number line to visualize adding and subtracting integers (signed numbers). Let's discuss addition first. To add signed numbers, move to the right on the number line when adding a positive number. Suppose we want to add −8 + 7. We start on the number line at −8 and count over 7 units to the right (since 7 is a positive number). The solution is −1.

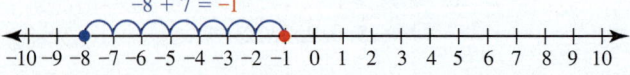

We move to the left on the number line when adding a negative number. Suppose we want to add −2 + (−5). We start at −2 on the number line and move over 5 units to the left because −5 is a negative number. The solution is −7.

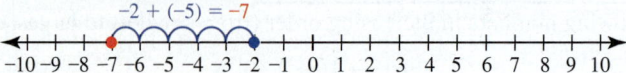

SECTION R.1 Operations with Integers

Drawing a number line each time we want to add or subtract signed numbers can be time-consuming. We usually add and subtract signed numbers using the following steps.

> **To Add or Subtract Integers**
> - **To add integers with the same sign:** Add the absolute values of the numbers. Attach the same sign of the numbers to the sum.
> - **To add integers with different signs:** Take the absolute value of each number. Subtract the smaller absolute value from the larger. Attach the sign of the number that is larger in absolute value.
> - **To subtract integers:** Change the sign of the second integer (reading left to right) and add as explained above.

> **What's That Mean?**
>
> **Sum**
>
> The *sum* is the result of performing an addition. In $5 + 7 = 12$, the sum is 12.

Example 8 Adding integers

Add the integers.

a. $-29 + 15$ **b.** $-13 + 18$

SOLUTION

a. These two integers have different signs.

Take the absolute value of each number.

$$|-29| = 29 \quad \text{and} \quad |15| = 15$$

Subtract the smaller absolute value from the larger.

$$29 - 15 = 14$$

Attach the sign of the number that is larger in absolute value.

The number that is larger in absolute value is -29, so the answer is negative.

$$-29 + 15 = -14$$

b. These two integers have different signs. Take the absolute value of each number and subtract the smaller absolute value from the larger. Subtract 13 from 18.

$$18 - 13 = 5$$

Since the number that is larger is $+18$, the answer is $+5$ or just 5.

PRACTICE PROBLEM FOR EXAMPLE 8

Add the integers.

a. $-42 + 8$ **b.** $-12 + 26$

Example 9 Subtracting integers

Subtract the integers.

a. $3 - 7$ **b.** $-5 - (-9)$

SOLUTION

a. These two integers are being subtracted.

Change the sign of the second integer and add.
$$3 - 7 = 3 + (-7)$$
Take the absolute value of each number.
$$|3| = 3 \quad \text{and} \quad |-7| = 7$$
Subtract the smaller absolute value from the larger.
$$7 - 3 = 4$$
Attach the sign of the number that is larger in absolute value.

The number that is larger in absolute value is -7, so the answer is negative.
$$3 + (-7) = -4$$

b. The two integers are being subtracted. Change the sign of the second integer (-9 to 9). Add the two numbers using the rule for adding integers.
$$-5 - (-9) = -5 + (9) = 4$$
The final answer is 4.

PRACTICE PROBLEM FOR EXAMPLE 9
Subtract the integers.

a. $8 - 15$ **b.** $7 - (-10)$

Example 10 — Rising and falling temperatures

a. If the temperature is $-5°F$ and the temperature rises by $8°F$, then what is the new temperature?

b. If the temperature is $-12°F$ and the temperature drops by $2°F$, then what is the new temperature?

SOLUTION

a. The temperature can be found by adding $-5 + 8$. These integers have different signs. The number with the larger absolute value is 8.

$8 - 5 = 3$ *Subtract 5 from 8, as the number with the larger absolute value is 8.*
$-5 + 8 = 3$ *The result is positive, as the number with the larger absolute value was positive 8.*

If the temperature is $-5°F$ and rises by $8°F$, the final temperature is $3°F$.

b. The temperature can be found by subtracting $-12 - 2$. To subtract two integers, change the sign of the second integer and add. Therefore, we compute
$$-12 + (-2) = -14$$
If the temperature is $-12°F$ and drops by $2°F$, the final temperature is $-14°F$.

PRACTICE PROBLEM FOR EXAMPLE 10
The highest point in the state of California is Mount Whitney, which has an elevation of 14,494 feet, and the lowest point in California is Death Valley, which is 282 feet below sea level.

SECTION R.1 Operations with Integers   11

Source: www.netstate.com/states/geography

a. Write the elevation of Mount Whitney as an integer. *Note*: A positive elevation will be above sea level.

b. Write the elevation of Death Valley as an integer. *Note*: A negative elevation will be below sea level.

c. Find the difference in elevation between the highest and lowest points in the state of California.

Absolute value can be used to find the distance between two points on a number line.

DEFINITION

Distance between Two Points on the Number Line   The distance between two points on a number line, a and b, can be found as $|b - a|$.

Example 11   Finding the distance between two points on a number line

Find the distance between -8 and -3.

SOLUTION
One way to find the distance between -8 and -3 is to graph the two points on a number line and count the spaces in between them.

Since there are five spaces in between -8 and -3, the distance between these two numbers is 5.

Another way to find the distance between these two points is to use the formula $|b - a|$. Let $a = -8$ and $b = -3$ and substitute into the formula.

$$\begin{aligned} |b - a| &= |-3 - (-8)| \\ &= |-3 + 8| \\ &= |5| \\ &= 5 \end{aligned}$$

Note that if we switch the order of the points, the formula still gives the correct answer. Let $a = -3$ and $b = -8$ and substitute into the formula.

$$\begin{aligned} |b - a| &= |-8 - (-3)| \\ &= |-8 + 3| \\ &= |-5| \\ &= 5 \end{aligned}$$

PRACTICE PROBLEM FOR EXAMPLE 11

Find the distance between -9 and -2.

The two other basic operations are multiplication and division. Recall that multiplication of natural numbers is shorthand for a repeated addition. So $3 \cdot 4 = 4 + 4 + 4 = 3 + 3 + 3 + 3$. So $3 \cdot 4$ is the same as adding 4 three times or adding 3 four times.

Likewise, division of natural numbers is a shortcut for a repeated subtraction. In $12 \div 3 = 4$, when we repeatedly subtract 3 from 12, we get

$$12 - 3 = 9$$
$$9 - 3 = 6$$
$$6 - 3 = 3$$
$$3 - 3 = 0$$

The number of times we subtract 3 from 12 is 4. Therefore, $12 \div 3 = 4$.

Writing multiplication and division problems as repeated additions and/or subtractions can be time-consuming. That is why people memorize multiplication tables. The following Concept Investigation will review the rules for multiplying signed numbers.

■ CONCEPT INVESTIGATION

What pattern of signs do you see?

1. Complete the following table.

Calculation	Result
$4 \cdot 4$	16
$4 \cdot 3$	
$4 \cdot 2$	
$4 \cdot 1$	
$4 \cdot 0$	

2. What happens to the result as the number you multiply 4 by goes down by 1 in each row?

3. When you multiply two positive numbers together, is the result a positive or negative number?

4. Use the pattern from part 1 to complete the following table.

Calculation	Result
$4 \cdot 2$	8
$4 \cdot 1$	
$4 \cdot 0$	
$4 \cdot (-1)$	
$4 \cdot (-2)$	
$4 \cdot (-3)$	
$4 \cdot (-4)$	

5. When a positive number and a negative number are multiplied together, is the result a positive or negative number?

6. Use a pattern to complete the following table.

Problem	Solution
$-4 \cdot 4$	-16
$-4 \cdot 3$	
$-4 \cdot 2$	
$-4 \cdot 1$	
$-4 \cdot 0$	
$-4 \cdot (-1)$	
$-4 \cdot (-2)$	
$-4 \cdot (-3)$	
$-4 \cdot (-4)$	

7. When a negative number and a negative number are multiplied together, is the result a positive or negative number?

To Multiply or Divide Signed Numbers

- **To multiply or divide two integers with the same sign:** Multiply or divide the absolute values of the numbers. The solution is always positive.
- **To multiply or divide two integers with different signs:** Multiply or divide the absolute values of the two numbers. The solution is always negative.

Example 12 Multiplying integers

Multiply the following integers.

a. $(-3)(7)$ b. $(-9)(-6)$

SOLUTION

a. These two integers have different signs.

Multiply the absolute values of the two numbers.

$$3 \cdot 7 = 21$$

The answer is negative, since the two integers have different signs.

$$(-3)(7) = -21$$

b. These two integers have the same sign.

Multiply the absolute values of the two numbers.

$$9 \cdot 6 = 54$$

The answer is positive, since the two integers have the same sign.

$$(-9)(-6) = 54$$

PRACTICE PROBLEM FOR EXAMPLE 12

Multiply the following integers.

a. $(-2)(4)$ b. $(-5)(-3)$

Example 13 Dividing integers

Divide the following integers.

a. $15 \div (-5)$ **b.** $(-48) \div 12$ **c.** $\dfrac{-27}{-3}$

SOLUTION

a. $15 \div (-5) = -3$

The answer is negative, since the two integers have different signs.

b. $(-48) \div 12 = -4$

The answer is negative, since the two integers have different signs.

c. $\dfrac{-27}{-3} = 9$

The answer is positive, since the two integers have the same sign.

PRACTICE PROBLEM FOR EXAMPLE 13

Divide the following integers.

a. $20 \div (-4)$ **b.** $\dfrac{-16}{-8}$ **c.** $(-27) \div 3$

Connecting the Concepts

Symbols that indicate division

There are several ways to show division. One way is with the $\div$ sign, such as $8 \div (-4)$.

Another way is with a fraction bar:

$$8 \div (-4) = \dfrac{8}{-4}.$$

Another way is using long division:

$8 \div (-4)$ means $-4\overline{)8}$.

What's That Mean?

Dividend, Divisor, and Quotient

The parts of a division problem have special names.

dividend $\div$ *divisor* = *quotient*

Written using a fraction bar, we have

$$\dfrac{\text{dividend}}{\text{divisor}} = \text{quotient}$$

Sometimes when a fraction bar is used, the parts are called the numerator and denominator.

$$\dfrac{\text{numerator}}{\text{denominator}}$$

When 0 is the dividend, such as $0 \div 16$ or $\dfrac{0}{16}$, the answer is 0. Division by 0 means that 0 is the divisor or is in the denominator. It is important to note that division by zero is **undefined** (sometimes mathematicians say the result of such an operation "does not exist" or "DNE"). We know that $16 \div 2 = 8$ or $\dfrac{16}{2} = 8$, since $8 \cdot 2 = 16$. Now suppose $16 \div 0 = c$ or $\dfrac{16}{0} = c$, where c is some real number. This means that $0 \cdot c = 16$, which is impossible. Any number multiplied by 0 is 0. Therefore, in math we say that division by zero is undefined.

> **DEFINITION**
> **Division by 0**   For any number a, $a \div 0$ is undefined.

■ Order of Operations

In arithmetic, we have the following four basic operations.

> **DEFINITIONS**
>
> **Terms**   Two (or more) quantities that are being added
> **Factors**   Two (or more) quantities that are multiplied
> **Basic Operations**   The basic operations used in arithmetic are as follows.
>
> 1. **Addition**   The result of an addition is called a **sum**.
> 2. **Subtraction**   The result of a subtraction is called a **difference**.
> 3. **Multiplication**   The result of a multiplication is called a **product**.
> 4. **Division**   The result of a division is called a **quotient**.

There are many times in a problem when we are faced with calculating more than one operation. For example, suppose we want to find the value of $4 \cdot 3 + 8$.

Janelle says to evaluate the multiplication first and then do the addition. Thus, we would get $4 \cdot 3 = 12$ and then compute $12 + 8 = 20$. But Curtis says to do the addition first and then the multiplication (working right to left). This would give us $3 + 8 = 11$ and then $4 \cdot 11 = 44$. This is not the same result at all.

$$4 \cdot 3 + 8 = ?$$

Janelle	Curtis
$(4 \cdot 3) + 8 = 12 + 8 = 20$	$4 \cdot (3 + 8) = 4 \cdot 11 = 44$

In mathematics, there cannot be two different answers, 20 and 44, for the same problem, $4 \cdot 3 + 8$.

To see which answer is correct, let's think about this problem in context. Suppose we buy four items for $3.00 each and one item for $8.00. How much did we spend? The four items at $3.00 cost $12.00, and then we add on the $8.00 for a total of $20.00. Thus, we know that 20 is the correct answer.

Because of this potential for confusion, mathematicians have come up with an order-of-operations agreement. This agreement will tell us in what order we should perform operations. *In arithmetic, terms are identified as the quantities that are being added or subtracted.* To determine the number of terms in an expression, look for the quantities that are separated by addition or subtraction symbols.

What's That Mean?

Evaluate

In mathematics, to *evaluate* something means to compute the numerical value.

Connecting the Concepts

Why does a subtraction symbol separate terms?

Recall that subtraction is the same as adding the opposite of the second number.

$$2 - 3 = 2 + (-3)$$

Therefore, 2 and -3 are the two terms being added.

Example 14 Identifying terms

Identify the terms in each expression.

a. $6 - 4 \cdot 5 + 10 \div 2$ **b.** $8 \div 4 - 16 + 9 \cdot 3 - 5$ **c.** $50 \cdot 2 \div 10$

SOLUTION

a. This expression has three terms.

$6 + (-4) \cdot 5 + 10 \div 2$ Rewrite the subtraction as adding the opposite
$(6) + ((-4) \cdot 5) + (10 \div 2)$ and circle the terms.

b. This expression has four terms.

$8 \div 4 + (-16) + 9 \cdot 3 + (-5)$ Rewrite the subtractions as adding the opposite.
$(8 \div 4) + (-16) + (9 \cdot 3) + (-5)$ Circle the terms.

c. This expression has one term. There are no additions or subtractions.

$(50 \cdot 2 \div 10)$

PRACTICE PROBLEM FOR EXAMPLE 14

Identify the terms in each expression.

a. $4 - 2 \cdot 6 + 8$ **b.** $3 \cdot 7 - 10 \div 2 \cdot 3 - 5 \cdot 4 + 1$ **c.** $-65 \div 5 \cdot 4$

Now that we have learned to identify terms, we can use the order-of-operations agreement.

16 CHAPTER R Review of Prealgebra

> **Steps to Use the Order-of-Operations Agreement**
> 1. Circle or underline the terms. (Remember: Terms are quantities that are added.)
> 2. Evaluate the operations (multiplications and divisions) inside of each term, working from left to right.
> 3. Add the simplified terms, working from left to right.

Example 15 Placing the operations in order

Simplify each expression.

a. $9 - 6 \cdot 3 + 2$ **b.** $48 \div 3 \cdot 4$

c. $7 \cdot 5 - 10 \div 2 + 5 \cdot 2$ **d.** $36 \div 9 \cdot 3 - 2 \cdot 4 + 7 \cdot 2 - 5$

SOLUTION
First circle the terms. Then working from left to right inside each term, do the multiplications and divisions. Do the additions and subtractions last, working from left to right.

a. This expression has three terms.

Step 1 Circle or underline the terms.

$$9 + (-6) \cdot 3 + 2 \quad \text{Rewrite subtraction as adding the opposite.}$$
$$⑨ + ⟮(-6) \cdot 3⟯ + ②$$

Step 2 Evaluate the operations inside of each term, working left to right.

$$= 9 + (-18) + 2$$

Step 3 Add the simplified terms, working left to right.

$$= -9 + 2$$
$$= -7$$

b. There are no additions or subtractions, so this expression has one term.

$$⟮48 \div 3 \cdot 4⟯ \quad \text{Circle the term.}$$
$$= 48 \div 3 \cdot 4 \quad \text{Inside the term, do the operations in}$$
$$= 16 \cdot 4 \quad \text{order, working from left to right.}$$
$$= 64$$

c. This expression has three terms.

$$7 \cdot 5 + (-10) \div 2 + 5 \cdot 2 \quad \text{Rewrite subtraction as adding the opposite.}$$
$$⟮7 \cdot 5⟯ + ⟮(-10) \div 2⟯ + ⟮5 \cdot 2⟯ \quad \text{Circle the terms.}$$
$$= 35 + (-5) + 10 \quad \text{Do the operations inside each term.}$$
$$= 30 + 10 \quad \text{Do additions, working from left to right.}$$
$$= 40$$

d. This expression has four terms.

$$36 \div 9 \cdot 3 + (-2) \cdot 4 + 7 \cdot 2 + (-5) \quad \text{Rewrite subtraction as adding the opposite.}$$
$$⟮36 \div 9 \cdot 3⟯ + ⟮(-2) \cdot 4⟯ + ⟮7 \cdot 2⟯ + ⟮(-5)⟯ \quad \text{Circle the terms.}$$
$$= ⟮4 \cdot 3⟯ + (-8) + 14 + (-5) \quad \text{Do the operations inside each term.}$$
$$= 12 + (-8) + 14 + (-5) \quad \text{Do the additions, working from left to right.}$$
$$= 4 + 14 + (-5)$$
$$= 18 + (-5)$$
$$= 13$$

> **■ What's That Mean?**
>
> **Simplify**
>
> In mathematics, to *simplify* something means to use arithmetic or algebraic operations to express the answer in the simplest possible form. In this chapter, we work with arithmetic. In later chapters, we will work with both arithmetic and algebraic operations.

SECTION R.1 Operations with Integers

PRACTICE PROBLEM FOR EXAMPLE 15
Simplify each of the following expressions.

a. $24 \div 2 \cdot 3$ **b.** $9 + 6 \cdot 5 - 8 \div 4$ **c.** $6 \div 3 + 4 \cdot 5 - 2 \cdot 2 + 24 \div 8$

R.1 Vocabulary Exercises

1. The set {0, 1, 2, 3,...} is called the set of _____ numbers.

2. The scale of a number line is the _____ between the tick marks on the number line.

3. _____ are two numbers with the same absolute value, but with different signs.

4. In an expression, _____ are the things that are added.

5. We use the _____ agreement to simplify expressions that have multiple operations.

6. In an expression, _____ are the things that are multiplied together.

7. The result of an addition problem is called a(n) _____.

8. The result of a multiplication problem is called a(n) _____.

9. Division by zero is _____.

R.1 Exercises

For Exercises 1 through 10, classify each number as a natural number, a whole number, or an integer. If a number belongs to more than one set, list all possible sets it belongs to.

1. 0
2. 6
3. −20
4. −7
5. 11
6. 0
7. −9
8. −3
9. 1
10. −1

For Exercises 11 through 20, determine the set(s), natural numbers, whole numbers, or integers to which all possible numbers described by the statement could belong.

11. The population of Los Angeles, California, each year

12. The number of people in a crowd at the beach

13. The population of Norfolk, Virginia

14. The number of children attending Sullivan Middle School graduation

15. The number of kittens in a local animal shelter

16. The number of puppies in a local animal shelter

17. The number of hours worked weekly by a Home Depot employee
18. The number of hours worked yearly by an auto worker
19. The daily high temperature in Missoula, Montana
20. The daily high temperature in Honolulu, Hawaii

For Exercises 21 through 26, list all numbers in each given set.

21. All integers between −5 and 1, including −5 and 1
22. All integers between −2 and 4, including −2 and 4
23. All natural numbers between −5 and 5, not including −5 and 5
24. All natural numbers between −6 and 4, not including −6 and 4
25. All whole numbers between −4 and 0, including −4 and 0
26. All whole numbers between −3 and 3, including −3 and 3

For Exercises 27 through 30, determine the scale of each number line.

27.
28.
29.
30.

For Exercises 31 through 36, graph each number on a number line like the one below.

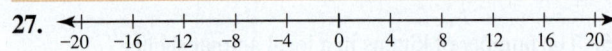

31. −5
32. −2
33. 1
34. 9
35. −4
36. −7

For Exercises 37 through 44, use an appropriate inequality symbol (<, >, ≤, or ≥) between each set of two numbers.

37. −5   −15
38. −6   0
39. −3   −3
40. 8   8
41. −4   −3
42. −9   −8
43. 0   −7
44. 0   −1

For Exercises 45 through 52, find the value of each absolute value expression. See Example 5.

45. $|-3|$
46. $|-11|$
47. $|8|$
48. $|0|$
49. $|16|$
50. $|9|$
51. $|-7|$
52. $|-15|$

For Exercises 53 through 58, place each list of numbers in increasing order (from smallest to largest).

53. $3, \ -\dfrac{3}{2}, \ -1, \ 2.95, \ 0.5, \ \dfrac{1}{3}$
54. $-2, \ -\dfrac{1}{4}, \ 5, \ -1.75, \ 3.5, \ -4.5$
55. $|-5|, \ -5, \ |-2|, \ -2$
56. $|8|, \ -8, \ |-0|, \ 0$
57. $|-4|, \ -3.5, \ |-5|, \ \dfrac{2}{3}, \ 4.2$
58. $-7, \ |-7|, \ \dfrac{5}{4}, \ |-1|, \ \dfrac{9}{3}, \ |-3|$

For Exercises 59 through 64, first rewrite each mathematical expression in sentence form and then find the value of the expressions.

59. $-(5)$
60. $-(-5)$
61. $-|-2|$
62. $-|-4|$
63. $-(-6)$
64. $-(-(-5))$

For Exercises 65 through 76, add or subtract the integers without a calculator.

65. $-3 + (-5)$
66. $-4 + (-9)$
67. $16 + (-7)$
68. $-8 + 15$
69. $-13 + 6$
70. $-20 + 12$
71. $-7 - 9$
72. $-6 - 13$
73. $-4 - (-8)$
74. $-7 - (-1)$
75. $-12 - (-11)$
76. $-17 - (-18)$

For Exercises 77 through 80, use the information given in each exercise to answer the questions.

77. At noon, it was $-23°F$ outside. The temperature drops 3 degrees in an hour. What is the new temperature?

78. At 10:00 A.M., it was $-6°F$, and the temperature rose 20 degrees by 1:00 P.M. that afternoon. What was the temperature at 1:00 P.M.?

79. Assume that depths below the ocean's surface are represented by negative values and heights above the ocean's surface are represented by positive values. Tom and Cassie are scuba diving in the Bahamas. They are at 30 feet below the surface of the ocean, and they dive another 20 feet. Now what is their depth?

80. The Badwater Ultramarathon is a 135-mile race that is run every year in California. The runners begin in Death Valley, at an elevation of 282 feet below sea level. They end up at the Mount Whitney Portal, which is 8360 feet above sea level. What is the total change in elevation that the runners experience in the race?

For Exercises 81 through 86, find the distance between the two points on the number line. See Example 11.

81. Find the distance between 7 and -2.
82. Find the distance between 5 and -6.
83. Find the distance between -1 and -9.
84. Find the distance between -8 and -22.
85. Find the distance between -16 and -35.
86. Find the distance between -26 and -5.

For Exercises 87 through 100, multiply or divide as indicated without a calculator.

87. $3(-2)$
88. $5(-6)$
89. $(-4)(5)$
90. $(-3)(15)$
91. $(-3)(-6)$
92. $(-5)(-9)$
93. $(-24) \div (-8)$
94. $(-12) \div (-3)$
95. $(30) \div (-5)$
96. $(27) \div (-9)$
97. $9 \div 0$
98. $-5 \div 0$
99. $0 \div 8$
100. $0 \div (-2)$

For Exercises 101 through 108, identify the terms in each expression.

101. $-3 + 5 \cdot 6$
102. $16 \cdot 3 + 9$
103. $-24 \div 3 \cdot 7$
104. $100 \cdot (-2) \div 5$
105. $3 + 22 \div 11 - 16 \div 4$
106. $42 \div 5 - (-6)7 + 12$
107. $92 \cdot (-8) + 26 \div 13 - 1 + 7(-3)$
108. $(-5) \cdot (-6) - 4 \div (-2) + 10 - 4(-2)$

For Exercises 109 through 118, identify the number of terms. Then simplify using the order-of-operations agreement.

109. $16 \div (-8) \cdot 3$
110. $(-25) \div 5 \cdot (-4)$
111. $-9 + 3 \cdot 5$
112. $6 \cdot (-2) + 18$
113. $9 + 24 \div 3 - 3 \cdot (-2)$
114. $56 \div (-8) + 3 \cdot 5 - (-7)$
115. $-14 \div 2 + 5 \cdot 4 \div 10 - (-1)$
116. $9 \cdot 4 \div (-6) - (-16) + 8 \cdot 2$
117. $100 \div 50 \cdot 3 - 2 \cdot (-7) + 4 \cdot 2 - (-2)$
118. $32 \div (-8) + 3 \cdot 7 - 9 - 48 \div 12 \cdot 5$

R.2 Operations with Fractions

LEARNING OBJECTIVES
- Find the prime factorization.
- Reduce a fraction to lowest terms.
- Find an equivalent fraction.
- Graph fractions on a number line.
- Add and subtract fractions.
- Multiply and divide fractions.
- Use the order-of-operations agreement with fractions.

Prime Numbers and Prime Factorization

Prime numbers are natural numbers greater than 1 that are evenly divisible only by themselves and 1. Natural numbers that are not prime are said to be **composite numbers**. Prime numbers are useful in several applications, including cryptography. Cryptography is the mathematics of encoding and decoding information. When we purchase an item over the Internet, our credit card information is most often encoded by using a procedure that uses prime numbers.

> **DEFINITION**
>
> **Prime Numbers** Prime numbers are natural numbers greater than 1 that are divisible only by themselves and 1. The prime numbers are the infinite set $\{2, 3, 5, 7, 11, 13, 17, 19, \ldots\}$.

In this course, we encounter prime numbers when we are *factoring* a number (or expression). We will use the following fact. Any natural number greater than 1 may be factored uniquely by using only prime numbers. This factorization is called the **prime factorization** of the number.

To find the prime factorization of a number means to write the number using only prime numbers as the factors. When we say that the factorization is unique, we mean that there is only one final answer.

What's That Mean?

Factor

A *factor* of a number divides the number evenly. For example, 3 is a factor of 6, since

$$6 \div 3 = 2$$

We say that 2 and 3 are factors of 6.

To *factor* a number means to write it as a product. For example, one factorization of 6 is $2 \cdot 3$, and another is $6 \cdot 1$.

Example 1 — Writing a number with prime factors only

Write the prime factorization of each number.

a. 12 b. 30 c. 48

SOLUTION

a. 12 — Think of any factorization of 12.
$= 3 \cdot 4$ — 4 is not prime, so continue to factor.
$= 3 \cdot 2 \cdot 2$ — Factor 4 into prime factors.

Note that in the last step, 4 is not a prime number, and we factor 4 into its prime factors, that is, we write $4 = 2 \cdot 2$.

We could start the problem by rewriting $12 = 6 \cdot 2$. Then factor 6 as $6 = 3 \cdot 2$.

12
$= 6 \cdot 2$ 6 is not prime, so we continue to factor.
$= 3 \cdot 2 \cdot 2$ Factor 6 into prime factors.

Notice that the final answer is the same.

b. 30   Think of any factorization of 30.
$= 5 \cdot 6$ Keep factoring until only prime factors remain.
$= 5 \cdot 2 \cdot 3$
$= 2 \cdot 3 \cdot 5$

c. Another way to factor is by using a factor tree.

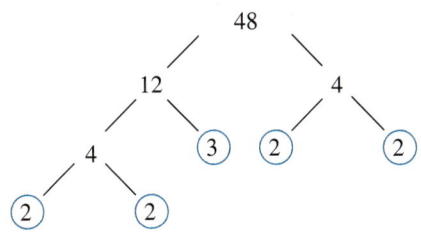

Only the factors at the bottom of each branch of the tree are part of the prime factorization.

$48 = 2 \cdot 2 \cdot 2 \cdot 2 \cdot 3$

PRACTICE PROBLEM FOR EXAMPLE 1

Write the prime factorization of each number.

a. 32   **b.** 10   **c.** 100

We use the idea of prime factorization to find the **greatest common factor (GCF)** for a list of numbers.

Steps to Find the Greatest Common Factor
1. Find the prime factorization of each number.
2. The greatest common factor is the product of the common prime factors.

Example 2 Finding the greatest common factor

Find the GCF for each set of numbers.

a. 8 and 20   **b.** 12 and 18

SOLUTION

a. Step 1 Find the prime factorization of each number.

$$8 = 2 \cdot 4 = 2 \cdot 2 \cdot 2$$
$$20 = 4 \cdot 5 = 2 \cdot 2 \cdot 5$$

Step 2 The greatest common factor is the product of the common prime factors.

$$\text{GCF is } 2 \cdot 2 = 4$$

b. Find the prime factorization of both 12 and 18.

$$12 = 4 \cdot 3 = 2 \cdot 2 \cdot 3$$
$$18 = 2 \cdot 9 = 2 \cdot 3 \cdot 3$$

What's That Mean?

Fraction

A *fraction* is often thought of as a part of a whole. People will say that something is "only a fraction of the cost," meaning that the object costs only part of what it should. A fraction can also refer to the breaking up of a whole. "The land was partitioned into fractions" means that the whole (the land) was broken into smaller parts (the fractions).

Sometimes a fraction will represent more than 1. In this case, the fraction is called an *improper fraction*.

The GCF is the product of the common factors. Therefore, the

GCF is $2 \cdot 3 = 6$

PRACTICE PROBLEM FOR EXAMPLE 2
Find the GCF for each set of numbers.

a. 4 and 22 **b.** 64 and 32

■ Simplifying Fractions and Equivalent Fractions

There are many times when we want to express the idea of a fractional part of something. For example, we may work only half a day and, therefore, receive half of our daily pay. The fraction $\frac{1}{2}$ represents half. In a fraction such as $\frac{1}{2}$, there are special names for the number on the top of the fraction and the number on the bottom.

> **DEFINITION**
>
> **Numerator and Denominator**   In the fraction $\frac{a}{b}$, the value of a is called the **numerator**. The value of b is called the **denominator**. The — symbol is called the fraction bar.
>
>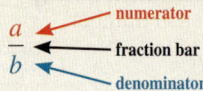

What's That Mean?

Numerator and Denominator

The *denominator* of a fraction tells us (denominates) how many equal-sized pieces the whole is divided into. For example, in the fraction $\frac{2}{4}$, the whole is divided up into four equal-sized pieces. The *numerator* of a fraction tells us (enumerates) how many of the equal-sized pieces to consider. In the fraction $\frac{2}{4}$, we count 2 of the 4 equal-sized pieces.

When we write a fraction such as $\frac{2}{4}$, it represents the shaded part of the circle.

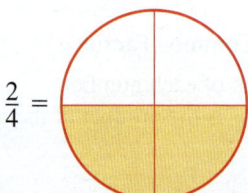

When we write the fraction $\frac{1}{2}$, it represents the shaded part of the circle. Because these two fractions represent the same portion of the circle, they are *equivalent fractions*. This means that $\frac{2}{4} = \frac{1}{2}$. The right side of this equation, $\frac{2}{4} = \frac{1}{2}$, is said to be **reduced to lowest terms.**

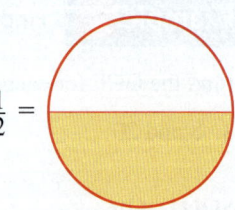

> **Steps to Reduce a Fraction to Lowest Terms**
>
> 1. Find the GCF for the numerator and denominator.
> 2. Factor the numerator and denominator, using the GCF as one of the factors.
> 3. Divide out the GCF from the numerator and denominator. The remaining fraction is the reduced or simplified form of the fraction.

SECTION R.2 Operations with Fractions 23

Example 3 — Reducing fractions to lowest terms

Reduce each fraction to lowest terms.

a. $\dfrac{9}{18}$ b. $\dfrac{16}{36}$ c. $\dfrac{25}{70}$

SOLUTION

a. **Step 1** Find the GCF for the numerator and denominator.

The GCF for 9 and 18 is 9.

Step 2 Factor the numerator and denominator, using the GCF as one of the factors.

$$\dfrac{9}{18} = \dfrac{9 \cdot 1}{9 \cdot 2}$$

Step 3 Divide out the GCF from the numerator and denominator.

$$\dfrac{\cancel{9} \cdot 1}{\cancel{9} \cdot 2} = \dfrac{1}{2}$$

b. The GCF for 16 and 36 is 4. Divide out 4 from the numerator and denominator of the fraction.

$$\dfrac{16}{36} = \dfrac{4 \cdot \cancel{4}}{9 \cdot \cancel{4}} = \dfrac{4}{9}$$

c. The GCF for 25 and 70 is 5. Divide out 5 from the numerator and denominator of the fraction.

$$\dfrac{25}{70} = \dfrac{5 \cdot \cancel{5}}{14 \cdot \cancel{5}} = \dfrac{5}{14}$$

PRACTICE PROBLEM FOR EXAMPLE 3

Reduce each fraction to lowest terms.

a. $\dfrac{3}{27}$ b. $\dfrac{21}{30}$ c. $\dfrac{32}{48}$

To reduce a fraction to lowest terms, divide the numerator and denominator by the GCF. To rewrite a fraction as an equivalent fraction, multiply the numerator and denominator by the same factor. We often use equivalent fractions to make two or more fractions have a common denominator.

> **Steps to Write an Equivalent Fraction**
>
> 1. When given the problem $\dfrac{a}{b} = \dfrac{?}{d}$, find the missing factor c such that $b \cdot c = d$.
>
> 2. Multiply the numerator and denominator by the missing factor c: $\dfrac{a \cdot c}{b \cdot c}$. This yields the equivalent fraction with a denominator of d.

Skill Connection

Fraction Bar

The fraction bar, such as the bar in the fraction $\dfrac{9}{18}$, may be thought of as a *division line*. This helps us to remember that fractions come from dividing integers.

What's That Mean?

Reduce to Lowest Terms

There is another common way to say *reduce a fraction to lowest terms*. It is to *simplify* the fraction to lowest terms. Sometimes the instructions will just say "reduce the fraction" or "simplify the fraction." The reduced or simplified answer is also known as *simplest form*. This means there are no common factors other than 1 in the numerator and denominator.

24 CHAPTER R Review of Prealgebra

Example 4 Writing an equivalent fraction

Rewrite each fraction as an equivalent fraction with the given denominator.

a. $\dfrac{2}{3} = \dfrac{?}{12}$ **b.** $\dfrac{5}{7} = \dfrac{?}{42}$

SOLUTION

a. Step 1 When given the problem $\dfrac{a}{b} = \dfrac{?}{d}$, find the missing factor c such that $b \cdot c = d$.

Comparing the denominators of 3 and 12, we see that the missing factor is 4, as $3 \cdot 4 = 12$.

Step 2 Multiply the numerator and denominator by the missing factor c.

$$\dfrac{2}{3} = \dfrac{2 \cdot 4}{3 \cdot 4} = \dfrac{8}{12}$$

b. Comparing the denominators of 7 and 42, we have that $7 \cdot 6 = 42$. Therefore, we multiply the numerator and denominator by 6.

$$\dfrac{5 \cdot 6}{7 \cdot 6} = \dfrac{30}{42}$$

PRACTICE PROBLEM FOR EXAMPLE 4

Rewrite each fraction as an equivalent fraction with the given denominator.

a. $\dfrac{6}{11} = \dfrac{?}{33}$ **b.** $\dfrac{3}{8} = \dfrac{?}{56}$

■ Fractions on Number Lines

To graph fractions on a number line, we must first determine the scale of the number line.

CONCEPT INVESTIGATION
What is the scale of that ruler?

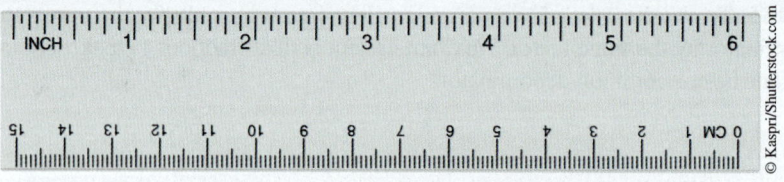

The ruler above has two measurement systems on it. The bottom side uses the metric system (in centimeters), and the top side uses the English system (in inches).

1. What is the scale for the metric (bottom side) of the ruler?
2. What is the scale for the English (top side) of the ruler?
3. Draw a point to show where $1\dfrac{5}{16}$ inches is located on the ruler.
4. Draw a point to show where $11\dfrac{7}{10}$ cm is located on the ruler.

Skill Connection

Scale

In Section R.1, we learned that the scale of a number line is the distance between the consistent and evenly spaced tick marks. Another way to find the scale is to count the spaces between integers. For example, if there are four spaces between 0 and 1, the scale is in fourths (or quarters).

Once the scale of a number line has been determined, fractions are graphed on number lines much as integers are. For example, the fraction $\frac{1}{2}$ is halfway between 0 and 1, so that is where the point is graphed on a number line.

Example 5 Graphing fractions on a number line

Answer each of the following questions.

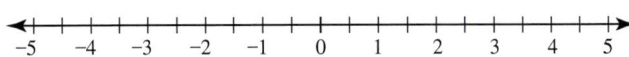

a. What is the scale of this number line?

b. Graph the numbers $\frac{1}{2}, -\frac{3}{2}, \frac{3}{4},$ and $-\frac{7}{2}.$

SOLUTION

a. The scale of the number line is 0.5 or $\frac{1}{2}$. This is because there are two even spaces between each whole number and the next whole number.

b. Graphing the points $\frac{1}{2}, -\frac{3}{2}, \frac{3}{4},$ and $-\frac{7}{2},$ gives the number line below.

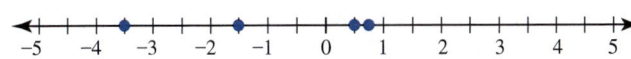

PRACTICE PROBLEM FOR EXAMPLE 5

Answer each of the following questions.

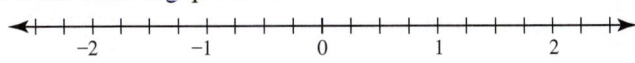

a. What is the scale of this number line?

b. Graph the numbers $\frac{1}{4}, -\frac{7}{4},$ and $\frac{5}{4}.$

■ Addition and Subtraction of Fractions

To add or subtract fractions that have the same denominator, just add or subtract the numerators and then reduce the answer to lowest terms.

> **Steps to Add or Subtract Fractions with Like (Common) Denominators**
> 1. Add (subtract) the fractions by adding (subtracting) the numerators.
> 2. Reduce the final answer to lowest terms (simplest form).

Example 6 Adding and subtracting fractions with like denominators

Add or subtract the fractions as shown. Reduce the answer to simplest form.

a. $\frac{5}{12} + \frac{7}{12}$ **b.** $2\frac{3}{8} - \frac{7}{8}$

What's That Mean?

Mixed Numbers and Improper Fractions

A *mixed number* has a whole number part and a fraction part, such as $1\frac{2}{3}$.

An *improper fraction* is a fraction in which the numerator is larger than the denominator, such as $\frac{10}{9}$.

Connecting the Concepts

Where does the negative sign go?

A negative sign in a fraction may go in one of three places:

$$-\frac{3}{2} = -\frac{3}{2} = \frac{3}{-2}$$

The first two forms are most commonly used: $-\frac{3}{2}$ or $-\frac{3}{2}$.

The last form, $\frac{3}{-2}$, is usually rewritten in one of the other two forms. Many people often do not "see" the negative sign in the denominator, and it gets lost.

Be careful to notice that $\frac{-3}{-2}$ is *not* the same as any of the above expressions.

SOLUTION

a. Step 1 Add (subtract) the fractions by adding (subtracting) the numerators.

$$\frac{5}{12} + \frac{7}{12} = \frac{12}{12}$$

Step 2 Reduce the final answer to lowest terms.

$$\frac{12}{12} = 1$$

b. $2\frac{3}{8} - \frac{7}{8}$ *Rewrite the mixed number as an improper fraction.*

$= \frac{19}{8} - \frac{7}{8}$ *Subtract the numerators.*

$= \frac{12}{8}$ *Reduce to lowest terms.*

$= \frac{3}{2} = 1\frac{1}{2}$

> **Skill Connection**
>
> **Convert a mixed number to an improper fraction**
>
> To convert a mixed number to an improper fraction, multiply the whole number part with the denominator and add on the numerator. Doing so gives the numerator of the improper fraction. The denominator will be the same.
>
> $1\frac{3}{4} = \frac{(4)(1) + 3}{4} = \frac{7}{4}$

PRACTICE PROBLEM FOR EXAMPLE 6

Add or subtract the fractions as indicated. Reduce the answer to simplest form.

a. $\frac{5}{6} + 2\frac{4}{6}$ **b.** $\frac{7}{15} - \frac{2}{15}$

Example 7   How much ribbon is left?

Florencia has $1\frac{1}{3}$ yards of ribbon for trim, and she uses $\frac{2}{3}$ yard. How much ribbon does she have left over?

SOLUTION

First convert the mixed number $1\frac{1}{3}$ to the equivalent improper fraction $\frac{4}{3}$. Then subtract the $\frac{2}{3}$ yard she used. Thus,

$$1\frac{1}{3} - \frac{2}{3} = \frac{4}{3} - \frac{2}{3}$$

$$= \frac{2}{3}$$

Florencia has $\frac{2}{3}$ yard of ribbon left over.

PRACTICE PROBLEM FOR EXAMPLE 7

Henry buys $\frac{1}{4}$ gallon of gas for his lawn mower and $\frac{1}{4}$ gallon of gas for his lawn trimmer. How much gas does he buy?

Adding and subtracting fractions with different denominators requires that each fraction be expressed with a common denominator.

SECTION R.2 Operations with Fractions

> **DEFINITION**
>
> **Least Common Denominator** The least common denominator is the smallest number that is a multiple of all the denominators. It is also known as the *least common multiple* of the denominators.

> **What's That Mean?**
>
> **Multiple**
>
> A *multiple* of a given number is a value into which the given number divides evenly. Multiples can also be thought of as the numbers that appear on the "times tables" for a given number. The multiples of 3 are 3, 6, 9, 12, 15, 18, ...

To find the least common denominator and then add or subtract fractions using the least common denominator, use the following steps.

> **Steps to Add or Subtract Fractions with Unlike Denominators**
>
> 1. Factor each denominator into its prime factorization.
> 2. The least common denominator is the product of all the different factors from step 1. If a factor occurs repeatedly (more than once), include the maximum number of times that factor is repeated.
> 3. Rewrite each fraction as an equivalent fraction with the denominator equal to the least common denominator.
> 4. Add or subtract the fractions by adding (subtracting) the numerators.
> 5. Reduce the final answer to lowest terms.

Example 8 Adding and subtracting with unlike denominators

Add or subtract the fractions. Reduce the answer to simplest form.

a. $\dfrac{5}{3} + \dfrac{5}{6}$ **b.** $\dfrac{11}{12} - \dfrac{3}{8}$

SOLUTION

a. Step 1 Factor each denominator into its prime factorization.

$$3 \text{ is prime, and } 6 = 3 \cdot 2.$$

Step 2 The LCD is the product of all the different factors from step 1.

$$\text{The LCD is } 3 \cdot 2 = 6.$$

Step 3 Rewrite each fraction as an equivalent fraction with the denominator equal to the LCD.

$$\dfrac{5}{3} + \dfrac{5}{6} = \dfrac{5 \cdot 2}{3 \cdot 2} + \dfrac{5}{6}$$

Step 4 Add or subtract the fractions by adding (subtracting) the numerators.

$$= \dfrac{10}{6} + \dfrac{5}{6} = \dfrac{15}{6}$$

Step 5 Reduce the final answer to lowest terms.

$$= \dfrac{5 \cdot 3}{2 \cdot 3} = \dfrac{5}{2}$$

b. The factorization of 12 is $12 = 3 \cdot 2 \cdot 2$, and that of 8 is $8 = 2 \cdot 2 \cdot 2$. The LCD is $3 \cdot 2 \cdot 2 \cdot 2 = 24$. Since 2 is a repeated factor, be sure to include 2 the maximum number of times it occurs (which is three).

$$\dfrac{11}{2} - \dfrac{3}{8} \qquad \text{Find a common denominator.}$$

$$= \dfrac{11 \cdot 2}{12 \cdot 2} - \dfrac{3 \cdot 3}{8 \cdot 3} \qquad \text{Rewrite each fraction over the common denominator.}$$

$$= \frac{22}{24} - \frac{9}{24}$$ Subtract the numerators.

$$= \frac{13}{24}$$ Reduce to lowest terms if possible.

Example 9 — How much paint is needed?

Lisa buys $1\frac{1}{2}$ quarts of blue paint for a home painting project and then buys an additional $2\frac{3}{4}$ quarts of yellow paint. How many quarts of paint does she buy?

SOLUTION
To find the total amount of paint Lisa bought, add the two amounts.

$$1\frac{1}{2} + 2\frac{3}{4} =$$ Change mixed numbers to improper fractions.

$$\frac{3}{2} + \frac{11}{4} =$$

$$\frac{3 \cdot 2}{2 \cdot 2} + \frac{11}{4} =$$ Rewrite fractions using common denominators.

$$\frac{6}{4} + \frac{11}{4} =$$

$$\frac{17}{4} = 4\frac{1}{4} \text{ quarts}$$ Add the numerators and reduce the final answer.

Lisa bought a total of $4\frac{1}{4}$ quarts of paint.

PRACTICE PROBLEM FOR EXAMPLE 9

Don bought 20 feet of molding to install in his living room. He used $12\frac{7}{8}$ feet. How much molding did he have left over?

■ Multiplication and Division of Fractions

We now review how to multiply or divide fractions. To do these operations, we need to define the **reciprocal** of a number.

> **DEFINITION**
>
> **Reciprocal** If a is a nonzero number, then $\frac{1}{a}$ is the reciprocal of a.

Informally, to find the reciprocal of a number means to invert, or flip, the fraction form of the number. The reciprocal of $\frac{2}{3}$ is $\frac{3}{2}$.

Skill Connection

The denominator of a whole number

The denominator of a whole number is 1. Therefore, $9 = \frac{9}{1}$.

SECTION R.2 Operations with Fractions   29

Example 10 Finding reciprocals

Find the reciprocal of the given number.

a. Find the reciprocal of -7.

b. Find the reciprocal of $\dfrac{1}{9}$.

c. Find the reciprocal of $-\dfrac{2}{3}$.

SOLUTION

To find a reciprocal, we invert the fraction.

a. To find the reciprocal of -7, recall that $-7 = -\dfrac{7}{1}$. Find the reciprocal, which is $-\dfrac{1}{7}$. Note that the reciprocal has the same sign as the original number.

b. The reciprocal of $\dfrac{1}{9}$ is $\dfrac{9}{1} = 9$.

c. The reciprocal of $-\dfrac{2}{3}$ is $-\dfrac{3}{2}$.

To multiply or divide fractions, we use the following steps.

Steps to Multiply or Divide Fractions

To multiply fractions:
1. Multiply the numerators together and multiply the denominators together (multiply across).
2. Reduce the final answer to lowest terms.

To divide fractions:
1. Invert (or find the reciprocal of) the second fraction. Change the quotient to a product.
2. Multiply the numerators together and multiply the denominators together (multiply across).
3. Reduce the final answer to lowest terms.

Connecting the Concepts

What are the different symbols used?

Multiplication can be written in the following ways:

$$3 \times 5 = 15$$
$$3 \cdot 5 = 15$$
$$(3)(5) = 15$$
$$3(5) = 15$$

In algebra, we do not generally use the form $3 \times 5 = 15$, writing the $\times$ to mean multiplication. Letters will be used for variables, and x is frequently used as a variable. In this book, we will often use the small dot $3 \cdot 5$ or parentheses $3(5)$ to represent multiplication.

Symbols used for division:

The following symbols all represent division.

$$6/3$$
$$3\overline{)6}$$
$$6 \div 3$$
$$\dfrac{6}{3}$$

In this book, we will use the last two forms, $6 \div 3$ or $\dfrac{6}{3}$.

Example 11 Multiplying and dividing fractions

Multiply or divide the fractions. Reduce the answers to simplest form.

a. $\dfrac{2}{3} \cdot \dfrac{6}{5}$ **b.** $\dfrac{4}{7} \div \dfrac{8}{35}$

SOLUTION

a. Step 1 Multiply the numerators together and multiply the denominators together (multiply across).

$$\dfrac{2}{3} \cdot \dfrac{6}{5} = \dfrac{12}{15}$$

Step 2 Reduce the final answer to lowest terms.

$$\frac{2 \cdot 2 \cdot \cancel{3}}{\cancel{3} \cdot 5} = \frac{4}{5}$$

b. Step 1 Invert the second fraction. Change the quotient to a product.

$$\frac{4}{7} \div \frac{8}{35} = \frac{4}{7} \cdot \frac{35}{8}$$

Step 2 Multiply the numerators together and multiply the denominators together (multiply across).

$$= \frac{4 \cdot 35}{7 \cdot 8}$$

Step 3 Reduce the final answer to lowest terms.

$$= \frac{\cancel{4} \cdot \cancel{7} \cdot 5}{\cancel{7} \cdot \cancel{4} \cdot 2}$$

$$= \frac{5}{2}$$

PRACTICE PROBLEM FOR EXAMPLE 11

Multiply or divide as indicated. Reduce the answers to simplest form.

a. $\dfrac{2}{3} \cdot \dfrac{12}{17}$ b. $\dfrac{5}{27} \div \dfrac{20}{9}$

■ Order of Operations

The order-of-operations agreement works the same when fractions are involved. What is different is that the rules for adding, subtracting, multiplying, and dividing fractions must also be used.

Example 12   Following the order-of-operations agreement with fractions

Identify the terms. Then simplify using the order-of-operations agreement.

a. $\dfrac{3}{2} + \dfrac{1}{2} \cdot \dfrac{6}{5}$ b. $\dfrac{4}{5} \div \dfrac{6}{5} - \dfrac{1}{4} + 2$

SOLUTION

a. There are two terms, $\left(\dfrac{3}{2}\right) + \left(\dfrac{1}{2} \cdot \dfrac{6}{5}\right)$. Start by simplifying inside each term.

$\left(\dfrac{3}{2}\right) + \left(\dfrac{1}{2} \cdot \dfrac{6}{5}\right)$ *Multiply the fractions in the second term.*

$= \dfrac{3}{2} + \dfrac{6}{10}$ *The LCD is 10.*

$= \dfrac{3 \cdot 5}{2 \cdot 5} + \dfrac{6}{10}$ *Rewrite each fraction over the LCD.*

$= \dfrac{15}{10} + \dfrac{6}{10}$ *Add the two fractions.*

$= \dfrac{21}{10}$ *This fraction is reduced.*

b. There are three terms, $\left(\dfrac{4}{5} \div \dfrac{6}{5}\right) + \left(-\dfrac{1}{4}\right) + 2$.

$\left(\dfrac{4}{5} \div \dfrac{6}{5}\right) + \left(-\dfrac{1}{4}\right) + 2$ *Simplify inside each term.*

$= \left(\dfrac{4}{5} \cdot \dfrac{5}{6}\right) + \left(-\dfrac{1}{4}\right) + 2$ *To divide fractions, invert the second fraction and multiply.*

$= \dfrac{4}{6} + \left(-\dfrac{1}{4}\right) + 2$ *Reduce the first term.*

$= \dfrac{2}{3} + \left(-\dfrac{1}{4}\right) + \dfrac{2}{1}$ *To add and subtract fractions, find the LCD. The LCD is 12.*

$= \dfrac{2 \cdot 4}{3 \cdot 4} + \left(-\dfrac{1 \cdot 3}{4 \cdot 3}\right) + \dfrac{2 \cdot 12}{1 \cdot 12}$ *Rewrite each fraction over the LCD.*

$= \dfrac{8}{12} + \left(-\dfrac{3}{12}\right) + \dfrac{24}{12}$ *Do additions in order from left to right.*

$= \dfrac{5}{12} + \dfrac{24}{12}$

$= \dfrac{29}{12}$ *This fraction is reduced to lowest terms.*

PRACTICE PROBLEM FOR EXAMPLE 12

Identify the terms. Then simplify, using the order-of-operations agreement.

a. $\dfrac{1}{3} \div \dfrac{5}{6} + \dfrac{5}{2}$ **b.** $7 - \dfrac{2}{5} \cdot \dfrac{15}{4} + \dfrac{1}{3}$

R.2 Vocabulary Exercises

1. _____ numbers are those natural numbers greater than 1 divisible only by themselves and 1.

2. Equivalent fractions represent the same _____.

3. Adding and subtracting fractions requires that the fractions have a(n) _____ denominator.

4. For $a \neq 0$, the _____ of a is $\dfrac{1}{a}$.

5. The _____ is the smallest number that is a multiple of all the denominators in a set of fractions.

6. The _____ is the product of the common prime factors of a set of numbers.

7. The number on the top of a fraction is called the _____, and the number on the bottom of a fraction is called the _____.

R.2 Exercises

For Exercises 1 through 10, find the prime factorization of each given number.

1. 16
2. 15
3. 34
4. 75
5. 42
6. 18
7. 64
8. 60
9. 70
10. 96

For Exercises 11 through 20, find the GCF for each set of numbers.

11. 12 and 15
12. 12 and 20
13. 60 and 15
14. 20 and 36
15. 9, 27, and 18
16. 21, 49, and 14
17. 48, 36, and 60
18. 32, 64, and 16
19. 51 and 27
20. 72 and 18

For Exercises 21 through 30, reduce each fraction to lowest terms.

21. $\dfrac{6}{15}$
22. $\dfrac{20}{25}$
23. $\dfrac{8}{42}$
24. $\dfrac{9}{33}$
25. $\dfrac{16}{48}$
26. $\dfrac{12}{60}$
27. $-\dfrac{9}{108}$
28. $-\dfrac{22}{121}$
29. $-5\dfrac{6}{21}$
30. $-9\dfrac{20}{36}$

For Exercises 31 through 40, rewrite each fraction as an equivalent fraction with the given denominator.

31. $\dfrac{2}{3} = \dfrac{?}{15}$
32. $\dfrac{4}{5} = \dfrac{?}{30}$
33. $\dfrac{5}{7} = \dfrac{?}{56}$
34. $\dfrac{3}{8} = \dfrac{?}{32}$
35. $-\dfrac{5}{16} = \dfrac{?}{32}$
36. $-\dfrac{2}{15} = \dfrac{?}{45}$
37. $4\dfrac{2}{9} = 4\dfrac{?}{45}$
38. $6\dfrac{7}{8} = 6\dfrac{?}{40}$
39. $5 = \dfrac{?}{3}$
40. $7 = \dfrac{?}{4}$

For Exercises 41 through 50, graph each value on the following number line.

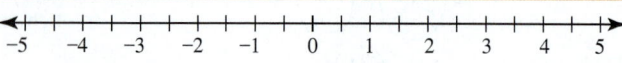

41. $\dfrac{3}{2}$
42. $\dfrac{5}{2}$
43. -3
44. -5
45. $-\dfrac{5}{2}$
46. $-\dfrac{7}{2}$
47. $\dfrac{1}{4}$
48. $\dfrac{5}{4}$
49. $-\dfrac{9}{4}$
50. $-\dfrac{7}{4}$

For Exercises 51 through 60, add or subtract the fractions with like denominators without a calculator. Reduce to simplest form.

51. $\dfrac{2}{3} + \dfrac{4}{3}$
52. $\dfrac{3}{8} + \dfrac{5}{8}$
53. $\dfrac{7}{5} - \dfrac{4}{5}$
54. $\dfrac{11}{14} - \dfrac{3}{14}$
55. $\dfrac{11}{15} - \dfrac{2}{15}$
56. $\dfrac{13}{24} + \dfrac{5}{24}$
57. $1\dfrac{1}{4} + 3\dfrac{3}{4}$
58. $4\dfrac{1}{5} + 7\dfrac{2}{5}$
59. $5\dfrac{2}{3} - 2\dfrac{1}{3}$
60. $8\dfrac{1}{6} - 3\dfrac{5}{6}$

For Exercises 61 through 70, add or subtract as indicated without a calculator. Reduce to simplest form.

61. $\dfrac{2}{3} + \dfrac{5}{6}$
62. $\dfrac{1}{4} + \dfrac{3}{8}$
63. $-\dfrac{4}{8} + \dfrac{1}{3}$
64. $-\dfrac{5}{6} + \dfrac{1}{4}$
65. $-\dfrac{2}{3} - \dfrac{3}{4}$
66. $-\dfrac{1}{5} - \dfrac{7}{10}$
67. $1\dfrac{1}{6} + 3\dfrac{2}{3}$
68. $2\dfrac{1}{3} + 1\dfrac{1}{2}$
69. $-2\dfrac{1}{4} - 3\dfrac{1}{2}$
70. $-1\dfrac{1}{5} - 3\dfrac{1}{2}$

For Exercises 71 through 76, add or subtract as indicated.

71. Sandy needs $2\frac{1}{4}$ gallons of ivory paint and $1\frac{1}{4}$ gallons of blue paint for a remodeling project. How many gallons of paint in total does she need?

72. James needs $\frac{2}{3}$ cup milk for one recipe and an additional $1\frac{1}{2}$ cups of milk for another recipe. How much milk does James need in all?

73. Dominique uses $\frac{1}{4}$ teaspoon of salt in one dish and $\frac{1}{2}$ teaspoon of salt in another dish. How much salt does Dominique use in all?

74. A carpenter needs $7\frac{3}{4}$ feet of molding for each side of a doorway and $3\frac{1}{4}$ feet for the top of the doorway. How many feet of molding does he need in all?

75. Ted needs $10\frac{3}{8}$ feet of lumber for a project, and he has 20 feet of lumber. How much will he have left over after his project?

76. Hanna needs $2\frac{1}{2}$ pounds of onions for a stew and $\frac{3}{4}$ pound of onions for soup. If she bought a 5-pound bag of onions, how many pounds will she have left over?

For Exercises 77 through 86, find each reciprocal.

77. -3
78. -5
79. $\frac{5}{3}$
80. $\frac{2}{7}$
81. $\frac{1}{9}$
82. $\frac{1}{12}$
83. 8
84. -2
85. 0
86. 1

For Exercises 87 through 98, multiply or divide as indicated. Reduce to simplest form.

87. $\frac{1}{4} \cdot 8$
88. $\frac{2}{3} \cdot 6$
89. $-\frac{3}{4} \cdot \left(-\frac{5}{9}\right)$
90. $-\frac{1}{6} \cdot \left(-\frac{12}{5}\right)$
91. $\frac{2}{3} \div \frac{14}{27}$
92. $\frac{3}{4} \div \frac{9}{16}$
93. $-100 \div (-20)$
94. $(-36) \div (-12)$
95. $\frac{4}{15} \div \left(-\frac{8}{45}\right)$
96. $-\frac{2}{5} \div \frac{4}{9}$
97. $-\frac{2}{9} \div (-12)$
98. $-\frac{3}{5} \div (15)$

For Exercises 99 through 108, identify the number of terms. Then simplify using the order-of-operations agreement.

99. $24 \div (-3) \cdot 2 + \frac{3}{2}$
100. $32 \cdot 2 \div (-8) + \frac{1}{3}$
101. $3 - \frac{1}{3} \div \frac{5}{6}$
102. $\frac{2}{5} \div \frac{7}{10} - 4$
103. $\frac{3}{8} \cdot \frac{4}{5} + \frac{1}{2} - 1$
104. $\frac{4}{15} \cdot \frac{3}{2} - \frac{1}{10} + 6$
105. $\frac{5}{8} \cdot \frac{4}{15} + \frac{1}{2} \cdot \frac{3}{2} + 2$
106. $\frac{3}{4} \cdot \frac{5}{9} + \frac{2}{3} \cdot \frac{7}{8} - 3$
107. $\frac{4}{9} \div \frac{5}{18} \cdot \frac{3}{2}$
108. $\frac{4}{7} \div \frac{3}{14} \cdot \frac{9}{16}$

R.3 Operations with Decimals

LEARNING OBJECTIVES
- Find the place value.
- Show the relationship between fractions and decimals.
- Graph decimals on a number line.
- Add and subtract decimals.
- Multiply and divide decimals.
- Use the order-of-operations agreement with decimals.
- Round decimals to a given place value.

■ Place Value

The system that we use to write numbers is called the *Hindu–Arabic* number system. It was developed in the arc of countries starting with India to the Middle East to Egypt to North Africa. This number system has base 10. It is an efficient way to write numbers because numbers are written using **place value.** This means that each place held by a number represents a multiple of ten. For example, the number 842 can be written as "8 hundreds, 4 tens, and 2 ones." The number 8 is in the hundreds place.

Many numbers involve decimals. For example, the average temperature of humans is 98.6°F. When a number is written with decimals, such as 345.1267, the part to the right of the decimal point also has place value. Some common place values are represented in the table at the right. Notice when the ending contains a *ths*, it represents a place value to the right of the decimal point.

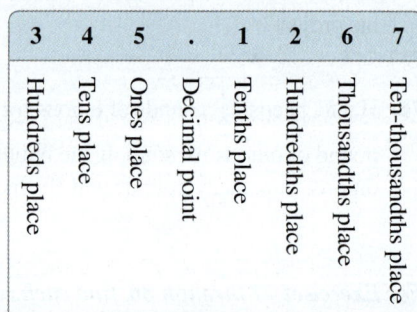

Example 1 Identifying place value

Identify the digit in each number that has the given place value.

a. The hundredths place: 6.3257

b. The thousandths place: 1.9108

SOLUTION

a. The digit in the hundredths place is 2. 6.3257

b. The digit in the thousandths place is 0. 1.9108

PRACTICE PROBLEM FOR EXAMPLE 1

Identify the digit in each number that has the given place value.

a. The tenths place: 897.351

b. The ten-thousandths place: 0.0025

■ Relationships between Fractions and Decimals

Finite and infinite repeating decimals may be written as fractions. To write a finite decimal as a fraction, identify the place value of the last nonzero digit. For example, the decimal 0.25 is read as "25 hundredths." To write 0.25 as a fraction, write it as $0.25 = \frac{25}{100} = \frac{1}{4}$. Recall that fractions should always be written in simplest form.

Example 2 — Writing a decimal in fraction form

Write each decimal in fraction form. Reduce all fractions to simplest form.

a. 1.6   **b.** 0.042   **c.** 2.67

SOLUTION

a. The number 1.6 is written in words as "one and 6 tenths." Writing 1.6 in fraction form gives

$$1.6 = 1\frac{6}{10}$$ Write the decimal part in fraction form.

$$= 1\frac{3}{5}$$ Reduce to lowest terms.

b. The number 0.042 is written in words as "forty-two thousandths." Writing 0.042 in fraction form gives

$$0.042 = \frac{42}{1000}$$ Write the decimal part in fraction form.

$$= \frac{42 \div 2}{1000 \div 2}$$ Reduce to lowest terms.

$$= \frac{21}{500}$$

c. The number 2.67 is written in words as "two and sixty-seven hundredths." Writing 2.67 in fraction form gives

$$2.67 = 2\frac{67}{100}$$ Write the decimal part in fraction form.

$$= 2\frac{67}{100}$$ This fraction is reduced.

■ Connecting the Concepts

What kind of decimal?

Decimals are classified as *finite* or *infinite*. A finite decimal has a number of nonzero decimal digits that come to an end. An infinite decimal has an unlimited number of nonzero decimal digits. Infinite decimals may be either repeating or nonrepeating. An example of a repeating infinite decimal is 0.333.... Here the three dots indicate the 3's continue on forever. In an infinite nonrepeating decimal, the decimal digits never settle into a pattern of repetition.

PRACTICE PROBLEM FOR EXAMPLE 2

Write each decimal in fraction form. Reduce all fractions to simplest form.

a. 2.46   **b.** 1.025   **c.** 10.9

36 CHAPTER R Review of Prealgebra

To write a fraction in decimal form, use long division. Remember that the fraction bar in $\frac{a}{b}$ indicates division. So $\frac{a}{b} = a \div b$. In a long division problem, all of the parts making up the problem have special names.

$$\frac{\text{dividend}}{\text{divisor}} = \text{quotient} + \frac{\text{remainder}}{\text{divisor}}$$

When writing a long division problem, recall that

$$\frac{\text{dividend}}{\text{divisor}} = \text{dividend} \div \text{divisor}$$

$$= \text{divisor})\overline{\text{dividend}}^{\text{quotient}}$$

Example 3 Writing a fraction as a decimal

Write each fraction in decimal form.

a. $\frac{4}{5}$ b. $\frac{3}{8}$ c. $\frac{1}{3}$

SOLUTION

a. To convert $\frac{4}{5}$ to decimal form, use long division. $\frac{4}{5}$ means $5)\overline{4}$. Since 4 is less than 5, add a decimal point and continue dividing.

$$\begin{array}{r} 0.8 \\ 5)\overline{4.0} \\ -4\,0 \\ \hline 0 \end{array}$$

Therefore, $\frac{4}{5} = 0.8$.

b. To convert $\frac{3}{8}$ to decimal form, use long division. $\frac{3}{8}$ means $8)\overline{3}$. Add a decimal point and continue dividing.

$$\begin{array}{r} 0.375 \\ 8)\overline{3.000} \\ -24 \\ \hline 60 \\ -56 \\ \hline 40 \\ -40 \\ \hline 0 \end{array}$$

Therefore, $\frac{3}{8} = 0.375$.

c. To convert $\frac{1}{3}$ to decimal form, use long division. $\frac{1}{3}$ means $3)\overline{1}$.

$$\begin{array}{r} 0.333\ldots \\ 3)\overline{1.000} \\ -9 \\ \hline 10 \\ -9 \\ \hline 10 \\ -9 \\ \hline 10 \end{array}$$

■ **Connecting the Concepts**

How is a repeating, infinite decimal written?

To show that a decimal repeats infinitely, we put a bar over the repeating part. Therefore, $0.333\ldots = 0.\overline{3}$, where the bar over the 3 indicates the 3's continue on forever. The infinite repeating decimal $0.125125125\ldots$ would be written as $0.\overline{125}$.

To show the decimal expansion continues in this repeating pattern, write
0.333 ... = 0.$\bar{3}$.

Therefore, $\frac{1}{3} = 0.\bar{3}$.

PRACTICE PROBLEM FOR EXAMPLE 3

Write each fraction in decimal form.

a. $\frac{2}{5}$
b. $\frac{3}{4}$
c. $\frac{1}{9}$

■ Graphing Decimals on a Number Line

To graph decimals on a number line, determine the scale of the number line. Then graph the number, using a dot to mark the location on the number line.

Example 4 Graphing the decimals on a number line

Answer each of the following questions.

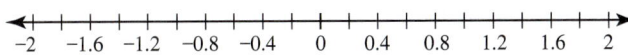

a. What is the scale of this number line?

b. Graph the numbers 1.6, −0.4, 0.7, and −1.5.

SOLUTION

a. The scale on this number line is 0.2 because that is how far the tick marks are apart.

b. 0.7 lies halfway between 0.6 and 0.8. The numbers 1.6 and −0.4 are located directly on this number line, so mark their locations with a dot. 0.7 lies halfway between 0.6 and 0.8. −1.5 lies halfway between −1.6 and −1.4. Then also mark the locations of 0.7 and −1.5 with dots.

−1.5 lies halfway between −1.4 and −1.6.

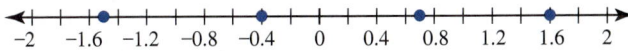

PRACTICE PROBLEM FOR EXAMPLE 4

Answer each of the following questions.

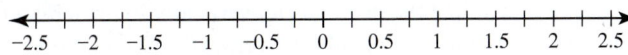

a. What is the scale of this number line?

b. Graph the numbers 1.4, −2.25, −0.2, and 1.9.

■ Rounding Decimals

Decimals are sometimes rounded to a given place value. Some people when grocery shopping keep a running tally in their head to estimate how much money they are spending. Instead of remembering that an item costs $2.99, many people will round that amount to $3.00.

To round a decimal, use the following steps.

> **Steps to Round a Decimal to a Given Place Value**
> 1. Underline the given place value in the number.
> 2. Look at the digit directly to the right of the underlined place.
>   a. If that digit is 0, 1, 2, 3, or 4, then delete all digits to the right of the underlined place. This is often called *rounding down*.
>   b. If that digit is 5, 6, 7, 8, or 9, then add 1 to the underlined place and delete all the digits to the right. This is often called *rounding up*.

Example 5 Rounding decimals

Round each decimal to the given place value.

a. 1.69 to the tenths place

b. 11.05 to the ones place

c. 0.3691 to the thousandths place

d. 16.895 to the hundredths place

SOLUTION

a. Step 1 Underline the tenths place in the number.

$$1.6\underline{6}9$$

Step 2 The next digit to the right is 9. Round up.

$$1.7$$

b. 1<u>1</u>.05   Underline the ones place. The next digit to the right is 0.
     11   Round down.

c. 0.36<u>9</u>1   Underline the thousandths place. The next digit to the right is 1.
     0.369   Round down.

d. 16.8<u>9</u>5   Underline the hundredths place. The next digit to the right is 5.
     16.90   Round up.

■ Addition and Subtraction of Decimals

To add or subtract decimals, line the decimals up vertically by matching place values. Starting with the rightmost digit, add or subtract vertically. Carry or borrow as needed.

Example 6 Adding and subtracting decimals

a. Add the decimals: $1.367 + 4.241$.

b. Subtract the decimals: $8.52 - 6.04$.

c. Emma bought 1 gallon of milk at $3.28 and a dozen eggs for $1.50. How much money did she spend?

SOLUTION

a. Write the problem vertically, carefully matching place values, and add.

$$\begin{array}{r} \overset{1}{1.3}67 \\ +4.241 \\ \hline 5.608 \end{array}$$

b. Write the problem vertically, carefully matching place values, and subtract.

$$\begin{array}{r} 8.\overset{4}{\cancel{5}}\overset{1}{2} \\ -6.04 \\ \hline 2.48 \end{array}$$

c. Add the amounts Emma spent on milk and eggs to get the total.

$$\begin{array}{r} 3.28 \\ +1.50 \\ \hline 4.78 \end{array}$$

Emma spent $4.78 on milk and eggs.

PRACTICE PROBLEM FOR EXAMPLE 6

a. Add the decimals: $4.891 + 6.099$.

b. Subtract the decimals: $9.73 - 2.57$.

c. Laurence bought a bottle of water for $1.69 and a box of cheese-flavored crackers for $0.99. He paid with a $10.00 bill. How much change did he receive?

■ Multiplication and Division of Decimals

When two decimals are multiplied, ignore the decimal points and multiply as if they were two integers. After that step, count the number of digits after the decimal point in both factors (numbers). Starting at the far right of the result, move the decimal point left the same number of places as the number of digits in both decimals.

Example 7 Multiplying decimals

a. Multiply: $2.35(1.2)$.

b. If you purchased two apps for $0.98 each, how much money did you spend?

SOLUTION

a.

$$\begin{array}{r} \overset{1}{2.3}5 \\ \times 1.2 \\ \hline \overset{1}{4}70 \\ +235 \\ \hline 2820 \end{array}$$

Ignore the decimal places.
Multiply as if two integers.

The numbers 2.3**5** and 1.**2** have a total of **three** digits after the decimal point. Move the decimal point from the right over **three** digits to get the solution.

2.820

b. You spent $2(0.98)$, or

$$\begin{array}{r} 0.98 \\ \times 2 \\ \hline 196 \end{array}$$

Ignore the decimal places.
Multiply as if two integers.

The numbers 0.98 and 2 have a total of *two* digits after the decimal point. Move the decimal point from the right over *two* digits to get the solution.

1.96

You spent a total of $1.96 on the two apps.

PRACTICE PROBLEM FOR EXAMPLE 7

a. Multiply: $4.21 \cdot 2.3$.

b. If you purchased three shirts for $12.99 each, how much money did you spend?

To divide decimals, use long division. Before starting long division, move the decimal point in the divisor (on the outside) all the way over to the right. Move the decimal point in the dividend (inside) the same number of places to the right. Then begin the long division process.

Example 8 Using long division of decimals

a. Divide: $4.25 \div 2.5$. Round to two decimal places if necessary.

b. Divide: $23.56 \div 2.4$. Round to two decimal places if necessary.

SOLUTION

a. $4.25 \div 2.5$ Write in long division form.

$2.5 \overline{)4.25}$ Move the decimal point in 2.5 and 4.25 one place to the right.

$$\begin{array}{r} 1.7 \\ 25\overline{)42.5} \\ -25 \\ \hline 175 \\ -175 \\ \hline 0 \end{array}$$ Perform long division.

The remainder is 0.

Since the remainder was 0, the numbers divided exactly. There is no need to round the answer. So $4.25 \div 2.5 = 1.7$.

b. $23.56 \div 2.4$ Write in long division form.

$2.4 \overline{)23.56}$ Move the decimal point in 2.4 and 23.56 one place to the right.

$$\begin{array}{r} 9.8166... \\ 24\overline{)235.600} \\ -216 \\ \hline 196 \\ -192 \\ \hline 40 \\ -24 \\ \hline 160 \\ -144 \\ \hline 160 \\ -144 \\ \hline 16 \end{array}$$ Perform long division.

The pattern will continue to repeat.

The quotient is 9.8166 . . . . Rounding to two decimal places gives the solution 9.82.

PRACTICE PROBLEM FOR EXAMPLE 8

a. Divide: 6.24 ÷ 2.6. Round to two decimal places if necessary.

b. Divide: 30.65 ÷ 1.6. Round to two decimal places if necessary.

■ Order-of-Operations with Decimals

The order-of-operations agreement is used the same way when decimals are involved. First identify the terms. Then simplify inside each term. Finally add and subtract in order, working from left to right.

Example 9 Applying the order-of-operations agreement to decimals

Identify the terms. Then simplify using the order-of-operations agreement.

a. $1.5 + 2.6 \cdot 8.4$ b. $7 \div 3.5 - 0.67 + 12$

SOLUTION

a. To simplify $1.5 + 2.6 \cdot 8.4$, first identify the terms. This expression has two terms.

$$(1.5) + (2.6 \cdot 8.4)$$

Then simplify the inside of each term and add.

$(1.5) + (2.6 \cdot 8.4)$ *Do the multiplication in the second term.*
$= 1.5 + 21.84$ *Add the two terms.*
$= 23.34$

b. Identify the terms. This expression has three terms.

$$(7 \div 3.5) + (-0.67) + (12)$$

Simplify the inside of each term and then add and subtract in order from left to right.

$(7 \div 3.5) + (-0.67) + (12)$ *Do the division in the first term.*
$= 2 + (-0.67) + 12$ *Do the additions, working left to right.*
$= 1.33 + 12$
$= 13.33$

PRACTICE PROBLEM FOR EXAMPLE 9

Identify the terms. Then simplify using the order-of-operations agreement.

a. $3.2 + 4.1 \cdot 2.5$ b. $6 \div 1.5 - 3.1 + 4$

R.3 Vocabulary Exercises

1. The digit 8 in the number 4.7981 is in the _____ place.

2. When rounding 5.67 to the tenths place, we round _____.

3. When rounding 5.61 to the tenths place, we round _____.

R.3 Exercises

For Exercises 1 through 10, identify the digit in each number that has the given place value.

1. The tenths place: 0.4732
2. The tenths place: 1.0943
3. The ten-thousandths place: 6.9154
4. The ten-thousandths place: 4.61507
5. The ones place: 7.25
6. The ones place: 8.91
7. The hundredths place: 6.2801
8. The hundredths place: 0.00541
9. The thousandths place: 8.9504
10. The thousandths place: 3.2974

For Exercises 11 through 20, write each decimal as a fraction.

11. 2.9
12. 3.7
13. 0.84
14. 0.42
15. 7.71
16. 5.23
17. 1.008
18. 2.014
19. 1.33
20. 2.67

For Exercises 21 through 30, write each fraction as a decimal. Write any repeating decimals with a bar over the repeating part.

21. $\frac{3}{5}$
22. $\frac{1}{5}$
23. $1\frac{1}{4}$
24. $3\frac{3}{4}$
25. $\frac{5}{12}$
26. $\frac{7}{12}$
27. $3\frac{1}{6}$
28. $4\frac{2}{3}$
29. $\frac{7}{8}$
30. $\frac{1}{8}$

For Exercises 31 through 40, graph each decimal on a number line like that below.

31. 0.2
32. 0.8
33. 1.7
34. 1.4
35. −0.9
36. −1.3
37. 1.3
38. 1.5
39. −1.6
40. −2.0

For Exercises 41 through 54, round each decimal to the given place value.

41. 8.25 to the tenths place
42. 6.27 to the tenths place
43. 9.207 to the hundredths place
44. 13.029 to the hundredths place
45. 205.69 to the tens place
46. 127.995 to the tens place
47. 0.12999 to the tenths place
48. 5.34997 to the tenths place
49. 0.99 to the ones place
50. 4.62 to the ones place
51. 4.32601 to the thousandths place
52. 9.99219 to the thousandths place
53. 6.51757 to the ten-thousandths place
54. 4.593217 to the ten-thousandths place

For Exercises 55 through 64, add or subtract the decimals as indicated without a calculator.

55. $2.75 + 3.01$
56. $6.14 + 9.73$
57. $8.68 - 1.39$
58. $10.701 - 3.334$
59. $0.2275 + 96.371$
60. $9.017 + 16.9027$
61. $9 - 3.67$
62. $16 - 2.994$
63. $9.75 - 3.001$
64. $16.2 - 5.447$

For Exercises 65 through 68, use the information in each exercise to answer the questions.

65. Michelle purchased textbooks for her two online courses. One book cost $149.95, and the other book cost $124.95. How much did the two books cost Michelle?

66. Jorge purchased textbooks for his math and reading courses. One book cost $120.95, and the other book cost $95.99. How much did the two books cost Jorge?

67. Rani purchased a soda for $4.00 and popcorn for $6.50. She paid with a $20 bill. How much change did she receive?

68. Ransi purchased a binder for $5.50 and a mechanical pencil for $2.50. He paid with a $20 bill. How much change did he receive?

For Exercises 69 through 74, multiply without a calculator.

69. $1.8 \cdot 3.01$
70. $2.01 \cdot 8.3$
71. $(3.5)(6.07)$
72. $(5.7)(1.24)$
73. $7(0.016)$
74. $(5.047)(8)$

SECTION R.3 Operations with Decimals 43

For Exercises 75 through 78, use the information given in each exercise to answer the questions.

75. If you purchased three apps for $0.99 apiece, how much money did you spend?

76. If you purchased six apps for $0.99 apiece, how much money did you spend?

77. Sam bought three candy bars on sale for $0.65 apiece. How much money did she spend?

78. Brogan bought two blouses for $23.75 apiece. How much did she spend?

For Exercises 79 through 86, divide the decimals without a calculator.

79. $8.2 \div 0.01$
80. $6.95 \div 0.01$
81. $0.12 \div 2$
82. $0.81 \div 9$
83. $19.32 \div 2.3$
84. $57.34 \div 6.1$
85. $0.01073 \div 0.29$
86. $0.09144 \div 0.36$

For Exercises 87 through 96, identify the number of terms. Then use the order-of-operations agreement to simplify each expression.

87. $4.2 + 1.5 \cdot 6.1$
88. $9.7 \cdot 2.3 + 7.1$
89. $4(8.95) - 16.23$
90. $7(1.02) - 21.54$
91. $4.6 + 8 \div 2.5 - 22.1$
92. $-3.7 + 10 \div 1.6 + 8.1$
93. $-12.3 + 5 \div 0.1 + 9.72$
94. $18.7 + 28 \div 0.1 - 14.25$
95. $24 \div 1.5 \cdot 8.4$
96. $20 \div 0.25 \cdot 0.61$

For Exercises 97 through 100, use the information given in each exercise to answer the questions.

97. Luc purchased one soda for $1.75, a bag of crackers for $0.99, and two packages of gum that cost $2.98 each.

   a. Write an expression that gives the *total* cost of Luc's purchase. Do not simplify.

   b. What two operations (addition, subtraction, multiplication, or division) are there in the expression?

   c. How many terms are in the expression?

   d. Which operation should be done first? Which operation should be done second?

   e. Find the total amount of Luc's purchase.

98. At the movies, Van purchased two sodas for $4.00 each, a bag of candy for $5.25, and two small buttered popcorns that cost $5.00 each.

   a. Write an expression that gives the *total* cost of Van's purchase. Do not simplify.

   b. What two operations (addition, subtraction, multiplication, or division) are there in the expression?

   c. How many terms are in the expression?

   d. Which operation should be done first? Which should be done second?

   e. Find the total amount of Van's purchase.

99. At the local taco shop, Kym purchased two sodas for $1.59 each and two taco meal deals for $3.99 each.

   a. Write an expression that gives the *total* cost of Kym's purchase. Do not simplify.

   b. What two operations (addition, subtraction, multiplication, or division) are there in the expression?

   c. How many terms are in the expression?

   d. Which operation should be done first? Which operation should be done second?

   e. Find the total amount of Kym's purchase.

100. Kwan purchased 5 gallons of gas for $3.95 a gallon and two bottles of water for $2.75 each.

   a. Write an expression that gives the *total* cost of Kwan's purchase. Do not simplify.

   b. What two operations (addition, subtraction, multiplication, or division) do you see in your expression?

   c. How many terms are in your expression?

   d. Which operation should be done first? Which operation should be done second?

   e. Find the total amount of Kwan's purchase.

R.4 Operations with Percents

LEARNING OBJECTIVES
- Interpret a percent.
- Convert between percents, decimals, and fractions.
- Solve problems involving percents.
- Find percent increase and decrease.

What Is a Percent?

Percent means "a part of hundred" or "per 100." *Per* in mathematics means to divide. So 1% means one part of 100, or $1\% = \dfrac{1}{100}$. The symbol % is the *percent sign*.

> **DEFINITION**
> **Percent** Percent means "per 100."
> $$1\% = \dfrac{1}{100}$$

Converting between Percents, Fractions, and Decimals

In the definition for percent, we can see a relationship between percents and fractions. Because $1\% = \dfrac{1}{100}$, we have that $p\% = \dfrac{p}{100}$. So $19\% = \dfrac{19}{100}$. Likewise, $25\% = \dfrac{25}{100} = \dfrac{1}{4}$. Always reduce fractions to lowest terms.

Example 1 Converting between percents and fractions

Convert each quantity as indicated. Write all fractions in lowest terms.

a. Write as a fraction: 40%.

b. Write as a percent: $\dfrac{9}{100}$.

c. Write as a percent: $\dfrac{3}{20}$.

SOLUTION

a.
$$40\% = \dfrac{40}{100} \quad \text{Rewrite as a fraction.}$$
$$= \dfrac{40 \div 20}{100 \div 20} \quad \text{Reduce fractions to lowest terms.}$$
$$= \dfrac{2}{5}$$

b. $\dfrac{9}{100} = 9\%$

c. This fraction is not written with a denominator of 100, so we cannot immediately write it as a percent. First rewrite the fraction with a denominator of 100. Then write it as a percent.

$$\frac{3}{20} = \frac{?}{100} \quad \text{Rewrite with a denominator of 100.}$$

$$\frac{3 \cdot 5}{20 \cdot 5} = \frac{15}{100}$$

$$\frac{3}{20} = 15\% \quad \text{Write as a percent.}$$

PRACTICE PROBLEM FOR EXAMPLE 1

Convert each quantity as indicated. Write all fractions in lowest terms.

a. Write as a fraction: 55%. b. Write as a percent: $\frac{84}{100}$. c. Write as a percent: $\frac{1}{5}$.

To Convert between Percents and Decimals

- **To convert a percent to a decimal:** Move the decimal point two places to the left and remove the % sign. This is the same as multiplying by $1\% = \frac{1}{100} = 0.01$ or dividing by 100.
- **To convert a decimal to a percent:** Move the decimal point two places to the right and add the % sign. This is the same as multiplying by 100.

Example 2 Converting between percents and decimals

Convert each quantity as indicated.

a. Write as a decimal: 18%. b. Write as a decimal: 0.5%.
c. Write as a percent: 0.125. d. Write as a percent: 4.01.

SOLUTION

a. Move the decimal point two places to the left. 18% = 0.18
b. Move the decimal point two places to the left. 0.5% = 0.005
c. Move the decimal point two places to the right and add the percent sign. 0.125 = 12.5%
d. Move the decimal point two places to the right and add the percent sign. 4.01 = 401%

PRACTICE PROBLEM FOR EXAMPLE 2

a. Write as a decimal: 8.25%. b. Write as a decimal: 0.36%.
c. Write as a percent: 0.22. d. Write as a percent: 110.

CHAPTER R Review of Prealgebra

Example 3 Converting between percents, decimals, and fractions

Complete the table by filling in the missing values.

Percent	Decimal	Fraction
	0.35	

SOLUTION

To convert the decimal 0.35 to a percent, move the decimal point two places to the right and add the percent sign. $0.35 = 35\%$.

To convert 0.35 to a fraction, first figure out the place value of the last digit, 5. The place value of the last digit, 5, is the hundredths place. Rewrite 0.35 as a fraction with a denominator of 100.

$$0.35 = \frac{35}{100} \qquad \text{Write the decimal part in fraction form.}$$

$$= \frac{35 \div 5}{100 \div 5} \qquad \text{Reduce to lowest terms.}$$

$$= \frac{7}{20}$$

So the completed table is as follows:

Percent	Decimal	Fraction
35%	0.35	$\frac{7}{20}$

■ Problem Solving with Percents

Many quantities in life are listed as percents, such as interest rates (APR of 5.25%) and discounts at sales (take 20 percent off). When calculating with percents in a problem, first convert the percents into decimals.

■ What's That Mean?

Of

In a mathematics problem, what the word *of* means depends on what comes before it. When a percent or fraction comes before the word *of* it translates as multiplication.

$$20\% \text{ of } 60 = 20\%(60)$$
$$\frac{1}{3} \text{ of } 90 = \frac{1}{3}(90)$$

If the word *of* follows another operation, it means that same operation.

 The sum of . . . = addition.
 The quotient of . . . = division.
 The product of . . . = multiplication.

Example 4 Calculating with percents

a. Find 25% of 42.

b. 55% of the students in a math class are female. There are 40 students in the class. How many are female?

c. You purchase an iPad for $499. Find the tax you paid if the sales tax rate in your area is 8%.

SOLUTION

a. First convert 25% to its decimal form. $25\% = 0.25$. "A percent of" means to multiply, so using the decimal form, calculate the answer.

$$25\% \cdot 42 = 0.25(42)$$
$$= 10.5$$

Therefore, 25% of 42 is 10.5.

b. This problem asks us to find 55% of 40.

$$55\% \cdot 40 = 0.55(40)$$
$$= 22$$

There are 22 female students in the class.

c. To find the tax, we multiply the price by the tax rate.

$$8\% \cdot 499 = 0.08(499)$$
$$= 39.92$$

The sales tax paid on the iPad was $39.92.

PRACTICE PROBLEM FOR EXAMPLE 4

a. Find 8% of 95.

b. 30% of the students in a woodworking class are female. There are 50 students in the class. How many are female?

c. You purchase a 65-inch curved 4K SUHD LED TV for $1600. Find the tax paid if the sales tax rate in your area is 4.5%.

■ Finding Percent Increase and Decrease

Suppose a house increases in value from $225,000 to $250,000. The house has increased in value by $25,000. We are often interested in describing that increase in value as a percent. This percent is called the *percent increase*. To do this, we use the following formula:

$$\text{percent increase} = \frac{\text{increase}}{\text{original amount}}$$

Example 5

A house increases in value from $225,000 to $250,000. What is the percent increase for the house?

SOLUTION

The original value of the house is $225,000. The increase in value of the house is $25,000. We don't know the percent increase yet.

Our formula is:

$$\text{percent increase} = \frac{\text{increase}}{\text{original amount}}$$

$$\text{percent increase} = \frac{25000}{225000}$$

$$\text{percent increase} \approx 0.11111$$

So, written as a percent, *percent increase* ≈ 11.1%.

The house increased in value by approximately 11.1%.

PRACTICE PROBLEM FOR EXAMPLE 5

A condo increases in value from $250,000 to $300,000. What is the percent increase for the condo?

In the retail world, the increase is called the *markup*, and the percent increase is called the *markup rate*. Markup problems are solved in a similar way to percent increase problems.

$$\text{markup} = (\text{markup rate}) \cdot (\text{original amount})$$

Example 6

A 0.5 carat diamond and platinum wedding band costs a jewelry store $531.42. The markup rate on the ring is 75%.

a. How much is the markup?

b. What is the retail price (that is, the price the wedding band will be sold for)?

SOLUTION

The original amount is the cost to the store, which is $531.42. The markup rate is 75% = 0.75. Substituting these two values into the markup formula yields

$$\text{markup} = (\text{markup rate}) \cdot (\text{original amount})$$
$$\text{markup} = (0.75)(531.42)$$
$$\text{markup} = 398.565$$

Since we are discussing money, we round the markup to two decimal places to get $398.57. The retail price is the total of the original amount and the markup.

$$\text{retail price} = \text{markup} + \text{original amount}$$
$$\text{retail price} = 398.57 + 531.42$$
$$\text{retail price} = 929.99$$

The 0.5 carat diamond and platinum wedding band retails for $929.99.

PRACTICE PROBLEM FOR EXAMPLE 6

The markup rate on cell phones is about 65%. Suppose a cell phone costs $200.00 to produce.

a. How much is the markup? **b.** What will the retail price be?

We can also use similar reasoning to find the percent decrease of a quantity. The formula becomes

$$\text{decrease} = (\text{percent decrease}) \cdot (\text{original amount})$$

In the retail world, the *decrease* is called the discount. The percent decrease is called the *markdown*.

What's That Mean?

Depreciation

Depreciation is a term used to describe the loss of value of a car (or other item) over time.

Example 7

The average depreciation rate for a new car is 19% in the first year. Suppose a new car is purchased for $27,950.

a. How much value will it lose in the first year?

b. How much is the car worth at the end of the first year?

SOLUTION

a. The original amount is the cost of the new car, which is $27,950. The percent decrease in the first year is 19%, or 0.19. Substituting these values into the formula yields

$$\text{decrease} = (\text{percent decrease}) \cdot (\text{original amount})$$
$$\text{decrease} = (0.19)(27950)$$
$$\text{decrease} = 5310.50$$

A new car purchased for $27,950 will decrease in value $5310.50 in the first year.

b. To find the value of the car at the end of the first year, we subtract the decrease from the original amount of the car.

$$\text{value} = \text{original amount} - \text{decrease}$$
$$\text{value} = 27950 - 5310.50$$
$$\text{value} = 22639.50$$

The new car is worth $22,639.50 at the end of the first year.

PRACTICE PROBLEM FOR EXAMPLE 7

A new car costs $33,275.00. It depreciates 19% in its first year.

a. How much value does the car lose in the first year?

b. How much is the car worth at the end of the first year?

R.4 Vocabulary Exercises

1. To find a *percent of* in mathematics, change the percent to a(n) _____ and multiply.

2. Percent means part of _____.

3. $\dfrac{\rule{1cm}{0.4pt}}{\rule{1cm}{0.4pt}} = \dfrac{\text{increase}}{\text{original amount}}$.

4. Markup = (_____) · (original amount).

R.4 Exercises

For Exercises 1 through 8, convert each value as indicated.

1. Convert 34% to a decimal.

2. Convert 5.1% to a decimal.

3. Convert $\dfrac{47}{100}$ to a percent.

4. Convert $\dfrac{9}{100}$ to a percent.

5. Convert 0.017 to a percent.

6. Convert 3.002 to a percent.

7. Convert $\dfrac{9}{25}$ to a percent.

8. Convert $\dfrac{3}{4}$ to a percent.

For Exercises 9 through 14, complete each table by filling in the missing values.

9.

Percent	Decimal	Fraction
	0.16	

10.

Percent	Decimal	Fraction
		$\dfrac{3}{5}$

11.

Percent	Decimal	Fraction
28.5%		

12.

Percent	Decimal	Fraction
		$\frac{7}{20}$

13.

Percent	Decimal	Fraction
0.12%		

14.

Percent	Decimal	Fraction
	0.067	

For Exercises 15 through 40, answer each problem as indicated.

15. Find 16% of 64.

16. Find 5% of 2.

17. Find 0.5% of 112.

18. Find 0.75% of 9.

19. 40% of the dogs in puppy class at an obedience school are Labrador retrievers. There are 25 dogs in the class. How many are Labrador retrievers?

20. 16% of the dogs in puppy class in an obedience school are American Eskimos. There are 25 dogs in the class. How many are American Eskimos?

21. 20% of the diners in a restaurant have a coupon. There are 30 diners in the restaurant. How many have a coupon?

22. 25% of the shoppers in a grocery store have coupons. There are 68 shoppers in the store. How many have coupons?

23. One serving of yellow split peas contains 48% of the recommended daily amount of 25 grams of dietary fiber. How many grams of dietary fiber are in one serving of yellow split peas?

24. One serving of lentils contains 6% of the recommended daily amount of 300 grams of total carbohydrates. How many grams of total carbohydrates are in one serving of lentils?

25. One serving of pasta sauce contains 25% of the recommended daily amount of 2400 milligrams of sodium. How many milligrams of sodium are in one serving of pasta sauce?

26. One serving of rice pilaf contains 27% of the recommended daily amount of 2400 milligrams of sodium. How many milligrams of sodium are in one serving of rice pilaf?

27. Brianna buys a purse for $65. Find the tax paid if the sales tax rate in her city is 6%.

28. Taria buys a cookware set for $270. Find the tax paid if the sales tax rate in her city is 7%.

29. Yerica buys a new set of tires for her car for $384. Find the sales tax if the tax rate in her city is 9.5%.

30. Jerrod needs $245 worth of supplies to fix up his bathroom at home. Find the tax Jerrod will pay if the sales tax rate in his city is 8.5%.

31. A home increases in value from $195,000 to $240,000. What is the percent increase for the house?

32. A home increases in value from $425,000 to $465,000. What is the percent increase for the house?

33. A home decreases in value from $259,000 to $245,000. What is the percent decrease for the house?

34. A home decreases in value from $520,000 to $495,000. What is the percent decrease for the house?

35. A princess cut 0.8 carat diamond necklace costs a jewelry store $1113.75. The markup rate is 75%. What is the retail price?

36. A golden South Sea pearl and diamond necklace costs a jewelry store $277.50. The markup rate is 72%. What is the retail price?

37. A new car sells for $22,259. It depreciates by 19% in its first year. How much is the car worth at the end of the first year?

38. A new car sells for $35,790. It depreciates by 18% in its first year. How much is the car worth at the end of the first year?

39. A shirt that originally retailed for $45 is marked down by 25%. What is the discounted price?

40. A pair of pants that originally retailed for $80 is marked down by 22%. What is the discounted price?

R.5 The Real Number System

LEARNING OBJECTIVES

- Define and categorize rational numbers.
- Define and categorize irrational numbers.
- Distinguish between rational and irrational numbers.
- Describe the sets that make up the real number system.
- Distinguish between exact and approximate answers.

■ Rational Numbers

In Section R.1, we studied natural numbers, whole numbers, and the integers. In Section R.2, we studied fractions. Fractions do not fit into the sets of numbers studied in Section R.1. After the set of integers, the number system expanded to include fractions. Some decimals may be represented as fractions.

Notice that all finite decimals may be expressed as fractions. For example, $0.25 = \frac{25}{100} = \frac{1}{4}$. Remember to write a decimal as a fraction means to read the place value of the rightmost decimal digit and use this value for the denominator of the fraction.

Infinite repeating decimals can also be represented as fractions. So $0.333333... = 0.\overline{3} = \frac{1}{3}$ is also a fraction.

When the set of integers is expanded to contain all fractions, the resulting set is called the rational numbers. Notice that the word *rational* has the word *ratio* in it. Because fractions are ratios of integers, the expanded set including fractions is called the **rational numbers**. Finite decimals and infinite repeating decimals are also fractions expressed in another form, so they are also rational numbers. It is also true that all integers can be written as fractions. Any whole number (or integer) can be written with a denominator of 1. For example, $5 = \frac{5}{1}$.

Consider your bank account balance. Suppose you have $158.92 in your bank account and you withdraw $20.00. You now have a total of $138.92 in your account. Your bank account balance is an example from the set of rational numbers, since it is a finite decimal. You can write $138.92 in fractional form by remembering that the name of the last decimal place in this number is the hundredths place. Therefore, $138.92 = \frac{13892}{100}$.

> ■ **What's That Mean?**
>
> **Ratio**
>
> A *ratio* is the quotient of one value divided by another, usually expressed as a fraction.

> **DEFINITION**
>
> **Rational Numbers** The rational numbers consist of the set of numbers that are *ratios* of integers. This category includes all of the integers, finite decimals, and infinite repeating decimals.

Example 1 — Which set does that number belong to?

Determine whether all possible numbers described by each statement belong to the set of integers as well as the set of rational numbers.

a. The average monthly charge for a cell phone

b. The number of minutes on a prepaid cell phone plan every month

c. The elevation of any location in California; note that the highest elevation in California is Mount Whitney at 14,494 feet above sea level, and the lowest elevation is in Death Valley, which is 282 feet below sea level

SOLUTION

a. The average monthly balance will often be a decimal number with two decimal digits, such as $55.92. Therefore, these values belong to the set of rational numbers.

b. Suppose a prepaid cell phone plan gives 100 minutes per month. The number 100 is an integer but can be considered both an integer and a rational number, since $100 = \frac{100}{1}$. Therefore, these values belong to both the integers and rational numbers.

c. The elevation of any location in California is a value between -282 feet and 14,494 feet. Elevations are typically measured to the nearest foot. Therefore, the elevations are integers and therefore also rational numbers.

PRACTICE PROBLEM FOR EXAMPLE 1

Determine whether all possible numbers described by each statement belong to the set of integers as well as the set of rational numbers.

a. The amount owed on a home electricity bill every month

b. The number of people who attend a jazz concert

c. The amount of time (in hours) that a college student exercises every week

Irrational Numbers

CONCEPT INVESTIGATION
What number is that?

Use your calculator to fill in the third column of the following table. Divide the circumference by the diameter and put the result in the third column. Carry out all results to four decimal digits of accuracy. Measure two additional circular objects as accurately as you can and add two rows to the table.

Circumference	Diameter	$\dfrac{\text{Circumference}}{\text{Diameter}}$
37.699	12	$\dfrac{37.699}{12} \approx 3.1416$
28″	8.91268″	
14 cm	4.45633 cm	
73.2200 m	23.3066 m	

What number does the last column always (approximately) come out to be? Can you think of another name for this number?

What's That Mean?

Circles

The *circumference* is the measure of the distance around a circle. The *diameter* is the measure of the distance from one side to the other of a circle through the center.

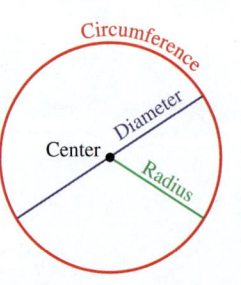

The number π is an example of an irrational number. We will encounter irrational numbers other than π in mathematics. The number $\sqrt{5}$ is also irrational. The first irrational number discovered was $\sqrt{2}$. It was discovered by the Greek mathematician Hippasus, who used geometric methods to prove that it was irrational.

> **DEFINITION**
> **Irrational Numbers** Irrational numbers can be expressed by infinite, nonrepeating decimals. These numbers cannot be written as a ratio of integers.

Because irrational numbers are infinite, nonrepeating decimals, we often approximate them using a calculator.

The Real Number System

Combining all rational numbers and all irrational numbers into one set results in what mathematicians call the **real number system**.

> **DEFINITION**
> **Real Number System** The real number system consists of the set of all rational and irrational numbers.

The following diagram is provided to help visualize the real number system. Notice that the set of natural numbers is contained in the set of whole numbers, the set of whole numbers is contained in the set of integers, and the set of integers is contained in the set of rational numbers. The irrational numbers are separate from the rational numbers, but these two sets together make up the entire real number system.

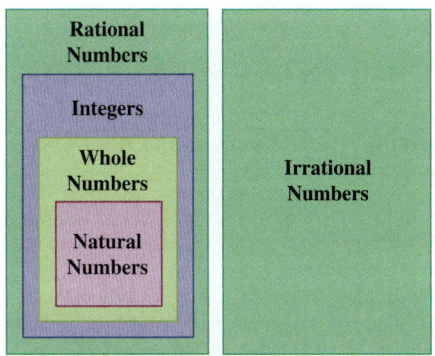

Real Number System

Example 2 Classifying numbers

Check all boxes that are appropriate for each number in the following table.

	Natural Number	Whole Number	Integer	Rational Number	Irrational Number	Real Number	Natural Number
-5							
$\dfrac{2}{5}$							
0							
$\sqrt{2}$							
0.29							

SOLUTION

−5 is an integer, a rational number, and a real number. So we will check all of those boxes.

$\frac{2}{5}$ is a rational number and a real number.

0 is a whole number, an integer, a rational number, and a real number.

$\sqrt{2}$ is an irrational number and a real number.

$0.29 = \frac{29}{100}$ is a rational number and a real number.

	Natural Number	Whole Number	Integer	Rational Number	Irrational Number	Real Number
−5			✓	✓		✓
$\frac{2}{5}$				✓		✓
0		✓	✓	✓		✓
$\sqrt{2}$					✓	✓
0.29				✓		✓

PRACTICE PROBLEM FOR EXAMPLE 2

Check all boxes that are appropriate for each number in the following table.

	Natural Number	Whole Number	Integer	Rational Number	Irrational Number	Real Number
8.3						
$\frac{-4}{3}$						
56						
π						
−15						

In Section R.3, we learned how to convert between fractions and decimals. We now look at those problems again, in the context of the real number system.

Example 3 Converting between fractions and decimals

a. Convert 0.635 to a fraction in simplest form. Is this number a rational number or an irrational number? Explain.

b. Convert $\frac{1}{6}$ to a decimal. Is this number a rational number or an irrational number? Explain.

SOLUTION

a. The number 0.635 can be written as

$$0.635 \quad \text{The last place value is thousandths.}$$

$$= \frac{635}{1000} \quad \text{Rewrite over 1000.}$$

$$= \frac{5 \cdot 127}{5 \cdot 200} \quad \text{Reduce to simplest form.}$$

$$= \frac{127}{200}$$

The number 0.635 is a rational number because it has a finite number of nonzero decimal digits and can be written as a ratio of integers.

b. The number $\frac{1}{6}$ can be written as

$$\frac{1}{6} \quad \text{Divide 6 into 1.}$$

$$\begin{array}{r} 0.166 \\ 6\overline{)1.000} \\ \underline{-6} \\ 40 \\ \underline{-36} \\ 40 \\ \underline{-36} \\ 4 \end{array}$$

Therefore, $\frac{1}{6} = 0.1666\ldots = 0.1\overline{6}$. The number $\frac{1}{6}$ is a rational number because it is a fraction. It can also be said that $\frac{1}{6}$ is rational because it has an infinite, repeating decimal expansion.

PRACTICE PROBLEM FOR EXAMPLE 3

a. Convert 0.255 to a fraction in simplest form. Is this number a rational number or an irrational number? Explain.

b. Convert $\frac{1}{9}$ to a decimal. Is this number a rational number or an irrational number? Explain.

■ Exact and Approximate Answers

It is important to pay careful attention to the difference between the *exact answer* and an *approximate answer*. The number π is an irrational number, which is approximated when used in problems. The value of π is $\pi = 3.1415926535\ldots$, where the decimal expansion continues indefinitely. When computing with π, many times students are told to use 3.14 as an approximation. Likewise, we have seen that $\sqrt{2}$ is an irrational number. The decimal expansion for $\sqrt{2}$ is $\sqrt{2} = 1.41421356\ldots$, where the decimal expansion continues indefinitely. An approximate value of $\sqrt{2}$ is 1.41. Whenever a number is rounded or truncated, the resulting value is an approximation of the original number. In this book, we will use only rounding.

Exact Value	An Approximate Value
π	3.14
$\sqrt{2}$	1.41

> **What's That Mean?**
>
> **Truncate**
>
> To *truncate* a number means to approximate the number by just cutting off the last decimal digits.
>
> For example, the fraction $\frac{2}{3}$ has the decimal expansion $\frac{2}{3} = 0.666\ldots$.
>
> *Truncating* $\frac{2}{3}$ to two decimal digits gives 0.66. *Rounding* $\frac{2}{3}$ to two decimal digits gives 0.67. Note that truncation and rounding can give different results. In this book, we use only rounding.

Approximate values for rational numbers can be found as well. Since $\frac{1}{3} = 0.333\ldots$, an approximation for $\frac{1}{3}$ is 0.3. Write $\frac{1}{3} \approx 0.3$.

Example 4 — Rounding to approximate values

Round each number to the given place value.

a. Round to the thousandths place: $\sqrt{5}$

b. Round to the tenths place: $\frac{1}{9}$

c. Round to the ten-thousandths place: π

SOLUTION

a. Using a calculator, $\sqrt{5} \approx 2.23606797\ldots$. To round to the thousandths place, underline the thousandths place: $2.23\underline{6}06797\ldots$. The digit to the right of the 6 in the thousandths place is 0, so round down.
Therefore, $\sqrt{5} \approx 2.236$.

b. Using a calculator, $\frac{1}{9} = 0.11111111\ldots$. Recall that to write a repeating decimal, put a bar over the repeating part: $\frac{1}{9} = 0.11111111\ldots = 0.\overline{1}$. To round to the tenths place, underline the tenths place: $0.\underline{1}1111111\ldots$. The digit to the right of the 1 in the tenths place is 1, so round down.
Therefore, $\frac{1}{9} \approx 0.1$.

c. Using a calculator, $\pi \approx 3.14159265\ldots$ To round to the ten-thousandths place, underline the ten-thousandths place: $3.141\underline{5}92654\ldots$. The digit to the right of the 5 in the ten-thousandths place is 9, so round up.
Therefore, $\pi \approx 3.1416$.

PRACTICE PROBLEM FOR EXAMPLE 4

Round each number to the given decimal place.

a. Round to the tenths place: $\sqrt{7}$

b. Round to the hundredths place: $\frac{2}{7}$

The grocery store is a place where approximations are often used. To help you compare prices on items, many grocery stores will provide the unit price for each item. The unit price is the cost for one unit of that item, such as $0.20 for 1 ounce of juice or $0.88 for 1 pound of oranges. To find a unit price, take the price for the item and divide by the number of units.

Example 5 — Unit prices

Find the unit price for each item. Round answers to the nearest cent.

a. A 10-ounce box of cereal for $5.49

b. A 12-ounce package of raisins for $6.99

Calculator Details

Square roots and pi

To approximate $\sqrt{5}$ using a calculator, look for the $\sqrt{}$ key. It is often above the x^2 key or near it. On older scientific calculators, enter 5 then the $\sqrt{}$ key. On newer calculators, enter $\sqrt{}$ then 5. If the $\sqrt{}$ is above the x^2 key, enter 2nd first then x^2 then 5. The 2nd function key does the operation above the key.

Many calculators have a π key. Ask your instructor for help if you cannot find it.

SOLUTION

a. To find the unit price, divide $5.49 by 10 ounces. To round to the nearest cent, divide out to the thousandths place.

$$\begin{array}{r} 0.549 \\ 10\overline{)5.490} \\ \underline{-50} \\ 49 \\ \underline{-40} \\ 90 \\ \underline{-90} \\ 0 \end{array}$$

The unit price for this box of cereal is approximately $0.55 per ounce.

b. Divide $6.99 by 12 ounces.

$$\begin{array}{r} 0.582 \\ 12\overline{)6.990} \\ \underline{-60} \\ 99 \\ \underline{-96} \\ 30 \\ \underline{-24} \\ 6 \end{array}$$

Divide out to the thousandths place.

Round to the nearest cent.

The unit price for these raisins is approximately $0.58 per ounce.

PRACTICE PROBLEM FOR EXAMPLE 5

Find the unit price for each item. Round answers to the nearest cent.

a. A 16-ounce package of bologna for $2.49

b. A 28-ounce jar of peanut butter for $8.92

R.5 Vocabulary Exercises

1. The _____ numbers are the set of numbers that are ratios of integers.

2. The number π is an example of a(n) _____ number.

3. The _____ number system consists of all rational and _____ numbers.

4. The smallest set of numbers in the real number system is the _____ numbers.

R.5 Exercises

For Exercises 1 and 2, check all boxes that are appropriate for each number in the table.

1.

	Whole Number	Integer	Rational Number	Irrational Number	Real Number
1.47					
$\frac{16}{49}$					
-7					
$\sqrt{5}$					
9					

2.

	Whole Number	Integer	Rational Number	Irrational Number	Real Number
$\frac{1}{9}$					
-21.3					
55					
π					
5.001					

For Exercises 3 through 14, determine the set(s), integers or rational numbers, to which all possible numbers described by the statement could belong.

3. The ratio of boys to girls in a calculus course
4. The ratio of girls to boys in a philosophy course
5. The ratio of flour to water in a bread recipe
6. The ratio of water to rice when cooking steamed rice
7. The daily high temperature in Milwaukee, Wisconsin, over the course of a year
8. The daily low temperature in Portland, Maine, over the course of a year
9. The amount owed on an electricity bill for a month
10. The amount owed on a cable TV bill for a month
11. The number of people who attend a college football game
12. The number of people who attend a college basketball game
13. The elevation of Death Valley, California. See Practice Problem for Example 10 in Section R.1 for the elevation
14. The elevation of Mount Everest, which is on the border of Nepal, Tibet, and China

For Exercises 15 through 24, list all the given numbers requested from each set.

15. All the rational numbers in the set $\left\{-3, 0, \pi, \frac{5}{2}, \sqrt{2}\right\}$
16. All the rational numbers in the set $\left\{-\frac{1}{2}, 0, 4, \frac{9}{4}, \pi\right\}$
17. All the irrational numbers in the set $\left\{-3, 0, \pi, \frac{5}{2}, \sqrt{2}\right\}$
18. All the irrational numbers in the set $\left\{-\frac{1}{2}, 0, 4, \frac{9}{4}, \pi\right\}$
19. All the whole numbers in the set $\left\{16, -5, \frac{9}{2}, \sqrt{5}, 0, 2\right\}$
20. All the whole numbers in the set $\{-3.5, -21, 5, \sqrt{2}, 10, 0.5\}$
21. All the integers in the set $\left\{-5.9, 8.6, \frac{7}{25}, 3, 111, \pi\right\}$
22. All the integers in the set $\left\{\frac{9}{77}, 8, \frac{3}{2}, -71, 4\right\}$
23. All the natural numbers in the set $\left\{-9, 0, \frac{2}{3}, 56, \sqrt{2}\right\}$
24. All the natural numbers in the set $\left\{\frac{4}{5}, \pi, 96, -4.2, 12\right\}$

For Exercises 25 through 32, convert each number to the requested form. Is the given number a rational number or an irrational number? Explain.

25. Convert $\frac{5}{8}$ to a decimal. Is $\frac{5}{8}$ a rational number or an irrational number? Explain.

26. Convert $\frac{3}{4}$ to a decimal. Is $\frac{3}{4}$ a rational number or an irrational number? Explain.

27. Convert 1.34 to a fraction in simplest form. Is 1.34 a rational number or an irrational number? Explain.

28. Convert 2.78 to a fraction in simplest form. Is 2.78 a rational number or an irrational number? Explain.

29. Convert $\frac{1}{11}$ to a decimal. Is $\frac{1}{11}$ a rational number or an irrational number? Explain.

30. Convert $\frac{1}{15}$ to a decimal. Is $\frac{1}{15}$ a rational number or an irrational number? Explain.

31. Convert 0.0133 to a fraction in simplest form. Is 0.0133 a rational number or an irrational number? Explain.

32. Convert 0.0527 to a fraction in simplest form. Is 0.0527 a rational number or an irrational number? Explain.

For Exercises 33 through 38, find a decimal approximation for each given number and round to the given place value. Is the given number a rational number or an irrational number? Explain.

33. Round $\frac{3}{11}$ to the thousandths place. Is $\frac{3}{11}$ a rational number or an irrational number? Explain.

34. Round $\frac{5}{21}$ to the ten-thousandths place. Is $\frac{5}{21}$ a rational number or an irrational number? Explain.

35. Round $\sqrt{11}$ to the ten-thousandths place. Is $\sqrt{11}$ a rational number or an irrational number? Explain.

36. Round $\sqrt{13}$ to the tenths place. Is $\sqrt{13}$ a rational number or an irrational number? Explain.

37. Round π to the millionths place. Is π a rational number or an irrational number? Explain.

38. Round π to the tenths place. Is π a rational number or an irrational number? Explain.

For Exercises 39 through 44, find the unit price for each item. Round answers to the nearest cent.

39. A 40-ounce container of mixed nuts for $26.99

40. A 5.25-ounce package of sunflower seeds for $1.67

41. A 16-ounce package of angel hair pasta for $1.42

42. A 16-ounce package of elbow macaroni for $1.50

43. A 24-ounce jar of marinara sauce for $2.59

44. A 24-ounce jar of pasta sauce for $3.75

For Exercises 45 through 56, identify the number of terms. Then use the order-of-operations agreement to simplify the expression.

45. $\frac{3}{5} \cdot \frac{1}{6} + \frac{1}{2}$

46. $\frac{4}{3} \div \frac{5}{6} - \frac{1}{2}$

47. $1.2 \cdot 2 - 6.2 \div 2 + 12.1$

48. $106 - 2.4 \div 4 + 6.9 \cdot 3$

49. $\frac{3}{5} + 1.5 \cdot 3$

50. $3.3 \cdot 5 - \frac{1}{8}$

51. $26.1 \div 6 - \frac{17}{5} \cdot \frac{1}{20}$

52. $\frac{1}{9} \cdot \frac{3}{4} + 11.25 \cdot 0.8$

53. $5 - 6.75 + \frac{2}{3} \div 4$

54. $\frac{6}{7} \div \frac{3}{14} + 8.95 - 12$

55. $\frac{1}{4} + \frac{3}{8} \cdot \frac{16}{21} - 4 \div \frac{2}{3}$

56. $\frac{1}{6} + \frac{3}{5} \cdot \frac{20}{21} - 5 \div \frac{10}{3}$

Chapter Summary

Section R.1 Operations with Integers

- The **natural numbers** are the set $\{1, 2, 3, \ldots\}$.
- The **whole numbers** are the set $\{0, 1, 2, 3, \ldots\}$.
- The **integers** are the set $\{\ldots, -2, -1, 0, 1, 2, \ldots\}$.
- A **number line** is used to visually represent numbers.

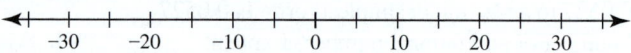

- The **scale** of a number line is the distance between the tick marks on the number line.
- Numbers have order, and both equalities and **inequalities** are used to represent the order relationship between numbers.
- The **absolute value** of a number is the distance between that number and 0. It is always nonnegative.
- **Opposites** have the same absolute value but differ only in sign.
- **Terms** are numbers or expressions that are added.
- **To add integers with the same sign:** Add the absolute values of the numbers. Attach the same sign of the numbers to the sum.
- **To add integers with different signs:** Take the absolute value of each number. Subtract the smaller absolute value from the larger. Attach the sign of the number that is larger in absolute value.
- **To subtract integers:** Change the sign of the second integer (reading left to right) and add as explained above.
- **To multiply or divide two integers with the same sign:** Multiply or divide the absolute values of the numbers. The solution is always positive.
- **To multiply or divide two integers with different signs:** Multiply or divide the absolute values of the two numbers. The solution is always negative.
- The **order-of-operations agreement** is used to compute the value of an expression that contains multiple operations.
  1. Circle or underline the terms. (Remember: Terms are quantities that are added.)
  2. Evaluate the operations (multiplications and divisions) inside of each term, working from left to right.
  3. Add the simplified terms, working from left to right.

Example 1 Place the appropriate inequality symbol, < or >, between the two numbers.

  a. $-3 \quad -7$

  b. $-6 \quad -(-6)$

SOLUTION

  a. $-3 > -7$

  b. $-6 < -(-6)$

Example 2

Find the absolute value of each of the following numbers.

a. $|9|$

b. $|-6|$

SOLUTION

a. 9

b. 6

Example 3

Identify the terms. Then simplify using the order-of-operations agreement.

a. $-12 + 3(-5)$

b. $7 + 5 \cdot 4 - 12 \div 2 \cdot 7$

SOLUTION

a. $\boxed{-12} + \boxed{3(-5)}$ There are two terms.
   $= -12 + (-15)$ Do the multiplication in the second term.
   $= -27$ Add the integers.

b. $\boxed{7} + \boxed{5 \cdot 4} + \boxed{(-12) \div 2 \cdot 7}$ There are three terms.
   $= 7 + 20 + (-6) \cdot 7$ Do the operations in the second and third terms, working from left to right.
   $= 7 + 20 + (-42)$ Add terms, working from left to right.
   $= 27 + (-42)$
   $= -15$

Section R.2   Operations with Fractions

- **Prime numbers** are natural numbers greater than 1 that are evenly divisible only by themselves and 1.
- To find the **prime factorization** of a number, first write any factorization of the number. Continue to factor all composite numbers until there is only a product of primes left.
- To find the **greatest common factor**, follow these steps:
  1. Find the prime factorization of each number.
  2. The greatest common factor is the product of the common prime factors.
- Fractions are parts of a whole number and should always be **reduced to lowest terms**.
- To **reduce a fraction to lowest terms**, follow these steps:
  1. Find the GCF for the numerator and denominator.
  2. Factor the numerator and denominator, using the GCF as one of the factors.
  3. Divide out the GCF from the numerator and denominator. The remaining fraction is the reduced or simplified form of the fraction.
- **Equivalent fractions** represent the same quantity: $\frac{1}{2} = \frac{5}{10}$.
- To write an **equivalent fraction**, follow these steps:
  1. When given the problem $\frac{a}{b} = \frac{?}{d}$, find the missing factors c such that $b \cdot c = d$.
  2. Multiply the numerator and denominator by the missing factor c: $\frac{a \cdot c}{b \cdot c}$. This yields the equivalent fraction with a denominator of d.

62 CHAPTER R Review of Prealgebra

- To **add or subtract fractions with like (common) denominators**, follow these steps:
    1. Add or subtract the fractions by adding (subtracting) the numerators.
    2. Reduce the final answer to lowest terms (simplest form).
- To **add or subtract fractions with unlike denominators**, follow these steps:
    1. Factor each denominator into its prime factorization.
    2. The least common denominator is the product of all the different factors from step 1.
        If a factor occurs repeatedly (more than once), include the maximum number of times that factor is repeated.
    3. Rewrite each fraction as an equivalent fraction with the denominator equal to the least common denominator.
    4. Add or subtract the fractions by adding (subtracting) the numerators.
    5. Reduce the final answer to lowest terms.
- To **multiply fractions**, follow these steps:
    1. Multiply the numerators together and multiply the denominators together (multiply across).
    2. Reduce the final answer to lowest terms.
- The **reciprocal** of a is $\frac{1}{a}$, if $a \neq 0$.
- To **divide fractions**, follow these steps:
    1. Invert (or find the reciprocal of) the second fraction. Change the quotient to a product.
    2. Multiply the numerators together and multiply the denominators together (multiply across).
    3. Reduce the final answer to lowest terms.
- The **order-of-operations agreement** is used with fractions the same way it is used with integers.

Example 4 Write the prime factorization of the number 24.

SOLUTION $24 = 2 \cdot 2 \cdot 2 \cdot 3$

Example 5 Add or subtract the following numbers.

a. $-\dfrac{5}{4} + \dfrac{3}{4}$ b. $\dfrac{2}{3} + \dfrac{1}{2}$

SOLUTION a. $-\dfrac{5}{4} + \dfrac{3}{4} = -\dfrac{2}{4} = -\dfrac{1}{2}$ b. $\dfrac{2}{3} + \dfrac{1}{2} = \dfrac{4}{6} + \dfrac{3}{6} = \dfrac{7}{6} = 1\dfrac{1}{6}$

Example 6 Multiply or divide the following fractions.

a. $\dfrac{4}{9} \cdot \dfrac{3}{8}$ b. $\dfrac{2}{3} \div \dfrac{5}{6}$

SOLUTION a. $\dfrac{4}{9} \cdot \dfrac{3}{8} = \dfrac{12}{72} = \dfrac{1}{6}$ b. $\dfrac{2}{3} \div \dfrac{5}{6} = \dfrac{2}{3} \cdot \dfrac{6}{5} = \dfrac{12}{15} = \dfrac{4}{5}$

Section R.3 Operations with Decimals

- The numbers to the right of the decimal point have **place value** that accords with a multiple of ten. The names of place values to the right of a decimal point will end with "ths."
- **Place value** can be used to convert decimals to fractions. Find the rightmost place value in the given decimal. Removing the decimal point, write the number over the rightmost place value to give the fractional form. Reduce to lowest terms.
- **Decimals can be added or subtracted** by lining them up vertically by place value. Then add or subtract, carrying or borrowing as necessary.
- **Decimals can be multiplied or divided** like whole numbers. Pay close attention to where the decimal point goes in the final answer.
- The **order-of-operations** agreement is used in the same way when decimals are involved.
- To **round a decimal to a given place value,** follow these steps:
    1. Underline the given place value in the number.
    2. Look at the digit directly to the right of the underlined place.
        a. If that digit is 0, 1, 2, 3, or 4, then delete all digits to the right of the underlined place. This is often called *rounding down*.
        b. If that digit is 5, 6, 7, 8, or 9, then add 1 to the underlined place and delete all the digits to the right. This is often called *rounding up*.

Example 7

Answer as indicated.

a. Find the digit in the thousandths place: 1067.52943.

b. Round the number to the tenths place: 3.259.

c. Find 7% of 65.

SOLUTION

a. The digit in the thousandths place is 9. 1067.52943.

b. 3.259   *The digit to the right is 5 or more: round up.*
   3.3

c. 7% of 65 = 0.07 · 65 = 4.55

Example 8

Add or subtract as indicated.

a. 8.20597 + 31.451   b. 16.0243 − 9.951

SOLUTION a. 8.20597 + 31.451 = 39.65697   b. 16.0243 − 9.951 = 6.0733

Example 9

Multiply or divide as indicated.

a. 3.6(2.7)   b. 4.34 ÷ 1.4

SOLUTION a. 3.6(2.7) = 9.72   b. 4.34 ÷ 1.4 = 3.1

Section R.4 Operations with Percents

- **Percent** means "a part of 100."
- To **convert a percent to fraction form**, put the percent over 100.
- To **convert a percent to a decimal:** Move the decimal point two places to the left and remove the % sign. This is the same as multiplying by $1\% = \dfrac{1}{100} = 0.001$ or dividing by 100.
- To **convert a decimal to a percent:** Move the decimal point two places to the right and add the % sign. This is the same as multiplying by 100.
- **When using a percent in a problem**, first convert to decimal form.
- Percents are also used to describe percent increase or decrease.

Example 10   Find 20% of 75.

SOLUTION   Convert 20% to its decimal form, which is 0.20. Then calculate the answer.

$$(0.20)(75) = 15$$

Example 11   A shirt that costs a clothing store $20 is marked up 25%. What is the retail cost?

SOLUTION   The markup is computed as follows.

$$0.25(20) = 5$$

The retail cost is therefore

$$20 + 5 = 25$$

The retail cost of this shirt is $25.

Section R.5 The Real Number System

- The **rational numbers** are the set of numbers that are ratios of integers. Natural numbers, whole numbers, integers, finite decimals, and infinite repeating decimals are all part of the rational numbers.
- The **irrational numbers** are the set of numbers that cannot be expressed as ratios of integers. Irrational numbers are infinite, nonrepeating decimals.
- The **real numbers** consist of the set of all rational and irrational numbers.
- **Approximate answers** can be found by rounding to a desired place value.

Example 12   List all the rational numbers in the set $\left\{-3, \dfrac{5}{6}, 0.22, \pi, 9\dfrac{2}{3}\right\}$.

SOLUTION   The rational numbers are $-3, \dfrac{5}{6}, 0.22,$ and $9\dfrac{2}{3}$.

Example 13   Find a decimal approximation for $\sqrt{85}$. Round your answer to the thousandths place.

SOLUTION   Using a calculator, we find $\sqrt{85} \approx 9.21954$, and rounding to the thousandths place gives $\sqrt{85} \approx 9.220$.

Chapter Review Exercises

At the end of each exercise, the section number in brackets indicates where the material is covered.

For Exercises 1 through 4, classify each number as a natural number, a whole number, or an integer. If a number belongs to more than one set, list all possible sets it belongs to.

1. a. -12 b. 0 c. 17 [R.1]
2. a. 9 b. -1 c. 0 [R.1]
3. The number of students in a chemistry lab [R.1]
4. The number of trees in a public park [R.1]
5. List all numbers in the set. All the whole numbers between -3 and 4, including -3 and 4. [R.1]
6. List all the numbers in the set. All the integers between -6 and -3, including -6 and -3. [R.1]
7. Determine the scale of the number line. [R.1]

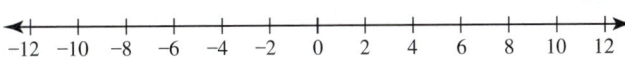

8. Determine the scale of the number line. [R.1]

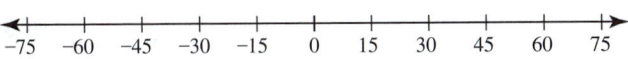

9. Plot the numbers on a number line.

$$-3, \frac{1}{2}, 0, -\frac{4}{2}, 4.25 \quad \text{[R.2]}$$

10. Plot the numbers on a number line.

$$2.5, -\frac{1}{3}, -2, \frac{5}{2}, -0.8 \quad \text{[R.2]}$$

11. Place the appropriate inequality symbol ($<$ or $>$) between the two numbers.

$$-3 \quad -5 \quad \text{[R.1]}$$

12. Place the appropriate inequality symbol ($<$ or $>$) between the two numbers.

$$-9 \quad -2 \quad \text{[R.1]}$$

13. Find the value of the absolute value expression:

$$|-9| \quad \text{[R.1]}$$

14. Find the value of the absolute value expression:

$$|-6.7| \quad \text{[R.1]}$$

15. Place the list of numbers in increasing order from smallest to largest.

$$3, -2, \frac{2}{5}, |-7| \quad \text{[R.2]}$$

16. Place the list of numbers in increasing order from smallest to largest.

$$-4, \left|-\frac{2}{3}\right|, 1.8, -0.1 \quad \text{[R.2]}$$

17. Rewrite the math expression in sentence form and then find the value: $-(-8)$ [R.1]

18. Rewrite the math expression in sentence form and then find the value: $-(-1)$ [R.1]

19. Add or subtract as indicated.
    a. $-22 + (-45)$ b. $-9 - 17$
    c. $-18 + 8$ d. $-16 - (-11)$ [R.1]

20. Add or subtract as indicated.
    a. $-13 + (-5)$ b. $50 - (-18)$
    c. $4 - 18$ d. $-19 + 12$ [R.1]

21. Multiply or divide as indicated.
    a. $-5(-10)(-3)$ b. $-2(4)(-1)$
    c. $64 \div (-4)$ d. $-55 \div (-11)$ [R.1]

22. Multiply or divide as indicated.
    a. $(-3)(-5)(2)$ b. $-(-3)(-6)$
    c. $27 \div (-9)$ d. $-24 \div (-3)$ [R.1]

23. Identify number of the terms in the expression:

$$-9 \cdot 4 \div 6 + 18 - 3 \quad \text{[R.1]}$$

24. Identify number of the terms in the expression:

$$-12 + 8 \div 2 - 1 \quad \text{[R.1]}$$

25. Identify the number of terms. Then simplify, using the order-of-operations agreement.

$$-48 \div (-12) \cdot 5 + 4 \cdot 3 - 7 \quad \text{[R.1]}$$

26. Identify the number of terms. Then simplify, using the order-of-operations agreement.

$$9 \cdot (-2) + 17 + 32 \div (-16) \cdot 3 \quad \text{[R.1]}$$

27. Find the prime factorization of 27. [R.2]

66 CHAPTER R Review of Prealgebra

28. Find the prime factorization of 45. [R.2]

29. Find the GCF for the set of numbers.

    8 and 28 [R.2]

30. Find the GCF for the set of numbers.

    15 and 45 [R.2]

31. Reduce the fraction to lowest terms.

    $$\frac{20}{44}$$ [R.2]

32. Reduce the fraction to lowest terms.

    $$\frac{12}{30}$$ [R.2]

33. Rewrite the fraction as an equivalent fraction with the given denominator.

    $$\frac{5}{8} = \frac{}{56}$$ [R.2]

34. Rewrite the fraction as an equivalent fraction with the given denominator.

    $$\frac{3}{7} = \frac{}{42}$$ [R.2]

35. Classify each number as an integer or a rational number or both.

    a. $\frac{3}{5}$ b. -5 c. 16.1 [R.4]

36. Classify each number as an integer or a rational number or both.

    a. -0.7 b. $-\frac{1}{9}$ c. 0 [R.4]

37. Add or subtract as indicated. Reduce the answers to lowest terms.

    a. $\frac{1}{6} + \frac{3}{6}$ b. $\frac{1}{6} + \frac{1}{4}$

    c. $\frac{1}{2} - \frac{2}{3}$ d. $3\frac{1}{2} - 1\frac{1}{3}$ [R.2]

38. Add or subtract as indicated. Reduce the answers to lowest terms.

    a. $\frac{5}{8} - \frac{1}{8}$ b. $\frac{1}{5} + \frac{1}{15}$

    c. $\frac{1}{4} - \frac{2}{5}$ d. $4\frac{1}{5} - 3\frac{1}{2}$ [R.2]

39. Find the reciprocal of each number.

    a. 6 b. $\frac{1}{4}$ [R.2]

40. Find the reciprocal of each number.

    a. -5 b. $\frac{3}{2}$ [R.2]

41. Multiply or divide as indicated. Reduce the answers to lowest terms.

    a. $\frac{2}{7} \cdot \frac{14}{22}$ b. $\frac{1}{5} \div 9$ c. $\frac{3}{8} \div \frac{9}{16}$ [R.2]

42. Multiply or divide as indicated. Reduce the answers to lowest terms.

    a. $\frac{4}{5} \cdot \frac{15}{28}$ b. $\frac{3}{7} \div 2$ c. $\frac{5}{6} \div \frac{10}{9}$ [R.2]

43. Identify the number of terms. Simplify, using the order-of-operations agreement.

    $$\frac{1}{3} \div \frac{2}{9} - 7$$ [R.2]

44. Identify the number of terms. Simplify, using the order-of-operations agreement.

    $$3 - \frac{4}{5} \cdot \frac{10}{3}$$ [R.2]

45. Identify the place value of the digit 9 in the number 2.497031. [R.3]

46. Identify the place value of the digit 5 in the number 8.26571. [R.3]

47. Write the decimal as a fraction in lowest terms.

    2.75 [R.3]

48. Write the decimal as a fraction in lowest terms.

    4.68 [R.3]

49. Write the fraction as a decimal. Put a bar over any repeating decimal part.

    $$\frac{4}{9}$$ [R.3]

50. Write the fraction as a decimal. Put a bar over any repeating decimal part.

    $$\frac{5}{8}$$ [R.3]

51. Add or subtract as indicated.

    a. $3.625 + 7.019$ b. $4.061 - 3.987$ [R.3]

52. Add or subtract as indicated.

    a. $4.073 + 16.998$ b. $5.072 - 3.826$ [R.3]

53. Multiply or divide as indicated.

    a. $(4.7)(1.03)$ b. $14.58 \div 5.4$ [R.3]

54. Multiply or divide as indicated.

    a. $(9.2)(3.8)$ b. $40.85 \div 4.3$ [R.3]

55. Identify the terms. Simplify, using the order-of-operations agreement.

    $$9.6 - 11.73 \div 5.1 - 3.2$$ [R.3]

56. Identify the terms. Simplify, using the order-of-operations agreement.
$$0.92 \div (-0.1) + 6.3 - 4.75$$ [R.3]

57. Round 0.01996 to the hundredths place. [R.3]

58. Round 5.409999 to the tenths place. [R.3]

59. Convert 8.75% to a decimal. [R.4]

60. Convert 5.45% to a decimal. [R.4]

61. Convert 28% to a fraction in lowest terms. [R.4]

62. Convert 45% to a fraction in lowest terms. [R.4]

63. Convert 0.0456 to a percent. [R.4]

64. Convert 0.0775 to a percent. [R.4]

65. Convert 1.905 to a percent. [R.4]

66. Convert 2.01 to a percent. [R.4]

67. Convert $\frac{4}{5}$ to a percent. [R.4]

68. Convert $\frac{3}{50}$ to a percent. [R.4]

69. Convert $1\frac{1}{5}$ to a percent. [R.4]

70. Convert $3\frac{1}{3}$ to a percent. [R.4]

71. Find 8.25% of 45. [R.4]

72. Find 3.45% of 9. [R.4]

73. Find 5.5% of 20. [R.4]

74. Find 10.2% of 16. [R.4]

75. Find the sales tax on a $235 purchase if the sales tax rate is 8.5%. [R.4]

76. Find the sales tax on a $130 purchase if the sales tax rate is 4.5%. [R.4]

77. A house increases in value from $195,000 to $276,000. What is the percent increase? [R.4]

78. A house decreases in value from $349,000 to $309,000. What is the percent decrease? [R.4]

79. A ring that costs a jewelry store $500 is marked up 75%. What is the retail price? [R.4]

80. A dress that costs a clothing store $35 is marked up 33%. What is the retail price? [R.4]

81. List all rational numbers in the set.
$$\left\{\frac{4}{3}, -16.9, \sqrt{5}, 0\right\}$$ [R.5]

82. List all rational numbers in the set.
$$\left\{-7.5, 2, \frac{5}{4}, \pi\right\}$$ [R.5]

83. List all irrational numbers in the set.
$$\left\{\frac{4}{3}, -16.9, \sqrt{5}, 0\right\}$$ [R.5]

84. List all irrational numbers in the set.
$$\left\{-7.5, 2, \frac{5}{4}, \pi\right\}$$ [R.5]

85. Find a decimal approximation for the number. Round to the hundredths place: $\frac{3}{7}$. [R.5]

86. Find a decimal approximation for the number. Round to the thousandths place: $\sqrt{21}$. [R.5]

For Exercises 87 through 96, use the order-of-operations agreement to simplify each expression.

87. $\frac{1}{2} \div 5 + 5.2 - 7.9$ [R.5]

88. $6 \div \frac{3}{5} - 5.69 + 8.4$ [R.5]

89. $\frac{3}{5} \div \frac{9}{10} - \frac{3}{2} + 4$ [R.5]

90. $7 - \frac{4}{7} \div \frac{1}{14} + \frac{1}{3}$ [R.5]

91. $0.35 + \frac{5}{4} \cdot \frac{3}{125}$ [R.5]

92. $1.27 - \frac{2}{10} \cdot \frac{3}{20}$ [R.5]

93. $44.2 \div 2 + 5 \cdot 7.1 - 5 \div \frac{5}{2}$ [R.5]

94. $25.15 \div 5 + 7 \div 0.2 - 6 \div 12$ [R.5]

95. $\frac{4}{5} \div \frac{8}{15} \cdot \frac{4}{9}$ [R.5]

96. $3 \div \frac{6}{11} \cdot \frac{2}{5}$ [R.5]

Chapter Test

This chapter test should take approximately one hour to complete. Read each question carefully. Show all of your work.

1. Check all boxes that apply for each number listed below.

	Whole Number	Integer	Rational Number	Irrational Number	Real Number
2.25					
$\frac{2}{3}$					
-6					
$\sqrt{2}$					
10					

2. Determine the scale of the number line.

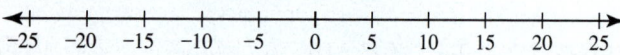

3. Place the numbers on a number line.
$$4, \quad -5.5, \quad \frac{7}{2}, \quad 0, \quad -0.9$$

4. Place the list of numbers in increasing order, from smallest to largest.
$$-4, \quad \frac{13}{5}, \quad -(-6.8), \quad |-11|$$

5. Use the appropriate inequality symbol ($<$ or $>$) between the two numbers.
   a. $-9 \quad -11$ b. $\frac{1}{3} \quad 0.3$

6. Find the value of each of the following.
   a. $-(-(-3))$ b. $|-6|$

7. Find the prime factorization of 36.

8. Add or subtract as indicated.
   a. $-5 - (-9)$ b. $-0.3 - 1.5 + 0.75$
   c. $-2 + (-8) - (-6)$

9. Multiply or divide as indicated.
   a. $(-4)(-6)(-2)$ b. $(-36) \div (-4)$
   c. $(32.2) \div (3.5)$

For Exercises 10 through 15, identify the number of terms. Then simplify each expression.

10. $-6 + 3 \cdot 4 - (-5) + 12 \div (-3)$

11. $24 \div 2 \cdot 3 + (-2) - (-1)$

12. $(-64) \div 16 \cdot 5 - 9$

13. $|-18| - (-5)(-7) + 36 \cdot 4$

14. $\frac{6}{5} \div \frac{2}{15} - \frac{1}{3} + 2$

15. $(1.2)(3.5) - 6.4 + 7$

16. Find the prime factorization of 50.

17. Find the GCF for 18 and 36.

18. Reduce to lowest terms. $\frac{48}{56}$

19. Write an equivalent fraction with the given denominator.
$$\frac{2}{7} = \frac{?}{63}$$

20. Add as indicated. Reduce the answer to lowest terms.
$$\frac{3}{5} + \frac{12}{5}$$

21. Subtract as indicated. Reduce the answer to lowest terms.
$$\frac{9}{5} - \frac{3}{4}$$

22. Multiply or divide as indicated. Reduce the answers to lowest terms.
   a. $-\frac{2}{5} \cdot \frac{15}{4}$ b. $\frac{3}{4} \div \left(-\frac{9}{16}\right)$

23. What is the place value of the digit 4 in the number 6.02478?

24. Round 5.02317 to the thousandths place.

25. Complete the table by filling in the missing values.

Percent	Decimal	Fraction
1.05%		

26. Find 9% of 206.

27. One tablespoon of ketchup has 16% of the recommended daily amount of 100 grams of sugar for adult women. How many grams of sugar are in one tablespoon of ketchup?

28. A house increases in value from $269,000 to $295,000. What is the percent increase?

Chapter Projects

What Kind of Number Is That?

Class Project
Four to five people in a group

What you will need
- Set of 20 index cards from your instructor
- Set of five containers from your instructor
- Response sheet to record the answers

Your instructor will give you five buckets labeled "natural numbers," "whole numbers," "integers," "rational numbers," and "irrational numbers." You will also receive a set of 20 index cards and a sheet of paper on which to record your results. Each card has a number written on it.

1. With your group members, decide which of the five buckets represents the smallest set to which the number belongs. Place the card in that bucket.
2. When your group is done with all 20 cards, write the answers on the sheet of paper your instructor has given you.

How Did the Ancients Record Their Counting?

Research Project
One or more people

What you will need
- Find information from a web search of the Ishango Bone.
- Follow the MLA Style Guide for all citations. If your school recommends another style guide, use it.

Research the Ishango Bone artifact online. Write a one-page paper on the Ishango Bone. Include the age, what kind of bone was used, what continent it was found on, and how the Ishango Bone is thought to have been used for counting and possibly arithmetic. Clearly state which websites you used as a part of your paper.

Pricing Your Trip to a Professional Sporting Event

Research Project
One or more people

What you will need
- Find data online to make your estimates.
- Follow the MLA Style Guide for all citations. If your school recommends another style guide, use it.

You are planning a trip for four friends and family to attend a professional sporting event. The cost of the event will be determined by a number of factors, such as the type of sporting event you are going to see, whether you have to travel out of town or will see a local event, and lodging. To come up with an estimate of the total cost of your trip, use the following steps. Do your research online and provide printouts of the cost estimates you used.

1. Select which professional sporting event you want to arrange the trip to see. Some possible selections are events of the National Football League (NFL), the National Basketball Association (NBA), the Major League Soccer Association (MLS), and Major League Baseball (MLB).
2. How much are the tickets per person? For four people?

3. Decide which city you want to see the event in. Is it your city, or do you need to arrange to travel out of town?

4. If you have to travel out of town, decide on the mode of travel. Will you arrange to fly and then rent a car? Will you travel by private car (drive)? Will you take a train and then a taxi? Will you take a bus and then a taxi? Estimate the cost of travel for four people. *Note:* If you are arranging to drive by private car, look up the Internal Revenue Service (IRS) mileage reimbursement rate and use that to estimate the driving cost.

5. Do you need to make reservations at a hotel? For how many rooms (at least two) and how many days?

6. If your trip requires travel, estimate the food cost per person per day.

7. How much does it cost to park at the stadium or arena if you are driving?

8. How much does it cost to eat and drink at the sporting event? Remember you are taking four people.

9. Add up all the costs above.

10. What is the cost per person?

Write a Review of a Section for Presentation

Research Project

One or more people

What you will need

- This presentation may be done in class or as an online video.
- If it is done as a video, post it on a website where it may be easily viewed by the class.
- You may want to use homework or other review problems from this book.
- Make it creative and fun.

Create a 5-minute review presentation of one section from this chapter. The format of the presentation can be a poster presentation, a blackboard presentation from notes, an online video, or a game format (for example, math jeopardy). The presentation should include the following:

1. Any important formulas in the chapter
2. Important skills in the chapter, backed up by examples
3. Explanation of terms used (for example, what are prime numbers?)
4. Common mistakes and how to avoid them

Building Blocks of Algebra

1.1 Exponents, Order of Operations, and Properties of Real Numbers

1.2 Algebra and Working with Variables

1.3 Simplifying Expressions

1.4 Graphs and the Rectangular Coordinate System

Many colleges charge students by credit hour (or unit) taken. To figure out how much tuition they owe for a semester (or quarter), the student must multiply the number of credit hours by the cost per credit hour. Then any additional fees, such as health fees or parking fees, must be added on. Algebra can be used to represent this calculation in a way that allows for different numbers of credit hours to be taken. In this chapter, we start to study the foundations of algebra.

1.1 Exponents, Order of Operations, and Properties of Real Numbers

LEARNING OBJECTIVES
- Evaluate exponents.
- Recognize scientific notation.
- Simplify using the order-of-operations agreement.
- Use the properties of real numbers.

Exponents

Multiplication of natural numbers is a way to write a repeated addition.

$$6 + 6 + 6 + 6 = 6 \cdot 4$$

A natural number exponent is a way to write a repeated multiplication of natural numbers.

$$2 \cdot 2 \cdot 2 \cdot 2 = 2^4$$

Here 2 is said to be the **base** (the number repeatedly multiplied), and 4 is said to be the **exponent** (the number of repeated multiplications).

> **DEFINITION**
>
> **Natural Number Exponent**   If a is a real number and n is a natural number, then
>
> $$a^n = \underbrace{a \cdot a \cdot a \cdot \ldots \cdot a}_{n}$$
>
> (a is multiplied by itself n times).
>
> a is called the **base**, and n is called the **exponent**.
>
>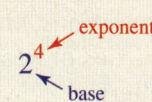

When working with natural number exponents, we say that $(2 \cdot 2 \cdot 2 \cdot 2)$ is in **expanded form** and (2^4) is in **exponential form**.

Example 1   Working with natural number exponents

Write each value as indicated.

a. Write 5^4 in expanded form.

b. Write $7 \cdot 7 \cdot 7 \cdot 7 \cdot 7 \cdot 7$ in exponential form.

c. Find the value of $(-3)^4$ without a calculator.

d. Find the value of 6^5 with a calculator.

SOLUTION

a. The base is 5, and the exponent of 4 tells us to multiply 5 by itself four times. The expanded form is $5 \cdot 5 \cdot 5 \cdot 5$.

b. The base is 7, and we are multiplying 7 by itself 6 times. So the exponent is 6. The exponential form is 7^6.

c. The base is -3, and we are multiplying -3 by itself four times.
$(-3) \cdot (-3) \cdot (-3) \cdot (-3) = 81$

d. Using a calculator, we have that $6^5 = 7776$.

PRACTICE PROBLEM FOR EXAMPLE 1

Write each value as indicated.

a. Write 16^5 in expanded form.

b. Write $29 \cdot 29 \cdot 29 \cdot 29$ in exponential form.

c. Find the value of $(-2)^5$ without a calculator.

d. Find the value of 5^6 with a calculator.

When calculating exponents, be very careful with negative signs. The expression -5^2 means the *opposite* of 5^2. Therefore,

$$-5^2 = -25$$

To raise a negative number to an exponent, the negative number *must be in parentheses*. The expression $(-5)^2$ means to square the number -5. Therefore,

$$(-5)^2 = (-5)(-5) = 25$$

Use the Concept Investigation to see what happens when we raise a number to the zero power.

■ CONCEPT INVESTIGATION

What happens when 0 is the exponent?

1. Fill in the following table, using your calculator.

2. What do you think that a number raised to the 0 power is in general?

3. Now try 0^0 in your calculator. Does this give the result you expected? Explain.

Exponential Expression	Calculator Value
4^0	1
3^0	
2^0	
1^0	

DEFINITION

Zero as an Exponent $a^0 = 1$ for all $a \neq 0$.

$$3^0 = 1 \quad (-8)^0 = 1$$

The expression 0^0 does not exist (or is said to be undefined).

Example 2 Exponents on negative numbers and zero as an exponent

Evaluate the following exponents.

a. -6^4 **b.** $(-8)^2$ **c.** 124^0 **d.** $(-2)^3$

SOLUTION

a. In the expression -6^4, we are finding the opposite of 6^4. $-6^4 = -1296$.

■ Calculator Details

Entering Exponents

On some scientific calculators, look for the key of the form x^y or y^x. To enter 4^3, you first enter 4, hit the exponent key, then enter the exponent 3.

You may then have to hit the = key to get a display of the answer, 64.

The exponent key on other scientific calculators looks like the ^ symbol. To enter 4^3, first enter 4, then ^ , and then the exponent of 3.

This will display the answer of 64.

■ Calculator Details

Squaring a Number

Most scientific calculators have an x^2 key. To find the value of 15^2 on a calculator, enter 15 then the x^2 key. To enter the square of a negative number, such as $(-11)^2$, depends on the programming of your calculator. To be safe, use the parentheses keys, usually located above the number 8 and 9 keys.

) then enter the x^2 key. The display should read 121.

■ Rules for Exponents x^2y^5

Zero as an Exponent

When any exponential expression with a base other than zero is raised to the power of zero, the expression will equal 1.

$$x^0 = 1 \quad (x) \neq 0$$

b. In the expression $(-8)^2$, the number -8 is being squared, so the result is positive. $(-8)^2 = (-8)(-8) = 64$.

c. Any nonzero number raised to the zero power is 1. Therefore, $124^0 = 1$.

d. The number -2 is being raised to the third power in the expression $(-2)^3$. Because there are three negative numbers multiplied together, the result is negative. $(-2)^3 = (-2)(-2)(-2) = -8$.

PRACTICE PROBLEM FOR EXAMPLE 2

Evaluate the following exponents.

a. $(-4)^2$ **b.** $(-5)^3$ **c.** 38^0 **d.** -7^2

■ Scientific Notation

When using positive exponents, we may encounter very large numbers. Large numbers are often displayed on a calculator in a special format. This format is a version of what is called **scientific notation.**

■ CONCEPT INVESTIGATION

What happened to my calculator display?

1. Use your calculator to fill in the following table.

Exponential Expression	Numerical Value
8^4	4096
8^5	
8^7	
8^{10}	
8^{13}	

2. What do you notice about the number your calculator displayed for 8^{13}?

Notice the appearance of the calculator display for the value of 8^{13}. Most calculators commonly show one of the three answers below. All three forms of the following calculator display have the same meaning; that is, the value of 8^{13}.

Input	Calculator Display	
8^{13}	5.497558139E11	Notice the E just before the 11.
8^{13}	5.497558139 ^{11}	Notice the space between the number and the exponent.
8^{13}	5.497558139 × 10^{11}	

A calculator display like those above means that the calculator has gone into *scientific notation* mode. The number is so large that it cannot be displayed all at once on the screen, so the calculator displays as much of the number as it can.

When a calculator displays 5.497558139E11, the number 11 to the right of the E represents the exponent on 10.

$$5.497558139E11 = 5.497558139 \times 10^{11}$$

When a calculator displays 5.497558139 ^{11}, it means

$$5.497558139\ ^{11} = 5.497558139 \times 10^{11}$$

$$5.497558139 \times 10^{11} = 549{,}755{,}813{,}900$$

Remember our number system has base 10. This means each place value is a **power** of 10. For example, 400 can be written as $400 = 4 \times 100$. Because $100 = 10^2$, 400 in scientific notation would be written as $400 = 4 \times 10^2$. A calculator may display this as 4E2 or 4 ^{2}. Be very careful to read 4 ^{2} correctly when there is a space between the 4 and the exponent of 2 on a calculator. It means 4×10^2, not 4 squared. The form that we use to write it by hand is 4×10^2. We say that 400 is in standard form and 4×10^2 is written using scientific notation.

> **What's That Mean?**
>
> **Power**
>
> Exponents are sometimes referred to as *powers*. In the expression 5^3, 3 is referred to as the *exponent* or the *power* of 5.

> **DEFINITION**
>
> **Scientific Notation** A number in the form $a \times 10^n$ where $1 \leq |a| < 10$ and n is an integer is said to be in scientific notation.

We now learn how to change scientific notation into standard form.

> **To Convert a Number Written in Scientific Notation to Standard Form**
>
> **The exponent on 10 is positive n:** Multiply by 10^n, which is equivalent to moving the decimal point n places to the right.

Example 3 Converting scientific notation to standard form

Write each of the following numbers in standard form.

a. 4.53×10^5 **b.** 9.32×10^7

SOLUTION

a. Multiply by 10^5, which is equivalent to moving the decimal point 5 spaces to the right.

$$4.53 \times 10^5 = 4.53 \cdot 100{,}000 = 453{,}000$$

Therefore, 4.53×10^5 has the standard form 453,000.

b. The exponent on 10 is 7. Move the decimal point seven places to the right.

$$9.32 \times 10^7 = 9.32 \cdot 10{,}000{,}000 = 93{,}200{,}000$$

Therefore, 9.32×10^7 has the standard form 93,200,000.

PRACTICE PROBLEM FOR EXAMPLE 3

Write each of the following numbers in standard form.

a. 1.02×10^6 **b.** 4.51×10^9

In Chapter 5, we will study scientific notation in more detail. For now, it is important to be able to read these numbers correctly from a calculator.

■ Order of Operations

In Chapter R, we studied four operations used to combine real numbers. Those operations were addition, subtraction, multiplication, and division. We have just learned a new way to combine numbers, namely, **exponentiation**. The list of basic operations now includes these five operations, which are used throughout algebra.

> **What's That Mean?**
>
> **Exponentiation**
>
> The operation of finding a power of a number is called *exponentiation*.

DEFINITIONS

Terms Two (or more) quantities that are added
Factors Two (or more) quantities that are multiplied
Basic Operations The basic operations used in algebra are as follows:

1. **Addition** The result of an addition is called a **sum.**
2. **Subtraction** The result of a subtraction is called a **difference.**
3. **Multiplication** The result of a multiplication is called a **product.**
4. **Division** The result of a division is called a **quotient.**
5. **Exponentiation** Exponential notation takes the form a^n, where a is called the **base** and n is the **exponent.**

The order-of-operations agreement must be updated to include this new operation, exponentiation. Recall that to use the order-of-operations agreement, terms are first identified. In an expression, terms are identified as the things that are added. Now we will see terms that include exponents.

Example 4 Identifying terms

Identify the terms in each expression.

a. $\dfrac{8}{4} - 16 + 9 \cdot 3 - 5^2$ **b.** $2^3 + 16 \div 4 \cdot 3$

SOLUTION

a. This expression has four terms. Remember that the terms are quantities that are added. When subtraction is involved we will rewrite the subtraction as addition and change the sign of the term following the plus sign. The terms are circled.

$$\dfrac{8}{4} - 16 + 9 \cdot 3 - 5^2$$

$$\dfrac{8}{4} + (-16) + 9 \cdot 3 + (-5^2)$$

$$\left(\dfrac{8}{4}\right) + (-16) + (9 \cdot 3) + (-5^2)$$

b. This expression has two terms. There are two terms separated by an addition symbol. The terms are circled.

$$(2^3) + (16 \div 4 \cdot 3)$$

PRACTICE PROBLEM FOR EXAMPLE 4

Identify the terms in each expression.

a. $3^4 - \dfrac{10}{2} + 5 \cdot 4$ **b.** $9^4 - 16 \cdot 3 + 9 - 6$

The order-of-operations agreement has been expanded to include exponents.

Steps to Use the Order-of-Operations Agreement

1. Evaluate any exponents inside each term, working from left to right.
2. Evaluate all of the operations (multiplications and divisions) inside of each term, working from left to right.
3. Add and subtract the terms, working from left to right.

Steps 2 and 3 of the order-of-operations agreement call for working on the multiplications and divisions in order from left to right, then working on the additions and subtractions in order from left to right. Think of evaluating an expression in the same order as English is read, from left to right, and it will seem natural to do the operations in that order.

SECTION 1.1 Exponents, Order of Operations, and Properties of Real Numbers 77

Example 5 Ordering the operations

Evaluate each expression.

a. $75 - 7^2$ **b.** $5^3 - 16 \div 8 - (-3)$ **c.** $4^3 \div 8 \cdot 7$

SOLUTION
Circle the terms to help focus on doing the correct operations.

a. Step 1 Evaluate any exponents inside each term, working left to right.

$75 - 7^2$
$= 75 + (-7^2)$ Rewrite the subtraction as addition.
$= \widehat{75} + \widehat{(-7^2)}$ Circle the terms.
$= 75 - 49$

Steps 2 and 3 Evaluate all of the operations (multiplications and divisions) inside of each term, working left to right. Add and subtract the terms, working from left to right.

$= 75 - 49$
$= 26$

b. This expression has three terms.

$5^3 - 16 \div 8 - (-3)$
$= 5^3 + (-16) \div 8 + 3$ Rewrite the subtraction as addition.
$= \widehat{5^3} + \widehat{(-16) \div 8} + \widehat{3}$ Circle the terms.
$= 125 + (-16) \div 8 + 3$ Do the exponent.
$= 125 + (-2) + 3$ Do the division.
$= 123 + 3$ Do the additions from left to right.
$= 126$

c. This expression has one term.

$\widehat{4^3 \div 8 \cdot 7}$
$= 64 \div 8 \cdot 7$ Do the exponent.
$= 8 \cdot 7$ Do the division.
$= 56$ Do the multiplication.

> **Calculator Details**
>
> **Order-of-Operations**
> Scientific calculators have been programmed with the order-of-operations agreement. If the expression is entered *exactly* the way it is written in the problem, the calculator will evaluate it correctly. For example, to enter $4^3 \div 8 \cdot 7$, enter
>
>
>
> The calculator will display the correct answer of 56.

PRACTICE PROBLEM FOR EXAMPLE 5

Evaluate each expression.

a. $6 + 3 \cdot 4 - 2 \div 2$ **b.** $8 \div 2 + 6 \cdot 3 - 3 \cdot 3 + 28 \div 4$ **c.** $5^2 - 16 + 9$

Now try to enter all of these expressions, exactly as they are written, all at once on a calculator. Do not hit Enter or = until the entire expression has been typed in. The results will be the same as those in Example 5. This is because calculators are programmed to evaluate expressions using the order-of-operations agreement.

More-complicated expressions involving grouping symbols, such as parentheses or absolute value bars, can appear. Therefore, the order-of-operations agreement has been expanded to include grouping symbols.

> **Connecting the Concepts**
>
> **What are grouping symbols?**
> Grouping symbols can include any of the following.
>
> **Parentheses:** $(9 - 16)$
> **Square brackets:** $[4 \cdot 12 - 10]$
> **Curly braces:** $\{2 \div 14 + 7\}$
> **Absolute value bars:** $|-2 + 5|$
> **Fraction bar:** $\dfrac{(6(-5) + 8)}{2}$
> **Square root:** $\sqrt{6 + 3}$

> **Steps to Perform the Order-of-Operations Agreement with Grouping Symbols**
>
> 1. Evaluate everything inside the grouping symbols first, inside to outside, working from left to right, using the basic order-of-operations agreement.
> 2. Evaluate all exponents, inside to outside, working from left to right.
> 3. Do all multiplications and divisions in the order they occur, working from left to right.
> 4. Do all additions and subtractions in the order they occur, working from left to right.

78   CHAPTER 1   Building Blocks of Algebra

Example 6 Evaluating expressions

Evaluate each expression using the order-of-operations agreement.

a. $8 \cdot 5 \div (12 - 10 \div 5) + 8 \cdot 4$

b. $(7 - 2)^3$

c. $(26 - 10 \cdot 2) \div 3 + \dfrac{18 - 3 \cdot 2}{2 + 1}$

d. $|17 - 4|^2 + \dfrac{10 \cdot 4}{2 \cdot 6 - 4} \cdot 3$

SOLUTION

a. Step 1 Evaluate everything inside the grouping symbols first, inside to outside, working from left to right, using the basic order-of-operations agreement.

$8 \cdot 5 \div (12 - 10 \div 5) + 8 \cdot 4$ *Do the operations inside the parentheses first.*

$= 8 \cdot 5 \div (12 - 2) + 8 \cdot 4$ *Divide inside the parentheses, then subtract.*

Steps 2 and 3 Evaluate all exponents, inside to outside, working from left to right. There are no exponents to evaluate for this expression. Do all multiplications and divisions in the order they occur, working from left to right.

$= 8 \cdot 5 \div 10 + 8 \cdot 4$

$= 40 \div 10 + 8 \cdot 4$

$= 4 + 32$

Step 4 Do all additions and subtractions in the order they occur, working from left to right.

$= 4 + 32$

$= 36$

b. $(7 - 2)^3$ *Do the operations inside the parentheses first.*

$= 5^3$ *Do the exponent second.*

$= 125$

c. The key to doing this problem correctly is to recall that a fraction bar is considered a grouping symbol. Work out the numerator and denominator separately. Then evaluate the division indicated by the fraction bar.

$(26 - 10 \cdot 2) \div 3 + \dfrac{18 - 3 \cdot 2}{2 + 1}$ *Evaluate inside the parentheses and the fraction bar first.*

$= (26 - 20) \div 3 + \dfrac{18 - 6}{3}$

$= 6 \div 3 + \dfrac{12}{3}$ *Do multiplications and divisions.*

$= 2 + 4$ *Do the addition.*

$= 6$

d. $|17 - 4|^2 + \dfrac{10 \cdot 4}{2 \cdot 6 - 4} \cdot 3$ *Evaluate within the absolute value and the fraction first.*

$= |13|^2 + \dfrac{40}{12 - 4} \cdot 3$ *Do the exponent next.*

$= 169 + \dfrac{40}{8} \cdot 3$ *Do the multiplications and divisions in order from left to right.*

$= 169 + 5 \cdot 3$

$= 169 + 15$ *Do the addition.*

$= 184$

What's That Mean?

PEMDAS

There is a handy mnemonic to help remember the order-of-operations agreement. It is PEMDAS, which can be remembered as Please Excuse My Dear Aunt Sally. The letters in PEMDAS stand for:

P = Parentheses (evaluate grouping symbols first, working from left to right)

E = Exponents (evaluate exponents second, working from left to right)

MD = multiplications and divisions (evaluate multiplications and divisions in the order in which they occur, working from left to right)

AS = additions and subtractions (evaluate additions and subtractions in the order in which they occur, working from left to right)

PRACTICE PROBLEM FOR EXAMPLE 6

Evaluate each expression using the order-of-operations agreement.

a. $16 \div (-8) \cdot 3 - (5 \cdot 3) + 12 \cdot 5$

b. $(-3 - 4)^2$

c. $|9 - 2 \cdot 3| + (-6) \div (-18 \div 3)$

d. $\dfrac{(-2)^2 + 4 \cdot 3}{7 - 2 \cdot 3} - [10 \div 2]$

■ Properties of Real Numbers

There are three very important properties about the operations of addition, subtraction, multiplication, and division. These properties allow us to perform operations on numbers quickly.

If the order in which an operation is performed does not matter (that is, does not change the result), then the operation is said to be **commutative.** For example, in the morning, most of us brush our hair and brush our teeth. It does not matter which brushing we perform first. This, mathematically, is a commutative operation.

Addition is commutative: $5 + 10 = 10 + 5$. The result is 15 regardless of which way 10 and 5 are added. The commutative properties state that real numbers can be added or multiplied in any order.

■ What's That Mean?

Property

In mathematics, the word *property* means an essential attribute or characteristic of something. This is different from the everyday use of the word *property*, which means something you own, such as land or a car.

> **COMMUTATIVE PROPERTY**
>
> If a and b are any real numbers, then the commutative properties are as follows:
>
> The **commutative property of addition** is $a + b = b + a$
>
> $$8 + 5 = 5 + 8$$
>
> The **commutative property of multiplication** is $a \cdot b = b \cdot a$
>
> $$7 \cdot 9 = 9 \cdot 7$$

■ What's That Mean?

Commutative

The word *commutative* comes from a French word that means "to switch or substitute." The commutative property says that we can switch the order of terms or factors in an operation and get the same result.

Suppose Natalie wants to buy a piece of candy for $0.90, a bag of chips for $1.25, and a soda for $1.75. If she adds the numbers in the given order using the order-of-operations agreement (PEMDAS) to compute $0.90 + 1.25 + 1.75$, she will first add $0.90 and $1.25, to get $2.15. Then Natalie has to add on $1.75 to get the final result of $3.90. But if she first adds $1.25 (chips) and $1.75 in her head to get $3.00 then adds on the additional $0.90 (candy), Natalie will get the same answer of $3.90. This alternative calculation can be written by using parentheses to group the cost for the chips and soda.

$$0.90 + (1.25 + 1.75)$$

It can be easier to combine the $1.25 and $1.75 to get $3.00 first then add on the $0.90 at the end. This is an example of the **associative property of addition.** The associative property of addition tells us that it is permissible to group the order of addition in any convenient way and get the same result.

> **ASSOCIATIVE PROPERTY**
>
> If a, b, and c are any real numbers, then the associative properties are as follows:
> The **associative property of addition** is $a + (b + c) = (a + b) + c$
>
> $$(0.90 + 1.25) + 1.75 = 0.90 + (1.25 + 1.75)$$
>
> The **associative property of multiplication** is $a \cdot (b \cdot c) = (a \cdot b) \cdot c$
>
> $$2(3 \cdot 5) = (2 \cdot 3) \cdot 5$$

Example 7   Identifying properties

Identify the property used in each statement.

a. $2 + 7 = 7 + 2$
b. $3(-2 \cdot 6) = (3 \cdot (-2)) \cdot 6$
c. $(3 + 5) + 10 = 3 + (5 + 10)$
d. $6(-41) = (-41) \cdot 6$

SOLUTION

a. $2 + 7 = 7 + 2$
This is an example of the commutative property of addition because the order of the terms has been changed.

b. $3(-2 \cdot 6) = (3 \cdot (-2)) \cdot 6$
This is an example of the associative property of multiplication because the order of the numbers was not changed, but how they are grouped was changed.

c. $(3 + 5) + 10 = 3 + (5 + 10)$
This is an example of the associative property of addition because the order of the numbers was not changed, but how they are grouped was changed.

d. $6(-41) = (-41) \cdot 6$
This is an example of the commutative property of multiplication because the order of the numbers was changed.

PRACTICE PROBLEM FOR EXAMPLE 7

Identify the property used in each statement.

a. $29 + (15 + 29) = (29 + 15) + 29$
b. $(-6) \cdot (-3) = (-3) \cdot (-6)$
c. $(5 \cdot 17) \cdot 3 = 5(17 \cdot 3)$
d. $-100 + 46 = 46 + (-100)$

CONCEPT INVESTIGATION

Which properties are not commutative?

1. Subtract the following integers. In the third column, note whether the results in the first column and second column are the same or different.

Subtract	Subtract	Same or Different?
$6 - 3 = 3$	$3 - 6 = -3$	Different
$10 - 5 =$	$5 - 10 =$	
$25 - 9 =$	$9 - 25 =$	
$8 - 9 =$	$9 - 8 =$	
$14 - 3 =$	$3 - 14 =$	

SECTION 1.1 Exponents, Order of Operations, and Properties of Real Numbers

2. Do you get the same answer when you reverse the order of subtraction?

3. Is subtraction commutative?

4. Perform the following division problems. In the third column, note whether the results in the first column and second column are the same or different.

Division	Division	Same or different?
$10 \div 2 = 5$	$2 \div 10 = 0.2$	Different
$16 \div 2 =$	$2 \div 16 =$	
$(-3) \div 1 =$	$1 \div (-3) =$	
$45 \div (-5) =$	$(-5) \div 45 =$	
$5 \div 2 =$	$2 \div 5 =$	

5. Do you get the same answer when you reverse the order of division?

6. Is division commutative?

Notice that both the commutative properties and associative properties are defined only for addition and multiplication. They do not work for subtraction and division. Order matters in problems involving these operations.

In the next Concept Investigation, we are going to explore another important property, called the **distributive property.** The goal is to understand when it is appropriate to use the distributive property.

CONCEPT INVESTIGATION
When is it okay to distribute?

1. Complete the following table using the first two rows as examples. In the third column, compare whether or not the results in the first two columns are the same or different. Make sure to use the order-of-operations agreement correctly.

		Same or Different?
$6(2 + 3)$ $= 6(5)$ $= 30$	$6 \cdot 2 + 6 \cdot 3$ $= 12 + 18$ $= 30$	Same
$4(5 + 6)$ $= 4(11)$ $= 44$	$4 \cdot 5 + 4 \cdot 6$ $= 20 + 24$ $= 44$	Same
$2(3 + 5)$	$2 \cdot 3 + 2 \cdot 5$	
$5(7 + 11)$	$5 \cdot 7 + 5 \cdot 11$	
$-2(8 - 4)$	$-2 \cdot 8 - (-2) \cdot 4$	

2. Do you always get the same value for columns 1 and 2?

3. Can you "distribute" a factor over the operation of addition or subtraction?

4. Complete the following table using the first two rows as examples. In the third column, compare whether or not the results in the first two columns are the same or different. Make sure you use the order-of-operations agreement correctly.

Calculator Details

Using Parentheses

To enter an expression such as $6(2 + 3)$ correctly on your calculator, enter it as it is written, including parentheses and the implied multiplication.

Your calculator will display 30.

If you do not enter the parentheses for the expression $6(2 + 3)$, the calculator will compute

The calculator will multiply the 6 and 2 and then add 3 to get the result 15.

To enter $6 \cdot 2 + 6 \cdot 3$, enter it exactly as written, including the operations.

The calculator will display 30.

		Same or Different?
$3(2 \cdot 5)$ $= 3(10)$ $= 30$	$(3 \cdot 2)(3 \cdot 5)$ $= 6 \cdot 15$ $= 90$	Different
$4(3 \cdot 2)$ $= 4(6)$ $= 24$	$(4 \cdot 3)(4 \cdot 2)$ $= (12)(8)$ $= 96$	Different
$-6(4 \div 2)$	$-6(4) \div -6(2)$	
$-3(-2 \cdot 5)$	$-3(-2) \cdot (-3)(5)$	
$5(9 \div 3)$	$(5 \cdot 9) \div (5 \cdot 3)$	

5. Do you always get the same value for columns 1 and 2?

6. Can you "distribute" a factor over the operation of multiplication or division?

In many problems, several operations are combined. The distributive property helps to simplify problems that combine multiplication and addition (or subtraction). Remember that the distributive property says that each term in the group must be multiplied.

> **DISTRIBUTIVE PROPERTY**
>
> If a, b, and c are real numbers, then the **distributive property** states
>
> $$a(b + c) = a \cdot b + a \cdot c$$
> $$a(b - c) = a \cdot b - a \cdot c$$

Skill Connection

Area of a Rectangle

The area of a rectangle with a length of l and a width of w is

$$l \cdot w$$

Example 8 — Using the distributive property

There are two bedrooms, side by side, that have to be carpeted. See the figure. This requires that the total square footage of both rooms be calculated.

a. Compute (separately) the areas of the two rooms. Find the total area by adding the areas of the two individual rooms.

b. Consider the two rooms to be one large room. First compute the total length (of both rooms). What is the width? Find the total area by multiplying the length times the width.

c. This is an example of which property?

SOLUTION

a. The area of the room on the right of the diagram is

$(16 \text{ ft})(10 \text{ ft}) = 160 \text{ ft}^2$

The area of the room on the left of the diagram is

$(12 \text{ ft})(10 \text{ ft}) = 120 \text{ ft}^2$

Therefore, the total area is

(16 ft)(10 ft) + (12 ft)(10 ft)
= 160 ft² + 120 ft²
= 280 ft²

b. The total length is 12 ft + 16 ft. The total area is

10 ft · (12 ft + 16 ft) = 10 ft · 28 ft = 280 ft²

c. Notice that we get the same total area using either method. This is an example of the distributive property.

Example 9 The distributive property versus the order-of-operations agreement

Evaluate the expression 7(9 + 3) as indicated.

a. Use the distributive property. **b.** Use the order-of-operations agreement.

c. Are the answers to parts a and b the same?

SOLUTION

a. Use the distributive property to evaluate.

$$7(9 + 3) = 7 \cdot 9 + 7 \cdot 3 \quad \text{Distribute the 7.}$$
$$= 63 + 21 \quad \text{Add.}$$
$$= 84$$

b. Use the order-of-operations agreement to evaluate the expression.

$$7(9 + 3) \quad \text{Simplify inside the parentheses first.}$$
$$= 7(12) \quad \text{Multiply.}$$
$$= 84$$

c. The answers to parts a and b are the same. It does not matter whether we use the distributive property first or follow the order of operations when we evaluate a problem like this one.

PRACTICE PROBLEM FOR EXAMPLE 9

Evaluate the expression 5(16 − 7) as indicated.

a. Use the distributive property. **b.** Use the order-of-operations agreement.

c. Are the answers to parts a and b the same?

Example 10 The distributive property

Evaluate the expression using the distributive property.

a. 4(3 + 6) − 10   **b.** −(4 − 11) + 34

SOLUTION

a. First distribute the 4 through the addition inside the parentheses.

$$4(3 + 6) - 10 = 4(3) + 4(6) - 10 \quad \text{Distribute the 4.}$$
$$= 12 + 24 - 10 \quad \text{Add and subtract.}$$
$$= 36 - 10$$
$$= 26$$

b. Because $-(4-11) = -1(4-11)$, distribute the negative 1 through the subtraction inside the parentheses. Watch your signs carefully.

$$-(4-11) + 34 = -(4) - (-11) + 34 \quad \text{Distribute the negative 1.}$$
$$= -4 + 11 + 34 \quad \text{Add.}$$
$$= 7 + 34$$
$$= 41$$

PRACTICE PROBLEM FOR EXAMPLE 10

Evaluate the expression using the distributive property.

a. $7(2+8) - 25$ **b.** $-(17-8) + 5$

1.1 Vocabulary Exercises

1. In the expression 5^3, 5 is the _____, and 3 is the _____.

2. _____ to the zero power is undefined; anything else to the zero power is _____.

3. The ability to group the order of addition in any convenient way is called _____.

4. The two or more quantities that are multiplied together are called _____.

5. We use _____ to write very large numbers in a more compact format.

6. The result of a multiplication is called a(n) _____. The result of a division is called a(n) _____.

7. The two or more quantities we are adding or subtracting are called _____.

8. We use the _____ multiply across addition or subtraction.

9. The result of an addition is called a(n) _____. The result of a subtraction is called a(n) _____.

10. Parentheses, absolute values, fraction bars, and square roots are examples of _____.

11. If the order in which an operation is performed does not matter, then the operation is _____.

12. Writing an _____ expression as repeated multiplication is the same as writing it in _____.

1.1 Exercises

1. Write in exponential form: $7 \cdot 7 \cdot 7 \cdot 7 \cdot 7$

2. Write in exponential form: $9 \cdot 9 \cdot 9 \cdot 9 \cdot 9 \cdot 9 \cdot 9$

3. Write in expanded form: 5^3

4. Write in expanded form: 3^4

5. Write in expanded form: $(-7)^3$

6. Write in expanded form: $(-1)^4$

7. Write in exponential form:
$(-3) \cdot (-3) \cdot (-3) \cdot (-3) \cdot (-3)$

8. Write in exponential form: $(-2)(-2)(-2)$

9. Write in exponential form: $\dfrac{2}{3} \cdot \dfrac{2}{3} \cdot \dfrac{2}{3}$

10. Write in exponential form: $\dfrac{4}{5} \cdot \dfrac{4}{5} \cdot \dfrac{4}{5} \cdot \dfrac{4}{5}$

11. Write in expanded form: $\left(\dfrac{5}{8}\right)^3$

12. Write in expanded form: $\left(\dfrac{6}{11}\right)^4$

SECTION 1.1 Exponents, Order of Operations, and Properties of Real Numbers

For Exercises 13 through 20, find each value without a calculator.

13. 2^3
14. 3^3
15. $(-2)^4$
16. $(-4)^2$
17. -5^2
18. -3^2
19. 15^0
20. 10^0

For Exercises 21 through 34, use a calculator to calculate each value of the exponential expressions.

21. 3^7
22. 8^5
23. $(-5)^2$
24. $(-3)^5$
25. -7^4
26. -5^4
27. $(-1)^3$
28. $(-1)^4$
29. $\left(\dfrac{3}{4}\right)^3$
30. $\left(\dfrac{1}{6}\right)^4$
31. $(0.35)^2$
32. $(1.42)^3$
33. -6^3
34. -4^3

For Exercises 35 through 42, convert each value in scientific notation to standard notation.

35. 1.034×10^6
36. 4.29×10^8
37. 9.3×10^7
38. 5.31×10^5
39. 4.001×10^9
40. 1.0095×10^8
41. -5.487×10^4
42. -1.295×10^{11}

For Exercises 43 through 52, identify the terms in each expression. Do not evaluate the expressions.

43. $4 \div 2 + 3 \cdot 27 - |-27|$
44. $15 + \dfrac{18}{3} - |-9 \cdot 4| - 22$
45. $8 \cdot 9 - 4 \cdot 16^2$
46. $9^3 - \dfrac{4 \cdot 3}{16}$
47. $\dfrac{3}{7} - \dfrac{1}{2} \div 8$
48. $\dfrac{6}{5} + \dfrac{5}{2} \cdot 6 - 12$
49. $(-4)^5 \cdot 2 \div 32$
50. $(-15)^2 \div 5 \cdot 7$
51. $(-64) \div 4 \div 2$
52. $81 \div (-9) \div 3$

For Exercises 53 through 66, evaluate each expression using the order-of-operations agreement.

53. $16 + 8 \div 2$
54. $9 - 18 \div 2$
55. $\dfrac{4}{5} - \dfrac{2}{3} \cdot \dfrac{9}{5}$
56. $8 \div \dfrac{16}{3} + \dfrac{4}{7}$
57. $3 \cdot 12 - 16 \div 4 + 1$
58. $4 \cdot 3 - (-7) + 16 \div 4$
59. $1 + \dfrac{7}{10} \cdot \dfrac{4}{5} - \dfrac{2}{5}$
60. $6 \div \dfrac{2}{5} + 4 \cdot \dfrac{3}{8} - 2$
61. $(-2)^3 + 4 \cdot (-6)$
62. $32 \div 8 + (-3)^2$
63. $-9 + 7 \cdot 3 - 15 \div 3 - 4^2 + (-8)$
64. $-(-6)^2 \cdot 3 + 16 \div (-4) - (-7)$
65. $-9 \cdot 3 + 5 - 15 \div 3 - 2^3$
66. $(-1)^4 + 9 \cdot (-2) - 6 \div 3$

For Exercises 67 through 80, evaluate each expression using the order-of-operations agreement. Pay attention to the grouping symbols.

67. $|16 - (3 \cdot 8)| + 9$
68. $4 + |-24 - 6|$
69. $\dfrac{4 \cdot 5 - 4}{6 - 4}$
70. $\dfrac{6 + 3 \cdot 5 - 7}{9 - 2}$
71. $3 \cdot \dfrac{-4 + 6 \cdot 8}{15 - 2^2} - 3^2$
72. $-5 + \dfrac{9 - 3}{2 \cdot 3} + (-2)^2$
73. $(18 - 2 \cdot 3)^3$
74. $10 + (-4 + 5 \cdot 2)^3$
75. $4 \div (-3 - 14 \div 2) - (-2)^2$
76. $-5 \cdot 4 - 2 + 3 \cdot 5^2 + |15 - 27|$
77. $(49 - 7 \cdot 2) \div 5 + \dfrac{30 - (-6) \cdot 2}{-4 + 2}$
78. $\dfrac{27 \cdot 2 + 3}{|3 - (6 \cdot 3)|}$
79. $|3 - 5 \cdot 6| + (-3 - 5)^2$
80. $(5 - 8)^3 + |-6 - 7 \cdot 2|$

For Exercises 81 through 92, identify the property used: the commutative property of addition or multiplication, the associative property of addition or multiplication, or the distributive property.

81. $3 + (2 + 5) = (3 + 2) + 5$
82. $(16 + 9) + 2 = 16 + (9 + 2)$

86 CHAPTER 1 Building Blocks of Algebra

83. $\frac{3}{5} \cdot 24 = 24 \cdot \frac{3}{5}$

84. $61\left(-\frac{5}{16}\right) = \left(-\frac{5}{16}\right) \cdot 61$

85. $-3(14 + 7) = -3 \cdot 14 + (-3) \cdot 7$

86. $2(5 + 3) = 2 \cdot 5 + 2 \cdot 3$

87. $(-11) + 19 = 19 + (-11)$

88. $3 + 9 = 9 + 3$

89. $3(9 - 2) = 3 \cdot 9 - 3 \cdot 2$

90. $-7(25 - 16) = -7 \cdot 25 - (-7) \cdot 16$

91. $(5 \cdot 3) \cdot 10 = 5(3 \cdot 10)$

92. $(9 \cdot 8)(-3) = 9(8(-3))$

93. The walls of a rectangular room need to be painted. This requires that the total square footage of the walls be calculated.

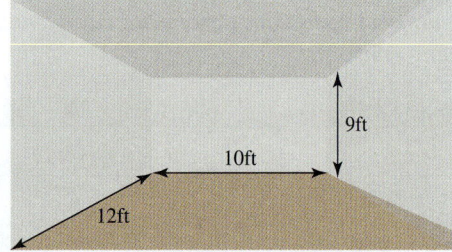

  a. Compute (separately) the areas of the four walls. Find the total area by adding the areas of the four walls.

  b. Consider the four walls as one long wall. First compute the total length (of all four walls). Find the total area by multiplying the total length times the height.

  c. Notice that parts a and b give the same answer. This is due to which property?

94. Consider that the room in Exercise 93 was 15 ft long by 12 ft wide and 10 ft tall.

  a. Compute (separately) the areas of the four walls. Find the total area by adding the areas of the four walls.

  b. Consider the four walls as one long wall. First compute the total length (of all four walls). Find the total area by multiplying the total length times the height.

  c. Notice that parts a and b give the same answer. This is due to which property?

95. A horse farmer wants to put new grass seed down in a few of her horse corrals. In order to purchase enough seed, she needs to compute the total area of the corrals.

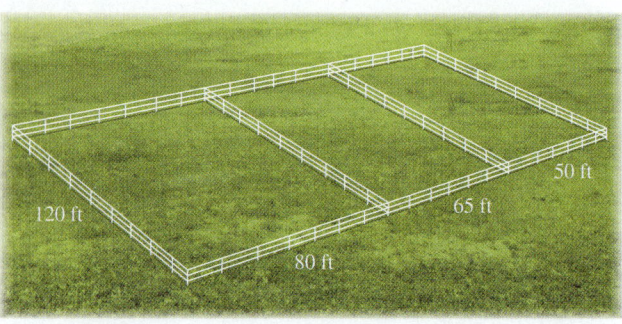

  a. Compute (separately) the areas of the three corrals. Find the total area by adding the areas together.

  b. Consider the three corrals as one long corral. First compute the total length (of all three). Find the total area by multiplying the total length times the width.

  c. Notice that parts a and b give the same answer. This is due to which property?

96. Consider that the corrals in Exercise 95 were 70 ft, 50 ft, and 40 ft long, and all were 75 ft wide. Find the total area of the corrals.

  a. Compute (separately) the areas of the three corrals. Find the total area by adding the areas together.

  b. Consider the three corrals as one long corral. First compute the total length (of all three). Find the total area by multiplying the total length times the width.

  c. Notice that parts a and b give the same answer. This is due to which property?

97. The owner of the room in Exercise 93 decided to install molding around the ceiling of the room. The molding chosen costs $2.18 per linear foot.

  a. Compute (separately) the cost for the molding for each wall. Find the total cost by adding the costs for each walls molding.

  b. Consider the walls as one long wall. First compute the total length of molding needed for the room. Find the total cost by multiplying the total length by the cost per linear foot.

  c. Notice that parts a and b give the same answer. This is due to which property?

98. The owner of the room in Exercise 94 decided to install molding around the ceiling of the room. The molding chosen costs $1.88 per linear foot.

  a. Compute (separately) the cost for the molding for each wall. Find the total cost by adding the costs for each walls molding.

  b. Consider the walls as one long wall. First compute the total length of molding needed for the room.

Find the total cost by multiplying the total length by the cost per linear foot.

  c. Notice that parts a and b give the same answer. This is due to which property?

99. The corrals in Exercise 95 also need the fencing replaced. Each linear foot of fencing costs $5.59.

  a. Compute (separately) the cost for the fencing per side needed. Find the total cost by adding the costs for each side of fence needed.

  b. Consider the sides of the corrals as one long fence. First compute the total length of fencing needed for the corrals. Find the total cost by multiplying the total length by the cost per linear foot.

  c. Notice that parts a and b give the same answer. This is due to which property?

100. The corrals in Exercise 96 also need the fencing replaced. Each linear foot of fencing costs $6.79.

  a. Compute (separately) the cost for the fencing per side needed. Find the total cost by adding the costs for each side of fence needed.

  b. Consider the sides of the corrals as one long fence. First compute the total length of fencing needed for the corrals. Find the total cost by multiplying the total length by the cost per linear foot.

  c. Notice that parts a and b give the same answer. This is due to which property?

101. Complete the following table by placing a check in each box that is true.

Operation	Commutative Property	Associative Property
Addition		
Subtraction		
Multiplication		
Division		

102. Complete the following table by answering true or false.

Equation	True or False?
$4(8 - 5) = 4 \cdot 8 - 4 \cdot 5$	
$2(3 \cdot 6) = 2 \cdot 3 \cdot 2 \cdot 6$	
$3(9 + 7) = 3 \cdot 9 + 3 \cdot 7$	
$8(4 \div 2) = 8 \cdot 4 \div 8 \cdot 2$	

For Exercises 103 through 112, use the distributive property to rewrite each expression. Do not evaluate the expressions.

103. $2(5 + 9)$
104. $2(8 + 7)$
105. $16(9 - 5)$
106. $9(8 - 10)$
107. $-2(8 + 7)$
108. $-2(5 + 10)$
109. $-2(7 - 16)$
110. $-1(-3 - 7)$
111. $\dfrac{7}{9}(11 + 15)$
112. $\dfrac{5}{8}(29 - 17)$

For Exercises 113 through 122, evaluate each expression using either the order-of-operations agreement or the distributive property.

113. $6(3 + 17)$
114. $8(14 + 9)$
115. $-5(6 + 27)$
116. $-7(15 + 3)$
117. $6(7 - 16) + 12$
118. $3(8 - 24) + 18$
119. $-(7 - 28) - 9$
120. $-(16 - 30) - 7$
121. $\dfrac{1}{2}(5 - 17)$
122. $\dfrac{3}{5}(-16 + 31)$

1.2 Algebra and Working with Variables

LEARNING OBJECTIVES
- Work with variables.
- Evaluate expressions.
- Convert units.
- Define variables.
- Translate sentences into expressions.
- Generate expressions from tables.

■ Constants and Variables

Rhind Papyrus

People have always wanted to solve problems that involved quantities that are unknown or that can vary. The need to solve these types of problems created the field of mathematics that we call *algebra*. In this course, we will study algebra, which is the branch of mathematics that solves problems involving quantities that are unknown or can change.

Algebra has been used since ancient times to solve problems. Ancient civilizations, such as those of Babylonia, Egypt, Greece, India, and China, stated problems using words and sentences and solved those problems using techniques that we will study. One example is to be found in the Rhind Papyrus (a paper document from Egypt, written in 1650 B.C.E.), which when translated states:

Divide 100 *loaves among* 10 *men including a boatman, a foreman, and a doorkeeper, who receive double portions. What is the share of each?*

Source: http://vmoc.museophile.com/algebra/

In this problem, the number of loaves for each man is unknown. One of the arithmetic operations involved is division. Like this problem, many algebra problems contain both arithmetic operations and an unknown quantity. Since algebra builds on arithmetic and arithmetic depends on numbers, the first topic we studied in Chapter R was sets of numbers.

In algebra, we use letters to represent quantities that can change, or vary. These letters are called **variables.** A value that cannot change is called a **constant.** Often the letters x, y, and z are used as variables, but be aware that any letter may be used. In application problems ("real-world" problems that come out of a situation), other letters that connect to the meaning of the quantity that is changing are often used. The letter o is avoided as a variable, though, because it can be confused with the number 0 (zero).

> **DEFINITIONS**
> **Variable** A variable is a symbol, usually a letter, used to represent a quantity that can change (vary).
> **Constant** A constant is a value that does not change.

Frequently variables are used in a problem in which a quantity is repeatedly computed. For example, suppose we are interested in computing your weekly salary over a period of 9 weeks. Since your pay is dependent on the number of hours you worked and the number of hours might vary each week, it is efficient to introduce a variable in this situation. Let the variable h represent the number of hours that you worked in a particular week.

Example 1 Varying the weekly pay

Kevin works for Patio Furniture Plus. He currently earns $11.25 an hour. His weekly pay can be computed as $11.25h$, where h is the number of hours he worked for a week. Here $11.25h$ means 11.25 multiplied by h.

a. If Kevin works 18 hours in a week, what is his weekly pay?

b. If Kevin works 29 hours in a week, what is his weekly pay?

SOLUTION

a. If Kevin works 18 hours in a week, substitute 18 for h to get

$$11.25(18) = 202.50$$

If Kevin works 18 *hours in a week, his weekly pay is* $202.50.

b. If Kevin works 29 hours in a week, then we can substitute 29 for h to get

$$11.25(29) = 326.25$$

If Kevin works 29 *hours in a week, his weekly pay is* $326.25.

PRACTICE PROBLEM FOR EXAMPLE 1

The Cruz family is planning a driving trip. They figure that with occasional rest stops, they will drive an average of 50 miles an hour. The distance they drive can be computed by $50t$, where t is the number of hours for which they drive.

a. What is the distance they will drive if they travel for 4.5 hours?

b. What is the distance they will drive if they travel for 7.25 hours?

In all application problems, write the answer using complete sentences, as in Example 1.

In Example 1, Kevin's weekly pay was given by $11.25h$. The number 11.25 (the multiplier of h) has a special name. It is called the **coefficient** of h.

> **DEFINITION**
>
> **Coefficient** The coefficient of a variable term is the numerical factor of the term.
>
> The coefficient of the term $4x$ is 4.

Connecting the Concepts

How can the product of a variable and constant be written?

A product of a variable and constant can be written in several ways.

$$11.25h$$
$$11.25 \cdot h$$
$$11.25(h)$$
$$(11.25)h$$
$$(11.25)(h)$$

The first way, $11.25h$, is the most common. When there is no operation symbol between a constant and a variable, the operation is a multiplication.

Example 2 — Managing the monthly budget

Rebecca gets paid monthly from her employer. The amount she makes each month depends on the number of hours she worked that month. Let d be the dollar amount of her paycheck for a month. Rebecca has to pay $550 a month for rent. After paying her rent, Rebecca has $d - 550$ left.

a. If Rebecca earns $1200 in a month, how much money does she have left after paying her rent?
b. If Rebecca earns $1450 in a month, how much money does she have left after paying her rent?

SOLUTION

a. If Rebecca earns $1200 in a month, then substitute in $d = 1200$ and get
$$1200 - 550 = 650$$
If Rebecca earns $1200 in a month, she will have $650 left after paying her rent.

b. If Rebecca earns $1450 in a month, then substitute in $d = 1450$ and get
$$1450 - 550 = 900$$
If Rebecca earns $1450 in a month, she will have $900 left after paying her rent.

■ Evaluating Expressions

In Examples 1 and 2, we substituted in a value for the variable and then simplified the remaining expression. This process is called **evaluating an expression.**

> **Steps to Evaluate an Expression**
> 1. Substitute the given value of the variable in the expression.
> 2. Simplify, using the order-of-operations agreement.

Example 3 — Evaluating an expression

Evaluate each of the variable expressions for the given value(s) of the variable.

a. The fees for taking c credits at Grossmont College are $19 + 46c$ dollars. Find the fees for taking 6 credits, 9 credits, and 12 credits.

b. The cost for renting a 29-foot U-Haul truck for 1 day is $29.95 + 0.79m$ dollars, where m is the number of miles driven. Find the cost for renting the truck if you drive 25 miles. Find the cost for renting the truck if you drive 72 miles.

SOLUTION

a. **Step 1** Substitute the given value of the variable into the expression.

To find the cost of taking 6 credits, substitute $c = 6$ into expression.
$$19 + 46(6)$$

Step 2 Simplify, using the order-of-operations agreement.
$$= 19 + 276$$
$$= 295$$

What's That Mean?

Credits and Units

Depending on which part of the United States a college is located in, students will refer to the weight or value of a class in credits or units. For an undergraduate student in the United States, 12 units or more is considered full-time enrollment. Formally, once a student successfully completes a course, units turn into credits.

The cost of taking 6 credits is $295.00.

To find the cost of taking 9 credits, substitute $c = 9$ into the expression and evaluate.
$$19 + 46(9) = 433$$

The cost of taking 9 credits is $433.00.

To find the cost of taking 12 credits, substitute $c = 12$ into the expression and evaluate.
$$19 + 46(12) = 571$$

The cost of taking 12 credits is $571.00.

b. To find the cost for renting the truck if you drive 25 miles, substitute $m = 25$ into the expression and evaluate.
$$29.95 + 0.79(25) = 49.70$$

The cost of renting the truck and driving 25 miles is $49.70.

To find the cost for renting the truck if you drive 72 miles, substitute $m = 72$ into the expression and evaluate.
$$29.95 + 0.79(72) = 86.83$$

The cost of renting the truck and driving 72 miles is $86.83.

Note that we round to the cents (hundredths) place, since the units are dollars.

PRACTICE PROBLEM FOR EXAMPLE 3

Evaluate each of the following expressions.

a. The cost in dollars of printing n pages at Palomar College is $0.10n$ when it costs $0.10 a page to print. Find the cost of printing 5, 10, and 25 pages.

b. The average home air conditioner uses $3500h$ watt-hours of electricity when it runs for h hours. Find the number of watt-hours of electricity used by the average home air conditioner in 4, 8, and 24 hours.

Source: http://michaelbluejay.com/electricity/cooling.html

Example 4 Evaluating algebraic expressions

Evaluate each of the following expressions for the given value of the variable.

a. $6n + 8$ for $n = -2$

b. $2y^2 - \dfrac{12}{y} + 4$ for $y = 3$

c. $-4(t \div 6) + (9 - 2)^2$ for $t = -18$

SOLUTION

a.
$6n + 8$ *Substitute in $n = -2$.*
$= 6(-2) + 8$ *Simplify.*
$= -12 + 8$
$= -4$

b. $2y^2 - \dfrac{12}{y} + 4$ *Substitute in $y = 3$ for every y in the expression.*

$= 2(3)^2 - \dfrac{12}{3} + 4$ *Simplify using the order-of-operations agreement.*

$$= 2(9) - 4 + 4$$
$$= 18 - 4 + 4$$
$$= 14 + 4$$
$$= 18$$

c. $-4(t \div 6) + (9 - 2)^2$ *Substitute in $t = -18$.*
$= -4[(-18) \div 6] + (9 - 2)^2$ *Add **brackets** to distinguish from **parentheses**.*
$= -4(-3) + 7^2$ *Simplify using the order-of-operations agreement.*
$= 12 + 49$
$= 61$

Connecting the Concepts

When do we use brackets?

In Example 4c, we can use either two sets of parentheses

$$((-18) \div 6)$$

or a pair of brackets and then parentheses:

$$[(-18) \div 6]$$

Often we use the second form, $[(-18) \div 6]$, because it is easier to read.

PRACTICE PROBLEM FOR EXAMPLE 4

Evaluate each of the following expressions for the given value of the variable.

a. $-3h + 12$ for $h = 7$
b. $-3x^2 + 6x + 1$ for $x = -2$
c. $|-t + 6| + 2t^2 - 1$ for $t = 5$

Unit Conversions

We often have to convert from one system of units to another. For example, suppose you take a trip to Mexico. As of this writing, $1.00 (U.S.) is equivalent to 18.8104 MXN (Mexican pesos) (*Source: www.XE.com*). You want to buy a gift for your mother on a trip to Mexico, and it costs 142 pesos. How many U.S. dollars is that equivalent to? To convert from one system of units (in this example, pesos) to another (U.S. dollars), we rely on the concept of a **unity fraction**.

© Gary/Fotolia

What's That Mean?

Units

The *units* of a quantity refer to the way the quantity is being measured. When we say that 5 meters has the units of meters, we mean that the quantity 5 is being measured in meters.

What's That Mean?

Unity Fraction

In fields such as chemistry and physics, unity fractions are called **conversion factors**.

> **DEFINITION**
>
> **Unity Fraction**   A unity fraction is a fraction that has units in the numerator and denominator and whose simplified value is equal to 1.

For example, 3 feet = 1 yard, so $\dfrac{3 \text{ feet}}{1 \text{ yard}}$ is a unity fraction (since the value of the fraction is 1). To convert units, multiply the given quantity by the appropriate unity fraction. Place the units we are converting to in the numerator and the units to change from in the denominator. Units can be divided out when multiplying fractions. Therefore, to convert 142 pesos to U.S. dollars, we multiply as follows:

$$142 \text{ pesos} \cdot \frac{1 \text{ dollar}}{18.8104 \text{ pesos}}$$

$$= \frac{142 \, \cancel{\text{pesos}}}{1} \cdot \frac{1 \text{ dollar}}{18.8104 \, \cancel{\text{pesos}}} \approx 7.55 \text{ dollars}$$

Steps to Convert Units
1. Find the appropriate unity fraction needed for the problem. Put the units that are wanted in the final answer in the numerator and the units to be converted from in the denominator.
2. Multiply the given value by the unity fraction. Divide out the common units.
3. Reduce the fraction to lowest terms.

Note: Common unit conversions are found at the back of the book.

What's That Mean?

Unit Abbreviations

Here is a list of abbreviations for commonly used units.

Length:
in. = inch
ft = foot
mi = mile
cm = centimeter
m = meter
km = kilometer

Weight:
lb = pound
kg = kilogram
g = gram

Capacity:
qt = quart
gal = gallon
oz = ounce
l = liter

Dry Measurement:
tsp = teaspoon
tbl = tablespoon

Area:
ft^2 = square feet
yd^2 = square yard

Example 5 Convert units

Convert units as indicated. Round to the hundredths place if necessary.

a. Convert 56 cm (centimeters) to inches (in.). *Note:* There are 2.54 cm in 1 in.

b. Convert 2640 ft (feet) to miles (mi). *Note:* There are 5280 ft in 1 mi.

c. Convert 80 qt (quarts) to gallons (gal). *Note:* There are 4 qt in 1 gal.

d. Convert 60 kg (kilograms) to pounds (lb). *Note:* There are approximately 2.2 lb in 1 kg.

SOLUTION

a. **Step 1** Find the appropriate unity fraction needed for the problem. Put the units that are wanted in the final answer in the numerator and the units to be converted from in the denominator. We want to convert from the units of centimeters, so the units of centimeters go into the denominator. We want to convert to the units of inches, so inches go in the numerator. There are 2.54 cm in 1 inch, so the unity fraction is $\dfrac{1 \text{ in}}{2.54 \text{ cm}}$.

Step 2. Multiply the given value by the unity fraction. Divide out the common units.

$$56 \text{ cm} \cdot \dfrac{1 \text{ in}}{2.54 \text{ cm}}$$

$$= \dfrac{56 \text{ cm}}{1} \cdot \dfrac{1 \text{ in}}{2.54 \text{ cm}}$$

Step 3 Reduce the fraction to lowest terms.

$$= \dfrac{56 \text{ in}}{2.54} \approx 22.05 \text{ in}$$

Therefore, 56 cm is equal to approximately 22.05 in.

b. First, form the unity fraction. We want to convert to the units of miles, so miles go into the numerator. We want to convert from the units of feet, so feet go into the denominator. There are 5280 ft in 1 mi, so the unity fraction is $\dfrac{1 \text{ mi}}{5280 \text{ ft}}$. Multiply 2640 ft by this unity fraction and simplify the expression as follows:

$$2640 \text{ ft} \cdot \dfrac{1 \text{ mi}}{5280 \text{ ft}}$$

$$= \dfrac{2640 \text{ ft}}{1} \cdot \dfrac{1 \text{ mi}}{5280 \text{ ft}} = 0.5 \text{ mi}$$

Therefore, 2640 ft is equal to 0.5 mi.

Connecting the Concepts

What's in the denominator?

A value that does not have a denominator can be rewritten as a fraction with a denominator of 1.

$$80 = \dfrac{80}{1}$$

c. There are 4 qt in 1 gal, so the unity fraction is $\dfrac{1 \text{ gal}}{4 \text{ qt}}$. Multiply 80 qt by this unity fraction and simplify the expression as follows:

$$80 \text{ qt} \cdot \dfrac{1 \text{ gal}}{4 \text{ qt}}$$

$$= \dfrac{80 \text{ qt}}{1} \cdot \dfrac{1 \text{ gal}}{4 \text{ qt}} = 20 \text{ gal}$$

Therefore, 80 qt is equal to 20 gal.

d. There are approximately 2.2 lb in 1 kg, so the unity fraction is $\dfrac{2.2 \text{ lb}}{1 \text{ kg}}$. Multiply 60 kg by this unity fraction and simplify the expression as follows:

$$60 \text{ kg} \cdot \dfrac{2.2 \text{ lb}}{1 \text{ kg}}$$

$$= \dfrac{60 \text{ kg}}{1} \cdot \dfrac{2.2 \text{ lb}}{1 \text{ kg}} = 132 \text{ lb}$$

Therefore, 60 kg is approximately equal to 132 lb.

PRACTICE PROBLEM FOR EXAMPLE 5

Convert units as indicated. Round to the hundredths place if necessary.

a. 500 feet (ft) to meters (m). There are approximately 3.28 ft in 1 m.

b. 42 oz (ounces) to grams (g). There are approximately 28.35 g in 1 oz.

c. 156 pounds (lb) to kilograms (kg). There are approximately 2.2 lb in 1 kg.

■ Defining Variables

A variable is a quantity that can change or is unknown. Defining variables and translating the problem statement into an expression allows us to evaluate the expression for any value of the variable. In an application problem that represents a situation, to define a variable means to write a statement of what the variable represents, including units.

Example 6 Defining the variables

Define variables for each problem situation below. This means to write a sentence that explains what the variable represents. Include the units of the variable.

a. Your weekly paycheck depends on how many hours you worked for that week.

b. You are driving to San Francisco from San Diego. How far you have driven since you left San Diego depends on your speed and how long you have been driving.

SOLUTION
Here are some reasonable definitions for the variables.

a. In this problem, there are two quantities that can change. One quantity is the number of hours worked for a given week, and the second is the amount of that week's paycheck. Let

$h =$ time worked each week in hours

$P =$ amount of the weekly paycheck in dollars

b. The three quantities that can change are the distance you have traveled, the speed of the car, and the number of hours spent driving. Let

$$d = \text{distance traveled in miles}$$
$$h = \text{time driven in hours}$$
$$s = \text{speed of the car in miles per hour (mph)}$$

PRACTICE PROBLEM FOR EXAMPLE 6

Define variables for each problem situation. Include the units of the variable.

a. The temperature depends on the time of day.

b. The cost of your tuition for the semester depends on how many credits you are taking.

■ Translating Sentences into Expressions

After defining variables for a problem situation, we then translate the problem into an algebraic expression. When translating applications into algebraic expressions, it helps to know how to translate some commonly used phrases.

Addition

Phrases Used for Addition	Translation
The sum of n and 4	$n + 4$
The total of 7 and x	$7 + x$
f increased by 16	$f + 16$
9 plus c	$9 + c$
100 more than t	$t + 100$
exceeds c by 50	$c + 50$
P added to 3	$3 + P$

Because addition is a commutative operation, we could have written any of the above in the opposite order. Thus, the statement *P added to 3* translates as both $3 + P$ and $P + 3$.

Subtraction

Phrases Used for Subtraction	Translation
92 minus y	$92 - y$
The difference of 24 and n	$24 - n$
9 subtracted from t	$t - 9$
W reduced by 8	$W - 8$
n decreased by 8	$n - 8$
100 less than x	$x - 100$

■ Connecting the Concepts

When do we reverse the order in subtraction?

Be very careful when translating *less than* and *subtracted from*. Notice that these are the only phrases we have seen so far for which you reverse the order of the terms. Ask yourself what number is being subtracted, and always put that number after the minus sign.

Subtraction is *not* a commutative operation, so the order does matter in the above statements. Remember that $5 - 3 = 2$ is different from $3 - 5 = -2$.

Multiplication

Phrases Used for Multiplication	Translation
The product of 6 and m	$6m$
4 times r	$4r$
Twice z	$2z$
Half of x	$\frac{1}{2}x$
30% of n	$0.30n$

Multiplication is a commutative operation, so order does not matter in the above translations. For this reason, $6m$ is equal to $m6$. However, it is traditional in algebra to write the number part first and the variable part last, so always write the final answer in the form $6m$. Recall that in the expression $6m$, 6 is called the coefficient of m.

Division

Phrases Used for Division	Translation
The quotient of n and 17	$\frac{n}{17}$
The ratio of 5 and x	$\frac{5}{x}$
Q divided by 12	$\frac{Q}{12}$

Division is *not* a commutative operation, so order matters in the above translations. Remember that $6 \div 3 = 2$ is different from $3 \div 6 = \frac{1}{2}$.

Example 7 Translating a sentence

Translate each of the following phrases into expressions. Note that the phrase "*a number*" translates as a variable.

a. Twice a number added to 6

b. The product of 3 and the difference of a number and 5

c. The quotient of 8 and a number

d. 9 less than a number

SOLUTION

Let $x =$ a number.

a. Twice a number means to multiply a number by 2. So "twice a number" translates as $2x$. Then we add this to 6. The sentence translates as $6 + 2x$ or $2x + 6$.

b. We are told to find the product of 3 and something. So we know we need to multiply 3 by something. That something is "the difference of a number and 5." This translates as $x - 5$. So the product of 3 and the difference of a number and 5 translates as $3(x - 5)$.

c. Quotient means to divide, and order matters. We divide 8 by a number. Therefore, the translation is $\dfrac{8}{x}$.

d. Remember that "less than" translates as a subtraction, but it is done in reverse order. So this translates as something minus 9. The "something" is a number, which is x. The translation is $x - 9$.

PRACTICE PROBLEM FOR EXAMPLE 7

Translate each of the following sentences into expressions. Note that the phrase "a number" translates as a variable.

a. 5 times a number

b. The sum of 4 times a number and 9

c. The ratio of 6 and a number

d. 16 subtracted from twice a number

Example 8 Defining variables and translating sentences

Define the variables. Translate each phrase into an expression.

a. As of 2017, the fees at Oregon Coast Community College in Newport, Oregon, are $99.00 per credit hour. Find an expression for the total cost of taking a variable number of units.

Source: www.occc.cc.or.us

b. As of 2017, the Internal Revenue Service set the standard mileage rate for the cost of operating a car for business at $0.535 a mile. Find an expression for the total cost of operating a car for a variable number of miles driven.

Source: www.irs.gov

c. 16 less than a number

SOLUTION

First, define the variables.

a. Let c = number of credit hours a student is registered for. The total cost is the expense of taking c credit hours. The cost per credit hour is $99.00, so the total cost of taking c credits is

$$99c \text{ dollars}$$

b. Let m = number of miles driven for business. The total cost of operating a car will be the expense of driving m miles. The cost per mile is $0.535, so the cost of driving m miles is

$$0.535m \text{ dollars}$$

c. Let x = an unknown number. The phrase translates as

$$x - 16$$

Remember that *less than* translates in expression form in reverse order from the sentence form.

98 CHAPTER 1 Building Blocks of Algebra

PRACTICE PROBLEM FOR EXAMPLE 8

Define the variables. Translate each phrase into an expression.

a. 29 subtracted from a number

b. At Grand Valley State University, an adult ticket to a basketball game costs $9. Find an expression for the total cost of buying a variable number of tickets.

c. The total of 150 and three times a number

In converting units, it makes sense to use a variable if we need to repeatedly convert the same units in a problem situation.

Example 9 Defining variables and converting units

Define the variables. Write expressions for each of the following unit conversions.

a. Convert a variable number of feet to miles. There are 5280 ft in 1 mi.

b. Convert a variable number of kilograms (kg) into pounds (lb). There are approximately 2.2 lb in 1 kg.

c. Convert a variable number of square feet (ft²) into square yards (yd²). There are 9 ft² in 1 yd².

SOLUTION

a. Let x = number of feet.

$$x \text{ ft} = \frac{x \text{ ft}}{1} \cdot \frac{1 \text{ mi}}{5280 \text{ ft}} = \frac{x}{5280} \text{ mi}$$

Notice that the units of the final output are in miles.

b. Let x = number of kilograms.

$$x \text{ kg} \approx \frac{x \text{ kg}}{1} \cdot \frac{2.2 \text{ lb}}{1 \text{ kg}} \approx 2.2x \text{ lb}$$

Notice that the units of the final answer are in pounds.

c. Let x = number of square feet.

$$x \text{ ft}^2 = \frac{x \text{ ft}^2}{1} \cdot \frac{1 \text{ yd}^2}{9 \text{ ft}^2} = \frac{x}{9} \text{ yd}^2$$

Notice that the units of the final answer are in square yards.

PRACTICE PROBLEM FOR EXAMPLE 9

Define the variables. Write expressions for each of the following unit conversions.

a. Convert a variable number of miles to kilometers. There are approximately 1.609 km (kilometers) in 1 mile.

b. Convert a variable number of teaspoons (tsp) to tablespoons (tbl). There are 3 teaspoons in 1 tablespoon.

Example 10 · Using units at home

You want to estimate the cost of installing new sod on your front lawn. You measure your front yard and find that it is 20 ft by 15 ft.

a. What is the area of your front yard? (Recall that the area of a rectangle is length times width.)

b. What are the units on your answer to part a?

c. Sod is sold by the square foot. The type of sod that you decide on, Marathon, costs $0.70 per square foot. How much will it cost you to buy the sod for your lawn?

Source: www.greenthumb.com

SOLUTION

a. To find the area of a rectangle, multiply length times width. The area is
$(20 \text{ feet})(15 \text{ feet}) = 300 \text{ feet}^2$.
The area of the front yard is 300 feet².

b. The units are square feet, or feet² = ft².

c. The cost will be
$$\left(\frac{300 \text{ ft}^2}{1}\right)\left(\frac{\$0.70}{1 \text{ ft}^2}\right) = \$210.00$$

The cost for 300 ft² of sod is $210.00.

Notice that when we multiply, the units of square feet divide out and the final units are dollars ($).

PRACTICE PROBLEM FOR EXAMPLE 10

You want to estimate the cost of installing laminate flooring in your living room. Your living room is 15 ft by 18 ft.

a. What is the area of your living room? (Recall that the area of a rectangle is length times width.)

b. What are the units on your answer in part a?

c. The laminate flooring that you decide on costs $2.38 per square foot. What is the total cost of flooring for your living room?

■ Generating Expressions from Input–Output Tables

Many times, a table may be used to help generate an expression for a problem situation. These tables are often called **input–output** tables. The *input*, or **independent** *variable*, may be thought of as the data or information that we are entering into the problem. The *output*, or **dependent** *variable*, is the information that results from, or is produced from, the problem. In a table, the input variable is placed in the left-hand column, and the output variable is placed in the right-hand column. In generating expressions from an Input–Output table, substitute enough values for the input variable so that a pattern emerges in the calculation of the output variable. Write out all the operations in the output column. Don't write just the simplified result.

■ What's That Mean?

Independent and Dependent

The independent variable is not changed or controlled by another variable. The dependent variable tends to be changed by the independent variable.

Steps to Generate Expressions from Input–Output Tables

1. Determine the input variable and the output variable. Put the input variable in the left column and output variable in the right column.
2. Select four or more reasonable values for the input variable. Think carefully about the problem situation. The last row of the input column should be the general variable used for the input (n, x, and so forth).
3. Calculate the value of the outputs for the chosen inputs. Make sure to write out all of the operations involved in computing the output.
4. Find the pattern used to calculate the outputs from the inputs.
5. Write the general algebraic expression in the last row.

Example 11 — Generating expressions from input–output tables

Use an input–output table to generate expressions for the following problem situations.

a. Convert h hours into days by completing the input–output table.

h = number of hours	d = number of days
12	
24	
36	
72	
h	

b. There is 1 inch of rain in 1 foot of snow. Convert inches of snow into inches of rain by completing the input–output table.

s = inches of snow	r = inches of rain
12	
24	
36	
42	
s	

c. Rex's Shipping Company boxes 1-quart cans of paint and 1-quart cans of glaze to ship to home improvement stores. Each shipping box can hold 24 cans. Determine how many cans of glaze (g) you can include in one shipping box if you ship 0, 1, 2, 3, 10, or p 1-quart cans of paint. Remember that one shipping box can hold only 24 cans.

p = cans of paint	g = cans of glaze
0	
1	
2	
3	
10	
p	

SECTION 1.2 Algebra and Working with Variables 101

SOLUTION

Notice in the solutions that the variables are defined and that the operations used to compute the output are clearly written out. Doing so lets us see the pattern from the inputs to the outputs. Once we know what operations and constants are used in the pattern, we can generalize the pattern to find the general variable expression.

a.

h = number of hours	d = number of days
12	$\frac{12}{24} = 0.5$
24	$\frac{24}{24} = 1$
36	$\frac{36}{24} = 1.5$
72	$\frac{72}{24} = 3$
h	$\frac{h}{24}$

Take the number of hours and divide it by 24 (there are 24 hours in a day) to get the number of days. The general variable expression is $\frac{h}{24}$.

b.

s = inches of snow	r = inches of rain
12	$\frac{12}{12} = 1$
24	$\frac{24}{12} = 2$
36	$\frac{36}{12} = 3$
42	$\frac{42}{12} = 3.5$
s	$\frac{s}{12}$

To get the number of inches of rain, divide the number of inches of snow by 12. The general variable expression is $\frac{s}{12}$.

c.

p = cans of paint	g = cans of glaze
0	$24 - 0 = 24$
1	$24 - 1 = 23$
2	$24 - 2 = 22$
3	$24 - 3 = 21$
10	$24 - 10 = 14$
p	$24 - p$

To get the number of cans of glaze, subtract the number of cans of paint from 24. The general variable expression is $24 - p$.

Example 12 Using tables to explore tuition costs

Northern Virginia Community College charges a fee of $163.15 per credit hour plus $14.10 in other mandatory student fees per credit hour.

a. Fill in the following input–output table with some reasonable inputs to determine an expression for the total cost (tuition) for one semester.

n = number of credits	T = total cost (tuition)
n	

b. Suppose that a student has $1800 budgeted for tuition for one semester. Approximately how many credits can the student take?

SOLUTION

a. Here are some reasonable inputs.

n = number of credits	T = total cost (tuition)
0	0
3	$163.15(3) + 14.10(3) = 531.75$
6	$163.15(6) + 14.10(6) = 1063.50$
12	$163.15(12) + 14.10(12) = 2127.00$
n	$163.15(n) + 14.10(n)$

b. Notice that 6 credits cost $1063.50 and 12 credits cost $2127.00. The number $1800 is between these two values and a bit closer to $1063.50. Extend the table to get a good estimate of how many credits a student can take for $1800.

n = number of credits	T = total cost of tuition (in dollars)
11	$163.15(11) + 14.10(11) = 1949.80$
10	$163.15(10) + 14.10(10) = 1772.50$

On the basis of this information, a student can enroll in 10 or fewer credits.

PRACTICE PROBLEM FOR EXAMPLE 12

Carla and Don are getting married and are trying to figure out how many people they can invite to their reception. They have determined that they have a maximum of $3500 to pay for the food at the reception. They have also decided on a no-host bar, so liquor costs will not be paid for by the bride and groom. The New Day Catering Co. charges $425.00 for the cake and $22.00 per person for the reception dinner.

a. Generate an expression for the reception costs per person for Carla and Don. Use an Input–Output table to help generate the expression.

b. Approximately how many people can Carla and Don invite?

1.2 Vocabulary Exercises

1. A symbol used to represent a quantity that can change is called a(n) _____.

2. A(n) _____ is a value that does not change.

3. A(n) _____ is used to convert units and has a simplified value equal to 1.

4. The process of substituting a value for the variable and then simplifying the remaining expression is called _____ the expression.

5. The numerical factor of a term is called the _____.

6. When defining variables, we should always include the _____ of the variable.

7. The words *sum*, *total*, and *plus* all represent _____.

8. When translating sentences involving subtraction into expressions, the phrases _____ and _____ require you to reverse the order of the terms.

9. The phrases *ratio of*, _____, and _____ represent division.

10. The output variable is also called the _____ variable. The input variable is also called the _____ variable.

1.2 Exercises

For Exercises 1 through 12, find the value of each expression for the given value of the variable.

1. Sarah's hourly wage is $14.75 working as a mathematics tutor. Her weekly pay can be computed as $14.75h$, where h is the number of hours she worked for a week. What would Sarah's weekly pay be if she worked 15 hours in a week?

2. Josh's hourly wage is $14.25 working as a biology lab assistant. His weekly pay can be computed as $14.25h$, where h is the number of hours he worked for a week. What would Josh's weekly pay be if he worked 13 hours in a week?

3. Thomas sells wrapping paper to raise money for his elementary school. The wrapping paper costs $8.00 per roll. The total revenue Thomas earns for his school can be computed as $8.00p$, where p is the number of rolls of wrapping paper he sells. How much revenue would Thomas earn if he sells 53 rolls of wrapping paper?

4. Brin sells Girl Scout cookies that cost $5.00 a box. The total revenue Brin earns for the Girl Scouts can be computed as $5.00b$, where b is the number of boxes she sells. How much revenue would Brin earn if she sells 400 boxes of cookies?

5. Marcus works in sales at an electronics store. His commission is 7% of his total sales for the week. His commission can be computed as $0.07s$, where s is his total weekly sales in dollars. What will Marcus's commission be if he sells $1045 worth of merchandise in a week?

6. Keesha works in sales at a jewelry store. Her commission is 6.75% of her total sales for the week. Her commission can be computed as $0.0675s$, where s is her total weekly sales in dollars. What will Keesha's commission be if she sells $2750 worth of jewelry in a week?

7. Vinny runs at a pace of 7.5 miles per hour. His total miles run for a day can be computed as $7.5h$, where h is the number of hours he runs that day. How many miles will Vinny run if he runs for 2.5 hours?

8. Frank rides his road bike at a pace of 18 miles per hour. His total miles for the day can be computed as $18h$, where h is the number of hours he rode his bike. How many miles will Frank ride if he rides for 4.25 hours?

9. Judy sells real estate and estimates that for each house she is listing, she spends $250.00 in advertising. The total amount she spends in advertising may be computed as $250l$, where l is the number of houses she is listing. How much does Judy estimate that she will spend on advertising if she is currently listing five houses?

10. Tomas sells real estate and estimates that for each property he is listing, he spends $550.00 in advertising. The total amount he spends in advertising may be computed as $550l$, where l is the number of houses he is listing. How much does Tomas estimate that he will spend on advertising if he is currently listing 12 houses?

11. Fusako is taking a road trip and drives at an average speed of 75 miles per hour. The total distance she travels in one day can be computed as $75h$, where h is the number of hours she drives for that day. How many miles does Fusako drive if she drives for 6.5 hours?

12. Hiroyoshi is taking a road trip and drives at an average speed of 68 miles per hour. The total distance he travels in one day can be computed as $68h$, where h is the number of hours he drives that day. How many miles does Hiroyoshi drive if he drives for 9.25 hours?

For Exercises 13 through 32, evaluate each expression for the given values of the variables.

13. $-6x + 9$ for $x = 4$
14. $15x - 24$ for $x = 3$
15. $\frac{2}{5}n + 7$ for $n = -10$
16. $-\frac{7}{4}n + 11$ for $n = -8$
17. $x^2 - 6x + 7$ for $x = -1$
18. $2t^2 - 3t + 11$ for $t = -2$
19. $-y^2 + 7$ for $y = -4$
20. $-n^2 + 2n$ for $n = -1$
21. $\frac{6y - 3}{2y + 8}$ for $y = 5$
22. $\frac{-2x + 9}{4x - 3}$ for $x = 5$
23. Evaluate $36x^2 + 4y^2$ for $x = -5$ and $y = -2$.
24. Evaluate $-16st + 7s - 2t$ for $s = 3$ and $t = 0$.
25. Evaluate $-22a - 8b$ for $a = 10$ and $b = -10$.
26. Evaluate $t + h$ for $t = \frac{1}{2}$ and $h = \frac{3}{5}$.
27. Evaluate $\frac{x - y}{x + 2y}$ for $x = 3$ and $y = -4$.
28. Evaluate $\frac{2a - b}{2a + b}$ for $a = -1$ and $b = 8$.
29. Evaluate $-9x + 51$ for $x = -6$.
30. Evaluate $2xy + y$ for $x = 3$ and $y = -3$.
31. Evaluate $2x^2 - 6x + 5$ for $x = 3$.
32. Evaluate $4 - 16c + 2t$ for $c = -3$ and $t = -9$.

For Exercises 33 through 50, convert units as indicated.

33. The unity fraction to convert feet to yards is $\frac{1 \text{ yd}}{3 \text{ ft}}$. Convert 22 feet to yards.

34. The unity fraction to convert miles per hour to feet per second is $\frac{5280 \text{ ft}}{3600 \text{ sec}}$. Convert 75 miles per hour to feet per second.

35. Convert 1057 feet to miles. *Note:* There are 5280 ft in 1 mi.

36. Convert 692 centimeters to meters. *Note:* There are 100 cm in 1 m.

37. As of this writing, 1.00 Japanese yen is equal to 0.00917 U.S. dollar. Convert 225 Japanese yen to U.S. dollars.

38. As of this writing, 1.00 euro is equal to 1.07255 U.S. dollar. Convert 149.00 euros to U.S. dollars.
    Source: www.XE.com

39. Convert 7.89 kilograms to pounds. *Note:* There are approximately 2.2 lb in 1 kg.

40. Convert 300 grams to ounces. *Note:* There are approximately 28.35 grams in 1 ounce.

41. Convert 15 gallons to liters. *Note:* There are approximately 3.785 l in 1 gal.

42. Convert 22 quarts to gallons. *Note:* There are 4 qt in 1 gal.

43. Convert 22,000 square feet to acres. *Note:* There are 43,560 ft² in 1 acre.

44. Convert 60 square feet to square yards. *Note:* There are 9 ft² in 1 yd².

45. Convert 601 million to thousands. *Hint:* 1 million = 1000 thousands.

46. Convert 10,011 thousands to millions.

47. Convert 5397 million to billions. *Hint:* 1 billion = 1000 million.

48. Convert 8.6 billion to millions.

49. Convert 9 trillion to billions. *Hint:* 1 trillion = 1000 billion.

50. Convert 8,375 billion to trillions.

51. Convert 2.5 gigabytes to megabytes. *Note:* There are 1024 MB in 1 GB.

52. Convert 12,000 gigabytes to terabytes. *Note:* There are 1024 GB in 1 TB.

For Exercises 53 through 62, define variables for each problem situation.

53. The amount owed on a monthly electricity bill depends on how many kilowatt-hours of electricity were used during that month.

54. The height of the ocean tide (in feet) at high tide depends on the phase of the moon.

55. The amount paid (in dollars) yearly in federal taxes depends on income (in dollars) for that year.

56. The amount of money (in dollars) a family spends a week on gas depends on the number of miles they drive that week.

57. The temperature outdoors (in degrees Fahrenheit) depends on the time of day.

58. The amount of revenue (in dollars) a theater makes depends on the number of tickets sold.

59. The amount of first-class postage you owe (in dollars) for a package depends on the weight of the package you are shipping (in pounds).

60. The amount (in milliliters) of IV fluids a hospitalized patient requires depends on the patient's weight (in kilograms).

61. The cost of attending a community college depends on how many credits a student enrolls in per semester.

62. The amount of money you earn weekly depends on the number of hours you worked that week.

For Exercises 63 through 82, translate each of the following phrases into an expression.

63. The sum of a number and 62

64. The sum of -19 and a number

65. 9 more than twice a number

66. Three times a number plus -7

67. The quotient of 16 and a number

68. Five times a number divided by 12

69. 7 less than $\frac{1}{3}$ times a number

70. 9 subtracted from $\frac{1}{2}$ a number

71. The difference between 8 times a number and 12

72. The difference between -7 times a number and 18

73. The product of -5 and the square of a number

74. The product of a number squared and -1

75. 16 increased by the quotient of a number and 4

76. The sum of 7 divided by a number and 3

77. The quotient of 6 and the difference of a number and 2

78. The quotient of a number minus 2 and 3

79. Half of a number added to 6

80. One-third of a number minus 16

81. The difference between 7 and twice a number

82. The difference between -3 times a number and 8

For Exercises 83 through 88, define each variable (include units). Then translate each sentence into an expression.

83. Cayuga County Community College (in upstate New York) charges part-time resident students (11 credit hours or less) $187.00 per credit plus a $10.00 per credit activity fee plus an $11.00 per credit technology fee plus a $20 wellness fee each semester. Write an expression that gives the cost of attending Cayuga County Community College for one semester.

    Source: www.cayuga-cc.edu

84. A car repair shop quotes a rate of $60.00 an hour for labor. Write an expression that gives the labor cost for a car repair.

85. A house-painting company quotes a price to paint a house. The quote includes the price of the paint, which is $475.00, plus labor, which costs $23.00 an hour. Write an expression that determines the cost of painting this house.

86. A local horse-boarding facility buys alfalfa hay to feed the horses. At the time of this writing, the average price of a small square bale of alfalfa hay in California is $8.06 a bale. The hay distributor also

charges a $75.00 delivery fee. Write an expression that determines the cost of purchasing and delivering the hay.

Source: www.hayexchange.com

87. As of this writing, the cost of first-class (letter) mail in the United States is $0.49 for the first ounce plus $0.21 per each additional ounce. Write an expression for the cost of mailing a first-class letter.

Source: www.usps.gov

88. A movie theater charges $16.50 a ticket for an adult ticket. Write an expression for the amount of money (revenue) the theater will take in if they sell t tickets.

For Exercises 89 through 98, first define each variable. Then write an expression for the unit conversion.

89. Convert a variable number of pounds to kilograms.

90. Convert a variable number of ounces to pounds.

91. Convert a variable number of miles to feet.

92. Convert a variable number of meters to feet.

93. Convert a variable number of liters to gallons.

94. Convert a variable number of quarts to gallons.

95. Convert a variable number of acres to square feet.

96. Convert a variable number of square yards to square feet.

97. Convert a variable number of seconds to minutes.

98. Convert a variable number of days to years.

For Exercises 99 through 102, use the information given in each exercise to answer the questions.

99. Bill wants to estimate the cost to recarpet his living room. Bill measures his living room and finds that its length is 20 feet and its width is 15 feet.

   a. Find the area of Bill's living room. What are the units?

   b. Carpet is sold by the square yard. Convert the area of Bill's living room to units of square yards. Round up to the nearest full square yard.

   c. The Berber carpet that Bill wants costs $23.49 per square yard. How much will the carpet cost?

   d. Bill is quoted a price of $2.75 per square yard for new padding. How much will it cost for the new padding?

   e. What will the total cost of the carpet and padding be?

100. JoAnn wants to estimate the cost to repaint her living room. JoAnn measures her living room and finds the floor dimensions are a length of 15 feet and width of 15 feet. The walls are 10 feet tall.

   a. Find the area of her living room walls. What are the units? *Hint:* You must find the total area of all four walls. For the purpose of the estimate, don't worry about doors or windows.

   b. The paint JoAnn wants covers 200 square feet per gallon. How many gallons of paint does JoAnn need to buy?

   c. The paint costs $25.97 per gallon. How much will it cost JoAnn to buy the paint?

101. You and some friends are going on a 7-night cruise. The cruise ship company allows each passenger to bring a total of only 75 pounds (lb) of luggage on board. You want to bring some clothing and toiletries for your trip. Let $c =$ amount in pounds of clothing you will pack, $l =$ weight of your suitcase in pounds, and $t =$ amount in pounds of toiletries that you will pack. The total weight of your luggage is

$$c + l + t$$

   a. Suppose that you take 50 lb of clothes and 10 lb of toiletries and that your suitcase weighs 12 lb. What is the total weight of your luggage?

   b. Suppose that you take 54.6 lb of clothes and 9.6 lb of toiletries and that your suitcase weighs 10 lb. What is the total weight of your luggage?

102. You want to go on a backpacking trip. In your backpack, you will have to carry all your food and equipment. Let $f =$ number of pounds of food you will pack and let $e =$ number of pounds of equipment (including the backpack) you will pack. The total weight you will have to carry is

$$f + e$$

   a. Suppose that you take 14 lb of food and 16 lb of equipment. What is the total weight you will have to carry?

b. Suppose that you take 12.6 lb of food and 18.9 lb of equipment. What is the total weight you will have to carry?

c. Suppose that you can pack a total of only 24 lb. Fill in the following table to determine various combinations of food and equipment you could take on the trip.

f	e	$f + e$
10		24
	15	24
13		24
	12	24

For Exercises 103 through 108, use input–output tables to generate expressions for each problem.

103. Carlos drives from San Diego to Los Angeles and back in a total time (round trip) of 4.5 hours. If he spends t hours driving from San Diego to Los Angeles, find an expression for the time it takes him to drive back.

Time to Drive from San Diego to Los Angeles	Time to Drive from Los Angeles to San Diego
1	
2	
2.5	
3	
t	

104. Sean rides his bike from his home to school daily. One day, it takes him a total time of 90 minutes for the round trip. If he spends m minutes riding his bike to school, find an expression for the time it takes him to ride from school to home.

Time to Ride Bike from Home to School	Time to Ride Bike from School to Home
20	
30	
40	
45	
m	

105. Rebecca has a total of $2500 to invest. She is going to invest part of the money in an account earning 3.5% interest and part of the money in a stock fund earning 7% interest. If Rebecca invests d dollars in the 3.5% account, find an expression for the amount of money she invests in the 7% account.

Amount Invested in the 3.5% Account	Amount Invested in the 7% Account
0	
500	
1000	
1500	
d	

106. Tom has 40 feet of lumber that he needs to cut into two pieces. If Tom cuts f feet of lumber for the first piece, find an expression for the remaining amount of lumber in the second piece.

First Piece of Lumber	Second Piece of Lumber
5	
10	
20	
35	
f	

107. Heather, owner of Cup to Go Coffee, wants 65 lb total of coffee in a coffee blend of Kenyan and Colombian. If Heather includes c lb of Colombian coffee in her blend, find an expression for the amount (in pounds) of Kenyan coffee she will need to include.

Pounds of Colombian Coffee	Pounds of Kenyan Coffee
0	
10	
20	
50	
c	

108. Anya, a chemist at Bio-chem, Inc., is mixing a 5% saline solution with a 7.5% saline solution for a total of 10 gallons in the mix. If Anya uses g gallons of the 5% saline solution, find an expression for the amount (in gallons) of the 7.5% solution she will need to include.

Amount of 5% Saline Solution	Amount of 7.5% Saline Solution
0	
2	
4	
6	
g	

For Exercises 109 through 112, use the information given in each exercise to answer the questions.

109. Sandy is throwing a 25th anniversary party for her parents. She is trying to figure out how many people she can invite to the party. She has budgeted $500.00 to pay for the food at the party. Sheila's Catering Company says they will charge Sandy $12.00 per person for the food.

   a. Generate an expression for the food cost for the party based on the number of people. Remember to use an input–output table to help generate the expression if needed.

   b. Approximately how many people can Sandy invite?

110. Gerry and Maureen want to throw a fifth birthday party for their daughter. They plan to serve take-out fried chicken, which will cost $1.49 per child. They also figure that the party favors will cost $3.50 per child. They have budgeted $90.00 for food and party favors.

   a. Generate an expression for the total cost of food and party favors, based on the number of children. Remember to use an input–output table to help generate the expression if needed.

   b. Approximately how many children can Gerry and Maureen invite?

111. Tricia is a professional event planner. She has been hired to set up a business reception at a conference. Tricia has been told that her budget for food is $1700.00. Jeff's Catering Affair has quoted Tricia $18.00 per person for drinks and $15.00 per person for appetizers.

   a. Generate an expression for the total food and drink cost based on the number of people attending. Remember to use an input–output table to help generate the expression if needed.

   b. Approximately how many people can attend the conference reception and stay within Tricia's budget?

112. Casey sets up paid phone interviews with industry professionals for his corporate clients. His clients hire him to ask industry professionals for their opinions about products and services. Casey is setting up interviews for a client who has given him a budget of $4000.00. Each interviewee will be paid $110.00 for participation.

   a. Generate an expression for the cost for all the interviews. Remember to use an input–output table to help generate the expression if needed.

   b. Approximately how many people can Casey have participate in the phone interview to stay within his $4000 budget?

1.3 Simplifying Expressions

LEARNING OBJECTIVES
- Recognize like terms.
- Use the addition and subtraction properties.
- Use the multiplication and distributive properties.
- Simplify expressions.

Like Terms

Terms are the building blocks of expressions. Expressions may involve the four main operations that we are already familiar with. In this section, we review the concept of a term and learn the basic operations for combining terms that are used throughout algebra.

> **DEFINITION**
>
> **Term** A constant (a number) or a product of a constant and a variable (or variables) is called a term.
>
> $$3 \qquad -4a \qquad \frac{2}{3}km^2 \qquad x$$
>
> *Note:* The numerical part of a term is called the **coefficient.**

We can combine more than one term together with the operations of addition or subtraction to create expressions. One of the more useful concepts in algebra is that of **like terms.** Two or more terms are "like" if they have the same variable part, where the variables are raised to the same exponents. Notice that the coefficients do not need to be the same for two terms to be "like."

What's That Mean?

Terms
Two phrases are used in discussing terms.
Constant Term: A constant is a number by itself.
Variable Term: A term with a variable

> **DEFINITION**
>
> **Like Terms** Two or more terms that have the same variable part, where the variables are raised to the same exponents, are called **like terms.**
>
> $3x$ and $5x$ are like terms
>
> $-25a^2bc^3$ and $10a^2bc^3$ are like terms

Example 1 Identifying like terms

Identify the number of terms. Then circle the like terms (if any) in each expression.

a. $5x^2 + 4x - 7$

b. $3a - 4b + 7a + 2$

c. $105 + 8m^2 + 7m - 4m^2 - 17$

d. $14x^2y + 3xy^2 + 4x^2y - 19y^2$

110　CHAPTER 1　Building Blocks of Algebra

SOLUTION

a. $5x^2 + 4x - 7$. There are three terms in this expression. There are no like terms in this expression. None of the terms in this expression have the same variable raised to the same exponent. The first two terms have the same variable x, but the variable x is raised to different powers.

b. ⟨$3a$⟩ $- 4b +$ ⟨$7a$⟩ $+ 2$. There are four terms in this expression. The like terms are $3a$ and $7a$ because they both have a raised to the first power.

c. ⟨105⟩ $+$ ⟨$8m^2$⟩ $+ 7m +$ ⟨$-4m^2$⟩ $+$ ⟨-17⟩. There are five terms in this expression. Both 105 and -17 are constant terms and are like terms. The terms $8m^2$ and $-4m^2$ are also like terms. Notice that the sign in front of a term is part of the term. This is why we have the term -17 and not 17 and $-4m^2$ not $4m^2$.

d. ⟨$14x^2y$⟩ $+ 3xy^2 +$ ⟨$4x^2y$⟩ $- 19y^2$. There are four terms in this expression. The like terms are $14x^2y$ and $4x^2y$. Notice that the variables are the same and that each is raised to the same power.

> **Connecting the Concepts**
>
> **Are signs included in a term?**
>
> In *algebra*, it is very important to include the sign of the coefficient in a term. In the expression
>
> $105 + 8m^2 + 7m - 4m^2 - 17$
>
> the two terms involving m^2 are $8m^2$ and $-4m^2$.

PRACTICE PROBLEM FOR EXAMPLE 1

Identify the number of terms. Then circle the like terms (if any) in each expression.

a. $-5x^2 + 7c + 4d^2 + 8c$ 　　 b. $45x^2 + 7x - 9$

c. $23g^2k + 45g^2k^2 - 16g^2k + 57g^2k^2$

■ Addition and Subtraction Properties

Recognizing like terms is so important because like terms are the basic components of addition and subtraction in algebra. To add or subtract two or more terms in an expression, they must be like terms. When adding or subtracting like terms, add or subtract their coefficients.

The distributive property that we learned in Section 1.1 validates why like terms can be added or subtracted.

$(2 + 3)x = 2x + 3x$ 　　 *The distributive property is used to rewrite*
$(2 + 3)x = 5x$ 　　　　　 *the problem. Combine the coefficients.*

Throughout this text, we will say to add or subtract like terms, but remember that this is justified by the distributive property. Just add or subtract the coefficients.

$2x + 3x = 5x$

> **Connecting the Concepts**
>
> **How to write a coefficient of ± 1**
>
> When the coefficient of a term is 1, we do not usually write the 1 in front of the variable. The coefficient is assumed to be 1 if no number is written. Therefore, $1x$ is written as x, and $1w^2xyz$ is written as w^2xyz.
> When the coefficient is -1, we have a similar situation. The 1 is not written, just the negative sign. $-1x$ would be written as $-x$, and $-1rt^3$ would be written as $-rt^3$.

> **ADDITION AND SUBTRACTION PROPERTY**
>
> Two or more terms can be added or subtracted only if they are like terms. Add or subtract the coefficients of the like terms. The variable part remains the same.
>
> $5x + 7x$ simplifies to $12x$ 　　 $9x^2 - 13x^2$ simplifies to $-4x^2$

> **Add Like Terms**
>
> Add or subtract like terms that have the same variables with the same exponents.
>
> $2x^2 + 7x + 9x^2 - 3x$
> $= 11x^2 + 4x$

Example 2 **Simplifying expressions by combining like terms**

Simplify each expression by combining like terms.

a. $12a - 3a$ 　　 b. $8m + 5n - 3m + 14n$

c. $7a + 10b - 15$ 　　 d. $4x^2 + 5x - x^2 + 9$

SECTION 1.3 Simplifying Expressions

SOLUTION $2x^2 + 3x^2$

a. $12a - 3a$ *Combine like terms by subtracting the coefficients.*
$= 9a$

b. $8m + 5n - 3m + 14n$ *Rearrange terms using the commutative property.*
$= 8m - 3m + 5n + 14n$ *Note that the negative sign in front of the $3m$ term goes with that term.*
$= 5m + 19n$ *Combine like terms by adding or subtracting the coefficients.*

c. $7a + 10b - 15$ *There are no like terms, so this expression is already simplified.*

d. $4x^2 + 5x - x^2 + 9$ *Rearrange terms using the commutative property.*
$= 4x^2 - x^2 + 5x + 9$ *Combine like terms.*
$= 3x^2 + 5x + 9$

Skill Connection

Commutative property of addition

The commutative property of addition was introduced in Section 1.1. In words, it says that we can add in any order. In symbols, we have that $a + b = b + a$. This property lets us rearrange the terms of an expression.

PRACTICE PROBLEM FOR EXAMPLE 2

Simplify each expression by combining like terms.

a. $8p + 5p$
b. $2w + 7 - 9w + 3$
c. $3x^2y + 4xy - 7xy + 15x^2y$
d. $8z^2 + 5x - 2z - 20$

It is **conventional** that the terms in an expression are ordered alphabetically, with descending exponents and with constant terms last. Writing terms with descending exponents means that the term with the highest power exponent should go first, the next highest power second, and so on. When simplifying an expression, write the final answer using this convention.

What's That Mean?

Conventional

In mathematics, when we use the word *conventional*, we mean that there is a general agreement on how something should be written or done.

Example 3   Writing the expression in conventional form

Write each expression in conventional form. This means that terms are ordered alphabetically, have descending exponents, and end with the constant terms.

a. $2b + 5a$
b. $-5x + 3x^2 + 6$
c. $3y - 3x + 11 + 6x^2$

SOLUTION

a. This expression is not in alphabetical order, reading left to right. Use the commutative property of addition to reorder the terms.

$$2b + 5a = 5a + 2b$$

b. This expression is not written with descending exponents, reading left to right. The term with the exponent of the highest power is $3x^2$, and it should go first. The term with the second highest power exponent is $-5x$, and it goes second. The only remaining term is the constant term 6, and it is placed last.

$$-5x + 3x^2 + 6$$
$$= 3x^2 - 5x + 6$$

c. This expression is not in alphabetical order. The x-terms are not in descending order. Rearranging the expression yields the answer below.

$$3y - 3x + 11 + 6x^2$$
$$= 6x^2 - 3x + 3y + 11$$

PRACTICE PROBLEM FOR EXAMPLE 3

Write each expression in conventional form. This means the terms are ordered alphabetically, have descending exponents, and end with the constant terms.

a. $-3d + 9c$ **b.** $16 - 2x + 3x^2$ **c.** $7 - 4b^2 - 2a + 3a^2$

Example 4 Combining like terms that represent cost

Debra is the captain of the Leffingwell Lions cheerleading squad. She wants to buy T-shirts for all the squad members. Debra called a local shirt company and was given the following information:

$13 per T-shirt

$6 per T-shirt for the school logo to be printed on the back of a shirt

$1.50 per letter for a name to be embroidered on the front of the shirt

These prices include all taxes.

Debra defined the following variables:

$s =$ Number of T-shirts ordered

$n =$ Number of letters in a squad member's name

a. Write an expression for the cost to buy s T-shirts.

b. Write an expression for the cost to print the logo on the back of s T-shirts.

c. Use the expressions from parts a and b to write an expression for the total cost to buy and print the logos on s T-shirts. Simplify the expression if possible.

d. Write an expression for the cost to embroider a name with n number of letters on the front of the T-shirt.

e. Does it make sense to add the expression found in part c to the expression in part d? Explain why or why not.

SOLUTION

a. The problem states that $s =$ number of T-shirts ordered.

It costs $13 to buy each shirt, so 13 times the number of shirts, $13s$, is an expression for the cost in dollars to buy the shirts. The expression is

$$13s$$

b. To print the logo on the back of the shirts, it costs $6 for each logo, so $6s$ is an expression for the cost in dollars to print the logos. The expression is

$$6s$$

c. Recall that *total* means to *add*. Adding these two expressions together gives us the expression $13s + 6s$. This expression represents the total cost to buy the shirts and print the logos. We can simplify this expression because both of these costs are per T-shirt and the terms are like terms. Therefore, combining $13s + 6s$ gives us the simplified expression $19s$ for the total cost in dollars to buy and print the logo on s T-shirts. The expression is

$$19s$$

d. The problem states that $n =$ number of letters. It costs $1.50 per letter for the embroidery, so a name with n letters will cost $1.50n$ dollars to be embroidered. The expression is

$$1.50n$$

e. It does not make sense to add these expressions together. We cannot add these two terms, since the $19 is a cost per T-shirt and the $1.50 is a cost per letter embroidered. If we did try to add these together, we would get $20.50, but the units do not match. $19s$ and $1.50n$ are not like terms and cannot be added together.

PRACTICE PROBLEM FOR EXAMPLE 4

Murray's Hitch installs trailer hitches on cars and trucks. When ordering supplies for the shop, Murray must order hitches as well as wiring kits and bolt kits. Murray's cost for these items are as follows:

$125 per standard hitch

$23 per wiring kit

$2 per bolt kit (Some hitches require two or three bolt kits to be installed properly.)

All of these prices include shipping and applicable taxes.

a. Write an expression for the cost to buy h standard hitches.

b. Write an expression for the cost to buy wiring kits for h hitches.

c. Use the expressions from parts a and b to write an expression for the total cost to buy h standard hitches and wiring kits. Simplify the expression if possible.

d. Write an expression for the cost to buy b bolt kits.

e. Does it make sense to add the expression found in part c to the expression in part d? Explain why or why not.

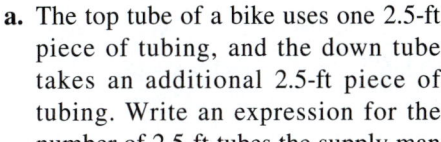

Example 5 Ordering parts to manufacture a bicycle

A custom bike shop produces bikes for small bike stores across the country. When an order is placed for bikes, the supply manager must order the right amount of tubing for the production process. To save money, the supply manager orders tubing in standard lengths of 2.5 ft.

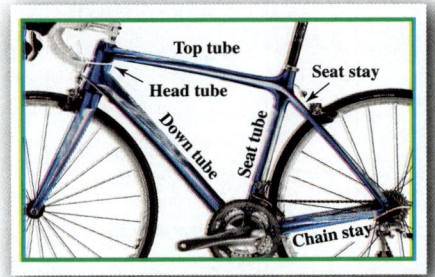

a. The top tube of a bike uses one 2.5-ft piece of tubing, and the down tube takes an additional 2.5-ft piece of tubing. Write an expression for the number of 2.5-ft tubes the supply manager should order to make the top tubes and down tubes for b bikes.

b. The seat tube and head tube of each bike can be made from the same 2.5-ft tube. Write an expression for the number of 2.5-ft tubes the supply manager should order to make the seat tubes and head tubes for b bikes.

c. Use the expressions from parts a and b to find an expression that will represent the total number of 2.5-ft pieces of tubing that need to be ordered for b bikes. Simplify the expression if possible.

114 CHAPTER 1 Building Blocks of Algebra

SOLUTION

a. Let $b =$ number of bikes ordered. Since each bike will need two tubes, one for the top tube and one for the down tube, the supply manager should order $2b$ 2.5-ft tubes for an order of b bikes.

b. Since the seat and head tubes can be made from one 2.5-ft tube, the supply manager needs to order one tube for each bike so $1b$ or b tubes should be ordered for an order of b bikes.

c. To find the total number of 2.5-ft tubes, the supply manager needs to add the order for the top and down tubes to the order for the seat and head tubes. Since the two expressions are both orders for 2.5-ft tubes, the supply manager can add them together and therefore should order $2b + b$ or $3b$ 2.5-ft tubes for the b bikes.

These last examples should help us see that to add two terms together, they must be like terms.

Multiplication and Distributive Properties

In multiplying a term or expression by a constant, we must know what to multiply together. Use the following Concept Investigation to determine the difference between multiplying one term by a constant and multiplying several terms by a constant.

CONCEPT INVESTIGATION
What do I multiply by?

Use the order-of-operations agreement or use your calculator to evaluate the expressions in each column, substituting in the values given. If you use your calculator, be sure to enter the expressions on your calculator in the same order as they appear in the columns below, working from left to right. Then determine which columns result in the same value and which result in a different value.

First consider multiplying a *single term* by a constant.

a.	$2(3xy)$	$6xy$	$(6x)(2y)$
$x = 5$ $y = 4$	$2(3(5)(4)) = 120$ Original expression	$6(5)(4) = 120$ Same value as the Original expression	$(6(5))(2(4)) = 240$ Different value from the Original expression
b.	$4(xy)$	$4xy$	$(4x)(4y)$
$x = 7$ $y = 3$	$4((7)(3)) = 84$ Original expression		
c.	$-4(3gh)$	$-12gh$	$(-12g)(-4h)$
$g = 2$ $h = 5$			

1. What column gave the same result as the original expression?

2. What does your answer to question 1 imply about what you should multiply together when multiplying a **single term** by a constant?

Now consider the result when multiplying the *several terms* by a constant.

Calculator Details

Evaluating Expressions

When evaluating an expression for different values of a variable, use parentheses around the value you are substituting in for the variable. Using parentheses helps to eliminate some calculator errors. This is especially important in substituting in negative values. To substitute in $x = -2$ in the expression $3x^2 + 1$, enter it as

$$3(-2)^2 + 1$$

SECTION 1.3 Simplifying Expressions   **115**

a. $x = 5$ $y = 4$	$2(3x + y)$ $2(3(5) + (4)) = 38$ Original expression	$6x + y$ $6(5) + (4) = 34$ Different value from the Original expression	$6x + 2y$ $6(5) + 2(4) = 38$ Same value as the Original expression
b. $x = 7$ $y = 3$	$4(5x + 3y)$	$20x + 3y$	$20x + 12y$
c. $g = 2$ $h = 5$	$-4(3g - 8h)$	$-12g - 8h$	$-12g + 32h$

3. What column gave the same result as the original expression?

4. What does the result in question 3 tell you about what to multiply together when multiplying **several terms** by a constant?

This Concept Investigation shows that in multiplying a **single** term by a constant, multiply the constant by the coefficient of the term. This rule is called the **multiplication property.**

> **MULTIPLICATION PROPERTY**
> When multiplying a single term by a constant, multiply the coefficient by the constant, and the variable parts stay the same.
> $$5(4x^2y) = 20x^2y$$

If there are several terms in the parentheses, we multiply the coefficient of each term by the constant. In Section 1.1, we learned how to use the distributive property to multiply expressions such as $a(b + c) = a \cdot b + a \cdot c$, where a, b, and c are real numbers. The distributive property is now extended to the case in which a *variable expression* is in the parentheses.

> **DISTRIBUTIVE PROPERTY**
> When multiplying a multiple-term expression by a constant, multiply the coefficient of each term by the constant, and the variable parts will stay the same.
> $$a(b + c) = a \cdot b + a \cdot c \quad \text{or} \quad a(b - c) = a \cdot b - a \cdot c$$
> $$3(2x - 7y) = 6x - 21y$$

Distributive Property $5(2x-7)$

Distribute across addition or subtraction.

$3(2x-4)$
$= 6x - 12$

$(5x+6)(2x-9)$
$= 10x^2 - 45x + 12x - 54$
$= 10x^2 - 33x - 54$

Example 6   Using the multiplication and distributive properties

State whether the multiplication property or the distributive property should be used. Then simplify the following expressions by multiplying.

a. $4(7x)$

b. $5(7x^2 + 8x - 3)$

c. $-\dfrac{2}{3}(9t)$

d. $-(m^2 - 6m + 3)$

SOLUTION

The property that is used depends on the number of terms in the parentheses. The multiplication property is used if only one term is in the parentheses. The distributive property is used if multiple terms are in the parentheses.

a. $4(7x)$ A single term. Multiplication property.
$= 28x$ Multiply the 4 by the coefficient 7.

b. $5(7x^2 + 8x - 3)$ Several terms. Distributive property.
$= 5(7x^2) + 5(8x) - 5(3)$ Distribute the 5 to all the terms, and multiply each coefficient.
$= 35x^2 + 40x - 15$

c. $-\dfrac{2}{3}(9t)$ A single term. Multiplication property.

$= -\dfrac{18}{3}t$ Multiply the coefficient.

$= -6t$ Reduce the fraction.

d. $-(m^2 - 6m + 3)$ A negative sign in front of the parenthesis is the same as multiplying by -1.
$= -1(m^2 - 6m + 3)$ Several terms. Distributive property.
$= -1(m^2) - 1(-6m) - 1(3)$ Distribute the -1 to each term.
$= -m^2 + 6m - 3$ Distributing the -1 changes the signs of all the terms enclosed in the parentheses.

PRACTICE PROBLEM FOR EXAMPLE 6

State whether the multiplication property or the distributive property should be used. Then simplify the following expressions by multiplying.

a. $-9(4m^3)$ **b.** $7(-4x + 6y)$

c. $-\dfrac{4}{5}(25c^2 + 5c - 10)$ **d.** $0.4(3x - 6)$

■ Simplifying Expressions

The properties discussed in this section can be used to simplify expressions that have a combination of operations. When simplifying more complicated expressions, follow the order-of-operations agreement. Therefore, simplify anything inside grouping symbols first, then multiply or divide, and finally add or subtract. Always keep in mind that only like terms can be added or subtracted.

Example 7 Simplifying expressions

Simplify the following expressions.

a. $-3m + 4n + 2(5m + 7)$

b. $0.5(3x - 8) + 2(x + 6)$

SOLUTION

a. Using the order-of-operations agreement, first consider the parentheses. The inside of the parentheses cannot be simplified because the terms are not like. There are no exponents, so multiply first by using the distributive property.

$-3m + 4n + 2(5m + 7)$ Multiply first by distributing the 2 to each term in the parentheses.
$= -3m + 4n + 10m + 14$
$= -3m + 10m + 4n + 14$ Rearrange terms using the commutative property. Combine like terms.
$= 7m + 4n + 14$

b. $0.5(3x - 8) + 2(x + 6)$ Multiply first by distributing the 0.5 and 2.
$= 1.5x - 4 + 2x + 12$
$= 1.5x + 2x - 4 + 12$ Rearrange the terms.
$= 3.5x + 8$ Combine like terms.

PRACTICE PROBLEM FOR EXAMPLE 7

Simplify the following expressions.

a. $5(4x - 9) + 8x + 20$

b. $1.5(2w + 7) + 3.5(4w - 6)$

When simplifying expressions using the distributive property, be careful to keep track of the signs involved. Whenever the distributive property is applied using a negative number, each sign inside the grouping symbol changes.

Example 8   Simplifying more complex expressions

Simplify the following expressions.

a. $6(3a^2c + 2a) - (5a^2c - 7a)$

b. $4(2x + 7y) + 3(5x - 4) - 2(-6x - 8)$

SOLUTION

a. $6(3a^2c + 2a) - (5a^2c - 7a)$ Recall that a negative sign in front of an expression or term is the same as a -1 coefficient.
$= 6(3a^2c + 2a) - 1(5a^2c - 7a)$ Multiply first by distributing the 6 and the -1.
$= 18a^2c + 12a - 5a^2c + 7a$
$= 18a^2c - 5a^2c + 12a + 7a$ The -1 changes the sign of both terms in the second parentheses.
$= 13a^2c + 19a$ Rearrange and combine like terms.

b. $4(2x + 7y) + 3(5x - 4) - 2(-6x - 8)$ Multiply first by distributing the 4, 3, and -2.
$= 8x + 28y + 15x - 12 + 12x + 16$
$= 8x + 15x + 12x + 28y - 12 + 16$ Rearrange the terms.
$= 35x + 28y + 4$ Combine like terms.

PRACTICE PROBLEM FOR EXAMPLE 8

Simplify the following expressions.

a. $(4x^2 + 5x) - (x^2 - 3x)$ **b.** $4(8t + 9u) - 3(6t - 5u)$

118 CHAPTER 1 Building Blocks of Algebra

1.3 Vocabulary Exercises

1. Use the _____ when multiplying multiple terms by a constant.
2. A(n) _____ is a constant or a product of a constant and a variable(s).
3. Only _____ terms can be combined by adding or subtracting.
4. The numeric part of a term is called the _____.
5. Use the _____ when multiplying a single term by a constant.
6. When simplifying expressions, be sure to use the _____ agreement.

1.3 Exercises

For Exercises 1 through 10, identify the number of terms and the like terms (if any) in each expression.

1. $2x + 8 + 7x$
2. $4p - 5p + 3$
3. $3 + 2h - 5 + 7h$
4. $4b + 5 - 6b + 2$
5. $3x^2 + 5x + 7x^2 + 3x$
6. $t^5 + 4t^3 + 3t^5 - t^3$
7. $2x^2 + 5x - 9$
8. $2n^4 + n^3 + 2n + 1$
9. $7xy + 8x - 2xy - 3y$
10. $4ab + 2a - 3b + 7ab$

For Exercises 11 through 20, simplify each expression by combining like terms.

11. $13x + 7x$
12. $7y + 20y$
13. $3p - 9p$
14. $10q - 16q$
15. $4x + 5y - 6x - y$
16. $a - 4b - 6a + 5b$
17. $-18n - 3 + 7n - 5$
18. $-3b + 11 - 7b - 9$
19. $3x^2 + 9x - 1 - 4x^2 + 2x - 5$
20. $5y^2 - 3y - 2 + 8y^2 - 7y - 9$

For Exercises 21 through 30, write each expression in conventional form. This means that terms are ordered alphabetically and have descending exponents and that constant terms are last in the expression.

21. $-8y + 3x$
22. $15b - 29a$
23. $7n - 8m + 12$
24. $9y + 7 - 8x$
25. $5 - x + 7x^2$
26. $8 - 6y - y^2$
27. $8 + 5x - 7x^5 - 3x^2$
28. $4y - 6 - y^3 - 2y^2$
29. $a - \frac{7}{5}b + 3a^2 - \frac{1}{2}a^6$
30. $3y - \frac{4}{5}x^2 + 5x + \frac{2}{5}x^4$

For Exercises 31 through 36, use the information given in each exercise to answer the questions.

31. Julie owns Children's Paradise Preschool and wants to order supplies for the children in the 4- to 5-year-old class for a project. Julie researches on the Internet and finds the following prices:

    Birdhouse kit: $4.19 each

    Paint-with-brush set: $1.39 each

   a. If Julie needs supplies for k children in the class, write an expression for the cost in dollars to order each child a birdhouse kit.

   b. Write an expression for the cost in dollars to order each child a paint-with-brush set.

   c. Use the expressions found in parts a and b to write an expression for the total cost in dollars to buy each child a birdhouse kit and a paint-with-brush set.

   d. Simplify the expression from part c if possible.

32. Ms. Binkinz is buying her first kindergarten class gifts for the end of the school year. She wants to order a book for each student that costs $1.25. Ms. Binkinz also wants to buy each student an educational toy that costs $2.35.

a. If Ms. Binkinz has *s* students in her kindergarten class, write an expression for the cost in dollars to buy each student a book.

b. Write an expression for the cost in dollars to buy each student an educational toy.

c. Use the expressions found in parts a and b to write an expression for the total cost in dollars to buy each student a book and educational toy.

d. Simplify the expression from part c if possible.

33. Julie, from Exercise 31, decides to have the 4- to 5-year-old children do another project and build pinewood derby cars. Julie again finds the prices for the supplies on the Internet:

   Pinewood derby complete car kit: $8.79 each
   Sandpaper and sealer: $1.19 each
   Paint-with-brush set: $1.39 each

a. If Julie needs supplies for *k* children in the class, write an expression for the cost to order each child a pinewood derby car kit.

b. Write an expression for the cost to order each child sandpaper and sealer.

c. Write an expression for the cost to order each child a paint-with-brush set.

d. Use the expressions found in parts a, b, and c to write an expression for the total cost to buy each child all the supplies needed for this project. Simplify the expression if possible.

34. A local Boy Scout troop is going to build model rockets to launch on their next adventure weekend. The scout leader looks up the following prices for the parts for each rocket they build: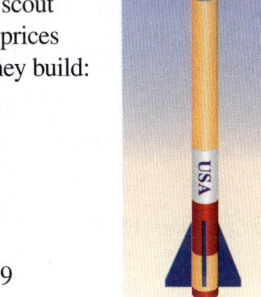

   BT-50 9″ body tube: $0.89
   Plastic nose cone: $1.39
   Fin kit: $1.49
   Engine mount kit: $5.29
   3 pack D engine pack: $7.79
   24″ parachute kit: $3.49

a. If the scout leader wants to make *r* rockets with the scouts, write an expression for the cost to buy the body tubes for the rockets.

b. Write expressions for the cost for each of the remaining parts listed.

c. Using the expressions found in parts a and b, write an expression for the total cost of buying the parts to make *r* rockets for the scouts. Simplify the expression if possible.

35. When doing his taxes, Steve discovers that he is in the 15% tax bracket for federal taxes and in the 8% tax bracket for his state.

a. If *I* is Steve's taxable income in dollars, write an expression for the amount Steve pays in federal taxes.

b. Write an expression for the amount Steve pays for his state taxes.

c. Use the expressions found in parts a and b to write an expression for the total amount Steve pays in both federal and state taxes. Simplify the expression if possible.

36. Perry looked at his first paycheck and noticed that money was withheld from his pay for taxes. Perry figured out the following withholdings:

   Federal tax: 17%
   Social Security tax: 6%
   Medicare tax: 1.5%
   State tax: 5%

a. If *T* is Perry's taxable income in dollars, write expressions for the amount of each tax that is taken out of his paycheck.

b. Use the expressions found in part a to write an expression for the total amount taken out of Perry's paycheck for taxes. Simplify the expression if possible.

For Exercises 37 through 50, simplify each expression by combining like terms.

37. $3x + 7 + 5x$

38. $9m + 4m + 8$

39. $7t + 9 - 3t + 5$

40. $4h - 8 - 9h + 6$

41. $2x^2 + 3x + 5x^2 + 6x$

42. $w^3 + 5w^3 + 3w^3 - w^2$

43. $4k^2 + 5k - 9$

44. $3g^5 + 5g^3 + 2g + 6$

45. $2xy + 5x - 3xy + 7y$

46. $8st + 5t^2 + 4st + 9s$

47. $5a^2b^3 + 4ab^2 + 8a^2b^2 + 2ab^2$

48. $-7n^3p^2 + 5n^2p + 4n^2p + 13n^3p^2$

49. $s^5t^3 + 4s^3t^5 + 7s^2t + 6s^5t^3 + 2s^2t$

50. $4xy + 2xz - xy + 3yz + 5xz$

CHAPTER 1 Building Blocks of Algebra

For Exercises 51 through 60:
 a. State whether the multiplication property or the distributive property will be used.
 b. Simplify each expression.

51. $2(3x + 7)$ **52.** $6(10m^2 + 2m)$

53. $5(2xy)$ **54.** $-3(4mn^2)$

55. $-4(7x^2 + 9x - 6)$

56. $7(3g^5 + 5g^3 + 2g - 8)$

57. $-5(4a^2bc)$ **58.** $6(9w^3)$

59. $-2(-3xy + 7y)$ **60.** $-3(12st + 5t^2 - 9s)$

For Exercises 61 through 70, a problem and a student's solution are given.
 a. Determine whether the student simplified correctly or incorrectly.
 b. If the student simplified the expression incorrectly, give the correct simplification.

61. $2(3x\,y)$ **62.** $5(2ab)$

> **Bill**
> $6x2y$

> **Evin**
> $10a5b$

63. $3(2k - 4)$ **64.** $4(3x + 5)$

> **Mandee**
> $6k - 4$

> **James**
> $12x + 20$

65. $7(3mn)$ **66.** $3(5xy)$

> **Marta**
> $21m21n$

> **Vanessa**
> $15xy$

67. $\frac{1}{5}(25ab)$ **68.** $\frac{1}{2}(10xy)$

> **Jennifer**
> $5ab$

> **Rita**
> $5x\frac{1}{2}y$

69. $\frac{2}{3}(6x + 9)$ **70.** $\frac{3}{4}(8y - 12)$

> **Joanne**
> $4x + 9$

> **Chad**
> $6y - 9$

For Exercises 71 through 96, simplify each expression.

71. $5(4x + 8) + 3x - 10$ **72.** $-2(5x - 4) + 10$

73. $-3(4g + 7) + 2g + 9$ **74.** $7(8d + 3) - 20d - 15$

75. $(2x + 5) - (5x - 7)$ **76.** $(5w + 9) - (-3w + 8)$

77. $(4k^2 + 5) - 3(2k^2 + 5k)$

78. $3(2a + 7) + (3a^2 + 5)$

79. $2xy + 5x + 2(3xy + 7)$

80. $-5(2st + 3) + 4st + 9$

81. $3(2x^2 + 3x) - 2(5x^2 + 6x)$

82. $4(3a^2 + 5a) - 7(6a^2 + 9a)$

83. $4(5w^2 + 6w - 3) + 3(2w^2 + 4w + 1)$

84. $2(3y^2 + 4y - 1) + 5(6y^2 + 8y + 3)$

85. $-2(4r^2 + 6r - 9) - 3(-5r^2 - 5r + 3)$

86. $-7(9x^2 - 5x + 1) - (-2x^2 + 7x - 2)$

87. $1.3(2t + 6) + 6.3t + 8$

88. $2.1(3b + 5) + 6.1b + 10$

89. $2.5(4h + 6) - 3.5(2h - 8)$

90. $3.2(4m - 3) - 6.3(2m + 8)$

91. $\frac{1}{2}(4x - 8) + 7x - 3$ **92.** $\frac{1}{2}(10 - 2x) - 7x + 12$

93. $\frac{1}{3}(9a - 6) + 8a - 14$ **94.** $\frac{1}{3}(12 - 3b) - 2b - 1$

95. $\frac{1}{5}(10a - 5b) + \frac{1}{2}(-4a + 12b)$

96. $\frac{1}{3}(6b - 15a) - \frac{1}{4}(16a - 8b)$

SECTION 1.3 Simplifying Expressions **121**

For Exercises 97 through 106, use the given figures to find expressions for the perimeter of the shapes. Simplify the expressions if possible. Note: Perimeter is the sum of the lengths of the sides.

97. Perimeter of a rectangle:

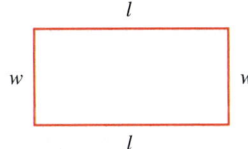

98. Perimeter of a square:

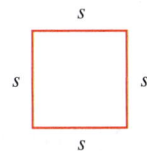

99.

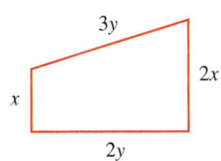

100.

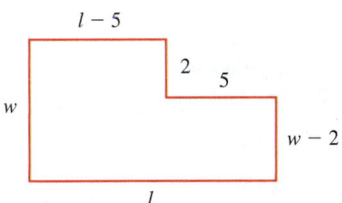

101. Perimeter of a rectangle:

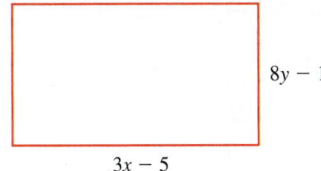

102. Perimeter of a rectangle:

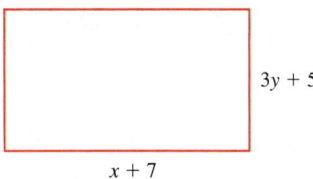

103. Perimeter of an equilateral triangle:

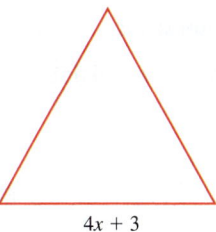

104. Perimeter of an equilateral triangle:

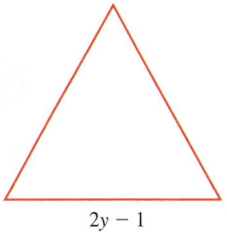

105. Perimeter of a square:

106. Perimeter of a square:

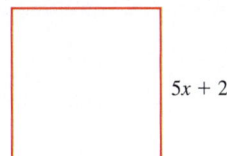

1.4 Graphs and the Rectangular Coordinate System

LEARNING OBJECTIVES

- Interpret data tables.
- Read and create bar graphs and scatterplots.
- Plot points on the rectangular coordinate system.
- Describe scale.
- Graph real-world data.

■ Data Tables

In Section 1.2, we saw that algebraic expressions may be derived from input–output tables. Tables are also used to organize and display data. In many fields, such as health organizations and corporations, information is often presented in tables.

Example 1 Reading a table

The following table gives data on the influenza season from the first 10 weeks in 2017. The first column of the table (the input) is the week number in 2017. The second column (the output) gives the percentage of positive isolates (tests) for the influenza virus from testing labs across the United States. This percentage means the number of people who actually had the flu out of those who were tested for the flu.

Week	Percentage of Positive Isolates
1	13.19
2	14.8
3	17.57
4	17.98
5	21.25
6	24.14
7	24.21
8	24.26
9	19.17
10	18.83

Source: Centers for Disease Control and Prevention (www.cdc.gov/flu).

a. What is the maximum percentage of positive isolates recorded? When does this occur?

b. What is the minimum percentage of positive isolates recorded? When does this occur?

SOLUTION

a. If we examine these data carefully, we can see that the percentage is increasing through week 8 and then starts to decrease.

The maximum percentage of positive isolates is 24.26%, and it occurs in week 8.

b. *The minimum percentage of positive isolates is 13.19%, and it occurs in week 1.*

PRACTICE PROBLEM FOR EXAMPLE 1

The following table gives the number of bachelor's degrees awarded in the United States to science and engineering majors for the years 2008 through 2012.

Year	Number of Science and Engineering Degrees
2008	496,168
2009	504,435
2010	525,374
2011	554,365
2012	589,330

Source: www.nsf.gov

a. What was the maximum number of bachelor's degrees awarded? When did this occur?

b. What was the minimum number of bachelor's degrees recorded? When did this occur?

■ Bar Graphs

Often data are easier for many people to examine if they are visually presented in a graph. The **horizontal axis,** sometimes called the *x*-axis, is where the input values are plotted. The **vertical axis,** sometimes called the *y*-axis, is where the output values are plotted. The height of each bar represents the value of the output for the corresponding value of the input. The bars are of uniform width and evenly spaced along the horizontal axis.

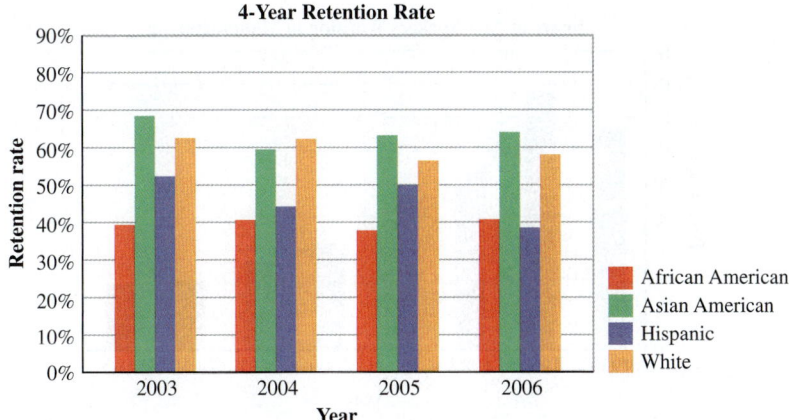

Example 2 — Reading a bar chart

The following graph is a bar chart giving the numbers of immigrants to the United States during the years 1820–2010.

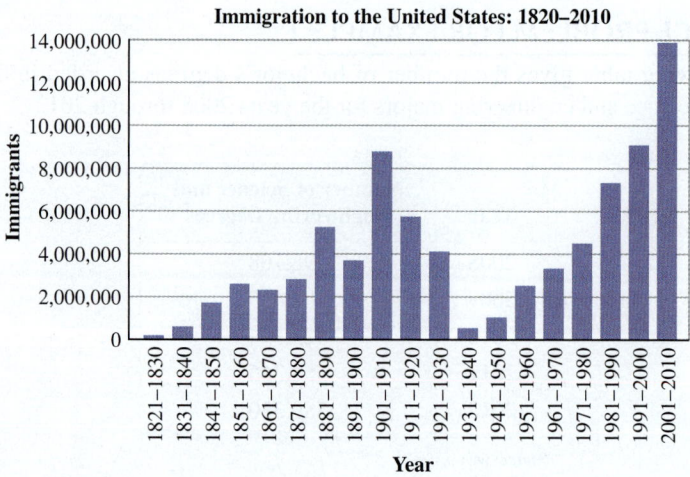

Source: Population Environment Balance.

a. For what years was the number of immigrants to the United States about 4,000,000 (four million)?

b. For what time period is the number of immigrants to the United States the largest?

SOLUTION

a. *For the years 1921–1930, the bar is about (just slightly above) 4,000,000.*

b. *The highest bar is for the years 2001–2010, so that is the time period when the number of immigrants to the United States was the largest.*

PRACTICE PROBLEM FOR EXAMPLE 2

The following graph is a bar chart showing the share of New Yorkers who work in manufacturing.

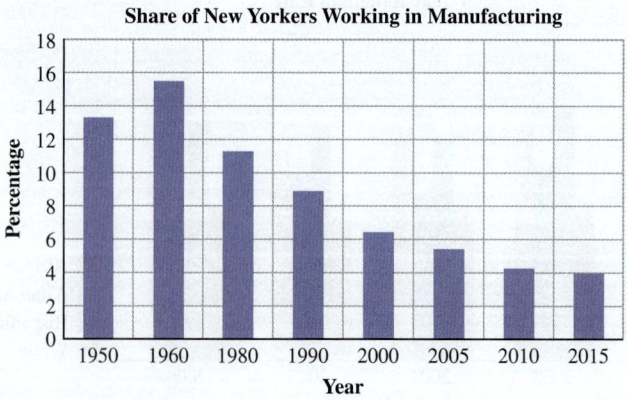

Source: U.S. Census Bureau.

a. What percentage of New Yorkers worked in manufacturing in 2010?

b. During what year did the highest percentage of New Yorkers work in manufacturing?

Steps to Create a Bar Graph

1. The input values will be plotted on the horizontal axis, and the output values will be plotted on the vertical axis.
2. Evenly and consistently space the horizontal axis so that it includes the input values (data).
3. Evenly and consistently space the vertical axis so that it includes the output values (data).
4. Draw vertical bars of uniform width. Center each bar over the appropriate input value. Draw the bars of height corresponding to the output value.

Example 3 Drawing a bar chart

Make a bar chart for the influenza data presented in Example 1 on page 122.

Week	Percentage of Positive Isolates
1	13.19
2	14.8
3	17.57
4	17.98
5	21.25
6	24.14
7	24.21
8	24.26
9	19.17
10	18.83

Source: Centers for Disease Control and Prevention (www.cdc.gov/flu).

SOLUTION

The input values are weeks 1 through 10, so we plot these values on the horizontal axis. The output values range from the smallest value of 13.19 to the largest value of 24.26. Thus, we will draw the vertical axis from 0 to 30 by counting by fives.

Notice that each bar is centered over the week to which it corresponds (the input value) and that the bars are evenly spaced along the horizontal axis. The heights of the bars represent the output value (the percent of positive isolates).

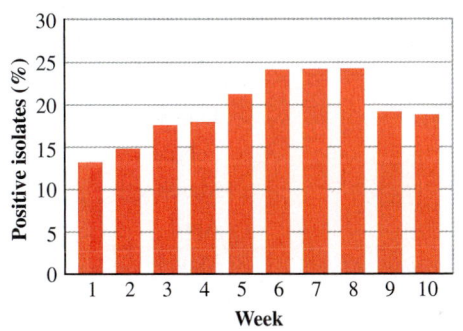

Example 4 Drawing a bar chart that includes negative values

The following table gives the net income for Volkswagen from 2012 to 2016. Draw a bar chart for this data set.

Year	Net Income (billions of dollars)
2012	21.7
2013	9.1
2014	10.9
2015	−1.6
2016	5.1

Source: www.marketwatch.com

SOLUTION

Notice that the output values take on both positive and negative values. When the output values take on negative values, draw the bar *upside down* under the horizontal axis. The horizontal axis is scaled by 1, which will accommodate the years 2012 through 2016. The vertical axis is scaled by 4. This will accommodate the smallest output value of −1.6 and the largest output value of 21.7.

126 CHAPTER 1 Building Blocks of Algebra

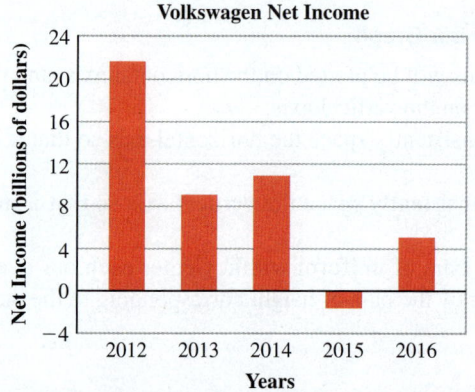

PRACTICE PROBLEM FOR EXAMPLE 4

The following table shows the average daily high temperature at Vostok Research Station in Antarctica. Draw a bar chart for this data set.

Month	Average High Temperature (°F)
January	−17°F
February	−38°F
March	−63°F
April	−78°F
May	−80°F

Source: Wikipedia.

■ Scatterplots

Another important type of graph is a scatterplot. The horizontal and vertical axes are scaled as they are for bar graphs. Instead of having a bar represent the height (or output value), use a large dot where the top of the bar would be.

Example 5 Drawing a scatterplot

Draw a scatterplot for Volkswagen data given in Example 4.

SOLUTION

A reasonable vertical axis for these data would range between −4 and 24, with a scale of 4. A negative value for Volkswagen's net income means that the company lost money that year and will be graphed below the horizontal axis. For a scatterplot, instead of using a bar to represent the height, use a large dot.

Year	Net Income (Billions of Dollars)
2012	21.7
2013	9.1
2014	10.9
2015	−1.6
2016	5.1

Source: www.marketwatch.com

SECTION 1.4 Graphs and the Rectangular Coordinate System 127

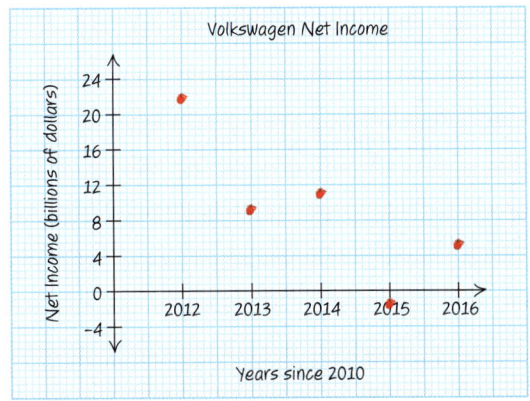

Month	Average High Temperature (°F)
January	−17°F
February	−38°F
March	−63°F
April	−78°F
May	−80°F

Source: Wikipedia.

PRACTICE PROBLEM FOR EXAMPLE 5
Draw a scatterplot for the Vostok Station data given in Practice Problem for Example 4.

▪ Rectangular Coordinate System

In Section 1.2, we learned how to generate relationships between two variables, an input and an output, from tables and expressions. For example, we developed the following table, which gave us the relationship between the number of inches of snow (s) and the number of inches of rain (r).

The data consist of values for the input variable (s) and values for the output variable (r). We write individual data items as **ordered pairs** (input, output) = (s, r). For example, the ordered pair (12, 1) for this equation means that $s = 12$ inches of snow converts to $r = 1$ inch of rain. The order matters in an ordered pair. We cannot swap the input and output variables. In math, we often use the variables x and y, and have ordered pairs (x, y).

s = inches of snow	r = inches of rain
12	$\frac{12}{12} = 1$
24	$\frac{24}{12} = 2$
36	$\frac{36}{12} = 3$
42	$\frac{42}{12} = 3.5$
s	$\frac{s}{12}$

> **DEFINITION**
> **Ordered Pair** An ordered pair is of the form (x, y), where x is the value of the input variable and y is the value of the output variable.

To graph data, we use the **rectangular coordinate system.** The input data are graphed on the horizontal axis (or x-axis), and the output data are graphed on the vertical axis (or y-axis). We graph each ordered pair as a point. The input, or x-coordinate, tells us how many units the point is located to the right (x is positive) or to the left (x is negative) of the y-axis. The output, or y-coordinate, tells us how many units the point is located above (y is positive) the x-axis, or below (y is negative) the x-axis. The point where the x-axis and y-axis cross is written as (0, 0) and is called the **origin.**

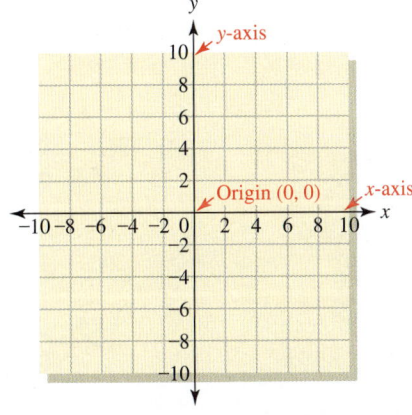

What's That Mean?

Cartesian Coordinate System
The rectangular coordinate system is also called the Cartesian coordinate system. It is named after the French mathematician, philosopher, and scientist René Descartes.

128 CHAPTER 1 Building Blocks of Algebra

Notice that on the graph above, the scale on the *x*-axis is 2. The scale on the *y*-axis is also 2.

> **What's That Mean?**
>
> **Coordinates**
>
> A point is described by an ordered pair (x, y). The values of x and y are called the **coordinates** of that point.

Example 6 Writing the ordered pairs shown on the graph

On the following graph, there are eight different points. Write the ordered pairs that give the coordinates of each point.

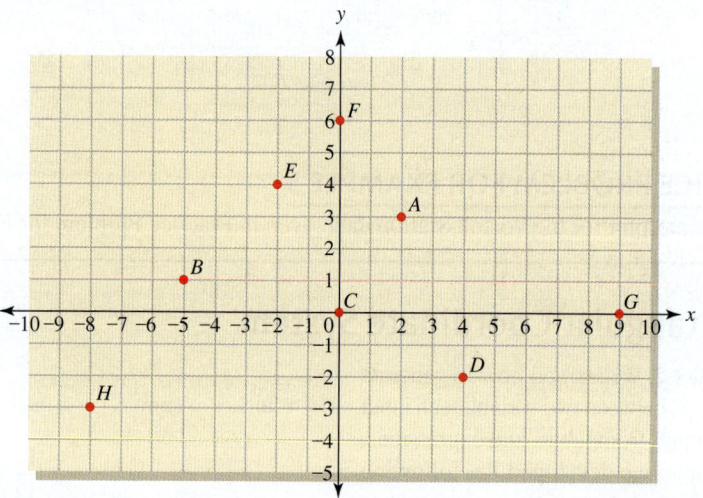

SOLUTION
To determine the coordinates of point *A*, we need to find the values of the *x*- and *y*-coordinates. Point *A* is located 2 units to the right of the *y*-axis, so the *x*-coordinate is $x = 2$. Point *A* is located 3 units above the *x*-axis, so the *y*-coordinate is $y = 3$. The coordinates of point *A* are (2, 3).

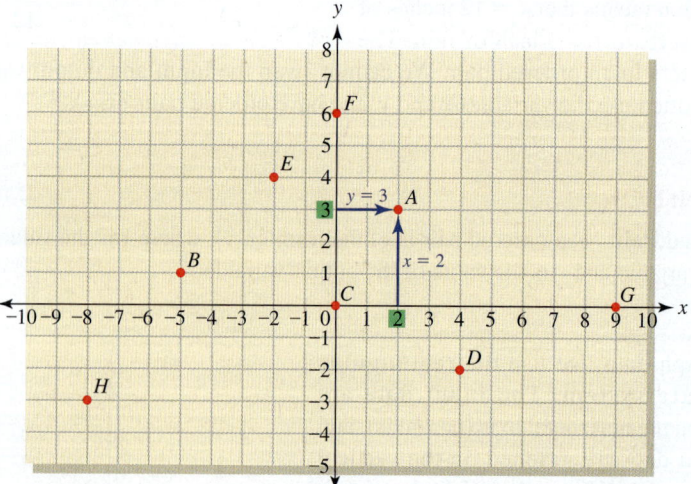

Point B is located 5 units to the left of the y-axis, so the x-coordinate is $x = -5$. Point B is located 1 unit above the x-axis, so the y-coordinate is $y = 1$. The coordinates of point B are $(-5, 1)$.

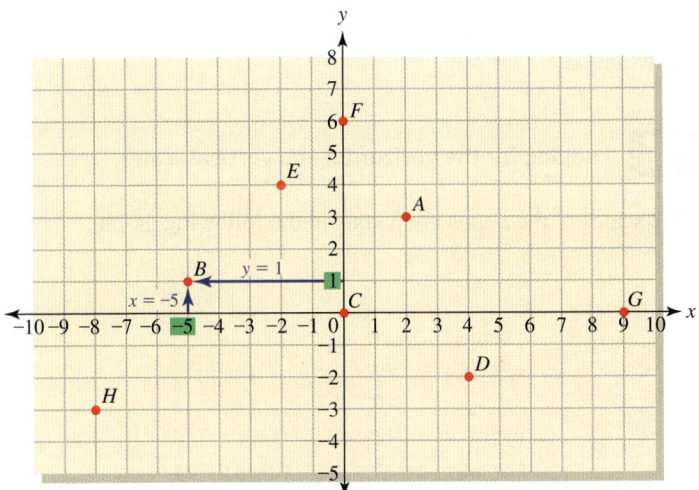

We find the coordinates of the rest of the points in a similar fashion.

Point C has coordinates $(0, 0)$: Point C is located at the origin.

Point D has coordinates $(4, -2)$.

Point E has coordinates $(-2, 4)$.

Point F has coordinates $(0, 6)$. Point F is located on the y-axis, which passes through $x = 0$.

Point G has coordinates $(9, 0)$. Point G is located on the x-axis, which passes through $y = 0$.

Point H has coordinates $(-8, -3)$.

PRACTICE PROBLEM FOR EXAMPLE 6

On the following graph, there are eight different points. Write the ordered pairs that give the coordinates of each point.

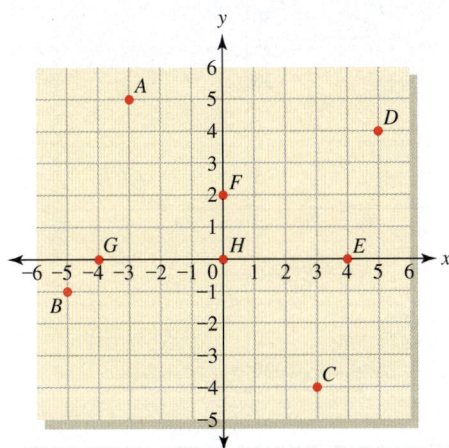

In Example 6, point D has coordinates (4, −2), and point E has coordinates (−2, 4). We can see from the graph that points D and E are not the same. This means that (4, −2) ≠ (−2, 4). This is why we call them *ordered* pairs. The order matters.

It is very important to pay attention to reading the scale from a graph. The scales on the axes are not always 1 and can be different from one another. This is demonstrated in the following example.

Example 7   Looking for the horizontal and vertical scales

Determine the scales of the *x*-axis and *y*-axis in the following graph.

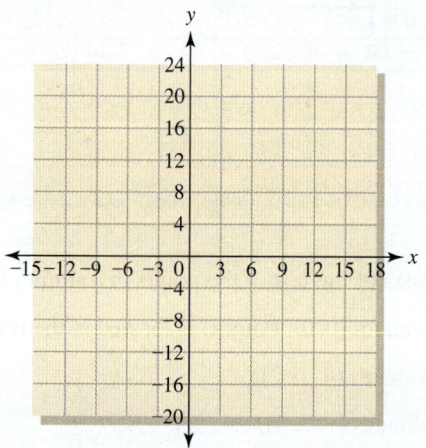

SOLUTION
The *x*-axis has a scale of 3 because the even spacing on the *x*-axis is by 3's. The *y*-axis has a scale of 4 because the even spacing on the *y*-axis is by 4's. Here the two axes have different scales from each other, but the scale on a single axis must be consistent.

PRACTICE PROBLEM FOR EXAMPLE 7
Determine the scales of the *x*-axis and *y*-axis in the following graph.

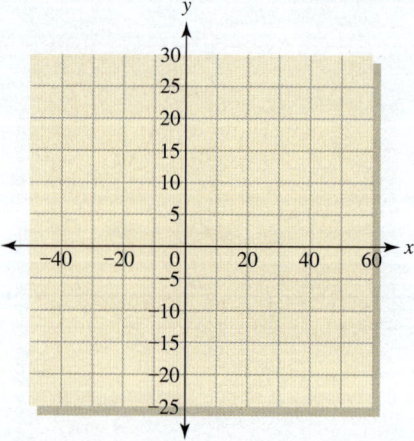

> **Steps to Create a Graph in the Rectangular Coordinate System**
> 1. Evenly and consistently space and scale the horizontal axis so that it includes the input values.
> 2. Evenly and consistently space and scale the vertical axis so that it includes the output values. *Note:* The scale of the vertical and horizontal axes may be different.
> 3. Plot each ordered pair (point) by first finding the intersection of the input value and the output value and then drawing a dot.

Skill Connection

When $x = 0$, the point is located on the *y*-axis. It has the form $(0, y)$. When $y = 0$, the point is located on the *x*-axis. It has the form $(x, 0)$.

Example 8 Graphing ordered pairs

Graph the following ordered pairs on the same axes. Clearly label and scale the axes.

a. $(-3, 4)$ b. $(5, -2)$ c. $(3, 3)$
d. $(-4, -1)$ e. $(2, 0)$ f. $(0, -3)$

SOLUTION

Step 1 Evenly and consistently space and scale the horizontal axis so that it includes the input values.

The input values range from -4 to 5. It is reasonable to scale the horizontal axis by 1's.

Step 2 Evenly and consistently space and scale the vertical axis so that it includes the input values.

The output values range from -3 to 4 so it will also be reasonable to scale the *y*-axis by 1's.

Step 3 Plot each ordered pair (point) by first finding the intersection of the input value and the output value and then drawing a dot.

The points are plotted on the following graph.

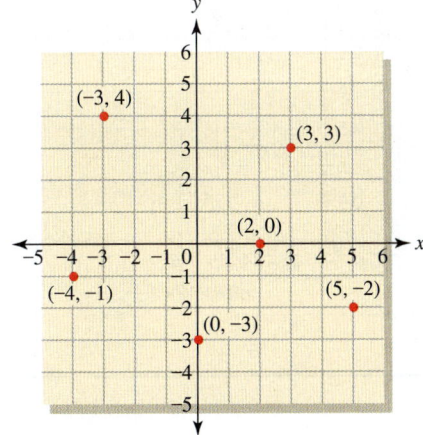

PRACTICE PROBLEM FOR EXAMPLE 8

Graph the following ordered pairs on the same axes. Clearly label and scale the axes.

a. $(2, 6)$ b. $(-5, 12)$ c. $(-1, 15)$
d. $(-6, -10)$ e. $(0, 20)$ f. $(-5, 0)$

132 CHAPTER 1 Building Blocks of Algebra

When graphing real data from a problem situation, consider the scale to be used on the axes. Remember that the scale on an axis must be consistent. This means that values on that particular axis must be evenly spaced. However, the horizontal axis and the vertical axis may have different scales.

Example 9 Graphing data

Graph the following ordered pairs on the same axes. Clearly label and scale your axes.

s = inches of snow	r = inches of rain
12	1
24	2
36	3
42	3.5

SOLUTION
To graph the data, first decide on the scale to use. For the horizontal axis, the data values are increasing by 12, so we plot starting at 0, with a scale of 12. The output values, which are plotted on the vertical axis, range from 1 to 3.5. So plotting from 0 to 5 with a scale of 0.5 should graph these values effectively.

When ordered pairs appear in a table, we pair them by row. In the first row, we are given that $s = 12$ and $r = 1$. This row represents the ordered pair $(s, r) = (12, 1)$.

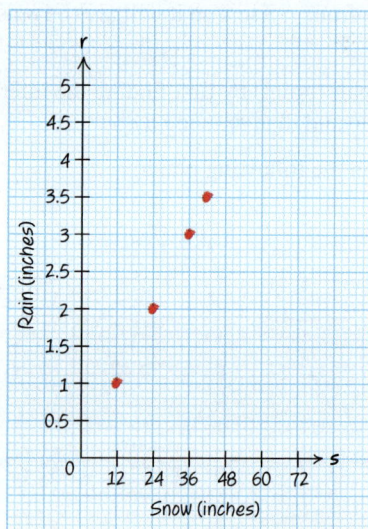

Sometimes the data presented to us do not start anywhere near the zero value of either axis. When we do not start graphing an axis at zero, we use a *zigzag* pattern to show there is a break in the numbering. This zigzag is called a *break* in the graph. This is shown in Example 10.

Example 10 Graphing data and adjusting axes

For the following data set, graph the points. Clearly scale and label the axes. The table lists the number of robberies reported in the Uniform Crime Reporting (UCR) Program database for the years 2005–2015.

Year	Robberies
2005	417,438
2006	447,403
2007	445,125
2008	443,574
2009	408,742
2010	369,089
2011	354,772
2012	355,051
2013	345,095
2014	322,905
2015	327,374

SOLUTION

We graph the input values along the horizontal axis. Since the years do not begin at zero but at 2005, we place a zigzag in the graph along the lower left corner to indicate the break in time. We start with the year 2004 and end at 2016, with a scale of 1 year. The vertical axis is where we graph the output data, which have a minimum value of 322,905 and a maximum value of 447,403. We put a zigzag in the graph to indicate that the vertical axis does not start at zero. We graph from 300,000 to 460,000, with a scale of 20,000. The scale of 20,000 was selected because it is easy to count by 20,000's and the data were spread out by several thousand.

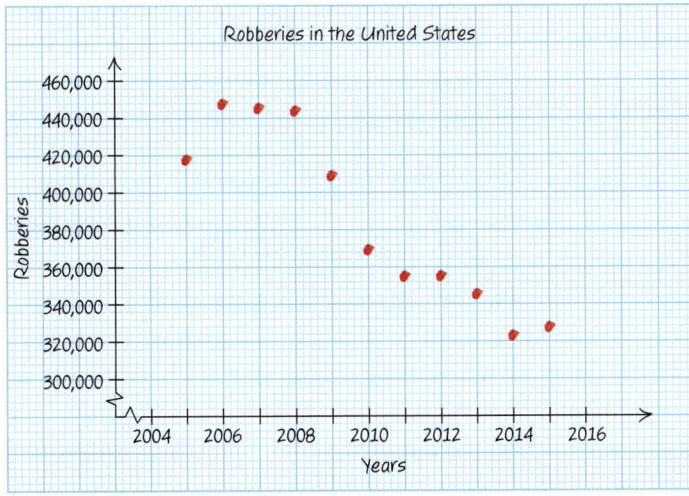

This graph is a bit different from the graphs we have drawn previously. First of all, there is a break in the graph near the origin to indicate that we are not starting to count at 0. The second difference is that the output values in the data table do not exactly match up with the values on the vertical axis. This is common with real-world data, and we often have to make a best estimate of where the point lies on the graph.

PRACTICE PROBLEM FOR EXAMPLE 10

The percentage of Americans who own stock are given in the table. Graph the data, clearly scale and label the axes.

Year	Share of Americans Who Own Stocks (%)
1989	31.90
1992	37.00
1995	40.50
1998	48.90
2001	53.00
2004	50.30
2007	53.20
2010	49.80
2013	58.42

1.4 Vocabulary Exercises

1. The input values are plotted on the _____.
2. The output values are plotted on the _____.
3. The height of each bar in a bar chart represents the value of the _____ for the corresponding _____ value.
4. Every point in the rectangular coordinate system should be written as a(n) _____.
5. The point (0, 0) is called the _____.
6. In the coordinate (2, 5), the 2 is the ___-value, and the 5 is the ___-value. The point will be 2 units to the _____ of the origin and 5 units _____ the origin.
7. Another name for the x-axis is _____.
8. Another name for the y-axis is _____.
9. A(n) _____ is placed in the graph to indicate a break in the values.
10. Each axis must have a consistent _____.

1.4 Exercises

For Exercises 1–4, use the data table given in each exercise to answer the questions.

1. The table lists the sales of iPhones and iPads for several quarters.

Quarter	iPhone (millions)	iPad (millions)
Q1 '14	51.03	26.04
Q2 '14	43.72	16.35
Q3 '14	35.2	13.28
Q4 '14	39.27	12.32
Q1 '15	74.47	21.42
Q2 '15	61.17	12.62
Q3 '15	47.53	10.93
Q4 '15	48.05	9.88
Q1 '16	74.78	16.12
Q2 '16	51.19	10.25
Q3 '16	40.4	9.95
Q4 '16	45.51	9.27
Q1 '17	78.29	13.08

Source: Apple Inc.

 a. What is the maximum number of iPads sold in a quarter? When did this happen?

 b. What is the minimum number of iPhones sold in a quarter? When did this happen?

2. The table lists the total revenue for scheduled freight air transportation in the United States.

Year	Revenue (millions of $)
2009	5795
2010	6441
2011	7078
2012	6671
2013	6233
2014	6146
2015	6042

Source: U.S. Census Bureau.

 a. What is the maximum revenue for scheduled freight air transportation? When did this happen?

 b. What is the minimum revenue for scheduled freight air transportation? When did this happen?

3. The world record times for the men's marathon are listed in the table.

Time	Athlete	Country	Date
2:08:18	Robert De Castella	Australia	6-Dec-81
2:08:05	Steve Jones	United Kingdom	21-Oct-84
2:07:12	Carlos Lopes	Portugal	20-Apr-85
2:06:50	Belayneh Dinsamo	Ethiopia	17-Apr-88
2:06:05	Ronaldo da Costa	Brazil	20-Sep-98
2:05:42	Khalid Khannouchi	Morocco	24-Oct-99
2:05:38	Khalid Khannouchi	United States	14-Apr-02
2:04:55	Paul Tergat	Kenya	28-Sep-03
2:04:26	Haile Gebrselassie	Ethiopia	30-Sep-07
2:03:59	Haile Gebrselassie	Ethiopia	28-Sep-08
2:03:38	Patrick Makau	Kenya	25-Sep-11
2:03:23	Wilson Kipsang	Kenya	29-Sep-13
2:02:57	Dennis Kimetto	Kenya	28-Sep-14

Source: https://en.wikipedia.org/wiki/Marathon_world_record_progression

 a. What is the fastest marathon time? In what year did this happen?

 b. What country has the most world record holders?

4. The world record times for the women's marathon are listed in the table.

Time	Athlete	Country	Date
2:21:06	Ingrid Kristiansen	Norway	21-Apr-85
2:20:47	Tegla Loroupe	Kenya	19-Apr-98
2:20:43	Tegla Loroupe	Kenya	26-Sep-99
2:19:46	Naoko Takahashi	Japan	30-Sep-01
2:18:47	Catherine Ndereba	Kenya	7-Oct-01
2:17:18	Paula Radcliffe	United Kingdom	13-Oct-02
2:15:25	Paula Radcliffe	United Kingdom	13-Apr-03
2:17:42	Paula Radcliffe	United Kingdom	17-Apr-05
2:17:01	Mary Jepkosgei Keitany	Kenya	23-Apr-17

Source: https://en.wikipedia.org/wiki/Marathon_world_record_progression

a. What is the fastest women's marathon time? In what year did this happen?

b. What country do the most world record holders come from?

For Exercises 5 through 8, use the graph given in each exercise to answer the questions.

5. The graph gives the percentage of adults in the United States who are considered obese for the years 2004–2014.

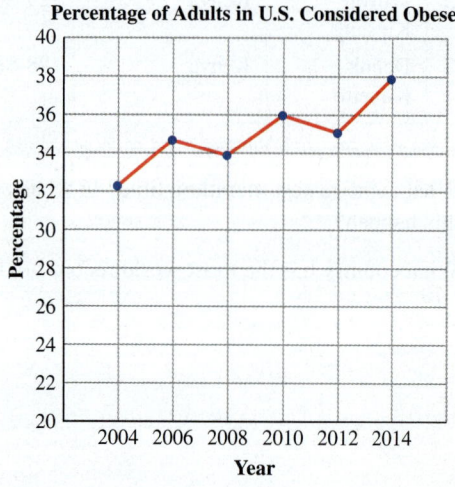

Source: www.cdc.gov

a. During which year(s) was the percentage the highest? Approximately what percentage of the adult population was obese that year?

b. During which year(s) was the percentage the lowest? Approximately what percentage of the adult population was obese that year?

6. The graph shows the number of named storms and hurricanes from 1960–2015 in the Atlantic Ocean. The hurricanes are in red, and the other named storms are in blue.

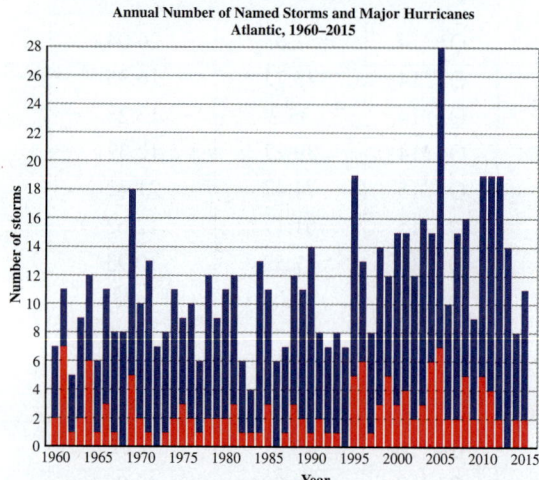

Source: Atlantic Oceanographic and Meteorological Laboratory, NOAA.

a. Which year(s) had the lowest number of Atlantic named storms and hurricanes? Approximately how many storms and hurricanes occurred during that year(s)?

b. Which year(s) had the highest number of Atlantic named storms and hurricanes? Approximately how many storms and hurricanes occurred during that year(s)?

7. The graph below shows sales of new homes in the United States from 1970 to 2016.

Source: U.S. Census Bureau.

a. In which year were new home sales at their lowest point? Approximately how many sales were made?

b. In which year were new home sales at their highest point? Approximately how many sales were made?

8. The graph shows gasoline prices in San Diego, California, during 2014–2017.

Source: http://www.sandiegogasprices.com/

a. During which month and year did San Diego have the lowest retail gas prices?

b. During which month and year did San Diego have the highest retail gas prices?

For Exercises 9 through 14, use the table given in each exercise to make a bar chart. Clearly label and scale the axes.

9. The table gives Gina's personal income over a 5-year period.

Year	Personal Income
2015	$27,000
2016	$29,000
2017	$30,500
2018	$31,000
2019	$31,500

10. The table lists the cost of attending Cayuga County Community College (in upstate New York), as determined by the number of units a resident part-time student enrolls in.

Number of Units	Cost (dollars)
3	624.00
6	1248.00
8	1664.00
9	1872.00
10	2080.00

11. The table lists the relationship between the number of years and the number of centuries. Note that a century is a 100-year time period.

Number of Years	Number of Centuries
100	1
150	1.5
200	2
250	2.5
300	3

12. The table lists the relationship between the number of centuries and the number of millennia. Note that a century is a 100-year time period and a millennium is a 1000-year time period.

Number of Centuries	Number of Millennia
10	1
15	1.5
20	2
25	2.5
30	3

13.

t	P
0	23
2	17
4	11
6	5
8	-7
10	-25

14.

x	y
0	-4
0.5	-1
1	5
1.5	6
2	3
2.5	1

For Exercises 15–20, use the data given in each exercise to create a scatterplot. Clearly label and scale the axes.

15. Use the table from Exercise 9.

16. Use the table from Exercise 10.

17.

x	y
-50	-20
-25	-10
0	0
10	4
40	16
50	20

18.

x	y
-2	-4
-1.5	-1
-1.0	5
-0.5	6
0	3
0.5	1

138   CHAPTER 1   Building Blocks of Algebra

19. The average high monthly temperature in San Diego, California, for the months of January through June is listed in the table. *Hint:* Let January = 1, February = 2, and so on.

Month	Average Monthly High Temperature
January	65.1°F
February	65°F
March	65.6°F
April	67.5°F
May	68.5°F
June	70.8°F

20. The average high monthly temperature in St. Paul, Minnesota, for the months of January through June is listed in the table.

Month	Average Monthly High Temperature
January	22.6°F
February	28.7°F
March	40.6°F
April	57.7°F
May	70°F
June	78.4°F

For Exercises 21 through 24, write the coordinates of each ordered pair shown on the graph.

21.
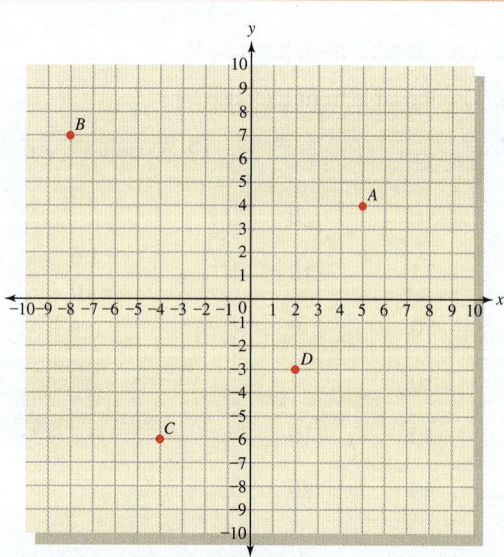

22.
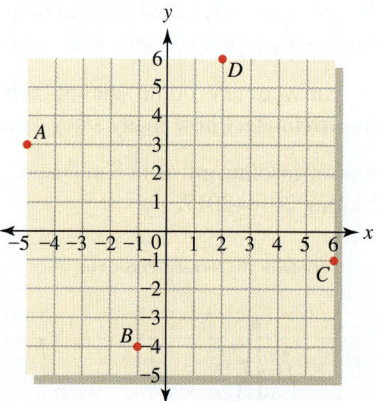

23.
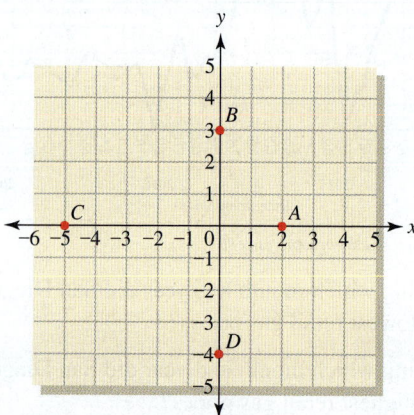

24.
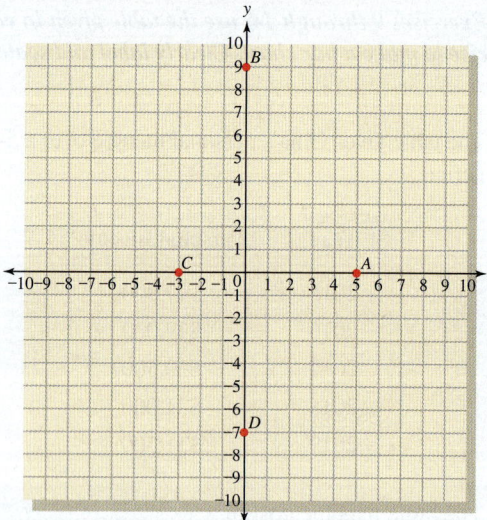

SECTION 1.4 Graphs and the Rectangular Coordinate System 139

For Exercises 25 through 28, state the scale on both the x- and y-axes.

25.

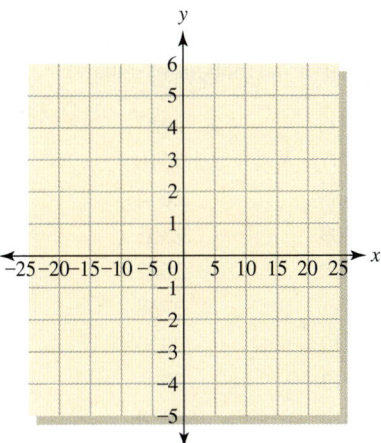

28.

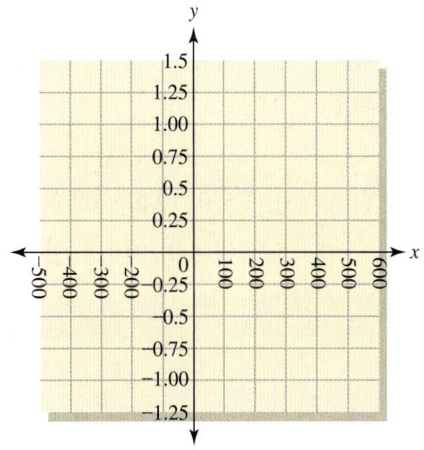

26.

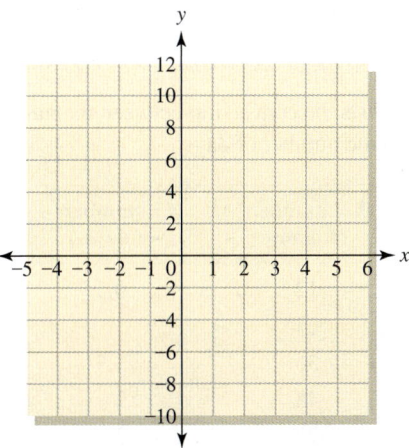

For Exercises 29 and 30, write the coordinates of each ordered pair shown on the graph.

29.

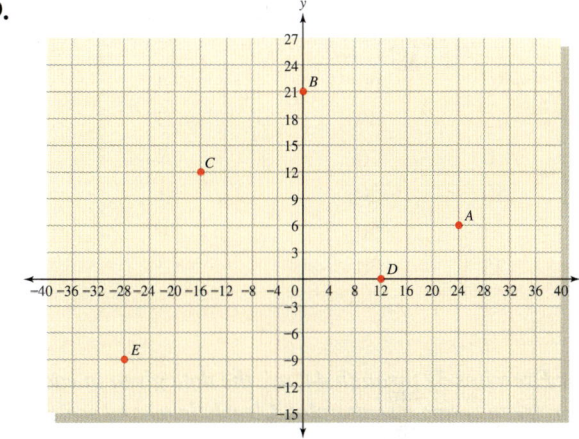

27.

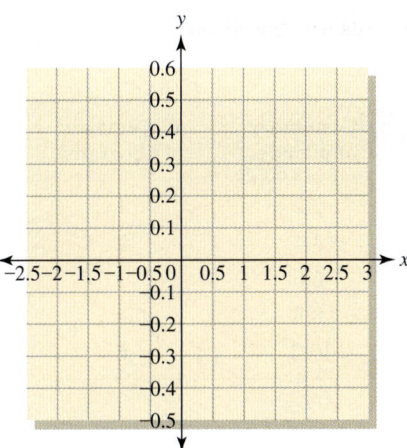

30.

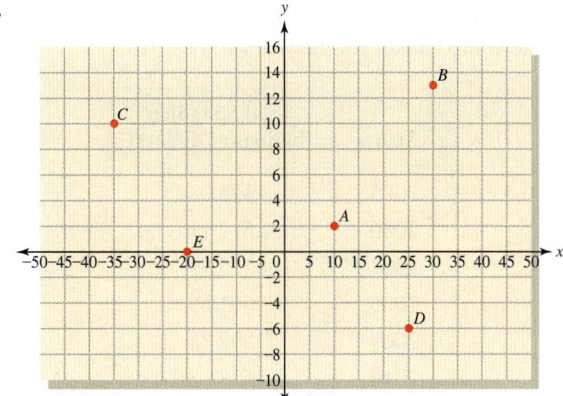

140 CHAPTER 1 Building Blocks of Algebra

For Exercises 31 through 38, graph the given ordered pairs on the same axes. Clearly label and scale the axes.

31. {(3, 2), (5, 4), (−2, 1), (−3, 6)}

32. {(−16, −8), (−8, −4), (0, 0), (8, 4)}

33. {(0.5, 1.0), (1.0, 0.5), (1.5, 0), (2.0, −0.5)}

34. {(−0.8, 3), (−0.4, 3.5), (0, 4.0), (0.4, 4.5)}

35.

x	y
0	−2
10	−4
20	16
30	−8
40	0
50	12

36.

x	y
−9	5
0	−10
3	15
6	−20
9	10
15	20

37.

x	y
−50	60
−25	40
0	10
50	−20
75	50
125	−30

38.

x	y
4	2.5
3	1.5
2	0
1	−2
0	−0.5
−3	1

For Exercises 39 through 44, use the data given in each exercise to create a scatterplot. Clearly label and scale the axes.

39. The table gives the total Christmas holiday spending in the United States for the years 2005–2016.

Year	Total Spending (billions of dollars)
2005	496.04
2006	512.11
2007	525.99
2008	501.59
2009	502.78
2010	528.93
2011	553.40
2012	567.77
2013	585.09
2014	613.27
2015	632.84
2016	655.87

Source: www.nrf.com

40. The table gives United States consumer credit for households and nonprofit organizations (in billions of dollars) for the years 2006–2016.

Year	Consumer Credit (billions of dollars)
2006	2457
2007	2610
2008	2644
2009	2555
2010	2647
2011	2758
2012	2920
2013	3096
2014	3318
2015	3536
2016	3765

Source: U.S. Department of the Treasury, Fiscal Service.

41. The table lists the conversion between the number of hours and the number of days.

h = number of hours	d = number of days
12	0.5
24	1
36	1.5
48	2
72	3

42. The table lists the conversion between pounds and kilograms.

P = number of pounds	K = number of kilograms
2.2	1
4.4	2
6.6	3
8.8	4
13.2	6

43. The table gives the U.S. federal debt (total amount owed) during the years 1990–2015.

Year	U.S. Debt (trillions of dollars)
1990	$3.36
1995	$4.99
2000	$5.66
2005	$8.17
2010	$14.03
2015	$18.92

Source: U.S. Department of the Treasury, Fiscal Service.

44. The table gives new privately owned housing starts for the year 2016. The units are in thousands.

Month	Number of New Housing Starts (thousands)
January	1128
February	1213
March	1113
April	1155
May	1128
June	1195
July	1218
August	1104
September	1052
October	1320
November	1149
December	1275

Source: www.census.gov/indicator

For Exercises 45 through 50, find the mistake in the student work shown.

45. Graph the point (3, 4).

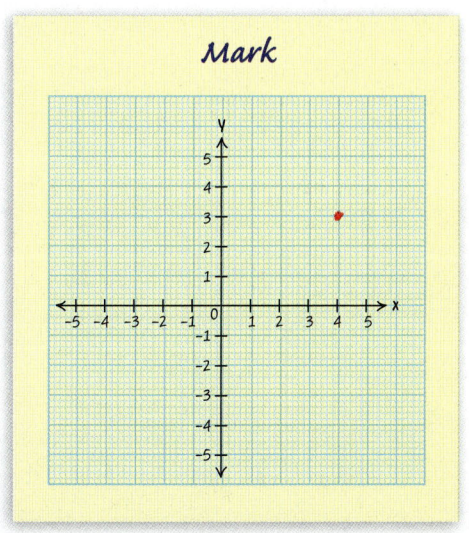

46. Graph the point $(-5, -2)$.

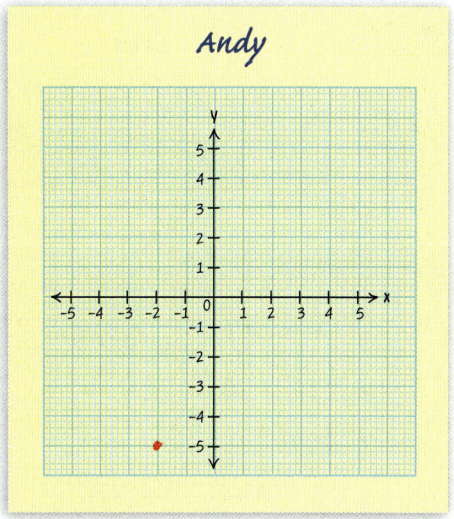

47. Graph the points in the table.

x	y
2	5
0	-2
5	-1
-3	-4

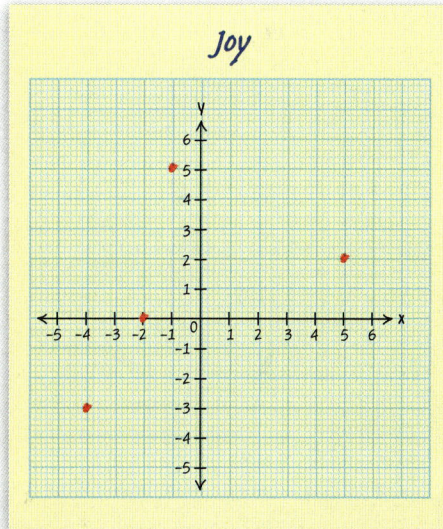

48. Graph the points in the table.

x	y
3	-1
6	0
-2	-4
4	1

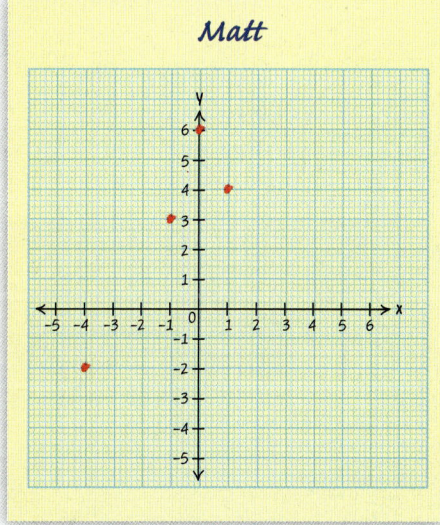

49. Graph the points in the table.

x	y
5	40
−20	−2

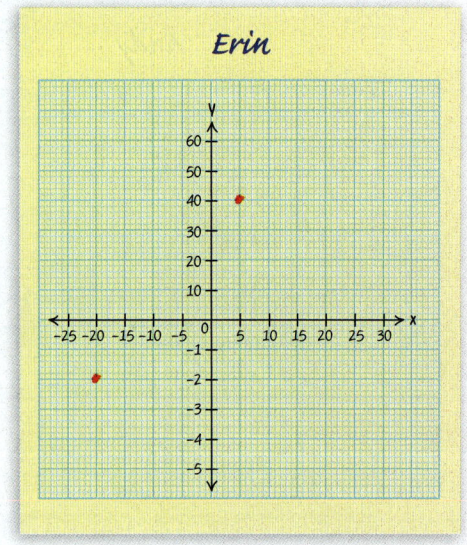

50. Graph the points in the table.

x	y
−4	50
0.5	−20

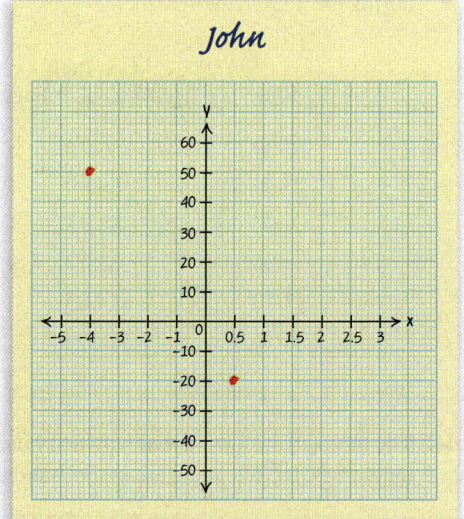

Chapter Summary

Section 1.1 Exponents, Order of Operations, and Properties of Real Numbers

- A **natural number** exponent is a way of writing a repeated multiplication. **Exponentiation** is another operation on real numbers.
- $a^0 = 1$ for all $a \neq 0$. 0^0 does not exist.
- **Scientific notation** is used to write extremely large or small numbers in a more compact form. For $1 \leq |a| < 10$, $a \times 10^n$ is written using scientific notation.
- To **convert a number written in scientific notation to standard form**, follow these steps: The exponent on 10 is positive n: Multiply by 10^n, which is equivalent to moving the decimal point n places to the right.
- The **basic arithmetic operation**s are addition, subtraction, multiplication, division, and exponentiation.
- In arithmetic, the two (or more) quantities being added or subtracted are called **terms.**
- The **order-of-operation**s agreement is used to simplify expressions with multiple operations. **PEMDAS** is an acronym to help memorize the order-of-operations agreement.
- To **use the order-of-operations agreement**, follow these steps:
  1. Evaluate any exponents inside each term, working from left to right.
  2. Evaluate all of the operations (multiplications and divisions) inside of each term, working from left to right.
  3. Add and subtract the terms, working from left to right.
- When **grouping symbol**s appear in an expression to be simplified, they must be evaluated first.
- To **use the order-of-operations agreement involving grouping symbols**, follow these steps:
  1. Evaluate everything inside the grouping symbols first, inside to outside, working from left to right, using the basic order-of-operations agreement.
  2. Evaluate all exponents, inside to outside, working from left to right.
  3. Do all multiplications and divisions in the order they occur, working from left to right.
  4. Do all additions and subtractions, in the order they occur, working from left to right.
- The **commutative property** tells us that we may add or multiply in any order.
- The **associative property** tells us that how we group terms (factors) in an addition (multiplication) problem does not change the result.
- The **distributive property** allows us to distribute the operation of multiplication over addition (or subtraction).

Example 1 Answer as indicated.

a. Write in exponential form: $(-5) \cdot (-5) \cdot (-5)$
b. Write in expanded form: 4^6
c. Find the value without a calculator: $(-2)^4$

144 CHAPTER 1 Building Blocks of Algebra

SOLUTION
a. $(-5) \cdot (-5) \cdot (-5) = (-5)^3$
b. $4^6 = 4 \cdot 4 \cdot 4 \cdot 4 \cdot 4 \cdot 4$
c. $(-2)^4 = 16$

Example 2 Simplify each expression using the order-of-operations agreement.

a. $-16 + \dfrac{12}{4} + 3 \cdot (-2) - 8$ b. $|4 - 2 \cdot 6| + (-4 - 3)^2$

SOLUTION
a. $-16 + \dfrac{12}{4} + 3 \cdot (-2) - 8 = -27$ b. $|4 - 2 \cdot 6| + (-4 - 3)^2 = 57$

Example 3 Identify the property used to rewrite the expression: the commutative property, the associative property, or the distributive property.

a. $3(5 - 4) = 3(5) - 3(4)$
b. $46 \cdot 2 = 2 \cdot 46$
c. $3 + (4 + 19) = (3 + 4) + 19$

SOLUTION
a. This is an example of the distributive property.
b. This is an example of the commutative property.
c. This is an example of the associative property.

Section 1.2 Algebra and Working with Variables

- **Variables** are used to represent an unknown quantity or a quantity that can change (vary).
- A **constant** is a quantity that cannot change.
- A **term** is a constant or the product of a constant and a variable(s). An **algebraic expression** is a combination of variables and constants under algebraic operations.
- To **evaluate an expression**, follow these steps:
  1. Substitute the given value of the variable into the expression.
  2. Simplify, using the order-of-operations agreement.
- To form a **unity fraction**, find a relationship between the two quantities. Put the units required in the final answer in the numerator, and the units that are being converted from in the denominator.
- To **convert units**, follow these steps:
  1. Find the appropriate unity fraction needed for the problem. Put the units that are wanted in the final answer in the numerator and the units to be converted from in the denominator.
  2. Multiply the given value by the unity fraction. Divide out the common units.
  3. Reduce the fraction to lowest terms.
- When **translating a sentence into an algebraic expression,** first **define** the variables. Then translate the phrases into an algebraic expression.
- To **generate expressions from input-output tables**, follow these steps:
  1. Determine the input variable and the output variable. Put the input variable I the left column and output variable in the right column.
  2. Select four or more reasonable values for the input variable. Think carefully about the problem situation. The last row of the input column should be the general variable used for the input.

CHAPTER 1 Summary   145

3. Calculate the value of the outputs for the chosen inputs. Make sure to write out all of the operations involved in computing the output.
4. Find the pattern used to calculate the outputs from the inputs.
5. Write the general algebraic expression in the last row.

Example 4

Convert units as indicated. Convert 25 kilograms to pounds. There are approximately 2.2 lb in 1 kg.

SOLUTION

$$25 \text{ kg} \cdot \frac{2.2 \text{ lb}}{1 \text{ kg}} \approx 55 \text{ lb}$$

Example 5

Evaluate the algebraic expression for the given value of the variable. $-3x^2 + 12x$ for $x = -2$.

SOLUTION

$$-3x^2 + 12x$$
$$= -3(-2)^2 + 12(-2) \quad \text{Substitute in } x = -2.$$
$$= -3(4) + (-24) \quad \text{Simplify.}$$
$$= -12 + (-24)$$
$$= -36$$

Example 6

Define the variables and translate into an expression.
A house-cleaning service quotes a rate of $15.00 an hour for labor. Write an expression that gives the labor cost for house cleaning.

SOLUTION

Let $h =$ number of hours worked cleaning a home. The total cost in dollars for labor will be $15h$.

Section 1.3   Simplifying Expressions

- A **term** is a constant, a variable, or the product of any number of constants and variables.
- The **coefficient** is the numerical factor of a term.
- Terms are called **like** if they have the same variables and those variables are raised to the same exponents.
- Only **like terms** can be added or subtracted. To add or subtract like terms, add or subtract their coefficients.
- When multiplying a single term by a constant, use the **multiplication property** to multiply the coefficient of the term by the constant.
- When multiplying several terms by a constant, use the **distributive property** to multiply the coefficient of each term by the constant.
- To **simplify** an expression means to use arithmetic or algebraic properties to rewrite the expression in simplest form. To simplify an expression, first multiply or distribute constants if necessary, and then combine like terms, all the while following the order of operations.

Example 7

Identify the number of terms in the expression. Simplify. $-3y^2 + 4y + 7 - y^2 + 9$.

SOLUTION

$$-3y^2 + 4y + 7 - y^2 + 9$$
$$= -3y^2 + 4y + 7 + (-y^2) + 9 \quad \text{Change subtraction to addition.}$$
$$= \boxed{-3y^2} + \boxed{4y} + \boxed{7} + \boxed{(-y^2)} + \boxed{9}$$

There are five terms in this expression. To simplify, add like terms.

$$-3y^2 + 4y + 7 + (-y^2) + 9 \qquad \text{Add } y^2 \text{ terms (like).}$$
$$= -4y^2 + 4y + 7 + 9 \qquad \text{Add 7 and 9 (like terms).}$$
$$= -4y^2 + 4y + 16$$

Example 8 Multiply each expression.

a. $-3(-2a)$ 	**b.** $-2(3x^2 - 5x + 1)$

SOLUTION
a. $-3(-2a) = 6a$ 	Use the multiplication property.
b. $-2(3x^2 - 5x + 1) = -6x^2 + 10x - 2$ 	Use the distributive property.

Example 9 Simplify the expression. $-2(4x - 3) - (-5x + 7)$.

SOLUTION
$$-2(4x - 3) - (-5x + 7)$$
$$= -8x + 6 + 5x - 7$$
$$= -3x - 1$$

Section 1.4 Graphs and the Rectangular Coordinate System

- **Data table**s are often used to organize and display data.
- **Graphs** are a visual way to look at tabular data or an equation in two variables.
- The most common kinds of graphs are **bar graphs** and **scatterplots**.
- To **create a bar graph**, follow these steps:
  1. The input values will be plotted on the horizontal axis, and the output values will be plotted on the vertical axis.
  2. Evenly and consistently space the horizontal axis so that it includes the input values (data).
  3. Evenly and consistently space the vertical axis so that it includes the output values (data).
  4. Draw vertical bars of uniform width. Center each bar over the appropriate input value. Draw the bars of height corresponding to the output value.
- The **input** (independent) variable is plotted on the horizontal or x-axis, and the **output** (dependent) variable is plotted on the vertical or y-axis.
- An **ordered pair** has the form (x, y), where x is the value of the input and y is the value of the output. The values of x and y are also called the **coordinates** of the ordered pair or point.
- To **create a graph in the rectangular coordinate system**, follow these steps:
  1. Evenly and consistently space and scale the horizontal axis so that it includes the input values.
  2. Evenly and consistently space and scale the vertical axis so that it includes the output values.
  3. Plot each ordered pair (point) by first finding the intersection of the input value and the output value and then drawing a dot.

CHAPTER 1   Review Exercises   **147**

Example 10   Graph the given ordered pairs on the same axes. Clearly label and scale the axes.
$(-6, 20)\ (-4, -20)\ (12, -50)\ (10, 40)\ (0, 12)$

SOLUTION

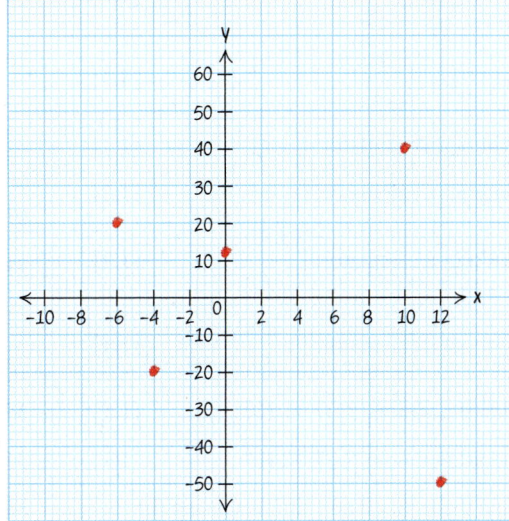

Example 11   Determine the scale on the x- and y-axes.

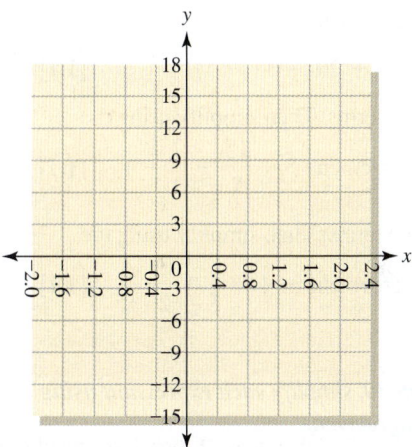

SOLUTION   The x-axis has a scale of 0.4 because the tick marks are evenly and consistently 0.4 unit apart. Likewise, the y-axis has a scale of 3.

Chapter Review Exercises

1. Write in exponential form: $6 \cdot 6 \cdot 6 \cdot 6 \cdot 6$ [1.1]

2. Write in exponential form: $8 \cdot 8 \cdot 8 \cdot 8$ [1.1]

3. Write in exponential form: $\left(\dfrac{3}{8}\right) \cdot \left(\dfrac{3}{8}\right) \cdot \left(\dfrac{3}{8}\right)$ [1.1]

4. Write in exponential form:

   $\left(-\dfrac{4}{5}\right) \cdot \left(-\dfrac{4}{5}\right) \cdot \left(-\dfrac{4}{5}\right) \cdot \left(-\dfrac{4}{5}\right) \cdot \left(-\dfrac{4}{5}\right)$ [1.1]

5. Write in expanded form: 5^6 [1.1]

148 CHAPTER 1 Building Blocks of Algebra

6. Write in expanded form: $(-2)^3$ [1.1]

7. Write in expanded form: $\left(-\dfrac{2}{5}\right)^4$ [1.1]

8. Write in expanded form: $\left(\dfrac{1}{11}\right)^3$ [1.1]

9. Find the value without a calculator: $(-3)^3$ [1.1]

10. Find the value without a calculator: $(-6)^2$ [1.1]

11. Find the value without a calculator: $\left(\dfrac{4}{9}\right)^2$ [1.1]

12. Find the value without a calculator: $\left(-\dfrac{2}{3}\right)^3$ [1.1]

13. Write using standard notation: 2.07×10^5 [1.1]

14. Write using standard notation: 5.67×10^6 [1.1]

15. Underline or circle the terms. Then simplify using the order-of-operations agreement:
$-3 + 4 \cdot (-2) + 24 \div 8 - 3^2 + (-2)$ [1.1]

16. Underline or circle the terms. Then simplify using the order-of-operations agreement:
$27 \div 3 - 5^2 + (-10) + 20 - 3 \cdot (-5)$ [1.1]

17. Underline or circle the terms. Then simplify using the order-of-operations agreement: $5 \div \dfrac{15}{7} + \dfrac{2}{3} - 1$ [1.1]

18. Underline or circle the terms. Then simplify using the order-of-operations agreement: $-4 - \dfrac{1}{5} + \dfrac{3}{2} \cdot \dfrac{4}{15}$ [1.1]

For Exercises 19 through 30, simplify each expression using the order-of-operations agreement.

19. $6 \div (-2) \cdot 5 + (-3)^2$ [1.1]

20. $(-2)^3 - 36 \div 9 \cdot 7$ [1.1]

21. $\dfrac{3}{2} \cdot \dfrac{4}{9} - 5 + \dfrac{2}{3}$ [1.1]

22. $-\dfrac{1}{4} \cdot \dfrac{12}{5} + 3 \div \dfrac{5}{2} - 1$ [1.1]

23. $-7 \cdot (-4) \div 2 + (-1)^4 + \dfrac{6 \cdot 3}{5 - 2}$ [1.1]

24. $\dfrac{8 - 5}{4 + 2} + 16 \cdot (-3) \div 12 + (-3)^2$ [1.1]

25. $-5 + 7 \cdot (-3) + 44 \div 2 - 6^2 + (-8)$ [1.1]

26. $(-5)^2 - 24 \div (-3) \cdot 2 - (-6 - 4) + 9$ [1.1]

27. $5 + \dfrac{22 - 4 \cdot 2}{16 - 3^2} - (-2)^2$ [1.1]

28. $\dfrac{-2 + 6 \cdot 2}{14 - 3^2} - (3 \cdot 2 - 7)$ [1.1]

29. $|-2 + 3 \cdot (-4)| + (-7)^2$ [1.1]

30. $|-8 + (-3)(-2)| + 3^3$ [1.1]

For Exercises 31 through 38, identify the property (commutative, associative, or distributive) used to rewrite each statement.

31. $(-2)(-8) = (-8)(-2)$ [1.1]

32. $-3 + (-8) = -8 + (-3)$ [1.1]

33. $-3(4 + 5) = -3 \cdot 4 + (-3) \cdot 5$ [1.1]

34. $2 \cdot (3 - 7) = 2 \cdot 3 - 2 \cdot 7$ [1.1]

35. $(-3) + (-4 + 9) = (-3 + (-4)) + 9$ [1.1]

36. $(5 \cdot 3)(-2) = 5 \cdot (3 \cdot (-2))$ [1.1]

37. $\dfrac{4}{5} + \dfrac{8}{9} = \dfrac{8}{9} + \dfrac{4}{5}$ [1.1]

38. $\dfrac{5}{7} \cdot \dfrac{3}{25} = \dfrac{3}{25} \cdot \dfrac{5}{7}$ [1.1]

39. Define the variables: The amount of money you earn in a month depends on the number of hours you work in a month. [1.2]

40. Define the variables: The amount of money (revenue) a local playhouse earns for a night depends on the number of tickets sold for that night. [1.2]

For Exercises 41 through 44, translate each sentence into a variable expression.

41. Six less than three times a number [1.2]

42. Eight more than five times a number [1.2]

43. One-half of the difference between a number and 4 [1.2]

44. The quotient of the sum of a number and 12, and 16 [1.2]

For Exercises 45 through 50, convert units as indicated. Round to the hundredths place if necessary.

45. Convert 9 inches to centimeters. [1.2]

46. Convert 55 kilometers to miles. [1.2]

47. Convert 185 pounds to kilograms. [1.2]

48. Convert 100 kilograms to pounds. [1.2]

49. Convert 55 grams to ounces. [1.2]

50. Convert 5 ounces to grams. [1.2]

For Exercises 51 through 60, evaluate each expression for the given value of the variable.

51. $-\dfrac{5h}{18} + 9$ for $h = -3$ [1.2]

52. $-4r + \dfrac{r}{2}$ for $r = 8$ [1.2]

53. $5x - 2y$ for $x = -3, y = -5$ [1.2]

54. $5y - 6x$ for $x = -2, y = -7$ [1.2]

55. $3xy + 2x - y$ for $x = -2, y = 3$ [1.2]

56. $l + k - 10$ for $l = 4, k = -3$ [1.2]

57. $0.25a + 1.75b$ for $a = 0.8, b = 4.6$ [1.2]

58. $0.5h - 2.5k$ for $h = 8, k = -10$ [1.2]

59. $15ab + c$ for $a = 2, b = -3, c = 4$ [1.2]

60. $a + b + c$ for $a = 3, b = 4, c = 5$ [1.2]

61. Wang Li is going to invest $4500.00 in two different accounts. He invests d dollars in the first account. Fill out the input-output table to write an expression for the amount of money he invests in the second account. [1.2]

Amount Invested in the First Account	Amount Invested in the Second Account
0	
1000	
2000	
3500	
d	

62. Colin has to cut a 20-foot board into two pieces. He cuts x feet of lumber in the first piece. Fill out the input-output table to write an expression for the length of the second piece. [1.2]

First Piece	Second Piece
0	
4	
8	
16	
x	

63. Convert 800 million to thousands. *Hint:* There are 1000 thousands in 1 million. [1.2]

64. Convert 700 billion into millions. *Hint:* There are 1000 millions in 1 billion. [1.2]

65. The table gives the number of apps in the iTunes App Store from 2008 to 2016. Use the given table to make a bar chart. Clearly label and scale the axes. [1.4]

Year	Available Apps (thousands)
2008	15
2009	46
2010	167
2011	356
2012	586
2013	721
2014	958
2015	1405
2016	1975

Source: Apple Inc.

66. Use the given table to make a scatterplot. Clearly label and scale the axes. [1.4]

x	y
-20	-3.5
-10	1
-5	4.5
0	-2.5
20	0.5
25	4.5

For Exercises 67 through 70, identify the number of terms and the like terms (if any) in each expression.

67. $-6p + 4 + 7p - 9$ [1.3]

68. $5x + 16 - \dfrac{2}{3}x + 12$ [1.3]

69. $-3x^2 + 2x - 4x^2 + 1$ [1.3]

70. $-\dfrac{2}{5}y + y^2 - 6y + 12$ [1.3]

71. Ms. Welch is buying her first grade class gifts for the end of the school year. She wants to order a book that costs $3.25 for each student. Ms. Welch also wants to buy each student a candy bar that costs $1.25. [1.3]
   a. If Ms. Welch has s students in her first grade class, write an expression for the cost in dollars to buy each student a book.
   b. Write an expression for the cost in dollars to buy each student a candy bar.
   c. Use the expressions you wrote from parts a and b to write an expression for the total cost in dollars to buy each student a book and a candy bar.
   d. Simplify the expression if possible.

150 CHAPTER 1 Building Blocks of Algebra

72. Yong Park has to pay $500 monthly for his part of the rent. He also pays $85.00 a week for groceries. [1.3]
    a. If Yong Park makes a monthly salary of *d* dollars, write an expression for how much money he has left after paying his rent.
    b. Write an expression for how much he pays for groceries each month. (Assume that there are 4 weeks in a month.)
    c. Use the expressions you wrote from parts a and b to write an expression for the total amount of money he has left over each month after he pays his rent and purchases groceries.
    d. Simplify the expression if possible.

For Exercises 73 through 80, simplify each expression by combining like terms.

73. $y^2 - 4y^3 + 3y^2 - 8y^3$ [1.3]
74. $-8a^2b + 3ab - 16ab + 12a^2b$ [1.3]
75. $-\dfrac{3}{4}x + 7 - \dfrac{1}{4}x - 12$ [1.3]
76. $-5 + \dfrac{2}{3}x - 7 + \dfrac{1}{3}x$ [1.3]
77. $1.25a - 6b - 7.5a + 8.1b$ [1.3]
78. $4.7x + 3.2y - 5x - 1.3y$ [1.3]
79. $5x^2 - 6x + 4 - 3x^2 + 7x + 8$ [1.3]
80. $12 - y + 3y^2 - 6y - 2y^2 + 7$ [1.3]

For Exercises 81 through 86:
  a. State whether the multiplication property or the distributive property can be used on the problem.
  b. Simplify each expression.

81. $16(-3ab^2)$ [1.3]
82. $-4(3xy)$ [1.3]
83. $-5(3x - 6y)$ [1.3]
84. $2(3a - 8b)$ [1.3]
85. $-(-13x)$ [1.3]
86. $-(5y - 7)$ [1.3]

For Exercises 87 through 96, simplify each expression.

87. $9 - 3(2x - 6)$ [1.3]
88. $-2(3x - 4) + 7x - 1$ [1.3]
89. $-(-2x + 3)$ [1.3]
90. $-(9y - 5)$ [1.3]
91. $(-2a + 3b) - (4a - 5b)$ [1.3]
92. $3x + 2y - 4(-x + 5y)$ [1.3]
93. $\dfrac{1}{2}(4x - 12) + 9x - 3$ [1.3]
94. $3h - 2k + \dfrac{1}{4}(4h - 16k)$ [1.3]
95. $-0.2(10x - 5) + 2x - 1$ [1.3]
96. $0.25(7x - 6y) - 1.25x + 9.8y$ [1.3]

For Exercises 97 and 98, use the given figures to find expressions for the perimeter of the shapes. Simplify the expressions if possible. **Note:** *The perimeter is the lengths of all the sides added together.*

97. [1.3]

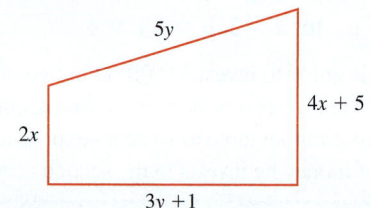

98. [1.3]

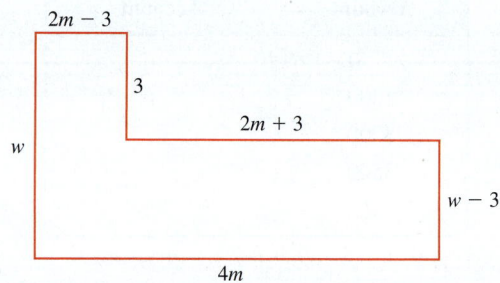

99. Write the coordinates of each ordered pair shown on the graph. [1.4]

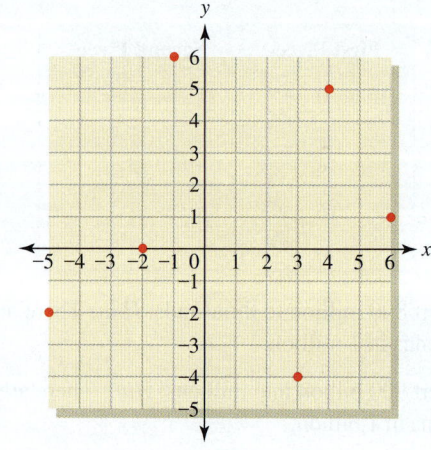

100. Write the coordinates of each ordered pair shown on the graph. [1.4]

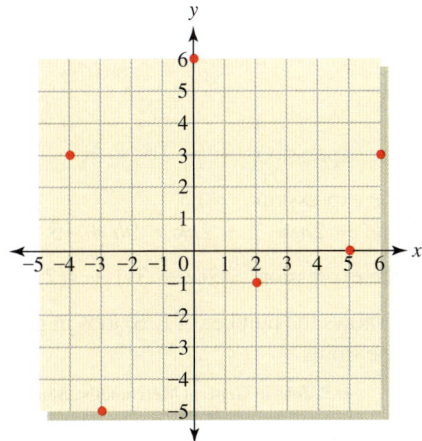

101. State the scales on both the *x*- and *y*-axes. [1.4]

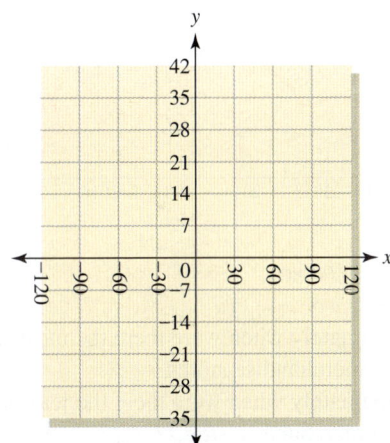

102. State the scales on both the *x*- and *y*-axes. [1.4]

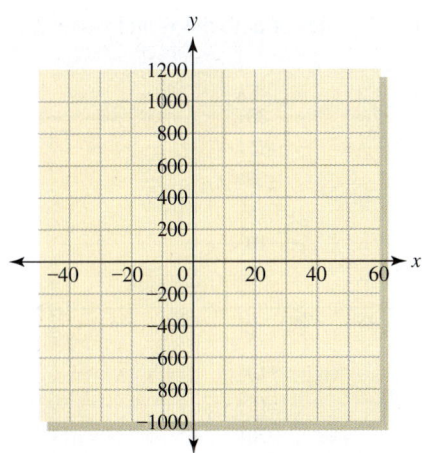

103. Graph the ordered pairs on the same set of axes. [1.4]

(4, −7), (−2, −5), (−3, 9), (1, 6), (0, −3), (−5, 0)

104. Graph the ordered pairs on the same set of axes. [1.4]

(−2, −5), (1, −1), (4, 1), (−9, 2), (0, 5), (−2, 0)

105. Use the given data to create a bar graph. Clearly label and scale the axes. [1.4]

State	Unemployment Rate as of Feb. 2017
Michigan	5.3%
Rhode Island	4.5%
California	5.0%
Texas	4.9%
South Carolina	4.4%
Ohio	5.1%
Illinois	5.4%
Nevada	4.9%

Source: www.bls.gov

106. Use the given data to create a bar graph. Clearly label and scale the axes. [1.4]

Year	U.S. National Unemployment Rate
2007	4.6%
2008	5.8%
2009	9.3%
2010	9.6%
2011	8.9%
2012	8.1%
2013	7.4%
2014	6.2%
2015	5.3%
2016	4.9%

Source: www.bls.gov

Chapter Test

This chapter test should take approximately one hour to complete. Read each question carefully. Show all of your work.

1. Write in exponential form: $3 \cdot 3 \cdot 3 \cdot 3 \cdot 3 \cdot 3$

2. Write in expanded form: $\left(-\dfrac{5}{8}\right)^3$

3. Find the value without a calculator: $(-2)^3$

4. Write in standard form: 4.299×10^7

For Exercises 5 through 8, identify the number of terms. Then simplify each expression.

5. $-6 + 3 \cdot 4 - (-5) + 12 \div (-3)$

6. $24 \div 2 \cdot 3 + (-2)^2 - (-1)$

7. $\dfrac{12 - 3 \cdot 2}{4 - 2} + 8 \div 4$

8. $-3 + |7 \cdot (-2)| - (-5) + 3 \cdot 6$

For Exercises 9 through 11, identify the property used to rewrite each expression: the commutative property of addition or multiplication, the associative property of addition or multiplication, or the distributive property.

9. $(-3 + 9) + (-6) = -3 + (9 + (-6))$

10. $4(59 - 11) = 4 \cdot 59 - 4 \cdot 11$

11. $72(-89) = (-89) \cdot 72$

12. Lucas's hourly wage is $13.75 working as an office assistant. His weekly salary can be computed as $13.75h$, where h is the number of hours he worked for a week. What would Lucas's weekly salary be if he worked 24 hours in a week?

13. Define the variables. Translate the following sentence into an expression.

    A plumber quotes a homeowner a price of $20.00 plus $65.00 an hour for labor. Write an expression that gives the cost to the homeowner for this plumber's services.

14. Convert 100 yards to meters. Note that $1 \text{ m} \approx 1.09$ yd.

15. Evaluate the expression for the given value of the variable.
    $x^2 + 2x - 1$ for $x = -3$

16. Simplify the expression: $-3(-2x^2 + 5x - 1)$

17. Simplify the expression:
    $-5hk + 3k^2 - 2h + 7 + 2hk + 9h + 15$

18. Simplify the expression: $-2(x + 3y) + 4(2y) + 5x$

19. The graph shows the number of U.S. jobs in general merchandise stores.

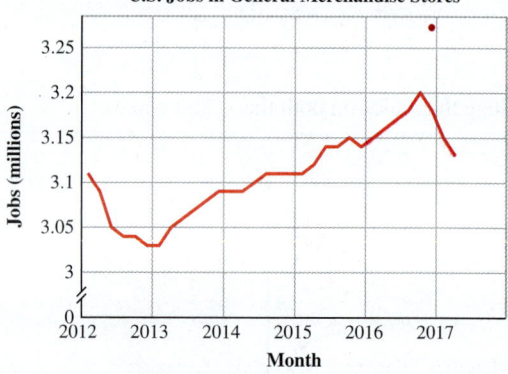

Source: www.bls.gov

a. Approximately when were there the most jobs in general merchandise stores?
b. Approximately when were there the least jobs in general merchandise stores?
c. What is the scale on the vertical axis?
d. What is the scale on the horizontal axis?

20. What are the scales of both the x- and y-axes?

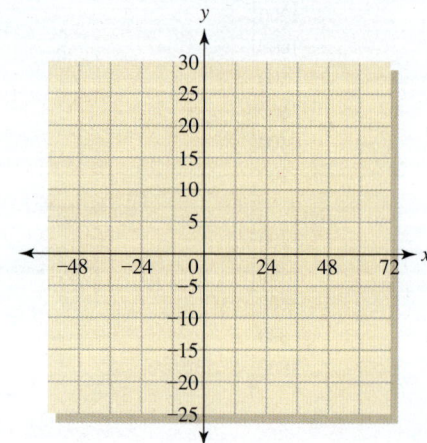

21. Graph the ordered pairs on the same axes. Clearly label and scale the axes.

x	y
−50	4.5
−30	2.5
−15	0.5
0	−1.0
20	−2.5
40	−3.0

22. Write the coordinates of each ordered pair on the graph.

Chapter Projects

■ How Many Miles Are in a Light Year?

Written Project
One or more people

What you will need
- A scientific calculator

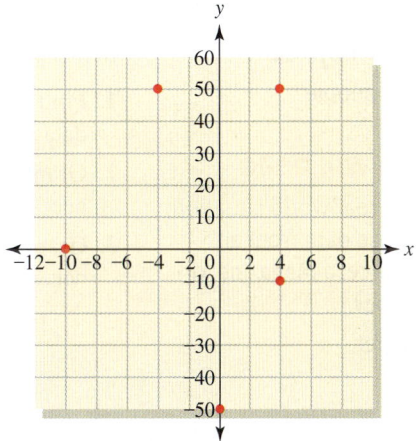

Light travels at the speed of 186,000 miles per second. The number of miles light travels in one year is called a *light-year*. To find out how many miles are in a light-year, use the following steps:

1. Convert the speed of light of 186,000 miles per second to miles per minute. *Hint:* To find the unity fraction, write down how many seconds are in 1 minute.
2. Convert the speed of light in miles per minute from step 1 to miles per hour. *Hint:* To find the unity fraction, write down how many minutes are in 1 hour.
3. Convert the speed of light in miles per hour from step 2 to miles per day. *Hint:* To find the unity fraction, write down how many hours are in 1 day.
4. Convert the speed of light in miles per day from step 3 to miles per year. *Hint:* To find the unity fraction, write down how many days are in 1 year. Assume it is a not a leap year.

■ Create a Bar Graph and a Scatterplot

Research Project
One or more people

What you will need
- Data for a real-world situation; you may use data from an online source, a magazine, or a newspaper
- Graph paper

In this project, you will find a data set and display it in various ways.

1. Find a data set, either from print media or an online source. Include a copy of the data set with your project. Make sure it contains at least seven points.
2. Graph the data set in a bar graph. Clearly label and scale the axes. Include any necessary breaks.
3. Graph the data set in a scatterplot. Clearly label and scale the axes. Include any necessary breaks.

■ Write a Review of a Section for Presentation

Research Project
One or more people

What you will need
- This presentation may be done in class or as an online video.
- If it is done as a video, post it on a website where it may be easily viewed by the class.
- You might want to use homework or other review problems from the book.
- Make it creative and fun.

Create a 5-minute review presentation of one section from Chapter 1. The format of the presentation can be a poster presentation, a blackboard presentation from notes, an online video, or a game format (for example, math jeopardy). The presentation should include the following:

1. Any important formulas in the chapter
2. Important skills in the chapter, backed up by examples
3. Explanation of terms used (for example, what is modeling?)
4. Common mistakes and how to avoid them

Linear Equations and Inequalities with One Variable

2

2.1 Addition and Subtraction Properties of Equality

2.2 Multiplication and Division Properties of Equality

2.3 Solving Equations with Variables on Both Sides

2.4 Solving and Graphing Linear Inequalities on a Number Line

When determining the number of school buses needed, school districts must take into account the capacity of each school bus. The *capacity* of the bus means how many students can ride on the bus. The capacity of a school bus seat is determined by dividing the width of the seat by 13 and rounding the result down to the nearest whole number. The school district can then multiply this result by the number of seats in the bus to determine how many students can be transported in each bus.

2.1 Addition and Subtraction Properties of Equality

LEARNING OBJECTIVES
- Recognize an equation and a solution.
- Use the addition and subtraction properties of equality.
- Solve literal equations.

Recognizing Equations and Their Solutions

In Chapter 1, we worked with applications using expressions that represented situations in real life. Recall that the expression $11.25h$ represented the weekly salary of an employee who is making \$11.25 an hour where h is the number of hours worked for that week. This expression can be used repeatedly to calculate different weekly salaries for any employee who makes \$11.25 an hour. Since the result of the calculation will change depending on what number of hours is substituted for the variable h, the salary could also be represented by a variable. Using the variable S for the salary will result in the equation $S = 11.25h$. This equation states that the salary S (the output) in dollars is equal to 11.25 times the number of hours, h (the input), that the employee worked that week. The equal sign ($=$) is used to indicate that the value of the two expressions is the same. A statement that two expressions are equal is called an **equation.**

> **DEFINITION**
>
> **Equation** A statement that two expressions are equal is called an **equation.**
>
> $$125 = 125 \qquad 5x + 7 = 205$$

Many applications in algebra generate equations. These equations can be used to determine the value of the output for a given value of the input. Using the equation $S = 11.25h$, we can evaluate this equation for different values of h, hours worked, to find the corresponding values of S, salary earned. If someone worked 30 hours in a week, substitute 30 for h to determine their salary.

$$S = 11.25(30)$$
$$S = 337.50$$

Someone who works 30 hours will earn a salary of \$337.50. This equation can be used to calculate the weekly salary for any number of hours that a person works. Any combination of hours and salary that makes this equation true is said to be a **solution** to the equation.

■ Connecting the Concepts

What is the difference between an expression and an equation?

The difference between an **expression** and an **equation** is important in understanding directions given in algebra.
An **expression** is any combination of terms.
$12xy^2 + 5x$ is an expression.
We simplify expressions.
An **equation** says that two expressions are equal.
An *equation will have an equal sign* ($=$).
$12xy^2 + 5x = 4xy + 8$ is an equation. *We solve equations.*

Example 1 Is it an expression or an equation?

Identify each of the following as an expression or an equation. Do not solve or simplify.

a. $5x - 2(4 + 3x)$
b. $-x + 3(4 - 2x) = 7$
c. $7 = 7$
d. $-x$

SECTION 2.1 Addition and Subtraction Properties of Equality   157

SOLUTION

a. This is an expression. There is no equal sign.
b. This is an equation. There is an equal sign.
c. This is an equation. There is an equal sign.
d. This is an expression. There is no equal sign.

PRACTICE PROBLEM FOR EXAMPLE 1

Identify each of the following as an expression or an equation. Do not solve or simplify.

a. $-9x = 7x$ b. $3y - 1$

DEFINITION

Solution   The solution(s) is (are) any value(s) of the variable(s) that make the equation true (when evaluated).
Note: The equation is true when both sides are equal.

For the equation $S = 11.25h$, the values $h = 20$ and $S = 225.00$ are a solution to the equation. When we substitute 225 for S and 20 for h, the two sides are equal, $225.00 = 11.25(20)$. In this case, these values represent a salary of $225.00 for 20 hours of work. To **check** whether certain values for the variables are solutions, substitute in the given value(s) of the variable(s) and determine whether both sides are equal.

Steps to Check a Solution

1. Substitute in the given value(s) of the variable(s).
2. Evaluate both sides of the equation. Simplify using the order-of-operations agreement. If both sides are equal (the equation is true), then that (those) value(s) of the variable(s) is a (are) solution(s) to the equation.

Example 2   Checking whether the given values are solutions

Determine whether or not the given values of the variables are solutions to the given equation.

a. $S = 11.25h$, where S is the salary in dollars (output) if h hours (input) are worked in a week; $h = 15$ and $S = 168.75$. Explain what these values mean in terms of this situation.
b. $d = 50t$, where d is the distance traveled in miles (output) when you have been driving for t hours (input); $t = 3$ and $d = 200$. Explain what these values mean in terms of this situation.
c. $3x + 25 = 55; x = 10$

SOLUTION

a. **Step 1** Substitute in the given value(s) of the variable(s).

$$S = 11.25h \quad \text{Original equation.}$$

$$168.75 \stackrel{?}{=} 11.25(15) \quad \text{Substitute 168.75 for S and 15 for h.}$$

Step 2 Evaluate both sides of the equation. Simplify using the order-of-operations agreement. If both sides are equal (the equation is true), then that (those) value(s) of the variable(s) is (are) solutions to the equation.

$$168.75 \stackrel{?}{=} 11.25(15)$$

$$168.75 = 168.75 \quad \text{This statement is true.}$$

What's That Mean?

$\stackrel{?}{=}$

The $\stackrel{?}{=}$ symbol will be used when checking solutions. It means that we are trying to determine whether or not the two sides of the equation are equal.

Because the final statement is true, $h = 15$ and $S = 168.75$ is a solution to the equation $S = 11.25h$.

If a person works 15 hours, then the person will earn a salary of $168.75 for the week.

b. $d = 50t$ *Original equation.*
  $200 \stackrel{?}{=} 50(3)$ *Substitute 200 for D and 3 for t.*
  $200 \stackrel{?}{=} 150$ *This statement is false.*
  $200 \neq 150$

Because the final results are not equal, $t = 3$ and $d = 200$ do not give a solution to the equation $d = 50t$. These values state that if a person drives for 3 hours, then they will travel 200 miles. These values do not make the equation true, so they do not make sense in this situation.

c. $3x + 25 = 55$ *Original equation.*
  $3(10) + 25 \stackrel{?}{=} 55$ *Substitute 10 for x.*
  $30 + 25 \stackrel{?}{=} 55$ *Simplify both sides.*
  $55 = 55$ *This statement is true.*

Because the final statement is true, $x = 10$ is a solution to the equation $3x + 25 = 55$.

PRACTICE PROBLEM FOR EXAMPLE 2

Determine whether or not the given values of the variables are solutions to the given equation.

a. $P = 25n$ if P is the profit in dollars (output) when n car batteries (input) are sold; $n = 120$ and $P = 4000$. Explain what these values mean in this situation.

b. $7x + 34 = 10x - 11$; $x = 15$

c. $4(x + 6) - 2x = x + 15$; $x = -10$

■ Addition and Subtraction Properties of Equality

We now study the process of solving equations. In solving an equation, the goal is to isolate the variable on one side of the equal sign. In this section, we will introduce some of the properties of equations that will be used throughout algebra.

An equation can be thought of as similar to an old-fashioned balance scale used to weigh objects. Someone wanting to weigh gold, for instance, would put the gold object in the pan on one side of the balance scale and then place known weights in the pan on the other side of the balance scale until the two pans were balanced. When the pans were balanced, there was an equal weight on both sides. The weight of the gold was equal to the standard weight in the pan on the other side.

An equation is similar in that both sides of an equation must represent the same *amount* or value. The equal sign represents the balance. This is easy to see with numbers.

$$5 = 5 \quad \text{or} \quad 2 + 8 = 10$$

If a variable is in the equation, it is the unknown value. Be very careful to keep the two "pans" in the "balance" of the equation true when solving for the variable. Consider the equation $x - 5 = 12$ and think of the equal sign as a balance.

> **■ What's That Mean?**
>
> **Solve an Equation**
>
> To solve an equation means to find the value(s) of the variable(s) that make the equation true.

SECTION 2.1 Addition and Subtraction Properties of Equality   159

Because this is an equation, the two pans are in balance—both sides are equal. If 5 is added to the left side, it becomes out of balance because equal "weight" must be added to both sides.

The scales are out of balance because 5 has been added to only the left side. When out of balance, the equation is no longer true. To keep the sides balanced, add 5 to the scale on the right side as well.

Now the pans are back in balance, and the equation is again true. Simplify the expressions on both sides of the scale and get

This last scale shows that $x = 17$. To check that this is the solution to the equation, substitute 17 for x in the original equation and make sure the equation is still true.

$x - 5 = 12$ *Original equation.*
$(17) - 5 \stackrel{?}{=} 12$ *Substitute 17 for x.*
$12 = 12$ *This statement is true.*

A balance scale is a conceptual way to think about the **addition property of equality.** A similar line of reasoning leads to the **subtraction property of equality.**

What's That Mean?

Mathematical Verbs

Simplify: Use arithmetic and basic algebra rules to make an expression simpler. *We simplify expressions.*

Evaluate: Substitute any given values for variables and simplify the resulting expression or equation.

Solve: Isolate the given variable in an equation on one side of the equal sign using the properties of equality. This will result in a value that the isolated variable is equal to. *We solve equations.*

ADDITION PROPERTY OF EQUALITY

When the same real number or algebraic expression is added to **both** sides of an equation, the solution to the equation is unchanged.

If $a = b$, then $a + c = b + c$

SUBTRACTION PROPERTY OF EQUALITY

When the same real number or algebraic expression is subtracted from **both** sides of an equation, the solution to the equation is unchanged.

If $a = b$, then $a - c = b - c$

■ **Connecting the Concepts**

Why do we check?

Checking an answer(s) is an important step in solving an equation. We are all capable of making mistakes. Common mistakes are adding incorrectly and giving a number the wrong sign.

Every time we solve an equation, we check the answers to be sure we did not make any mistakes.

These two properties are used to solve equations that involve addition or subtraction. When solving an equation, the goal is to *isolate* the variable (get the variable by itself) on one side of the equation. This allows us to determine what that variable is equal to. To use these properties, it is helpful to ask what operation is being performed on the variable and then use the reverse operation to *undo* that operation. To undo addition, subtract the same number from both sides of the equation. To undo subtraction, add the same number to both sides of the equation.

Undoing an Operation Using the Properties of Equality

- To undo an operation, use the reverse operation on both sides of the equal sign.
- To undo addition, use subtraction.
- To undo subtraction, use addition.

Note: Another way to think about undoing addition or subtraction is to "add the opposite."

Linear Equations

Addition and Subtraction Property of Equality

Use to isolate a variable term.

$3x + 7 = 20$
$-7\ -7$
$\overline{3x\ = 13}$

Example 3 Finding the total weight

Mark, one of the authors of this book, went on a hike to the top of Mount Whitney in California during the summer of 2005. He planned to be gone for 2 nights and 3 days and had to carry a backpack with all the equipment and food he needed on the hike. The total weight of the backpack depended on the weight of the food and equipment. The total weight can be represented by the equation

$$W = e + f$$

where W is the total weight in pounds of the backpack, e is the weight in pounds of the equipment, and f is the weight in pounds of the food. Use this equation to answer the following questions.

a. If Mark wanted the total weight of the backpack to be 25 pounds and his equipment weighed 15 pounds, how heavy could the food be?

b. If the total weight of the backpack was 30 pounds and the food weighed 12 pounds, how heavy could the equipment be?

SOLUTION

a. The total weight of the backpack is given as 25 pounds and the weight of the equipment as 15 pounds, so substitute these values into the given equation.

$W = e + f$ *Original equation.*

$(25) = (15) + f$ *Substitute 25 for W and 15 for e.*

To solve for f, isolate the variable f on one side of the equation. Because 15 is being added to f, undo adding 15 by subtracting 15 from both sides of the equation.

$$25 = 15 + \boxed{f}$$ *Identify the variable term.*
$$\underline{-15 \quad -15}$$ *Subtract 15 from both sides.*
$$10 = f$$ *f is isolated, so we have an answer.*

Check the answer to be sure that it works in the original equation and makes sense in the context of the problem.

$$W = e + f$$ *Substitute the values for each variable.*
$$(25) \stackrel{?}{=} (15) + (10)$$
$$25 = 25$$ *This statement is true, so our answer is mathematically correct.*

Carrying 15 pounds of equipment and 10 pounds of food gives the total weight of the backpack of 25 pounds. This is a correct solution. The final answer should be written in a complete sentence as follows:

Mark should carry 10 pounds of food with his 15 pounds of equipment to keep his backpack weight at 25 pounds.

b. The total weight of the backpack is 30 pounds, and the food weighs 12 pounds, so substitute these values into the given equation.

$$W = e + f$$ *Original equation.*
$$(30) = e + (12)$$ *Substitute 30 for W and 12 for f.*

To solve, we must isolate e. To do so, remove the term of 12 from the right side of the equation. Because 12 is being added to e, subtract 12 from both sides of the equation.

$$30 = \boxed{e} + 12$$ *Identify the variable term.*
$$\underline{-12 -12}$$ *Subtract 12 from both sides.*
$$18 = e$$ *e is isolated, so we have an answer.*

Check this answer in the original equation and determine whether it makes sense in the context of the problem.

$$W = e + f$$ *Substitute the values for each variable.*
$$(30) \stackrel{?}{=} (18) + (12)$$
$$30 = 30$$ *This statement is true, so our answer is mathematically correct.*

> **Skill Connection**
>
> **Checking a solution**
>
> Check the answer by substituting the value(s) found back into the *original* problem. Then simplify the equation and see whether it is true (both sides are equal). If the equation is true, then the solution checks.

Since carrying 12 pounds of equipment and 18 pounds of food gives us a total of 30 pounds, the solution checks. The final answer should be written in a complete sentence as follows:

Mark should carry 18 pounds of equipment with his 12 pounds of food to keep his backpack at 30 total pounds.

In Example 3, the solution to the equation has to be checked by substituting it back into the original equation. Something else to check is whether or not the solution makes sense in the context of the original information, units of measurement, and data given in the problem. Complete sentence answers are always required for application problems.

PRACTICE PROBLEM FOR EXAMPLE 3

Jennifer is a new college student and has to take 15 credits per semester to stay on her parents' health insurance plan as a full-time student. Jennifer wants to take a combination of general education classes (English, math, and science) and some electives (sports and art). If Jennifer does not want to take more than 15 credits a semester, she can represent the credits she takes using the equation

$$G + E = 15$$

162 CHAPTER 2 Linear Equations and Inequalities with One Variable

where G = the number of general education credits Jennifer takes and E = the number of elective credits Jennifer takes.

a. If Jennifer takes 11 general education credits this semester, how many elective credits should she take?

b. If Jennifer takes 2.5 credits of electives this semester, how many general education credits should she take?

The equations we are now going to study have a special form and name. They are called *linear equations in one variable*.

> **DEFINITION**
>
> **Linear Equation in One Variable** A linear equation in one variable may be written in the form $mx + b = 0$, where m and b are real numbers and $m \neq 0$.

Example 4 Solving equations using the addition and subtraction properties

Solve the equations. Check the answers.

a. $x + 20 = 55$ **b.** $t - 6 = 9$ **c.** $13.48 + k = 21.52$

SOLUTION

a. 20 is being added to x, so subtract 20 from both sides of the equation.

$$\begin{aligned}
\boxed{x} + 20 &= 55 & &\text{Identify the variable term.} \\
-20\;\;-20 & & &\text{Subtract 20 from both sides of the equation.} \\
x &= 35 & &\text{x is isolated. This is the answer.} \\
(35) + 20 &\stackrel{?}{=} 55 & &\text{Check the answer.} \\
55 &= 55 & &\text{The answer checks.}
\end{aligned}$$

b. 6 is being subtracted from t, so add 6 to both sides of the equation.

$$\begin{aligned}
\boxed{t} - 6 &= 9 & &\text{Identify the variable term.} \\
+6\;\;+6 & & &\text{Add 6 to both sides of the equation.} \\
t &= 15 & &\text{t is isolated. This is the answer.} \\
(15) - 6 &\stackrel{?}{=} 9 & &\text{Check the answer.} \\
9 &= 9 & &\text{The answer checks.}
\end{aligned}$$

c. 13.48 is being added to k, so subtract 13.48 from both sides of the equation.

$$\begin{aligned}
13.48 + \boxed{k} &= 21.52 & &\text{Identify the variable term.} \\
-13.48-13.48 & & &\text{Subtract 13.48 from both sides of the equation.} \\
k &= 8.04 & &\text{k is isolated. This is the answer.} \\
13.48 + (8.04) &\stackrel{?}{=} 21.52 & &\text{Check the answer.} \\
21.52 &= 21.52 & &\text{The answer checks.}
\end{aligned}$$

PRACTICE PROBLEM FOR EXAMPLE 4

Solve the equation. Check the answer.

a. $r + 9 = 24$ **b.** $p - 15.7 = 16.9$

Example 5   Solving an equation from business

The amount of profit a company makes is found by subtracting the company's costs from its revenue. This relationship can be written as the following equation:

$$P = R - C$$

where P is the **profit** in dollars, R is the **revenue** in dollars, and C is the **cost** in dollars for the company.

a. If a company has monthly revenue of $425,000 and monthly costs of $195,000, what is its profit?

b. If this company wants a profit of $21,000 for the month and has costs of $25,000, what revenue will this company have to generate?

SOLUTION

a. If the revenue is $425,000, then $R = 425000$. If the costs are $195,000, then $C = 195000$. Substituting these values into the equation $P = R - C$ yields the profit:

$$P = R - C$$
$$P = 425000 - 195000$$
$$P = 230000$$

A company with $425,000 in revenue and $195,000 in costs has $230,000 in profit.

b. If the profit is $21,000, then $P = 21000$. If the costs are $25,000, then set $C = 25000$. Substituting these values into the equation $P = R - C$ yields

$$P = R - C$$
$$21000 = R - 25000$$

To solve this equation for R, add 25,000 to both sides.

$$\begin{array}{ll} 21000 = R - 25000 & \\ \underline{+25000 \qquad +25000} & \text{Add 25,000 to both sides of the equation.} \\ 46000 = R & \\ 21000 \stackrel{?}{=} 46000 - 25000 & \text{Check the answer.} \\ 21000 = 21000 & \text{The answer checks.} \end{array}$$

If a company wants $21,000 profit and has $25,000 in costs, it must generate $46,000 of revenue.

> **What's That Mean?**
>
> **Profit, Revenue, and Cost**
>
> In business, revenue is the amount of money a company earns, such as from sales. The cost refers to both fixed costs (such as rent) and variable costs (such as electricity) that a company must pay. The profit a company makes is equal to the difference between the revenue and the cost.

PRACTICE PROBLEM FOR EXAMPLE 5

The amount of profit a company makes is found by subtracting the company's costs from its revenue. This relationship can be written as the following equation:

$$P = R - C$$

where P is the profit in dollars, R is the revenue in dollars, and C is the cost in dollars for the company.

a. If a company has monthly revenue of $65,000 and monthly costs of $22,500, what is its profit?

b. If this company wants a profit of $135,000 for the month and has costs of $75,000, what revenue will this company have to generate?

Example 6 Solving equations involving fractions

Solve the equations. Check the answers.

a. $n - \dfrac{2}{9} = \dfrac{7}{9}$ **b.** $x + \dfrac{4}{5} = \dfrac{7}{10}$

SOLUTION

a. The number $\dfrac{2}{9}$ is being subtracted from n, so add $\dfrac{2}{9}$ to both sides of the equation.

$$\begin{aligned} \boxed{n} - \dfrac{2}{9} &= \dfrac{7}{9} \qquad &\text{Identify the variable term.} \\ +\dfrac{2}{9} &+ \dfrac{2}{9} \qquad &\text{Add } \dfrac{2}{9} \text{ to both sides of the equation.} \\ n &= \dfrac{9}{9} \\ n &= 1 \qquad &\text{Simplify the fraction.} \\ (1) - \dfrac{2}{9} &\stackrel{?}{=} \dfrac{7}{9} \qquad &\text{Check the answer.} \\ \dfrac{9}{9} - \dfrac{2}{9} &\stackrel{?}{=} \dfrac{7}{9} \qquad &\text{Find like denominators.} \\ \dfrac{7}{9} &= \dfrac{7}{9} \qquad &\text{The answer checks.} \end{aligned}$$

b. The number $\dfrac{4}{5}$ is being added to x, so subtract $\dfrac{4}{5}$ from both sides of the equation.

$$\begin{aligned} \boxed{x} + \dfrac{4}{5} &= \dfrac{7}{10} \qquad &\text{Identify the variable term.} \\ -\dfrac{4}{5} &- \dfrac{4}{5} \qquad &\text{Subtract } \dfrac{4}{5} \text{ from both sides of the equation.} \\ x &= \dfrac{7}{10} - \dfrac{4}{5} \qquad &\text{Find like denominators to subtract the fractions.} \\ x &= \dfrac{7}{10} - \dfrac{8}{10} \\ x &= -\dfrac{1}{10} \\ \left(-\dfrac{1}{10}\right) + \dfrac{4}{5} &\stackrel{?}{=} \dfrac{7}{10} \qquad &\text{Check the answer.} \\ -\dfrac{1}{10} + \dfrac{8}{10} &\stackrel{?}{=} \dfrac{7}{10} \qquad &\text{Find like denominators.} \\ \dfrac{7}{10} &= \dfrac{7}{10} \qquad &\text{The answer checks.} \end{aligned}$$

> **Skill Connection**
>
> **Find the LCD**
>
> In Section R.2, we learned that finding the least common denominator, or LCD, involves two steps.
>
> 1. Find the prime factorization of each denominator.
> 2. The LCD equals the product of the distinct factors from step 1. If a factor occurs more than once, include the highest power of that factor in the LCD.

PRACTICE PROBLEM FOR EXAMPLE 6

Solve the equation. Check the answer. $x + \dfrac{2}{3} = \dfrac{5}{6}$

Exercise caution when isolating a variable that is not the leftmost term. For example, in the equation $-5 + x = 3$, add 5 to both sides because that will isolate x on the left side of the equation.

$$-5 + \boxed{x} = 3 \qquad \text{Identify the variable term.}$$
$$\underline{+5 \qquad\quad +5} \qquad \text{Add 5 to both sides of the equation.}$$
$$x = 8 \qquad \text{This is the answer.}$$

Remember to include the sign in front of the term when considering whether to use the addition or subtraction property to isolate the variable.

Example 7 Solving equations

Solve the equations. Check the answers.

a. $3 + t = 15$ **b.** $12 = -9 + w$ **c.** $-20 + x = -50$

SOLUTION

a.
$$3 + \boxed{t} = 15 \qquad \text{Identify the variable term.}$$
$$\underline{-3 \qquad\quad -3} \qquad \text{The 3 is positive, so subtract 3 from both sides.}$$
$$t = 12$$
$$3 + (12) \stackrel{?}{=} 15 \qquad \text{Check the answer.}$$
$$15 = 15 \qquad \text{The answer checks.}$$

b.
$$12 = -9 + \boxed{w} \qquad \text{Identify the variable term.}$$
$$\underline{+9 \quad +9} \qquad \text{The 9 is negative, so add 9 to both sides.}$$
$$21 = w$$
$$12 \stackrel{?}{=} -9 + (21) \qquad \text{Check the answer.}$$
$$12 = 12 \qquad \text{The answer checks.}$$

c.
$$-20 + \boxed{x} = -50 \qquad \text{Identify the variable term.}$$
$$\underline{+20 \qquad\quad +20} \qquad \text{Add 20 to both sides.}$$
$$x = -30$$
$$-20 + (-30) \stackrel{?}{=} -50 \qquad \text{Check the answer.}$$
$$-50 = -50 \qquad \text{The answer checks.}$$

PRACTICE PROBLEM FOR EXAMPLE 7

Solve the equation. Check the answer.

a. $7 + m = 20$ **b.** $-5 = -8 + c$

■ Solving Literal Equations

In many applications, equations or **formulas** are used to represent common calculations. Several examples from geometry are the formulas for perimeter, area, and volume. Some examples from science are the position of an object and velocity. A commonly used formula is $d = r \cdot t$. Here, d stands for distance, r stands for the rate (speed), and t stands for time. We use this formula to estimate travel distances and times. If a person drives 50 miles an hour for 2 hours, then the distance traveled is

$$d = r \cdot t$$
$$d = 50 \cdot 2$$
$$d = 100 \text{ miles}$$

166 CHAPTER 2 Linear Equations and Inequalities with One Variable

Several of these frequently used formulas are given at the back of the book. In mathematics, equations or formulas that have several variables are called **literal equations**.

> ### DEFINITIONS
> **Formula** An equation that expresses a general relationship between quantities is called a **formula**.
>
> **Literal Equation** An equation or formula that contains several variables is called a **literal equation**.

To solve a literal equation means to solve for only one of the variables. Solving literal equations is similar to solving other equations we have seen in this section. The properties of equality are used to isolate the variable we wish to solve for. When solving literal equations for a variable, we are finding a new formula with the variable being solved for isolated on one side of the equal sign.

Example 8 Solving literal equations

Solve the following literal equations for the indicated variables.

a. Sum of angle measures in a triangle: $A + B + C = 180$ for A.

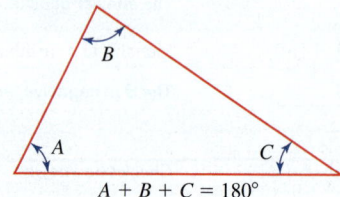

$A + B + C = 180°$

b. Slope-intercept form of a linear equation: $y = mx + b$ for b.
c. $H = r - p$ for r.

SOLUTION

a. To solve for A means to isolate A on one side of the equal sign. Since B is being added to A, remove B from the left side by using the reverse operation of subtraction.

$$\begin{array}{ll} \textcircled{A} + B + C = 180 & \text{Identify the variable term being solved for.} \\ \underline{-B -B} & \text{Subtract } B \text{ from both sides of the equation.} \\ \textcircled{A} + C = 180 - B \end{array}$$

To isolate A, move C from the left side by subtracting it from both sides of the equation.

$$\begin{array}{ll} \textcircled{A} + C = 180 - B & \\ \underline{-C -C} & \text{Subtract } C \text{ from both sides of the equation.} \\ A = 180 - B - C \end{array}$$

The variable A has been solved for. Notice that we have found a formula for A. This formula for A states that the measure of angle A of a triangle is 180 degrees minus the measures of the two other angles, B and C.

b. To solve for b, remove the expression mx from the right side of the equation. Since it is being added to b, use the subtraction property of equality to remove it.

$$y = mx + \boxed{b}$$ Identify the variable term being solved for.
$$\underline{-mx \quad -mx}$$ Subtract mx from both sides of the equation.
$$y - mx = b \text{ or } b = y - mx$$

This equation is now solved for b.

c. To solve, isolate the r on one side of the equal sign.

$$H = \boxed{r} - p$$ Identify the variable term being solved for.
$$\underline{+p \quad +p}$$ Add p to both sides of the equation.
$$H + p = r \text{ or } r = H + p$$

r is isolated on one side of the equal sign, so the literal equation has been solved for r.

■ Connecting the Concepts

How do I write the answer?

The two equations in Example 8b are equivalent: $y - mx = b$ or $b = y - mx$. Ask your instructor whether he or she prefers the variable being solved for on the left side of the equal sign. Many instructors do prefer the variable being isolated on the left.

PRACTICE PROBLEM FOR EXAMPLE 8

Solve the following literal equations for the indicated variables.

a. Weight of a backpack: $W = e + f$ for e.

b. $M + N - P = 200$ for N.

2.1 Vocabulary Exercises

1. A(n) _____ is a statement that two expressions are equal.

2. An example of a(n) _____ is $A = l \cdot w$.

3. The _____ of an equation is (are) value(s) of the variable(s) that make the equation true.

4. An equation is _____ when both sides are equal.

5. We _____ expressions.

6. We _____ equations.

2.1 Exercises

For Exercises 1 through 12, identify each problem as an expression or an equation.

1. $3x - 2$
2. $5 - 2a$
3. $b = 1$
4. $n = 0$
5. -6
6. $-2x$
7. $-5x^2 + 10 = 0$
8. $7 - 2y + y^2 = 2$
9. $2l + 2w = P$
10. $C = 2\pi r$
11. $x^2 + 4x - 9$
12. $-x^2 + 2x - 1$

For Exercises 13 and 14, answer the questions about each problem situation.

13. Joe has a job at a local hotel that pays $10.75 an hour. Joe's weekly salary can be represented by the equation $S = 10.75h$, where S is Joe's weekly salary in dollars when he works h hours a week.

   a. Use the equation to find Joe's salary if he works 20 hours in a week.

   b. Determine whether or not the following values are a solution to the equation $S = 376.25$ when $h = 35$. Explain what these values mean in this situation.

   c. Determine whether or not the following values are a solution to the equation $S = 260$ when $h = 25$. Explain what these values mean in this situation.

   d. Determine whether or not the following values are a solution to the equation $S = 430$ when $h = 40$. Explain what these values mean in this situation.

14. Danny works on the assembly line at a local golf club manufacturer. If it takes Danny 7 minutes to put together the parts for a golf club, then the amount of time it takes Danny to assemble g golf clubs can be represented by the equation $t = 7g$, where t is the total time in minutes it takes for him to assemble g golf clubs.

   a. Use the equation to find how long it will take Danny to assemble 45 golf clubs.

b. Determine whether or not the following values are a solution to the equation $t = 340$ when $g = 50$. Explain what these values mean in this situation.

c. Determine whether or not the following values are a solution to the equation $t = 329$ when $g = 47$. Explain what these values mean in this situation.

For Exercises 15 through 24:
a. Determine whether or not the given values of the variables are solutions to the equations.
b. Explain what the values mean in this situation.

15. $d = 60t$, where d is the distance traveled in miles when driving for t hours; $t = 4$ and $d = 240$.

16. $d = 65t$, where d is the distance traveled in miles when driving for t hours; $t = 30$ and $d = 1950$.

17. $C = 2\pi r$, where C is the circumference of a circle with radius r; $C = 37.7$ and $r = 6$. (Answers rounded to one decimal place.)

18. $C = 2\pi r$, where C is the circumference of a circle with radius r; $C = 163.4$ and $r = 26$. (Answers rounded to one decimal place.)

19. $F = 15L$, where F are the origination fees in dollars paid for a loan of L thousand dollars; $F = 4500$ and $L = 300$. (*Note:* An origination fee in a mortgage is the fee paid to the company originating your loan to cover costs associated with creating, processing, and closing your mortgage.)

20. $F = 20L$, where F are the origination fees in dollars paid for a loan of L thousand dollars; $F = 9000$ and $L = 450$.

21. $F = 10L$, where F are the origination fees in dollars paid for a loan of L thousand dollars; $F = 2000$ and $L = 225$.

22. $F = 12.5L$, where F are the origination fees in dollars paid for a loan of L thousand dollars; $F = 4500$ and $L = 400$.

23. $S = 77.8f$, where S is the number of cubic yards of sand needed to fill a sand volleyball court f feet deep; $S = 155.6$ and $f = 2$.

24. $S = 77.8f$, where S is the number of cubic yards of sand needed to fill a sand volleyball court f feet deep; $S = 200$ and $f = 3$.

For Exercises 25 through 44, determine whether or not the given value(s) of the variable(s) are solutions to the equations.

25. $2x + 8 = 30$; $x = 11$

26. $4g + 23 = 80$; $g = 10$

27. $7m + 15 = 10m$; $m = 8$

28. $-4y + 7 = -25$; $y = -6$

29. $\frac{3}{4}x - 13 = 8$; $x = 28$

30. $7 = -\frac{5}{3}y + \frac{1}{3}$; $y = -4$

31. $\frac{1}{2}x - 10 = -\frac{15}{2}$; $x = 5$

32. $-1 = \frac{2}{3}x + 1$; $x = -3$

33. $0.15x + 1.2 = 1.35$; $x = 1$

34. $-1.4x - 6 = -14.4$; $x - 5$

35. $0.5t + 3.2 = 1.7$; $t = 3$

36. $-6.1 + 3y = -10.3$; $y = 1.4$

37. $3x + 5y = 20$; $x = 5$ and $y = 1$

38. $4x - 9y = 30$; $x = 3$ and $y = -2$

39. $-x + 2y = 4$; $x = -4$ and $y = 0$

40. $-3x + 4y = -12$; $x = 0$ and $y = -3$

41. $25m + 15 = 10n + 40$; $m = 3$ and $n = 5$

42. $-3x + 7 = 4y + 9$; $x = 0$ and $y = -2$

43. $-5x + 3y = -15$; $x = 0$ and $y = 5$

44. $4x + 3y = -12$; $x = -4$ and $y = 1$

45. The maximum weight limit on the roof of a 2001 Volvo V70 station wagon is 220 lb. If you consider the weight of a cargo storage system and the luggage and other cargo, you can represent the maximum weight using the equation

$$s + l + c = 220$$

where s is the weight in pounds of the cargo storage system, l is the weight in pounds of the luggage, and c is the weight in pounds of any additional cargo.
Source: Volvo V70 2001 owners manual.

a. If the cargo storage system weighs 75 pounds and the luggage that you want to store on the roof weighs 100 pounds, how many pounds of other cargo can you store on the roof of the car?

b. If the cargo storage system weighs 75 pounds and the additional cargo you want to store on the roof weighs 60 pounds, how many pounds of luggage can you store on the roof of the car?

46. Alex has budgeted his expenses for the month and does not want to spend more than he makes at work. Alex needs to pay rent and other bills and wants some money for entertainment. Alex's total expenses can be represented by the equation

$$E = r + b + e$$

where E is the total monthly expenses in dollars, r is the amount he pays for rent in dollars, b is the amount of his other bills in dollars, and e is the amount of money he has for entertainment in dollars.

a. If Alex has $1200 this month for expenses and pays $600 in rent and $230 for other bills, how much does he have left for entertainment?

b. If Alex has only $1000 this month but still pays $600 for rent and needs $350 for other bills, how much does he have left for entertainment?

For Exercises 47 through 52, use the following information. The amount of profit a company makes is found by subtracting the company's costs from its revenue. This relationship can be written as the following equation:

$$P = R - C$$

where P is the profit in dollars, R is the revenue in dollars, and C is the cost in dollars for the company. Note: Revenue is the amount of money a company brings in from sales.

47. If a company has monthly revenue of $47,000 and monthly costs of $39,000 what is the profit?

48. If a company has monthly revenue of $180,000 and monthly costs of $165,000, what is the profit?

49. If a company has monthly revenue of $38,000 and monthly costs of $41,000, what is the profit?

50. If a company has monthly revenue of $430,000 and monthly costs of $500,000, what is the profit?

51. If a company wants a profit of $4000 for the month and has costs of $25,000, what revenue does this company have to generate?

52. If this company wants a profit of $11,000 for the month and has costs of $156,000, what revenue will this company have to generate?

For Exercises 53 through 82, solve each equation. Check the answers.

53. $x + 5 = 30$

54. $w + 12 = 28$

55. $x + 20.5 = 45$

56. $4.2 + t = -7.3$

57. $m - 14 = 5$

58. $k - 8 = 22$

59. $-4 + p = 23$

60. $-4 + y = 9$

61. $-1 + k = -16$

62. $-3 + w = -9$

63. $5 = -3 + x$

64. $-8 = -2 + r$

65. $a + 0.25 = 2.75$

66. $-1.35 + b = 7.60$

67. $\dfrac{3}{2} = \dfrac{1}{2} + x$

68. $\dfrac{5}{3} = -\dfrac{1}{3} + y$

69. $m + \dfrac{2}{3} = \dfrac{7}{3}$

70. $t - \dfrac{9}{5} = \dfrac{6}{5}$

71. $\dfrac{4}{9} + m = \dfrac{5}{9}$

72. $\dfrac{1}{3} + y = \dfrac{5}{3}$

73. $5 + x = 14$

74. $7 + z = 5$

75. $d - 1.75 = 6.5$

76. $-2.8 + k = 12.3$

77. $-1.3 + x = -4.7$

78. $13.5 = -7.8 + y$

79. $\dfrac{2}{5} + x = \dfrac{3}{7}$

80. $\dfrac{1}{2} + x = \dfrac{2}{3}$

81. $\dfrac{4}{5} + y = \dfrac{3}{10}$

82. $\dfrac{6}{11} + a = \dfrac{1}{22}$

For Exercises 83 through 100, solve each literal equation for the indicated variable.

83. Profit: $P = R - C$ for R.

84. Sale price: $S = P - D$ for P.

85. Maximum weight limit for the roof of a car: $s + l + c = 220$ for l.

86. Retail price: $R = C + M$ for C.

87. Sum of the measures of the angles in a right triangle: $A + B + 90 = 180°$ for A.

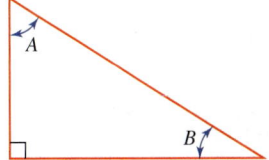

88. Sum of the angles in a quadrilateral:
$A + B + C + D = 360°$ for C.

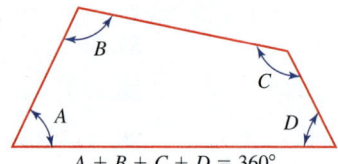

$A + B + C + D = 360°$

89. $T = 45 + M$ for M.
90. $g = 3f + h$ for h.
91. $x + y = -4$ for y.
92. $2x + y = 7$ for y.
93. $D + 10 = 3F + G - 20$ for G.
94. $5a - 8 = b + 2c + 15$ for b.
95. $t = -3u + w$ for w.
96. $y = -4x + z - 20$ for z.
97. $P = a + b + c$ for b.
98. $P = a + b + c + d$ for d.
99. $3y = x - 5$ for x.
100. $-4y = x + 9$ for x.

For Exercises 101 through 112, identify each problem as an expression or equation. Simplify any expressions and solve any equations.

101. $23 = x + 9$
102. $14 + m = 45$
103. $5x + 7 - 3x$
104. $8h - 3h + 14 + 2h$
105. $20a^2 + 6a - 3a^2 + 10$
106. $-11n^2 + 8 - 5n^2 + 7n$
107. $-14 + c = -10$
108. $-16 + y = -4$
109. $p - 14 = 20$
110. $r - 27 = 41$
111. $p - 14 + 20$
112. $r - 27 + 41$

For Exercises 113 through 120, use the formula $P = 4.99t + 146.4$. Here P represents the population of South America, in millions, t years since 1960.
Source: www.worldmetersinfo.org

113. One solution to the equation is $t = 35, P = 321.05$. Explain what these values mean in the context of the situation.

114. One solution to the equation is $t = 47, P = 380.93$. Explain what these values mean in the context of the situation.

115. One solution to the equation is $t = 58, P = 435.82$. Explain what these values mean in the context of the situation.

116. One solution to the equation is $t = 25, P = 271.15$. Explain what these values mean in the context of the situation.

117. Fill in each blank.

  a. The year 1985 corresponds to _____ years since 1960.

  b. The year 1985 corresponds to $t = $ _____.

  c. The population of South America in the year 1985 was about _____ million.

118. Fill in each blank.

  a. The year 1998 corresponds to _____ years since 1960.

  b. The year 1998 corresponds to $t = $ _____.

  c. The population of South America in the year 1998 was about _____ million.

119. Fill in each blank.

  a. The year 2016 corresponds to _____ years since 1960.

  b. The year 2016 corresponds to $t = $ _____.

  c. The population of South America in the year 2016 was about _____ million.

120. Fill in each blank.

  a. The year 2012 corresponds to _____ years since 1960.

  b. The year 2012 corresponds to $t = $ _____.

  c. The population of South America in the year 2012 was about _____ million.

2.2 Multiplication and Division Properties of Equality

LEARNING OBJECTIVES
- Use the multiplication and division properties of equality.
- Solve multiple-step equations.
- Generate equations from applications.
- Solve more literal equations.

Multiplication and Division Properties of Equality

In Section 2.1, we studied the addition and subtraction properties of equality and learned how to use them to solve some types of equations. In this section, two more properties of equality are presented. Keep in mind the visual example of scales in balance. Think of scales that are balanced: If we put twice as much weight on each side, then the two scales are still balanced.

If a scale is in balance and you double both sides, the scale will remain in balance.

This is the same as multiplying both sides of an equation by 2.

The same would be true if we took half of the weight away from both sides of the balance scale. It would still be in balance. These examples demonstrate that when multiplying or dividing both sides of an equation by the same nonzero real number, the equation will stay true.

This leads us to the **multiplication and division properties of equality.**

MULTIPLICATION PROPERTY OF EQUALITY
When **both** sides of an equation are multiplied by the same nonzero real number or nonzero algebraic expression, the solution to the equation is unchanged.

$$\text{If } a = b \text{ and } c \neq 0, \text{ then } a \cdot c = b \cdot c$$

DIVISION PROPERTY OF EQUALITY
When **both** sides of an equation are divided by the same nonzero real number or nonzero algebraic expression, the solution to the equation is unchanged.

$$\text{If } a = b \text{ and } c \neq 0, \text{ then } \frac{a}{c} = \frac{b}{c}$$

Connecting the Concepts

How do we show multiplication?

Multiplication can be shown in several ways. Multiplication is the only operation that does not need a separate symbol. When a number, variable, or parentheses are next to a variable or parentheses, it means to multiply.

$$9h = 9 \cdot h$$
$$2(3x + 5) = 6x + 10$$
$$(2)(4x) = 8x$$

From this definition, there is only one number by which we should not multiply or divide both sides of an equation. That number is zero. Division by zero is undefined.

172 CHAPTER 2 Linear Equations and Inequalities with One Variable

Likewise, both sides of an equation are not multiplied by 0. To see why, multiply both sides of the following equation by 0.

$$x + 5 = 7$$
$$0(x + 5) = 0(7) \quad \text{Multiply both sides by 0.}$$
$$0 = 0$$

Multiplying both sides by 0 only shows that $0 = 0$. This does not help solve the equation at all. Therefore, do not multiply both sides of an equation by 0.

Linear Equations

Addition and Subtraction Property of Equality

Use to isolate a variable term.

$$3x + 7 = 20$$
$$\underline{-7 \quad -7}$$
$$3x = 13$$

Multiplication and Division Property of Equality

Use to isolate a variable.

$$5x = 70$$
$$\frac{5x}{5} = \frac{70}{5}$$
$$x = 14$$

Example 1 Hourly pay

José has a job at a local Post and Pack store and gets paid about $12.00 an hour after taxes are taken out of his paycheck. José can calculate his weekly take-home salary using the equation

$$S = 12h$$

where S represents his weekly take-home salary in dollars and h is the number of hours he works that week.

a. Calculate José's weekly take-home salary if he works 15 hours.

b. How many hours does José have to work if he needs $240 to pay his car payment?

SOLUTION

a. We are given that José worked 15 hours that week, so substitute $h = 15$ and calculate his weekly take-home salary.

$$S = 12(15)$$
$$S = 180$$

José will take home about $180 for working 15 hours in a week.

b. In this situation, José needs $240 for his car payment, and we want to find out how many hours he has to work to earn that much. Since he wants his take-home salary to be $240, substitute $S = 240$, and solve for h to find the hours he has to work.

$$S = 12h$$
$$240 = 12h$$

Since h is being multiplied by 12, undo the multiplication by dividing both sides of the equation by 12.

$240 = 12h$	The variable term is by itself, so isolate the variable.
$\dfrac{240}{12} = \dfrac{12h}{12}$	Divide both sides by 12.
$20 = \dfrac{\cancel{12}h}{\cancel{12}}$	Reduce.
$20 = h$	

Check the answer by substituting both 240 and 20 into the equation and checking that the statement is true.

$240 \stackrel{?}{=} 12(20)$ The answer checks and seems like a
$240 = 240$ reasonable number of hours for this situation.

José has to work 20 hours this week to make enough money to pay his $240 car payment.

■ Connecting the Concepts

Should you answer an application using decimals?

When working with applications, read the problem to see if a decimal approximation is more appropriate. For example, if the problem is given in dollars, round the answer to two decimal places (the cents place).

SECTION 2.2 Multiplication and Division Properties of Equality 173

PRACTICE PROBLEM FOR EXAMPLE 1

Ann sells subscriptions to a local newspaper door to door. For every subscription she sells, she earns $6. The amount Ann gets paid for selling subscriptions can be represented by using the equation

$$P = 6s$$

where P is the amount Ann gets paid in dollars for selling s newspaper subscriptions.

a. How much does Ann get paid for 20 subscriptions?
b. Find how many subscriptions Ann must sell to make $240.

Example 2 Solving equations using the division property of equality

Solve the equations. Check the answers.

a. $5b = 140$ **b.** $14 = -3.5x$ **c.** $-x = 3.5$

SOLUTION

a. To solve for b, isolate the variable b on one side of the equation. In this case, we have to undo the multiplication by 5 to isolate b on the left side of the equation. To undo the multiplication, divide both sides by 5.

$5b = 140$ The variable term is already by itself, so isolate the variable.
$\dfrac{5b}{5} = \dfrac{140}{5}$ Isolate b by dividing both sides by 5.
$\dfrac{\cancel{5}b}{\cancel{5}} = 28$ Reduce.
$b = 28$
$5(28) \stackrel{?}{=} 140$ Check the answer.
$140 = 140$ The answer checks.

b. Since x is multiplied by -3.5, undo the multiplication by dividing both sides of the equation by -3.5.

$14 = -3.5x$ The variable term is already by itself, so isolate the variable.
$\dfrac{14}{-3.5} = \dfrac{-3.5x}{-3.5}$ Isolate x by dividing both sides by -3.5.
$-4 = x$
$14 \stackrel{?}{=} -3.5(-4)$ Check the answer.
$14 = 14$ The answer checks.

c. The variable term is $-x = -1 \cdot x$, so the coefficient of x is -1. To solve this equation, divide both sides by -1.

$-x = 3.5$ Rewrite the left side with a coefficient of -1.
$\dfrac{-1 \cdot x}{-1} = \dfrac{3.5}{-1}$ Divide both sides by -1.
$x = -3.5$
$-(-3.5) \stackrel{?}{=} 3.5$ Check the answer.
$3.5 = 3.5$ The answer checks.

> **Connecting the Concepts**
>
> **Which operations undo others?**
>
> Remember that the four properties of equality help us to isolate the variable by undoing operations that are on the same side of the equation as the variable.
>
> Use the table to help remember how operations undo each other.
>
> | Do | Undo |
> |---|---|
> | Add | Subtract |
> | Subtract | Add |
> | Multiply | Divide |
> | Divide | Multiply |

PRACTICE PROBLEM FOR EXAMPLE 2

Solve the equation. Check the answer.

a. $6d = 78$ **b.** $-7m = -45.5$ **c.** $8.42 = -n$

Example 3 — Solving equations with fractional coefficients

Solve the equation. Check the answer.

a. $\dfrac{w}{4} = 7$ **b.** $\dfrac{3}{8}y = -6$

SOLUTION

a. w is being divided by 4, so undo the division by multiplying both sides of the equation by 4.

$$\frac{w}{4} = 7 \qquad \text{The variable term is already by itself, so isolate the variable.}$$

$$4\left(\frac{w}{4}\right) = 4(7) \qquad \text{Isolate } w \text{ by multiplying both sides by 4.}$$

$$\cancel{4}\left(\frac{w}{\cancel{4}}\right) = 28 \qquad \text{Reduce.}$$

$$w = 28$$

$$\frac{28}{4} \stackrel{?}{=} 7 \qquad \text{Check the answer.}$$

$$7 = 7 \qquad \text{The answer checks.}$$

b. The coefficient of y is $\dfrac{3}{8}$. Undo multiplication by $\dfrac{3}{8}$ by dividing both sides of the equation by $\dfrac{3}{8}$.

$$\frac{3}{8}y = -6$$

$$\frac{\frac{3}{8}y}{\frac{3}{8}} = \frac{-6}{\frac{3}{8}} \qquad \text{Divide both sides by the coefficient.}$$

$$\frac{\cancel{\frac{3}{8}}y}{\cancel{\frac{3}{8}}} = -6 \cdot \frac{8}{3} \qquad \text{To divide by a fraction, invert and multiply by the reciprocal.}$$

$$y = -\overset{2}{\cancel{6}} \cdot \frac{8}{\cancel{3}} \qquad \text{Divide out like factors.}$$

$$y = -16$$

Notice that in the above problem, we divided both sides by the fraction $\dfrac{3}{8}$. Dividing by $\dfrac{3}{8}$ is equivalent to inverting $\dfrac{3}{8}$ and multiplying both sides by $\dfrac{8}{3}$.

■ Connecting the Concepts

How do we write a coefficient of $\dfrac{1}{2}$?

If a problem says to find one-half of a number, then it translates as $\dfrac{1}{2}x$. This can also be written as

$$\frac{1}{2}x = \frac{1}{2} \cdot x = \frac{1}{2} \cdot \frac{x}{1} = \frac{x}{2}$$

Likewise, $\dfrac{3}{8}y$ can be written as $\dfrac{3y}{8}$.

$$\frac{3}{8}y = -6$$

$$\frac{8}{3} \cdot \frac{3}{8}y = \frac{8}{3} \cdot (-6)$$ Multiply both sides by the reciprocal.

$$\frac{8}{3} \cdot \frac{3}{8}y = \frac{8}{3} \cdot (-\overset{2}{6})$$ Divide out like factors.

$$y = -16$$

To check, substitute $y = -16$ into the original problem.

$$\frac{3}{8}(-16) \overset{?}{=} -6$$

$$\frac{3}{8}(-\overset{2}{16}) \overset{?}{=} -6$$

$$-6 = -6$$ The answer checks.

PRACTICE PROBLEM FOR EXAMPLE 3

Solve the equations. Check the answers.

a. $\dfrac{n}{5} = -\dfrac{3}{10}$ **b.** $\dfrac{2}{3}x = -8$

Solving Multiple-Step Equations

Remember that to solve a linear equation in one variable, the variable must be isolated on one side of the equation. So far, we have solved only equations that involved one operation. One of the four properties of equality has been used to undo the operation and isolate the variable.

Many equations contain more than one operation. This means that we will use more than one property of equality to isolate the variable. In solving equations with more than one operation, a combination of the properties of equality studied so far will be applied. In general, first isolate the variable term of an equation and then isolate the variable itself. In the equation $3x + 9 = 21$, first isolate the variable term $3x$ by subtracting 9 from both sides and then isolate x by dividing both sides by 3.

$$\boxed{3x} + 9 = 21$$ Identify the variable term.

$$\underline{-9 \; -9}$$ Isolate the variable term by subtracting 9 from both sides of the equation.

$$3x = 12$$

$$\frac{3x}{3} = \frac{12}{3}$$ Now isolate the variable by dividing both sides by 3.

$$x = 4$$

$$3(4) + 9 \overset{?}{=} 21$$ Check the answer.

$$12 + 9 \overset{?}{=} 21$$

$$21 = 21$$ The answer checks.

When solving multiple-step problems, be sure to consider one step at a time. Right now, we have only four basic properties of equality to use. It is useful to circle the variable term and then work to isolate it on one side of the equation.

Connecting the Concepts

Why circle the variable term?

When solving an equation that has several operations to be undone, visualize a present that is in a box and has been gift-wrapped.

If you want to get to your present, you must open the package. You cannot get to the present without first taking off the wrapping paper and then opening the box. You cannot open the box first and then take off the wrapping paper. This is similar to an equation that has more than one operation involved. You should typically undo any operations that are farther away from the variable (the present) you are trying to isolate. This is why we often circle the variable term so that we can concentrate on isolating it first. Then work toward the variable by undoing one layer of operations at a time.

CHAPTER 2 Linear Equations and Inequalities with One Variable

Example 4 — Cost of T-shirts

The local YMCA wants to order T-shirts for kids who attend its summer day camps. After calling a few local shirt printers, the staff members find one that will charge a $350 setup fee for the design and $4 per shirt for the T-shirt with design printed on it. Using this information, the staff at the YMCA can estimate costs using the formula

$$C = 350 + 4s$$

where C represents the total cost in dollars for s T-shirts.

a. What will it cost the YMCA to purchase 100 T-shirts?

b. How many T-shirts can the YMCA purchase if the budget for T-shirts is $1250?

SOLUTION

a. Since the staff members want to purchase 100 T-shirts, $s = 100$. Substitute this in the equation and evaluate to find the cost.

$$C = 350 + 4(100) \quad \text{Substitute 100 for s.}$$
$$C = 350 + 400$$
$$C = 750$$

If the YMCA purchases 100 T-shirts, it will cost $750.

b. The YMCA has $1250 to spend on T-shirts, so we let $C = 1250$ and solve for s.

$$1250 = 350 + 4s \quad \text{Substitute in 1250 for C. Identify the variable term.}$$
$$\begin{array}{r} 1250 = 350 + 4s \\ -350 \quad -350 \end{array} \quad \text{Isolate the variable term by subtracting 350 from both sides.}$$
$$900 = 4s$$
$$\frac{900}{4} = \frac{4s}{4} \quad \text{Isolate s by dividing both sides by 4.}$$
$$225 = s$$
$$1250 \stackrel{?}{=} 350 + 4(225) \quad \text{Check the answer.}$$
$$1250 \stackrel{?}{=} 350 + 900$$
$$1250 = 1250 \quad \text{The answer checks.}$$

If the YMCA has $1250 to spend on T-shirts, the staff can purchase 225 shirts.

PRACTICE PROBLEM FOR EXAMPLE 4

The annual number of passenger-miles of travel on transit in the United States can be estimated by using the equation

$$M = 938.1t + 44260$$

where M represents annual number of passenger-miles of travel on transit in the United States, in millions, and t is time in years since 2000.

Source: https://www.rita.dot.gov

a. Estimate the annual number of passenger-miles on transit in the United States in 2018.

b. Find what year the number of passenger-miles on transit in the United States reached 58,000 million miles.

SECTION 2.2 Multiplication and Division Properties of Equality   177

> **Steps to Solving Linear Equations**
> 1. Simplify both sides of the equation independently, if needed, using the distributive property and combining like terms.
> 2. Isolate the variable term. Circle the variable term. Use the addition or subtraction properties of equality to isolate the variable term on one side of the equation.
> 3. Isolate the variable. Use the multiplication or division properties of equality to isolate the variable on one side of the equation.
> 4. Check the answer. Substitute the answer into the original equation and simplify both sides to check that they give the same result.

Example 5 Solving multiple-step equations

Solve the equations. Check the answers.

a. $3x + 8 = 44$
b. $75 = \dfrac{p}{2} - 25$
c. $84 = 6x - 12 - 10x$

SOLUTION

a. **Steps 1 and 2** Simplify both sides of the equation independently, if needed, using the distributive property and combining like terms. Isolate the variable term. Circle the variable term. Use the addition or subtraction properties of equality to isolate the variable term on one side of the equation.

$$\boxed{3x} + 8 = 44 \quad \text{Identify the variable term.}$$
$$\underline{-8 \quad -8} \quad \text{Isolate the variable term by subtracting 8 from}$$
$$3x = 36 \quad \text{both sides of the equation.}$$

Steps 3 Isolate the variable. Use the multiplication or division properties of equality to isolate the variable on one side of the equation.

$$\dfrac{3x}{3} = \dfrac{36}{3} \quad \text{Isolate x by dividing both sides by 3.}$$
$$x = 12$$

Steps 4 Check the answer. Substitute the answer into the original equation and simplify both sides to check that they give the same result.

$$3(12) + 8 \stackrel{?}{=} 44$$
$$36 + 8 \stackrel{?}{=} 44$$
$$44 = 44 \quad \text{The answer checks.}$$

b. First isolate the variable term on one side of the equal sign. Begin by adding 25 to both sides of the equation to undo the subtraction by 25. Then multiply both sides by 2 to undo the division.

$$75 = \boxed{\dfrac{p}{2}} - 25 \quad \text{Identify the variable term.}$$
$$\underline{+25 +25} \quad \text{Isolate the variable term by adding 25 to both sides.}$$
$$100 = \dfrac{p}{2}$$

$$2(100) = 2\left(\dfrac{p}{2}\right) \quad \text{Isolate p by multiplying both sides by 2.}$$
$$200 = p$$

$$75 \stackrel{?}{=} \dfrac{(200)}{2} - 25 \quad \text{Check the answer.}$$
$$75 \stackrel{?}{=} 100 - 25$$
$$75 = 75 \quad \text{The answer checks.}$$

c. In this equation, simplify the right side by combining like terms.

$$84 = 6x - 12 - 10x$$
$$84 = -4x - 12$$
$$+12 \quad\quad +12$$
$$96 = -4x$$
$$\frac{96}{-4} = \frac{-4x}{-4}$$
$$-24 = x$$

Simplify the right side by combining like terms.

Identify the variable term.

Isolate the variable term by adding 12 to both sides.

Isolate x by dividing both sides by −4.

$$84 \stackrel{?}{=} 6(-24) - 12 - 10(-24)$$
$$84 \stackrel{?}{=} -144 - 12 + 240$$
$$84 = 84$$

Check the answer.

The answer checks.

PRACTICE PROBLEM FOR EXAMPLE 5

Solve the equations. Check the answers.

a. $8b - 7 = 41$ **b.** $55 = \dfrac{x}{3} + 30$ **c.** $7x + 13 + 3x = 188$

■ Generating Equations from Applications

When a problem is given in a sentence form, first translate it into an algebraic equation and then solve it. The most common phrases used for the operations are reviewed in the What's That Mean? feature on this page. The phrases that represent an equal sign are in the following table.

Phrases	Translation
x is equal to 10	$x = 10$
15 is the same as n	$15 = n$
p is 4	$p = 4$
t was −8	$t = -8$

Any words or phrases that mean equal or the same will represent an equal sign. The words *is*, *are*, *were*, and *was* also represent an equal sign when translated. In some situations, we can use a known formula to write the equation needed to find the value of the variable.

Example 6 — Translating sentences into equations and solving

Translate each of the following sentences into an equation. Solve the equations. Check the answers.

a. Twice a number added to 10 is equal to 20.

b. The quotient of a number and 5 is 30.

c. Five times the sum of a number and 6 is equal to 40.

d. The perimeter of a rectangle is 40 inches, and its length is 12 inches. Find the rectangle's width.

■ What's That Mean?

Operations

Recall from Chapter 1 that the common phrases for addition, subtraction, multiplication, and division are as follows:

Addition:
sum
total
increased by
plus
more than
exceeds by
added to

Subtraction:
minus
difference
decreased by
less than
subtracted from
reduced by

Multiplication:
product
times
twice
doubled
tripled
half of

Division:
quotient
ratio
divided by

SECTION 2.2 Multiplication and Division Properties of Equality

SOLUTION

In these solutions, let $x =$ a number.

a. "Twice" means to multiply by 2, and "is equal to" translates as the equal sign. This yields the equation

$$10 + 2x = 20$$

Solve the equation.

$10 + \boxed{2x} = 20$	Identify the variable term.
$\underline{-10 \qquad -10}$	Isolate the variable term by subtracting 10 from both sides.
$2x = 10$	
$\dfrac{2x}{2} = \dfrac{10}{2}$	Isolate x by dividing both sides by 2.
$x = 5$	
$10 + 2(5) \stackrel{?}{=} 20$	Check the answer.
$20 = 20$	The answer checks.

> **Connecting the Concepts**
>
> **Does it matter which side of the equal sign the variable is on?**
>
> The equation $15 = n$ is the same as $n = 15$. It is okay to switch both sides of the equality.

b. *Quotient* represents division, and *is* represents the equal sign. Remember that the order of division is important. Read from left to right. Write the quotient in the same order as it is read. The equation translates as

$$\dfrac{x}{5} = 30$$

The variable term is by itself. To isolate x, multiply both sides of the equation by 5.

$\dfrac{x}{5} = 30$	
$5\left(\dfrac{x}{5}\right) = 5(30)$	Isolate x by multiplying both sides by 5.
$x = 150$	
$\dfrac{150}{5} \stackrel{?}{=} 30$	Check the answer.
$30 = 30$	The answer checks.

c. *Times* means to multiply, and *times the sum* means to multiply an addition that has already been done. To show the addition, include a set of parentheses around the sum. This grouping gives us the equation

$$5(x + 6) = 40$$

Solve the equation.

$5(x + 6) = 40$	Simplify the left side by distributing the 5.
$\boxed{5x} + 30 = 40$	Identify the variable term.
$\underline{-30 \quad -30}$	Isolate the variable term by subtracting 30 from both sides.
$5x = 10$	
$\dfrac{5x}{5} = \dfrac{10}{5}$	Isolate x by dividing both sides by 5.
$x = 2$	
$5((2) + 6) \stackrel{?}{=} 40$	Check the answer.
$5(8) \stackrel{?}{=} 40$	
$40 = 40$	The answer checks.

> **Connecting the Concepts**
>
> **Is order important when translating?**
>
> With addition and multiplication, the order in which you add or multiply is not important, since these operations are commutative. In subtracting or dividing, the order matters a great deal. Subtraction and division are not commutative. In subtracting, the phrases *less than* and *subtracted from* are translated in the reverse order of the way they are stated.
>
> 6 less than 20 is
>
> $20 - 6$
>
> 10 subtracted from 30 is
>
> $30 - 10$

Connecting the Concepts

Providing reasons for each step

When solving an equation, it is good practice to ask yourself *why* you are performing each step as you are working out the problem. In the examples, the blue annotations next to each step give you these reasons. Some exercises in this book will require you to provide reasons for each step. Use the annotations in the examples and the equation solving toolboxes in the back of the book as a guide. Be sure to state what step you are performing and why.

d. Recall that the perimeter of a rectangle can be calculated by using the formula

$$P = 2l + 2w$$

where P is the perimeter of the rectangle with length l and width w. Substitute $P = 40$ and $l = 12$.

$40 = 2(12) + 2w$	Substitute the given values and simplify.
$40 = 24 + 2w$	Identify the variable term.
$\underline{-24 \quad -24}$	Isolate the variable term by subtracting 24 from both sides.
$16 = 2w$	
$\dfrac{16}{2} = \dfrac{2w}{2}$	Isolate w by dividing both sides by 2.
$8 = w$	
$40 \stackrel{?}{=} 2(12) + 2(8)$	Check the answer.
$40 \stackrel{?}{=} 24 + 16$	
$40 = 40$	The answer checks.

A rectangle with a perimeter of 40 inches and length 12 inches has a width of 8 inches.

PRACTICE PROBLEM FOR EXAMPLE 6

Translate each of the following sentences into an equation. Solve the equations. Check the answers.

a. Half a number minus 13 is equal to 7.

b. The product of 3 and a number plus 8 is 23.

In an application problem, be sure to verify that the solution makes sense as well as checks in the equation. Remember that sometimes we round the answers to make them reasonable for the situation. For example, a solution that represents dollars should be rounded to the cents (hundredths) place. Any solution to an application problem should be written in a complete sentence using units.

Example 7 Translating application problems

Define the variables and translate each sentence into an equation. Solve the equations. Check the answers for accuracy and reasonableness.

a. The fees at Grossmont College are $46.00 per credit plus a $15.00 health fee. If Leticia paid $613.00 in fees this semester, how many credits did she take?

b. The cost of renting a 17-foot truck from U-Haul is $29.95 for one day and $0.79 per mile. If it cost Mandel $91.57 to rent the truck for one day, how many miles did he drive it?

SOLUTION

a. There are two quantities involved in this problem: the fees Leticia pays and the number of credits she takes. Define two variables to represent this situation.

$c =$ The number of credits taken this semester

$F =$ The fees paid in dollars to Grossmont College

Multiply by the per credit charge and add the health fee. The equation is

$$F = 46c + 15$$

Since Leticia paid $613.00 in fees this semester, $F = 613.00$. Substitute this value into the equation and solve for c (the number of credits).

$F = 46.00c + 15.00$ — Substitute 613 for F.
$613.00 = \boxed{46.00c} + 15.00$ — Identify the variable term.
$\underline{-15.00 \qquad\quad -15.00}$ — Isolate the variable term by subtracting 15 from both sides.
$598.00 = 46.00c$
$\dfrac{598.00}{46.00} = \dfrac{46.00c}{46.00}$ — Isolate c by dividing both sides by 46.00.
$13 = c$
$613.00 \stackrel{?}{=} 46.00(13) + 15$ — Check the answer.
$613.00 = 613.00$ — The answer checks.

Many students take between 6 and 15 credits a semester, so 13 credits is a reasonable answer. The quantity 13 checks in the equation.

Leticia paid $613.00 to take 13 credits at Grossmont College this semester.

b. $m =$ the miles the truck is driven.
$C =$ the cost in dollars to rent a 17-foot truck from U-Haul for one day.

Mandel must pay the one-day fee of $29.95 and a per mile charge of $0.79 per mile. The equation is

$$C = 29.95 + 0.79m$$

If it cost Mandel $91.57 to rent the truck for a day, then $C = 91.57$. Substitute this amount in the equation and solve for the number of miles driven.

$C = 29.95 + 0.79m$ — Substitute 91.57 for C.
$91.57 = 29.95 + \boxed{0.79m}$ — Identify the variable term.
$\underline{-29.95 \quad -29.95}$ — Isolate the variable term by subtracting 29.95 from both sides.
$61.62 = 0.79m$
$\dfrac{61.62}{0.79} = \dfrac{0.79m}{0.79}$ — Isolate m by dividing both sides by 0.79.
$78 = m$
$91.57 \stackrel{?}{=} 29.95 + 0.79(78)$ — Check the answer.
$91.57 = 91.57$ — The answer works.

If the truck rental cost Mandel $91.57, then he drove the truck 78 miles.

This is a reasonable amount of miles for a 1-day truck rental, and it checks in the equation.

PRACTICE PROBLEM FOR EXAMPLE 7

Define the variables and translate the sentence into an equation. Solve the equation. Check the answer for accuracy and reasonableness.

The value of a used backhoe loader decreases by an average of $5000 per year. A particular used backhoe loader is worth $61,000. After how many years will the backhoe loader be worth only $36,000?

■ **More on Solving Literal Equations**

Solving literal equations uses the same properties of equality that are used to isolate a variable in a formula or equation. Start by isolating the term that contains the variable to be solved for. Then isolate that variable. In this section, we assume that a variable expression does not equal zero so that division will not result in an undefined answer.

Skill Connection

Using tables to generate equations

In Section 1.2, we learned how to use input–output tables to generate expressions. In Example 7a, we can use a table as follows. The problem tells us that the fees are $46 a credit plus a $15 health fee. Using this information, we can generate a table for several values of the number of credits taken, c.

$c =$ no. of credits	$F =$ fees paid ($)
0	15
1	46 + 15
2	46 · 2 + 15
3	46 · 3 + 15
c	46 · c + 15

Therefore, from the last row, we have that $F = 46.00c + 15.00$.

182 CHAPTER 2 Linear Equations and Inequalities with One Variable

Example 8 — Solving literal equations for a variable

Solve each literal equation for the indicated variable.

a. Distance: $d = rt$ for r.

b. Slope-intercept form of a linear equation: $y = mx + b$ for x.

c. Perimeter of a rectangle: $P = 2l + 2w$ for w.

SOLUTION

a.
$d = rt$

$\dfrac{d}{t} = \dfrac{rt}{t}$ — Isolate r by dividing both sides by t.

$\dfrac{d}{t} = \dfrac{r\cancel{t}}{\cancel{t}}$ — Reduce.

$\dfrac{d}{t} = r$

b.
$y = \boxed{mx} + b$ — Identify the x-term.

$\underline{-b \qquad\quad -b}$ — Isolate the x-term by subtracting b from both sides.

$y - b = mx$ — Because y and b are not like terms, the subtraction on the left side cannot be simplified.

$\dfrac{y - b}{m} = \dfrac{mx}{m}$

$\dfrac{y - b}{m} = \dfrac{\cancel{m}x}{\cancel{m}}$ — Isolate x by dividing both sides by m and reduce.

$\dfrac{y - b}{m} = x$

c.
$P = 2l + \boxed{2w}$ — Identify the w-term.

$\underline{-2l \quad -2l}$ — Isolate the w-term by subtracting $2l$ from both sides.

$P - 2l = 2w$ — Because P and $2l$ are not like terms, the subtraction on the left side cannot be simplified.

$\dfrac{P - 2l}{2} = \dfrac{2w}{2}$ — Isolate w by dividing both sides by 2. Reduce.

$\dfrac{P - 2l}{2} = w$

> **Skill Connection**
>
> **Reducing fractions**
> When reducing a fraction, always be careful to divide out factors, not terms.
>
> | Correct | Wrong |
> |---|---|
> | $\dfrac{2 + 3}{4}$ | $\dfrac{\cancel{2} + 3}{^2\cancel{4}}$ |
> | $= \dfrac{5}{4}$ vs. $= \dfrac{3}{2}$ |
> | $= 1.25$ | $= 1.5$ |
>
> We can divide out like factors in the numerator and denominator (multiplication) but not through additions or subtractions.
>
> When the fraction contains only factors, we can multiply first or reduce the fraction first.
>
> | Correct | Also correct |
> |---|---|
> | $\dfrac{2 \cdot 3}{4}$ | $\dfrac{\cancel{2} \cdot 3}{^2\cancel{4}}$ |
> | $= \dfrac{6}{4}$ vs. $= \dfrac{3}{2}$ |
> | $= 1.5$ | $= 1.5$ |

PRACTICE PROBLEM FOR EXAMPLE 8

Solve each literal equation for the indicated variable.

a. Volume of a rectangular solid: $V = lwh$ solve for w.

b. Area of a trapezoid: $A = \dfrac{1}{2}h(b_1 + b_2)$ solve for h.

Example 9 — Solving more literal equations for a variable

Solve each equation for the indicated variable.

a. $-6x + 3y = -18$ for y.

b. $y - 2 = \dfrac{2}{3}(6x - 9)$ for x.

SECTION 2.2 Multiplication and Division Properties of Equality

SOLUTION

a. To solve the equation for y, isolate the variable y on one side of the equal sign.

$$-6x + 3y = -18 \quad \text{Isolate } y \text{ by adding } 6x \text{ to both sides.}$$
$$\underline{+6x \qquad\qquad +6x}$$
$$3y = -18 + 6x \quad \text{Rewrite with the } x\text{-term first.}$$
$$3y = 6x - 18$$
$$\frac{3y}{3} = \frac{6x}{3} - \frac{18}{3} \quad \text{Divide each term by 3 to solve for } y.$$
$$y = 2x - 6$$

b. To solve this equation for x, first multiply out the right side.

$$y - 2 = \frac{2}{3}(6x - 9) \quad \text{Multiply out the right side.}$$
$$y - 2 = \frac{2}{3}(\overset{2}{6}x) - \frac{2}{3}(\overset{3}{9}) \quad \text{Divide out like factors.}$$
$$y - 2 = 4x - 6$$
$$\underline{+6 \qquad +6} \quad \text{Add 6 to both sides to isolate the } x\text{-term.}$$
$$y + 4 = 4x$$
$$4x = y + 4 \quad \text{Rewrite the equation with the } x\text{-term on the left.}$$
$$\frac{4x}{4} = \frac{y}{4} + \frac{4}{4} \quad \text{Divide each term by 4 to solve for } x.$$
$$x = \frac{y}{4} + 1$$

PRACTICE PROBLEM FOR EXAMPLE 9

Solve each equation for the indicated variable.

a. $-5x + 15y = -45$ for y.

b. $y + 1 = \frac{4}{5}(-10x + 5)$ for x.

2.2 Vocabulary Exercises

1. The _____ property of equality allows us to multiply both sides of an equation by a nonzero number.

2. The _____ property of equality allows us to divide both sides of an equation by a nonzero number.

3. The phrase "x is four" translates as _____.

4. The phrase "five more than a number" translates as _____.

2.2 Exercises

For Exercises 1 through 22, solve each equation and check the answer. If the problem has decimals in it, then answer in decimal form. Otherwise, write any fractional answers as reduced fractions.

1. $3x = -81$
2. $4x = -48$
3. $-10y = -55$
4. $-6y = -27$
5. $7g = 91$
6. $14k = 84$
7. $-x = -9$
8. $17 = -n$
9. $42 = 2.8x$
10. $3.7m = 31.45$
11. $15.25 = 6.1n$
12. $49.68 = 5.4n$
13. $-6c = 42$
14. $-4x = 45$
15. $\dfrac{t}{5} = 14$
16. $\dfrac{p}{3} = 9$
17. $\dfrac{2}{3}x = -10$
18. $-\dfrac{4}{5}x = 8$
19. $\dfrac{5n}{7} = -\dfrac{2}{5}$
20. $\dfrac{3n}{4} = -\dfrac{9}{5}$
21. $-\dfrac{k}{3.5} = -4$
22. $-\dfrac{g}{7} = 4.6$

23. Victoria works part time as a legal assistant and makes $17 per hour after taxes. Victoria can calculate her monthly salary using the equation:
$$S = 17h$$
where S is the monthly salary in dollars that Victoria makes if she works h hours that month.

   a. Find Victoria's monthly salary if she works 80 hours.
   b. How many hours does Victoria have to work a month to make $2040 a month?
   c. How many hours does Victoria have to work a month if she only wants to earn $700 a month?

24. Emily is taking a driving trip across the country to go to college and plans to average a speed of about 60 miles per hour. The distance Emily can go in a day can be calculated by using the equation
$$d = 60t$$
where d is the distance in miles Emily can travel in a day that she drives for t hours.

   a. How far will Emily travel in a day if she drives for 12 hours?
   b. If Emily wants to drive about 450 miles a day, how many hours will she have to drive?
   c. How many hours will Emily have to drive to make it the approximately 2450 miles from Charlotte, North Carolina, to San Diego, California?

25. Pablo works for a discount mortgage broker who advertises home loans that cost only 2.0 points (2.0%) of the loan amount. No other fees will be charged. When a customer calls, Pablo uses the equation
$$F = 0.02L$$
where F is the fees in dollars the customer will be charged and L is the loan amount in dollars requested by the customer.

   a. If a customer calls and requests a loan for $250,000, how much will the fees be?
   b. If Pablo wants to keep the fees for a customer down to only $4000, what is the maximum loan amount he can suggest?

26. Rachel is applying for a home loan and wants to calculate the fees she will pay. The mortgage broker she is working with says that Rachel will pay 2.5 percent (2.5 points) of the amount of the loan in fees. With this information, Rachel can calculate the fees for her home loan using the equation
$$F = 0.025L$$
where F is the fee in dollars paid for a home loan of L dollars.

© Andy Dean Photography/Shutterstock.com.

   a. Find the fees Rachel will pay if her home loan is for $300,000.
   b. If Rachel has $5000 she is willing to pay in fees for her home loan, what is the most she can borrow from this broker?

27. The LEGO Group requires such accuracy in its production process that only 18 elements in every million produced do not meet the company's high quality standards. The number of elements that do not meet the quality standards of the company can be represented by the equation
$$D = 18b$$
where D is the number of bricks that do not meet the quality standards when b million bricks are produced.
Source: LEGO Company profile 2005. www.lego.com

a. If 500 million bricks are produced, how many will not meet the quality standards?

b. If LEGO makes about 40.8 million bricks a day, how many of those do not meet the quality standards?

c. If 21,600 bricks do not meet the company's standards, how many bricks did Lego produce?

28. The LEGO Group makes LEGO toys around the world. They make approximately 2.16 million items an hour 24 hours a day, 365 days a year. That means about 19 billion LEGO bricks and other components are made every year. A certain machine can make 50 bricks every 7 seconds. The number of bricks B that this machine can produce after m minutes can be represented by the equation

$$B = 429m$$

Source: LEGO Company profile 2005. www.lego.com

a. How many bricks can this machine make in 1 minute?

b. How many bricks can this machine make in 1 hour?

c. If the company has an order for 15 million bricks, how long will it take this machine to produce them?

29. If the average bus can seat 60 people, then the number of buses needed to transport a group of people can be calculated by using the formula

$$B = \frac{p}{60}$$

where B is the number of buses needed to transport a group of p people.

a. How many buses are needed to transport 300 people?

b. If a city has 350 buses, how many people can public transport handle at a time?

c. If a city has 600 buses, how many people can the city transport at a time?

30. The "rated maximum capacity" of a school bus seat gives the number of people who can sit on the seat. It is determined by dividing the width of the seat by the number 13. Using this calculation, the seat capacity can be determined by using the formula

$$C = \frac{w}{13}$$

where C is the rated maximum capacity in number of students when the bus seat is w inches wide.

Source: National Association of State Directors of Pupil Transportation Services. http://www.nasdpts.org/paperSeatCap.html

a. Use this formula to determine the rated maximum capacity of a bus seat that is 39 inches wide.

b. What is the smallest width a bus seat would have to be to have a maximum capacity of two people?

c. What is the smallest width a bus seat would have to be to have a maximum capacity of four people?

d. If a bus has 24 seats in it that are 39 inches wide, what is the total maximum capacity for the bus?

31. The amount of mortgage debt for multifamily homes in the United States can be found by using the formula

$$M = 75270.9t + 712058.6$$

where M is the amount in millions of dollars of multifamily home mortgages taken out in the United States during the tth year since 2010. (2010 is $t = 0$, 2011 is $t = 1$, 2015 is $t = 5$ etc.)

Source: www.federalreserve.gov

a. Find the total amount of mortgage debt for multifamily homes in the United States taken out in the year 2018.

b. Find what year the total amount of mortgage debt for multifamily homes in the United States was $1,464,767.6 million.

32. An online store shows the following chart for shipping charges within the United States.

Option	Delivery	Tracking	Flat Fee	Per Item
USPS	3–7 days	No	$2.99	$0.99
UPS Ground	3–5 days	Yes	$6.99	$1.99
UPS 2nd Day	2 days	Yes	$13.98	$0.00

The total cost to ship items from this store using the U.S. Postal Service (USPS) option can be found by using the formula

$$C = 2.99 + 0.99n$$

where C is the total cost in dollars for shipping n items from the online store using the USPS.

a. Find the cost to ship five items via USPS.

b. If the total shipping cost through the USPS is $5.96, how many items were shipped?

c. If the total shipping cost through the USPS is $12.89, how many items were shipped?

33. Business or industrial capital equipment includes items such as tractors, farm machinery, and manufacturing equipment. When business or industrial capital equipment is sold on eBay, the seller is charged an insertion fee of $20 for posting the item and a final value fee of 8% of the final price of the item. The total fees can be calculated by using the formula

$$f = 0.08p + 20$$

where f is the total fee in dollars charged by eBay for selling a business or industrial capital equipment item for a final price of p dollars.

Source: www.ebay.com

a. What are the total fees if the final price of a John Deere tractor is $17,900?

b. If someone was charged a total fee of $104 for selling a log splitter, what was the final price of the item on eBay?

c. If a company was charged a total fee of $1100 for selling a Kubota tractor, what was the final price for this item on eBay?

34. Fidelity Investments charges online customers commissions for buying stock options using the following formula:

$$C = 4.95 + 0.65n$$

where C is the commission in dollars when a customer orders n options contracts.

Source: www.Fidelity.com

a. Find the commission Fidelity Investments will charge if the customer orders 20 options contracts online.

b. If the total commission a customer paid for an online options order was $102.45, how many contracts did the customer order?

c. If the total commission a customer paid for an online options order was $66.70, how many contracts did the customer order?

35. When using Amazon payments service to process payments, zShops pays Amazon 0.30 plus 2.9% of the sale price, including shipping and handling. The total fee paid by the zShop for an Amazon payments transaction can be calculated by using the formula

$$F = 0.30 + 0.029p$$

where F is the fee paid by the zShop to Amazon when a transaction of p dollars is charged using the Amazon payments option.

Source: www.Amazon.com

a. What is the fee for a $100 transaction?

b. If the zShop is charged a $10.74 fee for an Amazon payments transaction, how much was the total transaction?

c. If the zShop is charged a $43.80 fee for an Amazon payments transaction, how much was the total transaction?

36. Maria is a realtor who charges a flat fee of $1500 plus 1.5% of the sales price of the home. The total commission that Maria earns on the sale of a home can be calculated by using the formula

$$C = 1500 + 0.015p$$

where C is the total commission in dollars from selling a home for a sales price of p dollars.

a. Find the commission on a home that sold for $325,000.

b. If the total commission paid was $5625, what was the sales price of the home?

c. If the total commission paid was $7875, what was the sales price of the home?

For Exercises 37 and 38, each linear equation is being solved. Complete the tables by filling in the missing algebraic steps to solve each equation and supply the missing reasons for the steps.

37. $-2.5c + 15 = -5$

Algebraic Step to Solve the Equation	Reason for Each Step
$-2.5c + 15 = -5$ $\underline{ -15 -15}$	
	Isolate the variable by using the division property of equality.

38. $-6.5h - 64 = -93.25$

Algebraic Step to Solve the Equation	Reason for Each Step
	Isolate the variable term by using the addition property of equality.
$-6.5h = -29.25$ $\dfrac{-6.5h}{-6.5} = \dfrac{-29.25}{-6.5}$ $h = 4.5$	

For Exercises 39 through 46, solve each equation and check the answer. Provide reasons for each step. Reduce all fractions. For problems given with decimals, round to two decimal places, if necessary.

39. $5x + 20 = 60$
40. $8m + 34 = 154$
41. $-3p + 50 = -85$
42. $-2q + 25 = -45$
43. $-2.3t + 20 = 26.9$
44. $21.2 = 4.8k + 50$
45. $20.1 = \dfrac{w}{8} + 17$
46. $-8.75 = -14 - \dfrac{b}{2}$

For Exercises 47 through 64, solve each equation and check the answer. Reduce all fractions. For problems given with decimals, round to two decimal places if necessary.

47. $48 = 12m - 30$
48. $-20 = 8d - 56$
49. $-251.8 = -8.6h + 14.8$
50. $-46.8 = -2.3y - 12.3$
51. $-134.0 = -4.7x - 16.5$
52. $4v - 52.2 = -3$
53. $-2n + 3.8 = 4.85$
54. $-5 + 2.5m = -16.25$
55. $7m + 8 = 31.8$
56. $6k - 56 = -11.6$
57. $-5z + 16 = 32$
58. $-4x + 26 = 84$
59. $\dfrac{x}{4} + 6 = 9$
60. $\dfrac{t}{3} + 14 = -8$
61. $-\dfrac{x}{16} + 5 = 2.5$
62. $\dfrac{s}{1.5} - 6 = -3$
63. $\dfrac{2x}{3} + \dfrac{1}{5} = \dfrac{4}{5}$
64. $-\dfrac{5x}{7} - \dfrac{2}{3} = \dfrac{5}{3}$

65. In Exercise 32, we were told that the online store charges a flat fee of $6.99 and an additional $1.99 per item to ship using UPS Ground.

  a. Define the variables. Write an equation for the cost to ship items purchased using UPS ground.

  b. What does it cost to ship 5 items UPS Ground?

  c. How many items can be shipped UPS Ground for $12.96?

66. Vanguard charges its Voyager Services customers commissions of $7.00 plus $1.00 per option contract for all option orders.

  Source: https://investor.vanguard.com

  a. Define the variables. Write an equation for the total commission when buying option contracts.

  b. Find the commission a Voyager Services customer will be charged if the customer orders 20 option contracts.

  c. If a Voyager Services customer is charged $92.00 in commissions for an online options order, how many option contracts did she order?

67. A local machine shop gets an order from a gardening supply manufacturer to create a crank for a portable composter. The production foreman determines that it will take about 6 hours to set up the manufacturing process and about 5 minutes per crank to manufacture.

  a. Define the variables. Write an equation for the total time in *minutes* it will take to set up and produce the cranks.

  b. How long will it take to set up and produce 50 cranks?

  c. If it takes the machine shop 1175 minutes to produce the cranks, how many did they make?

68. Richard copies EPs (Extended Play) for local bands. Each EP that he burns takes 12 minutes to create. The total time it takes for Richard to process an order for EPs depends on how many are requested.

  a. Define the variables. Write an equation for the total time it takes Richard to process an order of EPs.

  b. How long will it take Richard to create 250 EPs?

  c. If Richard has 4 hours today to spend creating EPs, how many can he get done? (*Hint*: Change 4 hours into minutes first.)

  d. If Richard spends 6 hours a day creating EPs and works 5 days a week, how many EPs can he make in a week?

For Exercises 69 through 86, define variables and translate each sentence into an equation. Then solve that equation. Check each answer for accuracy and reasonableness.

69. Three times a number is 45.

70. The sum of 20 and 4 times a number is equal to 44.

71. The perimeter of a rectangular garden is 56 feet, and its width is 8 feet. Find the length of the garden.

72. The perimeter of a volleyball court is 180 feet, and its width is 30 feet. Find the length of the volleyball court.

73. The area of a triangle is 40 square inches, and its base is 8 inches. Find the height of the triangle.

74. The area of a triangle is 75 square inches, and its height is 15 inches. Find the base of the triangle.

75. The quotient of a number and 7 is 13.

76. The quotient of a number and 5 added to 12 is 19.2.

77. Six plus the product of 7 and a number is equal to 62.

78. Four times the sum of a number and 3 is equal to 68.

79. One third of a number minus 8 is 28.

80. Half a number plus 18 is 15.
81. The difference between twice a number and 15 is 40.
82. The difference between three times a number and 7 is 17.
83. An unknown number subtracted from 8 is 19.
84. An unknown number subtracted from 36 is -5.
85. Seven less than twice a number is 11.
86. 8 less than three times a number is -23.

For Exercises 87 through 106, solve each literal equation for the indicated variable.

87. Perimeter of a rectangle: $P = 2l + 2w$ for l.
88. Circumference of a circle: $C = 2\pi r$ for r.
89. Volume of a cylinder: $V = \pi r^2 h$ for h.
90. Volume of a cone: $V = \frac{1}{3}\pi r^2 h$ for h.
91. Area of a triangle: $A = \frac{1}{2}bh$ for b.
92. Area of a trapezoid: $A = \frac{1}{2}h(b_1 + b_2)$ for h.
93. $W = 3t - 20$ for t.
94. $H = rt + 50$ for r.
95. $P = 2b + 2B$ for B.
96. $V = lwh$ for w.
97. $5x - 3y = -15$ for y.
98. $-3x + 4y = 24$ for y.
99. $x - 5y = 9$ for y.
100. $\frac{1}{3}x + 6y = -4$ for y.
101. $y - 2 = 3(x + 7)$ for x.
102. $y - 3 = -2(x - 5)$ for x.
103. $y + 1 = \frac{3}{4}(12x + 8)$ for x.
104. $y - 6 = \frac{1}{6}(12x + 6)$ for x.
105. Simple interest: $I = Prt$ for r.
106. Simple interest: $I = Prt$ for t.

For Exercises 107 through 116, identify each problem as an expression or equation. Simplify any expressions, and solve any equations.

107. $5a^2 + 6a - 8 + 12a + 3$
108. $10m^2 - 3n + 10 - 4m^2 + 4$
109. $2x + 7 = 21$
110. $-6x + 14 = -46$
111. $\frac{1}{3}r + 8 - \frac{5}{6}r = -2$
112. $\frac{3}{10}d + 15 - \frac{4}{5}d = -8$
113. $5n + 7 - 8n + 20m + 4$
114. $8w - 6 + 4v - 14w + 23$
115. $3.4y + 4 - 2.1y = 11.28$
116. $-4.6h - 8.7 + 1.2h = -32.84$

2.3 Solving Equations with Variables on Both Sides

LEARNING OBJECTIVES
- Solve equations with variables on both sides.
- Solve equations that contain fractions.
- Recognize equations that are identities or have no solution.
- Translate sentences into equations and solve.

Solving Equations with Variables on Both Sides

Some equations have the same variable appearing in terms on both sides of the equal sign. Simplify each side of the equation first. If a term has to be moved to the other side of the equation, use the addition or subtraction property of equality to move it. Then combine the like variable terms so that we can solve for the variable.

> **Steps to Solving Equations with Variables on Both Sides**
> 1. Simplify each side of the equation independently by performing any arithmetic and combining any like terms.
> 2. Move the variable terms to one side of the equation by using the addition or subtraction property of equality. Combine like terms.
> 3. Isolate the variable term by using the addition or subtraction property of equality.
> 4. Isolate the variable by using the multiplication or division property of equality.
> 5. Check the answer in the original problem.

Example 1 Solving equations with variables on both sides

Solve each equation. Check the answer.

a. $3(4x - 7) = 7x - 9 - x$ **b.** $0.25(3t + 6) = 1.25t - 3$
c. $2m + 15 - 3m = 28$

SOLUTION

a. Step 1 Simplify each side of the equation independently by performing any arithmetic and combining like terms.

$3(4x - 7) = 7x - 9 - x$ Distribute 3 and simplify both sides.
$12x - 21 = 6x - 9$

Step 2 Move the variable terms to one side of the equation by using the addition or subtraction property of equality. Combine like terms.

$12x - 21 = 6x - 9$ Combine like variable terms by subtracting 6x from both sides.
$\underline{-6x \qquad -6x}$
$6x - 21 = -9$

Skill Connection

Answer as a fraction or decimal.
A problem that is given with fractions is answered using fractions.
A problem that is given with decimals is answered using decimals.

Step 3 Isolate the variable term by using the addition or subtraction property of equality.

$$6x - 21 = -9$$
$$\underline{+21 +21}$$
$$6x = 12$$

Isolate the variable term by adding 21 to both sides.

Step 4 Isolate the variable by using the multiplication or division property of equality.

$$6x = 12$$
$$\frac{6x}{6} = \frac{12}{6}$$
$$x = 2$$

Isolate x by dividing both sides by 6.

Step 5 Check the answer in the original equation.

$$3(4(2) - 7) \stackrel{?}{=} 7(2) - 9 - (2)$$ Check the answer.
$$3(8 - 7) \stackrel{?}{=} 14 - 9 - 2$$
$$3(1) \stackrel{?}{=} 5 - 2$$
$$3 = 3$$ The answer checks.

b.
$$0.25(3t + 6) = 1.25t - 3$$ Simplify the left side by distributing 0.25.
$$0.75t + 1.5 = 1.25t - 3$$
$$\underline{-0.75t -0.75t}$$ Combine the like variable terms by subtracting 0.75t from both sides.
$$1.5 = 0.5t - 3$$
$$\underline{+3 +3}$$ Isolate the variable term by adding 3 to both sides.
$$4.5 = 0.5t$$
$$\frac{4.5}{0.5} = \frac{0.5t}{0.5}$$ Isolate the variable by dividing both sides by 0.5.
$$9 = t$$

$$0.25(3(9) + 6) \stackrel{?}{=} 1.25(9) - 3$$ Check the answer.
$$0.25(27 + 6) \stackrel{?}{=} 11.25 - 3$$
$$0.25(33) \stackrel{?}{=} 8.25$$
$$8.25 = 8.25$$ The answer checks.

c.
$$2m + 15 - 3m = 28$$ Combine like terms to simplify the left side.
$$-m + 15 = 28$$ Isolate the variable by subtracting 15 from both sides.
$$\underline{-15 -15}$$
$$-m = 13$$
$$\frac{-m}{-1} = \frac{13}{-1}$$ Isolate the variable by dividing both sides by -1.
$$m = -13$$

$$2(-13) + 15 - 3(-13) \stackrel{?}{=} 28$$ Check the answer.
$$-26 + 15 + 39 \stackrel{?}{=} 28$$
$$28 = 28$$ The answer checks.

PRACTICE PROBLEM FOR EXAMPLE 1

Solve each equation. Check the answers.

a. $4x + 80 = 2x + 40$ **b.** $2.5(6x + 8) = -4x + 59$ **c.** $20 = 7k - 9 - 8k$

Connecting the Concepts

What is the coefficient?

When working with variable terms, remember that the coefficient is the number in front of the variables.

$$2x$$

x has a coefficient of 2. If no number is written in front of the variable, then the coefficient is assumed to be 1.

$$x = 1 \cdot x$$

x has a coefficient of 1. When a term has a negative sign in front, the coefficient is -1.
x has a coefficient of -1.

$$-x = -1 \cdot x$$

When solving, if the coefficient is -1, we can isolate the variable by multiplying or dividing both sides by negative one. Be sure to watch the signs.

SECTION 2.3 Solving Equations with Variables on Both Sides 191

■ Solving Equations That Contain Fractions

When an equation involves fractions, we can approach the problem in a few ways. One way to solve an equation involving fractions is to use a common denominator whenever adding or subtracting fractions. Another way to solve an equation with fractions is to clear the equation of fractions at the beginning of the problem. This is done by multiplying both sides of the equation by a common denominator and using the distributive property. This results in a problem without fractions that can be solved. Either method is appropriate to use.

Example 2 Solving equations with fractions

Solve each equation. Check the answers.

a. $\dfrac{2}{3}x + \dfrac{1}{3} = 5$ **b.** $\dfrac{3}{5}k + 7 = 8 - \dfrac{1}{5}k$ **c.** $\dfrac{1}{3}(m + 4) = \dfrac{2}{5}m - 4$

SOLUTION

a. This equation starts with fractions that have a denominator of 3. Eliminate the fractions in the problem by multiplying both sides of the equation by 3.

$\dfrac{2}{3}x + \dfrac{1}{3} = 5$ Clear the fractions by multiplying both sides by the common denominator 3.

$3\left(\dfrac{2}{3}x + \dfrac{1}{3}\right) = 3(5)$

$3\left(\dfrac{2}{3}x\right) + 3\left(\dfrac{1}{3}\right) = 15$ Distribute the 3 and simplify.

$2x + 1 = 15$

$\quad\quad -1 \ -1$ Isolate the variable term by subtracting 1 from both sides.

$2x = 14$

$\dfrac{2x}{2} = \dfrac{14}{2}$ Isolate the variable by dividing both sides by 2.

$x = 7$

$\dfrac{2}{3}(7) + \dfrac{1}{3} \stackrel{?}{=} 5$ Check the answer.

$\dfrac{14}{3} + \dfrac{1}{3} \stackrel{?}{=} 5$

$\dfrac{15}{3} \stackrel{?}{=} 5$

$5 = 5$ The answer checks.

b. This equation has fractions, but this time, we leave the fractions and solve the equation.

$\dfrac{3}{5}k + 7 = 8 - \dfrac{1}{5}k$ Combine the like variable terms by adding $\dfrac{1}{5}k$ to both sides.

$+\dfrac{1}{5}k \quad\quad\quad +\dfrac{1}{5}k$

$\dfrac{4}{5}k + 7 = 8$ Isolate the variable term by subtracting 7 from both sides.

$\quad\quad -7 \ -7$

$\dfrac{4}{5}k = 1$

$$\frac{5}{4}\left(\frac{4}{5}k\right) = \frac{5}{4}(1)$$ Isolate k by multiplying both sides by $\frac{5}{4}$ (the reciprocal).

$$k = \frac{5}{4} = 1\frac{1}{4}$$ Write the answer as either a reduced improper fraction or a mixed number.

$$\frac{3}{5}\left(\frac{5}{4}\right) + 7 \stackrel{?}{=} 8 - \frac{1}{5}\left(\frac{5}{4}\right)$$ Check the answer.

$$\frac{3}{4} + 7 \stackrel{?}{=} 8 - \frac{1}{4}$$

$$\frac{3}{4} + \frac{28}{4} \stackrel{?}{=} \frac{32}{4} - \frac{1}{4}$$ Find common denominators.

$$\frac{31}{4} = \frac{31}{4}$$ The answer checks.

c. This equation starts with fractions. First simplify both sides and then clear the fractions by multiplying by the common denominator.

$$\frac{1}{3}(m + 4) = \frac{2}{5}m - 4$$ Simplify the left side by distributing the $\frac{1}{3}$.

$$\frac{1}{3}m + \frac{4}{3} = \frac{2}{5}m - 4$$

$$15\left(\frac{1}{3}m + \frac{4}{3}\right) = 15\left(\frac{2}{5}m - 4\right)$$ Multiply both sides by the common denominator 15.

$$\overset{5}{15}\left(\frac{1}{3}m\right) + \overset{5}{15}\left(\frac{4}{3}\right) = \overset{3}{15}\left(\frac{2}{5}m\right) - 15(4)$$ Distribute and simplify.

$$5m + 20 = 6m - 60$$ Combine the variable terms by subtracting $5m$ from both sides.
$$\underline{-5m \qquad\qquad -5m}$$
$$20 = m - 60$$
$$\underline{+60 \qquad +60}$$ Isolate the variable by adding 60 to both sides.
$$80 = m$$

$$\frac{1}{3}((80) + 4) \stackrel{?}{=} \frac{2}{5}(80) - 4$$ Check the answer in the original equation.

$$\frac{1}{3}(84) \stackrel{?}{=} 32 - 4$$

$$28 = 28$$ The answer checks.

PRACTICE PROBLEM FOR EXAMPLE 2
Solve each equation. Check the answers.

a. $\frac{3}{4}x + 7 = \frac{1}{4}x - 5$ **b.** $\frac{1}{2}(c + 3) = \frac{2}{5}c - \frac{3}{4}$

■ Equations That Are Identities or Have No Solution

When solving equations, we may run into two special situations. All of the equations we have solved so far have had one solution. Some equations can have many solutions, while other equations may have no solution at all. Use the following Concept Investigation to discover how these situations come about.

CONCEPT INVESTIGATION

What happened to the answer?

Solve each of the following equations for the variable, if possible. If this is not possible, state whether the equation is a true statement or a false statement.

1. $3 + 4 - x = 3x - 4x + 7$
2. $2x - 1 = 5(x + 1) - 3(x + 2)$
3. $-5x + 4 = -10x + 5x + 1$
4. $\frac{1}{2}(6y) = 3(y + 3)$

This Concept Investigation has revealed that we can get two other situations while solving equations, besides a solution for the variable. We can get an equation that is always true (such as $0 = 0$ or $x = x$), or we can get an equation that is always false (such as $0 = 9$).

The simple equation

$$x = x$$

is true for any value of x.

$$x = x \quad \text{if} \quad x = 5$$
$$5 = 5 \quad \text{Checks.}$$
$$x = x \quad \text{if} \quad x = -29$$
$$-29 = -29 \quad \text{Checks.}$$

so the solution set is all real numbers. An equation that has all real numbers as the solution set is called an **identity**.

> **DEFINITION**
>
> **Identity** An equation that has the solution set all real numbers is called an **identity**. An identity can often be simplified so that both sides are exactly the same (identical).

Some equations have **no solution.** The equation

$$x = x + 2$$

has no solution because there are no numbers that are equal to themselves plus 2. If we solve this equation, it will lead to a false statement.

$$x = x + 2$$
$$\underline{-x \ \ -x} \quad \text{Subtract } x \text{ from both sides of the equation.}$$
$$0 \stackrel{?}{=} 2 \quad \text{This is always a false statement. This equation has no solution.}$$

This equation simplified to a false statement, $0 = 2$, which has no variables left and is not true. The equation has no solutions.

It is common not to recognize at first that an equation is an identity or has no solution. In solving an equation, one of the following situations will occur at the very end.

$x = 3$ This is a solution.

$3 = 3$ This is a true statement without variables, so the solution set will be all real numbers.

$3 \stackrel{?}{=} 7$ This is a false statement without variables, so there is no solution.

What's That Mean?

All Real Numbers and No Solution

There are several ways to state the solution set of an identity or a problem with no solution. All real numbers can be represented by any of the following notations.

$$\mathbb{R}$$
$$(-\infty, \infty)$$
All real numbers

The solution set of an equation that has no solution can be written as

No solution

$\varnothing$ is the empty set

$\{\ \}$ is the empty set

Do not confuse the empty set symbol $\varnothing$ with the number zero. Zero is a number and is a possible solution to an equation. The empty set means that there are no solutions to the equation.

CHAPTER 2 Linear Equations and Inequalities with One Variable

> **Example 3** Recognizing equations that either are identities or have no solution

Solve. If the equation has no solution, write, "No solution." If it is an identity, write, "All real numbers."

a. $3x + 15 = 3(x + 6) - 3$ **b.** $4x + 6 - x = 2x + 5 + x$ **c.** $2x + 5 = 5$

SOLUTION

a. Solve as usual.

$3x + 15 = 3(x + 6) - 3$ Use the distributive property.
$3x + 15 = 3x + 18 - 3$
$3x + 15 = 3x + 15$ Combine the variable terms.
$\underline{-3x \qquad\quad -3x}$
$15 = 15$ This is a true statement (identity).

This equation simplifies to an identity, so the solution set is *all real numbers*.

b. $4x + 6 - x = 2x + 5 + x$ Simplify both sides.
$3x + 6 = 3x + 5$
$\underline{-3x \qquad -3x}$
$6 \stackrel{?}{=} 5$ This is a false statement.

This equation simplifies to a false statement, so there are *no solutions* to this equation.

c. $2x + 5 = 5$ Isolate the variable term by subtracting 5 from both sides.
$\underline{-5 \; -5}$
$2x = 0$
$\dfrac{2x}{2} = \dfrac{0}{2}$ Isolate the variable by dividing both sides by 2.
$x = 0$ Zero divided by 2 is zero, so we have a solution.
$2(0) + 5 \stackrel{?}{=} 5$ Check the answer.
$5 = 5$ The answer checks.

PRACTICE PROBLEM FOR EXAMPLE 3

Solve. If the equation has no solution, write, "No solution." If it is an identity, write, "All real numbers."

a. $4(2x + 3) - 8 = 8x + 6$

b. $2(x + 1) - 3 = -1 - 5x$

c. $5x + 7 - 3x = \dfrac{1}{2}(4x + 12) + 1$

Note that an equation with no solution is *not* the same as having the number zero as a solution. For instance, $x = 0$ can be a solution to an equation just as the amount of money in our checking account can be zero. "No solution" means that there are no values of the variable that are a solution to the equation.

SECTION 2.3 Solving Equations with Variables on Both Sides   195

■ Translating Sentences into Equations and Solving

Example 4 Translating and solving

Translate each of the following sentences into an equation and solve. Check the answers.
a. Three times a number is equal to twice the sum of the number and 5.
b. Half a number is equal to the number minus 14.

SOLUTION

Let x = a number.

a. "Three times a number" translates as $3x$. "Twice" means to multiply by two, but "twice the sum" means to multiply a sum by 2, so put the sum in parentheses. Using these translations yields the equation

$$3x = 2(x + 5)$$

Now solve this equation.

$3x = 2(x + 5)$ Distribute the 2 to simplify the right side.
$3x = 2x + 10$
$\underline{-2x \quad -2x}$ Combine the like variable terms
$x = 10$ by subtracting 2x from both sides.

$3(10) \stackrel{?}{=} 2((10) + 5)$ Check the answer.
$30 = 30$ The answer checks.

b. "Half a number" translates as $\frac{1}{2}x$, and "the number minus 14" means $x - 14$. Translating yields the equation

$$\frac{1}{2}x = x - 14$$

Now solve the equation.

$\frac{1}{2}x = x - 14$

$2\left(\frac{1}{2}x\right) = 2(x - 14)$ Clear the fraction by multiplying both sides by 2.
$x = 2x - 28$ Combine the like variable terms by subtracting
$\underline{-2x \quad -2x}$ 2x from both sides.
$-x = -28$
$-1(-x) = -1(-28)$ Multiply both sides by −1 to remove the
$x = 28$ negative in front of x and isolate the variable.

$\frac{1}{2}(28) \stackrel{?}{=} (28) - 14$ Check the answer.
$14 = 14$ The answer checks.

PRACTICE PROBLEM FOR EXAMPLE 4

Translate each of the following sentences into equations and solve. Check the answers.
a. $\frac{2}{3}$ of a number plus 8 equals -2 times the number plus 40.
b. A number plus twice the difference between the number and 10 is equal to 5 times the number.

Geometry is a subject in which many formulas and relationships are found that are used in everyday life. People in construction and manufacturing use many volume, area, and perimeter formulas to calculate costs and material needs. Architects and designers use some of the basic relationships from geometry to calculate and check different shapes that they design. We see area when buying tables (square feet) or volume when considering the capacity of a refrigerator (cubic feet). Most of the formulas we will use are found at the back of the book.

Example 5 Using a geometric formula

Given the following triangle, find the measure of each angle.

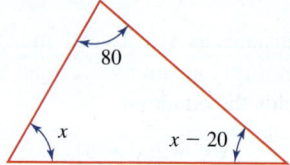

SOLUTION
The measures of the angles in a triangle must sum to 180 degrees. This yields the equation

$$(x) + (80) + (x - 20) = 180$$

Solve this equation.

$$x + 80 + x - 20 = 180 \quad \text{Simplify each side independently.}$$
$$2x + 60 = 180 \quad \text{Isolate the variable term by subtracting 60 from both sides.}$$
$$\underline{-60 \quad -60}$$
$$2x = 120$$
$$\frac{2x}{2} = \frac{120}{2} \quad \text{Isolate } x \text{ by dividing both sides by 2.}$$
$$x = 60$$
$$(60) + 80 + (60) - 20 \stackrel{?}{=} 180 \quad \text{Check the answer.}$$
$$180 = 180 \quad \text{The answer checks.}$$

Since $x = 60$, the angles measure 80°, 60°, and 60° − 20° = 40°.

PRACTICE PROBLEM FOR EXAMPLE 5

Given the following triangle, find the measure of each angle.

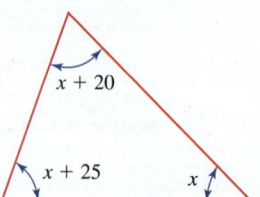

In some problems, it might seem that more than one variable should be defined. Because we have not introduced the process to solve for more than one variable, we will represent each problem using only one variable. Look at each problem carefully and determine what missing quantity is used to describe the other missing quantities. Be careful not to define variables for quantities that are already given in the problem. For example, if the problem states that gas costs $3.25 per gallon, don't define a variable for the cost of gas. Simply use 3.25 in that equation or calculation.

Example 6 Buying supplies

Amy, an elementary school teacher, wants to order supplies for her classroom. She plans to get each student an art kit each filled with a box of crayons, four colored pencils, and a watercolor paint set. It costs $2.50 for a box of crayons, $0.26 for each colored pencil, and $6.00 for each watercolor paint set. Amy wants to purchase the supplies to fill an art kit for each student plus three extra boxes of crayons and five extra paint sets. If Amy has a $304.62 budget for this year, for how many students can she purchase kits?

SOLUTION
We first consider how many of each item Amy wants to buy. We should not define different variables for each item. We will instead use the number of students to describe how many of each item she needs to purchase.

Let $s =$ The number of students in Amy's class

Amy should buy the following art supplies:

- $s + 3$ boxes of crayons, one for each student, plus the three extra she wanted
- $4s$ colored pencils, since she wanted four for each student with no extras
- $s + 5$ watercolor paint sets, one for each student and five extras

Now we multiply each of these expressions by how much they cost and then add those costs together to get the total of $304.62. Translating will give us the following costs as terms and then an equation.

- $2.50(s + 3)$ dollars for the crayons
- $0.26(4s)$ dollars for the colored pencils
- $6(s + 5)$ dollars for the watercolor paint sets
- The total cost will be all these costs added together. In this case, we have to get a total of 304.62, Amy's budget, so we write out the following equation:

$$2.50(s + 3) + 0.26(4s) + 6(s + 5) = 304.62$$

We can solve this equation by first simplifying the left side and then solving for s.

$2.50(s + 3) + 0.26(4s) + 6(s + 5) = 304.62$	Distribute to simplify.
$2.50s + 7.50 + 1.04s + 6s + 30 = 304.62$	
$9.54s + 37.50 = 304.62$	Isolate the variable term by subtracting 37.50 from both sides.
$ -37.50 -37.50$	
$9.54s = 267.12$	
$\dfrac{9.54s}{9.54} = \dfrac{267.12}{9.54}$	Isolate s by dividing both sides by 9.54.
$s = 28$	
$2.50((28) + 3) + 0.26(4(28)) + 6((28) + 5) \stackrel{?}{=} 304.62$	Check the answer.
$2.50(31) + 0.26(112) + 6(33) \stackrel{?}{=} 304.62$	
$304.62 = 304.62$	The answer checks.

With a budget of $304.62, Amy can purchase art kits for 28 students.

In Example 6, it is important to attack or approach the problem in parts so that we do not become overwhelmed by the amount of information presented. Read the problem carefully, and break it up into pieces to help build the equation, and then solve it.

PRACTICE PROBLEM FOR EXAMPLE 6

The Clarks are throwing a party and plan to serve their famous ribs, but they must stay within their budget of $500.00. For each person they invite, they will buy ribs, drinks, and paper goods. From past parties, the Clarks know that they will need about two drinks and an average of 1.5 sets of paper goods per person (the ribs can be spicy and messy). The ribs cost $5 per person. Each drink costs $0.50. Each set of paper goods costs $0.20. If they spend their entire budget, how many people can attend the Clarks' party?

2.3 Vocabulary Exercises

1. An equation that is a(n) _____ simplifies to a statement such as $5 = 5$.

2. The solution set to an equation that simplifies to a statement such as $5 = 0$ is _____.

2.3 Exercises

For Exercises 1 through 16, translate each sentence and solve. Check the answer.

1. Four times a number is equal to the number plus 33.

2. A number added to eight is equal to five times the number.

3. 5 times a number plus 8 is equal to 3 times the number minus 4.

4. 7 times a number plus 10 is equal to 4 times the number minus 18.

5. Twice a number is equal to 5 times the sum of the number and 6.

6. The product of 7 and a number is equal to 2 times the difference of the number and 5.

7. Twenty minus 3 times a number is equal to 4 times the number minus 22.

8. Twice the difference between a number and 6 is equal to three times the sum of the number and two.

9. Half of a number is equal to the number plus 5.

10. One third of a number plus 5 is equal to the number minus seven.

11. The quotient of a number and 2 added to 5 is equal to the number minus seven.

12. The quotient of a number and 3 is equal to the sum of five times the number and 28.

13. 8 less than a number is equal to five times the number plus 4.

14. 29 is equal to seven less than twice a number.

15. One fourth of a number minus 8 is equal to twice the number minus 29.

16. Half of a number plus 20 is equal to 3 times the number plus 5.

For Exercises 17 and 18, each linear equation is being solved. Complete the tables by filling in the missing algebraic steps to solve the equations and supply the missing reasons for the steps.

17. $2(x - 7) + 3 = -5(x + 2) - 8$

Algebraic Step to Solve the Equation	Reason for Each Step
$2(x - 7) + 3 = -5(x + 2) - 8$ $2x - 14 + 3 = -5x - 10 - 8$	
	Simplify each side by combining like terms.
$2x - 11 = -5x - 18$ $+5x +5x$	
	Use the addition property of equality to isolate the variable term.
$7x = -7$ $\dfrac{7x}{7} = \dfrac{-7}{7}$ $x = -1$	

18. $\frac{1}{3}x + 2 = \frac{2}{5}x - 4$

Algebraic Step to Solve the Equation	Reason for Each Step
	Use the multiplication property of equality to multiply by the LCD and clear all fractions.
$5x + 30 = 6x - 60$ $-5x \qquad -5x$	
$30 = x - 60$ $+60 \qquad +60$ $90 = x$	

For Exercises 19 through 30, solve each equation. Provide reasons for each step. Check the answer. If the equation is an identity, write the answer as "all real numbers." If there is no solution, write "no solution."

19. $3x + 12 = 7x - 28$ **20.** $6x + 20 = 10x - 36$

21. $12t - 50 = 4t + 14$ **22.** $7m - 30 = 3m - 10$

23. $4b + 36 = 9b + 66$ **24.** $15p - 20 = 4p - 64$

25. $2c + 5 + c = 3c - 4$ **26.** $5f - 7 = 8f + 2 - 3f$

27. $3(2x + 5) - 7 = 2(4x - 6)$

28. $5(x - 3) - 9 = 6(2x - 8) + 3$

29. $\frac{2}{3}x + 4 = 5x - 22$ **30.** $\frac{1}{4}x - 6 = \frac{2}{7}x - 5$

For Exercises 31 through 52, solve each equation. Check the answer. If the equation is an identity, write the answer as "all real numbers." If there is no solution, write "no solution."

31. $4t + 5 = 3t + 5 + t$

32. $2z + 8 - 5z = -3z + 8$

33. $3.5h = 2h - 12$ **34.** $4.8d = 20 = 7.3d$

35. $15 - x = 8$ **36.** $8x + 5 - 9x = 20$

37. $3h + 8 - 4h = 13$ **38.** $14 = 5m - 8 - 6m$

39. $2(3x + 5) = 4x + 22$ **40.** $3(8x - 9) = 30x - 99$

41. $5(2c + 3) - 8 = 3(4c + 2) - 2c$

42. $2(6w - 5) = 3(4w - 2) - 7$

43. $7d + 20 - 3d = 9d + 50$

44. $12r - 15 - 5r = 3r - 51$

45. $\frac{1}{2}x + 5 = 3x - 45$ **46.** $\frac{1}{2}z - 4 = 5z - 31$

47. $\frac{1}{3}x + 20 = 2x + 20$ **48.** $\frac{1}{7}h + 8 = 8$

49. $9y - 2.1 = 5y + 3.5$ **50.** $15y + 3.6 = 8y - 11.8$

51. $35t + 10.2 = 5t + 9.9$

52. $3w - 7.9 = -4.75 - 4w$

53. A triangle has angles with measures x, $2x$, and $5x$. Find the measure of each angle.

54. A triangle has angles with measures y, $3y$, and $6y$. Find the measure of each angle.

55. A rectangular yard has sides which measure $3x + 1$ feet and $5x$ feet. The perimeter is 66 feet. Find the length of each side of the yard.

56. A rectangular yard has sides which measure $4x + 2$ feet and $3x$ feet. The perimeter is 88 feet. Find the length of each side of the yard.

57. A square yard has sides which measure $3x - 2$ feet. The perimeter is 83 feet. Find the length of each side of the yard.

58. A square yard has sides which measure $6x - 3$ feet. The perimeter is 60 feet. Find the length of each side of the yard.

59. A rectangular lawn has sides which measure $x - 5$ feet and $4x$ feet. The perimeter is 100 feet. Find the length of each side of the lawn.

60. A rectangular lawn has sides which measure $x - 7$ feet and $3x$ feet. The perimeter is 58 feet. Find the length of each side of the lawn.

61. A triangle has angles with measures x, $2x$, and $2x$. Find the measure of each angle.

62. A triangle has angles with measures x, $2x$, and $3x$. Find the measure of each angle.

For Exercises 63 through 76, solve each equation. Check the answer. If the equation is an identity, write the answer as "all real numbers." If there is no solution, write "no solution."

63. $3(x - 0.5) + 1 = 3x - 0.5$

64. $2.1(x + 2) - 3 = 2.1x + 1.2$

65. $0.3x + 0.3 = 0.5x + 0.1 - 0.2x$

66. $0.9x - 0.2 = 0.5(3x + 1) - 0.6x$

67. $\frac{1}{2}r + 5 = \frac{7}{5}r + 5$ **68.** $\frac{1}{4}w + 50 = 3w + 50$

69. $\frac{3}{5}d - 1 = \frac{1}{10}(6d - 10)$ **70.** $\frac{2}{3}n + 6 = \frac{1}{3}(2n + 18)$

71. $\frac{1}{5}g + 7 = 2 + \frac{2}{5}g$ **72.** $\frac{2}{7}t - 5 = 15 - \frac{3}{7}t$

73. $\frac{1}{2}x + 7 = \frac{2}{5}x + 9$ **74.** $\frac{1}{3}v + 4 = \frac{1}{4}v + \frac{9}{2}$

75. $\frac{1}{2}x + 5 = \frac{1}{2}x + 3$ **76.** $\frac{1}{3}(3x + 6) = 2\left(\frac{1}{2}x + 4\right)$

200 CHAPTER 2 Linear Equations and Inequalities with One Variable

For Exercises 77 through 82, solve each geometry problem.

77. Given the following triangle, find the measure of each angle.

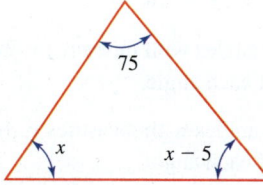

78. Given the following triangle, find the measure of each angle.

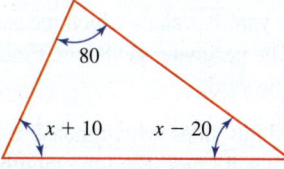

79. Given the following triangle, find the measure of each angle.

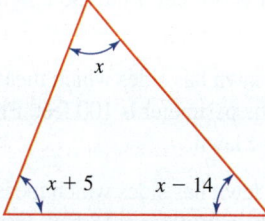

80. Given the following triangle, find the measure of each angle.

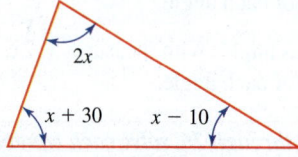

Geometry gives us the following property of quadrilaterals.

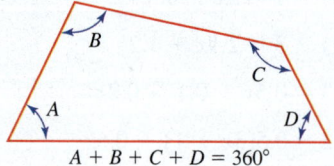

$A + B + C + D = 360°$

The measures of the four angles of a quadrilateral always add up to 360 degrees.

81. Given the following quadrilateral, find the measures of each angle.

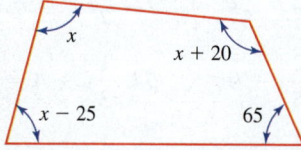

82. Given the following quadrilateral, find the measures of each angle.

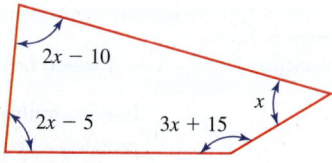

For Exercises 83 through 86, use the drawing of land plots below to answer each application problem. The figure is not drawn to scale.

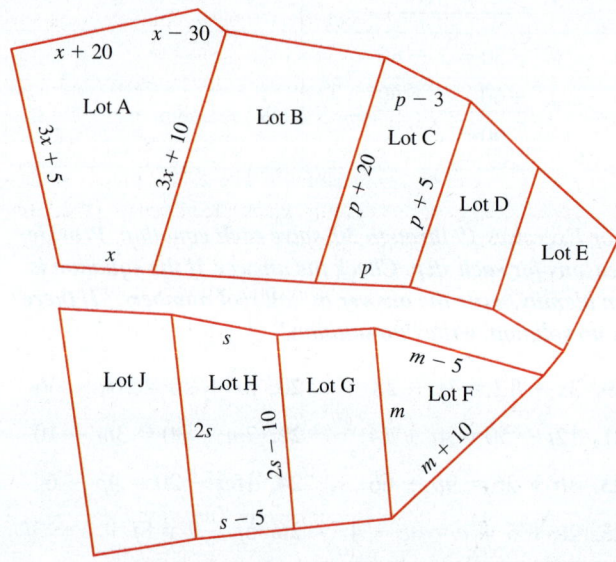

83. The perimeter of lot C above is 342 feet. Find the length of each side.

84. The perimeter of lot F is 335 feet. Find the length of each side.

85. The perimeter of lot H is 345 feet. Find the length of each side.

86. The perimeter of lot A is 500 feet. Find the length of each side.

87. A group of college students is taking a bike tour through Europe this summer. Carlotta is organizing the supplies for the trip. For each person, Carlotta must rent a bike, and she must also buy two water bottles and three food sacks for meals on the road. Carlotta calls the bike tour company and finds that each bike will cost $210 to rent, each water bottle is $2, and each food sack is $8.50. How many students can Carlotta supply if she has $7664.00 to spend?

88. Bill's EduToys manufactures educational toys for kids. For each Alphabet Car that Bill manufactures, he ships a few extra parts to reduce customer requests for replacements. For each Alphabet Car that Bill

manufactures, he wants to order four wheels, two axles, and five hubcaps. Each wheel costs him $1.00, each axle costs $0.39, and each hubcap costs $0.06. If Bill budgets $12,192 a year for these parts, how many Alphabet Cars can he build?

89. The North Oceanside Home Depot is sponsoring a Kids Workshop in which the kids are going to build wood desk calendars for their parents. Each kit requires 1 foot of a pine 2 × 2 to make the month and dates blocks, 1.5 feet of a pine 1 × 3 for the holder, and a set of nails to put it together with. For mistakes that happen, the Kids Workshop manager always wants about 8 feet of extra 2 × 2, an extra 10 feet of 1 × 3, and 12 extra sets of nails. The 2 × 2 pine costs $1.22 per foot, the 1 × 3 pine costs $1.34 per foot, and the sets of nails are $0.05 each.

a. Write an equation for the total amount spent on supplies for these calendars if k kids come to build the project.

b. Use your equation to find the cost if 75 kids come to build the project.

c. If the manager has $417.36 left in his budget, for how many projects can he buy the supplies?

90. Quality Landscaping Co. contracts to put front yard sprinkler systems into each home in a new development. For each system installed in a yard, Quality Landscaping calculates that the supplies required are twelve 10-foot PVC pipes, 9 sprinkler heads, and a set of miscellaneous PVC parts. The project manager decides that he should also order 15 extra 10-foot PVC pipes, 6 extra sprinkler heads, and 2 extra sets of miscellaneous PVC parts. Each 10-foot PVC pipe costs $1.50, each sprinkler costs $2.50, and each set of PVC parts costs $15.

a. Write an equation for the total the landscaping company will pay for the supplies needed to install sprinkler systems in y yards.

b. If the company installs 12 sprinkler systems during the first phase of the development, how much will it cost for the supplies?

c. If the manager budgets $6006 for the entire project, how many sprinkler systems can be installed?

For Exercises 91 through 98, use the following formulas:

$$I = rA$$

where I is the increase, r is the percent increase, and A is the original amount.

$$D = rA$$

where D is the decrease, r is the percent decrease, and A is the original amount.

$$M = rP$$

where M is the markup, r is the markup rate, and P is the original price.

91. A bookstore marks up each book 30% from the wholesale cost. If a book costs the bookstore $120 wholesale, what will the sales price be?

92. A grocery store marks up fruit 3% from the wholesale cost. If a pound of grapes costs the store $1.50 wholesale, what will the sales price be?

93. A store has an advertisement for 35% off the entire store. If the original price of a set of headphones is $20 what will the discount be? What will the sales price be?

94. A store gives military customers 15% off. If a service member comes in to purchase a storage container priced at $22, what will the discount be? What will the final price be?

95. A car is originally purchased for $31,900 and after 1 year, it is worth only $26,158. What is the percent decrease in the value of this car after 1 year?

96. A car is originally purchased for $15,500 and after 1 year, it is worth only $11,470. What is the percent decrease in the value of this car after 1 year?

97. A house was originally $456,000 and is now valued at $512,000. What is the percent increase in the value of this house?

98. A house was originally $345,000 and is now valued at $462,000. What is the percent increase in the value of this house?

For Exercises 99 through 108, identify each problem as an expression or equation. Simplify any expressions and solve any equations.

99. $5ab + 6a^2 - 4a + 12ab + 3$

100. $10mn^2 + 9n + 1 - 7mn^2 + 4$

101. $0.5x - 4 = 2x + 7$

102. $-0.25r + 12 = 4.5r - 15$

103. $\frac{1}{8}n + 12 = \frac{3}{24}n - 2$ 104. $\frac{3}{4}d + 14 = \frac{12}{16}d - 20$

105. $2x + 11 - 4x + 8$ 106. $7y^3 - 10y + 3y^2 + 2y$

107. $\frac{1}{2}(3x + 4) = \frac{1}{3}x + 20$

108. $\frac{1}{4}(5w + 2) = 11 - \frac{5}{18}w$

2.4 Solving and Graphing Linear Inequalities on a Number Line

LEARNING OBJECTIVES
- Solve inequalities.
- Write a solution set using interval notation.
- Graph a solution set on a number line.
- Solve compound inequalities.

Skill Connection

Inequality symbols

In Section R.1, we learned the inequality symbols.

Symbol	Meaning
>	Greater than
≥	Greater than or equal to
<	Less than
≤	Less than or equal to

Introduction to Inequalities

There are many situations in which quantities are less than or greater than another value. For example, we want our bills to be less than the money we earn. In school, students like their grades to be greater than the minimum to pass. For health reasons, it is desirable to have one's weight in the normal range, between underweight and overweight. Doctors want their patients to have less bad (LDL) cholesterol and more good (HDL) cholesterol. Many relationships such as these can be mathematically expressed as *inequalities*.

When working with inequalities, it is common to end up with a solution set that contains many possible solutions. Consider how many days we can rent a car and stay within our budget. The result will be anywhere from no days to the maximum number of days that will fit within our budget. To rent a bus for a school trip, the number of people who go on the bus must be less than or equal to the maximum occupancy of the bus. These situations have many possible answers that are mathematically correct as well as reasonable in the situation. The solution set to an inequality is an interval that includes all possible solutions.

Solving inequalities is much like solving equations, with one difference in the solving process. To discover this difference, we start by looking at the properties of equality and see whether they hold true for inequalities.

■ CONCEPT INVESTIGATION
Do the properties of equality work with inequalities?

Use the given inequality to test whether each property of equality is also true for inequalities. Test different kinds of numbers such as small, big, positive, and negative numbers. In this investigation, we will consider the resulting inequality true if the original inequality symbol is still the correct symbol to use.

a. **Addition Property**

The addition property of equality says that we can add the same real number to both sides of an equation and the equation will remain true. Fill in the table below to test if the property is true for inequalities.

	Add a Small Positive Number	Add a Large Positive Number	Add a Negative Number Close to Zero	Add a Negative Number Far from Zero	Add a Decimal or Fraction
5 > 2	5 + <u>4</u> > 2 + <u>4</u> <u>9</u> > <u>6</u>	5 + __ > 2 + __ __ > __	5 + __ > 2 + __ __ > __	5 + __ > 2 + __ __ > __	5 + __ > 2 + __ __ > __
True or False	True				

SECTION 2.4 Solving and Graphing Linear Inequalities on a Number Line   203

b. Subtraction Property
The subtraction property of equality says that we can subtract the same real number from both sides of an equation and the equation will remain true. Fill in the table below to test if the property is true for inequalities.

	Subtract a Small Positive Number	Subtract a Large Positive Number	Subtract a Negative Number Close to Zero	Subtract a Negative Number Far from Zero	Subtract a Decimal or Fraction
6 < 100	6 − 2 < 100 − 2 4 < 98	6 − __ < 100 − __ __ < __	6 − __ < 100 − __ __ < __	6 − __ < 100 − __ __ < __	6 − __ < 100 − __ __ < __
True or False	True				

c. Multiplication Property
The multiplication property says that we can multiply both sides of an equation by any nonzero real number and the equation will remain true. Fill in the following table to test if the property is true for inequalities.

	Multiply by a Small Positive Number	Multiply by a Large Positive Number	Multiply by a Negative Number Close to Zero	Multiply by a Negative Number Far from Zero	Multiply by a Negative Decimal or Fraction
3 < 5	2(3) < 2(5) 6 < 10	_(3) < _(5) _ < _	_(3) < _(5) _ < _	_(3) < _(5) _ < _	_(3) < _(5) _ < _
True or False	True				

d. Division Property
The division property says that we can divide both sides of an equation by any nonzero real number and the equation will remain true. (You might want to use decimals to compare the results.) Fill in the table below to test if the property is true for inequalities.

	Divide by a Small Positive Number	Divide by a Large Positive Number	Divide by a Negative Number Close to Zero	Divide by a Negative Number Far from Zero	Divide by a Negative Decimal or Fraction
40 > 30	$\frac{40}{2} > \frac{30}{2}$ 20 > 15	$\frac{40}{\_} > \frac{30}{\_}$ _ > _	$\frac{40}{\_} > \frac{30}{\_}$ _ > _	$\frac{40}{\_} > \frac{30}{\_}$ _ > _	$\frac{40}{\_} > \frac{30}{\_}$ _ > _
True or False	True				

Answer the following questions that refer to this Concept Investigation.

1. Which operations always keep the inequality true?
2. Which operations do not always keep the inequality true?
3. When do these operations fail to keep the inequality true? When do they still work?
4. When an inequality is *not* true, how can you change the inequality symbol to make the inequality true again?

This Concept Investigation can be summarized by the following properties for inequalities.

ADDITION PROPERTY OF INEQUALITIES

If the same real number or algebraic expression is added to both sides of an inequality, the solution to the inequality is unchanged.

If $a < b$, then $a + c < b + c$
If $a > b$, then $a + c > b + c$

SUBTRACTION PROPERTY OF INEQUALITIES

If the same real number or algebraic expression is subtracted from both sides of an inequality, the solution to the inequality is unchanged.

If $a < b$, then $a - c < b - c$
If $a > b$, then $a - c > b - c$

MULTIPLICATION PROPERTY OF INEQUALITIES

When both sides of an inequality are multiplied by a positive real number, the solution to the inequality is unchanged.

If $a < b$ and c is positive, then $a \cdot c < b \cdot c$
If $a > b$ and c is positive, then $a \cdot c > b \cdot c$

When both sides of an inequality are multiplied by a **negative** real number, the direction of the inequality symbol must be reversed.

If $a < b$ and c is negative, then $a \cdot c > b \cdot c$
If $a > b$ and c is negative, then $a \cdot c < b \cdot c$

DIVISION PROPERTY OF INEQUALITIES

When both sides of an inequality are divided by a positive real number, the solution to the inequality is unchanged.

If $a < b$ and c is positive, then $\dfrac{a}{c} < \dfrac{b}{c}$

If $a > b$ and c is positive, then $\dfrac{a}{c} > \dfrac{b}{c}$

When both sides of an inequality are divided by a **negative** real number, the direction of the inequality symbol must be reversed.

If $a < b$ and c is negative, then $\dfrac{a}{c} > \dfrac{b}{c}$

If $a > b$ and c is negative, then $\dfrac{a}{c} < \dfrac{b}{c}$

■ **What's That Mean?**

End Point

In the inequality $x \geq 2$, we call 2 the *end point*. In the inequality $x < -1$, we call -1 the *end point*.

■ Solving Inequalities

With these properties, we can solve many inequalities. We will solve inequalities using the properties of inequalities. Checking a solution for an inequality requires two steps. First, we check the **end point(s)** of the interval, and then we check the direction of the inequality symbol. The end point(s) is (are) the number(s) at which the solution set starts or ends.

SECTION 2.4 Solving and Graphing Linear Inequalities on a Number Line   205

Solving Inequalities
- Solve inequalities the same way as equations, except that when both sides are **multiplied or divided** by a negative number, *reverse* the inequality symbol.

Checking the Solution to an Inequality
- Substitute the end point in the inequality. Both sides should be equal (not unequal).
- Pick a number other than the end point from the solution set. Substitute this value in the original inequality. The original inequality should remain true.

Linear Inequalities

Addition and Subtraction Properties of Inequalities

Use to isolate a variable term.

$$4x + 5 < 21$$
$$\underline{-5 \quad -5}$$
$$4x < 16$$

Multiplication and Division Properties of Inequalities

Use to isolate a variable. When multiplying or dividing by a negative number, reverse the inequality symbol.

$$-3x < 21 \qquad 5x < 45$$
$$\frac{-3x}{-3} > \frac{21}{-3} \qquad \frac{5x}{5} < \frac{45}{5}$$
$$x > -7 \qquad x < 9$$

Example 1 Weekly salary

In Example 1 of Section 2.2, recall José, who worked for Post and Pack and earned $12 per hour. Let h be the number of hours José works in a week.

a. Write an inequality to show that José needs to earn **at least** $216 this week for his car payment.
b. Solve the inequality found in part a. Write the solution as a complete sentence.
c. Find the number of hours José must work to earn at least $350 a week.

SOLUTION

a. Because h is the number of hours José works a week, his salary is $12h$. His salary should be "at least $216," so we set it to be greater than or equal to 216. This gives us the inequality

$$12h \geq 216$$

b. Solve the inequality from part a in a similar way to solving an equation by dividing both sides by 12.

$$12h \geq 216$$

$$\frac{12h}{12} \geq \frac{216}{12}$$ 　　Divide both sides by 12.

$$h \geq 18$$ 　　This means that any values greater than or equal to 18 will be a solution. $h = 18$ is called the end point of the solution set.

$$12(18) \stackrel{?}{=} 216$$ 　　Check the end point 18 for equality.

$$216 = 216$$ 　　The end point checks.

$$12(20) \stackrel{?}{\geq} 216$$ 　　Now check a number greater than 18 to check the direction of the inequality symbol. The direction of the inequality works.

$$240 \geq 216$$

José has to work at least 18 hours a week to earn enough money to make his car payment.

c. José's salary needs to be at least $350, so we have the inequality

$$12h \geq 350$$

$$12h \geq 350$$ 　　Solve this inequality.

$$\frac{12h}{12} \geq \frac{350}{12}$$ 　　Divide both sides by 12.

$$h \geq \frac{350}{12} \approx 29.2$$ 　　The rounded approximation will help us to interpret the answer.

$$12\left(\frac{350}{12}\right) \stackrel{?}{=} 350$$ 　　Check the end point for equality.

What's That Mean?

At Least and At Most

"at least"

"I need *at least* $20 for the day," implies that $20 would be enough for the day but that more money would come in handy. Writing it as an inequality, we include the value 20 or more, so we use a symbol for greater than or equal to.

$$x \geq 20$$

"at most"

"Mom, I will need *at most* $300 for my trip," implies that less than $300 would probably be enough, but the trip could run as much as $300. Writing it as an inequality, we include the possibility that you will need $300 or less, which we express as a "less than or equal to" symbol.

$$x \leq 300$$

206 CHAPTER 2 Linear Equations and Inequalities with One Variable

Connecting the Concepts

Do we need to write the inequality symbol when solving inequalities?

When solving an inequality, keep writing the inequality symbol throughout the solving process. This is similar to solving an equation, in which we keep writing the equal sign.

$$350 = 350$$ The end point checks.

$$12(30) \stackrel{?}{\geq} 350$$ Check a number larger than 29.2 to check the direction of the inequality symbol. The direction of the inequality works.

$$360 \geq 350$$

José needs to work 29.2 or more hours a week to earn at least $350. This result means he could work 29.2, 30, 31, 32, or more hours a week and earn at least $350.

Example 2 — Solving inequalities

Solve the following inequalities. Check the answers.

a. $-2x + 8 \geq 20$ b. $4(x - 9) < 7x + 51$ c. $\dfrac{x}{3} + 9 \leq 4$

Skill Connection

The location of negative numbers on the number line

A number to the left of a number is less than that number. A number to the right of a number is greater than that number.

$-7 < -5$
-7 is less than -5 because it is to the left of -5 on the number line.

$-3 > -6$
-3 is greater than -6 because it is to the right of -6 on the number line.

SOLUTION

We solve inequalities in the same way that equations are solved, except remember to reverse the inequality symbol when multiplying or dividing both sides by a negative number.

a.
$-2x + 8 \geq 20$ Subtract 8 from both sides.
$\quad\;\; -8 \;\; -8$
$\overline{-2x \geq 12}$

$\dfrac{-2x}{-2} \leq \dfrac{12}{-2}$ Divide both sides by -2. When dividing by a negative, **reverse** the inequality symbol.

$x \leq -6$

$-2(-6) + 8 \stackrel{?}{=} 20$ Check the end point using $x = -6$.
$12 + 8 \stackrel{?}{=} 20$
$20 = 20$ The end point checks.

$-2(-7) + 8 \stackrel{?}{\geq} 20$ Check a number less than -6 ($x = -7$) to test the direction of the inequality symbol.
$14 + 8 \stackrel{?}{\geq} 20$
$22 \geq 20$ The direction of the inequality checks.

The solution set is $x \leq -6$.

b.
$4(x - 9) < 7x - 51$ Distribute the 4 to simplify.
$4x - 36 < 7x - 51$ Subtract $7x$ from both sides.
$\underline{-7x \qquad\; -7x}$
$-3x - 36 < -51$ Add 36 to both sides.
$\underline{\;+36 \;\; +36}$
$-3x < -15$

$\dfrac{-3x}{-3} > \dfrac{-15}{-3}$ Divide both sides by -3. When dividing by a negative, **reverse** the inequality symbol.

$x > 5$

$4((5) - 9) \stackrel{?}{=} 7(5) - 51$ Use $x = 5$ to check the value of the end point.
$4(-4) \stackrel{?}{=} 35 - 51$
$-16 = -16$ The end point checks.

$4((6) - 9) \stackrel{?}{<} 7(6) - 51$ Check a number larger than 5 ($x = 6$) to test the direction of the inequality symbol.
$4(-3) \stackrel{?}{<} 42 - 51$
$-12 < -9$ The direction of the inequality checks.

The solution set is $x > 5$.

c. $\dfrac{x}{3} + 9 \leq 4$

$3\left(\dfrac{x}{3} + 9\right) \leq 3(4)$ Multiply both sides by 3. Be sure to use the distributive property to divide out the denominator.

$x + 27 \leq 12$

$\underline{-27 \quad -27}$ Subtract 27 from both sides.

$x \leq -15$

$\dfrac{-15}{3} + 9 \stackrel{?}{=} 4$ Use $x = -15$ to check the value of the end point.

$-5 + 9 \stackrel{?}{=} 4$

$4 = 4$ The end point checks.

$\dfrac{-18}{3} + 9 \stackrel{?}{\leq} 4$ Check a number less than -15 ($x = -18$) to test the direction of the inequality symbol.

$-6 + 9 \stackrel{?}{\leq} 4$

$3 \leq 4$ The direction of the inequality checks.

The solution set is $x \leq -15$.

PRACTICE PROBLEM FOR EXAMPLE 2

Solve the following inequalities. Check the answers.

a. $-6x + 12 > 36$ **b.** $2(x + 5) \leq 7x + 75$

When solving inequalities, always remember to reverse the inequality symbol when multiplying or dividing by a negative number. Also, keep in mind that an answer that is negative does not reverse the symbol. Only the operation of multiplying or dividing by a negative number reverses the symbol.

Do Not Reverse the Inequality	Reverse the Inequality
$2x < -4$	$-2x < -4$
$\dfrac{2x}{2} < \dfrac{-4}{2}$	$\dfrac{-2x}{-2} > \dfrac{4}{-2}$
$x < -2$	$x > -2$

Students often find it simplest to interpret inequalities correctly if the **variable is on the left side** of the inequality symbol. Notice that all inequalities in this textbook are solved this way. It is not wrong for a variable to be on the right side of the inequality, but be careful with the direction of the inequality symbol.

$$x < 5 \text{ is the same as } 5 > x$$

The inequality symbol was reversed when expressions on the sides were switched. If x is less than 5, then 5 must be greater than x.

Example 3   Writing the variable on the left

Rewrite each inequality so that the variable appears on the left side. Solve the inequalities.

a. $16 > x$ **b.** $-4 \leq y$ **c.** $-15 < 5x$

SOLUTION

a. If $16 > x$, then x must be less than 16. This means that we can rewrite the inequality as

$$x < 16$$

b. If $-4 \leq y$, then y must be greater than or equal to -4. This means that we can rewrite the inequality as

$$y \geq -4$$

c. If $-15 < 5x$, then $5x$ must be greater than -15. Rewriting the inequality yields

$$5x > -15$$

Dividing both sides by 5 to solve the inequality for x yields

$$\frac{5x}{5} > \frac{-15}{5}$$

$$x > -3$$

Example 4 Solving inequalities that arise from applications

Alicia is on a budget and looking to control her family's energy usage. The natural gas that supplies her house has a baseline monthly fee of $1.02. The cost per therm of natural gas is $0.91. How many therms can Alicia's family use per month if she has $45.00 budgeted for the gas bill? *Note:* The units of measurement for natural gas are therms. A therm is equivalent to 100,000 BTUs (British thermal units), which is a measurement of heat energy.

a. Write an inequality to show that Alicia must keep her monthly gas bill to at most $45.

b. Solve the inequality from part a. Write the solution in a complete sentence.

SOLUTION

a. Let T be the number of therms that Alicia's family uses per month. The monthly gas charges are $1.02 + 0.91T$. The inequality that represents Alicia's family keeping the gas bill to at most $45 per month is

$$1.02 + 0.91T \leq 45$$

b. To solve the inequality, we isolate the term with T on the left side.

$$1.02 + 0.91T \leq 45 \quad \text{Subtract 1.02 from both sides.}$$
$$\underline{-1.02 \qquad \quad -1.02}$$
$$0.91T \leq 43.98$$
$$\frac{0.91T}{0.91} \leq \frac{43.98}{0.91} \quad \text{Divide both sides by 0.91 to isolate } T.$$
$$T \leq 48.329\ldots$$

Alicia's family can use at most approximately 48.3 therms of gas per month to keep her monthly bill less than $45.00.

PRACTICE PROBLEM FOR EXAMPLE 4

Marcela wants to try Zumba classes for her fitness routine. She finds a studio that offers the following deal for the first month: She can take the first class for $3.00, and after that, it is $5.00 per class. How many classes can Marcela take in her first month if she has budgeted a maximum of $58.00 to try out Zumba classes?

a. Write an inequality to show that Marcela must keep her first month's Zumba bill to at most $58.

b. Solve the inequality from part a. Write the solution in a complete sentence.

Interval Notation and Number Lines

The solution set for an inequality is usually a set of numbers that can be represented in several ways. So far in this section, we have worked all the examples using inequalities. Interval notation and number lines are also used to represent the solution set of an inequality. The inequality $x < 5$ can be represented in any of the following three ways.

In Words	Mathematical Representation
As an inequality	$x < 5$
Using interval notation	$(-\infty, 5)$
Graphed on a number line	number line from -10 to 10 with open parenthesis at 5, shaded to the left

Connecting the Concepts

Is there more than one way to write the solution set?

The solution set to an inequality may be written in one of three ways.

1. Inequality notation $x < -2$
2. Interval notation $(-\infty, -2)$
3. Number line

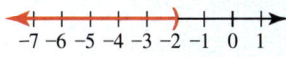

In interval notation, use either parentheses ( ) or brackets [ ] to indicate the ends of the interval. Parentheses are used when the end point is **not** included in the interval. When either a less than symbol or a greater than symbol is used in the inequality, parentheses are used in writing the solution set as an interval. If the end point is included in the interval, then a bracket is used. This happens when an "equal to" part is included with the inequality, such as "greater than or equal to."

The infinity symbol, ∞, is used to indicate that the interval does not stop in that direction and gets increasingly larger. Many intervals in this book will continue either to positive infinity or to negative infinity. Always use a parenthesis next to an infinity symbol. As an example, the inequality $x > -1$ is written as $(-1, \infty)$. Remember that we always write intervals starting with the left end point and ending with the right end point.

[left end point, right end point]

Number lines also use parentheses or brackets to indicate whether an end point of an interval is included or not. If an end point is *not* included, a parenthesis is used. If the end point is included in the interval, then a bracket is used. Some math instructors and textbooks will use open dots and closed dots with number lines instead of parentheses and brackets. If an interval goes to ∞ or $-\infty$, an arrowhead is used to indicate the line goes on forever. Following are some examples of the three types of notation being used.

Connecting the Concepts

Why don't we write $[-3, \infty]$?

A square bracket, [or], next to a number means that the interval will stop at and include that number in the interval. Infinity, ∞, is not a number on a number line. Since ∞ is not a value we ever get to on a number line, we never include it using square brackets. We always write a parenthesis symbol next to ∞.

	End Point Not Included	
Inequality <, >	$x < -3$	$x > -3$
Interval Notation Parentheses	$(-\infty, -3)$	$(-3, \infty)$
Number Line Parentheses	number line with) at -3, shaded left	number line with (at -3, shaded right

	End Point Included	
Inequality <, >	$x \leq -3$	$x \geq -3$
Interval Notation Bracket	$(-\infty, -3]$	$[-3, \infty)$
Number Line Bracket	number line with] at -3, shaded left	number line with [at -3, shaded right

Example 5 Writing the solution set using interval notation and graphing it on a number line

Solve the following inequalities. Write the solution set both using interval notation and on a number line.

a. $\dfrac{x}{7} > 3$ **b.** $-2x \geq -12$

SOLUTION

a.
$$\dfrac{x}{7} > 3$$

$$7\left(\dfrac{x}{7}\right) > 7(3) \quad \text{Multiply both sides by 7.}$$

$$x > 21 \quad \text{This is the solution set in inequality notation.}$$

$$\dfrac{21}{7} \stackrel{?}{=} 3 \quad \text{Use } x = 21 \text{ to check the value of the interval's end point.}$$

$$3 = 3 \quad \text{The end point checks.}$$

$$\dfrac{28}{7} \stackrel{?}{\geq} 3 \quad \text{Check a number greater than 21 to verify the direction of the inequality symbol. The direction of the inequality checks.}$$

$$4 > 3$$

Using interval notation, the solution set is written as $(21, \infty)$.

Use a parenthesis next to the 21 because the inequality does not have an "equal to" part. Always use a parenthesis next to the infinity symbol.

Using a number line:

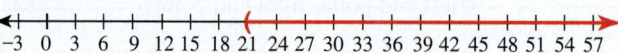

b.
$$-2x \geq -12$$

$$\dfrac{-2x}{-2} \leq \dfrac{-12}{-2} \quad \text{Divide both sides by } -2. \text{ Reverse the inequality symbol.}$$

$$x \leq 6 \quad \text{This is the solution set in inequality notation.}$$

$$-2(6) \stackrel{?}{=} -12 \quad \text{Use } x = 6 \text{ to check the value of the end point of the interval.}$$

$$-12 = -12 \quad \text{The end point checks.}$$

$$-2(5) \stackrel{?}{\geq} -12 \quad \text{Check a number less than 6 to verify the direction of the inequality symbol. The direction of the inequality checks.}$$

$$-10 \geq -12$$

Using interval notation, write the solution set as $(-\infty, 6]$.

Use a bracket next to the 6 because of the "equal to" part of the inequality symbol. This means that 6 is part of the solution set.

Using a number line:

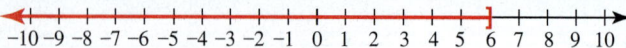

PRACTICE PROBLEM FOR EXAMPLE 5

Solve the following inequalities. Write the solution set both using interval notation and on a number line.

a. $-4x + 7 \leq 35$ **b.** $\dfrac{x}{2} > -4$

Example 6 — Graphing from the interval notation

Given the interval form of an inequality, graph the solution set on a number line. Then rewrite using inequality notation.

a. $[3, \infty)$ **b.** $(-\infty, -5)$

SOLUTION

a. The graph of this interval is given in the following graph.

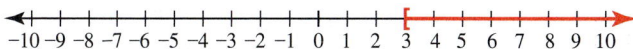

The values that are shown start at 3, include 3, and include all numbers greater than 3. This is equivalent to the inequality $x \geq 3$.

b. The graph of this interval is given in the following graph.

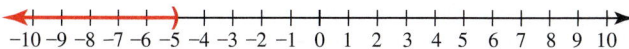

The values that are shown start at -5, do not include -5, and include all numbers less than -5. This is equivalent to the inequality $x < -5$.

PRACTICE PROBLEM FOR EXAMPLE 6

Given the interval form of an inequality, graph the solution set on a number line. Then rewrite using inequality notation.

a. $(-\infty, 7]$ **b.** $(-6, \infty)$

Compound Inequalities

In many areas of daily life, we face situations that require values to be within a certain range. A Boeing 787-9 Dreamliner, for example, can hold up to a maximum of 259 passengers (*Source: Boeing.com*).

Because it does not make sense to have a negative number of passengers, we say that the number of passengers must be between 0 and 259 inclusive. If we let $p =$ the number of passengers, we can write this range of values using inequality symbols as $0 \leq p \leq 259$. In this example, we interpret p as a whole number because it is representing the number of passengers on a jet. Because both 0 and 259 are possible values, include both end points for the set. This is an example of a **compound inequality** because it uses two inequalities together to represent the set. This set can be written in interval notation as $[0, 259]$ and as a number line as follows:

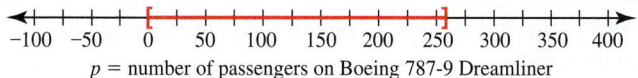

$p =$ number of passengers on Boeing 787-9 Dreamliner

DEFINITION

Compound Inequality When two inequalities are used together to show that a quantity lies between two values, the inequality is called a **compound inequality.**

$$5 < x < 20$$

means that x is between 5 and 20.

$$5 \leq x \leq 20$$

means that x is between 5 and 20, including the end points.
Note: Write the compound inequality from left to right and from the smaller value to the larger, just like with interval notation.

Example 7 Writing compound inequalities given in an application

Write a compound inequality for the following statements.

a. The number of passengers on a tour bus must not exceed 60.

b. The amount of open space on a certain computer hard drive is limited to 80 gigabytes.

c. To work correctly, an Aqua Star 250SX tankless water heater must have natural gas pressure between 5 and 14 inches water column (a unit of pressure).

SOLUTION

a. We cannot have a negative number of passengers on a tour bus, so we must have between 0 and 60 passengers. Since capacity cannot exceed 60, include the number 60 in our interval. Let $p =$ the number of passengers on the tour bus.

$$0 \leq p \leq 60$$

b. A computer hard drive cannot have a negative amount of open space, but it can be full. Include 0 to allow for no open space and 80 for the maximum open space. Let $s =$ the amount of open space in gigabytes on this computer hard drive.

$$0 \leq s \leq 80$$

c. Between 5 and 14 inches water column are required, and the problem does not indicate whether the pressure can equal 5 or 14. Therefore, we should not use "equal to" parts on the end points. Let $p =$ the gas pressure in inches of water column.

$$5 < p < 14$$

Note: $5 \leq p \leq 14$ might be valid, but since it was not explicitly stated in the problem that we include the end points, we will leave the solution as less than symbols.

PRACTICE PROBLEM FOR EXAMPLE 7

Write a compound inequality for the following statements.

a. The normal height for boys at 5 years of age is between 40 and 47 inches tall.
Source: FPnotebook.com (Family Practice Notebook, a Family Medicine Resource).

b. The Walt Disney Concert Hall in Los Angeles, California, can seat up to 2265 people.
Source: www.wdch.laphil.com

Keep in mind, when solving compound inequalities, that there are more than two sides to the inequality. We must be careful to perform operations on all sides of the inequalities to keep the entire inequality true. The goal is to isolate the variable in the *middle* of the compound inequality. For example, in the inequality $5 \leq p \leq 14$, the variable p is in the middle.

Example 8 Solving compound inequalities

Solve each compound inequality. Write the answer in the requested form.

a. $20 < 2x + 6 < 30$ Answer using a number line.

b. $-50 \leq -5x - 20 \leq 40$ Answer using interval notation.

c. $8 < \dfrac{x}{2} \leq 12$ Answer using a number line and interval notation.

SOLUTION

a. $20 < 2x + 6 < 30$ Isolate the x-term by subtracting 6 from all sides.
$\underline{-6 -6 -6}$
$14 < 2x < 24$
$\dfrac{14}{2} < \dfrac{2x}{2} < \dfrac{24}{2}$ Isolate the x by dividing all sides by 2.
$7 < x < 12$ This is the solution set in inequality notation.
$20 \stackrel{?}{=} 2(7) + 6$ Check the left end point using $x = 7$.
$20 = 20$ This end point checks.
$2(12) + 6 \stackrel{?}{=} 30$ Check the right end point using $x = 12$.
$30 = 30$ This end point checks.
$20 \stackrel{?}{<} 2(10) + 6 \stackrel{?}{<} 30$ Check the direction of the inequality symbols by using a number
$20 < 26 < 30$ between 7 and 12. The direction of the inequality symbols checks.

Since x is between 7 and 12 and there is no "equal to" part, we have a line from 7 to 12 with parentheses on both ends.

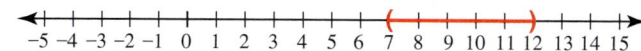

b. $-50 \leq -5x - 20 \leq 40$ Isolate the x-term by adding 20 to all sides.
$\underline{+20 +20 +20}$
$-30 \leq -5x \leq 60$
$\dfrac{-30}{-5} \geq \dfrac{-5x}{-5} \geq \dfrac{60}{-5}$ Isolate the x by dividing all the sides by -5.
    Reverse the inequality symbols.
$6 \geq x \geq -12$
or   Write compound inequalities from left to right and from small to large.
$-12 \leq x \leq 6$ This is the solution set in inequality notation.
$-50 \stackrel{?}{=} -5(6) - 20$ Check the right end point using $x = 6$.
$-50 = -50$ This end point checks.
$-5(-12) - 20 \stackrel{?}{=} 40$ Check the left end point using $x = -12$.
$40 = 40$ This end point checks.
$-50 \stackrel{?}{\leq} -5(0) - 20 \stackrel{?}{\leq} 40$ Check the direction of the inequality symbols using
$-50 \leq -20 \leq 40$ a number between -12 and 6. The direction of the inequality symbols checks.

Since both sides have "equal to" parts, we will use brackets for both sides. Always write intervals from lowest to highest numbers, so we get the interval $[-12, 6]$.

c. $8 < \dfrac{x}{2} \leq 12$

$2(8) < 2\left(\dfrac{x}{2}\right) \leq 2(12)$ Isolate the x by multiplying all sides by 2.

$16 < x \leq 24$
$8 \stackrel{?}{=} \dfrac{16}{2}$ Check the left end point using $x = 16$.
$8 = 8$ This end point checks.
$\dfrac{24}{2} \stackrel{?}{=} 12$ Check the right end point using $x = 24$.
$12 = 12$ This end point checks.
$8 \stackrel{?}{<} \dfrac{20}{2} \stackrel{?}{\leq} 12$ Check the direction of the inequality symbols using a number between 16 and 24.
$8 < 10 \leq 12$ The direction of the inequality symbols checks.

In interval notation, the solution set will have a parenthesis on the left side, since it is less than, and a bracket on the right side, since it has a less than or equal to sign.

(16, 24]

Using a number line, we have the following:

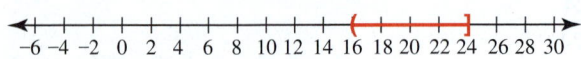

PRACTICE PROBLEM FOR EXAMPLE 8
Solve the following inequalities. Write the answer in the requested form.

a. $10 < 7x - 4 < 31$ Answer using interval notation.

b. $-8 \geq -\dfrac{x}{4} - 12 > -12$ Answer using a number line.

2.4 Vocabulary Exercises

1. To check the solution set of an inequality, check both the _____ and the direction of the inequality.

2. A(n) _____ inequality involves more than one inequality.

3. When graphing the solution set of an inequality, if $\leq$ or $\geq$ is involved, we draw a(n) _____ on the number line.

2.4 Exercises

For Exercises 1 through 22, solve each inequality. Check the answers. Write each solution set using inequality symbols.

1. $3x \geq 21$
2. $5w \geq 300$
3. $2d < 15$
4. $8z < 56$
5. $-\dfrac{1}{4}x \geq 48$
6. $-\dfrac{1}{3}y \leq 15$
7. $-a < -61$
8. $-x > -55$
9. $2g + 5 \leq 35$
10. $3x - 8 \geq 4$
11. $\dfrac{1}{5}m - 2 > -7$
12. $\dfrac{1}{5}x + 8 < -1$
13. $-2x + 9 \leq 25$
14. $4 - 3y > -13$
15. $4x + 7 > 6x + 3$
16. $3t - 14 < 4t - 20$
17. $-4x + 7 \geq 5x + 25$
18. $2 - 5y \leq y + 20$
19. $2(k + 5) - 3 < k + 18$
20. $4(z - 3) + 8 \geq 7z + 20$
21. $7(x - 2) + 1 > 8x - 6$
22. $9(x + 2) - 3 < 10x + 13$

For Exercises 23 through 34, solve each application problem. Round to the hundredths place if necessary.

23. Kati works for the school bookstore and gets paid $11 per hour. Let h be the number of hours Kati works per week.

    a. Write an inequality to show that Kati needs to earn at least $275 this week so that she can pay her rent.

    b. Solve the inequality you found in part a and write the solution as a complete sentence.

    c. Rewrite the inequality and find the number of hours Kati must work to earn at least $400 in a week. Write the solution as a complete sentence.

24. Tyler works for a local coffee shop and earns $11.25 per hour. Let *h* be the number of hours Tyler works per week.

    a. Write an inequality to show that Tyler must earn less than $225 per week if he wants to continue getting financial aid at his college.

    b. Solve the inequality you found in part a and write the solution as a complete sentence.

    c. Rewrite the inequality and find the number of hours Tyler has to work to earn at least $180 per week. Write the solution as a complete sentence.

25. Kevin wants to get the wood floors in his house professionally refinished. He calls Pro Floors, and they quote him a price of $25.00 for clean-up and $3.00 per square foot refinished. Let *s* be the number of square feet of wood floors that Kevin can have refinished.

   a. Write an inequality to show that Kevin must keep his budget for refinishing his hardwood floors to at most $5000.

   b. Solve the inequality from part a. Write the solution using a complete sentence.

   c. Rewrite the inequality and find out how many square feet he can have refinished if he needs to keep his bill to at most $4000.

26. Ricardo calls Quality Carpet Cleaning to get a quote on cleaning his carpets. They quote him a price of $100 to clean three rooms and then $40.00 for each room over three. Let *n* be the number of rooms over three that Ricardo needs to have cleaned.

   a. Write an inequality to show that Ricardo must keep his carpet cleaning bill to $220.00. (This is within his budget.)

   b. Solve the inequality from part a. Write the solution in a complete sentence.

27. Eddie needs to rent a flat-bed truck for a construction job. His local equipment rental company has a flat-bed truck for rent that costs $19.00 for the first 1.5 hours and $20.00 for each additional hour. Let *h* be the number of hours over 1.5 hours that Eddie uses the flat-bed truck.

   a. Write an inequality to show that Eddie must keep the flat-bed truck rental cost to at most $119.00.

   b. Solve the inequality from part a. Write the solution using a complete sentence.

   c. Rewrite the inequality and find out how many hours Eddie can rent the truck if he needs to keep his bill to at most $89.00.

28. Steven needs to rent a Bobcat for a construction job. His local equipment rental company quotes him a price of $19.00 for the first 1.5 hours plus $25.00 for each additional hour. Let *h* be the number of hours over 1.5 hours that Steven rents the Bobcat.

   a. Write an inequality to show that Steven must keep the Bobcat rental to at most $244.

   b. Solve the inequality from part a. Write the solution using a complete sentence.

   c. Rewrite the inequality and find out how many hours Steven can rent the Bobcat if he needs to keep his bill to at most $169.

29. Alicia calls Reliable Towing to tow her car after it broke down. Reliable Towing charges a $75.00 hook-up fee for towing and $3.00 a mile after the first 10 miles towed. Let *m* be the number of miles towed over 10 miles.

   a. Write an inequality to show that Alicia must keep her tow bill to at most $115.00. (This is within her budget.)

   b. Solve the inequality from part a. Write the solution in a complete sentence.

30. Amy called a local truck rental company and was told that it will charge her $20 for the day and $0.60 per mile to rent a truck. Let *m* be the number of miles Amy drives the truck in a day.

   a. Write an inequality to show that Amy has to keep the cost of her truck rental less than or equal to $95.

   b. Solve the inequality you wrote in part a and write the solution in a complete sentence.

   c. Rewrite the inequality and find how many miles Amy can drive the truck if she must limit the rental costs to at most $125. Write the solution as a complete sentence.

31. Solve the inequality

$$0.01t + 2.93 \geq 3$$

where *t* is years since 2010, to determine what year the population of Arkansas will be 3 million or more. Write the solution as a complete sentence.
Source: Model derived from data from St. Louis Federal Reserve.

32. Solve the inequality

$$0.09t + 19.41 \geq 20.3$$

where *t* is years since 2010, to determine what year the population of New York will be 20.3 million or more. Write the solution as a complete sentence.
Source: Model derived from data from U.S. Census Bureau.

33. Solve the inequality

$$0.26t + 18.84 \geq 21$$

where *t* is years since 2010, to determine what year the population of Florida will be 21 million or more. Write the solution as a complete sentence.
Source: Model derived from data from U.S. Census Bureau.

34. Solve the inequality

$$2p - 400 > 0$$

where *p* is the number of pieces of jewelry sold to find how many pieces of jewelry Juanita must sell to make a profit. Write the solution as a complete sentence.

For Exercises 35 through 44, rewrite each inequality so the variable appears on the left. Solve the inequality if necessary.

35. $4 > x$

36. $5 > y$

37. $-9 < n$

38. $-16 > m$

39. $0 \geq t$ **40.** $10 \leq t$
41. $6 > 2x$ **42.** $4 < -8x$
43. $5 \leq 2x + 1$ **44.** $-3 \geq 3x + 6$

For Exercises 45 through 62, solve each inequality. Check the answers. Write each solution set both using interval notation and on a number line.

45. $4x > 20$ **46.** $2x > 16$
47. $\dfrac{m}{3} < -2$ **48.** $\dfrac{h}{4} < 1.5$
49. $-3x \geq 12$ **50.** $-6x \geq -30$
51. $-\dfrac{2}{5}x < -\dfrac{1}{10}$ **52.** $-\dfrac{5y}{7} > \dfrac{3}{5}$
53. $-\dfrac{a}{3} \leq -4$ **54.** $-\dfrac{b}{5} \geq -11$
55. $2(k + 5) \leq k + 13$
56. $-(p + 4) + 5 \leq 3p + 9$
57. $2.5z + 8 \geq 3.5z - 4$
58. $-0.5x + 11 \geq -x - 4$
59. $2c + 7 \leq 6c + 17$
60. $-5z - 16 \leq 9z - 44$
61. $5w - 12 \geq 2(5w - 10) - 30$
62. $4(3g + 8) < 5(4g - 10) + 26$

For Exercises 63 through 70, given the interval form of an inequality, graph the solution set on a number line. Then rewrite using inequality notation.

63. $[-1, \infty)$ **64.** $[-7, \infty)$
65. $(-\infty, 9)$ **66.** $(-\infty, -12)$
67. $(69, \infty)$ **68.** $(42, \infty)$
69. $(-\infty, 16]$ **70.** $(-\infty, 0]$

For Exercises 71 through 80, write a compound inequality for each statement.

71. The maximum enrollment in a beginning algebra class on a college campus is 38. The class will be canceled if it does not have at least 16 students in it.

72. A Boeing 717-200 can hold a maximum of 106 passengers.
Source: www.boeing.com

73. School bus 12 has a capacity of at most 71 students.

74. Oakbridge, a Young Life camp in southern California, can accommodate up to 177 people.
Source: www.younglife.org

75. Another Young Life camp, Timber Wolf Lake, can accommodate up to 450 people. The minimum group size is 200.
Source: www.younglife.org

76. A local certified public accountant charges her clients between $150 and $500 to prepare tax returns depending on the complexity of the return.

77. At LEGOLAND California, to ride on some attractions, a child between 42 and 52 inches tall must have an adult ride with them.

78. College-level courses at Palomar College are numbered between 100 and 300, including 100 but not including 300.

79. The average adult has a normal resting heart rate of 66 to 100 beats per minute.

80. The average newborn baby has a normal resting heart rate of 100 to 160 beats per minute.

For Exercises 81 through 104, solve the given inequality. Write the solution set both using interval notation and on a number line.

81. $6 < 2x < 8$ **82.** $25 < 5x < 55$
83. $-7 < y - 1 < -3$ **84.** $-4 < b + 5 < 6$
85. $12 < 3x + 3 < 21$ **86.** $20 < 2x + 10 < 36$
87. $3 < \dfrac{k}{2} < 5$ **88.** $-2 < \dfrac{n}{4} < 3$
89. $40 \leq 6x + 4 \leq 64$
90. $-25 \leq 8x - 1 \leq 15$
91. $-1.5 \leq \dfrac{m}{6} \leq -0.5$
92. $3 \leq \dfrac{w}{4} + 2 \leq 5$
93. $-12 \leq \dfrac{1}{2}x - 2 \leq 0$ **94.** $0 \leq \dfrac{1}{3}x + 1 \leq 16$
95. $4 < \dfrac{x}{2} < 30$ **96.** $5 < \dfrac{x}{15} < 25$
97. $-7 < \dfrac{3x}{4} < -1$
98. $-25 < \dfrac{5x}{7} < -2$
99. $3.5 \leq 2x - 2 \leq 8.5$
100. $-12.3 \leq 6x + 4 \leq 6.6$
101. $-9 \leq 3v + 6 < 3$

102. $5 < 2x - 4 \leq 8$

103. $-7 < \dfrac{5a}{3} - 8 \leq 3$

104. $-3 < 4x - 9 \leq 17$

For Exercises 105 through 120, identify each problem as an expression, equation or inequality. Simplify each expression, and solve each inequality or equation. Write the solution set of each inequality using interval notation.

105. $2x + 7 > 41$

106. $6c - 10 < 32$

107. $4x + 7 - 12x$

108. $2m^2 + 3 - 4m - 7m^2$

109. $5h + 12 = 3h - 16$

110. $2(4n - 7) = -5n + 19$

111. $3a + 4 \geq 7a - 20$

112. $-5t - 14 \leq 3t + 2$

113. $\dfrac{2}{3}x + \dfrac{4}{3} = \dfrac{1}{3}x + 7$

114. $\dfrac{4}{7}d - \dfrac{2}{7} = 2d + \dfrac{3}{7}$

115. $-20 < 4a + 2 < 12$

116 $8 \leq \dfrac{y}{4} + 10 < 16$

117. $\dfrac{2}{3}b + 5 + \dfrac{4}{9}b - 2$

118. $\dfrac{3}{5}x - 8 + \dfrac{3}{10}x + \dfrac{1}{2}$

119. $4 < -\dfrac{g}{2} \leq 7$

120. $-8 \leq \dfrac{x}{3} \leq 4$

Chapter Summary

Section 2.1 — Addition and Subtraction Properties of Equality

- An **equation** is a statement that two expressions are equal. Equations have equal signs.
- **Solutions to an equation** are the values for the variables that make the equation true.
- **To check solutions** to an equation, use the following steps:
  1. Substitute in the given value(s) of the variables.
  2. Evaluate both sides of the equation. Simplify using the order-of-operations agreement. If both sides are equal then that (those) value(s) of the variable(s) is a (are) solution(s) of the equation.
- The **addition property of equality:** If the same real number or algebraic expression is added to **both** sides of an equation, the solution to the equation is not changed.
- The **subtraction property of equality:** If the same real number or algebraic expression is subtracted from **both** sides of an equation, the solution to the equation is not changed.
- To solve an equation for a variable, **isolate** the variable on one side of the equation.
- **To isolate a variable on one side** of the equation, use the reverse operation ("undo") to move any expressions being added to or subtracted from the variable.
- An equation that involves several variables or is a formula is called a **literal equation.** Solve literal equations using the same process as for solving other equations.

Example 1

Solve. Check the answer. $-8 + x = -16$.

SOLUTION

$$-8 + x = -16 \quad \text{Add 8 to both sides.}$$
$$+8 \qquad +8$$
$$x = -8$$
$$-8 + (-8) \stackrel{?}{=} -16 \quad \text{Check the answer.}$$
$$-16 = -16 \quad \text{The answer checks.}$$

Example 2

Solve. Check the answer. $x - \dfrac{2}{5} = \dfrac{1}{2}$.

SOLUTION

$$x - \frac{2}{5} = \frac{1}{2} \quad \text{Add } \frac{2}{5} \text{ to both sides.}$$
$$+\frac{2}{5} \quad +\frac{2}{5}$$
$$x = \frac{1}{2} + \frac{2}{5} \quad \text{The LCD is 10.}$$
$$x = \frac{5}{10} + \frac{4}{10} \quad \text{Rewrite fractions over LCD.}$$
$$x = \frac{9}{10}$$

CHAPTER 2 Summary 219

$$\frac{9}{10} - \frac{2}{5} \stackrel{?}{=} \frac{1}{2}$$ Check the answer.

$$\frac{9}{10} - \frac{4}{10} \stackrel{?}{=} \frac{1}{2}$$

$$\frac{5}{10} \stackrel{?}{=} \frac{1}{2}$$

$$\frac{1}{2} = \frac{1}{2}$$ The answer checks.

Example 3   Solve the literal equation for b: $2a + b = c$.

SOLUTION

$$2a + \boxed{b} = c$$ To isolate b, subtract $2a$ from both sides.
$$\underline{-2a \qquad -2a}$$
$$b = c - 2a$$

Section 2.2 Multiplication and Division Properties of Equality

- **The multiplication property of equality:** If *both* sides of an equation are multiplied by the same nonzero real number or nonzero algebraic expression, the solution to the equation is not changed.
- **The division property of equality:** If *both* sides of an equation are divided by the same nonzero real number or nonzero algebraic expression, the solution to the equation is not changed.
- **Simplify both sides of an equation before** isolating the variable.
- **When solving equations with several operations** involved, use the following steps:
  1. Simplify both sides of the equation independently, if needed, using the distributive property and combining like terms.
  2. Isolate the variable term. Circle the variable term. Use the addition or subtraction properties of equality to isolate the variable term on one side of the equation.
  3. Isolate the variable. Use the multiplication or division properties of equality to isolate the variable on one side of the equation.
  4. Check the answer. Substitute the answer into the original equation and simplify both sides to check that they give the same result.
- When **solving literal equations,** assume that all variables represent nonzero real numbers so that multiplying or dividing both sides of the equation by the variable will not make the equation undefined.

Example 4   Solve the equation for the variable. Check the answer. $-5d = -55$.

SOLUTION

$$\frac{-5d}{-5} = \frac{-55}{-5}$$ Divide both sides by -5.
$$d = 11$$
$$-5(11) \stackrel{?}{=} -55$$ Check the answer.
$$-55 = -55$$ The answer checks.

220 CHAPTER 2 Linear Equations and Inequalities with One Variable

Example 5 Solve the equation for the variable. Check the answer. $-3x + 8 + 9x = 44$.

SOLUTION

$-3x + 8 + 9x = 44$ Simplify the left side.

$6x + 8 = 44$ Subtract 8 from both sides.

$\underline{-8 -8}$

$6x = 36$

$\dfrac{6x}{6} = \dfrac{36}{6}$ Divide both sides by 6.

$x = 6$

$-3(6) + 8 + 9(6) \stackrel{?}{=} 44$ Check the answer.

$-18 + 8 + 54 \stackrel{?}{=} 44$

$44 = 44$ The answer checks.

Example 6 Translate the following sentences into equations and solve. Check the answer.
The difference between 8 and three times a number is 32.

SOLUTION The sentence translates as

$$8 - 3x = 32$$

Solving the equation yields

$8 - 3x = 32$ Subtract 8 from both sides.

$\underline{-8 -8}$

$\dfrac{-3x}{-3} = \dfrac{24}{-3}$ Divide both sides by -3.

$x = -8$

$8 - 3(-8) \stackrel{?}{=} 32$ Check the answer.

$8 + 24 \stackrel{?}{=} 32$

$32 = 32$ The answer checks.

Section 2.3   Solving Equations with Variables on Both Sides

- When **solving equations with variables on both sides**, follow these steps:
  1. Simplify each side of the equation by performing any arithmetic and combining any like terms.
  2. Move the variable terms to one side of the equation using the addition or subtraction property of equality. Combine like terms.
  3. Isolate the variable term by using the addition or subtraction property of equality.
  4. Isolate the variable by using the multiplication or division property of equality.
  5. Check the answer in the original problem.
- When **solving equations that contain fractions**, eliminate the fractions by multiplying both sides of the equation by the common denominator. Use the distributive property.
- An **identity** is an equation that is true for all possible values of the variable. Equations that are identities will have all real numbers as the solution set. Identities simplify to a statement that is always true, such as $0 = 0$.
- **Some equations have no solutions**. If an equation simplifies to a false statement, then it has no solution. An example of a false statement is $0 = 5$.
- When **translating sentences into equations**, use only one variable to describe the missing quantities.

CHAPTER 2 Summary 221

- **Read an application problem carefully,** breaking it up into smaller pieces to solve the entire problem.
- Some applications require us to use a **known formula or relationship.** Often these are geometry formulas. Many formulas are available at the back of the book.

Example 7 Solve for the variable. Check the answer. $5x - 4 = 3x + 8$.

SOLUTION

$$5x - 4 = 3x + 8$$ Combine x-terms by subtracting 3x from both sides.
$$\underline{-3x \qquad -3x}$$
$$2x - 4 = 8$$ Isolate the variable term by adding 4 to both sides.
$$\underline{+4 \ +4}$$
$$2x = 12$$
$$\frac{2x}{2} = \frac{12}{2}$$ Divide both sides by 2 to isolate x.
$$x = 6$$
$$5(6) - 4 \stackrel{?}{=} 3(6) + 8$$ Check the answer.
$$30 - 4 \stackrel{?}{=} 18 + 8$$
$$26 = 26$$ The answer checks.

Example 8 Solve for the variable. Check the answer. $\frac{1}{5}x + 4 = -3 + \frac{6}{5}x$.

SOLUTION

$$\frac{1}{5}x + 4 = -3 + \frac{6}{5}x$$

$$5\left(\frac{1}{5}x + 4\right) = 5\left(-3 + \frac{6}{5}x\right)$$ Eliminate the fractions by multiplying both sides by 5.

$$5\left(\frac{1}{5}x\right) + 20 = -15 + 5\left(\frac{6}{5}x\right)$$ Distribute and reduce.

$$x + 20 = -15 + 6x$$
$$\underline{-x \qquad\qquad -x}$$ Combine variable terms by subtracting x from both sides.
$$20 = -15 + 5x$$ Isolate the variable term by adding 15 to both sides.
$$\underline{+15 \ \ +15}$$
$$35 = 5x$$
$$\frac{35}{5} = \frac{5x}{5}$$ Isolate the variable by dividing both sides by 5.
$$7 = x$$

$$\frac{1}{5}(7) + 4 \stackrel{?}{=} -3 + \frac{6}{5}(7)$$ Check the answer.

$$\frac{7}{5} + 4 \stackrel{?}{=} -3 + \frac{42}{5}$$

$$\frac{7}{5} + \frac{20}{5} \stackrel{?}{=} -\frac{15}{5} + \frac{42}{5}$$

$$\frac{27}{5} = \frac{27}{5}$$ The answer checks.

Example 9 Solve. If the equation has no solution, answer "no solution." If it is an identity, answer "all real numbers."

a. $5(x - 2) = 5x - 10$ **b.** $4x + 3 = 4x - 5$

SOLUTION

a. $5(x - 2) = 5x - 10$ Use the distributive property to multiply.
$5x - 10 = 5x - 10$ Move the x-terms to the left.
$\underline{-5x \qquad\quad -5x}$
$-10 = -10$ This is a true statement.

Because this statement is true, the equation is an identity. The solution set is all real numbers.

b. $4x + 3 = 4x - 5$ Move the x-terms to the left.
$\underline{-4x \qquad\quad -4x}$
$3 = -5$ This is a false statement.

Because this equation is false, there is no solution.

Section 2.4   Solving and Graphing Linear Inequalities on a Number Line

- **Inequalities** will usually have a set of values that represent all possible solutions.
- **Inequalities are solved** by using the same properties as equations, except that we **reverse the inequality symbol when multiplying or dividing** both sides by a negative number.
- When writing using **interval notation,** use parentheses when the end point is not included in the interval. Use brackets when there is an "equal to" part and you want to include the end point in the interval.
- When **drawing a number line,** use parentheses to indicate that the end point is not included in the interval and brackets to indicate the "equal to" part that includes the end point of the interval.
- **Whenever infinity or negative infinity** is part of a solution set written in interval notation, you **must use a parenthesis** next to the infinity symbol. The symbol for infinity is ∞.
- **When drawing a number line, use arrowheads** on the end of the line to indicate that the set continues toward ∞ or $-\infty$.
- **Compound inequalities** involve more than one inequality.
- **When checking the solution to an inequality:**
  - Substitute the end point into the inequality. Both sides should be equal (not unequal).
  - Pick a number other than the end point from the solution set. Substitute this value in the original inequality. The original inequality should remain true.

Example 10 Solve the following inequality. Check the answer. $4 - 2x < -10$.

SOLUTION

$4 - 2x < -10$
$\underline{-4 \qquad\quad -4}$
$-2x < -14$
$\dfrac{-2x}{-2} > \dfrac{-14}{-2}$ **Reverse the inequality** when dividing by a negative number.
$x > 7$

$$4 - 2(7) \stackrel{?}{=} -10$$ Check the end point for equality.
$$-10 = -10$$ The end point checks.
$$4 - 2(8) \stackrel{?}{<} -10$$ Check the direction of the inequality symbol using a number greater than 7.
$$4 - 16 \stackrel{?}{<} -10$$
$$-12 < -10$$ The direction of the inequality checks.

Example 11 Solve the inequality and write the solution in interval notation and graph it on a number line. $-3x + 1 \geq -17$.

SOLUTION
$$-3x + 1 \geq -17$$
$$\underline{\; -1 \quad\; -1}$$
$$-3x \geq -18$$
$$\frac{-3x}{-3} \leq \frac{-18}{-3}$$ Reverse the inequality.
$$x \leq 6$$
$$-3(6) + 1 \stackrel{?}{=} -17$$ Check the end point for equality.
$$-18 + 1 \stackrel{?}{=} -17$$
$$-17 = -17$$ This end point checks.
$$-3(0) + 1 \stackrel{?}{\geq} -17$$ Check the direction of the inequality symbol using a number less than 6.
$$0 + 1 \stackrel{?}{\geq} -17$$
$$1 \geq -17$$ The direction of the inequality checks.

In interval notation, the solution is $(-\infty, 6]$.

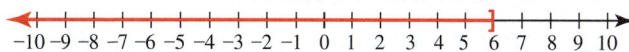

Example 12 Solve the inequality. Graph the solution on a number line. $-4 < 2x - 4 < 16$.

SOLUTION
$$-4 < 2x - 4 < 16$$ Add 4 to all sides.
$$\underline{+4 \qquad +4 \;\; +4}$$
$$0 < 2x < 20$$
$$\frac{0}{2} < \frac{2x}{2} < \frac{20}{2}$$ Divide all sides by 2.
$$0 < x < 10$$
$$-4 \stackrel{?}{=} 2(0) - 4$$ Check the left end point using $x = 0$.
$$-4 = -4$$ This end point checks.
$$2(10) - 4 \stackrel{?}{=} 16$$ Check the right end point using $x = 10$.
$$16 = 16$$ This end point checks.
$$-4 \stackrel{?}{<} 2(3) - 4 \stackrel{?}{<} 16$$ Check the direction of the inequality symbols by using a number between 0 and 10.
$$-4 \stackrel{?}{<} 6 - 4 \stackrel{?}{<} 16$$
$$-4 < 2 < 16$$ The direction of the inequality symbols checks.

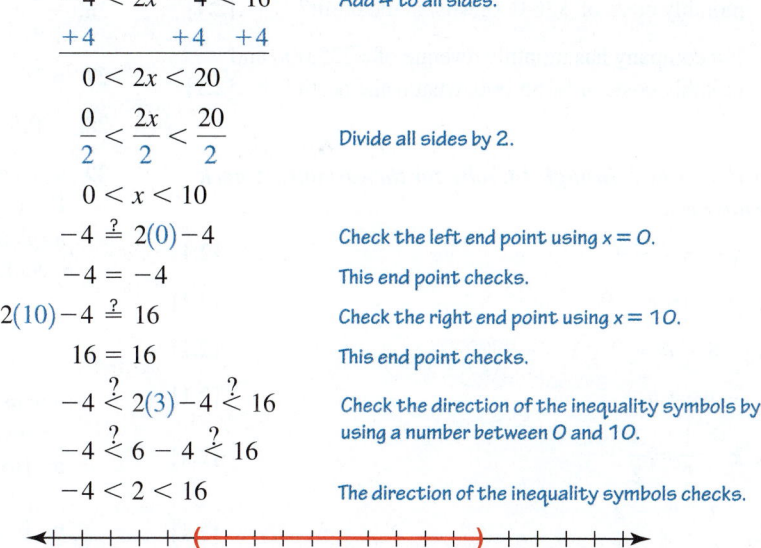

Chapter Review Exercises

For Exercises 1 and 2,
a. *Determine whether or not the given values of the variables are solutions to the equations.*
b. *Explain what the values mean in the problem situation.*

1. $d = 75t$, where d is the distance traveled in miles when driving for t hours; $t = 3.5$ and $d = 262.5$. [2.1]

2. $F = 11.5L$, where F are the origination fees in dollars paid for a loan of L thousand dollars; $F = 6900$ and $L = 600$. [2.1]

For Exercises 3 and 4, determine whether or not the given value(s) of the variable(s) are solutions to the equations.

3. $-2x + 4y = -8$; $x = -2$ and $y = 3$ [2.1]
4. $y = -3x + 5$; $x = -3$ and $y = 14$ [2.1]

For Exercises 5 and 6, use the following information.

The amount of profit a company makes is found by subtracting the company's costs from its revenue. This relationship can be written as the following equation:

$$P = R - C$$

where P is the profit in dollars, R is the revenue in dollars, and C is the cost in dollars for the company. Note: Revenue is the amount of money a company brings in from sales.

5. If a company has monthly revenue of $49,000 and monthly costs of $26,000, what is the profit? [2.1]

6. If a company has monthly revenue of $225,000 and monthly costs of $186,000, what is the profit? [2.1]

For Exercises 7 through 14, solve for the variables. Check the answers.

7. $x - 3 = -9$ [2.1]
8. $p - 16 = -9$ [2.1]
9. $-8 = a + 9$ [2.1]
10. $-16 = b + 7$ [2.1]
11. $k - \frac{1}{2} = \frac{3}{2}$ [2.1]
12. $-\frac{2}{5} + x = \frac{4}{5}$ [2.1]
13. $y - 9.8 = 4.6$ [2.1]
14. $0.02 + h = 1.04$ [2.1]

For Exercises 15 through 18, solve the given literal equations for the indicated variables.

15. $2p + q = 9$ for q [2.1]
16. $x - y = 2$ for x [2.1]
17. $8x - 24y = 48$ for y [2.2]
18. $3p - 5q = -30$ for p [2.2]

For Exercises 19 through 28, solve the equations for the variables. Check the answers. Round to the hundredths place, if necessary.

19. $7y = 84$ [2.2]
20. $4x = -64$ [2.2]
21. $-a = 7$ [2.2]
22. $-8 = -x$ [2.2]
23. $-2b = -22$ [2.2]
24. $-5y = 35$ [2.2]
25. $\frac{2}{3}x = -16$ [2.2]
26. $-\frac{3}{4}n = -9$ [2.2]
27. $6.2t = -8.4$ [2.2]
28. $-0.5k = -16.5$ [2.2]

29. The average short-haul jet airplane (such as a Boeing 737) can seat 126 people. The number of these airplanes needed to transport a group of people can be calculated by using the formula

$$J = \frac{p}{126}$$

where J is the number of airplanes needed to transport a group of p people. [2.2]
a. How many airplanes are needed to transport 520 people?
b. If an airline carrier has 15 planes, how many people can the carrier transport at a time?

30. Devora works as a medical assistant and makes $14 per hour after taxes. Devora can calculate her monthly salary using the equation

$$S = 14h$$

where S is the monthly salary in dollars that Devora makes if she works h hours that month. [2.2]
   a. Find Devora's monthly salary if she works 140 hours.
   b. How many hours does Devora have to work a month to make $2100 a month?
   c. How many hours does Devora have to work a month if she wants to earn only $1330 a month?

For Exercises 31 through 38, solve the equations. Check the answers.

31. $6x - 9 = -63$ [2.2]
32. $5y - 17 = 33$ [2.2]
33. $-2a + 9 = -5$ [2.2]
34. $-8k - 19 = 5$ [2.2]
35. $\dfrac{x}{2} - 1 = -11$ [2.2]
36. $-\dfrac{x}{3} + 2 = -4$ [2.2]
37. $-8.6 - 1.2x = 10.6$ [2.2]
38. $0.25y - 1 = -7$ [2.2]

For Exercises 39 through 46, define variables and translate each sentence into an equation. Then solve that equations and check the answers.

39. Nine times a number is 76. [2.2]
40. 12 is equal to the product of a number and -2. [2.2]
41. The sum of twice a number and 16 is -44. [2.2]
42. The sum of one-half a number and 4 is -18. [2.2]
43. The quotient of a number and 4 is 0.25. [2.2]
44. The quotient of a number and -3 is 0.75. [2.2]
45. The difference of 8 and 3 times a number is 17. [2.2]
46. Nine less than twice a number is 7. [2.2]

For Exercises 47 through 52, solve the literal equations for the indicated variables.

47. Solve $D = cma$ for m. [2.2]
48. Solve $V = lwh$ for h. [2.2]
49. Solve $C = \pi d$ for d. [2.2]
50. Solve $A = \dfrac{1}{2}bh$ for h. [2.2]
51. Solve $9x - 3y = -12$ for y. [2.2]
52. Solve $-8x + 4y = -24$ for y. [2.2]

For Exercises 53 and 54, translate each sentence and solve. Check the answers.

53. The sum of twice a number and 1 is equal to the difference between three times the number and 4. [2.2]
54. The sum of half a number and ten is equal to the difference between that number and 4. [2.2]

For Exercises 55 through 62, solve each equation. Provide reasons for each step. Check the answers. If the equation is an identity, write the answer as "all real numbers." If there is no solution, write "no solution."

55. $4x - 8 = -3x + 20$ [2.3]
56. $9y + 12 = 7y + 6$ [2.3]
57. $5(a + 5) = 12a - 13$ [2.3]
58. $11 + 7x + 7 = 5x - 2$ [2.3]
59. $4k + 1 = -5k + 3(3k - 1)$ [2.3]
60. $9h - 7(h + 1) = 2h + 9$ [2.3]
61. $-5x + 3 = 13 - 5(x + 2)$ [2.3]
62. $7y - 10 + 2(-3y + 2) = y - 6$ [2.3]

63. Given the following triangle, find the measure of each angle. [2.3]

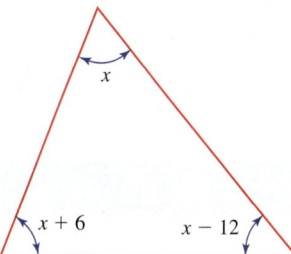

64. Given the following triangle, find the measure of each angle. [2.3]

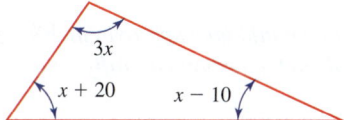

For Exercises 65 through 78, solve each inequality. Check the answers. Write each solution set in each of the following ways: using inequality symbols, using interval notation, and graphing the solution on a number line.

65. $4x + 1 > 49$ [2.4]
66. $3x - 2 \leq 19$ [2.4]
67. $-x \leq 7$ [2.4]
68. $-y \geq 3$ [2.4]
69. $-5y + 1 \geq -16$ [2.4]
70. $-6x - 7 < -3$ [2.4]
71. $-\dfrac{1}{4}y + 3 \leq -2$ [2.4]
72. $4 - \dfrac{2}{5}t > -8$ [2.4]
73. $-3x - 3 < -2x + 9$ [2.4]
74. $-8y - 1 > -3y + 9$ [2.4]
75. $-3 < 2x + 3 < 11$ [2.4]
76. $-1 < 5x - 6 < 24$ [2.4]
77. $-2 < \dfrac{x}{3} < 9$ [2.4]
78. $-8 < \dfrac{y}{2} + 11 < -5$ [2.4]

For Exercises 79 through 82, given the interval form of each inequality, graph each solution set on a number line. Then rewrite using inequality notation.

79. $[-3, \infty)$ [2.4]
80. $(5, \infty)$ [2.4]
81. $(-\infty, 6)$ [2.4]
82. $(-\infty, -12]$ [2.4]

For Exercises 81 and 82, rewrite each inequality with the variable on the left side.

83. $-6 > x$ [2.4]
84. $0 \leq y$ [2.4]

85. Solve the inequality [2.4]
$$80n - 2400 > 0$$
where n is the number of stained-glass windows that Kiano must sell to make a profit. Write the solution as a complete sentence.

86. Solve the inequality [2.4]
$$32.49d + 21.65 \leq 119.12$$
where d is the number of days Jessie can rent a car if she wants to spend only up to $119.12. Write the solution as a complete sentence.

87. Karyn is planning the reception for her daughter's wedding. It is a big event, and she is looking to hire a bartender. One bartender she contacted said the first 3 hours costs $135 and then it is $40 an hour after that. Let h be the number of hours after 3 hours that the bartender is working at the wedding. [2.4]
   a. Write an inequality to show that Karyn must keep the bartending bill at the wedding to at most $235 to stay within budget.
   b. Solve the inequality from part a. Write the solution in a complete sentence.

88. Trey purchases a flat-screen TV at a local electronics store. The store tells him that the minimum delivery charge is $50.00 plus $1.50 per mile. Let m be the number of miles that Trey lives from the store. [2.4]
   a. Write an inequality to show that Trey must keep his shipping charge to at most $65. (This is within his budget.)
   b. Solve the inequality from part a. Write the solution in a complete sentence.

Chapter Test

This chapter test should take approximately one hour to complete. Read each question carefully. Show all of your work.

1. Is $b = -2$ a solution of $5b + 8 = 2$?
2. Is $t = 4.2$ and $d = 294$ a solution to $d = 70t$, where d is the distance traveled in miles when driving for t hours?
3. Solve and check the answer: $-5 + x = -17$.
4. Solve the literal equation for the variable c. $P = a + b + c$.
5. Use the equation $P = R - C$, where P represents the profit, R represents the revenue, and C is the cost. If a company has monthly revenue of $250,000 and monthly costs of $88,000, what is the profit?
6. Solve and check the answer: $-5w = 450$.

7. Solve and check the answer: $-4p + 25 = -119$.

8. Solve and check the answer: $\dfrac{x}{3} + 7 = -5$.

9. Solve and check the answer. If the equation is an identity, write the answer as "all real numbers." If there is no solution, write "no solution." $4(x + 4) - 5 = 4x + 11$.

10. Solve and check the answer. If the equation is an identity, write the answer as "all real numbers." If there is no solution, write "no solution." $3x + 20 = 20 + 3(x - 5)$.

11. Define your variables and translate the problem into an equation. Solve the equation and check the answer. The perimeter of a rectangular building lot is 330 feet. The width of the lot is 40 feet. Find the length of the lot.

12. Solve the literal equation for the variable x. $y = mx + b$.

13. Translate the sentence into an equation and solve. The sum of four times a number and 7 is negative seventeen.

14. Solve the equation, provide reasons for each step, and check the answer. $10x - 15 = 7x - 9$.

15. Solve the equation, provide reasons for each step, and check the answer. $4x + 3 = 4(x - 1) + 7$.

16. A triangle has angles that measure x, $11x$, and $6x$ degrees. Find the measure of each angle of the triangle.

17. Rewrite the inequality with the variable on the left side. $-16 \leq x$.

18. Solve the inequality. Graph the solution set on a number line. $2x + 1 > 4x - 3$.

19. Solve the inequality. $5 \leq 2x - 7 \leq 15$.

20. Aiden works for a large retailer and earns $10.25 per hour. Let h be the number of hours Aiden works per week.
    a. Write an inequality that shows that Aiden needs to earn more than $287.00 a week to meet his expenses.
    b. Solve the inequality from part a and write the solution set using a complete sentence.

Chapter Projects

■ What Is That Pattern?

Written Project
One or more people

a. Look at the pattern in the following table.

$n = 1$	***
$n = 2$	*****
$n = 3$	*******
$n = 4$	*********
$n = 5$	***********
$n = 6$	
$n = 7$	

1. Using the pattern you observe, fill in the last two rows of the table, continuing the pattern.

2. Let n represent the row number. Write a general expression for the number of asterisks in the table in terms of n.

3. Suppose there are 107 asterisks in a row. In what row number does that occur? *Hint:* Find n using the expression you found in step 2.

b. Look at the pattern in the following table.

$n = 1$	**
$n = 2$	*****
$n = 3$	********
$n = 4$	***********
$n = 5$	
$n = 6$	
$n = 7$	

1. Using the pattern you observe, fill in the last three rows of the table, continuing the pattern.
2. Let n represent the row number. Write a general expression for the number of asterisks in the table in terms of n.
3. Suppose there are 107 asterisks in a row. In what row number does that occur? *Hint:* Find n using the expression you found in step 2.

■ Men's Boxing Weight Classes at the Summer Olympics

Research Project
One or more people

What you will need
- Data for this situation; you may use data from online, a magazine, or a newspaper

In this project, you will search online for the weight classes for men's Olympic boxing. Using that information, you will rewrite each weight class as an inequality.

1. Search online for a list of the weight classes for men's boxing at the Olympics. Print out the list and turn it in with your report.
2. Rewrite each weight class as an inequality. Make sure that you consider the end points of each weight class as part of the inequality.
3. List all weight classes in which a 65-kg man could participate.

■ Write a Review of a Section for Presentation

Research Project
One or more people

What you will need
- This presentation may be done in class or as an online video.
- If it is done as a video, post it on a website where it may be easily viewed by the class.
- You might want to use homework or other review problems from the book.
- Make it creative and fun.

Create a 5-minute review presentation of one section from Chapter 2. The format of the presentation can be a poster presentation, a blackboard presentation from notes, an online video, or a game format (for example, math jeopardy). The presentation should include the following:

a. Any important formulas in the chapter
b. Important skills in the chapter, backed up by examples
c. Explanation of terms used (for example, what is modeling?)
d. Common mistakes and how to avoid them

Cumulative Review — Chapters 1-2

1. Write in exponential form: $\dfrac{9}{4} \cdot \dfrac{9}{4} \cdot \dfrac{9}{4} \cdot \dfrac{9}{4} \cdot \dfrac{9}{4} \cdot \dfrac{9}{4}$.

2. Write in expanded form: $(-4)^3$.

3. Find the value without a calculator: $(-1)^5$.

4. Write using standard notation: 3.045×10^8.

5. Underline or circle the terms and simplify using the order-of-operations agreement:

$$\dfrac{84}{(10-6)} \cdot 3 - 2^2 - (-5) + 20 \div 4 \cdot (-3)$$

6. Simplify the expression using the order-of-operations agreement: $18 \div (-2) + |6 - 7 \cdot (-2)| + (-3)^2$.

7. Identify the property (commutative, associative, or distributive) used to rewrite the statement:

$(-16) + (3 - 7) = (3 - 7) + (-16)$

8. Define the variables: A local youth soccer club is selling candy bars. The amount of money (revenue) the club makes depends on the number of candy bars sold.

9. Translate the words into a variable expression: Seven less than 8 times a number.

10. Convert 95 feet to meters. There is 0.3048 meter in 1 foot.

11. Evaluate the expression for the given value of the variable: $-\dfrac{x}{3} + 7x$ for $x = -6$.

12. Lisa has a box of 50 dog biscuits she is going to divide up between her two dogs, Scout and Shadow. Fill in the input–output table to generate an expression for the problem situation.

Number of Biscuits for Scout	Number of Biscuits for Shadow
0	
5	
10	
15	
n	

13. Simplify: $7xy - 3x^2 + 6x - 9x^2 + 8xy - x$

14. Simplify: $6x - (3x - 5) + 8$

15. Simplify: $(3b - c) - (-b + 7c)$

16. Write the coordinates of each ordered pair shown on the graph.

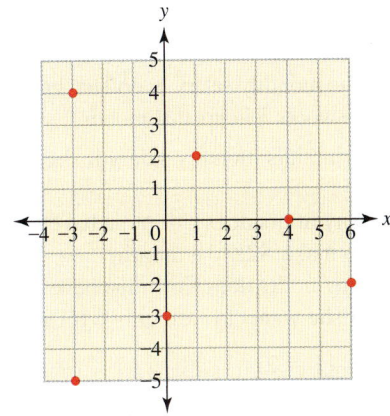

17. Graph the ordered pairs on the same set of axes: $(5, 6)\ (-1, -2)\ (7, -4)\ (-3, -3)\ (0, 4)\ (-1, 0)$.

18. The following table gives the world record progression for the men's long jump. Use the given data to create a scatterplot. Clearly label and scale the axes.

Year	Distance Traveled in Meters (athlete's name in parentheses)
1960	8.21 m (Ralph Boston, USA)
1961	8.24 m (Ralph Boston, USA)
1961	8.28 m (Ralph Boston, USA)
1962	8.31 m (Igor Ter-Ovanesyan, URS)
1964	8.31 m (Ralph Boston, USA)
1964	8.34 m (Ralph Boston, USA)
1965	8.34 m (Ralph Boston, USA)
1967	8.35 m (Igor Ter-Ovanesyan, URS)
1968	8.90 m (Bob Beamon, USA)
1991	8.95 m (Mike Powell, USA)

Source: www.track-and-field-jumpers.com
Prior to 1960, the world record was 8.13 m by Jesse Owens of the United States.

19. Are the values of the variables solutions to the equation? $4x - 7y = -6$: $x = -1.5$ and $y = 0$.

20. Solve: $-2 + y = -7$.

21. Solve: $x - \dfrac{1}{3} = \dfrac{5}{6}$.

22. Solve: $-0.051 + n = 4.006$.

23. Solve for the indicated variable: $3x - 5y = -30$ for y.

24. Solve: $-7x = -21$.

25. Solve: $\dfrac{2}{3}n = -8$.

26. Solve: $-5z = 7.95$.

27. Corinthia works as a dental assistant and makes $16 per hour after taxes. Corinthia can calculate her monthly salary using the equation:
$$S = 16h$$
where S is the monthly salary in dollars that Corinthia makes if she works h hours that month.
   a. Find Corinthia's monthly salary if she works 120 hours.
   b. How many hours does Corinthia have to work a month to make $2240 a month?
   c. How many hours does Corinthia have to work a month if she wants to earn only $1600 a month?

28. Solve: $8y - 11 = 21$.

29. Solve: $\dfrac{x}{5} - 4 = -9$.

30. Solve: $-1.2 + 3.7x = 15.45$.

31. Translate and solve: The sum of twice a number and 20 is 6.

32. Solve: $P = rgm$ for m.

33. Solve: $6x - 3y = -30$ for y.

34. Translate and solve: The sum of twice a number and 8 is equal to the difference between four times the number and 18.

35. Solve: $8y + 4 = 12y + 20$.

36. Solve: $7n + 8 = -3n + 2(4n - 12)$.

37. Solve: $5w - 6 + 3(-4w + 9) = w + 5$.

38. Given the following triangle, find the measure of each angle.

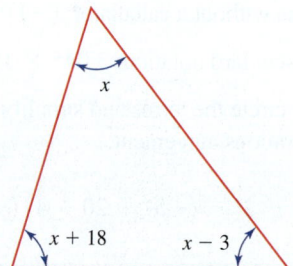

39. Solve the inequality: $6x + 1 > 49$.

40. Solve the inequality: $-\dfrac{1}{7}r + 5 \leq 9$.

41. Solve the inequality: $14 < 3x + 5 < 26$.

42. Solve the inequality: $10 - \dfrac{3}{8}t > -4$.

43. Write a compound inequality for the statement: The maximum enrollment in an English class on a college campus is 25. The class will be canceled if it does not have at least 15 students in it.

44. Write a compound inequality for the statement. At a state university, upper-division courses are labeled beginning with 200 to less than 400.

Linear Equations with Two Variables

3

3.1 Graphing Equations with Two Variables

3.2 Finding and Interpreting Slope

3.3 Slope-Intercept Form of Lines

3.4 Linear Equations and Their Graphs

3.5 Finding Equations of Lines

3.6 Modeling Linear Data

The relationship between the sighting of a flash of lightning and the distance the lightning is away from the observer can be represented by a linear equation. After seeing the lightning flash, count the seconds until the bang of thunder is heard. Divide this value by 5. The result will give an estimate of how many miles the lightning is away from you.

3.1 Graphing Equations with Two Variables

LEARNING OBJECTIVES

- Use tables to represent ordered pairs and data.
- Graph equations by plotting points.
- Graph nonlinear equations by plotting points.
- Graph vertical and horizontal lines.

Using Tables to Represent Ordered Pairs and Data

Many of the equations and problems we have studied so far have involved more than one variable. When the value of the input variable changes, the output value may change as well. In many problems, we are interested in more than one possible situation at a time. For example, if you work a part-time job, you might be interested in how much money (output) you will earn if you work different numbers of hours per week (input). A nutritionist might be interested in how different numbers of calories (input) affect a person's weight gain or loss (output).

In Chapter 1, we studied input–output tables relying on one variable, the input. We are now going to assign the output a variable as well because its value depends on the input. Recall that we sometimes call the input the *independent variable* and the output the *dependent variable*. We will now create input–output tables with two variables.

If Robin earns $11.50 per hour at her job as a receptionist, we could represent her weekly pay as

$$P = 11.50h$$

where P is the weekly pay in dollars when Robin works h hours per week. Because Robin may work a different number of hours each week, she may earn a different amount of money each week. In this situation, we would say that Robin's pay depends on how many hours she works. Therefore, P is the dependent variable, and h is the independent variable.

We can list some possibilities in a table of values such as the following.

h = hours worked	P = weekly pay (dollars)
1	11.50
5	57.50
20	230.00
30	345.00
40	460.00

We can easily fill a table such as this one by evaluating the given equation for several values of the input variable. In this case, we used the number of hours given in the table and calculated Robin's weekly pay using the given equation.

> **Skill Connection**
>
> **Input–Output Tables**
>
> In Section 1.2, we learned to put the input in the left column of the table and the output in the right column of the table. Recall that the input is the data entering a problem, and the output is the value being generated by the problem.

Example 1 Filling a table with data

Maria's monthly salary from her sales job includes $700 per month in base pay and 3% commission on all sales she makes. We can calculate Maria's monthly salary using the equation

$$M = 700 + 0.03s$$

where M is Maria's monthly salary in dollars when she makes s dollars in sales during the month. Use this equation to fill the following table with possible sales and salaries. Fill in the last two rows with any sales amounts you think are reasonable.

s = sales made (dollars)	M = monthly salary (dollars)
0	
1000	
5000	
10,000	
20,000	

SOLUTION

We are given some sales values to start with, so use the equation to calculate the monthly salary for the given sales amounts. In the last two rows, put any sales amounts that you think are reasonable for Maria in a month and then calculate the monthly salary using the equation.

s = sales made (dollars)	M = monthly salary (dollars)
0	700
1000	730
5000	850
10,000	1000
20,000	1300
50,000	2200
100,000	3700

Using the equation to find M

$M = 700 + 0.03(0)$
$M = 700$

$M = 700 + 0.03(1000)$
$M = 730$

$M = 700 + 0.03(5000)$
$M = 850$

$M = 700 + 0.03(10,000)$
$M = 1000$

$M = 700 + 0.03(20,000)$
$M = 1300$

$M = 700 + 0.03(50,000)$
$M = 2200$

$M = 700 + 0.03(100,000)$
$M = 3700$

The last two rows could have been filled with any sales amounts. The ones listed are just two possibilities.

PRACTICE PROBLEM FOR EXAMPLE 1

The Rocky Mountain White Water Rafting Company keeps track of costs for the rafting season using the equation

$$C = 50r + 4000$$

where C is the cost in dollars for the rafting season when r riders take a rafting trip during the season. Use this equation to fill the following table with possible riders and costs. Fill in the last two rows with other reasonable values for the number of riders.

r = riders	C = costs (dollars)
0	
50	
100	
150	
200	

Graphing Equations by Plotting Points

Problems can be presented with verbal statements, equations, or data tables. Often we would like to represent these problems visually using a graph. It is often easier to see trends in data by looking at a graph than by looking at a table or equation. This ability to gain information quickly from visual representations is what makes graphing important in mathematics.

We use the rectangular coordinate system first introduced in Section 1.4 to graph the information given by an equation or table of data. When considering a table of data, we have points, or coordinates, that can be plotted on a graph. Each table or equation that we have looked at so far has had an input and an output that are traditionally the x- and y-coordinates on the graph.

$$(\text{input, output}) = (x, y)$$

Looking at the equation $P = 11.50h$ and the table that represented Robin's weekly pay and hours worked, we have an input of hours worked and an output of weekly pay. The variable h is the **input variable** (sometimes called the independent variable). It will take the place of the x in each coordinate. The variable P is the **output variable** (sometimes called the dependent variable). It takes the place of the y in each coordinate. The table of data can be used to write coordinates for points on the graph.

h = hours worked	P = weekly pay (dollars)	Points on the graph (h, P)
1	11.50	(1, 11.50)
5	57.50	(5, 57.50)
20	230.00	(20, 230.00)
30	345.00	(30, 345.00)
40	460.00	(40, 460.00)

In Section 1.4, we learned that the order used in the coordinates is important. The input must be the first part of the coordinate, and the output is always the second part. The input of each coordinate represents the horizontal position, and the output of each coordinate represents the vertical position of the point. These points can now be plotted on a graph using the rectangular coordinate plane. Connect these points with a smooth curve. The line represents all of the possible combinations of hours worked and the associated weekly pay.

What's That Mean?

Input Variable and Output Variable

We have many names for the variables found in an equation or represented in a table or graph.

Input variable:
- input
- independent variable
- usually x

Output variable:
- output
- dependent variable
- usually y

Connecting the Concepts

What variables go on which axes?

Recall from Section 1.4 that we plot the input values on the horizontal axis and the output values on the vertical axis.

SECTION 3.1 Graphing Equations with Two Variables   235

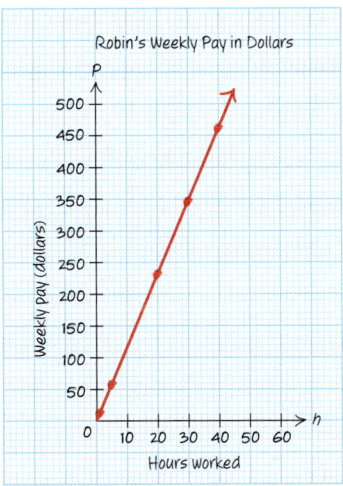

Skill Connection

Scale

Recall from Section 1.4 that we examine the data carefully to determine the scale on the axes.

In this discussion of Robin's pay, the horizontal scale is 10 because the input values range from 1 to 40. The vertical scale is 50 because the output values range from 11.50 to 460.00. A scale of 50 allows the vertical axis to include all the output values.

Steps to Sketch a Graph by Plotting Points

1. Create a table of points using the equation.
2. Plot the points on a set of axes. Clearly label and scale the axes.
3. Connect the points with a smooth curve.

Example 2 Creating a graph given a table

Use the table to write coordinates for points and graph them. Connect the points with a smooth curve. The equation is $y = 2x + 3$.

x	$y = 2x + 3$	(x, y)
-2	-1	
-1	1	
0	3	
1	5	
2	7	

SOLUTION

In this case, we are given an equation and a table already filled in with x- and y-values. Complete the table by writing coordinates using the given values. Then plot the points on a graph and connect them with a smooth curve.

Step 1 Complete the table.

x	$y = 2x + 3$	(x, y)
-2	-1	$(-2, -1)$
-1	1	$(-1, 1)$
0	3	$(0, 3)$
1	5	$(1, 5)$
2	7	$(2, 7)$

All these points lie on a straight line.

Steps 2 and 3 Plot the points and connect them with a smooth curve.

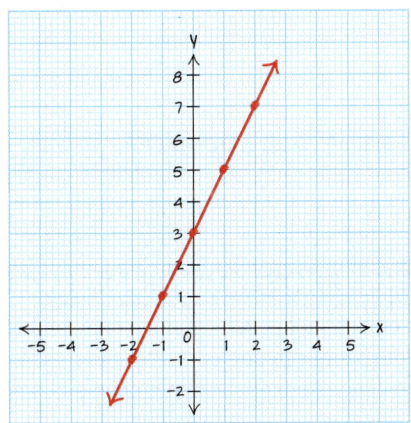

Connecting the Concepts

Is a straight line a smooth curve?

Mathematicians consider a straight line a smooth curve. The instructions for graphing will usually say to connect the points with a smooth curve. If the points follow a straight line, then draw a straight line. If the points do not follow a straight line, draw in a curve. Do not just connect the points with line segments.

236 CHAPTER 3 Linear Equations with Two Variables

PRACTICE PROBLEM FOR EXAMPLE 2

Use the table to write coordinates for points and graph them. Connect the points with a smooth curve. The equation is $y = -3x + 6$.

x	$y = -3x + 6$	(x, y)
-2	12	
-1	9	
0	6	
1	3	
2	0	

When building a table of points from an equation, we can choose any points we want. In most situations, it is best to pick at least one negative number, zero, and at least one positive number.

Example 3 — Generating a table to create a graph

Use the equation $y = x - 2$ to create a table of points and graph them. Connect the points with a smooth curve.

x	$y = x - 2$	(x, y)

SOLUTION

To fill in the table, select values for x and calculate y. We can choose any values for x. Select a few negative values, zero, and a few positive values that are easy to work with.

x	$y = x - 2$	(x, y)
-4	-6	$(-4, -6)$
-2	-4	$(-2, -4)$
0	-2	$(0, -2)$
2	0	$(2, 0)$
4	2	$(4, 2)$

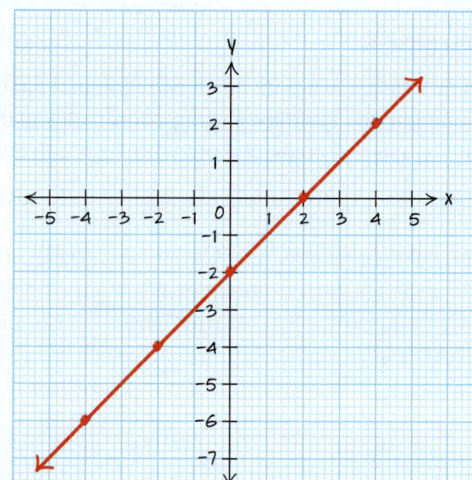

All these points lie on a straight line.

PRACTICE PROBLEM FOR EXAMPLE 3

Use the equation $y = 3x - 6$ to create a table of points and graph them. Connect the points with a smooth curve.

Skill Connection

Using graph paper

In algebra, we use graph paper to graph equations because we can draw a very accurate graph on it. Graph paper is readily available online.

SECTION 3.1 Graphing Equations with Two Variables 237

Example 4 Graphing Camille's weekly pay

Camille earns $10.75 per hour at her job and can represent her weekly pay with the equation $P = 10.75h$, where P is Camille's weekly pay in dollars when she works h hours.

a. Create a table of points that make sense in this situation.

b. Graph the points and connect the points with a smooth curve.

SOLUTION

a. The number of hours that Camille works is our input, so h is the independent variable and goes on the horizontal axis. The weekly pay in dollars is the output, so P is the dependent variable and goes on the vertical axis. Select different numbers of hours for Camille to work during a week and calculate the different weekly pay she would earn from each of these hours.

h	$P = 10.75h$	(h, P)
10	107.50	(10, 107.50)
15	161.25	(15, 161.25)
20	215.00	(20, 215.00)
30	322.50	(30, 322.50)
40	430.00	(40, 430.00)

b. Negative numbers of hours or a negative amount of pay does not make sense, so we will graph only positive values on both axes. In this situation, think about the scale that we should use on the axes. The horizontal axis should include values that go from zero to about 50, so we use a scale of 5. The vertical axis should include values from zero to about 550, so we use a scale of 50. These scales include all the values we found in the table. Label the axes with what each variable represents in the situation. The horizontal axis is labeled h for hours worked. The vertical axis is labeled P for weekly pay in dollars.

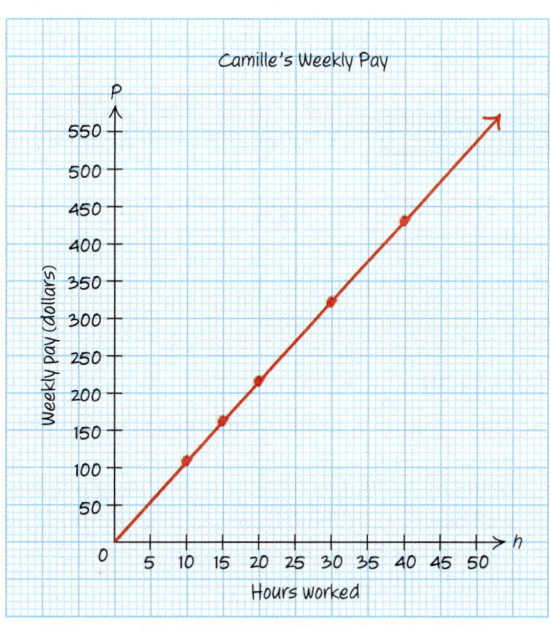

All the points on this line represent possible hours that Camille might work in a week and the weekly pay she earned. The table shows only a few specific examples, but the graph makes visible many more possibilities. Notice that the graph also shows the overall trend of Camille's pay increasing when she works more hours in a week.

Connecting the Concepts

How many values make a good display?

We generally place between 5 and 20 tick marks on each axis. Any more than 20 is too much detail to look at. Fewer than 5 may not give enough detail. Therefore, adjust the scale so that neither too many nor too few values sit on an axis. Be sure that all the input and output values will be included in the graph.

238 CHAPTER 3 Linear Equations with Two Variables

> ■ **Connecting the Concepts**
>
> **How many points do we graph?**
>
> In Examples 2 through 4, we plotted five points for each equation. For more complicated equations, such as those involving an x^2 or $|x|$ term, graph nine or more points. These additional points will help to make the graph more accurate.

PRACTICE PROBLEM FOR EXAMPLE 4

Over the course of an 8-week weight-loss diet, Philip's weight could be estimated by using the equation

$$W = 210 - 5n$$

where W is Philip's weight in pounds after n weeks on the diet.

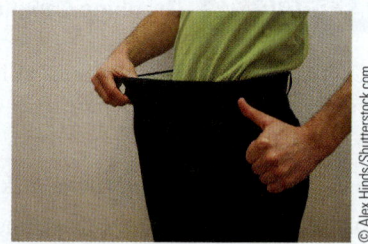

a. Create a table of points that make sense in this situation.

b. Graph the points and connect the points with a smooth curve.

All of the equations above graphed as straight lines. These graphs are called *linear*. An equation that generates a graph that is not a straight line is said to be *nonlinear*.

■ Graphing Nonlinear Equations by Plotting Points

When working with more complicated equations, we still graph them by creating a table of points, plotting the points, and connecting the points with a smooth curve. In these cases, plot additional points on the graph to be certain about the shape of the curve. Equations that do not graph as a straight line are called *nonlinear* equations.

Example 5 — Creating a nonlinear graph using a table

Use each equation to create a table of points and graph. Connect the points with a smooth curve.

a. $y = x^2 + 3$

b. $y = |x|$

SOLUTION

a. This equation is more complicated than previous ones because it includes a term with x^2. To make the graph more accurate, we find more points to plot for the graph. Begin by building a table of points using a few positive and negative x-values, as well as $x = 0$. Although we can pick any input values, we select x-values between -4 and 4 to make graphing easier. Then plot the points and connect them with a smooth curve.

x	$y = x^2 + 3$	(x, y)
-4	19	$(-4, 19)$
-3	12	$(-3, 12)$
-2	7	$(-2, 7)$
-1	4	$(-1, 4)$
0	3	$(0, 3)$
1	4	$(1, 4)$
2	7	$(2, 7)$
3	12	$(3, 12)$
4	19	$(4, 19)$

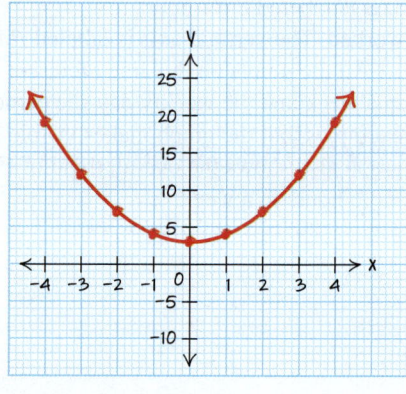

b. This equation includes an absolute value, so we generate a table of more points to plot for the graph. Build a table of points using some positive and negative *x*-values, as well as $x = 0$.

x	$y = |x|$	(x, y)
-5	5	$(-5, 5)$
-4	4	$(-4, 4)$
-3	3	$(-3, 3)$
-2	2	$(-2, 2)$
-1	1	$(-1, 1)$
0	0	$(0, 0)$
1	1	$(1, 1)$
2	2	$(2, 2)$
3	3	$(3, 3)$
4	4	$(4, 4)$
5	5	$(5, 5)$

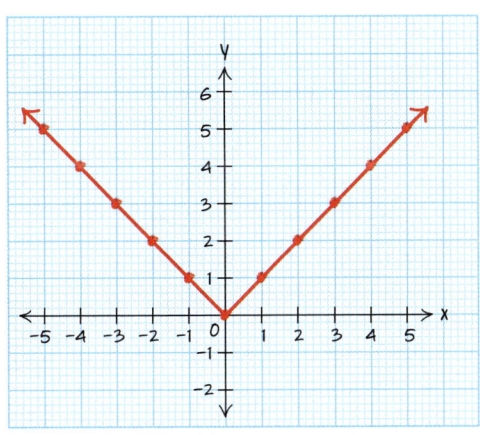

PRACTICE PROBLEM FOR EXAMPLE 5

Use each equation to create a table of points and graph. Connect the points with a smooth curve.

a. $y = x^2 - 5$

b. $y = |x + 4|$

Vertical and Horizontal Lines

All of the equations that we have graphed in this section have two variables. To create a graph for an equation with one variable, we generate a table of points. However, the table of points will be very simple. If $y = 5$ is the equation, then y will always be 5 no matter what value x is. Equations that have only one variable will result in **horizontal** or **vertical lines**.

> **DEFINITIONS**
>
> **Vertical Line**   The graph of an equation of the form $x = c$, where c is a real number, is a **vertical line.**
>
> **Horizontal Line**   The graph of an equation of the form $y = b$, where b is a real number, is a **horizontal line.**

What's That Mean?

Horizontal and Vertical

The word *horizontal* has the word *horizon* in it. This will help you remember that a horizontal line opens left and right, like the horizon. A vertical line opens straight up and down.

Example 6   Graphing horizontal and vertical lines

Use each equation to create a table of points and graph. Connect the points with a smooth curve.

a. $y = 5$

b. $x = -2$

SOLUTION

a. The value of *y* is constantly 5, so the values for *x* do not affect *y*. For any values of *x* selected, the value of *y* always equals 5.

x	y = 5	(x, y)
−3	5	(−3, 5)
−1	5	(−1, 5)
0	5	(0, 5)
1	5	(1, 5)
3	5	(3, 5)

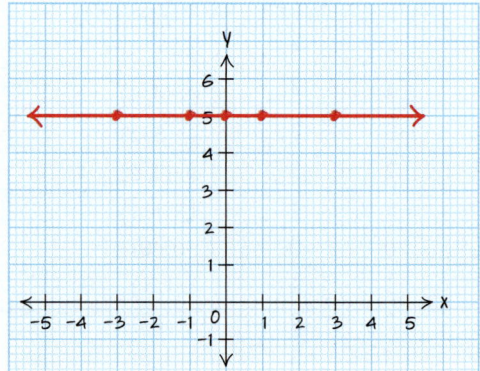

b. The value of *x* is always −2, so the values for *y* do not affect the value of *x*. Therefore, we can choose any values of *y* we want, and *x* equals −2.

x = −2	y	(x, y)
−2	−3	(−2, −3)
−2	−1	(−2, −1)
−2	1	(−2, 1)
−2	2	(−2, 2)
−2	4	(−2, 4)

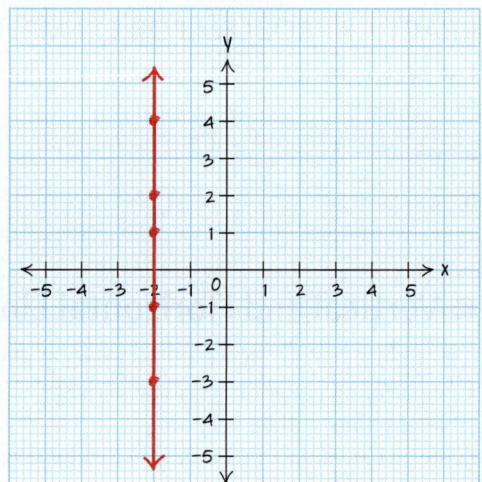

3.1 Vocabulary Exercises

1. _____ equations do not graph as a straight line.

2. _____ equations graph as a straight line.

3. The _____ variable is also known as the independent variable.

4. The _____ variable is also known as the dependent variable.

3.1 Exercises

For Exercises 1 through 8, fill in the given table. Complete the last two rows with values that are reasonable in the context of the given situation.

1. Martina earns a base salary of $300 per month and 5% commission on all sales she makes. Martina's monthly pay in dollars can be calculated by using the equation $P = 0.05s + 300$ if she sells s dollars of merchandise.

s = sales (dollars)	P = monthly pay (dollars)
0	
1000	
8000	

2. Maizie earns a base salary of $250 per month and 6% commission on all sales she makes. Maizie's monthly pay in dollars can be calculated by using the equation $P = 0.06s + 250$ if she sells s dollars of merchandise.

s = sales (dollars)	P = monthly pay (dollars)
0	
2000	
6000	

3. The total number of customers who visit a local sports equipment store can be estimated by using the equation $C = 8h + 2$, where h is the number of hours after opening for the day.

h = hours since opening	C = number of customers for the day
0	
1	
3	

4. The number of animals at the shelter has been growing since the beginning of the year. The number of animals can be calculated by using the equation $A = 4w + 20$, where w is the number of weeks since the beginning of the year.

w = weeks since the beginning of the year	A = animals in the shelter
0	
2	
6	

5. The cost in dollars to purchase one 2 × 4 stud is $3.50. The cost for framing a wall can be calculated by using the equation $C = 3.5n$ when n studs are purchased.

n = number of studs purchased	C = cost (dollars)
4	
6	
8	

6. The cost to purchase one plant in a 2-gallon container is $8.50. The cost to purchase plants in 2-gallon containers for a garden can be calculated by using the equation $C = 8.5n$ when n plants are purchased.

n = number of plants purchased	C = cost (dollars)
2	
4	
8	

7. The number of calories burned by a 190-pound person while bowling is 259 calories per hour. The number of calories burned by a 190-pound person while bowling for h hours can be estimated by the equation $C = 259h$.

h = hours of bowling	C = calories burned
1	
2	
3	

242 CHAPTER 3 Linear Equations with Two Variables

8. The number of calories burned by a 190-pound person while bicycling is 406 per hour. The number of calories burned by a 190-pound person while bicycling with a moderate effort for h hours can be estimated by the equation $C = 406h$.

h = hours of bicycling	C = calories burned
1	
2	
3	

For Exercises 9 through 16, graph the given set of points using graph paper. Clearly label and scale the axes. Connect the points with a smooth curve.

9. Graph the points in the table from Exercise 1 and connect them with a smooth curve.

10. Graph the points in the table from Exercise 2 and connect them with a smooth curve.

11. Graph the points in the table from Exercise 3 and connect them with a smooth curve.

12. Graph the points in the table from Exercise 4 and connect them with a smooth curve.

13. Graph the points in the table from Exercise 5 and connect them with a smooth curve.

14. Graph the points in the table from Exercise 6 and connect them with a smooth curve.

15. Graph the points in the table from Exercise 7 and connect them with a smooth curve.

16. Graph the points in the table from Exercise 8 and connect them with a smooth curve.

For Exercises 17 through 24, fill in the given table using the given equation. Graph the points. Connect the points with a smooth curve. Clearly label and scale the axes. Determine whether the graph is linear or nonlinear.

17.

t	$P = -8t + 17$
-4	
-2	
0	
2	
4	

18.

n	$C = 42n - 250$
-100	
-50	
0	
50	
100	

19.

x	$y = |x - 2|$
-6	
-3	
-1	
0	
2	
3	
4	
6	
10	

20.

x	$y = |5 - x|$
-16	
-8	
-2	
0	
5	
8	
10	
16	
20	

21.

s	$R = s^2 + 7$
-4	
-3	
-2	
-1	
0	
1	
2	
3	
4	

SECTION 3.1 Graphing Equations with Two Variables **243**

22.

s	$V = 3s^2$
−4	
−3	
−2	
−1	
0	
1	
2	
3	
4	

23.

x	$y = -2x^2 + 3$
−4	
−3	
−2	
−1	
0	
1	
2	
3	
4	

24.

x	$y = -x^2 + 1$
−4	
−3	
−2	
−1	
0	
1	
2	
3	
4	

For Exercises 25 through 42, use each given equation to create a table of five or more points and graph them. Connect the points with a smooth curve. Clearly label and scale the axes. Determine whether each graph is linear or nonlinear.

25. $y = 2x - 8$

26. $y = 3x - 9$

27. $y = 0.5x + 3$

28. $y = 0.25x + 1$

29. $y = -2x + 10$

30. $y = -5x + 15$

31. $y = -0.4x + 2$

32. $y = -0.75x + 4$

33. $y = 3x$

34. $y = 5x$

35. $y = -3x$

36. $y = -4x$

37. $y = \frac{1}{2}x + 3$

38. $y = \frac{1}{2}x - 4$

39. $y = \frac{2}{3}x - 5$

40. $y = \frac{3}{4}x + 1$

41. $y = -\frac{1}{5}x - 2$

42. $y = -\frac{1}{3}x + 4$

For Exercises 43 through 48, use the given information to answer the questions.

43. When lightning strikes, the time that it will take to hear thunder depends on the distance of the lightning strike from a location. The "flash to bang" rule says that the time delay is 5 seconds per mile. The time delay can be estimated by the equation $T = 5d$, where T is the time delay in seconds when lightning strikes d miles away. Create a table of points that make sense in this situation and graph them. Connect the points with a smooth curve.

44. The average American uses about 90 gallons of water per day. The amount of water the average person will use during d days can be estimated by using the equation $G = 90d$, where G is the total number of gallons of water the average person uses during d days. Create a table of points that make sense in this situation and graph them. Connect the points with a smooth curve.

45. The outside temperature can be estimated by using the equation $T = 75 - 3h$, where T is the outside temperature in degrees Fahrenheit h hours after 12 noon. Create a table of points that make sense in this situation and graph them. Connect the points with a smooth curve.

46. The distance that Marissa is from home after h hours of riding her bike can be estimated by using the equation $D = 8h + 2$, where D is her distance from home in miles. Create a table of points that make sense in this situation and graph them. Connect the points with a smooth curve.

47. The amount of grass still to be mowed at a park after m minutes of mowing can be estimated by using the equation $g = 24{,}000 - 100m$, where g is the number of square feet of grass still to be mowed. Create a table of points that make sense in this situation and graph them. Connect the points with a smooth curve.

48. The amount of money, m (in dollars), left in Choung's savings account w weeks after starting a home improvement project can be calculated by using the equation $m = 45{,}000 - 7500w$. Create a table of points that make sense in this situation and graph them. Connect the points with a smooth curve.

For Exercises 49 through 62, use each given equation to create a table of nine or more points and graph them. Connect the points with a smooth curve. Clearly label and scale the axes. Determine whether the graph is linear or nonlinear.

49. $y = x^2 + 1$
50. $y = x^2 + 5$
51. $y = x^2 - 4$
52. $y = x^2 - 2$
53. $y = (x + 3)^2$
54. $y = (x + 1)^2$
55. $y = (x - 2)^2$
56. $y = (x - 5)^2$
57. $y = |x| + 2$
58. $y = |x| - 3$
59. $y = |x + 2|$
60. $y = |x + 5|$
61. $y = |x - 3|$
62. $y = |x - 1|$

For Exercises 63 through 72, use each given equation to create a table of five or more points and graph them. Connect the points with a smooth curve. Clearly label and scale the axes. Determine whether the graph is linear or nonlinear.

63. $y = 6$
64. $y = 3$
65. $x = 7$
66. $x = 2$
67. $y = -3.5$
68. $y = -7$
69. $x = -1.5$
70. $x = -4$
71. $x = 0$
72. $y = 0$

For Exercises 73 through 76, simplify each expression using the order-of-operations agreement.

73. $30 \div (-15) \cdot 3^2 + (-2)^3 + |5 - 4(3)|$
74. $\dfrac{42}{15 - 2(4)} - (-5)^2 + 7 \cdot 4 \div 14$
75. $\dfrac{5}{2} \cdot \dfrac{4}{25} - \dfrac{2}{9} \cdot 3$
76. $\dfrac{2}{5} + 2 \cdot \dfrac{3}{10} - 4$

For Exercises 77 through 80, convert units as indicated. Round to the hundredths place if necessary. If needed see unit conversions in the back of the book.

77. Convert 12 centimeters to inches.
78. Convert 525 miles to kilometers.
79. Convert 185 pounds to kilograms.
80. Convert 95 kilograms to pounds.

For Exercises 81 through 84, simplify each expression.

81. $16x - 5(3x - 12)$
82. $-2(3 - 4t) - (t + 8)$
83. $-3(4n^2 - 2n + 7)$
84. $7(-6p^2 - 3p - 5)$

For Exercises 85 through 90, solve each equation. Provide reasons for each step. Check the answers. Round to the hundredths place if necessary.

85. $\dfrac{1}{2}t + \dfrac{1}{2} = -1$
86. $\dfrac{x}{15} - 1 = -\dfrac{2}{3}$
87. $0.4k - 2.2 = 3$
88. $-3.5x + 14 = -10.5$
89. $8(3 - x) + 7 = 5(2x - 1)$
90. $2(x - 5) = 14 - x$

For Exercises 91 and 92, solve each given literal equation for the indicated variables.

91. Solve $3x - 5y = 15$ for y.
92. Solve $-2x + 3y = -6$ for x.

For Exercises 93 and 94, solve each inequality. Write the solution set using inequality symbols, using interval notation, and graphing the solution set on a number line.

93. $-5x + 7 \leq 22$
94. $8 - 3x \geq 20$

3.2 Finding and Interpreting Slope

LEARNING OBJECTIVES
- Interpret graphs.
- Describe the rate of change.
- Calculate slope.
- Interpret slope.

Interpreting Graphs

When we examine graphs, there are several characteristics of the graph that give us information about the situation portrayed in the graph. The height of the line or curve tells us about the values of the outputs. The direction the curve goes in, reading from left to right, tells us about the kind of change that occurs over a unit change in input.

Consider a graph that compares the populations of two cities in the United States.

Which city has the larger population? Looking at the graph, we see that Cedar Rapids, Iowa, has a larger population than Fargo, North Dakota, over this time period. The line for Cedar Rapids is above the line for Fargo.

Which city's population is growing faster? Since the line representing Fargo is going up (reading from left to right) faster than the line for Cedar Rapids, we can tell that the population of Fargo is growing more rapidly than the population of Cedar Rapids.

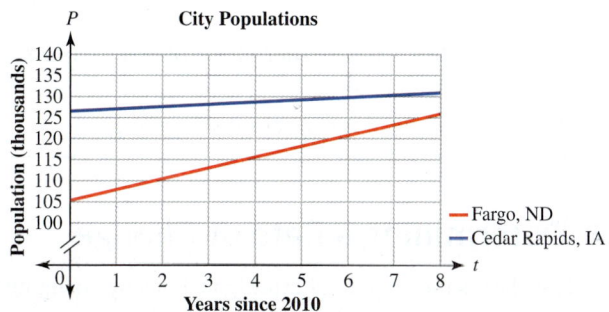

Source: Models derived from data found at the U.S. Census Bureau.

Connecting the Concepts

How do we read a graph?
Remember that in math, we read graphs from left to right. This is because the x-values increase from left to right.

Example 1 Interpreting graphs

Use the graph to answer the following questions.

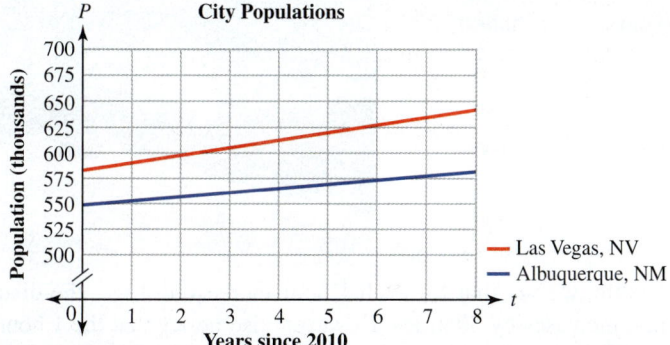

Source: Models derived from data found at the U.S. Census Bureau.

a. Which city has the smaller population? Explain your response.

b. Which city's population is growing faster? Explain your response.

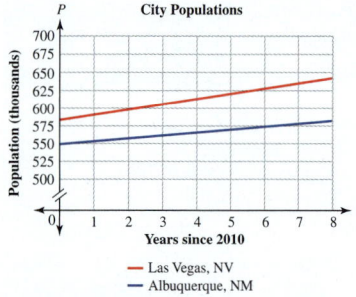

Source: Models derived from data found at the U.S. Census Bureau.

SOLUTION

a. Albuquerque, New Mexico, has a much smaller population than Las Vegas, Nevada. The graph shows the line representing Albuquerque as being much lower on the vertical axis.

b. Las Vegas, Nevada, is growing much more quickly than Albuquerque. The Las Vegas line is much steeper, so its population growth is greater.

PRACTICE PROBLEM FOR EXAMPLE 1

Use the graph to answer the following questions.

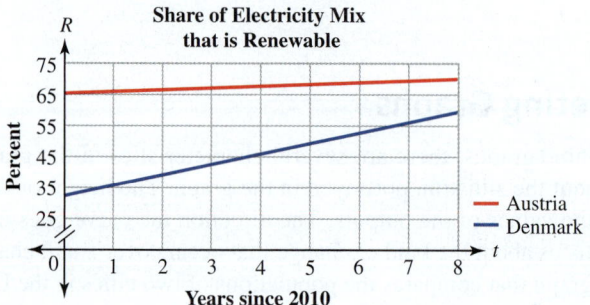

Source: Models derived from data found at Eurostat.

a. Which country has a higher share of renewable energy in the electricity mix? Explain how you know.

b. Which is growing at the faster rate? Explain.

■ Determining a Rate of Change

How fast is the amount of affordable housing changing?

How slowly is your salary increasing?

These are both questions about how quickly something is changing. In mathematics, how fast a quantity is changing is called a **rate of change.** Being able to measure change is a powerful tool that can help us to better understand a situation.

A familiar example of a rate of change is speed. When you are traveling in a car at 50 miles per hour, the distance from the starting place is increasing by 50 miles for every hour driven. A possible table of values of your driving distance from a starting location, traveling at an average speed of 50 mph, is as follows.

> ### ■ What's That Mean?
>
> **Rate**
>
> A *rate* is a ratio of two different kinds of quantities. The units for a rate are often written with the word *per*, which indicates the division of the two quantities.
>
> mph = miles per hour
>
> $$\frac{\text{miles}}{\text{hour}}$$
>
> m/s² = meters per second squared
>
> $$\frac{\text{meters}}{\text{second}^2}$$

t (hours)	Distance (miles)
0	0
1	50
2	100
3	150
4	200

From the table, we see that for each 1 hour increase in time, the distance in the second column increases by 50 miles. However, also notice that the 1 hour represents the *difference* between two consecutive inputs (time). The 50 miles also represents the *difference* between two consecutive outputs (distance). Recalling that a *difference* means to *subtract*, we write out this calculation by finding the changes using subtraction and then dividing the two values to get a speed in miles per hour.

SECTION 3.2 Finding and Interpreting Slope 247

t (hours)	Distance (miles)
0	0
1	50
2	100
3	150
4	200

Change in distance
$100 - 50 = 50$

Change in time
$2 - 1 = 1$

Divide the changes to find the rate of change, which is speed.

$\dfrac{50 \text{ miles}}{1 \text{ hour}} = 50 \text{ mph}$

Try again with two different rows.

$\dfrac{\text{change in distance}}{\text{change in time}} = \dfrac{200 - 150}{4 - 3} = \dfrac{50 \text{ miles}}{1 \text{ hour}} = 50 \text{ mph}$

Keep in mind that two quantities are compared in computing speed: the change in distance and the change in time. If we know the distance but do not know how long it took to travel that distance (that is, the time), then we cannot determine the speed. Because speed can be measured by using different units such as feet per second, miles per hour, meters per second, or kilometers per hour, be sure to use units when discussing speed or other rates of change.

This discussion of *speed* explains why in math and science, speed is often called the *rate*. Therefore, speed is the rate of change in a distance problem.

DEFINITION

Rate of Change The rate of change between two points is the change in outputs divided by the change in inputs.

$$\dfrac{\text{change in outputs}}{\text{change in inputs}}$$

Let's investigate the idea of speed a little more, using the following Concept Investigation.

■ CONCEPT INVESTIGATION

Are the speeds the same?

Use the tables to determine the average speed that each car was driving. Check more than one set of rows to verify any speed you find.

a.

t (hours)	Distance (miles)	Speed (mph)
1	60	
2	120	$\dfrac{120 - 60}{2 - 1} = \dfrac{60}{1} = 60$
3	180	
4	240	
5	300	

$\dfrac{\text{change in distance}}{\text{change in time}}$

248 CHAPTER 3 Linear Equations with Two Variables

What's That Mean?

Constant

In mathematics, the word *constant* is used to describe something that does not change.

In an expression or equation, we call numbers by themselves *constant terms*.

When we are discussing a rate of change, a *constant rate* means the rate is not changing.

b.

t (hours)	Distance (miles)	Speed (mph)
2	50	
3	75	
4	100	
6	150	
7	175	

c.

t (hours)	Distance (miles)	Speed (mph)
1	65	
2	100	
3	140	
4	200	
5	300	

Do all of these tables represent cars that are traveling at a **constant** (not changing) speed? If not, which table(s) represents constant speeds?

Using the table given before the preceding Concept Investigation and graphing the points yields the following graph.

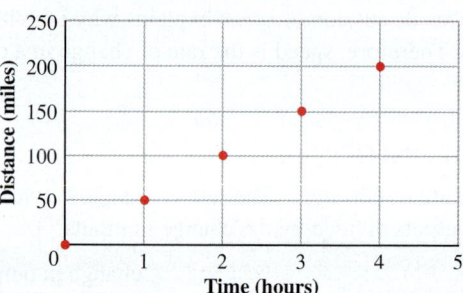

t (hours)	Distance (miles)
0	0
1	50
2	100
3	150
4	200

The distance is constantly changing as time increases. We represent this constant change by drawing a smooth curve through the points, as we did in Section 3.1. Connecting the points on this graph yields a straight line.

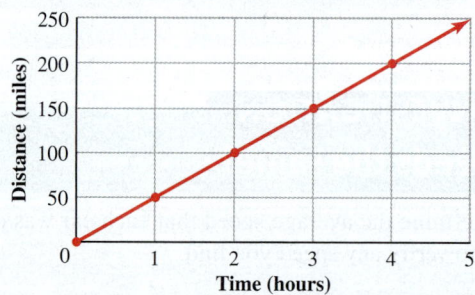

Now let's consider the data from the first Concept Investigation.

■ CONCEPT INVESTIGATION
Do these points make a line?

1. Plot the data from the following tables. Draw a smooth curve through the points.

a.

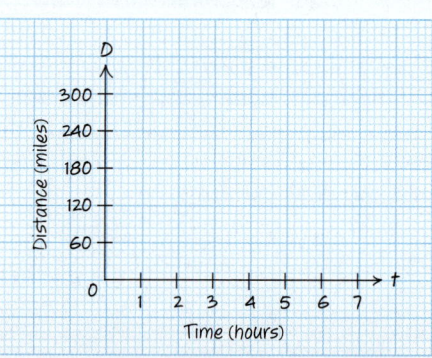

t (hours)	Distance (miles)
1	60
2	120
3	180
4	240
5	300

b.

t (hours)	Distance (miles)
2	50
3	75
4	100
6	150
7	175

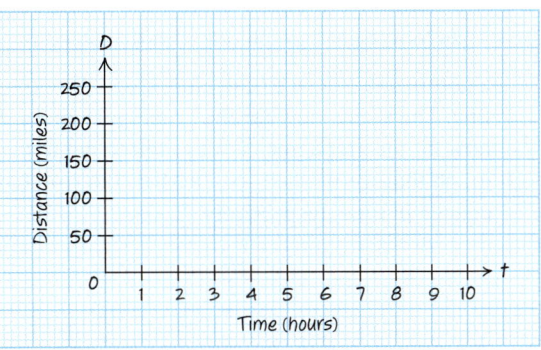

c.

t (hours)	Distance (miles)
1	65
2	100
3	140
4	200
5	300

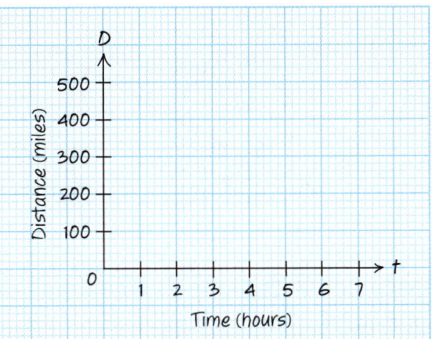

2. Connect the data points in each plot. Does each of the graphs look like a straight line? If not, which ones are not straight?

3. Do the two Concept Investigations in this section show a connection between the constant speed and the graph of a straight line? Explain your response.

The rate of change or speed in the preceding Concept Investigation is seen on the graph as the steepness of the line and whether the graph is rising or falling when read from left to right. In Example 1, we saw the rate of change of the populations.

In mathematics, the rate of change, steepness, and direction of the line are all called the **slope** of the line. An important feature of a linear graph is that the slope, or rate of change, is the same, or constant, for the entire line.

DEFINITION

Slope The slope of a line is the rate of change of a line.

$$\text{slope} = \frac{\text{change in outputs}}{\text{change in inputs}}$$

■ Calculating Slope

We now learn how to calculate the slope of a line. The basic components of rate of change are finding differences and then dividing them. The difference in outputs can be seen on a graph as the vertical change between the two points, reading from left to right. The vertical change is called the **rise**. The difference in inputs can be seen on the graph as horizontal change between the two points, reading from left to right. The horizontal change is called the **run**. Use one of the following descriptions of slope to remember the order correctly.

$$\text{slope} = \frac{\text{change in outputs}}{\text{change in inputs}} = \frac{\text{rise}}{\text{run}}$$

What's That Mean?

Rate of Change and Slope

In an everyday *application*, we often use the phrase *rate of change* to indicate how fast a quantity is changing. For example, when a car is traveling 65 miles per hour, the rate of change of distance (miles) per each unit of time (hour) is 65 mph.

When discussing a *linear equation*, we use the word *slope* to indicate the steepness of a line and whether the graph is rising or falling when read from left to right.

In looking at a graph, the idea that slope is $\frac{\text{rise}}{\text{run}}$ is usually the best way to understand and remember how to calculate slope. Working from left to right, pick two points on a line. The rise is how far the graph goes up or down to get to the next point. The run is how far the graph goes left to right to get to the next point. Using the distance and time graph that we worked with earlier, the rise and run are the change in distance and the change in time, respectively.

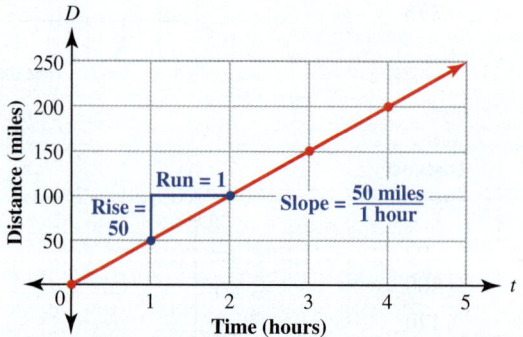

Steps to Calculate Slope Using a Graph

1. Pick two points on the line with coordinates that are easy to read.
2. Start with one point and draw a right triangle to the other point.
3. Determine the rise and run from one point to the other.
4. Divide the rise by the run.

$$\text{slope} = \frac{\text{change in outputs}}{\text{change in inputs}} = \frac{\text{rise}}{\text{run}}$$

Example 2 Using the graph to find slope

Use the graph to estimate the slope of the line. Make sure to work from left to right.

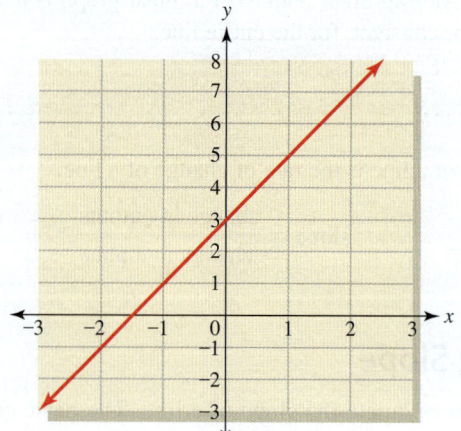

SOLUTION

Step 1 Pick two points on the line with coordinates that are easy to read.

When looking at a graph of a line, think of $\frac{\text{rise}}{\text{run}}$ as the template of how to calculate slope. First look for two points on the line that are easy to read. Usually points are

Skill Connection

Work left to right

In using a graph to find slope, it is best to work from left to right. Working from left to right reduces the number of possible sign errors that might be made.

easier to read when the line crosses an intersection on the grid. In this case, we chose the points (1, 5) and (2, 7).

Step 2 Start with one point and draw a right triangle to the other point.

Step 3 Determine the rise and run from one point to the other.

Looking at the graph, we see that we have a rise of 2 and a run of 1.

Step 4 Divide the rise by the run.

$$\text{slope} = \frac{\text{rise}}{\text{run}} = \frac{2}{1} = 2$$

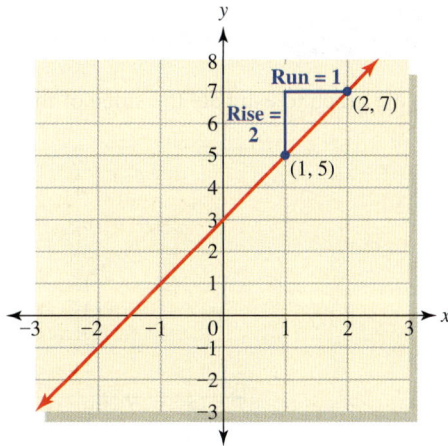

Notice that this slope is positive, which indicates that the line increases from left to right. Picking any other two points to calculate slope, yields the same value for slope. For example, the following graph uses the points $(-2, -1)$ and $(1, 5)$.

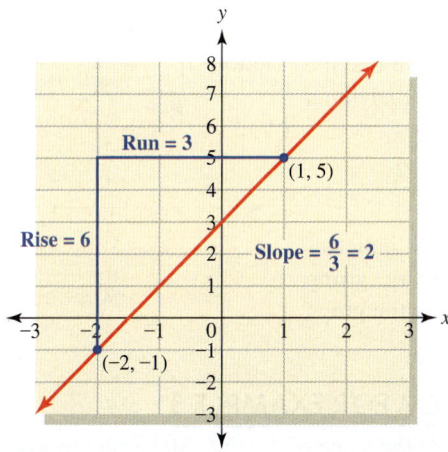

Therefore, any two points on the line can be used to calculate slope.

PRACTICE PROBLEM FOR EXAMPLE 2

Use the graph to estimate the slope of the line. Make sure to work from left to right.

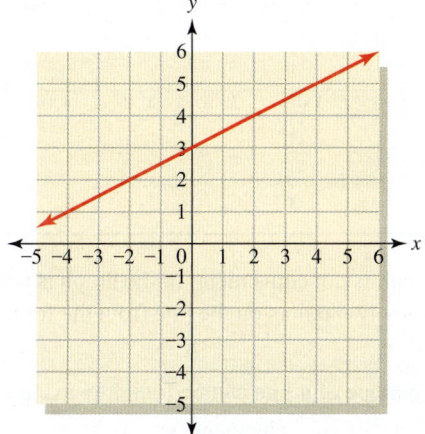

Example 3 The slope of decreasing lines

Use the graph to estimate the slope of the line. Make sure to work from left to right.

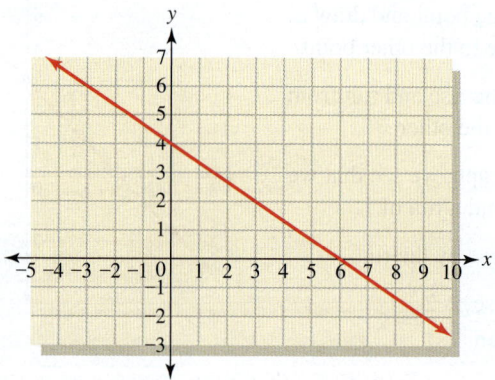

Connecting the Concepts

What is an increasing or decreasing line?

We say that a line **increases** if it goes up from left to right. A positive slope indicates an increasing line.

We say that a line **decreases** if it goes down from left to right. A negative slope indicates a decreasing line.

SOLUTION

Using the points $(-3, 6)$ and $(0, 4)$ gives a rise of -2 and a run of 3. In this case, the line is going down, so the rise is negative because the line is falling from left to right. Dividing the rise and run, yields the

$$\text{slope} = \frac{-2}{3} = -\frac{2}{3}$$

Note that the negative sign indicates that the line is decreasing, going downward, when reading the graph from left to right.

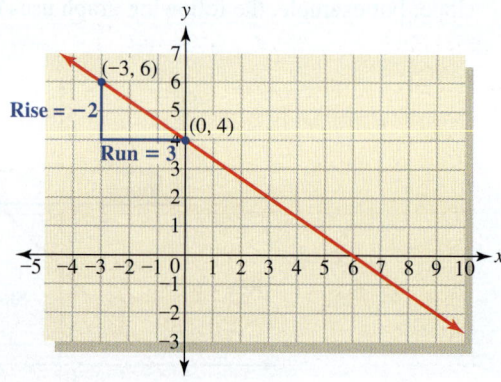

PRACTICE PROBLEM FOR EXAMPLE 3

Use the graph to estimate the slope of the line. Make sure to work from left to right.

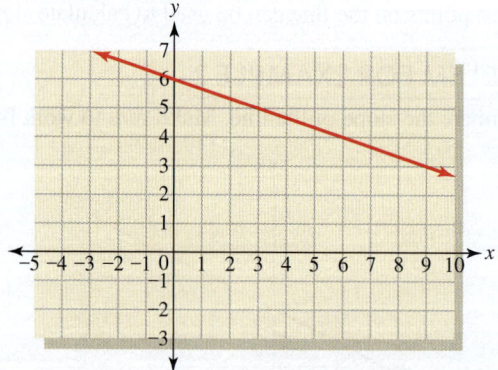

Sometimes using a graph is not convenient, or the graph is not given. In this case, we can find the slope using any two points on the line from the formula definition for slope.

> **Slope Formula** The slope of a line can be found by using the formula
> $$\text{slope} = \frac{\text{change in } y}{\text{change in } x} = \frac{y_2 - y_1}{x_2 - x_1}$$
> when (x_1, y_1) and (x_2, y_2) are any two points on the line.

SECTION 3.2 Finding and Interpreting Slope 253

Example 4 Using two points to find the slope

a. Find the slope of the line.

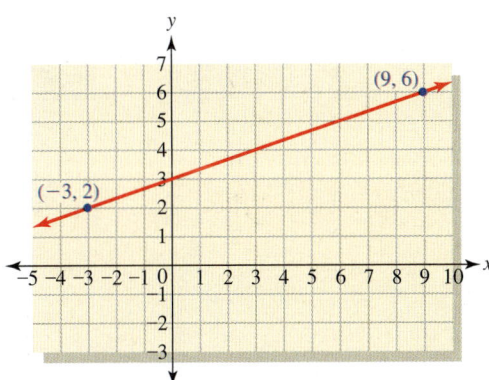

b. Find the slope of the line that goes through points (5, 7) and (8, 3).

SOLUTION

a. The points are given on the graph. Let $(x_1, y_1) = (-3, 2)$ and $(x_2, y_2) = (9, 6)$. Be careful with the negative x-value when subtracting.

$$\text{slope} = \frac{y_2 - y_1}{x_2 - x_1} = \frac{6 - 2}{9 - (-3)}$$

$$= \frac{4}{9 + 3} = \frac{4}{12} = \frac{1}{3}$$

The slope is positive, so the line should be increasing. From the graph, we see that the slope and graph agree.

b. The points are given as (5, 7) and (8, 3). Let $(x_1, y_1) = (5, 7)$ and $(x_2, y_2) = (8, 3)$. Substituting these points into the slope formula yields

$$\text{slope} = \frac{y_2 - y_1}{x_2 - x_1} = \frac{3 - 7}{8 - 5}$$

$$= \frac{-4}{3} = -\frac{4}{3}$$

This slope is negative, so the line should be decreasing. Graphing these two points and connecting them with a line yields a line that goes down from left to right.

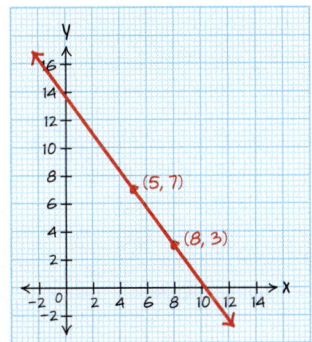

■ Connecting the Concepts

Does the slope formula give the same slope as $\frac{\text{rise}}{\text{run}}$?

In Example 4a, the slope formula gives the slope of $m = \frac{1}{3}$. Using the $\frac{\text{rise}}{\text{run}}$ method, we have that

$$m = \frac{\text{rise}}{\text{run}} = \frac{4}{12} = \frac{1}{3}$$

The slope is the same no matter which method is used to calculate it.

PRACTICE PROBLEM FOR EXAMPLE 4

Find the slope of the line that goes through the points (2, −5) and (7, 9).

Example 5 — Using any two points from a table to find slope

Use any two points from the table to find the slope of the line that passes through the points.

a.

x	y
1	10
3	18
7	34
8	38
10	46

b.

x	y
3	−1
3	0
3	1
3	2
3	5

c.

x	y
−2	1
−1	1
0	1
2	1

SOLUTION

a. Choose any two points in the table. Selecting the second and last points yields

$$\text{slope} = \frac{46 - 18}{10 - 3} = \frac{28}{7} = 4$$

Check this calculation by selecting two other points and substituting them into the slope formula.

$$\text{slope} = \frac{38 - 34}{8 - 7} = \frac{4}{1} = 4$$

b. Choose any two points in the table. Selecting the first and fourth points yields

$$\text{slope} = \frac{2 - (-1)}{3 - 3} = \frac{3}{0} \quad \text{Division by 0 is undefined.}$$

The slope of this line is undefined.

c. Choose any two points in the table. Selecting the first and second points yields

$$\text{slope} = \frac{1 - 1}{(-1) - (-2)} = \frac{0}{1} = 0$$

The slope of this line is 0.

PRACTICE PROBLEM FOR EXAMPLE 5

Use any two points from the table to find the slope of the line that passes through the points.

a.

x	y
0	24
2	23.5
8	22
10	21.5
20	19

b.

x	y
7	−3
7	0
7	8
7	11
7	15

In Example 5, we saw two special cases of slope. If we look at the rise and run of horizontal and vertical lines, we get the following.

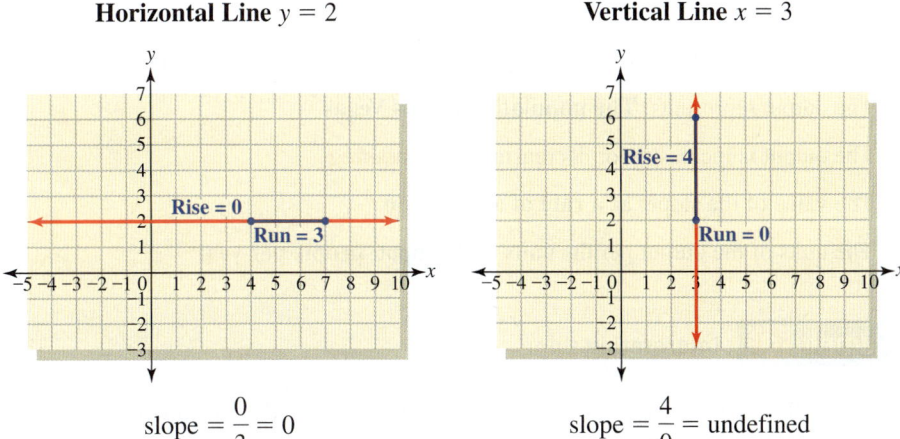

$$\text{slope} = \frac{0}{3} = 0 \qquad\qquad \text{slope} = \frac{4}{0} = \text{undefined}$$

These results are true for horizontal and vertical lines in general.

> **DEFINITIONS (TWO SPECIAL CASES)**
>
> **Slope of a Horizontal Line** The slope of a horizontal line is zero.
>
> **Slope of a Vertical Line** The slope of a vertical line is undefined.

■ Interpreting Slope

Recall Example 1 comparing the populations of Las Vegas and Albuquerque. In the following population graph, two points are given on each line, and the calculation of slope is carried out.

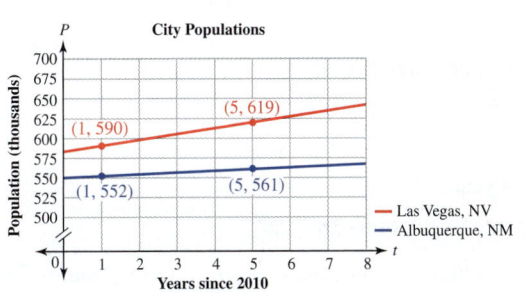

Source: Models derived from data at the U.S. Census Bureau.

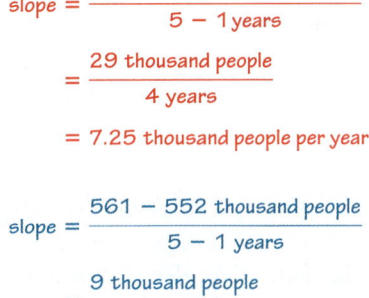

Connecting the Concepts

How can we convert units in thousands?

Recall from Section 1.2 that units are very important.

7.25 thousand people is the same as

7.25(1000) = 7250 people

The thousand can be in words or multiplied out to be represented in number form.

The population of Las Vegas is growing at a rate of about 7250 people per year, while the population of Albuquerque is growing at a rate of about 2250 people per year. Graphically, this correlates with the visual sense that the red line that represents Las Vegas is "steeper" than the blue line that represents Albuquerque.

When you are interpreting the slope in an application, it is important to include all the following information:

1. What the output represents
2. Whether the output is increasing or decreasing
3. The value of the slope
4. The units of the output variable per the units of the input variable

256 CHAPTER 3 Linear Equations with Two Variables

The explanation for the slope of the Las Vegas line above is:

> The population of Las Vegas **is increasing** at a rate of about 7250 **people per year.**

In this sentence, we have all four components of an interpretation for slope:

1. The output represents: **The population of Las Vegas**
2. The output is increasing or decreasing: **is increasing**
3. The value of the slope: **at a rate of about 7250**
4. The units of the output per the units of the input: **people per year**

Example 6 Interpreting slope

Calculate the slope of the lines in the given graph. Interpret the meaning of each slope in the context of the problem situation.

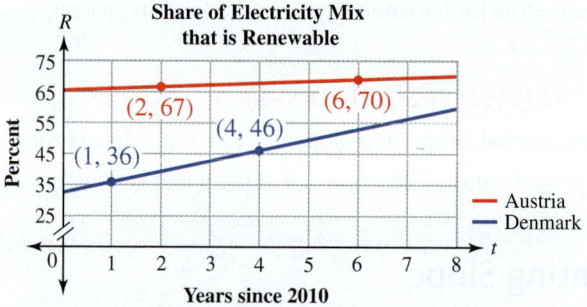

Source: Models derived from data found at Eurostat.

SOLUTION

Start with the Austria line. As we calculate the slope, include the units for the input and output values.

$$\text{slope} = \frac{70 - 67 \text{ percentage points}}{6 - 2 \text{ years}}$$

$$= \frac{3 \text{ percentage points}}{4 \text{ years}}$$

$$= 0.75 \text{ percentage points per year}.$$

The share of the electricity mix in Austria that is renewable is increasing by about 0.75 percentage points per year.

Calculating the slope for the Denmark line yields

$$\text{slope} = \frac{46 - 36 \text{ percentage points}}{4 - 1 \text{ years}}$$

$$= \frac{10 \text{ percentage points}}{3 \text{ years}}$$

$$\approx 3.33 \text{ percentage points per year}.$$

The share of the electricity mix in Denmark that is renewable is increasing by about 3.33 percentage points per year.

PRACTICE PROBLEM FOR EXAMPLE 6

Calculate the slope of the line in the given graph. Interpret the meaning of the slope in terms of the ticket sales.

■ Connecting the Concepts

Percents or percentage points?

When working with percentages, you may have some confusion between a percent increase or decrease and a change in percentage points.

If the percentage of people who smoke is currently 40% and there is a 5% increase in the number of people who smoke, we have the following:

original + increase = new amount
40 + (40 * 0.05) = 42

After a 5% increase, now 42% of the population smokes.

This represents a 2 *percentage point* increase from 40% to 42%. Therefore, a 5% increase is not always the same as a 5 percentage point increase.

In working with an application that involves an output measured as a percentage, the slope will be interpreted as a percentage point increase or decrease.

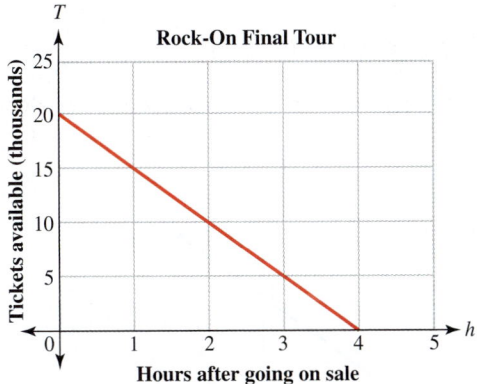

CONCEPT INVESTIGATION

Which line has the greater slope?

Consider the following graph.

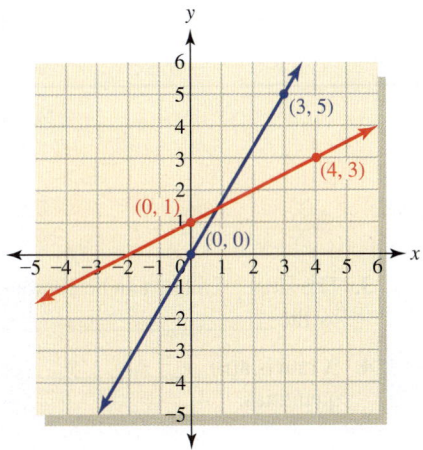

1. Compute the slope for the blue line.
2. Compute the slope for the red line.
3. Which line has the greater slope? (*Hint*: Compare the two numbers you got in parts a and b.)
4. Which line is steeper? Another way to ask this is which line is increasing faster? Read the graph from left to right.
5. How does the slope of the line relate to which line is steeper?

The preceding Concept Investigation shows that for lines with positive slope, the greater the slope, the steeper the line.

Example 7 Comparing slopes of lines

Examine the graph given below. Which line has the greater slope? Explain how you know.

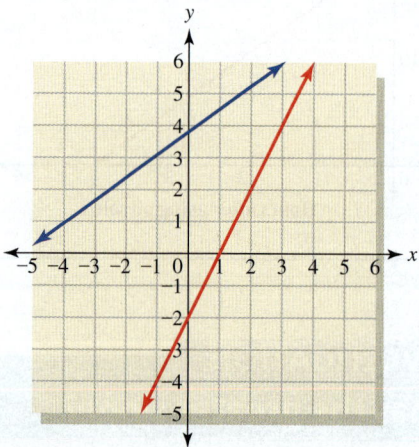

SOLUTION

The red line is steeper than the blue line, reading from left to right. Therefore, the red line has the greater slope.

3.2 Vocabulary Exercises

1. A line that goes up, reading left to right, is said to be a(n) _____ line.

2. A line that goes down, reading left to right, is said to be a(n) _____ line.

3. The slope of a line can be computed as the rise divided by the _____.

4. A rate is a(n) _____ of two different quantities.

3.2 Exercises

For Exercises 1 through 8, use the given graphs to answer the questions. Do not calculate any values.

1.

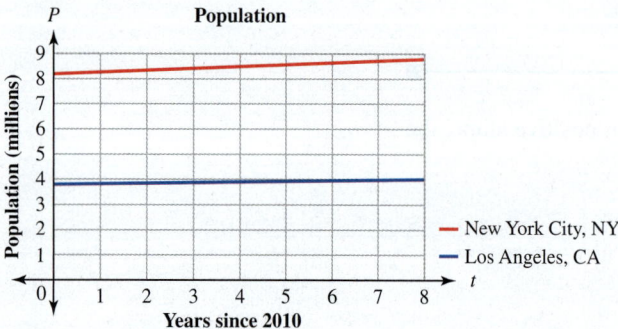

Source: Models derived from data at the U.S. Census Bureau.

a. Which city has the larger population during these years? Explain how you conclude that from the graph.

b. Which city is growing faster during these years? Explain how you conclude that from the graph.

2.

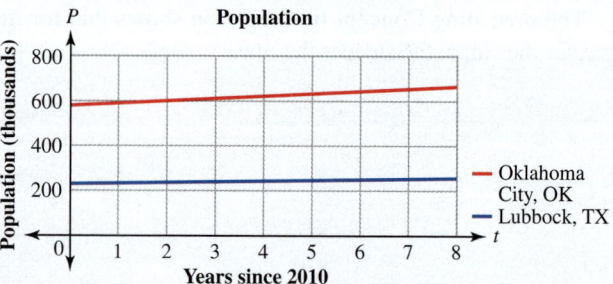

Source: Models derived from data at the U.S. Census Bureau.

a. Which city has the larger population during these years? Explain how you conclude that from the graph.

b. Which city is growing faster during these years? Explain how you conclude that from the graph.

3.

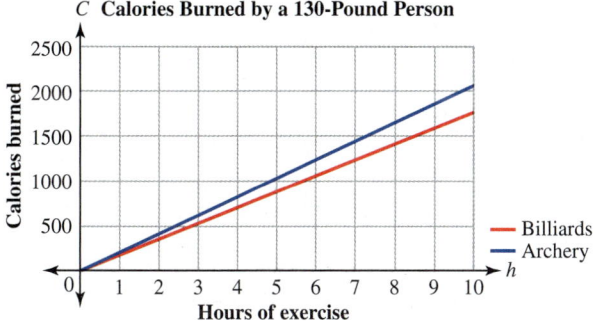

Source: Nutristrategy.com

Which kind of exercise burns more calories per hour? Explain how you conclude that from the graph.

4.

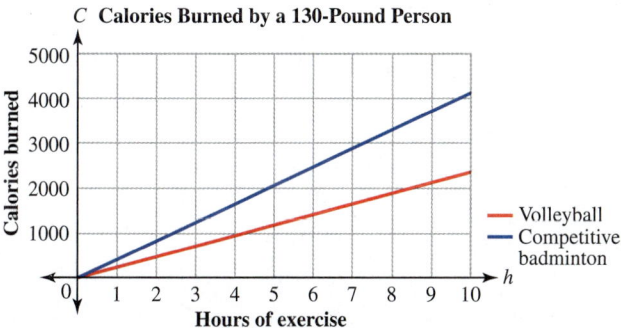

Source: Nutristrategy.com

Which kind of exercise burns more calories per hour? Explain how you conclude that from the graph.

5.

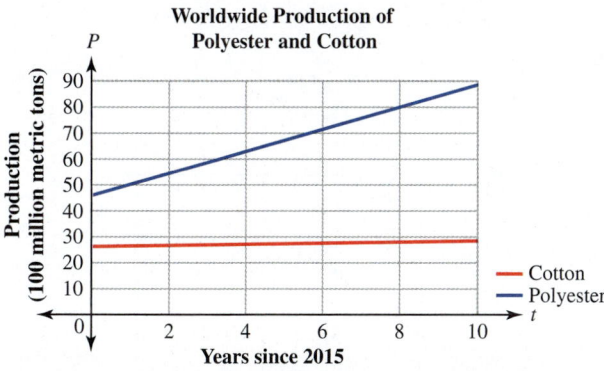

Source: Models derived from data found at Quartz.com

a. Which fiber has a higher production worldwide? Explain how you conclude that from the graph.

b. Which fiber's worldwide production is growing more slowly? Explain how you conclude that from the graph.

6.

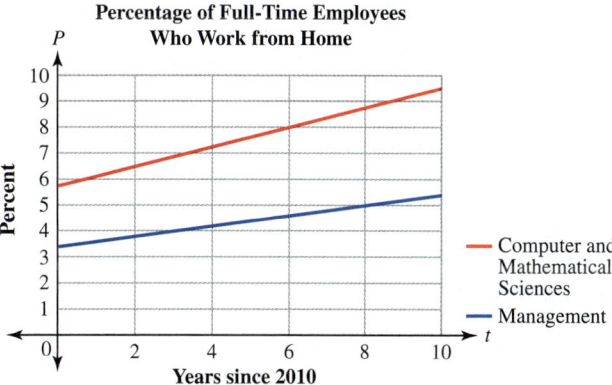

Source: Centers for Disease Control and Prevention.

a. Which job category has the highest percentage of full-time employees who work from home? Explain how you conclude that from the graph.

b. Which job category has the fastest percentage growth in work from home employees? Explain how you conclude that from the graph.

7.

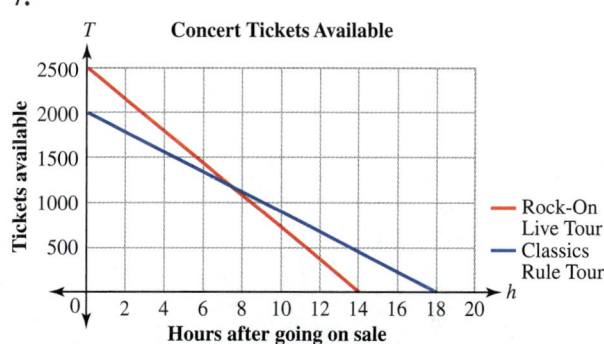

a. Which tour had more tickets available at the start of sales? Explain how you conclude that from the graph.

b. Which tour sold out first? Explain how you conclude that from the graph.

c. Which tour's available tickets decreased faster? Explain how you conclude that from the graph.

8.

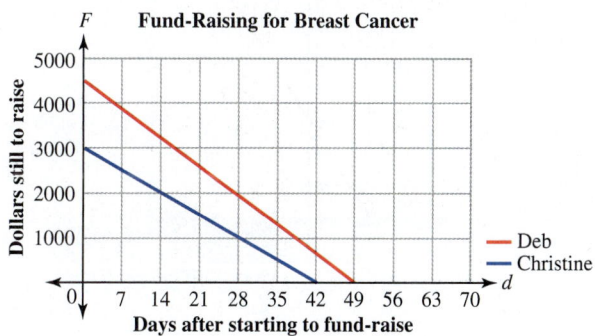

260 CHAPTER 3 Linear Equations with Two Variables

a. Who had the higher fund-raising goal? Explain how you conclude that from the graph.

b. Who finished fund-raising first? Explain how you conclude that from the graph.

c. Whose fund-raising came in at a faster rate? Explain how you conclude that from the graph.

For Exercises 9 through 22, find the slope of each line given in the graphs.

9.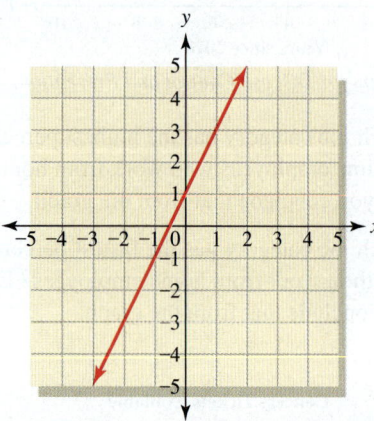

10.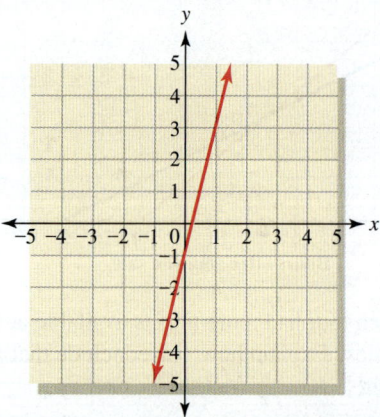

11.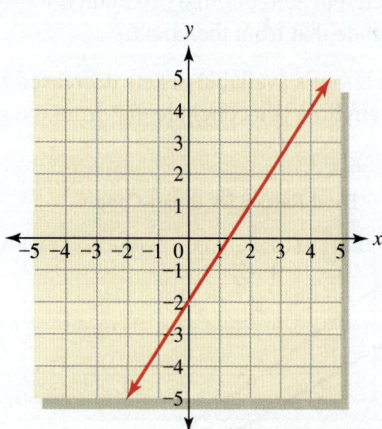

12.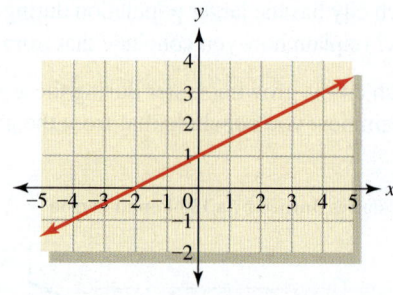

13.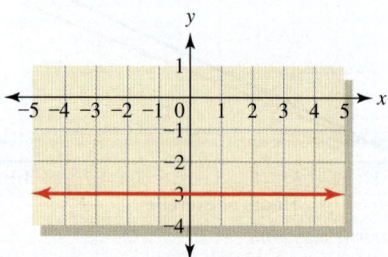

14.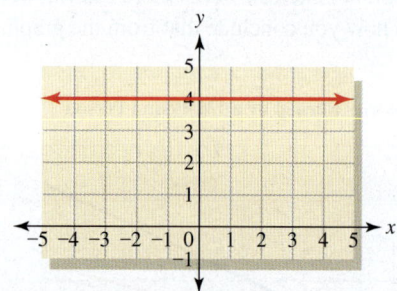

15.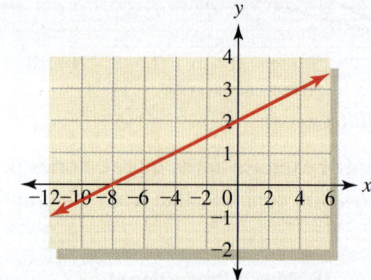

16.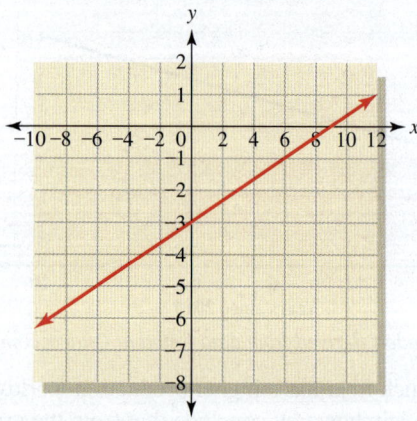

17.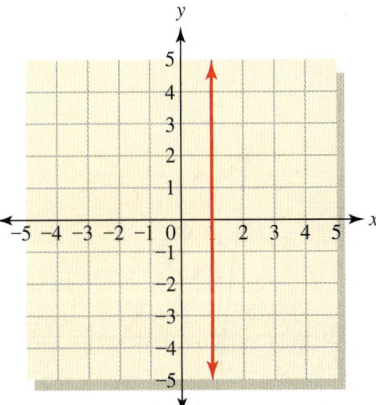

18.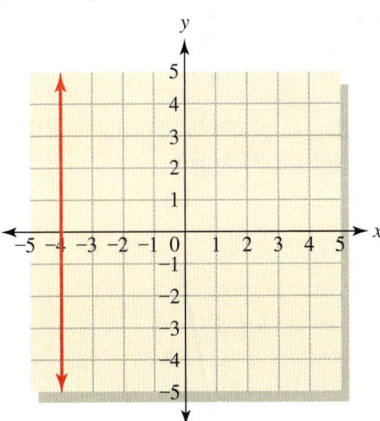

19.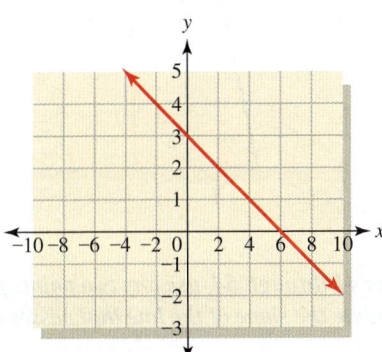

20.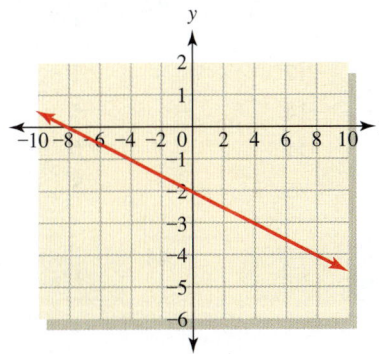

21.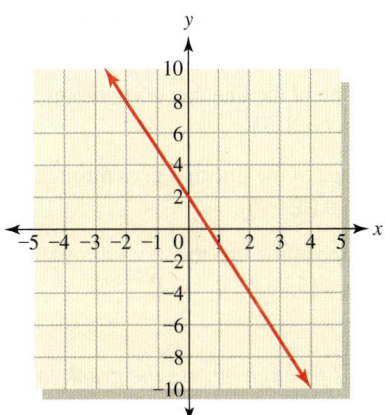

22.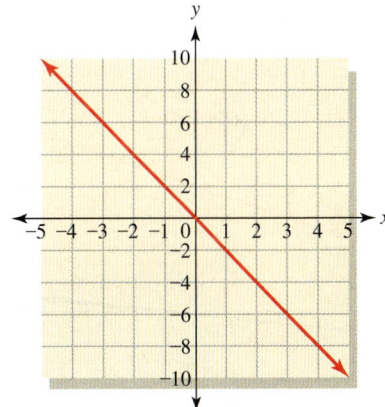

23. Find the slope of the line that goes through points (2, 5) and (6, 13).

24. Find the slope of the line that goes through points (3, 7) and (5, 17).

25. Find the slope of the line that goes through points (0, 2) and (4, 3).

26. Find the slope of the line that goes through points (1, 3) and (7, 5).

27. Find the slope of the line that goes through points $(-4, -7)$ and (2, 11).

28. Find the slope of the line that goes through points $(-2, -14)$ and $(1, -2)$.

29. Find the slope of the line that goes through points (4, 8) and (1, 8).

30. Find the slope of the line that goes through points $(-4, -3)$ and $(2, -3)$.

31. Find the slope of the line that goes through points $(-3, -7.5)$ and $(1, -5.5)$.

32. Find the slope of the line that goes through points $(-4, -9)$ and $(2, -7.5)$.

262 CHAPTER 3 Linear Equations with Two Variables

33. Find the slope of the line that goes through points (2, 5) and (2, 6).

34. Find the slope of the line that goes through points (−8, 9) and (−8, 7).

35. Find the slope of the line that goes through points (−3, 18) and (3, −6).

36. Find the slope of the line that goes through points (−5, 22) and (5, −8).

For Exercises 37 through 40, use the graph to answer the following questions.

37. Which line has slope closer to zero: the red line or the blue line? Explain.

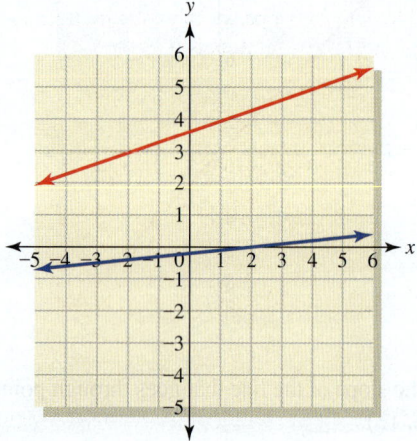

38. Which line has slope closer to zero: the red line or the blue line? Explain.

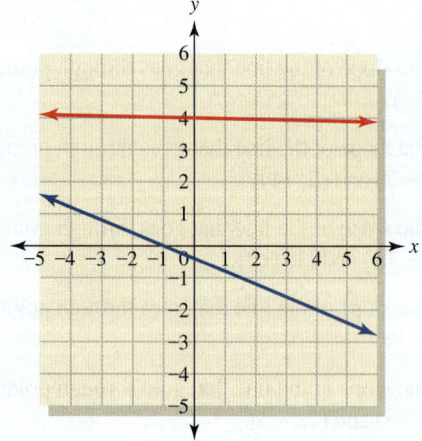

39. Which line has slope closer to being undefined: the red line or the blue line? Explain.

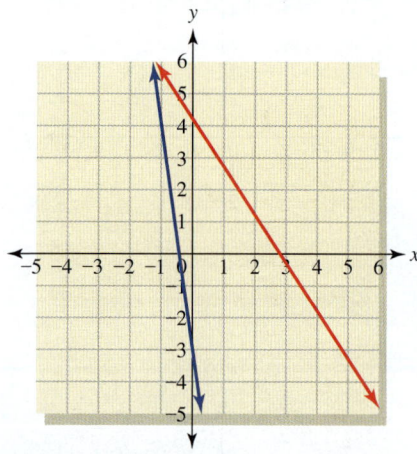

40. Which line has slope closer to being undefined: the red line or the blue line? Explain.

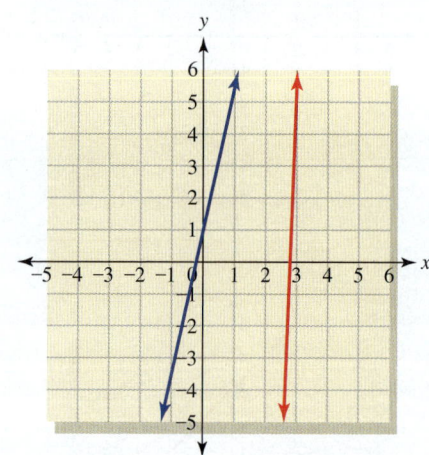

For Exercises 41 through 54, use any two points from each table to calculate the slope of the line that passes through the points.

41.

x	y
1	−17
2	−14
3	−11
5	−5

42.

x	y
1	−3
5	5
6	7
8	11

43.

x	y
2	−2
5	2.5
7	5.5
10	10

44.

x	y
1	−6.75
5	−1.75
7	0.75
10	4.5

SECTION 3.2 Finding and Interpreting Slope

45.

x	y
−4	2
−4	−3
−4	−7
−4	−12

46.

x	y
6	−2
6	−1
6	3
6	11

47.

x	y
3	8
9	10
12	11
15	12

48.

x	y
2	−7.5
6	−6.5
10	−5.5
18	−3.5

49.

x	y
1	2
3	−6
8	−26
11	−38

50.

x	y
2	−11
8	−29
11	−38
16	−53

51.

x	y
1	10
7	10
12	10
22	10

52.

x	y
−8	−2
−1	−2
3	−2
7	−2

53.

x	y
−3	9.5
−1	8.5
2	7
5	5.5

54.

x	y
−5	12.5
−2	11.6
0	11
5	9.5

55. **a.** Calculate the slope for the New York City, New York, line given in the graph in Exercise 1 on page 258. Use the points (4, 8.5) and (8, 8.8).

   b. Interpret the meaning of the slope you found in part a.

56. **a.** Calculate the slope for the Los Angeles, California, line given in the graph in Exercise 1 on page 258. Use the points (0, 3.8) and (8, 4).

   b. Interpret the meaning of the slope you found in part a.

57. **a.** Calculate the slope for the Oklahoma City, Oklahoma, line given in the graph in Exercise 2 on page 258. Use the points (0, 582) and (6, 640).

   b. Interpret the meaning of the slope you found in part a.

58. **a.** Calculate the slope for the Lubbock, Texas, line given in the graph in Exercise 2 on page 258. Use the points (1, 234) and (8, 257).

   b. Interpret the meaning of the slope you found in part a.

59. **a.** Calculate the slope for the archery line given in the graph in Exercise 3 on page 259.

   b. Interpret the meaning of the slope you found in part a.

60. **a.** Calculate the slope for the billiards line given in the graph in Exercise 3 on page 259.

   b. Interpret the meaning of the slope you found in part a.

61. **a.** Calculate the slope for the competitive badminton line given in the graph in Exercise 4 on page 259.

   b. Interpret the meaning of the slope you found in part a.

62. **a.** Calculate the slope for the volleyball line given in the graph in Exercise 4 on page 259.

   b. Interpret the meaning of the slope you found in part a.

63. **a.** Calculate the slope for the cotton line given in the graph in Exercise 5 on page 259. Use the points (4, 27.13) and (8, 28.01).

   b. Interpret the meaning of the slope you found in part a.

64. **a.** Calculate the slope for the Polyester line given in the graph in Exercise 5 on page 259. Use the points (0, 46) and (5, 67.35).

   b. Interpret the meaning of the slope you found in part a.

65. **a.** Calculate the slope for the Computer and Mathematical Sciences line given in the graph in Exercise 6 on page 259. Use the points (0, 5.8) and (10, 9.5).

   b. Interpret the meaning of the slope you found in part a.

66. **a.** Calculate the slope for the Management line given in the graph in Exercise 6 on page 259. Use the points (3, 4) and (8, 5).

   b. Interpret the meaning of the slope you found in part a.

67. **a.** Calculate the slope for the Rock-On Live Tour line given in the graph in Exercise 7 on page 259.

   b. Interpret the meaning of the slope you found in part a.

68. **a.** Calculate the slope for the Classics Rule Tour line given in the graph in Exercise 7 on page 259.

   b. Interpret the meaning of the slope you found in part a.

69. **a.** Calculate the slope for the Deb line given in the graph in Exercise 8 on page 259.

   b. Interpret the meaning of the slope you found in part a.

70. **a.** Calculate the slope for the Christine line given in the graph in Exercise 8 on page 259.

   b. Interpret the meaning of the slope you found in part a.

264 CHAPTER 3 Linear Equations with Two Variables

For Exercises 71 through 74, state which line in each graph (red or blue) has the greater slope. Do not calculate the slope. See Example 7 on page 259 for help. Explain how you know.

71.

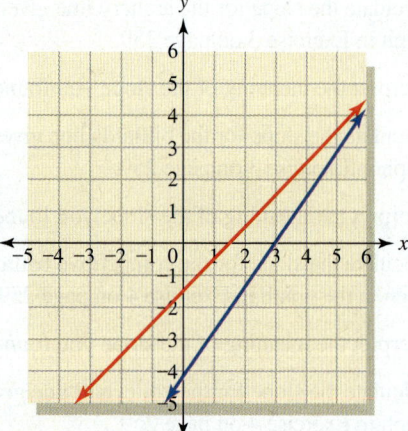

72.

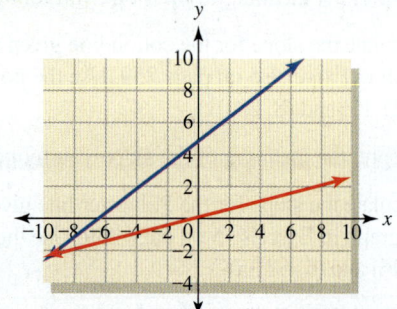

73.

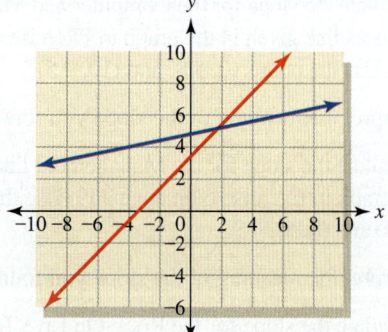

74.

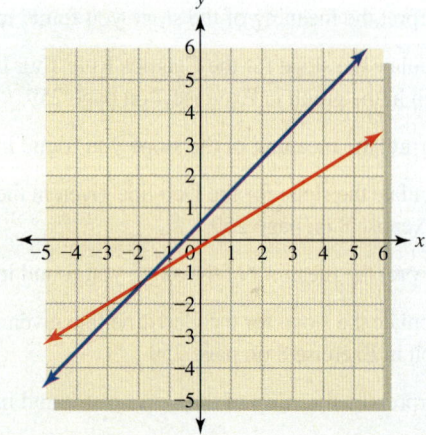

For Exercises 75 through 80, state whether the line in each graph is increasing, decreasing, or neither. Explain how you know.

75.

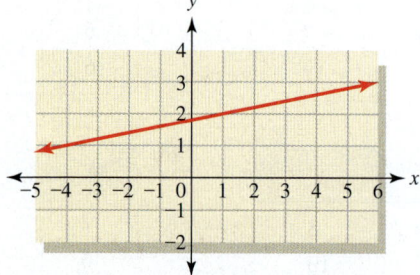

76.

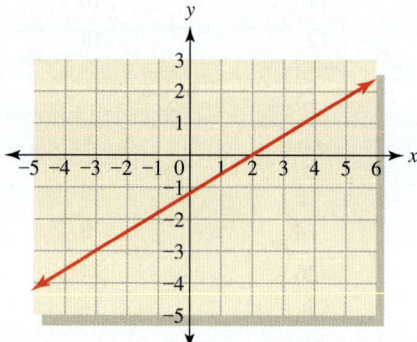

77.

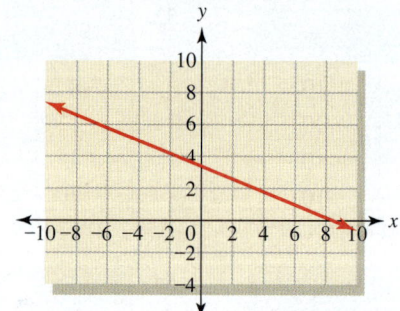

78.

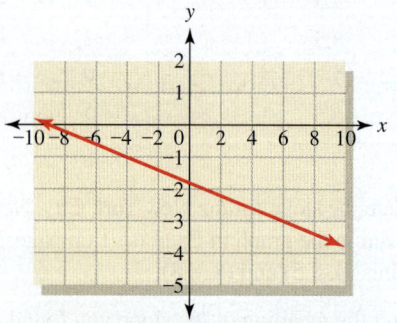

79.

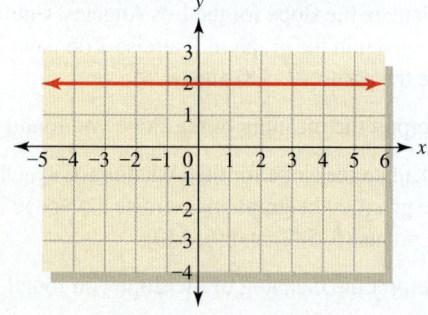

80.

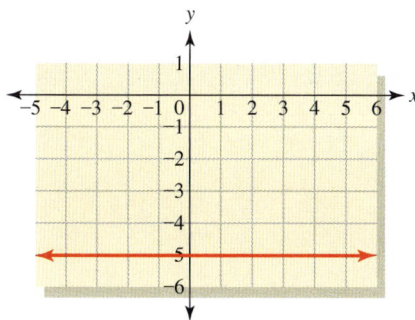

For Exercises 81 through 86, state whether each line is increasing, decreasing, or neither. Explain how you know. (Hint: Graph the line.)

81. $y = -3x + 19$

82. $y = -0.5x - 12$

83. $y = -\dfrac{5}{4}$

84. $y = -\dfrac{2}{3}$

85. $y = \dfrac{6}{5}x - 7$

86. $y = 8x - 21$

For Exercises 87 through 94, state whether each output quantity is increasing, decreasing, or neither.

87. Mallory's weekly pay P can be computed by the formula $P = 12.5h$, where h is the number of hours she works per week.

88. Jonas's weekly pay P can be computed by the formula $P = 11.75h$, where h is the number of hours he works per week.

89. The temperature T outside after 2:00 P.M. may be computed by $T = 89 - 3h$, where h is the number of hours after 2:00 P.M.

90. The temperature T outside after noon may be computed by $T = 65 - 2h$, where h is the number of hours after noon.

91. Patrick's weight W while dieting can be computed by the formula $W = 189 - 2t$, where t is the number of weeks that he has been on the diet.

92. Larissa's weight W while dieting can be computed by the formula $W = 165 - t$, where t is the number of weeks that she has been on the diet.

93. The distance D that Minh drives in h hours can be computed by the formula $D = 66h$.

94. The distance D that Trevor drives in h hours can be computed by the formula $D = 71h$.

For Exercises 95 and 96, answer the questions on the basis of the given information.

95. Use the graph to answer the following questions.

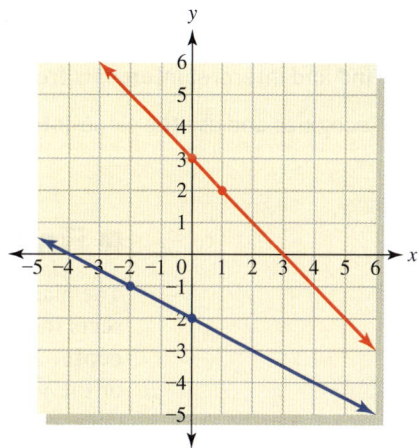

a. Which line appears to be decreasing more rapidly: the red line or the blue line?

b. Compute the slopes of both lines.

c. Look at your answers from parts a and b. Does the line that is decreasing more rapidly (part a) have a slope that is greater than or less than the slope of the other line (part b)?

96. Use the graph to answer the following questions.

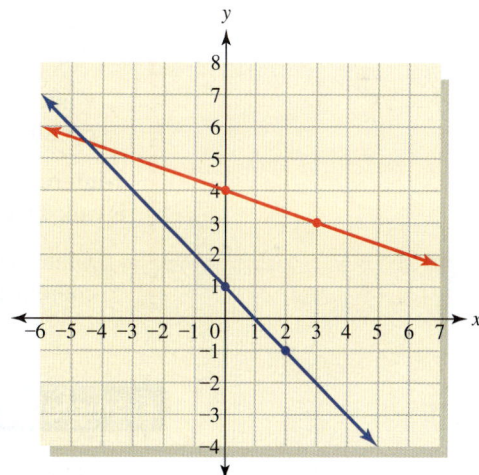

a. Which line appears to be decreasing more rapidly: the red line or the blue line?

b. Compute the slopes of both lines.

c. Look at your answers from parts a and b. Does the line that is decreasing more rapidly (part a) have a slope that is greater than or less than the slope of the other line (part b)?

3.3 Slope-Intercept Form of Lines

LEARNING OBJECTIVES
- Find and interpret intercepts from graphs.
- Find and interpret intercepts from equations.
- Recognize and use the slope-intercept form of a line.

■ Finding and Interpreting Intercepts from Graphs

The places where a curve intersects the axes are called **intercepts**. The vertical intercept, or the *y*-intercept, is where the curve touches the vertical axis or *y*-axis. The horizontal intercept, or the *x*-intercept is where the curve touches the horizontal axis or *x*-axis.

The *y*-intercept on this graph is the point (0, 4), and the *x*-intercept is the point (6, 0).

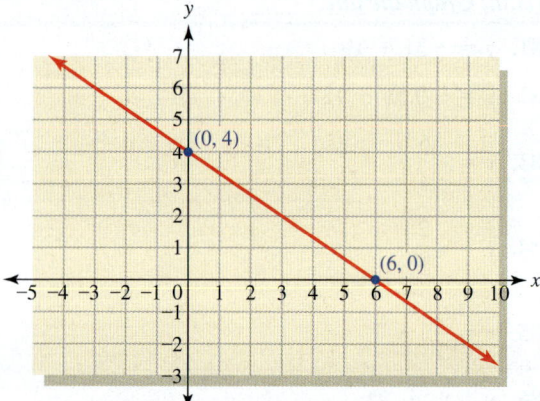

DEFINITIONS

y-intercept   The point where the curve touches the *y*-axis (vertical axis) is called the **y-intercept** or **vertical intercept**.

y-intercepts are points on the graph and should always be written as a coordinate point (0, *b*).

x-intercept   The point where the curve touches the *x*-axis (horizontal axis) is called the **x-intercept** or **horizontal intercept**.

x-intercepts are points on the graph and should always be written as a coordinate point (*c*, 0).

Example 1   Reading intercepts from graphs

a. Find the *x*-intercept and *y*-intercept of the following line.

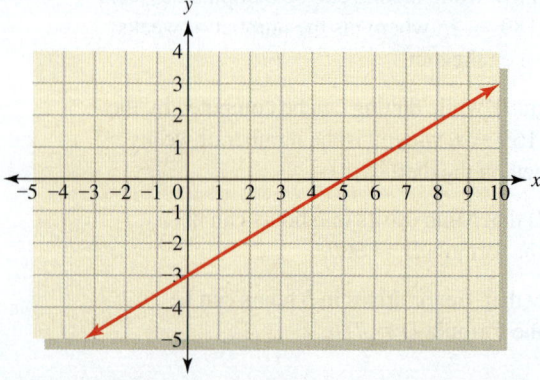

SECTION 3.3 Slope-Intercept Form of Lines   267

b. Find the *x*-intercepts and *y*-intercept of the following curve.

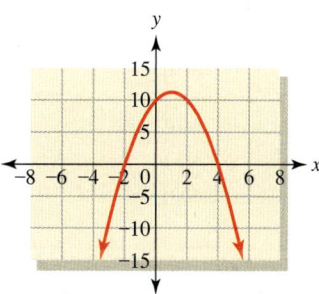

SOLUTION

a. Reading the graph, we find the following intercepts.

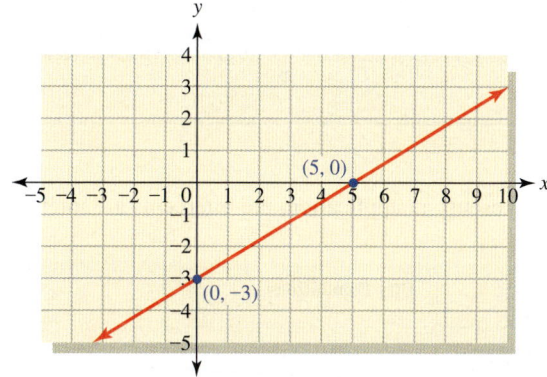

The *y*-intercept is the point $(0, -3)$, and the *x*-intercept is the point $(5, 0)$.

b. This curve has two *x*-intercepts and one *y*-intercept.

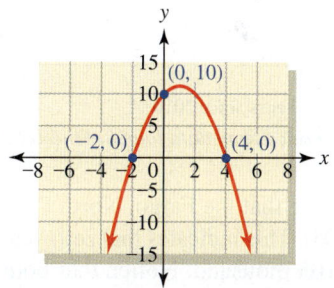

The *x*-intercepts are the points $(-2, 0)$ and $(4, 0)$, and the *y*-intercept is the point $(0, 10)$. We can see from this graph that a curve can have more than one *x*- or *y*-intercept.

PRACTICE PROBLEM FOR EXAMPLE 1

a. Find the *x*-intercept and *y*-intercept of the following line.

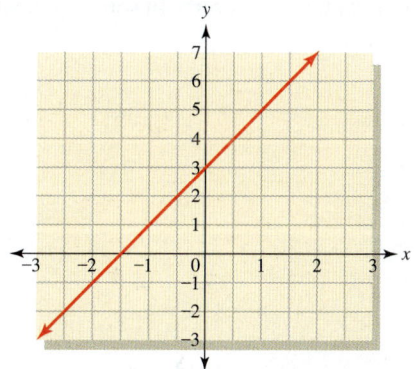

Connecting the Concepts

What are other ways to discuss intercepts?

The following words or phrases all mean the same thing.

y-intercept:
- Vertical intercept
- Can take on the name of the dependent variable that is represented on the vertical axis.

x-intercept:
- Horizontal intercept
- Can take on the name of the independent variable that is represented on the horizontal axis.

We use *horizontal* and *vertical* intercept when the variables in the problem are not *x* and *y*. This often occurs in application problems.

b. Find the *x*-intercepts and *y*-intercept of the following curve.

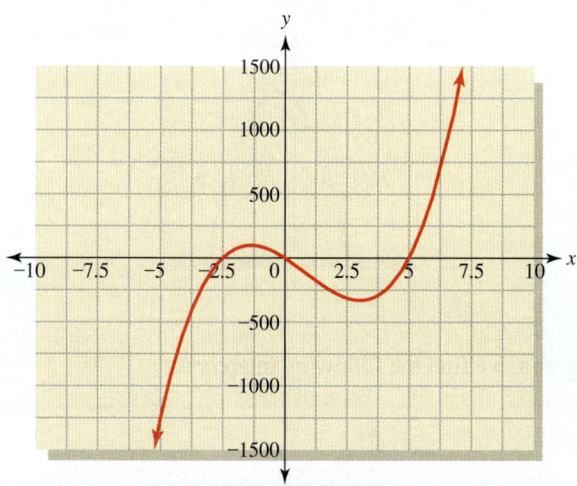

When a graph represents an application, the intercepts are interpreted in the context of the problem.

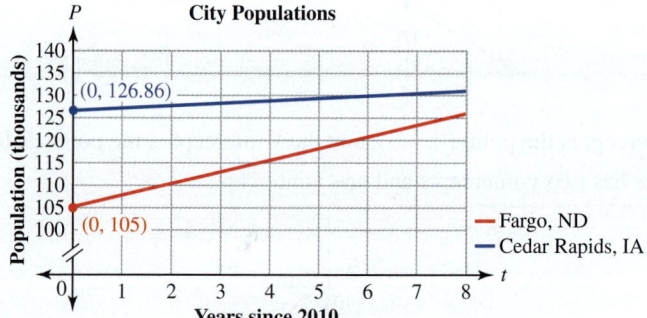

Source: Models derived from data found at the U.S. Census Bureau.

From this graph, we can see that the Fargo, North Dakota, line crosses the vertical axis at about the point (0, 105). This indicates that zero years since 2010, the year 2010, the population was about 105 thousand. Notice that both the zero and the 105 have meaning in the context of the situation.

The vertical intercept for the Cedar Rapids, Iowa, line is approximately (0, 127). The meaning of this point is that in the year 2010, the population was about 127 thousand. Both of these points are approximations, since reading an exact value from this graph is difficult.

The horizontal intercepts on the city population graph above cannot be seen on the graph and do not make sense in this situation. The horizontal intercept indicates the year in which the population of the city is zero. In some applications, one or both of the intercepts might not make sense.

SECTION 3.3 Slope-Intercept Form of Lines 269

Example 2 Interpreting intercepts from graphs

a. Estimate and interpret the vertical intercepts for the lines in this graph.

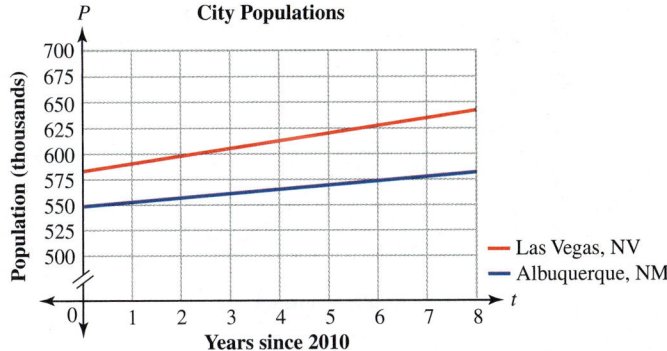

Source: Models derived from data found at the U.S. Census Bureau.

b. Estimate and interpret the vertical and horizontal intercepts for the line in this graph.

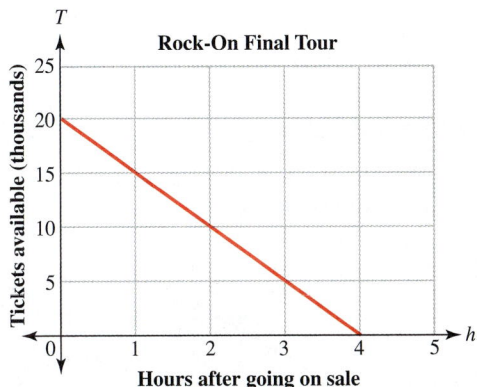

Source: Models derived from data found at the US Census Bureau.

SOLUTION

a. Estimate the vertical intercepts from this graph. Then interpret what the points represent in this situation.

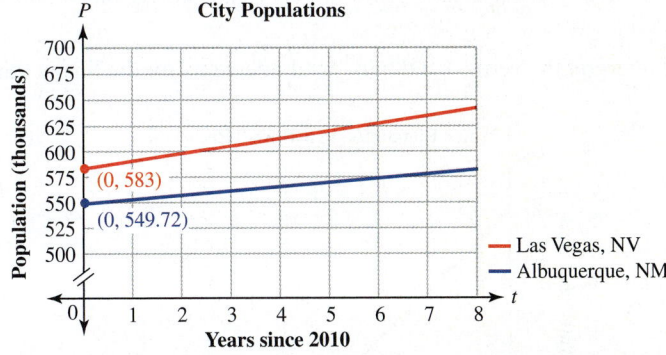

Source: Models derived from data found at the U.S. Census Bureau.

The vertical intercept for the Las Vegas, Nevada, line is approximately (0, 583). *The population of Las Vegas, NV, was about 583 thousand in 2010.*

The vertical intercept for the Albuquerque, New Mexico, line is approximately (0, 550).
The population of Albuquerque, NM, was about 550 thousand in 2010.

b. We first estimate both intercepts from the graph. Then interpret what they mean in the context of the situation.

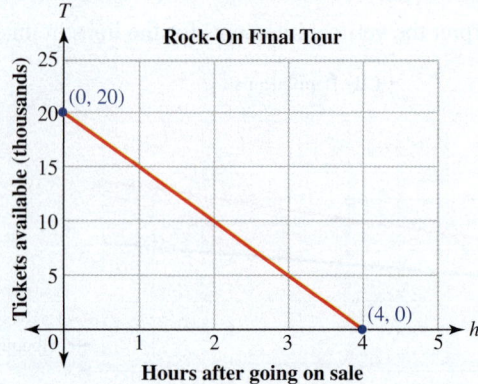

The vertical intercept is the point (0, 20).
When they started selling tickets, there were 20,000 tickets available for the Rock-On final tour.

The horizontal intercept is the point (4, 0).
Four hours after going on sale, no more tickets were available for the Rock-On final tour.

PRACTICE PROBLEM FOR EXAMPLE 2

a. Find and interpret the vertical intercept for the line in this graph.

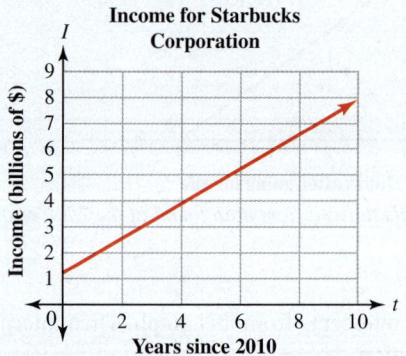

Source: Graph derived from data found at http://www.marketwatch.com

b. Find and interpret the vertical and horizontal intercepts for the line in this graph.

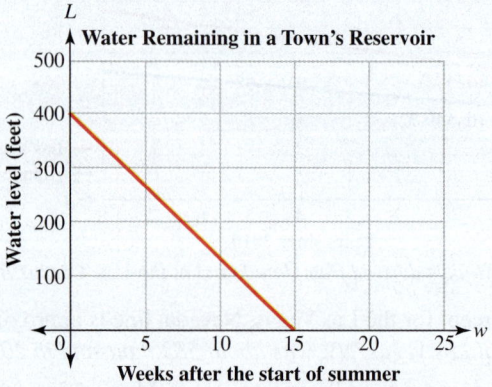

Finding and Interpreting Intercepts from Equations

When looking for x- and y-intercepts on a graph, notice that every intercept has a zero as part of the point. The *zero* is always the value of the *other* variable. Therefore, the y-intercept always has a zero for the x-value, and the x-intercept always has a zero for the y-value.

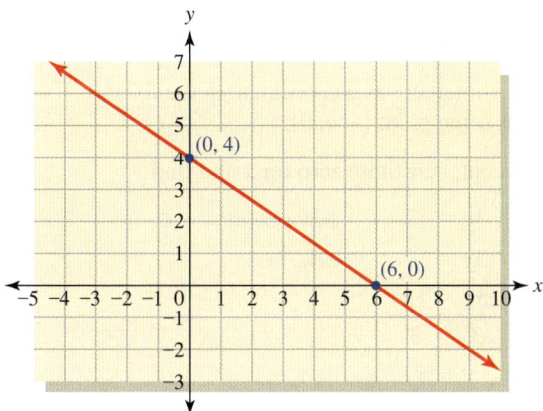

The fact that an intercept always has a zero coordinate allows us to find intercepts from equations. When given an equation, to find a y-intercept, substitute zero for x and solve for y. Remember that an intercept is a point on a graph, so it has a coordinate with both an x-value and a y-value.

> **Steps to Finding Intercepts from Equations**
>
> **y-intercept:**
> 1. Substitute zero for x, $x = 0$, and find the value of y.
> 2. Write the y-intercept as a coordinate in the form $(0, b)$.
>
> **x-intercept:**
> 1. Substitute zero for y, $y = 0$, and find the value of x.
> 2. Write the x-intercept as a coordinate in the form $(c, 0)$.

Example 3 Finding intercepts from equations

Find the x- and y-intercepts of the following equations. Write the intercepts as coordinates.

a. $y = 5x + 20$ **b.** $2x + 3y = 12$

SOLUTION

a. Step 1 Substitute zero for x, $x = 0$, and find the value of y.

$$y = 5(0) + 20 \quad \text{Substitute in } x = 0.$$
$$y = 20 \quad \text{Simplify.}$$

Step 2 Write the y-intercept as a coordinate in the form $(0, b)$.
The y-intercept is the point $(0, 20)$.

Step 1 Substitute zero for y, $y = 0$, and find the value of x.

$$0 = 5x + 20 \quad \text{Substitute in } y = 0.$$
$$\underline{-20 \quad\quad -20} \quad \text{Solve for x.}$$
$$-20 = 5x$$
$$\frac{-20}{5} = \frac{5x}{5}$$
$$-4 = x$$

Step 2 Write the x-intercept as a coordinate in the form $(c, 0)$.
The x-intercept is the point $(-4, 0)$.

b. To find the y-intercept, substitute zero for x and find y.

$$2(0) + 3y = 12 \quad \text{Substitute in } x = 0.$$
$$3y = 12 \quad \text{Solve for y.}$$
$$\frac{3y}{3} = \frac{12}{3}$$
$$y = 4$$

The y-intercept is the point $(0, 4)$.

To find the x-intercept, substitute zero for y and find x.

$$2x + 3(0) = 12 \quad \text{Substitute in } y = 0.$$
$$2x = 12 \quad \text{Solve for x.}$$
$$\frac{2x}{2} = \frac{12}{2}$$
$$x = 6$$

The x-intercept is the point $(6, 0)$.

PRACTICE PROBLEM FOR EXAMPLE 3

Find the x- and y-intercepts of the following equations. Write the intercepts as coordinates.

a. $y = -7x + 21$ **b.** $-4x + 7y = -56$

When working with an application, intercepts are found in the same way. Pay close attention to the meaning of the variables and use that information to understand the intercepts. Remember that each intercept has two parts. Include the meaning of both parts in the interpretation.

Example 4 Finding and interpreting intercepts from equations

Paulo started a diet plan two months ago to help him make weight for his next boxing match. Paulo's weight in pounds n weeks after starting the diet can be approximated by the equation

$$W = -1.5n + 175$$

a. Find and interpret the W-intercept for this equation.
b. Find and interpret the n-intercept.
c. Do the intercepts make sense in this situation?

SOLUTION

a. To find the W-intercept, substitute zero for n and find W.

$$W = -1.5(0) + 175 \quad \text{Substitute in } n = 0.$$
$$W = 175 \quad \text{Simplify.}$$

The W-intercept is the point (0, 175).
When the diet started, Paulo weighed 175 pounds.

b. To find the n-intercept, substitute zero for W and solve for n.

$$0 = -1.5n + 175 \quad \text{Substitute in } W = 0.$$
$$\underline{-175 \qquad\qquad -175} \quad \text{Solve for } n.$$
$$-175 = -1.5n$$
$$\frac{-175}{-1.5} = \frac{-1.5n}{-1.5}$$
$$116.7 \approx n$$

The n-intercept is the point (116.7, 0).
After about 117 weeks on the diet, Paulo will weigh zero pounds.

Since n is weeks after starting the diet, we round to a whole week.

c. The n-intercept does not make sense because Paulo cannot weigh zero pounds. The W-intercept seems reasonable, as 175 pounds is a fairly normal weight.

PRACTICE PROBLEM FOR EXAMPLE 4

Deb and Christine are walking in the 60-mile Breast Cancer 3-Day Walk and want to keep track of the number of miles they still have to walk. They will use the equation

$$D = -3h + 60$$

where D is the distance in miles they still have to walk after h hours of walking.

a. Find and interpret the D-intercept for this equation.

b. Find and interpret the h-intercept.

c. Do the intercepts make sense in this situation?

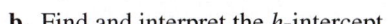

Slope-Intercept Form of a Line

CONCEPT INVESTIGATION

How are the y-intercepts and slopes connected to the equation?

1. Find the y-intercept for the following equations. Write it as a coordinate in the form (0, b).

    a. Equation: $y = 4x + 9$ y-intercept:

    b. Equation: $y = \frac{1}{2}x + 3$ y-intercept:

    c. Equation: $y = 1.25x - 2.5$ y-intercept:

2. What is the relationship between the y-intercept and the equation?

3. Find the slopes of the following equations using any two points given in each table.

a. Equation: $y = 4x + 9$ **Slope:**

x	$y = 4x + 9$
1	13
3	21
7	37
10	49

b. Equation: $y = \dfrac{1}{2}x + 3$ **Slope:**

x	$y = \dfrac{1}{2}x + 3$
2	4
6	6
10	8
16	11

4. What is the relationship between the slope and the equation?

All of the equations in the Concept Investigation are in what is called the **slope-intercept form of a line**. The slope-intercept form of a line is written as

$$y = mx + b$$

where m and b are real numbers. When an equation is written in this form, we can read the slope and the y-intercept of the line directly from the equation. In the first half of the Concept Investigation, we noticed that the point $(0, b)$ is the y-intercept for each line. In the second half of the Concept Investigation, we noticed that the slope for each equation is m, the coefficient of the x-term.

■ **Connecting the Concepts**

Are there other forms of lines?

In this section, we study the slope-intercept form of a line. There are several other forms of lines, which we will study in Sections 3.4 and 3.5.

DEFINITION

Slope-Intercept Form of a Line   The slope-intercept form of a line is

$$y = mx + b$$

where m and b are real numbers. The slope of the line is m, and the y-intercept is the point $(0, b)$.

Using this information, we can find the slope and y-intercept of any line that is in the slope-intercept form.

Example 5   Finding the slope and the intercepts from the slope-intercept form

Find the slope, y-intercept, and x-intercept for each of the following linear equations.

a. $y = \dfrac{2}{3}x - \dfrac{1}{3}$ **b.** $y = -0.5x + 7.5$

SOLUTION

a. This equation is in slope-intercept form, so the slope is the coefficient of the x-term, and the y-intercept is the point $(0, b)$.

The slope is $\frac{2}{3}$, and the y-intercept is $\left(0, -\frac{1}{3}\right)$.

To find the x-intercept, substitute zero for y and solve for x.

$$0 = \frac{2}{3}x - \frac{1}{3} \quad \text{Substitute in } y = 0.$$

$$\underline{+\frac{1}{3} \qquad\qquad +\frac{1}{3}} \quad \text{Solve for x.}$$

$$\frac{1}{2} = \frac{2}{3}x$$

$$\frac{3}{2}\left(\frac{1}{2}\right) = \frac{3}{2}\left(\frac{2}{3}x\right) \quad \text{Multiply both sides by the reciprocal } \frac{3}{2}.$$

$$\frac{1}{2} = x$$

The x-intercept is the point $\left(\frac{1}{2}, 0\right)$.

> **Skill Connection**
>
> **Dividing fractions**
>
> To divide by a fraction, we multiply by the reciprocal of that fraction.
>
> $$8 \div \frac{2}{5} =$$
>
> $$8 \cdot \frac{5}{2} =$$
>
> $$^4 8 \cdot \frac{5}{\cancel{2}} = 20$$

b. This equation is in slope-intercept form, so the slope is -0.5, and the y-intercept is $(0, 7.5)$.

To find the x-intercept, substitute zero for y and solve for x.

$$0 = -0.5x + 7.5$$

$$\underline{-7.5 \qquad\qquad -7.5}$$

$$-7.5 = -0.5x$$

$$\frac{-7.5}{-0.5} = \frac{-0.5x}{-0.5}$$

$$15 = x$$

The x-intercept is the point $(15, 0)$.

PRACTICE PROBLEM FOR EXAMPLE 5

Find the slope, y-intercept, and x-intercept for each of the following linear equations.

a. $y = -\frac{2}{3}x + 4$ **b.** $y = \frac{1}{4}x - \frac{2}{5}$

Example 6 Finding and interpreting slopes and intercepts from applications

The cost in dollars to have the carpets cleaned in your home can be calculated by using the equation

$$C = 50r + 25$$

where C is the cost in dollars to have the carpet in r rooms cleaned.

a. Find and interpret the C-intercept for this equation.

b. Find and interpret the slope of this equation.

SOLUTION

a. This equation is in slope-intercept form, so the C-intercept is $(0, 25)$.
If you have no carpet cleaned, the company will charge you $25.
This could be a base fee for coming out to the home.

b. The slope of this equation is 50. Since slope can be thought of as the change in outputs over the change in inputs, we can represent this slope as

$$50 = \frac{50 \text{ dollars}}{1 \text{ room}}$$

Interpret slope:
The cost for cleaning the carpets increases by $50 for each room cleaned.

PRACTICE PROBLEM FOR EXAMPLE 6

The time it takes to manufacture f bicycle frames for a certain type of bike can be calculated by using the equation

$$T = 0.5f + 2$$

where T is the time in hours it takes to produce f bicycle frames.

a. Find and interpret the T-intercept for this equation.

b. Find and interpret the slope of this equation.

Whenever a given equation is not in slope-intercept form, we cannot directly read the slope and y-intercept. Therefore, when an equation is not given in slope-intercept form, solve the equation for y to put the equation into slope-intercept form. Then we can read the slope and y-intercept directly from the equation.

> **Skill Connection**
>
> **Solutions to literal equations**
>
> Solving an equation for a variable is known as solving a literal equation. We studied this in Sections 2.1 and 2.2.

Example 7 — Finding slopes and intercepts from equations

Find the slope and y-intercept for each of the following linear equations.

a. $2x + y = 6$ **b.** $y + 5 = 4(x - 3)$ **c.** $2x - 5y = 3$

SOLUTION

a. This equation is not in slope-intercept form, so first put the equation into slope-intercept form.

$$\begin{aligned} 2x + y &= 6 \quad &\text{Solve for } y.\\ -2x & -2x \\ \hline y &= -2x + 6 \end{aligned}$$

Now that the equation is in slope-intercept form, we see that the slope is -2, and the y-intercept is $(0, 6)$.

b. This equation is not in slope-intercept form, so first put the equation into slope-intercept form.

$$\begin{aligned} y + 5 &= 4(x - 3) \quad &\text{Distribute on the right.}\\ y + 5 &= 4x - 12 \quad &\text{Solve for } y.\\ -5 & -5 \\ \hline y &= 4x - 17 \end{aligned}$$

Now that the equation is in slope-intercept form, we see that the slope is 4, and the y-intercept is $(0, -17)$.

c. This equation is not in slope-intercept form, so first put the equation into slope-intercept form.

$$2x - 5y = 3 \quad \text{Solve for y.}$$
$$\underline{-2x \qquad -2x}$$
$$-5y = -2x + 3$$
$$\frac{-5y}{-5} = \frac{-2x + 3}{-5} \quad \text{Divide each term by } -5.$$
$$y = \frac{2}{5}x - \frac{3}{5}$$

Now that the equation is in slope-intercept form, we see that the slope is $\frac{2}{5}$, and the y-intercept is $\left(0, -\frac{3}{5}\right)$.

PRACTICE PROBLEM FOR EXAMPLE 7

Find the slope and y-intercept of the given equations.

a. $2y = 8x + 5$ **b.** $y - 8 = \frac{2}{3}(x + 6)$

c. $4x + 3y = 1$

3.3 Vocabulary Exercises

1. The _____ is where the curve touches the vertical axis.

2. The _____ is where the curve touches the horizontal axis.

3. The _____ form of a line is $y = mx + b$.

4. In an equation of the form $y = mx + b$, the m represents the _____ of the line.

3.3 Exercises

For Exercises 1 through 10, use the graphs to find the x-intercept(s) and y-intercept for each line or curve.

1.

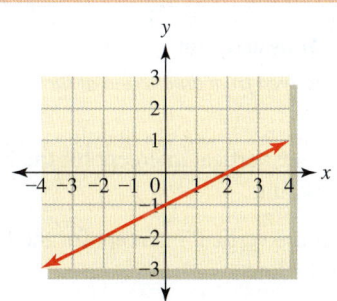

2.

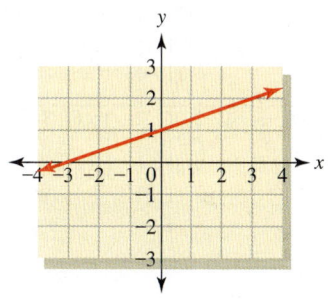

278 CHAPTER 3 Linear Equations with Two Variables

3.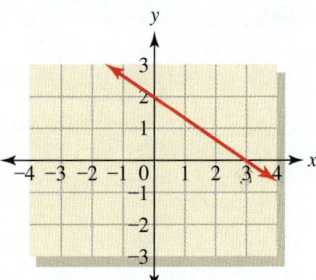

4.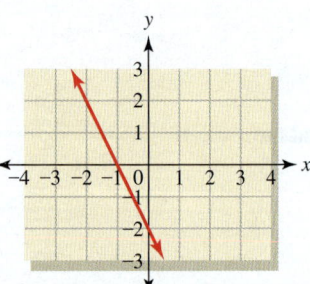

5.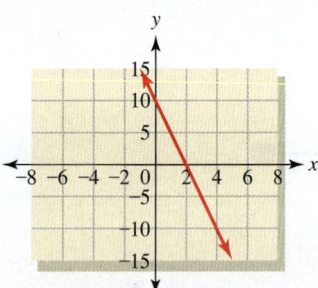

6.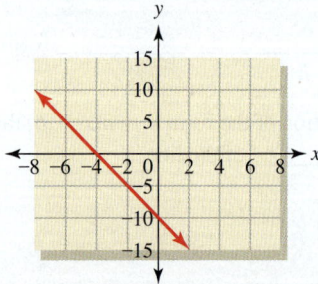

7.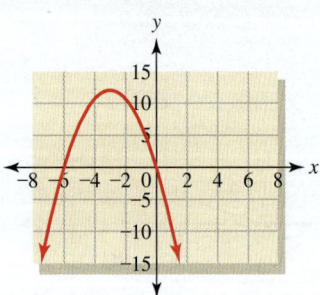

8.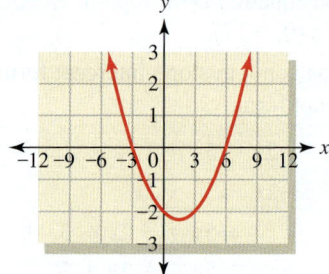

9.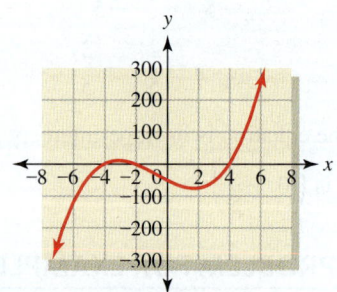

10.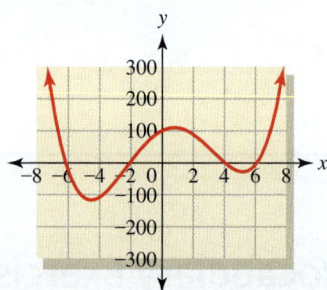

For Exercises 11 through 18, use the given information to answer the questions.

11.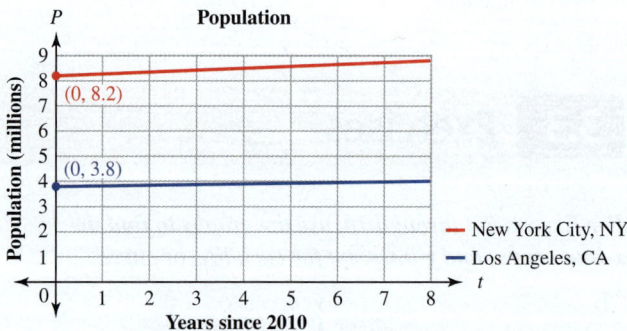

Source: Models derived from data found at the U.S. Census Bureau.

a. Interpret the vertical intercept for the New York City, New York, line.

b. Interpret the vertical intercept for the Los Angeles, California, line.

12.

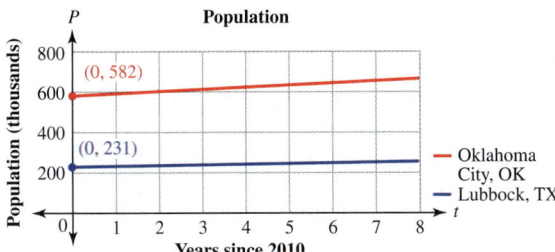

Source: Models derived from data found at the U.S. Census Bureau.

a. Interpret the vertical intercept for the Oklahoma City, Oklahoma, line.

b. Interpret the vertical intercept for the Lubbock, Texas, line.

13.

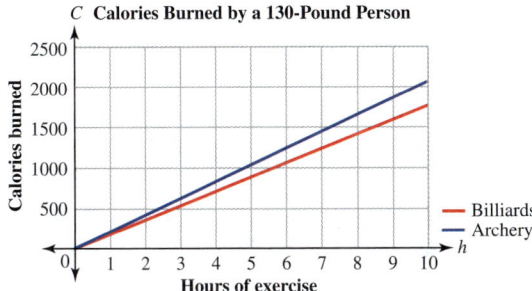

Source: Nutristrategy.com

Interpret the intercept for these lines.

14.

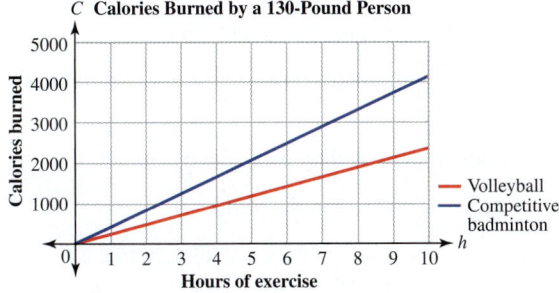

Source: Nutristrategy.com

Interpret the intercept for these lines.

15.

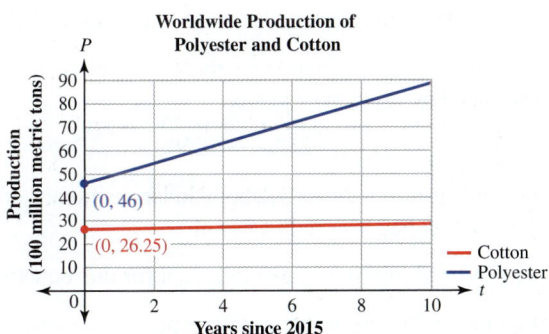

Source: Models derived from data found at Quartz.com

a. Interpret the vertical intercept for the cotton line.

b. Interpret the vertical intercept for the polyester line.

16.

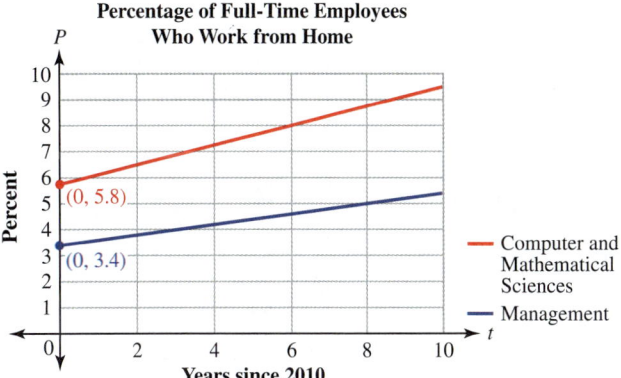

Source: Centers for Disease Control and Prevention.

a. Interpret the vertical intercept for the Computer and Mathematical Sciences line.

b. Interpret the vertical intercept for the Management line.

17.

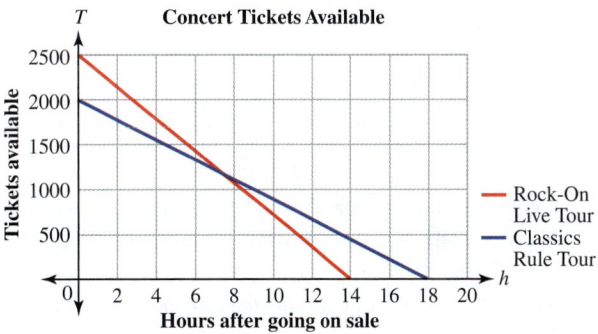

a. Interpret the vertical and horizontal intercepts for the Rock-On Live Tour line.

b. Interpret the vertical and horizontal intercepts for the Classics Rule Tour line.

18.

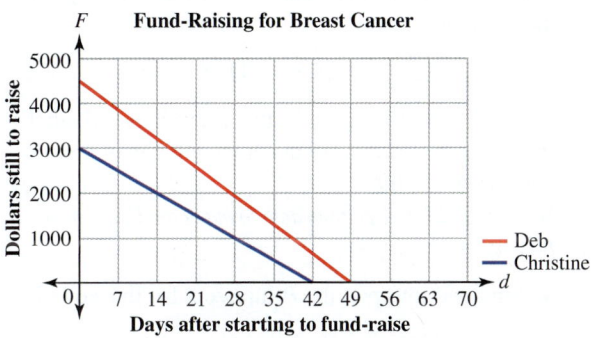

a. Interpret the vertical and horizontal intercepts for the line that represents Deb's fund-raising.

b. Interpret the vertical and horizontal intercepts for the line that represents Christine's fund-raising.

For Exercises 19 through 40, find the x-intercept and y-intercept of each given equation. Write intercepts as coordinates.

19. $y = 5x + 20$
20. $y = 2x + 12$
21. $y = 4x - 36$
22. $y = 3x - 12$
23. $y = \frac{1}{2}x + 3$
24. $y = \frac{1}{2}x - 5$
25. $y = \frac{2}{3}x + 6$
26. $y = \frac{4}{5}x + 10$
27. $y = 0.25x + 3.75$
28. $y = -1.4x + 2.8$
29. $y = x$
30. $y = -x$
31. $x + 5y = 15$
32. $x + 4y = 20$
33. $4x - 2y = 10$
34. $6x - 3y = 15$
35. $4x + 3y = 30$
36. $2x - 3y = 12$
37. $y = 6$
38. $y = -3$
39. $x = -5$
40. $x = 2$

For Exercises 41 through 48, use the given information to answer the questions.

41. The number of American Indians and Alaska Natives in the United States can be estimated by using the equation

$$P = 76.4t + 5058$$

where P is the number of American Indians and Alaska Natives alone or in combination in thousands t years since 2010.
Source: Model derived from data found at the U.S. Census Bureau.

   a. Find and interpret the P-intercept for this equation.

   b. Find and interpret the t-intercept.

   c. Do these intercepts make sense in this situation?

42. The number of Native Hawaiians and other Pacific Islanders in the United States can be estimated by using the equation

$$P = 28.5t + 1158$$

where P is the number of Native Hawaiians and other Pacific Islanders alone or in combination in thousands t years since 2000.
Source: Model derived from data found at the U.S. Census Bureau.

   a. Find and interpret the P-intercept for this equation.

   b. Find and interpret the t-intercept.

   c. Do these intercepts make sense in this situation?

43. The percentage of college enrollees that are White can be estimated by using the equation

$$P = -0.35t + 59$$

where P is the percentage of college enrollees that are White t years since 2015.
Source: National Center for Education Statistics.

   a. Find and interpret the P-intercept for this equation.

   b. Find and interpret the t-intercept.

   c. Do these intercepts make sense in this situation?

44. The percentage of college enrollees that are Black can be estimated by using the equation

$$P = 0.13t + 15.6$$

where P is the percentage of college enrollees that are Black t years since 2015.
Source: National Center for Education Statistics.

   a. Find and interpret the P-intercept for this equation.

   b. Find and interpret the t-intercept.

   c. Do these intercepts make sense in this situation?

45. The number of cars sold at Big Jim's Car Lot can be estimated by using the equation

$$C = -2w + 25$$

where C is the number of cars sold during week w of the year.

   a. Find and interpret the C-intercept for this equation.

   b. Find and interpret the w-intercept.

   c. Do these intercepts make sense in this situation?

46. The amount of time it takes to create personalized candy bars can be estimated by using the equation

$$T = 2c + 40$$

where T is the time in minutes it takes to make c personalized candy bars.

   a. Find and interpret the T-intercept for this equation.

   b. Find and interpret the c-intercept.

   c. Do these intercepts make sense in this situation?

47. The amount of credit card debt that Gabi has can be estimated by using the equation

$$D = -250m + 12500$$

where D is the credit card debt in dollars m months after she starts to pay her credit cards off.

   a. Find and interpret the D-intercept for this equation.

   b. Find and interpret the m-intercept.

   c. Do these intercepts make sense in this situation?

48. The amount of credit card debt that Svetlana has can be estimated by using the equation

$$D = -100m + 5400$$

where D is the credit card debt in dollars m months after starting to pay her credit cards off.

  a. Find and interpret the D-intercept for this equation.

  b. Find and interpret the m-intercept.

  c. Do these intercepts make sense in this situation?

For Exercises 49 through 64, find the slope, x-intercept, and y-intercept of the given equations.

49. $y = -3x + 27$

50. $y = -4x + 12$

51. $y = 8x - 16$

52. $y = 5x - 10$

53. $y = \frac{1}{2}x + 6$

54. $y = \frac{1}{2}x - 4$

55. $y = -\frac{1}{3}x + 5$

56. $y = -\frac{1}{5}x + 2$

57. $y = \frac{2}{3}x - 4$

58. $y = \frac{3}{4}x - 18$

59. $y = 1.5x - 3$

60. $y = 0.25x - 5$

61. $y = -5$

62. $y = 4$

63. $x = 6$

64. $x = -2$

65. Find and interpret the slope of the equation from Exercise 41.

66. Find and interpret the slope of the equation from Exercise 42.

67. Find and interpret the slope of the equation from Exercise 43.

68. Find and interpret the slope of the equation from Exercise 44.

69. Find and interpret the slope of the equation from Exercise 45.

70. Find and interpret the slope of the equation from Exercise 46.

71. The total revenue earned by Nike, Inc. can be estimated by the equation

$$R = 2.4t + 18.2$$

where R is the total revenue in billions of dollars t years since 2010.
Source: http://finance.google.com, accessed May 2017.

  a. Find and interpret the R-intercept for this equation.

  b. Find and interpret the slope of this equation.

72. The total gross profit earned by Nike, Inc. can be estimated by the equation

$$P = 1.36t + 7$$

where P is the total gross profit in billions of dollars t years since 2010.
Source: http://finance.google.com, accessed May 2017.

  a. Find and interpret the P-intercept for this equation.

  b. Find and interpret the slope of this equation.

For Exercises 73 through 86, find the slope and y-intercept of the given equations. (Hint: Put in $y = mx + b$ form as in Example 7.)

73. $3x + y = 18$

74. $2x + y = 4$

75. $3x + 2y = 6$

76. $4x + 5y = 10$

77. $2x - 7y = 21$

78. $5x - 6y = 42$

79. $x - 2y = -7$

80. $5x - 4y = -15$

81. $y + 3 = 2(x - 5)$

82. $y + 7 = 3(x + 4)$

83. $y - 2 = \frac{1}{2}(x + 20)$

84. $y - 6 = \frac{1}{5}(x + 55)$

85. $y - 7 = 0$

86. $y + 2 = 0$

For Exercises 87 through 90, solve each given literal equation for the indicated variable.

87. $A = \frac{1}{2}(b_1 + b_2)h$ for h

88. $A = \pi ab$ for b

89. $3x - 5y = 15$ for y

90. $-x + 4y = -12$ for y

For Exercises 91 through 94, convert units as indicated. Round to the hundredths place if necessary.

91. Convert 175 miles to kilometers.

92. Convert 625 kilometers to miles.

93. Convert 55 ounces to grams.

94. Convert 80 grams to ounces.

282 CHAPTER 3 Linear Equations with Two Variables

3.4 Linear Equations and Their Graphs

LEARNING OBJECTIVES
- Graph from slope-intercept form.
- Graph from general form.
- Describe a linear equation.
- Discern and graph parallel lines.
- Distinguish and graph perpendicular lines.

■ Graphing from Slope-Intercept Form

Now that we know how to find the slope and y-intercept of a line from an equation in slope-intercept form, $y = mx + b$, we want to use this information to make graphing lines easier. The y-intercept is a point on the line that is used as a starting point. The slope tells us in what direction to go from that point. The idea of slope as rise over run is the basis for finding other points on the line.

Skill Connection

Graph three points for a line

Although we need only two points to graph a line, we will graph three points as a check on our work. If the three points do not all lie on a straight line, then we will need to rework the problem.

Steps to Graphing Lines Using the Slope and y-Intercept

1. Determine the slope and y-intercept from the equation $y = mx + b$.
2. Plot the y-intercept $(0, b)$ on the graph.
3. Write the slope in fraction form. Use the slope to count the rise and run from the y-intercept to another point on the line. Repeat the rise and run process to find a third point. Work from left to right.
4. Draw a straight line through the points.

Example 1 Graphing lines using the slope and y-intercept

Graph the following linear equations using the slope and y-intercept. Graph at least three points for each line.

a. $y = \dfrac{2}{3}x + 1$ **b.** $y = -2x + 3$

SOLUTION

a. Step 1 Determine the slope and y-intercept from the equation $y = mx + b$.

From the equation $y = \dfrac{2}{3}x + 1$, we see that the slope is $m = \dfrac{2}{3} = \dfrac{\text{rise}}{\text{run}}$ and the y-intercept is the point $(0, 1)$.

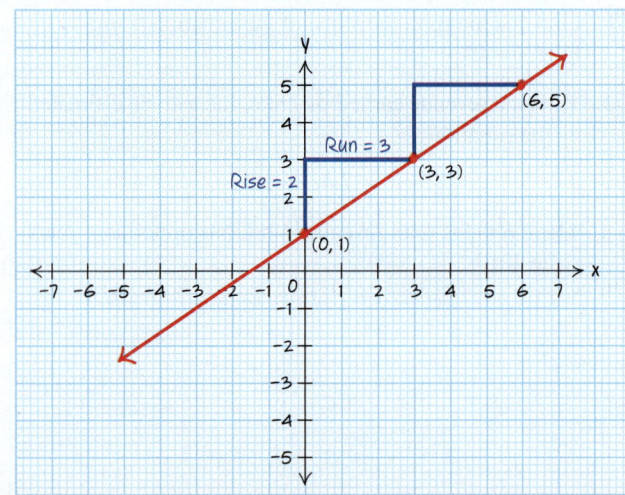

Step 2 Plot the y-intercept (0, b) on the graph.

Plot the y-intercept (0, 1).

Step 3 Use the slope to count the rise and run from the y-intercept to another point on the line. Repeat the rise and run process to find a third point.

Use the slope to rise 2 and run 3 from the y-intercept. Repeat the rise/run process again to find an additional point on the line.

Step 4 Draw a straight line through the points.

b. From the equation $y = -2x + 3$, we see that the slope is $-2 = \frac{-2}{1} = \frac{\text{rise}}{\text{run}}$ and the y-intercept is the point (0, 3). Start by plotting the y-intercept, (0, 3), and then use the slope to rise -2 and run 1 from the y-intercept. Since the rise is negative, we go down 2 instead of up. Repeating this process again yields another point on the graph.

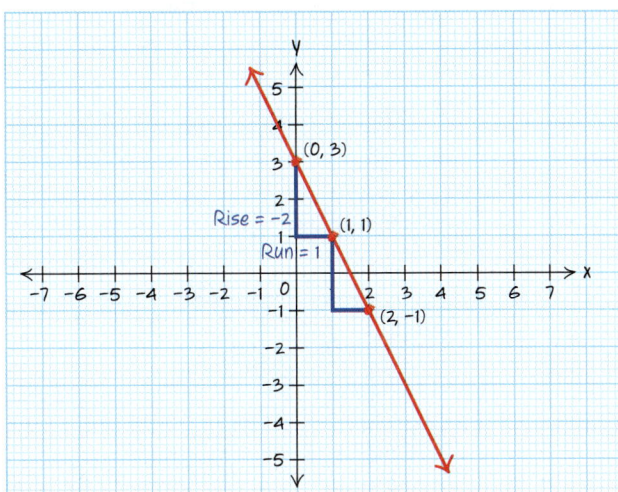

PRACTICE PROBLEM FOR EXAMPLE 1

Graph the following linear equations using the slope and y-intercept. Graph at least three points for each line.

a. $y = 3x - 4$ **b.** $y = -\frac{3}{4}x - 3$

Some linear equations are not already in slope-intercept form. When we are given an equation not in slope-intercept form, we start by putting the equation into slope-intercept form. We then determine the slope and the y-intercept and use this information to graph the line.

Example 2   Graphing lines from a different form

Graph the linear equation $y + 3 = \frac{1}{5}(x + 10)$ using the slope and y-intercept. Graph at least three points.

SOLUTION

The equation $y + 3 = \frac{1}{5}(x + 10)$ is not in slope-intercept form. First we solve for y and put the equation into slope-intercept form.

284 CHAPTER 3 Linear Equations with Two Variables

$$y + 3 = \frac{1}{5}(x + 10)$$ Distribute on the right side.

$$y + 3 = \frac{1}{5}x + 2$$ Solve for y.

$$ -3 \phantom{\frac{1}{5}x} -3$$

$$y = \frac{1}{5}x - 1$$

The slope is $\frac{1}{5}$, and the y-intercept is the point $(0, -1)$. First plot the y-intercept of $(0, -1)$ and then use the slope $m = \frac{1}{5} = \frac{\text{rise}}{\text{run}}$ to get the following graph.

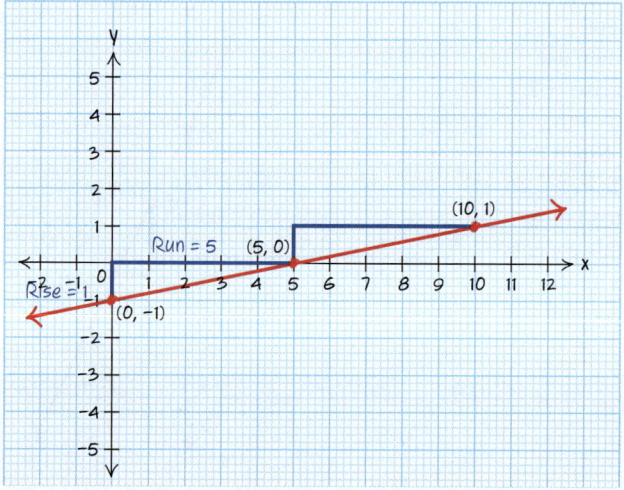

PRACTICE PROBLEM FOR EXAMPLE 2

Graph the linear equation $y - 8 = \frac{2}{3}(x - 6)$ using the slope and y-intercept. Graph at least three points.

Example 3 Graphing a cost equation

It costs $7 for admission to the local carnival and $1.50 for each game or ride ticket. The cost for someone who pays for admission and purchases t tickets can be found with the equation

$$C = 1.5t + 7$$

where C is the cost in dollars for admission and tickets at the local carnival when t tickets are purchased. Graph the cost equation.

SOLUTION

This equation is in slope-intercept form, so the slope is 1.5, and the vertical intercept is the point $(0, 7)$. With a decimal value of slope, it is easier to convert the decimal to a fraction to determine the rise and run.

$$m = 1.5 = \frac{15}{10} = \frac{3}{2} = \frac{\text{rise}}{\text{run}}$$

Starting at the vertical intercept, (0, 7), we rise up 3 and run over 2. In this case, the horizontal axis is scaled by 2's to include more numbers of tickets. We also scaled the vertical axis by 5's to include enough costs. When counting the rise, notice that each little box on the graph paper represents 1, so we rise 3 little boxes. When counting the run, each tick mark represents 2.

Skill Connection

Converting decimals to fractions

Recall from Section R.3 how to convert a decimal to a fraction. Remove the decimal and divide the number by the place value of the last decimal digit.

- Convert 1.5 to a fraction.

    Because the 5 is in the tenths place, divide by 10. This fraction can then be reduced.
    $$1.5 = \frac{15}{10} = \frac{3}{2}$$

- Convert 2.37 to a fraction.

    Because the 7 is in the hundredths place, divide by 100.
    $$2.37 = \frac{237}{100}$$

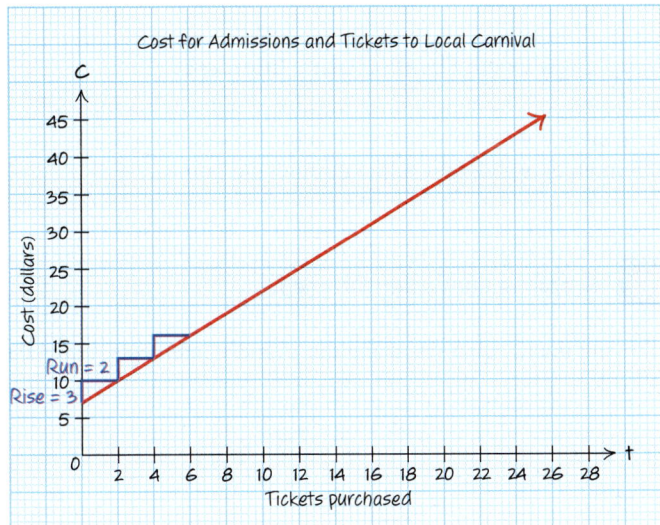

PRACTICE PROBLEM FOR EXAMPLE 3

Fatima earns $11.50 per hour at her part-time job. Her weekly pay can be found by using the equation $P = 11.50h$, where P is Fatima's weekly pay in dollars when she works h hours a week. Graph this equation.

■ Graphing from General Form

When an equation is not in slope-intercept form, we have a few options to graph it.

- Put the equation into slope-intercept form and graph it using the slope and y-intercept.
- Build a table of points, as in Section 3.1.
- Find the x- and y-intercepts for the equation and use those to graph the line.

Another form that linear equations are often written in is called the **general form of a line**. The general form is

$$Ax + By = C$$

where A, B, and C are real numbers and both A and B are not zero. One of the differences between the slope-intercept form and general form of a line is that all lines can be represented with the general form, but the slope-intercept form cannot represent vertical lines. We will see the general form of a line used extensively in Chapter 4.

Connecting the Concepts

How do we write a line in general form?

When we write a line in general form, the values of A, B, and C are usually written as integers. To put the equation $\frac{1}{2}x + y = -3$ in general form, we clear out the fraction. Both sides of the equation are multiplied by 2 (the least common denominator). This gives the equation

$$2\left(\frac{1}{2}x + y\right) = 2(-3)$$
$$x + 2y = -6$$

This last equation is in the general form of the line.

DEFINITION

The General Form of a Line   The general form of a line is

$$Ax + By = C$$

where A, B, and C are real numbers and both A and B are not zero.

When an equation is in general form, we cannot read the slope and y-intercept directly from the equation. Instead we find the x- and y-intercepts and use these points to draw the graph. Up until now, we have required at least three points to graph a line. Now that we have become more proficient at graphing lines, we use a fact from geometry that states that two points uniquely determine a line.

> **Steps to Graphing a Linear Equation Using Intercepts**
> 1. Find the y-intercept by substituting zero for x and solving for y.
> 2. Find the x-intercept by substituting zero for y and solving for x.
> 3. Plot both intercepts and draw a line through the points.

Example 4 Graphing lines using intercepts

Graph the following linear equations using the x- and y-intercepts. Label the intercepts.

a. $3x + 2y = 12$

b. $2x - 5y = 15$

SOLUTION

a. The equation $3x + 2y = 12$ is in general form, so find the intercepts and graph the line.

Step 1 Find the y-intercept by substituting zero for x and solving for y.

To find the y-intercept, substitute in $x = 0$ and solve for y.

$$3(0) + 2y = 12$$
$$2y = 12$$
$$y = 6$$

Therefore, the point $(0, 6)$ is the y-intercept.

Step 2 Find the x-intercept by substituting zero for y and solving for x.

To find the x-intercept, substitute in $y = 0$ and solve for x.

$$3x + 2(0) = 12$$
$$3x = 12$$
$$x = 4$$

Therefore, the point $(4, 0)$ is the x-intercept.

Step 3 Plot both intercepts and draw a line through the points.

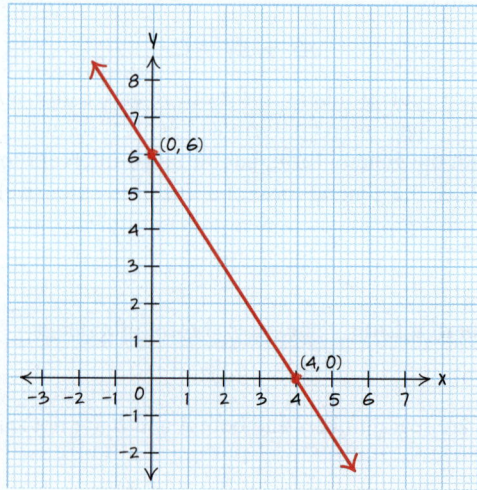

b. The equation $2x - 5y = 15$ is in general form, so find the intercepts and graph the line. To find the y-intercept, substitute in $x = 0$ and solve for y.

$$2(0) - 5y = 15$$
$$-5y = 15$$
$$y = -3$$

Therefore, the point $(0, -3)$ is the y-intercept. To find the x-intercept, substitute in $y = 0$ and solve for x.

$$2x - 5(0) = 15$$
$$2x = 15$$
$$x = 7.5$$

Therefore, the point $(7.5, 0)$ is the x-intercept. Use these two points to graph the line.

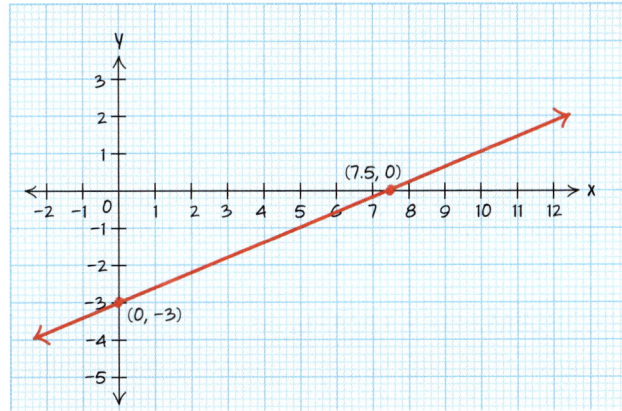

PRACTICE PROBLEM FOR EXAMPLE 4

Graph the following linear equations using the x- and y-intercepts. Label the intercepts.

a. $x + 5y = 20$ **b.** $3x - 4y = 8$

Example 5 Graphing lines using intercepts

To stay within budget, Antonio must not spend more than $300 on incentives for the next sales meeting. Antonio plans to buy a combination of $30 and $15 gift cards. If Antonio spends the entire $300 budget, the possible number of each kind of gift card can be found by using the equation $30T + 15F = 300$. Here T is the number of $30 gift cards and F is the number of $15 gift cards purchased.

a. Graph the given equation.

b. Using the graph, determine how many $15 gift cards can be purchased if five $30 gift cards are purchased.

SOLUTION

a. Begin by finding the intercepts for the graph. In this case, either variable can be used on the horizontal or vertical axis. We select T on the horizontal axis and F on the vertical axis.

$$30T + 15F = 300$$
$$30(0) + 15F = 300$$
$$15F = 300$$
$$F = 20$$
$$(T, F) = (0, 20) \quad \text{The vertical intercept}$$

$$30T + 15(0) = 300$$
$$30T = 300$$
$$T = 10$$
$$(T, F) = (10, 0) \quad \text{The horizontal intercept}$$

Using these intercepts, we get the following graph.

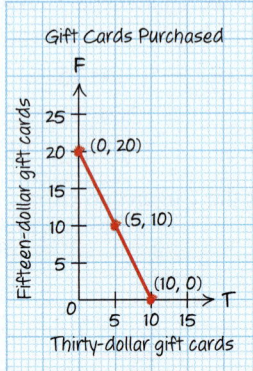

b. On the graph, we can see that the point (5, 10) is on the line.
This point means that if five $30 cards are purchased, then ten $15 cards can be bought.

PRACTICE PROBLEM FOR EXAMPLE 5

Amanda has two credit cards, one charging 15% interest and the other charging 20% interest. Amanda paid $1135 in interest this year. The possible balances on each credit card can be found by using the equation $0.15A + 0.20B = 1135$, where A is the amount in dollars owed on the credit card that charges 15% interest and B is the amount in dollars owed on the credit card that charges 20% interest.

a. Graph the given equation.

b. Using the graph, estimate the amount Amanda owes on the card charging 20% if she owes $4500 on the card charging 15%.

There are several ways to graph a line given in general form. One is by using the intercepts, as we just learned. Another way is to create a table of values and graph the points, as in Section 3.1. Yet a third way is to rewrite the equation of the line as $y = mx + b$ and then graph using the slope and y-intercept.

Example 6 — Converting general form into slope-intercept form for graphing

Graph the line $12x - 3y = -6$ by completing the following steps.

a. Put the equation $12x - 3y = -6$ into $y = mx + b$ form.

b. Graph the line using the slope and y-intercept.

SOLUTION

a. Solve the equation $12x - 3y = -6$ for y.

$$12x - 3y = -6$$
$$\underline{-12x \qquad\qquad -12x} \qquad \text{Subtract } 12x \text{ from both sides.}$$
$$-3y = -12x - 6$$
$$\frac{-3y}{-3} = \frac{-12x}{-3} - \frac{6}{-3} \qquad \text{Divide each term by } -3 \text{ on both sides.}$$
$$y = 4x + 2$$

b. The equation of the line $y = 4x + 2$ is now in $y = mx + b$ form. We see that the y-intercept is the point $(0, 2)$ and the slope is $m = 4 = \frac{4}{1}$. Graphing the line yields the following graph.

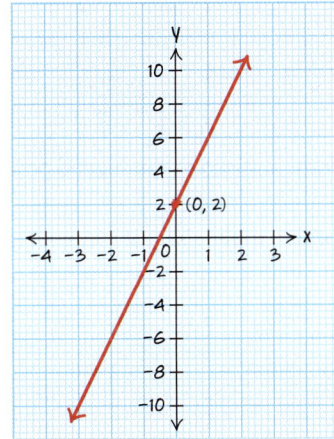

■ Recognizing a Linear Equation

When working with equations, it is helpful to recognize which equations are linear. The two forms of **linear equations** that we have studied so far give us some information about what a linear equation will or will not contain.

$$y = mx + b \quad \text{and} \quad Ax + By = C$$

Both have variables that have an exponent of 1 and are not multiplied or divided by each other. Any equation of two variables that can be simplified into one of these forms is linear.

DEFINITION

Linear Equation An equation that can be put into one of the following forms is called a **linear equation**.

$$y = mx + b \quad \text{and} \quad Ax + By = C$$

where m, b, A, B, and C are real numbers and both A and B are not zero. The graph of a linear equation will be a straight line.

Note: The exponents on x and y in a linear equation can be only 1.

■ Connecting the Concepts

What is not in a linear equation?
Linear equations have x and y raised only to the first power. The variables x and y cannot be inside an absolute value, under a square root, or in a denominator. The variables x and y are in separate terms, never combined into the same term.

Example 7 — Determining linear equations

Determine whether the following equations are linear equations. If the equation is linear, put it into slope-intercept or general form.

a. $y + 3 = 2(x + 5)$ **b.** $y = 3x^2 + 5$

c. $3y - 4x = -5x + 7$ **d.** $5xy = 30$

SOLUTION

a. This equation can be simplified to slope-intercept form, so it is a linear equation.

$$y + 3 = 2(x + 5)$$
$$y + 3 = 2x + 10$$
$$\underline{ -3 -3}$$
$$y = 2x + 7$$

b. This equation has the variable x raised to the second power, so this is not a linear equation.

c. This equation can be simplified to the general form, so it is a linear equation.

$$3y - 4x = -5x + 7$$
$$-4x + 3y = -5x + 7$$
$$\underline{+5x +5x}$$
$$x + 3y = 7$$

d. This equation has the variables x and y *multiplied* together, so this is not a linear equation.

PRACTICE PROBLEM FOR EXAMPLE 7

Determine whether the following equations are linear equations. If the equation is linear, put it into slope-intercept or general form.

a. $y - 8 = 4(x - 2) - 7$ **b.** $20 = 2xy + 7$

■ Parallel and Perpendicular Lines

■ CONCEPT INVESTIGATION

Are those lines parallel?

1. Graph the two lines on the same axes.

$y = -2x + 3$

$y = -2x - 4$

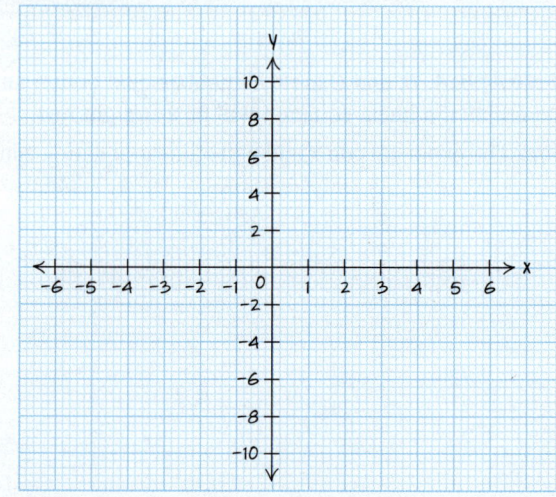

SECTION 3.4 Linear Equations and Their Graphs   291

2. Are the two lines in part 1 parallel?
3. Find the slopes of the two lines in part 1. Are the slopes the same or different?
4. Graph the two lines on the same axes.

What's That Mean?

Parallel

In mathematics, two lines are said to be *parallel* if they extend in the same direction(s) and are always the same distance apart. In music, two melodies are parallel if they keep the same distance apart in musical pitch.

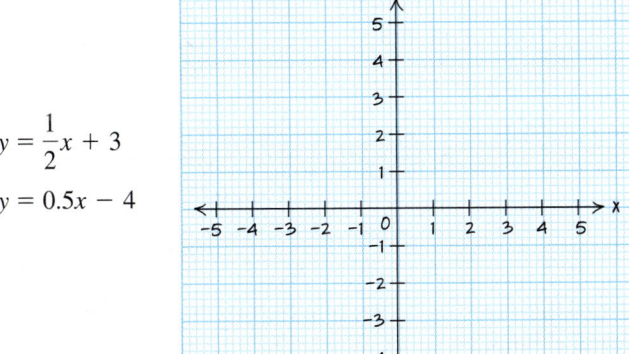

$y = \dfrac{1}{2}x + 3$

$y = 0.5x - 4$

5. Are the two lines in part 4 parallel?
6. Find the slopes of the two lines in part 4. Are the slopes the same or different?
7. Make a conjecture about the slopes of parallel lines.

In the preceding Concept Investigation, we saw that the lines graphed were parallel and had the same slope. In fact, this is always true.

DEFINITION

Parallel Lines Two lines that have the same slope are parallel. If the slope of the first line is m_1 and the slope of the second line is m_2, then the two lines are parallel when $m_1 = m_2$.

Example 8 Determine whether the lines are parallel

Determine whether the two lines are parallel.

a. $y = -2x + 10$
   $6x + 3y = -18$

b. $y = 5x + 1$
   $40x + 8y = 3$

SOLUTION

a. Parallel lines have the same slope. The first line, $y = -2x + 10$, is in slope-intercept form. The slope is $m = -2$.
The second line, $6x + 3y = -18$, is in general form. To find the slope, solve this equation for y to put it in slope-intercept form.

$$6x + 3y = -18$$
$$\underline{-6x \qquad\qquad -6x}$$ Isolate y by subtracting $6x$ from both sides.
$$3y = -6x - 18$$
$$\dfrac{3y}{3} = \dfrac{-6x}{3} - \dfrac{18}{3}$$ Divide both sides by 3 to solve for y.
$$y = -2x - 6$$ This is in slope-intercept form.

The slope of this second line is $m = -2$. Because the two lines have the same slope, they are parallel lines.

b. Parallel lines have the same slope. The first line, $y = 5x + 1$, is in slope-intercept form. The slope is $m = 5$.

The second line, $40x + 8y = 3$, is in general form. To find the slope, solve this equation for y to put it in slope-intercept form.

$$40x + 8y = 3$$
$$\underline{-40x \qquad\qquad -40x}$$ Isolate y by subtracting 40x from both sides.
$$8y = -40x + 3$$
$$\frac{8y}{8} = \frac{-40x}{8} + \frac{3}{8}$$ Divide both sides by 8 to solve for y.
$$y = -5x + \frac{3}{8}$$ This is in slope-intercept form.

The slope of this second line is $m = -5$. Because the two lines have different slope, they are not parallel lines.

PRACTICE PROBLEM FOR EXAMPLE 8

Determine whether the two lines are parallel.

a. $y = 5x + 3$

$-20x + 4y = 1$

b. $y = -\frac{3}{5}x + 8$

$3x - 5y = 5$

Another type of special lines to be aware of are **perpendicular lines**. Perpendicular lines meet at 90°, or right, angles.

> ### DEFINITION
> **Perpendicular Lines** Two lines that have slopes that are opposite reciprocals of each other are said to be perpendicular. If the slope of the first line is m_1 and the slope of the second line is m_2, then the two lines are perpendicular when $m_1 = -\frac{1}{m_2}$.

Example 9 Determining whether lines are perpendicular

Determine whether the two lines are perpendicular.

a. $y = 3x + 7$

$x + 3y = 15$

b. $y = 4x - 2$

$x - 4y = -24$

Skill Connection

Opposite reciprocals

Recall that a reciprocal means to "flip" a fraction over. The reciprocal of $\frac{3}{5}$ is $\frac{5}{3}$. To find an opposite reciprocal, flip the fraction over *and* change the sign.

The opposite reciprocal of $\frac{3}{5}$ is $-\frac{5}{3}$.

SOLUTION

a. Perpendicular lines have slopes that are opposite reciprocals. The first line, $y = 3x + 7$, is in slope-intercept form. The slope is $m = 3$.

The second line, $x + 3y = 15$, is in general form. To find the slope, we solve this equation for y to put it in slope-intercept form.

$$x + 3y = 15$$
$$\underline{-x \qquad\qquad -x}$$ Isolate y by subtracting x from both sides.
$$3y = -x + 15$$
$$\frac{3y}{3} = \frac{-x}{3} + \frac{15}{3}$$ Divide both sides by 3 to solve for y.

$$y = \frac{-x}{3} + 5 \qquad \text{This is in slope-intercept form.}$$

$$y = -\frac{1}{3}x + 5 \qquad \text{The coefficient on } x \text{ is } -\frac{1}{3}.$$

The slope of this second line is $m = -\frac{1}{3}$. Take this slope, $m = -\frac{1}{3}$, and find the opposite reciprocal.

$$-\frac{1}{3} \qquad \text{Flip over the fraction and change the sign.}$$

$$-\left(-\frac{3}{1}\right) = \frac{3}{1} = 3$$

Because this yields the slope of the first line, 3, the two lines are perpendicular.

b. Perpendicular lines have slopes that are opposite reciprocals. The first line, $y = 4x - 2$, is in slope-intercept form. The slope is $m = 4$.
The second line, $x - 4y = -24$, is in general form. To find the slope, we solve this equation for y to put it in slope-intercept form.

$$x - 4y = -24 \qquad \text{Isolate } y \text{ by subtracting } x \text{ from both sides.}$$

$$\underline{-x -x}$$

$$-4y = -x - 24$$

$$\frac{-4y}{-4} = \frac{-x}{-4} - \frac{24}{-4} \qquad \text{Divide both sides by } -4 \text{ to solve for } y.$$

$$y = \frac{x}{4} + 6 \qquad \text{This is in slope-intercept form.}$$

$$y = \frac{1}{4}x + 6 \qquad \text{The coefficient on } x \text{ is } \frac{1}{4}.$$

The slope of this second line is $m = \frac{1}{4}$. The opposite reciprocal of $\frac{1}{4}$ is $-\frac{4}{1} = -4$.
Since this does not equal the slope of the first line, $m = 4$, the two lines are not perpendicular.

PRACTICE PROBLEM FOR EXAMPLE 9

Determine whether the two lines are perpendicular.

a. $y = \frac{5}{2}x + 1$

$2x - 5y = 35$

b. $y = x$

$x + y = 1$

3.4 Vocabulary Exercises

1. $Ax + By = C$ is called the _____ of the equation of a line.

2. Parallel lines have the _____ slope.

3. Perpendicular lines have slopes that are _____ reciprocals.

294 CHAPTER 3   Linear Equations with Two Variables

3.4 Exercises

For Exercises 1 through 4, match the equations to the given graphs.

1. $y = 2x + 3$
   $y = -2x + 3$
   $y = 2x - 3$

   Graph A 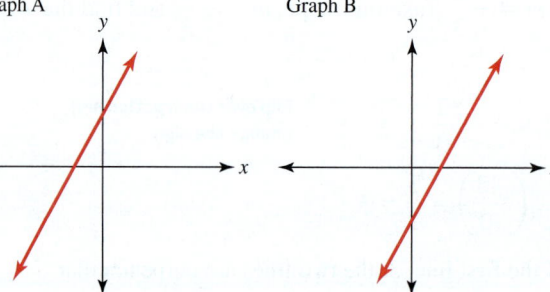 Graph B   Graph C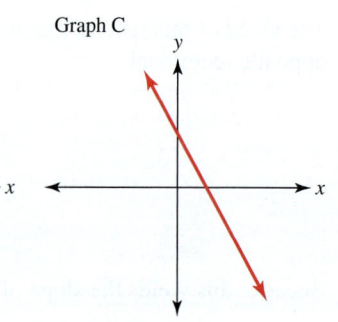

2. $y = -3x + 5$
   $y = 3x - 5$
   $y = -3x - 5$

   Graph A   Graph B 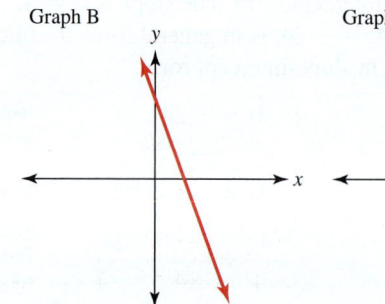 Graph C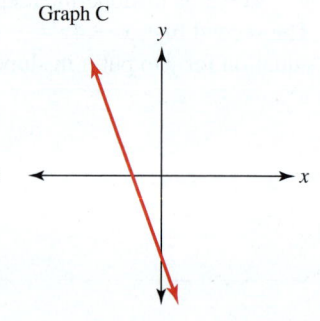

3. $y = \dfrac{1}{2}x + 2$
   $y = \dfrac{1}{2}x - 3$
   $y = \dfrac{1}{2}x + 5$

   Graph A 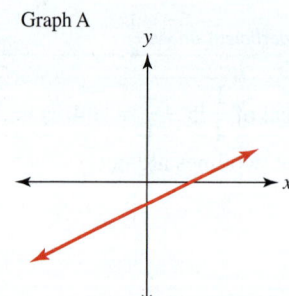 Graph B 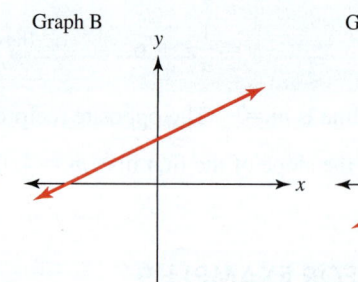 Graph C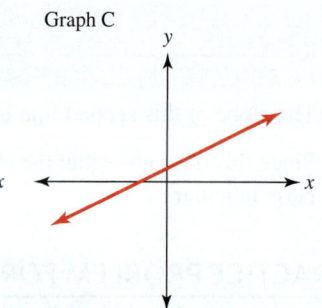

4. $y = -\dfrac{2}{3}x + 1$
   $y = -\dfrac{2}{3}x + 4$
   $y = -\dfrac{2}{3}x - 6$

   Graph A 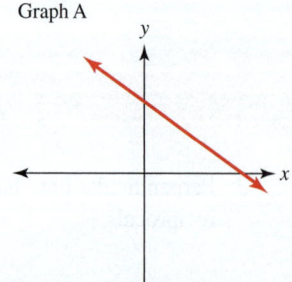 Graph B 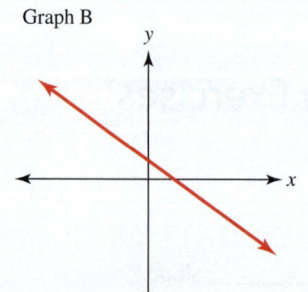 Graph C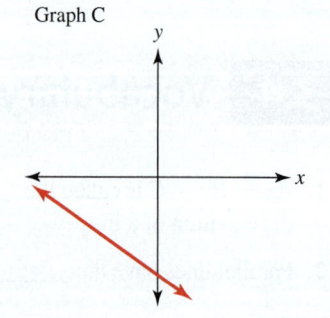

SECTION 3.4 Linear Equations and Their Graphs   295

For Exercises 5 through 22, graph each equation using the slope and y-intercept. Graph at least three points for each line. Clearly label and scale the axes.

5. $y = \frac{1}{2}x - 3$
6. $y = \frac{2}{5}x - 4$
7. $y = \frac{1}{3}x + 2$
8. $y = \frac{2}{3}x + 1$
9. $y = -\frac{3}{5}x + 6$
10. $y = -\frac{3}{4}x + 5$
11. $y = 3x - 8$
12. $y = 4x - 10$
13. $y = -2x + 4$
14. $y = -3x + 7$
15. $y + 5 = \frac{1}{2}(x + 26)$
16. $y + 3 = \frac{1}{3}(x + 15)$
17. $y + 6 = -2(x + 1)$
18. $y + 4 = -3(x + 2)$
19. $5y = 4x - 30$
20. $2y = 3x - 8$
21. $-4y = 3x - 12$
22. $-y = x + 3$

For Exercises 23 through 30, an equation is given with a student's graph. Each graph has a mistake in it. Explain what the student did wrong when graphing the given equation.

23. $y = \frac{1}{2}x + 2$

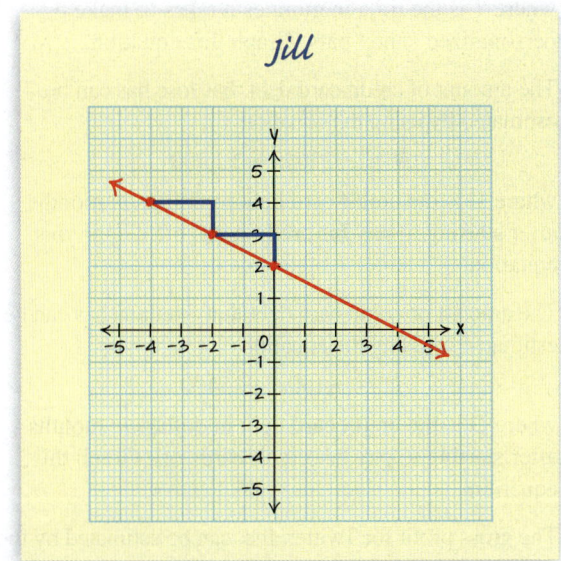

Jill

24. $y = \frac{2}{3}x + 1$

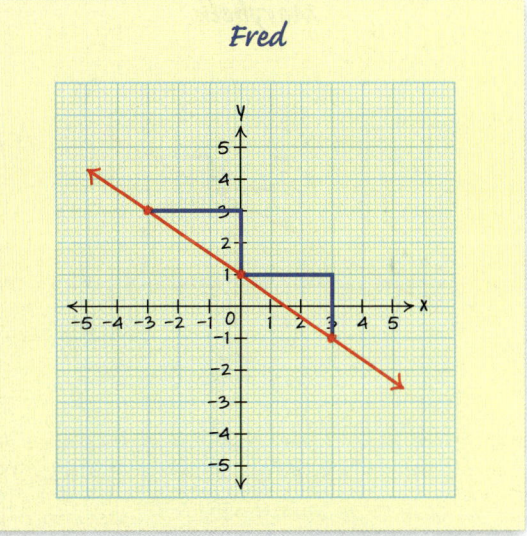

Fred

25. $y = -2x - 3$

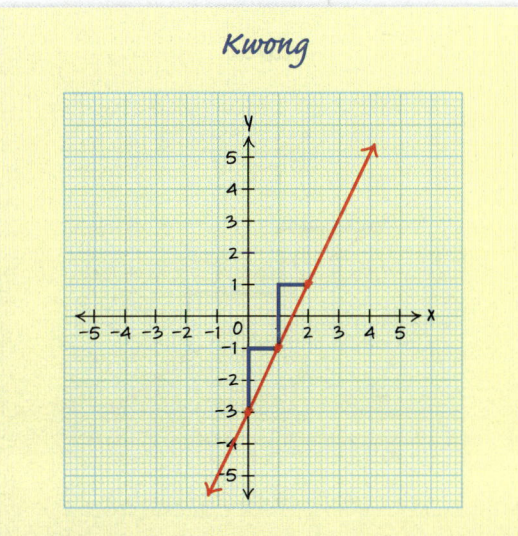

Kwong

26. $y = -\frac{1}{2}x - 1$

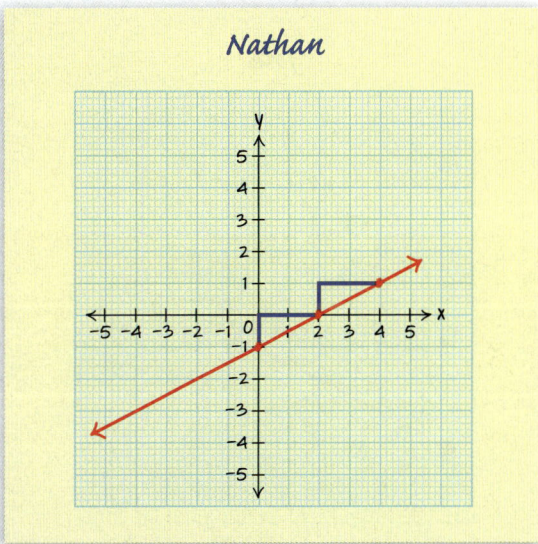

Nathan

296 CHAPTER 3 Linear Equations with Two Variables

27. $y = 2x + 2$

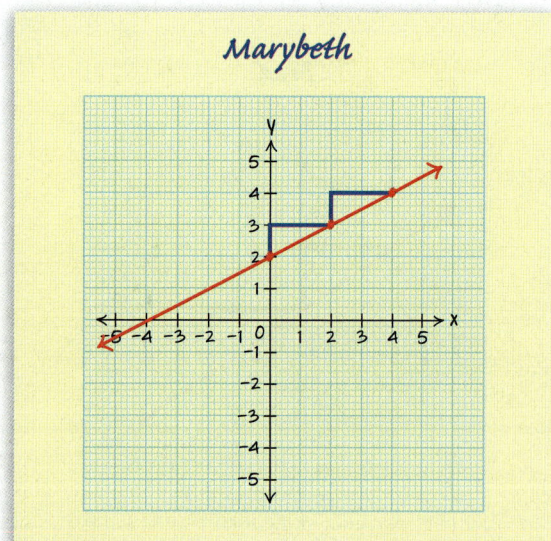
Marybeth

28. $y = -\dfrac{3}{2}x + 1$

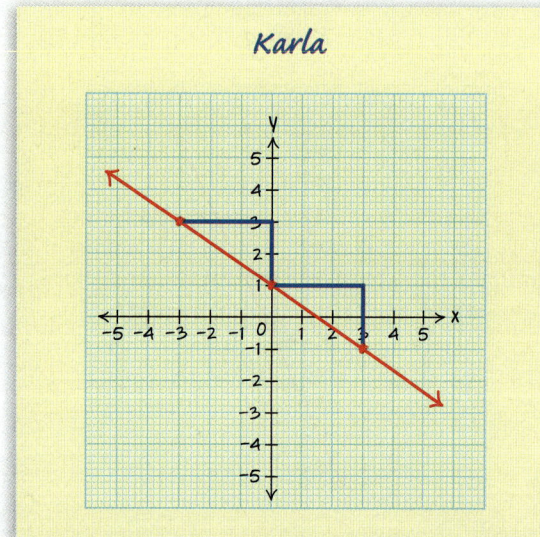
Karla

29. $y = \dfrac{1}{2}x + 2$

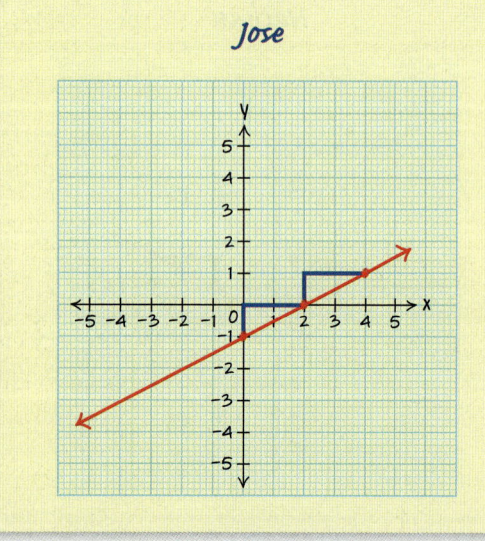
Jose

30. $y = -\dfrac{2}{3}x + 3$

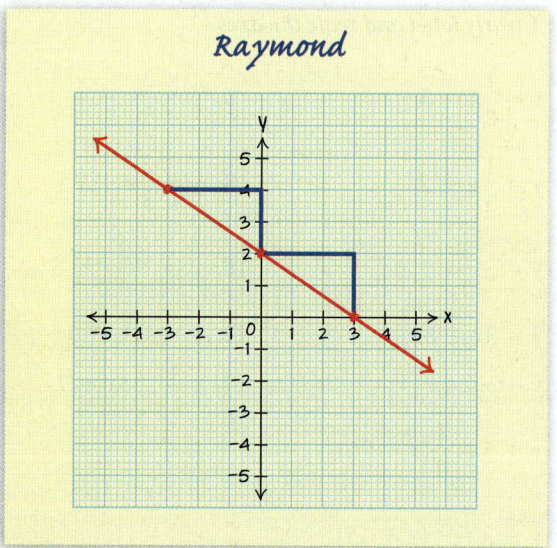
Raymond

For Exercises 31 through 40, use the given problem situations to graph the equations. Clearly label and scale the axes.

31. The number of cars sold at Big Jim's Car Lot can be estimated by using the equation
$$C = -2w + 25$$
where C is the number of cars sold during week w of the year. Graph this equation.

32. The amount of time it takes to create personalized candy bars can be estimated by using the equation
$$T = 2c + 40$$
where T is the time in minutes it takes to make c personalized candy bars. Graph this equation.

33. The amount of credit card debt that Jose has can be estimated by using the equation
$$D = -250m + 12{,}500$$
where D is the credit card debt in dollars m months after starting to pay his credit cards off. Graph this equation.

34. The amount of credit card debt that Svetlana has can be estimated by using the equation
$$D = -100m + 5400$$
where D is the credit card debt in dollars m months after starting to pay her credit cards off. Graph this equation.

35. The gross profit for Twitter, Inc. can be estimated by the equation
$$P = 413t + 1317$$
where P is the gross profit in millions of dollars t years since 2015. Graph this equation.

Source: http://finance.google.com, accessed May 2017.

36. The cost for Hamid to make EPs for his new garage band can be calculated by using the equation
$$C = 1.25n + 100$$
where C is the cost in dollars to make n EPs. Graph this equation.

37. The total revenue for Deere and Company, the makers of John Deere tractors and other agricultural equipment, can be estimated by using the equation
$$R = -4t + 51$$
where R is the revenue in billions of dollars t years since 2010. Graph this equation.

Source: http://finance.google.com, accessed May 2017.

38. The total revenue for Twitter, Inc. can be estimated by using the equation
$$R = 641t + 2025$$
where R is the revenue in millions of dollars t years since 2015. Graph this equation.

Source: http://finance.google.com, accessed May 2017.

39. The total revenue for Papa John's International, Inc. can be estimated by the equation
$$R = 86.4t + 1208$$
where R is the total revenue in millions of dollars t years since 2010. Graph this equation.

Source: http://finance.google.com, accessed May 2017.

40. The total revenue for McDonalds Corporation can be estimated by the equation
$$R = -1.25t + 32$$
where P is the gross profit in billions of dollars t years since 2010. Graph this equation.

Source: http://finance.google.com, accessed May 2017.

For Exercises 41 through 50, graph the given linear equations using the x- and y-intercepts. Clearly label and scale the axes. Label the intercepts.

41. $2x + 5y = 20$
42. $3x + 4y = 24$
43. $x + 2y = 6$
44. $x + 3y = 12$
45. $4x - 2y = 10$
46. $5x - 7y = 14$
47. $4x - 5y = 15$
48. $3x - 8y = 12$
49. $10x + 35y = 210$
50. $15x - 25y = 300$

For Exercises 51 through 54, use the given information to answer the questions.

51. The number of lawn mowers and tillers that a factory can produce is limited by the amount of time available. It takes 2 hours to make a lawn mower and 3 hours to make a tiller. If there are 80 hours a week when the production line can be run, the number of lawn mowers and tillers that can be produced can be found by using the equation
$$2m + 3t = 80$$
where m is the number of lawn mowers and t is the number of tillers that the plant can produce.

   a. Graph the given equation.

   b. Using the graph, determine how many lawn mowers the factory can produce if the workers make 12 tillers that week.

52. The number of sets of 20-inch wheels and 24-inch wheels that a custom shop can produce is limited by the amount of time available. It takes 4.5 hours to make a set of 20-inch wheels and 6 hours to make a set of 24-inch wheels. If there are 36 hours a week to run the production equipment, the number of sets of 20-inch wheels and 24-inch wheels that can be produced can be found by using the equation
$$4.5t + 6f = 36$$
where t is the number of 20-inch wheel sets and f is the number of 24-inch wheel sets that the shop can produce.

   a. Graph the given equation.

   b. Using the graph, determine how many 24-inch wheel sets the shop can produce if the workers make four sets of 20-inch wheel sets that week.

53. Maggie has two credit cards, one charging 16% interest and the other charging 24% interest. Maggie paid $800 in interest this year. The possible balances on each credit card can be found by using the equation
$$0.16A + 0.24B = 800$$
where A is the amount in dollars owed on the credit card that charges 16% interest and B is the amount in dollars owed on the credit card that charges 24% interest.

a. Graph the given equation.

b. Interpret the *A*-intercept of the problem situation.

54. Irina has two credit cards, one charging 12% interest and the other charging 17% interest. Irina paid $1204 in interest this year. The possible balances on each credit card can be found by using the equation

$$0.12A + 0.17B = 1204$$

where *A* is the amount in dollars owed on the credit card that charges 12% interest and *B* is the amount in dollars owed on the credit card that charges 17% interest.

a. Graph the given equation.

b. Interpret the *B*-intercept of the problem situation.

For Exercises 55 through 70, determine whether the given equations are linear. If the equation is linear, put it into slope-intercept form.

55. $2y + 7 = 3x + 21$
56. $3y - 6 = 4x + 8$
57. $12 = 3xy + 9$
58. $8 = -2xy - 10$
59. $y = |3x + 1|$
60. $y = |2x - 4|$
61. $2x + 3 = 4(y + 8)$
62. $5x - 8 = 7(2 - y)$
63. $y = 3x^2 + 5$
64. $y = -2x^2 + 10$
65. $y = \dfrac{5}{x} + 2$
66. $y = \dfrac{2}{x} - 7$
67. $3x + 8 = 4y - 9$
68. $x + 5 = 8 - 7y$
69. $y = x^3 + 2$
70. $y = -x^3 + 1$

For Exercises 71 through 82, decide whether you would graph the given equation by using the slope and y-intercept, by using the x- and y-intercepts, or by building a table of points to plot. Do not graph these equations. There are several possible correct answers.

71. $y = \dfrac{2}{9}x + 5$
72. $y = \dfrac{7}{3}x - 8$
73. $2x + 8y = 93$
74. $5x - 3y = 24$
75. $y = 2x^2 + 5$
76. $y = -x^2 + 3$
77. $y + 5 = 2(x - 9)$
78. $y - 7 = -3(x + 8)$
79. $y = 2.45x - 3.75$
80. $3.5y = 1.78x - 4.26$
81. $y = |x - 4|$
82. $y = |7 - x|$

For Exercises 83 through 94, determine whether the lines are parallel, perpendicular, or neither. Do not graph the equations.

83. $y = -3x + 5$
    $y = -3x$
84. $y = \dfrac{5}{2}x - 1$
    $y = 2.5x + 6$
85. $y = -7x + 2$
    $y = \dfrac{1}{7}x - 44$
86. $y = x - 17$
    $y = -x + 9$
87. $y = \dfrac{2}{5}x + 3.5$
    $y = \dfrac{5}{2}x - 3.5$
88. $y = 8x - 12$
    $y = \dfrac{1}{8}x + 11$
89. $y = \dfrac{1}{2}x - 7$
    $6x + 3y = 27$
90. $4x + 3y = 15$
    $y = \dfrac{3}{4}x - 2$
91. $2x - 12y = 12$
    $y = \dfrac{1}{6}x + 2$
92. $-4x + 5y = 20$
    $y = \dfrac{4}{5}x - 3$
93. $x - 6y = 12$
    $y = 6x + 7$
94. $y = \dfrac{3}{8}x - 4$
    $8x - 3y = -3$

For Exercises 95 through 98, solve each equation. Provide reasons for each step. Check the answers. Round to the hundredths place if necessary.

95. $-7t + 11 = -10$
96. $-\dfrac{x}{2} + 4 = \dfrac{1}{3}$
97. $2(3x - 4) + 5 = -2x + 5$
98. $-(4x - 7) + 6 = -2x + 1$

For Exercises 99 and 100, solve each inequality. Write the solution sets using inequality symbols, using interval notation, and graphing the solution set on a number line.

99. $-2x + 6 \leq -4$
100. $3 - 3x \geq 21$

For Exercises 101 and 102, answer each problem as indicated. Round to two decimal places if necessary.

101. Find 89% of 110.
102. Find 15% of 205.

SECTION 3.5 Finding Equations of Lines 299

3.5 Finding Equations of Lines

LEARNING OBJECTIVES
- Find equations of lines using slope-intercept form.
- Solve applications.
- Find equations of lines using the point-slope form.
- Find equations of parallel and perpendicular lines.

■ Finding Equations of Lines Using Slope-Intercept Form

In Section 3.3, we saw that $y = mx + b$ is the equation of a line in slope-intercept form. We will now learn how to use the slope-intercept form of a line to write the equation of a line. To do so, we must determine the value of the slope, m, and the y-intercept $(0, b)$.

Example 1 Finding the equation of a line from a graph with a known y-intercept

Use the graph to find the equation of the line.

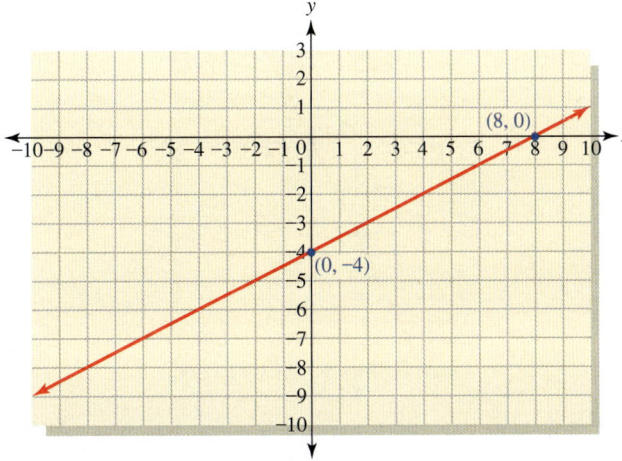

SOLUTION
We have to find the slope, m, and the y-intercept, $(0, b)$, of this line. To find the slope, select two points on the line and use the definition of slope. Recall that one form of the definition of slope is $m = \dfrac{\text{rise}}{\text{run}}$. Two points on the line are $(0, -4)$ and $(8, 0)$.

Therefore, the slope is

$$m = \frac{\text{rise}}{\text{run}} = \frac{4}{8} = \frac{1}{2}$$

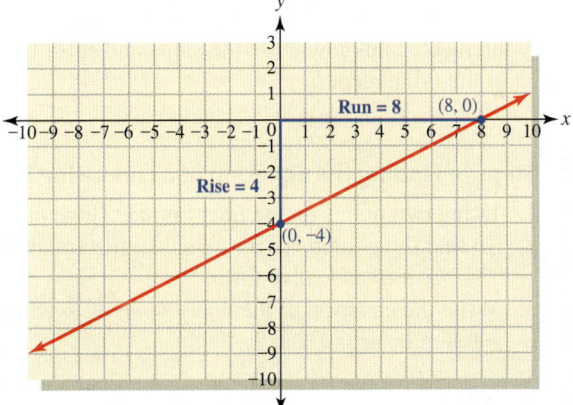

■ Connecting the Concepts

Why do we find slope working from left to right?

Recall that we read graphs from left to right. When we find the slope from a graph, we also read the graph from left to right. A graph that goes up from left to right will have a positive slope. A graph that goes down from left to right will have a negative slope.

Substitute this value of slope into the slope-intercept form of a line.

$$y = mx + b$$

$$y = \frac{1}{2}x + b \qquad \text{Substitute in } m = \frac{1}{2}.$$

The point where the graph crosses the y-axis is the y-intercept $(0, b)$. That point is $(0, -4)$, so $b = -4$. Substituting this value into our equation yields

$$y = \frac{1}{2}x + b \qquad \text{The linear equation}$$

$$y = \frac{1}{2}x + (-4) \qquad \text{Substitute in } b = -4.$$

$$y = \frac{1}{2}x - 4 \qquad \text{Simplify.}$$

Therefore, the equation of the line is $y = \frac{1}{2}x - 4$.

PRACTICE PROBLEM FOR EXAMPLE 1

Use the following graph to find the equation of the line.

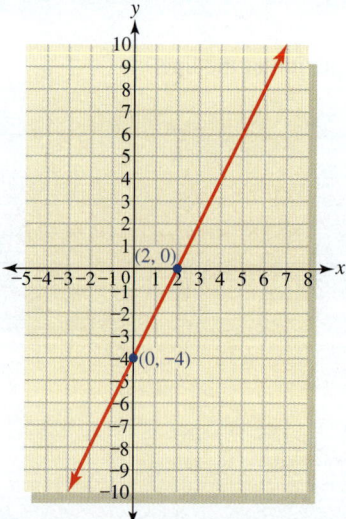

In Example 1, the y-intercept was found by inspection from the graph. However, we cannot always determine the y-intercept exactly by inspection. In problems in which the y-intercept is unknown, use the slope and substitute any point on the line to find the value of the y-intercept b.

> **Steps to Finding the Equation of a Line Using the Slope-Intercept Form**
>
> 1. Using two points on the line, find the slope m.
> 2. To find the y-intercept $(0, b)$, substitute any point on the line in for x and y in $y = mx + b$ and solve for b.
> 3. Write the equation of the line $y = mx + b$ with the specific values for m and b found in steps 1 and 2.
>
> *Note: x and y remain as variables in the final answer.*

SECTION 3.5 Finding Equations of Lines **301**

Example 2 — Finding the equation of a line from a graph with an unknown y-intercept

Use the graph to find the equation of the line.

a.

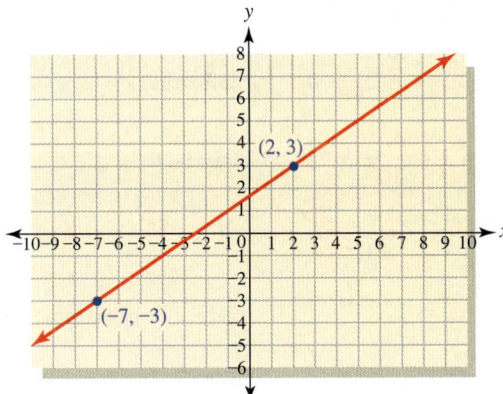

b.

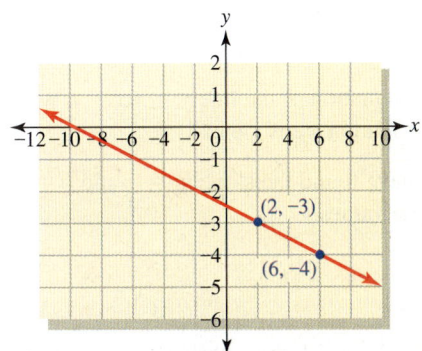

SOLUTION

a. Step 1 Using two points on the line, find the slope m.

Two points on the graph are $(-7, -3)$ and $(2, 3)$. To find the slope, we use the $m = \dfrac{\text{rise}}{\text{run}}$ definition.

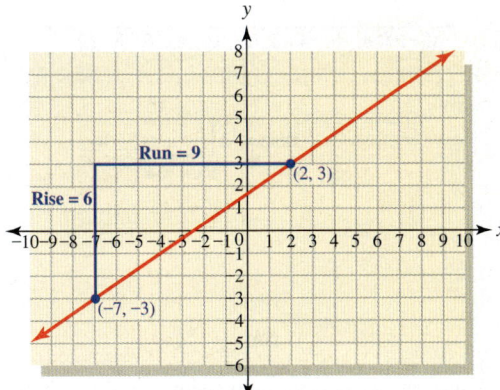

Therefore, the slope is

$$m = \frac{\text{rise}}{\text{run}} = \frac{6}{9} = \frac{2}{3}$$

Substitute $m = \dfrac{2}{3}$ into the slope-intercept form of a line.

$$y = mx + b$$

$$y = \frac{2}{3}x + b$$

Step 2 To find the y-intercept $(0, b)$, substitute any point on the line in for x and y into $y = mx + b$ and solve for b.

Selecting $(2, 3)$ and substituting it into the equation yields

$$y = \frac{2}{3}x + b$$

$$3 = \frac{2}{3}(2) + b \quad \text{Substitute } x = 2, y = 3, \text{ and simplify.}$$

$$3 = \frac{4}{3} + b \quad \text{Solve for } b.$$

$$\underline{-\frac{4}{3} \quad -\frac{4}{3}}$$

$$3 - \frac{4}{3} = b \quad \text{Find like denominators.}$$

$$\frac{9}{3} - \frac{4}{3} = b$$

$$\frac{5}{3} = b$$

Step 3 Write the equation of the line $y = mx + b$ with the specific values for m and b found in steps 1 and 2.

Now that we have found the slope $m = \frac{2}{3}$ and $b = \frac{5}{3}$, write the equation of the line using these values.

$$y = \frac{2}{3}x + \frac{5}{3}$$

b. Two points on the graph are $(2, -3)$ and $(6, -4)$. To find the slope, use the $m = \dfrac{\text{rise}}{\text{run}}$ definition.

■ Connecting the Concepts

When is it okay to use decimals?

In Example 2b, the slope of the line is

$$m = -\frac{1}{4}$$

Since $-\dfrac{1}{4} = -0.25$ is an *exact* decimal representation, it would be correct to use the decimal form for the slope.

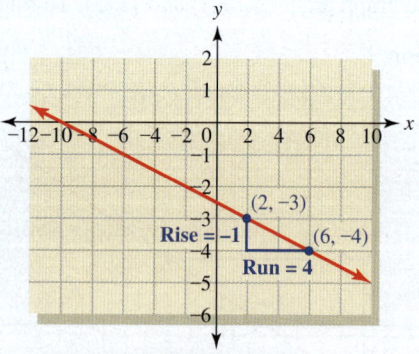

$$m = \frac{\text{rise}}{\text{run}} = -\frac{1}{4}$$

Substitute $m = -\dfrac{1}{4}$ into the slope-intercept form of a line.

$$y = mx + b$$

$$y = -\frac{1}{4}x + b$$

SECTION 3.5 Finding Equations of Lines   **303**

To find the *y*-intercept, $(0, b)$, substitute either point into the equation. Selecting $(2, -3)$ and substituting it into the equation yields

$$y = -\frac{1}{4}x + b$$

$$-3 = -\frac{1}{4}(2) + b \quad \text{Substitute } x = 2, y = -3, \text{and simplify.}$$

$$-3 = -\frac{1}{2} + b \quad \text{Solve for } b.$$

$$\underline{+\frac{1}{2} \quad +\frac{1}{2}}$$

$$-3 + \frac{1}{2} = b$$

$$-\frac{6}{2} + \frac{1}{2} = b$$

$$-\frac{5}{2} = b \quad \text{Find like denominators.}$$

Now that we have found the slope $m = -\frac{1}{4}$ and $b = -\frac{5}{2}$, write the equation of the line using these values.

$$y = -\frac{1}{4}x - \frac{5}{2}$$

PRACTICE PROBLEM FOR EXAMPLE 2

Use the graph to find the equation of the line.

a.

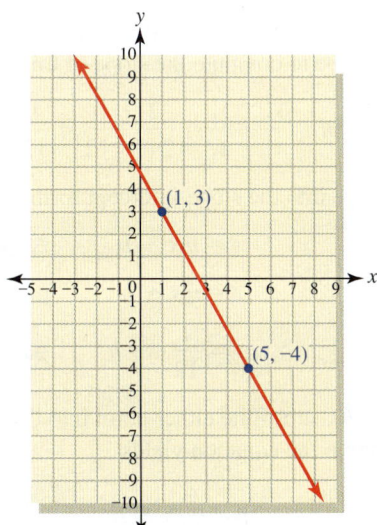

b.

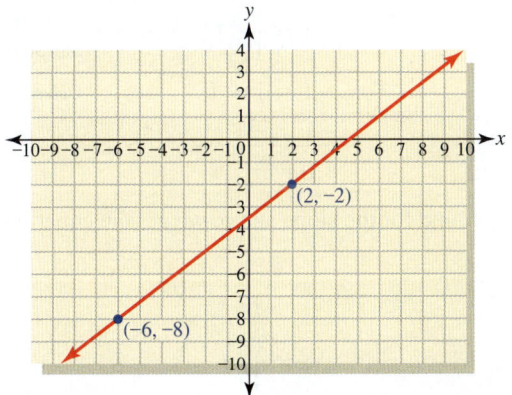

Finding Equations of Lines from Applications

If we know that the relationship between the variables in an application is linear, then we can use any two points to find the value of the slope and y-intercept. Using these, we can generate the equation of the line.

Example 3   Inches of rain from snow

In Section 1.2, we examined the relationship in the conversion between the inches of snowfall into the inches of rainfall. There is a linear relationship between the inches of snow, s, and the inches of rain, r. Find the equation of the line that passes through these points.

s = inches of snow	r = inches of rain
12	$\frac{12}{12} = 1$
24	$\frac{24}{12} = 2$
36	$\frac{36}{12} = 3$
42	$\frac{42}{12} = 3.5$

SOLUTION

Two points on the line are (12, 1) and (24, 2). Because we were not given the graph, use the formula definition of slope to compute the slope of this line.

$$m = \frac{y_2 - y_1}{x_2 - x_1} = \frac{2 - 1}{24 - 12} = \frac{1}{12}$$

Substitute this value of slope into the slope-intercept form of a line.

$y = mx + b$ Slope-intercept form of a line

$r = ms + b$ Use the variables from the problem.

$r = \frac{1}{12}s + b$ Substitute $m = \frac{1}{12}$.

To find the y-intercept, $(0, b)$, substitute either point into the equation. Selecting (12, 1) and substituting it into the equation yields

$r = \frac{1}{12}s + b$

$1 = \frac{1}{12}(12) + b$ Substitute $s = 12$, $r = 1$, and simplify.

$1 = 1 + b$ Solve for b.

$\underline{-1 \quad -1}$

$0 = b$

Using the values we found for the slope and b, we get the equation of the line

$$r = \frac{1}{12}s + 0$$

$$r = \frac{1}{12}s$$

PRACTICE PROBLEM FOR EXAMPLE 3

The conversion between the number of hours, h, and the number of days, d, is given in the following table. A linear relationship exists between the variables h and d. Find the equation of the line that passes through these points.

h = number of hours	d = number of days
12	$\frac{12}{24} = 0.5$
24	$\frac{24}{24} = 1$
36	$\frac{36}{24} = 1.5$
72	$\frac{72}{24} = 3$

In some applications, the two points are given in the wording of the problem. We have to read the problem carefully to find the relationship between the two variables. These relationships can be written as points and used to find the slope and the value of b. We can then write the equation of the line that describes the linear relationship between the variables.

Example 4 Converting from Celsius to Fahrenheit

Two methods of measuring temperature are degrees Fahrenheit (°F) and degrees Celsius (°C). Water freezes at 32°F, which is the same as 0°C. Water boils at 212°F, which is equal to 100°C. The two quantities (°F and °C) have a linear relationship.

a. Find the equation of the line that describes this relationship. Let the input C represent degrees Celsius and let the output F represent degrees Fahrenheit.

b. Interpret the slope of the linear equation.

c. Find and interpret the intercepts of the linear equation.

d. Find F when $C = 20$. Interpret the solution in the context of measuring temperature.

SOLUTION

a. First define the variables. The input variable C represents the degrees Celsius. The output variable F represents the degrees Fahrenheit. Two points on the line are $(C, F) = (0, 32)$ and $(C, F) = (100, 212)$. Notice that we could switch the input and output in this situation, but we will select the input variable C and the output variable F. Therefore, in the slope-intercept form of a line,

$$y = mx + b$$

we will have

$$F = mC + b$$

The slope of this line can be calculated by using the slope formula.

$$m = \frac{y_2 - y_1}{x_2 - x_1} \quad \text{We are using } x = C \text{ and } y = F.$$

$$m = \frac{F_2 - F_1}{C_2 - C_1} \quad \begin{array}{l}(C_1, F_1) = (0, 32)\\ (C_2, F_2) = (100, 212)\end{array}$$

$$m = \frac{F_2 - F_1}{C_2 - C_1} = \frac{212 - 32}{100 - 0} = \frac{180}{100} = \frac{9}{5}$$

Substituting $m = \dfrac{9}{5}$ into $F = mC + b$ yields

$$F = \dfrac{9}{5}C + b$$

To determine the value of b, substitute in either point. Selecting $(C, F) = (0, 32)$ and substituting these values in yields

$$F = \dfrac{9}{5}C + b$$

$$32 = \dfrac{9}{5}(0) + b \qquad \textbf{F = 32, C = 0}$$

$$32 = 0 + b$$

$$32 = b$$

The point $(0, 32)$ represents the vertical intercept. If we notice that the vertical intercept is given, we can set $b = 32$ without substituting any points into the equation. Using the slope $m = \dfrac{9}{5}$ and $b = 32$, we get the equation of the line.

$$F = \dfrac{9}{5}C + 32$$

b. The slope is $m = \dfrac{9}{5}$. To interpret the slope, recall that

$$m = \dfrac{9}{5} = \dfrac{\text{change in } F}{\text{change in } C}$$

Attaching units to the numerator and denominator will allow us to interpret the slope in terms of the problem.

$$m = \dfrac{9 \text{ degrees Fahrenheit}}{5 \text{ degrees Celsius}} = 1.8 \text{ degrees Fahrenheit per degree Celsius}$$

The temperature in degrees Fahrenheit increases 9 degrees Fahrenheit for an increase of 5 degrees Celsius. In other words, the temperature increases 1.8 degrees Fahrenheit for each degree Celsius increase.

c. One point on the line is $(C, F) = (0, 32)$. This point is the F-intercept because $C = 0$. The interpretation of this point is that it is $32°F$ when it is $0°C$. To find the C-intercept, let $F = 0$ and solve for C.

$$F = \dfrac{9}{5}C + 32$$

$$0 = \dfrac{9}{5}C + 32 \qquad \textbf{Substitute F = 0.}$$

$$\underline{-32 \qquad\qquad -32} \qquad \textbf{Solve for C.}$$

$$-32 = \dfrac{9}{5}C$$

$$\dfrac{5}{9}(-32) = \dfrac{5}{9}\left(\dfrac{9}{5}C\right) \qquad \textbf{Multiply both sides by the reciprocal.}$$

$$-\dfrac{160}{9} = C$$

$$-17.8 \approx C$$

$0°F$ is equivalent to approximately $-17.8°C$.

d. When $C = 20$, we have

$$F = \frac{9}{5}(20) + 32 \quad \text{Substitute in } C = 20.$$
$$F = 36 + 32$$
$$F = 68$$

20°C is equivalent to 68°F.

Example 5 Manufacturing toys

A toy manufacturer is currently producing 2000 toys a week. Beginning in October, the manufacturer will begin to increase production each week by 500 more toys during each week of the Christmas production season. A linear relationship exists between the week, w, and the number of toys produced, P. Here $w = 1$ is the first week in October, $w = 2$ is the second week in October, and so on.

a. Find the equation of the line that describes this relationship. Let the input w represent the week number beginning with the first week in October and let the output P represent the production level of toys in week w.

b. Interpret the slope of the linear equation.

c. Find and interpret the intercepts of the linear equation.

d. Find P when $w = 5$. Interpret the solution in terms of toy production.

SOLUTION

a. In this problem, the toy production is being increased by 500 toys per week. This rate is how fast the production is changing and, therefore, is also the slope of the line. The production of toys depends on the week, so the input is w, and the output is P. Substituting $m = 500$ into $P = mw + b$ yields

$$P = 500w + b$$

When $w = 0$, the plant is producing $P = 2000$ toys. To determine b, we can substitute these values into the linear equation to get

$$P = 500w + b \quad w = 0, P = 2000$$
$$2000 = 500(0) + b$$
$$2000 = b \quad \text{Solve for } b.$$

Therefore, the linear equation is

$$P = 500w + 2000$$

b. The slope (from above) is $m = 500$. To interpret the slope, attach the proper units to the numerator and denominator.

$$m = \frac{500}{1} = \frac{\text{change in } P}{\text{change in } w}$$
$$m = \frac{500 \text{ toys}}{1 \text{ week}}$$

The plant is producing 500 more toys per week.

c. From part a, we have that when $w = 0$, the plant is producing $P = 2000$ toys. This is the *P*-intercept, the point $(0, 2000)$.

At the start of the Christmas production season, the plant is producing 2000 toys a week.

d.
$$P = 500(5) + 2000 \quad \text{Substitute in } w = 5.$$
$$P = 4500$$

During the fifth week of the Christmas production season, the plant is producing 4500 toys in that week.

PRACTICE PROBLEM FOR EXAMPLE 5

a. There is a "flash to bang" method for computing the distance that lightning is from you. When you see lightning (the flash), count the number of seconds until you hear the thunder (the bang). Divide the number of seconds by 5 to get the number of miles that the lightning is from you. Let the input s be the number of seconds you count from flash to bang and the output d represent the distance in miles the lightning is from you. There is a linear relationship between s and d. Find the equation of the line that describes this relationship.
Source: http://www.srh.noaa.gov/jetstream/lightning/lightning_safety.htm.

b. Kaito plans to lose weight on his diet. He currently weighs 195 pounds (week 0). He would like to lose 1 pound a week for w weeks. Let the input w represent the week number that he has been on his diet. Let the output N represent the number of pounds that Kaito weighs that week. There is a linear relationship between w and N. Find the equation of the line that describes this relationship.

■ Finding Equations of Lines Using Point-Slope Form

We have been finding equations of lines using the slope-intercept form, $y = mx + b$. From the formula for slope, we can get another form for the equation of a line called the **point-slope form.**

> **DEFINITION**
>
> **The Point-Slope Form of the Equation of a Line** Given the slope of a line, m, and a point on the line, (x_1, y_1), the point-slope form of the line is
>
> $$y - y_1 = m(x - x_1)$$

Example 6 Finding equations using the point-slope form

a. Find the equation of the line with slope $m = -\dfrac{2}{5}$ that passes through the point $(5, 1)$. Write the final answer in slope-intercept form, $y = mx + b$.

b. Find the equation of the line passing through the points $(3, -4)$ and $(-3, -8)$. Write the final answer in slope-intercept form, $y = mx + b$.

SOLUTION

a. Notice that we have been given a point on the line and the slope. This is a situation in which the point-slope form of a line works efficiently. We have that $m = -\dfrac{2}{5}$ and $(x_1, y_1) = (5, 1)$. Substituting these values into the point-slope form of a line yields

$$(x_1, y_1) = (5, 1)$$

$$y - 1 = -\frac{2}{5}(x - 5) \qquad \text{Substitute } m = -\frac{2}{5} \text{ and } (x_1, y_1) = (5, 1).$$

$$y - 1 = -\frac{2}{5}x - \left(-\frac{2}{5}\right)(5) \qquad \text{Distribute and simplify.}$$

$$y - 1 = -\frac{2}{5}x + 2 \qquad \text{Solve for } y.$$

$$\underline{+1 \qquad\qquad\quad +1}$$

$$y = -\frac{2}{5}x + 3$$

The equation of the line is $y = -\frac{2}{5}x + 3$.

b. As before, first determine the slope of the line.

$$m = \frac{y_2 - y_1}{x_2 - x_1} = \frac{-8 - (-4)}{-3 - 3}$$

$$m = \frac{-8 + 4}{-6} = \frac{-4}{-6} = \frac{2}{3}$$

Substituting this value in for the slope in the point-slope form yields

$$y - y_1 = \frac{2}{3}(x - x_1)$$

We can choose either of the two given points to substitute in for (x_1, y_1). Choosing $(x_1, y_1) = (3, -4)$ and substituting this into our equation, we get

$$y - (-4) = \frac{2}{3}(x - 3) \qquad \text{Substitute in } (x_1, y_1) = (3, -4).$$

$$y + 4 = \frac{2}{3}(x - 3) \qquad \text{Distribute and simplify.}$$

$$y + 4 = \frac{2}{3}x - \frac{2}{3}(3)$$

$$y + 4 = \frac{2}{3}x - 2 \qquad \text{Solve for } y.$$

$$\underline{-4 \qquad\qquad\quad -4}$$

$$y = \frac{2}{3}x - 6$$

The equation of the line is $y = \frac{2}{3}x - 6$.

PRACTICE PROBLEM FOR EXAMPLE 6

a. Find the equation of the line passing through the points $(-2, 5)$ and $(2, 3)$. Put the final answer in slope-intercept form, $y = mx + b$.

b. Find the equation of the line with slope $m = \frac{1}{6}$ that passes through the point $(6, -1)$. Put the final answer in slope-intercept form, $y = mx + b$.

310 CHAPTER 3 Linear Equations with Two Variables

■ Finding Equations of Parallel and Perpendicular Lines

We can use the techniques of this section to find the equation of parallel and perpendicular lines.

Example 7 Finding the equation of a line parallel to a given line

Find the equation of the line parallel to $y = -\frac{3}{5}x + 1$ passing through the point (5, 3).

SOLUTION
To find the equation of a line, we need both the slope and a point. Parallel lines have the same slope. Therefore, since the slope of the given line is $m = -\frac{3}{5}$, the slope of the line for which we are trying to find the equation is also $m = -\frac{3}{5}$. Substituting this into slope-intercept form yields

$$y = mx + b \qquad \text{Substitute in } m = -\frac{3}{5}.$$
$$y = -\frac{3}{5}x + b$$

To find the value of b, substitute in the given point (5, 3) and solve for b.

$$y = -\frac{3}{5}x + b \qquad \text{Substitute in the given point (5, 3).}$$
$$3 = -\frac{3}{5}(5) + b \qquad \text{Simplify the right side.}$$
$$3 = -3 + b \qquad \text{Solve for } b.$$
$$\underline{+3 \quad +3}$$
$$6 = b$$
$$y = -\frac{3}{5}x + 6 \qquad \text{The equation of the line}$$

PRACTICE PROBLEM FOR EXAMPLE 7

Find the equation of the line parallel to $y = -\frac{1}{2}x - 17$ passing through the point $(-6, 1)$.

Example 8 Finding the equation of a line perpendicular to a given line

Find the equation of the line perpendicular to $y = -3x + 12$ passing through the point $(-9, 2)$.

SOLUTION
To find the equation of a line, we need both the slope and a point. Perpendicular lines have slopes that are opposite reciprocals. The slope of the given line is $m = -3$. The slope can be written as

$$-3 = -\frac{3}{1}$$

The opposite reciprocal is

$$-\left(-\frac{1}{3}\right) = \frac{1}{3}$$

The slope of the line we are trying to find is $m = \dfrac{1}{3}$.

Substituting this in slope-intercept form yields

$y = mx + b$ Substitute in $m = \dfrac{1}{3}$.

$y = \dfrac{1}{3}x + b$

To find the value of b, substitute in the given point $(-9, 2)$ and solve for b.

$y = \dfrac{1}{3}x + b$ Substitute in the given point $(-9, 2)$.

$2 = \dfrac{1}{3}(-9) + b$ Simplify the right side.

$2 = -3 + b$ Solve for b.

$\underline{+3 \quad +3}$

$5 = b$

$y = \dfrac{1}{3}x + 5$ The equation of the line

PRACTICE PROBLEM FOR EXAMPLE 8

Find the equation of the line perpendicular to $y = \dfrac{3}{7}x - 1$ passing through the point $(6, -1)$.

3.5 Vocabulary Exercises

1. $y - y_1 = m(x - x_1)$ is called the _____ of the equation of a line.

2. Finding the slope of a line requires _____ points.

3. The slope of a line can be found from the *graph* using the slope formula or the _____ definition.

3.5 Exercises

For Exercises 1 through 10, use the graph to find the equation of the line. Put the answer in $y = mx + b$ form.

1.

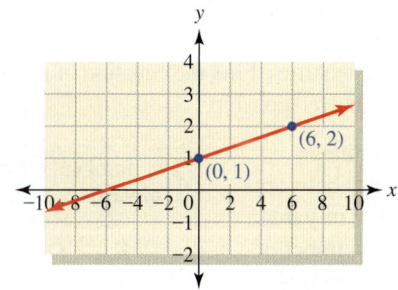

2.

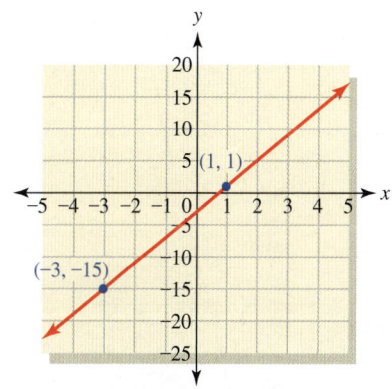

312 CHAPTER 3 Linear Equations with Two Variables

3.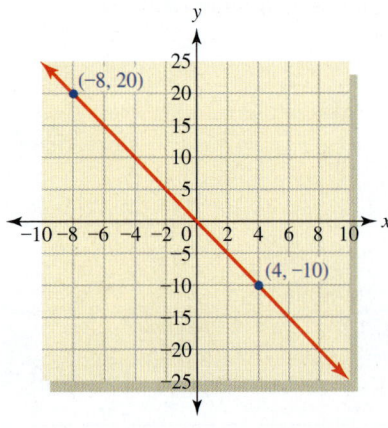

4.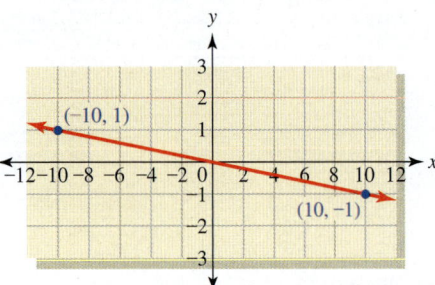

5.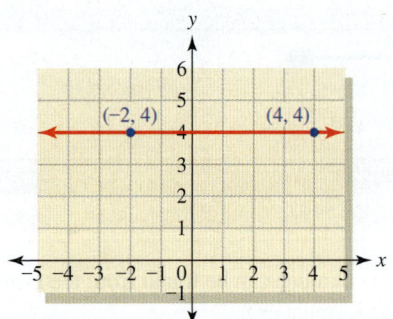

6.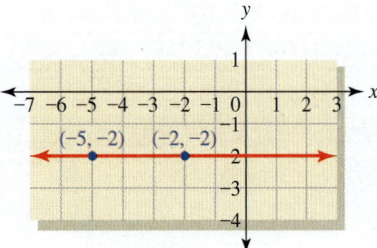

7.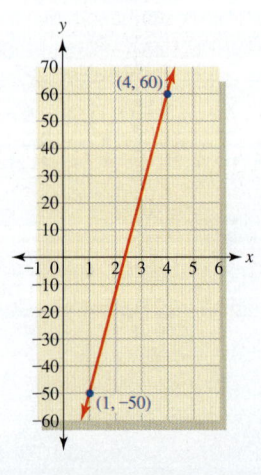

8.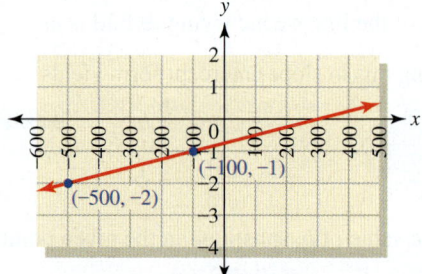

9.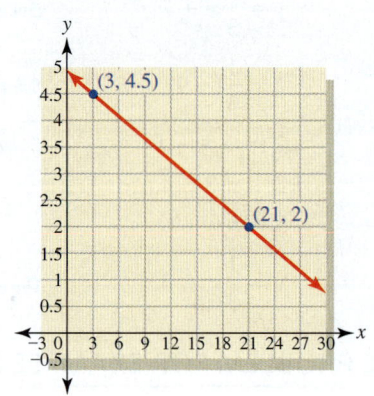

10.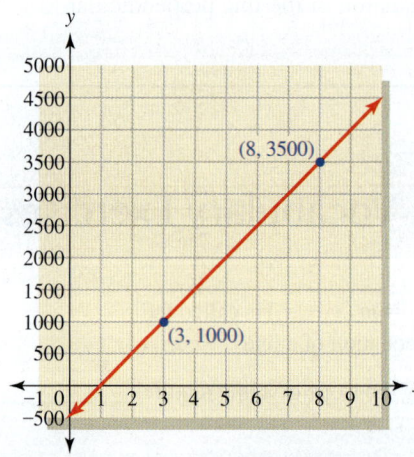

11. There is a linear relationship between the number of years, y, and the number of centuries, c. The accompanying table lists some conversions between the number of years and the number of centuries. Find the equation of the line that describes this relationship.

y = number of years	c = number of centuries
0	0
100	1
150	1.5
200	2

12. There is a linear relationship between the number of yards, y, and the number of feet, f. The accompanying table lists some conversions between the number of yards

y = number of yards	f = number of feet
0	0
1	3
2	6
3	9

and the number of feet. Find the equation of the line that describes this relationship.

13. The temperature drops from 1:00 P.M. onward. Let the input t represent the time of the day and let the output F represent the temperature in degrees Fahrenheit. At 1:00 P.M. ($t = 1$), the temperature is $F = 75$. At 3:00 P.M. ($t = 3$), the temperature is $F = 71$. Find the linear relationship between t and F.

14. The temperature increases from 6:00 A.M. onward. Let the input t represent the time of the day and let the output F represent the temperature in degrees Fahrenheit. At 6:00 A.M. ($t = 6$), the temperature is $F = 54$. At 10:00 A.M. ($t = 10$), the temperature is $F = 68$. Find the linear relationship between t and F.

15. Irina goes on a diet to lose weight. Let the input t represent the week of her diet and the output W represent Irina's weight in pounds (lb). At the start of her diet ($t = 0$), Irina weighed 165 pounds. After 4 weeks on her diet, she weighed 157 pounds. Find the linear relationship between t and W.

16. Peter goes on a diet to lose weight. Let the input t represent the week of his diet and let the output W represent Peter's weight in pounds. At the start of his diet ($t = 0$), Peter weighed 225 pounds. After 6 weeks on his diet, he weighed 207 pounds. Find the linear relationship between t and W.

For Exercises 17 through 22, find and interpret the slope in the context of each given problem.

17. Interpret the slope in terms of the problem situation for Exercise 11.

18. Interpret the slope in terms of the problem situation for Exercise 12.

19. Interpret the slope in terms of the problem situation for Exercise 13.

20. Interpret the slope in terms of the problem situation for Exercise 14.

21. Interpret the slope in terms of the problem situation for Exercise 15.

22. Interpret the slope in terms of the problem situation for Exercise 16.

For Exercises 23 through 28, find and interpret the intercepts in the context of each given problem.

23. Find and interpret the intercepts in terms of the problem situation for Exercise 11.

24. Find and interpret the intercepts in terms of the problem situation for Exercise 12.

25. Find and interpret the intercepts in terms of the problem situation for Exercise 13.

26. Find and interpret the intercepts in terms of the problem situation for Exercise 14.

27. Find and interpret the intercepts in terms of the problem situation for Exercise 15.

28. Find and interpret the intercepts in terms of the problem situation for Exercise 16.

29. The 2017 Toyota Prius has average combined mileage as given in the following table. Let the input g represent the number of gallons of gas used and let the output d represent the distance in (combined) miles driven.

   a. Find the equation of the line that describes this relationship.

   b. Interpret the slope in terms of the problem situation.

g = number of gallons of gas	d = number of miles driven
2	104
3	156
4	208
5	260

   Source: Data derived from information found at www.toyota.com

   c. Interpret the intercepts in terms of the problem situation.

30. The 2017 Honda Civic has average city mileage as given in the following table. Let the input g represent the number of gallons of gas used and let the output d represent the distance in (city) miles driven.

   a. Find the equation of the line that describes this relationship.

   b. Interpret the slope in terms of the problem situation.

g = number of gallons of gas	d = number of miles driven
2	56
3	84
4	112
5	140

   Source: Data derived from information found at www.automobiles.honda.com

   c. Interpret the intercepts in terms of the problem situation.

31. A local gym is advertising memberships that cost $45.00 a month plus an initial $125 sign-up fee. Let the input m represent the number of months and let the output c represent the cost. Find the linear relationship between m and c.

  a. Find the equation of the line that describes this relationship.

  b. Interpret the slope in terms of the problem situation.

  c. Interpret the intercepts in terms of the problem situation.

32. A community college charges $46.00 per unit plus a $50.00 health fee. Let the input u represent the number of units enrolled in and let the output T represent the tuition (charges) for that semester.

  a. Find the equation of the line that describes this relationship.

  b. Interpret the slope in terms of the problem situation.

  c. Interpret the intercepts in terms of the problem situation.

For Exercises 33 through 42, determine the equation of the line that passes through the given points. Put the final answers in $y = mx + b$ form.

33. Points: $(0, -9)$ and $(4, -5)$

34. Points: $(0, 6)$ and $(3, 15)$

35. Points: $(-3, -1)$ and $(0, 1)$

36. Points: $(0, -2)$ and $(-5, -3)$

37. Points: $(1, -6)$ and $(-1, 2)$

38. Points: $(-2, -3)$ and $(3, -13)$

39. Points: $(3, 5)$ and $(-4, 5)$

40. Points: $(6, -1)$ and $(3, -1)$

41. Points: $(6, 3)$ and $(7, 6)$

42. Points: $(-3, 6)$ and $(-1, 2)$

For Exercises 43 through 58, determine the equation of the line by using the point-slope form. Put the final answer in $y = mx + b$ form.

43. $m = -2$ and point $(0, -7)$

44. $m = -4$ and point $(0, 9)$

45. $m = 6$ and point $(2, 3)$

46. $m = 5$ and point $(3, -1)$

47. $m = \dfrac{2}{3}$ and point $(6, 4)$

48. $m = \dfrac{1}{2}$ and point $(-8, -1)$

49. $m = 0$ and point $(-3, 5)$

50. $m = 0$ and point $(4, 6)$

51. Points: $(1, 5)$ and $(-1, -1)$

52. Points: $(2, 3)$ and $(4, 11)$

53. Points: $(-1, 7)$ and $(3, -5)$

54. Points: $(-1, 6)$ and $(1, -4)$

55. Points: $(2, 2)$ and $(-2, 0)$

56. Points: $(4, -1)$ and $(8, -4)$

57. Points: $(3, -6)$ and $(-2, -6)$

58. Points: $(0, 18)$ and $(-5, 18)$

For Exercises 59 through 66, rewrite each linear equation in the requested form.

59. Write in general form: $y = -\dfrac{2}{3}x + 7$.

60. Write in general form: $y = \dfrac{2}{5}x - 3$.

61. Write in slope-intercept form: $-3x + 4y = -24$.

62. Write in slope-intercept form: $7x - 14y = 21$.

63. Write in slope-intercept form: $y - 3 = -6(x + 11)$.

64. Write in slope-intercept form: $y + 9 = -4(x - 5)$.

65. Write in general form: $y - 2 = 7(x + 1)$.

66. Write in general form: $y + 12 = -\dfrac{1}{2}(x + 16)$.

For Exercises 67 through 70, write the equation of the line through the given points. (Hint: Graph the two points and connect them with a line. Review Section 3.1 if necessary.)

67. Points: $(5, 15)$ and $(5, 3)$

68. Points: $(7, 3)$ and $(7, -3)$

69. Points: $(0, -6)$ and $(0, -10)$

70. Points: $(-12, 5)$ and $(-12, 0)$

For Exercises 71 through 78, find the equation of the requested line. See Examples 7 and 8 for help.

71. Find the equation of the line parallel to $y = 6x - 10$ passing through the point $(-7, 0)$.

72. Find the equation of the line parallel to $y = -11x + 3$ passing through the point $(0, -5)$.

73. Find the equation of the line parallel to $x - 5y = 9$ passing through the point $(-10, 2)$.

74. Find the equation of the line parallel to $2x + 3y = -1$ passing through the point $(-6, -1)$.

75. Find the equation of the line perpendicular to $y = \frac{3}{8}x - 15$ passing through the point $(-12, 4)$.

76. Find the equation of the line perpendicular to $y = \frac{6}{7}x - 9$ passing through the point $(-6, 5)$.

77. Find the equation of the line perpendicular to $-5x + 4y = -9$ passing through the point $(-5, 1)$.

78. Find the equation of the line perpendicular to $-2x + 9y = -7$ passing through the point $(-4, 1)$.

For Exercises 79 through 86, find the equation of the line using the given information. (Hint: Draw the graph to help yourself visualize the situation.)

79. Find the equation of the line parallel to $y = 4$ passing through the point $(-3, 6)$.

80. Find the equation of the line parallel to $y = 0$ passing through the point $(5, -3)$.

81. Find the equation of the line perpendicular to $x = 3$ passing through the point $(2, 5)$.

82. Find the equation of the line perpendicular to $x = -1$ passing through the point $(0, 2)$.

83. Find the equation of the line parallel to $x = -5$ passing through the point $(1, -3)$.

84. Find the equation of the line parallel to $x = 4$ passing through the point $(6, -1)$.

85. Find the equation of the line perpendicular to $y = -3$ passing through the point $(5, 2)$.

86. Find the equation of the line perpendicular to $y = 7$ passing through the point $(0, 4)$.

For Exercises 87 through 90, evaluate each expression using the order-of-operations agreement.

87. $(24 - 6) \div 9 + 4^2 - |3 - 2(5)| + 8$

88. $(-5)^2 - 6\left(\frac{2}{3}\right) - (5 - 8)$

89. $\frac{7 - (-3)}{8 - (-2)^2} + \frac{3}{5} \div \frac{1}{15}$

90. $\frac{3(5) - 5}{-12 + 7} - (-3)^2(3 - 10) + |-16 + 3(-2)| - (-1)$

For Exercises 91 through 94, translate each of the following sentences into an expression.

91. 9 less than 5 times a number

92. The sum of a square of a number and 18

93. The quotient of the square of a number and 4

94. The difference between 5 times a number and 7

For Exercises 95 through 98, solve each equation. Provide reasons for each step. Check the answers. Round to the hundredths place if necessary.

95. $4 - 2(3 - t) = 5(2t - 1) + 7$

96. $-2(15 - n) + 6 = 1 - (3n - 5)$

97. $\frac{1}{2}(x - 7) + 3 = \frac{1}{3}x + 1$

98. $-2 + \frac{1}{3}(x - 2) = \frac{1}{4}(2x + 1) - 1$

For Exercises 99 and 100, complete each table by filling in the missing values.

99.

Percent	Decimal	Fraction
12.5%		

100.

Percent	Decimal	Fraction
		$\frac{2}{25}$

For Exercises 101 and 102, solve each inequality. Check the answers. Write each solution set in each of the following ways: using inequality symbols, using interval notation, and graphing the solution on a number line.

101. $-5 < 2x - 1 < 7$

102. $3 < 4x - 5 < 11$

316 CHAPTER 3 Linear Equations with Two Variables

3.6 Modeling Linear Data

LEARNING OBJECTIVES
- Find a linear model for real data.
- Find linear models.
- Use linear models to make estimates.
- Determine model breakdown.

We now turn our attention to the concept of modeling. Observations in the real world are often collected and organized in a table. These observations are called *data*. Often it is easier to see what is happening if the data are graphed. The shape the data take can help us to see that a quantity is going up or down (a trend). Many times, finding an equation that approximately fits the data points and can be used to make estimates is very helpful.

The process of finding the equation that fits the data is called **modeling**, and the equation is called a **model**. For example, if the data points follow a linear pattern, the equation of a line that closely fits the data points is called a *linear model*.

■ Finding a Graphical Model for Linear Data

So far in Chapter 3, we have studied lines in detail. We have learned about slope, the x- and y-intercepts, the forms of equations of a line, and how to graph lines. We now apply that knowledge to model some real-world situations. In particular, we are interested in data sets that, when graphed, have the data points more or less lying in a straight line pattern.

Recall from Section 1.4 that a *scatterplot* is a dot plot of data from a table. The input (or independent) variable is often in the left column of the data table. The output (or dependent) variable is often in the right column of the data table. The output variable *depends* on the input variable. The input variable is graphed on the x-axis (horizontal axis), and the output variable is graphed on the y-axis (vertical axis).

Example 1 Do the data lie in a straight line?

The annual amounts that Apple, Inc., spent on research and development are listed in the following table.

Year	Annual Expenditures on Research and Development (billions of dollars)
2012	3.38
2013	4.48
2014	6.04
2015	8.07
2016	10.05

Source: www.marketwatch.com

a. Create a scatterplot for the data in the table.

b. Do the data lie more or less in a straight line?

SECTION 3.6 Modeling Linear Data 317

SOLUTION

a. To create a scatterplot, first look at the data carefully and determine a good scale for the data. The input values begin at 2012 and end at 2016. Therefore, scale the horizontal axis by 1 year. However, because the first input value is 2012, we do not want to start graphing on the horizontal axis at 0. We include a *break* in the graph (a zigzag) to indicate that we skipped the values between 0 and 2010. The output values begin at 3.38 and end at 10.05. Start graphing at 0, end at 12, and scale by 1.

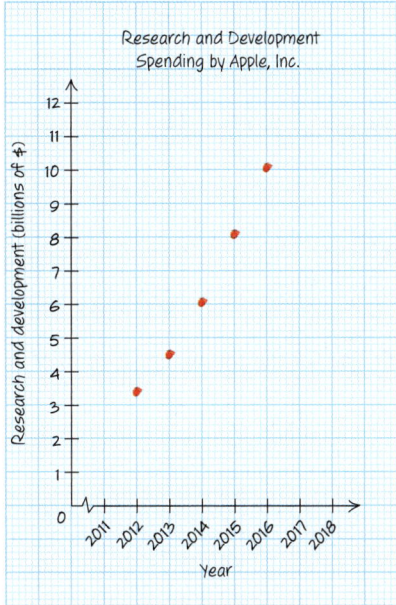

b. The points do more or less lie in a straight line.

Connecting the Concepts

What are units of billions of dollars?

In Example 1, we see that in the year 2012, Apple, Inc., spent approximately $3.38 billion on research and development. The units are billions of dollars. To convert the units from billions of dollars to dollars, multiply 3.38 by 1 billion.

$3.38 \cdot 1$ billion
$= 3.38 \cdot 1{,}000{,}000{,}000$
$= 3{,}380{,}000{,}000$

The data in Example 1 are said to be **linearly related**, since the data points lie more or less on a straight line. We want to find the line that fits the data best visually. This is called the *eyeball best-fit line*. To find this line, imagine drawing several lines on the scatterplot. Look for the line that comes as close to all of the points as possible on the scatterplot. The points that the line misses should be approximately equally spread out above and below the line.

CONCEPT INVESTIGATION
Finding an eyeball best-fit line

1. For the given scatterplot, decide which line better fits the data. Explain why you selected the line you did.

Graph 1

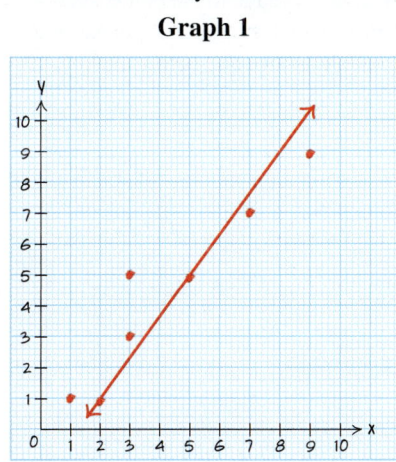

Graph 2

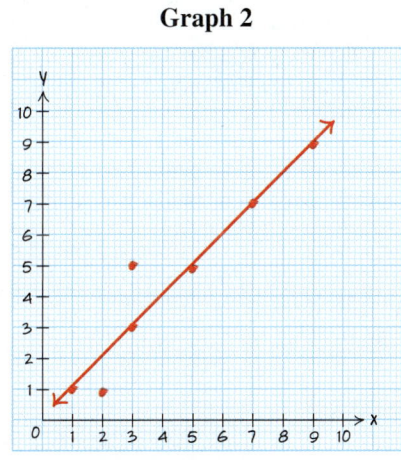

Connecting the Concepts

When the line does not fit, what do we do?

If an eyeball best-fit line does not fit the data well, pick two other points and redraw the line. The goals are to have the line go through as many data points as possible and to have an equal number of missed data points above and below the line. You can use the eyeball best-fit line to make estimations and predictions. A line that does not fit can lead to faulty conclusions about the data.

318 CHAPTER 3 Linear Equations with Two Variables

2. For the given scatterplot, decide which line better fits the data. Explain why you selected the line you did.

Graph 1

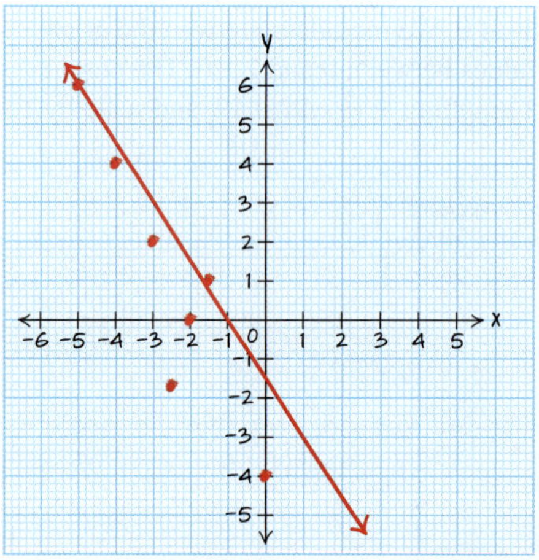

Graph 2

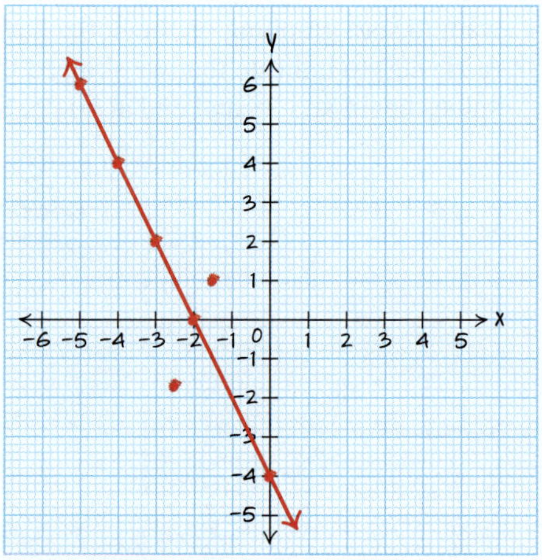

3. Draw an eyeball best-fit line for the following scatterplot.

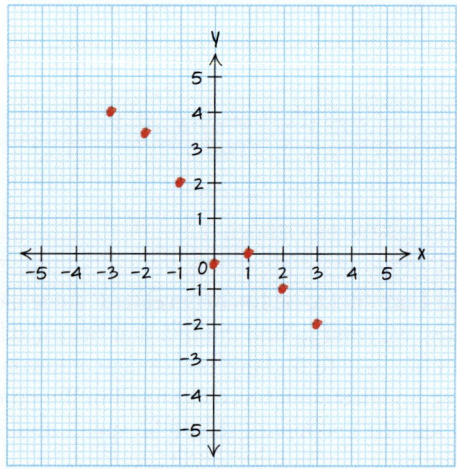

Steps to Find an Eyeball Best-Fit Line

1. Graph the data on a scatterplot.
2. Using a straightedge (a small clear ruler is best), fit the edge to the points as closely as possible. If the data do not completely lie on a straight line, make sure that the points the line misses are spread out equally above and below the line.
3. Draw the line on the scatterplot.

Example 2 Draw an eyeball best-fit line

For the following scatterplot, draw an eyeball best-fit line.

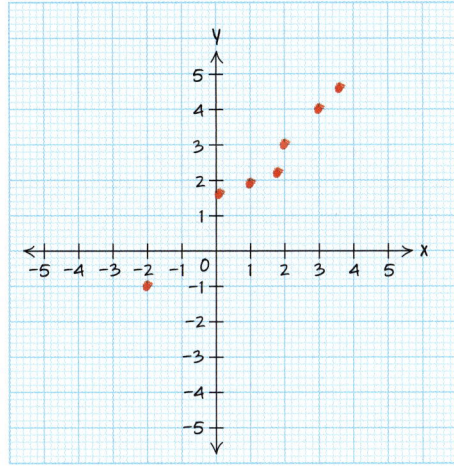

SOLUTION

Step 1 Graph the data on a scatterplot.
The first step has been completed, as the data were given on a scatterplot.

Step 2 Using a straightedge, fit the edge to the points as closely as possible. If the data do not completely lie on a straight line, make sure that the points the line misses are spread out equally above and below the line.

Step 3 Draw the line on the scatterplot.
The best-fit line misses two points. One missed point is above the line, and one is below the line.

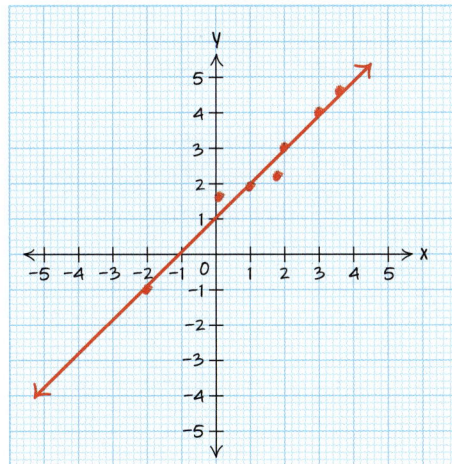

PRACTICE PROBLEM FOR EXAMPLE 2

For the following scatterplot, draw an eyeball best-fit line.

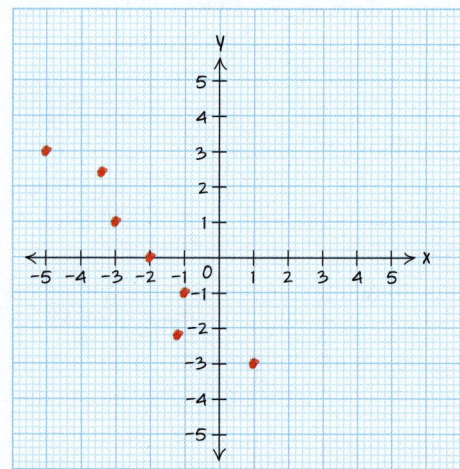

In Example 2, the data did not have a real-world connection. An eyeball best-fit line can also be drawn for data that do have a real-world connection. We now look at the Apple, Inc. data that were originally presented in Example 1 and find an eyeball best-fit line.

Example 3 An eyeball best-fit line for real data

Draw an eyeball best-fit line for the Apple, Inc. data that were presented in Example 1. The annual amounts spend on research and development for Apple, Inc. are listed in the following table.

Year	Annual Expenditures on Research and Development (billions of dollars)
2012	3.38
2013	4.48
2014	6.04
2015	8.07
2016	10.05

Source: www.marketwatch.com

SOLUTION
Drawing an eyeball best-fit line on the data that we graphed in Example 1 yields the following graph.

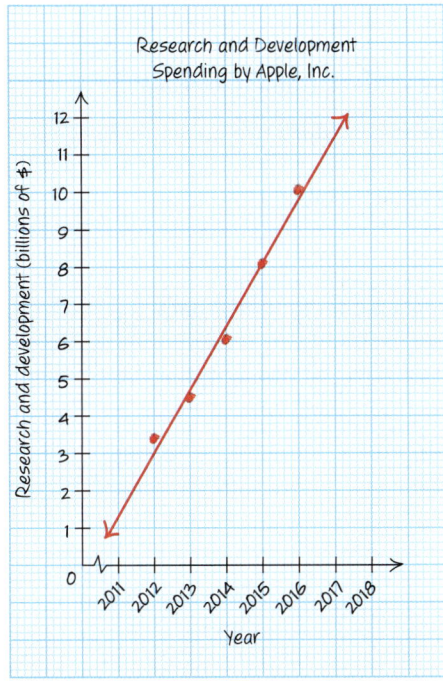

PRACTICE PROBLEM FOR EXAMPLE 3
The Bureau of Labor Statistics reports the following data for percentage of workers covered by unions in the United States. Draw an eyeball best-fit line for the data. (*Hint*: Scale the vertical axis by 0.2.)

Year	Percentage of Workers Represented by Unions
2008	13.7
2010	13.1
2012	12.5
2014	12.3
2016	11.9

Source: bls.data.gov

■ Finding a Linear Model

After learning how to find an eyeball best-fit line for a data set, we now learn how to come up with the equation of that line. In Section 3.3, the slope-intercept form of a line, $y = mx + b$, was introduced. We now use this form of a line to come up with the equation of the eyeball best-fit line. The equation of the eyeball best-fit line has a special name: It is called the **model**.

> **DEFINITION**
> **Model** The **model** for a linear data set is the equation of the line which represents the eyeball best-fit for the data.

What's That Mean?

Model

A **model** is a representation of something. When finding linear models (equations of lines), we are finding equations that are representations of the real data. The real data and the model (equation) will not be exactly the same in most cases.

Skill Connection

How to find the equation of a line

In Section 3.3, we learned to find the equation of a line. The formula $y = mx + b$ requires that we first find the slope using the slope formula, $m = \dfrac{y_2 - y_1}{x_2 - x_1}$. After the slope has been found, substitute in either point, and solve for b, the y-intercept. Then write the equation of the line $y = mx + b$.

To find the model for a data set, use the following steps.

> **Steps to Find a Linear Model for a Data Set**
> 1. Graph the data, and draw an eyeball best-fit line.
> 2. Find two points *on* the eyeball best-fit line. Use these two points to find the equation of the line. This equation is the *model*.
>
> *Note:* The model depends on which two points are selected to find the eyeball best-fit line. Answers may vary. The slopes of each model should be approximately equal to one another.

Example 4 Finding a model for linear data

For each data set, graph the data and find an eyeball best-fit line. Then find the equation of that line (the model).

a.

x	y
−2	−6.5
0	−5.5
1	−4.25
2	−3.5
3	−2.5
4	−2

b.

x	y
−6	8
−3	6
0	5
3	2
6	−1
9	−2

SOLUTION

a. Step 1 Graph the data and draw an eyeball best-fit line.

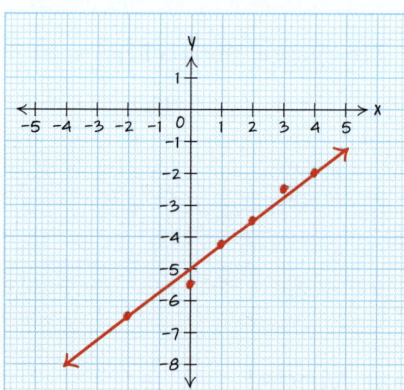

Connecting the Concepts

Does slope give trends in data?

In Section 3.2, we learned that slope can be interpreted as a rate of change. The slope of a model is a rate of change and can give us information about trends. When the slope is positive, the model is estimating an upward trend. When the slope is negative, the model is estimating a downward trend.

Step 2 Find two points on the eyeball best-fit line. Use these two points to find the equation of the line.

Two points on the line are $(-2, -6.5)$ and $(4, -2)$. Substituting $(x_1, y_1) = (-2, -6.5)$ and $(x_2, y_2) = (4, -2)$ in the slope formula yields

$$m = \frac{y_2 - y_1}{x_2 - x_1}$$

$$m = \frac{-2 - (-6.5)}{4 - (-2)}$$

$$m = \frac{4.5}{6} = \frac{3}{4}$$

Substituting the value of the slope into the equation of a line yields

$$y = mx + b$$
$$y = \frac{3}{4}x + b$$

To find the value of b, substitute either point into the equation and solve for b. Substituting in the point $(x_2, y_2) = (4, -2)$ yields

$$y = mx + b$$
$$y = \frac{3}{4}x + b \qquad \text{Substitute in } x, y = (4, -2).$$
$$-2 = \frac{3}{4}(4) + b$$
$$-2 = 3 + b \qquad \text{Solve for } b.$$
$$\underline{-3 \ -3}$$
$$-5 = b$$

$$y = \frac{3}{4}x - 5 \qquad \text{Write the equation of the line.}$$

The *model* is $y = \frac{3}{4}x - 5$. The slope of this model is positive, so the model is estimating an upward or positive trend.

b. Graphing the points and drawing an eyeball best-fit line, we get the following graph.

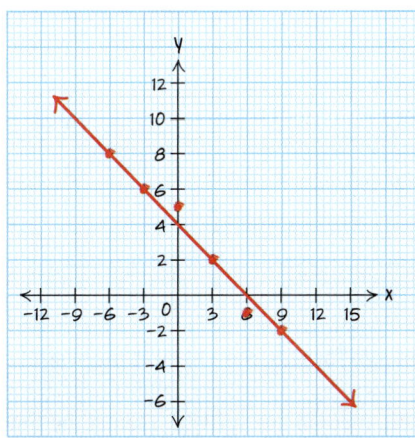

To find the equation of the model, which is the same as the eyeball best-fit line, use two points on the line. Two points on the line are $(-3, 6)$ and $(3, 2)$. Setting $(x_1, y_1) = (-3, 6)$ and $(x_2, y_2) = (3, 2)$ in the slope formula yields

$$m = \frac{y_2 - y_1}{x_2 - x_1}$$
$$m = \frac{2 - 6}{3 - (-3)}$$
$$m = \frac{-4}{6} = -\frac{2}{3}$$

We can also use the point-slope formula to find the equation of a line. Substituting the value of the slope into the point-slope form of a line yields

$$y - y_1 = m(x - x_1)$$
$$y - y_1 = -\frac{2}{3}(x - x_1)$$

■ **Connecting the Concepts**

Does an eyeball best-fit line have to go through two data points?

No, an eyeball best-fit line does not have to go through two data points. It may be easier to compute the equation of the model if it does, though.

324 CHAPTER 3 Linear Equations with Two Variables

We can substitute either point into the equation. Substituting in the point $(x_1, y_1) = (3, 2)$ yields

$$y - y_1 = m(x - x_1)$$
$$y - y_1 = -\frac{2}{3}(x - x_1)$$
$$y - 2 = -\frac{2}{3}(x - 3) \quad \text{Substitute in } (x_1, y_1) = (3, 2).$$
$$y - 2 = -\frac{2}{3}x + 2 \quad \text{Solve for y.}$$
$$\underline{+2 \qquad\qquad +2}$$
$$y = -\frac{2}{3}x + 4 \quad \text{Write the equation of the line.}$$

The model is $y = -\frac{2}{3}x + 4$. The slope of this model is negative, so the model is estimating a downward or negative trend.

PRACTICE PROBLEM FOR EXAMPLE 4

For each data set, graph the data, and draw an eyeball best-fit line. Then find the equation of the model for that line.

a.

x	y
−10	−11
−8	−9.5
−5	−9
0	−7
2	−7
5	−5

b.

x	y
−6	3
−3	2.5
0	1
3	−0.5
6	−1
9	−2

Example 5   Median age for U.S. men at first marriage

The data in the table list the median age for U.S. men at the time of their first marriage.

Year	Median Age at First Marriage (Men)
1970	23.2
1980	24.7
1990	26.1
2000	26.8
2010	28.2
2015	29.2

Source: U.S. Census Bureau, Current Population Survey.

a. Define the variables.
b. Graph the data on a scatterplot and find an eyeball best-fit line.
c. Find the equation of the linear model.

SOLUTION

a. Let t = years since 1950 (the input value) and let M = median age at first marriage for men (the output value). This means that the ordered pairs will have the form (t, M).

b. To graph the data on a scatterplot, first determine the scales of the axes. The input data (year) start at 20 (1970) and end at 65 (2015). Scale the horizontal axis by 5-year increments, beginning with 0 and ending with 70. The vertical axis will be scaled by 0.5, beginning at 22 and ending at 30. This will include the range of output values, which begin at 23.2 and end at 29.2. Include a break in the graph to show that the vertical axis is not starting at 0.

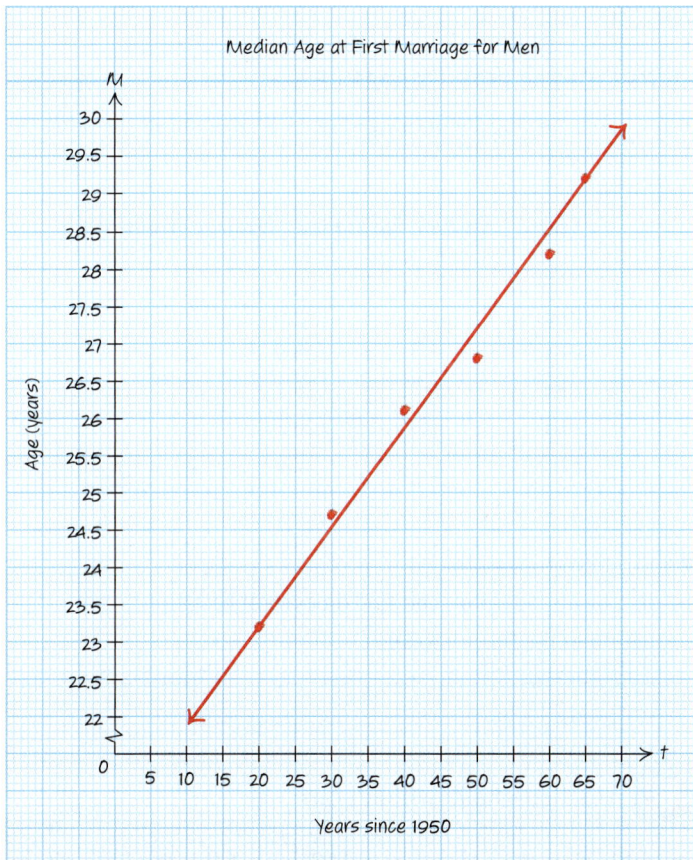

What's That Mean?

M versus m

The lowercase m represents the slope in the equation of the line. Therefore, we do not often use lowercase m as a variable when working with linear models. However, sometimes we represent a variable by an uppercase M. In Example 5, the M represents the median age at first marriage for men.

The eyeball best-fit line that was drawn has one point above the line and two points below the line.

c. To find the equation of the linear model, find two points on the line. Two points on the line are (20, 23.2) and (65, 29.2). Setting $(t_1, M_1) = (20, 23.2)$ and $(t_2, M_2) = (65, 29.2)$ in the slope formula yields

$$m = \frac{M_2 - M_1}{t_2 - t_1}$$

$$m = \frac{29.2 - 23.2}{65 - 20}$$

$$m = \frac{6}{45} \approx 0.133$$

Substituting the value of the slope into the equation of a line yields

$$M = mt + b$$
$$M = 0.133t + b$$

To find the value of b, substitute either point into the equation, and solve for b. Substituting in the point $(t_2, M_2) = (65, 29.2)$ yields

$$M = mt + b$$
$$M = 0.133t + b \quad \text{Substitute in } (t, M) = (65, 29.2).$$
$$29.2 = 0.133(65)$$
$$29.2 = 8.645 + b \quad \text{Solve for } b.$$
$$\underline{-8.645 \quad -8.645}$$
$$20.555 = b$$
$$M = 0.133t + 20.555$$

The model is $M = 0.133t + 20.555$.

PRACTICE PROBLEM FOR EXAMPLE 5

The data in the table are the median age for U.S. women at the time of their first marriage in certain years.

Year	Median Age at First Marriage (Women)
1970	20.8
1980	22.0
1990	23.9
2000	25.1
2010	26.1
2015	27.1

Source: U.S. Census Bureau, Current Population Survey.

a. Define the variables.
b. Graph the data on a scatterplot, and find an eyeball best-fit line.
c. Find the equation of the linear model.

Example 6 U.S. households without cars

The following table gives the percentage of U.S. households that have no car during the years 1980–2015.

Year	U.S. Households with No Car (%)
1980	13
1990	12
2000	10.3
2010	8.9
2015	9.1

Source: U.S. Census Bureau.

a. Define the variables.
b. Graph the data on a scatterplot, and find an eyeball best-fit line.
c. Find the equation of the linear model.

SECTION 3.6 Modeling Linear Data **327**

SOLUTION

a. Let t = years since 2000 and let H represent the percentage of U.S. households with no car.

b. To graph the data on a scatterplot, first determine the scales of the axes. The input data (year) starts at -20 (1980) and ends at 15 (2015). Scale the horizontal axis by 5-year increments, beginning with -30 and ending with 25. The vertical axis is scaled by 1, beginning at 6 and ending at 16. This includes the range of output values, which begin at 8.9 and end at 13. Include a break in the graph to show that we are not starting the axes at 0.

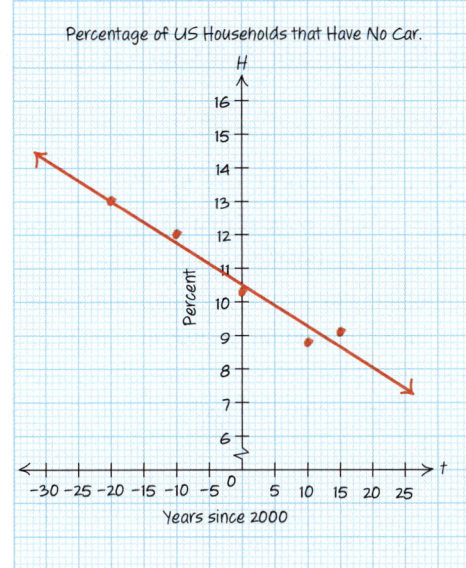

c. To find the equation of the linear model, we find two points on the line. Two points on the line are $(-20, 13)$ and approximately $(12, 9)$. Setting $(t_1, H_1) = (-20, 13)$ and $(t_2, H_2) = (12, 9)$ in the slope formula, we have that the slope is

$$m = \frac{H_2 - H_1}{t_2 - t_1}$$

$$m = \frac{9 - 13}{12 - (-20)}$$

$$m = \frac{-4}{32} = -0.125$$

Substituting the value of the slope into the equation of a line yields

$$H = mt + b$$
$$H = -0.125t + b$$

To find the value of b, substitute either point into the equation, and solve for b. Substituting in the point $(t_2, H_2) = (12, 9)$ yields

$H = mt + b$
$H = -0.125t + b$ Substitute in $(t, H) = (12, 9)$.
$9 = -0.125(12) + b$
$9 = -1.5 + b$ Solve for b.
$+1.5 \quad +1.5$
$\overline{10.5 = b}$
$H = -0.125t + 10.5$

The model is $H = -0.125t + 10.5$.

What's That Mean?

Estimate, approximate, predict, and forecast

When we use a model to come up with a value in the past, we say that we are *estimating* or *approximating* that past value. When we use a model to come up with a value in the future, we say that we are *predicting* or *forecasting* that future value.

Using a Linear Model to Make Estimates

We have learned how to find an eyeball best-fit line and the equation of the corresponding line (the model). The model is generated by the data, which are the observed values in the table. Now we want to use the model to help to investigate other values that are not given in the table. Models can help us make reasonable estimates or forecasts about values that do not appear in the table. The model can be used to *estimate* an output value for a given input value that does not appear in the table but falls within the scope of the input values in the table. The model can also be used to *forecast* an output value for an input value that falls beyond the scope of the input values in the table.

Example 7 — Using the median age for U.S. men at first marriage model

In Example 5, the model for the median age of U.S. men at first marriage was determined to be

$$M = 0.133t + 20.555$$

where M is the median age of U.S. men at first marriage t years since 1950. Use the model to answer the following questions.

a. Using the model, estimate the median age of men at the time of their first marriage in the year 1985. If necessary, round the answer to two decimal places.

b. Using the model, forecast the median age of men at the time of their first marriage in the year 2020. If necessary, round the answer to two decimal places.

SOLUTION

a. To estimate the median age at first marriage for men in the year 1985, substitute $t = 35$ into the model and evaluate to find M.

$M = 0.133t + 20.555$ *Substitute in $t = 35$.*
$M = 0.133(35) + 20.555$
$M \approx 25.21$ *Round to two decimal places.*

The model approximates that in 1985, the median age at first marriage for men in the U.S. was 25.21 years.

b. To forecast the median age at first marriage for men in the year 2020, substitute $t = 70$ into the model, and evaluate to find M.

$M = 0.133t + 20.555$ *Substitute in $t = 70$.*
$M = 0.133(70) + 20.555$
$M \approx 29.87$ *Round to two decimal places.*

The model forecasts that in 2020, the median age at first marriage for a man in the U.S. will be 29.87 years.

Connecting the Concepts

What are interpolation and extrapolation?

Models can *estimate* an output value for an input value that is not given in the table but lies within the given data values. Statisticians call the process of estimation *interpolation*.

Forecasting an output value for a data value that falls outside of the table is called *extrapolation* by statisticians.

PRACTICE PROBLEM FOR EXAMPLE 7

Use the model found in Practice Problem for Example 5 for the median age of U.S. women at first marriage to answer the following questions.

a. Using the model, estimate the median age at the time of their first marriage for women in the year 1985. If necessary, round the answer to two decimal places.

b. Using the model, forecast the median age at the time of their first marriage for women in the year 2020. If necessary, round the answer to two decimal places.

Determining Model Breakdown

In Example 7, we saw how we can use a model to make estimates of values that are not in the data table. For example, in the year 2020, the estimated median age at first marriage for men is about 29.87 years. This seems to be a reasonable value. However, sometimes the prediction that we make is either not possible or not reasonable. In these cases, **model breakdown** is said to have happened.

CONCEPT INVESTIGATION

When does model breakdown happen?

A model for the percentage of U.S. households that have no car was found in Example 6 to be $H = -0.125t + 10.5$. Here, t is years since 2000, and H is the percentage of U.S. households who have no car.

1. Use the model to predict the percentage of U.S. households who have no car in the year 2010. Is this is a reasonable value for the year 2010? Explain your response.
2. Use the model to predict the percentage of U.S. households who have no car in the year 2050. Is this is a reasonable value for the year 2050? Explain your response.
3. Use the model to predict the percentage of U.S. households who have no car in the year 2100. Is this is a reasonable value for the year 2100? Explain your response.

Connecting the Concepts

Does the model break down, or does it lack data?

Models can break down when projecting future and past values, as will be discussed in Example 8. Models can also be limited in their usefulness when the data set that was used to generate the model is too small or leaves out important data values. These concepts are covered in later courses, such as statistics.

This Concept Investigation shows why it is often not safe to go too far outside of the given data set when using a model to make predictions. The model gives output values that do not make sense. This is what is meant by model breakdown.

Example 8 Model breakdown when finding life expectancy

The following table gives the life expectancy for Americans during the years 1940–2010 (both sexes).

Year	Life Expectancy
1940	62.9
1950	68.2
1960	69.7
1970	70.8
1980	73.7
1990	75.4
2000	77
2010	78.7

Source: www.infoplease.com

a. Define the variables.
b. Graph the data on a scatterplot, and find an eyeball best-fit line.
c. Find the equation of the linear model.
d. Using the model, estimate the life expectancy for Americans in the year 1930. If necessary, round the answer to two decimal places. Is this answer reasonable?
e. Using the model, estimate the life expectancy for Americans in the year 2100. If necessary, round the answer to two decimal places. Is this answer reasonable?

SOLUTION

a. Let t = years since 1900 (the input value) and let L = life expectancy in years (the output value). This means that the ordered pairs have the form (t, L).

b. To graph the data on a scatterplot, we first determine the scales of the axes. The input data (year) starts at 40 (1940) and ends at 110 (2010). We scale the horizontal axis by 10-year increments, beginning with 30 and ending with 120. The vertical axis will be scaled by 2, beginning at 56 and ending at 80. This includes the range of output values, which begin at 62.9 and end at 78.7. We include breaks in the graph to show that we are not starting the axes at 0.

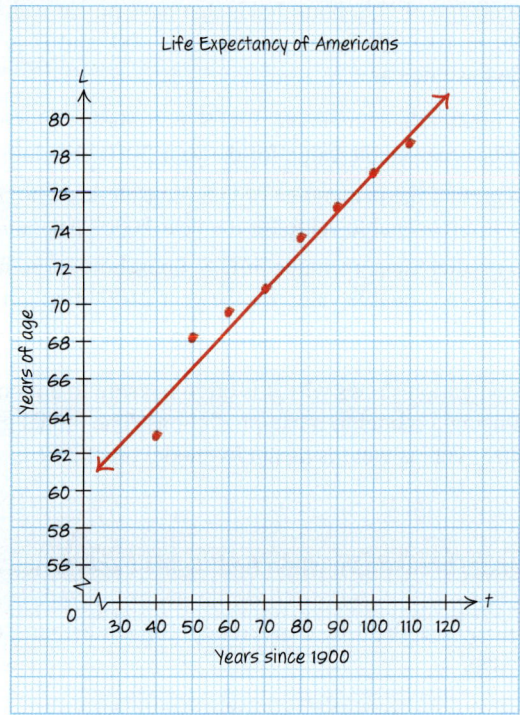

c. To find the equation of the linear model, find two points on the line. Two points on the line are (70, 70.8) and (100, 77). Substitute $(t_1, L_1) = (70, 70.8)$ and $(t_2, L_2) = (100, 77)$ in the slope formula to determine the slope.

$$m = \frac{L_2 - L_1}{t_2 - t_1}$$

$$m = \frac{77 - 70.8}{100 - 70}$$

$$m = \frac{6.2}{30} \approx 0.21$$

Substituting the value of the slope into the equation of a line yields

$$L = mt + b$$
$$L = 0.21t + b$$

To find the value of b, substitute either point into the equation, and solve for b. Substituting in the point $(t_2, L_2) = (100, 77)$ yields

$$L = mt + b$$
$$L = 0.21t + b \qquad \text{Substitute in } (t, L) = (100, 77).$$
$$77 = 0.21(100) + b$$
$$77 = 21 + b \qquad \text{Solve for } b.$$
$$\underline{-21 \quad -21}$$
$$56 = b$$
$$L = 0.21t + 56$$

The model is $L = 0.21t + 56$.

d. To estimate the life expectancy for Americans in the year 1930, substitute $t = 30$ into the model, and evaluate to find L.

$$L = 0.21t + 56 \qquad \text{Substitute in } t = 30.$$
$$L = 0.21(30) + 56$$
$$L \approx 62.3 \qquad \text{Round to two decimal places.}$$

In 1930, the model approximates that the life expectancy for Americans is 62.3 years. This answer seems reasonable.

e. To estimate the life expectancy for Americans in the year 2100, substitute $t = 200$ into the model, and evaluate to find L.

$$L = 0.21t + 56 \qquad \text{Substitute in } t = 200.$$
$$L = 0.21(200) + 56$$
$$L = 98.0$$

The model approximates that the life expectancy for Americans in the year 2100 is 98 years.
We have no idea whether it will be possible for the average person to live to 98 years by the year 2100. The year 2100 is far outside of our data input values that go between 1940 and 2010. Therefore, this is a case of model breakdown.

Example 9 Finding and interpreting values using a given model

Use the model found in Example 5 for the median age for men at first marriage in the United States and answer the questions based on the model. If the solution does not make sense, then write the answer as model breakdown. The model is $M = 0.133t + 20.555$, where t is years since 1950 and M is the median age for men at first marriage.

a. Find the slope of the model and explain the value of the slope in terms of the problem. Does the slope predict a positive or a negative trend?

b. Find the horizontal intercept. Explain the value of the horizontal intercept in terms of the problem.

c. Find the vertical intercept. Explain the value of the vertical intercept in terms of the problem.

SOLUTION

a. The slope of the model is 0.133.
Including units, the slope is

$$\frac{0.133}{1} = \frac{0.133 \text{ year in men's median age at first marriage}}{1 \text{ year in time}}$$

The slope means that for every year increase in time, the median age at first marriage for men has increased by about 0.133 year.

Since the slope is positive, the trend is that as time increases, the median age for men at the time of their first marriage is going up.

b. To find the horizontal intercept, substitute $M = 0$ into the model, and solve for t.

$$0 = 0.133t + 20.555 \quad \text{Substitute in } M = 0.$$
$$0 = 0.133t + 20.555 \quad \text{Solve for } t.$$
$$\underline{-20.555 \qquad\qquad -20.555}$$
$$-20.555 = 0.133$$
$$\frac{-20.555}{0.133} = \frac{0.133}{0.133}$$
$$-154.55 \approx t$$
$$1950 - 154.55 = 1795.45$$

This means that in about the year 1795, the median age for men at first marriage was 0.

This makes no sense, so this is a case of model breakdown.

c. To find the vertical intercept, substitute $t = 0$ into the model, and solve for M.

$$M = 0.133(0) + 20.555 \quad \text{Substitute in } t = 0.$$
$$M = 20.555 \quad \text{Solve for } M.$$

This means that in the year 1950, the median age for men at first marriage was 20.555 years.

PRACTICE PROBLEM FOR EXAMPLE 9

Use the model found in the Practice Problem for Example 5 for the median age for women at first marriage in the United States and answer the questions based on the model. If the solution does not make sense, then write the answer as model breakdown.

a. Find the slope of the model and explain the value of the slope in terms of the problem. Does the slope predict a positive or a negative trend?

b. Find the horizontal intercept. Explain the value of the horizontal intercept in terms of the problem.

c. Find the vertical intercept. Explain the value of the vertical intercept in terms of the problem.

3.6 Vocabulary Exercises

1. Linear data creates a(n) _____ pattern in the scatterplot.

2. A linear model is the _____ of the eyeball best-fit line.

3. _____ occurs when a model returns a value that does not make sense in the context of the problem situation.

3.6 Exercises

For Exercises 1 through 8, draw a scatterplot for each data set. Clearly label and scale the axes. Think carefully about scaling the axes. Don't forget to include a break when necessary. Are the data approximately linear?

1.

x	y
−7	−30
−4	−16
−1	0
2	14
5	25

2.

x	y
−5	−21
−3	−9
−1	3
1	15
3	27

3.

x	y
−2	−9
−1	−14
0	−7
2	−1
4	20

4.

x	y
−3	−21
−1	−19
1	−13
3	0
5	32

5. The population of Nebraska is given in the following table.

Year	Population (millions)
2010	1.83
2011	1.84
2012	1.86
2013	1.87
2014	1.88
2015	1.89
2016	1.91

Source: U.S. Census Bureau.

6. The population of Kentucky is given in the following table.

Year	Population (millions)
2010	4.35
2011	4.37
2012	4.38
2013	4.40
2014	4.41
2015	4.42
2016	4.44

Source: U.S. Census Bureau.

7. The number of monthly active Facebook users is given in the following table.

Year	Users (billions)
2010	0.61
2011	0.85
2012	1.06
2013	1.23
2014	1.39
2015	1.55
2016	1.86

Source: Facebook.

8. The rate to send a domestic letter in the United States, per ounce, for the years 1968 to 1981 is given in the following table.

Year	Postal Rate (cents)
1968	6
1971	8
1974	10
1975	13
1978	15
1981	18

Source: www.usps.com/postalhistory/

334 CHAPTER 3 Linear Equations with Two Variables

For Exercises 9 through 12, decide which line better fits the data.

9.   Graph 1   Graph 2

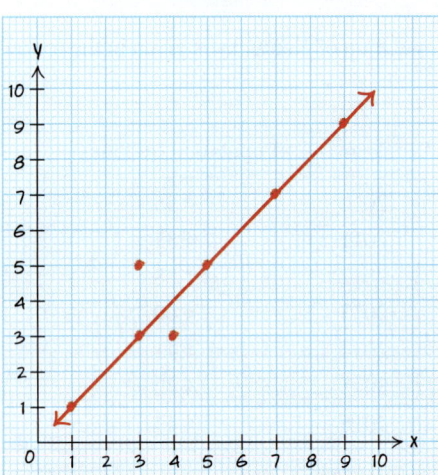 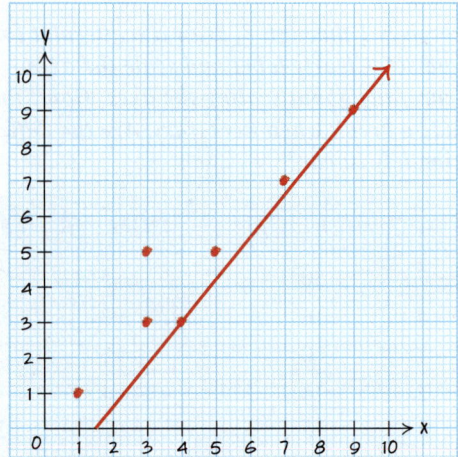

10.   Graph 1   Graph 2

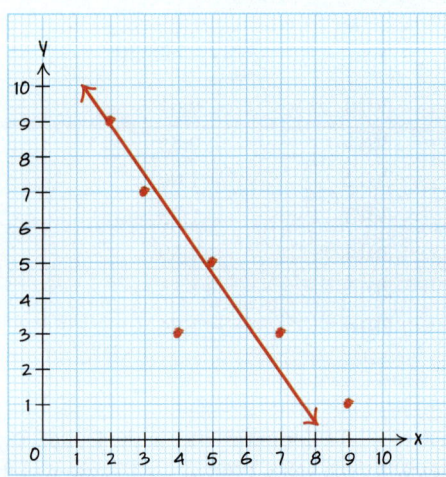 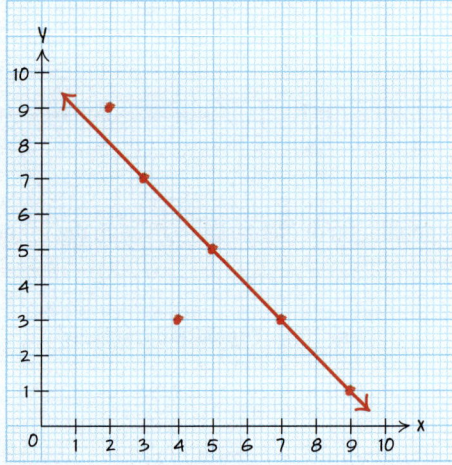

11.   Graph 1   Graph 2

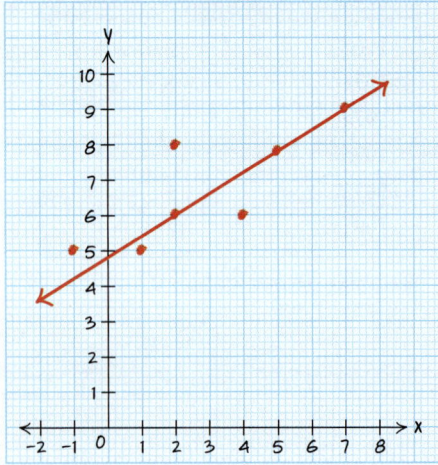 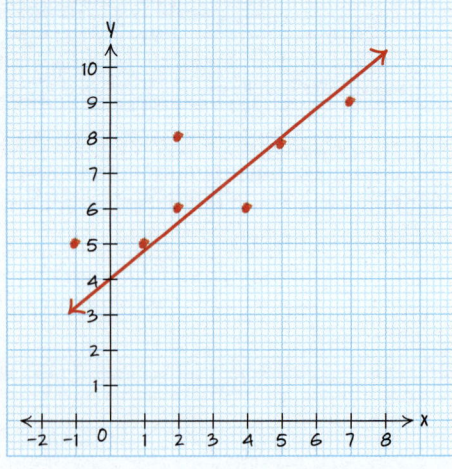

SECTION 3.6 Modeling Linear Data **335**

12.

Graph 1

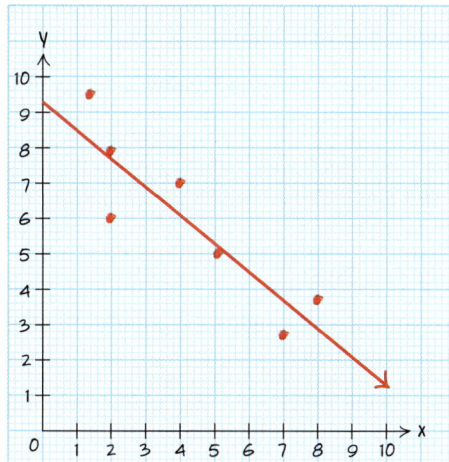

Graph 2

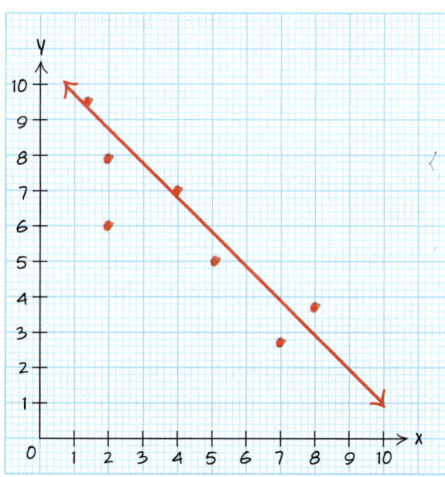

For Exercises 13 through 18, do each of the following.

a. Draw a scatterplot for each data set. Clearly label and scale the axes. Think carefully about scaling the axes. Don't forget to include a break when necessary. Are the data approximately linear?

b. Draw an eyeball best-fit line on the scatterplot. Be sure to clearly draw the line through two data points.

c. Use the two points you labeled on the graph for the best-fit line to find the model for the line. Remember that the model is the equation of the line passing through the data. Write the equation of the line in $y = mx + b$ form.

13.

x	y
−3	12
−1	7
1	4
3	0
5	−5

14.

x	y
−4	35
−2	30
0	20
2	8
4	0

15.

x	y
−10	9
−5	8
−1	7
5	6
10	4

16.

x	y
−6	8
−3	6
0	5
3	2
6	−1

17.

x	y
12	9
15	7
18	6
21	5
24	3

18.

x	y
21	47
28	45
35	44
42	43
49	40

For Exercises 19 through 22, do each of the following.

a. Define the variables.

b. Draw a scatterplot for each data set. Clearly label and scale the axes. Think carefully about scaling the axes. Don't forget to include a break when necessary. Are the data approximately linear?

c. Draw an eyeball best-fit line on the scatterplot. Be sure to clearly draw the line through two data points.

d. Use the two points you labeled on the graph for the best-fit line to find the model for the line. Remember that the model is the equation of the line passing through the data. Write the equation of the line in slope-intercept form.

19. The number of immigrants in the United States as a percentage of the total population is given in the following table. Round the slope and y-intercept to three decimal places.

Year	Immigrants as a Percentage of U.S. Population
1970	4.7
1980	6.2
1990	7.9
2000	11.1
2010	12.9
2015	13.3

Source: www.cis.org

336 CHAPTER 3 Linear Equations with Two Variables

20. The population of Nevada is given in the following table. Round the slope and y-intercept to four decimal places.

Year	Population (millions)
2011	2.72
2012	2.75
2013	2.79
2014	2.83
2015	2.88
2016	2.94

Source: U.S. Census Bureau.

21. The following table shows the age of a person and the target heart rate during exercise (50% of maximum).

Age	Heart Rate (beats per minute)
20	100
30	95
40	90
50	85
60	80
70	75

Source: www.americanheart.org

22. The following table shows the percentage of the world's population connected to the Internet. Round the slope to three decimal places.

Year	Percentage of the World's Population Connected to the Internet
2010	29.15
2011	31.73
2012	34.95
2013	37.42
2014	40.65
2015	44.00

Source: World Bank.

For Exercises 23 through 26, use each model for the application problem and answer the questions based on the model.

23. Use the model for the data on the number of immigrants in the United States as a percentage of the population found in Exercise 19 to answer the questions.

    a. Using the model, estimate the percentage of the population who were immigrants in the United States in the year 1995.

    b. Using the model, predict the percentage of the population who will be immigrants in the United States in the year 2020.

    c. Interpret the slope of the model in terms of the problem.

24. Use the model for the data on the population of Nevada found in Exercise 20 to answer the questions.

    a. Using the model, estimate the population of Nevada in the year 2010.

    b. Using the model, predict the population of Nevada in the year 2018.

    c. Interpret the slope of the model in terms of the problem.

25. Use the model for the data on the age versus target heart rate found in Exercise 21 to answer the questions.

    a. Using the model, estimate the target heart rate of a 25-year-old.

    b. Using the model, predict the target heart rate of an 80-year-old.

    c. Interpret the slope of the model in terms of the problem.

26. Use the model for the data on the percentage of the world's population connected to the Internet found in Exercise 22 to answer the questions.

    a. Using the model, predict the percentage of the world's population connected to the Internet in the year 2018.

    b. Using the model, predict the percentage of the world's population connected to the Internet in the year 2020.

    c. Interpret the slope of the model in terms of the problem.

For Exercises 27 through 30, find an equation for a model of each given data set, and answer the questions based on the model.

27. To find a model for the unemployment data in California in the year 2016, use the following steps.

Month	Unemployment Rate
Jan.	5.7
March	5.6
May	5.5
July	5.4
Sept.	5.3

Source: www.bls.gov

a. Relabel the first column of the data set with 1 = January, 2 = February, 3 = March, and so on. This is sometimes called *adjusting the data*.

Month	Unemployment Rate
1(= January)	5.7
	5.6
	5.5
	5.4
	5.3

b. Define the variables.

c. Draw the scatterplot and the best-fit line through two points.

d. Using the two points through which the best-fit line passed, find the model. Remember that the model is the equation of the line through those two points.

e. Using the model, predict the unemployment rate in December 2016. Round to one decimal place if necessary. (*Hint:* December corresponds to what number?)

28. To find a model for the unemployment data for the state of Michigan in the year 2014, use the following steps.

Month	Unemployment Rate
Jan.	7.9
Feb.	7.7
March	7.4
April	7.0
May	6.7
June	6.4

Source: www.bls.gov

a. Relabel the first column of the data set with 1 = January, 2 = February, 3 = March, and so on. This is sometimes called *adjusting the data*.

Month	Unemployment Rate
1	7.9
	7.7
	7.4
	7.0
	6.7
	6.4

b. Define the variables.

c. Draw the scatterplot and the best-fit line through two points.

d. Using the two points through which the best-fit line passed, find the model. Remember that the model is the equation of the line through those two points.

e. Using the model, estimate the unemployment rate in December 2014. (*Hint:* December corresponds to what number?)

29. The annual revenues for JetBlue Airways Corporation are listed in the following table.

Year	Annual Revenue (millions of dollars)
2011	4504
2012	4982
2013	5441
2014	5817
2015	6416
2016	6632

Source: http://finance.google.com, accessed in May 2017.

a. Define the variables.

b. Find an equation of a model.

c. Interpret the slope of the model in terms of the problem. Is the model predicting a positive or a negative trend for the revenue of JetBlue?

d. Use the model to approximate the revenue for JetBlue in the year 2018.

30. The Bureau of Labor Statistics reports the following data for the percentage of employed workers who are represented by unions.

Year	Percentage of Workers Represented by Unions
2008	13.7
2010	13.1
2012	12.5
2014	12.3
2016	11.9

Source: bls.data.gov

a. Define the variables.

b. Find the equation of the model. (*Hint:* Use the eyeball best-fit line drawn in the Practice Problem for Example 3.)

c. Interpret the slope of the model in terms of the problem. Is the model predicting a positive or a negative trend for the percentage of workers represented by unions in the United States?

d. Use the model to approximate the percentage of workers who were represented by unions in the year 2020.

For Exercises 31 through 36, evaluate each model for the given value of the variable. Determine whether the solution makes sense in the context of the problem. If the solution does not make sense, then write the answer as model breakdown.

31. Smoking among 12th grade students in the United States is declining. A linear model for the percentage of 12th grade students who smoke is $P = -1.22t + 11.8$. Here, P is the percentage of 12th graders who smoke, and t is years since 2010 (time).

Source: U.S. Department of Health and Human Services.

a. Use the model to predict the percentage of 12th grade students who smoke in the year 2018. Is this answer reasonable? Explain.

b. Use the model to predict the percentage of 12th grade students who will smoke in the year 2020. Is this answer reasonable? Explain.

32. Smoking among tenth grade students in the U.S. is declining. A linear model for the percentage of tenth grade students who smoke is $P = -0.68t + 6.2$. Here, P is the percentage of tenth grade students who smoke, and t is years since 2010 (time).

Source: U.S. Department of Health and Human Services.

a. Use the model to predict the percentage of tenth grade students who smoke in the year 2018. Is this reasonable? Explain.

b. Use the model to predict the percentage of tenth grade students who will smoke in the year 2020. Is this reasonable? Explain.

33. High cholesterol among adults aged 20 and over is decreasing in the United States. A linear model for the percentage of U.S. adults aged 20 and over who have high cholesterol is given by $P = -0.525t + 13.75$. Here P is the percentage of U.S. adults who have high cholesterol, and t is years since 2010 (time).

Source: www.cdc.gov/nchs/

a. Use the model to approximate the percentage of U.S. adults who had high cholesterol in the year 2010. Is this answer reasonable? Explain.

b. Use the model to approximate the percentage of U.S. adults who had high cholesterol in the year 2020. Is this answer reasonable? Explain.

34. The number of U.S. adults aged 20 and over who have diabetes is increasing. The percentage of U.S. adults aged 20 and older who have diabetes is given by $P = 0.14t + 12.16$. Here P is the percentage of U.S. adults who have diabetes, and t is years since 2010 (time).

Source: www.cdc.gov/nchs/

a. Use the model to approximate the percentage of U.S. adults who had diabetes in the year 1920. Is this answer reasonable? Explain.

b. Use the model to approximate the percentage of U.S. adults who will have diabetes in the year 2020. Is this answer reasonable? Explain.

35. The estimated number of mobile payment transactions in the United States is given by $M = 50.903t - 40.455$. Here M is the number of mobile payment transactions in billions, and t is years since 2015 (time).

Source: Statista Digital Market Outlook.

a. Use the model to estimate the number of mobile payment transactions in the United States in the year 2000. Is this number reasonable? Explain.

b. Use the model to estimate the number of mobile payment transactions in the United States in the year 2020. Is this number reasonable? Explain.

36. The number of Facebook mobile daily active users is given by $F = 0.182t - 0.73$. Here F is the number of Facebook mobile daily active users in billions, and t is years since 2015 (time).

Source: Facebook.

a. Use the model to estimate the number of Facebook mobile daily active users in the year 2000. Is this number reasonable? Explain.

b. Use the model to estimate the number of Facebook mobile daily active users in the year 2020. Is this number reasonable? Explain.

For Exercises 37 through 42, use the model for the application problem, and answer the questions based on the model. If the solution does not make sense, then write the answer as model breakdown.

37. Smoking among 12th grade students in the United States is declining. A linear model for the percentage of 12th grade students who smoke is $P = -1.22t + 11.8$. Here, P is the percentage of 12th graders who smoke, and t is years since 2010 (time).

Source: U.S. Department of Health and Human Services.

a. Find the slope of the model and explain the value of the slope in terms of the problem. Does the slope predict a positive or a negative trend?

b. Find the horizontal intercept. Explain the value of the horizontal intercept in terms of the problem.

c. Find the vertical intercept. Explain the value of the vertical intercept in terms of the problem.

38. Smoking among tenth grade students in the United States is declining. A linear model for the percentage of tenth grade students who smoke is $P = -0.68t + 6.2$. Here, P is the percentage of tenth grade students who smoke, and t is years since 2010 (time).

Source: U.S. Department of Health and Human Services.

a. Find the slope of the model and explain the value of the slope in terms of the problem. Does the slope predict a positive or a negative trend?

b. Find the horizontal intercept. Explain the value of the horizontal intercept in terms of the problem.

c. Find the vertical intercept. Explain the value of the vertical intercept in terms of the problem.

39. High cholesterol among adults aged 20 and over is decreasing in the United States. A linear model for the percentage of U.S. adults aged 20 and over who have high cholesterol is given by $P = -0.525t + 13.75$. Here P is the percentage of U.S. adults who have high cholesterol, and t is years since 2010 (time).

Source: www.cdc.gov/nchs/

a. Find the slope of the model and explain the value of the slope in terms of the problem. Does the slope predict a positive or a negative trend?

b. Find the horizontal intercept. Explain the value of the horizontal intercept in terms of the problem.

c. Find the vertical intercept. Explain the value of the vertical intercept in terms of the problem.

40. The number of U.S. adults aged 20 and over who have diabetes is increasing. The percentage of U.S. adults aged 20 and older who have diabetes is given by $P = 0.14t + 12.16$. Here P is the percentage of U.S. adults who have diabetes, and t is years since 2010 (time).

Source: www.cdc.gov/nchs/

a. Find the slope of the model and explain the value of the slope in terms of the problem. Does the slope predict a positive or a negative trend?

b. Find the horizontal intercept. Explain the value of the horizontal intercept in terms of the problem.

c. Find the vertical intercept. Explain the value of the vertical intercept in terms of the problem.

41. The estimated number of mobile payment transactions in the United States is given by $M = 50.903t - 40.455$. Here M is the number of mobile payment transactions in billions, and t is years since 2015 (time).

Source: Statista Digital Market Outlook.

a. Find the slope of the model and explain the value of the slope in terms of the problem. Does the slope predict a positive or a negative trend?

b. Find the horizontal intercept. Explain the value of the horizontal intercept in terms of the problem.

c. Find the vertical intercept. Explain the value of the vertical intercept in terms of the problem.

42. The number of Facebook mobile daily active users is given by $F = 0.182t - 0.73$. Here F is the number of Facebook mobile daily active users in billions, and t is years since 2015 (time).

Source: Facebook.

a. Find the slope of the model and explain the value of the slope in terms of the problem. Does the slope predict a positive or a negative trend?

b. Find the horizontal intercept. Explain the value of the horizontal intercept in terms of the problem.

c. Find the vertical intercept. Explain the value of the vertical intercept in terms of the problem.

Chapter Summary

Section 3.1 — Graphing Equations with Two Variables

- **Tables** can be used to represent data or ordered pairs.
- **Equations** with two variables may be **graphed** by generating a table of points and then plotting the points and connecting them with a smooth curve.
- Both the *x*- and *y*-axes should be **scaled** so that all points in the table are displayed on the graph. Each axis should have 5 to 20 values listed.
- To **graph by plotting points** follow these steps:
  1. Create a table of points using the equation.
  2. Plot the points on a set of axes. Clearly label and scale the axes.
  3. Connect the points with a smooth curve.
- To **graph simpler equations in two variables,** graph five or more points.
- To **graph more complicated equations in two variables,** such as those with an exponent of 2 on one of the variables or with absolute values, graph nine or more points.
- A **vertical line** is the graph of an equation of the form $x = c$.
- A **horizontal line** is the graph of an equation of the form $y = b$.

Example 1

Use the table to write coordinates for points and then graph them. Connect the points with a smooth curve. The equation is $y = -3x + 5$.

x	$y = -3x + 5$	(x, y)
-3	14	
-2	11	
-1	8	
1	2	
2	-1	

SOLUTION

x	$y = -3x + 5$	(x, y)
-3	14	$(-3, 14)$
-2	11	$(-2, 11)$
-1	8	$(-1, 8)$
1	2	$(1, 2)$
2	-1	$(2, -1)$

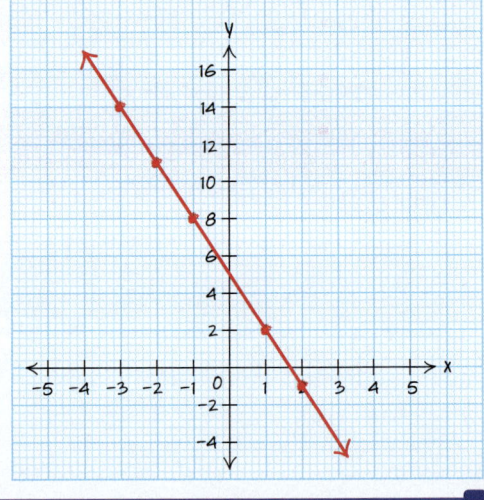

CHAPTER 3 Summary 341

Example 2 Use each equation to create a table of points and then graph them. Connect the points with a smooth curve. The equation is $y = |x - 1|$.

SOLUTION

x	$y = |x-1|$	(x, y)
-5	6	$(-5, 6)$
-4	5	$(-4, 5)$
-3	4	$(-3, 4)$
-2	3	$(-2, 3)$
-1	2	$(-1, 2)$
0	1	$(0, 1)$
1	0	$(1, 0)$
2	1	$(2, 1)$
3	2	$(3, 2)$
4	3	$(4, 3)$
5	4	$(5, 4)$

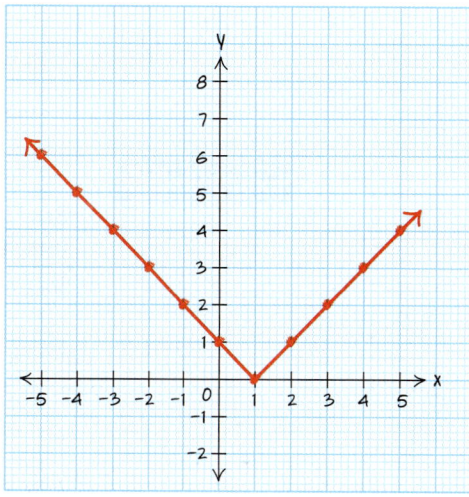

Section 3.2 Finding and Interpreting Slope

- **Graphs** give information about the output values from the height of the graph.
- **Graphs** are read from left to right.
- **The rate of change** between two points is the change in the outputs divided by the change in the inputs.
- **Speed** is the rate of change of a distance (output) versus time (input) graph.
- The rate of change of a linear graph is called the **slope of the line**.
- To find the **slope of a line using a graph** follow these steps:
  1. Pick two points on the line with coordinates that are easy to read.
  2. Start with one point and draw a right triangle to the other point.
  3. Determine the rise and run from one point to the other.
  4. Divide the rise by the run.

$$\text{slope} = \frac{\text{rise}}{\text{run}}$$

- **A line is said to be increasing** if it goes up, reading from left to right. A line is increasing when its slope is positive.
- **A line is said to be decreasing** if it goes down, reading from left to right. A line is decreasing when its slope is negative.
- The **slope of a line between two points may be found by using the formula**

$$\text{slope} = \frac{\text{change in } y}{\text{change in } x} = \frac{y_2 - y_1}{x_2 - x_1}$$

  where (x_1, y_1) and (x_2, y_2) are any two points on the line.
- To **interpret slope in the context of an application,** first compute the slope and then attach the units of the output to the numerator and the units of the input to the denominator. Then write the slope as a rate, including the units.

Example 3 Find the slope of the line that passes through the two points $(-3, -1)$ and $(4, 6)$.

SOLUTION Using the slope formula, we find that the slope of the line is 1.

$$m = \frac{6 - (-1)}{4 - (-3)} = \frac{7}{7} = 1$$

Example 4 The population of San Diego, California, in 2010 was 1,307,420. The population of San Diego in 2016 was 1,394,928. Find and interpret the slope of the line between these two points.

Source: Wikipedia.

SOLUTION Using the slope formula, we find that the slope of the line is about 14584.7.

$$m = \frac{1394928 - 1307420}{2016 - 2010} = \frac{87508}{6} \approx \frac{14584.7 \text{ people}}{1 \text{ year}}$$

This means that the population of San Diego is increasing at the rate of 14,584.7 people per year.

Section 3.3 Slope-Intercept Form of Lines

- The **y-intercept** is the point(s) where the curve touches the *y*-axis. It is written in the form $(0, b)$.
- The **x-intercept** is the point(s) where the curve touches the *x*-axis. It is written in the form $(c, 0)$.
- **Interpreting intercepts in a problem** may or may not make sense in the problem situation.
- **To find the y-intercept,** follow these steps:
  1. Substitute zero for *x*, $x = 0$, and find the value of *y*.
  2. Write the *y*-intercept as a coordinate in the form $(0, b)$.
- **To find the x-intercept,** follow these steps:
  1. Substitute zero for *y*, $y = 0$, and find the value of *x*.
  2. Write the *x*-intercept as a coordinate in the form $(c, 0)$.
- **The slope-intercept form** of a line is $y = mx + b$, where *m* and *b* are real numbers. The slope of the line is *m*, and the *y*-intercept is the point $(0, b)$.

Example 5 Find the *x*- and *y*-intercepts of the equation $-2x + 5y = 20$.

SOLUTION To find the *x*-intercept, let $y = 0$, and solve for *x*.

$$-2x + 5(0) = 20 \quad \text{Substitute in } y = 0.$$

$$\frac{-2x}{-2} = \frac{20}{-2} \quad \text{Solve for } x.$$

$$x = -10$$

The *x*-intercept is the point $(-10, 0)$.

To find the y-intercept, let $x = 0$, and solve for y.

$$-2(0) + 5y = 20 \quad \text{Substitute in } x = 0.$$
$$\frac{5y}{5} = \frac{20}{5} \quad \text{Solve for } y.$$
$$y = 4$$

The y-intercept is the point $(0, 4)$.

Example 6 Find the slope and y-intercept of the line $y = -\frac{1}{2}x + 7$.

SOLUTION This line is in the form $y = mx + b$. Read the slope from the equation, $m = -\frac{1}{2}$. The value of b is $b = 7$. Therefore, the y-intercept is the point $(0, 7)$.

Section 3.4 Linear Equations and Their Graphs

- **To graph a line using the slope and y-intercept**, follow these steps:
  1. Determine the slope and y-intercept from the equation $y = mx + b$.
  2. Plot the y-intercept $(0, b)$ on the graph.
  3. Write the slope in fraction form. Use the slope to count the rise and run from the y-intercept to another point on the line. Repeat the rise and run process to find a third point. Work from left to right.
  4. Draw a straight line through the points.
- The **general form of the equation of a line** is $Ax + By = C$, where A, B, and C are real numbers, and both A and B are not zero.
- **To graph a line using the intercepts,** follow these steps:
  1. Find the y-intercept by substituting zero for x and solving for y.
  2. Find the x-intercept by substituting zero for y and solving for x.
  3. Plot both intercepts and draw a line through the points.
- A **linear equation in two variables** is an equation that may be put into either form
$$y = mx + b \quad \text{or} \quad Ax + By = C$$
- Two lines are **parallel** if they have the same slope: $m_1 = m_2$.
- Two lines are **perpendicular** if their slopes are opposite reciprocals: $m_1 = -\frac{1}{m_2}$.

Example 7 Graph the line using the slope and y-intercept.
$$y = \frac{2}{3}x - 5$$

SOLUTION The y-intercept of this line is the point $(0, -5)$. The slope is $m = \frac{2}{3} = \frac{\text{rise}}{\text{run}}$. First graph the y-intercept $(0, -5)$ and then use the slope to get two more points on the line.

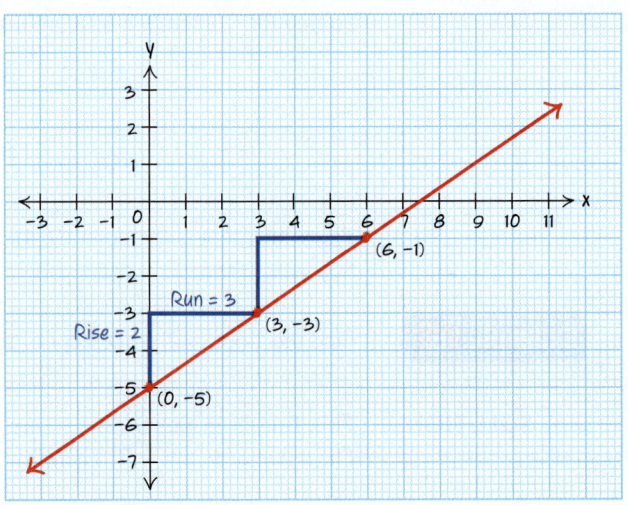

Example 8 Graph the line by finding and graphing the x- and y-intercepts.

$$-3x + 4y = -12$$

SOLUTION To find the x-intercept, let $y = 0$ and solve for x.

$$-3x + 4(0) = -12 \qquad \text{Let } y = 0.$$

$$\frac{-3x}{-3} = \frac{-12}{-3} \qquad \text{Solve for } x.$$

$$x = 4$$

Therefore, the x-intercept is the point $(4, 0)$.
To find the y-intercept, let $x = 0$ and solve for y.

$$-3(0) + 4y = -12 \qquad \text{Let } x = 0.$$

$$\frac{4y}{4} = \frac{-12}{4} \qquad \text{Solve for } y.$$

$$y = -3$$

Therefore, the y-intercept is the point $(0, -3)$.
Graphing these two points and connecting them with a line, we get the following graph.

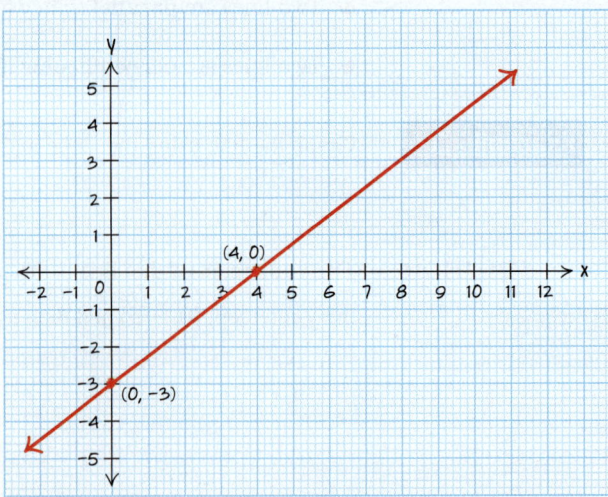

Example 9 Determine whether each of the following equations is linear. If it is linear, put it in $y = mx + b$ form.

   a. $y = x^2 + 6$ **b.** $-2x + y = 7$

SOLUTION

a. This equation is NOT linear. The variable x has an exponent of 2, which is not a characteristic of a linear equation.

b. This equation is linear. It is in general form, $Ax + By = C$. To put it in slope-intercept form, we must solve this equation for y. Remember that to solve for y means to isolate y on one side of the equal sign.

$$-2x + y = 7$$
$$\underline{+2x \qquad\quad +2x}$$
$$y = 2x + 7$$

Section 3.5 Finding Equations of Lines

- **To find the equation of a line using the slope-intercept form**, use these steps:
    1. Using two points on the line, find the slope m.
    2. To find the y-intercept $(0, b)$, substitute any point on the line in for x and y in $y = mx + b$ and solve for b.
    3. Write the equation of the line $y = mx + b$ with the specific values for m and b found in steps 1 and 2.
- **The point-slope form of a line** is $y - y_1 = m(x - x_1)$, where (x_1, y_1) is any point on the line and m is the slope of the line.

Example 10 Find the equation of the line that passes through the points $(5, 1)$ and $(-5, -3)$.

SOLUTION The slope of the line is

$$m = \frac{-3 - 1}{-5 - 5} = \frac{-4}{-10} = \frac{2}{5}$$

Substituting this result into $y = mx + b$ yields $y = \frac{2}{5}x + b$. To find the value of b, substitute either point into the equation for x and y and solve for b.

$$1 = \frac{2}{5}(5) + b \qquad \text{Substitute in } (x, y) = (5, 1).$$
$$1 = 2 + b$$
$$\underline{-2 \;\; -2}$$
$$-1 = b$$

Therefore, the equation of the line is $y = \frac{2}{5}x - 1$.

346 CHAPTER 3 Linear Equations with Two Variables

Example 11 Using point-slope form, find the equation of the line passing through the points $(1, -3)$ and $(-2, -15)$. Put the final answer in slope-intercept form, $y = mx + b$.

SOLUTION The slope of the line is

$$m = \frac{-15 - (-3)}{-2 - 1} = \frac{-12}{-3} = 4$$

Substituting the slope and the point $(x_1, y_1) = (1, -3)$ into the point-slope form yields

$$y - (-3) = 4(x - 1)$$
$$y + 3 = 4x - 4$$
$$\underline{-3 \qquad -3}$$
$$y = 4x - 7$$

Section 3.6 Modeling Linear Data

- **Scatterplots** can be used to look for patterns in data.
- **Linear data** will create a linear pattern when graphed in a scatterplot.
- **To find an eyeball best-fit line**, follow these steps:
  1. Graph the data on a scatterplot.
  2. Using a straightedge (a small clear ruler is best), fit the edge to the points as closely as possible. If the data do not completely lie on a straight line, make sure that the points the line misses are spread out equally above and below the line.
  3. Draw the line on the scatterplot.
- **To find a linear model for a data set**, follow these steps:
  1. Graph the data, and draw an eyeball best-fit line.
  2. Find two points on the eyeball best-fit line. Use these two points to find the equation of the line. This equation is the *model*.
- **Estimates or predictions** of values may be made with the linear model.
- **Model breakdown** occurs when a linear model returns a value that does not make sense in the context of a problem.

Example 12 The population of North Carolina is given in the following table.

Year	Population (millions)
2008	9.31
2010	9.56
2012	9.75
2014	9.93
2016	10.15

Source: www.census.gov

a. Draw a scatterplot for the data. Clearly label and scale the axes. Do the data look linear?

b. Draw an eyeball best-fit line on the scatterplot.

SOLUTION   a. A scatterplot follows. The arrangement of the data looks linear.

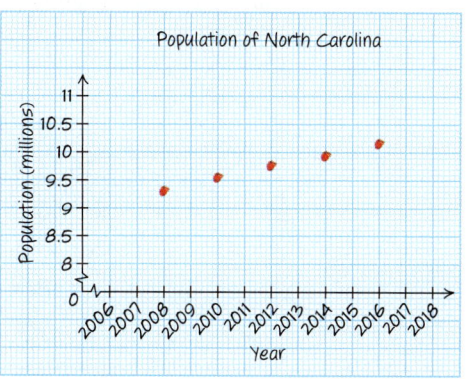

b. The eyeball best-fit line is drawn on the following graph.

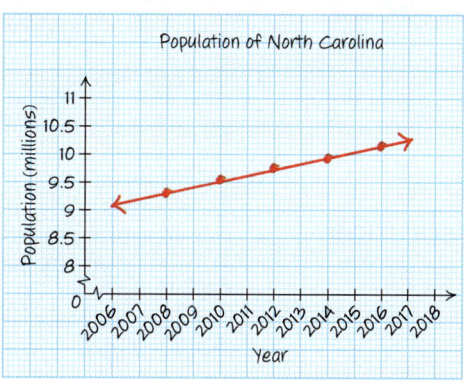

Example 13   Define the variables and find the equation of the eyeball best-fit line in Example 1.

SOLUTION   Let t represent the year and let P represent the population of North Carolina in millions. To find the equation of the linear model, find two points on the line. Two points on the line are (2008, 9.31) and (2016, 10.15). Setting $(t_1, P_1) = (2008, 9.31)$ and $(t_2, P_2) = (2016, 10.15)$ in the slope formula, we have that

$$m = \frac{P_2 - P_1}{t_2 - t_1}$$

$$m = \frac{10.15 - 9.31}{2016 - 2008}$$

$$m = \frac{0.84}{8} = 0.105$$

Substituting the value of the slope into the equation of a line yields

$$P = mt + b$$
$$P = 0.105t + b$$

348 CHAPTER 3 Linear Equations with Two Variables

To find the value of b, substitute either point into the equation, and solve for b. Substituting in the point $(t_2, P_2) = (2016, 10.15)$ yields

$$P = mt + b$$
$$P = 0.105t + b \qquad \text{Substitute in } (t, P) = (2016, 10.15).$$
$$10.15 = 0.105(2016) + b$$
$$10.15 = 211.68 + b \qquad \text{Solve for } b.$$
$$\underline{-211.68 \quad -211.68}$$
$$-201.53 = b$$
$$P = 0.105t - 201.53$$

The model is $P = 0.105t - 201.53$.

Example 14 Use the model in Example 13 to answer the following questions. Determine whether the answers are reasonable.

a. Find and interpret the slope of the model in terms of the problem.

b. Use the model to estimate the population of North Carolina in the year 2020.

c. Use the model to estimate the population of North Carolina in the year 1910.

SOLUTION

a. The slope of the model $P = 0.105t - 201.53$ is $m = 0.105$. Including units, the slope is $m = \dfrac{0.105 \text{ million}}{1 \text{ year}}$.

The population of North Carolina is increasing at the rate of 0.105 million people per year.

b. Substitute in $t = 2020$ to find the value of P.

$$P = 0.105t - 201.53 \qquad \text{Substitute in } t = 2020.$$
$$P = 0.105(2020) - 201.53$$
$$P = 10.57$$

The model estimates the population of North Carolina in 2020 as 10.57 million people. This seems reasonable considering current trends.

c. Substitute in $t = 1910$ to find the value of P.

$$P = 0.105t - 201.53 \qquad \text{Substitute in } t = 1910.$$
$$P = 0.105(1910) - 201.53$$
$$P = -0.98$$

The model estimates the population of North Carolina in 1910 was -0.98 million people. Since there cannot be a negative number of people in a state, this is a case of model breakdown.

Chapter Review Exercises

In Exercises 1 and 2, fill in the given tables. Complete the last two rows of each table with values that you think are reasonable in the problem situation.

1. The number of calories burned by a 160-pound person while walking at the rate of 2.5 mph for h hours can be estimated by the equation $C = 87h$. [3.1]

h = hours of walking	C = calories burned
1	
2	
3	

2. The cost to purchase a plant in a 1-gallon container for a garden can be calculated by using the equation $C = 5.25n$ when n plants are purchased. [3.1]

n = number of plants purchased	C = cost (dollars)
2	
4	
8	

3. Graph the points from Exercise 1 and connect them with a smooth curve. [3.1]

4. Graph the points from Exercise 2 and connect them with a smooth curve. [3.1]

For Exercises 5 through 12, use each given equation to create a table of five or more points. Graph the points. Clearly label and scale the axes. Connect the points with a smooth curve.

5. $y = -3x + 8$
6. $y = -5x - 3$ [3.1]
7. $y = \frac{3}{4}x + 1$
8. $y = -\frac{1}{5}x - 6$ [3.1]
9. $y = 5$
10. $y = -4$ [3.1]
11. $x = -3$
12. $x = 5$ [3.1]

For Exercises 13 through 16, use each given equation to create a table of nine or more points. Graph the points. Clearly label and scale the axes. Connect the points with a smooth curve.

13. $y = x^2 + 2$
14. $y = -x^2 + 5$ [3.1]
15. $y = |9 - 3x|$
16. $y = |x + 2|$ [3.1]

17. [3.2]

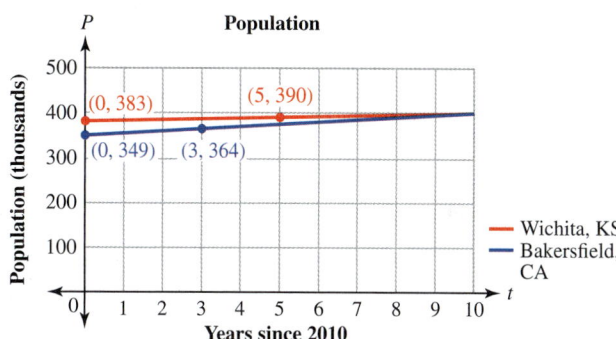

Source: Models derived from data found at the U.S. Census Bureau.

a. Which city has the larger population during these years? Explain your response.
b. Which city is growing faster during these years? Explain your response.

18. [3.2]

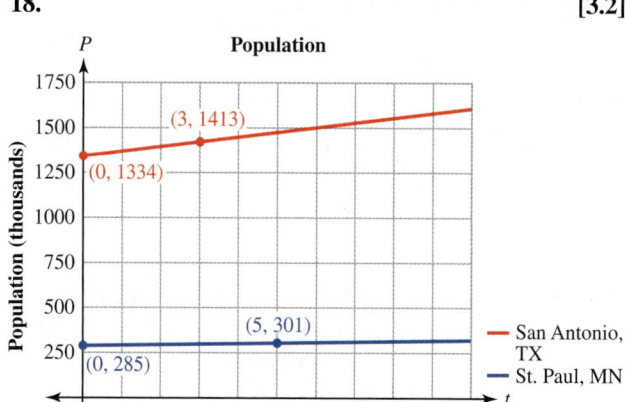

Source: Models derived from data found at the U.S. Census Bureau.

a. Which city has the larger population during these years? Explain your response.
b. Which city is growing faster during these years? Explain your response.

For Exercises 19 and 20, find the slope of the line given in each graph.

19. [3.2]

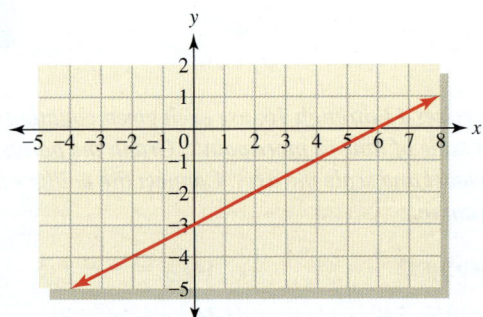

20. [3.2]

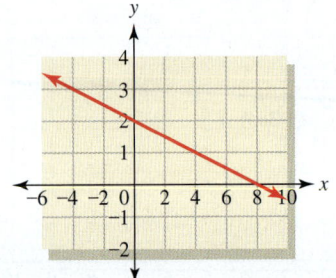

21. Find the slope of the line that goes through the points $(-3, 4)$ and $(5, -2)$. [3.2]

22. Find the slope of the line that goes through the points $(-8, 2)$ and $(0, 5)$. [3.2]

23. Use two points from the table to find the slope of the line that passes through the points. [3.2]

x	y
−10	−11
−5	−9
0	−7
5	−5

24. Use two points from the table to find the slope of the line that passes through the points. [3.2]

x	y
−6	−6
−3	−5
0	−4
3	−3

25. Calculate the slope for Bakersfield, California, given in Exercise 17. Interpret the meaning of the slope in terms of the problem situation. [3.2]

26. Calculate the slope for San Antonio, Texas, given in Exercise 18. Interpret the meaning of the slope in terms of the problem situation. [3.2]

27. Use the graph to find the x-intercept(s) and y-intercept of the curve. [3.3]

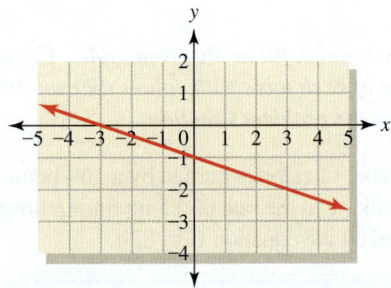

28. Use the graph to find the x-intercept(s) and y-intercept of the curve. [3.3]

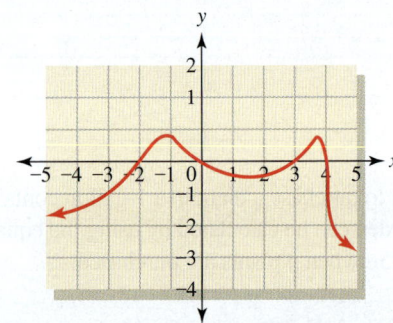

29. a. Interpret the vertical intercept for the Bakersfield line in Exercise 17 in terms of the problem situation.

b. Interpret the vertical intercept for the Wichita line in Exercise 17 in terms of the problem situation. [3.3]

30. [3.3]

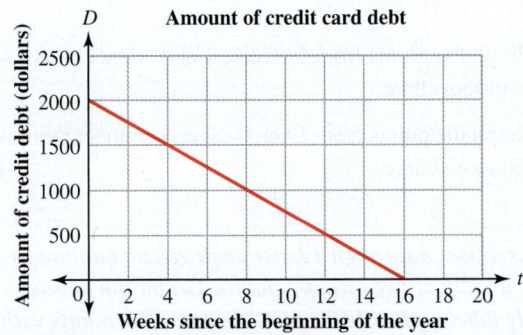

a. Interpret the vertical intercept for the line in terms of the problem situation.

b. Interpret the horizontal intercept for the line in terms of the problem situation.

31. Find the x- and y-intercepts of the equation $y = -\frac{3}{4}x + 3$. [3.3]

32. Find the *x*- and *y*-intercepts of the equation
$-6x + 3y = -18$. [3.3]

33. The time it takes to hand-paint toy dolls can be estimated by the equation
$$T = 1.5d + 1$$
where *T* is the time in hours it takes to paint *d* dolls.
a. Find and interpret the *T*-intercept for this equation in terms of the problem situation.
b. Find and interpret the *d*-intercept for this equation in terms of the problem situation. [3.3]

34. The cost to hand-paint custom toy dolls can be estimated by the equation
$$C = 25d + 10$$
where *C* is the cost in dollars to paint *d* dolls.
a. Find and interpret the *C*-intercept for this equation in terms of the problem situation.
b. Find and interpret the *d*-intercept for this equation in terms of the problem situation. [3.3]

35. Find the *x*-intercept, *y*-intercept, and slope of the line:
$y = -3x + 9$. [3.3]

36. Find the *x*-intercept, *y*-intercept, and slope of the line:
$y = 2x + 12$. [3.3]

37. Find the *x*-intercept, *y*-intercept, and slope of the line:
$y = 1.005x - 7.112$. [3.3]

38. Find the *x*-intercept, *y*-intercept, and slope of the line:
$y = -0.25x + 9.75$. [3.3]

39. Find the *x*-intercept, *y*-intercept, and slope of the line:
$4x + 8y = -16$. [3.3]

40. Find the *x*-intercept, *y*-intercept, and slope of the line:
$-x - 3y = 12$. [3.3]

41. Find the *x*-intercept, *y*-intercept, and slope of the line:
$4x - 5y = -20$. [3.3]

42. Find the *x*-intercept, *y*-intercept, and slope of the line:
$2x - 10y = -20$. [3.3]

43. The population of Alaska can be estimated by using the equation
$$P = 4.7t + 714$$
where *P* is the population of Alaska in thousands *t* years since 2010. [3.3]

Source: Models derived from data found at the U.S. Census Bureau.
a. Find the slope for this equation, and interpret it in this situation.
b. Find the *P*-intercept and interpret it in this situation.

44. The cost to produce sets of clamps to be used on conveyor belts can be estimated by using the equation
$$C = 2.5s + 200$$
where *C* is the cost in dollars for *s* sets of clamps. [3.3]
a. Find the slope for this equation and interpret it in this situation.
b. Find the *C*-intercept and interpret it in this situation.

For Exercises 45 through 54, graph each equation using the slope and y-intercept. Graph at least three points. Clearly label and scale the axes.

45. $y = \frac{1}{4}x - 5$ **46.** $y = \frac{2}{3}x - 6$ [3.4]

47. $y = -\frac{1}{6}x + 2$ **48.** $y = -\frac{2}{7}x + 3$ [3.4]

49. $y = 4x - 9$ **50.** $y = -4x + 8$ [3.4]

51. $y + 3 = \frac{1}{2}(x + 12)$ **52.** $y - 5 = -3(x + 4)$ [3.4]

53. $3y = 2x - 30$ **54.** $-6y = 3x - 18$ [3.4]

For Exercises 55 through 58, graph the given linear equations using the x- and y-intercepts. Clearly label and scale the axes. Label the intercepts.

55. $3x + 4y = 28$ **56.** $2x + 5y = 30$ [3.4]

57. $x + 3y = 9$ **58.** $-x + 6y = 24$ [3.4]

For Exercises 59 and 60, use the graphs to find the equations of the lines. Put the answers in the slope-intercept form (y = mx + b form).

59. [3.5]

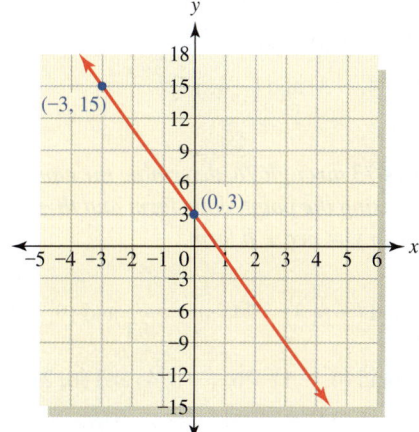

60. [3.5]

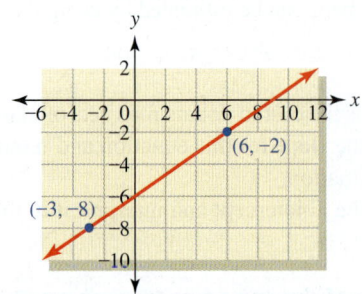

For Exercises 61 and 62, find the equation of the lines that describe the given linear relationships.

61. The number of calories burned every hour by a 160- to 170-pound person while running is 852. (*Note:* This is at a pace of 8 mph.) [3.5]

62. The number of calories a 160- to 170-pound person burns each hour while golfing using a handcart is 264. [3.5]

63. Interpret the slope of the line in Exercise 61 in terms of the problem situation. [3.5]

64. Interpret the slope of the line in Exercise 62 in terms of the problem situation. [3.5]

For Exercises 65 through 72, determine the equations of the lines that pass through the given points. If possible, put the final answers in y = mx + b form.

65. Points $(6, 2)$ and $(11, -3)$ [3.5]

66. Points $(6, 1)$ and $(9, -1)$ [3.5]

67. Points $(-10, 4)$ and $(0, 5)$ [3.5]

68. Points $(-2, -8)$ and $(2, -6)$ [3.5]

69. Points $(3, -11)$ and $(-2, -11)$ [3.5]

70. Points $(-10, 12)$ and $(16, 12)$ [3.5]

71. Points $(1, -8)$ and $(1, 7)$ [3.5]

72. Points $(-4, -3)$ and $(-4, 17)$ [3.5]

For Exercises 73 through 76, determine the equation of each line by using the point-slope form and then putting the final answer in y = mx + b form.

73. $m = \dfrac{1}{2}$ and $(-6, 3)$ **74.** $m = -\dfrac{1}{3}$ and $(9, -1)$ [3.5]

75. $(-4, 5)$ and $(4, -1)$ **76.** $(-6, 4)$ and $(6, 2)$ [3.5]

For Exercises 77 through 80, determine whether each pair of lines are parallel, perpendicular, or neither.

77. $y = -3x + 7$ **78.** $y = \dfrac{1}{3}x - 13$ [3.5]

$6x + 2y = 11$ $5x - 15y = 19$

79. $5x + 2y = 6$ **80.** $y = -3x + 9$ [3.5]

$y = \dfrac{2}{5}x - 1$ $x - 3y = -15$

For Exercises 81 through 84, find the equation of each requested line.

81. Find the equation of the line parallel to $-x + 8y = 1$ passing through the point $(-8, 3)$. [3.5]

82. Find the equation of the line parallel to $5x - 2y = 7$ passing through the point $(10, -1)$. [3.5]

83. Find the equation of the line perpendicular to $y = -\dfrac{4}{3}x + 1$ passing through the point $(12, -5)$. [3.5]

84. Find the equation of the line perpendicular to $y = -\dfrac{x}{2} + 16$ passing through the point $(-3, -7)$. [3.5]

For Exercises 85 through 92, put the equation of each line in the requested form.

85. Put in general form: $y = -\dfrac{5}{6}x + 7$. [3.5]

86. Put in general form: $y = -\dfrac{4}{5}x - 9$. [3.5]

87. Put in slope-intercept form: $-6x + 5y = -30$. [3.5]

88. Put in slope-intercept form: $5x - 10y = -100$. [3.5]

89. Put in general form: $y - 1 = -2(x + 4)$. [3.5]

90. Put in general form: $y + 3 = -7(x - 5)$. [3.5]

91. Put in slope-intercept form: $y + 4 = \dfrac{2}{3}(x - 6)$. [3.5]

92. Put in slope-intercept form: $y - 7 = -\dfrac{5}{7}(x + 14)$. [3.5]

For Exercises 93 through 96, draw a scatterplot for each data set. Clearly label and scale the axes, including a break when necessary. Do the data show a linear pattern?

93. [3.6]

x	y
−6	8
−3	6
0	5
3	1
6	0

94. [3.6]

x	y
20	−1
22	−0.8
24	−0.6
26	−0.4
28	−0.2

95. In a study of 161 children in Kalama, Egypt, the average heights in centimeters versus the ages in months of the children are listed in the following table. [3.6]

Age (months)	Average Height (cm)
18	76.1
19	77
20	78.1
21	78.2
22	78.8
23	79.7
24	79.9
25	81.1

Source: lib.stat.cmu.edu

96. The following table gives the conversion between a cat's age and the corresponding human age. [3.6]

Cat's Age (years)	Human's Age (years)
1	15
2	24
3	28
4	32
5	36
6	40

Source: cats.about.com

For Exercises 97 and 98, decide which line better fits the data graphed in each scatterplot.

97. Graph 1 [3.6]

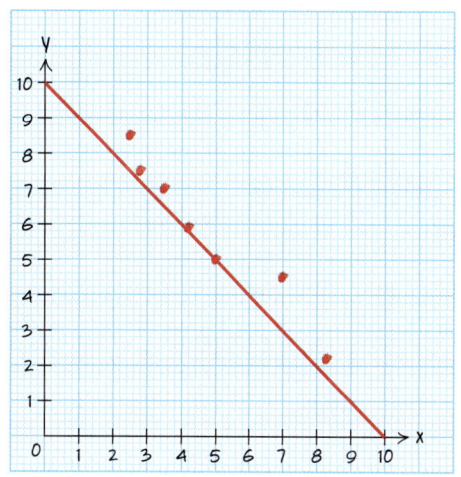

Graph 2

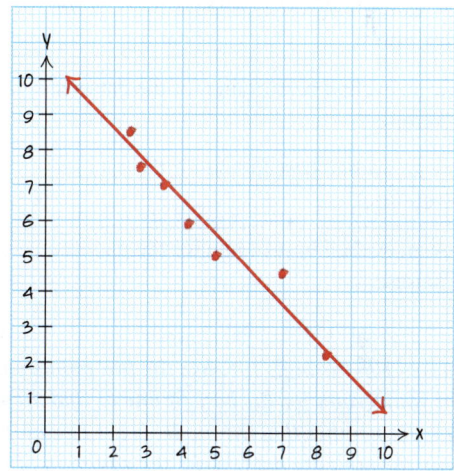

98. Graph 1 [3.6]

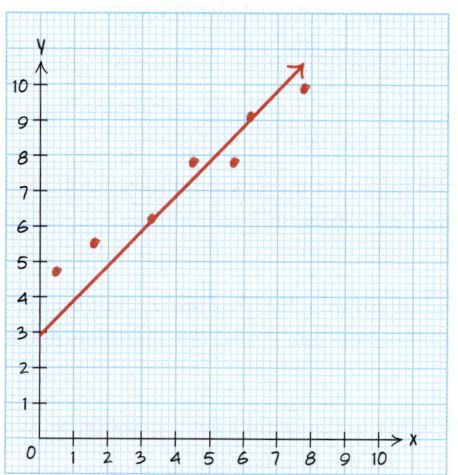

Graph 2

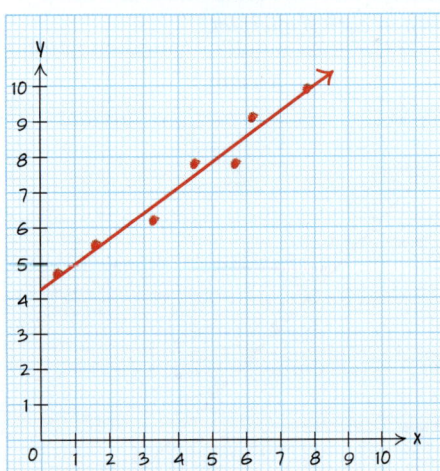

For Exercises 99 through 102, draw an eyeball best-fit line for each data set.

99. Draw an eyeball best-fit line for the data set in Exercise 1. [3.6]

100. Draw an eyeball best-fit line for the data set in Exercise 2. [3.6]

101. Draw an eyeball best-fit line for the data set in Exercise 3. [3.6]

102. Draw an eyeball best-fit line for the data set in Exercise 4. [3.6]

For Exercises 103 through 106, use the two points you labeled on your graph for the best-fit line to find the model for the line. Remember that the model is the equation of the line passing through the data. Write the equation of the line in y = mx + b form.

103. Find a model for the data set in Exercise 1. [3.6]

104. Find a model for the data set in Exercise 2. [3.6]

105. Find a model for the data set in Exercise 3. Make sure you define your variables. [3.6]

106. Find a model for the data set in Exercise 4. Make sure you define your variables. [3.6]

For Exercises 107 and 108, find the model for each application problem and answer the questions based on that model.

107. The following table gives the conversion between a horse's age and a human's age. [3.6]

Horse's Age (years)	Human's Age (years)
1	6.5
2	13
3	18
4	20.5
5	24.5
7	28
10	35.5

Source: iceryder.net

a. Define the variables.
b. Graph the data on a scatterplot. Draw an eyeball best-fit line.
c. Find a linear model for the data.
d. A 25-year-old horse is how old in human years?
e. A 4-year-old human correlates to what age of a horse? Does this answer make sense?

108. The following table gives the population of Montana. [3.6]

Year	Population (thousands)
2008	976
2010	991
2012	1005
2014	1023
2016	1043

Source: U.S. Census Bureau.

a. Define the variables.
b. Graph the data on a scatterplot. Draw an eyeball best-fit line.
c. Find a linear model for the data.
d. What does the model estimate the population of Montana to be in the year 2020? Is this answer reasonable? Explain.
e. What does the model estimate the population of Montana to be in the year 1860? Is this answer reasonable? Explain.

Chapter Test

This chapter test should take approximately one hour to complete. Read each question carefully. Show all of your work.

1. For the given equation, create a table of five or more points and graph them. Connect the points with a smooth curve. $y = -5x + 2$.

2. For the given equation, create a table of nine or more points and graph them. Connect the points with a smooth curve. $y = x^2 - 1$.

3. Find the slope of the line that goes through the points $(3, 7)$ and $(-4, 0)$.

4. In 1999, there were 66,043 children in the United States diagnosed with autism. In 2003, there were 141,022 children in the United States with autism. These values correspond to the points (1999, 66,043) and (2003, 141,022). Find the slope of the line between these points. Interpret the slope in terms of the problem situation.
   Source: http://fightingautism.org/idea/autism.php?

5. Find the x- and y-intercepts of the equation $-2x + 3y = -6$.

6.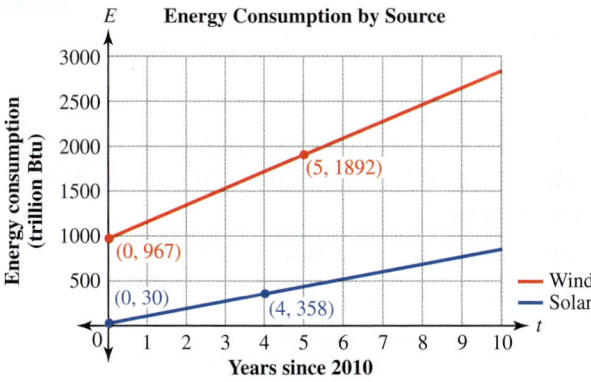
   Source: Models derived from data found at U.S. Energy Information Administration.

   a. Interpret the vertical intercept for the line representing electricity generation by wind.
   b. Interpret the vertical intercept for the line representing electricity generation by Solar.

7. Find the slope and the x- and y-intercepts of the equation $y = \frac{3}{5}x - \frac{2}{5}$.

8. Graph the line using the slope and y-intercept. Graph at least three points.
$$y = -\frac{1}{2}x + 7$$

9. Ahmed earns $12.00 an hour as a tutor. His wages can be computed by using the equation $W = 12h$, where h represents the number of hours Ahmed works and W represents his wages. Graph this line by graphing at least three points.

10. Graph the line by finding the x- and y-intercepts and plotting them. $4x + y = -7$.

11. Determine whether each equation is linear. If the equation is linear, put it in $y = mx + b$ form.
    a. $y = 6 - x^2$
    b. $y - 2 = 3(x + 4)$

12. Find the equation of the line that passes through the points $(1, -2)$ and $(-2, -17)$. Put the final answer in $y = mx + b$ form.

13. Use the graph to find the equation of the line. Put the final answer in $y = mx + b$ form.

    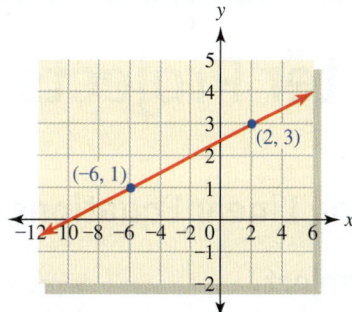

14. Paul goes on a diet to lose weight. Let the input t represent the week of his diet and let the output W represent Paul's weight in pounds. At the start of his diet ($t = 0$), Paul weighed 235 pounds. After 5 weeks on his diet, he weighed 210 pounds. Find the linear relationship between t and W.

15. Interpret the slope of the line in Problem 14 in terms of the problem situation (weight loss).

16. Use point-slope form to determine the equation of the line that passes through the points. Put your final answer in $y = mx + b$ form. The points are $(4, -2)$ and $(-8, -11)$.

356   CHAPTER 3   Linear Equations with Two Variables

17. Are the lines parallel, perpendicular, or neither?
$$y = -7x + 11$$
$$2x - 14y = 3$$

18. Find the equation of the line parallel to $y = \frac{2}{5}x + 9$ passing through the point $(-5, -3)$.

19. The following table gives the comparison between a dog's age and a human's age.

Dog's Age (years)	Human's Age (years)
1	15
2	24
3	28
4	32
5	36
7	44
10	56

Source: www.dogbreedinfo.com

a. Define the variables.
b. Graph the data on a scatterplot. Does the scatterplot look linear?
c. Draw the eyeball best-fit line.
d. Find a linear model for the data. Make sure to define the variables.
e. An 8-year-old dog corresponds to what human age? Is the answer reasonable? Explain.
f. A 5-year-old human corresponds to what age of dog? Is the answer reasonable? Explain.

20. The population of West Virginia is given in the following table.

Year	Population (millions)
2012	1.857
2013	1.853
2014	1.849
2015	1.841
2016	1.831

Source: U.S. Census Bureau.

a. Define the variables.
b. Graph the data on a scatterplot. Does the scatterplot look linear?
c. Draw the eyeball best-fit line.
d. Find a linear model for the data. Make sure to define the variables.
e. Use the model to predict the population of West Virginia in the year 2020. Is the answer reasonable? Explain.

Chapter Projects

■ Solving Linear Equations Using Tables

Written Project
One or more people

Suppose we want to solve the linear equation $5x + 6 = 12 - x$ using tables. We generate two tables using the following steps.

1. Let y = left side of the equation and let y = right side of the equation.
2. For each equation from step 1, fill in the table.

x	$y = 5x + 6$
-1	
0	
1	
2	
3	

x	$y = 12 - x$
-1	
0	
1	
2	
3	

3. For which point (x, y) are the two tables the same?
4. To solve $5x + 6 = 12 - x$, we are trying to find x. What is the value of x from step 3?

How Old Is the Grand Canyon?

Written Project
One or more people

The Grand Canyon is a steep canyon located in the state of Arizona. It is about 277 miles long and is anywhere from 4 to 18 miles wide. The Grand Canyon has been cut into the landscape by the Colorado River, which is gradually eroding the rocks. The central region of the canyon is about 5000 feet deep. Measurements show that the Colorado River is cutting into the canyon at the approximate rate of 1 inch deeper every 80 years. Suppose the river has been cutting (eroding) the canyon at a constant rate.
Source: Dr. Steve Spear, Palomar College.

1. How deep is the central part of the Grand Canyon?
2. What is the rate of erosion by the Colorado River?
3. How long did it take for the Colorado River to erode the current depth of the canyon?
4. How old (approximately) is the Grand Canyon?

Write a Review of a Section for Presentation

Research Project
One or more people

What you will need
- This presentation may be done in class or as an online video.
- If it is done as a video, post it on a website where it may be easily viewed by the class.
- You may want to use homework or other review problems from the book.
- Make it creative and fun.

Create a 5-minute review presentation of one section from Chapter 3. The format of the presentation can be a poster presentation, a blackboard presentation from notes, an online video, or a game format (e.g., math jeopardy). The presentation should include the following:

a. Any important formulas in the chapter
b. Important skills in the chapter, backed up by examples
c. Explanation of terms used (for example, what is slope?)
d. Common mistakes and how to avoid them

Systems of Linear Equations

- **4.1** Identifying Systems of Linear Equations
- **4.2** Solving Systems Using the Substitution Method
- **4.3** Solving Systems Using the Elimination Method
- **4.4** Solving Linear Inequalities in Two Variables Graphically
- **4.5** Systems of Linear Inequalities

Students in chemistry and nursing often have to mix solutions containing different amounts of saline or acid. They have to be careful that the mixture contains the desired amount of saline or acid after mixing. To determine how much of each type of solution to mix together, algebra can be used to solve a system of equations.

359

4.1 Identifying Systems of Linear Equations

LEARNING OBJECTIVES
- Identify a system of equations.
- Interpret solutions to systems of equations.
- Solve systems graphically.

Introduction to Systems of Equations

In Chapters 2 and 3, we studied solving linear equations and graphing linear equations. In this chapter, we are interested in solving a **system of linear equations**. A system of linear equations is a set of two (or more) linear equations. Linear systems arise in many applications in which we want to compare quantities. For example, a consumer might want to compare two cellular phone plans and determine, on the basis of the number of minutes used, the point at which the two plans are equal in cost. In general, the goal is to determine an input and output that satisfy both equations. This point is the solution of the system.

> **DEFINITION**
>
> **System of Linear Equations** A **system of linear equations** is a set of two (or more) linear equations.

What's That Mean?

System

In different fields, the meaning of the word *system* may vary. A biological system is an arrangement of many parts that work together. An example is the circulatory system. The parts include the heart, the veins, and the arteries. In mathematics, the term *system* has a slightly different meaning. A system of equations is a set of simultaneous equations. This means that two or more equations are being considered at the same time.

Consider the following two linear equations:

$$y = 2x$$
$$y = -3x + 10$$

These two equations are an example of a system of linear equations. A table of values for each of these two lines is given below.

x	$y = 2x$
-1	-2
0	0
1	2
2	4
3	6

x	$y = -3x + 10$
-1	13
0	10
1	7
2	4
3	1

The goal is to determine the input and output that satisfy both equations. From these two tables, we see that the two lines have one input and output pair that are the same: (2, 4). The point (2, 4) is called the **solution of the system of equations**.

> **DEFINITION**
>
> **Solution to a System of Linear Equations** A **solution to a system of linear equations** is one (or more) point that is a solution to every linear equation in the system.
>
> *Note:* The solution to a system of linear equations must be written as the ordered pair (x, y). The values of *both* variables must be found.

SECTION 4.1 Identifying Systems of Linear Equations **361**

Example 1   Comparing rental fees

Grace is moving to Seattle, and she has to rent a truck. She compares the rental costs for two companies. For a small truck, Vince's Truck Rentals quotes her $19.99 a day plus a one-time fee of $15.00. Gina's Rent-a-Truck quotes her $24.99 a day plus a one-time fee of $5.00.

a. Find equations for the cost for each of the two companies for renting a small truck for n days. Let V be the cost in dollars of renting a truck from Vince's, and let G be the cost in dollars of renting a truck from Gina's.

b. Fill in the following tables for Vince's Truck Rentals and Gina's Rent-a-Truck.

n = number of days	V = total cost of rental from Vince's
0	
1	
2	
3	
4	

n = number of days	G = total cost of rental from Gina's
0	
1	
2	
3	
4	

c. After how many days is the rental cost (approximately) the same?

d. If Grace has to rent a truck for 4 days, which company should she choose?

SOLUTION

a. First find the equation for the cost of renting from Vince's. The cost of renting a truck is $15.00 in fees plus $19.99 a day. The equation for the cost of renting from Vince's is

$$V = 15 + 19.99n$$

for a rental of n days.

Find the equation for the cost for renting a truck from Gina's Rent-a-Truck. The cost of renting the truck is $5.00 plus $24.99 a day. The equation for the cost of renting a truck from Gina's is

$$G = 5 + 24.99n$$

for a rental of n days.

b. Use the equations from part a to fill in the tables.

n = number of days	V = total cost of rental from Vince's
0	$15.00
1	$34.99
2	$54.98
3	$74.97
4	$94.96

n = number of days	G = total cost of rental from Gina's
0	$5.00
1	$29.99
2	$54.98
3	$79.97
4	$104.96

c. Remember that the solution to a system of equations is the point at which the two equations are equal. For 2 days, the rental cost is the same to rent a truck from

■ **Connecting the Concepts**

How do we estimate a reasonable value?

In Example 1, Grace is interested in renting a truck. To make a quick estimate of the cost, she should round the values to make the calculations easy. For example, for Vince's Truck Rentals, the quote of $19.99 a day rounds to $20.00. If Grace needs to rent a truck for 3 days, she can estimate $3 \cdot 20 = 60$ and then add on the $15.00 fee to get $75.00.

Vince's Truck Rentals as to rent a truck from Gina's Rent-a-Truck. The rental cost is $54.98 for both companies. The solution is 2 days and a cost of $54.98.

d. *It costs $94.96 to rent from Vince's for 4 days and $104.96 to rent from Gina's. So for a 4-day rental, Vince's Truck Rentals is cheaper.*

PRACTICE PROBLEM FOR EXAMPLE 1

Angelica makes belts and handbags, which she sells to local clothing boutiques. A major cost that concerns her is shipping. She must send the items to her customers using a local delivery service. Her customers currently all live in the same city that she does but are located more than 15 miles away from her studio. The first local shipping company she contacts, Ship Quick, quotes her a base rate of $55.00 plus $2.30 per mile for delivery beyond 15 miles. The second local shipping company, In Town Shipping, quotes Angelica $50.00 plus $2.50 per mile for delivery beyond 15 miles.

a. Find equations for the cost (in dollars) for shipping a box m miles beyond 15 miles. Let S = cost in dollars of shipping with Ship Quick, and let I = cost in dollars of shipping with In Town Shipping.

b. Fill in the following tables for Ship Quick and In Town Shipping.

m = number of miles over 15 miles	S = cost of shipping for Ship Quick
0	
5	
10	
15	
20	
25	
30	

m = number of miles over 15 miles	I = cost of shipping for In Town Shipping
0	
5	
10	
15	
20	
25	
30	

c. After how many miles over 15 miles is the shipping cost the same for both companies? That is, find the solution to the system of equations.

d. If Angelica has to ship a package 20 miles (i.e., 5 miles beyond 15 miles), which company should she use?

In Example 1, we are interested in finding the point at which the number of days and cost were the same for Vince's and Gina's truck rentals. In more formal language, we are trying to find what value of the input (n) makes both outputs (cost) the same for the two equations. The solution to this system is n = 2 days and cost = $54.98. Notice that this point is a solution that works in both linear equations.

To *check* a solution of a system, substitute the coordinates of the point into *both* original equations and determine whether it makes both equations true (or satisfies both equations).

Example 2 — Checking solutions of a system of equations

Check the solution of each system of equations.

a. Is $(1, 1)$ a solution of the system?

$$5x - 2y = 3$$
$$-20x + 8y = -12$$

b. The solution to Example 1 was $(2, 54.98)$. The equations are below. Check the solution.

$$V = 15 + 19.99n \quad G = 5 + 24.99n$$

c. Is $(0, -3)$ a solution of the system?

$$3x - 2y = 6$$
$$-2x + y = 8$$

SOLUTION

a. Substitute $x = 1$ and $y = 1$ into both equations to see whether they both are true.

First Equation		Second Equation
$5(1) - 2(1) \stackrel{?}{=} 3$	Substitute the values for x and y.	$-20(1) + 8(1) \stackrel{?}{=} -12$
$5 - 2 \stackrel{?}{=} 3$		$-20 + 8 \stackrel{?}{=} -12$
$3 = 3$	Both statements are true.	$-12 = -12$

Therefore, the point $(1, 1)$ is a solution of the system.

b. Substituting $n = 2$, $V = 54.98$, and $G = 54.98$ into Vince's and Gina's cost equations yields the following:

Vince's Cost Equation		Gina's Cost Equation
$54.98 \stackrel{?}{=} 15 + 19.99(2)$		$54.98 \stackrel{?}{=} 5 + 24.99(2)$
$54.98 = 54.98$	Both statements are true.	$54.98 = 54.98$

The point $(2, 54.98)$ is a solution of the system.

c. Substitute $x = 0$ and $y = -3$ into both equations to see whether they both are true.

First Equation		Second Equation
$3(0) - 2(-3) \stackrel{?}{=} 6$	Substitute in the values for x and y.	$-2(0) + (-3) \stackrel{?}{=} 8$
$0 - (-6) \stackrel{?}{=} 6$		$0 - 3 \stackrel{?}{=} 8$
$6 = 6$	Only one statement is true.	$-3 = 8$ False.

The point is on the first line but not on the second. Since $(0, -3)$ does not make the second equation a true statement, it is *not* a solution of the system.

PRACTICE PROBLEM FOR EXAMPLE 2

Check the solution of each system of equations.

a. Is $(-2, -2)$ a solution of the system?

$$x + y = -4$$
$$y = \frac{5}{2}x + 3$$

b. Is (25, 112.50) a solution of the system?

$$C = 55 + 2.30m$$
$$C = 50 + 2.50m$$

c. Is (4, 9) a solution of the system?

$$2x - y = -1$$
$$-3x + 7y = 4$$

Solving Systems Graphically

We are now going to look at some systems graphically and determine the solution using the graph. Recall that the solution to a system is a point that satisfies both linear equations. Graphically this point lies on both lines: It is the *intersection point* of both lines.

At the start of this section, we examined the following two linear equations.

$$y = 2x$$
$$y = -3x + 10$$

We found, using a table of values, that a solution to this system is (2, 4). Graphing both lines on the same axes yields the accompanying graph.

The two lines intersect at (2, 4), which is the same point that we found using the tables. Therefore, when solving a linear system graphically, we are looking for the point(s) where the lines intersect.

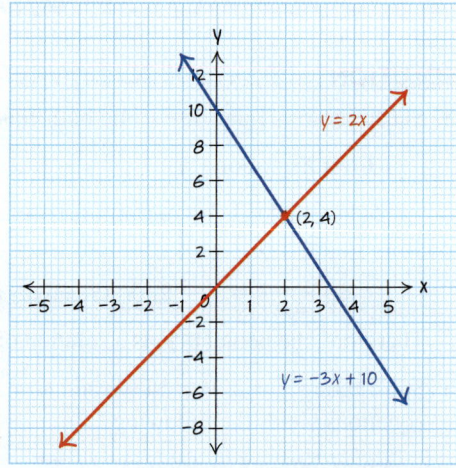

Example 3 — Graphical solutions from real data

The types of nonhousing debt balances in the United States for the years 2004 through 2016 is given in the following graph. Use this graph to answer the questions below. *Note:* HE revolving debt refers to housing equity lines of credit.

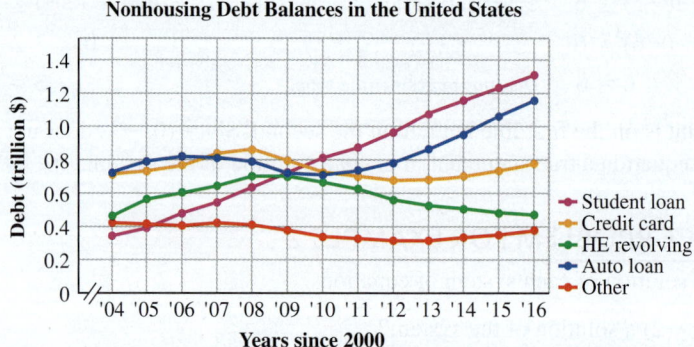

Source: Center for Microeconomic Data.

a. During which year(s) was credit card debt equal to auto loan debt? Estimate the debt at the point(s) of intersection.

b. During which year(s) was HE revolving debt equal to student loan debt? Estimate the debt at the point(s) of intersection.

SOLUTION

Using the graph above, we see the following.

a. Credit card debt was equal to auto loan debt midway between 2006 and 2007, and again halfway through the years of 2010 and 2011 (see red arrows on the graph below). From the graph, we estimate the debt at the intersection point between 2006 and 2007 to be approximately $0.8 trillion. From the graph, we estimate the debt at the intersection point between 2010 and 2011 to be approximately $0.7 trillion.

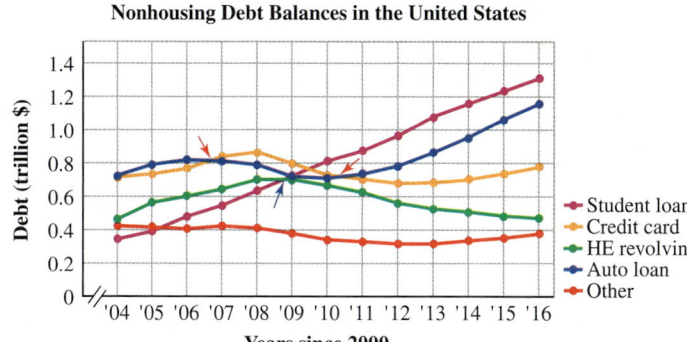

b. HE revolving debt equals student loan debt in about 2009 (see the blue arrow on the graph above). From the graph, we estimate the debt at the intersection point to be approximately $0.7 trillion.

PRACTICE PROBLEM FOR EXAMPLE 3

Using the graph in Example 3, answer the following questions about the nonhousing debt balances in the United States.

a. During which years did student loan debt equal credit card debt? Estimate the debt at the point(s) of intersection.

b. During which years did student loan debt equal other debt? Estimate the debt at the point(s) of intersection.

In Example 3, we used the graphs to determine the solution(s) to a system. Typically in these problems, there are just two lines, and we are looking for the intersection point. In the next example, we use a graph to find the intersection point.

Example 4 Reading the solution to a system from a graph

Use the graph to determine the solution to each system of equations.

a.

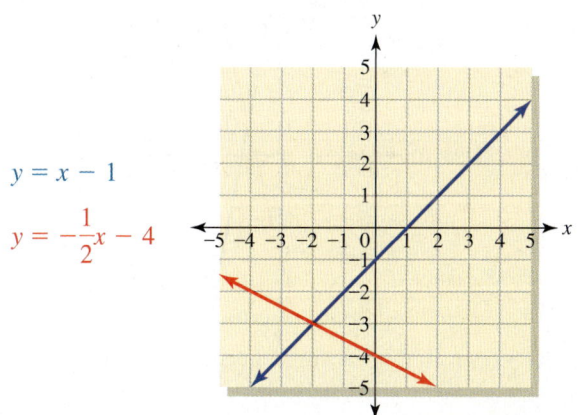

$y = x - 1$

$y = -\dfrac{1}{2}x - 4$

b.

$y = -5x + 6$

$y = 4x - 3$

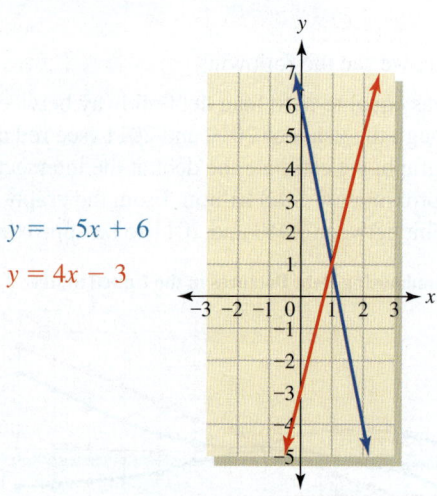

SOLUTION

a. To find the solution to this system, look for the point of intersection of the two lines. Looking at the graph, we see that the point of intersection seems to be at $x = -2$ and $y = -3$ or, as an ordered pair, $(-2, -3)$.

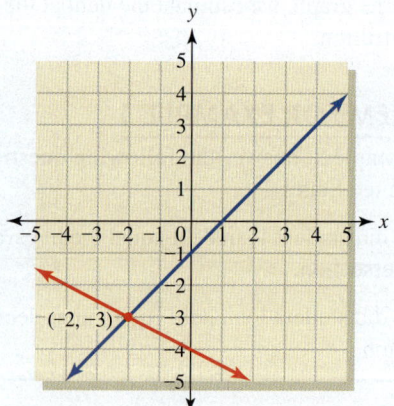

b. To find the solution to this system, look for the point of intersection of the two lines. When we look at the graph, it is clear that the two lines intersect with the x-coordinate $x = 1$. The y-coordinate is $y = 1$. The solution is $(1, 1)$.

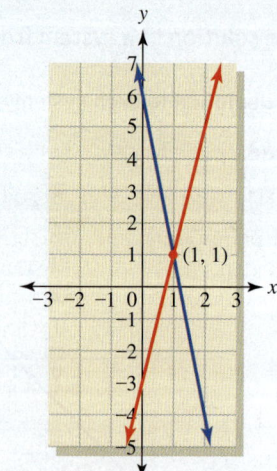

PRACTICE PROBLEM FOR EXAMPLE 4

Use the graph to determine the solution to the system of equations.

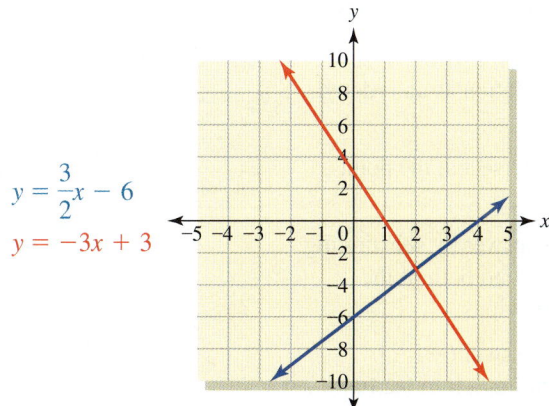

In Example 4, we were given the graph, and our task was to examine the graph and find the intersection point. Sometimes the graph is not given. In that case, first graph the system of equations and then use that graph to find the solution. To solve a system accurately using a graph, it is important to use graph paper and to draw the lines using a straightedge or ruler.

> **Steps to Solving a System of Equations Graphically**
> 1. On graph paper, carefully graph both lines on the same pair of axes.
>   a. Scale the axes so that both lines can be graphed simultaneously.
>   b. Use a ruler to draw in the lines.
> 2. Find the point of intersection. This is the solution to the system.
> 3. Write the solution as an ordered pair (x, y).
> 4. Check the answer by substituting the solution into *both* of the original equations.

Example 5 Solving a system graphically

Solve the system graphically. Check the solution.

$$y = 2x + 3$$
$$2x + y = -1$$

SOLUTION

Step 1 On graph paper, carefully graph both lines on the same pair of axes.

To begin, carefully graph both lines. The first line has a y-intercept of $(0, 3)$, so graph the point $(0, 3)$. The slope is $2 = \frac{2}{1} = \frac{\text{rise}}{\text{run}}$, so from the point $(0, 3)$, go up 2 and over 1 to get to another point on the line, which is $(1, 5)$. Draw in the line.

The second line is not in slope-intercept form. To put this line in slope-intercept form, solve for y.

$$\begin{array}{rl} 2x + y = & -1 \\ -2x & -2x \\ \hline y = & -2x - 1 \end{array}$$ Isolate y by subtracting $2x$ from both sides.

This is in $y = mx + b$ form.

The y-intercept is $(0, -1)$, and the slope is $-2 = \frac{-2}{1} = \frac{\text{rise}}{\text{run}}$. Start at the point $(0, -1)$ and go down 2 units and over 1 unit to get to another point on the line, $(1, -3)$. Draw in the line.

Skill Connection

Slope-intercept form

The slope-intercept form of a line is $y = mx + b$, where m is the slope and $(0, b)$ is the y-intercept.

Step 2 Find the point of intersection. This is the solution to the system.

Looking at the graph, the two lines intersect at $x = -1$ and $y = 1$.

Step 3 Write the solution as an ordered pair (x, y).

The two lines intersect at the point $(-1, 1)$.

Step 4 Check the answer by substituting the solution into both of the original equations.

Now check the answer $(-1, 1)$. Start by substituting it into the first equation.

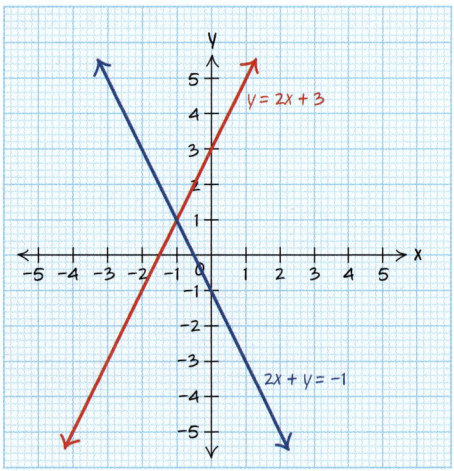

First Equation		Second Equation
$y = 2x + 3$		$y = -2x - 1$
$1 \stackrel{?}{=} 2(-1) + 3$	Substitute in $x = -1$ and $y = 1$.	$1 \stackrel{?}{=} -2(-1) - 1$
$1 = 1$	Both statements are true.	$1 = 1$

Therefore, the point $(-1, 1)$ is a solution to the system of linear equations.

PRACTICE PROBLEM FOR EXAMPLE 5

Solve the system graphically. Check the solution.

$$y = -2x - 16$$
$$y = 4x - 4$$

■ Types of Systems

In the following Concept Investigation, we explore how many solutions a system of linear equations can have. Remember that we are trying to determine the number of intersection points that two lines can have.

CONCEPT INVESTIGATION
How many answers are there?

Graph each system on the axes provided and answer the questions.

a. Graphically determine the solution to the system

$$y = 8x - 4$$
$$y = -2x + 6$$

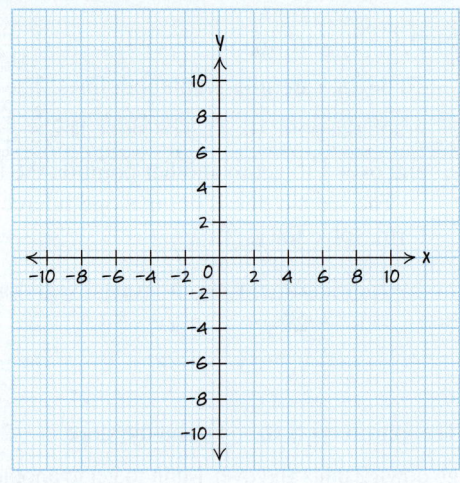

How many solutions does this system have?

b. Graphically determine the solution to the system

$$y = -3x + 5$$
$$y = -3x - 3$$

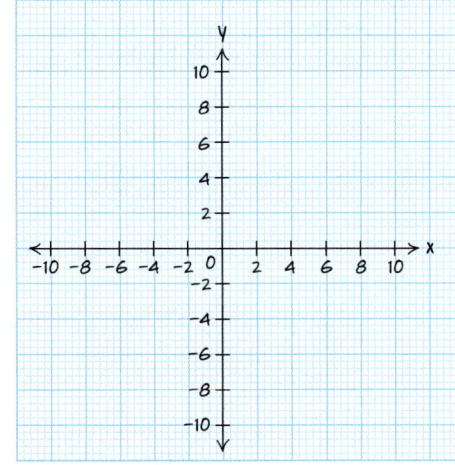

How many solutions does this system have?

c. Graphically determine the solution to the system

$$6x + 2y = 10$$
$$y = -3x + 5$$

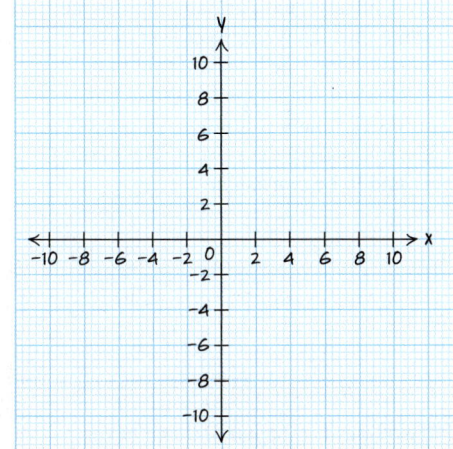

How many solutions does this system have?

In the Concept Investigation, we saw that two lines can intersect at exactly one point, never intersect, or intersect everywhere on the line, as they are the same line. The following terms are used to describe the number of solutions a system of equations has.

> **DEFINITIONS**
>
> **Consistent System** A system of linear equations is said to be consistent if it has at least one solution. This means that the two lines can intersect at exactly one point or intersect at infinitely many points on the same line.
>
> **Inconsistent System** A system of linear equations is said to be inconsistent if it has no solutions.

We summarize this new terminology and the number of solutions to linear systems in the following box.

370 CHAPTER 4 Systems of Linear Equations

Interpreting the Number of Solutions to a System of Linear Equations	Graphs
Two lines can intersect at exactly one point. This type of system has one solution, and it is written as a point (x, y). The system is said to be *consistent*.	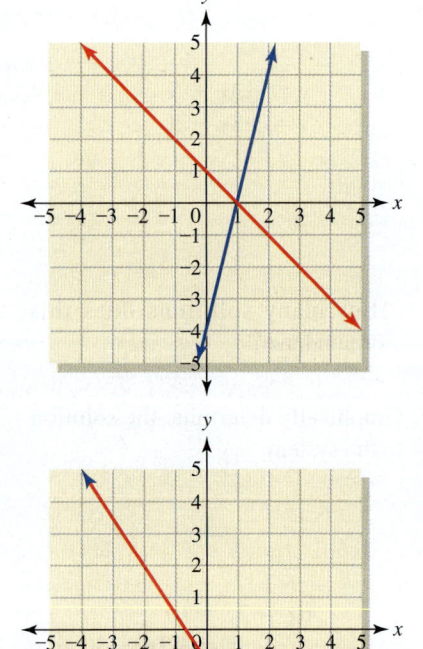
Two linear equations can graph as the same line. They intersect everywhere on that line. The lines are said to be *dependent*. This type of system has infinitely many solutions. All points on the line satisfy the system. The system is said to be *consistent*.	
Two lines can be parallel and never intersect. This type of system has no solutions. The system is said to be *inconsistent*.	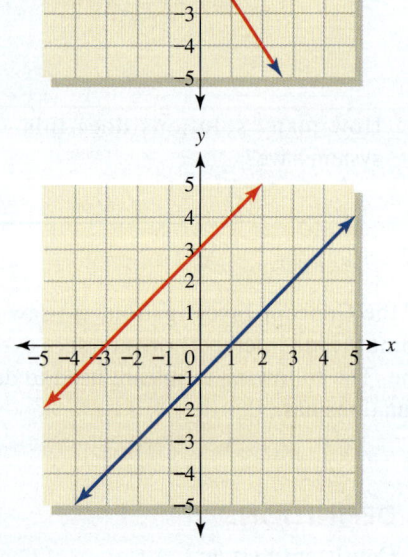

Example 6 — Consistent or inconsistent systems?

Solve each system graphically. Identify each system as consistent or inconsistent. Clearly label and scale the axes. Check the solution(s), when possible, by substituting into the original problem.

a. $4x - 3y = 12$

$y = \dfrac{4}{3}x + 5$

b. $y = \dfrac{3}{2}x - 4$

$y = -\dfrac{x}{4} + 3$

c. $2x + 5y = 25$

$y = -\dfrac{2}{5}x + 5$

SECTION 4.1 Identifying Systems of Linear Equations 371

SOLUTION

a. The equation of the first line is $4x - 3y = 12$. This equation is in general form. To graph it requires two points on the line. For this particular line, it is fairly easy to find the two intercepts.

x-intercept	y-intercept
Let $y = 0$	Let $x = 0$
$4x - 3(0) = 12$	$4(0) - 3y = 12$
$4x = 12$	$-3y = 12$
$\dfrac{4x}{4} = \dfrac{12}{4}$	$\dfrac{-3y}{-3} = \dfrac{12}{-3}$
$x = 3$	$y = -4$
x-intercept: $(3, 0)$	y-intercept: $(0, -4)$

Graph the line using these two points. The equation of the second line is $y = \dfrac{4}{3}x + 5$. The second line is given in slope-intercept form. The y-intercept is the point $(0, 5)$, and we can use the slope to find another point on the graph. The slope is $m = \dfrac{4}{3} = \dfrac{\text{rise}}{\text{run}}$. The graph looks like the following:

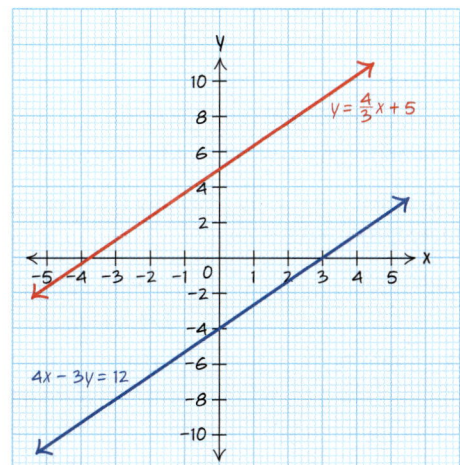

From the graph, it appears that the lines are parallel. To determine whether the lines are parallel, find and compare the slopes of both lines. The first line, $4x - 3y = 12$, has intercepts $(3, 0)$ and $(0, -4)$. Using the slope formula, we find the slope as follows:

$$m = \dfrac{-4 - 0}{0 - 3} = \dfrac{-4}{-3} = \dfrac{4}{3}$$

The second line is given in slope-intercept form, $y = \dfrac{4}{3}x + 5$, and the slope is $\dfrac{4}{3}$. Because the two slopes are equal but the y-intercepts are different, the lines are parallel.

There are *no solutions* to this system of equations. *The system is inconsistent.*

b. Both lines are given in the slope-intercept form. To graph them, use the y-intercept and then use the slope to find an additional point on the line.

$y = \dfrac{3}{2}x - 4$	$y = -\dfrac{x}{4} + 3$
Slope: $m = \dfrac{3}{2}$	Slope: $m = -\dfrac{1}{4}$
y-intercept: $(0, -4)$	y-intercept: $(0, 3)$

■ **Connecting the Concepts**

How do we use slope to find parallel lines?

Recall from Chapter 3 that two lines are parallel if they have the same slope but different y-intercepts. Therefore, to determine whether two lines are parallel, find their slopes and check to see whether they are equal. To find the slope, either put each line in the $y = mx + b$ form or select two points on the line and use the slope formula,

$$m = \dfrac{y_2 - y_1}{x_2 - x_1}$$

Graphing both lines carefully on the same axes yields the following graph.

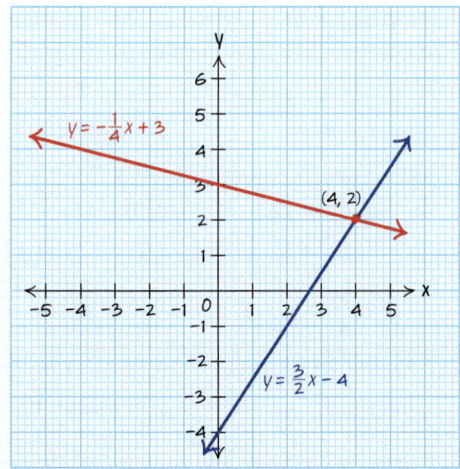

The two lines intersect at the point (4, 2). Check the solution (4, 2) by substituting $x = 4$ and $y = 2$ into both equations to see that they are true.

First Equation

$2 \stackrel{?}{=} \dfrac{3}{2}(4) - 4$ Substitute in $x = 4$, $y = 2$.

$2 \stackrel{?}{=} 6 - 4$

$2 = 2$ Both equations are true.

Second Equation

$2 \stackrel{?}{=} -\dfrac{(4)}{4} + 3$

$2 \stackrel{?}{=} -1 + 3$

$2 = 2$

The solution checks in both equations. *The system is consistent.*

c. The equation of the first line is $2x + 5y = 25$ and is in general form. We will graph this line using its intercepts.

x-intercept
Let $y = 0$
$2x + 5(0) = 25$
$2x = 25$
$x = \dfrac{25}{5} = 12.5$
x-intercept: (12.5, 0)

y-intercept
Let $x = 0$
$2(0) + 5y = 25$
$5y = 25$
$y = 5$
y-intercept: (0, 5)

The equation of the second line is $y = -\dfrac{2}{5}x + 5$ and is in slope-intercept form ($y = mx + b$). The y-intercept is (0, 5), and the slope is $m = -\dfrac{2}{5} = \dfrac{\text{rise}}{\text{run}}$.

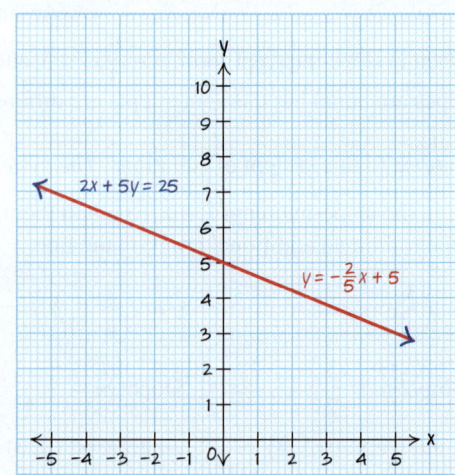

The two linear equations graph as the same line. Therefore, the two equations intersect everywhere, so there are *infinitely many solutions. This system is consistent.*

To show that these two lines are really the same line, solve the first equation for y.

$$2x + 5y = 25$$
$$\underline{-2x \qquad\qquad -2x} \qquad \text{Subtract 2x from both sides.}$$
$$5y = -2x + 25$$
$$\frac{5y}{5} = \frac{-2x}{5} + \frac{25}{5} \qquad \text{Divide both sides by 5 and simplify.}$$
$$y = -\frac{2}{5}x + 5$$

This is equal to the second linear equation, $y = -\frac{2}{5}x + 5$.

When solving a system of equations that result from an application problem, there is an additional first step of determining the two linear equations that result from the problem. Then proceed as before by graphing the two lines carefully and using the graph to determine the point(s) of intersection (if any).

Example 7 Comparing salaries

Renee is considering two part-time jobs in sales. One job, at Electronics-4-U, offers her $140 a week plus a 4% commission on her weekly sales. The other job, at Clothes to Go, offers her a weekly base salary of $130 plus a 5% commission on her weekly sales.

What's That Mean?

Commission

Sales jobs are sometimes paid on a base salary plus a *commission*. The commission is a percentage of the sales that is paid back to the salesperson. If a salesperson sells $1000.00 and earns a 5% commission, the pay would be $0.05 \cdot 1000 = 50$, or $50.00.

a. Determine the equations for Renee's weekly salary for each of the two jobs.

b. Graph the two lines on graph paper. Think carefully about how to scale the axes.

c. What is the solution to the system of equations? Interpret the solution in terms of her salary and weekly sales.

d. If Renee estimates that her weekly sales averages $550, which job should she take?

SOLUTION

a. Let w be the weekly sales in dollars, and let S be the weekly salary in dollars. The equation for Renee's weekly salary at Electronics-4-U is

$$S = 140 + 0.04w$$

Make sure to write the percentage as a decimal in the problem. The equation for Renee's weekly salary at Clothes to Go is given by

$$S = 130 + 0.05w$$

b. Notice that the y-intercepts are (0, 140) and (0, 130). A reasonable starting scale for the vertical axis would be from 0 to 200 by an increment of 20. Since the input is w, which represents weekly sales, a reasonable scale for the horizontal axis would be from 0 to 2500, with an increment of 500. One way to graph these two lines is to generate a table of values for each equation.

w	Electronics-4-U $S = 140 + 0.04w$
0	140
1000	180
2000	220

w	Clothes to Go $S = 130 + 0.05w$
0	130
1000	180
2000	230

This is an estimate of what her weekly sales in clothing or electronics would be.

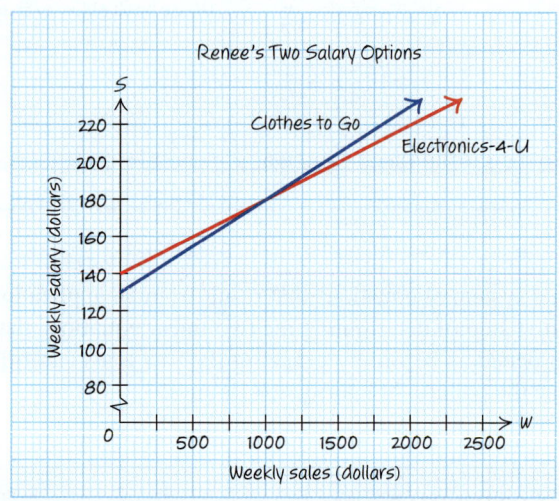

c. From the graph, the two lines intersect at (1000, 180).

This means that if Renee's weekly sales are $1000, her weekly salary will be $180.00 at both companies.

d. Substitute 550 into w in each equation to find the weekly salaries.

Electronics-4-U	Clothes to Go
$S = 140 + 0.04w$	$S = 130 + 0.05w$
$S = 140 + 0.04(550)$	$S = 130 + 0.05(550)$
$S = 162.00$	$S = 157.50$

Since she earns more at Electronics-4-U, she should work there.

Skill Connection

Set the scale

Recall from Chapter R that the consistent and even spacing along an axis is called the *scale*. To scale an axis, find the smallest and largest values to be plotted along that axis. Round these values as needed, remembering that it is easy to count in multiples of twos, fives, and tens. Then decide on the scale, aiming for about ten tick marks on the axis.

PRACTICE PROBLEM FOR EXAMPLE 7

You are moving and want to store some of your things temporarily. You contact two local storage companies and get quotes on a 10-foot by 10-foot storage space. Help-U-Stor gives you a quote of $160.00 per month with a one-time $10.00 administration fee. Local Storage Solutions quotes you $150.00 per month with a one-time $30.00 administration fee.

a. Determine the equations for the cost of renting a 10-foot × 10-foot storage unit for each of the two companies. Let m be the number of months and let C be the total monthly rental cost in dollars.

b. Graph the two lines on graph paper. Think carefully about how to scale the axes.

c. What is the solution to the system of equations? Interpret the solution in terms of the cost of renting the two units.

d. If you have to rent the storage space for 3 months, which company should you store your things with?

4.1 Vocabulary Exercises

1. A(n) _____ of equations is a set of two or more linear equations.

2. A(n) _____ to a system of equations is a point that satisfies both equations in the system.

3. Graphically the solution to a system of linear equations is the point(s) where the two lines _____.

4. There are _____ possibilities for the solution set to a system of equations.

5. A system of equations that has at least one solution is called a(n) _____ system.

6. A system of equations that has no solutions is called a(n) _____ system.

4.1 Exercises

For Exercises 1 through 8, use the table of values to solve the systems.

1.

x	$y = -\frac{2}{5}x + 4$
-7	6.8
-6	6.4
-5	6
-4	5.6
-3	5.2

x	$y = -\frac{8}{5}x - 2$
-7	9.2
-6	7.6
-5	6
-4	4.4
-3	2.8

2.

x	$y = -x + 1$
-5	6
-4	5
-3	4
-2	3
-1	2

x	$y = \frac{1}{3}x + 5$
-5	$3\frac{1}{3}$
-4	$3\frac{2}{3}$
-3	4
-2	$4\frac{1}{3}$
-1	$4\frac{2}{3}$

3.

x	$y = 5.5x - 5$
-2	-16
-1	-10.5
0	-5
1	0.5
2	6

x	$y = -1.5x + 9$
-2	12
-1	10.5
0	9
1	7.5
2	6

4.

x	$y = 2x - 1$
-1	-3
0	-1
1	1
2	3
3	5

x	$y = -x + 8$
-1	9
0	8
1	7
2	6
3	5

5.

x	$y = \frac{1}{2}x + 3$
-1	2.5
0	3
1	3.5
2	4
3	4.5

x	$y = \frac{3}{2}x + 3$
-1	1.5
0	3
1	4.5
2	6
3	7.5

6.

x	$y = -2x + 7$
-2	11
0	7
2	3
4	-1
6	-5

x	$y = 3x + 7$
-2	1
0	7
2	13
4	19
6	25

7.

x	$y = -x + 10$
6	4
7	3
8	2
9	1
10	0

x	$y = x - 8$
6	-2
7	-1
8	0
9	1
10	2

8.

x	$y = -3x + 2$
-1	5
0	2
1	-1
2	-4
3	-7

x	$y = 3x - 10$
-1	-13
0	-10
1	-7
2	-4
3	-1

For Exercises 9 through 20, determine whether or not the given points are solutions of the systems.

9.
$$C = 3x - 11$$
$$C = -2x + 14$$

376 CHAPTER 4 Systems of Linear Equations

a. Is $(x, C) = (-3, 2)$ a solution of the system?
b. Is $(x, C) = (5, 4)$ a solution of the system?

10.
$$y = 4x$$
$$y = -7x - 22$$

a. Is $(x, y) = (1, 4)$ a solution of the system?
b. Is $(x, y) = (-2, -8)$ a solution of the system?

11.
$$5a + 3b = -15$$
$$a + 3b = 3$$

a. Is $(a, b) = (0, 3)$ a solution of the system?
b. Is $(a, b) = (3, 0)$ a solution of the system?

12.
$$5t - 3P = -33$$
$$t + 2P = -4$$

a. Is $(t, P) = (-5, 1)$ a solution of the system?
b. Is $(t, P) = (1, -6)$ a solution of the system?

13.
$$-12x + 3y = -24$$
$$4x - y = 8$$

a. Is $x = 0, y = -8$ a solution of the system?
b. Is $x = 1, y = -4$ a solution of the system?

14.
$$-3c + d = 5$$
$$6c - 2d = -10$$

a. Is $c = -\frac{5}{3}, d = 0$ a solution of the system?
b. Is $c = 0, d = 5$ a solution of the system?

15.
$$-2x + y = -12$$
$$y = 2x + 16$$

a. Is $x = 6, y = 0$ a solution of the system?
b. Is $x = -8, y = 0$ a solution of the system?

16.
$$s + 3r = 3$$
$$3r = -s - 9$$

a. Is $s = 0, r = 1$ a solution of the system?
b. Is $s = 3, r = 1$ a solution of the system?

17.
$$x = 3$$
$$y = -1$$

a. Is $(3, 1)$ a solution of the system?
b. Is $(3, -1)$ a solution of the system?

18.
$$x = 0$$
$$y = -7$$

a. Is $(0, 0)$ a solution of the system?
b. Is $(0, -7)$ a solution of the system?

19.
$$y = -5x + 6$$
$$5x + y = 6$$

a. Is $(1, 1)$ a solution of the system?
b. Is $(2, -4)$ a solution of the system?

20.
$$y = \frac{1}{3}x - 4$$
$$x - 3y = 12$$

a. Is $(3, -3)$ a solution of the system?
b. Is $(0, -4)$ a solution of the system?

For Exercises 21 and 22, use the given graph to answer the questions. The graph below shows the death rates (percentage) involving CHD (which is short for coronary heart disease) organized by serum cholesterol level in various countries.

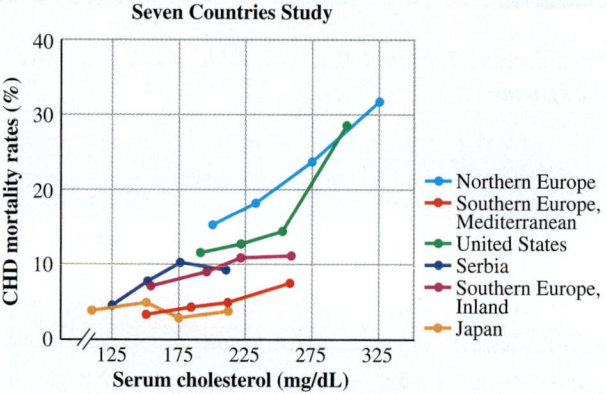

Seven Countries Study

21. For which serum cholesterol level are the CHD mortality rates for Serbia and Japan equal?

22. For which serum cholesterol level are the CHD mortality rates for Northern Europe and the United States equal?

For Exercises 23 through 26, fill in the given tables. Answer the corresponding questions.

23. Your band is interested in renting one of two music recording studios. Music by You charges $100 plus an additional $60 per hour. Bling Music Studio charges $150 plus an additional $50 per hour.

a. Define the variables.
Write equations for the cost of renting each studio.

b. Fill in each table below.

Hours	Cost for Music by You
1	
2	
3	
4	
5	
6	
7	

Hours	Cost for Bling Music Studio
1	
2	
3	
4	
5	
6	
7	

c. Examine the tables to find the number of hours that makes the cost at both studios equal.

d. If you need the studio for 7 hours, which studio will be cheaper for you to rent?

24. You are interested in comparing two video-on-demand services on the Internet. VideoNow charges $15.00 per month plus $3.00 per movie. MoviesPlus charges $5.00 a month plus $8.00 per movie.

   a. Define the variables.
   Write equations that give the monthly cost of downloading movies from both companies.

   b. Fill in each table.

Number of Movies Downloaded	Monthly Cost for VideoNow
0	
1	
2	
3	
4	
5	

Number of Movies Downloaded	Monthly Cost for MoviesPlus
0	
1	
2	
3	
4	
5	

   c. Examine the tables to find the number of movies downloaded that makes the monthly cost of the two services equal.

   d. If you download an average of five movies a month, which service will be cheaper for you to use?

25. You want to join a fitness club and are considering two options in your area. The first club, Fitness Plus, charges a start-up fee of $100.00 plus $15.00 per month. The second club, Rock's Gym, charges an initial fee of $80.00 plus $20.00 per month.

   a. Define the variables.
   Write equations for the cost of joining each gym.

   b. Fill in each table.

Number of Months	Cost of Fitness Plus
0	
1	
2	
3	
4	
5	
6	

Number of Months	Cost of Rock's Gym
0	
1	
2	
3	
4	
5	
6	

   c. After how many months will the cost of joining the two fitness clubs be equal?

   d. If you plan on joining for 6 months, which fitness club will be cheaper for you to use?

26. The cost to a retailer for each small bottle of water she sells is $0.37. She sells each small bottle for $1.50.

   a. Define the variables.
   Write equations for the cost that the retailer pays to purchase the water and the revenue the retailer takes in from selling the water.

   b. Fill in each table below.

Number of Small Bottles of Water	Cost to Retailer
0	
2	
4	
6	
8	
10	
12	

Number of Bottles of Water	Revenue Retailer Earns from Sale of Water
0	
2	
4	
6	
8	
10	
12	

   c. How many bottles of water does the retailer have to sell for her cost to equal her revenue?

   d. Does your answer to part c make sense in the context of this problem? Explain.

For Exercises 27 through 34, use the graphs to determine the solutions to the systems of equations.

27.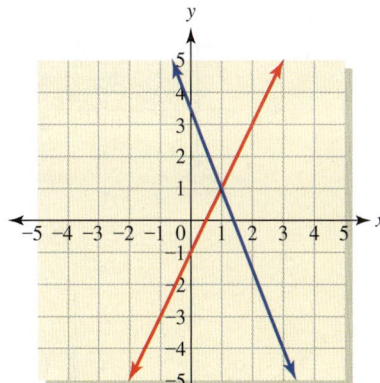

28.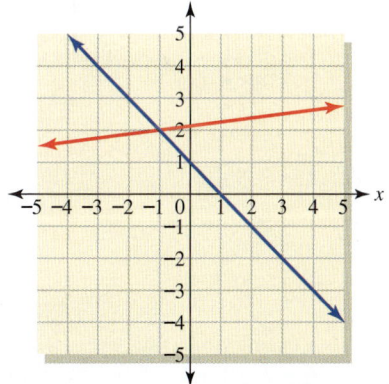

378 CHAPTER 4 Systems of Linear Equations

29.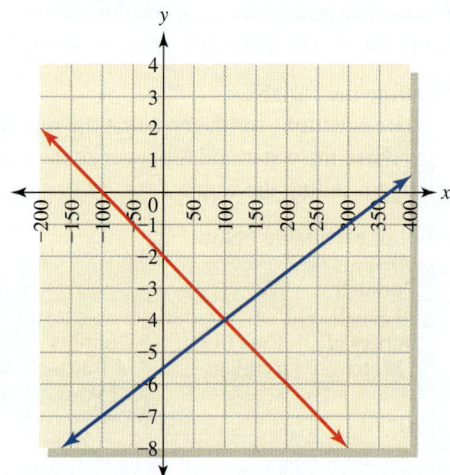

30.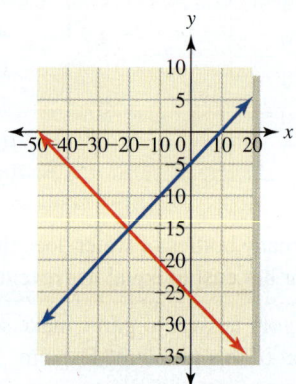

31.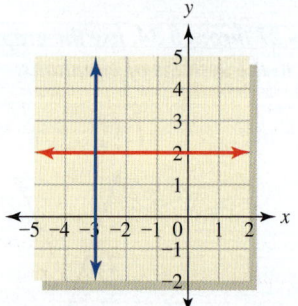

32.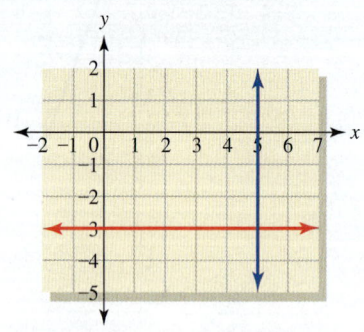

33.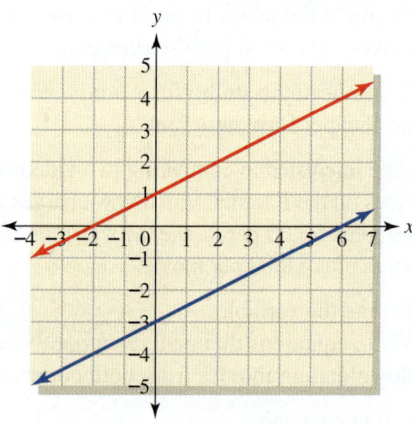

34.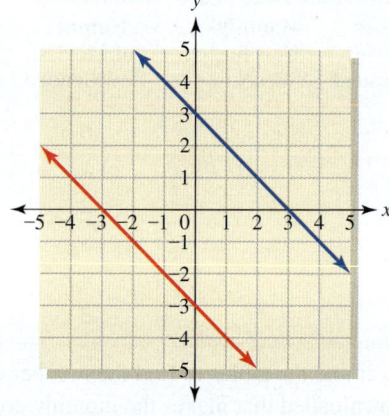

For Exercises 35 through 38, use the given information to answer the questions.

35. Ayesha is moving and gets quotes from two truck rental companies for a cargo van. The first company quotes her a 1-day rental price of $19.95 plus $0.79 per mile. The second company quotes a 1-day rental price of $21.95 plus $0.59 per mile. The system of equations that describes the cost C for a 1-day rental given the number of miles m driven is as follows:

$$C = 19.95 + 0.79m$$
$$C = 21.95 + 0.59m$$

  a. Is $(m, C) = (10, 27.85)$ a solution of the system?
  b. Is $(m, C) = (15, 31.80)$ a solution of the system?

36. Amir is moving and gets two quotes from two rental companies for a 14-foot truck. The first company quotes him a 1-day rate of $29.95 plus $0.79 per mile. The second company quotes the 1-day rental price of $34.95 plus $0.59 per mile. The system of equations that describes the cost C for a 1-day rental given the number of miles m driven is as follows:

$$C = 29.95 + 0.79m$$
$$C = 34.95 + 0.59m$$

  a. Is $(m, C) = (30, 52.65)$ a solution of the system?
  b. Is $(m, C) = (25, 49.70)$ a solution of the system?

37. Verena is considering taking one of two jobs. The first job pays a base salary of $9.00 an hour plus 3% commission on the dollar amount of sales. The second job pays a base salary of $10.00 an hour plus 1% commission on the dollar amount of sales. The system of equations that describes the hourly salary S earned given the dollar amount of sales x is as follows:

$$S = 9.00 + 0.03x$$
$$S = 10.00 + 0.01x$$

a. Is $(x, S) = (45, 10.35)$ a solution of the system?

b. Is $(x, S) = (50, 10.50)$ a solution of the system?

38. Alex is considering taking one of two jobs. The first job pays a base salary of $9.50 an hour plus 2% commission on the dollar amount of sales. The second job pays a base salary of $10.50 an hour plus 1% commission on the dollar amount of sales. The system of equations that describes the hourly salary S earned given the dollar amount of sales x is as follows:

$$S = 9.50 + 0.02x$$
$$S = 10.50 + 0.01x$$

a. Is $(x, S) = (100, 11.50)$ a solution of the system?

b. Is $(x, S) = (75, 11)$ a solution of the system?

For Exercises 39 through 62, graphically determine the solution to each system of equations. Identify each system as consistent or inconsistent. Clearly label and scale all graphs. Check the solutions by substituting into the original systems.

39. $y = 3x - 6$
$y = -6x + 3$

40. $y = -5x + 10$
$y = 10x - 5$

41. $y = 4x - 20$
$y = -2x + 10$

42. $y = -\frac{1}{2}x + 6$
$y = 2x - 4$

43. $y = \frac{1}{2}x - 1$
$-x + 2y = -2$

44. $y = \frac{1}{6}x + 2$
$y = \frac{1}{6}x - 4$

45. $x + y = 1$
$x + y = -5$

46. $4x + y = 7$
$4x + y = -1$

47. $-2x + y = 4$
$4x + y = -2$

48. $-x + 2y = 2$
$3x + 4y = -16$

49. $-9x + y = 6$
$2x + y = -5$

50. $2x + 3y = 21$
$-4x + 3y = 3$

51. $y = 5$
$x = -4$

52. $x = 3$
$y = -1$

53. $x = -9$
$y = -2.5$

54. $y = -3.8$
$x = 5.7$

55. $y = -\frac{1}{3}x + 2$
$2x + 6y = 4$

56. $\frac{x}{3} + y = 7$
$-3x + y = -3$

57. $-3x + 5y = 7$
$y = \frac{3}{5}x + 2$

58. $y = \frac{x}{3}$
$y = x + 2$

59. $y = \frac{4}{5}x + 3$
$5y = 4x + 15$

60. $x - 4y = 12$
$y = \frac{1}{4}x - 3$

61. $y = \frac{2}{3}x + 3$
$y = -2x + 19$

62. $y = -\frac{1}{2}x + 7$
$x + 2y = 14$

For Exercises 63 through 66, solve the application problems graphically. Clearly label and scale all graphs. Check each solution by substituting into the original system. Recall that the break-even point is the point at which revenue equals cost.

63. Jason's mobile auto repair has the following weekly revenue and cost equations for his business:

$$R = 75h$$
$$C = 50h + 300$$

where R represents the weekly revenue in dollars and C represents the weekly costs in dollars when he does h hours of repair work a week. Find the break-even point for Jason's mobile auto repair.

64. Gourmet Apples sells chocolate- and caramel-covered apples at local farmers' markets. Their weekly revenue and cost equations are

$$R = 15a$$
$$C = 3a + 600$$

where R represents the weekly revenue in dollars and C represents the weekly costs in dollars when Gourmet Apples sells a apples a week. Find the break-even point for Gourmet Apples.

65. Hummus R Us sells specialty hummus products at small grocery stores. Their weekly revenue and cost equations are

$$R = 6n$$
$$C = 2n + 460$$

where R represents the weekly revenue in dollars and C represents the weekly costs in dollars when Hummus R Us sells n hummus products a week. Find the break-even point for Hummus R Us.

66. Eve knits scarves to sell at craft fairs. Her monthly revenue and cost equations are

$$R = 40n$$
$$C = 25n + 300$$

where R represents the monthly revenue in dollars and C represents the monthly cost in dollars when Eve sells n scarves a month. Find the break-even point for Eve.

67. Janelle wants to hire a handyman for 1 day to do some repairs to her condo. A handyman named Mike advertises an hourly rate of $60.00 plus a $10.00 travel fee. Josh, another handyman, advertises an hourly rate of $55.00 plus a $15.00 travel fee.

  a. Find an equation for the cost of hiring Mike for 1 day.

  b. Find an equation for the cost of hiring Josh for 1 day.

  c. Carefully graph both lines on the same graph. Determine the point at which the cost of hiring Mike is equal to the cost of hiring Josh.

68. One plumber, Rocco, advertises that his labor rate is $65.00 an hour plus a $20.00 travel fee. A second plumber, Marco, advertises his labor rate at $60.00 an hour plus a $25.00 travel fee.

  a. Find an equation for the cost of hiring Rocco for 1 day.

  b. Find an equation for the cost of hiring Marco for 1 day.

  c. Carefully graph both lines on the same graph. Determine the point at which the cost of hiring Rocco is equal to the cost of hiring Marco.

69. An oven and microwave combination costs $1200.00. Purchased individually, the oven costs twice as much as the microwave. Find the cost of the oven and microwave.

70. A washer and dryer combination costs $1300. If each is bought separately, the dryer costs $\frac{7}{6}$ times as much as the washer. Find the cost of the washer and dryer.

71. The perimeter of a rectangle is 24 centimeters. The length is three times the width. Find the length and the width of the rectangle.

72. The perimeter of a rectangle is 18 inches. The length is twice the width. Find the length and the width of the rectangle.

73. The local high school track team is holding a luncheon fund-raiser. They rented space in a nearby hotel for $255.00 for one afternoon. The catering company tells them that it will cost $5.75 per person for lunch. They plan to sell tickets to the fund-raiser for $10.00 per person.

  a. Find an equation for the total cost. This will include both the cost of renting the space and the cost of providing lunch.

  b. Find an equation for the total revenue. This will be all the money they take in for their fund-raiser.

  c. Carefully graph both lines on the same graph. Determine the break-even point, where cost = revenue.

  d. How many tickets do they have to sell to break even?

74. The Women's Charity League wants to hold a charity fund-raiser. They plan on a dinner and dancing. A local church has offered them space for free. The League plans to spend $9.50 per person for dinner. The band they hired is charging $715 for the evening. The League plans to sell tickets to the fund-raiser for $16.00 per person.

  a. Find an equation for the total cost. This will include both the cost of hiring the band and the cost of dinner.

  b. Find an equation for the total revenue. This will include all the money the League takes in for their fund-raiser.

  c. Carefully graph both lines on the same graph. Determine the break-even point, where the cost = revenue.

  d. How many tickets do they have to sell to break even?

For Exercises 75 through 84, determine whether each pair of lines is parallel, perpendicular, or neither. See Section 3.4 to review as needed.

75. $y = -3x + 5$

$y = -3x$

76. $y = \frac{2}{3}x + 7$

$y = \frac{2}{3}x - 1$

77. $y = -7x + 16$

$y = -\frac{1}{7}x - 24$

78. $y = \frac{3}{4}x + 5$

$y = \frac{4}{3}x - 12$

79. $y = -\frac{5}{6}x + 7$

$y = \frac{6}{5}x - 14$

80. $y = -9x + 71$

$y = \frac{1}{9}x + 19$

81. $x + y = 3$

$y = -x - 8$

82. $x - 2y = -4$

$y = \frac{1}{2}x + 9$

83. $2x + 5y = 5$

$6x + 15y = -3$

84. $x + y = -9$

$x = 15 - y$

For Exercises 85 through 90, determine how many solutions there are to each system. You do NOT have to find the solution to the system.

85.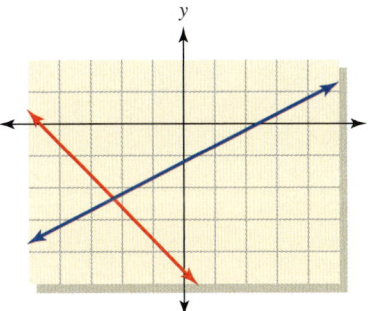

86.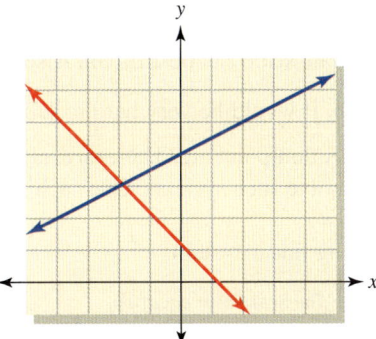

87.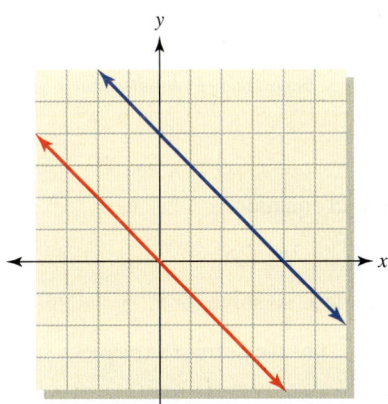

88.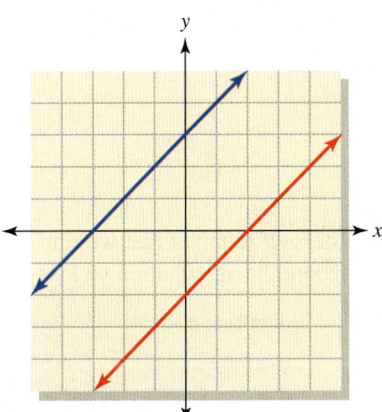

89.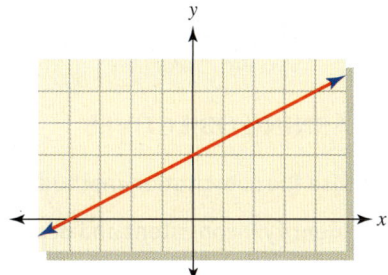

90.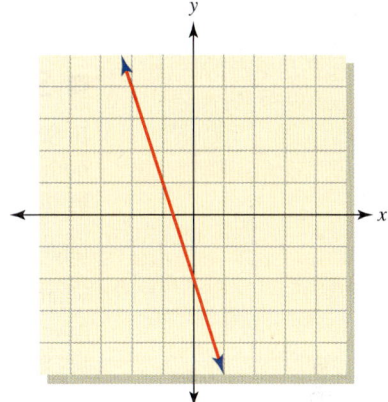

For Exercises 91 through 96, solve the equation for the indicated variable.

91. $5x - 3y = -15$ for y.

92. $-x + 2y = -12$ for y.

93. $\frac{x}{2} - 3y = 7$ for x.

94. $\frac{2x}{3} + y = 4$ for x.

95. $C = 2\pi r$ for r.

96. $A = \frac{1}{2}(b_1 + b_2)h$ for h.

For Exercises 97 through 100, convert units as indicated. Round to two decimal places.

97. Convert 0.5 tablespoons to milliliters. There are about 14.79 milliliters in one U.S. tablespoon.

98. Convert 20.5 milliliters to tablespoons. There are about 14.79 milliliters in one U.S. tablespoon.

99. Convert 4.6 cups to liters. There are about 0.24 liters to one cup.

100. Convert 16.9 liters to cups. There are about 0.24 liters to one cup.

4.2 Solving Systems Using the Substitution Method

LEARNING OBJECTIVES

- Use the substitution method.
- Identify inconsistent and dependent systems.
- Solve practical applications of systems of linear equations.

■ Substitution Method

In Section 4.1, we learned that a system of linear equations can be solved by graphing both lines and determining the point of intersection (the solution). However, graphing by hand can be slow, and finding an accurate scale can be difficult. Also, if the two lines intersected at a point such as $\left(\frac{1}{3}, -\frac{3}{8}\right)$, it would be very hard to tell from the graph what the actual coordinates of the intersection point are. Therefore, algebraic methods have been developed to more efficiently solve such systems. One such method is called the **substitution method.**

Suppose we want to solve the following linear system.

$2x + y = 7$ *The first linear equation.*
$x - y = 5$ *The second linear equation.*

Solve the first equation for the variable y.

$$\begin{aligned} 2x + y &= 7 \\ -2x & -2x \\ \hline y &= 7 - 2x \end{aligned}$$ *Solve the first equation for y.*

Substitute $7 - 2x$ in for y in the second equation.

$$x - (7 - 2x) = 5$$

This equation contains only the variable x, so solve for x.

$x - (7 - 2x) = 5$ *Use the distributive property.*
$x - 7 + 2x = 5$ *Combine like terms.*
$3x - 7 = 5$ *Solve for x.*
$\underline{ +7 \ +7}$
$3x = 12$
$\dfrac{3x}{3} = \dfrac{12}{3}$
$x = 4$

To determine where these two lines intersect requires that we find both the x and y coordinates. Substituting $x = 4$ into the second linear equation yields

$x - y = 5$ *Substitute in x = 4.*
$4 - y = 5$ *Solve for y.*
$\underline{-4 \ -4}$
$-y = 1$
$y = -1$

Therefore, the solution is (4, −1). To check, substitute $x = 4$ and $y = -1$ into both original equations.

Equation 1		Equation 2
$2(4) + (-1) \stackrel{?}{=} 7$	Substitute in $x = 4, y = -1$.	$4 - (-1) \stackrel{?}{=} 5$
$8 + (-1) \stackrel{?}{=} 7$		$4 + 1 \stackrel{?}{=} 5$
$7 = 7$	Both equations are true.	$5 = 5$

To determine where two lines intersect requires that we find a solution that is an ordered pair (x, y). The steps involved in solving a system using the substitution method are as follows.

Steps to Solve a System of Equations Using the Substitution Method

1. Solve one of the two equations for one of the variables. (*Hint:* Whenever possible, solve for a variable with a coefficient of ±1 to avoid introducing fractions.)
2. Substitute the expression from step 1 in for the value of the variable in the other equation. Use parentheses.
3. Simplify the equation in step 2 and solve for the remaining variable.
4. Substitute the answer from step 3 into either of the original equations and solve for the remaining variable.
5. Check the solution by substituting into both original equations.

Skill Connection

Solving for a variable

In step 1, we are told to solve one of the equations for a variable. Review Sections 2.1 and 2.2 on solving literal equations if you need help. In steps 3 and 4, we find *both* values of the variables because two lines intersect at a point (x, y).

Systems of Equations Tools

Substitution Method
Use when a variable is isolated or can be easily isolated.

$$y = 6x + 8$$
$$2x + 4y = 30$$
$$2x + 4(6x + 8) = 30$$

Example 1 Trying out the substitution method

Solve each of the following linear systems using the substitution method. Check the solution.

a. $7x - 2y = 4$
   $y = x + 3$

b. $4x - y = -4$
   $3x - 2y = 12$

SOLUTION

a. **Step 1** Solve one of the two equations for one of the variables.

The second equation, $y = x + 3$, already has y isolated.

Step 2 Substitute the expression from step 1 in for the value of the variable in the other equation. Use parentheses.

Substitute $x + 3$ for y into the first equation.

| $7x - 2y = 4$ | Substitute in $y = x + 3$ for y. |
| $7x - 2(x + 3) = 4$ | Distribute. |

Step 3 Simplify the equation in step 2 and solve for the remaining variable.

$$7x - 2x - 6 = 4 \quad \text{Simplify.}$$
$$5x - 6 = 4$$
$$\underline{+6 \; +6}$$
$$5x = 10 \quad \text{Solve for } x.$$
$$x = 2$$

Step 4 Substitute the answer from step 3 into either of the original equations and solve for the remaining variable.

$$y = x + 3$$
$$y = (2) + 3 \qquad \text{Substitute in } x = 2 \text{ to find } y.$$
$$y = 5$$
$$(x, y) = (2, 5) \qquad \text{This is the solution.}$$

Step 5 Check the solution by substituting into both original equations.

To check, substitute the point $(x, y) = (2, 5)$ into both equations. Check $(x, y) = (2, 5)$ in both equations.

Equation 1	Equation 2
$7x - 2y = 4$	$y = x + 3$
$7(2) - 2(5) \stackrel{?}{=} 4$	$5 \stackrel{?}{=} (2) + 3$
$14 - 10 \stackrel{?}{=} 4$	$5 = 5$
$4 = 4$	

Since both equations are true, the solution checks.

b. To avoid introducing fractions into the problem, solve the first equation for y, since it has a coefficient of -1.

$$4x - y = -4 \qquad \text{Solve the first equation for } y.$$
$$\underline{+y \quad +y}$$
$$4x = -4 + y \qquad \text{Add 4 to both sides.}$$
$$\underline{+4 \quad +4}$$
$$4x + 4 = y$$
$$y = 4x + 4$$

Substituting $4x + 4$ in for y in the second equation results in

$$3x - 2(4x + 4) = 12 \qquad \text{Substitute in } y = 4x + 4.$$
$$3x - 8x - 8 = 12 \qquad \text{Distribute and simplify.}$$
$$-5x - 8 = 12$$
$$\underline{+8 \quad +8} \qquad \text{Add 8 to both sides.}$$
$$-5x = 20$$
$$\frac{-5x}{-5} = \frac{20}{-5} \qquad \text{Divide both sides by } -5.$$
$$x = -4$$

Now find the value for y. Substituting $x = -4$ into the first equation yields

$$4(-4) - y = -4 \qquad \text{Substitute in } x = -4.$$
$$-16 - y = -4$$
$$\underline{+16 \qquad +16} \qquad \text{Add 16 to both sides.}$$
$$-y = 12$$
$$\frac{-y}{-1} = \frac{12}{-1} \qquad \text{Divide both sides by } -1.$$
$$y = -12$$

The solution is the point $(-4, -12)$.
Check $(x, y) = (-4, -12)$ in both equations.

Equation 1	Equation 2
$4x - y = -4$	$3x - 2y = 12$
$4(-4) - (-12) \stackrel{?}{=} -4$	$3(-4) - 2(-12) \stackrel{?}{=} 12$
$-16 + 12 \stackrel{?}{=} -4$	$-12 + 24 \stackrel{?}{=} 12$
$-4 = -4$	$12 = 12$

Since both equations are true, the solution checks.

PRACTICE PROBLEM FOR EXAMPLE 1

Solve each linear system using the substitution method. Check the solutions.

a. $x - 2y = 7$
   $-3x + y = 24$

b. $3x - 6y = 12$
   $-2x + 8y = 10$

Example 2 Using the substitution method to solve the truck rental problem

In Example 1 of Section 4.1, we generated the equations for Vince's Truck Rentals and Gina's Rent-a-Truck. The cost in dollars for renting a truck from Vince's was given by

$$V = 15 + 19.99n$$

where n is the number of days the truck is rented. The cost in dollars for renting a truck from Gina's was given by

$$G = 5 + 24.99n$$

Use the substitution method to solve this system.

SOLUTION

Notice that in both equations, the variable representing cost (V or G) has been isolated. Since we are trying to determine where the cost of renting from Vince's equals the cost of renting from Gina's, we can *set the two costs equal* and solve for n. Setting the two costs equal is the same as setting these two equations equal.

$V = G$	Set V equal to G.
$15 + 19.99n = 5 + 24.99n$	
$ -5 -5$	Subtract 5 from both sides.
$10 + 19.99n = 24.99n$	
$10 + 19.99n = 24.99n$	
$ -19.99n -19.99n$	Subtract 19.99n from both sides.
$10 = 5n$	
$\dfrac{10}{5} = \dfrac{5n}{5}$	Divide both sides by 5.
$n = 2$	

Substituting $n = 2$ into the two equations yields the following:

Vince's Cost	Gina's Cost
$V = 15 + 19.99(2)$	$G = 5 + 24.99(2)$
$V = 54.98$	$G = 54.98$

So the cost for renting a truck from Vince's is the same as the cost of renting a truck from Gina's when a truck is rented for 2 days.

> **What's That Mean?**
>
> **Set the Two Equations Equal**
>
> Suppose we have two equations that are both solved for the output variable y, such as
>
> $$y = -3x + 4$$
>
> and
>
> $$y = 2x - 5$$
>
> Since y is isolated in both equations, we can take the equation
>
> $$y = -3x + 4$$
>
> and substitute in for y the expression $2x - 5$.
> This yields
>
> $$2x - 5 = -3x + 4$$
>
> This is often called *setting the two equations equal*.

PRACTICE PROBLEM FOR EXAMPLE 2

In Practice Problem 7 of Section 4.1, we generated equations for the cost of renting a 10-foot by 10-foot storage space. The cost C in dollars for Help-U-Stor was

$$C = 160m + 10$$

and the cost in dollars for Local Storage Solutions was

$$C = 150m + 30$$

where m is the number of months. Solve this system using the substitution method.

■ Inconsistent and Consistent Systems

When using the substitution method, we typically don't look at the graphs of the two lines in the system. Therefore, we cannot see if the two lines are parallel (no solution) or are the same line (infinitely many solutions). In the next example, we look at how the substitution method behaves when there are no solutions or infinitely many solutions.

Example 3 — Using the substitution method to explore solutions

Solve each of the following systems using the substitution method. If possible, check the solutions.

a. $2x + 5y = 40$

$y = -\dfrac{2}{5}x - 3$

b. $-3x + 4y = 4$

$6x - 8y = -8$

SOLUTION

a. $2x + 5y = 40$

$y = -\dfrac{2}{5}x - 3$

Notice that the second equation has been solved for y. Substitute $-\dfrac{2}{5}x - 3$ in for y in the first equation.

$2x + 5\left(-\dfrac{2}{5}x - 3\right) = 40$ Substitute into the first equation.

$2x + 5\left(-\dfrac{2}{5}x\right) - 15 = 40$ Distribute and simplify.

$2x - 2x - 15 = 40$ The variable terms add to zero.

$-15 \stackrel{?}{=} 40$ This statement is false.

Since this is a false statement ($-15 = 40$), there are *no solutions* to the system. The lines are parallel. This is similar to Section 2.3, in which we saw that some equations have no solutions. To prove that these two lines are parallel, show that their slopes are equal. The second line has slope $m = -\dfrac{2}{5}$, since this line is given in slope-intercept form. To find the slope of the first line, rewrite this equation in slope-intercept form.

SECTION 4.2 Solving Systems Using the Substitution Method   387

$$2x + 5y = 40$$
$$\underline{-2x \qquad\qquad -2x}$$
$$5y = 40 - 2x \qquad \text{Switch the order of the terms.}$$
$$5y = -2x + 40$$
$$\frac{5y}{5} = \frac{-2}{5}x + \frac{40}{5} \qquad \text{Divide both sides by 5.}$$
$$y = -\frac{2}{5}x + 8$$

The slope of both lines is $m = -\frac{2}{5}$. Because the slopes are equal and the y-intercepts are different, the lines are parallel. This is an example of the inconsistent case. The graphs of these two lines are shown on the following graph. They confirm that the lines are indeed parallel and thus never intersect. The lines being parallel confirms that the answer to the system is no solution.

> **Connecting the Concepts**
>
> **How many possibilities are there for the solution?**
>
> In Section 4.1, we learned that there are three possibilities for the number of solutions to a system of two linear equations in two unknowns. The possibilities are a single solution, no solutions (parallel lines), and infinitely many solutions (the same line).

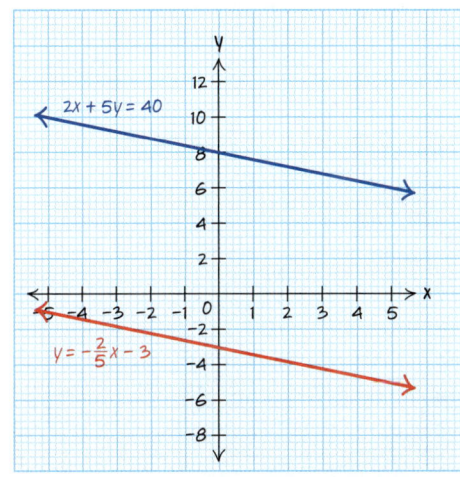

The system has no solutions.

b. Neither equation has a variable with coefficient ± 1, so there is no way to avoid coefficients that are in fraction form. Solving the first equation for y yields

$$-3x + 4y = 4$$
$$\underline{+3x \qquad\qquad +3x} \qquad \text{Add 3x to both sides.}$$
$$4y = 3x + 4$$
$$\frac{4y}{4} = \frac{3x}{4} + \frac{4}{4} \qquad \text{Divide both sides by 4.}$$
$$y = \frac{3}{4}x + 1$$

Substituting $\frac{3}{4}x + 1$ for y into the second equation yields

$$6x - 8\left(\frac{3}{4}x + 1\right) = -8 \qquad \text{Distribute and simplify.}$$
$$6x - 8\left(\frac{3}{4}x\right) - 8 = -8$$
$$6x - 6x - 8 = -8 \qquad \text{The variable terms add to zero.}$$
$$-8 = -8 \qquad \text{This statement is true.}$$

Because this equation is always true ($-8 = -8$), there are infinitely many solutions. These two equations graph as the same line. This is an example of the dependent case.

To prove that these really are the same lines, put both lines in slope-intercept form. The first equation was put into slope-intercept form above, with

$$y = \frac{3}{4}x + 1$$

Rewriting the second line in slope-intercept form yields

$$6x - 8y = -8$$
$$\underline{-6x \qquad\quad -6x} \qquad \text{Subtract 6x from both sides.}$$
$$-8y = -6x - 8$$
$$\frac{-8y}{-8} = \frac{-6x}{-8} - \frac{8}{-8} \qquad \text{Divide both sides by } -8 \text{ to solve for } y.$$
$$y = \frac{3}{4}x + 1$$

The graphs of these two lines are on the following graph. We have confirmed that the lines are the same and intersect at each point on the line.

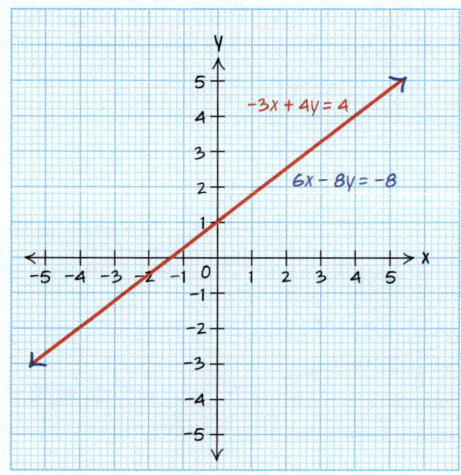

PRACTICE PROBLEM FOR EXAMPLE 3

Solve each of the systems using the substitution method. If possible, check the solutions.

a. $10x - 15y = 30$
$-2x + 3y = -6$

b. $7x + y = 10$
$-14x - 2y = 10$

> **Interpreting Results From the Substitution Method**
> - There is a **unique solution** (x, y) to the system when a value can be determined for both x and y. This is also known as a **consistent** system. Graphically this means the two lines intersect at exactly one point.
> - There are **infinitely many solutions** to the system when all the **variables add to zero** and a true statement remains (for example, $-8 = -8$). This is also known as a **consistent** system, and the lines are **dependent**. Graphically this means the two lines graph as the same line and intersect at all points on that line.
> - There are **no solutions** to the system when all the **variables add to zero** and a **false statement** remains (for example, $0 = -1$). This is also known as an **inconsistent** system. Graphically this means the two lines are parallel: There are no points of intersection.

Practical Applications of Systems of Linear Equations

We have looked at various applications that result in a system of equations. These systems have been solved by using the graphical method and the substitution method. A variety of other applications require that we use a formula before we can set up and solve the linear system. The first formula that we are going to study is the *distance formula*.

> **DISTANCE FORMULA**
>
> $$d = r \cdot t, \quad \text{or} \quad \text{distance} = \text{rate} \cdot \text{time}$$

This formula applies to situations in which the *rate*, which we commonly refer to as the *speed*, is constant. Often, the distance formula shows up in problems involving airplanes, cars, and ships. To set up these problems correctly, it is important to understand the relationship between the rate (speed) of the plane or ship and the rate (speed) of the wind or current. We examine the rate in the next example.

Example 4 Finding the rate

Answer each of the following questions.

a. A small plane can fly at a rate of 100 mph in still air (without any wind). Currently, the wind speed is 15 mph. Find the rate of the plane if the wind is a headwind (the plane is flying into the wind, so the wind is slowing the plane down).

b. A boat can travel at the rate of 9 knots in still water. The rate of the current is 4 knots. Find the rate of the boat if the boat is traveling downstream (with the current, so the current is speeding the boat up).

SOLUTION

a. With a headwind, the wind is slowing the plane down. Therefore, subtract the rate of the wind from the rate of the plane. The rate of the plane with the headwind is $100 - 15 = 85$, or 85 mph.

b. When the boat is traveling downstream, the current is pushing the boat, so its rate is faster. Therefore, add the rate of the current to the rate of the boat. The rate of the boat traveling downstream is $9 + 4 = 13$, or 13 knots.

PRACTICE PROBLEM FOR EXAMPLE 4

Answer each of the following questions.

a. A passenger jet can fly at a rate of 480 mph in still air. Currently the wind speed is 25 mph. Find the rate of the plane if the wind is a tailwind (the plane is flying with the wind, so the wind is speeding the plane up).

b. A ship can travel at the rate of 15 mph in still water. The rate of the current is 4 mph. Find the rate of the ship if the ship is traveling upstream (against the current, so the current is slowing the ship down).

Interpreting the Rate Correctly

- Let s represent the rate of the plane in still air and let w represent the rate of the wind. When a plane is flying with a **headwind** (against the wind, or into the wind), its rate is $s - w$. When a plane is flying with a **tailwind** (with the wind), its rate is $s + w$.

- Let b represent the rate of the ship (boat) in still water and c represent the rate of the current. When a boat is traveling **downstream** (with the current), the rate of the ship (boat) is $b + c$. When a boat is traveling **upstream** (against the current), the rate of the ship (boat) is $b - c$.

Now we apply these ideas to some problems.

Example 5 — Rate of a plane

A small plane can fly 150 miles with a headwind in 2.5 hours. It makes the return trip (with a tailwind) in 1.5 hours. Find the rate of the plane in still air and the rate of the wind.

SOLUTION

First write two equations to describe what is happening in this problem. The two situations are the plane flying against a headwind and the plane flying with a tailwind.

Let s represent the rate of the plane in still air, in miles per hour, and let w represent the rate of the wind in miles per hour.

Headwind: A headwind slows the plane down, so the plane's rate with a headwind is $s - w$. Using the formula $d = r \cdot t$ with the given distance $d = 150$ and time $t = 2.5$ yields

$$150 = (s - w) \cdot 2.5$$

Tailwind: A tailwind speeds the plane up, so the plane's rate with a tailwind is $s + w$. Using the formula $d = r \cdot t$, with the given distance $d = 150$ and time $t = 1.5$ yields

$$150 = (s + w) \cdot 1.5$$

These two equations form a system of equations that we will solve using the substitution method.

$$150 = (s - w) \cdot 2.5$$
$$150 = (s + w) \cdot 1.5$$

Solving the first equation for s may be done by first dividing both sides by 2.5.

$$150 = (s - w) \cdot 2.5$$
$$\frac{150}{2.5} = \frac{(s - w) \cdot 2.5}{2.5} \quad \text{Divide both sides by 2.5.}$$
$$60 = s - w$$
$$\underline{+w \qquad +w} \quad \text{Add } w \text{ to both sides to solve for } s.$$
$$60 + w = s$$

or

$$s = 60 + w$$

Connecting the Concepts

How fast is reasonable?

Interpreting whether a solution is *reasonable* to a plane or ship (boat) problem can be difficult if you are not a pilot or a navigator. Small planes typically fly from 75 to 150 miles per hour. Commercial jets often fly at speeds near 500 mph. Usually only military jets fly at speeds greater than 600 mph. Reasonable wind speeds are 5 to 74 miles per hour. Wind speeds of over 74 miles per hour are hurricane-force winds. Only the most violent tornadoes will have wind speeds of over 200 mph. Boats and ships do not, in general, travel very fast, as the drag of the water is hard to overcome. Boats and ships typically travel at 10 to 25 knots (nautical miles per hour). In these problems, current rates are from 3 to 20 knots. Here a *boat* refers to a craft of less than 200 feet, and a *ship* refers to a craft over 200 feet.

Substitute $60 + w$ into the s in the second equation to get

$$150 = [(60 + w) + w] \cdot 1.5$$
$$\frac{150}{1.5} = \frac{[(60 + w) + w] \cdot 1.5}{1.5} \quad \text{Divide both sides by 1.5.}$$
$$100 = 60 + 2w$$
$$\underline{-60 \quad -60} \quad \text{Subtract 60 from both sides.}$$
$$40 = 2w$$
$$\frac{40}{2} = \frac{2w}{2} \quad \text{Divide both sides by 2 to isolate } w.$$
$$w = 20$$

Substituting $w = 20$ into the first equation yields

$$150 = (s - 20) \cdot 2.5$$
$$\frac{150}{2.5} = \frac{(s - 20) \cdot 2.5}{2.5} \quad \text{Divide both sides by 2.5.}$$
$$60 = s - 20$$
$$\underline{+20 \quad +20} \quad \text{Add 20 to both sides to solve for } s.$$
$$80 = s$$

Check the solution.
Substituting $s = 80$ and $w = 20$ into both equations yields

$150 \stackrel{?}{=} (80 - 20) \cdot 2.5$ Substitute in $s = 80$ and $w = 20$. $150 \stackrel{?}{=} (80 + 20) \cdot 1.5$
$150 \stackrel{?}{=} 60 \cdot 2.5$ Simplify. $150 \stackrel{?}{=} 100 \cdot 1.5$
$150 = 150$ Both equations are true. $150 = 150$

The rate of the plane in still air is 80 mph, and the rate of the wind is 20 mph.

PRACTICE PROBLEM FOR EXAMPLE 5
A paddleboat travels 20 miles downstream in 6 hours. It takes 8 hours to make the return trip upstream. Find the rate of the paddleboat in still water and the rate of the current.

Another type of problem that we study involves integers. Recall that the integers are the whole numbers and their opposites, which is the set $\{\ldots, -3, -2, -1, 0, 1, 2, 3, \ldots\}$.

Example 6   Finding the numbers

The sum of two numbers is 100. One number is 8 less than 3 times the other number. Find the two numbers.

SOLUTION
The two unknowns are two numbers. Let x represent one number and let y represent the other number. The first sentence translates as

$$x + y = 100$$

In Section 1.2, we learned not only that *less than* means to subtract but also that we subtract in the opposite order in which the sentence is written. Thus, the second sentence translates as

$$x = 3y - 8$$

The system of equations is

$$x + y = 100$$
$$x = 3y - 8$$

Since the second equation has already been solved for x, substitute $3y - 8$ directly into the first equation.

$(3y - 8) + y = 100$ Substitute $3y - 8$ in for x in the first equation.

$4y - 8 = 100$ Add 8 to both sides.

$ +8 +8$

$4y = 108$

$\dfrac{4y}{4} = \dfrac{108}{4}$ Divide both sides by 4 to solve for y.

$y = 27$

The first integer is 27. To find the second integer, substitute 27 for y in one of the equations.

$x = 3(27) - 8$

$x = 73$ Substitute $y = 27$ into the second equation and evaluate.

To check the work, substitute these values into both equations.

Equation 1		Equation 2
$73 + 27 \stackrel{?}{=} 100$	Substitute in $x = 73$ and $y = 27$.	$73 \stackrel{?}{=} 3(27) - 8$
$100 = 100$	Both equations are true.	$73 = 73$

The solution checks in both equations.

PRACTICE PROBLEM FOR EXAMPLE 6

The difference between a first integer and a second integer is 54. Three times the first integer is equal to the sum of the second integer and 6. Find the two integers.

Another application that gives rise to systems of linear equations comes from mixture problems. These problems often include computing the **value** of a quantity. For example, if apples are selling at $0.75 per pound and we purchase 3 pounds, the value (worth) is $3 \cdot 0.75 = 2.25$, or $2.25. The amount purchased is 3 pounds, and we say that the unit price (or price per pound) is $0.75.

> **VALUE FORMULA**
>
> The value of a quantity is the monetary worth of that quantity. To compute the value, we use the formula
>
> $$\text{value} = \text{amount} \cdot \text{unit price}$$

When two different quantities are mixed or blended, such as a mix of almonds and peanuts, the *value of the mix equals the sum of the values of the individual quantities.* The unit price of the mix is a number in between the unit price of each quantity.

Example 7   Unit price

Suppose a store has 20 pounds of almonds priced at $5.00 per pound (the unit price). The store also has 30 pounds of peanuts that cost $2.50 per pound (the unit price). The store wants to mix together the almonds and the peanuts.

a. Find the value of the almonds and the value of the peanuts.

b. Find the unit price of the mix.

SOLUTION

a. The value of the almonds is $20 \cdot \$5.00 = \$100.00$. The value of the peanuts is $30 \cdot \$2.50 = \$75.00$.

b. The value of the almonds plus the value of the peanuts equals the value of the mix. Since the unit price of the mix is unknown, we let n represent the unit price of the mix in dollars per pound. The mix contains 50 pounds of nuts.

$$20 \cdot \$5.00 + 30 \cdot \$2.50 = 50 \cdot n$$

Solving this equation for n yields

$$100 + 75 = 50n$$
$$175 = 50n$$
$$\frac{175}{50} = \frac{50n}{50}$$
$$3.50 = n$$

Therefore, the unit price of the mix is $3.50 per pound. Notice that this price is a number in between the unit price of almonds ($5.00 per pound) and the unit price of peanuts ($2.50 per pound).

PRACTICE PROBLEM FOR EXAMPLE 7

Suppose you have 15 pounds of shelled walnuts priced at $7.49 per pound (the unit price). You also have 20 pounds of hazelnuts that cost $8.49 per pound (the unit price). You want to mix together the almonds and the hazelnuts.

a. Find the value of the walnuts and the value of the hazelnuts.

b. Find the unit price of the mix.

Example 8 Coffee mix

A local coffee shop is making 30 pounds of its house blend of coffee by mixing two types of coffee. The first type is Colombian, and it costs $14.95 per pound. The second type is Guatemalan, and it costs $16.95 per pound. The blend is priced at $15.50 per pound.

Find how many pounds of each type of coffee the barista should use in the shop's blend.

SOLUTION

Let C represent the pounds of the Colombian coffee used and let G represent the pounds of the Guatemalan coffee used. Since the shop is making 30 pounds of the blend, we have that

$$C + G = 30$$

Compute the value of each individual type of coffee and of the blend. The value of the Colombian coffee is

$$C \cdot 14.95 = 14.95C$$

The value of the Guatemalan coffee is

$$G \cdot 16.95 = 16.95G$$

The value of the blend is the 30 pounds of mix times the unit price for the mix, $15.50.

$$30 \cdot 15.50 = 465$$

The sum of the individual values is equal to the value of the blend, so the second equation is

$$14.95C + 16.95G = 465$$

394 CHAPTER 4 Systems of Linear Equations

Therefore, the two equations in two unknowns are

$$C + G = 30$$
$$14.95C + 16.95G = 465$$

Solving the first equation for C yields

$$C + G = 30 \qquad \text{Subtract } G \text{ from both sides.}$$
$$\underline{-G \quad -G}$$
$$C = 30 - G$$

Substituting this expression into the second equation gives the following equation to solve for G.

$$14.95(30 - G) + 16.95G = 465 \qquad \text{Substitute in } C = 30 - G.$$
$$448.50 - 14.95G + 16.95G = 465 \qquad \text{Simplify.}$$
$$448.50 + 2.00G = 465$$
$$\underline{-448.50 \qquad\qquad -448.50} \qquad \text{Subtract 448.50 from both sides.}$$
$$2.00G = 16.50$$
$$\frac{2.00G}{2.00} = \frac{16.50}{2.00} \qquad \text{Divide both sides by 2.00 to isolate } G.$$
$$G = 8.25$$

The coffee shop needs to use 8.25 pounds of the Guatemalan coffee. To find the amount of Colombian coffee, substitute $G = 8.25$ into the first equation.

$$C + 8.25 = 30 \qquad \text{Substitute into the first equation to find } C.$$
$$\underline{-8.25 \quad -8.25}$$
$$C = 21.75$$

The coffee shop has to use 21.75 pounds of the Colombian coffee and 8.25 pounds of the Guatemalan coffee to make the 30 pounds of the house blend.

PRACTICE PROBLEM FOR EXAMPLE 8

Organic Foods is making a custom nut blend. They are making a 50-pound mix of hazelnuts and almonds. Hazelnuts are priced at $5.65 a pound, and almonds are priced at $7.95 a pound. The mix will sell for $6.50 a pound. Find the amount of hazelnuts and almonds that Organic Foods should include.

4.2 Vocabulary Exercises

1. The substitution method is used to algebraically _____ a system of linear equations.
2. The _____ formula is $d = r \cdot t$.
3. The _____ of a quantity is the amount multiplied by the unit price.
4. When flying with a(n) _____, the plane speed will be reduced.
5. When flying with a(n) _____, the plane speed will be increased.

4.2 Exercises

For Exercises 1 through 6, solve each equation for the requested variable.

1. Solve $3x + y = -4$ for y.
2. Solve $-5x + y = 7$ for y.
3. Solve $x + 4y = -3$ for x.
4. Solve $x - 3y = -16$ for x.
5. Solve $5x - 10y = 30$ for y.
6. Solve $-3x + 4y = 12$ for x.

SECTION 4.2 Solving Systems Using the Substitution Method

For Exercises 7 through 12, select the equation in each set containing a variable with coefficient 1 or −1. Then solve that equation for that variable. Do not solve the system. See Example 2 if you need help getting started.

7. $5x - 6y = 9$
   $x + 3y = 7$

8. $2x - y = 17$
   $5x - 2y = -4$

9. $-3a + 12b = 24$
   $-a + 9b = 6$

10. $\frac{2}{3}a - b = 4$
    $-\frac{1}{2}a + 3b = 5$

11. $-6x + 4y = 7$
    $y = -2x + 1$

12. $y = -5x + 14$
    $3x + 5y = -2$

For Exercises 13 through 22, solve the systems of equations using the substitution method. Provide reasons for each step. Check the solution in both of the original equations.

13. $x = 3y - 1$
    $2x + y = 12$

14. $y = 3x - 5$
    $2x + y = 4$

15. $x = -5y + 1$
    $-3x + 2y = 31$

16. $y = -0.5x + 6$
    $2x - y = 8$

17. $2x + y = 10$
    $-6x + y = -5$

18. $x + 3y = 3$
    $3x - y = 6$

19. $y = -6x + 5$
    $y = 4x - 15$

20. $y = \frac{5}{3}x + 4$
    $y = -\frac{2}{3}x - 3$

21. $x = y$
    $2x + y = -6$

22. $y = x$
    $y = -3x + 20$

For Exercises 23 through 36, solve each system using the substitution method.

23. The sum of two numbers is 65. One number is 12 times the other number. Find the two numbers.
    a. Define the variables.
    b. Determine the two equations.
    c. Solve the system.

24. A first number minus a second number is −9. The first number is four times the second number. Find the two numbers.
    a. Define the variables.
    b. Determine the two equations.
    c. Solve the system.

25. The sum of two numbers is 25. One number is four times the other. Find the two numbers.

26. The sum of two numbers is 48. One number is three times the other. Find the two numbers.

27. The sum of two numbers is 12. One number is five times the other. Find the two numbers.

28. The sum of two numbers is 20. One number is nine times the other. Find the two numbers.

29. A first number minus a second number is −7. Four times the first number added to 1 is equal to the second number. Find the two numbers.

30. The sum of two numbers is 22. The first number is equal to twice the second number subtracted from 9. Find the two numbers.

31. The sum of two numbers is 12. One number is equal to the quotient of the other number and 2. Find the two numbers.

32. Twice the first number added to 6 is equal to the second number. The difference between the first number and the second number is 17. Find the two numbers.

33. Ethan is considering two part-time jobs in sales. One job, at a mobile-phone cart, offers him $100 per week plus a 5% commission on his weekly sales. The other job, at a tire store, offers him a weekly base pay of $120 plus a 4% commission on his weekly sales.
    a. Define the variables.
    b. Determine the equations for Ethan's weekly pay for each of the two jobs.
    c. Find the solution to the system. Interpret the solution in terms of Ethan's weekly pay.

34. Maya is considering two part-time jobs in clothing sales. One job, at Value Fashion, offers her $250 per week plus a 3% commission on her weekly sales. The other job, at Amy's Boutique, offers her a weekly base pay of $150 plus a 5% commission on her weekly sales.
    a. Define the variables.
    b. Determine the equations for Maya's weekly pay for each of the two jobs.
    c. Find the solution to the system. Interpret the solution in terms of Maya's weekly pay.

35. Marcus is working two jobs. One job pays him $10.25 an hour, and the other job pays him $11.50 an hour. The total hours he can work each week at both jobs is 20. Marcus would like to earn $220.00 weekly from both jobs.
    a. Define the variables.
    b. Determine the system of equations described in the problem.
    c. Find the solution to the system. Interpret the solution in terms of Marcus's salary and jobs.

36. Amir is working two jobs. The first job pays him $10.00 an hour, and the second job pays him $11.75 an hour. The total number of hours Amir works each week at both jobs is 22. He would like to earn $241.00 a week working both jobs.

   a. Define the variables.

   b. Determine the system of equations described in the problem.

   c. Find the solution to the system. Interpret the solution in terms of Amir's salary and jobs.

For Exercises 37 through 64, solve each system of equations using the substitution method. Check the solutions in both of the original equations. If a system is inconsistent, answer "No solution." If a set of lines are dependent, answer "Infinitely many solutions."

37. $y = 2x - 1$
    $y = -4x + 5$

38. $y = -x + 9$
    $y = 4x + 6$

39. $y = 4x + 5$
    $y = -\frac{1}{2}x - 4$

40. $y = -\frac{4}{5}x - 2$
    $y = \frac{1}{5}x - 7$

41. $-x + 3y = 1$
    $2x + y = 5$

42. $3x - y = -4$
    $x + 2y = -6$

43. $x + 2y = 3$
    $2x - y = 4$

44. $4x + y = -2$
    $x - 2y = -5$

45. $3x + 2y = 6$
    $y = 2.5x - 4$

46. $-x + 6y = 4$
    $y = -0.5x + 2$

47. $x + 5y = 6$
    $y = 1$

48. $y = 2$
    $3x + 2y = -9$

49. $5x + 2y = -4$
    $-x + y = 5$

50. $x - 2y = 0$
    $-5x + 8y = 8$

51. $-3x + 4y = 12$
    $-\frac{3}{4}x + y = -1$

52. $-8x - 2y = 1$
    $4x + y = 2$

53. $5x - 10y = 15$
    $x = 2y + 3$

54. $-\frac{1}{4}x + 5y = \frac{1}{2}$
    $x - 20y = -2$

55. $x + 3y = 18$
    $2x + 6y = -12$

56. $x - y = 9$
    $y = x - 2$

57. $-x + 3y = 6$
    $3x + y = -5$

58. $\frac{3}{7}x - 7y = 35$
    $x = -14$

59. $x + 1 = -9$
    $y - 3 = -3$

60. $x - 5 = -12$
    $y + 6 = 3.5$

61. $y = 7x + 1$
    $14x - 2y = -2$

62. $4x + y = 5$
    $y = -4x + 5$

63. $x - \frac{1}{4}y = 9$
    $2x + y = 1$

64. $6x + 3y = 5$
    $y = \frac{3}{2}x - \frac{4}{3}$

For Exercises 65 through 78, answer each question.

65. The rate of a plane in still air is 120 mph, and the rate of the wind is 15 mph. Find the rate of the plane if the plane is flying into a headwind.

66. The rate of a plane in still air is 550 mph, and the rate of the wind is 25 mph. Find the rate of the plane if the plane is flying with a tailwind.

67. The rate of a plane in still air is 100 miles per hour, and the rate of the wind is 5 mph. Find the rate of the plane if the plane is flying with a tailwind.

68. The rate of a plane in still air is 340 miles per hour, and the rate of the wind is 25 miles per hour. Find the rate of the plane if the plane is flying into a headwind.

69. A boat travels at 7 mph in still water, and the current is flowing at a rate of 1.5 mph. Find the rate of the boat if it is traveling downstream.

70. A boat travels at 11 mph in still water, and the current has a rate of 3.5 mph. Find the rate of the boat if it is traveling downstream.

71. A boat travels at 9 mph in still water, and the current flows at a rate of 2 mph. Find the rate of the boat if it is traveling upstream.

72. A boat travels at 12 mph in still water, and the current has a rate of 1.5 mph. Find the rate of the boat if it is traveling upstream.

73. Oolong tea sells for $57.95 per pound.

   a. Find the value of 5.5 pounds of tea.

   b. Find the value of t pounds of tea.

74. Cashews sell for $19.30 per pound.

   a. Find the value of 26 pounds of cashews.

   b. Find the value of c pounds of cashews.

75. French Roast coffee sells for $16.95 per pound.

   a. Find the value of 26 pounds of coffee.

   b. Find the value of c pounds of coffee.

76. Dried cherries sell for $11.99 per pound.

   a. Find the value of 2.5 pounds of dried cherries.

   b. Find the value of b pounds of dried cherries.

77. The total value of 50 pounds of coffee is $797.50. What is the unit price? Recall that the unit price is the cost per pound.

78. The total value of 5 pounds of Earl Grey tea is $199.75. What is the unit cost? Recall that the unit cost is the cost per pound.

For Exercises 79 through 90, solve each system using the substitution method.

79. A boat travels 50 miles downstream in 2.5 hours. It takes 5 hours to make the return trip upstream. Find the rate of the boat in still water and the rate of the current.

   a. Define the variables.

   b. Determine the two equations.

   c. Solve the system.

80. A plane can fly 600 miles in 4 hours with a tailwind. It makes the return trip (with a headwind) in 6 hours. Find the rate of the plane in still air and the rate of the wind.

   a. Define the variables.

   b. Determine the two equations.

   c. Solve the system.

81. A plane can fly 1275 miles in 3.4 hours with a headwind. It makes the return trip (with a tailwind) in 3 hours. Find the rate of the plane in still air and the rate of the wind.

   a. Define the variables.

   b. Determine the two equations.

   c. Solve the system.

82. A boat travels 30 miles downstream in 3 hours. It takes 5 hours to make the return trip upstream. Find the rate of the boat in still water and the rate of the current.

   a. Define the variables.

   b. Determine the two equations.

   c. Solve the system.

83. Julie wants to make a 50-pound mix of tea by blending two types of tea. The first type is Ceylon, which costs $23.00 per pound, and the second type is black currant tea, which costs $27.00 per pound. Julie wants to price the blend at $25.00 per pound. How many pounds of each type of tea does Julie need to use in her blend?

   a. Define the variables.

   b. Determine the two equations.

   c. Solve the system.

84. Michelle wants to make 10 gallons of punch for sale at her café. She is going to mix sparkling water and lemonade. The sparkling water costs $4.50 a gallon, and the lemonade costs $2.50 a gallon. Michelle would like to price the punch at $3.70 a gallon. Find how many gallons of sparkling water and lemonade Michelle should buy.

   a. Define the variables.

   b. Determine the two equations.

   c. Solve the system.

85. Graciela is making a 25-pound mix of dried papaya and crystallized ginger for her health food store. The dried papaya costs $3.80 per pound, and the crystallized ginger costs $9.50 per pound. Graciela would like to price the mix at $7.22 per pound. Find how many pounds of dried papaya and crystallized ginger Graciela has to put in the mix.

   a. Define the variables.

   b. Determine the two equations.

   c. Solve the system.

86. Andy is making a 55-pound mix of Costa Rican coffee and Colombian coffee for his cafe. The Costa Rican coffee costs $10.95 per pound, and the Colombian coffee costs $12.95 per pound. Andy would like to price the blend at $12.50 per pound. Find how many pounds of each type of coffee Andy should put in the mix.

   a. Define the variables.

   b. Determine the two equations.

   c. Solve the system.

87. Rita's Auto Detail has the following weekly revenue and cost equations for her business:

$$R = 150n$$
$$C = 100n + 1400$$

where R represents the weekly revenue in dollars and C represents the weekly costs in dollars when she details n cars a week. Find the break-even point for Rita's Auto Detail.

88. Walz Caps has the following weekly revenue and cost equations from selling cycling caps:

$$R = 25h$$
$$C = 9h + 1200$$

where R represents the weekly revenue in dollars and C represents the weekly costs in dollars when the company sells h cycling caps a week. Find the break-even point for Walz Caps.

89. Mobile Style has the following monthly revenue and cost equations from selling mobile phone accessories at a mall kiosk:

$$R = 24.50n$$
$$C = 8.50n + 4300$$

where R represents the monthly revenue in dollars and C represents the monthly costs in dollars when Mobile Style sells n accessories. Find the break-even point for Mobile Style.

90. Tracy's Old Time Foods sells organic canned goods at renaissance fairs. Tracy has the following revenue and cost equations for her business:

$$R = 8n$$
$$C = 4.5n + 680$$

where R represents the revenue in dollars and C represents the costs in dollars when Tracy sells n jars of canned goods at a local fair. Find the break-even point for Tracy's Old Time Foods.

For Exercises 91 through 94, answer the following questions.

91. Consider the system of equations:

$$-3x + 5y = 15$$
$$y = \frac{3}{5}x + 3$$

a. Solve this system using the substitution method. How many solutions to this system are there?

b. Graph the two lines on the same axes.

c. Two points on this line are (0, 3) and (5, 6). Are these points solutions to this system?

d. Two points that are not on this line are (0, 0) and (1, 0). Are these points solutions to this system?

e. If there are infinitely many solutions, are the solutions points on the line or points not on the line?

92. Consider the system of equations:

$$4x - 3y = 24$$
$$y = \frac{4}{3}x - 8$$

a. Solve this system using the substitution method. How many solutions to this system are there?

b. Graph the two lines on the same axes.

c. Two points on this line are (0, −8) and (12, 8). Are these points solutions to this system?

d. Two points that are not on this line are (0, 0) and (2, 4). Are these points solutions to this system?

e. If there are infinitely many solutions, are the solutions points on the line or points not on the line?

93. Consider the following system of equations:

$$y = -6x + 8$$
$$18x + 3y = -21$$

a. Solve this system using the substitution method. How many solutions to this system are there?

b. Graph the two lines on the same axes.

c. Two points on the first line, $y = -6x + 8$, are (0, 8) and (1, 2). Are these points solutions to this system?

d. Two points on the second line, $18x + 3y = -21$, are (0, −7) and (1, −13). Are these points solutions to this system?

e. Explain why there are no solutions to this system.

94. Consider the following system of equations:

$$y = -7x + 4$$
$$21x + 3y = -6$$

a. Solve this system using the substitution method. How many solutions to this system are there?

b. Graph the two lines on the same axes.

c. Two points on the first line, $y = -7x + 4$ are (0, 4) and (1, −3). Are these points solutions to this system?

d. Two points on the second line, $21x + 3y = -6$, are (0, −2) and (−1, 5). Are these points solutions to this system?

e. Explain why there are no solutions to this system.

For Exercises 95 through 98, the students were asked to solve the given system using the substitution method. Read over the student's work and find the mistake.

95. $y = 4x - 8$
$2x - y = 6$

Matthew

$y = 4x - 8$
$2x - y = 6$
$2x - (4x - 8) = 6$
$2x - 4x - 8 = 6$
$-2x - 8 = 6$
$\quad\quad +8 \quad +8$
$\overline{-2x = 14}$
$x = -7$
$y = 4(-7) - 8 = -28 - 8 = -36$
The solution is (−7, −36).

96. $x = -3y + 4$
$-x + y = 5$

> **Maya**
> $x = -3y + 4$
> $-x + y = 5$
> $-(-3y + 4) + y = 5$
> $3y - 4 + y = 5$
> $-2y + 4 = 5$
> $ -4 -4$
> $\overline{-2y = 1}$
> $y = -\dfrac{1}{2}$
> $x = -3y + 4$
> $x = -3\left(-\dfrac{1}{2}\right) + 4$
> $x = \dfrac{3}{2} + 4 = \dfrac{3}{2} + \dfrac{8}{2} = \dfrac{11}{2}$
> The solution is $\left(\dfrac{11}{2}, -\dfrac{1}{2}\right)$.

97. $3x + 4y = 11$
$-5x + 2y = -1$

> **Zachary**
> $3x + 4y = 11$
> $-5x + 2y = -1$
> $3x = -4y + 11$
> $-5x + 2y = -1$
> $-5(-4y + 11) + 2y = -1$
> $20y - 55 + 2y = -1$
> $22y = 54$
> $y = \dfrac{54}{22} = \dfrac{27}{11}$
> $3x = -4\left(\dfrac{27}{11}\right) + 11 = \dfrac{13}{11}$
> The solution is $\left(\dfrac{13}{33}, \dfrac{27}{11}\right)$.

98. $-2x + 5y = 18$
$-6x - 9y = 6$

> **Sophia**
> $-2x + 5y = 18$
> $-6x - 9y = 6$
> $5y = 2x + 18$
> $-6x - 9(2x + 18) = 6$
> $-6x - 18x - 162 = 6$
> $-24x - 162 = 6$
> $-24x = 168$
> $x = -7$
> $5y = 2(-7) + 18 = -14 + 18 = 4$
> The solution is $\left(-7, \dfrac{4}{5}\right)$.

For Exercises 99 through 104, answer each of the following questions.

99. If solving a system of equations using the substitution method results in $0 = 4$, is the system said to be consistent or inconsistent?

100. If solving a system of equations using the substitution method results in $-1 = -1$, is the system said to be consistent or inconsistent?

101. If solving a system of equations by graphing results in two parallel lines, is the system said to be consistent or inconsistent?

102. If solving a system of equations by graphing results in one line, is the system said to be consistent or inconsistent?

103. If solving a system of equations by graphing results in two intersecting lines, is the system said to be consistent or inconsistent?

104. If solving a system of equations using the substitution method results in the solution (2, 3), is the system said to be consistent or inconsistent?

4.3 Solving Systems Using the Elimination Method

LEARNING OBJECTIVES

- Use the elimination method.
- Solve practical applications of systems of linear equations.
- Decide whether to use substitution or elimination.

Using the Elimination Method

There is another algebraic method that is used to solve systems of linear equations. It is called the **elimination** method. This method is the one that computers use to solve linear systems because it is easy to code and is very efficient.

Suppose we want to solve the following linear system:

$$x - y = 10$$
$$x + y = 22$$

Because the two sides of the first equation are equal, we can use the idea of a scale to show that the two sides are in balance.

What's That Mean?

Elimination, Addition, or Gaussian Elimination

The *elimination* method goes by various names. Sometimes it is called the *addition* method because the two equations are added together. Sometimes it is called the *Gaussian elimination* method. Karl Gauss was a very influential 19th-century mathematician who developed this method to solve a system.

Recall the addition property of equality, which states that we may add the same quantity to both sides of an equation. For example, we could add 22 to both sides of the first equation.

$$x - y = 10$$
$$\underline{+22 \quad\quad +22}$$
$$x - y + 22 = 32$$

Adding 22 did not help us to solve the system. However, the second equation tells us that 22 is the same as $x + y$. Therefore, we can add $x + y$ on the left side and 22 on the right side of the first equation.

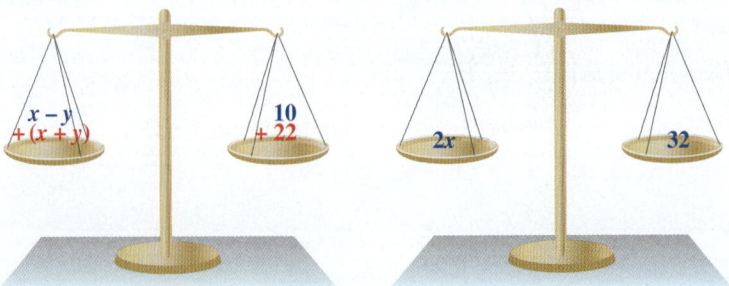

Algebraically we follow the same steps. Add the two equations vertically. The second equation, $x + y = 22$, states that $x + y$ is the same amount as 22 because the two sides are equal. Therefore, when the second equation is added to the first equation, we are adding the same quantity to both sides.

$$x - y = 10$$
$$x + y = 22$$
$$2x = 32 \quad \text{The variable } y \text{ has been eliminated. Solve for } x.$$
$$\frac{2x}{2} = \frac{32}{2}$$
$$x = 16$$

> **Connecting the Concepts**
>
> **What is the addition property of equality?**
>
> Recall from Chapter 2 that we studied the addition property of equality, which states that the same value can be added to both sides of an equation. This is the property we are using when we add two equations vertically.

Now that we know that $x = 16$, substitute this result into either one of the original equations to solve for y.

$$16 + y = 22 \quad \text{Substitute in } x = 16.$$
$$16 + y = 22$$
$$\underline{-16 \quad -16} \quad \text{Subtract 16 from both sides to solve for } y.$$
$$y = 6$$

The solution to the system is the point $(16, 6)$. To check our solution, we substitute the values $x = 16$ and $y = 6$ into both original equations.

Equation 1
$16 - 6 \stackrel{?}{=} 10$ Substitute $x = 16$, $y = 6$ into both equations.
$10 = 10$ Both statements are true. The solution checks.

Equation 2
$16 + 6 \stackrel{?}{=} 22$
$22 = 22$

The whole key to the elimination method is to have the coefficients on one variable (in both equations) that are opposites. This means the coefficients have the same absolute value but different signs. Then when we add the two equations vertically, one variable sums to 0 and is "eliminated."

> **Connecting the Concepts**
>
> **What is an opposite?**
>
> Recall from Chapter R that opposites are two numbers that are the same in absolute value but differ in sign. For example, 6 and -6 are opposites.

> **Steps to Solve a System of Equations Using the Elimination Method**
>
> 1. Put the equations of both lines in general form ($Ax + By = C$), if necessary.
> 2. If necessary, multiply both sides of one (or both) equations so that the coefficients of one of the variables are opposites.
> 3. Add the two equations vertically. At this point in the process, one of the variables should be eliminated.
> 4. Solve the equation for the remaining variable.
> 5. Substitute the value of the known variable from step 4 into either of the original equations and solve for the other variable. Remember that we are trying to find where two lines intersect, so we must determine the values of *both* variables.
> 6. Check the solution by substituting into *both* original equations.

> **Systems of Equations Tools**
>
> **Substitution Method**
>
> Use when a variable is isolated or can be easily isolated.
>
> $$y = 6x + 8$$
> $$2x + 4y = 30$$
> $$2x + 4(6x + 8) = 30$$
>
> **Elimination Method**
>
> Use when the equations are in general form.
>
> $$5x - 7y = 42$$
> $$2x + 4y = 30$$
> $$4(5x - 7y) = 4(42)$$
> $$7(2x + 4y) = 7(30)$$
> $$20x - 28y = 168$$
> $$\underline{14x + 28y = 210}$$
> $$34x = 378$$

Example 1   **Putting the elimination method to work**

Solve the system using the elimination method. Check the solution.

$$-2x + y = 16$$
$$2x + 3y = 8$$

SOLUTION

Step 1 Put the equations of both lines in general form, if necessary.

The given equations are both already in general form.

Step 2 If necessary, multiply both sides of one (or both) equations so that the coefficients of one of the variables are opposites.

The coefficients of x are -2 and 2; that is, the coefficients are opposites.

Step 3 Add the two equations vertically. At this point in the process, one of the variables should be eliminated.

Add the two equations vertically by adding like terms. The x-variable should be eliminated after this step.

$$-2x + y = 16$$
$$2x + 3y = 8$$
$$\overline{4y = 24}$$

Add the two equations vertically.

Step 4 Solve the equation for the remaining variable.

$$\frac{4y}{4} = \frac{24}{4}$$ Divide both sides by 4 to solve for y.
$$y = 6$$

Step 5 Substitute the answer from step 4 into either of the original equations and solve for the other variable.

Now that we know that $y = 6$, substitute this value into either of the original equations to find the value of x. Substituting into the second equation yields

$$2x + 3(6) = 8 \quad \text{Substitute in } y = 6.$$
$$2x + 18 = 8 \quad \text{Subtract 18 from both sides.}$$
$$\underline{-18 \; -18}$$
$$2x = -10$$
$$\frac{2x}{2} = \frac{-10}{2} \quad \text{Divide both sides by 2.}$$
$$x = -5$$

Step 6 Check the solution by substituting into both original equations.

Check the solution $(-5, 6)$ by substituting $x = -5$, $y = 6$ into both of the original equations.

Equation 1	Equation 2
$-2(-5) + 6 \stackrel{?}{=} 16$	$2(-5) + 3(6) \stackrel{?}{=} 8$
$10 + 6 \stackrel{?}{=} 16$	$-10 + 18 \stackrel{?}{=} 8$
$16 = 16$	$8 = 8$

Both equations check, so the solution is $(-5, 6)$.

PRACTICE PROBLEM FOR EXAMPLE 1

Solve the system using the elimination method. Check the solution.

$$x - 2y = 6$$
$$5x + 2y = 18$$

> **Skill Connection**
>
> **Adding vertically**
> To add vertically, line up the like terms vertically. Then add the like terms.
> To add
>
> $$(2x + 4) + (3x + 9)$$
>
> vertically, we write as follows.
>
> $$2x + 4$$
> $$\underline{+3x + 9}$$
> $$5x + 13$$

Example 2 Finding opposites to use the elimination method

Solve each system using the elimination method. Check the solution.

a. $2x + y = 11$
$-x + 4y = 8$

b. $2x - 3y = 7$
$-3x + 4y = -10$

SECTION 4.3 Solving Systems Using the Elimination Method 403

SOLUTION

a. First notice that neither of the coefficients of x or y is an opposite.

$$2x + y = 11 \quad \text{The original linear system}$$
$$-x + 4y = 8$$

To eliminate the variable x, multiply both sides of the second equation by 2 so that the coefficients of x are 2 and -2. Use the distributive property on the left side.

$$2x + y = 11 \qquad\qquad 2x + y = 11$$
$$2(-x + 4y) = 2(8) \longrightarrow -2x + 8y = 16$$

Now that the coefficients of x are opposites, add the two equations vertically to eliminate x.

$$2x + y = 11$$
$$\underline{-2x + 8y = 16} \quad \text{Add the two equations.}$$
$$9y = 27$$

With the variable x eliminated, we can now solve for y.

$$\frac{9y}{9} = \frac{27}{9} \quad \text{Divide by 9 and solve for } y.$$
$$y = 3$$

Substitute $y = 3$ into either of the original equations and solve for x.

$$2x + (3) = 11 \quad \text{Substitute } y = 3 \text{ into the first equation, and solve for } x.$$
$$2x = 8$$
$$\frac{2x}{2} = \frac{8}{2}$$
$$x = 4$$

The solution to this system is the point $(4, 3)$. Check by substituting $x = 4$ and $y = 3$ into both original equations.

Equation 1	Equation 2
$2(4) + 3 \stackrel{?}{=} 11$	$-(4) + 4(3) \stackrel{?}{=} 8$
$8 + 3 \stackrel{?}{=} 11$	$-4 + 12 \stackrel{?}{=} 8$
$11 = 11$	$8 = 8$

Both equations check.

b. $2x - 3y = 7 \quad \text{The original linear system}$
$-3x + 4y = -10$

Considering the coefficients of x, the least common multiple of 2 and 3 is 6. To get coefficients of 6 and -6 for the variable x, multiply the first equation by 3 and the second by 2.

$$3(2x - 3y) = 3(7) \qquad\qquad 6x - 9y = 21$$
$$2(-3x + 4y) = 2(-10) \longrightarrow -6x + 8y = -20$$

Add the two equations to eliminate the variable x.

$$6x - 9y = 21$$
$$\underline{-6x + 8y = -20} \quad \text{Add the two equations vertically.}$$
$$-y = 1$$
$$\frac{-y}{-1} = \frac{1}{-1} \quad \text{Solve for } y.$$
$$y = -1$$

> **Skill Connection**
>
> **Find the least common multiple**
> To get the coefficients of one of the variables to be opposites, we first find the least common multiple of the coefficients. Recall that the least common multiple is the smallest number that is a multiple of the given numbers.

Substitute $y = -1$ into the first equation and solve for x.

$$2x - 3(-1) = 7$$
$$2x + 3 = 7$$
$$\underline{-3 -3}$$
$$2x = 4$$
$$\frac{2x}{2} = \frac{4}{2}$$
$$x = 2$$

The solution is $(2, -1)$. Check the solution of $(2, -1)$ by substituting $x = 2$ and $y = -1$ back in both original equations.

Equation 1	Equation 2
$2(2) - 3(-1) \stackrel{?}{=} 7$	$-3(2) + 4(-1) \stackrel{?}{=} -10$
$4 + 3 \stackrel{?}{=} 7$	$-6 + -4 \stackrel{?}{=} -10$
$7 = 7$	$-10 = -10$

Because $(2, -1)$ satisfies both original equations, it is the solution to the system.

PRACTICE PROBLEM FOR EXAMPLE 2

Solve each system using the elimination method. Check the solutions.

a. $-3x + 5y = 30$
 $x + 5y = 10$

b. $2x + 3y = 18$
 $6x + 9y = -9$

In Example 2, the variable x was eliminated first. The variable y could instead have been eliminated in the first step, and the final answer would have been the same. The solution is not affected by which variable is eliminated first.

Example 3 Switching to the general form first

Solve the system using the elimination method. Check the solutions.

$$-2x + 3y = 0$$
$$3y = -4x - 18$$

SOLUTION

$$-2x + 3y = 0 \qquad \text{The original linear system}$$
$$3y = -4x - 18$$

The second equation is *not* in general form. Notice that x and y are not on the same side of the equation. The first step is to put the second equation in general form. We do this by moving the x-term on the right side of the equation to the left side of the equation.

$$3y = -4x - 18 \qquad \text{Add } 4x \text{ to both sides.}$$
$$\underline{+4x +4x}$$
$$4x + 3y = -18$$

The general form of the system is

$$-2x + 3y = 0$$
$$4x + 3y = -18$$

The coefficients of the y variable are both 3. Multiply the first equation by -1 to get the coefficients of the y variables to be opposites (-3 and 3).

$$-1(-2x + 3y) = -1(0) \qquad\qquad 2x - 3y = 0$$
$$4x + 3y = -18 \qquad\longrightarrow\qquad 4x + 3y = -18$$

Now that the coefficients of y are opposites, add the two equations vertically to eliminate y.

$$\begin{aligned} 2x - 3y &= 0 \\ 4x + 3y &= -18 \end{aligned}$$ Add the two equations vertically.

$$6x = -18$$

$$\frac{6x}{6} = \frac{-18}{6}$$ Divide both sides by 6 to solve for x.

$$x = -3$$

Now that we have found the value of x, we also have to find the value of y. Substituting $x = -3$ into the first equation and solving for y yields

$$2(-3) - 3y = 0 \qquad \text{Substitute in } x = -3.$$
$$-6 - 3y = 0 \qquad \text{Add 6 to both sides to isolate y.}$$
$$+6 \qquad +6$$
$$-3y = 6$$
$$\frac{-3y}{-3} = \frac{6}{-3} \qquad \text{Divide both sides by } -3 \text{ to solve for y.}$$
$$y = -2$$

The solution is the point $(-3, -2)$.
Check the solution $(-3, -2)$ by substituting $x = -3$ and $y = -2$ into both original equations.

Equation 1   **Equation 2**

$$2(-3) - 3(-2) \stackrel{?}{=} 0 \qquad\qquad 4(-3) + 3(-2) \stackrel{?}{=} -18$$
$$-6 - (-6) \stackrel{?}{=} 0 \qquad\qquad -12 + (-6) \stackrel{?}{=} -18$$
$$0 = 0 \qquad\qquad -18 = -18$$

Because $(-3, -2)$ satisfies both original equations, it is the solution to the system.

PRACTICE PROBLEM FOR EXAMPLE 3

Solve the system using the elimination method. Check the solution.

$$-2x + y = -3$$
$$4y = -x + 24$$

■ More Practical Applications of Systems of Linear Equations

The elimination method can be used to solve a variety of applications.

Example 4 Purchasing school supplies

Siria buys two spiral notebooks and one USB drive at the campus bookstore for $25.60. Two days later, Siria buys three more spiral notebooks and two USB drives at the campus bookstore for $48.40. How much does each notebook and USB drive cost individually?

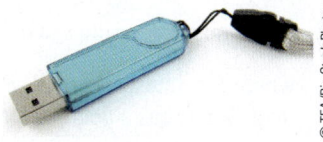

SOLUTION

First, define the variables. Let

$$n = \text{Cost of each spiral notebook in dollars}$$
$$b = \text{Cost of each USB drive in dollars}$$

Siria first purchased two spiral notebooks and one USB drive for $25.60. This purchase translates into the equation

$$2n + b = 25.60$$

On her second trip to the campus bookstore, Siria bought three spiral notebooks and two more USB drives for $48.40. This translates into the equation

$$3n + 2b = 48.40$$

The system of equations to solve is

$$2n + b = 25.60$$
$$3n + 2b = 48.40$$

To eliminate b, multiply the first equation by -2 to get the coefficients of b to be opposites.

$$-2(2n + b) = -2(25.60) \qquad\qquad -4n - 2b = -51.20$$
$$3n + 2b = 48.40 \qquad\longrightarrow\qquad 3n + 2b = 48.40$$

$$-4n - 2b = -51.20$$
$$\underline{3n + 2b = 48.40}\qquad \text{Add the two equations vertically.}$$
$$-1n = -2.80$$
$$-1(-1n) = -1(-2.80)\qquad\text{Multiply both sides by }-1.$$
$$n = 2.80$$

To solve for b, substitute the value for n into the first equation and solve.

$$2n + b = 25.60$$
$$2(2.80) + b = 25.60\qquad\text{Substitute 2.80 for }n\text{ and solve for }b.$$
$$5.60 + b = 25.60$$
$$\underline{-5.60 \qquad\quad -5.60}$$
$$b = 20.00$$

Each spiral notebook costs $2.80, and each USB drive costs $20.00.

PRACTICE PROBLEM FOR EXAMPLE 4

Mario is teaching a cake-decorating class at a culinary school. He is buying supplies for his new class. He buys six food-coloring kits and eight cake-decorating kits for $246.00. Later some more students add his class, and he has to purchase three more food-coloring kits and three cake-decorating kits for $99.00. How much does each food-coloring kit and each cake-decorating kit cost?

Now we will examine two other types of applications that involve percentages and are commonly encountered. The first type of problem involves simple interest earned on an investment. To calculate simple interest, we use the following formula.

SIMPLE INTEREST FORMULA

$$I = P \cdot r \cdot t$$

where I = interest earned, P = principal (original investment), r = annual interest rate (as a decimal), and t = time (in years).

What's That Mean?

Simple and Compound Interest

Simple interest refers to interest that is paid only at the end of the time period. For example, if you put $1000 into an account earning 5% interest for 1 year, at the end of the year, you will be paid $50.00 in simple interest. *Compound interest* is interest that is paid at set intervals of time, such as daily, monthly, or quarterly, until the end of the time period.

SECTION 4.3 Solving Systems Using the Elimination Method 407

Example 5 · Computing simple interest

Compute the simple interest earned on a $2500 investment earning a 4.5% annual interest rate over a 2-year time period.

SOLUTION
The principal P is the original investment, so $P = 2500$. The interest rate r has to be converted to decimal form. So

$r = 4.5\%$ *Move the decimal point two places to the left.*
$r = 0.045$

The time t is two years, so $t = 2$. Substituting these values into the formula yields

$I = P \cdot r \cdot t$
$I = (2500)(0.045)(2)$ *Substitute in the given values and multiply.*
$I = 225.00$

This account earns $225.00 in interest in 2 years.

PRACTICE PROBLEM FOR EXAMPLE 5
Compute the simple interest earned on a $15,000 investment earning a 4% annual interest rate over a 10-year period.

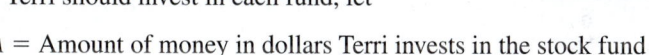

The simple interest formula appears in investment problems. When a person has some money to invest, an investment advisor will encourage the person to split the money into several accounts. These accounts include stock accounts, bond accounts, and cash accounts.

People invest money in stocks and bonds because stocks and bonds may earn more than the money would in a bank account. The more risk an investor takes on, the more the potential return on the investment (the money earned). Financial advisors will typically advise that money be split between stocks and bonds to minimize risk.

Skill Connection

Percents

To convert a percentage to decimal form, move the decimal point two places to the left. Add zeros if necessary. For example,

$9\% = 0.09$

Example 6 · Investing in stock and bond funds

Terri receives $20,000.00 from the estate of her great-aunt Mildred. Terri goes to a financial advisor, who tells her to put part of her money in a stock fund that is currently returning 10% interest and part of her money in a bond fund that is currently returning 6.5% interest. How much money should Terri put in each fund if she wants to earn $1500.00 a year in interest?

SOLUTION
We are concerned with two situations in this problem: the amount of money invested in the funds and the amount of interest earned in the funds. Because the question is how much money Terri should invest in each fund, let

A = Amount of money in dollars Terri invests in the stock fund

B = Amount of money in dollars Terri invests in the bond fund

Because she has a total of $20,000, we have that

$A + B = 20{,}000$

This equation describes one situation we were concerned with: the amount of money invested in the funds.

What's That Mean?

Risk and Investments

Bank accounts, such as checking, savings, and certificate of deposits (CDs), are often federally insured. If a bank declares bankruptcy, the U.S. government will pay the account holders back the value of their account up to $250,000.

Stock accounts and bond accounts are not insured, meaning that if the company whose stocks or bonds are purchased declares bankruptcy, the investor loses money. Bonds are considered less risky than stocks.

What's That Mean?

Stocks versus Bonds

Do you know the difference between a stock and a bond? When you buy *stock* in a company, you are purchasing part ownership in that company. When you buy a *bond* from a company, you are lending that company money, which the company promises to pay back with interest. So think of a bond as lending a company your money and a stock as purchasing part of a company.

Skill Connection

Rounding

In a problem that involves money, round to two decimal places. This is the cents place.

The second situation involves the amount of interest earned in the funds. The total interest Terri wants to earn yearly is $1500. Remember that in math, *total* means to *add*. Add the interest she earns on the stock fund to the interest she earns on the bond fund. The total yearly interest of these two funds is $1500.00. The amounts invested, A and B, are the original investments, or the principal, in each account. Since she wants to earn $1500.00 in one year, $t = 1$.

$$I = P \cdot r \cdot t$$
$$I = A \cdot (0.10) \cdot 1 = 0.10A \qquad I = B \cdot (0.065) \cdot 1 = 0.065B$$

Adding the two interest amounts together to get the total interest, we get the equation

$$0.10A + 0.065B = 1500 \qquad \text{Total the interest to get \$1500.00.}$$

So the system of equations we have to solve is

$$A + B = 20{,}000 \qquad \text{The original system}$$
$$0.10A + 0.065B = 1500$$

Neither the coefficient of A nor the coefficient of B is an opposite. Multiply both sides of the first equation by -0.10 to make the coefficients of A opposites.

$$-0.10(A + B) = -0.10(20{,}000) \quad \longrightarrow \quad -0.10A - 0.10B = -2000$$
$$0.10A + 0.065B = 1500 \qquad\qquad\qquad\qquad 0.10A + 0.065B = 1500$$

$$-0.10A - 0.10B = -2000 \qquad \text{Add the two equations.}$$
$$\underline{0.10A + 0.065B = 1500}$$
$$-0.035B = -500$$
$$\frac{-0.035B}{-0.035} = \frac{-500}{-0.035} \qquad \text{Solve for } B.$$
$$B \approx 14285.71 \qquad \text{Round to two decimal places.}$$

Substitute the value of B into the first equation and solve for A.

$$A + (14285.71) = 20000$$
$$A = 5714.29$$

Notice that the solution is rounded to the hundredths place (cents place) because we are talking about money in dollars.

Terri should invest $5714.29 in the stock fund (which is riskier) and $14,285.71 in the bond fund (less risk) to generate $1500.00 in interest annually.

PRACTICE PROBLEM FOR EXAMPLE 6

Russ receives an inheritance of $45,000 from his grandparents. His financial advisor wants him to invest his inheritance in a stock fund that is currently returning 9.5% interest and a bond fund that is currently earning 6.5% interest. Russ wants to earn $4000 a year in interest from his investments. How much should he invest in each account?

The next type of application we are going to look at involves percent mixture problems. These problems are similar to interest problems in that we will multiply a percent (as a decimal) by an amount.

> **Amount of a Given Substance in a Solution Formula** To find the amount of a given substance in a solution, use the formula
>
> $$\text{given substance} = \text{percent} \cdot \text{amount of solution}$$

When using a percent in a problem, always convert it to the decimal form first.

Example 7 Computing a quantity in a solution

Sara is taking a chemistry lab class and has 5 liters of a 70% alcohol solution. How much alcohol is in the solution?

SOLUTION
The amount of solution is 5 liters. The percent must be converted to its decimal form before using it in the formula: 70% = 0.70. Therefore,

$$\text{amount of alcohol} = 0.70 \cdot 5 = 3.50$$

There are 3.50 liters of alcohol in the solution.

We now apply this formula to a system of equations.

Example 8 Setting up a mixture problem

James needs to mix a 40% saline solution (a salt and water solution is called *saline*) with a 20% saline solution to make a 25% saline solution. He wants a total of 16 liters in the mix. How many liters of the 40% saline solution and 20% saline solution should he use?

SOLUTION
There are two situations to examine in this problem. The total amount of solution should be 16 liters, and the amount of salt in the mix (solution) should be 25%. Because the question is how many liters of each solution James should include, let

$$A = \text{Amount of the 40\% saline solution (in liters)}$$
$$B = \text{Amount of the 20\% saline solution (in liters)}$$

James wants a total of 16 liters. Remember that *total* means to add. We have our first equation, which is

$$A + B = 16$$

Total (add) the amount of salt in the 40% solution to the amount of salt in the 20% solution to get the total amount of salt in the 25% solution.

$$\text{amount of salt in the 40\% solution} = 0.40 \cdot A = 0.40A$$
$$\text{amount of salt in the 20\% solution} = 0.20 \cdot B = 0.20B$$
$$\text{amount of salt in the final mix of 25\% solution} = 0.25(16) = 4$$

$$0.40A + 0.20B = 4 \qquad \text{Total the two amounts of salt.}$$

So the system of equations is

$$A + B = 16 \qquad \text{The original linear system}$$
$$0.40A + 0.20B = 4$$

Neither of the coefficients of A or B is an opposite. Multiply both sides of the first equation by -0.20 so that the coefficients of A are opposites.

$$-0.20(A + B) = -0.20(16) \qquad \longrightarrow \qquad -0.20A - 0.20B = -3.2$$
$$0.40A + 0.20B = 4 \qquad\qquad\qquad\qquad\quad 0.40A + 0.20B = 4$$

Add the two equations to eliminate B.

$$-0.20A - 0.20B = -3.2$$
$$\underline{0.40A + 0.20B = 4} \qquad \text{Add the equations vertically.}$$
$$0.20A = 0.8$$
$$\frac{0.20A}{0.20} = \frac{0.80}{0.20} \qquad \text{Solve the system for } A.$$
$$A = 4$$

Substitute the value for A back into the first equation to solve for B.

$$4 + B = 16$$
$$B = 12$$

James needs to use 4 liters of the 40% saline solution and 12 liters of the 20% saline solution to make 16 liters of a 25% saline solution.

PRACTICE PROBLEM FOR EXAMPLE 8

A dairy wants to mix whole milk, which has a 4% fat content, with skim milk (a 1% fat content) to make 500 gallons of 2% milk. How many gallons of whole milk and skim milk should be used?

■ Substitution or Elimination?

There are many times in an application problem that using one method over the other makes the solving process a bit quicker. In the next example, we will see how using the substitution method might be a bit faster.

Example 9 Comparing angles in a geometry problem

Two angles are supplementary. The measure of the first angle is 3 times as large as the measure of the second angle. Find the degree measures of the two angles.

SOLUTION

Let

a = Measure of the first angle in degrees
b = Measure of the second angle in degrees

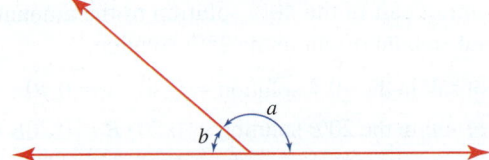

The statement that the two angles are supplementary translates as

$$a + b = 180$$

We are told that the first angle is 3 times as large as the second. This sentence translates as

$$a = 3b$$

The two equations in two unknowns are

$$a + b = 180$$
$$a = 3b$$

Substitution would be the quickest method to solve this system because the second equation has already been solved for a. Substituting the second equation into the first and solving for b yields

$$3b + b = 180$$ Substitute $a = 3b$ into the first
$$4b = 180$$ equation and solve for b.
$$\frac{4b}{4} = \frac{180}{4}$$
$$b = 45$$

What's That Mean?

Supplementary and Complementary

In geometry, two angles are said to be *supplementary* if their sum is 180°. Two angles are said to be *complementary* if their sum is 90°.

Skill Connection

Elimination

Solving Example 9 using the elimination method is done below.

$$a + b = 180$$
$$a = 3b$$

$$a + b = 180$$
$$a - 3b = 0$$

$$a + b = 180$$
$$\underline{-a + 3b = 0}$$
$$4b = 180$$

$$\frac{4b}{4} = \frac{180}{4}$$

$$b = 45$$
$$a = 3b = 3(45) = 135$$

To solve using elimination required rewriting the second equation, multiplying the second equation by -1, and then adding the two equations. In this case, the substitution method was more efficient.

Substitute the value for *b* into the second equation and solve for *a*.

$$a = 3(45)$$
$$a = 135$$

The first angle has measure 135°, and the second angle has measure 45°.

PRACTICE PROBLEM FOR EXAMPLE 9
Two angles are complementary. The measure of the second angle is four times the measure of the first angle. Find the degree measure of the two angles.

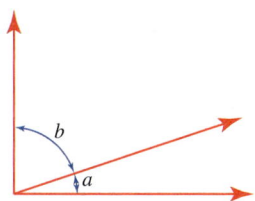

Example 10 Choosing substitution or choosing elimination?

Decide whether to solve the system using the substitution method or the elimination method. Do *not* solve the system but do justify your answer.

a. $y = 6x - 4$
   $2x + 3y = 7$

b. $x + 2y = 7$
   $3x + 4y = -8$

c. $-2x + 3y = 5$
   $4x + 5y = -9$

SOLUTION

a. The first equation is $y = 6x - 4$. Because *y* has already been solved for, this system can be efficiently solved by using the substitution method.

b. The first equation is $x + 2y = 7$. The coefficient of *x* is 1, so *x* will be easy to solve for. The system can be solved by using the substitution method. Notice that both equations are in general form, so the system can also be solved using the elimination method. Either method is reasonable.

c. Neither variable in either equation has a coefficient of ±1. The coefficient of *x* in the first equation is −2, and the coefficient of *x* in the second equation is +4. If the first equation is multiplied by 2, the coefficients of *x* will be opposites. This system can be efficiently solved by using the elimination method.

■ Connecting the Concepts

Which method should be used to solve a system?

To solve a linear system algebraically, we can use either the substitution method or the elimination method. When a variable has already been solved for or a variable has a coefficient of ±1, the substitution method often is used because it can be faster. When the system is given in general form ($Ax + By = C$) and either a variable has coefficients that are opposites or there is no coefficient that is ±1, the elimination method is often used.

4.3   Vocabulary Exercises

1. The _____ method is used to algebraically find the solution to a system of linear equations.

2. When using the elimination method, both equations are first put into _____ form.

3. The _____ formula is $I = Prt$.

4. Two angles are said to be _____ if their sum is 180°.

5. Two angles are said to be _____ if their sum is 90°.

4.3 Exercises

For Exercises 1 through 8, solve each system using the elimination method. Provide reasons for each step. Check the solutions.

1. $2x - y = 16$
   $3x + y = 9$

2. $-3x + y = 9$
   $5x - y = 15$

3. $-x + 6y = -3$
   $x - 2y = -5$

4. $x + 3y = 8$
   $-x + 4y = 6$

5. $-4x + 2y = 12$
   $8x - 2y = -40$

6. $3x + y = -13$
   $3x - y = 1$

7. $4x + 2y = 8$
   $4x - y = -4$

8. $x + 3y = 6$
   $3x + 3y = -18$

For Exercises 9 through 24, solve each system using the elimination method. Check the solutions.

9. $4x + 3y = -5$
   $2x - y = 5$

10. $3x + y = 9$
    $-x + 5y = -35$

11. $3x - y = -1$
    $-6x + 2y = 2$

12. $5x - 15y = 45$
    $-x + 3y = -9$

13. $3x - 2y = -14$
    $2x + 3y = 8$

14. $3x + 8y = -7$
    $5x + 3y = -22$

15. $13x + 2y = -40$
    $10x - 3y = 1$

16. $3x - 5y = 16$
    $5x - 4y = 31$

17. $10x + 3y = 22$
    $3x - 2y = 24$

18. $7x - 8y = 12$
    $2x + 5y = -33$

19. $6x + 8y = -8$
    $3x + 4y = 36$

20. $x - y = 7$
    $-3x + 3y = 10$

21. $7x + 2y = -2$
    $-21x - 6y = 6$

22. $2x + 5y = 20$
    $-8x - 20y = -80$

23. $-6x + 2y = -16$
    $3x - y = -2$

24. $4x - 3y = 21$
    $8x - 6y = -3$

For Exercises 25 through 30, solve each system using the elimination method.

25. Two integers sum to 95. The difference of 4 times the first and the second number is 15. Find the two integers.
    a. Define the variables.
    b. Determine the two equations.
    c. Solve the system using the elimination method. Check the solution.

26. A first number minus a second number is 5. The sum of twice the first number and 3 times the second number is 45. Find the two numbers.
    a. Define the variables.
    b. Determine the two equations.
    c. Solve the system using the elimination method. Check the solution.

27. Hunter is the coach of a youth soccer team. Hunter purchases eight T-shirts and six water bottles for $186 to give out at an upcoming youth soccer tournament. Later she purchases ten more T-shirts and eight water bottles for $237 as additional players sign up. How much does each T-shirt and water bottle cost?
    a. Define the variables.
    b. Determine the two equations.
    c. Solve the system using the elimination method. Check the solution.

28. Gary is organizing a charity dinner and is purchasing door prizes. He buys three coffee mugs and two pen sets for $29.50. As additional attendees sign up for the event, he purchases an additional three coffee mugs and four pen sets for $42.50. How much does each coffee mug and pen set cost?
    a. Define the variables.
    b. Determine the two equations.
    c. Solve the system using the elimination method. Check the solution.

29. Martinique, a busy mother of four, purchases three gallons of milk and four loaves of French bread for $23.40. The next week, she purchases two gallons of milk and three loaves of French bread for $16.35. How much does each gallon of milk and loaf of French bread cost?
    a. Define the variables.
    b. Determine the two equations.
    c. Solve the system using the elimination method. Check the solution.

30. Jordan is planting his spring garden. He purchases three packages of seeds plus two tomato plants for $11.75. The next week, he purchases an additional four packages of seeds plus one more tomato plant for $11.80. How much does each packet of seeds and each tomato plant cost?

© iStock.com/fotokostic

a. Define the variables.

b. Determine the two equations.

c. Solve the system using the elimination method. Check the solution.

For Exercises 31 through 50, solve each system using the elimination method. Check the solutions.

31. $-2x + 5y = 10$
    $2x = 5y + 25$

32. $4x + 5y = -20$
    $5y = -4x + 15$

33. $x + 2y = 18$
    $-3x + 2y = 2$

34. $-2x + 3y = -21$
    $x + 3y = -12$

35. $4x - 3y = -24$
    $x + y = 1$

36. $x + y = 3$
    $6x - 5y = 20$

37. $4x - 5y = -15$
    $2x + y = -11$

38. $11x - 6y = 24$
    $x + 3y = 27$

39. $-14x + 35y = 21$
    $2x - 5y = -3$

40. $12x - 4y = 4$
    $-3x + y = -1$

41. $3x + 5y = 30$
    $y = -\dfrac{3}{5}x + 6$

42. $y = -\dfrac{1}{2}x + 7$
    $x + 2y = 14$

43. $2x - 2y = -14$
    $4x - y = -1$

44. $5x + 4y = 0$
    $-5x + 2y = 30$

45. $y = \dfrac{5}{2}x - 3$
    $5x - 2y = 3$

46. $12x - 2y = 2$
    $-6x + y = 7$

47. $3x - y = 7$
    $-9x + 3y = -21$

48. $\dfrac{3}{4}x + y = -6$
    $-3x - 4y = 24$

49. $2x - 2y = 10$
    $-x + y = 8$

50. $-2x - 3y = 3$
    $2x + 3y = 12$

For Exercises 51 through 58, decide whether to solve each system using the substitution method or elimination method. Do not solve the systems but do justify your answers.

51. $3x + 4y = 9$
    $2x - 4y = -3$

52. $2x + 7y = -3$
    $-2x - 3y = 4$

53. $3x - 5y = 7$
    $y = 6x + 1$

54. $-6 + 3y = x$
    $4x + 9y = 8$

55. $y = 2x - 5$
    $y = x + 7$

56. $y = -4x + 5$
    $y = -2x - 3$

57. $x - y = 6$
    $x + y = 3$

58. $2x - 7y = 4$
    $-3x + 7y = 2$

For Exercises 59 through 74, answer each question.

59. Tomas has $500.00 to invest in a simple interest account for 4 years. The interest rate is 4.25%. Find the amount of interest he earns.

60. Tess has $12,500.00 to invest in a simple interest account for 2 years. The interest rate is 4.75%. Find the amount of interest she earns.

61. Steven has $8720.00 to invest in a simple interest account for 3 years. The interest rate is 4.25%. Find the amount of interest he earns.

62. Alejandro has $5000.00 to invest in a simple interest account for 1 year. The interest rate is 3.75%. Find the amount of interest he earns.

63. A simple saline solution is 8.5% sodium. Find the amount of sodium in 10 liters of saline.

64. A mouthwash contains 8% alcohol. Find the amount of alcohol in 16 ounces of mouthwash.

65. Christy is working with an 8% sucrose solution. Find the amount of sucrose in 8 liters of the solution.

66. A juice drink contains 12% real fruit juice. Find the amount of real fruit juice in 8 ounces of the drink.

67. Two angles are complementary. One angle has a measure of 25°. Find the measure of the other angle.

68. Two angles are complementary. One angle has a measure of 64°. Find the measure of the other angle.

69. Two angles are supplementary. One angle has a measure of 110°. Find the measure of the other angle.

70. Two angles are supplementary. One angle has a measure of 125°. Find the measure of the other angle.

71. Teresa has $10,000 to invest. She puts part of her money in a stock account and the rest in a bond account. If Teresa puts n dollars in the stock account, write an expression for the remainder that she puts in the bond account.

72. Nikolai has $4000 to invest. He puts part of his money in a stock account and the rest in a bond account. If Nikolai puts n dollars in the stock account, write an expression for the remainder that he puts in the bond account.

73. Alexis has $3800 to invest. He puts part of his money in a CD and the rest in a money market account. If Alexis puts n dollars in the CD, write an expression for the remainder that he puts in the money market.

74. Lupe has $1500 to invest. She puts part of her money in a certificate of deposit and the rest in a money market account. If Lupe puts n dollars in the CD account, write an expression for the remainder that she puts in the money market account.

For Exercises 75 through 94, set up the system of equations and solve using either the substitution method or elimination method.

75. A small plane flying with the wind travels 375 miles in 3 hours. On the return trip, flying against the wind, the same plane takes 5 hours to travel 375 miles. Find the rate of the plane in still air and the rate of the wind.

  a. Define the variables.

  b. Determine the two equations.

  c. Solve the system. Check the solution.

76. A plane flying with the wind travels 4400 miles in 10 hours. On the return trip, flying against the wind, the same plane takes 11 hours to travel 4400 miles. Find the rate of the plane in still air and the rate of the wind.

  a. Define the variables.

  b. Determine the two equations.

  c. Solve the system. Check the solution.

77. A boat travels with the current 60 miles in 3 hours. It takes the boat 6 hours to travel the same 60 miles against the current on the return trip. Find the rate of the boat in still water and the rate of the current.

  a. Define the variables.

  b. Determine the two equations.

  c. Solve the system. Check the solution.

78. A boat travels with the current 130 miles in 5 hours. Against the current, it takes the boat 13 hours to travel the same 130 miles on its return trip. Find the rate of the boat in still water and the rate of the current.

  a. Define the variables.

  b. Determine the two equations.

  c. Solve the system. Check the solution.

79. To make healthier bread dough, a baker wants to blend whole-wheat and white flour. He has a total of 50 pounds of whole-wheat and white flour in the mix. The whole-wheat flour costs $4.25 per pound, and the white flour costs $3.75 per pound. The mix costs $4.00 per pound. Find the amount of each type of flour that he should include in the mix.

  a. Define the variables.

  b. Determine the two equations.

  c. Solve the system. Check the solution.

80. To make a batch of trail mix, a health food store wants to blend walnuts and yogurt-covered raisins. There is a total of 50 pounds of walnuts and raisins in the mix. The walnuts cost $7.49 per pound, and the yogurt-covered raisins cost $3.49 per pound. The total value of the trail mix is $249.50. Find the amount of walnuts and yogurt-covered raisins that they have to include in the mix.

  a. Define the variables.

  b. Determine the two equations.

  c. Solve the system. Check the solution.

81. Renata is working two jobs. Her first job pays $10.25 an hour, and her second job pays $11.50 an hour. She would like to work a total of 25 hours a week and earn $275.00 a week. Determine how many hours a week Renata should work at each job.

  a. Define the variables.

  b. Determine the two equations.

  c. Solve the system. Check the solution.

82. Gayle is working two jobs. Her first job pays $11.00 an hour, and her second job pays $12.00 an hour. She would like to work a total of 20 hours a week and earn $232.00 a week. Determine how many hours a week Gayle should work at each job.

  a. Define the variables.

  b. Determine the two equations.

  c. Solve the system. Check the solution.

83. Cate has $4000.00 to invest. An investment advisor tells her to put part of her money in a CD earning 4% interest and the rest in a bond fund earning 5% interest. Cate would like to earn $185.00 in interest every year. How much should she invest in each account?

84. Dianna has $25,000 to invest. An investment advisor tells her to put part of her money in a bond fund at 5% interest and the rest in a stock fund earning 7.5% interest. Dianna would like to earn $1625.00 in interest every year. How much should she invest in each account?

85. Michael has two credit cards. One card is a MasterCard with an annual percentage rate (APR) of 18%, and the other is a store credit card with an APR of 22%. The total Michael owes is $3500.00, and he paid $730.00 this year in interest. How much does Michael owe on each card?

86. Renaldo has two credit cards. One card is a MasterCard with an annual percentage rate (APR) of 15%, and the other is a store credit card with an APR of 20%. The total Renaldo owes is $4500.00, and he paid $750.00 this year in interest. How much does Renaldo owe on each card?

87. Greg needs to mix a 15% hydrochloric acid solution with 20% hydrochloric acid solution to make an 18.75% hydrochloric acid solution. He wants a total of 20 liters in the mix. How many liters of the 15% solution and the 20% solution should he use?

88. Elizabeth wants to mix a 25% saline solution with a 40% saline solution to make a 30% saline solution. She needs a total of 12 ounces in the mix. How many ounces of the 25% saline solution and the 40% saline solution should she use?

89. Christine has to mix a 30% antiseptic alcohol solution with a 50% antiseptic alcohol solution to make a 42% antiseptic alcohol solution. She needs a total of 100 liters in the mix. How many liters of the 30% antiseptic alcohol solution and the 50% antiseptic alcohol solution should she use?

90. Eddie has to mix 5% butterfat milk with 2% butterfat milk to make 3% butterfat milk. He needs a total of 120 gallons in the mix. How many gallons of the 5% butterfat milk and the 2% butterfat milk should he use?

91. Two angles are complementary. The second angle measures 3 degrees more than twice the measure of the first angle. Find the degree measures of both angles.

92. Two angles are complementary. The second angle measures 2 degrees more than three times the measure of the first angle. Find the degree measures of both angles.

93. Two angles are supplementary. The second angle measures 5 degrees less than four times the measure of the first angle. Find the degree measures of both angles.

94. Two angles are supplementary. The second angle measures 4 degrees less than three times the measure of the first angle. Find the degree measures of both angles.

For Exercises 95 through 98, read each student's work carefully and find the errors.

95. Solve the system:

$$x + y = 5$$
$$x + 2y = 6$$

Micah
$x + y = 5$
$x + 2y = 6$
$3y = 11$
$y = \dfrac{11}{3}$
$x + \dfrac{11}{3} = 5$
$x = 5 - \dfrac{11}{3} = \dfrac{15}{3} - \dfrac{11}{3} = \dfrac{4}{3}$
The solution is $\left(\dfrac{4}{3}, \dfrac{11}{3}\right)$.

96. Solve the system:

$$3x + y = 6$$
$$2x + y = -1$$

Kristine
$3x + y = 6$
$2x + y = -1$
$5x = 5$
$x = 1$
$3(1) + y = 6$
$3 + y = 6$
$y = 3$
The solution is $(1, 3)$.

97. Solve the system:

$$2x - 3y = 15$$
$$x + 5y = -2$$

Aaron
$2x - 3y = 15$
$x + 5y = -2$
$2x - 3y = 15$
$-2(x + 5y) = -2$
$2x - 3y = 15$
$-2x - 10y = -2$
$-13y = 13$
$y = -1$
$x + 5(-1) = -2$
$x - 5 = -2$
$x = 5 + -2 = 3$
The solution is $(3, -1)$.

98. Solve the system:

$$4x + y = 8$$
$$2x + y = 6$$

Angel
$4x + y = 8$
$2x + y = 6$
$4x + y = 8$
$-1(2x + y) = 6$
$4x + y = 8$
$-2x - y = 6$
$2x = 14$
$x = 7$
$2(7) + y = 6$
$14 + y = 6$
$y = -8$
The solution is $(7, -8)$.

4.4 Solving Linear Inequalities in Two Variables Graphically

LEARNING OBJECTIVES

- Find the solution set for linear inequalities in two variables.
- Graph vertical and horizontal inequalities.

Linear Inequalities in Two Variables

In some situations, we want to come up with a solution to a problem, but the situation has some constraints. For example, a couple may be getting married and would like to know how many people they can invite to their wedding yet stay within their budget. The constraint on this situation is the amount of money the couple has budgeted for guests at their wedding.

What's That Mean?

Constraint

In general, a *constraint* on a problem is a condition that limits the possible solutions to the problem. Constraints give rise to inequalities.

> **DEFINITION**
>
> **Linear Inequality in Two Variables** A linear inequality in two variables is of the form $Ax + By < C$, where A, B, and C are real numbers with neither A nor B as zero. The inequality symbols $>$, $\leq$, and $\geq$ may also be used.

We will now examine what the solution set to a linear inequality looks like.

Example 1 The wedding planner

Arlene is a wedding consultant. Her clients, Sara and Alex, are getting married and have budgeted $1200 for dinner at the reception. Sara and Alex want to serve their guests a choice of a chicken dinner or a salmon dinner. The catering company that Arlene uses quotes the cost of chicken dinners at $6.50 a plate and the salmon at $8.50 a plate. The total cost of the reception meal is $6.50C + 8.50S$, where C represents the number of chicken dinners served and S represents the number of salmon dinners served. Since Sara and Alex want to spend a maximum of $1200 on the reception meal, Arlene wants the total cost to be less than or equal to $1200 or, mathematically,

$$6.50C + 8.50S \leq 1200$$

Given this constraint on their budget, fill in the table below to determine whether the given possible combination of chicken and salmon dinners is within Sara and Alex's budget. Fill in the last two rows with different combinations of your choice.

Number of Chicken Dinners	Number of Salmon Dinners	Cost ≤ 1200?
80	75	
29	119	
80	85	

SOLUTION

Number of Chicken Dinners	Number of Salmon Dinners	Cost ≤ 1200?	
80	75	1157.50	6.50(80) + 8.50(75)
29	119	1200	= 1157.5 ≤ 1200
80	85	1242.50 Not a solution!	6.50(80) + 8.50(85) = 1242.50 ≥ 1200
75	75	1125	
80	80	1200	

Notice that the third row in the table does not check because the cost exceeds $1200.00 (our constraint).

From this example, we see there are many solutions to a linear inequality.

DEFINITION

Solution Set to a Linear Inequality in Two Variables The solution set to a linear inequality in two variables is the set of all points (x, y) that make the inequality true.

Example 2 Checking solutions

Determine whether each given point is in the solution set of the linear inequality.

$$y > 3x - 5$$

a. $(1, 1)$ **b.** $(0, -5)$
c. $(0, 0)$ **d.** $(2, -6)$

SOLUTION

To see whether a given point is in the solution set, substitute the coordinates of that point into the inequality and see whether it makes the inequality true. If the inequality is true, then the point is in the solution set. If the inequality is false, then the point is *not* in the solution set.

a.
$(x, y) = (1, 1)$

$1 \overset{?}{>} 3(1) - 5$ Substitute in $x = 1, y = 1$.

$1 > -2$ This statement is true.

The point $(1, 1)$ is in the solution set.

b.
$(x, y) = (0, -5)$

$-5 \overset{?}{>} 3(0) - 5$ Substitute in $x = 0, y = -5$.

$-5 \overset{?}{>} -5$ This statement is false.

Since -5 is not strictly greater than -5 (it is equal), the point $(0, -5)$ is not in the solution set.

c.
$(x, y) = (0, 0)$

$0 \overset{?}{>} 3(0) - 5$ Substitute in $x = 0, y = 0$.

$0 > -5$ This statement is true.

The point $(0, 0)$ is in the solution set.

d.
$$(x, y) = (2, -6)$$
$$-6 \overset{?}{>} 3(2) - 5 \quad \text{Substitute in } x = 2, y = -6.$$
$$-6 \overset{?}{>} 1 \quad \text{This statement is false.}$$
The point $(2, -6)$ is not in the solution set.

PRACTICE PROBLEM FOR EXAMPLE 2

Determine whether the given point is in the solution set of the linear inequality.
$$y < -x + 7$$

a. $(-2, 1)$ **b.** $(0, 0)$

c. $(0, 7)$ **d.** $(8, 2)$

In this Concept Investigation, we explore how many solutions a linear inequality in two variables can have.

CONCEPT INVESTIGATION
How many solutions are there?

Consider the following linear inequality:
$$y \leq -2x + 5$$

1. Check to see whether the following points are in the solution set by substituting the values into the inequality.

   $(0, 5) \quad (2, -1) \quad (4, 4) \quad (0, 2) \quad (3, 0) \quad (-2, 3)$

2. Plot the points from part 1 that are in the solution set on the following graph. Plot only the points that made the inequality true.

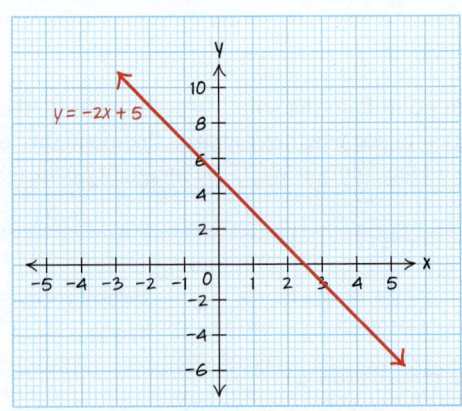

3. List three other points in the solution set. Graph these points on the graph in part 2.

4. Are there other points in the solution set besides the points you have already plotted? How many points do you think there are in the solution set?

From this Concept Investigation, we see that there are a lot of points in the solution set. In fact, the solution set contains infinitely many points. For this particular inequality, all the points below the line are in the solution set. Because we cannot possibly list

SECTION 4.4 Solving Linear Inequalities in Two Variables Graphically   419

them all, we shade the graph to indicate that all of those points are solutions to the linear inequality.

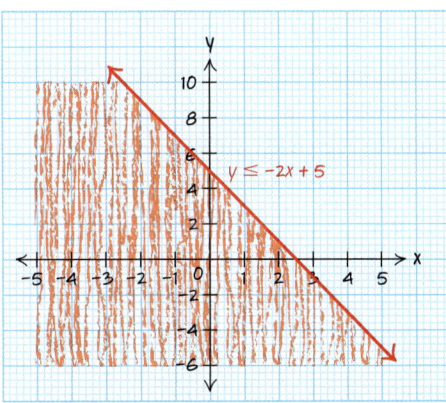

The solution set to a linear inequality in two variables always consists of all the points on one side of the line. When graphing the solution set, we want to determine which side of the line to shade. Use the following steps to determine which side of the line is the solution set and thus should be shaded.

Steps to Find the Solution Set of a Linear Inequality in Two Variables
1. Select a test point that is not on the line.
2. Substitute the test point into the inequality. Determine whether it makes the inequality true or false.
3. Shade in the side of the line that consists of the points that make the inequality true.

Example 3 Determine which side should be shaded

Consider the accompanying graph of the line $y = \frac{2}{3}x - 1$. Shade the side of the line that represents the solution set for the inequality $y \geq \frac{2}{3}x - 1$.

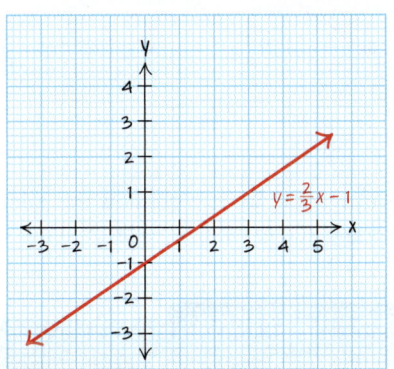

■ **Connecting the Concepts**

Why do we shade part of the graph?

In Chapter 2, we learned how to solve a linear inequality in one variable algebraically, and we examined how to graph the solution on a number line. Remember that the solution consisted of infinitely many values, and we represented the solution set by shading in an interval on a number line. To solve a linear inequality in two variables graphically, we use the xy-plane because we have two variables. Just as in Chapter 2, the solution will consist of infinitely many points, and we now shade in regions on the xy-plane instead of an interval on a number line.

SOLUTION

Step 1 Select a test point that is not on the line.

Pick a test point that is not on the line, say, (3, 0).

Step 2 Substitute the test point into the inequality. Determine whether it makes the inequality true or false.

Substituting (3, 0) into the inequality yields

$$y \geq \frac{2}{3}x - 1 \qquad \text{The test point is } (x, y) = (3, 0).$$

$$0 \stackrel{?}{\geq} \frac{2}{3}(3) - 1 \qquad \text{Substitute in } x = 3, y = 0.$$

$$0 \stackrel{?}{\geq} 1 \qquad \text{This statement is false.}$$

The point (3, 0) does not satisfy the inequality.

Step 3 Shade in the side of the line that consists of the points that make the inequality true.

This point lies below the line, so we will not shade below the line. Therefore, we should shade above the line. All points above the line make the inequality true. To confirm this, pick the point (0, 2), which lies above the line. Substituting (0, 2) into the inequality yields

$$y \geq \frac{2}{3}x - 1 \qquad \text{The test point is } (x, y) = (0, 2).$$

$$2 \stackrel{?}{\geq} \frac{2}{3}(0) - 1 \qquad \text{Substitute in } x = 0, y = 2.$$

$$2 \geq -1 \qquad \text{This statement is true.}$$

The graph of the solution set of the inequality is as follows.

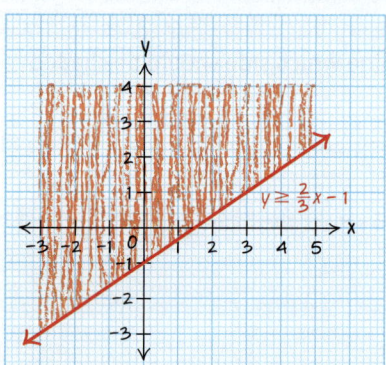

In the next Concept Investigation, we explore how the solution set is different when we use a $<$ or $>$ symbol compared to using a $\leq$ or $\geq$ symbol.

■ CONCEPT INVESTIGATION

Should we include the line?

1. In the table, the first column contains a list of the points on the line $y = -2x + 5$. Test whether these points are in the solution set of the inequalities given in the second and third column. The first row has been done as an example.

SECTION 4.4 Solving Linear Inequalities in Two Variables Graphically   421

Points on the line (x, y)	$y \leq -2x + 5$	$y < -2x + 5$
$(0, 5)$	$5 \stackrel{?}{\leq} -2(0) + 5$ Substitute in $x = 0, y = 5.$ $5 \leq 5$ True This point is in the solution set.	$5 \stackrel{?}{<} -2(0) + 5$ Substitute in $x = 0, y = 5.$ $5 < 5$ False This point is not in the solution set.
$(1, 3)$		
$(2, 1)$		
$(3, -1)$		
$(4, -3)$		

Connecting the Concepts

When is an inequality true?
To determine whether an inequality is true, first simplify both sides of the inequality separately. Then see whether the inequality is a correct statement. If so, the inequality is true.

2. All of the points in the first column are on the line. Are these points part of the solution set of $y \leq -2x + 5$? That is, do these points make this inequality true?

3. Should the line $y = -2x + 5$ be included in the graph of the solution set for the inequality $y \leq -2x + 5$?

4. All of the points in the first column are on the line. Are these points part of the solution set of $y < -2x + 5$? That is, do these points make this inequality true?

5. Should the line $y = -2x + 5$ be included in the graph of the solution set for the inequality $y < -2x + 5$?

When the points on the line are included in the solution set, the line itself is part of the solution set and is represented by drawing a solid line. When the points on the line are not included in the solution set, we represent that on the graph with a dashed line.

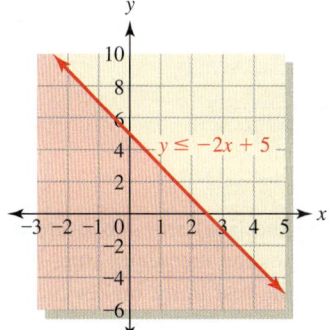
$y \leq -2x + 5$

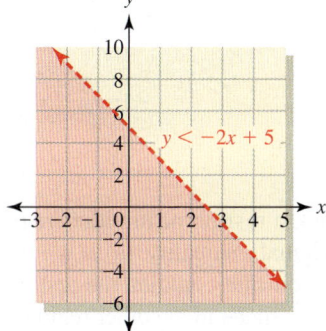
$y < -2x + 5$

Putting these concepts together yields the following steps for graphing the solution set to a linear inequality in two variables.

Steps for Graphing Linear Inequalities in Two Variables

1. Graph the line.
2. If the inequality is of the form $<$ or $>$, draw in a dashed line. If the inequality is of the form $\leq$ or $\geq$, draw in a solid line.
3. The line separates the plane into two regions, above and below the line. Pick a test point in either region (not on the line) and substitute it into the original inequality.
   a. If the test point makes the inequality true, shade in the region that contains the test point.
   b. If the test point makes the inequality false, shade in the other region (the one that does not contain the test point).

Example 4 Graphing inequalities

Graph the solution set of each inequality.

a. $-3x + 6y \leq 18$ **b.** $-2y > 4x - 9$

SOLUTION

a. Step 1 Graph the line.

Graph the line $-3x + 6y = 18$ by finding the intercepts.

x-intercept	y-intercept
Let $y = 0$	Let $x = 0$
$-3x + 6(0) = 18$	$-3(0) + 6y = 18$
$-3x = 18$	$6y = 18$
$\dfrac{-3x}{-3} = \dfrac{18}{-3}$	$\dfrac{6y}{6} = \dfrac{18}{6}$
$x = -6$	$y = 3$
x-intercept: $(-6, 0)$	y-intercept: $(0, 3)$

Step 2 If the inequality is of the form $<$ or $>$, draw a dashed line. If the inequality is of the form $\leq$ or $\geq$, draw a solid line.

Since the original inequality symbol was $\leq$, draw in a solid line.

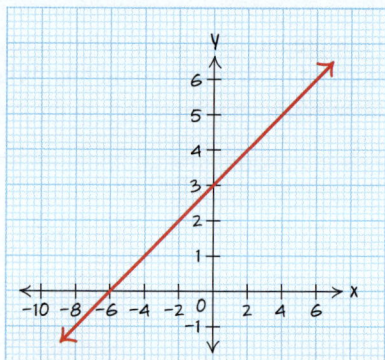

Step 3 Pick a test point in either region (not on the line) and substitute it into the original inequality. (a) If the test point makes the inequality true, shade in the region that contains the test point. (b) If the test point makes the inequality false, shade in the other region.

Pick any test point that is not on the line, for example, $(0, 4)$. Substituting $x = 0$, $y = 4$ into the original inequality yields

$-3(0) + 6(4) \stackrel{?}{\leq} 18$ Substitute in $x = 0, y = 4$.

$0 + 24 \stackrel{?}{\leq} 18$

$24 \stackrel{?}{\leq} 18$ This is a false statement.

Because this is not true, we do not shade in the region containing the point $(0, 4)$. Shade in the region below the line.

Skill Connection

Graphing lines

To graph a line, use any of the following techniques from Chapter 3.

1. Generate a table of values. Plot the points and connect with a line.
2. Put the equation of the line in slope-intercept form. Graph using the y-intercept and slope.
3. Find the x- and y-intercepts. Plot the points and connect with a line.

SECTION 4.4 Solving Linear Inequalities in Two Variables Graphically   423

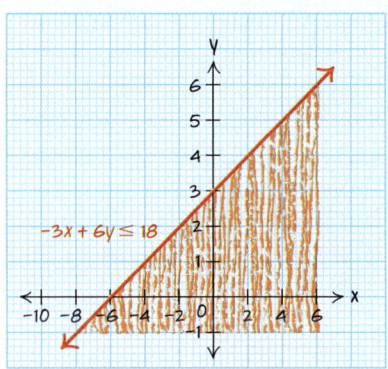

Connecting the Concepts

Where are the solutions?

All of the points in the shaded region are solutions to the linear inequality. All of these points satisfy the original linear inequality.

b. If we solve the inequality $-2y > 4x - 9$ for y first, it will be easier to determine how to graph the inequality and what region to shade. Solving for y yields

$$-2y > 4x - 9 \quad \text{The original inequality}$$

$$\frac{-2y}{-2} < \frac{4x}{-2} - \frac{9}{-2} \quad \text{Divide both sides by } -2 \text{ to isolate } y.\\ \text{Reverse the inequality symbol.}$$

Remember that when we divide by a negative number, we must reverse the inequality.

$$y < -2x + \frac{9}{2}$$

What's That Mean?

Strict Inequality

Inequalities that do not contain an equal to part are called *strict inequalities*. The symbols $>$ and $<$ are strict inequalities.

To graph the line $y = -2x + \frac{9}{2}$, plot the y-intercept $\left(0, \frac{9}{2}\right)$ and use the slope $m = \frac{-2}{1} = \frac{\text{rise}}{\text{run}}$ to find another point on the graph. Since this is a strict inequality ($<$), graph a dashed line as below.

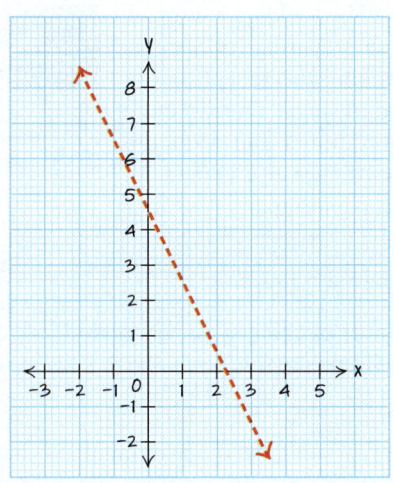

Substitute in a test point to see which region should be shaded. This test point cannot be on the line but should instead be in either of the two regions above the line and below the line.

Select (0, 0), which is below the line. Substitute (0, 0) into the original inequality.

$$-2(0) \stackrel{?}{>} 4(0) - 9 \quad \text{Substitute in } x = 0, y = 0.$$

$$0 \stackrel{?}{>} -9 \quad \text{Simplify.}$$

$$0 > -9 \quad \text{This is a true statement.}$$

Since this statement is true, we shaded the region below the line that contains the point (0, 0).

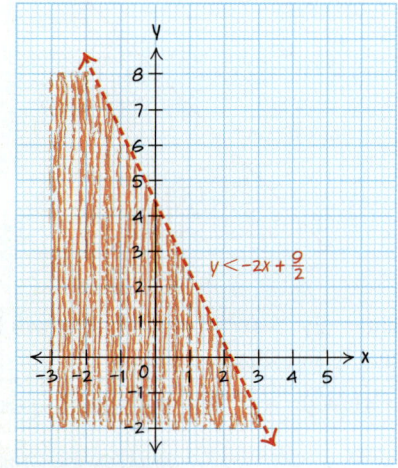

PRACTICE PROBLEM FOR EXAMPLE 4

Graph the solution set of each inequality.

a. $-5x + 4y \geq -20$

b. $y > -\dfrac{4}{3}x$

Example 5 The wedding planner graphed

In Example 1, we discussed the wedding consultant, Arlene. One of her tasks in preparing for Sara and Alex's wedding was planning the reception dinner. Sara and Alex wanted to spend a maximum of $1200.00 on the reception dinner and wanted to serve a choice of a chicken dinner or a salmon dinner. Arlene determined that the inequality describing this situation was

$$6.50C + 8.50S \leq 12.00$$

Here C represents the number of chicken dinners served, and S represents the number of salmon dinners served. Graph the solution set to this linear inequality.

SOLUTION

To graph the solution set, first put C on the horizontal axis and S on the vertical axis. The choice of axis is not significant in this problem. We could choose to graph C on the vertical axis and S on the horizontal axis, but for now, we will do it the other way.

Graph the line $6.50C + 8.50S = 1200$. To graph the line, graph two points on the line. Two easy points to find are the intercepts.

S-intercept
Let $C = 0$
$6.50(0) + 8.50S = 1200$
$\dfrac{8.50S}{8.50} = \dfrac{1200}{8.50}$
$S \approx 141.18$
S-intercept: $(0, 141.18)$

C-intercept
Let $S = 0$
$6.50C + 8.50(0) = 1200$
$\dfrac{6.50C}{6.50} = \dfrac{1200}{6.50}$
$C \approx 184.62$
C-intercept: $(184.62, 0)$

The graph of the line looks like the following.

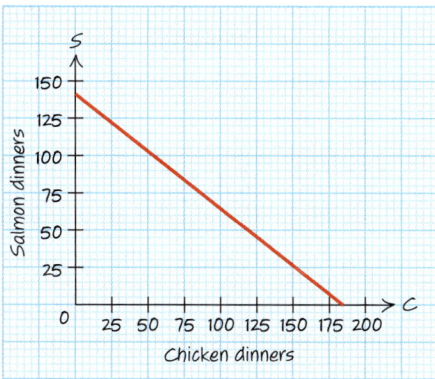

All points on the line are part of the solution set, since those points make the cost equal to $1200.00. We are graphing only positive values because negative chicken or salmon dinners does not make sense.

Notice that the solutions we discussed in Example 1 lie below the line. In fact, selecting any point below the line makes the inequality true. To show graphically that all points below the line make the inequality true, shade the half-plane in the first quadrant below the line.

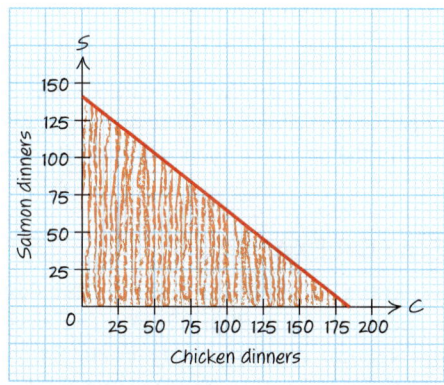

Example 6 Meeting living expenses

Ricardo works two part-time jobs. He has to earn more than $450 a week to pay his bills. His first job pays $11.75 an hour, and his second job pays $12.50 an hour. Write an inequality that expresses this relationship between his total weekly earnings and his goal (minimum) earning per week. Then graph the solution set for this inequality.

SOLUTION

Let a represents the number of hours per week Ricardo works at the first job, and let b represents the number of hours per week Ricardo works at the second job. His weekly earnings from both jobs is $11.75a + 12.50b$. He has to earn more than $450 per week to meet his income goals. So his weekly earnings are greater than $450, or

$$11.75a + 12.50b > 450$$

Since the variables a and b represent the number of hours he works in a week, it makes sense that both a and b must be nonnegative numbers.

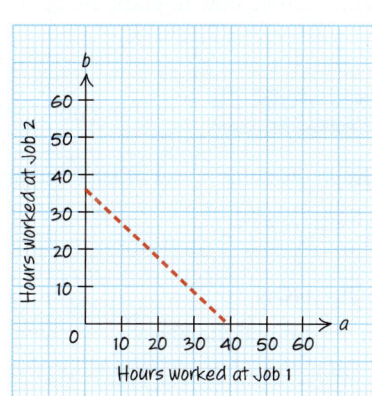

426 CHAPTER 4 Systems of Linear Equations

What's That Mean?

Quadrant

The xy-coordinate plane is divided into four *quadrants* by the axes. They are labeled counterclockwise, as below.

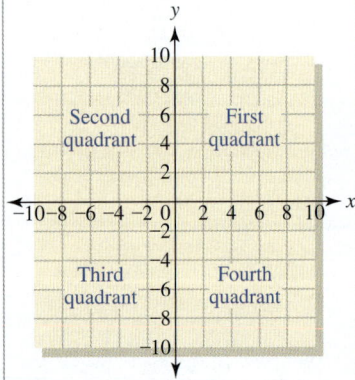

We are looking only for the part of the solution set that lies in the first quadrant. Graph the line $11.75a + 12.50b = 450$ in the first quadrant. Graphing a on the horizontal axis and b on the vertical axis yields the following graph.

Selecting a point that is not on the line, for instance, $(0, 0)$, and substituting it into the original inequality yields

$$11.75(0) + 12.50(0) \stackrel{?}{>} 450 \quad \text{Substitute in } a = 0, b = 0.$$

$$0 \stackrel{?}{>} 450 \quad \text{This is a false statement.}$$

Therefore, $(0, 0)$ is not in the solution set. Shade in the region above the line.

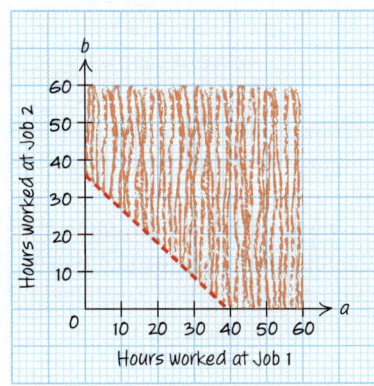

The point $(20, 25)$ lies in the solution set. So if Ricardo works 20 hours a week at his first job and 25 hours a week at his second job, he can earn more than $450.00 a week. Any combination of hours in the shaded region will earn Ricardo more than $450.00 a week.

PRACTICE PROBLEM FOR EXAMPLE 6

Kelli wants to buy two kinds of candy to bring to her daughter's scout troop meeting. She wants to spend no more than $20.00. The first type of candy costs $2.25 a pound, and the second type of candy costs $4.25 a pound. Write an inequality that expresses the relationship between the total cost of the candy and the maximum Kelli wants to spend on candy ($20.00). Then graph the solution set for this inequality.

Connecting the Concepts

What is the solution set for horizontal and vertical lines?

Since the value of y gives the vertical coordinate, when we graph the solution set to a $y > b$ problem, we shade above the line. Likewise, when graphing the solution set to a $y < b$ problem, we shade below the line. The value of x gives the horizontal coordinate. We shade to the right of the line when graphing the solution set of an $x > c$ problem and to the left of the line for an $x < c$ problem.

■ Graphing Vertical and Horizontal Inequalities

In Chapter 3, we learned how to graph vertical and horizontal lines. Recall that the graph of a horizontal line has the form $y = b$, where b is any real number. The graph of a vertical line has the form $x = c$, where c is any real number. In the next example, we will examine how to graph linear inequalities involving horizontal and vertical lines.

Example 7 Graphing horizontal and vertical inequalities

Graph the solution set of each inequality.

a. $y > -2$ **b.** $x \leq 3.5$

SOLUTION

a. First graph the line $y = -2$. $y = -2$ is of the form $y = b$, so this equation will graph as a horizontal line through $y = -2$. This line should be dashed, since the inequality does not have an equal to part. Selecting the test point to be (0, 0) and substituting it into the original inequality, we get

$$0 \stackrel{?}{>} -2 \quad \text{Substitute in } y = 0.$$
$$0 > -2 \quad \text{This is a true statement.}$$

Since this is a true statement, we shade in the region above the line $y = -2$. This shaded region contains the point (0, 0).

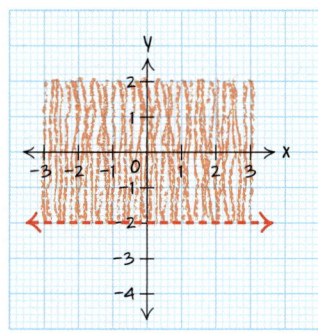

b. The graph of $x = 3.5$ is the graph of a vertical line. Select (0, 0) as a test point, since it is not on the line. Substituting it into the original inequality yields

$$0 \stackrel{?}{\leq} 3.5 \quad \text{Substitute in } x = 0.$$
$$0 \leq 3.5 \quad \text{This is a true statement.}$$

Since this is a true statement, we shade in the part of the plane that contains (0, 0).

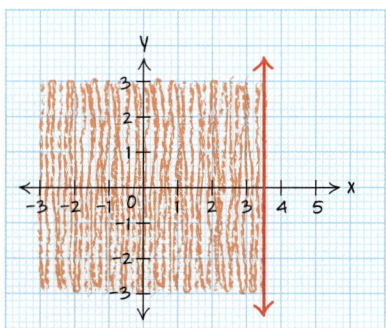

PRACTICE PROBLEM FOR EXAMPLE 7

Graph the solution set of each inequality.

a. $x > 1$ **b.** $y \leq 4$

4.4 Vocabulary Exercises

1. A linear inequality has one of the symbols _____ in it instead of an equal sign.

2. The solution set to a linear inequality is an entire _____ on the graph.

3. The solution set of a linear inequality is drawn with a(n) _____ line when the inequality symbols is $<$ or $>$.

4. The solution set of a linear inequality is drawn with a(n) _____ line when the inequality symbols is $\geq$ or $\leq$.

428 CHAPTER 4 Systems of Linear Equations

4.4 Exercises

For Exercises 1 through 10, determine whether each given point is in the solution set of the linear inequality.

1. $y < 2x - 3$
   a. $(5, 0)$
   b. $(6, 0)$

2. $y \geq -\dfrac{1}{4}x - 7$
   a. $(4, 6)$
   b. $(0, 0)$

3. $x \leq -5$
   a. $(-2, 1)$
   b. $(6, 6)$

4. $x > 7.9$
   a. $(8, 1)$
   b. $(-2, 60)$

5. $y > 33$
   a. $(0, 0)$
   b. $(-3, 58)$

6. $y \leq -1.5$
   a. $(3, 0)$
   b. $(-4, -7)$

7. $-4x + y \leq 4$
   a. $(0, 7)$
   b. $(1, -2)$

8. $2x - 3y \geq 2$
   a. $(0, -6)$
   b. $(-1, 5)$

9. $x > -6.5$
   a. $(1, 1)$
   b. $(-8, 10)$

10. $y < 9$
    a. $(-2, 9)$
    b. $(8, 10)$

For Exercises 11 through 18, use each given inequality to answer the following questions: (a) Would you graph the solution using a solid line or a dashed line? (b) Is the given point in the solution set? Note: Do not graph the inequality.

11. Inequality: $y < \dfrac{3}{5}x + 8$
    Point: $(0, 8)$

12. Inequality: $y > -6x + 12$
    Point: $(2, 0)$

13. Inequality: $x \leq -5$
    Point: $(-5, 2)$

14. Inequality: $y \geq -6.4$
    Point: $(0, -6.4)$

15. Inequality: $-4x + y \leq 4$
    Point: $(1, 8)$

16. Inequality: $2x - 3y \geq 2$
    Point: $(-2, -2)$

17. Inequality: $x - 6y > 9$
    Point: $(9, 0)$

18. Inequality: $-3x - 5y < 2$
    Point: $(6, -4)$

For Exercises 19 through 28:
a. *Select a test point.*
b. *Substitute it into the inequality.*
c. *Shade in the correct region to finish the graph of the inequality.*

19. $y < 3x - 3$

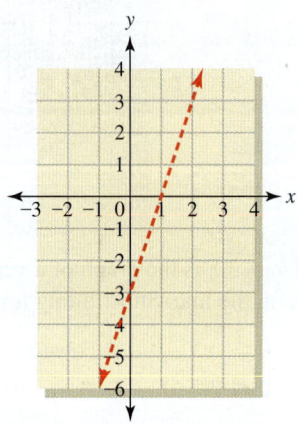

20. $y > -3x + 5$

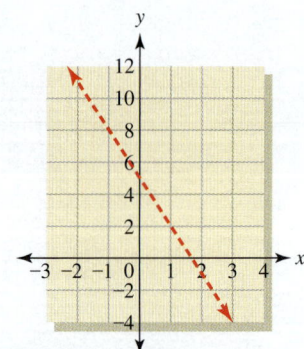

21. $4x - 3y \geq 12$

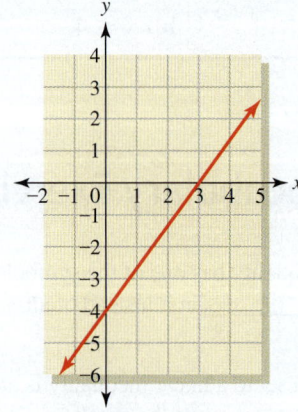

22. $-5x + 4y \leq 20$

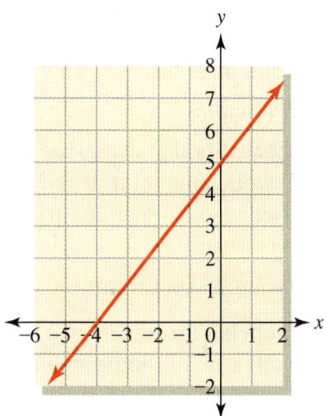

23. $-2y < 3x$

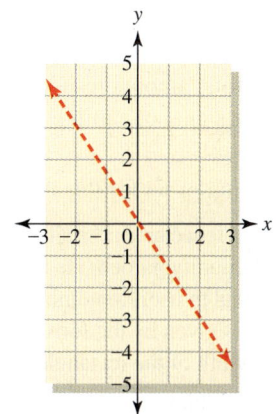

24. $x > 3y$

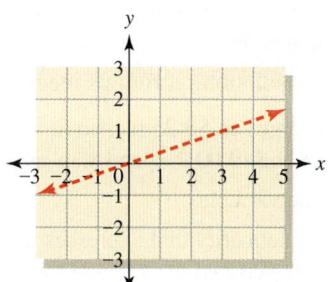

25. $x > 2$

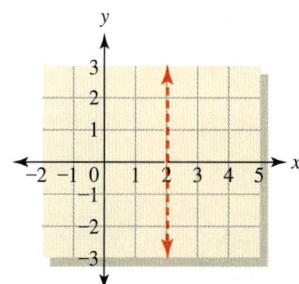

26. $x \leq -1$

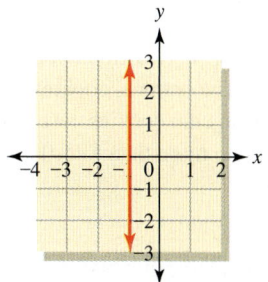

27. $y \geq -2$

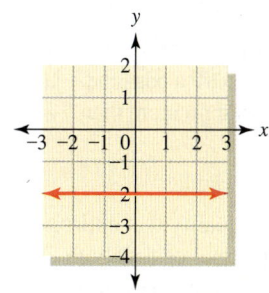

28. $y < 1$

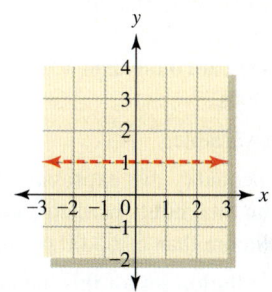

For Exercises 29 through 46, graph the solution set of each linear inequality. Clearly label and scale the axes.

29. $y < -2x + 6$ **30.** $y > -5x + 8$

31. $y \geq 3x - 7$ **32.** $y \leq 5x - 6$

33. $y \leq \frac{2}{3}x - 4$ **34.** $y \geq -\frac{1}{2}x + 1$

35. $y < -\frac{1}{4}x + 3$ **36.** $y < \frac{2}{5}x - 7$

37. $4x + 2y \leq 8$ **38.** $x + 3y \geq 6$

39. $4x + 3y < -12$ **40.** $3x + y > 9$

41. $x > -2$ **42.** $x \leq 5$

43. $y \geq 3.5$ **44.** $y < -1$

45. $x \geq 0$ **46.** $y < 0$

430 CHAPTER 4 Systems of Linear Equations

For Exercises 47 through 54, use the given information to answer the questions.

47. Jane is asked to bring two kinds of cookies to a Boy Scout meeting. The oatmeal cookies cost $4.00 a box, and the chocolate chip cookies cost $5.50 a box. She wants to spend no more than $25.00 on both kinds of cookies.
    a. Define the variables.
    b. Write an inequality that expresses the relationship between the total cost of both cookies and the maximum Jane wants to spend on cookies ($25.00).
    c. Graph the solution set for this inequality.

48. Joe has been asked to bring chips and salsa to a party. The chips cost $3.99 a bag, and a jar of salsa costs $3.29. He wants to spend no more than $20.00 on both chips and salsa.
    a. Define the variables.
    b. Write an inequality that expresses the relationship between the total cost of both chips and salsa and the maximum that Joe wants to spend ($20.00).
    c. Graph the solution set for this inequality.

49. Shahab works two jobs. The first job pays him $12.00 an hour, and the second job pays him $10.00 an hour. Shahab wants to earn more than $360.00 a week working both jobs.
    a. Define the variables.
    b. Write an inequality that expresses the relationship between the total hours worked and Shahab's goal of earning more than $360.00 per week.
    c. Graph the solution set for this inequality.

50. Tanya works two jobs. The first job pays her $11.50 an hour, and the second job pays $10.50 an hour. She has to earn more than $500.00 a week working both jobs.
    a. Define the variables.
    b. Write an inequality that expresses the relationship between the total hours Tanya works a week and her goal of earning more than $500.00 a week.
    c. Graph the solution set for this inequality.

51. Francisco has split his investment into two accounts. The first account pays 5% simple interest a year, and the second account pays 6% simple interest a year. Francisco would like to earn more than $500 a year in interest.
    a. Define the variables.
    b. Write an inequality that expresses the relationship between the total amount invested in both accounts and Francisco's goal of earning more than $500 in interest every year.
    c. Graph the solution set for this inequality.

52. LaDonna has split her investment into two accounts. The first account pays 3% simple interest a year, and the second account pays 5% simple interest every year. LaDonna would like to earn more than $1200 a year in interest.
    a. Define the variables.
    b. Write an inequality that expresses the relationship between the total amount invested in both accounts and LaDonna's goal of earning more than $1200.00 a year in interest.
    c. Graph the solution set for this inequality.

53. Paul is building a fence around a site that will become a rectangular community garden. He has at most 200 feet of fencing material to use. Remember that the perimeter P of a rectangle is given by the formula $P = 2w + 2l$, where w is the width of the rectangle and l is the length.
    a. Define the variables.
    b. Write an inequality that expresses the relationship between the total amount of fencing actually used and fact that Paul has at most 200 feet of fencing material available.
    c. Graph the solution set for this inequality.

54. A taxicab charges an $8.25 pickup fee plus $3.25 a mile. Maria has only $40.00 to spend on a cab.
    a. Define the variables.
    b. Write an inequality that expresses the relationship between the cost of the cab and the maximum amount Maria has to spend.
    c. Graph the solution set for this inequality.

For Exercises 55 through 60, match the inequalities with the graphs.

55.

1. $y \geq 2$
2. $y > \frac{1}{2}x + 2$

Graph A

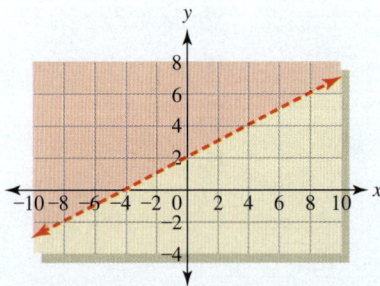

Graph B

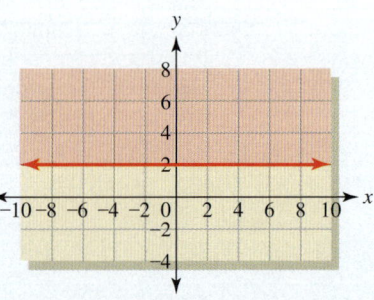

56.

1. $y \geq x - 4$
2. $x \leq 6$

Graph A
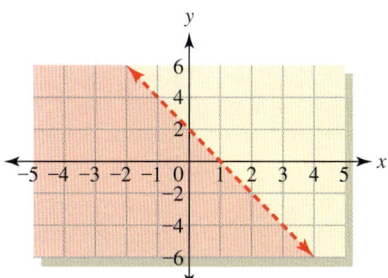

Graph B
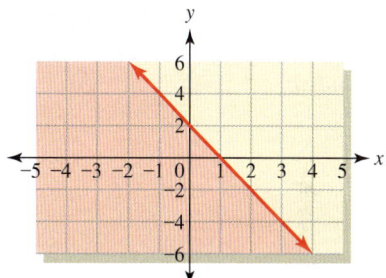

57.

1. $y < -2x + 2$
2. $y \leq -2x + 2$

Graph A
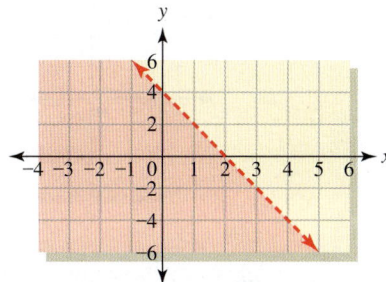

Graph B
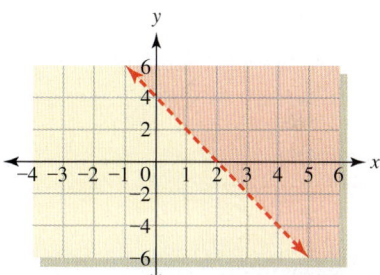

58.

1. $y < -2x + 4$
2. $y > -2x + 4$

Graph A
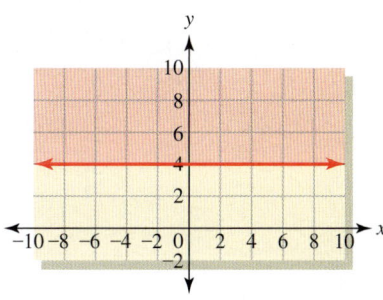

Graph B
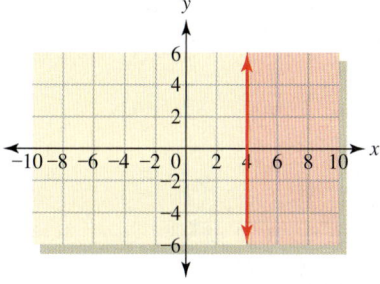

59.

1. $y \geq 4$
2. $x \geq 4$

Graph A

Graph B

432 CHAPTER 4 Systems of Linear Equations

60.

Graph A

Graph B

1. $x \geq 0$
2. $y \geq 0$

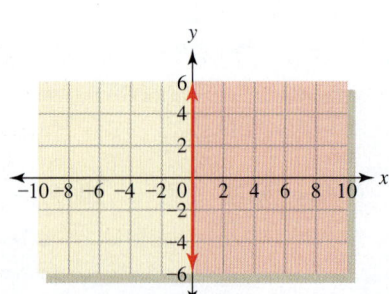
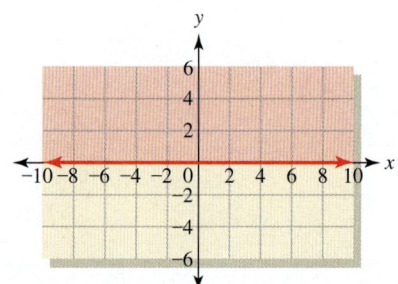

For Exercises 61 through 68, write the linear inequalities for the given graphs. Use the following steps: (a) Find the equation of the line shown in the graph. (b) Determine whether to use < or >, or ≤ or ≥. This depends on whether the line is solid or dashed. (c) Determine the direction of the inequality symbol.

61.

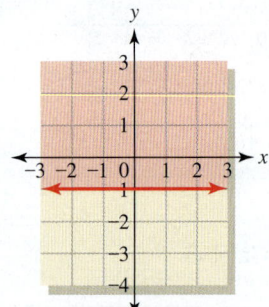

62.

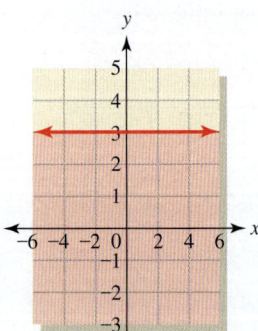

63.

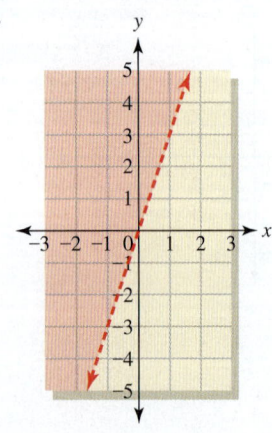

64.

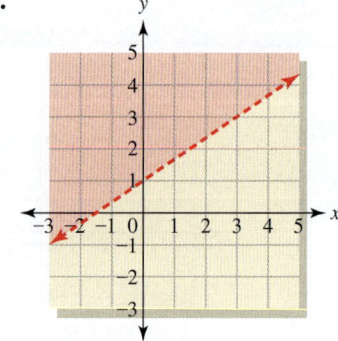

65.

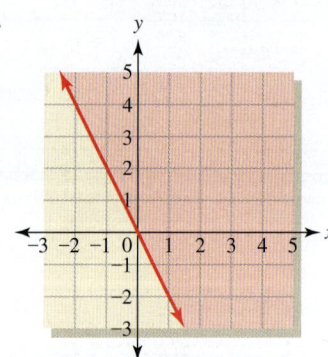

66.

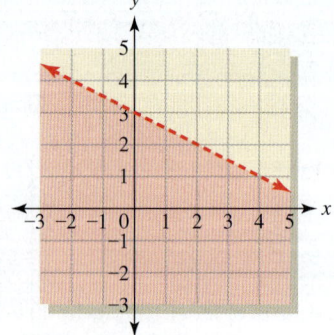

67.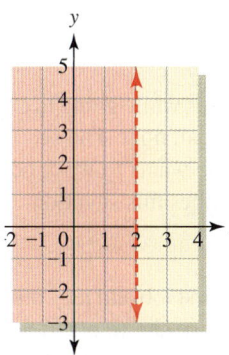

68.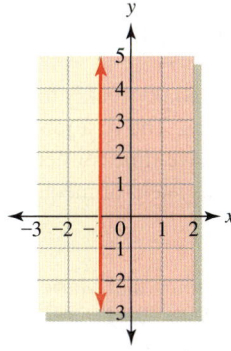

For Exercises 69 through 76, read each problem and determine whether it is an example of an equality or inequality. Do not solve.

69. Pavel wants to earn at least $150.00 in interest a year.

70. Philippa wants to earn $250.00 in interest a year.

71. Naveed wants to work 20 hours a week.

72. Parmida wants to work at most 25 hours a week.

73. Chato wants to study at least 15 hours a week.

74. Zhina wants to earn a 90% on her English essay.

75. Mailin wants to earn at least a 95% on her math test.

76. Sheng wants to earn $14.50 an hour working at his job.

For Exercises 77 through 90, answer as indicated.

77. Find the slope and y-intercept of the line $-3x + 8y = -24$.

78. Find the slope and y-intercept of the line $\frac{2}{3}x - 6y = -4$.

79. Write the equation of the line passing through the points $(-5, 3)$ and $(10, -3)$.

80. Write the equation of the line passing through the points $(-8, -11)$ and $(4, -2)$.

81. Simplify: $-81 \div 27 \cdot (-3) + (-2)^2 - |4 - 12|$.

82. Simplify: $-125 \div (5)^2 + \frac{25 - 16}{5 - 2} - |8 - 4(5)|$.

83. 65% of the class is female. There are 40 students in the class. How many students are male?

84. 55% of the class is male. There are 60 students in the class. How many students are female?

85. Convert units as indicated. Convert 490 miles to kilometers. Round to two decimal places if necessary.

86. Convert units as indicated. Convert 760 kilometers to miles. Round to two decimal places if necessary.

87. Solve for the variable. Check the answer. $2(2x - 1) + 5 = -21 - 2x$.

88. Solve for the variable. Check the answer. $-3y + 5(y - 4) + 1 = 18 - y$.

89. Translate the sentence into an equation and solve. The product of 5 and a number plus 12 is 10.

90. Translate the sentence into an equation and solve. The quotient of a twice a number and 5 is 4.

4.5 Systems of Linear Inequalities

LEARNING OBJECTIVE

- Find the solution set to a system of linear inequalities.

In Section 4.4, we learned that linear inequalities may be used to solve various applications problems that involve a constraint. In this section, we will look at systems of linear inequalities. Systems will apply to problems that involve more than one constraint.

> **DEFINITION**
>
> **System of Linear Inequalities** A system of linear inequalities is a set of two or more linear inequalities.

Example 1 — Writing inequalities

Marielle has to earn at least $250.00 weekly to meet her expenses while going to college. She works at two part-time jobs. The first job pays her $11.25 an hour. The second job pays her $11.70 an hour. To have time for her studies, Marielle wants to work no more than 25 hours a week. Write a system of inequalities that describes this situation.

SOLUTION

Let x represent the number of hours that Marielle works at her first job and let y represent the number of hours that Marielle works at her second job. Since she wants to work no more than 25 hours a week, we have the first inequality $x + y \leq 25$. Her total earnings from both jobs should be at least $250.00. Therefore, we have the second inequality $11.25x + 11.70y \geq 250.00$. The system of inequalities that describes this situation is

$$x + y \leq 25$$
$$11.25x + 11.70y \geq 250.00$$

Here both x and y are greater than or equal to zero because both variables refer to the number of hours worked weekly.

In Section 4.1, we saw that the solution point to a system of linear equations had to satisfy both equations. The solution set to a system of linear inequalities will consist of all the points that satisfy *both* inequalities. In the Concept Investigation, we will examine whether a point is in the solution set of a system of linear inequalities.

CONCEPT INVESTIGATION

What's in the solution set?

Consider the following system of linear inequalities.

$$y \geq -x + 1$$
$$y \geq 2x - 2$$

Answer each of the following questions. Remember that if a point is in the solution set, it must make *both* inequalities true.

1. Is the point (0, 0) in the solution set?
2. Is the point (3, 0) in the solution set?
3. Is the point (1, 4) in the solution set?
4. The graph of the system is given below. Which shaded region represents the solution set for the system of linear inequalities?

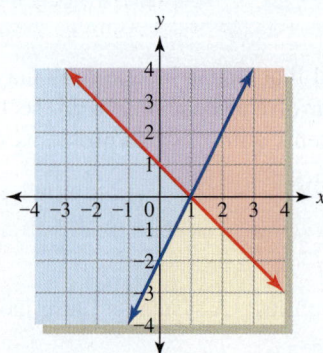

In Section 4.4, we learned how to find the solution set for a single linear inequality in two variables graphically. We now apply the same approach to finding the solution set for a system of inequalities graphically.

> **Steps to Determine the Solution Set to a System of Linear Inequalities**
> 1. Graph the first inequality as a solid or dashed line and shade the appropriate region (see Section 4.4).
> 2. On the same grid, graph the second inequality as a solid or dashed line and shade in the appropriate region.
> 3. The **solution set** to the system is the **intersection** of the two shaded areas from steps 1 and 2. Carefully shade in the solution set darker to distinguish it.

Example 2 Working to meet expenses

In Example 1, we discussed how Marielle has to earn at least $250.00 weekly to meet her expenses while going to college. She works two part-time jobs. The first job pays her $11.25 an hour. The second job pays her $11.70 an hour. To have time for her studies, Marielle wants to work no more than 25 hours a week. The system of linear inequalities that describes this situation (from Example 1) is

$$x + y \leq 25$$
$$11.25x + 11.70y \geq 250.00$$

Here x represents the number of hours that Marielle works weekly at her first job, and y represents the number of hours that Marielle works weekly at her second job. Since both x and y refer to the number of hours worked a week, they both take on positive values (or 0) only.

a. Graphically solve the system of linear inequalities.

b. Check the solution set with a sample test point. Interpret a sample point in the solution set in terms of Marielle's problem.

SOLUTION
a. Step 1 Graph the first inequality as a solid or dashed line, and shade the appropriate region.

The first inequality is $x + y \leq 25$. This line has an x-intercept of $(25, 0)$ and y-intercept of $(0, 25)$.

Graph this inequality as a solid line because it has a $\leq$ and therefore includes the equal to part.

Use $(0, 0)$ as a test point to determine whether to shade above or below the line.

$$0 + 0 \stackrel{?}{\leq} 25$$
$$0 \leq 25$$

Since (0, 0) is in the solution set, shade below the line.

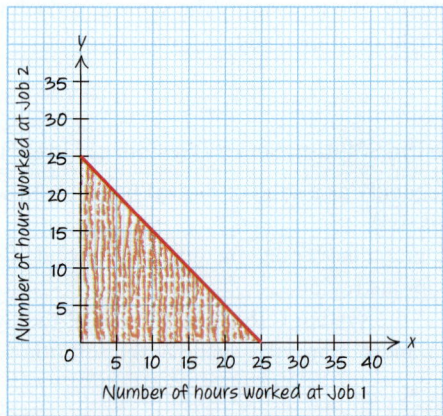

Step 2 On the same grid, graph the second inequality as a solid or dashed line and shade the appropriate region.

The second inequality is $11.25x + 11.70y \geq 250$. This line has an *x*-intercept of (22.22, 0), and *y*-intercept of (0, 21.37). Graph this inequality as a solid line because it has a $\geq$ symbol.

Since (0, 0) is not on the line, we can substitute it in the original inequality as a test point.

$$11.25(0) + 11.70(0) \stackrel{?}{\geq} 250 \quad \text{Substitute in } x = 0, y = 0.$$
$$0 \stackrel{?}{\geq} 250$$
$$0 \geq 250 \quad \text{This is a false statement.}$$

Since this is a false statement, shade in the half-plane that does *not* include the point (0, 0), that is, above the line.

Step 3 The solution set to the system is the intersection of the two shaded areas from steps 1 and 2.

The graph of the system of inequalities is as follows.

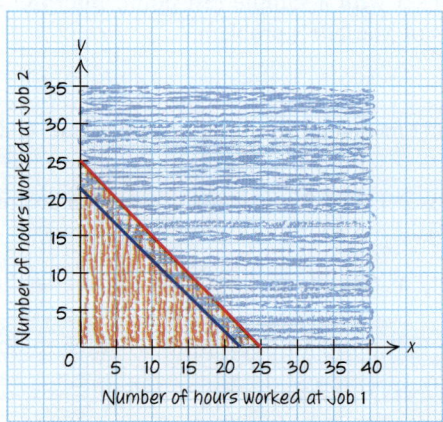

b. As a check on our work, we see that the point (5, 18) is in the shaded region. Substitute $x = 5$ and $y = 18$ into the original system.

$$5 + 18 \stackrel{?}{\leq} 25 \quad \text{Substitute in } x = 5, y = 18. \quad 11.25(5) + 11.70(18) \stackrel{?}{\geq} 250$$
$$23 \leq 25 \quad \text{These are true statements.} \quad 266.85 \geq 250$$

Since these are both true statements, the point (5, 18) is a solution of the system. This means that one possible solution is that Marielle can work 5 hours a week at her first job and 18 hours a week at her second job, earn more than $250.00 for the week, and work less than 25 hours weekly.

What's That Mean?

Intersection

In mathematics, an *intersection* means the set of elements that are *in common* to two or more sets.

SECTION 4.5 Systems of Linear Inequalities   437

PRACTICE PROBLEM FOR EXAMPLE 2

Tim's Cabinet Shop makes kitchen cabinets and bathroom cabinets for local builders. Kitchen cabinets take 8 hours of labor to make, and bathroom cabinets take 3 hours of labor to make. To stay profitable, the shop has to produce at least 60 cabinets a week. The total labor available at the shop is a maximum of 500 hours per week.

a. Write a system of linear inequalities that describes this situation.

b. Graphically solve the system of linear inequalities.

c. Check the solution set with a sample test point. Interpret a sample point in the solution set in terms of the problem.

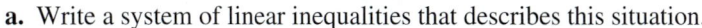

Example 3 Graphing systems of inequalities

Graph the solution set of each system of inequalities.

a. $y \leq -3x + 6$
   $y \geq 2x - 4$

b. $10x - 16y \geq -80$
   $y < 3$

SOLUTION

a. Notice that both inequalities are of the form $\geq$ or $\leq$, so both will be graphed as solid lines. While graphing the first line $y \leq -3x + 6$, notice that it is in $y = mx + b$ form. Graph it by using the y-intercept of $(0, 6)$ and the slope $m = -\frac{3}{1}$ to get to another point on the graph, $(1, 3)$. Selecting $(0, 0)$ as a test point and substituting it into the inequality, we get

$$0 \stackrel{?}{\leq} -3(0) + 6 \qquad \text{Substitute in } x = 0, y = 0.$$
$$0 \leq 6 \qquad \text{This is a true statement.}$$

Shade the region including $(0, 0)$, which is below the line.

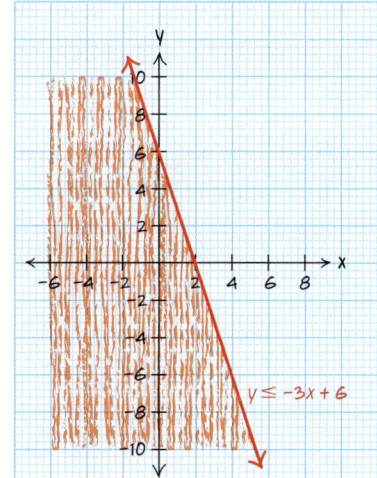

To graph the second inequality, $y \geq 2x - 4$, first graph the line by using the slope and the y-intercept. The y-intercept is the point $(0, -4)$, and the slope is $m = \frac{2}{1} = \frac{\text{rise}}{\text{run}}$. Starting from the y-intercept and using the slope, we get to another point on the line, $(1, -2)$. Select $(0, 0)$ as a test point because it is not on the line. Substituting $(0, 0)$ into the inequality yields

$$0 \stackrel{?}{\geq} 2(0) - 4 \qquad \text{Substitute in } x = 0, y = 0.$$
$$0 \geq -4 \qquad \text{This is a true statement.}$$

Therefore, we shade above the line.

After graphing both inequalities on the same axes, we shade the overlapping region (the intersection). This produces the following graph.

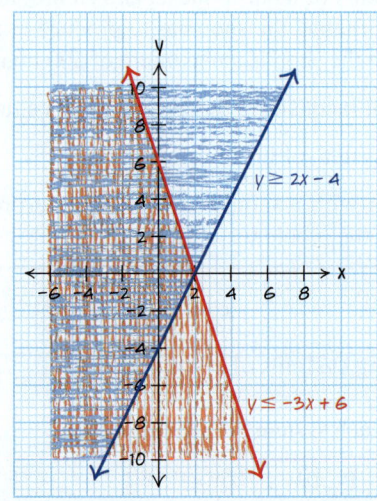

Note that the solution set consists of all points in the overlapping shaded region.

As a check of our work, we see that the point $(-2, 0)$ is in the solution set. Substituting $x = -2$ and $y = 0$ into both original inequalities yields

Inequality 1

$0 \stackrel{?}{\leq} -3(-2) + 6$ Substitute in $x = -2, y = 0$.

$0 \stackrel{?}{\leq} 6 + 6$

$0 \leq 12$ These are true statements.

Inequality 2

$0 \stackrel{?}{\geq} 2(-2) - 4$

$0 \stackrel{?}{\geq} -4 - 4$

$0 \geq -8$

b. The first inequality is $10x - 16y \geq -80$. Begin by graphing the line $10x - 16y = -80$. This line has an x-intercept of $(-8, 0)$, and y-intercept of $(0, 5)$. Since the point $(0, 0)$ is not on the line, use it as a test point.

$10(0) - 16(0) \stackrel{?}{\geq} -80$ Substitute in $x = 0, y = 0$.

$0 \geq -80$ This is a true statement.

Since this statement is true, shade the half-plane containing $(0, 0)$, that is, below the line.

The second inequality is $y < 3$. Therefore, using a dashed line, graph the horizontal line $y = 3$, and shade below the line.

The graph of the solution set of the system of inequalities is the intersection of the two solution sets and has been shaded in the following graph.

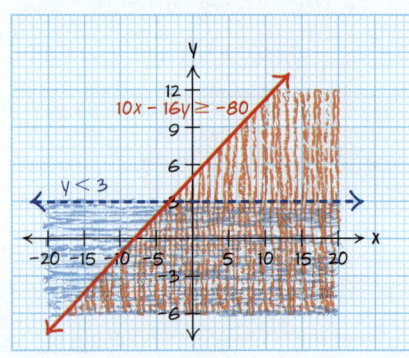

Connecting the Concepts

What's true about points in the solution set?

Remember that a point in the solution set must make *both* inequalities true.

Skill Connection

Horizontal Inequalities

In Section 4.4, we learned that since the value of y gives the vertical coordinate, when graphing the solution set to a $y \geq b$ problem, we shade above the line. Likewise, when graphing the solution set to a $y \leq b$ problem, we shade below the line.

To check our work, we see that the point (0, 0) is in the solution set. Substituting $x = 0$ and $y = 0$ into both original inequalities yields

Inequality 1	**Inequality 2**
$10(0) - 16(0) \stackrel{?}{\geq} -80$	$0 < 3$
$0 \geq -80$ True	$0 < 3$ True

PRACTICE PROBLEM FOR EXAMPLE 3

Graph the solution set for each system of inequalities.

a. $y < -\dfrac{5}{2}x + 8$
$y \geq 4x - 6$

b. $2x - 3y \geq 21$
$y < -3$

Some of the applications that we previously looked at in systems of linear equations may be more realistic when stated as a system of inequalities. One such type of application comes from investment problems.

Example 4 Applying constraints to an investment

Mitsuko recently won a lottery game and has a maximum of $50,000.00 to invest. Her financial advisor discusses investment options with her, and Mitsuko decides to take on less risky investments. Her financial advisor tells her to put part of the money in a CD at the bank earning 2.5% simple interest and part of the money in a bond fund earning 4% interest. Mitsuko's yearly goal is to earn a total of at least $1200.00 in interest from both accounts.

a. Write a system of linear inequalities that describes this situation.

b. Graphically solve the system of linear inequalities.

c. Check the solution set. Interpret a sample point from the solution set in terms of Mitsuko's investments.

SOLUTION

a. Let b represent the amount of money (in dollars) invested in bonds and let c represent the amount of money (in dollars) invested in a CD. Since Mitsuko has a maximum of $50,000.00 to invest, our first inequality is

$$b + c \leq 50000.00$$

The minimum total interest Mitsuko would like to earn every year is $1200.00. We have to use the formula $I = P \cdot r \cdot t$, with $t = 1$ year. The simple interest earned on bonds added to the simple interest earned on the CDs should be at least $1200.00 per year.

$$0.04b + 0.025c \geq 1200$$

Therefore, the system of inequalities that describes this situation is

$$b + c \leq 50000$$
$$0.04b + 0.025c \geq 1200$$

Note that since b and c represent the amount of money (in dollars) invested, we have that $b \geq 0$ and $c \geq 0$.

What's That Mean?

Certificate of Deposit (CD) and Money Market Accounts

A CD is a *certificate of deposit*. CDs are like savings accounts in that they are federally insured. How a CD differs from a savings account is that the money must be left in for a fixed amount of time (usually 3 months, 6 months, or longer) for a fixed interest rate. The interest rates for CDs are usually higher than those for standard savings accounts.

A *money market account* is a savings account that is not federally insured. Money market accounts earn higher interest than most CDs and savings accounts.

b. Graph b on the horizontal axis and c on the vertical axis. Since Mitsuko could invest at most the entire $50,000 in bonds or the entire $50,000 in CDs, we scale both axes from 0 to 60,000 by 10,000's. The line $b + c = 50000$ can be graphed by finding the intercepts. This graph appears below. Draw a solid line because it is an inequality of the form $\leq$ and then pick (10000, 20000) as a test point that is not on the line. Substituting these values into the original inequality yields

$$10000 + 20000 \stackrel{?}{\leq} 50000 \qquad \text{Substitute in } b = 10000, c = 20000.$$
$$30000 \leq 50000$$

Since this is a true statement, shade the half-plane that contains the test point (10000, 20000). Shade below the line.

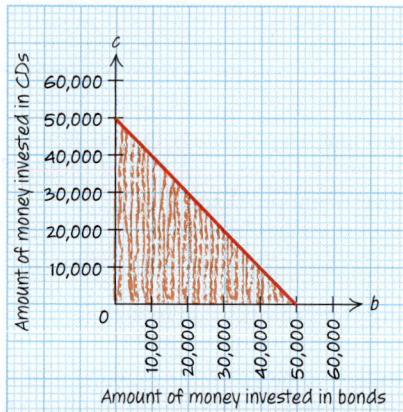

Now graph the second inequality on the same graph as the first. The second inequality is $0.04b + 0.025c \geq 1200$. To graph the line $0.04b + 0.025c = 1200$, plot the intercepts. This line has a b-intercept of (30000, 0) and c-intercept of (0, 48000). Since the origin (0, 0) is not on the line, we can use this as a test point. Substituting (0, 0) into the original inequality yields

$$0.04(0) + 0.025(0) \stackrel{?}{\geq} 1200 \qquad \text{Substitute in } b = 0, c = 0.$$
$$0 \geq 1200 \qquad \text{This is a false statement.}$$

Because this is a false statement, shade above the line.

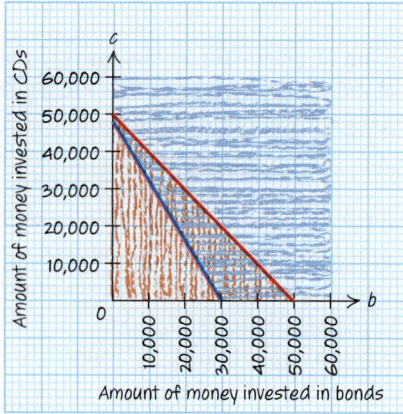

The overlapping region is between the two lines, and that region is the solution set.

c. As a check on our work, select a point in the solution set and determine whether it satisfies both inequalities. The point (20000, 20000) is in the solution set (shaded region). Checking yields

$20000 + 20000 \stackrel{?}{\leq} 50000$ Check both inequalities. $0.04(20000) + 0.025(20000) \stackrel{?}{\geq} 1200$

$40000 \leq 50000$ Both inequalities check. $1300 \geq 1200$

Therefore, this check of our work is valid. The point (20,000, 20,000) means that if Mitsuko invests $20,000 in the bond fund and invests $20,000 in CDs, she will earn over $1200 a year in interest, and her total investment will be less than or equal to $50,000.

PRACTICE PROBLEM FOR EXAMPLE 4

Members of the local high school track team are having a fund-raiser. They have only up to 1000 candy bars to sell. Two types of candy bars are for sale. The Nutter Bar costs $1.50 per bar, and the Peanut Bar costs $2.50 per bar. The team would like to raise at least $2000.00 from this sale.

a. Write a system of linear inequalities that describes this situation.

b. Graphically solve the system of linear inequalities.

c. Check the solution set. Interpret a sample point from the solution set in terms of candy bar sales for the track team.

4.5 Vocabulary Exercises

1. The solution set to a system of linear inequalities is the _____ of the solution sets of the two inequalities.

2. Graphically, we _____ in the intersection of the solution sets of the two inequalities to show the solution set to the system of linear inequalities.

4.5 Exercises

For Exercises 1 through 8, test to determine whether the given points are in the solution set of the systems of inequalities.

1. $x + y < 1$
   $x - y > 2$
   a. $(0, 0)$
   b. $(0, -5)$

2. $y \geq x - 3$
   $y \leq -x + 2$
   a. $(0, 0)$
   b. $(-1, 7)$

3. $y > 2x$
   $y > -3x + 2$
   a. $(0, 0)$
   b. $(6, -3)$

4. $y \leq x - 4$
   $y > -2x$
   a. $(0, 0)$
   b. $(0, -6)$

5. $3x - 4y \leq 12$
   $x + 2y > 6$
   a. $(4, 5)$
   b. $(0, 0)$

6. $y < -2x + 4$
   $5x + 10y \geq 20$
   a. $(0, 0)$
   b. $(-2, 6)$

7. $x < 1$
   $y > 2$
   a. $(0, 0)$
   b. $(0, 6)$

8. $x \geq -3$
   $y \leq -5$
   a. $(0, 0)$
   b. $(2, -6)$

For Exercises 9 through 14, determine whether the given points are in the solutions set of the systems of inequalities. Note: The blue region is the solution set for the blue line, and the light pink region is the solution set for the red line. The purple region is the overlap of these two solution sets.

9.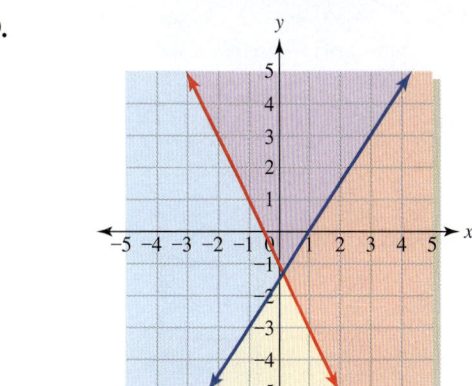

a. $(0, 0)$ b. $(5, 1)$

442 CHAPTER 4 Systems of Linear Equations

10.

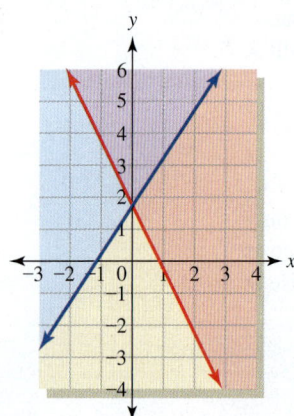

a. (0, 3)   b. (3, −1)

11.

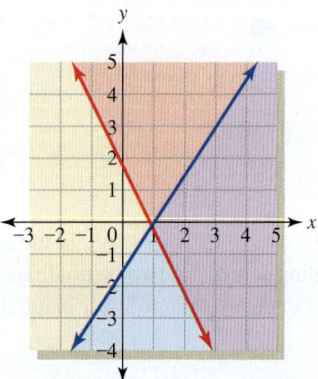

a. (5, 0)   b. (1, 4)

12.

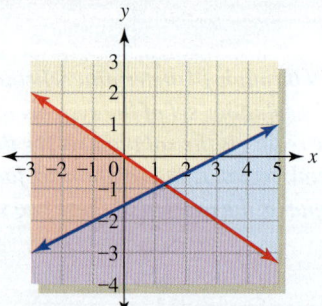

a. (4, −1)   b. (1, −3)

13.

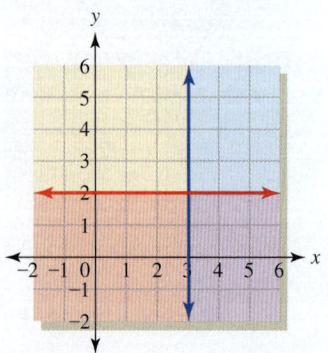

a. (4, 1)   b. (4, 0)

14.

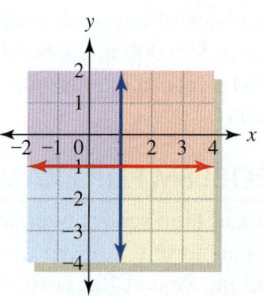

a. (−1, 1)   b. (0, 0)

For Exercises 15 through 30, graph the solution set for each system of inequalities. Check the solution set with a sample test point.

15. $y > -x + 1$
    $y < 2x + 2$

16. $y > x + 4$
    $y < -2x$

17. $y \geq -\dfrac{1}{2}x + 5$
    $y > 4x$

18. $y > 5x$
    $y \leq -2x + 3$

19. $2x + y \geq 4$
    $-x + y \leq 6$

20. $4x + 3y \leq 12$
    $x + 2y \geq 2$

21. $x - 3y < 12$
    $2x - 3y > -3$

22. $x - 5y < 10$
    $2x + 3y > 0$

23. $x > 5$
    $y < 1$

24. $x < -2$
    $y < 4$

25. $x \geq 3$
    $y < 2x + 1$

26. $y \geq 1$
    $y < -3x$

27. $y > 0$
    $x > 0$

28. $y > 0$
    $x < 0$

29. $y < 0$
    $x < 0$

30. $y < 0$
    $x > 0$

For Exercises 31 through 38, solve each system of linear inequalities as indicated.

31. Samuel has to earn more than $175 a week to meet his living expenses while going to college. He works at two part-time jobs. The first job pays $11.00 an hour, and the second job pays $11.50 an hour. To have time to attend college, Samuel wants to work no more than 20 hours a week.

    a. Write a system of linear inequalities that describes this situation.

    b. Graphically solve the system of linear inequalities.

c. Check the solution set with a sample test point. Interpret a sample point from the solution set in terms of Samuel's work schedule.

32. Daniella wants to earn more than $275 a week to meet her living expenses while going to school. She works at two part-time jobs. The first job pays $10.75 an hour, and the second job pays $11.25 an hour. To have time to attend college, Daniella wants to work no more than 25 hours a week.

   a. Write a system of linear inequalities that describes this situation.

   b. Graphically solve the system of linear inequalities.

   c. Check the solution set with a sample test point. Interpret a sample point from the solution set in terms of Daniella's work schedule.

33. Darren has at most $5000.00 to invest in stocks. He has two stock funds in mind. The first stock fund returns 8% a year and is very stable. The second, more risky stock fund may return 10% a year. Darren would like to earn more than $100 a year in simple interest.

   a. Write a system of linear inequalities that describes this situation.

   b. Graphically solve the system of linear inequalities.

   c. Check the solution set with a sample test point. Interpret a sample point from the solution set in terms of Darren's investments.

34. Donna has at most $10,000.00 to invest in bonds. She has two bond funds in mind. The first bond fund returns 5% a year and is very stable. The second, more risky bond fund may return 6.5% a year. Donna would like to earn more than $120 a year in simple interest.

   a. Write a system of linear inequalities that describes this situation.

   b. Graphically solve the system of linear inequalities.

   c. Check the solution set with a sample test point. Interpret a sample point from the solution set in terms of Donna's investments.

35. A toy manufacturer is making two toys for the Christmas season. The manufacturing plant works c hours per week producing cars and t hours per week producing trucks. The total maximum number of production hours the plant can operate per week is 80. However, the plant must stay in operation at least 40 hours a week to pay its employees.

   a. Write a system of linear inequalities that describes this situation.

   b. Graphically solve the system of linear inequalities.

   c. Check the solution set with a sample test point. Interpret a sample point from the solution set in terms of the hours spent each week producing cars and trucks.

36. Francesca is a handbag designer and has two handbags in production for fall: a tote bag and a hobo bag. The manufacturing plant with which she has contracted spends t hours per week producing tote bags and spends h hours per week producing hobo bags. The plant can work at most 100 production hours per week and needs to work at least 60 production hours per week to pay the employees.

   a. Write a system of linear inequalities that describes this situation.

   b. Graphically solve the system of linear inequalities.

   c. Check the solution set with a sample test point. Interpret a sample point from the solution set in terms of the hours spent each week producing tote bags and hobo bags.

37. A university has at most 55,000 tickets to sell to its football games. There are two kinds of tickets: student tickets and general admission tickets. The university would like to sell no more than 10,000 student tickets.

   a. Write a system of linear inequalities that describes this situation.

   b. Graphically solve the system of linear inequalities.

   c. Check the solution set with a sample test point. Interpret a sample point from the solution set in terms of student tickets and general admission tickets sold.

38. A local high school has at most 3000 tickets to sell for its football games. There are two kinds of tickets: student tickets and general admission tickets. The high school would like to sell no more than 1500 student tickets.

   a. Write a system of linear inequalities that describes this situation.

   b. Graphically solve the system of linear inequalities.

   c. Check the solution set with a sample test point. Interpret a sample point from the solution set in terms of student tickets and general admission tickets sold.

CHAPTER 4 Systems of Linear Equations

For Exercises 39 through 44, fill in each blank with one of the following vocabulary words.

- *Inequality*
- *Equality*
- *Union*
- *Solid*
- *Linear inequalities*
- *Dashed*
- *Intersection*

39. $<$ is an example of a(n) _____ symbol.

40. A system of linear inequalities has two or more _____.

41. The solution set to a system of linear inequalities is the _____ of the solution sets of the individual inequalities in the system.

42. To graph an inequality of the form $\leq$, a(n) _____ line is used.

43. To graph an inequality of the form $>$, a(n) _____ line is used.

44. $=$ is an example of a(n) _____ symbol.

45. Solve the system.
$$2x + y = 7$$
$$y = -3x + 4$$

46. Solve the system.
$$4y - x = 9$$
$$x = 2y - 3$$

47. Solve the system.
$$x - 2y = 11$$
$$3x + 2y = -3$$

48. Solve the system.
$$-7x + y = 9$$
$$7x - 5y = 15$$

49. Is $(-3, 4)$ in the solution set of the linear inequality?
$$2x - 3y < 5$$

50. Is $(-2, -1)$ in the solution set of the linear inequality?
$$y \geq -2x + 4$$

51. Is $(10, 1)$ in the solution set of the system of linear inequalities?
$$y > -3x + 6$$
$$y < 2x - 4$$

52. Is $(6, -1)$ in the solution set of the system of linear inequalities?
$$y < -2x + 1$$
$$x > 0$$

53. Solve the system.
$$y < -\frac{2}{3}x + 6$$
$$y > \frac{1}{4}x - 10$$

54. Solve the system.
$$y \geq -2x + 1$$
$$y < 3x - 9$$

55. Solve the system.
$$2x + 7y > 21$$
$$4x - 3y > 15$$

56. Solve the system.
$$3x - 5y \leq 30$$
$$8x - 4y \geq -24$$

Chapter Summary

Section 4.1   Identifying Systems of Linear Equations

- A **system of equations** is a set of two (or more) linear equations.
- A **solution** to a system of equations is an ordered pair, (x, y), that satisfies both linear equations in the system.
- To **check** the solution to a system of equations, substitute the x-value and the y-value into both original equations. If both equations are true, then the solution checks.
- To **solve a system graphically,** use the following steps:
  1. On graph paper, carefully graph both lines on the same pair of axes.
     a. Scale the axes so both lines can be graphed simultaneously.
     b. Use a ruler to draw in the lines.
  2. Find the point of intersection of the system. This is the solution to the system.
  3. Write the solution as an ordered pair, (x, y).
  4. Check the answer by substituting the solution into *both* of the original equations.
- There are **three possibilities** for the solution set to a system of equations:
  1. There is one solution because the two lines intersect at exactly one point.
  2. There is no solution because the two lines are parallel and never intersect.
  3. There are infinitely many solutions consisting of all points on the line because the two lines are in fact the same line.

Example 1   Use the graph to determine the solution to the system of equations.

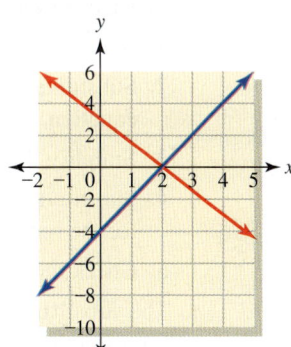

SOLUTION   The two lines intersect at the point $(2, 0)$. The solution to the system is the point $(2, 0)$.

Example 2   Graphically determine the solution to the system of equations. Check the solution.

$$y = \frac{1}{2}x$$
$$y = -2x + 5$$

446 CHAPTER 4 Systems of Linear Equations

SOLUTION Graphing the two lines on the same axes yields the graph below.

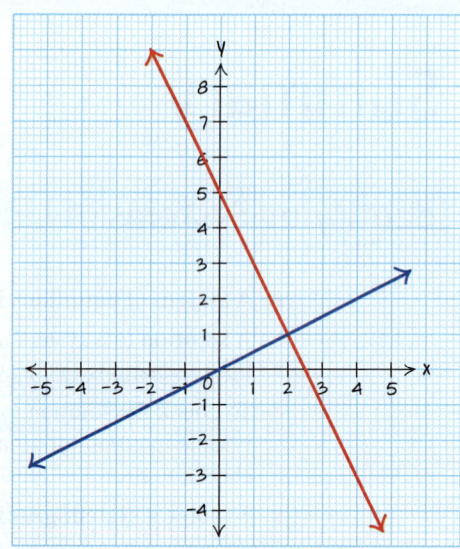

From the graph, we see the two lines intersect at the point $(x, y) = (2, 1)$. Checking the solution yields

Equation 1

$1 \stackrel{?}{=} \frac{1}{2}(2)$ Substitute in $x = 2$, $y = 1$.

$1 \stackrel{?}{=} 1$

$1 = 1$ The point checks in both equations.

Equation 2

$1 \stackrel{?}{=} -2(2) + 5$

$1 \stackrel{?}{=} -4 + 5$

$1 = 1$

Section 4.2 Solving Systems Using the Substitution Method

- The **substitution method** is an algebraic method of solving a system of equations.
- To solve a system using the **substitution method** follow these steps:
  1. Solve one of the two equations for one of the variables.
  2. Substitute the expression from step 1 in for the value of the variable in the other equation.
  3. Simplify the equation in step 2 and solve for the remaining variable.
  4. Substitute the answer from step 3 into either of the original equations and solve for the remaining variable.
  5. Check the solution by substituting into both original equations.
- The **substitution method** can determine whether a system has a single ordered pair for a solution, infinitely many solutions, or no solutions.
- The **distance formula** is $d = r \cdot t$. Here, d refers to the distance, r to the rate (or speed), and t to the time an object has traveled.
- The **value formula** is value = amount · unit price.

Example 3 Solve the system using the substitution method. Check the solution.

$$2x - 3y = 3$$
$$x - y = 2$$

SOLUTION The original system is

$$2x - 3y = 3$$
$$x - y = 2$$

CHAPTER 4 Summary 447

The second equation has a coefficient of 1 for the *x*-variable. Solving for *x* in the second equation yields

$$x - y = 2$$
$$\underline{+y \quad +y}$$
$$x = 2 + y$$

Substituting this into the first equation and simplifying, we get

$$2(2 + y) - 3y = 3 \qquad \text{Distribute and simplify.}$$
$$4 + 2y - 3y = 3$$
$$4 - y = 3 \qquad \text{Subtract 4 from both sides.}$$
$$\underline{-4 \qquad\qquad -4}$$
$$-y = -1 \qquad \text{Multiply both sides by } -1.$$
$$y = 1$$

Solve for *x*.

$$x = 2 + y \qquad \text{Substitute in } y = 1.$$
$$x = 2 + 1$$
$$x = 3$$

The solution is $(x, y) = (3, 1)$.

Checking $(x, y) = (3, 1)$ in both original equations yields

Equation 1	**Equation 2**
$2(3) - 3(1) \stackrel{?}{=} 3$ Substitute in $x = 3, y = 1$.	$3 - 1 \stackrel{?}{=} 2$
$6 - 3 \stackrel{?}{=} 3$	$2 \stackrel{?}{=} 2$
$3 = 3$	$2 = 2$

The solution checks in both equations.

Example 4 A boat traveled 30 miles downstream in 2 hours. It takes 3 hours to make the return trip upstream. Find the rate of the boat in still water and the rate of the current.

  a. Define the variables.
  b. Determine the two equations.
  c. Solve the system.

SOLUTION

a. Let *b* represent the rate of the boat in still water, in miles per hour, and *c* represent the rate of the current in miles per hour.

b. The two equations are below.

Recall that $r \cdot t = d$.

$$(b + c) \cdot 2 = 30 \qquad \text{Boat traveling with the current.}$$
$$(b - c) \cdot 3 = 30 \qquad \text{Boat traveling against the current.}$$

c.
$$\frac{(b + c) \cdot 2}{2} = \frac{30}{2} \qquad \text{Divide both sides of the first equation by 2.}$$
$$b + c = 15 \qquad \text{Solve for } b.$$
$$\underline{-c \quad -c}$$
$$b = 15 - c$$

Substituting this value in the second equation and simplifying, we get

$$(b - c) \cdot 3 = 30 \quad \text{Distribute and simplify.}$$
$$3b - 3c = 30 \quad \text{Substitute in } b = 15 - c.$$
$$3(15 - c) - 3c = 30 \quad \text{Distribute and simplify.}$$
$$45 - 3c - 3c = 30$$
$$45 - 6c = 30 \quad \text{Solve for } c.$$
$$\underline{-45 \qquad -45}$$
$$\frac{-6c}{-6} = \frac{-15}{-6} \quad \text{Reduce fractions.}$$
$$c = \frac{15}{6}$$
$$c = \frac{5}{2} = 2.5$$
$$b = 15 - c \quad \text{Find the value of } b.$$
$$b = 15 - 2.5$$
$$b = 12.5$$

The boat is traveling at 12.5 mph, and the current is 2.5 mph.

Section 4.3 Solving Systems Using the Elimination Method

- The **elimination method** is another algebraic method of solving a system of linear equations.
  1. Put the equations of both lines in general form ($Ax + By + C$), if necessary.
  2. If necessary, multiply both sides of one (or both) equation(s) so that the coefficients of one of the variables are opposites.
  3. Add the two equations vertically.
  4. Solve the equation for the remaining variable.
  5. Substitute the value of the known variable from step 4 into either of the original equations and solve for the other variable.
  6. Check the solution by substituting into *both* original equations.
- The **elimination method** can determine whether a system has a single ordered pair for a solution, infinitely many solutions, or no solution.
- The **simple interest** formula is
$$I = P \cdot r \cdot t$$
  Here I represents the interest earned, P is the principal, r is the interest rate, and t is the time in years.
- **The amount of a quantity in a solution** formula is
$$\text{given substance} = \text{percent} \cdot \text{amount of solution}$$
- Remember to write percents in decimal form when used in a problem.

Example 5 Solve each system using the elimination method. Check the solutions.

a. $4x - 3y = 21$
$x + 3y = 9$

b. $4x - 6y = -6$
$-2x + 3y = -21$

SOLUTION a. Because the coefficients of y are opposites (3 and -3), add the two equations vertically to eliminate y.

$$4x - 3y = 21$$
$$x + 3y = 9$$
$$\frac{5x}{5} = \frac{30}{5}$$
$$x = 6$$

Substituting this value in the second equation to find the value of y yields

$$(6) + 3y = 9$$
$$-6 \quad\quad -6$$
$$\frac{3y}{3} = \frac{3}{3}$$
$$y = 1$$

The solution is the point $(x, y) = (6, 1)$. Check the solution in both original equations.

Equation 1	Equation 2
$4(6) - 3(1) \stackrel{?}{=} 21$	$6 + 3(1) \stackrel{?}{=} 9$
$24 - 3 \stackrel{?}{=} 21$	$6 + 3 \stackrel{?}{=} 9$
$21 = 21$	$9 = 9$

The solution checks.

b. Examining the two equations, we notice that neither variable has coefficients that are opposites. Multiply the second equation by 2 to get the coefficients of x to be opposites.

$$4x - 6y = -6 \quad\quad\quad\quad 4x - 6y = -6$$
$$2(-2x + 3y) = 2(-21) \quad\quad -4x + 6y = -42$$

$$4x - 6y = -6$$
$$\underline{-4x + 6y = -42} \quad \text{Add the two equations.}$$
$$0 = -48 \quad \text{This statement is false.}$$

Because $0 \neq -48$, the system has no solution.

Section 4.4 Solving Linear Inequalities in Two Variables Graphically

- A **linear inequality** is an inequality of the form $Ax + By < C$. Here the symbol $<$ may be any one of the following inequality symbols: $<$, $>$, $\leq$, or $\geq$.
- Many applications involve inequalities. We are often interested in when one quantity is greater than or less than another.
- We **solve linear inequalities** in two variables graphically by the following steps:
  1. Graph the line.
  2. If the inequality is of the form $<$ or $>$, draw in a dashed line. If the inequality is of the form $\leq$ or $\geq$, draw in a solid line.
  3. The line separates the plane into two regions, above and below the line. Pick a test point in either region (not on the line) and substitute it into the original inequality.
     a. If the test point makes the inequality true, shade in the region that contains the test point.
     b. If the test point makes the inequality false, shade in the other region (the one that does not contain the test point).

450 CHAPTER 4 Systems of Linear Equations

- The **solution set** is an entire region on the graph, not a single value.
- When multiplying or dividing both sides of an inequality by a negative number, reverse the inequality symbol.

Example 6 Graph the solution set of the linear inequality.

$$x - 4y < 8$$

SOLUTION First graph the line $x - 4y = 8$ by plotting the intercepts. Because the inequality involves a $<$ symbol, graph a dashed line to indicate that the points on the line are not part of the solution set. Then pick a test point that is not on the line. Pick $(0, 0)$. Substitute this point into the original inequality and see whether it makes the inequality true.

$x - 4y < 8$ ⠀⠀⠀Substitute in $(x, y) = (0, 0)$.
$0 - 4(0) \overset{?}{<} 8$
$0 < 8$ ⠀⠀⠀This is a true statement.

The point $(0, 0)$ is in the solution set. Shade above the line.

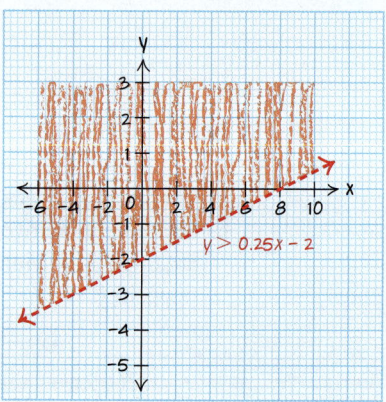

Example 7 Graph the solution set of the linear inequality.

$$x \geq -3$$

SOLUTION Graph the line $x = -3$. This is a vertical line. Because the inequality involves a $\geq$ symbol, draw a solid line. Pick a test point that is not on the line, for instance, $(0, 0)$. Substitute $(0, 0)$ into the original inequality and see whether the inequality is true.

$x \geq -3$ ⠀⠀⠀Substitute in $(x, y) = (0, 0)$.
$0 \geq -3$ ⠀⠀⠀This is a true statement.

The point $(0, 0)$ is in the solution set. Shade in the part of the graph to the right of the line.

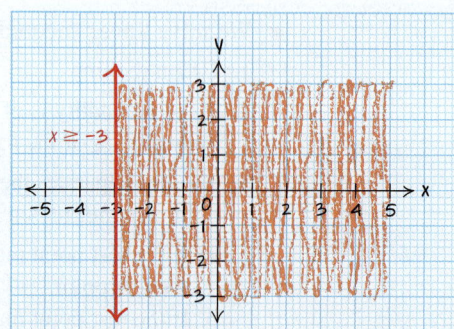

CHAPTER 4   Summary   451

Section 4.5 Systems of Linear Inequalities

- A **system of inequalities** is a set of two (or more) linear inequalities. Many applications that result in systems are more logically written as systems of inequalities.
- The **steps** to solving a system of linear inequalities are:
  1. Graph the first inequality as a solid or dashed line and shade the appropriate region (see Section 4.4).
  2. On the same grid, graph the second inequality as a solid or dashed line and shade in the appropriate region.
  3. The **solution set** to the system is the **intersection** of the two shaded areas from steps 1 and 2. Carefully shade in the solution set darker to distinguish it.

Example 8   Solve the system of linear inequalities graphically.

$$y \geq -x + 4$$
$$y \leq x - 3$$

SOLUTION   Graph both lines on the same axes. Both lines are in slope-intercept form and will graph as solid lines because of the equal to part of the inequality. The first line, $y = -x + 4$, has a y-intercept of $(0, 4)$ and a slope $m = -1$. The second line, $y = x - 3$, has a y-intercept of $(0, -3)$ and a slope $m = 1$. To determine which side of the graph to shade, pick $(0, 0)$ as a test point, as neither lines passes through the origin. Substituting $(0, 0)$ into both equations yields

$$y \geq -x + 4$$
$$y \leq x - 3 \qquad \text{Substitute in } x = 0, y = 0.$$

Inequality 1   **Inequality 2**

$0 \stackrel{?}{\geq} -(0) + 4$ $0 \stackrel{?}{\leq} (0) - 3$

$0 \stackrel{?}{\geq} 0 + 4$ $0 \stackrel{?}{\leq} -3$

$0 \stackrel{?}{\geq} 4$ False $0 \stackrel{?}{\leq} -3$ False

Shade in the part of the graph that does not contain $(0, 0)$ for both lines. The solution set for the system of inequalities is the overlapping area.

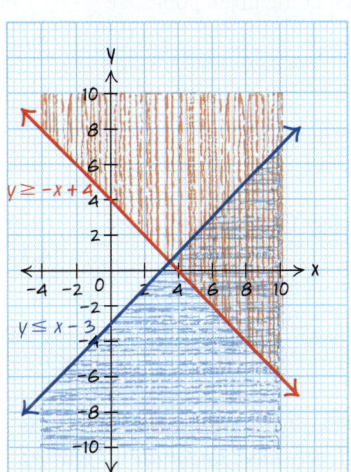

Chapter Review Exercises

1. Use the graph to determine the solution to the system of equations. [4.1]

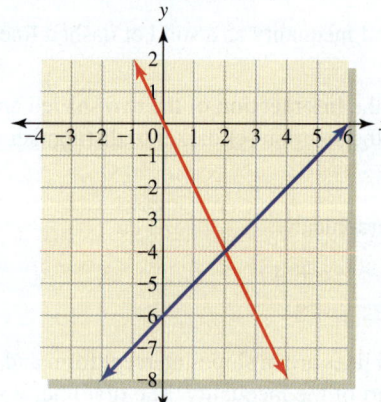

2. Use the graph to determine the solution to the system of equations. [4.1]

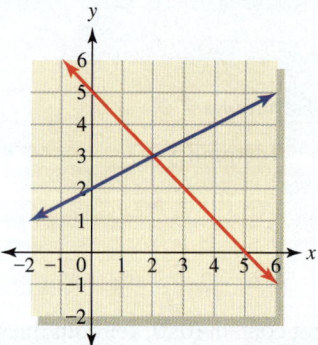

3. Check to determine whether the given points are solutions of the system. [4.1]

$$y = -\frac{4}{3}x$$
$$y = x - 7$$

   a. $(6, -8)$ b. $(3, -4)$

4. Check to determine whether the given points are solutions of the system. [4.1]

$$y = x + 6$$
$$y = -2x$$

   a. $(-2, 4)$ b. $(2, -4)$

For Exercises 5 and 6, graphically determine the solution to each system of equations. Clearly label and scale all graphs. Check the solutions.

5. $y = 5x$
   $y = 2x - 3$

6. $y = -\frac{5}{3}x - 3$
   $y = x + 5$ [4.1]

For Exercises 7 through 12, solve each system using the substitution method. Check the solutions.

7. $-3x + y = 1$
   $x + y = 13$

8. $2x + y = -4$
   $3x + 2y = 7$ [4.2]

9. $y = -\frac{2}{3}x + 1$
   $4x - 6y = -6$

10. $x = 4y + 12$
    $-2x + 8y = -24$ [4.2]

11. $6x + 2y = 14$
    $3x + y = 1$

12. $5y = 2x + 5$
    $y = \frac{2}{5}x - 5$ [4.2]

13. A cyclist can ride her bike 48 miles with a tailwind in 2 hours. It takes her 3 hours to make the return trip. Find the rate of the cyclist (with no wind) and the rate of the wind. [4.2]

14. A cyclist can ride his bike 45 miles with a tailwind in 3 hours. It takes him 5 hours to make the return trip. Find the rate of the cyclist (with no wind) and the rate of the wind. [4.2]

15. A small plane can fly with a tailwind 240 miles in 2 hours. It makes the return trip with a headwind in 3 hours. Find the rate of the plane in still air and the rate of the wind. [4.2]

16. A boat travels 28 miles downstream in 2 hours. It takes 4 hours to make the return trip upstream. Find the rate of the boat in still water and the rate of the current. [4.2]

17. The difference of two numbers is 8. Twice the first number added to the second number is 37. Find the two numbers. [4.2]

18. The difference of two numbers is 20. Three times the first number plus the second is 16. Find the two numbers. [4.2]

19. The sum of two numbers is 52. One number is three times the other. Find the two numbers. [4.2]

20. The sum of two numbers is 28. One number is six times the other. Find the two numbers. [4.2]

For Exercises 21 through 28, solve each system using the elimination method. Check the solutions.

21. $x - 2y = 7$
    $3x + 2y = -3$

22. $5x + 7y = 27$
    $-5x - y = 9$ [4.3]

23. $3x + 2y = 16$
    $10x - 4y = 0$

24. $2x + y = -7$
    $-x + 3y = 0$ [4.3]

25. $3x + y = 5$
    $3x + y = -8$

26. $6x + y = -5$
    $-6x - y = -1$ [4.3]

27. $x + 2y = 3$
    $-2x - 4y = -6$

28. $-2x + 3y = 18$
    $4x - 6y = -36$ [4.3]

29. Decide whether you would solve the system using the substitution method or elimination method. Do *not* solve the system, but do justify your answer. [4.3]

    $-2x + 5y = -16$
    $2x - 3y = -12$

30. Decide whether you would solve the system using the substitution method or elimination method. Do *not* solve the system, but do justify your answer. [4.3]

    $x - 4y = 9$
    $12x - 7y = -10$

31. Anne is a professional dog trainer. For her new puppy class, she has purchased four boxes of jerky treats and two leashes for $65.00. Because more students signed up for the class, she later purchased three more boxes of jerky treats and three leashes for $73.50. How much does each box of jerky treats and each leash cost? [4.3]

32. Christian is a piano teacher. For his new beginning students, he has purchased three large bags of candy and one metronome for $67.75. Because more students signed up for lessons, he later purchased two more bags of candy and two metronomes for $78.50. How much does each bag of candy and each metronome cost? [4.3]

33. Lance has $2000.00 to invest. An investment advisor tells him to put part of his money in a bond fund earning 3% interest and part in a stock fund earning 4% interest. Lance wants to earn $65.00 in simple interest yearly. How much should he invest in each account? [4.3]

34. Ming-Yue has $5000.00 to invest. An investment planner tells her to put part of her money in a bond fund earning 3% interest and part in a stock fund earning 5% interest. Ming-Yue wants to earn $184.00 in simple interest yearly. How much should she invest in each account? [4.3]

35. Two angles are supplementary. One angle is 5 times the measure of the other angle. Find the measures of the two angles. [4.3]

36. Two angles are supplementary. One angle is 4 times the second angle added to 5. Find the measure of the two angles. [4.3]

37. Two angles are complementary. One angle is 9 times the measure of the other angle. Find the measures of the two angles. [4.3]

38. Two angles are complementary. One angle is twice the second angle. Find the measures of the two angles. [4.3]

39. Tarin wants to mix a 5% saline solution with a 30% saline solution to make a 25% saline solution. He needs a total of 20 ounces in the mix. How many ounces of the 5% saline solution and the 30% saline solution should he use? [4.3]

40. Shelby wants to mix a 6% fruit juice solution with a 12% fruit juice solution to make a 10% fruit juice solution. He needs a total of 20 gallons in the mix. How many gallons of the 6% solution and the 12% solution should he use? Round to the nearest hundredths place if necessary. [4.3]

41. Inna needs to mix raisins and peanuts for a trail mix. She wants to make a total of 10 pounds of the mix. The peanuts cost $2.00 per pound, and the raisins cost $3.60 per pound. The total value of the trail mix is $26.40. Find the amount of peanuts and raisins she needs to include in the mix. [4.3]

42. Lorraine mixes two types of coffee for a blend for her café. The French Roast coffee costs $9.95 per pound, and the Guatemalan coffee costs $12.95 per pound. Lorraine would like to price the blend at $10.95 per pound. If she makes 20 pounds of the blend, how many pounds of each type of coffee does she need to add? [4.3]

43. Determine whether the given point is in the solution set of the linear inequality. [4.4]

    $y < -3x + 5$

    a. $(0, 5)$ b. $(-2, -1)$

44. Determine whether the given point is in the solution set of the linear inequality. [4.4]

    $y \geq \frac{2}{3}x - 1$

    a. $(0, -1)$ b. $(-3, 3)$

45. Shade in the correct region to graph the solution set of the inequality. [4.4]

    $y > -\frac{2}{3}x + 4$

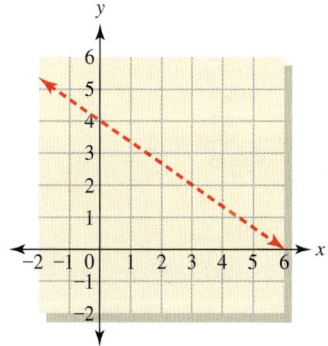

454 CHAPTER 4 Systems of Linear Equations

46. Shade in the correct region to graph the solution set of the inequality. [4.4]

$$y < \frac{1}{4}x - 3$$

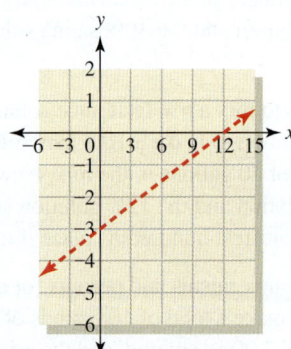

47. Solve the linear inequality graphically. [4.4]

$$y < -3x + 7$$

48. Solve the linear inequality graphically. [4.4]

$$y > \frac{1}{2}x - 1$$

49. Solve the linear inequality graphically. [4.4]

$$2x - y \leq 1$$

50. Solve the linear inequality graphically. [4.4]

$$x + 2y \geq -4$$

51. Solve the linear inequality graphically. [4.4]

$$x < -3$$

52. Solve the linear inequality graphically. [4.4]

$$x \geq 2.5$$

53. Solve the linear inequality graphically. [4.4]

$$y \leq -2$$

54. Solve the linear inequality graphically. [4.4]

$$y > 0$$

55. Zita works two jobs. The first job pays her $10.75 an hour, and the second job pays her $11.25 an hour. Zita wants to earn more than $500.00 a week working both jobs. [4.4]
 a. Define the variables.
 b. Write an inequality that expresses the relationship between the total hours worked and Zita's goal of earning more than $500.00 a week.
 c. Graph the solution set for this inequality.

56. Aaron works two jobs. The first job pays him $11.00 an hour, and the second job pays him $11.50 an hour. Aaron wants to earn more than $450.00 a week working both jobs. [4.4]
 a. Define the variables.
 b. Write an inequality that expresses the relationship between the total hours worked and Aaron's goal of earning more than $450.00 a week.
 c. Graph the solution set for this inequality.

For Exercises 57 and 58, match the inequalities with the graphs. [4.4]

57.

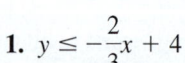

1. $y \leq -\frac{2}{3}x + 4$

2. $y < -\frac{1}{3}x + 6$

Graph A

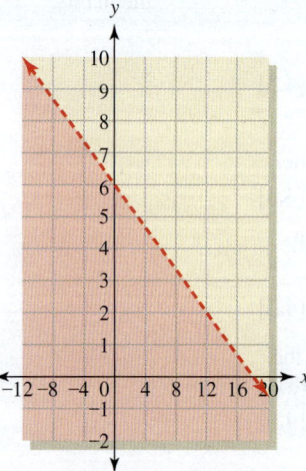

Graph B

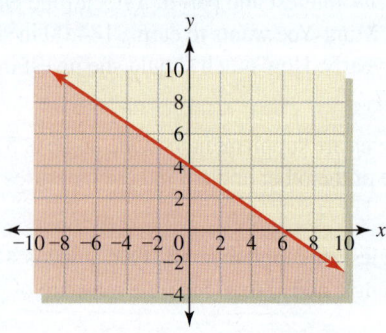

58.

Graph A

Graph B

1. $2x - 4y > 12$
2. $6x - 3y < 15$

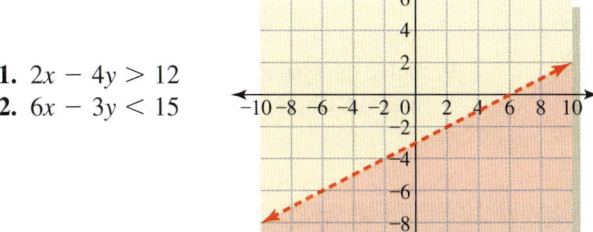
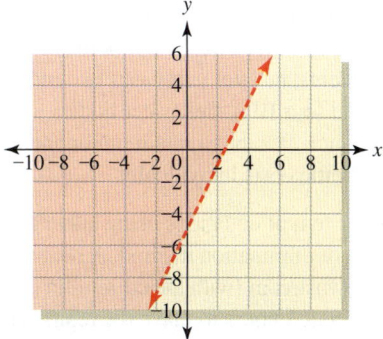

59. Are the given points in the solution set of the system of inequalities? **[4.5]**

$$x - y \geq 2$$
$$2x + y < 1$$

a. $(0, 0)$ **b.** $(0, -4)$

60. Are the given points in the solution set of the system of inequalities? **[4.5]**

$$y \geq 3x - 1$$
$$y \leq 2$$

a. $(3, 1)$ **b.** $(-3, 1)$

61. Solve the system of linear inequalities graphically. **[4.5]**

$$y > -2x + 3$$
$$y \leq x - 4$$

62. Solve the system of linear inequalities graphically. **[4.5]**

$$y \geq 2x - 5$$
$$y < -x + 1$$

63. Darius has at most $6000 to invest in stocks. He has two stock funds in mind. The first fund pays 5% simple interest a year, and the second, riskier fund pays 7% simple interest a year. Darius would like to earn more than $200 a year in simple interest. **[4.5]**

  a. Write a system of linear inequalities that describes this situation.
  b. Graphically solve the system of linear inequalities.
  c. Check the solution set with a sample test point. Interpret a sample point in the solution set in terms of the investments.

64. Leila has at most $7500.00 to invest in stock funds. She has two stock funds in mind. The first fund pays 7% simple interest a year, and the second, riskier fund pays 8% simple interest a year. Leila would like to earn more than $110.00 a year in simple interest. **[4.5]**

  a. Write a system of linear inequalities that describes this situation.
  b. Graphically solve the system of linear inequalities.
  c. Check the solution set with a sample test point. Interpret a sample point in the solution set in terms of the investments.

Chapter Test

This chapter test should take approximately one hour to complete. Read each question carefully. Show all of your work.

1. Graphically determine the solution to the system of equations. Identify the system as consistent or inconsistent. Clearly label and scale all graphs. Check the solution.

$$y = 3x - 1$$
$$y = -2x + 4$$

2. Solve for the variable y: $7x - 3y = -21$.

3. Solve the system of equations using the substitution method. Check the solution.

$$-2x + y = 1$$
$$2x - 3y = 5$$

4. Solve the system of equations using the elimination method. Check the solution.

$$4x - 5y = 10$$
$$-2x + 5y = 0$$

5. Solve the system of equations using either the substitution or elimination method. Rami is a personal trainer. As gifts for his clients, he buys three heart rate monitors and five BPA-free water bottles for $159.00. For other clients, he later purchases two more heart rate monitors and four BPA-free water bottles for $114.00. Find how much each heart rate monitor and each BPA-free water bottle costs individually.

6. Solve the system of equations using either the substitution or elimination method. A boat travels 24 miles downstream (with the current) in 2 hours. It takes the boat 3 hours to make the return trip upstream (against the current). Find the rate of the boat in still water and the rate of the current.

7. Solve the system of equations using either the substitution or elimination method. Laurent has $2000.00 to invest. His financial advisor recommends that he put part of his investment in a bond fund that returns 5% simple interest yearly and a stock fund that returns 7% simple interest yearly. Laurent would like to earn $124.00 in interest a year. How much should he invest in each account?

8. Solve the system of equations using the substitution method. Check the solution.

$$-x + 5y = 15$$
$$2x - 10y = 10$$

9. Solve the system of equations using the elimination method. Check the solution.

$$-2x - 8y = -8$$
$$x + 4y = 4$$

10. Solve the linear inequality graphically.

$$y < -4x + 7$$

11. Solve the linear inequality graphically.

$$x > 5$$

12. Solve the system of linear inequalities graphically.

$$y \geq x + 2$$
$$y < -x + 5$$

Chapter Projects

■ Writing an Infinite, Repeating Decimal as a Fraction

Written Project
One or more people

In this project, you will learn how to convert an infinite, repeating decimal to a fraction.

Example: Consider the infinite, repeating decimal 0.252525....

1. Let $n = 0.252525\ldots$.
2. This decimal repeats after the *hundredths* place.
3. Because the decimal repeats after the *hundredths* place, multiply both sides by 100.

$$100n = 25.252525\ldots$$

4. Subtract the equation from step 1 from the equation from step 3.

$$100n = 25.252525\ldots$$
$$-\quad n = 0.252525\ldots$$
$$99n = 25$$
$$\frac{99n}{99} = \frac{25}{99} \quad \text{Solve for } n.$$
$$n = \frac{25}{99}$$

Problem: Convert the infinite, repeating decimal to fraction form.

$$0.717171\ldots$$

1. Set the decimal equal to n.
2. After what place value does the decimal start to repeat?
3. Multiply both sides of the equation from step 1 by _____ because the decimal repeats after the _____ place.
4. Subtract the equation from step 1 from the equation from step 3.
5. Solve for n.

What Is Your Target Heart Rate?

Written Project
One or more people

When you are exercising, it is recommended that you exercise within your target heart rate zone. Let $x =$ your age and let $y =$ target heart rate. Doctors use the following formulas to compute the target heart rate zone for adults (age 20 and older).

1. To compute the lower bound, the target heart rate is equal to 50% of 220 minus the age. Write this equation using x and y. Simplify the equation as $y = mx + b$.
2. To compute the upper bound, the target heart rate is equal to 85% of 220 minus the age. Write this equation using x and y. Simplify the equation as $y = mx + b$.
3. Doctors recommend that people exercise with their target heart rate zone found as follows:
   a. The target heart rate should be greater than or equal to 50% of 220 minus the age. Use the equation from step 1 to write this inequality.
   b. The target heart rate should be less than or equal to 85% of 220 minus the age. Use the equation from step 2 to write this inequality.
4. Graph the two inequalities you found in step 3 on the same axes. Put a break in the x-axis to start the ages at 20.
5. Shade in the solution set.
6. What is the target heart rate zone for your age?

Write a Review of a Section for Presentation

Research Project
One or more people

What you will need

- This presentation may be done in class or as an online video.
- If it is done as a video, post it on a website where it may be easily viewed by the class.
- You might want to use homework or other review problems from the book.
- Make it creative and fun.

Create a 5-minute review presentation of one section from Chapter 4. The format of the presentation can be a poster presentation, a blackboard presentation from notes, an online video, or a game format (for example, math jeopardy). The presentation should include the following:

- Any important formulas in the chapter
- Important skills in the chapter, backed up by examples
- Explanation of terms used (for example, what is *modeling*?)
- Common mistakes and how to avoid them

Cumulative Review — Chapters 1-4

1. Write in exponential form:
$$\left(-\frac{1}{5}\right)\cdot\left(-\frac{1}{5}\right)\cdot\left(-\frac{1}{5}\right)\cdot\left(-\frac{1}{5}\right).$$

2. Find the value without a calculator: $(-2)^4$.

3. Write using standard notation: 6.103×10^6.

4. Underline or circle the terms and simplify using the order-of-operations agreement:
$$\left|-\frac{12}{11-7}\right| + 3^3 - (-1-8) + 18 \div 3 \cdot (9)$$

5. Define the variables: The amount of commission in dollars that a salesperson earns depends on the amount of sales in dollars she makes.

6. Translate the sentence into a variable expression: three times the sum of nine and a number.

7. Convert 12 meters to feet. There is 1 foot in 0.3048 meter. Round to the hundredths place.

8. Evaluate the expression for the given value of the variable: $-x^2 + 3x - 1$ for $x = -2$.

9. Simplify: $3(5x-1) - 4(3x-7)$.

10. Simplify: $-(-(-5x))$.

11. Graph the ordered pairs on the same set of axes:
$(-20, 4)\ (15, 3)\ (10, -1)\ (-10, -10)\ (25, 0)$

12. Are the values of the variables solutions to the equation?
$$4x + 3y = -12:\ x = 0 \text{ and } y = -4$$

13. Solve: $-\dfrac{3}{4}x = 18$.

14. Solve for the indicated variable: $-9x - 3y = 27$ for y.

15. Solve: $3(x-1) - 2(x+2) = -3 + \dfrac{1}{3}x$.

16. Daniel works as an office specialist and makes \$13 per hour after taxes. Daniel can calculate his monthly salary using the equation
$$S = 13h$$
where S is the monthly salary in dollars that Daniel makes if he works h hours that month.

  a. Find Daniel's monthly salary if he works 110 hours.

  b. How many hours does Daniel have to work a month to make \$1820 a month?

17. Solve: $12x + 9 = 5x + 26.5$

18. Translate and solve: The sum of eight and three times a number is equal to the difference of twice the number and negative six.

19. Solve: $9y - 8 = 6y + 2 + 3(y+1) - 10$.

20. Solve the inequality: $-5x + 2 > 37$.

21. Complete each part below.

  a. Fill in the table using the given equation.

x	$y = -\dfrac{3}{5}x - 7$
-5	
0	
5	
15	
20	

  b. Graph the line by plotting the points in the table. Clearly label and scale the axes.

  c. Is the graph linear or nonlinear?

22. Use the equation $y = -\dfrac{2}{3}x + 4$ to create a table of five or more points and graph them. Clearly label and scale the axes. Connect the points with a smooth curve. Is the graph linear or nonlinear?

23. The average American eats about 4000 calories per day. The amount of calories the average person will eat after d days can be estimated by using the equation $C = 4000d$, where C is the total number of calories the average person eats over the course of d days. Create a table of points that make sense in this situation and graph them. Connect the points with a smooth curve.

24. Use the equation $y = x^2 - 2$ to create a table of nine or more points and graph them. Clearly label and scale the axes. Connect the points with a smooth curve. Is the graph linear or nonlinear?

25. Use the equation $y = -5$ to create a table of five or more points and graph them. Clearly label and scale the axes. Connect the points with a smooth curve. Is the graph linear or nonlinear?

26. Use the equation $x = 6$ to create a table of five or more points and graph them. Clearly label and scale the axes. Connect the points with a smooth curve. Is the graph linear or nonlinear?

27. Find the slope of the line given in the graph.

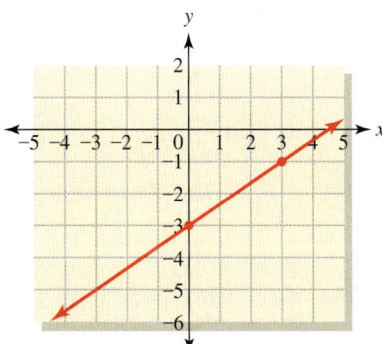

28. Find the slope of the line that goes through points $(-2, 5)$ and $(3, -7)$.

29. Use two points from the table to find the slope of the line that passes through the points.

x	y
−5	11
0	9
5	7
10	5

30. Use two points from the table to find the slope of the line that passes through the points.

x	y
5	−7
7	−7
8	−7
10	−7

31. State which line, the red line or the blue line, has the greater slope. Explain your answer.

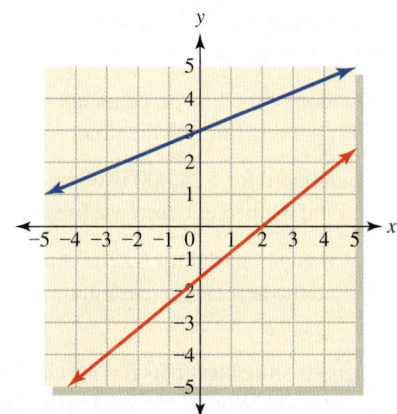

32. Explain whether the line is increasing, decreasing, or neither.

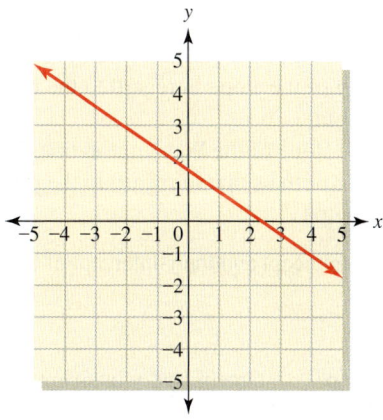

33. Is the line increasing, decreasing, or neither? Explain how you know.

$$y = 3x - 5$$

34. Use the graph to find the *x*- and *y*-intercepts. Write the intercepts as ordered pairs (x, y).

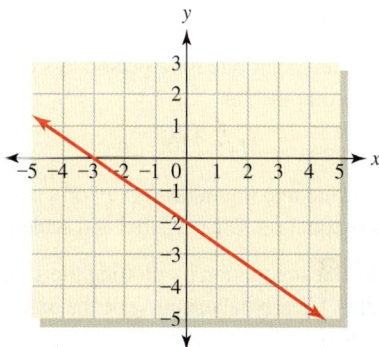

35. Use the following graph to answer the questions.

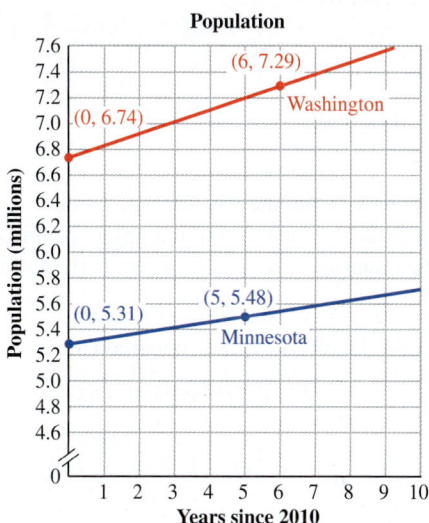

Source: U.S. Census Bureau, Population Division.

a. Interpret the vertical intercept for the Minnesota line.
b. Interpret the vertical intercept for the Washington line.
c. Find the slope of the Minnesota line. Explain the meaning of the slope in regard to the population of Minnesota.
d. Find the slope of the Washington line. Explain the meaning of the slope in regard to the population of Washington.

36. Find the x-intercept and the y-intercept of the equation $y = -\dfrac{3}{5}x + 15$.

37. Find the slope, x-intercept, and y-intercept of the equation $-4x + 8y = -12$.

38. Graph the equation $y = -\dfrac{3}{2}x + 5$ using the slope and the y-intercept. Graph at least three points. Clearly label and scale the axes.

39. Graph the equation $2x - 4y = 3$ using the slope and the y-intercept. Graph at least three points. Clearly label and scale the axes.

40. The maximum heart rate for a person less than 30 years old can be computed as
$$M = 220 - a$$
where M is the maximum heart rate for a person age a. Graph this equation. Clearly label and scale the axes.

41. Graph the linear equation by plotting the x- and y-intercepts. Clearly label and scale the axes. $3x - 2y = 6$.

42. Determine which of the following equations are linear. Explain your reasoning.
a. $x + 10y = -3$
b. $y = x^2 - 1$
c. $y = |3x - 6|$
d. $y = -2x + 17$

43. Determine whether the two lines are parallel, perpendicular, or neither. Do not graph the lines. The two lines are $y = -5x + 12$ and $10x + 2y = 6$.

44. Determine whether the two lines are parallel, perpendicular, or neither. Do not graph the lines. The two lines are $5x + 2y = 6$ and $y = \dfrac{2}{5}x + 1$.

45. Use the graph to find the equation of the line. Put the answer in $y = mx + b$ form.

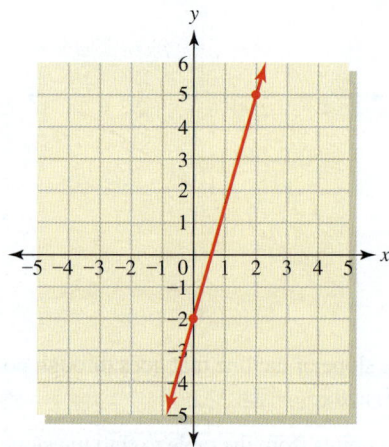

46. Maribel goes on a diet to lose weight. Let the input t represent the week of her diet, and let the output W represent Maribel's weight in pounds. At the start of her diet ($t = 0$), Maribel weighed 155 pounds. After 6 weeks on her diet, she weighed 130 pounds. Find the linear relationship between t and W.

47. Determine the equation of the line that passes through the points $(-5, 11)$ and $(5, 7)$. Put the answer in $y = mx + b$ form.

48. Determine the equation of the line by using point-slope form. The line has slope $m = \dfrac{2}{3}$ and passes through the point $(-6, 0)$. Put the final answer in $y = mx + b$ form.

49. Determine the equation of the line that passes through the points $(2, -5)$ and $(-3, -5)$. Put the answer in $y = mx + b$ form.

50. Write the equation of the line through the given points. The points are $(3, -5)$ and $(3, 2)$. (*Hint:* Graph the two points if necessary.)

51. Find the equation of the line parallel to $y = -5x + 7$, passing through the point $(-10, 3)$.

52. Find the equation of the line perpendicular to $y = 3x - 11$, passing through the point $(6, -4)$.

53. A new car purchased for $24,000 is worth only $19,400 after 1 year. Find the percent decrease in value for this car.

54. A house that was purchased for $220,000 is now worth $510,000. Find the percent increase in the value of this house.

55. A sporting goods store buys bats for $60 and has a markup rate of 40%. What is the markup and the sales price for these bats?

CHAPTER 4 Cumulative Review

56. Worldwide sales of makeup are rising. The projected worldwide sales of makeup are given in the table.

Year	Makeup Sales (billion $)
2014	56.9
2015	58.1
2016	59.5
2017	61.2
2018	63.0
2019	64.8

Source: Euromonitor.

  a. Define the variables.
  b. Graph the data on a scatterplot and find an eyeball best-fit line.
  c. Find the equation of the linear model.
  d. Using the model, estimate worldwide makeup sales in 2025.
  e. Find the slope of the model and explain the value of the slope in terms of the problem.

57. $T = 12c + 20$ represents the time in minutes it takes to decorate c Christmas ornaments. Find T when $c = 30$, and explain its meaning in terms of the problem situation.

58. $R = 13.00t$ represents the revenue a movie theater takes in for selling t tickets at $13.00 each. Find R when $t = 68$, and explain its meaning in terms of the problem situation.

59. Solve the system of equations using the table of values.

x	$y = -4x + 2$
-3	14
-2	10
-1	6
0	2
1	-2

x	$y = 3x + 9$
-3	0
-2	3
-1	6
0	9
1	12

60. Determine whether each point is a solution of the system.

$$y = -2x + 3$$
$$y = 4x - 3$$

  a. Is $(x, y) = (2, -1)$ a solution of the system?
  b. Is $(x, y) = (1, 1)$ a solution of the system?

61. Use the graph to determine the solution to the system.

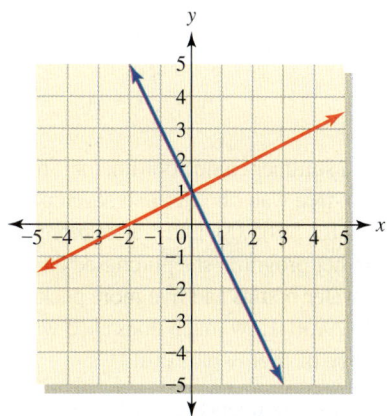

62. Martin is considering taking one of two jobs. The first job pays a base salary of $9.25 an hour plus 2% commission on the dollar amount of sales. The second job pays a base salary of $10.50 an hour plus 1% commission on the dollar amount of sales. The two equations that describe the hourly salary S earned given the dollar amount of sales x are given below.

$$S = 9.25 + 0.02x$$
$$S = 10.50 + 0.01x$$

  a. Is $(x, S) = (100, 11.25)$ a solution of the system?
  b. Is $(x, S) = (125, 11.75)$ a solution of the system?

63. Graphically determine the solution of the system. Clearly label and scale the axes. Check the solution by substituting it into both original equations.

$$y = -2x + 5$$
$$y = 2x - 3$$

64. Graphically determine the solution of the system. Clearly label and scale the axes. Check the solution by substituting it into both original equations.

$$y = 3x - 4$$
$$y = 2$$

65. Solve the system using the substitution method. Check the solution in both original equations.

$$5x + 4y = 4$$
$$2x + y = -2$$

66. Solve the system using the substitution method. Check the solution in both original equations.

$$-4x + y = 7$$
$$x - y = 8$$

67. Solve the system using the substitution method. Check the solution in both original equations.

$$-2x + 3y = 6$$
$$y = \frac{2}{3}x - 4$$

68. Haleigh is a summer camp coordinator. She purchased fifteen BPA-free water bottles and seven packages of energy bars for $265.00. Later, after more children enrolled in her program, she purchased an additional nine BPA-free water bottles and five more packages of energy bars for $179.00. How much does each water bottle and each package of energy bars cost?

69. The sum of two numbers is 32. The first number is equal to the difference between the second number and 6. Find the two numbers.

70. A plane can fly 450 miles in 3 hours with a tailwind. It makes the return trip (with a headwind) in 4.5 hours. Find the rate of the plane in still air and the rate of the wind.

71. Francis wants to make a 5-pound spice mix by blending two types of spices. The first spice is turmeric, which costs $8.90 per pound. The second spice is ground coriander, which costs $11.20 per pound. Francis wants to price the mix at $9.82 per pound. How many pounds of each type of spice does Francis need to use in his mix?

72. Solve the system using the elimination method. Check the solution in both original equations.

$$3x + y = 7$$
$$2x + y = 3$$

73. Solve the system using the elimination method. Check the solution in both original equations.

$$5x - 2y = 3$$
$$7x + 3y = -19$$

74. Solve the system using the elimination method. Check the solution in both original equations.

$$2x + 3y = 18$$
$$-4x - 6y = -36$$

75. Allyson is organizing PTA gifts for the teachers at her children's elementary school. She buys ten blank large scrapbooks and eight sticker sets for $280.00. Later she purchases an additional four blank large scrapbooks and six sticker sets for $147.00. How much does each scrapbook and sticker set cost?

76. Two angles are supplementary. The second angle measures five times the measure of the first angle. Find the degree measures of both angles.

77. Eric has $12,000.00 to invest. An investment advisor tells him to put part of his money in a CD earning 3% interest and the rest in a bond fund earning 5% interest. Eric would like to earn $440.00 in interest every year. How much should he invest in each account?

78. Determine whether the point (3, 10) is a solution of the linear inequality $3x - y < 7$.

79. Graph the solution set of the linear inequality $-2x + 3y > 6$.

80. Graph the solution set of the linear inequality $3x - 10y \leq 20$.

81. Mai has been asked to bring appetizers to a party. She calls a local restaurant, and they give her the following prices. The spring rolls cost $3.99, and chicken gyoza is $4.25. She wants to spend no more than $20.00 on both appetizers.

  a. Define the variables.

  b. Write an inequality that expresses the relationship between the total cost of both spring rolls and chicken gyoza and the maximum that Mai wants to spend ($20.00).

  c. Graph the solution set for this inequality.

82. Determine whether the point $(-2, 1)$ is in the solution set of the system of linear inequalities.

$$y > -5$$
$$4x + y < 0$$

83. Graph the solution set for each system of inequalities. Check the solution set with a sample test point.

$$y \leq 3x - 6$$
$$2x + y > 4$$

Exponents and Polynomials

5

5.1 Rules for Exponents

5.2 Negative Exponents and Scientific Notation

5.3 Adding and Subtracting Polynomials

5.4 Multiplying Polynomials

5.5 Dividing Polynomials

In astronomy, many large numbers are used because of the vast distances in space. To write these numbers in a more compact notation, scientists use scientific notation. The Andromeda Galaxy is the nearest spiral galaxy to the Earth. It is about 2,500,000 light-years away from Earth. Astronomers also say that it is about 2.5×10^6 light-years away. Here the extremely large number of light-years, 2.5×10^6, is written more compactly in scientific notation. A light-year is the distance light travels in space in one year. In kilometers, a light-year is 9.46×10^{12} km. In miles, a light-year is 5.88×10^{12} m.

5.1 Rules for Exponents

LEARNING OBJECTIVES
- Use the product rule for exponents.
- Use the quotient rule for exponents.
- Use the power rule for exponents.
- Work with powers of products and quotients.

When we encounter expressions that have exponents, relying on some rules helps us do our work. The rules for exponents that we learn in this section will help us to simplify expressions and do some more complicated algebra.

In this section, we assume that variables do not take on values that result in division by 0. Recall from Section R.1 that division by 0 is undefined.

First we review how to compute the values of some exponential expressions.

Example 1 Simplifying exponential expressions

Simplify each exponential expression.

a. 3^4 **b.** $(-3)^2$ **c.** -3^2 **d.** $(-x)^2$

SOLUTION

a. $3^4 = 3 \cdot 3 \cdot 3 \cdot 3 = 81$

b. $(-3)^2 = (-3)(-3) = 9$

c. -3^2 This means the opposite of 3^2.
$= -(3 \cdot 3)$
$= -9$

d. $(-x)^2$ Recall that a negative multiplied by a negative is positive.
$(-x)(-x)$
$= x^2$

PRACTICE PROBLEM FOR EXAMPLE 1

Simplify each exponential expression.

a. 2^5 **b.** $(-2)^4$ **c.** -2^4 **d.** $(-a)^3$

■ Product Rule for Exponents

Recall from Section 1.1 that natural number exponents represent repeated multiplication and that an exponential expression has a base that is raised to an exponent. Expanding on the idea of repeated multiplication, we can discover the rules for exponents. Use the Concept Investigation to look at the way we multiply expressions that contain exponents.

■ CONCEPT INVESTIGATION
What does multiplication do to the exponents?

1. Fill in the following table. First rewrite the original expression into expanded form. Then simplify the expanded form into an exponential.

Original Expression	Expanded Form	Simplified Exponential
$5^2 \cdot 5^4$	$5 \cdot 5 \cdot 5 \cdot 5 \cdot 5 \cdot 5$	5^6
$3^4 \cdot 3^3$		
$x^2 \cdot x^5$		
$y \cdot y^3$		

■ Calculator Details

Entering Exponents

Recall that when you are entering exponents into a scientific calculator, you look for the key of the form .

Enter 2^3 on a scientific calculator with the following keystrokes:

[2] [x^y] [3]

You may then have to hit the [=] key to get the answer, 8. The exponent key on some scientific calculators and graphing calculators looks like the ^ symbol.

2^3 is entered as

Hit the [ENTER] key last to display the answer of 8.

SECTION 5.1 Rules for Exponents 465

2. Use a calculator to fill in the following table.

Calculate using the order of operations.	Simplify using the pattern observed in part 1. Then calculate.
$2^3 \cdot 2^2 = \ 8 \cdot 4 \ = 32$	$2^3 \cdot 2^2 = \ 2^5 \ = 32$
$3^4 \cdot 3^2 = \ \ \ =$	$3^4 \cdot 3^2 = \ \ \ =$
$5^2 \cdot 5^3 = \ \ \ =$	$5^2 \cdot 5^3 = \ \ \ =$

3. What happens to the exponents in the original expressions when you write the simplified exponential form?

When we have a product of exponential expressions with the same base, we add the exponents. Adding exponents in this way is called the **product rule for exponents**. It is crucial that the bases be the same, or this rule for exponents does not apply.

> **Product Rule for Exponents**
>
> $$x^m x^n = x^{m+n}$$
>
> When multiplying exponential expressions that have the same base, add exponents, keeping the same base.
>
> $$x^4 x^3 = x^{4+3} = x^7$$

Example 2 Using the product rule for exponents to simplify

Simplify each expression, using the product rule for exponents.

a. $x^2 x^9$ b. $5^2 \cdot 5^4$ c. $y^3 y$

SOLUTION

a. This problem, $x^2 x^9$, is a product of exponential expressions with the same base (x), so add the exponents. Note that the base stays the same.

$x^2 x^9$
$= x^{2+9}$ Keep the base and add the exponents.
$= x^{11}$

b. This problem, $5^2 \cdot 5^4$, is a product of exponential expressions with the same base (5), so add the exponents, and the base stays the same.

$5^2 \cdot 5^4$
$= 5^{2+4}$ Keep the base and add the exponents.
$= 5^6$ Since the base is a constant, calculate the value.
$= 15625$

c. This problem, $y^3 y$, is a product of exponential expressions with the same base (y), so add the exponents. The y in the second part of the expression has an exponent of 1. Write the 1 when using the exponent rules because it helps to reduce the number of mistakes.

$y^3 y$
$= y^3 y^1$ Rewrite y as y^1.
$= y^{3+1}$ Keep the base and add the exponents.
$= y^4$

Connecting the Concepts

What operations change the exponents?

Avoid using the product rule for exponents with addition or subtraction problems.

Recall from Section 1.3 that we need like terms to add or subtract. The expression $x^2 + x^5$ cannot be simplified because these are not like terms.

The expression $x^2 \cdot x^5$ simplifies using the product rule for exponents. $x^2 \cdot x^5 = x^7$

Rules for Exponents

Product Rule for Exponents
When multiplying exponential expressions with the same base, add the exponents.
$$x^2 x^5 = x^7$$

Connecting the Concepts

Why does x have an exponent of 1?

A variable with no written exponent has an exponent of 1. The exponent of 1 is not written down. Therefore, $x = x^1$.

To make sense of this, look at the following pattern:

$x^3 = x \cdot x \cdot x$
$x^2 = x \cdot x$

So,

$x^1 = x$

466 CHAPTER 5 Exponents and Polynomials

PRACTICE PROBLEM FOR EXAMPLE 2

Simplify each expression, using the product rule for exponents.
a. $x^5 x^6$ **b.** $t^3 t$ **c.** $7^4 7^5$

When working with products, use the commutative property to group like bases and coefficients. Then multiply the coefficients and use the product rule for exponents to simplify the bases.

Example 3 — Simplifying products with constants

Simplify each expression.
a. $(3x^4 y^7)(5x^2 y^3)$ **b.** $(-4m^2 n)(8m^8 n^3)$

SOLUTION

a. Use the commutative property to group like bases and to multiply the coefficients. Then add exponents, using the product rule for exponents.

$(3x^4 y^7)(5x^2 y^3)$ — Use the commutative property to group like bases.
$= 3 \cdot 5 \cdot x^4 x^2 \cdot y^7 y^3$ — Multiply the coefficients. Keep the bases and add exponents, using the product rule for exponents.
$= 15 x^6 y^{10}$

b. Again use the commutative property and the product rule for exponents to simplify this product.

$(-4m^2 n)(8m^8 n^3)$ — Use the commutative property to group like bases. Use $n = n^1$.
$= -4 \cdot 8 \cdot m^2 m^8 \cdot n^1 n^3$ — Multiply the coefficients. Keep the bases and add exponents, using the product rule for exponents.
$= -32 m^{10} n^4$

PRACTICE PROBLEM FOR EXAMPLE 3

Simplify each expression.
a. $(5x^6 y^9)(6x^5 y^2)$ **b.** $(-9a^4 b^3)(4a^2 b^6)$

■ Quotient Rule for Exponents

Now we examine what to do with exponents when a quotient is involved. Use the next Concept Investigation to look at the way in which we divide exponential expressions.

■ CONCEPT INVESTIGATION

What does division do to the exponents?

1. Fill in the following table by first expanding the original expression into expanded form and then simplifying the expanded form into an exponential.

Original Expression	Expanded Form	Simplified Exponential
$\dfrac{3^5}{3^2}$	$\dfrac{3 \cdot 3 \cdot 3 \cdot 3 \cdot 3}{3 \cdot 3}$	$\dfrac{\cancel{3} \cdot \cancel{3} \cdot 3 \cdot 3 \cdot 3}{\cancel{3}_1 \cdot \cancel{3}_1} = 3 \cdot 3 \cdot 3 = 3^3$
$\dfrac{5^6}{5^2}$		
$\dfrac{x^3}{x}$		
$\dfrac{y^5}{y^3}$		

Skill Connection

Reducing fractions

We simplify fractions by dividing out like factors.

$\dfrac{4}{6} = \dfrac{\cancel{2} \cdot 2}{\cancel{2} \cdot 3} = \dfrac{2}{3}$

$\dfrac{x^3}{x^2} = \dfrac{\cancel{x} \cdot \cancel{x} \cdot x}{\cancel{x} \cdot \cancel{x}} = \dfrac{x}{1} = x$

2. What happens to the exponents in the original expressions when you write the simplified exponential form?
_____

From this Concept Investigation, we see that we subtract the exponents when dividing exponential expressions with the same base. Subtracting exponents in this way is called the **quotient rule for exponents**. It is required that the bases be the same, or this rule for exponents will not apply.

> **Quotient Rule for Exponents**
>
> $$\frac{x^m}{x^n} = x^{m-n}$$
>
> When dividing exponential expressions that have the same base, subtract the exponents, keeping the same base. Note that the order of subtraction is top exponent minus bottom exponent.
>
> $$\frac{x^7}{x^3} = x^{7-3} = x^4$$

Example 4 Using the quotient rule for exponents to simplify

Simplify each expression, using the quotient rule for exponents.

a. $\dfrac{x^7}{x^5}$ **b.** $\dfrac{2^8}{2^3}$ **c.** $\dfrac{n^5}{n}$

SOLUTION

a. This is a quotient of exponential expressions with the same base (x), so subtract the exponents.

$$\frac{x^7}{x^5}$$
$$= x^{7-5} \qquad \text{Keep the base and subtract the exponents.}$$
$$= x^2$$

b. This is a quotient of exponential expressions with the same base (2), so subtract the exponents, and the base will stay the same.

$$\frac{2^8}{2^3}$$
$$= 2^{8-3} \qquad \text{Keep the base and subtract the exponents.}$$
$$= 2^5 \qquad \text{Since the base is a number, calculate the value.}$$
$$= 32$$

c. This is also a quotient of exponential expressions with the same base (n), so subtract the exponents. Write the 1 as the exponent for the n in the denominator, to avoid mental arithmetic mistakes.

$$\frac{n^5}{n}$$
$$= \frac{n^5}{n^1}$$
$$= n^{5-1} \qquad \text{Keep the base and subtract the exponents.}$$
$$= n^4$$

Rules for Exponents

Quotient Rule for Exponents
When dividing exponential expressions with the same base, subtract the exponents.

$$\frac{x^5}{x^2} = x^{5-2} = x^3$$

PRACTICE PROBLEM FOR EXAMPLE 4

Simplify each expression, using the quotient rule for exponents.

a. $\dfrac{x^9}{x^2}$ b. $\dfrac{4^9}{4^3}$

Remember that the product and quotient rules for exponents apply only when the bases are the same. It is also important to note that the base does not change when these rules are used.

Example 5 — Combining the product and quotient rules for exponents to simplify

Simplify each expression, using the product or quotient rule for exponents or both.

a. $\dfrac{x^8 y^4}{x^3 y}$ b. $\dfrac{a^5 a^3 b^6}{a^2 b}$ c. $\dfrac{x^7 y^2}{x^4 y} \cdot \dfrac{x^5 y^7}{x^2 y^3}$

SOLUTION $\boxed{x^2 y^5}$

a. Since this expression involves a quotient, use the quotient rule for exponents and subtract exponents of the common bases.

$$\dfrac{x^8 y^4}{x^3 y}$$
$$= x^{8-3} y^{4-1} \quad \text{Keep the same bases and subtract exponents.}$$
$$= x^5 y^3$$

b. PEMDAS indicates that we first simplify the numerator. Then use the quotient rule for exponents.

$$\dfrac{a^5 a^3 b^6}{a^2 b} \quad \text{Using PEMDAS, simplify the numerator first.}$$
$$= \dfrac{a^8 b^6}{a^2 b} \quad \text{Add the exponents of the like bases in the numerator.}$$
$$= a^{8-2} b^{6-1} \quad \text{Keep the bases and subtract exponents.}$$
$$= a^6 b^5$$

c. This expression is the product of two quotients, so first use the quotient rule for exponents. Then use the product rule for exponents.

$$\dfrac{x^7 y^2}{x^4 y} \cdot \dfrac{x^5 y^7}{x^2 y^3}$$
$$= \dfrac{x^3 y}{1} \cdot \dfrac{x^3 y^4}{1} \quad \text{Subtract the exponents with the same bases.}$$
$$= x^3 x^3 \cdot y y^4 \quad \text{Regroup like bases and add the exponents.}$$
$$= x^6 y^5$$

■ Connecting the Concepts

Why do we simplify the numerator and denominator first?

A fraction bar is a grouping symbol. It acts much like parentheses around the numerator and denominator.

$$\dfrac{a^5 a^3 b^6}{a^2 b} = \dfrac{(a^5 a^3 b^6)}{(a^2 b)}$$

When PEMDAS, the order-of-operations agreement, is used, if there are operations in the numerator and denominator, they must be done first, before the division.

PRACTICE PROBLEM FOR EXAMPLE 5

Simplify each expression, using the product or quotient rule for exponents or both.

a. $\dfrac{a^3 a^5 b^7}{a^2 a^4 b^2}$ b. $\dfrac{x^5 y^6}{x^2 y^4} \cdot \dfrac{x^8 y^4}{x y^3}$

Power Rule for Exponents

Another rule for exponents that is used to simplify exponential expressions is the **power rule for exponents**. This rule allows us to simplify exponential expressions that are raised to another power. Considering the expression $(5^2)^3$, there is a base of 5^2 that is raised to the power of 3. This expression can be rewritten and simplified by using the product rule for exponents.

$$(5^2)^3$$
$$= 5^2 \cdot 5^2 \cdot 5^2 \quad \text{Rewrite using the definition of exponents.}$$
$$= 5^{2+2+2} \quad \text{Keep the base and add the exponents.}$$
$$= 5^6$$

Instead of using the product rule for exponents, we multiply the two powers together.

$$(5^2)^3 = 5^{2 \cdot 3} \quad \text{Keep the base and multiply the exponents.}$$
$$= 5^6$$

Power Rule for Exponents

$$(x^m)^n = x^{mn}$$

When raising an exponential expression to another power, multiply the exponents, keeping the same base.

$$(x^4)^7 = x^{28}$$

Example 6 Using the power rule for exponents to simplify

Simplify each expression, using the power rule for exponents.

a. $(d^3)^5$ **b.** $(3^2)^4$

SOLUTION

a. As this is an exponential expression raised to another power, we use the power rule for exponents and multiply the exponents.

$$(d^3)^5 = d^{3 \cdot 5} \quad \text{Keep the base and multiply the exponents.}$$
$$= d^{15}$$

b. This is an exponential expression raised to another power, so we use the power rule for exponents and multiply the exponents.

$$(3^2)^4 = 3^{2 \cdot 4} \quad \text{Keep the base and multiply the exponents.}$$
$$= 3^8 \quad \text{Since the base is a constant, calculate the value.}$$
$$= 6561$$

PRACTICE PROBLEM FOR EXAMPLE 6

Simplify each expression, using the power rule for exponents.

a. $(c^4)^6$ **b.** $(8^3)^2$

Rules for Exponents

Power Rule for Exponents

When raising an exponential expression to another power, multiply the exponents.

$$(x^5)^3 = x^{15}$$

470 CHAPTER 5 Exponents and Polynomials

■ Powers of Products and Quotients

The power rule for exponents can be extended to **powers of products and quotients**. When a product is raised to a power, the power can be applied to each factor. If a factor is already raised to an exponent, the power rule for exponents is then used to combine the two powers. Do not apply exponents over the operations of addition or subtraction. Exponents can be applied only over the operations of multiplication or division.

> **Powers of Products and Quotients**
>
> $$(xy)^m = x^m y^m \qquad \left(\frac{x}{y}\right)^m = \frac{x^m}{y^m}$$
>
> When an expression is raised to a power, that power can be applied over multiplication or division.
>
> $$(xy)^7 = x^7 y^7 \qquad \left(\frac{x}{y}\right)^4 = \frac{x^4}{y^4}$$
>
> *Note*: Exponents cannot be applied over the operations of addition or subtraction.

In the expression $(2x)^3$, we have two factors, 2 and x, that are raised to a power of 3. Using the powers of products rule, we have that

$$(2x)^3 = (2 \cdot x)^3 = 2^3 \cdot x^3 = 8x^3$$

We could also rewrite the expression using repeated multiplication and then simplify using the product rule for exponents.

$$(2x)^3 = (2x)(2x)(2x) = 8x^3$$

Example 7 Simplifying powers of products and quotients

Simplify each expression.

a. $(-2a)^3$ **b.** $(3a^3 b^6)^2$ **c.** $\left(\dfrac{x^2}{y^3}\right)^5$

SOLUTION

a. The power of 3 can be applied over the product by using the powers of products rule.

$(-2a)^3$ Apply the power of 3 over each factor.
$= (-2)^3 (a)^3$ Simplify $(-2)^3$.
$= -8a^3$

b. The power of 2 can be applied over the product by using the powers of products rule.

$(3a^3 b^6)^2$
$= (3)^2 (a^3)^2 (b^6)^2$ Apply the exponent of 2 to each factor.
$= 9a^6 b^{12}$ Multiply exponents, using the power rule for exponents.

Skill Connection

Applying exponents

With exponent rules, the exponents can be applied over multiplication and division but not over addition or subtraction. It is true that

$$(x \cdot y)^5 = x^5 y^5$$

In general,

$$(x + y)^5 \neq x^5 + y^5$$

■ Connecting the Concepts

Where is the multiplication?

When we look at a problem involving exponents, it can help to write out the multiplication. For example,

$$3x^2 = 3 \cdot x^2$$

When we put in the multiplication symbol, we see that the exponent applies only to the variable x.

The expression $(3x)^2$ is different. Rewriting with a product yields

$$(3x)^2 = (3 \cdot x)^2$$

Here the exponent will apply to *both factors*, 3 and x.

$$(3x)^2 = (3 \cdot x)^2$$
$$= 3^2 \cdot x^2 = 9x^2$$

Rules for Exponents

Powers of Product and Quotients

When raising an expression to a power, apply the power over multiplication or division.

$$(xy)^3 = x^3 y^3 \qquad \left(\frac{x}{y}\right)^5 = \frac{x^5}{y^5}$$

c. The power of 5 can be applied over the quotient by using the powers of quotients rule.

$$\left(\frac{x^2}{y^3}\right)^5$$

$$= \frac{(x^2)^5}{(y^3)^5}$$ Apply the exponent of 5 over the quotient.

$$= \frac{x^{10}}{y^{15}}$$ Multiply the exponents, using the power rule for exponents.

PRACTICE PROBLEM FOR EXAMPLE 7

Simplify each expression.

a. $(-5a)^4$ b. $(4ac^4)^3$ c. $\left(\frac{a^2}{b^5}\right)^2$

Summary of Exponent Rules

1. $x^m x^n = x^{m+n}$ 2. $\dfrac{x^m}{x^n} = x^{m-n}$ 3. $(x^m)^n = x^{mn}$

4. $(xy)^m = x^m y^m$ 5. $\left(\dfrac{x}{y}\right)^m = \dfrac{x^m}{y^m}$

We can use each of these rules to simplify expressions. In some expressions, we need to use more than one of the rules to simplify completely. In many cases, it is easiest to simplify an expression within parentheses before applying the exponents over a multiplication or division.

Example 8 Simplifying by using the rules for exponents

Simplify each expression.

a. $\dfrac{(2x^4)^3}{(3xy^5)^2}$ b. $\left(\dfrac{-2x^6 y^4}{x^2 y^3}\right)^5$

c. $(-3a^3 b^2)^4 (5a^2 b)^2$ d. $\dfrac{(3x)^{12}}{(3x)^4}$

SOLUTION $x^2 y^5$

a. This quotient has two expressions raised to additional powers. Apply the powers over the products and then simplify.

$$\frac{(2x^4)^3}{(3xy^5)^2}$$

$$= \frac{(2)^3 (x^4)^3}{(3)^2 (x)^2 (y^5)^2}$$ Apply the exponents, using the powers of products rule.

$$= \frac{8x^{12}}{9x^2 y^{10}}$$ Multiply the exponents, using the power rule for exponents.

$$= \frac{8x^{10}}{9y^{10}}$$ Subtract the exponents of x, using the quotient rule for exponents.

> **Skill Connection**
>
> **Negative signs and exponents**
> When applying exponents, watch for negative signs. When we apply exponents to a negative number, the negative must be included in parentheses.
>
> $$(-3)^2 = 9$$
>
> But
>
> $$-3^2 = -9$$
>
> The second operation squares only the 3 and not the negative sign.

b. This quotient is being raised to an additional power. First simplify the quotient, using the quotient rule for exponents. Then apply the exponent.

$$\left(\frac{-2x^6y^4}{x^2y^3}\right)^5$$ Simplify inside the parentheses first (PEMDAS).

$$= (-2x^4y)^5$$ Subtract exponents, using the quotient rule for exponents.

$$= (-2)^5(x^4)^5(y)^5$$ Apply the exponents to each factor.

$$= -32x^{20}y^5$$ Multiply the exponents, using the power rule for exponents. Be careful with the -2 raised to the power of 5.

c. There are two products raised to additional powers, so we apply the exponents and use the power and product rules for exponents to simplify.

$$(-3a^3b^2)^4(5a^2b)^2$$ Apply the exponents.

$$= (-3)^4(a^3)^4(b^2)^4(5)^2(a^2)^2(b)^2$$ Be sure the -3 is raised to the fourth power.

$$= 81a^{12}b^8 \cdot 25a^4b^2$$ Multiply the exponents, using the power rule for exponents.

$$= 81 \cdot 25 \cdot a^{12}a^4 \cdot b^8b^2$$ Group like bases, using the commutative property.

$$= 2025a^{16}b^{10}$$ Add exponents, using the product rule for exponents.

d. This is a quotient of exponential expressions with the same base ($3x$), so subtract the exponents, using the quotient rule for exponents.

$$\frac{(3x)^{12}}{(3x)^4}$$

$$= (3x)^{12-4}$$ Keep the base and subtract the exponents.

$$= (3x)^8$$

$$= 3^8x^8$$ Apply the exponents.

$$= 6561x^8$$

PRACTICE PROBLEM FOR EXAMPLE 8
Simplify each expression.

a. $\left(\dfrac{-3m^7n^5}{m^2n^3}\right)^4$

b. $\dfrac{(5x^3)^2}{(2x^2y^4)^3}$

c. $(-2a^5b)^3(6a^3b^5)^2$

d. $(4m)^5(4m)^8$

5.1 Vocabulary Exercises

1. Natural number exponents represent _____ multiplication.

2. When multiplying exponential expressions that have the same base, _____ exponents.

3. When dividing exponential expressions that have the same base, _____ exponents.

4. When raising an exponential expression to another power, _____ the exponents.

5. When an expression is raised to a power, that power can apply over the operations of _____ and _____.

6. Applying exponents does NOT work over the operations _____ and _____.

7. The exponent rules in this section do NOT apply to the operations of _____ or _____.

5.1 Exercises

For Exercises 1 through 34, simplify each exponential expression. List the exponent rule or other rule that was used to simplify.

1. $(-2)^2$
2. -2^2
3. $(-2)^3$
4. -2^3
5. -5^2
6. $(-5)^2$
7. -1^3
8. $(-1)^3$
9. $x^7 x^2$
10. $y^3 y^5$
11. $7^3 \cdot 7^5$
12. $3^4 \cdot 3^2$
13. $5w^4 \cdot 3w^8$
14. $7g^2 \cdot 2g^9$
15. $(5xy^2)(8x^5y^3)$
16. $(6a^5b^4)(2ab^2)$
17. $(-3m^3n^4)(5m^7n^3)$
18. $(-8c^5d^9)(4c^2d^5)$
19. $(-6x^2y)(3x^4y^3)$
20. $(-9mn^5)(2mn^2)$
21. $(6x)^5(6x)^3$
22. $(2a)^5(2a)^3$
23. $(7b)^2 \cdot (7b)^4$
24. $(5p)^6 \cdot (5p)^4$
25. $(-4m)^3(-4m)^2$
26. $(-3y^2)^5(-3y^2)^6$
27. $\dfrac{x^{12}}{x^5}$
28. $\dfrac{h^{15}}{h^{11}}$
29. $\dfrac{20k^{17}}{4k^3}$
30. $\dfrac{45x^6}{3x}$
31. $(x^2)^5$
32. $(d^4)^3$
33. $(a^2b^6)^5$
34. $(x^6y^5)^4$

For Exercises 35 through 44, determine which student in each pair simplified each expression incorrectly. Explain what the student did wrong.

35. $x^5 x^2$

Elizabeth	David
x^{10}	x^7

36. $(x^3)^5$

Kaitlyn	Daisy
x^8	x^{15}

37. $\dfrac{20x^5}{4x^2}$

Alexa	Elba
$5x^3$	$16x^3$

38. $\dfrac{x^8}{x^4}$

Yasmin	Keegan
x^2	x^4

39. $(4x^3)^2$

Jonathan	Luis
$16x^6$	$8x^6$

40. $(5x^4)^3$

Kaila	Hannah
$15x^{12}$	$125x^{12}$

41. $(-3x)^2$

Yomaira	Steven
$-9x^2$	$9x^2$

474 CHAPTER 5 Exponents and Polynomials

42. $(-2x^3)^4$

Danaya	Sabrina
$16x^{12}$	$-16x^{12}$

43. $\dfrac{5^5}{5^2}$

Andrew	Joel
1	125

44. $\dfrac{6^3}{6}$

Lydia	Eleene
1	36

For Exercises 45 through 74, simplify each exponential expression. List the exponent rule or other rule that was used to simplify.

45. $\dfrac{x^2 x^5 y^5}{x^4 y^2}$

46. $\dfrac{a^6 a^2 b^5}{a^4 a^3 b}$

47. $\dfrac{m^2 m^3 n^2}{mn}$

48. $\dfrac{c^4 c^7 d^3 d^2}{c^3 c d^4}$

49. $\dfrac{30 x^5 y^2}{14 x y^2}$

50. $\dfrac{-16 a^3 b^5}{6 a^3 b}$

51. $\dfrac{x^4}{25} \cdot \dfrac{5}{x^2}$

52. $\dfrac{-12}{n} \cdot \dfrac{n^5}{3}$

53. $\dfrac{a^3 b}{a^2 b} \cdot \dfrac{b^3}{a}$

54. $\dfrac{xy^5}{xy^2} \cdot \dfrac{x^4 y^3}{xy}$

55. $\dfrac{16 a^2 b^6}{ab^3} \cdot \dfrac{a^3 b^2}{48 ab^2}$

56. $\dfrac{-9 x^3 y^7}{x^2 y^2} \cdot \dfrac{xy^3}{24 xy^2}$

57. $(3c^2 d^5)^2$

58. $(5xy^4)^3$

59. $(-5a^7 b^5)^2$

60. $(-2m^2 n^8)^4$

61. $\left(\dfrac{12 x^5 y^4}{2x^2 y}\right)^2$

62. $\left(\dfrac{24 a^7 b^5}{8 a^3 b^2}\right)^3$

63. $\left(\dfrac{-26 x^3 y^5}{6xy}\right)^2$

64. $\left(\dfrac{-18 m^3 n^2}{4 m^5 n}\right)^2$

65. $(3xy^2)^2 (5x^3 y^7)$

66. $(2a^5 b^3)^4 (5a^2 b)^2$

67. $(4m^4 n^3)^2 (3mn^5)^3$

68. $(7p^7 r^2)^2 (2p^4 r^6)^3$

69. $(-3a^2)^4 (2ab^3)^3$

70. $(-5m^2)^3 (4m^3 n)^2$

71. $\dfrac{(2x^3 y)^4}{(5xy^2)^2}$

72. $\dfrac{(6a^3 b^5)^2}{(3ab^2)^3}$

73. $\dfrac{(-4x^5 y^3)^2}{(2x^2 y)^3}$

74. $\dfrac{(-2g^4 h^3)^4}{(5g^3 h^2)^3}$

For Exercises 75 through 82, examine each student's work, and find the mistake that the student made. Hint: Remember to use the order-of-operations agreement.

75.
Elias
$-2^2 = 4$

76.
Reggie
$-3^2 = 9$

77.
Beverly
$3^2 \cdot 2 = 6^2$

78.
Eliseo
$4 \cdot 5^3 = 20^3$

79.
Alexis
$(2+3)^2 = 2^2 + 3^2 = 4 + 9 = 13$

80.
Oscar
$(7-5)^2 = 7^2 - 5^2 = 49 - 25 = 24$

81.
Victoria
$\dfrac{9^4}{3} = 3^4$

82.
Cooper
$\dfrac{4^3}{2} = 2^3$

83. **a.** Can the expression $(x+3)^2 (x+3)^5$ be simplified to $(x+3)^7$? If so, what rule for exponents allows this? Explain.

   b. Can the exponent 7 in the expression $(x+3)^7$ be applied to x and 3 to get $x^7 + 3^7$? If so, what rule for exponents allows this? Explain.

84. a. Can the expression $(x - 6)^3(x - 6)^7$ be simplified to $(x - 6)^{10}$? If so, what rule for exponents allows this? Explain.

b. Can the exponent 10 in the expression $(x - 6)^{10}$ be applied to x and 6 to get $x^{10} - 6^{10}$? If so, what rule for exponents allows this? Explain.

85. Can the expression $x^2 + x^3$ be simplified? Explain.

86. Can the expression $x^7 + x^4$ be simplified? Explain.

For Exercises 87 through 94, find an expression for each geometric figure using the numbers given.

87. Find an expression for the area of the given square. ($A = s^2$, that is, area = side²)

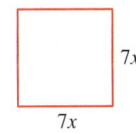

88. Find an expression for the area of the given square. ($A = s^2$, that is, area = side²)

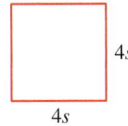

89. Find an expression for the volume of the given cube. ($V = s^3$, that is, volume = side³)

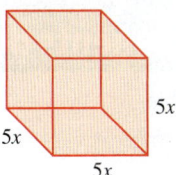

90. Find an expression for the volume of the given cube. ($V = s^3$, that is, volume = side³)

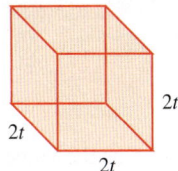

91. Find an expression for the area of the given circle. ($A = \pi r^2$)

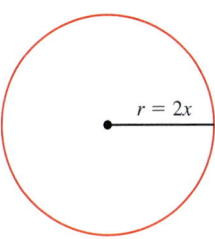

92. Find an expression for the area of the given circle. ($A = \pi r^2$)

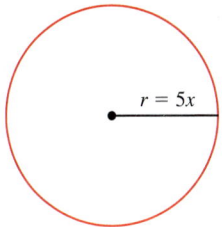

93. Find an expression for the area of the given triangle. $\left(A = \dfrac{1}{2}bh\right)$

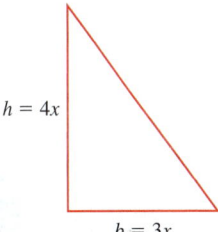

94. Find an expression for the area of the given triangle. $\left(A = \dfrac{1}{2}bh\right)$

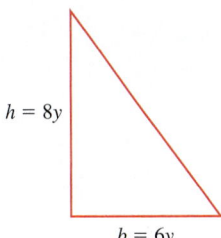

476 CHAPTER 5 Exponents and Polynomials

5.2 Negative Exponents and Scientific Notation

LEARNING OBJECTIVES
- Explain negative exponents.
- Use scientific notation in calculations.

■ Negative Exponents

When working with exponential expressions using the rules for exponents, we may end up with negative exponents. The most common way this occurs is when we are using the quotient rule for exponents, as it involves subtracting the exponents.

$$\frac{x^2}{x^5} = x^{2-5} = x^{-3}$$

Here we subtracted exponents and got an exponent of -3. Another way to simplify this same expression is to write out the exponents as repeated multiplication. Then divide out common factors from the numerator and denominator.

$$\frac{x^2}{x^5} = \frac{{}^1\cancel{x}\,{}^1\cancel{x}}{{}_1\cancel{x}\,{}_1\cancel{x}\,xxx} = \frac{1}{xxx} = \frac{1}{x^3}$$

Both of these methods must give the same result, Therefore,

$$x^{-3} = \frac{1}{x^3}$$

Let's look at this pattern by using numbers in a Concept Investigation.

■ CONCEPT INVESTIGATION
What are those negative exponents doing?

1. Fill in the missing values in the following table.

n	2^n
5	$2^5 = 32$
4	$2^4 = 16$
3	
2	
1	

2. What operation happens in the right-hand column as the value of n decreases by 1?

3. Use this pattern to continue the table for a few more values of n. Write your answers as fractions if necessary.

n	2^n
2	$2^2 = 4$
1	$2^1 = 2$
0	
-1	
-2	
-3	

4. In the last three rows of the table, rewrite the denominators of the fractions as powers of 2 (for instance, $\frac{1}{4} = \frac{1}{2^2}$).

What's That Mean?

Reciprocal

Remember that a *reciprocal* is a word associated with fractions and means to invert the fraction.

The reciprocal of $\frac{2}{3}$ is $\frac{3}{2}$.

The reciprocal of $\frac{1}{4}$ is 4.

The reciprocal of 5 is $\frac{1}{5}$.

From this investigation and the previous discussion, we see that a negative exponent is equivalent to the **reciprocal** with a positive exponent. In general, it is preferable to have only positive exponents in the answer. Use the reciprocal to change any negative exponents to positive ones.

Negative Exponents

$$x^{-n} = \frac{1}{x^n} \quad \text{and} \quad \frac{1}{x^{-n}} = x^n$$

When raising a base to a negative exponent, find the reciprocal of the base. Raise the base to the absolute value of the exponent.

$$x^{-4} = \frac{1}{x^4} \qquad \left(\frac{1}{2}\right)^{-1} = 2 \qquad \frac{2}{x^{-3}} = 2x^3$$

Example 1 — Working with negative exponents

Rewrite without negative exponents.

a. x^{-5} b. 3^{-2} c. $\frac{1}{y^{-3}}$ d. $5x^{-2}$

SOLUTION

a. The exponent is negative, so take the reciprocal of the base x with an exponent of 5.

$$x^{-5} = \frac{1}{x^5}$$

b. The exponent is negative, so take the reciprocal of the base 3 with an exponent of 2. Then calculate a value, as the base is a constant.

$$3^{-2} = \frac{1}{3^2} = \frac{1}{9}$$

c. This negative exponent is in the denominator. When taking the reciprocal of the base, the base moves to the numerator.

$$\frac{1}{y^{-3}} = y^3$$

d. Notice that 5 is not raised to an exponent. It is the coefficient of x^{-2}. The rule for negative exponents applies only to x^{-2}.

$$5x^{-2} = 5 \cdot x^{-2} \quad \text{Rewrite as a product.}$$
$$= 5 \cdot \frac{1}{x^2} \quad \text{Apply the rule for negative exponents to } x^{-2}.$$
$$= \frac{5}{x^2} \quad \text{Simplify.}$$

Rules for Exponents

Negative Exponents

When raising a base to a negative exponent, raise the reciprocal of that base to the absolute value of the exponent.

$$y^{-5} = \frac{1}{y^5}$$

Connecting the Concepts

Base or coefficient?

In the expression $5x^2$, there is a hidden product. $5x^2 = 5 \cdot x^2$, so 5 is *not* being raised to the second power. The exponent applies only to the variable.

In the expression $(5x)^2$, the base is $5x$, and that entire base is being raised to the second power. The exponent applies to both 5 and x.

PRACTICE PROBLEM FOR EXAMPLE 1

Rewrite without negative exponents.

a. b^{-6} b. $7h^{-2}$ c. $4x^{-3}$

All of the rules for exponents that we learned in the last section apply to negative exponents. When we use any of the rules for exponents, be careful to watch the signs to avoid calculation errors. When using the quotient rule for exponents, the resulting exponential always ends up in the numerator of the fraction. If the exponent is negative, find the reciprocal of the base.

$$\frac{20x^2}{14x^7}$$ Reduce the constants and subtract the exponents, using the quotient rule for exponents. The result is in the numerator.

$$= \frac{10x^{2-7}}{7}$$

$$= \frac{10x^{-5}}{7}$$ Take the reciprocal to rewrite without a negative exponent.

$$= \frac{10}{7x^5}$$

When simplifying an exponential expression, the final answer must contain no negative exponents.

> **What's That Mean?**
>
> **Simplify an Exponential Expression**
>
> To *simplify an exponential expression*, use the rules for exponents to combine exponents with the same base. Then use the rule for negative exponents to make sure the final answer has only positive exponents.

Example 2 Simplifying exponential expressions

Simplify. Write the final answer using positive exponents only.

a. $\dfrac{10x^3}{12x^5}$ b. $7x^2y^{-4}$ c. $\dfrac{9x^{-2}y^5}{8xy^{-3}}$ d. $\dfrac{-5a^2b^{-5}}{2a^7b}$

SOLUTION x^2y^5

a. The expression has a quotient, so use the quotient rule for exponents.

$$\frac{10x^3}{12x^5}$$ Reduce the constants and subtract the exponents using the quotient rule for exponents. The result is in the numerator.

$$= \frac{5x^{3-5}}{6}$$

$$= \frac{5x^{-2}}{6}$$ Take the reciprocal to rewrite without negative exponents.

$$= \frac{5}{6x^2}$$ Positive exponents only in the final answer.

b. Notice the variable y is raised to a negative exponent. Apply the rule for negative exponents to rewrite y with a positive exponent. The rest of the expression does not contain negative exponents.

$$7x^2y^{-4}$$ Rewrite y^{-4} without negative exponents.

$$= \frac{7x^2}{y^4}$$ The answer contains only positive exponents.

c. The expression has a quotient, so use the quotient rule for exponents. Write the final answer using only positive exponents.

$$\frac{9x^{-2}y^5}{8xy^{-3}}$$ Subtract the exponents, using the quotient rule for exponents.

$$=\frac{9x^{-2-1}y^{5-(-3)}}{8}$$ The result of the quotient rule for exponents is in the numerator.

$$=\frac{9x^{-3}y^8}{8}$$ Use the negative exponent rule to rewrite x^{-3}.

$$=\frac{9y^8}{8x^3}$$ The final answer has positive exponents only.

d. This expression has a quotient, so use the quotient rule for exponents. Then apply the rule for negative exponents to write the final answer with positive exponents only. Note that -5 is a coefficient. It does not have a negative exponent, so it remains unchanged in the numerator.

$$\frac{-5a^2b^{-5}}{2a^7b}$$ Subtract the exponents, using the quotient rule for exponents.

$$=\frac{-5a^{2-7}b^{-5-1}}{2}$$ The result of the quotient rule for exponents is in the numerator.

$$=\frac{-5a^{-5}b^{-6}}{2}$$ Take the reciprocal to rewrite without negative exponents.

$$=\frac{-5}{2a^5b^6}$$ The final answer has positive exponents only.

> **Connecting the Concepts**
>
> **Negative number or negative exponent?**
>
> In the expression $-5x^{-2}$, only the base x has a negative exponent. Therefore, the expression simplifies as $-5x^{-2} = \frac{-5}{x^2}$. Note that negative numbers do not move in the final answer.
>
> In the expression $(-5x)^{-2}$, the base is $-5x$, and the negative exponent applies to the entire base. Therefore, the expression simplifies as
> $$(-5x)^{-2} = \frac{1}{(-5x)^2} = \frac{1}{25x^2}.$$

PRACTICE PROBLEM FOR EXAMPLE 2

Simplify. Write the final answer using positive exponents only.

a. $\dfrac{35x^4}{14x^{10}}$ b. $-12x^{-4}y^3$ c. $\dfrac{5a^{-5}b^4}{7a^{-2}b^{-1}}$

■ Using Scientific Notation in Calculations

Every day people use very large and very small numbers. For example, countries spend billions of dollars, and engineers study structures built with nanotubes.

The names that are given to numbers are represented by powers of 10. Both the American naming system and the metric system are based on powers of 10. The following table lists several powers of 10 and the American name and metric prefix for each.

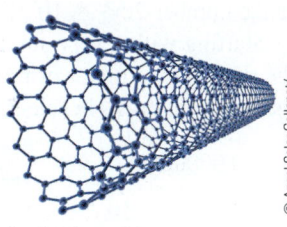

Drawing of a nanotube

American Name	Power of 10	Metric Prefix	American Name	Power of 10	Metric Prefix
1 trillion	10^{12}	tera- (T-)	1 tenth	10^{-1}	deci- (d-)
1 billion	10^9	giga- (G-)	1 hundredth	10^{-2}	centi- (c-)
1 million	10^6	mega- (M-)	1 thousandth	10^{-3}	milli- (m-)
1 thousand	10^3	kilo- (k-)	1 millionth	10^{-6}	micro- (μ-)
1 hundred	10^2	hecto- (h-)	1 billionth	10^{-9}	nano- (n-)

Connecting the Concepts

How can we convert from scientific notation?

In Section 1.1, we learned how to read a number in scientific notation. A number in the form $a \times 10^n$, where $1 \leq |a| < 10$ and n is an integer, is said to be in scientific notation.

When we convert a number from scientific notation to standard form, the sign of the exponent n determines which way the decimal point moves.

When the exponent on 10 is positive, multiply by 10^n, or move the decimal point n places to the right.

When the exponent of 10 is negative, multiply by 10^{-n}, or move the decimal point n places to the left.

What's That Mean?

Digit

The *digits* are the numbers 0, 1, 2, 3, 4, 5, 6, 7, 8, and 9. The nonzero digits are the integers between 1 and 9, including 1 and 9.

The state of California has a 2017 proposed budget of about $124,000,000,000. This number is usually represented by using words—as $124 billion—to make it easier to read. Using a power of 10 to write this number makes it easier to do calculations.

$124,000,000,000   *This is in standard form.*
$= $124 billion   *The number written in words.*
$= 124×10^9 *One billion is 10^9.*
$= 1.24×10^{11} *Scientific notation has the decimal point to the right of the leftmost digit.*

When working with very large or very small numbers, we often write the numbers using scientific notation to make them easier to work with. In Section 1.1, we introduced scientific notation and what the notation represents. Now we practice changing numbers from standard form to scientific notation. Then we will learn how to do calculations with numbers in scientific notation.

Because scientific notation uses powers of 10 as a shorter way to write some numbers, we can use the powers of 10 to move the decimal place left or right. Remember that a positive power of 10 results in a large number, and a negative power of 10 results in a small number.

$$10^8 = 100,000,000$$

$$10^{-5} = \frac{1}{10^5} = \frac{1}{100,000} = 0.00001$$

Starting with a large number in standard notation, move the decimal place to the left until it is to the right of the first (leftmost) **digit**. The number of times the decimal place is moved is equal to the exponent on 10 in scientific notation.

25,400,000   *The decimal point is not shown, so it is on the far right.*
25,400,000.   *Move the decimal point seven places to the left so that it is to the right of the first digit.*
$= 2.54 \times 10^7$ *This is the number written using scientific notation.*

Check this answer by multiplying it out.

$= 2.54 \times 10,000,000$ *Substitute in $10^7 = 10,000,000$.*
$= 25,400,000$ *The original number. It checks.*

In the number 2.54×10^7, 2.54 is the *coefficient*. The number 7 is the exponent on 10.

Starting with a very small number in standard notation, move the decimal place to the right. The number of times the decimal place is moved is equal to the *opposite* of the exponent on 10 in scientific notation.

0.00000843   *The decimal place needs to move six places to the right.*
   The decimal point should be to the right of the first nonzero digit.
$= 8.43 \times 10^{-6}$ *This was a very small number, so the exponent of 10 is negative.*

Example 3   Converting numbers into scientific notation

Convert the following numbers into scientific notation.

a. *Voyager* I was launched in September 1977. As of this writing, it is about 20,627,000,000 km from Earth.

b. The total national debt of the United States as of this writing was about $19,846,000,000,000.

c. The size of a plant cell is 0.00001276 meter.

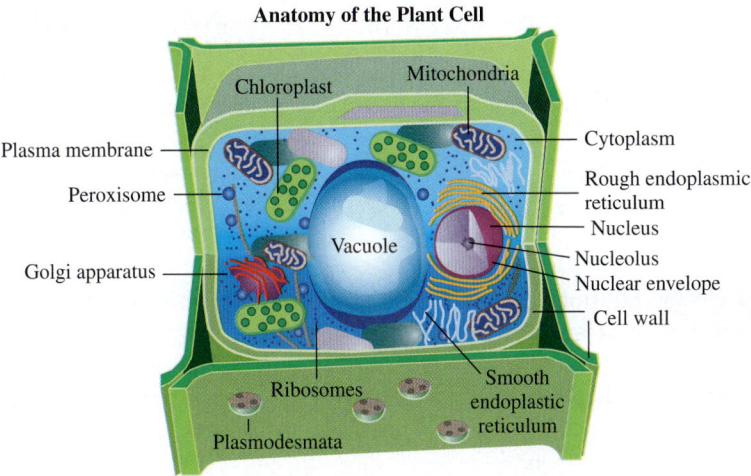

d. The diameter of a single-walled nanotube is 0.000000001 meter.

SOLUTION

a. This is a large number, so we move the decimal place to the left. The exponent of 10 will be positive.

$$20{,}627{,}000{,}000 \text{ km} = 2.0627 \times 10^{10}$$

Voyager I is about 2.0627×10^{10} km from Earth.

b. This is a large number, so we move the decimal place to the left. The exponent of 10 will be positive.

$$19{,}846{,}000{,}000{,}000 = 1.9846 \times 10^{13}$$

The United States national debt was about 1.9846×10^{13} dollars.

c. This is a small number, so we move the decimal place to the right. The exponent of 10 will be negative.

$$0.00001276 \text{ meter} = 1.276 \times 10^{-5} \text{ meter}$$

The size of a plant cell is about 1.276×10^{-5} meters.

d. This is a small number, so we move the decimal place to the right. The exponent of 10 will be negative.

$$0.000000001 \text{ meter} = 1.0 \times 10^{-9} \text{ meter}$$

The diameter of a single-walled nanotube is 1.0×10^{-9} meters.

PRACTICE PROBLEM FOR EXAMPLE 3

Convert the following numbers into scientific notation.

a. As of this writing, Mark Zuckerberg's Facebook stock was worth approximately $72,000,000,000.

b. 0.00000002 second.

Now we use the rules for exponents to do calculations involving scientific notation. When multiplying or dividing numbers in scientific notation, first multiply or divide the coefficients. Then apply the rules for exponents on the powers of 10.

When done, adjust the powers of 10 to make sure the absolute value of the coefficient is still greater than or equal to 1 and less than 10. If a coefficient is too large, move the decimal point to the left until the coefficient is less than 10. Moving the decimal point increases the exponent of 10.

482 CHAPTER 5 Exponents and Polynomials

Skill Connection

Working with powers of 10
To write 457.2×10^4 quickly in scientific notation, move the decimal point two places to the left. This will add 2 to the exponent on 10. Therefore,

$$457.2 \times 10^4 = 4.572 \times 10^6$$

Likewise to rewrite 0.00058×10^{11}, move the decimal place four places to the right. This will subtract 4 from the exponent on 10. Therefore,

$$0.00058 \times 10^{11} = 5.8 \times 10^7$$

$457.2 \times 10^4 = (4.572 \times 10^2) \times 10^4$ — Rewrite the coefficient in proper scientific notation.

$= 4.572 \times 10^2 \times 10^4 = 4.572 \times 10^6$ — Use the product rule to add the exponents of 10.

If the coefficient is less than 1, move the decimal place to the right until the coefficient is equal to or greater than 1. Moving the decimal point decreases the exponent of 10.

$0.00058 \times 10^{11} = (5.8 \times 10^{-4}) \times 10^{11}$ — Rewrite the coefficient in proper scientific notation.

$5.8 \times 10^{-4} \times 10^{11} = 5.8 \times 10^7$ — Use the product rule to add the exponents of 10.

Example 4 Multiplication and division of a number written in scientific notation

Perform the indicated operation. Write the answer in scientific notation.

a. $8(4.0 \times 10^{21})$ **b.** $\dfrac{1.2 \times 10^{18}}{4}$

SOLUTION

a. Multiply the coefficient 4.0 by 8 and adjust the exponent of 10 if needed.

$8(4.0 \times 10^{21})$
$= 8(4.0) \times 10^{21}$ — Multiply the coefficient by 8.
$= 32.0 \times 10^{21}$ — The new coefficient is more than 10, so we need to adjust the exponent.
$= (3.2 \times 10^1) \times 10^{21}$
$= 3.2 \times 10^{22}$ — The exponent increases by 1, and the decimal moves 1 place to the left.

b. Divide the coefficient by 4 and adjust the exponent of 10 if needed.

$\dfrac{1.2 \times 10^{18}}{4}$

$= \dfrac{1.2}{4} \times 10^{18}$ — Divide the coefficient by 4.

$= 0.3 \times 10^{18}$ — The new coefficient is less than 1, so we will adjust the exponent.
$= (3 \times 10^{-1}) \times 10^{18}$
$= 3 \times 10^{17}$ — The exponent decreases by 1, and the decimal moves 1 place to the right.

PRACTICE PROBLEM FOR EXAMPLE 4

Perform the indicated operation. Write the answer in scientific notation.

a. $4(1.25 \times 10^8)$ **b.** $25(3.4 \times 10^{13})$ **c.** $\dfrac{7.4 \times 10^8}{1480}$

Example 5 Charge of a particle

The elementary charge, the **electric charge** carried by a single proton, is approximately 1.602×10^{-19} coulomb. Find the electric charge of 500 protons.

SOLUTION
To find the electric charge of 500 protons, we multiply the elementary charge by 500.

$500(1.602 \times 10^{-19})$
$= 500(1.602) \times 10^{-19}$ — Multiply the coefficient by 500.

$= 801 \times 10^{-19}$ The new coefficient is more than 10, so adjust the exponent.
$= (8.01 \times 10^2) \times 10^{-19}$
$= 8.01 \times 10^{-17}$ The decimal moves two places to the left, so the exponent increases by 2.

The electric charge of 500 protons is 8.01×10^{-17} coulomb.

PRACTICE PROBLEM FOR EXAMPLE 5
An astronomical unit (au) equals the average distance from the Sun's center to the center of the Earth. An au is equal to 1.496×10^8 km. The average distance of Uranus from the Sun is about 19.19 au. Find the average distance of Uranus from the Sun in kilometers.

> **What's That Mean?**
> **The Electric Charge of a Proton**
> Atoms are made up of protons, neutrons, and electrons.
>
>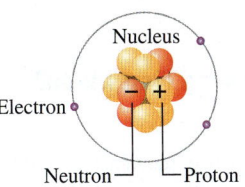
>
> The protons and neutrons are located in the nucleus at the center of the atom. The electrons circle the nucleus in shells. Protons are positively charged. Electrons are negatively charged. Neutrons have no (neutral) charge.

Example 6 — Multiplying and dividing numbers in scientific notation

Perform the indicated operation. Write the answer in scientific notation.

a. $(5.0 \times 10^4)(3.5 \times 10^{12})$ **b.** $(7.2 \times 10^{-6})(3.0 \times 10^{-4})$ **c.** $\dfrac{2.5 \times 10^{11}}{8.0 \times 10^8}$

SOLUTION

a. Multiply the coefficients. Add the exponents of the 10's, using the product rule for exponents. Finally, adjust the exponent of 10 if needed.

$(5.0 \times 10^4)(3.5 \times 10^{12})$

$= 5.0(3.5) \times 10^{4+12}$ Multiply the coefficients and add the exponents.
$= 17.5 \times 10^{16}$ The new coefficient is bigger than 10, so adjust the exponent.
$= (1.75 \times 10^1) \times 10^{16}$
$= 1.75 \times 10^{17}$ The exponent increases by 1, and the decimal moves one place to the left.

b. Multiply the coefficients. Add the exponents, using the product rule for exponents. Adjust the exponent if needed.

$(7.2 \times 10^{-6})(3.0 \times 10^{-4})$

$= 7.2(3.0) \times 10^{-6+(-4)}$ Multiply the coefficients and add the exponents.
$= 21.6 \times 10^{-10}$ The coefficient is bigger than 10, so adjust the exponent.
$= (2.16 \times 10^1) \times 10^{-10}$
$= 2.16 \times 10^{-9}$ The exponent increases by 1, and the decimal moves one place to the left.

c. Divide the coefficients. Subtract the exponents, using the quotient rule for exponents. Adjust the exponent if needed.

$\dfrac{2.5 \times 10^{11}}{8.0 \times 10^8}$

$= \dfrac{2.5}{8.0} \times 10^{11-8}$ Divide the coefficients and subtract the exponents.
$= 0.3125 \times 10^3$ The new coefficient is less than 1, so adjust the exponent.
$= (3.125 \times 10^{-1}) \times 10^3$
$= 3.125 \times 10^2$ The exponent decreases by 1, and the decimal moves 1 place to the right.

PRACTICE PROBLEM FOR EXAMPLE 6
Perform the indicated operation. Write the answer in scientific notation.

a. $(8.4 \times 10^6)(7.0 \times 10^8)$ **b.** $(1.5 \times 10^{-9})(4.0 \times 10^{-17})$ **c.** $\dfrac{6.4 \times 10^{15}}{2.0 \times 10^4}$

484 CHAPTER 5 Exponents and Polynomials

Example 7 Comparing the masses of planets

The mass of the planet Uranus is estimated at 8.68×10^{25} kg, and the mass of the Earth is 5.98×10^{24} kg. Divide the mass of Uranus by the mass of the Earth to find how much larger the mass of Uranus is than that of the Earth.

SOLUTION

$$\frac{8.68 \times 10^{25} \text{ kg}}{5.98 \times 10^{24} \text{ kg}}$$
$$\approx 1.45 \times 10$$
$$\approx 14.5$$

The mass of Uranus is 14.5 times the mass of the Earth.

PRACTICE PROBLEM FOR EXAMPLE 7

The mass of the planet Jupiter is estimated at 1.898×10^{27} kg, and the mass of the Earth is 5.98×10^{24} kg. Divide the mass of Jupiter by the mass of the Earth to find how much larger the mass of Jupiter is than that of the Earth.

What's That Mean?

Mass

In science, the *mass* of an object refers to the amount of matter it contains. The mass of an object has weight when it is in a gravitational field. We also use the word *mass* to describe very large objects (massive). We also say that a large grouping of things forms a mass, such as a mass rally.

5.2 Vocabulary Exercises

1. Scientific notation is a compact way to write very _____ or very _____ numbers.

2. To simplify an expression with a negative exponent, find the _____ of the base and make the exponent _____.

3. For a number in the form $a \times 10^n$ to be in proper scientific notation, the value of a must be _____ $\leq |a| <$ _____.

4. A number in proper scientific notation $a \times 10^n$ represents a large number when n is _____.

5. A number in proper scientific notation $a \times 10^n$ represents a small number when n is _____.

5.2 Exercises

For Exercises 1 through 20, rewrite using positive exponents.

1. x^{-3}
2. x^{-5}
3. 3^{-2}
4. 7^{-2}
5. 2^{-5}
6. 10^{-3}
7. $4d^{-6}$
8. $23m^{-1}$
9. $-5a^{-4}$
10. $-8m^{-7}$
11. $\dfrac{5x^2}{2y^{-3}}$
12. $\dfrac{12a^{-3}}{8b^{-4}}$
13. $\dfrac{1}{5x^{-4}}$
14. $\dfrac{1}{2a^{-3}}$
15. $\dfrac{a^{-3}}{2}$
16. $\dfrac{b^{-4}}{6}$
17. $\dfrac{3^{-2}}{x}$
18. $\dfrac{5^{-3}}{y^2}$
19. $\dfrac{x^{-2}}{y^{-3}}$
20. $\dfrac{x^{-1}}{y^{-4}}$

For Exercises 21 through 28, the students were asked to simplify. Explain what each student did wrong. Then simplify the expression correctly.

21. $\dfrac{-2x}{y}$

22. $\dfrac{-5a^2}{b^2}$

Julia
$$\dfrac{-2x}{y} = \dfrac{x}{2y}$$

Adelynne
$$\dfrac{-5a^2}{b^2} = \dfrac{a^2}{5b^2}$$

SECTION 5.2 Negative Exponents and Scientific Notation **485**

23. $42a^{-1}$

Frida
$42a^{-1} = \dfrac{1}{42a}$

24. $6b^{-1}$

Mike
$6b^{-1} = \dfrac{1}{6b}$

25. $(5x)^{-2}$

Christopher
$(5x)^{-2} = 5x^{-2} = \dfrac{5}{x^2}$

26. $(7y^2)^{-3}$

Helen
$(7y^2)^{-3} = 7y^{-6} = \dfrac{7}{y^6}$

27. $(-3x)^{-2}$

Madalyn
$(-3x)^{-2} = (-3)^2 x^{-2} = \dfrac{9}{x^2}$

28. $(-4y)^{-3}$

Nicholas
$(-4y)^{-3} = 4^3 y^{-3} = \dfrac{64}{y^3}$

For Exercises 29 through 50, simplify each expression. List the exponent rule(s) used. Write the final answers using positive exponents only.

29. $\dfrac{x^3}{x^7}$

30. $\dfrac{a^4}{a^{10}}$

31. $\dfrac{m^2 n}{m^5 n^2}$

32. $\dfrac{g^2 h^2}{g h^5}$

33. $\dfrac{10 x^5 y^7}{6 x^{11} y^2}$

34. $\dfrac{8 a^2 b^4}{24 a^5 b^2}$

35. $\dfrac{4 x^3 y^{-2}}{x^5 y^6}$

36. $\dfrac{6 a^4 b^{-3}}{a b^2}$

37. $\dfrac{24 m^{-3} n^5}{10 m^2 n^{-3}}$

38. $\dfrac{50 x^{-7} y^4}{20 x^6 y^{-2}}$

39. $\dfrac{-15 x^2 y^{-4}}{6 x y^3}$

40. $\dfrac{-32 a^4 b^{-3}}{12 a b^2}$

41. $(2x^{-2})^3$

42. $(3x^{-3})^4$

43. $(-5x^{-3})^2$

44. $(-3a^{-7})^2$

45. $(-3a^{-4})^2(-6a^3 b^{-3})$

46. $(-2x^{-3})^4(-5xy^{-1})$

47. $\left(\dfrac{-4x^5 y^3}{6x^7 y^{-1}}\right)^2$

48. $\left(\dfrac{-15 a^3 b^2}{35 a^{-2} b^4}\right)^2$

49. $\left(\dfrac{a}{b^3}\right)^{-2}$

50. $\left(\dfrac{x^2}{y^3}\right)^{-3}$

For Exercises 51 through 56, the students were asked to simplify. Explain what each student did wrong. Then simplify the expression correctly.

51. $\dfrac{x^3}{x^{-5}}$

Joseph
$\dfrac{x^3}{x^{-5}} = x^{3-5} = x^{-2} = \dfrac{1}{x^2}$

52. $\dfrac{a^8}{a^{-2}}$

Jonathan
$\dfrac{a^8}{a^{-2}} = a^{8-2} = a^6$

53. $\dfrac{-2x^{-4}}{16 x^3}$

Mackenzie
$\dfrac{-2x^{-4}}{16 x^3} = \dfrac{-2x^{-4-3}}{16} = 8x^{-7} = \dfrac{8}{x^7}$

54. $\dfrac{3y^3}{-9y^5}$

Wyatt
$\dfrac{3y^3}{-9y^5} = \dfrac{y^{3-5}}{-3} = 3y^{-2} = \dfrac{3}{y^2}$

55. $(x^3 y)^{-2}$

Tristan
$(x^3 y)^{-2} = x^3 y^{-2} = \dfrac{x^3}{y^2}$

486 CHAPTER 5 Exponents and Polynomials

56. $(x^{-2}y)^{-3}$

> *Miguel*
>
> $(x^{-2}y)^{-3} = x^{-5}y^{-3} = \dfrac{1}{x^5 y^3}$

For Exercises 57 through 64, convert the following numbers from scientific notation to standard form.

57. 6.23×10^8
58. 1.457×10^{11}
59. The mass of the Earth: 5.98×10^{24} kg.
60. The diameter of the Earth: 1.28×10^4 km.
61. 3.745×10^{-9}
62. 2.7×10^{-6}
63. An electron's mass: 1.67×10^{-27} kg.
64. The electrical charge of one proton: 1.602×10^{-19} coulomb.

For Exercises 65 through 76, convert the following numbers from standard form to scientific notation.

65. 62,000,000,000
66. 3,540,000,000,000
67. 0.0000043
68. 0.0000000082
69. The 2017 first-quarter Alphabet (Google) revenue was $24,750,000,000.
    Source: qz.com
70. The total revenue in 2016 for Apple, Inc. was $215,640,000,000.
    Source: www.statistica.com
71. The 2017 first-quarter U.S. gross domestic product (GDP) was $19,027,600,000.
    Source: St. Louis Federal Reserve.
72. The 2014 GDP of the world was $107,500,000,000,000.
    Source: CIA World Factbook.
73. The average width of a human hair: 0.00007 meter.

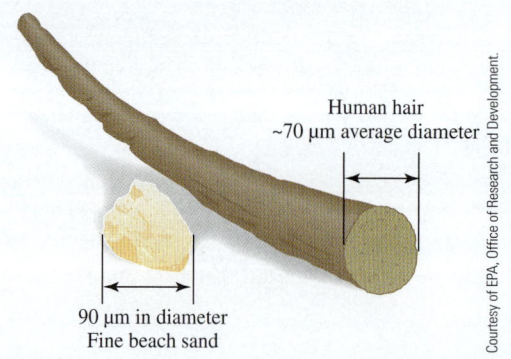
Human hair ~70 μm average diameter
90 μm in diameter Fine beach sand

74. The depth of pits where data on compact discs are stored: 0.0000001 meter.
75. The width of pits where data on compact discs are stored: 0.0000005 meter.
76. The diameter of a standard atom: 0.00000000006 meter.

For Exercises 77 through 100, perform the indicated operations. Write the answers in scientific notation.

77. $2(4.2 \times 10^{11})$
78. $3(1.25 \times 10^7)$
79. $5(1.5 \times 10^{-12})$
80. $7(1.2 \times 10^{-4})$
81. $4(6.3 \times 10^{13})$
82. $6(3.5 \times 10^8)$
83. $12(3.0 \times 10^{-7})$
84. $20(2.5 \times 10^{-14})$
85. $\dfrac{2.5 \times 10^8}{5}$
86. $\dfrac{3.5 \times 10^{15}}{5}$
87. $\dfrac{1.4 \times 10^{15}}{7}$
88. $\dfrac{1.6 \times 10^{20}}{4}$
89. $25(6.0 \times 10^{23})$
90. $30(8.0 \times 10^{41})$
91. $50(7.2 \times 10^{-24})$
92. $40(8.3 \times 10^{-34})$
93. $\dfrac{7.2 \times 10^{30}}{1200}$
94. $\dfrac{6.8 \times 10^{18}}{400}$
95. $\dfrac{7.2 \times 10^{-14}}{3000}$
96. $\dfrac{9.6 \times 10^{-45}}{15{,}000}$
97. $-3(4.5 \times 10^{11})$
98. $-6(2.5 \times 10^{26})$
99. $5(-3.6 \times 10^{23})$
100. $8(-4.5 \times 10^{15})$

For Exercises 101 through 110, use the following information to answer the questions. Write the answers using scientific notation.

Mass of the Earth	5.98×10^{24} kg
Astronomical unit (au): average distance from the center of the Sun to the center of the Earth	1.496×10^8 km

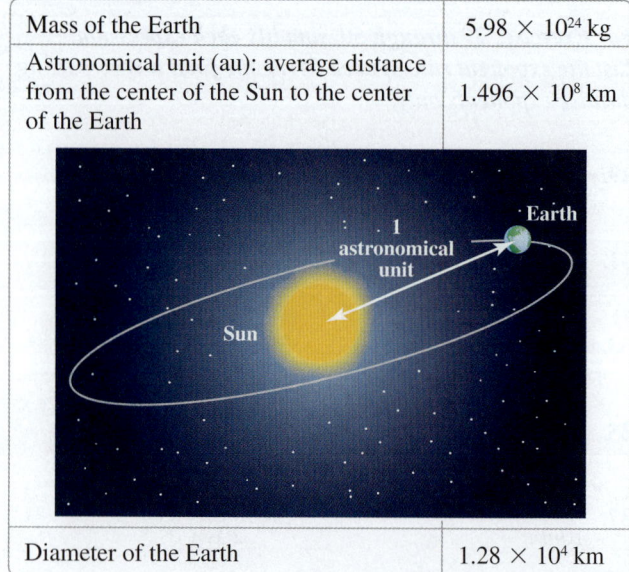

Diameter of the Earth	1.28×10^4 km

101. The mass of Saturn is about 95 times the mass of the Earth. Find the mass of Saturn in kilograms.

102. The mass of Neptune is about 17 times the mass of the Earth. Find the mass of Neptune in kilograms.

103. The mass of Mercury is about 0.055 times the mass of the Earth. Find the mass of Mercury in kilograms.

104. The mass of Mars is about 0.107 times the mass of the Earth. Find the mass of Mars in kilograms.

105. The average distance from Pluto to the Sun is approximately 39.5 au. Find the average distance in kilometers.

106. The average distance from Neptune to the Sun is approximately 30 au. Find the average distance in kilometers.

107. The average distance from Mars to the Sun is approximately 1.5 au. Find the average distance in kilometers.

108. The average distance from Venus to the Sun is approximately 0.72 au. Find the average distance in kilometers.

109. The diameter of Mars is about half the diameter of the Earth. Find the diameter of Mars.

110. The diameter of Mercury is about one-third the diameter of the Earth. Find the diameter of Mercury.

For Exercises 111 through 116, use the following information to answer the questions. Write the answers using scientific notation.

Elementary charge: the electrical charge of one proton	1.602×10^{-19} coulomb
An electron's mass	1.67×10^{-27} kg

111. Find the electric charge of 200 protons.

112. Find the electric charge of 1300 protons.

113. Find the mass of 500 electrons in kilograms.

114. Find the mass of 800 electrons in kilograms.

115. Find the mass of 3 million electrons in kilograms.

116. Find the mass of 7 million electrons in kilograms.

For Exercises 117 through 130, perform the indicated operations. Write the answers in scientific notation.

117. $(2.0 \times 10^{13})(4.2 \times 10^{11})$

118. $(3.0 \times 10^{8})(1.2 \times 10^{6})$

119. $(2.5 \times 10^{17})(1.5 \times 10^{-12})$

120. $(4.0 \times 10^{-14})(3.5 \times 10^{-6})$

121. $(6.0 \times 10^{20})(6.3 \times 10^{13})$

122. $(8.5 \times 10^{12})(3.5 \times 10^{8})$

123. $(7.1 \times 10^{-8})(3.0 \times 10^{-7})$

124. $(9.0 \times 10^{-11})(2.5 \times 10^{-14})$

125. $\dfrac{2.5 \times 10^{8}}{1.25 \times 10^{4}}$

126. $\dfrac{3.5 \times 10^{15}}{5.0 \times 10^{10}}$

127. $\dfrac{2.7 \times 10^{15}}{3.0 \times 10^{28}}$

128. $\dfrac{8.4 \times 10^{20}}{4.2 \times 10^{35}}$

129. $\dfrac{7.2 \times 10^{30}}{1.5 \times 10^{11}}$

130. $\dfrac{6.8 \times 10^{18}}{2.4 \times 10^{7}}$

131. The mass of the planet Venus is estimated at 4.87×10^{24} kg. Divide the mass of Venus by the mass of the Earth to find how much smaller the mass of Venus is than that of the Earth.

132. The mass of Pluto is estimated at 1.25×10^{22} kg. Divide the mass of Pluto by the mass of the Earth to find how much smaller the mass of Pluto is than that of the Earth.

133. The diameter of Jupiter is about 1.428×10^{5} km. Divide the diameter of Jupiter by the diameter of the Earth to find how much bigger the diameter of Jupiter is than that of the Earth.

134. The diameter of Saturn is about 1.2×10^{5} km. Divide the diameter of Saturn by the diameter of the Earth to find how much bigger the diameter of Saturn is than that of the Earth.

135. The approximate U.S. national debt in May 2017 was 1.98×10^{13} dollars. The estimated population of the United States was 3.26×10^{8} people. Divide the national debt by the number of people in the United States to estimate each person's share of the debt.

136. The approximate U.S. national debt in May 2017 was 1.98×10^{13} dollars. The estimated number of working-age people (15 to 64 years old) in the United States was 2.05×10^{8} people. Divide the national debt by the number of working-age people in the United States to estimate each working-age person's share of the debt.

488 CHAPTER 5 Exponents and Polynomials

5.3 Adding and Subtracting Polynomials

LEARNING OBJECTIVES
- Use the terminology to describe polynomials.
- Add and subtract polynomials.

■ The Terminology of Polynomials

Recall from Section 1.1 that a *term* is a constant or a product of a constant and a variable or variables. Terms are the basic building blocks of expressions called **polynomials**. When a combination of terms is added or subtracted together, different types of polynomials are created. Some polynomials have special names such as **monomial, binomial,** and **trinomial**.

> **DEFINITIONS**
>
> **Polynomial** Any combination of terms that are added or subtracted together is called a **polynomial**. The powers of all variables in a polynomial must be positive integers.
>
> **Monomial** A polynomial with one term is called a **monomial**.
>
> **Binomial** A polynomial with two terms is called a **binomial**.
>
> **Trinomial** A polynomial with three terms is called a **trinomial**.
>
> Any polynomial with more than three terms is simply called a polynomial.
>
> | Monomial | Binomial | Trinomial |
> |---|---|---|
> | 78 | $5x + 2$ | $x^2 + 2x + 5$ |
> | $5x^2$ | $8m^2 + 75$ | $7a^2 + 2ab - 6b^2$ |
> | $9a^3b^2c$ | $4xy^3 - 24y$ | $8x^3yz^2 - 5yz + 20$ |

Example 1 Naming terms and polynomials

Identify the terms in each of the following polynomials. Label the polynomial as a monomial, binomial, or trinomial.

a. $3x^2 + 4x - 8$ **b.** $5z + 12$ **c.** $22x^2y$

SOLUTION

a. This polynomial has three terms: $3x^2$, $4x$, and -8. This polynomial is called a trinomial.

b. This polynomial has two terms: $5z$ and 12. This polynomial is called a binomial.

c. This polynomial has one term: $22x^2y$. This polynomial is called a monomial.

PRACTICE PROBLEM FOR EXAMPLE 1

Identify the terms in each of the following polynomials. Label the polynomial as a monomial, binomial, or trinomial.

a. $5x^2 + 7x$ **b.** $4a^3b^2 - 2ab + b$

Another way in which we describe a term or polynomial is by using the exponents of the variables. The **degree** of a single-variable polynomial is the highest power of the exponents on that variable.

SECTION 5.3 Adding and Subtracting Polynomials

DEFINITION

For a polynomial of a single variable:

Degree of a Term The degree of a term is equal to the exponent on the variable in that term. A constant term has degree zero.

The degree of $4x^2$ is 2.

Degree of a Polynomial The degree of a polynomial is equal to the highest exponent on the variable in the polynomial.

The degree of $-7x^3 + 5x^2 + x - 4$ is 3.

What's That Mean?

Lead Coefficient

Recall from Chapter 1 that the *coefficient* of a variable term is the numerical factor of the term. The *lead coefficient* of a polynomial is the coefficient of the highest-degree term. In the polynomial $3x^2 + 4x - 8$, the lead coefficient is 3. The highest-degree term is $3x^2$ and 3 is the coefficient of that term.

To find the degree of a polynomial of one variable, it is very helpful if the polynomial is written in *descending order*. This means the first term (leftmost term) has the highest exponent on the variable, the next term has the second highest exponent, and so on. To write a polynomial in descending order, use the commutative property of addition.

Not in Descending Order		Descending Order
$4x + 2x^2 - 5$	$=$	$2x^2 + 4x - 5$

Example 2   Determining the degree of a polynomial of a single variable

For the given polynomials, list the degree of the polynomial. Rewrite each polynomial in descending order.

a. $-x^2 + 5x - 1$ b. $12 - 3x - 5x^3 + x^4$ c. $-\dfrac{1}{3}x + 1$

SOLUTION

a. The highest exponent on x in this polynomial is 2. So the degree of $-x^2 + 5x - 1$ is 2. This polynomial is already in descending order.

b. First rewrite this polynomial in descending order. So $12 - 3x - 5x^3 + x^4 = x^4 - 5x^3 - 3x + 12$. The highest exponent on x is 4. The degree of this polynomial is 4.

c. The polynomial $-\dfrac{1}{3}x + 1$ has degree 1. Recall that $x = x^1$. This polynomial is already in descending order.

PRACTICE PROBLEM FOR EXAMPLE 2

For the given polynomials, list the degree of the polynomial. Rewrite each polynomial in descending order.

a. $-x^2 + 4$ b. $x^2 + x - 4 + 3x^3$ c. $-7x + 5$

Some polynomials have more than one variable. The degree is determined by looking at the exponents, but now we have to consider exponents on all the variables. The degree of a term is found by adding the exponents of all the variables in the term. The degree of a polynomial is the same as that of the highest-degree term in the polynomial.

DEFINITION

For a polynomial of more than one variable:

Degree of a Term The degree of a term is the sum of all the exponents of the variables in the term.

The degree of $3x^2y^3$ is 5.

> **DEFINITION**
>
> **For a polynomial of more than one variable:**
>
> **Degree of a Polynomial** The degree of a polynomial is equal to the degree of the highest-degree term.
>
> $$\text{The degree of } 4xy - 7x^3y^3 \text{ is } 6$$

Example 3   Determining the degree of terms and polynomials

For the given polynomials, list the degree of each term. Then find the degree of each entire polynomial.

a. $5x^2y + 4x^3y^2 + 15x$ **b.** $3x^3y^2 + 4x^2y^5 - 2xy$ **c.** 5

SOLUTION

a. This polynomial has three terms. Recall that a variable such as y has an exponent of 1: $y = y^1$. Therefore, the first term is $5x^2y = 5x^2y^1$ and has degree 3. The second term is $4x^3y^2$ and has degree 5. The last term is $15x$ and has degree 1. The polynomial has degree 5.

b. This polynomial has three terms. The term $3x^3y^2$ has degree 5, $4x^2y^5$ has degree 7, and $-2xy$ has degree 2. The polynomial has degree 7.

c. This polynomial has degree 0. To see why, recall from Chapter 1 that $x^0 = 1$ (for $x \neq 0$). So we can write $5 = 5x^0$. Therefore, the degree is 0.

PRACTICE PROBLEM FOR EXAMPLE 3

For the given polynomials, list the degree of each term. Then find the degree of each entire polynomial.

a. $4x^5 - 7xy^3 + 3x^2 + 8$ **b.** $5a^4b^2 - 7a^3b^6 - 2ab^3$

In Chapters 1 through 4, we worked mostly with expressions and equations of degree 1. We used the word *linear* to describe many different expressions and equations that have degree 1. A linear equation of the form $y = mx + b$, where m and b are constants, is considered an equation of degree 1. A polynomial with one variable and degree 1 is also called a *linear polynomial*. The word *linear* is used to describe more than one object in mathematics. The following table lists several different things that we call *linear* and gives some of their characteristics.

Name	Degree	What We Do	Examples
Linear equation in one variable	Degree 1	Solve	$2x + 5 = 7(x - 1)$
Linear equation in two variables	Degree 1	Graph, solve	$y = 2x + 5$
Linear expression/polynomial in one or more variables	Degree 1	Simplify, evaluate	$2(x + 5) - 7x$ $5x + 7y - 8$

Notice that all of these linear equations or expressions have degree 1. If an equation or expression has degree higher than 1, it is not linear. Such expressions or equations are called *nonlinear*.

Example 4 Identifying linear expressions and equations

Classify the following as linear equation in one variable, linear equation in two variables, linear expression in one or more variables, or nonlinear. Explain each classification.

a. $2x + 7$ **b.** $y = 2x^2 + 5x - 8$ **c.** $2x + 5y + 81$

d. $y = \dfrac{2}{3}x + 7$ **e.** $5xy + 7 = 20$

SOLUTION

a. There is no equal sign, so it is an expression, not an equation. $2x + 7$ is a polynomial in one variable with degree 1, so it is a linear expression in one variable.

b. It is an equation because there is an equal sign. $y = 2x^2 + 5x - 8$ is an equation that has a second-degree term, so it is nonlinear.

c. It is an expression because there is no equal sign. $2x + 5y + 81$ is a polynomial in two variables with degree 1, so it is a linear expression in two variables.

d. It is an equation because there is an equal sign. $y = \dfrac{2}{3}x + 7$ is an equation in two variables that has degree 1, so it is a linear equation in two variables.

e. It is an equation because there is an equal sign. $5xy + 7 = 20$ is an equation in two variables that has a second degree term ($5xy$). Because the degree is not 1, it is nonlinear. Be careful to look at the degree of the term, not just the exponent of the variables.

PRACTICE PROBLEM FOR EXAMPLE 4

Classify the following as linear equation in one variable, linear equation in two variables, linear expression in one or more variables, or nonlinear. Explain each classification.

a. $2x + 7y = 20$ **b.** $3x + 7(4x - 12)$ **c.** $y = 2x^3 + 5x - 7$

Adding and Subtracting Polynomials

Recall that to add or subtract expressions, combine like terms.

$$(-3x + 8) + (-4x - 2)$$
$$= -7x + 6$$

We also combine like terms when adding or subtracting polynomials. To add or subtract polynomials, simply add or subtract the like terms from each polynomial.

> **Adding Polynomials**
> To add two or more polynomials, combine like terms.

Example 5 Adding polynomials

Add the following polynomials.

a. $(5x^2 + 2x + 10) + (7x^2 + 4x + 3)$

b. $(7a^2 + 8ab - 9b^2) + (4a^2 - 3ab + 6b^2)$

c. Find the sum of $4x^2 + 5$ and $2x^2 - 6x + 3$.

SOLUTION

a. Use the associative property to group the like terms. Add the like terms together.

$(5x^2 + 2x + 10) + (7x^2 + 4x + 3)$
$= (5x^2 + 7x^2) + (2x + 4x) + (10 + 3)$ Group, using the associative property.
$= 12x^2 + 6x + 13$ Add like terms.

b. Use the associative property to group like terms. Add the like terms together.

$(7a^2 + 8ab - 9b^2) + (4a^2 - 3ab + 6b^2)$
$= (7a^2 + 4a^2) + (8ab + (-3ab)) + (-9b^2 + 6b^2)$ Group, using the associative property.
$= 11a^2 + 5ab - 3b^2$ Add like terms.

c. To find the sum means to add the two quantities. This translates as $(4x^2 + 5) + (2x^2 - 6x + 3)$. Group the like terms together and add like terms.

$(4x^2 + 5) + (2x^2 - 6x + 3)$
$= (4x^2 + 2x^2) + (-6x) + (5 + 3)$
$= 6x^2 - 6x + 8$

In this case, the term $-6x$ does not have a corresponding like term, so it remains the same in the final result.

PRACTICE PROBLEM FOR EXAMPLE 5

Add the following polynomials.

a. $(4x^2 - 9x + 7) + (2x^2 + 3x + 4)$ **b.** $(2a^2 - 5b^2) + (3a^2 + 8ab + 12b^2)$

When subtracting polynomials, be cautious with the signs of each term. Start each subtraction problem by using the distributive property to distribute the sign of the second polynomial. Then combine like terms.

> **Subtracting Polynomials**
>
> To subtract polynomials, first use the distributive property to distribute the sign of the second polynomial. Then combine like terms.

Example 6   Subtracting polynomials

Subtract the following polynomials.

a. $(8x^2 + 6x + 10) - (2x^2 + x - 7)$ **b.** $(3y^2 - 2y - 4) - (6y^2 + 2y - 6)$

c. Subtract $-6x^2 - 5x + 7$ from $8x^2 + 2$.

SOLUTION

a. To subtract these polynomials, use the distributive property to distribute the sign of the second polynomial. Then combine like terms.

$(8x^2 + 6x + 10) - (2x^2 + x - 7)$ Use the distributive property.
$= 8x^2 + 6x + 10 - 2x^2 - x + 7$
$= (8x^2 - 2x^2) + (6x - x) + (10 + 7)$ Group and combine like terms.
$= 6x^2 + 5x + 17$

SECTION 5.3 Adding and Subtracting Polynomials 493

b. To subtract these polynomials, use the distributive property to distribute the sign of the second polynomial. Then combine like terms.

$(3y^2 - 2y - 4) - (6y^2 + 2y - 6)$ Use the distributive property.
$= 3y^2 - 2y - 4 - 6y^2 - 2y + 6$
$= -3y^2 - 4y + 2$ Combine like terms.

c. To subtract $-6x^2 - 5x + 7$ from $8x^2 + 2$, first place parentheses around each polynomial. This problem translates as $(8x^2 + 2) - (-6x^2 - 5x + 7)$.

$(8x^2 + 2) - (-6x^2 - 5x + 7)$ Use the distributive property.
$= 8x^2 + 2 + 6x^2 + 5x - 7$
$= 14x^2 + 5x - 5$ Combine like terms.

Connecting the Concepts

Why do we change the signs?

When we subtract polynomials such as $(x + 5) - (2x - 4)$, we change the signs of the second polynomial first. This is due to the distributive property. Examining the second polynomial, we have

$-(2x - 4)$
$= -1(2x - 4)$
$= -2x + 4$

Notice that the signs of this polynomial have been changed.

PRACTICE PROBLEM FOR EXAMPLE 6

Subtract the following polynomials.

a. $(12x^2 + 7x + 4) - (3x^2 + 5x - 6)$ b. Subtract $8y^2 + 3y - 16$ from $11y^2 - 12$.

In some applications, we are asked to find a *total* or *difference* of certain quantities. When these quantities are represented by polynomials, we add or subtract the polynomials. Doing so gives a new polynomial that represents the total or difference.

Skill Connection

Combining like terms

When we add or subtract like terms, we add or subtract only the coefficients of like terms. Therefore,

$(2x + 5) + (3x - 4) = 5x + 1$

When adding or subtracting like terms, we never change the exponents on the variables. The exponent rules apply only to products and quotients, not sums and differences.

Example 7 Total cost of a baby shower

Christine and Theresa are throwing a baby shower for their friend Wendy. Christine is buying all of the food for the party and knows that the cost depends on how many people attend the baby shower. Theresa is buying the decorations and party favors for each person who attends. Christine estimates the cost of the food to be $10.5n + 25$ dollars when n people attend the party. Theresa estimates that the decorations (including party favors) will cost $7.5n + 45$ dollars when n people attend the party.

a. Find a new polynomial that gives the total cost of food and decorations.

b. Use the new polynomial to find the total cost of the party if 12 people attend the baby shower.

SOLUTION

a. To find the total cost, add the two polynomials together.

cost for food + cost for decorations
$= (10.5n + 25) + (7.5n + 45)$
$= 18n + 70$ Add the like terms.

The total cost of food and decorations is $18n + 70$ dollars.

b. If 12 people attend the baby shower, n is 12. Substitute $n = 12$ into the new polynomial, and simplify.

$18n + 70$ Substitute in $n = 12$.
$= 18(12) + 70$
$= 216 + 70$
$= 286$

The total cost for 12 people to attend the baby shower is $286.

What's That Mean?

Total and Difference

Total in algebra means to add. *To find a difference* means to subtract.

494 CHAPTER 5 Exponents and Polynomials

PRACTICE PROBLEM FOR EXAMPLE 7

Ms. Fiske is ordering supplies for art projects for her second-grade class. The total cost for the supplies depends on the number of children she has in her class. After some research, Ms. Fiske determines that the total cost can be represented by $12c + 30$ dollars when c children are in her class.

a. Ms. Fiske decides not to purchase supplies for her winter projects that cost $3c + 5$ dollars. Find a new polynomial that represents the cost for the art projects without the winter projects.

b. How much will the supplies cost for the art projects without the winter project if she has 18 children in her class?

Example 8 Measuring angles

Write a polynomial that represents the total measurement of the two angles.

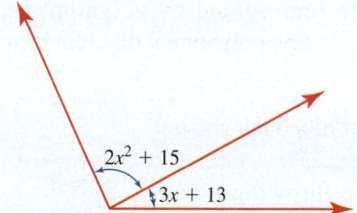

SOLUTION

To find the total measurement of both angles, add the two polynomials together.

$$(2x^2 + 15) + (3x + 13)$$
$$= 2x^2 + 3x + 28$$

The total measurement of both angles is $2x^2 + 3x + 28$ degrees.

5.3 Vocabulary Exercises

1. A monomial has _____ term(s).

2. A trinomial has _____ term(s).

3. $3x - 5y$ is an example of a(n) _____.

4. The degree of a polynomial is equal to the degree of the _____ term.

5. $-6x + 2y$ is an example of a linear _____.

6. $-6x + 2y = 1$ is an example of a linear _____.

5.3 Exercises

For Exercises 1 through 14, identify the terms in each of the polynomials. Label each polynomial as a monomial, binomial, or trinomial.

1. $x^2 + 2x + 5$

2. $5x^2 - 3x + 7$

3. $4x + 8$

4. $-6x + 3$

5. $8x^3y$

6. $25a^2b^2$

7. $7m^2 - 9$

8. $-4d^3 + 7d$

9. $14x^3y + 7x^2y^2 - 9xy$

10. $5a^8b^2 - 3a^4b^3 + 2ab^2$

11. $2m^7n^3 - 4m^2n^5$
12. $35g^4h - 17gh^7$
13. $-45mn^2 + 56$
14. $-12a^5b^2c + 2$

For Exercises 15 through 24, state the number of terms of each polynomial. Rewrite each polynomial in descending order. Then find the degree of each polynomial of a single variable.

15. $\frac{3}{2}x - 12$
16. $0.05x + 9.25$
17. $-6x^2 + 3x - 9$
18. $x^2 + 4x + 4$
19. -45
20. 9
21. $4y + 7 - y^3 + 9y^2$
22. $-3y^2 + 12y - 3 + y^4$
23. $9b - 3b^8 + 2b^7$
24. $2a^3 - 5a^2 + a^6$

For Exercises 25 through 34, find the degree of each term and the degree of each entire polynomial.

25. $-6xy^2$
26. $95x^3y$
27. $3a + 2ab - 6b^2$
28. $-ab^4 + 2a^3b^3 - 6a^7$
29. $1 - x^3y^3$
30. $x^2 + y^2 + 9$
31. $-24x^2yz^3 + 6z - 7y + 5$
32. $9x^2z^2 + 9x^2y^2 + 3yx + 7z^3$
33. -29
34. 0

For Exercises 35 through 48, classify each of the following as linear equation in one variable, linear equation in two variables, linear expression in one or more variables, or not linear. Explain each classification.

35. $x^2 + 2x + 5$
36. $5x^2 - 3x + 7$
37. $4x + 8$
38. $-6x + 3$
39. $y = 4x + 3$
40. $y = -8x + 5$
41. $5x - (2x + 8) = 4x$
42. $-4d + 75d = 5d + 2$
43. $4xy + 7x - 9xy$
44. $10ab - 32a + 2ab$
45. $y = 2x^2 + 3x - 20$
46. $y = -7x^2 + 5$
47. $2x + 7y = 68$
48. $7x - 6y = 12$

For Exercises 49 through 60, add or subtract the given polynomials.

49. $(3x + 5) + (4x + 9)$
50. $(6a + 8) + (2a + 4)$
51. $(t^2 + 5t - 9) + (4t^2 + 3t + 4)$
52. $(3x^2 + 2x - 8) + (5x^2 + 8x + 12)$
53. Find the sum of $-5m^2 + 6m - 3$ and $-4m^2 + 7m - 6$.
54. Find the sum of $-9y^2 - 4y + 7$ and $5y^2 + 12y - 4$.
55. $(6x + 9) - (4x + 3)$
56. $(8p + 14) - (2p + 5)$
57. $(n^2 + 8n - 6) - (n^2 - 5n + 3)$
58. $(x^2 - 4x - 8) - (x^2 + 6x - 11)$
59. Subtract $3h^2 - 5h + 8$ from $5h^2 - 7h + 12$.
60. Subtract $5b^2 - 4b + 5$ from $3b^2 + 8b - 9$.

For Exercises 61 through 66, explain what each student did wrong. Then simplify the expressions correctly.

61. $2x + x$

Reina
$2x + x = 2x^2$

62. $4y + 5y$

Michael
$4y + 5y = 9y^2$

63. $(3x - 2) - (4x + 5)$

Fari
$(3x - 2) - (4x + 5)$
$= 3x - 2 - 4x + 5$
$= -x + 3$

64. $(-2a + 3b) - (a - 3b)$

Ali
$(-2a + 3b) - (a - 3b)$
$= -2a + 3b - a - 3b$
$= -3a$

65. $(x^2 + 4x - 8) - (2x^2 - 5x)$

Anya
$(x^2 + 4x - 8) - (2x^2 - 5x)$
$= x^2 + 4x - 8 - 2x^2 - 5x$
$= -x^2 - x - 8$

66. $(y^2 - y + 4) - (-y^2 + 3y - 2)$

Ivan
$(y^2 - y + 4) - (-y^2 + 3y - 2)$
$= y^2 - y + 4 + y^2 + 3y - 2$
$= 2y^2 + 2y + 2$

496 CHAPTER 5 Exponents and Polynomials

For Exercises 67 through 70, use the following information.

Julio's Pumpkin Patch is trying to prepare for school tour season and has come up with the following expressions to calculate the cost for k students to participate in different activities.

Activity	Cost (dollars)
Life of a pumpkin demo	$1.5k + 5$
Petting zoo	$2k + 10$
Maize maze	$0.75k$
Tractor ride	$1.25k + 5$
Take home a pumpkin	$4.5k$

67. **a.** Write an expression for the total cost if a class wants to see the life of a pumpkin demo and visit the petting zoo.

    b. Use the expression from part a to determine the cost for 33 students to see the life of a pumpkin demo and visit the petting zoo.

68. **a.** Write an expression for the total cost if a class wants to visit the petting zoo, go through the maize maze, and take a tractor ride.

    b. Use the expression from part a to determine the cost for 25 students to visit the petting zoo, go through the maize maze, and take a tractor ride.

69. **a.** Write an expression for the total cost if a class wants to see the life of a pumpkin demo, take a tractor ride, and take home a pumpkin.

    b. Use the expression from part a to determine the cost for 60 students to see the life of a pumpkin demo, take a tractor ride, and take home a pumpkin.

70. **a.** Write an expression for the total cost if a class wants to do all the activities.

    b. Use the expression from part a to determine the cost for 32 students to do all the activities.

For Exercises 71 through 74, use the following information.

The Grizzlies crew of the local YMCA is starting a new year and deciding what gear to buy and what activities to attend. They have the following expressions to calculate the cost for m members to purchase each piece of gear or participate in each activity.

Gear	Cost (dollars)
T-shirt	$7.5m + 55$
Vest	$8m + 15$
Hat	$10.5m + 25$
Activity	**Cost (dollars)**
Dos Picos campout	$30m + 100$
Bowling and beach cleanup	$15m$
Pinewood Derby	$10m + 30$

71. **a.** Write an expression for the total cost if the Grizzlies want to purchase all the gear.

    b. Use the expression from part a to determine the cost for 15 members to purchase all the gear.

72. **a.** Write an expression for the total cost if the Grizzlies want to go on all the activities.

    b. Use the expression from part a to determine the cost for 15 members to go on all the activities.

73. **a.** Write an expression for the total cost if the Grizzlies want to purchase T-shirts and go on the Dos Picos campout.

    b. Use the expression from part a to determine the cost for 12 members to purchase T-shirts and go on the Dos Picos campout.

74. **a.** Write an expression for the total cost if the Grizzlies want to purchase vests and hats and participate in the Pinewood Derby.

    b. Use the expression from part a to determine the cost for 12 members to purchase vests and hats and participate in the Pinewood Derby.

75. The ski club is planning a week-long ski trip for its members. The total cost in dollars for a group of m members can be found by using the expression

    $$275m + 500$$

    To reduce costs, the club is thinking about making the trip shorter and going for only 5 days. If they do this, they can reduce their costs by $125m + 150$ dollars.

    a. Find an expression for the total cost for the 5-day trip.

    b. Use your expression from part a to determine the cost for 30 ski club members to go on the 5-day trip.

76. The total weight of a backpack depends on the number of days the hiking trip will include and what type of weather a hiker has to prepare for. A hiker has found that for a cold-weather hike lasting d days, the backpack's weight in pounds can be estimated by using the expression

    $$2.5d + 30$$

If the trip is during warm weather, the hiker can carry lighter clothing and less food. This reduces the backpack's weight by about $0.5d + 7$ pounds.

a. Write an expression for the weight of the hiker's backpack if he is going on a warm-weather trip.

b. Use the expression from part a to determine the weight of the backpack for a 5-day warm-weather trip.

77. Write a polynomial that represents the total measurement of the two angles.

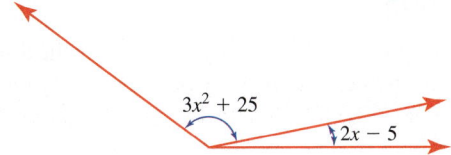

78. Write a polynomial that represents the total measurement of the two angles.

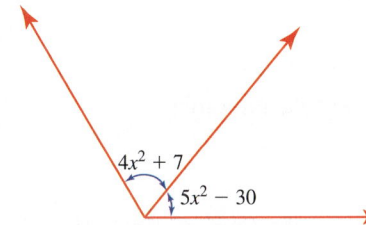

79. Write a polynomial for the perimeter of the following figure.

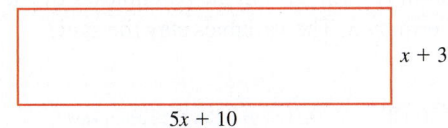

80. Write a polynomial for the perimeter of the following figure.

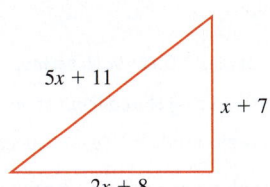

For Exercises 81 through 96, add or subtract the given polynomials. Simplify.

81. $(4x^2 + 21x) + (3x^2 + 5x)$

82. $(7x^2 + 21) + (8x^2 + 16)$

83. $(a^2 + 5ab - 8b) + (4a^2 - 9ab + 6b)$

84. $(3m^2 + 8mn - 12n) + (-9m^2 + 2mn + 14n)$

85. $(-2t^2 + 6t - 14) + (-8t^2 - 6)$

86. $(-8x^2 + 20) + (15x - 6)$

87. $(6a + 11b) - (18a + 5b)$

88. $(7x + 30y) - (4x + 9y)$

89. $(m^2 + 2mn - n) - (7m^2 - 11mn + 3n)$

90. $(20x^2 - 15xy - 17y) - (-11x^2 + 8xy - 15y)$

91. $(45a^2 + 17a) - (8a + 24)$

92. $(8h^2 + 3) - (7h - 9)$

93. $(7x^2 + 8x - 9) + (4x + 10) - (8x^2 - 3x + 4)$

94. $(2a^2 - 3ab + 8b) - (8ab + 2b) + (10a^2 + ab)$

95. $(-5m^3 + 3m^2 + 8m) + (7m^3 + 9m - 4) - (6m^2 - 3m + 2)$

96. $(-p^5 + 8p^3 - 12p) - (4p^3 + 3p^2 + 10) - (5p^2 - 4p)$

For Exercises 97 through 100, answer as indicated.

97. Graph the line. Clearly label and scale the axes.
$$-2x + 5y = -10$$

98. Graph the line. Clearly label and scale the axes.
$$3x - 4y = 8$$

99. Solve the system. Check the solution.
$$2x + y = 9$$
$$5x - y = 26$$

100. Solve the system. Check the solution.
$$5x + y = -17$$
$$-2x + y = 4$$

For Exercises 101 through 104, use the model $P = 0.457t - 13.162$, where P is the percentage share of 25- to 29-year-old employees in the United States with a bachelor's degree or higher t years since 1900.

Source: Model based on data from www.statista.com

101. Find the percentage share of 25- to 29-year-old employees in the United States with a bachelor's degree or higher in 1995.

102. Find when the percentage share of 25- to 29-year-old employees in the United States with a bachelor's degree or higher was 40%.

103. Find and interpret the slope of this model.

104. Find and interpret the y-intercept of this model.

5.4 Multiplying Polynomials

LEARNING OBJECTIVES
- Multiply polynomials.
- Explain FOIL, the handy acronym.
- Work with special products.

Distributive Property

Distribute terms across addition or subtraction.

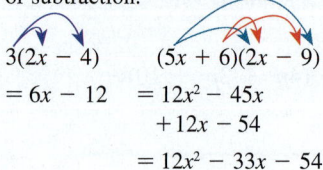

Multiplying Polynomials

In Chapter 1, we multiplied expressions (polynomials) by a constant, using the distributive property. Now we expand the distributive property to multiply polynomials by monomials and other polynomials.

> **Multiplying a Polynomial by a Monomial**
> Multiply each term of the polynomial by the monomial.

Example 1 Multiplying a polynomial by a monomial

Find each product and simplify.

a. $3(4x^2 + 8x - 2)$ **b.** $5a(3a + 8)$ **c.** $7x(2x^2 - 3x + 4)$

SOLUTION

a. Use the distributive property to multiply each term in the parentheses by the 3. Multiply the coefficients of each term by 3. The variables stay the same.

$3(4x^2 + 8x - 2)$
$= 3(4x^2) + 3(8x) + 3(-2)$ Distribute the 3 to each term.
$= 12x^2 + 24x - 6$

b. Use the distributive property to multiply each term in the parentheses by $5a$.

$5a(3a + 8)$
$= 5a(3a) + 5a(8)$ Distribute $5a$ to both terms.
$= (5 \cdot 3)(a \cdot a) + (5 \cdot 8)a$ Group using the commutative property.
$= 15a^2 + 40a$ Use the product rule for exponents and simplify.

c. Use the distributive property to multiply each term in the parentheses by $7x$.

$7x(2x^2 - 3x + 4)$
$= 7x(2x^2) + 7x(-3x) + 7x(4)$ Distribute $7x$ to all three terms.
$= (7 \cdot 2)(x \cdot x^2) + (7 \cdot (-3))(x \cdot x) + (7 \cdot 4)x$ Group using the commutative property.
$= 14x^3 - 21x^2 + 28x$ Use the product rule for exponents and simplify.

Skill Connection

Product rule for exponents
In Section 5.1, we learned the product rule for exponents.
$$x^n \cdot x^m = x^{n+m}$$

PRACTICE PROBLEM FOR EXAMPLE 1
Find each product and simplify.

a. $6x(4x - 8)$ **b.** $4a(5a^2 - 6a + 8)$

When multiplying a polynomial by more than one term, we extend the distributive property so that each term is distributed correctly. This can be done in a few ways, which are explored below.

Problem: Multiply $(2x + 5)(3x^2 + 7x - 4)$.
Option 1: First distribute the binomial to each term.

$$(2x + 5)(3x^2 + 7x - 4)$$
$$= (2x + 5)(3x^2) + (2x + 5)(7x) + (2x + 5)(-4)$$

Then distribute within each binomial.

$$= 2x(3x^2) + 5(3x^2) + 2x(7x) + 5(7x) + 2x(-4) + 5(-4)$$
$$= 6x^3 + 15x^2 + 14x^2 + 35x - 8x - 20 \quad \text{Combine like terms.}$$
$$= 6x^3 + 29x^2 + 27x - 20$$

Problem: Multiply $(2x + 5)(3x^2 + 7x - 4)$.
Option 2: Distribute each term of the binomial through each of the terms in the parentheses.

$$(2x + 5)(3x^2 + 7x - 4)$$
$$= 2x(3x^2 + 7x - 4) + 5(3x^2 + 7x - 4)$$
$$= 2x(3x^2) + 2x(7x) + 2x(-4) + 5(3x^2) + 5(7x) + 5(-4)$$
$$= 6x^3 + 14x^2 - 8x + 15x^2 + 35x - 20 \quad \text{Combine like terms.}$$
$$= 6x^3 + 29x^2 + 27x - 20$$

Both of these methods work when we are multiplying two polynomials. Looking at option 2 carefully, we see the same products as option 1, but the terms are arranged differently. If there are more terms in the first polynomial, distribute each term in the first polynomial to each term in the second polynomial.

> **Multiplying Polynomials by Polynomials**
> Multiply each term of one polynomial by each term of the other polynomial, simplify each term, and then combine like terms.

Example 2 Multiplying a polynomial by a polynomial

Find each product and simplify.

a. $(5x - 7)(2x^2 + 6x - 3)$ **b.** $(5a^3b + 2ab)(3a^2 + 4b - 11)$

SOLUTION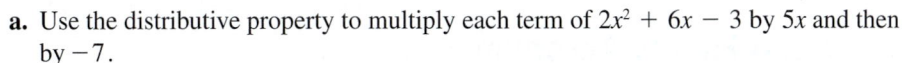

a. Use the distributive property to multiply each term of $2x^2 + 6x - 3$ by $5x$ and then by -7.

$$(5x - 7)(2x^2 + 6x - 3) \quad \text{Distribute } (5x - 7) \text{ to the terms.}$$
$$= (5x - 7)(2x^2) + (5x - 7)(6x) + (5x - 7)(-3)$$
$$= 5x(2x^2) - 7(2x^2) + 5x(6x) - 7(6x) + 5x(-3) - 7(-3)$$
$$= 10x^3 - 14x^2 + 30x^2 - 42x - 15x + 21 \quad \text{Combine like terms.}$$
$$= 10x^3 + 16x^2 - 57x + 21$$

Skill Connection

Commutative property of multiplication

The commutative property of multiplication allows us to multiply in any order. In part a of Example 3, we multiplied the factors in order, working left to right. In part b of Example 3, we multiplied the two binomials together first, then multiplied by the lead factor $-7a$. Many people like to multiply the two binomials together first, as in Example 3b, because it is the harder part of the problem and the numbers are not yet so large. Then they use the distributive property to finish.

b. Use the distributive property to multiply each term of $3a^2 + 4b - 11$ by $5a^3b$ and then by $2ab$.

$(5a^3b + 2ab)(3a^2 + 4b - 11)$ *Use the distributive property.*
$= 5a^3b(3a^2) + 5a^3b(4b) + 5a^3b(-11) + 2ab(3a^2) + 2ab(4b) + 2ab(-11)$
$= 15a^5b + 20a^3b^2 - 55a^3b + 6a^3b + 8ab^2 - 22ab$ *Combine like terms.*
$= 15a^5b + 20a^3b^2 - 49a^3b + 8ab^2 - 22ab$

PRACTICE PROBLEM FOR EXAMPLE 2

Find each product and simplify.

a. $(5x + 7)(4x^2 - 9x + 2)$ **b.** $(4x^4y^2 - 5x^2y)(3x^2 + 2xy - 5y^2)$

If more than two polynomials are multiplied together, start by multiplying any two of them together. Then multiply that result by the other polynomial.

Example 3 Multiplying more than two polynomials

Find each product and simplify.

a. $5x(2x + 8)(3x + 4)$ **b.** $-7a(2a + 3b)(5a - 2b)$

SOLUTION $5(2x-7)$

a. Multiply $5x$ by $(2x + 8)$ and then multiply the result by $(3x + 4)$.

$5x(2x + 8)(3x + 4)$ *Multiply the first two polynomials.*
$= (10x^2 + 40x)(3x + 4)$ *Multiply the result by the last polynomial.*
$= 30x^3 + 40x^2 + 120x^2 + 160x$ *Combine like terms.*
$= 30x^3 + 160x^2 + 160x$

b. Multiply the two binomials together and then multiply by $-7a$.

$-7a(2a + 3b)(5a - 2b)$ *Multiply the two binomials together.*
$= -7a(10a^2 - 4ab + 15ab - 6b^2)$ *Combine like terms.*
$= -7a(10a^2 + 11ab - 6b^2)$ *Multiply by $-7a$.*
$= -70a^3 - 77a^2b + 42ab^2$

PRACTICE PROBLEM FOR EXAMPLE 3

Find the product and simplify.

$$-4x(2x + 5y)(3x - 8y)$$

Connecting the Concepts

How are FOIL and the distributive property related?

The distributive property can be used to multiply out any two or more polynomials. In the case of two binomials, we often use FOIL.

FOIL starts by taking the first term and distributes it through the second binomial. FOIL then takes the second term and distributes it through the second binomial. This gives the same result as the distributive property but with a nice way to remember it.

■ FOIL: A Handy Acronym

One way to help us remember the distributive property when multiplying two binomials is the memory-jogging word **FOIL**. FOIL stands for First, Outer, Inner, and Last terms. Using FOIL helps us to remember which terms we are supposed to multiply together.

First terms $(2x + 7)(3x + 4)$
Outer terms $(2x + 7)(3x + 4)$
Inner terms $(2x + 7)(3x + 4)$
Last terms $(2x + 7)(3x + 4)$

When we are multiplying two binomials, FOIL looks like the following:

$$(2x + 7)(3x + 4)$$

First   Outer   Inner   Last
$$= 2x(3x) + (2x)(4) + 7(3x) + 7(4)$$
$$= 6x^2 + 8x + 21x + 28 \quad \text{Simplify by combining like terms.}$$
$$= 6x^2 + 29x + 28$$

> **Multiplying Two Binomials**
> When multiplying two binomials, use FOIL.
>
> First + Outer + Inner + Last
>
> Then simplify, by combining like terms, if possible.

Example 4 Using FOIL

Find each product and simplify.

a. $(x + 3)(x + 5)$ **b.** $(4x - 3)(5x + 8)$ **c.** $(2a + 3b)(5a - 4b)$

SOLUTION

a. Since this is the product of two binomials, use FOIL.

$$(x + 3)(x + 5)$$
$$= x(x) + x(5) + 3(x) + 3(5) \qquad \text{Use FOIL to multiply.}$$
$$= x^2 + 5x + 3x + 15 \qquad \text{Combine like terms to simplify.}$$
$$= x^2 + 8x + 15$$

b. $(4x - 3)(5x + 8)$
$$= 4x(5x) + 4x(8) - 3(5x) - 3(8) \qquad \text{Use FOIL to multiply.}$$
$$= 20x^2 + 32x - 15x - 24 \qquad \text{Combine like terms to simplify.}$$
$$= 20x^2 + 17x - 24$$

c. $(2a + 3b)(5a - 4b)$
$$= 2a(5a) + 2a(-4b) + 3b(5a) + 3b(-4b) \qquad \text{Use FOIL to multiply.}$$
$$= 10a^2 - 8ab + 15ab - 12b^2 \qquad \text{Combine like terms to simplify.}$$
$$= 10a^2 + 7ab - 12b^2$$

PRACTICE PROBLEM FOR EXAMPLE 4

Find each product and simplify.

a. $(6x - 2)(5x + 3)$ **b.** $(4x + 5y)(3x - 6y)$

Example 5 The rising costs of salad

The wholesale cost in dollars of c cases of local organic lettuce can be estimated by using the expression $7.5c + 12$. After a large winter storm, the price doubled. Write an expression for the wholesale cost of organic lettuce after the storm.

SOLUTION

Since the price doubled after the storm, multiply the cost expression by 2.

$$2(7.5c + 12)$$
$$15c + 24$$

The cost in dollars of c cases of local organic lettuce after the storm can be estimated by using the expression $15c + 24$.

Example 6 We eat lots of cheese

The average amount of cheese consumed per person in the United States t years since 2010 can be estimated by using the expression

$0.29t + 32.7$ pounds per person

Source: Model derived from data at www.statistia.com

The population of the United States in millions can be estimated by using the expression

$2.32t + 309.32$ million people t years since 2010

Source: Model derived from data at the U.S. Census Bureau.

a. Write a new expression to represent the total amount of cheese consumed by the U.S population.

b. Use the expression found in part a to estimate the amount of cheese consumed by the U.S. population in 2020.

SOLUTION

a. The expression given, $0.29t + 32.7$, provides the average amount of cheese consumed per person, so multiply by the number of people to get the total.

$(0.29t + 32.7)(2.32t + 309.32)$
$= 0.6728t^2 + 89.7028t + 75.864t + 10114.764$ *Multiply using the distributive property.*
$= 0.6728t^2 + 165.5668t + 10114.764$

Checking the units yields

$$\frac{\text{pounds}}{\text{person}} \cdot \text{million people} = \text{million pounds}$$

The person and people divide out to give units of million pounds.
The total amount of cheese consumed by Americans can be estimated as $0.6728t^2 + 165.5668t + 10114.764$ million pounds.

b. The year 2020 is 10 years since 2010, so $t = 10$. Substitute 10 for t and evaluate.

$0.6728(10)^2 + 165.5668(10) + 10114.764$ *Evaluate the expression.*
$= 67.28 + 1655.668 + 10114.764$ *Simplify.*
$= 11837.712$

In 2020, Americans will consume about 11,837.712 million pounds of cheese.

PRACTICE PROBLEM FOR EXAMPLE 6

The average amount of fresh fruit consumed per person in the United States t years since 2010 can be estimated by using the expression

$$2.16t + 127.98 \text{ pounds per person}$$

Source: Model derived from data at www.statistia.com

The population of the United States in millions t years since 2010 be estimated by using the expression

$$2.32t + 309.32 \text{ million people.}$$

Source: Model derived from data at the U.S. Census Bureau.

a. Write a new expression to represent the total amount of fresh fruit consumed by the U.S population.
b. Use the expression found in part a to find the amount of fresh fruit that will be consumed by the U.S. population in 2020.

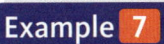

Example 7 Finding area

Write a polynomial that represents the area of the following rectangle.

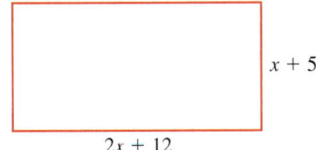

SOLUTION

To find the area of a rectangle, multiply the length by the width.

length · width
$(2x + 12)(x + 5)$
$= 2x^2 + 10x + 12x + 60$ Multiply, using the distributive property.
$= 2x^2 + 22x + 60$ Combine like terms.

The area of this rectangle can be represented by the polynomial $2x^2 + 22x + 60$.

PRACTICE PROBLEM FOR EXAMPLE 7

Write a polynomial that represents the area of the following rectangle.

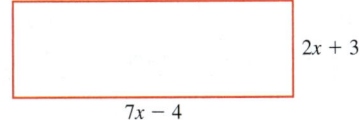

■ Special Products

In some special cases, products will follow a handy pattern that can be used to reduce the amount of work that needs to be done. One of these cases occurs when a binomial is squared or a binomial is multiplied by itself. Recall from Section 5.1 that we cannot apply exponents over addition or subtraction. When a sum or difference is being raised

■ Connecting the Concepts

What makes these products special?

When we square a binomial or multiply the sum and difference of two terms, the result forms two patterns. These two patterns, perfect square trinomials and the difference of two squares, are useful to recognize.

These formulas allow us to perform the product more quickly as well as to undo the product more easily.

to a power, multiply out the expression. The following pattern may help when squaring binomials.

$(a + b)^2$ *There is an addition sign. The exponent of 2 cannot be applied over the sum.*

$= (a + b)(a + b)$ *Squaring is the same as multiplying the base by itself.*

$= a^2 + ab + ab + b^2$ *Multiply out using the distributive property.*

$= a^2 + 2ab + b^2$ *Combine like terms.*

By looking at the final expression, we see that the first term is squared, plus twice the product of the two terms, plus the second term squared. This pattern also works for $(a - b)^2$.

> **Squaring a Binomial**
>
> To square a binomial, the result is the first term squared, plus twice the product of the two terms, plus the second term squared.
>
> $(a + b)^2 = a^2 + 2ab + b^2$ $(a - b)^2 = a^2 - 2ab + b^2$
>
> The resulting trinomials are called *perfect square trinomials*.

Example 8   Squaring binomials

Find each product and simplify.

a. $(x + 3)^2$

b. $(5x - 6)^2$

c. $(7n^2 + 4m)^2$

d. $\left(\dfrac{2}{3}x + y\right)^2$

SOLUTION

a. This is a binomial squared, so use the following formula: the first term squared, plus twice the product of the two terms, plus the second term squared.

$(x + 3)^2$ Use $a = x$ and $b = 3$.

$= x^2 + 2(x)(3) + 3^2$

$= x^2 + 6x + 9$

b. This is a binomial squared, so use the same formula.

$(5x - 6)^2$ Use $a = 5x$ and $b = 6$.

$= (5x)^2 - 2(5x)(6) + 6^2$

$= 25x^2 - 60x + 36$

c. $(7n^2 + 4m)^2$ Use $a = 7n^2$ and $b = 4m$.

$= (7n^2)^2 + 2(7n^2)(4m) + (4m)^2$

$= 49n^4 + 56n^2m + 16m^2$

d. $\left(\dfrac{2}{3}x + y\right)^2$ Use $a = \dfrac{2}{3}x$ and $b = y$.

$= \left(\dfrac{2}{3}x\right)^2 + 2\left(\dfrac{2}{3}x\right)(y) + y^2$

$= \dfrac{4}{9}x^2 + \dfrac{4}{3}xy + y^2$

PRACTICE PROBLEM FOR EXAMPLE 8

Find each product and simplify.

a. $(3x - 8)^2$
b. $(3x^2 + 5y)^2$
c. $\left(\dfrac{t}{4} - 5\right)^2$

It is important to look inside the parentheses first and determine what operation is being used before squaring. We examine two different operations in the next example.

Example 9 Identify operations to multiply polynomials

Identify the expression inside the parentheses as a product or a sum. Multiply each expression and simplify.

a. $(5x)^2$
b. $(5 + x)^2$

SOLUTION

a. The expression inside the parentheses is a *product*. Recall that $(5x)^2 = (5 \cdot x)^2$. Since exponent rules apply to powers of products, we multiply as follows.

$(5x)^2$
$= (5 \cdot x)^2$ Use the powers of products exponent rule.
$= 5^2 \cdot x^2$
$= 25x^2$

b. The expression inside the parentheses is a *sum*. This is an example of squaring a binomial (two terms).

$(5 + x)^2$
$= 5^2 + 2(5)(x) + x^2$ Apply the squaring a binomial formula.
$= 25 + 10x + x^2$
$= x^2 + 10x + 25$ Write the final answer in descending order.

> **■ Connecting the Concepts**
>
> **Why is $(x + 5)^2$ not equal to $x^2 + 5^2$?**
>
> Recall that exponent rules apply to products and quotients. In Example 9a, we simplify $(5x)^2$ using exponent rules because a product, $(5 \cdot x)^2$, is involved. Exponent rules do not apply to sums and differences. We cannot use an exponent rule on $(x + 5)^2$ because this expression involves a sum. It is multiplied out, as in Example 9b.

PRACTICE PROBLEM FOR EXAMPLE 9

Identify the expression inside the parentheses as a product or a sum. Multiply each expression and simplify.

a. $(8t)^2$
b. $(8 + t)^2$

Another pattern that is useful is the product of the sum and difference of the same two terms.

$(a + b)(a - b)$ Use the distributive property to multiply the binomials.
$= a^2 - ab + ab - b^2$ The middle two terms add to 0.
$= a^2 - b^2$

The middle terms add to zero. We are left with the difference of the two terms squared.

> **Multiplying the Sum and Difference of Two Terms**
>
> When multiplying the sum and difference of two terms, the result is the first term squared minus the second term squared.
>
> $$(a + b)(a - b) = a^2 - b^2$$
>
> The result is called the *difference of two squares*.

Example 10 — Multiplying the sum and difference of two terms

Find each product and simplify.

a. $(x + 3)(x - 3)$ **b.** $(4x - 7)(4x + 7)$ **c.** $(8t^3 + 5)(8t^3 - 5)$

SOLUTION

a. This is a product of the sum and difference of the two terms x and 3. The result follows the pattern of the first term squared minus the second term squared.

$$(x + 3)(x - 3)$$
$$= (x)^2 - 3^2$$
$$= x^2 - 9$$

b. This is the product of the sum and difference of the two terms $4x$ and 7.

$$(4x - 7)(4x + 7)$$
$$= (4x)^2 - (7)^2$$
$$= 16x^2 - 49$$

c. This is the product of the sum and difference of the two terms $8t^3$ and 5.

$$(8t^3 + 5)(8t^3 - 5)$$
$$= (8t^3)^2 - (5)^2$$
$$= 64t^6 - 25$$

PRACTICE PROBLEM FOR EXAMPLE 10

Find each product and simplify.

a. $(2x + 1)(2x - 1)$ **b.** $(x + 5y)(x - 5y)$

Example 11 — Combining operations

Perform the indicated operations and simplify.

a. $5x - 1 + (x - 3)(2x + 9)$ **b.** $(x - 4)^2 - (x + 5)^2$

SOLUTION

a. In this problem, $5x - 1 + (x - 3)(2x + 9)$, the multiplication is done before the addition. Recall the order-of-operations agreement, PEMDAS, from Chapter 1.

$5x - 1 + (x - 3)(2x + 9)$	Multiply the two binomials first.
$= 5x - 1 + (2x^2 + 9x - 6x - 27)$	Simplify the product in parentheses.
$= 5x - 1 + 2x^2 + 3x - 27$	Combine like terms.
$= 2x^2 + 8x - 28$	

b. In this problem, $(x - 4)^2 - (x + 5)^2$, the two binomials must be squared first before subtracting.

$(x - 4)^2 - (x + 5)^2$	Write out the square as a product.
$= (x - 4)(x - 4) - (x + 5)(x + 5)$	Multiply. Keep the result in parentheses.
$= (x^2 - 4x - 4x + 16) - (x^2 + 5x + 5x + 25)$	Simplify each product.
$= (x^2 - 8x + 16) - (x^2 + 10x + 25)$	Distribute the negative sign.
$= x^2 - 8x + 16 - x^2 - 10x - 25$	Combine like terms.
$= -18x - 9$	

5.4 Vocabulary Exercises

1. To multiply a polynomial by a monomial, use the _____ property.

2. The acronym _____ is used to help multiply two binomials.

5.4 Exercises

For Exercises 1 through 12, find each product and simplify.

1. $5(2x + 7)$
2. $8(3x + 4)$
3. $2x(7x + 6)$
4. $5x(3x - 8)$
5. $-3(4x + 2)$
6. $-4(5x + 1)$
7. $6(2x^2 + 3x - 5)$
8. $3(4x^2 - 5x + 8)$
9. $-7(x^2 + 4x - 3)$
10. $-5(3x^2 - 4x + 2)$
11. $3x(5x^2 - 3x + 4)$
12. $6x(2x^2 + 3x - 4)$

13. The number of screws required to hold down a lid on an industrial container can be represented by $3P$, where P is the perimeter of the lid in inches. After an accident, the number of screws required was doubled. Write a new expression to represent the number of screws needed to hold down a lid.

14. The cost for gasoline in dollars on a road trip depends on the number of miles traveled and can be represented by $0.1m$. Recently, the cost for gas has doubled. Write a new expression to represent the cost for gas in dollars for a road trip.

For Exercises 15 through 24, find each product and simplify.

15. $(x + 5)(x^2 + 2x - 3)$
16. $(x - 4)(x^2 + 7x + 5)$
17. $(x - 2)(5x^2 - 3x - 8)$
18. $(x - 6)(3x^2 + 5x - 9)$
19. $(2x + 7)(4x^2 - 6x + 3)$
20. $(3x - 5)(2x^2 + 7x + 6)$
21. $(4x - 3)(4x^2 - 6x - 3)$
22. $(5x - 7)(x^2 - 2x - 8)$
23. $(x^2 + 3)(2x^2 + 3x - 4)$
24. $(2x^2 - 6)(3x^2 + 4x - 2)$

For Exercises 25 through 34, find each product using FOIL. Simplify.

25. $(x + 3)(x + 5)$
26. $(x + 4)(x + 8)$
27. $(x + 6)(x - 2)$
28. $(x - 7)(x + 5)$
29. $(a + 2)(a + 5)$
30. $(m - 8)(m + 6)$
31. $(m + n)(2m + n)$
32. $(a + b)(6a - b)$
33. $(x^2 + 3)(5x - 4)$
34. $(2x^2 - 6)(3x + 7)$

35. The average amount of high-fructose corn syrup consumed per person in the United States t years since 2010 can be estimated by using the expression

$$-1.17t + 48.04 \text{ pounds per person}$$

Source: Model dervied from data at www.statista.com

The population of the United States in millions t years since 2010 can be estimated by using the expression

$$2.32t + 309.32 \text{ million people}$$

Source: Model derived from data found at the U.S. Census Bureau.

a. Write a new expression to represent the total amount of high-fructose corn syrup consumed by people in the United States.

b. Use the expression found in part a to find the total amount of high-fructose corn syrup that will be consumed by Americans in 2020.

36. The average amount of soft drinks consumed per person in the United States t years since 2010 can be estimated by using the expression

$$-1.01t + 45.54 \text{ gallons per person.}$$

Source: Model dervied from data at www.statista.com

The population of the United States in millions t years since 2010 can be estimated by using the expression

$$2.32t + 309.32 \text{ million people.}$$

Source: Model derived from data found at the U.S. Census Bureau.

a. Write a new expression to represent the total amount of soft drinks consumed by people in the United States.

b. Use the expression found in part a to find the amount of soft drinks that will be consumed by Americans in 2022.

37. The average income in thousands of dollars for a family in the United States t years since 2010 can be estimated by using the expression

$$1.22t + 61.73$$

The number of families in millions in the United States t years since 2010 can be estimated by using the expression

$$0.42t + 79.66$$

Source: Models derived from data found at www.deptofnumbers.com

a. Write the average income of a family in the United States in 2018.

b. Find the number of families in the United States in 2018.

c. Use the given expressions to find a new expression for the total income of all families in the United States.

d. Use the expression found in part c to find the total income of all families in the United States in 2020.

38. The average amount of red meat and poultry consumed per person in the United States t years since 2010 can be estimated by using the expression

$$5.12t + 183.14 \text{ pounds per person}$$

Source: Model derived from data found at The National Chicken Council.

The population of the United States in millions t years since 2010 can be estimated by using the expression

$$2.32t + 309.32$$

Source: Model derived from data found at the U.S. Census Bureau.

a. Find the average amount of red meat and poultry consumed by people in the United States in 2018.

b. Use the given expressions to write a new expression for the total amount of red meat and poultry consumed in the United States.

c. Use the expression found in part b to find the total amount of red meat and poultry consumed in the United States in 2020.

39. Write a polynomial that represents the area of the following rectangle.

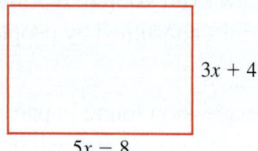

$3x + 4$

$5x - 8$

40. Write a polynomial that represents the area of the following rectangle.

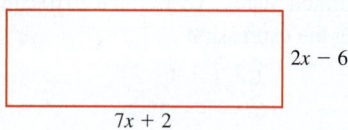

$2x - 6$

$7x + 2$

For Exercises 41 through 68, find each special product using a formula and simplify.

41. $(x + 6)^2$ 42. $(x + 7)^2$
43. $(x - 3)^2$ 44. $(x - 4)^2$
45. $(3x + 5)^2$ 46. $(2x + 8)^2$
47. $(5x - 3)^2$ 48. $(7x - 2)^2$
49. $(2a + 7b)^2$ 50. $(5m - 3n)^2$
51. $(3x^2 + 5)^2$ 52. $(6x^2 + 2)^2$
53. $(4d^3 - 5c)^2$ 54. $(8x^3 + 7y)^2$
55. $(x + 5)(x - 5)$ 56. $(x + 7)(x - 7)$
57. $(2a + 5)(2a - 5)$ 58. $(6p - 1)(6p + 1)$
59. $\left(\dfrac{x}{2} + 1\right)\left(\dfrac{x}{2} - 1\right)$ 60. $\left(5y - \dfrac{2}{5}\right)\left(5y + \dfrac{2}{5}\right)$
61. $\left(\dfrac{4}{5}x - 3\right)^2$ 62. $\left(\dfrac{1}{6}y + 5\right)^2$
63. $(3a + 2b)(3a - 2b)$ 64. $(4m - 6n)(4m + 6n)$
65. $(x^2 + 5)(x^2 - 5)$ 66. $(m^2 + 4)(m^2 - 4)$
67. $(2a^5 + 3b)(2a^5 - 3b)$ 68. $(4m^3 - 6n^2)(4m^3 + 6n^2)$

For Exercises 69 through 78, identify each expression in the parentheses as a product or a sum. Multiply the expression and simplify.

69. $(-2x)^2$ 70. $(-2 + x)^2$
71. $(y + 6)^2$ 72. $(6y)^2$
73. $(x + y)^2$ 74. $(xy)^2$
75. $\left(\dfrac{1}{2}n\right)^2$ 76. $\left(n + \dfrac{1}{2}\right)^2$
77. $\left(\dfrac{1}{5}a\right)^2$ 78. $\left(\dfrac{1}{5} + a\right)^2$

For Exercises 79 through 84, the students were asked to find each product and simplify. Explain what each student did wrong. Then simplify each expression correctly.

79. $(x + 4)(x - 3)$ 80. $(x - 2)(x + 5)$

Tristan

$(x + 4)(x - 3)$
$= x - 3x + 4x - 12$
$= 2x - 12$

Miguel

$(x - 2)(x + 5)$
$= x + 5x - 2x - 10$
$= 4x - 10$

81. $(x + 4)^2$
82. $(y - 2)^2$

Dolores
$(x + 4)^2$
$= x^2 + 16$

Pablo
$(y - 2)^2$
$= y^2 - 4$

83. $\left(\dfrac{2}{3}a - 1\right)^2$
84. $\left(\dfrac{y}{2} + 3\right)^2$

Tyler
$\left(\dfrac{2}{3}a - 1\right)^2$
$= \dfrac{2}{3}a^2 - 1$

Rebekah
$\left(\dfrac{y}{2} + 3\right)^2$
$= \dfrac{y^2}{4} + 9$

For Exercises 85 through 94, find each product using any method and simplify.

85. $5x(x + 7)(x + 3)$
86. $3x(x + 6)(x + 2)$
87. $2x(3x + 4)(4x - 7)$
88. $6a(2a - 8)(3a + 5)$
89. $7a(a + 2)(a - 2)$
90. $8m(m - 8)(m + 8)$
91. $7h(h + 5)^2$
92. $2x(x + 6)^2$
93. $8x^2(5x - 4)^2$
94. $2x^2(3x + 7)^2$

For Exercises 95 through 106, perform the indicated operations and simplify.

95. $2x + 5 + (x + 3)(x - 7)$
96. $7x^2 + 3x + (2x + 4)(x - 3)$
97. $5x^2 + 10x - (2x + 4)(x + 3)$
98. $8a^2 - 20a - (4a + 1)(2a - 6)$
99. $-(3x - 7)(x + 1) + 4x^2 - 7x$
100. $-(x - 6)(2x + 4) - x^2 + 9x$
101. $(x + 4)^2 - (x - 3)^2$
102. $(2m + 3)^2 - (m - 5)^2$
103. $(2x + 5)(2x - 5) + (3x^2 + 6x - 8)$
104. $(3x + 7)(3x - 7) + (4x^2 + 2x - 15)$
105. $(2t + 7)(3t - 5) - 4(7t^2 + 3t - 5)$
106. $(a + 4)(6a - 3) - 5(2a^2 + 4a - 7)$

5.5 Dividing Polynomials

LEARNING OBJECTIVES
- Divide a polynomial by a monomial.
- Divide a polynomial by a polynomial using long division.

So far in this chapter, polynomials have been added, subtracted, and multiplied. We now learn how to divide polynomials. We expand the quotient rule for exponents, from Section 5.1, to divide polynomials by monomials. We then learn how to divide a polynomial by another polynomial.

■ Dividing a Polynomial by a Monomial

When a polynomial is divided by a monomial, it is easiest to separate the polynomial by terms and then divide each term by the monomial.

Dividing a Polynomial by a Monomial
To divide a polynomial by a monomial, divide each term in the numerator (polynomial) by the denominator (monomial). Then simplify each term.

Note: A monomial has one term.

510 CHAPTER 5 Exponents and Polynomials

Example 1 — Dividing a polynomial by a monomial

Divide and simplify.

a. $\dfrac{15x^2 + 12x}{6x}$ b. $\dfrac{8a^3 - 20a^2}{5a^2}$

SOLUTION

a. This is a polynomial divided by a monomial. Separate the terms of the polynomial and divide each term by the denominator.

$$\dfrac{15x^2 + 12x}{6x}$$

$$= \dfrac{15x^{2\,1}}{6\cancel{x}} + \dfrac{12\cancel{x}}{6\cancel{x}} \qquad \text{Separate each term and use the exponent rules to reduce each fraction.}$$

$$= \dfrac{5x}{2} + 2$$

b. This is a polynomial divided by a monomial. Separate the terms of the polynomial and divide each term by the denominator.

$$\dfrac{8a^3 - 20a^2}{5a^2}$$

$$= \dfrac{8a^{3\,1}}{5\cancel{a^2}} - \dfrac{20\cancel{a^2}}{5\cancel{a^2}} \qquad \text{Separate each term and use the exponent rules to reduce each fraction.}$$

$$= \dfrac{8a}{5} - 4$$

PRACTICE PROBLEM FOR EXAMPLE 1

Divide and simplify.

a. $\dfrac{30x^2 + 4x}{5x}$ b. $\dfrac{24m^4 - 36m}{8m}$

Example 2 — Dividing more complicated polynomials by a monomial

Divide and simplify.

a. $\dfrac{35x^2 + 14x - 21}{7x}$ b. $\dfrac{20a^3b^2 + 15a^2b^2 - 30ab}{5ab^2}$

SOLUTION

a. This is a polynomial divided by a monomial. Separate the terms of the polynomial and divide each term by the denominator.

$$\dfrac{35x^2 + 14x - 21}{7x}$$

$$= \dfrac{35x^{2\,1}}{7\cancel{x}} + \dfrac{14\cancel{x}}{7\cancel{x}} - \dfrac{21}{7x} \qquad \text{Separate each term and use the exponent rules to reduce each fraction.}$$

$$= 5x + 2 - \dfrac{3}{x} \qquad \text{Keep the answer as separate terms.}$$

b. This is a polynomial divided by a monomial. Separate the terms of the polynomial and divide each term by the denominator.

$$\frac{20a^3b^2 + 15a^2b^2 - 30ab}{5ab^2}$$

$$= \frac{20a^{3\,2}b^2}{5ab^2} + \frac{15a^{2\,1}b^2}{5ab^2} - \frac{30ab}{5ab^{2\,1}} \quad \text{Separate each term and use the exponent rules to reduce each fraction.}$$

$$= 4a^2 + 3a - \frac{6}{b} \quad \text{Keep the answer as separate terms.}$$

PRACTICE PROBLEM FOR EXAMPLE 2

Divide and simplify.

a. $\dfrac{40x^2 + 12x - 72}{4x}$ **b.** $\dfrac{60m^4n^3 - 51m^3n^2 + 30m^2n}{3m^2n}$

Dividing a Polynomial by a Polynomial Using Long Division

When dividing a polynomial by a polynomial with more than one term, we use a method that is similar to long division of natural numbers. Let's review long division of natural numbers. To divide $\dfrac{583}{6}$, put the numerator on the inside of the long division symbol and the denominator on the outside. So write $\dfrac{583}{6}$ as $6\overline{)583}$. Here 583 is called the *dividend* (numerator), and 6 is called the *divisor* (denominator). Performing long division yields

$$\begin{array}{r} 97\ R1 \\ 6\overline{)583} \\ -54 \\ \hline 43 \\ -42 \\ \hline 1 \end{array}$$

The quotient is 97, and the remainder is 1. At each step, we divide and then subtract until we are left with something less than the divisor. We use a similar algorithm to divide a polynomial by a polynomial.

> **Division of a Polynomial by a Polynomial**
>
> To divide a polynomial by a polynomial that has more than one term, use long division. Continue dividing until the degree of the remainder is less than the degree of the divisor.

When dividing by a polynomial, pay attention to the first term of the divisor.

What's That Mean?

Dividend and Divisor

The numerator and denominator of a fraction are called the *dividend* and *divisor* when using long division. In the fraction $\dfrac{583}{6}$, 583 is the numerator and dividend. The 6 is the denominator and divisor.

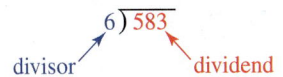

divisor   dividend

Example 3 Long division of polynomials

Divide and check the results.

a. $(x^2 + 12x + 35) \div (x + 5)$ **b.** $(6x^2 + 17x - 45) \div (3x - 5)$

CHAPTER 5 Exponents and Polynomials

SOLUTION

a. $(x^2 + 12x + 35) \div (x + 5)$

The divisor is $x + 5$, which has two terms. Use long division, since we are dividing by a binomial.

The first term in the divisor $(x + 5)$ is x. Divide the first term in the dividend (numerator) by x: $\dfrac{x^2}{x} = x$. This yields the first term in the quotient. Multiply the first term in the quotient by the divisor. Write this below the dividend and subtract.

$$\begin{array}{r} x \\ x + 5 \overline{) x^2 + 12x + 35} \\ -(x^2 + 5x) \end{array} \qquad \begin{array}{r} x \\ x + 5 \overline{) x^2 + 12x + 35} \\ -x^2 - 5x \\ \hline 7x + 35 \end{array}$$

To subtract, distribute the negative sign and combine like terms. Bring down $+35$.

Continue by dividing $7x$ by x: $\dfrac{7x}{x} = 7$. This yields the second term in the quotient. Multiply the second term in the quotient by the divisor. Subtract this from the dividend.

$$\begin{array}{r} x + 7 \\ x + 5 \overline{) x^2 + 12x + 35} \\ -x^2 - 5x \\ \hline 7x + 35 \\ -(7x + 35) \end{array} \qquad \begin{array}{r} x + 7 \\ x + 5 \overline{) x^2 + 12x + 35} \\ -x^2 - 5x \\ \hline 7x + 35 \\ -7x - 35 \\ \hline 0 \end{array}$$

To subtract, distribute the negative sign and combine like terms. The remainder is 0, so we are done.

The quotient is $x + 7$.

Check this answer by multiplying it by the original divisor.

$(x + 5)(x + 7)$
$= x^2 + 7x + 5x + 35$ Multiply out using FOIL.
$= x^2 + 12x + 35$ This is the original dividend, so the answer is correct.

b. $(6x^2 + 17x - 45) \div (3x - 5)$

The denominator, $3x - 5$, has two terms. Since we are dividing this polynomial by a binomial, we use long division.

The first term in the divisor $(3x - 5)$ is $3x$. Divide the first term in the dividend (numerator) by $3x$: $\dfrac{6x^2}{3x} = 2x$. This yields the first term in the quotient. Multiply $2x$ by the divisor. Subtract this result from the dividend.

$$\begin{array}{r} 2x \\ 3x - 5 \overline{) 6x^2 + 17x - 45} \\ -(6x^2 - 10x) \end{array} \qquad \begin{array}{r} 2x \\ 3x - 5 \overline{) 6x^2 + 17x - 45} \\ -6x^2 + 10x \\ \hline 27x - 45 \end{array}$$

To subtract, distribute the negative sign and combine like terms. Bring down -45.

Continue by dividing $27x$ by $3x$: $\dfrac{27x}{3x} = 9$. This yields the second term in the quotient. Multiply 9 by the divisor. Subtract this result from the dividend.

$$\begin{array}{r} 2x + 9 \\ 3x - 5 \overline{\smash{)}6x^2 + 17x - 45} \\ -6x^2 + 10x \\ \hline 27x - 45 \\ -(27x - 45) \end{array} \quad \begin{array}{r} 2x + 9 \\ 3x - 5 \overline{\smash{)}6x^2 + 17x - 45} \\ -6x^2 + 10x \\ \hline 27x - 45 \\ -27x + 45 \\ \hline 0 \end{array}$$ To subtract, distribute the negative sign and combine like terms. The remainder is 0, so we are done.

The quotient is $2x + 9$.

Check this using multiplication.

$(3x - 5)(2x + 9)$
$= 6x^2 + 27x - 10x - 45$ Multiply out using FOIL.
$= 6x^2 + 17x - 45$ This is the original dividend, so the answer is correct.

PRACTICE PROBLEM FOR EXAMPLE 3

Divide and check the results.

a. $(x^2 + 12x + 32) \div (x + 4)$ **b.** $(20x^2 - 9x - 18) \div (4x + 3)$

Example 4 Finding the width from the area

Find an expression that represents the width of this rectangle.

area = $3x^2 + 14x - 24$, $3x - 4$, ?

SOLUTION

Since the area of a rectangle is found by multiplying the length by the width, we undo this by dividing. Therefore, we need to divide the area by $3x - 4$.

The first term in the divisor $(3x - 4)$ is $3x$. Divide the first term in the dividend (numerator) by $3x$: $\dfrac{3x^2}{3x} = x$. This yields the first term in the quotient.

$$\begin{array}{r} x \\ 3x - 4 \overline{\smash{)}3x^2 + 14x - 24} \\ -(3x^2 - 4x) \end{array} \quad \begin{array}{r} x \\ 3x - 4 \overline{\smash{)}3x^2 + 14x - 24} \\ -3x^2 + 4x \\ \hline 18x - 24 \end{array}$$ To subtract, distribute the negative sign and combine like terms. Bring down -24.

Continue by dividing $18x$ by $3x$: $\dfrac{18x}{3x} = 6$. This gives the second term in the quotient.

$$\begin{array}{r} x + 6 \\ 3x - 4 \overline{\smash{)}3x^2 + 14x - 24} \\ -3x^2 + 4x \\ \hline 18x - 24 \\ -(18x - 24) \end{array} \quad \begin{array}{r} x + 6 \\ 3x - 4 \overline{\smash{)}3x^2 + 14x - 24} \\ -3x^2 + 4x \\ \hline 18x - 24 \\ -18x + 24 \\ \hline 0 \end{array}$$ To subtract, distribute the negative sign and combine like terms. The remainder is 0, so we are done.

Therefore, the width can be represented by the expression $x + 6$.

514 CHAPTER 5 Exponents and Polynomials

PRACTICE PROBLEM FOR EXAMPLE 4
Find an expression that represents the length of this rectangle.

| area = $4x^2 - 18x - 10$ | $x - 5$ |

?

As we have seen in the first numerical example, a remainder can be left after we have divided all the terms of the polynomial. When this happens, write the remainder over the dividend as a fraction. When using long division with polynomials, stop dividing when the degree of the remainder is less than the degree of the divisor.

Example 5   Long division of polynomials with a remainder

Divide and check the results.

a. $(7x^2 - 53x + 30) \div (x - 8)$

b. $(6x^3 + 14x^2 + 7x + 10) \div (x + 2)$

■ **Connecting the Concepts**

When do we stop dividing?

When using long division with integers, we stop dividing when the remainder is less than the divisor.
    This means that in long division with polynomials, we stop dividing *when the degree of the reminder is less than the degree of the divisor.*

■ **Skill Connection**

Write the solution to a division problem

We write the solution to a division problem in the following manner.

$$\text{quotient} + \frac{\text{remainder}}{\text{divisor}}$$

SOLUTION

a. $(7x^2 - 53x + 30) \div (x - 8)$

The divisor is $x - 8$, which has two terms. Use long division.

The first term in the divisor $(x - 8)$ is x. Divide the first term in the dividend (numerator) by x: $\dfrac{7x^2}{x} = 7x$. This yields the first term in the quotient.

$$\begin{array}{r} 7x \\ x-8{\overline{\smash{\big)}\,7x^2 - 53x + 30}} \\ -(7x^2 - 56x) \end{array}$$

$$\begin{array}{r} 7x \\ x-8{\overline{\smash{\big)}\,7x^2 - 53x + 30}} \\ -7x^2 + 56x \\ \hline 3x + 30 \end{array}$$

To subtract, distribute the negative sign and combine like terms. Bring down +30.

Continue dividing $3x$ by x: $\dfrac{3x}{x} = 3$. This gives the second term in the quotient.

$$\begin{array}{r} 7x + 3 \\ x-8{\overline{\smash{\big)}\,7x^2 - 53x + 30}} \\ -7x^2 + 56x \\ \hline 3x + 30 \\ -(3x - 24) \end{array}$$

$$\begin{array}{r} 7x + 3 \\ x-8{\overline{\smash{\big)}\,7x^2 - 53x + 30}} \\ -7x^2 + 56x \\ \hline 3x + 30 \\ -3x + 24 \\ \hline 54 \end{array}$$

To subtract, distribute the negative sign and combine like terms. The remainder is 54. The degree of 54 is less than the degree of $x - 8$, so we are done.

The solution is $7x + 3 + \dfrac{54}{x - 8}$.

Check by multiplying the quotient by the divisor and adding the remainder to the result.

$(x - 8)(7x + 3) + 54$

$= 7x^2 + 3x - 56x - 24 + 54$ Multiply using FOIL and add the remainder.

$= 7x^2 - 53x + 30$ This is the original dividend, so the answer checks.

b. $(6x^3 + 14x^2 + 7x + 10) \div (x + 2)$

The divisor is $x + 2$, which has two terms. Use long division.

The first term in the divisor $(x + 2)$ is x. Divide the first term in the dividend (numerator) by x: $\dfrac{6x^3}{x} = 6x^2$. This yields the first term in the quotient.

$$\begin{array}{r} 6x^2 \\ x + 2 \overline{\smash{)}6x^3 + 14x^2 + 7x + 10} \\ -(6x^3 + 12x^2) \end{array}$$

$$\begin{array}{r} 6x^2 \\ x + 2 \overline{\smash{)}6x^3 + 14x^2 + 7x + 10} \\ \underline{-6x^3 - 12x^2} \\ 2x^2 + 7x \end{array}$$

To subtract, distribute the negative sign and combine like terms. Bring down $+7x$.

Continue dividing $2x^2$ by x: $\dfrac{2x^2}{x} = 2x$. This gives the second term in the quotient.

$$\begin{array}{r} 6x^2 + 2x \\ x + 2 \overline{\smash{)}6x^3 + 14x^2 + 7x + 10} \\ \underline{-6x^3 - 12x^2} \\ 2x^2 + 7x \\ -(2x^2 + 4x) \end{array}$$

$$\begin{array}{r} 6x^2 + 2x \\ x + 2 \overline{\smash{)}6x^3 + 14x^2 + 7x + 10} \\ \underline{-6x^3 - 12x^2} \\ 2x^2 + 7x \\ \underline{-2x^2 - 4x} \\ 3x + 10 \end{array}$$

To subtract, distribute the negative sign and combine like terms. The remainder is $3x$. Since $3x$ is not less in degree than $x + 2$, we keep dividing. Bring down the $+10$.

Continue dividing $3x$ by x: $\dfrac{3x}{x} = 3$. This gives the third term in the quotient.

$$\begin{array}{r} 6x^2 + 2x + 3 \\ x + 2 \overline{\smash{)}6x^3 + 14x^2 + 7x + 10} \\ \underline{-6x^3 - 12x^2} \\ 2x^2 + 7x \\ \underline{-2x^2 - 4x} \\ 3x + 10 \\ -(3x + 6) \end{array}$$

$$\begin{array}{r} 6x^2 + 2x + 3 \\ x + 2 \overline{\smash{)}6x^3 + 14x^2 + 7x + 10} \\ \underline{-6x^3 - 12x^2} \\ 2x^2 + 7x \\ \underline{-2x^2 - 4x} \\ 3x + 10 \\ \underline{-3x - 6} \\ 4 \end{array}$$

To subtract, distribute the negative sign and combine like terms. The remainder is 4. Since 4 is less in degree than $x + 2$, we stop dividing.

The solution is $6x^2 + 2x + 3 + \dfrac{4}{x + 2}$.

Check by multiplying the quotient by the divisor and adding the remainder to the result.

$(x + 2)(6x^2 + 2x + 3) + 4$
$= 6x^3 + 2x^2 + 3x + 12x^2 + 4x + 6 + 4$
$= 6x^3 + 14x^2 + 7x + 10$ This is the original dividend, so the answer checks.

PRACTICE PROBLEM FOR EXAMPLE 5

Divide and check the results.

a. $(8x^2 - 19x - 19) \div (x - 3)$

b. $(3x^3 - 11x^2 - 27x + 37) \div (x + 3)$

In each of these examples, the terms of the dividend (numerator) have been in descending order, and no middle terms were missing. In some cases, the order of the

Skill Connection

Descending order

A polynomial is in descending order when the highest power of the variable is in the leftmost term and the powers decrease on the terms moving left to right. The polynomial $-2x + 5 + 7x^2$ is not in descending order. Rearranging the terms yields $7x^2 - 2x + 5$, which is in descending order.

---

terms has to be rearranged into descending order to make the division algorithm work. Missing terms also cause confusion, so any missing terms are represented by a term with a coefficient of zero. This allows the subtraction steps to be done more clearly.

Example 6   Long division of polynomials with missing terms

Divide and check the results.

a. $(5x^3 + 8 + 16x^2) \div (x + 3)$ **b.** $(8x^3 + 1) \div (2x + 1)$

SOLUTION

a. $(5x^3 + 8 + 16x^2) \div (x + 3)$

The terms in this polynomial are not in descending order, so we rearrange the terms.

$$5x^3 + 16x^2 + 8$$

The first-degree term, x, is missing, so we add the term $0x$ to take the place of the missing first-degree term.

$$5x^3 + 16x^2 + 0x + 8$$

Now use long division.

The first term in the divisor $(x + 3)$ is x. Divide the first term in the dividend (numerator) by x: $\dfrac{5x^3}{x} = 5x^2$. This yields the first term in the quotient.

$$\begin{array}{r} 5x^2 \\ x+3\overline{)5x^3 + 16x^2 + 0x + 8} \\ -(5x^3 + 15x^2) \end{array}$$

$$\begin{array}{r} 5x^2 \\ x+3\overline{)5x^3 + 16x^2 + 0x + 8} \\ -5x^3 - 15x^2 \\ \hline x^2 + 0x \end{array}$$

To subtract, distribute the negative sign and combine like terms. Bring down $+ 0x$.

Continue dividing x^2 by x: $\dfrac{x^2}{x} = x$. This gives the second term in the quotient.

$$\begin{array}{r} 5x^2 + x \\ x+3\overline{)5x^3 + 16x^2 + 0x + 8} \\ -5x^3 - 15x^2 \\ \hline x^2 + 0x \\ -(x^2 + 3x) \end{array}$$

$$\begin{array}{r} 5x^2 + x \\ x+3\overline{)5x^3 + 16x^2 + 0x + 8} \\ -5x^3 - 15x^2 \\ \hline x^2 + 0x \\ -x^2 - 3x \\ \hline -3x + 8 \end{array}$$

To subtract, distribute the negative sign and combine like terms. Bring down the $+8$.

Continue dividing $-3x$ by x: $\dfrac{-3x}{x} = -3$. This gives the third term in the quotient.

$$\begin{array}{r} 5x^2 + x - 3 \\ x+3\overline{)5x^3 + 16x^2 + 0x + 8} \\ -5x^3 - 15x^2 \\ \hline x^2 + 0x \\ -x^2 - 3x \\ \hline -3x + 8 \\ -(-3x - 9) \end{array}$$

$$\begin{array}{r} 5x^2 + x - 3 \\ x+3\overline{)5x^3 + 16x^2 + 0x + 8} \\ -5x^3 - 15x^2 \\ \hline x^2 + 0x \\ -x^2 - 3x \\ \hline -3x + 8 \\ 3x + 9 \\ \hline +17 \end{array}$$

To subtract, distribute the negative sign and combine like terms. The remainder is 17. Since the degree of 17 is less than the degree of $x + 3$, we are done.

The solution is $5x^2 + x - 3 + \dfrac{17}{x + 3}$.

Check by multiplying the quotient by the divisor and adding the remainder to the result.

$(x + 3)(5x^2 + x - 3) + 17$ Multiply and add the remainder.

$= 5x^3 + x^2 - 3x + 15x^2 + 3x - 9 + 17$

$= 5x^3 + 16x^2 + 8$ This is the original dividend, so the answer checks.

b. $(8x^3 + 1) \div (2x + 1)$

This polynomial is missing the second- and first-degree terms, so we add the terms $0x^2$ and $0x$.

$$8x^3 + 0x^2 + 0x + 1$$

Now use long division. The first term in the divisor $(2x + 1)$ is $2x$. Divide the first term in the dividend (numerator) by $2x$: $\dfrac{8x^3}{2x} = 4x^2$. This yields the first term in the quotient.

$$\begin{array}{r} 4x^2 \\ 2x + 1\overline{\smash{)}8x^3 + 0x^2 + 0x + 1} \\ -(8x^3 + 4x^2) \end{array}$$

$$\begin{array}{r} 4x^2 \\ 2x + 1\overline{\smash{)}8x^3 + 0x^2 + 0x + 1} \\ -8x^3 - 4x^2 \\ \hline -4x^2 + 0x \end{array}$$

To subtract, distribute the negative sign and combine like terms. Bring down $+0x$.

Continue dividing $-4x^2$ by $2x$: $\dfrac{-4x^2}{2x} = -2x$. This gives the second term in the quotient.

$$\begin{array}{r} 4x^2 - 2x \\ 2x + 1\overline{\smash{)}8x^3 + 0x^2 + 0x + 1} \\ -8x^3 - 4x^2 \\ \hline -4x^2 + 0x \\ -(-4x^2 - 2x) \end{array}$$

$$\begin{array}{r} 4x^2 - 2x \\ 2x + 1\overline{\smash{)}8x^3 + 0x^2 + 0x + 1} \\ -8x^3 - 4x^2 \\ \hline -4x^2 + 0x \\ 4x^2 + 2x \\ \hline 2x + 1 \end{array}$$

To subtract, distribute the negative sign and combine like terms. Bring down the $+1$.

Continue dividing $2x$ by $2x$: $\dfrac{2x}{2x} = 1$. This gives the third term in the quotient.

$$\begin{array}{r} 4x^2 - 2x + 1 \\ 2x + 1\overline{\smash{)}8x^3 + 0x^2 + 0x + 1} \\ -8x^3 - 4x^2 \\ \hline -4x^2 + 0x \\ 4x^2 + 2x \\ \hline 2x + 1 \\ -(2x + 1) \end{array}$$

$$\begin{array}{r} 4x^2 - 2x + 1 \\ 2x + 1\overline{\smash{)}8x^3 + 0x^2 + 0x + 1} \\ -8x^3 - 4x^2 \\ \hline -4x^2 + 0x \\ 4x^2 + 2x \\ \hline 2x + 1 \\ -2x - 1 \\ \hline 0 \end{array}$$

To subtract, distribute the negative sign and combine like terms. The remainder is 0, so we are done.

The solution is $4x^2 - 2x + 1$.

Check using multiplication.

$(2x + 1)(4x^2 - 2x + 1)$

$= 8x^3 - 4x^2 + 2x + 4x^2 - 2x + 1$

$= 8x^3 + 1$ This is the original dividend, so the answer is correct.

518 CHAPTER 5 Exponents and Polynomials

PRACTICE PROBLEM FOR EXAMPLE 6
Divide and check the results.

a. $(4x^3 + 7 - 3x) \div (x + 2)$

b. $(x^3 - 27) \div (x - 3)$

5.5 Vocabulary Exercises

1. To divide a polynomial by a monomial, divide each _____ of the polynomial by the monomial.

2. To divide a polynomial by a polynomial, use _____ division.

5.5 Exercises

For Exercises 1 through 10, warm up by identifying each divisor as a monomial or a polynomial of more than one term. Do not perform any calculations.

1. $\dfrac{-6x + 5}{3}$

2. $\dfrac{-4x + 12}{8}$

3. $\dfrac{7b + 4}{3b}$

4. $\dfrac{-9a + 16}{-2a}$

5. $\dfrac{y^2 + 6y + 9}{y + 3}$

6. $\dfrac{y^2 - 14y + 49}{y + 7}$

7. $\dfrac{-x^3 + 4x^2 + 6}{x^3 - 5x + 1}$

8. $\dfrac{3x^4 + 5x^3 - 2x - 1}{x^2 + x + 2}$

9. $\dfrac{8a^3 + 2a^2 - 6a + a}{2a - 1}$

10. $\dfrac{-5b^5 + 3b^4 + b^2 + 7}{4b - 3}$

For Exercises 11 through 26, divide the following expressions.

11. $\dfrac{12x^3 + 16x^2 - 8x}{2x}$

12. $\dfrac{15a^4 - 20a^2 + 30a}{5a}$

13. $\dfrac{30m^4 - 24m^3 + 72m^2}{6m^2}$

14. $\dfrac{14b^5 + 28b^3 - 56b^2}{7b^2}$

15. $\dfrac{2x^5 + 7x^3 - 15x^2 + 6x}{4x}$

16. $\dfrac{3a^4 - 18a^2 + 20a}{6a}$

17. $\dfrac{21h^5 + 12h^3 - 2h^2}{7h^2}$

18. $\dfrac{20c^4 - 15c^3 + 14c^2}{4c^2}$

19. $\dfrac{25x^3 + 30x^2 - 5x}{5x^2}$

20. $\dfrac{24a^3 + 18a^2 - 12a}{6a^2}$

21. $\dfrac{24d^5 - 21d^3 + 14d}{14d^2}$

22. $\dfrac{36y^4 + 20y^2 - 15y}{12y^2}$

23. $\dfrac{20x^2 - 10xy + 15y^2}{5xy}$

24. $\dfrac{10a^2 - 12ab + 4b^2}{2ab}$

25. $\dfrac{18m^4n^3 - 20m^2n^2 + 15mn}{6mn^2}$

26. $\dfrac{33x^6y^4 + 45x^4y^3 - 16x^2y^2}{9x^2y^2}$

For Exercises 27 through 32, the students were asked to simplify the given expressions. Check the results to see whether they are correct or incorrect. See the check for Example 3 for help if needed.

27. $\dfrac{3x - 6}{x - 2}$

Samuel
$\dfrac{3x - 6}{x - 2} = 3$

28. $\dfrac{-4x - 10}{2x + 5}$

Santiago
$\dfrac{-4x - 10}{2x + 5} = -2$

29. $\dfrac{x^2 + x - 6}{x - 2}$

Jacine
$\dfrac{x^2 + x - 6}{x - 2} = x + 3$

30. $\dfrac{x^2 + x - 20}{x + 5}$

Sterling
$$\dfrac{x^2 + x - 20}{x + 5} = x - 4$$

31. $\dfrac{x^3 - 1}{x - 1}$

Lyria
$$\dfrac{x^3 - 1}{x - 1} = x^2 + x + 1$$

32. $\dfrac{y^3 + 1}{y + 1}$

Olivia
$$\dfrac{y^3 + 1}{y + 1} = y^2 - y + 1$$

For Exercises 33 through 42, divide using long division. Check the results.

33. $(x^2 + 11x + 18) \div (x + 2)$

34. $(m^2 + 11m + 24) \div (m + 3)$

35. $(x^2 - 3x - 28) \div (x + 4)$

36. $(a^2 - 6a - 55) \div (a + 5)$

37. $(m^2 - 7m + 12) \div (m - 3)$

38. $(d^2 - 14d + 48) \div (d - 6)$

39. $(14x^2 + 41x + 15) \div (7x + 3)$

40. $(24a^2 + 25a + 6) \div (3a + 2)$

41. $(30d^2 - 62d + 28) \div (6d - 4)$

42. $(36c^2 - 77c + 40) \div (9c - 8)$

For Exercises 43 through 46, divide to find the length of each rectangle. See Example 4 for help if needed.

43. Write an expression that represents the length of this rectangle.

area = $2x^2 + 9x - 35$ $2x - 5$

?

44. Write an expression that represents the length of this rectangle.

area = $5x^2 - 44x + 32$ $x - 8$

?

45. Write an expression that represents the length of this rectangle.

area = $14a^2 + 5a - 24$ $2a + 3$

?

46. Write an expression that represents the length of this rectangle.

area = $12b^2 - 34b + 14$ $3b - 7$

?

For Exercises 47 through 60, divide using long division. Check the results. See Example 5 for help if needed.

47. $(x^2 + 5x - 29) \div (x - 4)$

48. $(a^2 - 3a - 31) \div (a - 5)$

49. $(b^2 - 10b + 13) \div (b - 3)$

50. $(m^2 - 13m + 26) \div (m - 4)$

51. $(2x^2 + 13x + 9) \div (2x + 5)$

52. $(7x^2 + 44x + 8) \div (7x + 2)$

53. $(12a^2 + 13a - 9) \div (4a + 7)$

54. $(18m^2 + 18m - 10) \div (3m + 5)$

55. $(15x^2 - 41x + 8) \div (5x - 2)$

56. $(16y^2 - 46y + 6) \div (2y - 5)$

57. $(2x^3 + 7x^2 - x - 21) \div (2x - 3)$

58. $(3x^3 + 19x^2 - 52x + 20) \div (3x - 5)$

59. $(10a^3 + 26a^2 - 13a - 15) \div (5a + 3)$

60. $(6c^3 + 2c^2 - 28c - 16) \div (2c + 4)$

For Exercises 61 through 66, first find any missing terms in the numerator and write them with 0 as a coefficient. Then use long division to divide. Check the results. See Example 6 for help if needed.

61. $(5x^3 + 2x + 12) \div (x - 3)$

62. $(8y^3 - 6y + 14) \div (2y - 5)$

63. $(a^3 + 8) \div (a + 2)$

64. $(m^3 + 27) \div (m + 3)$

65. $(8x^3 - 27) \div (2x - 3)$

66. $(27y^3 - 64) \div (3y - 4)$

For Exercises 67 through 70, simplify each expression using the order-of-operations agreement.

67. $\left(-\dfrac{2}{3}\right)^3 \div \dfrac{4}{3} - \dfrac{5}{3}$

68. $\dfrac{5}{4} - \dfrac{1}{3} + \left(\dfrac{13 - 8}{7 - 2^2}\right)^2$

69. $4 \cdot (-10)^2 \div 5 - |8 - 9(-6)|$

70. $(5 - 3(4))^2 - (-1)^3$

For Exercises 71 through 74, convert units as indicated. Round to the hundredths place if necessary.

71. Convert 1024 bits per second to kilobits per second. There is 1 bit per second in 0.001 kilobit per second.

72. Convert 12,048 kilobits per second to bits per second. There is 1 bit per second in 0.001 kilobit per second.

73. In 2014, the U.S. Library of Congress said it had collected 525 terabytes of archive data. Convert 525 terabytes to gigabytes. There are 1000 gigabytes in 1 terabyte.

74. In 2016, CERN (the European Organization for Nuclear Research) posted 300 terabytes of its data from the LHC (Large Hadron Collider) for other researchers to use. Convert 300 terabytes to gigabytes. There are 1000 gigabytes in 1 terabyte.

For Exercises 75 through 78, simplify each expression.

75. $\dfrac{-5x^4y^2}{10xy^3}$

76. $\dfrac{8xy^4}{24x^3y}$

77. $(-3nm^2)^3(-4n^2m)$

78. $(-7st^4)(-8t^2)^2$

For Exercises 79 through 82, solve each equation. Provide reasons for each step. Check the answers.

79. $t + 7 = -2t + 5$

80. $-p + 6 = 5p - 8$

81. $4(2n + 1) + 1 = 3(n + 3)$

82. $2(3y + 4) + 1 = -4(y - 2)$

For Exercises 83 and 84, find the percent increase or decrease as indicated. Round to the hundredths place if necessary.

83. A retailer purchases a shirt for $4.00 and sells it for $16.99. What is the percent increase?

84. A pair of jeans goes on sale from the regular price of $120.00 to $64.00. What is the percent decrease?

For Exercises 85 through 86, find the equation of each line using the given information.

85. Find the equation of the line parallel to $5x - 3y = -18$, passing through the point $(-3, 7)$.

86. Find the equation of the line parallel to $2x + 3y = 21$, passing through the point $(6, -1)$.

For Exercises 87 through 90, graph each line. Clearly label and scale the axes.

87. $x - 5 = 0$

88. $-x + 4 = 0$

89. $y + 2.5 = 0$

90. $-y - 3.75 = 0$

Chapter Summary

Section 5.1 Rules for Exponents

- The exponent rules are as follows:

  Product rule for exponents: $x^m x^n = x^{m+n}$ Quotient rule for exponents: $\dfrac{x^m}{x^n} = x^{m-n}$

  Power rule for exponents: $(x^m)^n = x^{mn}$ Powers of products: $(xy)^m = x^m y^m$

  Powers of quotients: $\left(\dfrac{x}{y}\right)^m = \dfrac{x^m}{y^m}$

Example 1 Simplify the following exponential expressions.

   a. $(3m^5 n)(7mn^2)$ b. $(5x^3 y^2)^2$ c. $\left(\dfrac{20a^5 b^6}{14ab^4}\right)^3$

SOLUTION

a. This is a product, so multiply the coefficients and use the product rule for exponents.

$(3m^5 n)(7mn^2)$
$= 3 \cdot 7 \cdot m^5 m \cdot n n^2$ *Multiply the constants and combine the like bases.*
$= 21 m^6 n^3$ *Add the exponents, using the product rule for exponents.*

b. This is an exponential expression being raised to a power. Apply the exponents over the product. Then use the power rule for exponents.

$(5x^3 y^2)^2$
$= (5)^2 (x^3)^2 (y^2)^2$ *Apply the exponents over the product.*
$= 25 x^6 y^4$ *Multiply the exponents, using the power rule for exponents.*

c. Simplify the quotient first and then apply the power through the remaining quotient.

$\left(\dfrac{20a^5 b^6}{14ab^4}\right)^3$ *Simplify inside the parentheses.*

$= \left(\dfrac{10 a^{5-1} b^{6-4}}{7}\right)^3$ *Subtract exponents, using the quotient rule for exponents.*

$= \left(\dfrac{10 a^4 b^2}{7}\right)^3$

$= \dfrac{(10)^3 (a^4)^3 (b^2)^3}{(7)^3}$ *Apply the exponent of 3.*

$= \dfrac{1000 a^{12} b^6}{343}$ *Multiply the exponents, using the power rule for exponents.*

Section 5.2 Negative Exponents and Scientific Notation

- When a base is raised to a **negative exponent,** find the reciprocal of that base raised to the absolute value of the exponent.

$$x^{-n} = \frac{1}{x^n}$$

- All the rules for exponents from Section 5.1 also apply to negative exponents.
- When simplifying exponential expressions, do not leave any negative exponents in the result.
- A number in the form $a \times 10^n$, where $1 \leq |a| < 10$ and n is an integer, is said to be in **scientific notation.**
- When doing calculations involving numbers in scientific notation, use the rules for exponents to simply the result.

Example 2 Simplify the following exponential expressions.

a. $7x^{-3}y$ **b.** $\dfrac{12a^3b^{-4}c^5}{15ab^3c^{-2}}$

SOLUTION **a.** To eliminate the negative exponent, we find the reciprocal of the base x and change the exponent to positive 3.

$7x^{-3}y$ Find the reciprocal of the base x.

$= \dfrac{7y}{x^3}$ The exponent is now positive.

b. This is a quotient, so use the quotient rule to subtract exponents. Then find the reciprocal of any bases with negative exponents.

$\dfrac{12a^3b^{-4}c^5}{15ab^3c^{-2}}$ Reduce the constants.

$= \dfrac{4a^2b^{-7}c^7}{5}$ Subtract the exponents using the quotient rule for exponents.

$= \dfrac{4a^2c^7}{5b^7}$ Simplify any negative exponents by finding the reciprocal of the base.

Example 3 Perform each indicated operation. Write the answers in scientific notation.

a. $8(4.5 \times 10^6)$ **b.** $(3.5 \times 10^4)(2.0 \times 10^{10})$ **c.** $\dfrac{1.2 \times 10^{18}}{4.0 \times 10^4}$

SOLUTION **a.** Multiply 8 by 4.5 and adjust the exponent of 10 to put the result in scientific notation.

$8(4.5 \times 10^6)$

$= 8 \cdot 4.5 \times 10^6$ Multiply the 8 by 4.5.

$= 36.0 \times 10^6$ The coefficient is bigger than 10, so adjust the exponent of 10.

$= 3.60 \times 10^7$ Simplify the decimal one place to the left and add 1 to the exponent.

b. Multiply the coefficients, and add the exponents, using the product rule for exponents.

$(3.5 \times 10^4)(2.0 \times 10^{10})$

$= 3.5 \cdot 2.0 \times 10^4 \times 10^{10}$

$= 7.0 \times 10^{14}$ Add the exponents using the product rule for exponents.

c. Divide the coefficients, and subtract the exponents, using the quotient rule for exponents. Adjust the power of 10 to put the result in standard scientific notation.

$$\frac{1.2 \times 10^{18}}{4.0 \times 10^{4}}$$

$$= 0.3 \times 10^{14} \quad \text{Divide the coefficients and subtract the exponents.}$$

$$= 3.0 \times 10^{13} \quad \text{Move the decimal one place to the right and subtract 1 from the exponent.}$$

Section 5.3 Adding and Subtracting Polynomials

- Any combination of terms with variables raised to positive integer exponents that are added or subtracted together is called a **polynomial.**
- Some special names for certain polynomials include **monomial, binomial,** and **trinomial.**
- The **degree** of a term is the sum of all the exponents for the variables in the term.
- The **degree of a polynomial** is the same as the degree of the highest-degree term.
- To **add polynomials,** combine the like terms from each polynomial.
- To **subtract polynomials,** change the sign of each term being subtracted and then combine like terms.

Example 4 Add or subtract the given polynomials.

a. $(3x^2 + 5x + 9) + (7x^2 - 13x + 2)$ **b.** $(4a^2 + 2b - 17) - (a^2 - 5b + 4)$

SOLUTION **a.** Add like terms.

$$(3x^2 + 5x + 9) + (7x^2 - 13x + 2)$$
$$= 3x^2 + 7x^2 + 5x - 13x + 9 + 2 \quad \text{Combine like terms.}$$
$$= 10x^2 - 8x + 11$$

b. To subtract, change the sign of each term in the second polynomial and combine like terms.

$$(4a^2 + 2b - 17) - (a^2 - 5b + 4)$$
$$= 4a^2 + 2b - 17 - a^2 + 5b - 4 \quad \text{Change the sign of each term being subtracted.}$$
$$= 4a^2 - a^2 + 2b + 5b - 17 - 4 \quad \text{Combine like terms.}$$
$$= 3a^2 + 7b - 21$$

Section 5.4 Multiplying Polynomials

- To **multiply a polynomial by a monomial,** multiply each term of the polynomial by the monomial.
- To **multiply two polynomials,** multiply each term of the first polynomial by each term of the second polynomial. Simplify by combining like terms.
- When multiplying two binomials, use the acronym **FOIL** to remember the distributive property.

$$\text{First} + \text{Outer} + \text{Inner} + \text{Last}$$

- When **squaring a binomial,** the result is a **perfect square trinomial.**

$$(a + b)^2 = a^2 + 2ab + b^2 \qquad (a - b)^2 = a^2 - 2ab + b^2$$

- When **multiplying the sum and difference of two terms,** the result is the **difference of two squares.**

$$(a + b)(a - b) = a^2 - b^2$$

Example 5 Multiply the following polynomials.

a. $(x + 7)(3x + 8)$
b. $(2x + 5)(3x^2 + 6x - 8)$
c. $(5a + 3)^2$
d. $(6m + n)(6m - n)$

SOLUTION **a.** Find the product.

$$(x + 7)(3x + 8) \qquad \text{Use the distributive property to multiply.}$$
$$= 3x^2 + 8x + 21x + 56 \qquad \text{Combine like terms.}$$
$$= 3x^2 + 29x + 56$$

b. Distribute each term from the first polynomial into each term of the second polynomial.

$$(2x + 5)(3x^2 + 6x - 8) \qquad \text{Distribute each term.}$$
$$= 2x(3x^2) + 2x(6x) + 2x(-8) + 5(3x^2) + 5(6x) + 5(-8)$$
$$= 6x^3 + 12x^2 - 16x + 15x^2 + 30x - 40 \qquad \text{Combine like terms.}$$
$$= 6x^3 + 27x^2 + 14x - 40$$

c. We are squaring a binomial, so the result is a perfect square trinomial.

$$(5a + 3)^2 = 25a^2 + 30a + 9 \qquad \text{Use the perfect square trinomial pattern.}$$

d. This is the product of the sum and difference of the same two terms. The result is the difference of two squares.

$$(6m + n)(6m - n) = 36m^2 - n^2 \qquad \text{Use the difference of two squares pattern.}$$

Section 5.5 Dividing Polynomials

- When **dividing a polynomial by a monomial,** separate the polynomial into terms. Divide each term by the monomial.
- When **dividing a polynomial by another polynomial,** use the long division algorithm.
- When using long division, make sure the terms are in descending order and that no terms are missing.

Example 6 Divide the following polynomials.

a. $\dfrac{15x^2 + 5x + 8}{5x}$
b. $(x^3 - 8x + 7) \div (x + 2)$

SOLUTION **a.** Since we are dividing by a monomial, separate the polynomial, and divide each term by the monomial.

$$\frac{15x^2 + 5x + 8}{5x}$$

$$= \frac{15x^2}{5x} + \frac{5x}{5x} + \frac{8}{5x} \qquad \text{Separate each term and simplify.}$$

$$= 3x + 1 + \frac{8}{5x}$$

b. Use long division. The second-degree term is missing, so insert $0x^2$ to hold its place in the division. The first term in the divisor $(x + 2)$ is x. Divide the first term in the dividend (numerator) by x: $\frac{x^3}{x} = x^2$. This yields the first term in the quotient.

$\,x^2$ $x + 2\overline{)x^3 + 0x^2 - 8x + 7}$ $\underline{-(x^3 + 2x^2)}$	$\,x^2$ $x + 2\overline{)x^3 + 0x^2 - 8x + 7}$ $\underline{-x^3 - 2x^2}$ $-2x^2 - 8x$	To subtract, distribute the negative sign and combine like terms. Bring down $-8x$.

Continue dividing $-2x^2$ by x: $\frac{-2x^2}{x} = -2x$. This gives the second term in the quotient.

$\,x^2 - 2x$ $x + 2\overline{)x^3 + 0x^2 - 8x + 7}$ $\underline{-x^3 - 2x^2}$ $-2x^2 - 8x$ $\underline{-(-2x^2 - 4x)}$	$\,x^2 - 2x$ $x + 2\overline{)x^3 + 0x^2 - 8x + 7}$ $\underline{-x^3 - 2x^2}$ $-2x^2 - 8x$ $\underline{2x^2 + 4x}$ $-4x + 7$	To subtract, distribute the negative sign and combine like terms. Bring down $+7$.

Continue dividing $-4x$ by x: $\frac{-4x}{x} = -4$. This gives the third term in the quotient.

$\,x^2 - 2x - 4$ $x + 2\overline{)x^3 + 0x^2 - 8x + 7}$ $\underline{-x^3 - 2x^2}$ $-2x^2 - 8x$ $\underline{2x^2 + 4x}$ $-4x + 7$ $\underline{-(-4x - 8)}$	$\,x^2 - 2x - 4$ $x + 2\overline{)x^3 + 0x^2 - 8x + 7}$ $\underline{-x^3 - 2x^2}$ $-2x^2 - 8x$ $\underline{2x^2 + 4x}$ $-4x + 7$ $\underline{4x + 8}$ 15	To subtract, distribute the negative sign and combine like terms. The remainder is 15, so we can stop dividing.

Therefore, $(x^3 - 8x + 7) \div (x + 2) = x^2 - 2x - 4 + \dfrac{15}{x + 2}$.

Chapter Review Exercises

For Exercises 1 through 20, simplify each exponential expression. Write the final answers using positive exponents only.

1. $4x^5 \cdot 7x^3$
2. $-3y^4 \cdot 8y$ [5.1]
3. $(3x^4y^3)(8x^5y^2)$
4. $(-4x^2y^7)(7xy^3)$ [5.1]
5. $\dfrac{-4x^5y^3}{16x^4y}$
6. $\dfrac{7x^7y^3}{-14xy^3}$ [5.1]
7. $\dfrac{c^5c^9d^5d^7}{c^2cd^8}$
8. $\dfrac{25a^3ab^6}{50a^5b^8}$ [5.1]
9. $(5a^2b^5)^3$
10. $(-2x^2y^3)^2$ [5.1]
11. $\left(\dfrac{-6a^3b^4}{12a^2b}\right)^3$
12. $\left(\dfrac{20m^5n^3}{14m^2n}\right)^4$ [5.1]
13. $(-7x^2y^5)^3(3x^3y)^2$
14. $(6x^3y^2)^2(-5x^3y)^4$ [5.1]
15. $\dfrac{-15x^3y^9}{6x^8y^5}$
16. $\dfrac{10a^{10}b^3}{30a^7b^5}$ [5.2]
17. $\dfrac{-9x^8y^{-4}}{x^{12}y^9}$
18. $\dfrac{-12m^5n^{-7}}{mn^3}$ [5.2]
19. $\dfrac{c^4d^{-5}}{c^{-2}d^3}$
20. $\dfrac{y^{-3}z^8}{y^{-7}z^{-5}}$ [5.2]

For Exercises 21 through 28, convert each number as indicated.

21. Convert 52,600,000,000 to scientific notation. [5.2]
22. Convert 65,120,000 to scientific notation. [5.2]
23. Convert the following to scientific notation. The average distance from Neptune to the Sun is about 4,550,000,000 km. [5.2]
24. Convert the following to scientific notation. The width of one helix of a DNA molecule is 0.000000002 meter. [5.2]
25. Convert 6.024×10^{-6} to standard form. [5.2]
26. Convert 4.3×10^7 to standard form. [5.2]
27. Convert to standard form. The width of a flu virus is 3.0×10^{-7} meter. [5.2]
28. Convert the following to standard form. The approximate width of a carbon nanotube shaft of one of the first nanomotors created is 7.5×10^{-9} meter. [5.2]

For Exercises 29 through 34, perform the indicated operations. Write the answers using scientific notation.

29. $40(3.5 \times 10^{18})$ [5.2]
30. $-30(5.7 \times 10^{-8})$ [5.2]
31. $\dfrac{1.5 \times 10^{-12}}{3.0 \times 10^8}$ [5.2]
32. $\dfrac{6.2 \times 10^{20}}{2.0 \times 10^4}$ [5.2]
33. $(2.5 \times 10^{10})(6.0 \times 10^8)$ [5.2]
34. $(4.0 \times 10^{-12})(3.6 \times 10^{-5})$ [5.2]
35. For the polynomial $8x^2 + 5x + 10$, state the number of terms and find the degree of the entire polynomial. [5.3]
36. For the polynomial $5x + 8$, state the number of terms and find the degree of the entire polynomial. [5.3]
37. For the polynomial $-5a^3b + 6a^5 - 7b^2 + 8$, determine the degree of each term and of the entire polynomial. [5.3]
38. For the polynomial $20a^2b^3c + 18ab^2c^2 - 9bc^8$, determine the degree of each term and of the entire polynomial. [5.3]
39. Is $7x - 16y = -23$ a linear equation? Explain your response. [5.3]
40. Is $y = 3x - 8$ a linear equation? Explain your response. [5.3]
41. Is $y = 3x^2 + 10$ a linear equation? Explain your response. [5.3]
42. Is $y = 25 - x^2$ a linear equation? Explain your response. [5.3]
43. Find a polynomial for the perimeter of the following figure. [5.3]

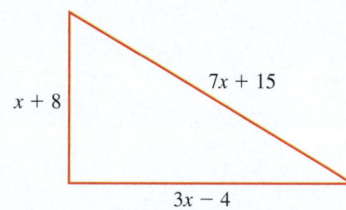

44. Find a polynomial for the perimeter of the following figure. [5.3]

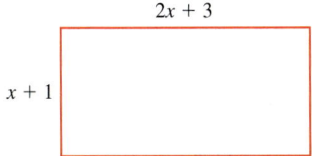

45. The number of students attending a math class after w weeks of the semester can be represented by $42 - 0.5w$. The number of students attending an English class after w weeks of the semester can be represented by $25 - 0.4w$. Find a polynomial for how many more students attend the math class than the English class. [5.3]

46. A group of students is planning a spring break trip. They find out that airfare will be $350 per student plus $100 a day for food and lodging. [5.3]
  a. Find the cost for one student to go on the trip for 5 days.
  b. How much will it cost for n students to go on the trip for 5 days?

For Exercises 47 through 50, add or subtract the given polynomials.

47. $(5a^2 + 6a - 12) + (3a^2 - 10a + 5)$ [5.3]

48. $(3x^2 - 8x + 14) + (7x^2 + 3x - 2)$ [5.3]

49. $(5x^2 - 8x + 2) - (2x^2 - 8)$ [5.3]

50. $(8n^2 - 7n - 5) - (15n^2 - 4n)$ [5.3]

51. Find a polynomial that represents the area of the following rectangle. [5.4]

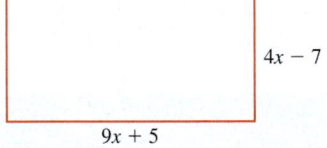

52. The price of gasoline in dollars per gallon, w weeks into the year, can be represented by $2.15 + 0.04w$. The number of gallons used by the Vista Unified School District during week w of the year can be represented by $150 + 5w$. Find a polynomial that represents the cost for the gasoline used by Vista Unified School District during week w of the year. [5.4]

For Exercises 53 through 64, multiply the given polynomials and simplify.

53. $5x(7x + 8)$ [5.4]

54. $-8x(7x^2 + 2x - 4)$ [5.4]

55. $(2x + 5)(3x - 6)$ [5.4]

56. $(4x - 7)(2x - 3)$ [5.4]

57. $(3x - 7)(3x + 7)$ [5.4]

58. $(4a + 5b)(4a - 5b)$ [5.4]

59. $(x + 2)(x^2 - 3x + 7)$ [5.4]

60. $(x - 4)(x^2 + 5x - 3)$ [5.4]

61. $(3x + 7)^2$ [5.4]

62. $(6x - 9)^2$ [5.4]

63. $5m(m + 4)(3m - 8)$ [5.4]

64. $-2a(3a + 5)(5a - 8)$ [5.4]

65. Find an expression that represents the length of this rectangle. [5.5]

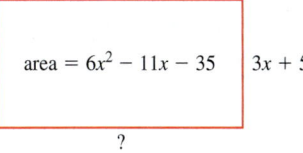

66. Find an expression that represents the length of this rectangle. [5.5]

area = $7x^2 - 31x + 12$ | $x - 4$
?

For Exercises 67 through 72, divide the following expressions and simplify.

67. $\dfrac{20x^2 + 6x - 8}{2x}$ [5.5]

68. $\dfrac{15a^2b^3 - 18ab^2 + 9b}{3ab}$ [5.5]

69. $(x^2 + 4x - 21) \div (x - 3)$ [5.5]

70. $(6a^2 + 26a + 28) \div (2a + 4)$ [5.5]

71. $(m^2 - 8m + 20) \div (m - 3)$ [5.5]

72. $(8p^3 + 125) \div (2p + 5)$ [5.5]

Chapter Test

This chapter test should take approximately one hour to complete. Read each question carefully. Show all of your work.

For Exercises 1 through 4, simplify each exponential expression. Write the answers using positive exponents only.

1. $(5x^3y^5)(4x^7y^2)$
2. $\dfrac{m^3m^4n^2n^4}{m^5n^5}$
3. $(2a^{-4}b^3)^3$
4. $\dfrac{20x^{-3}y^4}{14x^5y^{-3}}$

5. Convert the following to scientific notation: The diameter of an average red blood cell is 0.000007 meter.
6. Convert the following to standard form: The average distance from the Sun to Pluto is roughly 5.9×10^9 km.

For Exercises 7 through 9, simplify each exponential expression. Write the answers using scientific notation.

7. $5(7.3 \times 10^9)$
8. $(3.0 \times 10^7)(4.5 \times 10^4)$
9. $\dfrac{6.5 \times 10^{-8}}{5.0 \times 10^6}$

For Exercises 10 through 18, perform the indicated operations.

10. $(5x^2 + 2x + 8) + (3x - 12)$
11. $(7a^2 - 8a + 10) - (4a^2 + 5a - 3)$
12. $7x(2x^2 + 3x - 7)$
13. $(2x + 5)(2x - 5)$
14. $(3x - 4)(4x^2 + 2x + 5)$
15. $(6x - 7)^2$
16. $\dfrac{4m^2 + 6m - 8}{2m}$
17. $(2d^2 + 7d - 15) \div (d + 5)$
18. $(3x^3 + 7x^2 - 18) \div (x + 2)$

For Exercises 19 and 20, use the given rectangle to find the desired expression.

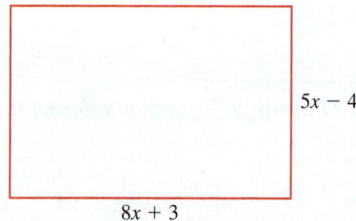

19. Write an expression for the area of the rectangle. Simplify.
20. Write an expression for the perimeter of the rectangle. Simplify.

Chapter Projects

■ Using the Box Method to Multiply Binomials

Written Project
One or more people

To represent multiplying the binomials $(x + 5)(x + 3)$ through a drawing, we illustrate the idea of finding the area of a rectangle (box).

The binomial $x + 5$ is represented by the length of the rectangle. The top of the rectangle has two lengths: one of length x and one of length 5. Adding these two lengths yields $x + 5$. The width of the rectangle represents the binomial $x + 3$. Recall that the area of a rectangle is length · width. Finding the area of each of the four smaller rectangles yields the following:

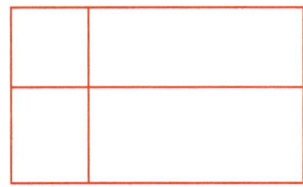

To find the area of the entire (big) rectangle, add up the areas of the four smaller rectangles.

$$x^2 + 5x + 3x + 15$$
$$= x^2 + 8x + 15 \quad \text{Simplify by adding the two like terms.}$$

Therefore, $(x + 5)(x + 3) = x^2 + 8x + 15$.

Problem: Multiply $(x + 3)(x + 7)$ using the box method.

1. Using the given binomials, $x + 3$ and $x + 7$, set the dimensions of the rectangle below.

2. Find the areas of the four smaller rectangles.
3. Find the area of the large rectangle. Recall that the area of the large rectangle is equal to the sum of the areas of the four smaller rectangles. Simplify.
4. Therefore, what does $(x + 3)(x + 7)$ equal?

■ Why Does That Exponent Rule Work?

Written Project

One or more people

In this project, you will explore why the power of products exponent rule works.

Power of Products Exponent Rule: $(a \cdot b)^n = a^n \cdot b^n$

Consider the product $(3x)^4$. This is equivalent to $(3x)^4 = (3 \cdot x)^4$.

1. Rewrite $(3x)^4$ in *expanded* form. Recall that the expanded form of 2^3 is $2 \cdot 2 \cdot 2$.
2. Use the commutative property of multiplication to rewrite the expanded form of $(3x)^4$ with like factors. The commutative property of multiplication justifies $(2a)(2a) = 2 \cdot 2 \cdot a \cdot a$.
3. Rewrite your expression from step 2 in exponential form.
4. Use the power of products rule to rewrite the original expression $(3x)^4$.
5. Are the answers to steps 3 and 4 the same?

Write a Review of a Section for Presentation

Research Project

One or more people

What you will need

- This presentation may be done in class or as an online video.
- If it is done as a video, post it on a website where it may be easily viewed by the class.
- You might want to use homework or other review problems from the book.
- Make it creative and fun.

Create a 5-minute review presentation of one section from Chapter 5. The format of the presentation can be a poster presentation, a blackboard presentation from notes, an online video, or a game format (e.g., math jeopardy). The presentation should include the following:

1. Any important formulas in the chapter
2. Important skills in the chapter, backed up by examples
3. Explanation of terms used (for example, what is *modeling*?)
4. Common mistakes and how to avoid them

Factoring

6

6.1 What It Means to Factor
6.2 Factoring Trinomials
6.3 Factoring Special Forms
6.4 Solving Quadratic Equations by Factoring

Students of business and management analyze profit, revenue, and cost equations. These equations are often *quadratic equations*. For example, an appliance manufacturer may want to know how many appliances they have to sell in order to maintain a certain level of average profit per appliance. In this chapter, we begin our study of how to factor quadratic expressions and solve quadratic equations.

6.1 What It Means to Factor

LEARNING OBJECTIVES
- Find the greatest common factor.
- Factor out the greatest common factor.
- Factor by grouping.
- Explain how to factor completely.

In Section R.2, we discussed *factoring* numbers. Recall that when asked to factor a number, that means to rewrite that number as a product. For example, the prime factors of 10 are 5 and 2 because $5 \cdot 2 = 10$. *Factoring* 10 means to rewrite it as the product (multiplication) of its factors, 5 and 2.

Instructions	What to do
Multiply 5 and 2.	Given 5 and 2, multiply them to get 10. $5 \cdot 2 = 10$
Factor 10.	Given 10, find the factors that multiply to 10. $10 = 5 \cdot 2$

> **DEFINITIONS**
>
> **Factor an Integer**   To factor an integer means to rewrite it as the product of other integers that, when multiplied together, result in the original integer.
>
> **Factor a Polynomial**   To factor a polynomial means to rewrite the polynomial expression as the product of simpler polynomials that, when multiplied together, result in the original polynomial.

When factoring an expression, it is best to begin by looking for the **greatest common factor**, or GCF, which helps to break down all the terms.

> **DEFINITION**
>
> **Greatest Common Factor (GCF) of Integers or Monomials**   The greatest common factor, or GCF, of a set of integers or monomials is the largest factor that is shared in common with all of the elements of the set.

To find the GCF of a set of integers or monomials, use the following steps.

> **Steps to Find the GCF of a Set of Integers or Monomials**
> 1. Find the prime factorization of each integer or coefficient. Write repeated factors in exponential form.
> 2. Circle the *common* factors of all the elements in the set.
> 3. The GCF equals the product of the common factors, each raised to the *lowest power* that appears.

Example 1 Finding the GCF of a set of integers or monomials

For each set of monomials, find the GCF.

a. $-8x^2, 24x^3$ **b.** $-5ab^2, 15ab, 20a^2b$

SOLUTION

a. Step 1 Find the prime factorization of each integer or coefficient. Write repeated factors in exponential form.

The two monomials are $-8x^2$ and $24x^3$. Factoring each term yields

$$-8x^2 = -1 \cdot 2 \cdot 4 \cdot x^2 = -1 \cdot 2 \cdot 2 \cdot 2 \cdot x^2 = -1 \cdot 2^3 \cdot x^2$$
$$24x^3 = 4 \cdot 6 \cdot x^3 = 2 \cdot 2 \cdot 2 \cdot 3 \cdot x^3 = 2^3 \cdot 3 \cdot x^3$$

Step 2 Circle the common factors of all the elements in the set.

$$-8x^2 = -1 \cdot \boxed{2^3 \cdot x^2}$$
$$24x^3 = \boxed{2^3} \cdot 3 \cdot \boxed{x^3}$$

Step 3 The GCF equals the product of the common factors, each raised to the lowest power that appears.

Therefore, the GCF is

$$2^3 \cdot x^2 = 8x^2$$

Notice in this example that both terms have a factor of x and that the GCF has in it the lowest power of x that appeared in both terms, namely, x^2.

b. The three monomials are $-5ab^2$, $15ab$, and $20a^2b$. Factoring each monomial yields

$$-5ab^2 = -1 \cdot 5 \cdot a \cdot b^2$$
$$15ab = 3 \cdot 5 \cdot a \cdot b$$
$$20a^2b = 4 \cdot 5 \cdot a^2 \cdot b = 2 \cdot 2 \cdot 5 \cdot a^2 \cdot b = 2^2 \cdot 5 \cdot a^2 \cdot b$$

Circling the common factors in these factorizations yields

$$-5ab^2 = -1 \cdot \boxed{5 \cdot a \cdot b^2}$$
$$15ab = 3 \cdot \boxed{5 \cdot a \cdot b}$$
$$20a^2b = 4 \cdot 5 \cdot a^2 \cdot b = 2 \cdot 2 \cdot 5 \cdot a^2 \cdot b = 2^2 \cdot \boxed{5 \cdot a^2 \cdot b}$$

Selecting the lowest powers on the common factors yields the GCF $5ab$.

PRACTICE PROBLEM FOR EXAMPLE 1

For each set of integers or monomials, find the GCF.

a. $-65, 20, 35$ **b.** $12y^2, 40y$ **c.** $9xy^3, -27x^2y, 30x^2y^2$

DEFINITION

Greatest Common Factor (GCF) of a Polynomial   The greatest common factor, or GCF, of a polynomial expression is the largest factor shared in common with all of the terms.

To find the GCF of a polynomial expression, we rely on the following steps.

> **Steps to Find the GCF of a Polynomial Expression**
> 1. Factor each term completely. Factor each integer into its prime factors.
> 2. The GCF equals the product of the common factors from each term, each raised to the lowest power that appears.

Example 2 Finding the GCF of a polynomial

For each polynomial expression, find the GCF.

a. $4x^2 + 16x$ **b.** $-x - 7$

SOLUTION

a. Step 1 Factor each term completely. Factor each integer into its prime factors.

The two terms are $4x^2$ and $16x$. Factoring each term yields

$$4x^2 = 2 \cdot 2 \cdot x \cdot x = 2^2 \cdot x^2$$
$$16x = 2 \cdot 2 \cdot 2 \cdot 2 \cdot x = 2^4 \cdot x$$

Step 2 The GCF equals the product of the common factors from each term, each raised to the lowest power that appears.

GCF is $2^2 \cdot x$, or $4x$.

b. The two terms are $-x$ and -7. Factoring each term yields

$$-x = -1 \cdot x$$
$$-7 = -1 \cdot 7$$

Therefore, the GCF is -1.

PRACTICE PROBLEM FOR EXAMPLE 2

For each polynomial expression, find the GCF.

a. $8x^4 - 16x^2$ **b.** $-25y^3 - 50y^2$

Example 3 Finding the GCF of more complicated expressions

For each polynomial expression, find the GCF.

a. $48ab + 16bc - 32b$ **b.** $3x^3 - 6x - 9x^2$ **c.** $25mn^2 - 50m^2n$

SOLUTION

a. There are three terms in this expression: $48ab$, $16bc$, and $-32b$. Factoring each term yields

$$48ab = 2 \cdot 2 \cdot 2 \cdot 2 \cdot 3 \cdot a \cdot b = 2^4 \cdot 3 \cdot a \cdot b$$
$$16bc = 2 \cdot 2 \cdot 2 \cdot 2 \cdot b \cdot c = 2^4 \cdot b \cdot c$$
$$-32b = -1 \cdot 2 \cdot 2 \cdot 2 \cdot 2 \cdot 2 \cdot b = -1 \cdot 2^5 \cdot b$$

Therefore, the GCF is $2^4 \cdot b$, or $16b$.

b. There are three terms in this expression: $3x^3$, $-6x$, and $-9x^2$. Factoring each term yields

$$3x^3 = 3 \cdot x^3$$
$$-6x = -1 \cdot 2 \cdot 3 \cdot x$$
$$-9x^2 = -1 \cdot 3^2 \cdot x^2$$

Therefore, the GCF is $3x$. Notice that x is a variable common to all terms and the GCF contains the lowest power of x that appears, namely, x.

c. There are two terms in this expression: $25mn^2$ and $-50m^2n$. There are two variables in this expression, so check whether both variables are common to all terms. Then select the lowest power of each variable that appears. Factoring each term yields

$$25mn^2 = 5^2 \cdot m \cdot n^2$$
$$-50m^2n = -1 \cdot 2 \cdot 5^2 \cdot m^2 \cdot n$$

Because both variables are common to all terms, we have to include both in the GCF. Therefore, the GCF is $5^2 \cdot m \cdot n$, or $25mn$.

PRACTICE PROBLEM FOR EXAMPLE 3

For each polynomial expression, find the GCF.

a. $18xy^2 + 24x^2y - 9xy$ **b.** $40p^2q^2 - 30pq$

Factoring Out the Greatest Common Factor

We now are going to learn how to factor polynomial expressions. In Chapter 5, we learned how to multiply polynomials. As we saw in the examples above, multiplication and factorization are closely related. Because factoring and multiplication are closely related, we begin our study of factoring by returning to the first method we learned to multiply polynomial expressions: the distributive property.

Multiply: $4(3x - 5) = 4 \cdot 3x - 4 \cdot 5$ Use the distributive property.
$ = 12x - 20$

When instructed to factor $12x - 20$, begin with the product and work backwards to rewrite it as the multiplication of factors.

Factor: $12x - 20 = 4 \cdot 3x - 4 \cdot 5$ The GCF is 4. Rewrite each term as a product.
$ = 4(3x - 5)$ Factor out the GCF. Recall that $4(3x - 5) = 4 \cdot (3x - 5)$.

The form $4(3x - 5)$ is called the **factored form** of this polynomial. Notice how the original polynomial expression, $12x - 20$, has been broken down into a product of simpler polynomial expressions, $4(3x - 5)$.

CONCEPT INVESTIGATION
Are they the same?

Consider the two polynomial expressions $15x - 25$ and $5(3x - 5)$. Substitute in each of the following values of x, and evaluate each of the expressions. Then answer the question of whether the two expressions evaluated for a particular value of x are the same or different.

536 CHAPTER 6 Factoring

The first row has been completed as an example to get you started.

x	Original Polynomial Expression $15x - 25$	Factored Form $5(3x - 5)$	Same or Different?
1	$15(1) - 25 = 15 - 25 = -10$	$5(3(1) - 5) = 5(3 - 5) = 5(-2) = -10$	Same
5			
0			
−2			
−3			

1. Do you think that the two expressions represent the same quantity? Explain.
2. Multiply out the factored form, $5(3x - 5)$. Do you get the original polynomial expression back?

From this Concept Investigation, we see that a polynomial expression and its factored form are equal. The factored form has been broken down into simpler polynomials.
To factor the GCF from a polynomial expression, use the following steps.

> **Steps to Factor the GCF from a Polynomial Expression**
> 1. Determine the GCF of all the terms.
> 2. Rewrite each term as the product of the GCF and the remaining factors.
> 3. Factor out the GCF using the distributive property.
>
> *Note:* Check the result by multiplying the expressions out. Make sure the signs are correct in step 3.

After factoring out the GCF, there will be the same number of terms inside the parentheses as in the original expression.

Factoring Polynomials with Three Terms $(x+2)(x-5)$

GCF First

Factor out the GCF first, if there is one.

$6x^2 + 9x - 15$
$= 3(2x^2 + 3x - 5)$

Example 4 Factoring out the GCF

For each expression, factor out the GCF. Check the result.

a. $8x - 12$ b. $3x^3 - 12x^2 - 6x$ c. $9a^2 + 9a$ d. $15ab^2 - 25ab$

SOLUTION $(x+2)(x-5)$

a. **Step 1** Determine the GCF of all the terms.

The GCF for $8x - 12$ is 4.

Step 2 Rewrite each term as the product of the GCF and the remaining factors.

As we factor out the GCF, note that there are two terms in this expression. This means that two terms will be in the parentheses.

$8x - 12 = 4 \cdot 2 \cdot x - 4 \cdot 3$ *The GCF is 4. Rewrite each term.*

Step 3 Factor out the GCF using the distributive property.

$= 4(2x - 3)$ *Factored form.*

To check the result, multiply the expression using the distributive property.

$$\text{Check:} \quad 4(2x - 3) = 4 \cdot 2x - 4 \cdot 3$$
$$= 8x - 12$$

b. The GCF for $3x^3 - 12x^2 - 6x$ is $3x$. As we factor out the GCF, note that there are three terms in this expression. This means that there will be three terms in the parentheses.

$$3x^3 - 12x^2 - 6x = (3x) \cdot x^2 - (3x) \cdot 4x - (3x) \cdot 2 \quad \text{The GCF is } 3x. \text{ Rewrite each term.}$$
$$= 3x(x^2 - 4x - 2) \quad \text{Factored form.}$$

To check the result, multiply using the distributive property. Remember that when multiplying factors with the same base, add the exponents.

$$\text{Check:} \quad 3x(x^2 - 4x - 2) = 3x(x^2) - 3x(4x) - 3x(2)$$
$$= 3x^3 - 12x^2 - 6x$$

c. The GCF for $9a^2 + 9a$ is $9a$. To check this answer, let's factor both terms and look at what is in common.

$$9a^2 = 3 \cdot 3 \cdot a^2 = (3^2 \cdot a^2)$$
$$9a = 3 \cdot 3 \cdot a = (3^2 \cdot a)$$

Both terms have 3^2 and a factor of a in common. The lowest power of a that appears in both terms is 1, so the GCF is $3^2 a = 9a$.

There are two terms in this expression, so there are two terms in the parentheses. The second term has only a 1 remaining, because the GCF is $9a$ and $9a = 9a \cdot 1$. When we factor out the GCF, there will be a 1 in the place of the second term.

$$9a^2 + 9a = (9a) \cdot a + (9a) \cdot 1 \quad \text{The GCF is } 9a. \text{ Rewrite each term.}$$
$$= 9a(a + 1) \quad \text{Factored form.}$$

To check the result, multiply using the distributive property.

$$\text{Check:} \quad 9a(a + 1) = 9a(a) + 9a(1)$$
$$= 9a^2 + 9a$$

d. The GCF for $15ab^2 - 25ab$ is $5ab$. Factor out the GCF. There are two terms in this expression, so there are two terms in the parentheses.

$$15ab^2 - 25ab = (5ab) \cdot 3b - (5ab) \cdot 5 \quad \text{The GCF is } 5ab. \text{ Rewrite each term.}$$
$$= 5ab(3b - 5) \quad \text{Factored form.}$$

To check the result, multiply using the distributive property.

$$\text{Check:} \quad 5ab(3b - 5) = (5ab) \cdot 3b - (5ab) \cdot 5$$
$$= 15ab^2 - 25ab$$

PRACTICE PROBLEM FOR EXAMPLE 4

For each expression, factor out the GCF. Check the result.

a. $5x - 35$ **b.** $2y^3 - 6y^2 + 12y^4$ **c.** $6x^3 - 2x^2 + 2x$

In a factored polynomial, it is common practice to give the coefficient of the highest-degree term a positive sign. To follow this convention, if the highest-degree term is negative, include a factor of -1 in the GCF. When factoring out a negative number, *reverse* the signs of all of the terms. The practice of writing polynomials with the highest-degree coefficient positive makes comparing expressions easier.

■ **Skill Connection**

Simplifying exponents

To check a factorization, remember the Product Rule for Exponents, $x^n \cdot x^m = x^{n+m}$. When we are multiplying expressions that have the same base, add the exponents to simplify the expression.

■ **Connecting the Concepts**

What happens when we factor everything out of a term?

When factoring out the GCF, we are dividing each term by the GCF. When a term is the GCF without any other factors, factoring out the GCF will result in a 1 in place of that term.

$$8x^2y + 4xy \quad \text{GCF is } 4xy.$$
$$= 4xy(2x + 1)$$

Factoring out the GCF is like dividing each term by $4xy$.

$$\frac{8x^2y}{4xy} = 2x \quad \frac{4xy}{4xy} = 1$$

Since the second term was the GCF, it was replaced by a 1 during factoring.

Example 5 Factoring out a negative number

For each polynomial, factor out the GCF. Make sure the coefficient of the highest-degree term is positive after factoring.

a. $-x + 7$ **b.** $-x - 7$ **c.** $-5x + 10$

SOLUTION $(x+2)(x-5)$

a. The two terms are $-x$ and 7. There are no common factors between $-x$ and 7. However, because the leading term is negative, $-x$, factor out -1. The GCF is -1.

$$-x + 7 = (-1) \cdot x + (-1) \cdot (-7) \quad \text{The GCF is } -1. \text{ Rewrite each term.}$$
$$= -1(x + (-7)) \quad \text{Factored form.}$$
$$= -(x - 7) \quad \text{Simplify.}$$

Inside the parentheses, all of the signs have been changed from the original expression. To check the result, multiply out the factored form.

Check: $-(x - 7) = -1 \cdot x - (-1) \cdot 7$
$$= -x + 7$$

b. The two terms, $-x$ and -7, have no common factors other than -1. So the GCF is -1.

$$-x - 7 = (-1) \cdot x + (-1) \cdot (7) \quad \text{The GCF is } -1. \text{ Rewrite each term.}$$
$$= -1(x + 7) \quad \text{Factored form.}$$
$$= -(x + 7) \quad \text{Simplify.}$$

Notice that when a negative sign has been factored out, all of the signs on the terms inside the parentheses have been changed when compared to the original expression.

To check the result, multiply out the factored form.

Check: $-(x + 7) = -1 \cdot x + (-1) \cdot 7$
$$= -x - 7$$

c. These two terms have a 5 in common. The highest-degree term, x, has a negative coefficient. Factor out a -5.

$$-5x + 10 = -5(x - 2)$$

To check the result, multiply out the factored form.

Check: $-5(x - 2) = -5 \cdot x - (-5) \cdot 2$
$$= -5x + 10$$

PRACTICE PROBLEM FOR EXAMPLE 5

For each polynomial, factor out the GCF. Make sure the coefficient of the highest-degree term is positive after factoring.

a. $-x^2 - 5$ **b.** $-y + 16$ **c.** $-16n + 48$

In some expressions, the GCF may contain more than one term. We still factor out the GCF in the same way as before, even though the GCF looks more complicated.

SECTION 6.1 What It Means to Factor 539

Example 6 Factoring out a GCF that has terms

For each expression, factor out the GCF. Check the result.

a. $x(y + 2) - 2(y + 2)$ **b.** $3x(x - 2) + 4(x - 2)$

SOLUTION $(x+2)(x-5)$

a. The original polynomial has two terms. They are circled below. Looking carefully at these two terms, we see that they have a common factor of $(y + 2)$.

$\boxed{x(y + 2)} - \boxed{2(y + 2)}$ The two terms are circled.
$x(y + 2) - 2(y + 2)$ The GCF is $y + 2$.

Factor out the GCF and rewrite the expression in factored form.

$x(y + 2) - 2(y + 2)$ Factor the GCF out of each term.
$(y + 2)(x - 2)$ This is factored form. It is a product.

To check the result, multiply out the factored form.

Check: $(y + 2)(x - 2) = xy - 2y + 2x - 4$ Use FOIL.

This does not match the original problem. Multiply out the original problem to see whether they are equivalent expressions.

$x(y + 2) - 2(y + 2) = xy + 2x - 2y - 4$
$ = xy - 2y + 2x - 4$ Rewrite using the commutative property of addition.

The expressions are the same, so the answer is correct.

b. The original polynomial has two terms. They are circled below. Looking carefully at these two terms, we see that the common factor is $(x - 2)$.

$\boxed{3x(x - 2)} + \boxed{4(x - 2)}$ The two terms are circled.
$3x(x - 2) + 4(x - 2)$ The GCF is $x - 2$.

Factor out the GCF and rewrite the expression in factored form.

$3x(x - 2) + 4(x - 2)$ Factor the GCF out of each term.
$(x - 2)(3x + 4)$ This is factored form. It is a product.

To check the result, multiply out the factored form.

Check: $(x - 2)(3x + 4) = 3x^2 + 4x - 6x - 8$ Use FOIL and combine like terms.
$ = 3x^2 - 2x - 8$

Once again, this does not match the original problem. Let's multiply out the original problem to see whether the two expressions are the same.

$3x(x - 2) + 4(x - 2) = 3x^2 - 6x + 4x - 8$ Multiply and combine like terms.
$ = 3x^2 - 2x - 8$

The expressions are the same, so the answer is correct.

PRACTICE PROBLEM FOR EXAMPLE 6

For each expression, factor out the GCF. Check the result.

a. $a(2 - b) - 3(2 - b)$ **b.** $x(x + 1) - 5(x + 1)$

Factoring by Grouping

Another useful factoring technique is called **factoring by grouping.** This technique is used on expressions that have four terms. The key idea is to group the first and second terms and factor them, and group the third and fourth terms and factor them. Finish up by factoring out the GCF from the remaining expression. Think of grouping the terms as separating terms two at a time.

$4x^2 + 4x + 5x + 5$
$= (4x^2 + 4x) + (5x + 5)$ Group the first two terms and the last two terms.
$= 4x(x + 1) + 5(x + 1)$ Factor the first group and the second group.
$= (x + 1)(4x + 5)$ Factor out the GCF.

Steps to Factor by Grouping

1. Check to determine whether or not there are four terms.
2. Factor out any GCF of all four terms, if needed.
3. Group the first two terms and the last two terms.
4. Factor the first two terms. Factor the last two terms.
5. Factor out the GCF of the remaining expression.

Note: Check the result by multiplying the expressions out.

Factoring Polynomials with Four Terms

GCF First

Factor out the GCF first, if there is one.
$6x^2 + 4x - 8x + 10$
$= 2(3x^2 + 2x - 4x + 5)$

Factor by Grouping
$12x^2 + 8x - 21x - 14$
$= (4x - 7)(3x + 2)$

Example 7 — Factor by grouping

Factor each expression. Check the result.

a. $3x^2 - 6x + 5x - 10$ **b.** $2xy + 8x - y - 4$ **c.** $6a^2 + 3ab - 4ab - 2b^2$

SOLUTION

a. Step 1 Check to determine whether or not there are four terms.

This expression has four terms, so we will try to factor by grouping.

Step 2 Factor out any GCF of all four terms if needed.

There is no GCF for these four terms.

Step 3 Group the first two terms and the last two terms.

$(3x^2 - 6x) + (5x - 10)$

Step 4 Factor the first two terms. Factor the last two terms.

$= 3x(x - 2) + 5(x - 2)$

Step 5 Factor out the GCF of the remaining expression.

Factor out $(x - 2)$, which is common to both groups.

$= (x - 2)(3x + 5)$

To check the result, multiply it out.

Check: $(x - 2)(3x + 5) = 3x^2 + 5x - 6x - 10$ Use FOIL.
$= 3x^2 - x - 10$ Combine like terms.

Connecting the Concepts

Why can factors appear in a different order?

Sometimes, when we factor and check our answer, the factors are in a different order from the answer. Remember that we can multiply in any order, using the commutative property of multiplication. So

$(x + 1)(4x - 5)$
$= (4x - 5)(x + 1)$

This does not look like the original problem. Simplify the original problem to see whether it is the same.

$$3x^2 - 6x + 5x - 10 \quad \text{Combine like terms.}$$
$$= 3x^2 - x - 10$$

The expressions are the same, so the factorization is correct.

b. The given expression has four terms, so try to factor by grouping. When grouping, we initially put a plus sign between the groups. Then we factor the GCF of -1 out of the second group.

$(2xy + 8x) + (-y - 4)$ *Group the first two terms and the last two terms.*
$= 2x(y + 4) - 1(y + 4)$ *To factor out a negative, change both signs.*
$= (y + 4)(2x + (-1))$ *Factor out $(y + 4)$, which is common to both groups.*
$= (y + 4)(2x - 1)$ *Simplify.*

To check the result, multiply it out.

Check: $(y + 4)(2x - 1) = 2xy - y + 8x - 4$ *Use FOIL.*
$ = 2xy + 8x - y - 4$ *Use the commutative property to rewrite.*

c. The given expression has four terms, so try to factor by grouping. Always put a plus sign between the groups.

$(6a^2 + 3ab) + (-4ab - 2b^2)$ *Group the first two terms and the last two terms.*
$= 3a(2a + b) - 2b(2a + b)$ *To factor out a negative, change both signs.*
$= (2a + b)(3a - 2b)$ *Factor out the $(2a + b)$, which is common to both groups.*

To check the result, multiply it out and see whether it equals the original expression.

Check: $(2a + b)(3a - 2b) = 6a^2 - 4ab + 3ab - 2b^2$ *Use FOIL. Combine like terms.*
$ = 6a^2 - ab - 2b^2$

This is not the original problem, so simplify the original problem to complete the check.

$$6a^2 + 3ab - 4ab - 2b^2 \quad \text{Combine like terms.}$$
$$= 6a^2 - ab - 2b^2$$

PRACTICE PROBLEM FOR EXAMPLE 7

Factor each expression. Check the result.

a. $2x^2 + 14x - 5x - 35$ **b.** $6a^2 - 3ab + 2ab - b^2$

■ Factoring Completely

When factoring expressions, we factor the expressions as much as possible. This is called **factoring completely.** Let's explore this idea in the following Concept Investigation.

CONCEPT INVESTIGATION

Who factored completely?

Two students are given the following problem to factor:

$$2x^2 + 10x + 4x + 20$$

Using factoring by grouping, because the expression has four terms, the two students came up with different answers.

Teresa	Sarah
$2(x + 2)(x + 5)$	$(2x + 4)(x + 5)$

1. Multiply out both students' work. Do their factorizations match the original problem?

2. Which student factored completely? By factoring completely, we mean which of the two students broke the problem down into more factors?

3. What did the other student, the one who did not factor completely, forget to do? Do you see a missing factor?

Remember that one way to think of factoring is to break down a polynomial into the product of simpler polynomials. When asked to factor a polynomial completely, factor it as much as possible. The following strategy is used to factor a polynomial expression completely.

Tools to Completely Factor a Polynomial Expression
1. Factor out the GCF first, if there is one.
2. If there are four terms, try factoring by grouping.

Note: We will continue to add new tools throughout Chapter 6. All of the tools we use for factoring are also listed in the Factoring Toolbox in the back of the book.

Example 8 Factoring completely

Completely factor each expression. Check the result.

a. $2xy - 6x^2$ **b.** $2x^2 - 6x + 8x - 24$

SOLUTION $(x+2)(x-5)$

a. This expression has a GCF of $2x$. Factoring out the GCF first yields

$$2xy - 6x^2 = 2x(y - 3x)$$

As the expression in the parentheses does not have four terms, we cannot use grouping. We have finished factoring.

To check, multiply out the factored form.

Check: $2x(y - 3x) = 2xy - 6x^2$

This is the same expression as the original problem, so the factorization is correct.

b. This expression has a GCF of 2. First factor out the GCF.

$$2x^2 - 6x + 8x - 24 = 2(x^2 - 3x + 4x - 12)$$

The expression in the parentheses has four terms, so try to factor it further using grouping.

$$2(x^2 - 3x + 4x - 12) = 2[(x^2 - 3x) + (4x - 12)]$$
$$= 2[x(x - 3) + 4(x - 3)] \quad \text{Do not lose the 2.}$$
$$= 2(x - 3)(x + 4)$$

The last expression is written as a product (it is factored). None of the individual factors can be "broken down" any further. We are done factoring.

To check, multiply out the factored form.

Check: $2(x + 4)(x - 3) = (2x + 8)(x - 3)$ Multiply left to right.
$$= 2x^2 - 6x + 8x - 24 \quad \text{Use FOIL. Then add like terms.}$$
$$= 2x^2 + 2x - 24$$

This does not equal the original expression, so simplify the original expression to complete our check.

$$2x^2 - 6x + 8x - 24 = 2x^2 + 2x - 24$$

The expressions are the same, so the factorization is correct.

PRACTICE PROBLEM FOR EXAMPLE 8
Completely factor each expression. Check the result.

a. $16y^2 - 48y$ **b.** $3x^2 + 15x - 3x - 15$

6.1 Vocabulary Exercises

1. Factoring means to rewrite an expression as a(n) _____.

2. When finding the GCF, we include common factors raised to the _____ power that appears.

3. Factoring out the GCF involves using the _____ property.

4. When a polynomial expression has _____ terms, we try to factor by grouping.

6.1 Exercises

For Exercises 1 through 10, find the GCF for each set of integers or monomials.

1. 3, 12, 24

2. 4, 16, 40

3. 16, 24, 48

4. 14, 28, 35

5. $15a, 30a^2, 9a$

6. $22x, 55x^3$

7. $27y^2, 54y^4$

8. $100p, 120p^3, -60p^5$

9. $15xy^3, 45x^2y^2, 16xy$

10. $3ab, 21ab^3, 10a^2b^3$

For Exercises 11 through 20, find the GCF for each polynomial expression.

11. $9x + 21$

12. $-4y + 36$

13. $5ab - 35a^2$

14. $6xy - 18x^2y$

15. $-3x - 9x^2$

16. $-7ab - 49b$

17. $-2ab^5 + 16ab^2 - 28ab$

18. $x^2y^3 - x^3y^2$

19. $30x^2y + 9xyz - 15xz$

20. $7abc^3 - 14ab^2c + 35abc$

For Exercises 21 through 44, for each polynomial expression, factor out the GCF. Check the result.

21. $10x + 55$ **22.** $9x + 21$

23. $6x + 2$ **24.** $15a + 5$

25. $-15y - 20y^2$ **26.** $-12a - 24a^2$

27. $48a^4 - 16a^2$ **28.** $36x^3 - 9x^2$

29. $-6x - 6y$ **30.** $-7a - 14b$

31. $12p^6 - 6p^3 + 3p$ **32.** $7x^6 - 14x^2 + 49x$

33. $48a^4b^2 - 16a^2b^4$ **34.** $-15x^3y + 20xy^2$

35. $2(x + 4) + y(x + 4)$ **36.** $6(x + 8) + y(x + 8)$

37. $9x(x - 2) + 5(x - 2)$ **38.** $5n(n - 3) - 7(n - 3)$

39. $2x(x + 8) - (x + 8)$ **40.** $4x(2x - 5) + (2x - 5)$

41. $3(y + 5) - 2y(y + 5)$

42. $6(x - 4) - 5x(x - 4)$

43. $-3x(5y + 7) + 2(5y + 7)$

44. $2a(3b + 1) - 9(3b + 1)$

For Exercises 45 through 50, answer each question with one of the following properties:

Commutative property of multiplication
Commutative property of addition
Associative property
Distributive property

45. What property justifies that the two expressions are equal?
$$(x - 3)(x + 1) = (x + 1)(x - 3)$$

46. What property justifies that the two expressions are equal?
$$(2y + 7)(4x - 1) = (4x - 1)(2y + 7)$$

47. What property justifies that the two expressions are equal?
$$(-3x + 1)(x + 2) = (1 - 3x)(x + 2)$$

48. What property justifies that the two expressions are equal?
$$(3 - a)(a + 5) = (-a + 3)(a + 5)$$

49. Seth says that the factorization of $-6x - 12$ is $-6(x + 2)$. Sonia says that the factorization is $6(-x - 2)$. Why are both of them correct? Which is the more conventional answer?

50. Alison says that the factorization of $-2y - 10$ is $-2(y + 5)$. Aaron says that the factorization is $2(-y - 5)$. Why are both of them correct? Which is the more conventional answer?

For Exercises 51 through 58, factor so that each lead coefficient is positive. See Example 5 for help if needed.

51. $-9x + 27$ **52.** $-3y + 42$

53. $-y - 2$ **54.** $-t - 9$

55. $5 - x$ **56.** $3 - n$

57. $-x^2 + 7$ **58.** $-t^2 + 11$

For Exercises 59 through 76, factor each polynomial expression by grouping. Check the result.

59. $x^2 + 5x + 4x + 20$ **60.** $x^2 + 7x + 3x + 21$

61. $x^2 + 7x + 11x + 77$ **62.** $x^2 + 5x + 9x + 45$

63. $x^2 - 2x + 6x - 12$ **64.** $x^2 - 4x + 9x - 36$

65. $x^2 + 8x - 3x - 24$ **66.** $x^2 + 9x - 6x - 54$

67. $x^2y + 8x - 3xy - 24$ **68.** $x^2y + 6xy - 5x - 30$

69. $2mn - 2m - 5n + 5$ **70.** $3gh - 15g - h + 5$

71. $3ab - 2a - 9b + 6$ **72.** $4xy - 3y + 4x - 3$

73. $2xy - y + 6x - 3$ **74.** $3ab - 12a - 2b + 8$

75. $2a^2 - 6ab + ab - 3b^2$ **76.** $5x^2 + 10xy + 3xy + 6y^2$

For Exercises 77 through 82, determine which student factored the given polynomial expression completely.

77. $6x^2 - 12x$

Javier	Joey
$3x(2x - 4)$	$6x(x - 2)$

78. $8a^2 - 24a$

Abraham	Geni
$8a(a - 3)$	$4a(2a - 6)$

79. $2x^2 - 2x + 10x - 10$

Abi	Noah
$2(x + 5)(x - 1)$	$(2x + 10)(x - 1)$

80. $3x^2 - 6x - 9x + 18$

Osvaldo	Alexandra
$(x-3)(3x-6)$	$3(x-3)(x-2)$

81. $-x + 16$

Kyla	Lucia
$-(x-16)$	$1(-x+16)$

82. $-2x + 6$

Samantha	Melanie
$-2(x-3)$	$2(-x+3)$

For Exercises 83 through 94, factor each polynomial expression completely. Check the result.

83. $6ab^3 - 12a^2b^2 + 18b^2$
84. $5a^4b - 15a^3b^2 + 35a^3$
85. $-6x^2 + 3x$
86. $-2y^2 - 8y$
87. $-17p + 34q$
88. $19h - 38k^2$
89. $81x^3 - 27x$
90. $-8x^5 + 12x$
91. $2p^2 + 6p + 4p + 12$
92. $3p^2 + 3p + 15p + 15$
93. $6n^2 - 12n - 6n + 12$
94. $10n^2 - 15n - 40n + 60$

For Exercises 95 through 100, find the error in each student's solution.

95. Factor: $x(x + 2) + 3(x + 2)$

Dylan
$x^2 + 5x + 6$

96. Factor: $n(n + 7) - 3(n + 7)$

Brian
$n^2 + 4n - 21$

97. Factor: $6x^2 - 8xy + 2x$

Maryam
$2x(3x - 4y)$

98. Factor: $6xy - 3y^2 + 3y$

Krista
$3y(2x - y)$

99. Multiply: $5(n - 4) + m(n - 4)$

Diego
$(5 + m)(n - 4)$

100. Multiply: $x(x - 1) - (x - 1)$

Ethan
$(x - 1)^2$

101. Karlee needs to mix a 20% hydrochloric acid solution with a 12% hydrochloric acid solution to make an 18% hydrochloric acid solution. She wants a total of 10 liters in the mix. How many liters of the 20% solution and of the 12% solutions should she use?

102. Miranda has $250,000 to invest into two accounts. One account is a bond fund that pays 3% simple interest. The second account is a stock fund that pays 4.5% simple interest. Miranda needs $9323.50 per year to pay for her medical insurance. Find how much Miranda should invest in each account to earn enough interest to pay her medical insurance.

103. Christopher gets a base salary of $100 per week plus 3% commission on the value of all sales made.

    a. Write an inequality to show that Christopher needs to earn at least $200 per week to pay his bills.

    b. Solve the inequality from part a. Write the solution in a complete sentence.

104. Jeremy is having a security system installed. The company charges $50 for basic installation plus an additional $15 for each door and window sensor installed beyond the front door.

    a. Write an inequality to show that Jeremy does not want to spend more than $100 for installation.

    b. Solve the inequality from part a. Write the solution in a complete sentence.

6.2 Factoring Trinomials

LEARNING OBJECTIVES

- Factor trinomials of the form $x^2 + bx + c$ by inspection.
- Factor trinomials of the form $ax^2 + bx + c$.
- Factor a trinomial completely.

■ Factoring Trinomials of the Form $x^2 + bx + c$ by Inspection

Recall that a polynomial with two terms is called a *binomial*, and a polynomial with three terms is called a *trinomial*. For example, $2x - 5$ is a binomial, and $x^2 + 2x + 4$ is a trinomial. In Chapter 5, we learned how to perform operations on polynomials, such as how to multiply polynomial expressions. The distributive property (FOIL) is used to multiply two binomials.

$$(x + 5)(x - 3) = x^2 - 3x + 5x - 15 \quad \text{Use FOIL to multiply the expression.}$$
$$= x^2 + 2x - 15 \quad \text{Simplify by combining like terms.}$$

We now want to reverse this process. In this section, we start with a trinomial such as $x^2 + 2x - 15$ and will rewrite it as the product of two binomials, in this case, $(x + 5)(x - 3)$. The first type of trinomial that we will learn to factor has the standard form $x^2 + bx + c$. What is important to recognize about this form is that the lead coefficient on x^2 is 1, b is the coefficient of the x-term, and c is the constant term.

Example 1 Identifying b and c

For each of the following polynomial expressions, state the value of b (the coefficient of x) and c (the constant term). Make sure the polynomial is in descending order first.

a. $x^2 - 2x + 9$ **b.** $-x - 7 + x^2$

SOLUTION

a. Matching the given expression, $x^2 - 2x + 9$, with the standard form $x^2 + bx + c$ yields

$$x^2 - 2x + 9 \quad \text{Matching x-terms, we see that } b = -2.$$
$$x^2 + bx + c \quad \text{Matching constant terms, we see that } c = 9.$$

b. This expression is not in descending order. First rewrite the expression in descending order.

$$-x - 7 + x^2 = x^2 - x - 7 \quad \text{Rearrange terms.}$$

Now we can match our expression in descending order with the standard form to determine the values of b and c.

$$x^2 - 1x - 7 \quad \text{Matching x-coefficients, we see that } b = -1.$$
$$x^2 + bx + c \quad \text{Matching constant terms, we see that } c = -7.$$

Be careful to remember that $-x = -1 \cdot x$; this is why $b = -1$ in the above expression. Likewise, with the constant term, include the sign in front of 7 when you find the value of c. This is why $c = -7$.

PRACTICE PROBLEM FOR EXAMPLE 1

For each of the following polynomial expressions, state the value of b (the coefficient of x) and c (the constant term). Make sure the polynomial is in descending order first.

a. $x^2 - 9x - 12$ b. $-5 - 3x + x^2$

The process of factoring $x^2 + bx + c$ may be thought of as the opposite of multiplying out two binomials using the distributive property. Let's examine multiplication for some clues on how we should proceed. Multiplying the binomials $(x + 2)(x + 5)$ looks like the following:

$(x + 2)(x + 5) = x^2 + 2x + 5x + 10$ Use FOIL.
$\qquad\qquad\qquad = x^2 + (2 + 5)x + 10$ Apply the distributive property to the x-terms.
$\qquad\qquad\qquad = x^2 + 7x + 10$ Simplify.

To go backward, proceed as follows:

$x^2 + 7x + 10 = x^2 + (2 + 5)x + 10$
$\qquad\qquad\quad = x^2 + 2x + 5x + 10$
$\qquad\qquad\quad = (x + 2)(x + 5)$

For the expression $x^2 + 7x + 10$, we see that $b = 7$ and $c = 10$. Starting with the value of $c = 10$, we find the two factors of 10, namely, 5 and 2, that add to $b = 7$. It is these two values, 5 and 2, that are in the factored final answer $(x + 2)(x + 5)$. We generalize this process to factor a trinomial of the form $x^2 + bx + c$ over the integers. This process of finding factors of c that add to b is often called **factoring by inspection** or **trial and error**.

> **Steps to Factor a Trinomial of the Form $x^2 + bx + c$**
>
> 1. Identify the values of b and c.
> 2. Find the factors of c that add to b.
> 3. Rewrite as the product of two binomials of the form $(x + ?)(x + ?)$, where the question marks are the two factors (including their signs) from step 2.
>
> *Note:* To check the result, multiply out the factors from step 3. The product should match the original trinomial, $x^2 + bx + c$.

Example 2 Factoring trinomials of the form $x^2 + bx + c$

Factor each trinomial into a product of two binomials. Check the result.

a. $x^2 + 9x + 18$ b. $x^2 - 9x + 14$ c. $2x - 24 + x^2$

SOLUTION (x+2)(x-5)

a. **Step 1** Identify the values of b and c.

Examining the trinomial $x^2 + 9x + 18$, we see that $b = 9$ and $c = 18$.

Step 2 Find the factors of c that add to b.

Look for the pair of numbers that multiplies to $c = 18$ and adds to $b = 9$.

$c = 18$	$b = 9$
$1 \cdot 18 = 18$	$1 + 18 = 19 \neq 9$
$2 \cdot 9 = 18$	$2 + 9 = 11 \neq 9$
$3 \cdot 6 = 18$	$3 + 6 = 9$

Factoring Polynomials with Three Terms   (x+2)(x-5)

GCF First
Factor out the GCF first, if there is one.

$6x^2 + 9x - 15$
$= 3(2x^2 + 3x - 5)$

Coefficient of x^2 is 1
Trial and error or AC method

$x^2 + 7x + 12$
$= (x + 3)(x + 4)$

The correct pair of numbers that we are looking for is 3 and 6.

Step 3 Rewrite as the product of two binomials of the form $(x + ?)(x + ?)$, where the question marks are the two factors (including their signs) from step 2.

$$(x + 3)(x + 6)$$

To check the result, multiply out the factorization $(x + 3)(x + 6)$.

$(x + 3)(x + 6) = x^2 + 6x + 3x + 18$ Multiply, using FOIL.
$ = x^2 + 9x + 18$ This is the original expression.

b. Examining the trinomial $x^2 - 9x + 14$, we see that $b = -9$ and $c = 14$. Look for the pair of numbers that multiplies to $c = 14$ and adds to $b = -9$.

$c = 14$	$b = -9$
$1 \cdot 14 = 14$	$1 + 14 = 15 \neq -9$
$2 \cdot 7 = 14$	$2 + 7 = 9 \neq -9$
$-1 \cdot (-14) = 14$	$-1 + (-14) = -15 \neq -9$
$-2 \cdot (-7) = 14$	$-2 + (-7) = -9$

The positive factors of c will not add to a negative number. Therefore, we consider two negative factors that will multiply together to give us a positive 14 and add to -9. The correct pair of numbers that we are looking for is -2 and -7. Therefore, the expression $x^2 - 9x + 14$ factors as

$$(x - 2)(x - 7)$$

To check the result, multiply out the factorization $(x - 2)(x - 7)$ to see whether we get the original polynomial back.

$(x - 2)(x - 7) = x^2 - 7x - 2x + 14$ Multiply, using FOIL. Combine like terms.
$ = x^2 - 9x + 14$ This is the original expression.

c. Before we identify b and c, put this trinomial $2x - 24 + x^2$ in descending order.

$$2x - 24 + x^2 = x^2 + 2x - 24$$

We see that $b = 2$ and $c = -24$. Look for the pair of numbers that multiplies to $c = -24$ and adds to $b = 2$. Because $c = -24$, look for factors that have opposite signs. The only way we can multiply two numbers together and get a negative is if one is a positive number and the other is a negative number.

$c = -24$	$b = 2$
$-2 \cdot 12 = -24$	$-2 + 12 = 10 \neq 2$
$2 \cdot (-12) = -24$	$2 + (-12) = -10 \neq 2$
$3 \cdot (-8) = -24$	$3 + (-8) = -5 \neq 2$
$-3 \cdot 8 = -24$	$-3 + 8 = 5 \neq 2$
$4 \cdot (-6) = -24$	$4 + (-6) = -2 \neq 2$ The sign is wrong. Switch
$-4 \cdot 6 = -24$	$-4 + 6 = 2$ signs on the factors.

The correct pair of numbers we are looking for is -4 and 6. Therefore, the expression $x^2 + 2x - 24$ factors as

$$(x - 4)(x + 6)$$

To check the result, we multiply out the factorization $(x - 4)(x + 6)$ to see whether we get the original polynomial back.

$(x - 4)(x + 6) = x^2 + 6x - 4x - 24$ Multiply, using FOIL. Combine like terms.
$ = x^2 + 2x - 24$ Rearrange the terms.
$ = 2x - 24 + x^2$ This is the original polynomial.

PRACTICE PROBLEM FOR EXAMPLE 2

Factor each trinomial into a product of two binomials. Check the result.

a. $x^2 + 11x + 24$ b. $-35 + 2x + x^2$ c. $x^2 - 13x + 36$

Factoring Trinomials of the Form $ax^2 + bx + c$

Now that we have learned how to factor trinomials in which the coefficient of x^2 equals 1, we move on to the case in which the coefficient of x^2 is a number other than positive 1. The standard form for a single-variable trinomial is $ax^2 + bx + c$, and it is important to correctly identify the values of a, b, and c.

Example 3 Finding a, b, and c

For each of the following polynomial expressions, state the value of a (the coefficient of x^2), b (the coefficient of x), and c (the constant term). Make sure the polynomial is in descending order first.

a. $-2x^2 + 5x - 9$ b. $-x^2 + 7x$ c. $-3 + 5x^2 - 6x$

SOLUTION

a. Match the given expression, $-2x^2 + 5x - 9$, with the standard form $ax^2 + bx + c$. Then equate (or match) coefficients.

$$-2x^2 + 5x - 9$$
$$ax^2 + bx + c$$

Matching x^2-terms, we see that $a = -2$, $b = 5$, and $c = -9$.

b. Match the given expression, $-x^2 + 7x$, with the standard form $ax^2 + bx + c$. Then equate (or match) coefficients. The coefficient of $-x^2$ is -1 because it can be rewritten $-1 \cdot x^2$.

$$-x^2 + 7x$$
$$-1 \cdot x^2 + 7x + 0$$
$$ax^2 + bx + c$$

There is no constant term. It is zero.
$a = -1, b = 7$, and $c = 0$.

c. The original expression, $-3 + 5x^2 - 6x$, is not in descending order. First rearrange the expression to put the terms in descending order.

$$-3 + 5x^2 - 6x = 5x^2 - 6x - 3$$

Now match the given expression, $5x^2 - 6x - 3$, with the standard form $ax^2 + bx + c$. Then equate (or match) coefficients.

$$5x^2 - 6x - 3$$
$$ax^2 + bx + c$$

Matching x^2-terms, we see that $a = 5$, $b = -6$, and $c = -3$.

PRACTICE PROBLEM FOR EXAMPLE 3

For each of the following polynomial expressions, state the value of a (the coefficient of x^2), b (the coefficient of x), and c (the constant term). Make sure the polynomial is in descending order first.

a. $-3x^2 - 16$ b. $-4x - x^2 + 3$ c. $16 - 4x^2 + 8x$

To factor a polynomial of the form $ax^2 + bx + c$, where $a \neq 1$, we will use a technique based on factoring by grouping. To understand this technique, let's multiply out two binomials and look carefully at how to reverse the steps.

$(2x + 1)(x + 3) = 2x^2 + 6x + x + 3$ — Use FOIL to multiply.
$= 2x^2 + (6 + 1)x + 3$ — Use the distributive property on the x-terms.
$= 2x^2 + 7x + 3$ — Simplify.

Now working backward with this line of reasoning yields

$2x^2 + 7x + 3 = 2x^2 + (6 + 1)x + 3$ — Begin with the trinomial.
$= 2x^2 + 6x + x + 3$ — Rewrite so that it can be factored by grouping.
$= 2x(x + 3) + 1(x + 3)$ — Four terms—factor by grouping.
$= (x + 3)(2x + 1)$

What's new? The example now has a middle step of rewriting the trinomial that can be factored by grouping. For this problem, $a = 2$, $b = 7$, and $c = 3$. If we multiply a and c, the result is $a \cdot c = 2 \cdot 3 = 6$. Factors of 6 that add to $b = 7$ are 6 and 1. These are the coefficients that are used on the x-terms in the middle step to "rewrite so it can be factored by grouping." We now look for factors of $a \cdot c$ that add to b. Then we factor by grouping.

> **Steps to Factor a Trinomial of the Form $ax^2 + bx + c$**
> 1. Factor out the GCF.
> 2. Find the product $a \cdot c$.
> 3. Find the factors of $a \cdot c$ that add to b.
> 4. Rewrite the middle x-term into the sum of two terms, with the factors from step 3 as coefficients of x.
> 5. Factor by grouping.
>
> *Note:* Check the result by multiplying out the expression. This method is often called the **AC method of factoring.**

Factoring Polynomials with Three Terms $(x+2)(x-5)$

GCF First
Factor out the GCF first, if there is one.

$6x^2 + 9x - 15$
$= 3(2x^2 + 3x - 5)$

Coefficient of x^2 is not 1
AC method or trial and error

$12x^2 + 8x - 15$
$= (2x + 3)(6x - 5)$

Example 4 Factoring trinomials of the form $ax^2 + bx + c$

Factor each trinomial into the product of two binomials. Check the result.

a. $3x^2 + 13x + 4$ **b.** $3x^2 - 10x + 8$ **c.** $18x - 5 + 8x^2$

SOLUTION $(x+2)(x-5)$

a. Step 1 Factor out the GCF.

There is no GCF to factor out.

Step 2 Find the product of $a \cdot c$.

Examining the trinomial $3x^2 + 13x + 4$, we see that $a = 3$, $b = 13$, and $c = 4$. Multiplying $a \cdot c$ yields $a \cdot c = 3 \cdot 4 = 12$.

Step 3 Find the factors of $a \cdot c$ that add to b.

List the factors of 12 and look for the pair that adds to $b = 13$.

$\dfrac{a \cdot c = 12}{1 \cdot 12 = 12} \qquad \dfrac{b = 13}{1 + 12 = 13}$ — The first factors that we tried worked.

The first factors that we tried worked, so we do not need to continue finding other factors of 12.

Step 4 Rewrite the middle x-term into the sum of two terms, with the factors from step 3 as coefficients of x.

Rewrite the middle x-term using 1 and 12 as coefficients of x.

$$3x^2 + 13x + 4 = 3x^2 + 12x + x + 4$$

Step 5 Factor by grouping.

$$\begin{aligned}3x^2 + 13x + 4 &= 3x^2 + 12x + x + 4\\ &= (3x^2 + 12x) + (x + 4)\\ &= 3x(x + 4) + 1(x + 4)\\ &= (x + 4)(3x + 1)\end{aligned}$$

To check, multiply out the factorization $(x + 4)(3x + 1)$ to see whether we get the original expression back.

$$\begin{aligned}(x + 4)(3x + 1)& \quad \text{Multiply, using FOIL.}\\ = 3x^2 + x + 12x + 4& \quad \text{Combine like terms.}\\ = 3x^2 + 13x + 4& \quad \text{This is the original polynomial.}\end{aligned}$$

Notice that the factor 12 from step 3 is not visible in the final factored form.

b. From the trinomial $3x^2 - 10x + 8$, we have that $a = 3$, $b = -10$, and $c = 8$. Multiplying $a \cdot c$ yields $a \cdot c = 3 \cdot 8 = 24$. List the factors of 24 and look for the pair that adds to $b = -10$.

$a \cdot c = 24$	$b = -10$	
$3 \cdot 8 = 24$	$3 + 8 = 11 \neq -10$	
$2 \cdot 12 = 24$	$2 + 12 = 14 \neq -10$	
$6 \cdot 4 = 24$	$6 + 4 = 10 \neq -10$	The sign is wrong. Make both factors negative.
$(-6)(-4) = 24$	$(-6) + (-4) = -10$	

> **Connecting the Concepts**
>
> **When do we use the AC method?**
>
> The AC method of factoring can also be used when factoring a trinomial of the form
>
> $$x^2 + bx + c$$
>
> In this case, $a = 1$, so when we multiply ac, we get c.
> The AC method may not be faster than factoring by inspection, but it can be used for both types of trinomials.

Rewrite the middle x-term, using -6 and -4 as coefficients of x, and factor by grouping.

$$\begin{aligned}3x^2 - 10x + 8 &= 3x^2 - 6x - 4x + 8 & &\text{When grouping, put a plus between the}\\ &= (3x^2 - 6x) + (-4x + 8) & &\text{groups and keep the negative sign with}\\ & & &\text{the 4x.}\\ &= 3x(x - 2) - 4(x - 2) & &\text{Factor out a negative if the lead}\\ & & &\text{coefficient is negative.}\\ &= (x - 2)(3x - 4) & &\text{Factor out the common factor of } x - 2.\end{aligned}$$

To check, multiply out the factorization $(x - 2)(3x - 4)$ to see whether we get the original expression back.

$$\begin{aligned}(x - 2)(3x - 4)& \quad \text{Multiply, using FOIL.}\\ = 3x^2 - 4x - 6x + 8& \quad \text{Combine like terms.}\\ = 3x^2 - 10x + 8& \quad \text{This is the original polynomial.}\end{aligned}$$

c. The original expression, $18x - 5 + 8x^2$, is not in descending order. First rewrite the expression so that the terms are in descending order.

$$18x - 5 + 8x^2 = 8x^2 + 18x - 5$$

We have that $a = 8$, $b = 18$, and $c = -5$. Multiplying $a \cdot c$ yields $a \cdot c = 8 \cdot (-5) = -40$. List the factors of -40 and look for the pair that adds to $b = 18$.

$a \cdot c = -40$	$b = 18$	
$4 \cdot (-10) = -40$	$4 + (-10) = -6 \neq 18$	
$5 \cdot (-8) = -40$	$5 + (-8) = -3 \neq 18$	
$2 \cdot (-20) = -40$	$2 + (-20) = -18 \neq 18$	Wrong sign. Change signs on the factors.
$(-2) \cdot 20 = -40$	$(-2) + 20 = 18$	

552 CHAPTER 6 Factoring

Rewrite the middle x-term, using -2 and 20 as coefficients of x, and factor by grouping.

$$8x^2 + 18x - 5 = 8x^2 - 2x + 20x - 5$$
$$= (8x^2 - 2x) + (20x - 5)$$
$$= 2x(4x - 1) + 5(4x - 1)$$
$$= (4x - 1)(2x + 5)$$

To check, multiply out the factorization $(4x - 1)(2x + 5)$ to see whether we get the original expression back.

$(4x - 1)(2x + 5)$	Multiply, using FOIL.
$= 8x^2 + 20x - 2x - 5$	Combine like terms.
$= 8x^2 + 18x - 5$	Rearrange the terms.
$= 18x - 5 + 8x^2$	This is the original polynomial.

PRACTICE PROBLEM FOR EXAMPLE 4

Factor each trinomial into the product of two binomials. Check the result.

a. $3x^2 + 23x + 14$ **b.** $2x^2 + 5x - 42$ **c.** $10 - 11x + 3x^2$

Some polynomials cannot be factored (using the rational numbers). These polynomials are called **prime** polynomials. When using the AC method of factoring, we will not be able to find factors of ac that add up to b in step 3.

Example 5   Finding a prime polynomial

Factor each trinomial into the product of two binomials. Check the result.

a. $x^2 + 3x + 1$ **b.** $3x^2 + 5x + 4$

SOLUTION

a. The coefficient of x^2 is 1, so find the factors of $c = 1$ that add to $b = 3$.

$c = 1$	$b = 3$
$1 \cdot 1 = 1$	$1 + 1 = 2 \neq 3$
$(-1) \cdot (-1) = 1$	$(-1) + (-1) = -2 \neq 3$

As there are no other integer factors of $c = 1$, and none of the pairs of factors in the above list sum to $b = 3$, this polynomial cannot be factored. The polynomial is prime.

b. The coefficient of x^2 is not 1. It is 3, so use the AC method of factoring. Multiplying $a \cdot c$ yields $a \cdot c = 3 \cdot 4 = 12$ and $b = 5$.

$a \cdot c = 12$	$b = 5$
$1 \cdot 12 = 12$	$1 + 12 = 13 \neq 5$
$(-1) \cdot (-12) = 12$	$(-1) + (-12) = -13 \neq 5$
$2 \cdot 6 = 12$	$2 + 6 = 8 \neq 5$
$(-2) \cdot (-6) = 12$	$(-2) + (-6) = -8 \neq 5$
$3 \cdot 4 = 12$	$3 + 4 = 7 \neq 5$
$(-3) \cdot (-4) = 12$	$(-3) + (-4) = -7 \neq 5$

Because none of the factors of 12 add to 5, this polynomial cannot be factored. The polynomial is prime.

What's That Mean?

Prime

Recall from Chapter R that a *prime* number has only itself and 1 as factors. For example, 5 is prime because the only whole number factors of 5 are 1 and 5. A polynomial is *prime* when it has only itself and 1 as rational number factors.

SECTION 6.2 Factoring Trinomials 553

More Techniques to Factor Completely

We now add more factoring techniques to our list for factoring polynomials.

> **Tools to Completely Factor a Polynomial Expression**
> 1. Factor out the GCF first, if there is one.
> 2. If there are four terms, try factoring by grouping.
> 3. If there are three terms and the coefficient of x^2 is 1, factor by finding the factors of c that sum to b, and factor as the product of two binomials. If the coefficient of x^2 is not 1, then use the AC method.
>
> *Note:* We will continue to add new tools throughout Chapter 6. All of the tools we use for factoring are also listed in the Factoring Toolbox in the back of the book.

Example 6 Factoring completely

Factor each polynomial completely.

a. $12x^2 + 8xy - 6xy - 4y^2$ **b.** $3x^2 + 6x - 105$ **c.** $-6m^2n + 28mn + 10n$

SOLUTION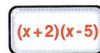

a. Look for a GCF in this polynomial. The coefficient of each term has a common factor of 2, so begin by factoring out the GCF of 2.

$$12x^2 + 8xy - 6xy - 4y^2 = 2(6x^2 + 4xy - 3xy - 2y^2)$$

Inside the parentheses, the polynomial has four terms, so we will try to factor that expression by grouping.

$$\begin{aligned}12x^2 + 8xy - 6xy - 4y^2 &= 2(6x^2 + 4xy - 3xy - 2y^2)\\ &= 2[(6x^2 + 4xy) + (-3xy - 2y^2)]\\ &= 2[2x(3x + 2y) - y(3x + 2y)]\\ &= 2(3x + 2y)(2x - y)\end{aligned}$$

None of these expressions in the parentheses can be factored any more, so we are finished factoring. The answer is

$$12x^2 + 8xy - 6xy - 4y^2 = 2(3x + 2y)(2x - y)$$

b. Look for a GCF in this polynomial. The coefficient of each term has a common factor of 3, so begin by factoring out the GCF of 3.

$$3x^2 + 6x - 105 = 3(x^2 + 2x - 35)$$

The trinomial inside the parentheses has a coefficient of 1 on x^2, so we look for factors of $c = -35$ that sum to $b = 2$. $7(-5) = -35$ and $7 + (-5) = 2$. The factors that we need are 7 and -5.

$$\begin{aligned}3x^2 + 6x - 105 &= 3(x^2 + 2x - 35)\\ &= 3(x + 7)(x - 5)\end{aligned}$$

c. This polynomial has a GCF that differs slightly. Each term is divisible by 2, so 2 is part of the GCF. However, each term also has the variable n in common. The lowest power of n that appears is 1, so we will factor out the GCF of $2n$.

$$-6m^2n + 28mn + 10n = 2n(-3m^2 + 14m + 5)$$

The coefficient on m^2 is -3. It is much easier to factor a trinomial if the coefficient on the squared term is positive. Factor out -1 to change the sign

> **Skill Connection**
>
> **Divisibility rules**
> When looking for a GCF, we can use some divisibility rules.
>   An integer is *divisible by* 2 if it ends in an even number (0, 2, 4, 6, 8).
>   An integer is *divisible by* 3 if the sum of its digits is divisible by 3 (105 is divisible by 3 because $1 + 0 + 5 = 6$, and 6 is divisible by 3).
>   A number is *divisible by* 5 if it ends in 0 or 5.

on the squared term. Doing so changes the signs on all of the terms inside the parentheses.

$$-6m^2n + 28mn + 10n = 2n(-3m^2 + 14m + 5)$$
$$= -2n(3m^2 - 14m - 5)$$

The coefficient of m^2 is 3, so use the AC method to factor it. Multiplying $a \cdot c$ yields $a \cdot c = 3 \cdot (-5) = -15$ and $b = -14$.

$a \cdot c = -15$	$b = -14$
$-1 \cdot 15 = -15$	$-1 + 15 = 14 \neq -14$
$1 \cdot (-15) = -15$	$1 + (-15) = -14$

Using the factors 1 and -15 to break up the middle term yields

$$-6m^2n + 28mn + 10n = -2n(3m^2 - 14m - 5)$$
$$= -2n(3m^2 - 15m + m - 5)$$
$$= -2n[(3m^2 - 15m) + (m - 5)]$$
$$= -2n[3m(m - 5) + 1(m - 5)]$$
$$= -2n(m - 5)(3m + 1)$$

PRACTICE PROBLEM FOR EXAMPLE 6
Factor each polynomial completely.

a. $12m^2 - 4mn + 6mn - 2n^2$ **b.** $-3t^2 - 18t + 21$ **c.** $2x^2y - 2x - 10y$

6.2 Vocabulary Exercises

1. A binomial has _____ terms.
2. A trinomial has _____ terms.
3. A monomial has _____ terms.
4. A polynomial has _____ terms.
5. In the equation $3 \cdot 7 = 21$, 3 and 7 are called _____ of 21.
6. In the equation $x \cdot y = xy$, x and y are called the _____ of xy.
7. In the equation $3 + 7 = 10$, 10 is known as the _____ of 3 and 7.
8. In the equation $9 + 7 = 16$, 16 is known as the _____ of 9 and 7.
9. What does it mean to factor a polynomial?
10. What does it mean to factor a number?
11. What property or properties justify the following equation: $-2x + 7 + x^2 = x^2 - 2x + 7$?
12. What property or properties justify the following equation: $(x - 3)(x + 6) = (x + 6)(x - 3)$?
13. A polynomial is considered _____ if it cannot be factored.
14. When factoring a trinomial with a lead coefficient of 1, use the _____ method to factor.
15. When factoring a trinomial with a lead coefficient other than 1, use the _____ method to factor.

6.2 Exercises

For Exercises 1 through 16, identify a, b, and c for each trinomial. Hint: Make sure the polynomial is in descending order first.

1. $x^2 - 6x + 5$
2. $x^2 + 5x - 16$
3. $y^2 - 6y - 1$
4. $t^2 - 2t - 1$
5. $x^2 - 7$
6. $x^2 + 18$
7. $m^2 - 2m$
8. $k^2 - 29k$
9. $2x^2 + 6x - 8$
10. $4x^2 - 8x + 16$
11. $-y^2 - 6y + 7$
12. $-x^2 + 6x - 9$

13. $5 - 3m - 6m^2$
14. $-8n - 9n^2 + 1$
15. $16 - 2x + x^2$
16. $-9 + 7y - y^2$

For Exercises 17 through 26, factor each trinomial. Check the result using multiplication. If the trinomial cannot be factored over the integers, then write that it is prime.

17. $x^2 + 6x + 5$
18. $x^2 + 5x + 6$
19. $x^2 + 3x - 18$
20. $x^2 + 3x - 28$
21. $x^2 - 3x - 40$
22. $x^2 - 7x - 18$
23. $x^2 - 15x + 56$
24. $x^2 - 9x + 20$
25. $x^2 - 6x + 2$
26. $x^2 + 8x - 3$

For Exercises 27 through 40, factor each trinomial. Check the result using multiplication. If the trinomial cannot be factored over the integers, then write that it is prime.

27. $2x^2 + 7x + 6$
28. $5x^2 + 16x + 3$
29. $2x^2 + 9x + 1$
30. $3x^2 - 15x - 2$
31. $7x^2 + 41x - 6$
32. $3x^2 + x - 2$
33. $3x^2 - 19x + 6$
34. $5x^2 - 11x + 2$
35. $4x^2 - 17x + 15$
36. $9x^2 - 55x + 6$
37. $-x^2 + x + 6$
38. $-x^2 + 4x + 5$
39. $-2x^2 + 4x - 3$
40. $3x^2 + 5x + 8$

For Exercises 41 through 60, state the number of terms in the polynomial. Then factor each polynomial completely. Check the result using multiplication.

41. $4xy - 8y^2$
42. $18x^2 - 21x^3$
43. $6a^2b - 12ab + 2b$
44. $-42xy^2 - 35xy + 7x$
45. $3x^2 - 6x - 72$
46. $2x^2 - 28x + 90$
47. $15a^2 - 15ab + 5ab - 5b^2$
48. $-6a^2 + 3ab - 12ab + 6b^2$
49. $12x^2 + 21x - 6$
50. $70x^2 - 90x + 20$
51. $t^3 - t^2 - 30t$
52. $x^3 + 10x^2 + 21x$
53. $2a^3 - 18a^2 + 40a$
54. $3m^3 - 15m^2 - 42m$
55. $24c^3 - 4c^2 - 160c$
56. $60d^3 - 5d^2 + 175d$
57. $x^2y - 16xy + 63y$
58. $a^2 - 2ab - 35b^2$
59. $10m^2 + 3mn - 27n^2$
60. $20x^2 - 39xy + 18y^2$

For Exercises 61 through 64, solve each inequality. Check the answers. Write each solution set in each of the following ways: using inequality symbols, using interval notation, and graphing the solution on a number line.

61. $\frac{2}{3}x + 8 > 24$
62. $\frac{3}{5}p - 4 \leq 26$
63. $2(x + 3) \geq 5x - 11$
64. $-6x - 7 < 24 + 2x$

For Exercises 65 through 68, graph the solution set of each linear inequality. Clearly label and scale the axes.

65. $y < \frac{2}{3}x + 1$
66. $y \geq -\frac{3}{5}x + 8$
67. $-5x + 4y > 24$
68. $4x + 6y \leq -30$

69. Find the equation of the line perpendicular to $y = \frac{2}{3}x + 4$, passing through the point $(2, 7)$.

70. Find the equation of the line parallel to $y = \frac{4}{3}x - 6$, passing through the point $(-4, -3)$.

For Exercises 71 through 74, solve each system of equations. Provide reasons for each step. Check each solution in both original equations.

71. $2x + 4y = 10$
    $-3x + 8y = 34$

72. $5x - 4y = 16$
    $2x + 12y = 3$

73. $y = 2x - 9$
    $-6x + 3y = 20$

74. $x = \frac{1}{3}y + 4$
    $-3x + y = -12$

For Exercises 75 through 78, simplify each exponential expression. Write the answers using positive exponents only.

75. $(x^2y^4)(5xy^5)^2$
76. $\frac{14a^2b^5}{21ab^{-2}}$
77. $(3x^7y^{-3})^2$
78. $(3x + 5)^2$

6.3 Factoring Special Forms

LEARNING OBJECTIVES

- Recognize and factor a difference of squares.
- Factor perfect square trinomials.
- Factor polynomials completely.

What's That Mean?

Perfect Square

We say that 9 is a *perfect square* because $9 = 3^2$. Some perfect squares are as follows:

Perfect Square	Explanation
1	$1 = 1^2$
4	$4 = 2^2$
9	$9 = 3^2$
16	$16 = 4^2$
25	$25 = 5^2$
36	$36 = 6^2$
49	$49 = 7^2$
64	$64 = 8^2$
81	$81 = 9^2$
100	$100 = 10^2$
121	$121 = 11^2$
144	$144 = 12^2$

■ Difference of Squares

There are a few special forms of polynomials and their factorizations that we now examine. The first form is called a **difference of squares.**

> **DEFINITION**
>
> **Difference of Squares** A polynomial expression is said to be a **difference of squares** if it has the form $a^2 - b^2$.

To recognize whether an expression is a difference of squares, make sure that the first term and last term are both **perfect squares** and that the two terms are subtracted. Recall that *subtracting* two terms is also called the *difference* between two terms. That is why we call these expressions a difference of squares.

Example 1 Recognizing a difference of squares

Determine whether or not each polynomial expression is a difference of squares.

a. $x^2 - 9$ **b.** $k^2 + 16$ **c.** $25t^2 - 1$ **d.** $x^2 - 9x - 16$

SOLUTION

a. The expression $x^2 - 9$ is a difference of squares. The first term is x^2, which is a perfect square. The second term is 9, which is 3^2. The two terms are being subtracted. Therefore, this is a difference of squares.

b. This expression is *not* a difference of squares. This is a sum, not a difference. A difference involves subtraction.

c. This expression is a difference of squares. The first term is $25t^2$. Since $(5t)^2 = 5^2 \cdot t^2 = 25t^2$, the first term is a perfect square. The second term is 1, which is 1^2. The two terms are being subtracted, so this is a difference of squares.

d. This is *not* a difference of squares. There are three terms, not two, so it does not fit the form in the definition, which is $a^2 - b^2$.

Now we learn how to factor a difference of squares. Let's go back to what we know about multiplying out polynomials and look for a pattern.

SECTION 6.3 Factoring Special Forms

CONCEPT INVESTIGATION
What form is it?

Multiply out and simplify each of the following polynomials.

a. $(x + 3)(x + 3)$ **b.** $(x - 3)(x + 3)$
c. $(x - 3)(x - 3)$ **d.** $(x - 4)(x + 4)$

Which parts resulted in a difference of squares? What do you notice about the factors on these parts?

To factor a difference of squares, use the following steps.

Steps to Factor a Difference of Squares

1. Find the values of a and b.
2. The expression factors as $a^2 - b^2 = (a - b)(a + b)$.

Note: Recall that $(a - b)(a + b) = (a + b)(a - b)$ because multiplication is commutative. It is all right to write the product with the factors in either order. Check the result by multiplying out the expression.

Example 2 Factoring a difference of squares

First, determine whether each polynomial expression is a difference of squares. Then factor each polynomial expression.

a. $x^2 - 25$ **b.** $4y^2 - 49$

SOLUTION

a. Step 1 Find the values of a and b.

This expression is a difference of squares. The first and second terms are perfect squares. To find the values of a and b, we equate the given expression with the general form. We match as follows:

$x^2 - 25$ *Original expression*
$a^2 - b^2$ *General form*

Since $(x)^2 = x^2$ and $5^2 = 25$, rewrite the original expression as

$(x)^2 - (5)^2$ *Rewrite the original expression.*
$a^2 - b^2$ *Match with the general form.*

Matching first term to first term and second term to second term yields

$a = x$ and $b = 5$

Step 2 The expression factors as $a^2 - b^2 = (a + b)(a - b)$.

In the factorization $a^2 - b^2 = (a - b)(a + b)$, substitute in $a = x$ and $b = 5$. The factorization is

$x^2 - 25 = (x - 5)(x + 5)$

Check the solution by multiplying out the expression.

$(x - 5)(x + 5) = x^2 + 5x - 5x - 25$
$ = x^2 - 25$

Factoring Polynomials with Two Terms

GCF First
Factor out the GCF first, if there is one.

$6x^2 + 4x$
$= 2x(3x + 2)$

Difference of Squares
Use the difference of squares pattern.

$a^2 - b^2 = (a + b)(a - b)$

$x^2 - 25$
$= (x + 5)(x - 5)$

b. This expression is a difference of squares. The first and second terms are perfect squares. To find the values of a and b, we will equate the given expression with the general form. We match as follows:

$$4y^2 - 49 \quad \text{Original expression}$$
$$a^2 - b^2 \quad \text{General form}$$

Since $(2y)^2 = 2^2 y^2 = 4y^2$ and $7^2 = 49$, rewrite the original expression as

$$(2y)^2 - (7)^2 \quad \text{Rewrite the original expression.}$$
$$a^2 - b^2 \quad \text{Match with the general form.}$$

Matching first term to first term and second term to second term yields

$$a = 2y \quad \text{and} \quad b = 7$$

In the factorization $a^2 - b^2 = (a - b)(a + b)$, substitute in $a = 2y$ and $b = 7$. The factorization is

$$4y^2 - 49 = (2y - 7)(2y + 7)$$

Check the solution by multiplying out the expression.

$$(2y - 7)(2y + 7) = 4y^2 + 14y - 14y - 49$$
$$= 4y^2 - 49$$

PRACTICE PROBLEM FOR EXAMPLE 2

First determine whether each polynomial expression is a difference of squares. Then factor each polynomial expression.

a. $p^2 - 100$ **b.** $9x^2 - 64$

Perfect Square Trinomials

Recall that a number is a perfect square if it is the square of a natural number. For example, 81 is a perfect square because $9^2 = 81$. A monomial can also be a perfect square. We learned that $4y^2$ is a perfect square because $(2y)^2 = 2^2 y^2 = 4y^2$. A polynomial expression can also be a perfect square. For example, the trinomial $x^2 + 4x + 4$ is a perfect square because $x^2 + 4x + 4 = (x + 2)^2$. To see why this is so, let's expand $(x + 2)^2$.

$$(x + 2)^2 = (x + 2)(x + 2) \quad \text{Rewrite as a product.}$$
$$= x^2 + 2x + 2x + 4 \quad \text{Use FOIL to multiply.}$$
$$= x^2 + 4x + 4 \quad \text{Add like terms.}$$

These **perfect square trinomials** have a special form. If we learn how to recognize a perfect square trinomial, we can use a shortcut to factor it.

> **DEFINITION**
>
> **Perfect Square Trinomial** A trinomial of the form $a^2 + 2ab + b^2$ or $a^2 - 2ab + b^2$ is a perfect square trinomial.
>
> *Note:* In both forms, the middle term is twice (a times b).

SECTION 6.3 Factoring Special Forms   559

Example 3 — Recognizing perfect square trinomials

Determine whether or not each polynomial expression is a perfect square trinomial.
a. $x^2 + 6x + 9$ **b.** $x^2 + 12x + 16$ **c.** $4x^2 - 20x + 25$

SOLUTION

a. Compare $x^2 + 6x + 9$ with the standard form given in the definition. Since the given expression involves an addition sign in front of the second term, match it with the first form $a^2 + 2ab + b^2$.

$x^2 + 6x + 9$	The original trinomial
$a^2 + 2ab + b^2$	General form
$a = x$ and $b = 3$	Compare the first and last terms. Recall that $3^2 = 9$.
$6x \stackrel{?}{=} 2ab$	Check the middle term.
$6x \stackrel{?}{=} 2(x)(3)$	Substitute in $a = x$ and $b = 3$.
$6x = 6x$	It checks.

Therefore, $x^2 + 6x + 9$ is a perfect square trinomial.

b. Compare $x^2 + 12x + 16$ with the standard form given in the definition. Since the given expression involves an addition sign in front of the second term, match it with the first form, $a^2 + 2ab + b^2$.

$x^2 + 12x + 16$	The original trinomial.
$a^2 + 2ab + b^2$	General form.
$a = x$ and $b = 4$	Compare the first and last terms. Recall that $4^2 = 16$.
$12x \stackrel{?}{=} 2ab$	Check the middle term.
$12x \stackrel{?}{=} 2(x)(4)$	Substitute in $a = x$ and $b = 4$.
$12x \neq 8x$	It does not check.

Therefore, $x^2 + 12x + 16$ is *not* a perfect square trinomial.

c. Compare $4x^2 - 20x + 25$ with the standard form given in the definition. Since the given expression involves a subtraction sign in front of the second term, match it with the second form, $a^2 - 2ab + b^2$.

$4x^2 - 20x + 25$	The original trinomial.
$a^2 - 2ab + b^2$	General form.
$a = 2x$ and $b = 5$	Compare the first and last terms. $(2x)^2 = 4x^2$ and $5^2 = 25$.
$20x \stackrel{?}{=} 2ab$	Check the middle term.
$20x \stackrel{?}{=} 2(2x)(5)$	Substitute in $a = 2x$ and $b = 5$.
$20x = 20x$	It checks.

Therefore, $4x^2 - 20x + 25$ is a perfect square trinomial.

PRACTICE PROBLEM FOR EXAMPLE 3

Determine whether each polynomial expression is a perfect square trinomial or not.

a. $x^2 - 4x + 4$ **b.** $9x^2 + 30x + 25$ **c.** $x^2 - 7x + 81$

Now that we can recognize a perfect square trinomial, we can use this information to factor it. Let's multiply out the following expressions and see what results.

$(a + b)^2 = (a + b)(a + b)$ Multiply out, using FOIL. $(a - b)^2 = (a - b)(a - b)$
$ = a^2 + ab + ab + b^2$ $ = a^2 - ab - ab + b^2$
$ = a^2 + 2ab + b^2$ These are perfect square trinomials. $= a^2 - 2ab + b^2$

560 CHAPTER 6 Factoring

> **Factoring Polynomials with Three Terms**
>
> **GCF First**
> Factor out the GCF first, if there is one.
> $$6x^2 + 9x - 15$$
> $$= 3(2x^2 + 3x - 5)$$
>
> **Perfect Square Trinomial**
> Use the patterns.
> $$a^2 + 2ab + b^2 = (a + b)^2$$
> $$a^2 - 2ab + b^2 = (a - b)^2$$
> $$x^2 + 6x + 9 = (x + 3)^2$$
> $$x^2 - 10x + 25 = (x - 5)^2$$

Looking at the process backwards, we see that the two perfect square trinomials factor as
$$a^2 + 2ab + b^2 = (a + b)^2 \qquad a^2 - 2ab + b^2 = (a - b)^2$$

> **Steps to Factor a Perfect Square Trinomial**
> 1. Determine whether or not the trinomial has the form $a^2 + 2ab + b^2$ or $a^2 - 2ab + b^2$. (Is it a perfect square trinomial?) If not, use the AC method of factoring.
> 2. If it is a perfect square trinomial, determine a and b.
> 3. Factor into one of the following forms, depending on the sign before the second term.
> $$a^2 + 2ab + b^2 = (a + b)^2$$
> $$a^2 - 2ab + b^2 = (a - b)^2$$
>
> *Note:* Pay careful attention to the sign before the second term. Check the result by multiplying out the expression.

> ■ **Connecting the Concepts**
>
> **Which factoring method should we use?**
>
> If we don't recognize a perfect square trinomial, it can still be factored. Perfect square trinomials are just a special type of trinomial. Therefore, we can still factor it by inspection or using the AC method.

Example 4 Factoring a perfect square trinomial

Factor each of the following trinomials.

a. $x^2 + 10x + 25$ **b.** $x^2 - 8x + 16$

SOLUTION

a. Step 1 Determine whether or not the trinomial has the form $a^2 + 2ab + b^2$ or $a^2 - 2ab + b^2$.

Determine whether or not $x^2 + 10x + 25$ is a perfect square trinomial. Since the given expression involves an addition sign in front of the second term, we compare it to the first form, $a^2 + 2ab + b^2$.

$x^2 + 10x + 25$	The original trinomial
$a^2 + 2ab + b^2$	General form
$a = x$ and $b = 5$	Compare the first and last terms.
$10x \stackrel{?}{=} 2ab$	Check the middle term.
$10x \stackrel{?}{=} 2(x)(5)$	Substitute in $a = x$ and $b = 5$.
$10x = 10x$	It checks.

Step 2 If it is a perfect square trinomial, determine a and b.

Since $x^2 + 10x + 25$ is a perfect square trinomial, we see that $a = x$ and $b = 5$.

Step 3 Factor into one of the following forms, depending on the sign before the second term. $a^2 + 2ab + b^2 = (a + b)^2$ or $a^2 - 2ab + b^2 = (a - b)^2$.

The sign in front of the second term is positive, so it factors like the form $a^2 + 2ab + b^2 = (a + b)^2$, or
$$x^2 + 10x + 25 = (x + 5)^2$$

b. Determine whether or not $x^2 - 8x + 16$ is a perfect square trinomial. The given expression involves a subtraction sign in front of the second term, so we compare it to the second form, $a^2 - 2ab + b^2$.

SECTION 6.3 Factoring Special Forms 561

$x^2 - 8x + 16$ The original trinomial
$a^2 - 2ab + b^2$ General form
$a = x$ and $b = 4$ Compare the first and last terms.
$8x \stackrel{?}{=} 2ab$ Check the middle term.
$8x \stackrel{?}{=} 2(x)(4)$ Substitute in $a = x$ and $b = 4$.
$8x = 8x$ It checks.

Because $x^2 - 8x + 16$ is a perfect square trinomial with $a = x$ and $b = 4$ and the sign in front of the second term is negative, it factors like the form $a^2 - 2ab + b^2 = (a - b)^2$, or

$$x^2 - 8x + 16 = (x - 4)^2$$

PRACTICE PROBLEM FOR EXAMPLE 4

Factor each of the following trinomials.

a. $x^2 + 16x + 64$ **b.** $9x^2 - 6x + 1$

■ Summary of Factoring Tools

Now we will summarize our list of how to factor a polynomial expression using the factoring tools below. All of these tools are also listed in the Factoring Toolbox in the back of the book.

Factoring Polynomials with Two Terms (x+2)(x-5)

GCF First
Factor out the GCF first, if there is one.
$6x^2 + 4x = 2x(3x + 2)$

Difference of Squares
Use the difference of squares pattern.
$a^2 - b^2 = (a + b)(a - b)$
$x^2 - 25 = (x + 5)(x - 5)$

Factoring Polynomials with Four Terms (x+2)(x-5)

GCF First
Factor out the GCF first, if there is one.
$6x^2 + 4x - 8x + 10$
$= 2(3x^2 + 2x - 4x + 5)$

Factor by Grouping
$12x^2 + 8x - 21x - 14$
$= (4x - 7)(3x + 2)$

Factoring Polynomials with Three Terms (x+2)(x-5)

GCF First
Factor out the GCF first, if there is one.
$6x^2 + 9x - 15 = 3(2x^2 + 3x - 5)$

Coefficient of x^2 is 1
Trial and error or AC method
$x^2 + 7x + 12 = (x + 3)(x + 4)$

Coefficient of x^2 is not 1
AC method or trial and error.
$12x^2 + 8x - 15 = (2x + 3)(6x - 5)$

Perfect Square Trinomial
Use the patterns
$a^2 + 2ab + b^2 = (a + b)^2$
$a^2 - 2ab + b^2 = (a - b)^2$
$x^2 + 6x + 9 = (x + 3)^2$
$x^2 - 10x + 25 = (x - 5)^2$

Example 5 Factoring completely

Factor each of the following polynomial expressions completely.

a. $6y^2 - 24$ **b.** $4xy + 6x + 4y + 6$ **c.** $12k^3 - 60k^2 + 75k$

SOLUTION

a. Look for the GCF of $6y^2 - 24$. The terms have a common factor of 6, so we factor out the GCF of 6.

$$6y^2 - 24 = 6 \cdot y^2 - 6 \cdot 4$$
$$= 6(y^2 - 4)$$

562 CHAPTER 6 Factoring

The expression in parentheses, $y^2 - 4$, has two terms that are being subtracted. Both terms are perfect squares, so this is a difference of squares.

$$6y^2 - 24 = 6(y^2 - 4) \qquad y^2 - 4 = a^2 - b^2 = (a-b)(a+b)$$
$$= 6(y-2)(y+2) \qquad a = y \text{ and } b = 2$$

The factorization is $6y^2 - 24 = 6(y-2)(y+2)$.

b. Look for the GCF of $4xy + 6x + 4y + 6$. All of the terms have a common factor of 2, so factor out the GCF of 2.

$$4xy + 6x + 4y + 6 = 2(2xy + 3x + 2y + 3)$$

The expression in the parentheses, $2xy + 3x + 2y + 3$, has four terms, so we try to factor it by grouping.

$$4xy + 6x + 4y + 6 = 2(2xy + 3x + 2y + 3) \qquad \text{Group the first two and last two terms.}$$
$$= 2[(2xy + 3x) + (2y + 3)] \qquad \text{The first two terms have } x \text{ in common.}$$
$$= 2[x(2y + 3) + 1(2y + 3)] \qquad \text{The last two terms have only 1 in common.}$$
$$= 2(2y + 3)(x + 1)$$

The factorization is $4xy + 6x + 4y + 6 = 2(2y + 3)(x + 1)$.

c. The polynomial $12k^3 - 60k^2 + 75k$ has a GCF of $3k$. Factoring out the GCF of $3k$ yields

$$12k^3 - 60k^2 + 75k = 3k(4k^2 - 20k + 25)$$

The expression inside the parentheses, $4k^2 - 20k + 25$, has three terms. Let's check to see whether it is a perfect square trinomial. If not, then we will factor it using the AC method, because the lead coefficient is not 1.

$4k^2 - 20k + 25$	The original trinomial
$a^2 - 2ab + b^2$	General form
$a = 2k$ and $b = 5$	Compare the first and last terms.
$20k \stackrel{?}{=} 2ab$	Check the middle term.
$20k \stackrel{?}{=} 2(2k)(5)$	Substitute in $a = 2k$ and $b = 5$.
$20k = 20k$	It checks.

Since $4k^2 - 20k + 25$ is a perfect square trinomial with $a = 2k$ and $b = 5$ and the sign in front of the second term is negative, it factors like the form $a^2 - 2ab + b^2 = (a-b)^2$, or

$$4k^2 - 20k + 25 = (2k - 5)^2$$

The original polynomial factors as $12k^3 - 60k^2 + 75k = 3k(2k - 5)^2$.

PRACTICE PROBLEM FOR EXAMPLE 5

Factor each of the following polynomial expressions completely.

a. $15pq - 60p + 10q - 40$ **b.** $3x^2 - 432$ **c.** $-6a^3 - 21a^2 + 12a$

6.3 Vocabulary Exercises

1. A polynomial that factors to the form $(a + b)^2$ or $(a - b)^2$ is called a(n) _____.

2. A polynomial of the form $a^2 - b^2$ is called _____.

3. When factoring, the number of _____ is a key indicator of what factoring tool should be used.

6.3 Exercises

For Exercises 1 through 8, determine the number of terms in each expression.

1. $-12x^2 + 2$
2. $16x + 8y$
3. $-6ab^2$
4. $9a^2b^2c$
5. $4x + 8 - 2y - 4$
6. $7 - 2x + 8xy - 3y$
7. $2x^2 + 6x - 7$
8. $10 + 2y - x^2$

For Exercises 9 through 18, determine whether or not each expression is a difference of squares. Do not factor the expressions.

9. $v^2 - 144$
10. $x^2 - 1$
11. $81 - y^2$
12. $121 - v^2$
13. $9 + p^2$
14. $64 + x^2$
15. $100x^2 - 49$
16. $25 - 36y^2$
17. $x^2 + 1$
18. $y^2 + 25$

For Exercises 19 through 34, factor each polynomial expression if possible. Check the results.

19. $x^2 - 144$
20. $r^2 - 4$
21. $1 - y^2$
22. $121 - v^2$
23. $9 + p^2$
24. $64 + x^2$
25. $100x^2 - 49$
26. $25 - 36y^2$
27. $x^2 + 1$
28. $y^2 + 25$
29. $81 - a^2$
30. $36 - p^2$
31. $16x^2 - 49$
32. $100m^2 - 1$
33. $9t^2 - 4$
34. $25p^2 - 81$

For Exercises 35 through 44, determine whether or not each expression is a perfect square trinomial. Do not factor the expressions.

35. $a^2 + 12a + 36$
36. $y^2 + 2y + 1$
37. $25x^2 - 10x + 1$
38. $16a^2 - 24a + 9$
39. $9y^2 + 72y + 4$
40. $x^2 - 50x + 25$
41. $x^2 - 6x + 12$
42. $y^2 + 7y + 16$
43. $a^2 - 8ab + 16b^2$
44. $4x^2 + 4xy + y^2$

For Exercises 45 through 62, factor each polynomial expression. If a polynomial cannot be factored, write the answer as prime.

45. $x^2 + 2x + 1$
46. $x^2 + 14x + 49$
47. $y^2 - 8y + 16$
48. $y^2 - 18y + 81$
49. $9p^2 - 6p + 1$
50. $9p^2 + 12p + 4$
51. $t^2 + 5t + 6$
52. $a^2 + 11a + 10$
53. $6x^2 + 7x - 20$
54. $25m^2 - 20m - 21$
55. $4x^2 + 20xy + 25y^2$
56. $36a^2 - 12ab + b^2$
57. $9x^2 + 24xy + 16y^2$
58. $a^2 + 6ab + 9b^2$
59. $-t^2 + 6t - 9$
60. $-x^2 + 10x - 25$
61. $y^2 + 9$
62. $x^2 + 25$

For Exercises 63 through 90, factor each of the following polynomial expressions completely. Remember to factor out the GCF first. If a polynomial cannot be factored, write the answer as prime.

63. $3x^2 - 75$
64. $4 - 144x^2$
65. $17xy - 34x^2$
66. $9ab + 45b^2$
67. $5x^2 - 125$
68. $3x^2 - 48$
69. $4pr + 12p - 2r - 6$
70. $3ab - 2a + 15b - 10$
71. $6x^2 - 27x - 15$
72. $10x^2 - 64x - 42$
73. $-18a^2 - 12a - 2$
74. $-9b^2 + 48b - 64$
75. $5x^2 + 20$
76. $18a^2 + 32$
77. $3t^2 + 6t + 3$
78. $18a^2 - 12a + 2$
79. $5xy^2 + 40xy + 80x$
80. $100m^2n - 40mn + 4n$
81. $5x^2 - 20x - 105$
82. $7x^2 - 42x + 56$
83. $-20x^2 + 46x - 24$
84. $18x^2 + 51x + 15$
85. $3a^3 + 7a^2 - 20a$
86. $5n^3 + 21n^2 + 4n$
87. $\dfrac{x^2}{4} - 25$
88. $\dfrac{x^2}{9} - 4$
89. $x^2 - \dfrac{25}{16}$
90. $x^2 - \dfrac{9}{49}$

For Exercises 91 through 94, graph the equations. Clearly label and scale the axes.

91. $y = 0.25x + 5$
92. $y = -0.2x - 3$
93. $-\dfrac{1}{2}x + 5y = 10$
94. $\dfrac{3}{4}x - 2y = 12$

For Exercises 95 through 100, perform the indicated operations.

95. $(2x + 5)(-3x + 8)$
96. $(x + 6)(-4x - 7)$
97. $(x^2 + 5x - 24) \div (x - 3)$
98. $(6x^2 + x - 35) \div (3x - 7)$
99. $(9x - 8) - (2x - 3)$
100. $(x^2 + 5x - 24) - (4x - 18)$

6.4 Solving Quadratic Equations by Factoring

LEARNING OBJECTIVES
- Recognize a quadratic equation.
- Use the zero-product property.
- Solve quadratic equations by factoring.

■ Recognizing a Quadratic Equation

An important use of algebra is to solve equations. Factoring can help us to solve certain types of equations, such as quadratic equations.

> **DEFINITION**
>
> **Quadratic Equation**   A **quadratic equation** has the form $ax^2 + bx + c = 0$, where a, b, and c are real numbers and $a \neq 0$. This is called the **standard form of a quadratic equation**.
>
> *Note:* A quadratic equation is a polynomial equation with the highest power on the variable equal to 2 (degree is two).

> ■ **Connecting the Concepts**
>
> **What is a linear equation?**
>
> In Section 2.1, we defined a *linear equation* as an equation that can be put in the form $mx + b = 0$. Notice that a linear equation is a polynomial equation with the highest power on the variable equal to 1 (degree is 1).

It is important to be able to distinguish between a linear equation and a quadratic equation. The methods we use to solve the two types of equations are different. Both linear and quadratic equations have polynomials on one or both sides of the equal sign. A linear equation has degree 1. A quadratic equation has degree 2. Recall from Chapter 5 that the highest power on the variable in a single-variable polynomial is the degree of the polynomial. Quadratic equations have degree equal to 2 and are also called *second-degree* equations.

Example 1   Types of equations

For each equation, identify if it is a linear equation or a quadratic equation.

a. $-7 + 14x = 0$ **b.** $x^2 - 6x + 9 = 0$ **c.** $9 - y^2 = 4$

SOLUTION

a. This equation can be rewritten as follows:

$$-7 + 14x = 0$$ *Rewrite the equation using the commutative property of addition.*
$$14x - 7 = 0$$

The equation has been written in the form $mx + b = 0$, so this equation is linear. Another way to recognize a linear equation is to note it is a polynomial of degree 1.

b. This equation is quadratic because it is in the form $ax^2 + bx + c = 0$.

$$x^2 - 6x + 9 = 0$$ *Given equation*
$$ax^2 + bx + c = 0$$ *Standard form of a quadratic equation*

Another way to recognize a quadratic equation is by noting it is a polynomial of degree 2.

c. Rewrite this equation to put it in descending order, which means that the highest-degree term should go first (reading from left to right).

$$9 - y^2 = 4 \qquad \text{\color{blue}Rewrite the equation.}$$
$$-y^2 + 9 = 4 \qquad \text{\color{blue}Set the right side equal to 0.}$$
$$\underline{ -4 \quad -4}$$
$$-y^2 + 5 = 0$$

The equation has been put into the standard form for a quadratic equation. Therefore, this is a quadratic equation. The highest degree on the variable is 2.

PRACTICE PROBLEM FOR EXAMPLE 1

For each equation, identify if it is a linear equation or a quadratic equation.

a. $-2x + x^2 = 6$ **b.** $4 - 8y = 17$

■ Zero-Product Property

Now that we can recognize the difference between linear and quadratic equations, we learn how to solve a quadratic equation. We studied solving linear equations in Chapter 2. Before we learn how to solve a quadratic equation, we will investigate an important property that will be used when solving quadratic equations. This property is called the **zero-product property**.

> **DEFINITION**
>
> **Zero-Product Property**   The zero-product property states that if any two numbers are multiplied together to get zero, then one or both of the two numbers must be zero. In symbols:
>
> If $a \cdot b = 0$, then either $a = 0$ or $b = 0$ or both are equal to 0.

■ CONCEPT INVESTIGATION

Why is it called the zero-product property?

1. In the zero-product property, does one side have to be 0? We will explore this idea. If two numbers multiplied together equal 1, does that mean that either or both of the numbers are 1?

In symbols, if $a \cdot b = 1$, then is $a = 1$ or $b = 1$, or are both equal to 1?

To test if the product of two numbers is 1 implies one or both of the numbers must equal 1, fill in the following table with some values that multiply to 1. The first two rows have been done for you to get you started.

$a \cdot b = 1$	a	b	$a \stackrel{?}{=} 1$	$b \stackrel{?}{=} 1$	Both a and b Equal to 1?
$1 \cdot 1 = 1$	$a = 1$	$b = 1$	Yes	Yes	Yes
$\frac{1}{2} \cdot 2 = 1$	$a = \frac{1}{2}$	$b = 2$	No	No	No
$4 \cdot 0.25 = 1$					
$-5 \cdot \left(-\frac{1}{5}\right) = 1$					

2. So if we multiply two numbers together and get 1, must either or both of the numbers always equal 1?

3. Does the zero-product property mean that one side must equal 0 for the rest of the property to make sense?

4. Now we will explore the operation involved in the zero-product property. The operation is a product, a multiplication. What happens if we change the product (multiplication) to another operation, for instance, a sum (addition)?

In symbols, if $a + b = 0$, does $a = 0$ or $b = 0$, or do both equal 0?

To test if the sum of two numbers is 0 implies one or both of the numbers must equal 0, fill in the following table with values that sum (add) to 0. Then check to see whether that means $a = 0$ or $b = 0$ or both equal 0. The first two rows have been done for you to get you started.

$a + b = 0$	a	b	$a \stackrel{?}{=} 0$	$b \stackrel{?}{=} 0$	Both a and b Equal to 0?
$0 + 0 = 0$	$a = 0$	$b = 0$	Yes	Yes	Yes
$1 + (-1) = 0$	$a = 1$	$b = -1$	No	No	No
$-3 + 3 = 0$					
$6 + (-6) = 0$					

5. If we add two numbers together and get 0, must one or both of the numbers equal 0?

6. Do you think that for the zero-product property to work, the operation must be multiplication?

We now see how the zero-product property helps us solve equations.

Example 2 Using the zero-product property to solve equations

Solve each equation, using the zero-product property if possible. Check the answer(s).

a. $3x = 0$ **b.** $4(x + 2) = 0$ **c.** $(x - 3) \cdot x = 0$ **d.** $3 + x = 0$

SOLUTION

a. To solve this equation, use the zero-product property because it is a product equal to zero.

$$3x = 0 \quad \text{Rewrite as a product.}$$
$$3 \cdot x = 0 \quad \text{Use the zero-product property.}$$
$$3 \neq 0 \quad \text{or} \quad x = 0 \quad \text{Since } 3 \neq 0, \text{ we have that } x = 0.$$

Therefore, the answer is $x = 0$. To check the answer, substitute $x = 0$ into the original equation and see whether both sides are equal.

$$3(0) \stackrel{?}{=} 0 \quad \text{Substitute in } x = 0.$$
$$0 = 0 \quad \text{The answer checks.}$$

b. First rewrite the left side as a product. Then solve, using the zero-product property.

$$4(x + 2) = 0 \quad \text{The left side is a product equal to zero.}$$
$$4 = 0 \quad \text{or} \quad x + 2 = 0 \quad \text{Apply the zero-product property.}$$
$$4 \neq 0 \quad \text{so} \quad x + 2 = 0$$
$$x + 2 = 0 \quad \text{Solve for } x.$$
$$\underline{-2 \quad -2}$$
$$x = -2$$

Therefore, the answer is $x = -2$. To check the answer, substitute $x = -2$ into the original equation and see whether both sides are equal.

$$4((-2) + 2) \stackrel{?}{=} 0 \quad \text{Substitute in } x = -2.$$
$$4(0) \stackrel{?}{=} 0$$
$$0 = 0 \quad \text{The answer checks.}$$

c. This equation is written with a product, so solve using the zero-product property.

$$(x - 3) \cdot x = 0 \quad \text{The left side is a product equal to zero.}$$
$$x - 3 = 0 \quad \text{or} \quad x = 0 \quad \text{Apply the zero-product property.}$$
$$x - 3 = 0 \quad \text{or} \quad x = 0 \quad \text{The second equation is solved; } x = 0.$$
$$\underline{+3 \quad +3} \quad \text{Solve the first equation for x.}$$
$$x = 3$$

The two answers are $x = 3$ or $x = 0$. Check both of these answers in the original problem.

$$\begin{array}{cc} \text{Check } x = 3 & \text{Check } x = 0 \\ (3 - 3) \cdot 3 \stackrel{?}{=} 0 & (0 - 3) \cdot 0 \stackrel{?}{=} 0 \\ (0) \cdot 3 \stackrel{?}{=} 0 & (-3) \cdot 0 \stackrel{?}{=} 0 \\ 0 = 0 \quad \text{Both answers check.} & 0 = 0 \end{array}$$

d. We cannot use the zero-product property to solve this equation because no product is involved. This is a linear equation. Solve it as in Chapter 2.

$$3 + x = 0$$
$$\underline{-3 \quad -3}$$
$$x = -3$$

Checking $x = -3$ in the original equation yields

$$3 + (-3) \stackrel{?}{=} 0$$
$$0 = 0 \quad \text{The answer checks.}$$

PRACTICE PROBLEM FOR EXAMPLE 2

Solve each equation, using the zero-product property if possible. Check the answer(s).

a. $5(x - 6) = 0$ **b.** $y(y - 1) = 0$ **c.** $y + y - 1 = 0$

Solving Quadratic Equations by Factoring

In part c of Example 2, we solved the equation $(x - 3) \cdot x = 0$. Using the distributive property, multiply the left side of this equation out to see whether it is a linear equation or a quadratic equation.

$$(x - 3) \cdot x = 0 \quad \text{The original equation. It is in factored form.}$$
$$x^2 - 3x = 0 \quad \text{Multiply out the left side.}$$

This is a quadratic equation. We solved the quadratic equation using the factored form. To solve a quadratic equation, first set it equal to 0. Then factor it and use the zero-product property to solve for the variable.

568 CHAPTER 6 Factoring

> **Steps to Solve a Quadratic Equation by Factoring**
> 1. Put the equation into the standard form of a quadratic equation, $ax^2 + bx + c = 0$.
> 2. Factor the quadratic expression.
> 3. Using the zero-product property, set each factor equal to 0, and solve for the variable.
> 4. Check all answers in the *original* equation.
>
> *Note:* One side of the equation must be equal to 0. If one side of the equation is not 0, use the addition property of equality to set one side equal to 0.

Quadratic Equations

Factoring

Use when the quadratic equation has small coefficients that factor easily. Set the quadratic equal to zero, factor and then use the zero factor property to write two or more simpler equations.

$$x^2 + 7x + 10 = 0$$
$$(x + 5)(x + 2) = 0$$
$$x + 5 = 0 \quad or \quad x + 2 = 0$$

Example 3 Solving a quadratic equation

Solve each quadratic equation. Check all answers.

a. $x^2 + 4x - 5 = 0$ **b.** $p^2 - 25 = 0$

c. $2y^2 + 7y = 4$

SOLUTION

a. Step 1 Put the equation into the standard form of a quadratic equation, $ax^2 + bx + c = 0$.

This quadratic equation is in the standard form of a quadratic equation.

Step 2 Factor the quadratic expression.

Because the lead coefficient, the coefficient on x^2, is 1, we use factoring by inspection. Look for the factors of $c = -5$ that sum to $b = 4$.

$$\begin{array}{ll} c = -5 & b = 4 \\ 1(-5) = -5 & 1 + (-5) = -4 \neq 4 \\ -1(5) = -5 & -1 + 5 = 4 = b \end{array}$$

Therefore, the factorization is

$$x^2 + 4x - 5 = 0$$
$$(x + 5)(x - 1) = 0$$

Step 3 Using the zero-product property, set each factor equal to 0, and solve for the variable.

$$(x + 5)(x - 1) = 0 \qquad \text{Apply the zero-product property.}$$
$$\begin{array}{ll} x + 5 = 0 & x - 1 = 0 \qquad \text{Solve both equations for x.} \\ \underline{-5 \quad -5} & \underline{+1 \quad +1} \\ x = -5 & x = 1 \end{array}$$

Step 4 Check all the answers in the original equation.

$$x^2 + 4x - 5 = 0$$

$$\begin{array}{cc} x = -5 & x = 1 \\ (-5)^2 + 4(-5) - 5 \stackrel{?}{=} 0 & (1)^2 + 4(1) - 5 \stackrel{?}{=} 0 \\ 25 - 20 - 5 \stackrel{?}{=} 0 & 1 + 4 - 5 \stackrel{?}{=} 0 \\ 0 = 0 & 0 = 0 \end{array}$$

Both answers check.

b. Skip the first step because this quadratic equation is set equal to 0. Following step 2, factor the left side. Looking at this equation carefully, notice that the left side is a difference of squares.

$$p^2 - 25 = 0$$
$$(p)^2 - 5^2 = 0$$

Using the difference of squares formula,

$$a^2 - b^2 = (a - b)(a + b) \quad \text{with} \quad a = p \text{ and } b = 5$$

the left side factors as

$$p^2 - 25 = 0$$
$$(p)^2 - 5^2 = 0$$
$$(p - 5)(p + 5) = 0$$

Use the zero-product property to solve for the variable p.

$$(p - 5)(p + 5) = 0$$
$$p - 5 = 0 \qquad p + 5 = 0$$
$$\underline{+5 \quad +5} \qquad \underline{-5 \quad -5}$$
$$p = 5 \qquad p = -5$$

Check the answers in the original problem.

$$\underline{p = -5} \qquad\qquad \underline{p = 5}$$
$$(-5)^2 - 25 \stackrel{?}{=} 0 \qquad (5)^2 - 25 \stackrel{?}{=} 0$$
$$25 - 25 \stackrel{?}{=} 0 \qquad 25 - 25 \stackrel{?}{=} 0$$
$$0 = 0 \quad \text{Both answers check.} \quad 0 = 0$$

c. Set the right side equal to 0.

$$2y^2 + 7y = 4 \qquad \text{Subtract 4 from both sides.}$$
$$\underline{\quad -4 \quad -4}$$
$$2y^2 + 7y - 4 = 0 \qquad \text{Now there is a 0 on one side.}$$

Factor the left side using the AC method.

$$2y^2 + 7y - 4 = 0$$
$$(2y - 1)(y + 4) = 0$$

Use the zero-product property to solve the quadratic equation.

$$(2y - 1)(y + 4) = 0$$
$$2y - 1 = 0 \qquad y + 4 = 0$$
$$\underline{+1 \quad +1} \qquad \underline{-4 \quad -4}$$
$$\qquad\qquad\qquad y = -4$$
$$\frac{2y}{2} = \frac{1}{2}$$
$$y = \frac{1}{2}$$

The answers are $y = \dfrac{1}{2}$ and $y = -4$. Check both answers in the original equation.

$$y = -4$$
$$2(-4)^2 + 7(-4) - 4 \stackrel{?}{=} 0$$
$$2(16) - 28 - 4 \stackrel{?}{=} 0$$
$$32 - 28 - 4 \stackrel{?}{=} 0$$
$$0 = 0 \quad \text{Both answers check.}$$

$$y = \dfrac{1}{2}$$
$$2\left(\dfrac{1}{2}\right)^2 + 7\left(\dfrac{1}{2}\right) - 4 \stackrel{?}{=} 0$$
$$2\left(\dfrac{1}{4}\right) + \dfrac{7}{2} - 4 \stackrel{?}{=} 0$$
$$\dfrac{1}{2} + \dfrac{7}{2} - 4 \stackrel{?}{=} 0$$
$$0 = 0$$

PRACTICE PROBLEM FOR EXAMPLE 3

Solve each quadratic equation. Check all answers.

a. $6d^2 + 11d + 3 = 0$ **b.** $t^2 - 144 = 0$ **c.** $2n^2 - 11n = 6$

Example 4 Casino revenue

A local casino's profit P in millions of dollars can be estimated using the equation $P = t^2 + 4t + 15$, where t is years since 2010.

a. Use the equation to estimate the casino's profit in 2020.

b. Find when the casino's profit was or will be $92 million.

SOLUTION

a. The year 2020 can be represented as $t = 10$, so we will substitute 10 for t and calculate the profit.

$$P = (10)^2 + 4(10) + 15 \quad \text{Substitute 10 for } t \text{ and calculate.}$$
$$P = 155$$

Therefore, in 2020, the casino's profit will be approximately $155 million.

b. Since we want to know the year or years when the profit will be $92 million, substitute 92 for P and solve for t.

$$92 = t^2 + 4t + 15 \quad \text{Substitute 92 for } P \text{ and solve.}$$
$$\underline{-92 \qquad\qquad -92} \quad \text{Set the quadratic equal to zero. Factor.}$$
$$0 = t^2 + 4t - 77 \quad \text{The factors of } -77 \text{ that add up to 4 are } -7 \text{ and } 11.$$
$$0 = (t - 7)(t + 11)$$
$$t - 7 = 0 \qquad t + 11 = 0 \quad \text{Set each factor equal to zero and solve.}$$
$$\underline{+7 \quad +7} \qquad \underline{-11 \quad -11}$$
$$t = 7 \qquad\qquad t = -11$$

Check these answers by substituting them into the original equation.

$$t = 7$$
$$92 \stackrel{?}{=} (7)^2 + 4(7) + 15$$
$$92 \stackrel{?}{=} 49 + 28 + 15$$
$$92 = 92 \quad \text{Both answers check.}$$

$$t = -11$$
$$92 \stackrel{?}{=} (-11)^2 + 4(-11) + 15$$
$$92 \stackrel{?}{=} 121 - 44 + 15$$
$$92 = 92$$

The variable *t* represents years since 2010.

$t = 7$ converts to the year $2010 + 7 = 2017$

$t = -11$ converts to the year $2010 + (-11) = 1999$

The casino's profit was $92 million in about 1999 and 2017.

PRACTICE PROBLEM FOR EXAMPLE 4

JT's Fishing Supplies profit *P* in hundreds of dollars from selling *n* hundred fishing kits can be estimated by the equation $P = n^2 + 6n + 10$.

a. Use the equation to estimate the profit from selling 400 fishing kits.

b. Find how many fishing kits JT's Fishing Supplies needs to sell to make $6500 in profit.

Example 5   Solving polynomial equations by factoring

Solve each polynomial equation by factoring. Check the answer(s).

a. $(x - 10)(x - 2) = -15$ **b.** $2x^3 - 4x^2 - 30x = 0$ **c.** $n(n - 3) = 10$

SOLUTION

a. The equation $(x - 10)(x - 2) = -15$ is not equal to 0. First multiply out the left side and then rewrite the equation to get a 0 on the right side of the equal sign.

$(x - 10)(x - 2) = -15$ Multiply out the left side, using FOIL.

$x^2 - 2x - 10x + 20 = -15$ Simplify the left side.

$x^2 - 12x + 20 = -15$ Add 15 to both sides to set one side equal to 0.

$ +15 +15$

$x^2 - 12x + 35 = 0$

Now the quadratic is in standard form, $ax^2 + bx + c = 0$. Since $a = 1$, factor using inspection.

$$x^2 - 12x + 35 = 0$$
$$(x - 5)(x - 7) = 0$$

Apply the zero-product property to solve this quadratic equation.

$$(x - 5)(x - 7) = 0$$

$x - 5 = 0$ $x - 7 = 0$

$ +5 +5$ $ +7 +7$

$x = 5$ $x = 7$

Check the answers in the original equation, $(x - 10)(x - 2) = -15$.

$x = 5$	$x = 7$
$(5 - 10)(5 - 2) \stackrel{?}{=} -15$	$(7 - 10)(7 - 2) \stackrel{?}{=} -15$
$(-5)(3) \stackrel{?}{=} -15$	$(-3)(5) \stackrel{?}{=} -15$
$-15 = -15$ Both answers check.	$-15 = -15$

The answers are $x = 5$ and $x = 7$.

> **Connecting the Concepts**
>
> **How do we solve polynomial equations of degree higher than 2?**
>
> In Example 5b, we are solving a polynomial equation of degree 3. Sometimes we can solve these equations by setting one side equal to 0 and then factoring. The zero-product property applies to any product that is equal to 0.

b. The equation $2x^3 - 4x^2 - 30x = 0$ is *not* a quadratic equation. The highest-degree term is 3. But it is set equal to 0, so factor the left side.

$$2x^3 - 4x^2 - 30x = 0$$
$$2x(x^2 - 2x - 15) = 0 \quad \text{Factor out the GCF first.}$$
$$2x(x - 5)(x + 3) = 0 \quad \text{Factor the quadratic } x^2 - 2x - 15.$$

Now that the left side is factored completely, apply the zero-product property.

$$2x(x - 5)(x + 3) = 0 \quad \text{Set each factor equal to 0 and solve.}$$

$2x = 0$	$x - 5 = 0$	$x + 3 = 0$
$\dfrac{2x}{2} = \dfrac{0}{2}$	$+5 \quad +5$	$-3 \quad -3$
$x = 0$	$x = 5$	$x = -3$

Check each answer in the original equation, $2x^3 - 4x^2 - 30x = 0$.

$$2x^3 - 4x^2 - 30x = 0$$

$\underline{x = 0}$
$$2(0)^3 - 4(0)^2 - 30(0) \stackrel{?}{=} 0$$
$$0 = 0$$

$\underline{x = 5}$
$$2(5)^3 - 4(5)^2 - 30(5) \stackrel{?}{=} 0$$
$$2(125) - 4(25) - 150 \stackrel{?}{=} 0$$
$$250 - 100 - 150 \stackrel{?}{=} 0$$
$$0 = 0$$

$\underline{x = -3}$
$$2(-3)^3 - 4(-3)^2 - 30(-3) \stackrel{?}{=} 0$$
$$2(-27) - 4(9) + 90 \stackrel{?}{=} 0$$
$$-54 - 36 + 90 \stackrel{?}{=} 0$$
$$0 = 0$$

All the answers check.

c. This is a quadratic equation, but it is not in standard form.

$$n(n - 3) = 10 \quad \text{Multiply out the left side.}$$
$$n^2 - 3n = 10 \quad \text{Subtract 10 from both sides.}$$
$$\underline{-10 \quad -10}$$
$$n^2 - 3n - 10 = 0 \quad \text{This is in standard form.}$$

Now the quadratic is in standard form, $ax^2 + bx + c = 0$. Factor the left side using inspection and apply the zero-product property to solve.

$$n^2 - 3n - 10 = 0$$
$$(n - 5)(n + 2) = 0$$

$n - 5 = 0$	$n + 2 = 0$
$+5 \quad +5$	$-2 \quad -2$
$n = 5$	$n = -2$

Check the answers in the original equation, $n(n - 3) = 10$.

$\underline{n = 5}$
$$5(5 - 3) \stackrel{?}{=} 10$$
$$5(2) \stackrel{?}{=} 10$$
$$10 = 10$$

$\underline{n = -2}$
$$-2(-2 - 3) \stackrel{?}{=} 10$$
$$-2(-5) \stackrel{?}{=} 10$$
$$10 = 10$$

Both answers check.

Therefore, the answers are $n = 5$ and $n = -2$.

PRACTICE PROBLEM FOR EXAMPLE 5

Solve each polynomial equation by factoring. Check the answer(s).

a. $20x^2 + 3x = 2$ **b.** $3x^3 + 6x^2 - 105x = 0$

6.4 Vocabulary Exercises

1. A quadratic equation is a polynomial equation with degree _____.

2. An equation of the form $ax^2 + bx + c = 0$ is called the _____ of a quadratic equation.

3. When two factors are multiplied together to get 0, the _____ states that one or both of the two must equal 0.

6.4 Exercises

For Exercises 1 through 14, determine whether each equation is linear, quadratic, or neither. Do not solve the equations.

1. $2x + 5 = 3$
2. $7t + 2 = 11$
3. $5x^3 + 7x^2 - 8 = 0$
4. $x^3 + 6x = 9$
5. $x^2 + 2x - 4 = 0$
6. $3x^2 - 6x + 7 = 0$
7. $4x^2 + 3x - 2 = 0$
8. $7b^2 - 5b + 6 = 0$
9. $6x = 12$
10. $3(x + 1) = -12$
11. $5(x + 2) = 3x - 8$
12. $-4(x + 1) = x$
13. $5 + 2y = 4y^2$
14. $7 = 8h^2 + 3h$

For Exercises 15 through 28, solve each equation using the zero-product property. Check the answers.

15. $5(x + 3) = 0$
16. $8(x - 6) = 0$
17. $-4(2n - 5) = 0$
18. $-2(3p + 1) = 0$
19. $a(a - 8) = 0$
20. $m(m - 4) = 0$
21. $y(2y + 7) = 0$
22. $x(3x + 4) = 0$
23. $3t(2t + 5) = 0$
24. $2h(5h + 12) = 0$
25. $(x + 5)(x - 6) = 0$
26. $5(x + 2)(x - 3) = 0$
27. $4p(2p + 3)(p - 9) = 0$
28. $2k(5k - 4)(2k + 3) = 0$

For Exercises 29 through 46, solve each equation by factoring. Check the answers.

29. $x^2 + 5x + 6 = 0$
30. $t^2 + 8t + 15 = 0$
31. $a^2 + 7a + 12 = 0$
32. $m^2 + 11m + 18 = 0$
33. $-x^2 - 2x + 3 = 0$
34. $-x^2 - 3x + 10 = 0$
35. $t^2 - 7t + 12 = 0$
36. $h^2 - 6h + 5 = 0$
37. $2x^2 + 6x + 4 = 0$
38. $3x^2 - 27x + 60 = 0$
39. $2x^2 + 9x - 5 = 0$
40. $6x^2 - 20x - 16 = 0$
41. $3x^2 - 10x - 8 = 0$
42. $5x^2 + 11x - 12 = 0$
43. $6x^2 + 23x + 20 = 0$
44. $7y^2 + 44y + 12 = 0$
45. $24a^2 + 25a + 6 = 0$
46. $15k^2 - 31k + 14 = 0$

47. An appliance manufacturer's average profit per appliance P in dollars when a appliances are sold in a single order can be estimated by the equation $P = -4a^2 + 80a$.

    a. Estimate the average profit per appliance for this company if it sells four appliances in a single order.

    b. Find the number of appliances the company needs to sell to have an average profit of $300 per appliance. (*Hint*: Factor out a negative with the GCF.)

48. A company's average cost per item C in dollars when n items are produced in a day can be estimated by the equation $C = -6n^2 + 162n$.

    a. Estimate the average cost per item when five items are produced in a day.

    b. Find the number of items produced in a day that results in the average cost per item being $840.

49. A plane manufacturer's average profit per plane P in millions of dollars when n planes are ordered by an airline at one time can be estimated by the equation $P = -n^2 + 16n$.

    a. Estimate the average profit per plane if an airline orders three planes.

    b. Find how many planes must be sold to an airline for the average profit per plane to be $60 million.

50. A customized paint manufacturer's average profit per quart P in dollars when q quarts of paint are mixed for one order can be estimated by the equation $P = -7q^2 + 105q$.

a. Find the average profit per quart for this company when 2 quarts are ordered.

b. Find how many quarts must be ordered at a time for the average profit per quart to be $350.

51. A rectangular hallway rug is 10 feet longer than it is wide. The total area of the rug is 24 square feet. Find the dimensions of the rug.

52. A rectangular dining room has a total area of 96 square feet. The dining room is 4 feet longer than it is wide. Find the dimensions of the room.

53. The area of a rectangle is 144 square centimeters. The length is 7 centimeters longer than the width. Find the length and the width of the rectangle.

54. The area of a rectangle is 84 square centimeters. The length is 5 centimeters longer than the width. Find the length and the width of the rectangle.

55. The area of a triangle is 30 square inches. The height is 4 inches less than the base. Find the base and height of the triangle.

56. The area of a triangle is 60 square inches. The height is 7 inches less than the base. Find the base and height of the triangle.

57. The height h of a ball in feet t seconds after being thrown from a roof can be estimated by the equation $h = -16t^2 + 32t + 48$. Find when the ball will hit the ground (height equal to zero).

58. The height h of a ball in feet t seconds after being thrown from a roof can be estimated by the equation $h = -16t^2 + 64t + 80$. Find when the ball will hit the ground (height equal to zero).

59. The height h of a ball in feet t seconds after being thrown from a roof can be estimated by the equation $h = -16t^2 + 64t + 80$. Find when the ball will reach a height of 128 feet.

60. The height h of a ball in feet t seconds after being thrown from a roof can be estimated by the equation $h = -16t^2 + 32t + 48$. Find when the ball will reach a height of 60 feet.

For Exercises 61 through 84, solve the equations. Check the answers.

61. $x^2 + 4x = -3$
62. $t^2 + 10t = -24$
63. $2a^2 + 9 = 9a$
64. $4m^2 + 5 = 21m$
65. $x(x - 5) = 14$
66. $x(x - 7) = -12$
67. $n(n + 9) = -20$
68. $y(y + 11) = -24$
69. $(x + 3)(x - 2) = 6$
70. $(x + 4)(x + 3) = 2$
71. $(y + 4)(y + 2) = 24$
72. $(n - 8)(n - 4) = 5$
73. $t^2 - 25 = 0$
74. $h^2 - 36 = 0$
75. $g^2 = 100$
76. $w^2 = 49$
77. $6x^2 - 22x - 120 = 0$
78. $15n^2 + 80n + 25 = 0$
79. $t^3 - 9t = 0$
80. $x^3 - 4x = 0$
81. $2x^3 + 14x^2 + 24x = 0$
82. $36x^3 - 3x^2 - 105x = 0$
83. $8t^3 - 24t^2 = 40t^2 - 120t$
84. $6a^3 - 100a = 6a^2 + 20a$

For Exercises 85 through 100, (a) determine for each whether you are given an expression or an equation, and (b) solve the equation or factor the expression.

85. $x^2 + 16x + 64$
86. $x^2 + 16x + 64 = 0$
87. $4n^2 - 25 = 0$
88. $4n^2 - 25$
89. $a^2 + 8a + 7$
90. $3m^2 + 4m - 7$
91. $4x^3 + 4x^2 - 24x = 0$
92. $t(t + 3) = 40$
93. $4m^2 - 4m = 3$
94. $3p^3 - 18p^2 - 21p$
95. $40b^2 + 38b - 15$
96. $x^2 - 14x + 49$
97. $2m^2 - 50 = 0$
98. $5x^3 - 6x^2 + x = 0$
99. $h^2 + 100$
100. $6w^2 + 5w = -1$

Chapter Summary

Section 6.1 What It Means to Factor

- The **greatest common factor,** or **GCF,** of a polynomial expression is the largest factor shared in common with all the terms.
- Steps to **find the GCF of a set of integers or monomials**
  1. Find the prime factorization of each integer or coefficient. Write repeated factors in exponential form.
  2. Circle the *common* factors of all the elements in the set.
  3. The GCF equals the product of the common factors, each raised to the *lowest power* that appears.
- Steps to **find the GCF of a polynomial expression**
  1. Factor each term completely. Factor each integer into its prime factors.
  2. The GCF equals the product of the common factors from each term, each raised to the lowest power that appears.
- Steps to **factor the GCF** from a polynomial expression:
  1. Determine the GCF of all the terms.
  2. Rewrite each term as the product of the GCF and the remaining factors.
  3. Factor out the GCF using the distributive property.
- Steps to **factor by grouping:**
  1. Check to determine whether or not there are four terms.
  2. Factor out any GCF of all four terms, if needed.
  3. Group the first two terms and last two terms.
  4. Factor out the GCF of the first two terms. Factor the last two terms.
  5. Factor out the GCF of the remaining expression.
- To check the result from factoring, multiply. The answer should equal the original expression.

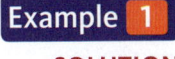

Factor out the GCF from the polynomial $24x^3y^2 - 32x^2y$.

SOLUTION   Find the prime factorization of each term. Then factor out the GCF.

$24x^3y^2 - 32x^2y$ Factor the coefficients into primes.
$(2^3) \cdot 3(x^3)(y^2) - (2^5)(x^2)(y)$ Look for common factors.
GCF is 2^3x^2y
$(2^3x^2y) \cdot 3xy - (2^3x^2y) \cdot 2^2$ Rewrite each term.
$2^3x^2y(3xy - 2^2)$ Factor the GCF out of the parentheses.
$8x^2y(3xy - 4)$

Example 2 Factor the polynomial $2a^2 - 5a + 4ab - 10b$ by grouping.

SOLUTION The GCF of all the terms in this expression is 1. There are four terms, so group the first two terms and last two terms, and factor.

$2a^2 - 5a + 4ab - 10b$
$(2a^2 - 5a) + (4ab - 10b)$ Group the first two terms and the last two terms.
$a(2a - 5) + 2b(2a - 5)$ Factor out the GCF from each group.
$(2a - 5)(a + 2b)$ Factor out the GCF from the remaining expression.

To check the result, multiply it out, using the distributive property.

$(2a - 5)(a + 2b)$ Multiply, using FOIL.
$= 2a^2 + 4ab - 5a - 10b$ Rearrange the terms.
$= 2a^2 - 5a + 4ab - 10b$ The answer checks.

Section 6.2 Factoring Trinomials

- Steps to **factor a trinomial of the form $x^2 + bx + c$ by inspection**:
  1. Identify the values of b and c.
  2. Find the factors of c that add to b.
  3. Rewrite as the product of two binomials of the form $(x + ?)(x + ?)$, where the question marks are the two factors (including their signs) from step 2.
- Steps to **factor a trinomial of the form $ax^2 + bx + c$ using the AC method**:
  1. Factor out the GCF.
  2. Find the product $a \cdot c$.
  3. Find the factors of $a \cdot c$ that add to b.
  4. Rewrite the middle x-term into the sum of two terms, with the factors from step 3 as coefficients of x.
  5. Factor by grouping.

Example 3 Factor the following polynomials. Check the result.

a. $m^2 - 5m - 24$ **b.** $8t^2 - 26t + 15$

SOLUTION **a.** Since $a = 1$, find factors of c that add up to b.

$c = -24$ $\quad\quad\quad\quad b = -5$
$1 \cdot (-24) = -24 \quad\quad 1 + (-24) = -23 \neq -5$
$2 \cdot (-12) = -24 \quad\quad 2 + (-12) = -10 \neq -5$
$3 \cdot (-8) = -24 \quad\quad 3 + (-8) = -5$

The correct pair of numbers is 3 and -8. The expression $m^2 - 5m - 24$ factors as

$(m + 3)(m - 8)$.

We check the result by multiplying out the factors.

$(m + 3)(m - 8)$ Multiply, using FOIL.
$= m^2 - 8m + 3m - 24$ Combine like terms.
$= m^2 - 5m - 24$ The answer checks.

b. This expression is of the form $ax^2 + bx + c$, so we use the AC method.

$$ac = (8)(15) = 120 \qquad\qquad b = -26$$

$$-1 \cdot (-120) = 120 \qquad\qquad -1 + (-120) = -121 \neq -26$$
$$-2 \cdot (-60) = 120 \qquad\qquad -2 + (-60) = -62 \neq -26$$
$$-3 \cdot (-40) = 120 \qquad\qquad -3 + (-40) = -43 \neq -26$$
$$-4 \cdot (-30) = 120 \qquad\qquad -4 + (-30) = -34 \neq -26$$
$$-5 \cdot (-24) = 120 \qquad\qquad -5 + (-24) = -29 \neq -26$$
$$-6 \cdot (-20) = 120 \qquad\qquad -6 + (-20) = -26$$

The correct pair of numbers we are looking for is -6 and -20. Rewrite the bx-term and factor by grouping.

$$8t^2 - 26t + 15$$
$$8t^2 - 20t - 6t + 15 \qquad \text{Rewrite the middle term.}$$
$$(8t^2 - 20t) + (-6t + 15) \qquad \text{Group and factor.}$$
$$4t(2t - 5) - 3(2t - 5)$$
$$(2t - 5)(4t - 3)$$

Check this answer by multiplying out the factors.

$$(2t - 5)(4t - 3)$$
$$= 8t^2 - 6t - 20t + 15$$
$$= 8t^2 - 26t + 15 \qquad \text{The answer checks.}$$

Section 6.3   Factoring Special Forms

- A **difference of squares** is of the form $a^2 - b^2$.
- Steps to **factor a difference of squares**:
  1. Find the values of a and b.
  2. The expression factors as $a^2 - b^2 = (a - b)(a + b)$.
- Steps to **factor a perfect square trinomial**:
  1. Determine whether or not the trinomial has the form $a^2 + 2ab + b^2$ or $a^2 - 2ab + b^2$. If not, use the AC method of factoring.
  2. If it is a perfect square trinomial, determine a and b.
  3. Factor into one of the following forms, depending on the sign before the second term.

$$a^2 + 2ab + b^2 = (a + b)^2$$
$$a^2 - 2ab + b^2 = (a - b)^2$$

Example 4   Factor the following polynomials. Check the result.

a. $x^2 - 64$ **b.** $9m^2 - 24m + 16$

SOLUTION   **a.** This expression is the difference of squares, x^2 and 8^2, so it factors as

$$(x - 8)(x + 8).$$

We can check this answer by multiplying out the factors.

$$(x - 8)(x + 8)$$
$$x^2 + 8x - 8x - 64$$
$$x^2 - 64 \quad \text{The answer checks.}$$

b. First check to see if this expression is a perfect square trinomial.

$9m^2 - 24m + 16$	Original expression
$a^2 - 2ab + b^2$	Standard form
$a = 3m$ and $b = 4$	Compare the first and last terms.
$24m \stackrel{?}{=} 2ab$	Check the middle term.
$24m \stackrel{?}{=} 2(3m)(4)$	
$24m = 24m$	It checks as a perfect square trinomial.

Since the expression is a perfect square trinomial of the form $a^2 - 2ab + b^2 = (a - b)^2$, with $a = 3m$ and $b = 4$, it factors as

$$(3m - 4)^2$$

Check this by multiplying out the factors.

$$(3m - 4)^2$$
$$= (3m - 4)(3m - 4)$$
$$= 9m^2 - 12m - 12m + 16$$
$$= 9m^2 - 24m + 16 \quad \text{The answer checks.}$$

Section 6.4 Solving Quadratic Equations by Factoring

- The **standard form of a quadratic equation** is $ax^2 + bx + c = 0$, where a, b, and c are real numbers and $a \neq 0$.
- The **zero-product property** states that if any two numbers are multiplied together and equal zero, one or both of the two numbers must be zero. In symbols: If $a \cdot b = 0$, then either $a = 0$ or $b = 0$, or both are equal to 0.
- Steps to **solve a quadratic equation by factoring**:
  1. Put the equation into the standard form of a quadratic equation, $ax^2 + bx + c = 0$.
  2. Factor the quadratic expression.
  3. Using the zero-product property, set each factor equal to 0 and solve for the variable.
  4. Check all answers in the original equation.

Example 5 Solve the equation $2k^2 - 9k = 35$ by factoring. Check the answers.

SOLUTION Set the equation equal to zero and then factor using the AC method.

$2k^2 - 9k = 35$
$\underline{-35 \quad\quad -35}$
$2k^2 - 9k - 35 = 0$
$2k^2 - 14k + 5k - 35 = 0$
$(2k^2 - 14k) + (5k - 35) = 0$
$2k(k - 7) + 5(k - 7) = 0$

$ac = -70$
$1(-70) = -70$
$2(-35) = -70$
$5(-14) = -70$

$b = -9$
$1 + (-70) = -69 \neq -9$
$2 + (-35) = -33 \neq -9$
$5 + (-14) = -9$

$$(k - 7)(2k + 5) = 0$$

$k - 7 = 0$ or $2k + 5 = 0$

$k = 7$ or $k = -\dfrac{5}{2}$

Check these answers by substituting the values into the original equation.

$\underline{k = 7}$ $\underline{k = -\dfrac{5}{2}}$

$2(7)^2 - 9(7) \stackrel{?}{=} 35$ $2\left(-\dfrac{5}{2}\right)^2 - 9\left(-\dfrac{5}{2}\right) \stackrel{?}{=} 35$

$2(49) - 63 \stackrel{?}{=} 35$ $2\left(\dfrac{25}{4}\right) + \left(\dfrac{45}{2}\right) \stackrel{?}{=} 35$

$35 = 35$ $\dfrac{25}{2} + \dfrac{45}{2} \stackrel{?}{=} 35$

Both answers check. $35 = 35$

Example 6   The height h of a ball in feet t seconds after being hit can be estimated by the equation $h = -16t^2 + 48t + 3$. Find when the ball will be at a height of 35 feet.

SOLUTION   Since we want the time when the ball reaches a height of 35 feet, substitute 35 for h and solve for t.

$$35 = -16t^2 + 48t + 3$$
$$\underline{-35 -35}$$
$$0 = -16t^2 + 48t - 32$$
$$\dfrac{0}{-16} = \dfrac{-16t^2 + 48t - 32}{-16} \quad \text{Divide both sides by the GCF.}$$
$$0 = t^2 - 3t + 2 \quad \text{Solve by factoring.}$$
$$0 = (t - 2)(t - 1)$$

$t - 2 = 0$ or $t - 1 = 0$

$t = 2$ or $t = 1$

Check the answers in the original equation.

$\underline{t = 2}$ $\underline{t = 1}$

$35 \stackrel{?}{=} -16(2)^2 + 48(2) + 3$ $35 \stackrel{?}{=} -16(1)^2 + 48(1) + 3$

$35 \stackrel{?}{=} -16(4) + 96 + 3$ $35 \stackrel{?}{=} -16 + 48 + 3$

$35 = 35$ **Both answers check.** $35 = 35$

Therefore, the ball will be at a height of 35 feet at 1 second after being hit and again at 2 seconds after being hit.

Chapter Review Exercises

For Exercises 1 through 4, for each polynomial expression, factor out the GCF. Check the results.

1. $12x^5 + 10x^3 - 6x^2$ [6.1]
2. $3a^3b^2 + 6ab^2 - 9b^2$ [6.1]
3. $2a(3a + 5) - 7(3a + 5)$ [6.1]
4. $2y(x + 3) - 5(x + 3)$ [6.1]

For Exercises 5 through 8, factor each polynomial expression by grouping. Check the results.

5. $4x^2 + 8x + 3x + 6$ [6.1]
6. $2t^2 - 12t + 7t - 42$ [6.1]
7. $5m^2 + 10mn - 7m - 14n$ [6.1]
8. $6xy + 10x - 3y - 5$ [6.1]

For Exercises 9 through 34, factor each polynomial expression completely. If it cannot be factored, answer that it is prime.

9. $a^2 + 10a + 24$
10. $m^2 + 16m + 64$ [6.2]
11. $x^2 - 9x + 20$
12. $h^2 - 13h + 36$ [6.2]
13. $x^2 - 4x - 21$
14. $y^2 - 7y - 18$ [6.2]
15. $2x^2 + 13x + 15$
16. $3a^2 + 19a + 6$ [6.2]
17. $5x^2 + 13x - 6$
18. $3b^2 + 17b - 28$ [6.2]
19. $4x^2 - 27x - 40$
20. $9y^2 - 24y + 16$ [6.2]
21. $-10b^2 + 19b - 6$
22. $-2t^2 + 15t - 25$ [6.2]
23. $g^3 - 8g^2 + 15g$
24. $6x^2 + 7xy - 20y^2$ [6.2]
25. $n^2 - 81$
26. $t^2 - 64$ [6.3]
27. $9a^2 - 100$
28. $4x^2 - 49$ [6.3]
29. $b^2 + 36$
30. $x^2 + 16$ [6.3]
31. $3x^2 - 27x - 7x + 63$ [6.3]
32. $4x^2y + 8y - 5x^2 - 10$ [6.3]
33. $6abc - 24ac + 3bc - 12c$ [6.3]
34. $4xyz + 2xz - 24yz - 12z$ [6.3]

For Exercises 35 through 52, solve each equation. Check the answer(s) in the original equation.

35. $(3x - 5)(x + 7) = 0$ [6.4]
36. $(2x - 9)(x - 3) = 0$ [6.4]
37. $y(4y - 6) = 0$ [6.4]
38. $x(3x - 1) = 0$ [6.4]
39. $x^2 + 3x - 28 = 0$ [6.4]
40. $n^2 - 3n - 54 = 0$ [6.4]
41. $2x^2 + 9x - 35 = 0$ [6.4]
42. $3y^2 + 11y - 4 = 0$ [6.4]
43. $m^2 + 4m = 32$ [6.4]
44. $n^2 + 2n = 35$ [6.4]
45. $a^2 - 81 = 0$ [6.4]
46. $x^2 - 49 = 0$ [6.4]
47. $3y^2 - 12 = 0$ [6.4]
48. $4a^2 - 36 = 0$ [6.4]
49. $2t^3 + 2t^2 - 60t = 0$ [6.4]
50. $2x^3 - 7x^2 - 15x = 0$ [6.4]
51. $x(x + 5) = 36$ [6.4]
52. $x(3x + 2) = 1$ [6.4]

53. The area of a rectangle is 36 square centimeters. The length of the rectangle is one more than twice the width. Find the length and the width of the rectangle. [6.4]

54. RC Toys, Inc.'s profit P in thousands of dollars from selling t thousand toy cars can be estimated by the equation $P = -2t^2 + 34t$. [6.4]
    a. Estimate the profit for RC Toys, Inc. if they sell 8000 toys.
    b. Find the number of toys RC Toys, Inc. needs to sell to earn $120 thousand profit.

Chapter Test

This chapter test should take approximately one hour to complete. Read each question carefully. Show all of your work.

For Exercises 1 through 13, factor each polynomial expression completely.

1. $x^2 + 2x - 24$
2. $a^2 - 14a + 48$
3. $6x^2 - 9x - 60$
4. $4x^2 + 25$
5. $2x^2 - 3x + 30$
6. $x^2 - 100$
7. $25m^2 - 1$
8. $6x^2 + 7x - 3$
9. $4y^2 + 20y + 25$
10. $4t^3 + 5t^2 - 21t$
11. $2x^3 - 6x^2 - 8x$
12. $5ab - 10a + 2b - 4$
13. $2m(n + 8) - 3(n + 8)$

For Exercises 14 through 18, solve each equation.

14. $t^2 + 13t + 30 = 0$
15. $9w^2 - 36 = 0$
16. $x^3 + 2x^2 = 48x$
17. $6x^2 - 22x = 40$
18. $5x^2 - 28x + 16 = -16$
19. XY Wireless profit P in thousands of dollars from selling n thousand Bluetooth earpieces can be estimated by the equation $P = -2n^2 + 48n$.
    a. Estimate the profit for XY Wireless if they sell 5000 earpieces.
    b. Find the number of Bluetooth earpieces XY Wireless must sell in order to earn $160 thousand profit.
20. The area of a rectangle is 63 square meters. The length of the rectangle is 5 less than twice the width. Find the length and width of the rectangle.

Chapter Projects

What Does a Perfect Square Trinomial Look Like?

Use the following diagram to answer the questions.

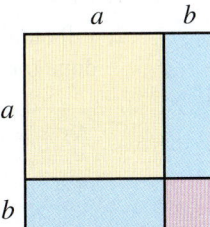

First consider the smaller squares and rectangles.

1. What expression represents the length of one side of the yellow square?
2. What expression represents the area of the yellow square?
3. What expressions represent the length and width of a blue rectangle?

4. What expression represents the area of a blue rectangle?
5. What expression represents the length of one side of the red square?
6. What expression represents the area of the red square?
7. Fill in each part of the diagram with the expressions that represent the area of each part.

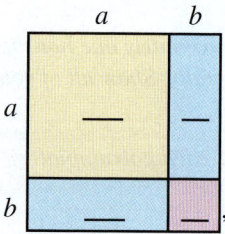

Now consider the square made up of all the smaller pieces.

8. What expression represents a side of the large square?
9. What expression represents the area of the large square (write it in the form of s^2, where s = side)?
10. Write an expression for the *total* of the areas for all the smaller parts of the square.
11. Since the area of the large square is equal to the total of the areas for all the smaller parts of the square, write an equation showing this relationship.

What Is That Quadratic Pattern?

Written Project

One or more people

Consider the following pattern of rectangles.

| $n = 1$ | $n = 2$ | $n = 3$ |

1. Draw the next two rectangles, one for $n = 4$ and one for $n = 5$.
2. Fill out the remaining five rows of the following table. The first row has been done for you to get you started.

N	Width	Length	Area
1	2	3	$2 \cdot 3 = 6$
2			
3			
4			
5			
N			

3. In the last row of the table, find the general pattern for the width and the length. Multiply the width and the length to find the general formula for the area.
4. Multiply out the area expression for the last row using FOIL.
5. What is the width of the 20th rectangle in this pattern? What is the length of the 20th rectangle in this pattern? What is the area of the 20th rectangle in this pattern?

■ Write a Review of a Section for Presentation

Research Project
One or more people

What you will need
- This presentation may be done in class or as an online video.
- If it is done as a video, post it on a website where it may be easily viewed by the class.
- You may want to use homework or other review problems from the book.
- Make it creative and fun.

Create a 5-minute review presentation of one section from Chapter 6. The format of the presentation can be a poster presentation, a blackboard presentation from notes, an online video, or a game format (e.g., math jeopardy). The presentation should include the following:

- Any important formulas in the chapter
- Important skills in the chapter, backed up by examples
- Explanation of terms used (e.g., what is *modeling*?)
- Common mistakes and how to avoid them

Cumulative Review — Chapters 1-6

1. Write in exponential form: $(2x) \cdot (2x) \cdot (2x) \cdot (2x) \cdot (2x)$.

2. Find the value without a calculator: $\left(-\dfrac{1}{2}\right)^3$.

3. Underline or circle the terms, and simplify using the order-of-operations agreement:
$$24 \div 8 \cdot (-5) + (-3)^3 - 6 \div (-2) + |-1 - 4(5)|$$

4. Evaluate the expression for the given value of the variable:
$$x^3 + 2x^2 - 3x + 7 \quad \text{for } x = -1$$

5. Simplify: $-3(4a + 2b) - (3a - 7b)$.

6. Simplify: $-(5x - 6x) + 3(x + 2)$.

7. Solve for y: $6x - 5y = 35$.

8. Solve: $-16 = \dfrac{4}{5}y$.

9. Solve for the indicated value of the variable: $P = a + b + c$ for b

10. Solve: $5(n + 1) - 6 = 2n + 1$.

11. Helena is driving to another city to visit her friends. She plans to average a speed of about 65 miles per hour. The distance Helena can travel in one day can be calculated by using the equation
$$D = 65t$$
where D is the distance she travels in miles if she drives t hours.
   a. Find the distance Helena travels if she drives for 6.5 hours.
   b. How many hours does Helena have to drive to travel 455 miles?

12. Solve: $\dfrac{1}{2}x + 3 = \dfrac{13}{4}x + 11$.

13. Solve the inequality: $-2(x - 3) + 6 < -8$.

14. Complete each part below.
   a. Fill in the table using the given equation.

x	$y = \dfrac{2}{3}x - 4$
-6	
-3	
0	
3	
6	

   b. Graph the points. Clearly label and scale the axes.
   c. Is the graph linear or nonlinear?

15. An exercise physiologist estimates that a man burns about 124 calories per mile while running. Let $C = 124m$, where C is the total number of calories a man burns while running m miles. Create a table of points that make sense in this situation and graph them. Connect the points with a smooth curve.

16. Use the equation $y = x^2 + 3$ to create a table of nine or more points and graph them. Connect the points with a smooth curve. Clearly label and scale the axes. Is the graph linear or nonlinear?

17. Find the slope of the line that goes through points $(4, -3)$ and $(-2, 9)$.

18. Find the x-intercept and the y-intercept of the equation $-3x + 8y = -24$. Graph the line using the intercepts. Clearly label and scale the axes.

19. Find the slope, x-intercept, and y-intercept of the line $2x + 7y = 14$.

20. Graph the line $y = -\dfrac{3}{5}x - 2$ using the slope and y-intercept. Graph at least three points. Clearly label and scale the axes.

21. Graph the line. Clearly label and scale the axes.
$$-\dfrac{2}{3}x + \dfrac{1}{6}y = -2$$

22. Determine whether or not the two lines are parallel, perpendicular, or neither. Do not graph the lines. The two lines are $y = \dfrac{4}{5}x - 3$ and $4x - 5y = -1$.

23. Use the graph to find the equation of the line. Put the answer in $y = mx + b$ form.

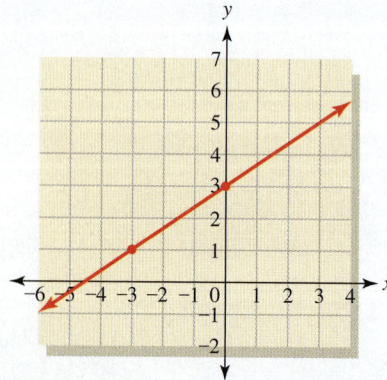

584

24. Jon is a rock climber. An online source tells him that he will burn about 600 calories per hour while ascending. Let the input h represent the number of hours Jon is ascending and let the output C represent the number of calories Jon burns. Find the linear relationship between h and C.

25. Determine the equation of the line that passes through the points $(-6, 21)$ and $(4, -4)$. Put the answer in $y = mx + b$ form.

26. Determine the equation of the line by using point-slope form. The line has slope $m = 0$ and passes through the point $(-3, 4)$. Put the final answer in $y = mx + b$ form.

27. Determine the equation of the line that passes through the points $(2, -4)$ and $(-4, 17)$. Put the answer in $y = mx + b$ form.

28. Find the equation of the line that is parallel to $y = \dfrac{x}{2} + 3$ and passes through the point $(-4, -5)$.

29. Find the equation of the line that is perpendicular to $y = \dfrac{3x}{5} - 16$ and passes through the point $(-3, 7)$.

30. A store buys pet toys for $4 each and sells them for $7.50 each. What is the markup rate?

31. A new computer sells for $1100 and one year later is worth only $300. Find the percent decrease in value.

32. A dress that sells for $120 is on sale for 30% off. Find the discount and sales price for the dress.

33. You buy a used car for $8000. The DMV (Department of Motor Vehicles) charges you 7.5% tax on the purchase of the car. Find the total tax paid for the car.

34. Solve the system of equations using the table of values.

x	$y = -\dfrac{4}{5}x - 1$
-10	7
-5	3
0	-1
5	-5
10	-9

x	$y = x + 8$
-10	-2
-5	3
0	8
5	13
10	18

35. Determine whether or not each point is a solution of the system.

$$y = -\dfrac{5}{7}x + 3$$
$$y = \dfrac{4}{7}x - 6$$

   a. Is $(x, y) = (-7, 8)$ a solution of the system?
   b. Is $(x, y) = (7, -2)$ a solution of the system?

36. Graphically determine the solution of the system. Clearly label and scale the axes. Check the solution by substituting it into both original equations.

$$y = x - 5$$
$$y = -6x + 2$$

37. Solve the system, using the substitution method. Check the solution in both original equations.

$$y = -\dfrac{3}{5}x + 2$$
$$3x + 5y = 2$$

38. Solve the system, using the substitution method. Check the solution in both original equations.

$$2x + y = 2$$
$$7x + 2y = -17$$

39. Darrion is a coordinator for a youth basketball league. He purchased 12 T-shirts and 7 basketballs for $220.00. Later, after more kids enrolled in his league, he purchased an additional 10 T-shirts and 8 more basketballs for $218.00. How much does each T-shirt and each basketball cost?

40. A boat travels downstream 40 miles in 4 hours. It makes the return trip upstream in 5 hours. Find the rate of the boat and the rate of the current.

41. Emma is making a 40-pound mix of dried papaya and dried pineapple for her organic food store. The dried papaya costs $7.99 per pound, and the dried pineapple costs $11.99 per pound. Emma would like to price the mix at $9.49 per pound. Find how many pounds of dried papaya and dried pineapple Emma has to put in the mix.

42. Solve the system using the elimination method. Check the solution in both original equations.

$$3x + 2y = 8$$
$$-2x + y = -17$$

43. Solve the system using the elimination method. Check the solution in both original equations.

$$3x + 4y = 7$$
$$-6x - 8y = 4$$

44. Two angles are complementary. The second angle measures 20 degrees more than the measure of the first angle. Find the degree measures of both angles.

45. Seamus has $10,000.00 to invest. An investment advisor tells him to put part of his money in a CD earning 3% interest and the rest in a bond fund earning 3.5% interest. Seamus would like to earn $340.00 in interest every year. How much should he invest in each account?

46. Determine whether or not the point (2, 4) is a solution of the linear inequality $6x - 9y \geq 10$.

47. Graph the solution set of the linear inequality $y < -\dfrac{2}{3}x + 7$.

48. Determine whether or not the point $(-5, 3)$ is in the solution set of the system of linear inequalities.
$$x - y < 3$$
$$y \geq 3x + 4$$

49. Graph the solution set for the system of inequalities. Check the solution set with a sample test point.
$$y > -2x + 1$$
$$-3x + 2y < 6$$

50. Simplify: $(-3)^2$.

51. Simplify: -3^2.

52. Simplify.
  a. $x^3 \cdot x^5$
  b. $3a^4 \cdot 4a$
  c. $(-3ab^3)(-5a^4b)$

53. Simplify.
  a. $n^5 \cdot n$
  b. $5t^3 \cdot 7t^2$
  c. $(9x^2y)(-3xy^6)$

54. Simplify.
  a. $(3x^4)^3$
  b. $(-2x^2)^3$
  c. $(2x^3y^4)^2$

55. Simplify.
  a. $(5y^4)^2$
  b. $(-3y^2)^5$
  c. $(4a^5b^2)^3$

56. Simplify.
  a. $(-7x)(3x^2)^2$
  b. $(-3xy)^2(-2y^3)^3$

57. Simplify.
  a. $(8n)(-2n^6)^3$
  b. $(-xy^4)^5(-5x^2y)^2$

58. Simplify.
  a. $\dfrac{x^9}{x^3}$
  b. $-\dfrac{8a^5}{16a}$
  c. $\dfrac{x^3xy^5}{x^2y^3}$

59. Simplify.
  a. $\dfrac{y^3}{y}$
  b. $-\dfrac{25y^4}{5y^3}$
  c. $\dfrac{a^4bb^6}{a^2b^3}$

60. Simplify: $\left(-\dfrac{15x^4y^2}{50x^3y}\right)^2$.

61. Simplify: $\left(-\dfrac{49n^4m^3}{7nm^2}\right)^3$.

62. Find an expression for the area of the square. ($A = s^2$ or area = side^2.)

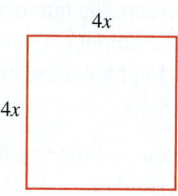

63. Find an expression for the area of the circle. ($A = \pi r^2$.)

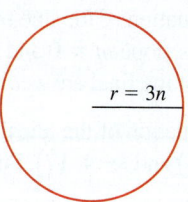

64. Rewrite each expression using positive exponents.
  a. $5t^{-2}$
  b. $\dfrac{-2}{x^{-3}}$

65. Rewrite each expression using positive exponents.
  a. $6n^{-5}$
  b. $\dfrac{-7}{y^{-4}}$

66. Simplify. Write the final answers using positive exponents only.
  a. $\dfrac{x^4y}{x^2y^2}$
  b. $\dfrac{-3x^{-3}y^3}{9xy^2}$
  c. $(-3n^{-3})^2$
  d. $\left(\dfrac{x^2}{y}\right)^{-2}$

67. Simplify. Write the final answers using positive exponents only.
  a. $\dfrac{r^5t^2}{r^6t^{-2}}$
  b. $\dfrac{-16n^5n}{24n^7m^{-2}}$
  c. $(-4m^{-2})^3$
  d. $\left(\dfrac{x}{y^4}\right)^{-3}$

68. Write each number using scientific notation.
  a. 895,100,000,000
  b. 0.000413

69. Write each number using scientific notation.
  a. 648,700,000
  b. 0.00000692

70. Convert each number in scientific notation to standard notation.
  a. 6.75×10^7
  b. 3.58×10^{-5}

71. Convert each number in scientific notation to standard notation.
  a. 4.01×10^8
  b. 8.43×10^{-4}

72. A computer can execute 1 billion instructions per second. Write this number using scientific notation.

73. A transistor gate for a microcomputer has length 2.5×10^{-9} meter (known as 2.5 nanometers). Write this number using standard notation.

74. Perform the indicated operations. Write the answers in scientific notation.
    a. $9(3.2 \times 10^8)$
    b. $\dfrac{6.8 \times 10^{12}}{4}$
    c. $34(2.5 \times 10^{-9})$
    d. $(6.2 \times 10^{13})(4.3 \times 10^8)$
    e. $\dfrac{3.4 \times 10^{12}}{1.7 \times 10^4}$

75. Identify the number of terms and find the degree of the polynomial: $-8x^3 + 3x^2 - 7x + 12$.

76. Identify the number of terms and find the degree of the polynomial: $x^2 - 6x$.

77. Find the degree of each term and of the polynomial: $4xy^2 - xy + 2y^4 + 9x$.

78. Find the degree of each term and of the polynomial: $8a^5b + 3a^3 - 6b^5 + 5$.

79. Add: $(x^2 - x - 4) + (-2x^2 - 3x + 1)$.

80. Add: $(9y - 7) + (2y^2 - 3y - 5)$.

81. Subtract: $(-6x^2 + 3x - 7) - (-2x + 9)$.

82. Subtract: $(5a^2 - 2a + 7) - 3(4a^2 - 11)$.

83. Multiply: $-2(5n - 9)$.

84. Multiply: $-3(-2n + 1)$.

85. Multiply and simplify.
    a. $(x - 4)(x + 7)$
    b. $(3x + 1)(2x - 5)$
    c. $(x + 2)(x^2 - 3x + 4)$
    d. $(n + 4)^2$
    e. $(2x + 1)(2x - 1)$

86. Multiply and simplify.
    a. $(y + 9)(y - 3)$
    b. $(4x + 7)(-2x + 1)$
    c. $(x - 1)(x^2 + 4x - 2)$
    d. $(n - 3)^2$
    e. $(3x + 5)(3x - 5)$

87. Perform the indicated operations and simplify: $5x - 4 - 2(6x + 8)$

88. Perform the indicated operations and simplify: $-4y + 7 - 3(2y - 5)$

89. Divide and simplify: $\dfrac{-6n^2 + 12n - 15}{3n}$.

90. Divide and simplify: $\dfrac{-4y^2 + 8y - 16}{4y}$.

91. Divide using long division: $(2x^2 + 7x - 4) \div (x + 4)$.

92. Divide using long division: $(3x^2 - 8x - 3) \div (x - 3)$.

93. Factor out the GCF.
    a. $-12a^2b + 3ab^2 - 30ab$
    b. $2n(n + 3) - 1(n + 3)$

94. Factor out the GCF.
    a. $16x^3y^3 + 48x^2y^2$
    b. $-3m(m - 4) + 2(m - 4)$

95. Factor by grouping: $3ab - 12a + b - 4$.

96. Factor by grouping: $2xy + 12x - 5y - 30$.

97. Factor the polynomial completely: $3x^2 - 15x + 3x - 15$.

98. Factor the polynomial completely: $2y^2 - 4y - 14y + 28$.

99. Factor the trinomials.
    a. $x^2 + x - 30$
    b. $-x^2 - x + 6$
    c. $4x^2 + 4x - 3$

100. Factor the trinomials.
    a. $x^2 - 5x - 36$
    b. $-x^2 + x + 20$
    c. $3x^2 - 10x - 3$

101. Factor the polynomials completely.
    a. $2n^3 - 6n^2 - 20n$
    b. $10x^2 + 5xy + 30xy + 15y^2$

102. Factor the polynomials completely.
    a. $3t^3 - 45t^2 + 162t$
    b. $6x^2 + 3xy - 42xy - 21y^2$

103. Factor the polynomials.
    a. $t^2 - 49$
    b. $9 - 4y^2$
    c. $4x^2 + 20x + 25$
    d. $a^2 + 36$
    e. $5x^2 - 45$

104. Factor the polynomials.
    a. $y^2 - 36$
    b. $1 - 25n^2$
    c. $9y^2 - 12y + 4$
    d. $b^2 + 16$
    e. $3y^2 - 48$

105. Determine whether the following equations are linear, quadratic, or neither. Do not solve the equations.
- **a.** $x^2 - 4x = 8$
- **b.** $x^3 + x^2 - 2x + 1 = 0$
- **c.** $3(x + 2) - 1 = 0$

106. Determine whether the following equations are linear, quadratic, or neither. Do not solve the equations.
- **a.** $5x - 2(x + 1) = 4$
- **b.** $x(x - 5) = 0$
- **c.** $1 + x^4 = 0$

107. Solve each quadratic equation by factoring.
- **a.** $x^2 + 12x + 32 = 0$
- **b.** $2x^2 + 7x - 15 = 0$
- **c.** $x(x - 2) = 35$
- **d.** $y^2 - 49 = 0$

108. Solve each quadratic equation by factoring.
- **a.** $x^2 + 10x + 21 = 0$
- **b.** $4x^2 - 23x - 6 = 0$
- **c.** $x(x - 1) = 6$
- **d.** $n^2 - 25 = 0$

Rational Expressions and Equations

7

- **7.1** The Basics of Rational Expressions and Equations
- **7.2** Multiplication and Division of Rational Expressions
- **7.3** Addition and Subtraction of Rational Expressions
- **7.4** Solving Rational Equations
- **7.5** Proportions, Similar Triangles, and Variation

In image processing, photographs must often be made larger or smaller from the original to fit a certain space. This sizing, which is known as *scaling*, can be analyzed mathematically using proportions. In this chapter, we develop the tools for rational expressions and equations that allow us to solve applications such as scaling.

7.1 The Basics of Rational Expressions and Equations

LEARNING OBJECTIVES
- Evaluate rational expressions and equations.
- Find excluded values.
- Simplify rational expressions.

Evaluating Rational Expressions and Equations

In this chapter, we examine many of the topics from Chapters 5 and 6 and combine them to study **rational expressions** and **equations**. Rational expressions result from dividing two polynomials. Rational expressions show up in many different applications, including finding averages.

> **What's That Mean?**
>
> **Rational**
>
> The word *ratio* in mathematics refers to a quotient of two mathematical elements. A *rational number* is a ratio of integers, with a nonzero denominator. A *rational expression* is a ratio of polynomials. This is different from everyday language, in which saying that someone is rational means that the person is logical and sensible.

> **DEFINITIONS**
>
> **Rational Expression**   An expression of the form
> $$\frac{P}{Q}$$
> where P and Q are polynomials and $Q \neq 0$ is called a rational expression.
>
> **Rational Equation**   An equation that contains one or more rational expressions is called a rational equation.
>
> | Rational Expression | Rational Equation |
> |---|---|
> | $\dfrac{2x + 5}{x - 4}$ | $\dfrac{4x - 7}{3x + 2} = 5$ |

Example 1 Cost per person for a ski trip

A group of students is going to rent a cabin that costs $1000 for a weekend of skiing, and they plan to pay equal shares. The students can find the cost per person by using the equation

$$c = \frac{1000}{n}$$

where c is the cost in dollars per person for n students to go to the cabin for the weekend.

a. Find the cost per person if five students stay in the cabin.

b. When the cabin is full, students will pay the lowest cost per person. If the cabin can hold up to 20 people, what is the lowest cost per person?

SOLUTION

a. If five people stay in the cabin, substitute $n = 5$, and calculate the per-person cost.

$$c = \frac{1000}{5} = 200$$

If five students stay at the cabin, they will each have to pay $200.

b. If the maximum number of people, 20, stay in the cabin, the lowest cost per person can be found by substituting $n = 20$ and calculating the per person cost.

$$c = \frac{1000}{20} = 50$$

The lowest cost per person to rent the cabin is $50 when 20 students go to the cabin.

PRACTICE PROBLEM FOR EXAMPLE 1

For a recent field trip, the school had to pay $600 for the bus and $10 per student for admission to the natural history museum. The cost per student for a class to take this field trip can be calculated by using the equation

$$c = \frac{600 + 10n}{n}$$

where c is the cost in dollars per person for n students to go on the field trip.

a. Find the cost per student if 20 students go on the field trip.

b. If the bus can hold up to 60 students, what is the lowest cost per student?

Evaluating an expression means substituting a given value for the variable into the expression and simplifying. When evaluating rational expressions, be careful to follow the order-of-operations agreement.

Example 2 Evaluating rational expressions

Evaluate each rational expression for the given value.

a. $\dfrac{x + 5}{x - 4}$; $x = 7$ **b.** $\dfrac{a^2 + 5}{a^2 - 7a + 12}$; $a = 4$ **c.** $\dfrac{t^2 + 2t - 3}{2t + 6}$; $t = 5$

SOLUTION

a. Substitute in $x = 7$ and simplify.

$$\frac{(7) + 5}{(7) - 4} \qquad \text{Substitute in } x = 7.$$

$$= \frac{12}{3} = 4 \qquad \text{Simplify.}$$

b. Substitute in $a = 4$ and simplify.

$$\frac{(4)^2 + 5}{(4)^2 - 7(4) + 12} \qquad \text{Substitute in } a = 4.$$

$$= \frac{16 + 5}{16 - 28 + 12} \qquad \text{Simplify the numerator and denominator separately.}$$

$$= \frac{21}{0} \qquad \text{This expression is undefined, we cannot divide by zero.}$$

When this rational expression is evaluated for $a = 4$, the denominator equals zero. Therefore, this expression is undefined when $a = 4$.

c. Substitute in $t = 5$ and simplify.

$$\frac{(5)^2 + 2(5) - 3}{2(5) + 6} \qquad \text{Substitute in } t = 5.$$

$$= \frac{25 + 10 - 3}{10 + 6} \qquad \text{Simplify.}$$

$$= \frac{32}{16} = 2$$

> **Skill Connection**
>
> **Order-of-operations agreement with grouping symbols**
>
> A fraction bar is considered a grouping symbol when using the order-of-operations agreement. This is seen in Example 2, where we first evaluated the numerator and denominator separately. Then we performed the division indicated by the fraction bar.

CHAPTER 7 Rational Expressions and Equations

PRACTICE PROBLEM FOR EXAMPLE 2
Evaluate each rational expression for the given value.

a. $\dfrac{x + 10}{x - 6}$; $x = 2$ **b.** $\dfrac{m - 8}{m^2 + 3m - 4}$; $m = 10$ **c.** $\dfrac{h^2 + 4h - 8}{2h - 12}$; $h = 6$

■ Excluded Values

Connecting the Concepts

What makes fractions undefined?

Remember from Section R.1 that division by zero is undefined, but dividing into zero is zero.

$\dfrac{5}{0}$ is undefined, but $\dfrac{0}{5} = 0$

The denominator being zero makes the expression undefined. This is why we exclude values for variables that make the denominator zero.

Notice that the definition for a rational expression has a restriction that the polynomial in the denominator cannot equal zero. This restriction is important because it avoids the problem of division by zero, which is undefined. When working with rational expressions, we find any values that make the denominator of a rational expression zero and *exclude* them. Values that make the denominator of a rational expression or equation zero are called **excluded values**.

> **DEFINITION**
>
> **Excluded Values** Any number that makes the denominator of a rational expression equal zero is called an excluded value. These values will be excluded from the possible values of the variable so that division by zero will not occur.

To find the excluded values for a rational expression, find the values that make the denominator equal zero.

> **Steps to Finding Excluded Values for a Rational Expression**
> 1. Set the denominator equal to zero, and solve for the variable.
> 2. Check the solution(s) in the original expression to verify the expression is undefined.

Example 3   Finding excluded values

Find the values of the variables that must be excluded for each rational expression.

a. $\dfrac{x + 5}{x - 4}$ **b.** $\dfrac{a^2 + 5}{a^2 - 7a + 12}$ **c.** $\dfrac{t^2 + 2t - 3}{6}$

SOLUTION

a. **Step 1** Set the denominator equal to zero, and solve for the variable.

$$\dfrac{x + 5}{x - 4}$$

$x - 4 = 0$ Set the denominator equal to zero and solve.

$x = 4$ This is the excluded value.

Step 2 Check the solution(s) in the original expression to verify the expression is undefined.

$\dfrac{(4) + 5}{(4) - 4} = \dfrac{9}{0}$ undefined   This value is an excluded value.

b. To find the excluded values, set the denominator equal to zero and solve. This denominator is a quadratic expression. Solve the quadratic equation by factoring.

$$\frac{a^2 + 5}{a^2 - 7a + 12}$$

$a^2 - 7a + 12 = 0$	*Set the denominator equal to zero and solve.*
$(a - 4)(a - 3) = 0$	*Solve by factoring.*
$a - 4 = 0$ or $a - 3 = 0$	
$a = 4$ or $a = 3$	*These are the excluded values.*

To check, substitute in $a = 4$.

$$\frac{(4)^2 + 5}{(4)^2 - 7(4) + 12} = \frac{16 + 5}{16 - 28 + 12} = \frac{21}{0} \text{ undefined}$$

To check, substitute in $a = 3$.

$$\frac{(3)^2 + 5}{(3)^2 - 7(3) + 12} = \frac{9 + 5}{9 - 21 + 12} = \frac{14}{0} \text{ undefined}$$

Therefore, both $a = 4$ and $a = 3$ are excluded values for this expression.

c. Set the denominator equal to zero and solve.

$$\frac{t^2 + 2t - 3}{6}$$

$6 \stackrel{?}{=} 0$ *Set the denominator equal to zero and solve. This has no solution.*

Therefore, there are *no* excluded values for this expression.

Calculator Details

Division by 0

Your calculator will display a message to alert you that division by zero is undefined. Depending on the type of calculator you have, the screen may read

"Error" or

"Error; Divide by 0."

PRACTICE PROBLEM FOR EXAMPLE 3

Find the value(s) of the variable that must be excluded for each rational expression.

a. $\dfrac{x + 7}{x + 3}$ **b.** $\dfrac{2t - 8}{t^2 - 3t - 10}$ **c.** $\dfrac{3a - 12}{5}$

■ Simplifying Rational Expressions

Simplifying rational expressions makes them easier to work with. A rational expression is a fraction, so simplifying is similar to simplifying numerical fractions. Use the following Concept Investigation to discover the similarities and differences between simplifying numerical fractions and rational expressions.

■ CONCEPT INVESTIGATION
Can I divide out that term or factor?

1. Simplify the following fractions without a calculator.

   a. $\dfrac{40}{24}$ **b.** $\dfrac{1200}{360}$

2. Follow the order-of-operations agreement to simplify the following fractions.

   a. $\dfrac{6 + 14}{4 + 1}$ **b.** $\dfrac{2 + 13}{2 + 4}$

Skill Connection

Reducing numerical fractions

When reducing numerical fractions, first find the prime factorization and then divide out any common factors.

$$\frac{100}{30} = \frac{5 \cdot 5 \cdot 2 \cdot 2}{5 \cdot 3 \cdot 2}$$

$$= \frac{\cancel{5} \cdot 5 \cdot \cancel{2} \cdot 2}{\cancel{5} \cdot 3 \cdot \cancel{2}}$$

$$= \frac{5 \cdot 2}{3} = \frac{10}{3}$$

Operations contained in grouping symbols should be calculated first. The division bar is one of the grouping symbols, so we perform any operations in the numerator and denominator before dividing.

$$\frac{5 + 7}{1 + 3} = \frac{12}{4}$$

$$= \frac{3 \cdot \cancel{2} \cdot \cancel{2}}{\cancel{2} \cdot \cancel{2}} = 3$$

What's That Mean?

Terms and Factors

In algebra, a *term* is something that is added or subtracted. The expression $3x + 4$ has two terms: $3x$ and 4. A *factor* is something that is multiplied. The expression $5x$ has a single term, and it has two factors 5 and x.

Simplifying Rationals

Factor the numerator and denominator, then divide out any common factors.

$$\frac{5\cancel{(x+2)}}{(x+3)\cancel{(x+2)}}$$

$$= \frac{5}{x+3}$$

---

3. Evaluate the following rational expressions for the given values.

   a. $\dfrac{x + 8}{x + 2}$; $x = 6, 10,$ and 25 b. $\dfrac{m + 12}{m - 3}$; $m = 5, 8,$ and 30

4. Dividing out the terms of x in the first fraction from part 3 yields

$$\frac{\cancel{x} + 8}{\cancel{x} + 2} = \frac{8}{2} = 4$$

   Does this result of 4 agree with any of the values calculated in part 3a?

5. What does your answer to question 4 tell us about dividing out common *terms* in a fraction?

6. Evaluate the following rational expressions for the given values.

   a. $\dfrac{2x(x + 3)}{5(x + 3)}$; $x = 4, 8,$ and 10 b. $\dfrac{(w - 4)(w + 2)}{(w - 4)(w - 3)}$; $w = 5, 10,$ and 16

7. Dividing out the common factor of $x + 3$ in the fraction from part 6a yields

$$\frac{2x\cancel{(x+3)}}{5\cancel{(x+3)}} = \frac{2x}{5}$$

   Evaluate this new fraction with the values $x = 4, 8,$ and 10. Do these values agree with the values you calculated in part 6a?

8. What does your answer to question 7 tell us about dividing out common *factors* in a fraction?

Warning: When working on the second fraction (part 2b above), some students might mistakenly divide out the terms of 2 and then reduce the fraction. Doing so does not give the same result. $\dfrac{13}{4}$ does not equal $\dfrac{5}{2}$.

Correct: Simplify numerator and denominator then divide.

$$\frac{2 + 13}{2 + 4} = \frac{15}{6} = \frac{5}{2}$$

Incorrect: Divide out the 2's first.

$$\frac{\cancel{2} + 13}{\cancel{2} + 4} = \frac{13}{4}$$

The results are not the same, so both processes cannot be correct.

$$\frac{13}{4} \neq \frac{5}{2}$$

The Concept Investigation above illustrates that *common terms cannot be divided out*. Because common factors can be divided out, to simplify rational expressions, first factor the numerator and denominator. Then divide out common factors. This results in a simpler form than the original expression.

> **Steps to Simplifying Rational Expressions**
>
> 1. Factor the numerator and denominator (if needed).
> 2. Divide out any common factors in the numerator and denominator.
>
> *Note:* Leave the numerator and denominator in factored form.

SECTION 7.1 The Basics of Rational Expressions and Equations   595

Example 4 Simplifying rational expressions

Simplify the following rational expressions.

a. $\dfrac{5x(x+2)}{2(x+2)}$ b. $\dfrac{(y+2)(y-8)}{(y-8)(y+5)}$ c. $\dfrac{h+3}{h^2+5h+6}$

SOLUTION $\boxed{\dfrac{x+2}{x-5}}$

a. **Step 1** Factor the numerator and denominator (if needed).

  Both the numerator and denominator are already factored.

  Step 2 Divide out any common factors in the numerator and denominator.

  $\dfrac{5x\cancel{(x+2)}}{2\cancel{(x+2)}}$ Divide out the common factor of $(x+2)$.

  $= \dfrac{5x}{2}$

b. Both the numerator and denominator are already factored, so divide out the common factors.

  $\dfrac{(y+2)\cancel{(y-8)}}{\cancel{(y-8)}(y+5)}$ Divide out the common factor of $(y-8)$.

  $= \dfrac{y+2}{y+5}$ This is in simplest form.

Do not be tempted to divide out the y's in the numerator and denominator in this last fraction. Common terms cannot be divided out. Only common factors can be divided out.

c. The denominator is not factored. First factor the denominator and then divide out any common factors.

  $\dfrac{h+3}{h^2+5h+6}$

  $= \dfrac{(h+3)}{(h+3)(h+2)}$ Factor the denominator.

  $= \dfrac{\overset{1}{\cancel{(h+3)}}}{\cancel{(h+3)}(h+2)}$ Divide out the common factor of $h+3$.

  $= \dfrac{1}{h+2}$ This is in simplest form.

In the numerator, $h+3$, h and 3 have only a common factor of 1. Therefore, $h+3$ factors as $h+3 = 1(h+3)$. When the common factor of $h+3$ is divided out in the numerator and denominator, a 1 remains in the numerator.

■ **Connecting the Concepts**

Are the original expression and simplified form equal?

In Example 4a, we simplified the original rational expression $\dfrac{5x(x+2)}{2(x+2)}$ to $\dfrac{5x}{2}$. These two expressions are equivalent everywhere except at the excluded value $x = -2$.

■ **What's That Mean?**

Reducing a Fraction to Lowest Terms

To *reduce a fraction to lowest terms* means to divide out common factors until the numerator and denominator have no common factors other than 1. Reducing a fraction to lowest terms is also called *simplifying* a fraction.

PRACTICE PROBLEM FOR EXAMPLE 4

Simplify the following rational expressions.

a. $\dfrac{(a+3)(a-4)}{(a+6)(a+3)}$ b. $\dfrac{x+2}{x^2-3x-10}$

Always check that both the numerator and denominator are factored completely before simplifying. Not factoring completely may prevent a rational expression from being reduced to simplest form.

596 CHAPTER 7 Rational Expressions and Equations

Example 5 Simplifying rational expressions

Simplify the following rational expressions.

a. $\dfrac{5 + (x + 7)}{(x - 3)(x + 7)}$ b. $\dfrac{t^2 - 4t + 4}{(2 - t)(t + 3)}$ c. $\dfrac{6m + 3}{2m^2 + 11m + 5}$

SOLUTION

a. The denominator is factored, but the numerator needs to be simplified first and then factored.

$$\dfrac{5 + (x + 7)}{(x - 3)(x + 7)}$$

$$= \dfrac{x + 12}{(x - 3)(x + 7)}$$

The numerator cannot be factored because x and 12 have no common factors. This rational expression cannot be simplified any further.

b. Start by factoring the numerator.

$$\dfrac{t^2 - 4t + 4}{(2 - t)(t + 3)}$$

$= \dfrac{(t - 2)(t - 2)}{(2 - t)(t + 3)}$ Factor the numerator and compare factors. The factor $(2 - t)$ is the opposite of the $(t - 2)$.

$= \dfrac{(t - 2)(t - 2)}{-1(t - 2)(t + 3)}$ Factor out -1 from $(2 - t)$.

$= \dfrac{\cancel{(t - 2)}(t - 2)}{-1\cancel{(t - 2)}(t + 3)}$ Divide out the common factor of $(t - 2)$.

$= -1\dfrac{(t - 2)}{(t + 3)}$ The -1 is usually placed in front.

$= -\dfrac{t - 2}{t + 3}$ This is in simplest form.

c. Factor both the numerator and the denominator. Factor out the GCF first.

$$\dfrac{6m + 3}{2m^2 + 11m + 5}$$

$= \dfrac{3(2m + 1)}{(2m + 1)(m + 5)}$ Factor the numerator and denominator.

$= \dfrac{3\cancel{(2m + 1)}}{\cancel{(2m + 1)}(m + 5)}$ Divide out the common factor of $(2m + 1)$.

$= \dfrac{3}{m + 5}$ This is in simplest form.

PRACTICE PROBLEM FOR EXAMPLE 5

Simplify the following rational expressions.

a. $\dfrac{(x + 5)(4 - x)}{x^2 + 6x - 40}$ b. $\dfrac{5y + 20}{y^2 + 8y + 16}$

Skill Connection

Factor out a negative sign

Recall from Section 6.1 that factoring a -1 changes the signs in an expression. In Example 5b, factoring out -1 is done as follows.

$(2 - t)$
$= -1(-2 + t)$
$= -1(t - 2)$

What's That Mean?

The GCF

In Section 6.1, we learned that GCF stands for *greatest common factor*. It is the largest factor that is common to all terms in an expression.

7.1 Vocabulary Exercises

1. In math, a ratio is a(n) _____ of two expressions.

2. A rational expression is a ratio of two _____.

3. To reduce a fraction to lowest terms is the same as _____ a fraction.

4. An excluded value is any number that makes the _____ of a rational expression equal to 0.

5. When reducing a fraction we can only divide out common _____.

7.1 Exercises

For Exercises 1 through 10, evaluate each rational expression for the given value.

1. $\dfrac{x+6}{x-3}$ for $x = 12$

2. $\dfrac{m-2}{m+10}$ for $m = 8$

3. $\dfrac{2a+5}{3a+4}$ for $a = 10$

4. $\dfrac{4h+2}{3h+5}$ for $h = 2$

5. $\dfrac{x-1}{x^2+2x+5}$ for $x = 5$

6. $\dfrac{t+3}{t^2+3t+7}$ for $t = 4$

7. $\dfrac{r^2+6}{r^2-5r+6}$ for $r = 3$

8. $\dfrac{5p+7}{p^2-8p+12}$ for $p = 6$

9. $\dfrac{b^2-6b-7}{3b+6}$ for $b = -1$

10. $\dfrac{n^2-5n+4}{2n+8}$ for $n = 4$

For Exercises 11 through 18, use the information given in each exercise to answer the questions.

11. Some employees of a local company want to honor a friend who is retiring by buying an engraved watch as a retirement gift. The watch and engraving will cost $250. The cost per person depends on the number of people who agree to help pay. If n employees agree to pay equal shares, the cost per person in dollars can be found by using the equation

$$c = \dfrac{250}{n}$$

where c is the cost per person in dollars.

a. Find the per-person cost if 10 people agree to help pay for the gift.

b. If there are 50 employees at the company besides the retiree, what is the lowest possible cost per person?

12. The lump-sum prize for the state lottery is $5 million. If p people pitch in for tickets and split the prize money evenly, their potential winnings per person can be found by using the equation

$$w = \dfrac{5000000}{p}$$

where w is each person's winnings in dollars.

a. Find the winnings per person if 20 people pool their ticket money.

b. If the group limits the number of people who can pitch in for tickets to 100, what is the least they could each win? (Assume that the group wins the lottery, which rarely happens.)

13. The cost per person for a college debate team to travel to the national championships can be found by using the equation

$$c = \dfrac{400 + 25p}{p}$$

where c is the cost per person in dollars when p people go on the trip.

a. What is the cost per person if eight people go on the trip?

b. If the trip is limited to 25 people, what is the lowest cost per person?

14. The Wake Tech Eagles basketball program is having T-shirts made to raise funds for their awards banquet. The cost can be found by using the equation

$$c = \frac{150 + 7t}{t}$$

where c is the cost per T-shirt in dollars when t T-shirts are made.

a. What is the cost per T-shirt if 50 T-shirts are made?

b. What is the cost per T-shirt if 200 T-shirts are made?

15. Wally's Custom Surfboards can make up to 10 surfboards a week. The company's cost to produce each board can be found by using the equation

$$c = \frac{800 + 100b}{b}$$

where c is the cost per surfboard in dollars when b boards are made in a week.

a. What does each surfboard cost if eight surfboards are made in a week?

b. What is the lowest cost per surfboard for a week?

16. The El Camino Country Club charges $1200 for room rental plus $45 per person for a banquet. The cost per person can be found by using the equation

$$c = \frac{1200 + 45n}{n}$$

where c is the cost per person in dollars when n people attend the banquet.

a. What is the cost per person if 30 people attend the banquet?

b. What is the cost per person if 100 people attend the banquet?

17. Spending per person by the state of Texas on public assistance can be found by using the equation

$$s = \frac{4014t + 22847}{0.466t + 25.08}$$

where s is the spending per person by the state of Texas on public assistance in dollars t years since 2010. Use the equation to find per person spending by the state of Texas on public assistance in 2020.

Source: Equation derived from data found at www.usgovernmentspending.com

18. The state of New York's spending per person can be found by using the equation

$$s = \frac{5919.8t + 114666.4}{0.027t + 19.6}$$

where s is the spending per person by the state of New York in dollars t years since 2010. Use the equation to find spending per person by the state of New York in 2020.

Source: Equation derived from data found at www.openbudget.ny.gov

For Exercises 19 through 36, find the value(s) of the variables that must be excluded for each rational expression.

19. $\dfrac{5x + 6}{x}$ **20.** $\dfrac{2a - 9}{a}$ **21.** $\dfrac{x + 6}{x - 3}$

22. $\dfrac{m - 2}{m + 10}$ **23.** $\dfrac{2a + 5}{3a + 4}$ **24.** $\dfrac{4h + 2}{3h + 5}$

25. $\dfrac{x^2 + 4}{7}$ **26.** $\dfrac{-3y + 1}{5}$

27. $\dfrac{x + 5}{(x - 3)(x + 2)}$ **28.** $\dfrac{y - 6}{(y + 6)(y + 7)}$

29. $\dfrac{25}{(h + 4)(h - 8)}$ **30.** $\dfrac{150}{4(b - 3)(b + 9)}$

31. $\dfrac{r^2 + 6}{r^2 - 5r + 6}$ **32.** $\dfrac{5p + 7}{p^2 - 8p + 12}$

33. $\dfrac{n - 5}{n^2 + 4n + 4}$ **34.** $\dfrac{g + 7}{g^2 - 6g + 9}$

35. $\dfrac{b^2 + 5b - 7}{3b + 6}$ **36.** $\dfrac{n^2 - 5n + 4}{2n - 8}$

For Exercises 37 through 48, simplify each rational expression.

37. $\dfrac{4(x - 3)}{x(x - 3)}$ **38.** $\dfrac{2(a - 7)}{a(a - 7)}$

39. $\dfrac{7n(n + 6)}{n + 6}$ **40.** $\dfrac{2m(m - 2)}{m - 2}$

41. $\dfrac{a + 8}{(a + 5)(a + 8)}$ **42.** $\dfrac{h + 2}{(h + 5)(h + 2)}$

43. $\dfrac{4x(x + 2)}{(x - 3)(x + 2)}$ **44.** $\dfrac{7y(y + 6)}{(y + 6)(y + 7)}$

45. $\dfrac{(h + 6)(h - 8)}{(h + 4)(h - 8)}$ **46.** $\dfrac{(b - 3)(b + 2)}{(b - 3)(b + 9)}$

47. $\dfrac{r - 3}{r^2 - 9}$ **48.** $\dfrac{p + 7}{p^2 - 49}$

For Exercises 49 through 56, use the information given in each exercise to answer the questions.

49. An experienced landscaper takes 4 hours to complete the mowing, trimming, edging, and blowing for a certain property. The experienced landscaper and his assistant working together complete the job in T hours.

$$T = \frac{4x}{4 + x}$$

Here x is the time in hours it takes the assistant to do the job alone.

  a. How long will it take the two landscapers working together if the assistant can complete the job alone in 6 hours?

  b. How long will it take the two landscapers working together if the assistant can complete the job alone in 5.5 hours?

50. An experienced carpenter takes 2 hours to install a door in a house. The experienced carpenter and his assistant working together complete the job in T hours.

$$T = \frac{2x}{2 + x}$$

Here x is the time in hours it takes the assistant to do the job alone.

  a. How long will it take the two carpenters working together if the assistant can complete the job alone in 4.5 hours?

  b. How long will it take the two carpenters working together if the assistant can complete the job alone in 3.5 hours?

51. An experienced administrative assistant takes 5 hours to prepare a budget report. Working together, the new office specialist and the administrative assistant complete the job in T hours.

$$T = \frac{5x}{5 + x}$$

Here x is the time in hours it takes the office specialist to do the job alone.

  a. How long will it take the two staff members working together if the office specialist can complete the job alone in 8 hours?

  b. How long will it take the two staff members working together if the office specialist can complete the job alone in 6.5 hours?

52. An experienced bookkeeper takes 5.5 hours to balance the books for a small business. When the bookkeeper works with an apprentice, the two complete the job in time t (in hours).

$$t = \frac{5.5x}{5.5 + x}$$

Here x is the time in hours it takes the apprentice to do the job alone.

  a. How long will it take the two working together if the apprentice can complete the job alone in 8 hours?

  b. How long will it take the two working together if the apprentice can complete the job alone in 7.5 hours?

53. Jack drives to a city 100 miles away at x mph. He returns at y mph for the same 100 miles. If he averages 60 miles per hour for the total trip, then the two speeds are related by

$$y = \frac{30x}{x - 30}$$

  a. If Jack drives at $x = 70$ mph to the city, what is his speed on the return trip?

  b. If Jack drives at $x = 55$ mph to the city, what is his speed on the return trip?

54. Jocelyn drives to a city 100 miles away at x mph. She returns at y mph for the same 100 miles. If she averages 72 miles per hour for the total trip, then the two speeds are related by

$$y = \frac{36x}{x - 36}$$

  a. If Jocelyn drives at $x = 70$ mph to the city, what is her speed on the return trip?

  b. If Jocelyn drives at $x = 55$ mph to the city, what is her speed on the return trip?

55. A pilot flies to a city 1000 miles away at x mph. She returns at y mph for the same 1000 miles. If she averages 450 miles per hour for the total trip, then the two speeds are related by

$$y = \frac{225x}{x - 225}$$

  a. If the pilot flies at $x = 500$ mph to the city, what is her speed on the return trip?

  b. If the pilot flies at $x = 425$ mph to the city, what is her speed on the return trip?

56. A pilot flies to a city 100 miles away at x mph. He returns at y mph for the same 100 miles. If he averages 200 miles per hour for the total trip, then the two speeds are related by

$$y = \frac{100x}{x - 100}$$

  a. If the pilot flies at $x = 175$ mph to the city, what is his speed on the return trip?

  b. If the pilot flies at $x = 225$ mph to the city, what is his speed on the return trip?

For Exercises 57 through 62, find the error in each student's work. Then correctly simplify the rational expressions.

57. Simplify: $\dfrac{x+2}{x+4}$

Ruben
$\dfrac{x+2}{x+4} = \dfrac{2}{4} = \dfrac{1}{2}$

58. Simplify: $\dfrac{x-14}{x-2}$

Ashley
$\dfrac{x - \overset{7}{\cancel{14}}}{x - \cancel{2}} = -7$

59. Simplify: $\dfrac{x}{x+7}$

Ciara
$\dfrac{\overset{1}{\cancel{x}}}{\cancel{x}+7} = \dfrac{1}{7}$

60. Simplify: $\dfrac{x+9}{x}$

Omar
$\dfrac{\cancel{x}+9}{\cancel{x}} = 9$

61. Simplify: $\dfrac{x^2-4}{x-2}$

Celina
$\dfrac{\overset{x}{\cancel{x^2}}-\overset{2}{\cancel{4}}}{x-2} = x-2$

62. Simplify: $\dfrac{x^2-25}{x-5}$

David
$\dfrac{\overset{2}{\cancel{x^2}}-\overset{5}{\cancel{25}}}{x-5} = x-5$

For Exercises 63 through 74, simplify each rational expression.

63. $\dfrac{n+2}{n^2+4n+4}$

64. $\dfrac{g-3}{g^2-6g+9}$

65. $\dfrac{2b+8}{b^2+9b+20}$

66. $\dfrac{5n+10}{n^2+7n+10}$

67. $\dfrac{8+(x+7)}{(x+2)(x+7)}$

68. $\dfrac{4+(m-3)}{(m-3)(m+5)}$

69. $\dfrac{5+(a-3)}{(a+2)(a-3)}$

70. $\dfrac{7+(a-8)}{(a-1)(a-8)}$

71. $\dfrac{(r+2)(r-9)}{r^2-5r-14}$

72. $\dfrac{(z-4)(z-6)}{z^2-12z+36}$

73. $\dfrac{y^2+11y-12}{(y-1)(y-11)}$

74. $\dfrac{b^2-3b-40}{(b-8)(b+3)}$

For Exercises 75 through 84, factor out -1 from the expression. Then rewrite the expression in descending order.

75. $1-x$

76. $3-y$

77. $4-y$

78. $8-x$

79. $5-2x-3x^2$

80. $7+x-x^2$

81. $3x-8-2x^2$

82. $-4x+5-x^2$

83. $4-8x^2$

84. $9-n^2$

For Exercises 85 through 100, factor any quadratic expressions first. Then, factor out -1 from a factor, if needed. Simplify each rational expression.

85. $\dfrac{15-3x}{x-5}$

86. $\dfrac{28-4x}{x-7}$

87. $\dfrac{y-1}{1-y}$

88. $\dfrac{9-w}{w-9}$

89. $\dfrac{4t+20}{3(t+5)}$

90. $\dfrac{8t-28}{2(2t-7)}$

91. $\dfrac{w^2-49}{(w+3)(7-w)}$

92. $\dfrac{x^2-1}{(x+6)(1-x)}$

93. $\dfrac{a+5}{(a+5)(a+4)}$

94. $\dfrac{h-3}{h^2-9}$

95. $\dfrac{5-x}{x^2-10x+25}$

96. $\dfrac{3-p}{p^2+p-12}$

97. $\dfrac{m^2+5m-6}{3m+18}$

98. $\dfrac{h^2-25}{h^2-10h+25}$

99. $\dfrac{12-3t}{t^2-10t+24}$

100. $\dfrac{t^2+t-30}{10-2t}$

For Exercises 101 through 108, multiply or divide as indicated. Reduce to simplest form.

101. $\dfrac{2}{3} \cdot \dfrac{4}{5}$

102. $\dfrac{4}{7} \cdot \dfrac{2}{5}$

103. $-\dfrac{12}{7} \cdot \dfrac{21}{30}$

104. $-\dfrac{6}{14} \cdot \dfrac{10}{8}$

105. $\dfrac{4}{3} \div \dfrac{4}{5}$

106. $\dfrac{2}{7} \div \dfrac{3}{14}$

107. $-\dfrac{8}{15} \div \dfrac{6}{11}$

108. $-\dfrac{10}{14} \div \dfrac{12}{21}$

7.2 Multiplication and Division of Rational Expressions

LEARNING OBJECTIVES
- Multiply rational expressions.
- Expand unit conversions.
- Divide rational expressions.
- Simplify complex fractions.

■ Multiplying Rational Expressions

A rational expression is a ratio of polynomials, and we have learned how to simplify them. Now we will learn how to multiply and divide rational expressions. To multiply rational expressions, use the same process as to multiply numerical fractions.

> **MULTIPLYING RATIONAL EXPRESSIONS**
>
> $$\frac{a}{b} \cdot \frac{c}{d} = \frac{a \cdot c}{b \cdot d} \quad \text{where} \quad b \neq 0, d \neq 0$$

Recall that when multiplying numerical fractions, multiply the numerators together and the denominators together. It is often helpful to simplify the fractions before multiplying them together. Consider the following two options for multiplying the same two numerical fractions.

Option 1	Option 2
$\dfrac{20}{18} \cdot \dfrac{15}{70}$	$\dfrac{20}{18} \cdot \dfrac{15}{70}$
$= \dfrac{300}{1260}$ Multiply first.	Reduce first.
Now reduce.	$= \dfrac{5 \cdot 2 \cdot 2}{3 \cdot 3 \cdot 2} \cdot \dfrac{5 \cdot 3}{7 \cdot 5 \cdot 2}$
$= \dfrac{5 \cdot 5 \cdot 3 \cdot 2 \cdot 2}{7 \cdot 5 \cdot 3 \cdot 3 \cdot 2 \cdot 2}$	$= \dfrac{5 \cdot \cancel{2} \cdot \cancel{2}}{3 \cdot \cancel{3} \cdot \cancel{2}} \cdot \dfrac{\cancel{5} \cdot \cancel{3}}{7 \cdot \cancel{5} \cdot \cancel{2}}$
$= \dfrac{5 \cdot 5 \cdot \cancel{3} \cdot \cancel{2} \cdot \cancel{2}}{7 \cdot \cancel{5} \cdot \cancel{3} \cdot 3 \cdot \cancel{2} \cdot \cancel{2}}$	$= \dfrac{5}{3} \cdot \dfrac{1}{7}$ Now multiply.
$= \dfrac{5}{21}$	$= \dfrac{5}{21}$

Both of these options are valid. We will use option 2, as the numbers do not get as large. These same steps work for rational expressions. When multiplying, first factor the rational expressions, divide out any common factors and then multiply the remaining numerators and denominators.

602 CHAPTER 7 Rational Expressions and Equations

> **What's That Mean?**
>
> **Rewrite as a Single Fraction**
>
> This means to multiply the remaining numerators together and multiply the remaining denominators together. Leave the result in factored form.

> **Steps to Multiplying Rational Expressions**
> 1. Factor the numerator and denominator of each fraction (if needed).
> 2. Divide out any common factors.
> 3. Rewrite as a single fraction.
>
> *Note:* Leave the result in factored form.

Example 1 — Multiplying rational expressions

Multiply the following rational expressions.

a. $\dfrac{x+5}{x+2} \cdot \dfrac{x-3}{x+5}$

b. $\dfrac{a+8}{a-5} \cdot \dfrac{a+7}{a-9}$

SOLUTION

a. **Step 1** Factor the numerator and denominator of each fraction (if needed).

These expressions are already factored.

Step 2 Divide out any common factors.

$$\dfrac{\cancel{x+5}}{x+2} \cdot \dfrac{x-3}{\cancel{x+5}} \qquad \text{Divide out the common factor.}$$

Step 3 Rewrite as a single fraction.

$$= \dfrac{x-3}{x+2}$$

b. The expressions are already factored. There are no common factors. Rewrite as a single fraction. Leave the answer factored.

$$\dfrac{a+8}{a-5} \cdot \dfrac{a+7}{a-9} \qquad \text{There are no common factors, so rewrite as a single fraction.}$$

$$= \dfrac{(a+8)(a+7)}{(a-5)(a-9)} \qquad \text{Leave the result in factored form.}$$

PRACTICE PROBLEM FOR EXAMPLE 1

Multiply the following rational expressions.

a. $\dfrac{8}{26} \cdot \dfrac{35}{10}$

b. $\dfrac{x-6}{x-4} \cdot \dfrac{x-4}{x-7}$

If the expressions get more complicated, concentrate on factoring one fraction at a time. That way, the numerators and denominators of all fractions will get factored.

Example 2 — Multiplying rational expressions

Multiply the following rational expressions.

a. $\dfrac{t+5}{t+3} \cdot \dfrac{2-t}{t^2-4}$

b. $\dfrac{x^2+6x+9}{x^2+8x+15} \cdot \dfrac{x^2+9x+20}{x^2-6x+8}$

SECTION 7.2 Multiplication and Division of Rational Expressions 603

SOLUTION

a. The second denominator is the difference of two squares. Factor it and then multiply.

$$\frac{t+5}{t+3} \cdot \frac{2-t}{t^2-4}$$

$$= \frac{t+5}{t+3} \cdot \frac{2-t}{(t-2)(t+2)} \qquad \text{Factor the denominator.}$$

$$= \frac{t+5}{t+3} \cdot \frac{-1(t-2)}{(t-2)(t+2)} \qquad \text{Factor out } -1 \text{ from } (2-t).$$

$$= \frac{t+5}{t+3} \cdot \frac{-1\cancel{(t-2)}}{\cancel{(t-2)}(t+2)} \qquad \text{Divide out the common factor.}$$

$$= \frac{-(t+5)}{(t+3)(t+2)} \qquad \text{Rewrite as a single fraction.}$$

b. Start by factoring each fraction completely.

$$\frac{x^2+6x+9}{x^2+8x+15} \cdot \frac{x^2+9x+20}{x^2-6x+8}$$

$$= \frac{(x+3)(x+3)}{(x+3)(x+5)} \cdot \frac{(x+5)(x+4)}{(x-4)(x-2)} \qquad \text{Factor.}$$

$$= \frac{\cancel{(x+3)}(x+3)}{\cancel{(x+3)}\cancel{(x+5)}} \cdot \frac{\cancel{(x+5)}(x+4)}{(x-4)(x-2)} \qquad \text{Divide out the common factors.}$$

$$= \frac{(x+3)(x+4)}{(x-4)(x-2)} \qquad \text{Rewrite as a single fraction.}$$

PRACTICE PROBLEM FOR EXAMPLE 2

Multiply the following rational expressions.

a. $\dfrac{a+11}{(8-a)(a+9)} \cdot \dfrac{a^2-64}{a^2+8a-33}$ **b.** $\dfrac{x^2-8x+15}{x^2-3x-10} \cdot \dfrac{x^2+8x+12}{x^2-6x-27}$

■ Expanding Unit Conversions

Unit conversions were previously discussed in Section 1.2. Just like rational expressions, common units can be divided out in the same way as common factors.

Example 3 Converting milligrams per kilogram to milligrams per pound

The correct dosage of a child's drug is 22 milligrams per kilogram. Find the drug dosage in milligrams per pound.

SOLUTION
The dosage is given in milligrams per kilogram, so we need to change the kilograms to pounds. Because there are about 2.2 pounds in 1 kilogram, use a unity fraction to convert the units.

$$\frac{22 \text{ mg}}{1 \text{ kg}}$$

$$\approx \frac{22 \text{ mg}}{1 \text{ kg}} \cdot \frac{1 \text{ kg}}{2.2 \text{ lb}} \qquad \text{Multiply by the unity fraction and divide out the common units.}$$

$$\approx \frac{22 \text{ mg}}{2.2 \text{ lb}} \approx 10 \frac{\text{mg}}{\text{lb}} \qquad \text{Reduce the fraction.}$$

The child's drug dosage is 10 milligrams per pound.

> **Skill Connection**
>
> **Converting units**
> Recall the definition of a unity fraction and the steps to convert units in Section 1.2.
>   A unity fraction is a fraction that has units in the numerator and denominator and a simplified value equal to 1.
>
> $$\frac{3 \text{ feet}}{1 \text{ yard}}$$
>
> **Steps to convert units:**
> 1. Determine the appropriate unity fraction needed for the problem.
> 2. Multiply the given value by the unity fraction. Divide out units as needed.
> 3. Simplify the final fraction.

PRACTICE PROBLEM FOR EXAMPLE 3

A box of cereal is priced at $3 per pound. Convert this price per pound (lb) to dollars per ounce (oz).

Example 4 — Converting miles per hour to feet per minute

A car is traveling at 90 mph. What is the car's speed in feet per minute?

SOLUTION

The speed is given in miles per hour. We have to change miles to feet and the hours to minutes. Do these conversions one at a time to make it easier to keep track of them. See the back of the book for common unit conversions.

$$\frac{90 \text{ miles}}{1 \text{ hour}}$$

$$= \frac{90 \text{ miles}}{1 \text{ hour}} \cdot \frac{5280 \text{ feet}}{1 \text{ mile}}$$

There are 5280 feet in 1 mile. Use this to make the unity fraction. Divide out the common units of miles.

$$= \frac{475200 \text{ feet}}{1 \text{ hour}}$$

$$\frac{475200 \text{ feet}}{1 \text{ hour}}$$

$$= \frac{475200 \text{ feet}}{1 \text{ hour}} \cdot \frac{1 \text{ hour}}{60 \text{ minutes}}$$

There are 60 minutes in 1 hour. Use this to make the unity fraction. Divide out the common units of hours.

$$= \frac{7920 \text{ feet}}{1 \text{ minute}}$$

The car is traveling at 7920 feet per minute.

PRACTICE PROBLEM FOR EXAMPLE 4

A car is traveling at 6600 feet per minute. What is the car's speed in meters per second?

Example 5 — Treating a patient with heart problems

A doctor has a 100-kg (220-lb) patient who is being treated for a heart attack and low blood pressure. The doctor prescribes 200 micrograms (μg) per minute of dopamine to raise the patient's blood pressure. The dopamine is to be added to an intravenous (IV) bag and administered through an IV pump.

a. The IV pump flow rate is measured in milligrams per hour. Convert the dopamine dosage to milligrams per hour.

b. Use the result from part a to calculate the number of milligrams of dopamine the patient will receive in a 24-hour period.

SOLUTION

a. The dopamine dosage is given in micrograms per minute. First convert the dopamine dosage to micrograms per hour. Convert to milligrams per hour.

$$\frac{200 \; \mu g}{1 \; min}$$

$$= \frac{200 \; \mu g}{1 \; \cancel{min}} \cdot \frac{60 \; \cancel{min}}{1 \; hr} \qquad \text{Multiply by the unity fraction to convert to hours.}$$

$$= \frac{12000 \; \mu g}{1 \; hr}$$

$$= \frac{12000 \; \cancel{\mu g}}{1 \; hr} \cdot \frac{1 \; mg}{1000 \; \cancel{\mu g}} \qquad \text{Multiply by the unity fraction to convert to milligrams.}$$

$$= \frac{12 \; mg}{1 \; hr} = 12 \; \text{milligrams per hour}$$

The patient's dosage should be 12 milligrams per hour.

b. We want the amount of dopamine the patient will receive in 24 hours, so multiply the result from part a by 24.

$$\frac{12 \; mg}{1 \; \cancel{hr}} \cdot 24 \; \cancel{hr} = 288 \; mg$$

The patient will receive 288 mg in a 24-hour period.

PRACTICE PROBLEM FOR EXAMPLE 5

A work crew is laying a new asphalt road. The machine they use can lay asphalt at a rate of 24 meters per hour.

a. Convert the rate into miles per hour.

b. Using the result from part a, how many miles of asphalt can the crew lay in an 8-hour work day?

Connecting the Concepts

Why do health care professionals convert units?

Prescription medications are often given in dose per kilogram. Kilograms are a measurement of a person's mass in the metric system. In the United States, weights are often measured in pounds. Health care workers have to be able to convert between pounds and kilograms to ensure that a patient is getting the correct dose. Also, when a drug must be added to an intravenous pump for a patient who is hospitalized, unit conversions are done to ensure that the patient receives the correct dose over a given time span.

■ Dividing Rational Expressions

Divide rational expressions in the same way as numerical fractions.

DIVIDING RATIONAL EXPRESSIONS

$$\frac{a}{b} \div \frac{c}{d} = \frac{a}{b} \cdot \frac{d}{c} \quad \text{where} \quad b \neq 0, c \neq 0, d \neq 0$$

When dividing two rational expressions, multiply the first fraction by the reciprocal of the second fraction.

Steps to Divide Rational Expressions

1. Multiply by the reciprocal of the second fraction.
2. Factor the numerator and denominator of each fraction (if needed).
3. Divide out any common factors.
4. Rewrite as a single fraction.

Note: Leave the result in factored form.

606 CHAPTER 7 Rational Expressions and Equations

Example 6 Dividing rational expressions

Divide the following rational expressions.

a. $\dfrac{t+5}{t+3} \div \dfrac{t+5}{t-4}$ b. $\dfrac{12x^2}{45y} \div \dfrac{9x}{30y}$

SOLUTION

a. Steps 1 and 2 Multiply by the reciprocal of the second fraction. Factor the numerator and denominator of each fraction.

$$\dfrac{t+5}{t+3} \div \dfrac{t+5}{t-4}$$

$$= \dfrac{t+5}{t+3} \cdot \dfrac{t-4}{t+5}$$

Step 3 Divide out any common factors.

$$= \dfrac{\cancel{t+5}}{t+3} \cdot \dfrac{t-4}{\cancel{t+5}}$$

Step 4 Rewrite as a single fraction.

$$= \dfrac{t-4}{t+3}$$

b. $\dfrac{12x^2}{45y} \div \dfrac{9x}{30y}$

$= \dfrac{12x^2}{45y} \cdot \dfrac{30y}{9x}$ Find the reciprocal of the second fraction.

$= \dfrac{3 \cdot 2 \cdot 2 \cdot x^2}{5 \cdot 3 \cdot 3 \cdot \cancel{y}} \cdot \dfrac{5 \cdot 3 \cdot 2 \cdot \cancel{y}}{3 \cdot 3 \cdot x}$ Factor and divide out the common factors.

$= \dfrac{8x^2}{9x}$ Simplify $\dfrac{x^2}{x} = x^{2-1} = x$.

$= \dfrac{8x}{9}$

PRACTICE PROBLEM FOR EXAMPLE 6

Divide the following rational expressions.

a. $\dfrac{6}{5} \div \dfrac{8}{15}$ b. $\dfrac{20x^2}{27y^3} \div \dfrac{4x^2}{9y}$ c. $\dfrac{y-3}{y+2} \div \dfrac{y-7}{y+2}$

Example 7 Dividing rational expressions

Divide the following rational expressions.

a. $\dfrac{(h+6)(h+2)}{(h-9)(h+2)} \div \dfrac{(h-8)(h+6)}{(h-9)(h-4)}$ b. $\dfrac{x^2-25}{x^2-5x-14} \div \dfrac{x^2-8x+15}{x^2-6x-16}$

SOLUTION

a. Both numerator and denominator are factored, so find the reciprocal of the second fraction and multiply.

$$\frac{(h+6)(h+2)}{(h-9)(h+2)} \div \frac{(h-8)(h+6)}{(h-9)(h-4)}$$

$$= \frac{(h+6)(h+2)}{(h-9)(h+2)} \cdot \frac{(h-9)(h-4)}{(h-8)(h+6)} \qquad \text{Find the reciprocal of the second fraction.}$$

$$= \frac{\cancel{(h+6)}\cancel{(h+2)}}{\cancel{(h-9)}\cancel{(h+2)}} \cdot \frac{\cancel{(h-9)}(h-4)}{(h-8)\cancel{(h+6)}} \qquad \text{Divide out common factors.}$$

$$= \frac{h-4}{h-8}$$

b. Find the reciprocal of the second fraction. Factor and multiply.

$$\frac{x^2-25}{x^2-5x-14} \div \frac{x^2-8x+15}{x^2-6x-16}$$

$$= \frac{x^2-25}{x^2-5x-14} \cdot \frac{x^2-6x-16}{x^2-8x+15}$$

$$= \frac{(x+5)(x-5)}{(x+2)(x-7)} \cdot \frac{(x+2)(x-8)}{(x-3)(x-5)}$$

$$= \frac{(x+5)\cancel{(x-5)}}{\cancel{(x+2)}(x-7)} \cdot \frac{\cancel{(x+2)}(x-8)}{(x-3)\cancel{(x-5)}}$$

$$= \frac{(x+5)(x-8)}{(x-7)(x-3)}$$

PRACTICE PROBLEM FOR EXAMPLE 7

Divide the following rational expressions.

a. $\dfrac{(b+10)(b+14)}{(b-15)(b+11)} \div \dfrac{(b+14)(b+6)}{(b-8)(b-15)}$ **b.** $\dfrac{x^2+2x-24}{x^2-8x+16} \div \dfrac{x^2+11x+30}{x^2-2x-8}$

■ Basics of Complex Fractions

In Chapter R, we saw that there are several ways to write a division problem. The problem $8 \div 4$ may also be written as $\dfrac{8}{4}$. Sometimes problems involving dividing rational expressions are written by using this second format. The problem $\dfrac{3x-6}{x+1} \div \dfrac{x-2}{4x+4}$ can be written by using the fraction bar to indicate division.

$$\frac{3x-6}{x+1} \div \frac{x-2}{4x+4} \text{ may be written as } \frac{\dfrac{3x-6}{x+1}}{\dfrac{x-2}{4x+4}}$$

It is helpful to put parentheses around the two fractions involved. Thus,

$$\left(\frac{3x-6}{x+1}\right) \div \left(\frac{x-2}{4x+4}\right) \text{ may be written as } \frac{\left(\dfrac{3x-6}{x+1}\right)}{\left(\dfrac{x-2}{4x+4}\right)}$$

This second form, namely, $\dfrac{\text{fraction}}{\text{fraction}}$, is often called a *complex fraction*.

Example 8 — Simplifying a complex fraction

Simplify the complex fraction.

$$\dfrac{\dfrac{3x-6}{x+1}}{\dfrac{x-2}{4x+4}}$$

SOLUTION

Recall that the fraction bar is another symbol that indicates division. Divide these expressions.

$$\dfrac{\dfrac{3x-6}{x+1}}{\dfrac{x-2}{4x+4}}$$

$$= \dfrac{3x-6}{x+1} \div \dfrac{x-2}{4x+4} \qquad \text{Rewrite with equivalent} \div \text{sign.}$$

$$= \dfrac{3x-6}{x+1} \cdot \dfrac{4x+4}{x-2} \qquad \text{Find the reciprocal of the second fraction.}$$

$$= \dfrac{3(x-2)}{(x+1)} \cdot \dfrac{4(x+1)}{(x-2)} \qquad \text{Factor.}$$

$$= \dfrac{3\cancel{(x-2)}}{\cancel{(x+1)}} \cdot \dfrac{4\cancel{(x+1)}}{\cancel{(x-2)}} \qquad \text{Divide out common factors.}$$

$$= \dfrac{3 \cdot 4}{1}$$

$$= 12$$

PRACTICE PROBLEM FOR EXAMPLE 8

Simplify the complex fraction.

$$\dfrac{\dfrac{x+6}{10x-2}}{\dfrac{x+4}{5x-1}}$$

7.2 Vocabulary Exercises

1. When multiplying rational expressions, first _____ the numerator and denominator.

2. When simplifying rational expressions, divide out _____.

3. To divide rational expressions, find the _____ of the second fraction and multiply.

4. A(n) _____ fraction has fraction(s) in the numerator and/or denominator.

7.2 Exercises

For Exercises 1 through 34, multiply the rational expressions. Simplify.

1. $\dfrac{12}{25} \cdot \dfrac{40}{9}$

2. $\dfrac{30}{16} \cdot \dfrac{24}{20}$

3. $\dfrac{-120}{230} \cdot \dfrac{45}{150}$

4. $\dfrac{140}{315} \cdot \left(-\dfrac{210}{425}\right)$

5. $\dfrac{3x}{4y^2} \cdot \dfrac{6y}{15x}$

6. $\dfrac{7y^3}{9x} \cdot \dfrac{18x^3}{14y}$

7. $\dfrac{-8mn^2}{25m^2} \cdot \dfrac{5mn}{4n^2}$

8. $\dfrac{-6hk}{11h^2} \cdot \dfrac{55k}{9h}$

9. $\dfrac{x+5}{x-8} \cdot \dfrac{x-8}{x+2}$

10. $\dfrac{x-3}{x-7} \cdot \dfrac{x-6}{x-3}$

11. $\dfrac{a+7}{a+2} \cdot \dfrac{a+3}{a-5}$

12. $\dfrac{h-2}{h+1} \cdot \dfrac{h+8}{h-4}$

13. $\dfrac{(x+2)(x-3)}{(x-3)(x+5)} \cdot \dfrac{x+5}{x-4}$

14. $\dfrac{(x-7)(x+7)}{(x+3)(x-6)} \cdot \dfrac{x+3}{x+7}$

15. $\dfrac{(m+2)(m-5)}{(m+3)(m+2)} \cdot \dfrac{m+3}{m+9}$

16. $\dfrac{(n+1)(n-2)}{(n+3)(n+4)} \cdot \dfrac{n+4}{n+1}$

17. $\dfrac{(h+5)(h+7)}{(h+4)(h+5)} \cdot \dfrac{(h+4)(h+3)}{(h+2)(h+7)}$

18. $\dfrac{(g-4)(g+11)}{(g-2)(g+13)} \cdot \dfrac{(g+11)(g-2)}{(g+15)(g-4)}$

19. $\dfrac{b+6}{(b+3)(b+4)} \cdot \dfrac{b+4}{b^2-36}$

20. $\dfrac{p+5}{(p-2)(p+9)} \cdot \dfrac{p^2-81}{p+5}$

21. $\dfrac{(k+2)(k-5)}{k+3} \cdot \dfrac{k+4}{k^2+6k+8}$

22. $\dfrac{n^2-5n-14}{(n-3)(n-7)} \cdot \dfrac{n-8}{n+2}$

23. $\dfrac{h+10}{h-3} \cdot \dfrac{(h+4)(h-1)}{h^2+9h-10}$

24. $\dfrac{(y-8)(y-6)}{y^2-12y+32} \cdot \dfrac{(y+9)(y-4)}{y+2}$

25. $\dfrac{x^2+7x+10}{x^2+8x+15} \cdot \dfrac{x^2+9x+18}{x^2+3x+2}$

26. $\dfrac{m^2-5m-36}{m^2-10m+16} \cdot \dfrac{m^2-9m+14}{m^2-16m+63}$

27. $\dfrac{t^2-t-12}{t^2-11t+30} \cdot \dfrac{t^2-3t-10}{t^2+5t+6}$

28. $\dfrac{b^2+14b+48}{b^2+11b+28} \cdot \dfrac{b^2+10b+21}{b^2+11b+24}$

29. $\dfrac{x-2}{20(x-3)} \cdot \dfrac{5x-15}{x^2-4}$

30. $\dfrac{x^2-16}{3x+12} \cdot \dfrac{2x-10}{x^2-25}$

31. $\dfrac{2-x}{x^2-1} \cdot \dfrac{4x+4}{x^2-4}$

32. $\dfrac{x^2-36}{x^2-25} \cdot \dfrac{9x-45}{6-x}$

33. $\dfrac{x^2-49}{x^2+2x-8} \cdot \dfrac{3x-6}{x^2+7x}$

34. $\dfrac{10x-80}{x^2-100} \cdot \dfrac{x+10}{x^2-8x}$

For Exercises 35 through 54, use the information given in each exercise to answer the questions. Round to the tenths place if necessary.

35. The recommended dosage of a child's drug is 30 milligrams per kilogram. Find the drug dosage in milligrams per pound.

36. A child's prescribed drug dosage is 20 milligrams per kilogram. Find the drug dosage in milligrams per pound.

37. An adult's prescribed drug dosage is 45 milligrams per kilogram. Find the drug dosage in milligrams per pound.

38. An adult's prescribed drug dosage is 40 milligrams per kilogram. Find the drug dosage in milligrams per pound.

39. A 5-pound bag of rice is priced at $0.90 per pound. Convert this price per pound (lb) to dollars per ounce (oz) so that it can be compared to the price of other sized bags. (Round to two decimal places.)

40. A 1-pound box of crackers is priced at $2.79. Convert this price per pound (lb) to dollars per ounce (oz) so that it can be compared to the price of different-sized boxes. (Round to two decimal places.)

41. A 1-pound jar of peanut butter is priced at $3.85. Convert this price per pound (lb) to dollars per ounce (oz) so that it can be compared to the price of other jars of peanut butter. (Round to two decimal places.)

42. A 1-pound jar of boysenberry preserves is priced at $11.68. Convert this price per pound (lb) to dollars per ounce (oz) so that it can be compared to the price of other containers of preserves. (Round to two decimal places.)

43. A car is traveling at 60 mph. What is the car's speed in kilometers per hour?

44. A cyclist is traveling at 20 mph. What is the cyclist's speed in kilometers per hour?

45. A car is traveling at 50 mph. What is the car's speed in feet per minute?

46. A cyclist is traveling at 22 mph. What is the cyclist's speed in feet per minute?

47. A car is traveling at 4400 feet per minute. What is the car's speed in meters per second?

48. A cyclist is traveling at 1936 feet per minute. What is the cyclist's speed in meters per second?

49. A doctor prescribes 150 micrograms (μg) per minute of dopamine to raise a patient's blood pressure. The dopamine is to be added to an IV bag and administered through an IV pump.

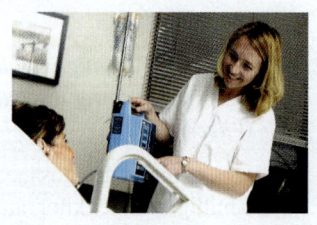

  a. The IV pump flow rate is measured in milligrams per hour. Convert the dosage to milligrams per hour.

  b. Use the result from part a to calculate the number of milligrams of dopamine that the patient will receive in a 24-hour period.

50. A doctor prescribes 200 milligrams of ampicillin per kilogram of body weight to be given four times a day to a sick child.

  a. Convert the dosage to milligrams per pound.

  b. The child weighs 60 lb. Use the result from part a to calculate the number of milligrams of ampicillin that the child will receive in one dose. (Round to the nearest whole number.)

  c. Use the result from part b to find the number of milligrams of ampicillin the child will receive in a day.

51. A cyclist can average 40 km/hr on a training ride.

  a. Convert the rate into miles per hour.

  b. Using the result from part a, how many miles can the cyclist ride during a 6-hour ride?

52. A 155-lb cross-country skier will burn approximately 985 calories per hour during a race. How many calories will the skier burn during a 90-minute race?

Source: Information found at www.nutristrategy.com

53. The Solartaxi is a Swiss-made solar car that has traveled around the world. The car can travel at a maximum speed of 90 km/h.

Source: www.solartaxi.com

  a. Convert this speed to miles per hour.

  b. If traveling at its maximum speed, how many miles can the Solartaxi travel in 5 hours?

54. The Solartaxi consumes 8 kilowatt-hours of electricity per 100 kilometers (8 kWh/100 km). How many kilowatt-hours will the car use when traveling 500 miles? (*Hint*: First convert miles to kilometers.)

Source: www.solartaxi.com

For Exercises 55 through 76, divide the rational expressions.

55. $\dfrac{24}{15} \div \dfrac{20}{9}$

56. $\dfrac{76}{110} \div \dfrac{16}{20}$

57. $\dfrac{85}{22} \div \dfrac{30}{55}$

58. $\dfrac{5}{8} \div \dfrac{25}{2}$

59. $\dfrac{5x}{4y} \div \dfrac{55x^3}{16y}$

60. $\dfrac{7y^2}{16x} \div \dfrac{21y}{20x^2}$

61. $\dfrac{-4m^3}{5n} \div \dfrac{12m}{25n^2}$

62. $\dfrac{8p}{15q} \div \dfrac{20p^2}{9q^3}$

63. $\dfrac{x+5}{x-3} \div \dfrac{x+5}{x+4}$

64. $\dfrac{x-1}{x-5} \div \dfrac{x-1}{x-9}$

65. $\dfrac{a+3}{a+4} \div \dfrac{a+1}{a+3}$

66. $\dfrac{h-6}{h+4} \div \dfrac{h+2}{h-6}$

67. $\dfrac{(x+2)(x-3)}{(x-3)(x+5)} \div \dfrac{x+2}{x-7}$

68. $\dfrac{(t-7)(t+7)}{(t+3)(t-6)} \div \dfrac{t-7}{t+1}$

69. $\dfrac{(n+2)(n-5)}{(n+8)(n+2)} \div \dfrac{n+3}{n+8}$

70. $\dfrac{(c+1)(c-2)}{(c+2)(c+4)} \div \dfrac{c+1}{c+2}$

71. $\dfrac{(h+8)(h+9)}{(h+4)(h+7)} \div \dfrac{(h+2)(h+8)}{(h+4)(h+9)}$

72. $\dfrac{(g-4)(g+12)}{(g-2)(g+13)} \div \dfrac{(g+12)(g-2)}{(g+13)(g-4)}$

73. $\dfrac{b+7}{(b-7)(b+5)} \div \dfrac{b+4}{b^2-49}$

74. $\dfrac{p-4}{(p-7)(p+4)} \div \dfrac{p^2-16}{p-7}$

75. $\dfrac{k^2+8k-9}{k^2-3k-28} \div \dfrac{k^2+5k-6}{k^2+6k+8}$

76. $\dfrac{n^2-5n-14}{n^2-4n-21} \div \dfrac{n^2+7n+10}{n^2-10n+21}$

For Exercises 77 through 82, simplify each complex fraction. See Example 8 for help if needed.

77. $\dfrac{\frac{4x}{9y^2}}{\frac{-6x^2}{27y}}$

78. $\dfrac{\frac{-20k^3}{3h}}{\frac{5k}{12h^2}}$

79. $\dfrac{\frac{5x-15}{6x}}{\frac{2x-6}{12x}}$

80. $\dfrac{\frac{3x-9}{8y}}{\frac{4x-12}{24y}}$

81. $\dfrac{\frac{x+5}{x^2-9}}{\frac{5x+25}{x+3}}$

82. $\dfrac{\frac{1-x}{2x+5}}{\frac{x^2-1}{6x+15}}$

For Exercises 83 through 100, perform the indicated operations.

83. $\dfrac{-9n^3}{5m} \cdot \dfrac{25m^3}{6n}$

84. $\dfrac{8n}{15p^2} \cdot \left(-\dfrac{45p}{16n^3}\right)$

85. $\dfrac{x+4}{x-8} \cdot \dfrac{x+7}{2x+8}$

86. $\dfrac{3x+12}{x+1} \cdot \dfrac{x+1}{x+4}$

87. $\dfrac{x^2-9}{12} \cdot \dfrac{28}{3-x}$

88. $\dfrac{-48}{4-y} \cdot \dfrac{y^2-16}{16}$

89. $\dfrac{-9a}{5b} \div \dfrac{21a^2}{20b^2}$

90. $\dfrac{8pq}{3q^2} \div \left(-\dfrac{24p}{5q}\right)$

91. $\dfrac{z+2}{z-6} \div \dfrac{z+5}{z-6}$

92. $\dfrac{t+11}{t-8} \div \dfrac{t+9}{t-8}$

93. $\dfrac{5x+10}{x-3} \div \dfrac{x+2}{6x-18}$

94. $\dfrac{h-1}{2h-18} \div \dfrac{-7h+7}{h-9}$

95. $\dfrac{a^2-9}{a-3} \div \dfrac{a+3}{a+2}$

96. $\dfrac{h-6}{h^2-36} \div \dfrac{h+5}{h-6}$

97. $\dfrac{a^2+6a+9}{a^2-9} \cdot \dfrac{a^2-5a-24}{a^2-64}$

98. $\dfrac{g^2-25}{g^2+11g+30} \cdot \dfrac{g^2-100}{g^2-12g+20}$

99. $\dfrac{b^2+10b+25}{b^2+8b+16} \div \dfrac{b^2-25}{b^2-16}$

100. $\dfrac{p^2-12p+27}{p^2-13p+36} \div \dfrac{p^2-6p+9}{p^2-8p+16}$

For Exercises 101 through 108, add or subtract as indicated. Reduce to simplest forms.

101. $\dfrac{2}{15} + \dfrac{4}{15}$

102. $\dfrac{4}{13} + \dfrac{5}{13}$

103. $\dfrac{12}{30} - \dfrac{4}{30}$

104. $\dfrac{8}{14} - \dfrac{2}{14}$

105. $\dfrac{1}{3} + \dfrac{3}{14}$

106. $\dfrac{2}{15} + \dfrac{8}{14}$

107. $\dfrac{8}{15} - \dfrac{7}{10}$

108. $\dfrac{10}{14} - \dfrac{12}{21}$

612 CHAPTER 7 Rational Expressions and Equations

7.3 Addition and Subtraction of Rational Expressions

LEARNING OBJECTIVES
- Add and subtract rational expressions with common denominators.
- Find the least common denominator (LCD).
- Add and subtract rational expressions with unlike denominators.
- Simplify complex fractions.

Adding and subtracting rational expressions is similar to adding and subtracting numerical fractions. When the fractions have common denominators, add or subtract the numerators and then simplify. When fractions do not have common denominators, first rewrite the fractions with a common denominator and then simplify.

■ Adding and Subtracting Rational Expressions with Common Denominators

We begin by adding and subtracting rational expressions with common denominators. Use the following Concept Investigation to review the procedure for adding two numerical fractions with common denominators.

▌CONCEPT INVESTIGATION
What do I add or subtract?

Skill Connection

Adding and subtracting numerical fractions with common denominators

To add or subtract numerical fractions with common denominators, add or subtract the numerators and keep the same denominator. Then reduce the fraction if possible. So

$$\frac{2}{3} + \frac{4}{3} = \frac{6}{3}$$
$$= \frac{2}{1}$$
$$= 2$$

and

$$\frac{5}{8} - \frac{3}{8} = \frac{2}{8}$$
$$= \frac{1}{4}$$

Adding and subtracting fractions were covered in Section R.2. Review that section as needed.

1. Choose which of the given diagrams correctly depicts the result of the given addition problem: $\frac{1}{2} + \frac{1}{2}$.

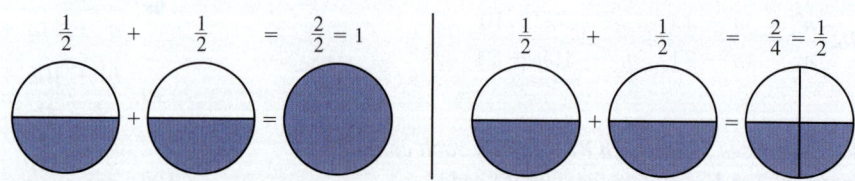

2. Using the diagram you chose in part 1, determine what parts of the fractions you added.
   a. Denominators
   b. Numerators
   c. Denominators and numerators

3. Choose which of the given diagrams correctly depicts the result of the given subtraction problem: $\frac{2}{3} - \frac{1}{3}$.

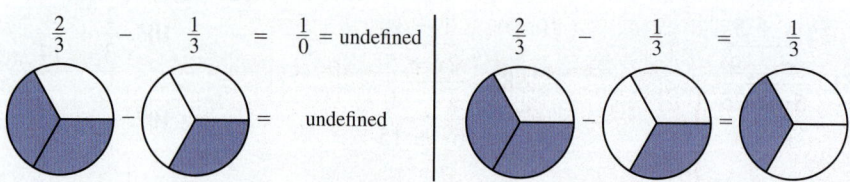

4. Using the diagram you chose in part 3, determine what parts of the fractions you subtracted.
   a. Denominators
   b. Numerators
   c. Denominators and numerators

These diagrams help us visualize adding or subtracting fractions with a common denominator.

> **ADDING OR SUBTRACTING RATIONAL EXPRESSIONS WITH COMMON DENOMINATORS**
>
> $$\frac{a}{b} + \frac{c}{b} = \frac{a+c}{b} \quad \text{and} \quad \frac{a}{b} - \frac{c}{b} = \frac{a-c}{b}$$
>
> where $b \neq 0$
>
> *Note:* The denominator stays the same.

The procedure for numerical fractions is similar to the procedure for adding and subtracting rational expressions.

> **Steps to Adding Rational Expressions with Common Denominators**
> 1. Check that all fractions have common denominators.
> 2. Add the numerators. Write the sum with the common denominator.
> 3. Factor the numerator and reduce (if possible).

Example 1 Adding rational expressions with common denominators

Add the following rational expressions. Simplify.

a. $\dfrac{2x}{x+5} + \dfrac{3}{x+5}$ b. $\dfrac{h+2}{(h+3)(h+5)} + \dfrac{h+4}{(h+3)(h+5)}$

SOLUTION

a. **Step 1** Check that all fractions have common denominators.

The denominators are the same.

$$\frac{2x}{x+5} + \frac{3}{x+5}$$

Step 2 Add the numerators. Write the sum with the common denominator.

$$= \frac{2x+3}{x+5}$$

Step 3 Factor the numerator and reduce (if possible).

The numerator does not factor. This is the answer.

b. The denominators are the same, so add the numerators.

$$\frac{h+2}{(h+3)(h+5)} + \frac{h+4}{(h+3)(h+5)}$$

$$= \frac{h+2+h+4}{(h+3)(h+5)} \quad \text{Add the numerators. Write with the common denominator.}$$

$$= \frac{2h+6}{(h+3)(h+5)}$$

$$= \frac{2(h+3)}{(h+3)(h+5)} \quad \text{Factor the numerator and reduce.}$$

$$= \frac{2\cancel{(h+3)}}{\cancel{(h+3)}(h+5)}$$

$$= \frac{2}{h+5}$$

PRACTICE PROBLEM FOR EXAMPLE 1

Add the following rational expressions.

a. $\dfrac{3a}{5b^2} + \dfrac{7}{5b^2}$ **b.** $\dfrac{4x}{x-3} + \dfrac{5}{x-3}$ **c.** $\dfrac{2m}{(m+5)(m-1)} + \dfrac{-2}{(m+5)(m-1)}$

Subtracting rational expressions is similar to adding, but be sure to subtract the entire numerator of the second fraction. A common mistake is to not use the distributive property to distribute the minus sign.

> **Steps to Subtracting Rational Expressions**
>
> 1. Check that all fractions have common denominators.
> 2. Subtract the numerators. Write the difference with the common denominator.
> 3. Factor the numerator and reduce (if possible).
>
> *Note:* When subtracting rational expressions, be sure to use the distributive property to distribute the minus sign to the entire numerator of the second fraction.

Example 2 Subtracting rational expressions with common denominators

Subtract the following rational expressions. Simplify.

a. $\dfrac{7m}{m+2} - \dfrac{4m+1}{m+2}$ **b.** $\dfrac{b}{8a} - \dfrac{3}{8a}$ **c.** $\dfrac{4z-11}{(z+1)(z-5)} - \dfrac{2z-1}{(z+1)(z-5)}$

SOLUTION

a. Step 1 Check that all fractions have common denominators.

These fractions have common denominators.

$$\frac{7m}{m+2} - \frac{4m+1}{m+2}$$

Step 2 Subtract the numerators. Write the difference with the common denominator.

$$= \frac{7m}{m+2} - \frac{(4m+1)}{m+2} \qquad \text{Use parentheses around the second numerator.}$$

$$= \frac{7m - (4m+1)}{m+2} \qquad \text{Rewrite as a single fraction.}$$

$$= \frac{7m - 1(4m+1)}{m+2} \qquad \text{Recall that } -(4m+1) = -1(4m+1).$$

$$= \frac{7m - 4m - 1}{m+2} \qquad \text{Use the distributive property.}$$

$$= \frac{3m - 1}{m+2} \qquad \text{Combine like terms in the numerator.}$$

Step 3 Factor the numerator and reduce (if possible).

The numerator does not factor. This fraction is simplified.

b. These fractions have common denominators, so subtract the numerators and keep the common denominator.

$$\frac{b}{8a} - \frac{3}{8a}$$

$$= \frac{b - 3}{8a} \qquad \text{Subtract the numerators. Write with the common denominator.}$$

c. Again these fractions have common denominators, so subtract the numerators. Write the difference over the common denominator.

$$\frac{4z - 11}{(z+1)(z-5)} - \frac{2z - 1}{(z+1)(z-5)}$$

$$= \frac{4z - 11 - (2z - 1)}{(z+1)(z-5)} \qquad \text{Use parentheses around the second numerator.}$$

$$= \frac{4z - 11 - 2z + 1}{(z+1)(z-5)} \qquad \text{Use the distributive property.}$$

$$= \frac{2z - 10}{(z+1)(z-5)} \qquad \text{Combine like terms.}$$

$$= \frac{2(z - 5)}{(z+1)(z-5)} \qquad \text{Factor the numerator and reduce.}$$

$$= \frac{2\cancel{(z-5)}}{(z+1)\cancel{(z-5)}}$$

$$= \frac{2}{z+1}$$

PRACTICE PROBLEM FOR EXAMPLE 2

Subtract the following rational expressions. Simplify.

a. $\dfrac{7x}{3y^2} - \dfrac{5}{3y^2}$ **b.** $\dfrac{5a}{a+3} - \dfrac{2a-9}{a+3}$ **c.** $\dfrac{5x+2}{(x+3)(x-4)} - \dfrac{x-6}{(x+3)(x-4)}$

■ Finding the Least Common Denominator (LCD)

To add and subtract rational expressions with unlike denominators, we first have to learn how to find the least common denominator (LCD) of rational expressions. The process will be like the process for numerical fractions because denominators must be completely factored to find the LCD.

616 CHAPTER 7 Rational Expressions and Equations

> **Steps to Finding the Least Common Denominator (LCD) for Rational Expressions**
>
> 1. Factor the denominator of each fraction (if needed). If the denominators involve numerical factors, find the prime factorization.
> 2. The LCD includes each distinct factor. Include the highest power of each factor in the LCD.
> 3. The LCD is the product of the factors from step 2. Leave the LCD in factored form.

Skill Connection

Prime factorization

Remember from Section R.2 that we can find the prime factorization for a natural number greater than 1.
Start with any factorization of the number. Keep factoring until only prime numbers remain.

$$120 = 12 \cdot 10$$
$$120 = 3 \cdot 4 \cdot 5 \cdot 2$$
$$120 = 3 \cdot 2 \cdot 2 \cdot 5 \cdot 2$$
$$120 = 5 \cdot 3 \cdot 2^3$$

Example 3 Finding the LCD with monomial denominators

Find the least common denominator for the following set of fractions.

$$\frac{7}{5xy^2} \qquad \frac{8}{xyz}$$

SOLUTION

Step 1 Factor the denominator of each fraction (if needed).

The denominators are already factored.

Step 2 The LCD includes each distinct factor. Include the highest power of each factor in the LCD.

$$\frac{7}{5xy^2} \qquad \frac{8}{xyz}$$

Step 3 The LCD is the product of the factors from step 2. Leave the LCD in factored form.

The LCD is $5xy^2z$.

PRACTICE PROBLEM FOR EXAMPLE 3

Find the least common denominator for the following set of fractions.

$$\frac{3}{40m^2n} \qquad \frac{8}{75mn}$$

When the denominators are polynomial expressions, be careful to include the correct factors for the LCD. Let's find the LCD for the two fractions

$$\frac{5}{x+3} \qquad \frac{7}{x+6}$$

Both denominators are factored. Therefore, both factors of $x + 3$ and $x + 6$ are included in the LCD. The LCD is $(x + 3)(x + 6)$. When working with polynomial denominators, always factor first.

Example 4 Finding the LCD

Find the least common denominator for each of the following sets of fractions.

a. $\dfrac{2x}{x+5} \qquad \dfrac{5}{x-7}$

b. $\dfrac{2}{(n+5)(n-3)} \qquad \dfrac{n-9}{(n+4)(n-3)}$

c. $\dfrac{h+2}{h^2 - 3h - 28}$ $\dfrac{5h}{h^2 - 4h - 21}$

SOLUTION

a. The denominators are already factored, so include the highest power of each factor for the LCD.

$$\dfrac{2x}{x+5} \qquad \dfrac{5}{x-7}$$

The LCD is $(x+5)(x-7)$.

b. The denominators are already factored. Include the highest power of each factor in the LCD. Note that one factor of $n-3$ is included because it occurs in each fraction only once.

$$\dfrac{2}{(n+5)(n-3)} \qquad \dfrac{n-9}{(n+4)(n-3)}$$

The LCD is $(n+5)(n-3)(n+4)$.

c. Factor the denominators before deciding which factors to include in the LCD.

$$\dfrac{h+2}{h^2 - 3h - 28} \qquad \dfrac{5h}{h^2 - 4h - 21}$$

$$\dfrac{h+2}{(h+4)(h-7)} \qquad \dfrac{5h}{(h+3)(h-7)} \qquad \text{Factor the denominators.}$$

The LCD is $(h+4)(h-7)(h+3)$.

PRACTICE PROBLEM FOR EXAMPLE 4

Find the least common denominator for each of the following pairs of fractions.

a. $\dfrac{k}{(k+1)(k-12)} \qquad \dfrac{7}{(k+1)(k-3)}$

b. $\dfrac{t+3}{t^2 - 9t + 20} \qquad \dfrac{2t}{t^2 + 4t - 32}$

When writing fractions in terms of the LCD, rewrite each fraction to include any missing factors from the LCD. To do so, multiply both the numerator and denominator by the missing factor(s).

Steps to Rewrite Rational Expressions with the LCD

1. Find the least common denominator.
2. Compare the denominator with the LCD to determine which factors are missing.
3. Multiply the numerator and denominator of the fraction by the missing factors.

Example 5 Rewriting rational expressions with the LCD

Find the LCD for each of the following sets of two fractions, and rewrite each fraction with the LCD.

a. $\dfrac{2x}{15y^2 z} \qquad \dfrac{3}{40xy^3}$

b. $\dfrac{4m}{m+2} \qquad \dfrac{3}{m-1}$

Rewriting Rationals with a Common Denominator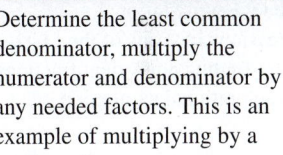

Determine the least common denominator, multiply the numerator and denominator by any needed factors. This is an example of multiplying by a version of one.

Rational 1   Rational 2
$\dfrac{x+6}{(x+2)(x-5)} \qquad \dfrac{5}{(x-5)(x+3)}$

LCD $= (x-5)(x+3)(x+2)$

Rational 1
$\dfrac{(x+3)}{(x+3)} \cdot \dfrac{x+6}{(x+2)(x-5)}$

$= \dfrac{(x+3)(x+6)}{(x-5)(x+3)(x+2)}$

Rational 2
$\dfrac{5}{(x-5)(x+3)} \cdot \dfrac{(x+2)}{(x+2)}$

$= \dfrac{5(x+2)}{(x-5)(x+3)(x+2)}$

618 CHAPTER 7 Rational Expressions and Equations

SOLUTION $\boxed{\frac{x+1}{x+1}}$

a. Step 1 Find the least common denominator.

Factor each denominator to find the LCD.

$$\frac{2x}{15y^2z} \qquad \frac{3}{40xy^3}$$

$$= \frac{2x}{5 \cdot 3y^2z} \qquad = \frac{3}{5 \cdot 2^3xy^3} \qquad \textit{Factor the denominators and find the LCD.}$$

The LCD is $5 \cdot 3 \cdot 2^3xy^3z = 120xy^3z$.

Steps 2 and 3 Compare the denominator with the LCD to determine which factors are missing. Multiply the numerator and denominator of the fraction by the missing factors.

$$\frac{2x}{5 \cdot 3y^2z} \cdot \frac{2^3xy}{2^3xy} \qquad \frac{3}{5 \cdot 2^3xy^3} \cdot \frac{3z}{3z}$$

$$= \frac{16x^2y}{120xy^3z} \qquad = \frac{9z}{120xy^3z}$$

b. Find the LCD. Multiply each fraction by the missing factors.

$$\frac{4m}{m+2} \qquad \frac{3}{m-1}$$

The LCD is $(m+2)(m-1)$.

Compare each denominator to the LCD to find the missing factors. Then multiply both the numerator and denominator by the missing factors.

$$\frac{4m}{m+2} \cdot \frac{(m-1)}{(m-1)} \qquad \frac{3}{m-1} \cdot \frac{(m+2)}{(m+2)}$$

$$= \frac{4m(m-1)}{(m-1)(m+2)} \qquad = \frac{3(m+2)}{(m-1)(m+2)}$$

$$= \frac{4m^2 - 4m}{(m-1)(m+2)} \qquad = \frac{3m+6}{(m-1)(m+2)}$$

PRACTICE PROBLEM FOR EXAMPLE 5

Find the LCD for each of the following sets of two fractions and rewrite each fraction with the LCD.

a. $\dfrac{4a}{25bc^2} \quad \dfrac{7}{30abc^3}$ **b.** $\dfrac{8x}{x+6} \quad \dfrac{5}{x-2}$

> **Connecting the Concepts**
>
> **Are the fractions equivalent?**
> After simplifying a fraction, we obtain a fraction that is equivalent to the original. This is also true if we multiply both the numerator and denominator of a fraction by the same nonzero number.
>
> $$\frac{20}{30} = \frac{2 \cdot \cancel{10}}{3 \cdot \cancel{10}} = \frac{2}{3}$$
>
> $$\frac{5}{7} = \frac{5 \cdot 3}{7 \cdot 3} = \frac{15}{21}$$
>
> When we write rational expressions in terms of the LCD, we are finding equivalent fractions for each rational expression.

Example 6 Rewriting rational expressions with the LCD

Find the LCD for each of the following sets of two fractions. Rewrite each fraction with the LCD.

a. $\dfrac{2}{(n+3)(n-4)} \quad \dfrac{n-2}{(n-4)(n-7)}$ **b.** $\dfrac{h+3}{h^2-3h-40} \quad \dfrac{4h}{h^2-10h+16}$

SOLUTION $\boxed{\frac{x+1}{x+1}}$

a. The denominators are already factored. Find the LCD and multiply each fraction's numerator and denominator by the missing factors.

SECTION 7.3 Addition and Subtraction of Rational Expressions

$$\frac{2}{(n+3)(n-4)} \qquad \frac{n-2}{(n-4)(n-7)}$$

The LCD is $(n+3)(n-4)(n-7)$.

$$\frac{2}{(n+3)(n-4)} \cdot \frac{(n-7)}{(n-7)} \qquad \frac{n-2}{(n-4)(n-7)} \cdot \frac{(n+3)}{(n+3)}$$

$$= \frac{2(n-7)}{(n+3)(n-4)(n-7)} \qquad = \frac{(n-2)(n+3)}{(n+3)(n-4)(n-7)}$$

$$= \frac{2n-14}{(n+3)(n-4)(n-7)} \qquad = \frac{n^2+n-6}{(n+3)(n-4)(n-7)}$$

b. Factor each denominator and find the LCD.

$$\frac{h+3}{h^2-3h-40} \qquad \frac{4h}{h^2-10h+16}$$

$$\frac{h+3}{(h+5)(h-8)} \qquad \frac{4h}{(h-8)(h-2)}$$

The LCD is $(h+5)(h-8)(h-2)$.

$$\frac{h+3}{(h+5)(h-8)} \cdot \frac{(h-2)}{(h-2)} \qquad \frac{4h}{(h-8)(h-2)} \cdot \frac{(h+5)}{(h+5)}$$

$$= \frac{(h-2)(h+3)}{(h+5)(h-8)(h-2)} \qquad = \frac{4h(h+5)}{(h+5)(h-8)(h-2)}$$

$$= \frac{h^2+h-6}{(h+5)(h-8)(h-2)} \qquad = \frac{4h^2+20h}{(h+5)(h-8)(h-2)}$$

Connecting the Concepts

When writing with an LCD, what are we multiplying by?

To rewrite the fraction $\frac{3}{5}$ with a denominator of 20, we multiply $\frac{3}{5}$ by $\frac{4}{4}$:

$$\frac{3}{5} \cdot \frac{4}{4} = \frac{12}{20}$$

Multiplying by $\frac{4}{4}$ is the same as multiplying by 1 because $\frac{4}{4} = 1$. Multiplying an expression by 1 does not change the expression. Likewise, when we multiply

$$\frac{2}{(n+3)(n-4)} \text{ by } \frac{(n-7)}{(n-7)}$$

we are just multiplying by 1.

PRACTICE PROBLEM FOR EXAMPLE 6

Find the LCD for each of the following set of fractions. Rewrite each fraction with the LCD.

$$\frac{t-5}{t^2+6t+9} \qquad \frac{6t}{t^2-t-12}$$

■ Adding and Subtracting Rational Expressions with Unlike Denominators

The process for adding and subtracting rational expressions that have unlike denominators is much like the process for adding and subtracting numerical fractions. First find a common denominator and rewrite each fraction with the common denominator. Then add or subtract the numerators. Finally, factor the numerator and reduce if possible.

> **Steps to Adding or Subtracting Rational Expressions**
> 1. Find the LCD.
> 2. Rewrite all fractions with the LCD.
> 3. Add or subtract the numerators. Write the sum or difference with the common denominator.
> 4. Factor the numerator and reduce (if possible).
>
> *Note:* When subtracting rational expressions, be sure to use the distributive property to distribute the minus sign to the entire numerator of the second fraction.

Skill Connection

Adding and subtracting numerical fractions with different denominators

Find the LCD and rewrite each fraction with the LCD. Next add or subtract the numerators, and keep the same denominator. Reduce the fraction if possible. To add $\frac{2}{5} + \frac{4}{3}$, find the LCD, which is $5 \cdot 3 = 15$. Rewriting each fraction with a denominator of 15 and adding, we get

$$\frac{2}{5} + \frac{4}{3} = \frac{2}{5} \cdot \frac{3}{3} + \frac{4}{3} \cdot \frac{5}{5}$$

$$= \frac{6}{15} + \frac{20}{15}$$

$$= \frac{26}{15}$$

620 CHAPTER 7 Rational Expressions and Equations

Example 7 — Adding and subtracting rational expressions with unlike denominators

Add or subtract the following rational expressions. Reduce the final answer.

a. $\dfrac{3}{xy} + \dfrac{2}{5y^2}$

b. $\dfrac{2x}{x+4} + \dfrac{5}{x-3}$

c. $\dfrac{3h+4}{h+2} - 7$

d. $\dfrac{5}{(n+1)(n-2)} + \dfrac{n+3}{(n+4)(n-2)}$

SOLUTION

a. Step 1 Find the LCD.

$$\dfrac{3}{xy} + \dfrac{2}{5y^2}$$

$$= \dfrac{3}{xy} + \dfrac{2}{5y^2} \qquad \text{The LCD is } 5xy^2.$$

Step 2 Rewrite all fractions with the LCD.

$$= \dfrac{5y}{5y} \cdot \dfrac{3}{xy} + \dfrac{2}{5y^2} \cdot \dfrac{x}{x}$$

$$= \dfrac{15y}{5xy^2} + \dfrac{2x}{5xy^2} \qquad \text{Multiply out the numerators.}$$

Step 3 Add or subtract the numerators. Write the sum or difference with the common denominator.

$$= \dfrac{15y + 2x}{5xy^2}$$

Step 4 Factor the numerator and reduce (if possible).

$$= \dfrac{2x + 15y}{5xy^2} \qquad \text{These are not like terms. This is simplified.}$$

b. Find a common denominator and then add.

$$\dfrac{2x}{x+4} + \dfrac{5}{x-3}$$

$$= \dfrac{2x}{x+4} + \dfrac{5}{x-3} \qquad \text{The LCD is } (x+4)(x-3).$$

$$= \dfrac{(x-3)}{(x-3)} \cdot \dfrac{2x}{x+4} + \dfrac{5}{x-3} \cdot \dfrac{(x+4)}{(x+4)} \qquad \text{Rewrite fractions with the LCD.}$$

$$= \dfrac{2x(x-3)}{(x-3)(x+4)} + \dfrac{5(x+4)}{(x-3)(x+4)} \qquad \text{Multiply out the numerators.}$$

$$= \dfrac{2x^2 - 6x}{(x-3)(x+4)} + \dfrac{5x + 20}{(x-3)(x+4)}$$

$$= \dfrac{2x^2 - 6x + 5x + 20}{(x-3)(x+4)} \qquad \text{Add the numerators and keep the common denominator.}$$

$$= \dfrac{2x^2 - x + 20}{(x-3)(x+4)} \qquad \text{The numerator does not factor.}$$

c. First write the number 7 as a fraction with a denominator of 1. Then multiply the numerator and denominator by $h + 2$ to get a common denominator.

$$\frac{3h + 4}{h + 2} - 7 \qquad \text{Rewrite } 7 = \frac{7}{1}.$$

$$= \frac{3h + 4}{h + 2} - \frac{7}{1} \qquad \text{The LCD is } (h + 2).$$

$$= \frac{3h + 4}{h + 2} - \frac{7}{1} \cdot \frac{(h + 2)}{(h + 2)} \qquad \text{Rewrite fractions with the LCD.}$$

$$= \frac{3h + 4}{h + 2} - \frac{7h + 14}{h + 2} \qquad \text{Subtract the numerators.}$$

$$= \frac{3h + 4 - (7h + 14)}{h + 2} \qquad \text{Use the distributive property.}$$

$$= \frac{3h + 4 - 7h - 14}{h + 2} \qquad \text{Combine like terms.}$$

$$= \frac{-4h - 10}{h + 2} \qquad \text{Factor the numerator.}$$

$$= \frac{-2(2h + 5)}{h + 2} \qquad \text{The expression does not reduce further.}$$

d. Find the LCD and then add.

$$\frac{5}{(n + 1)(n - 2)} + \frac{n + 3}{(n + 4)(n - 2)}$$

$$= \frac{5}{(n + 1)(n - 2)} + \frac{n + 3}{(n + 4)(n - 2)} \qquad \text{The LCD is } (n + 1)(n - 2)(n + 4).$$

$$= \frac{(n + 4)}{(n + 4)} \cdot \frac{5}{(n + 1)(n - 2)} + \frac{n + 3}{(n + 4)(n - 2)} \cdot \frac{(n + 1)}{(n + 1)} \qquad \text{Rewrite with the LCD.}$$

$$= \frac{5(n + 4)}{(n + 4)(n + 1)(n - 2)} + \frac{(n + 3)(n + 1)}{(n + 4)(n + 1)(n - 2)}$$

$$= \frac{5n + 20}{(n + 4)(n + 1)(n - 2)} + \frac{n^2 + 4n + 3}{(n + 4)(n + 1)(n - 2)} \qquad \text{Multiply out the numerators.}$$

$$= \frac{5n + 20 + n^2 + 4n + 3}{(n + 4)(n + 1)(n - 2)} \qquad \begin{array}{l}\text{Add like terms to simplify}\\ \text{the numerator.}\end{array}$$

$$= \frac{n^2 + 9n + 23}{(n + 4)(n + 1)(n - 2)} \qquad \begin{array}{l}\text{The expression does not}\\ \text{reduce further.}\end{array}$$

PRACTICE PROBLEM FOR EXAMPLE 7

Add or subtract the following rational expressions. Reduce the final answer.

a. $\dfrac{6}{x + 7} + 3x$ **b.** $\dfrac{5m}{m + 4} - \dfrac{7}{m - 6}$ **c.** $\dfrac{2}{(a + 3)(a + 2)} + \dfrac{a + 4}{(a + 3)(a + 5)}$

622 CHAPTER 7 Rational Expressions and Equations

CONCEPT INVESTIGATION
Are they equivalent expressions?

Consider the expressions $\frac{1}{a} + \frac{1}{b}$ and $\frac{a+b}{ab}$.

1. Fill in the following table by substituting in the given values of a and b. The first row has been done for you.

a	b	Simplify $\frac{1}{a} + \frac{1}{b}$	Simplify $\frac{a+b}{ab}$	Expressions Same or Different?
2	3	$\frac{1}{2} + \frac{1}{3}$ The LCD is 6. $= \frac{1}{2} \cdot \frac{3}{3} + \frac{1}{3} \cdot \frac{2}{2}$ Rewrite each term with the LCD. $= \frac{3}{6} + \frac{2}{6}$ Add. $= \frac{5}{6}$	$\frac{2+3}{2 \cdot 3}$ $= \frac{5}{6}$	Same $\frac{5}{6} = \frac{5}{6}$
4	5			
7	2			
5	3			

2. On the basis of the pattern you see in the table above, do you think that $\frac{1}{a} + \frac{1}{b}$ and $\frac{a+b}{ab}$ are equivalent?

3. Which expression was easier to compute with in part 1: $\frac{1}{a} + \frac{1}{b}$ or $\frac{a+b}{ab}$?

4. To show that these two expressions are equivalent, fill in the last steps.

$$\frac{a+b}{ab}$$ To divide a polynomial by a monomial, divide each term in the numerator by the denominator.

$$= \frac{a}{ab} + \frac{b}{ab}$$

$$= \frac{?}{?} + \frac{?}{?}$$ Simplify each fraction.

$$= \frac{?}{?} + \frac{?}{?}$$ Use the commutative property of addition.

This Concept Investigation illustrated how adding rational expressions results in an expression that is easier to evaluate.

Example 8 Evaluating the sum of two rational expressions

Angel and Maggie are grading exams. Angel can grade the exams for a class in x hours. Angel gets $\frac{1}{x}$ of the exams graded per hour. Maggie can grade the same exams for a class in y hours. Maggie gets $\frac{1}{y}$ of the exams graded per hour. Working together, the two graders can get $\frac{1}{x} + \frac{1}{y}$ of the exams graded per hour.

a. Rewrite the expression $\dfrac{1}{x} + \dfrac{1}{y}$ by finding a common denominator and adding.

b. Using the expression found in part a, if Angel can grade the exams in 4 hours and Maggie can grade the exams in 6 hours, find the *rate* at which they can grade the exams working together.

SOLUTION $\boxed{\tfrac{x+1}{x+1}}$

a. The LCD for $\dfrac{1}{x} + \dfrac{1}{y}$ is xy. Rewriting each fraction over the LCD and adding the numerators, we get

$$\dfrac{1}{x} + \dfrac{1}{y} \qquad \text{The LCD is } xy.$$

$$= \dfrac{1}{x} \cdot \dfrac{y}{y} + \dfrac{1}{y} \cdot \dfrac{x}{x} \qquad \text{Rewrite each fraction with the LCD.}$$

$$= \dfrac{y}{xy} + \dfrac{x}{xy} \qquad \text{Add the numerators.}$$

$$= \dfrac{y + x}{xy} \qquad \text{Use the commutative property of addition.}$$

$$= \dfrac{x + y}{xy}$$

b. If Angel can grade the exams in 4 hours, let $x = 4$. If Maggie can grade the exams in 6 hours, let $y = 6$. The rate when they work together is

$$\dfrac{x + y}{xy}$$

$$= \dfrac{4 + 6}{4 \cdot 6} \qquad \text{Substitute in } x = 4 \text{ and } y = 6.$$

$$= \dfrac{10}{24} \qquad \text{Simplify the fraction.}$$

$$= \dfrac{5}{12}$$

Working together, Angel and Maggie can grade at a rate of $\dfrac{5}{12}$ *of the exams per hour.*

PRACTICE PROBLEM FOR EXAMPLE 8

Jackson and Noah are working on the budget for their corporation. Jackson can create a budget in a hours. Jackson gets $\dfrac{1}{a}$ of the budget done per hour. Noah can create the same budget in b hours. Noah gets $\dfrac{1}{b}$ of the budget done per hour. Working together, the two employees can get $\dfrac{1}{a} + \dfrac{1}{b}$ of the budget done per hour.

a. Rewrite the expression $\dfrac{1}{a} + \dfrac{1}{b}$ by finding a common denominator and adding.

b. Using the expression found in part a, if Jackson can create the budget in 10 hours and Noah can create the budget in 15 hours, find the *rate* at which they can create the budget working together.

624 CHAPTER 7 Rational Expressions and Equations

Example 9 — Adding and subtracting rational expressions

Add or subtract the following rational expressions. Reduce the final answers.

a. $\dfrac{2a + 3}{a - 2} + \dfrac{7}{2 - a}$

b. $\dfrac{x}{x + 6} - \dfrac{x - 2}{x^2 + 9x + 18}$

c. $\dfrac{2}{h^2 - 4} + \dfrac{h + 4}{h^2 + 8h + 12}$

SOLUTION $\frac{x+1}{x+1}$

a. Find the LCD and then add. In this expression, the denominators look very similar. Note that the terms are reversed in the subtraction. Factor out -1 from the second denominator.

$\dfrac{2a + 3}{a - 2} + \dfrac{7}{2 - a}$

$= \dfrac{2a + 3}{a - 2} + \dfrac{7}{-1(a - 2)}$ Factor a -1 from the second denominator.

$= \dfrac{2a + 3}{a - 2} + \dfrac{-7}{a - 2}$ Move the -1 to the numerator.

$= \dfrac{2a + 3 + (-7)}{a - 2}$ Add the numerators.

$= \dfrac{2a - 4}{a - 2}$

$= \dfrac{2(a - 2)}{a - 2}$ Factor the numerator and reduce.

$= \dfrac{2\cancel{(a - 2)}}{\cancel{a - 2}}$

$= 2$

b. Factor the second denominator to find the LCD. Then subtract and reduce if possible.

$\dfrac{x}{x + 6} - \dfrac{x - 2}{x^2 + 9x + 18}$

$= \dfrac{x}{x + 6} - \dfrac{x - 2}{(x + 6)(x + 3)}$ Factor the denominator.

$= \dfrac{(x + 3)}{(x + 3)} \cdot \dfrac{x}{x + 6} - \dfrac{x - 2}{(x + 6)(x + 3)}$ Find the LCD.

$= \dfrac{x(x + 3)}{(x + 6)(x + 3)} - \dfrac{x - 2}{(x + 6)(x + 3)}$

$= \dfrac{x^2 + 3x}{(x + 6)(x + 3)} - \dfrac{x - 2}{(x + 6)(x + 3)}$

$= \dfrac{x^2 + 3x - (x - 2)}{(x + 6)(x + 3)}$ Subtract the numerators.

$= \dfrac{x^2 + 3x - x + 2}{(x + 6)(x + 3)}$ Use the distributive property. Combine like terms.

$= \dfrac{x^2 + 2x + 2}{(x + 6)(x + 3)}$ The numerator does not factor. This is simplest form.

SECTION 7.3 Addition and Subtraction of Rational Expressions

c. Factor the denominators to find the least common denominator. Then add and reduce if possible.

$\dfrac{2}{h^2 - 4} + \dfrac{h + 4}{h^2 + 8h + 12}$ Factor the denominators.

$= \dfrac{2}{(h - 2)(h + 2)} + \dfrac{h + 4}{(h + 6)(h + 2)}$ The LCD is $(h - 2)(h + 2)(h + 6)$.

$= \dfrac{(h + 6)}{(h + 6)} \cdot \dfrac{2}{(h - 2)(h + 2)} + \dfrac{h + 4}{(h + 6)(h + 2)} \cdot \dfrac{(h - 2)}{(h - 2)}$ Rewrite over the LCD.

$= \dfrac{2(h + 6)}{(h + 6)(h - 2)(h + 2)} + \dfrac{(h + 4)(h - 2)}{(h + 6)(h - 2)(h + 2)}$

$= \dfrac{2h + 12}{(h + 6)(h - 2)(h + 2)} + \dfrac{h^2 + 2h - 8}{(h + 6)(h - 2)(h + 2)}$ Multiply out the numerators.

$= \dfrac{2h + 12 + h^2 + 2h - 8}{(h + 6)(h - 2)(h + 2)}$ Add the numerators.

$= \dfrac{h^2 + 4h + 4}{(h + 6)(h - 2)(h + 2)}$

$= \dfrac{(h + 2)(h + 2)}{(h + 6)(h - 2)(h + 2)}$ Factor the numerator.

$= \dfrac{(h + 2)\cancel{(h + 2)}}{(h + 6)(h - 2)\cancel{(h + 2)}}$ Divide out the common factors.

$= \dfrac{h + 2}{(h + 6)(h - 2)}$

PRACTICE PROBLEM FOR EXAMPLE 9

Add or subtract the following rational expressions. Reduce the final answers.

a. $\dfrac{8}{5 - z} - \dfrac{2z}{z - 5}$ **b.** $\dfrac{-2}{n^2 - 16} + \dfrac{n - 2}{n^2 - 5n + 4}$

■ Simplifying Complex Fractions

In Section 7.2, we learned the basics of *complex* fractions—fractions in which the numerators and denominators themselves contain fractions. When the numerator and denominator contain only one term each, simplify as in Section 7.2 by dividing.

Example 10   Simplifying a complex fraction using division

Simplify: $\dfrac{\dfrac{4x - 20}{x + 4}}{\dfrac{x - 5}{x^2 - 16}}$.

SOLUTION

$\dfrac{\dfrac{4x - 20}{x + 4}}{\dfrac{x - 5}{x^2 - 16}}$

626 CHAPTER 7 Rational Expressions and Equations

$$= \frac{4x - 20}{x + 4} \div \frac{x - 5}{x^2 - 16}$$ Rewrite as a division problem.

$$= \frac{4x - 20}{x + 4} \cdot \frac{x^2 - 16}{x - 5}$$ Invert the second fraction and multiply.

$$= \frac{4(x - 5)}{(x + 4)} \cdot \frac{(x - 4)(x + 4)}{x - 5}$$ Factor.

$$= \frac{4\cancel{(x - 5)}}{\cancel{(x + 4)}} \cdot \frac{(x - 4)\cancel{(x + 4)}}{\cancel{(x - 5)}}$$ Divide out like factors.

$$= 4(x - 4) = 4x - 16$$

Some complex fractions have multiple terms, some of which are fractions, in the numerator and denominator. These problems can look like this:

$$\frac{1 + \dfrac{1}{x}}{2 - \dfrac{3}{x}}$$

Because there are multiple terms in the numerator and denominator and some of the terms are fractions, we cannot divide this problem out as we did in Example 10. To simplify complex fractions that have multiple terms in the numerator and denominator requires the following steps.

> **Complex Fractions**
>
> Determine the least common denominator of all the fractions inside the rational expression. Multiply the numerator and denominator of the rational expression by the LCD. This is an example of multiplying by a version of one.
>
> $$\frac{2 + \dfrac{3}{x}}{1 - \dfrac{7}{x^2}}$$
>
> $$\text{LCD} = x^2 = \frac{x^2}{1}$$
>
> $$\frac{\left(2 + \dfrac{3}{x}\right)}{\left(1 - \dfrac{7}{x^2}\right)} \cdot \frac{\dfrac{x^2}{1}}{\dfrac{x^2}{1}}$$
>
> $$\frac{2x^2 + \dfrac{3}{x} \dfrac{x^2}{1}}{x^2 - \dfrac{7}{x^2} \dfrac{x^2}{1}}$$
>
> $$\frac{2x^2 + 3x}{x^2 - 7}$$

Steps to Simplify Complex Fractions

1. Find the LCD for the entire expression. Include in the LCD all factors in the denominators of both the numerator and the denominator.
2. Multiply the numerator and denominator by the LCD.
3. Factor and reduce (if possible).

Example 11 Simplifying complex fractions

Simplify each complex fraction. Reduce to lowest terms.

a. $$\dfrac{\dfrac{2}{3} - \dfrac{1}{6}}{\dfrac{1}{2} + \dfrac{5}{6}}$$ b. $$\dfrac{1 + \dfrac{1}{x}}{2 - \dfrac{3}{x}}$$ c. $$\dfrac{1 - \dfrac{6}{y}}{1 - \dfrac{36}{y^2}}$$

SOLUTION

a. Step 1 Find the LCD for the entire expression. Include in the LCD all factors in the denominators of both the numerator and the denominator.

The only denominators that appear are 3, 6, and 2. Therefore, the LCD is 6, which can be written as $6 = \dfrac{6}{1}$.

$$\dfrac{\dfrac{2}{3} - \dfrac{1}{6}}{\dfrac{1}{2} + \dfrac{5}{6}} \qquad \text{The LCD is } 6 = \dfrac{6}{1}.$$

Step 2 Multiply the numerator and denominator by the LCD.

$$= \frac{\left(\frac{2}{3} - \frac{1}{6}\right) \cdot \frac{6}{1}}{\left(\frac{1}{2} + \frac{5}{6}\right) \cdot \frac{6}{1}}$$

$$= \frac{\frac{2}{3} \cdot \frac{6}{1} - \frac{1}{6} \cdot \frac{6}{1}}{\frac{1}{2} \cdot \frac{6}{1} + \frac{5}{6} \cdot \frac{6}{1}} \qquad \text{Use the distributive property.}$$

Step 3 Factor and reduce if possible.

$$= \frac{4 - 1}{3 + 5} \qquad \text{Simplify. Reduce if possible.}$$

$$= \frac{3}{8}$$

b. In this case, the only denominator that appears in all terms is x, so the LCD is x.

$$\frac{1 + \frac{1}{x}}{2 - \frac{3}{x}} \qquad \text{The LCD is } x = \frac{x}{1}.$$

$$= \frac{\left(1 + \frac{1}{x}\right) \cdot \frac{x}{1}}{\left(2 - \frac{3}{x}\right) \cdot \frac{x}{1}} \qquad \text{Multiply the numerator and denominator by the LCD.}$$

$$= \frac{1 \cdot \frac{x}{1} + \frac{1}{x} \cdot \frac{x}{1}}{2 \cdot \frac{x}{1} - \frac{3}{x} \cdot \frac{x}{1}} \qquad \text{Use the distributive property.}$$

$$= \frac{x + 1}{2x - 3} \qquad \text{Simplify. Reduce if possible.}$$

c. Find the LCD. In this case, the denominators that appear are y and y^2, so the LCD is y^2.

$$\frac{1 - \frac{6}{y}}{1 - \frac{36}{y^2}} \qquad \text{The LCD is } y^2 = \frac{y^2}{1}.$$

$$= \frac{\left(1 - \frac{6}{y}\right) \cdot \frac{y^2}{1}}{\left(1 - \frac{36}{y^2}\right) \cdot \frac{y^2}{1}} \qquad \text{Multiply the numerator and denominator by the LCD.}$$

$$= \frac{1 \cdot \frac{y^2}{1} - \frac{6}{y} \cdot \frac{y^2}{1}}{1 \cdot \frac{y^2}{1} - \frac{36}{y^2} \cdot \frac{y^2}{1}} \qquad \text{Use the distributive property.}$$

$$= \frac{y^2 - 6y}{y^2 - 36} \qquad \text{Simplify.}$$

628 CHAPTER 7 Rational Expressions and Equations

$$= \frac{y(y-6)}{(y-6)(y+6)}$$ *Factor. Divide out common factors.*

$$= \frac{y}{y+6}$$

PRACTICE PROBLEM FOR EXAMPLE 11
Simplify each complex fraction. Reduce to lowest terms.

a. $\dfrac{\dfrac{3}{10} - \dfrac{1}{5}}{\dfrac{1}{2} + \dfrac{7}{10}}$

b. $\dfrac{\dfrac{5}{x} - 1}{\dfrac{25}{x^2} - 1}$

7.3 Vocabulary Exercises

1. To add or subtract rational expressions requires _____ denominators.

2. To find the LCD, include the _____ power of each factor to be included in the LCD.

3. When adding or subtracting rational expressions, add or subtract the _____ and keep the common _____.

4. When subtracting rational expressions, be sure to distribute the _____ to the entire second numerator.

7.3 Exercises

For Exercises 1 through 20, add or subtract the rational expressions. Reduce the final answers.

1. $\dfrac{4}{7c^2} + \dfrac{6}{7c^2}$

2. $\dfrac{8}{3x} + \dfrac{2}{3x}$

3. $\dfrac{16}{5a^2} - \dfrac{1}{5a^2}$

4. $\dfrac{9}{2m^2} - \dfrac{3}{2m^2}$

5. $\dfrac{2x}{x+5} + \dfrac{3}{x+5}$

6. $\dfrac{3t}{t-7} + \dfrac{4}{t-7}$

7. $\dfrac{4h}{h-9} - \dfrac{3}{h-9}$

8. $\dfrac{2a}{a+3} - \dfrac{7}{a+3}$

9. $\dfrac{10}{x+2} + \dfrac{5x}{x+2}$

10. $\dfrac{20}{c+2} + \dfrac{10c}{c+2}$

11. $\dfrac{4a^2}{2a-1} - \dfrac{2a}{2a-1}$

12. $\dfrac{3h^2}{h-4} - \dfrac{12h}{h-4}$

13. $\dfrac{h+3}{(h+4)(h+7)} + \dfrac{4}{(h+4)(h+7)}$

14. $\dfrac{x+2}{(x+4)(x-6)} + \dfrac{2}{(x+4)(x-6)}$

15. $\dfrac{z+7}{(z+2)(z-3)} - \dfrac{5}{(z+2)(z-3)}$

16. $\dfrac{c-2}{(c-6)(c-4)} - \dfrac{4}{(c-6)(c-4)}$

17. $\dfrac{3x+8}{(x+2)(2x+5)} - \dfrac{x+3}{(x+2)(2x+5)}$

18. $\dfrac{4m-9}{(m-3)(m-7)} - \dfrac{2m+5}{(m-3)(m-7)}$

19. $\dfrac{7a-4}{(2a+3)(a-4)} - \dfrac{5a-8}{(2a+3)(a-4)}$

20. $\dfrac{6t-11}{(t+3)(4t+5)} - \dfrac{2t-3}{(t+3)(4t+5)}$

For Exercises 21 through 34, find the LCD for each of set of fractions.

21. $\dfrac{3}{50} \quad \dfrac{8}{15}$

22. $\dfrac{9}{40} \quad \dfrac{2}{75}$

23. $\dfrac{2}{7xy^3} \quad \dfrac{5}{yz^2}$

24. $\dfrac{10}{3a^2b} \quad \dfrac{11}{abc}$

SECTION 7.3 Addition and Subtraction of Rational Expressions

25. $\dfrac{4h}{h-9}$ $\dfrac{3}{h-6}$

26. $\dfrac{2a}{a+3}$ $\dfrac{7}{a+5}$

27. $\dfrac{5}{2a-6}$ $\dfrac{2}{a-3}$

28. $\dfrac{8}{3h-12}$ $\dfrac{1}{h-4}$

29. $\dfrac{5}{(h+4)(h+7)}$ $\dfrac{4}{(h+4)(h+2)}$

30. $\dfrac{3}{(x+4)(x-2)}$ $\dfrac{9}{(x+4)(x-6)}$

31. $\dfrac{z+7}{(z+2)(z-3)}$ $\dfrac{z+4}{(z-5)(z-3)}$

32. $\dfrac{r+2}{(r-5)(r-3)}$ $\dfrac{r-4}{(r-5)(r-7)}$

33. $\dfrac{5}{m^2+8m-20}$ $\dfrac{3}{m^2+3m-10}$

34. $\dfrac{z+5}{z^2+4z-12}$ $\dfrac{z-3}{z^2-5z+6}$

For Exercises 35 through 48, find the LCD for each set of fractions. Rewrite each fraction with the LCD.

35. $\dfrac{3}{56}$ $\dfrac{4}{21}$

36. $\dfrac{5}{42}$ $\dfrac{3}{28}$

37. $\dfrac{7n}{20m^3}$ $\dfrac{5}{18mn^2}$

38. $\dfrac{3s}{26t^2}$ $\dfrac{7}{12s^2t}$

39. $\dfrac{10}{x+4}$ $\dfrac{5x}{x+2}$

40. $\dfrac{20}{c+7}$ $\dfrac{10c}{c-3}$

41. $\dfrac{5}{7a-14}$ $\dfrac{9}{a-2}$

42. $\dfrac{3}{4h-28}$ $\dfrac{8}{h-7}$

43. $\dfrac{z+4}{(z+5)(z-2)}$ $\dfrac{z+1}{(z-8)(z-2)}$

44. $\dfrac{r+7}{(r-2)(r-1)}$ $\dfrac{r-5}{(r-1)(r-3)}$

45. $\dfrac{2x}{x^2-9}$ $\dfrac{x}{x^2+5x+6}$

46. $\dfrac{3a}{a^2-100}$ $\dfrac{2a}{a^2-11a+10}$

47. $\dfrac{15}{m^2+6m-16}$ $\dfrac{3}{m^2+7m-8}$

48. $\dfrac{z+1}{z^2+4z+4}$ $\dfrac{z-6}{z^2-3z-10}$

For Exercises 49 through 70, add or subtract the rational expressions. Reduce the final answers.

49. $\dfrac{8}{45}+\dfrac{6}{15}$

50. $\dfrac{9}{50}-\dfrac{2}{35}$

51. $\dfrac{5}{6xy}+\dfrac{8}{yz^2}$

52. $\dfrac{10c}{3a^2b^2}+\dfrac{4}{abc}$

53. $\dfrac{4n}{9m^3}-\dfrac{5}{15mn^2}$

54. $\dfrac{5s}{14t^2}-\dfrac{7}{32s^2t}$

55. $\dfrac{4h}{h-9}+\dfrac{3}{h-6}$

56. $\dfrac{2a}{a+3}+\dfrac{7}{a+5}$

57. $\dfrac{4a-14}{a-5}+\dfrac{6}{5-a}$

58. $\dfrac{3x+8}{x-4}+\dfrac{20}{4-x}$

59. $\dfrac{3b-1}{b-7}+\dfrac{6}{7-b}$

60. $\dfrac{2v+5}{4-v}+\dfrac{9}{v-4}$

61. $\dfrac{5}{4a-8}+\dfrac{3}{a-2}$

62. $\dfrac{22}{3h-12}-\dfrac{4}{h-4}$

63. $\dfrac{x}{x+4}-\dfrac{x-3}{x^2+9x+20}$

64. $\dfrac{z}{z-2}-\dfrac{5z-2}{z^2-9z+14}$

65. $\dfrac{m}{m+2}-\dfrac{2m-15}{m^2+12m+20}$

66. $\dfrac{t}{t-6}-\dfrac{5t-6}{t^2-8t+12}$

67. $\dfrac{2x}{x^2-9}+\dfrac{x}{x^2+5x+6}$

68. $\dfrac{3a}{a^2-100}+\dfrac{2a}{a^2-11a+10}$

69. $\dfrac{5}{m^2+8m-20}+\dfrac{3}{m^2+3m-10}$

70. $\dfrac{z+5}{z^2+4z-12}+\dfrac{z-3}{z^2-5z+6}$

For Exercises 71 through 74, use the information given in each exercise to answer the questions. See Example 8 for help if needed.

71. Adam can mow the lawn in a hours. Adam gets $\dfrac{1}{a}$ of the job done per hour. Matthew can mow the lawn in b hours. Matthew gets $\dfrac{1}{b}$ of the job done per hour. Working

together, the two brothers can get $\frac{1}{a}+\frac{1}{b}$ of the yard mowed per hour.

a. Rewrite the expression $\frac{1}{a}+\frac{1}{b}$ by finding a common denominator and adding.

b. Use the expression found in part a to find the rate it takes the two brothers working together to mow the lawn if Adam can mow the lawn in 3 hours and Matthew can mow the lawn in 2 hours.

72. Barbara can paint a mural in x hours. Barbara gets $\frac{1}{x}$ of the job done per hour. Christy can paint the same mural in y hours. Christy gets $\frac{1}{y}$ of the job done per hour. Working together, the two sisters can get $\frac{1}{x}+\frac{1}{y}$ of the mural painted per hour.

a. Rewrite the expression $\frac{1}{x}+\frac{1}{y}$ by finding a common denominator and adding.

b. Use the expression found in part a to find the rate it takes the two sisters working together to paint the mural if Barbara can paint the mural in 22 hours and Christy can paint the mural in 20 hours.

73. Luis can file college applications for the fall semester in m days. Luis can get $\frac{1}{m}$ of the job done per day. Veronica can file the same applications in n days. Veronica can get $\frac{1}{n}$ of the job done per day. Working together, they get $\frac{1}{m}+\frac{1}{n}$ of the applications filed per day.

a. Rewrite the expression $\frac{1}{m}+\frac{1}{n}$ by finding a common denominator and adding.

b. Find the rate it takes the two employees to file fall applications if Luis can get the job done in 4 days and Veronica can get the job done in 5 days.

74. Stefan can prepare a tax return in h hours. Stefan gets $\frac{1}{h}$ of the job done per hour. Jelena can prepare the same tax return in k hours. Jelena gets $\frac{1}{k}$ of the job done per hour. Working together, they get $\frac{1}{h}+\frac{1}{k}$ of the tax return done per hour.

a. Rewrite the expression $\frac{1}{h}+\frac{1}{k}$ by finding a common denominator and adding.

b. Find the rate it takes the two employees working together to prepare the tax return if Stefan can prepare the return in 9 hours and Jelena can prepare the return in 7.5 hours.

For Exercises 75 through 86, perform the indicated operations. Reduce the final answers.

75. $\dfrac{6}{x+3}+\dfrac{5}{x+1}$

76. $\dfrac{4}{a+2}-\dfrac{3}{a-5}$

77. $\dfrac{4x}{x+3}\cdot\dfrac{8}{x+2}$

78. $\dfrac{2x+6}{x+5}\div\dfrac{5x+15}{x-2}$

79. $\dfrac{n^2-9}{n^2+6n+9}\div\dfrac{2n-6}{n+5}$

80. $\dfrac{4}{t+5}-\dfrac{5t}{t^2+7t+10}$

81. $\dfrac{8h}{h-7}+\dfrac{6}{h-4}$

82. $\dfrac{4b}{b+8}\cdot\dfrac{7}{b+2}$

83. $\dfrac{7n-15}{n-3}+\dfrac{6}{3-n}$

84. $\dfrac{5y+2}{y-4}+\dfrac{11}{4-y}$

85. $\dfrac{c^2+3c+2}{c-1}\cdot\dfrac{5c+3}{c+2}$

86. $\dfrac{4x^2-25}{6x+15}\div\dfrac{2x-5}{x+4}$

For Exercises 87 through 100, simplify each complex fraction.

87. $\dfrac{\dfrac{2x+1}{x-5}}{\dfrac{4x+2}{3x-15}}$

88. $\dfrac{\dfrac{x-7}{x+9}}{\dfrac{x^2-49}{x^2-81}}$

89. $\dfrac{\dfrac{a+2}{3b}}{\dfrac{a^2-4}{7b^2}}$

90. $\dfrac{\dfrac{5y}{z^2-9}}{\dfrac{8y^3}{2z-6}}$

91. $\dfrac{\dfrac{3}{4}-\dfrac{1}{6}}{\dfrac{1}{12}+\dfrac{1}{3}}$

92. $\dfrac{\dfrac{5}{8}+\dfrac{1}{6}}{\dfrac{5}{4}-\dfrac{1}{3}}$

93. $\dfrac{\dfrac{3}{a}-4}{\dfrac{5}{a}+7}$

94. $\dfrac{\dfrac{5}{b}-1}{\dfrac{7}{b}-3}$

95. $\dfrac{\dfrac{4}{x}-16}{\dfrac{8}{x^2}-12}$

96. $\dfrac{20+\dfrac{5}{x}}{5-\dfrac{30}{x^2}}$

97. $\dfrac{8p + \dfrac{3}{p}}{8p + \dfrac{3}{p}}$

98. $\dfrac{7z - \dfrac{5}{z^2}}{7z - \dfrac{5}{z^2}}$

99. $\dfrac{\dfrac{4}{x} + \dfrac{5}{y}}{\dfrac{6}{x} - \dfrac{7}{y}}$

100. $\dfrac{\dfrac{3}{a^2} - \dfrac{1}{b^2}}{\dfrac{4}{a} + \dfrac{2}{b}}$

For Exercises 101 and 102, use the information given in the exercises to answer the questions.

101. A Little League baseball team and the team's two coaches are traveling to the state playoffs. The average cost for food for n players and two coaches is given by

$$\dfrac{80n + 400}{n + 2}$$

The average cost for transportation and lodging for the n players and two coaches is given by

$$\dfrac{150n + 1200}{n + 2}$$

a. If 20 players go on the trip, how much is the average cost for food?

b. If 20 players go on the trip, how much is the average cost for transportation and lodging?

c. Add the two expressions for average cost for food and average cost for transportation and lodging. Simplify.

d. Use the expression from part c to find the total average cost for all expenses if 25 players go on the trip.

102. A girls' soccer team and the team's three coaches are traveling to the state playoffs. The average cost for food for n players and three coaches is given by

$$\dfrac{55n + 450}{n + 3}$$

The average cost for transportation and lodging for the n players and three coaches is given by

$$\dfrac{200n + 1650}{n + 3}$$

a. If 18 players go on the trip, how much is the average cost for food?

b. If 18 players go on the trip, how much is the average cost for transportation and lodging?

c. Add the two expressions for average cost for food and average cost for transportation and lodging. Simplify.

d. Use the expression from part c to find the total average cost for all expenses if 20 players go on the trip.

7.4 Solving Rational Equations

LEARNING OBJECTIVES
- Solve rational equations.
- Set up and solve shared work problems.

Solving Rational Equations

In Section 7.1, we evaluated rational equations and discussed excluded values for rational expressions. Now we learn to solve rational equations using some of the tools introduced so far in this chapter. To solve a rational equation, multiply both sides of the equation by the LCD of all the rational expressions in the equation. Recall that the multiplication property of equality justifies multiplying both sides by the LCD. This multiplication will eliminate all the denominators and will result in a simpler equation to solve.

Steps to Solving Rational Equations
1. Find any excluded values by setting the denominators equal to zero.
2. Find the LCD of all the rational expressions.
3. Multiply both sides of the equation by the LCD and simplify.
4. Solve the remaining equation.
5. Check all solutions. Remove any that are excluded values.

Example 1 — Cost per person for a ski trip

Recall from Section 7.1 that a group of students is going to rent a cabin that costs $1000 for a weekend of snowboarding. They plan to pay equal shares. The students can find the cost per person using the equation

$$c = \frac{1000}{n}$$

where c is the cost in dollars per person for n students to go to the cabin for the weekend.

a. Find the cost per person if 10 students stay in the cabin.

b. If the students want to spend only $50 each, how many students must share the cabin?

SOLUTION

a. The number of students is 10, so substitute $n = 10$ and calculate the cost c.

$$c = \frac{1000}{10} = 100$$

If 10 students go to the cabin, it will cost each of them $100.

b. The students want to spend only $50 each. Therefore, substitute $c = 50$ for the cost and solve for n.

Step 1 Find any excluded values by setting the denominator equal to zero.

Because the denominator of the rational expression would be zero when $n = 0$, exclude this value from the possible solutions.

Step 2 Find the LCD of all the rational expressions.

$$50 = \frac{1000}{n} \qquad \text{The LCD is } n.$$

Step 3 Multiply both sides of the equation by the LCD and simplify.

$$n(50) = \left(\frac{1000}{\not n}\right) \cdot \not n$$

$$50n = 1000$$

Step 4 Solve the remaining equation.

$$\frac{50n}{50} = \frac{1000}{50}$$

$$n = 20$$

Step 5 Check all solutions. Remove any that are excluded values.

$$c = \frac{1000}{n}$$

$$50 \stackrel{?}{=} \frac{1000}{20}$$

$$50 = 50$$

For the cost per student to be $50, 20 students must share the cabin.

When we started to solve this equation, we noted that we must exclude zero as a possible solution because it would make the denominator of the rational expression 0. In this situation, having zero students go on the ski trip would also not make sense and would also have been excluded as a possible solution.

SECTION 7.4 Solving Rational Equations   633

PRACTICE PROBLEM FOR EXAMPLE 1

For a recent field trip to the Blue Ridge Parkway Institute and Museum, a school had to pay $650 for the bus and $8 per student for admission. The cost per student for a class to take this field trip can be calculated by using the equation

$$c = \frac{650 + 8n}{n}$$

where c is the cost in dollars per student for n students to go on the field trip.

a. Find the cost per student if 40 students go on the field trip.

b. If the school wants the cost per student to be $21, how many students must go on the field trip?

In Example 1b, we had to solve the equation $50 = \frac{1000}{n}$ to determine the number of students who had to sign up for the ski trip so that the cost could be kept to $50.00 per student. This is an example of a practical reason for solving a rational equation. In the next example, we further study solving rational equations.

Example 2 Solving rational equations with common denominators

Solve the following rational equations. Check the solutions.

a. $5 + \dfrac{12}{x} = 11 - \dfrac{6}{x}$

b. $\dfrac{-5}{m-4} = 3 - \dfrac{5}{m-4}$

c. $\dfrac{15}{a-4} + 5 = a - 1$

SOLUTION

a. To find the excluded values, set the denominators equal to 0. The only value that will make the denominators 0 is $x = 0$, so exclude $x = 0$ from the possible solutions. The LCD is x, so multiply both sides of the equation by x and solve the remaining equation.

$5 + \dfrac{12}{x} = 11 - \dfrac{6}{x}$ The LCD is x. Recall that $x = \dfrac{x}{1}$.

$\dfrac{x}{1}\left(5 + \dfrac{12}{x}\right) = \left(11 - \dfrac{6}{x}\right) \cdot \dfrac{x}{1}$ Multiply both sides of the equation by the LCD.

$5 \cdot \dfrac{x}{1} + \dfrac{x}{1}\left(\dfrac{12}{x}\right) = 11 \cdot \dfrac{x}{1} - \dfrac{x}{1}\left(\dfrac{6}{x}\right)$ Use the distributive property.

$5x + \dfrac{x}{1}\left(\dfrac{12}{x}\right) = 11x - \dfrac{x}{1}\left(\dfrac{6}{x}\right)$ Reduce and simplify.

$5x + 12 = 11x - 6$ Solve the remaining equation.

$-5x -5x$

$12 = 6x - 6$

$+6 +6$

$18 = 6x$

$\dfrac{18}{6} = \dfrac{6x}{6}$

$3 = x$

Rational Equations

Multiply by the LCD

Multiply both sides of the equation by the least common denominator. This will clear all fractions from the equation.

$$\frac{3}{x+1} = \frac{2}{x}$$

$$(x+1)(x) \cdot \frac{3}{x+1} = \frac{2}{x} \cdot (x+1)(x)$$

$$3x = 2x + 2$$

Skill Connection

Finding excluded values

In Section 7.1, we learned that to find the excluded values for a rational expression or equation means to find those values of the variable that make the denominator zero. Therefore, set the denominators equal to 0 and solve for the variable. Then exclude those values.

Check this solution by substituting it into the original equation.

$$5 + \frac{12}{(3)} \stackrel{?}{=} 11 - \frac{6}{(3)}$$

$$5 + 4 \stackrel{?}{=} 11 - 2$$

$$9 = 9$$

The solution checks and is not an excluded value. So $x = 3$ is the solution.

b. To find the excluded values, set the denominators equal to 0. Setting $m - 4 = 0$ and solving for m gives the only excluded value of $m = 4$. The LCD is $m - 4$, so multiply both sides of the equation by $m - 4$ and solve the resulting equation.

$\dfrac{-5}{m-4} = 3 - \dfrac{5}{m-4}$ The LCD is $m - 4$. Recall that $m - 4 = \dfrac{m-4}{1}$.

$\dfrac{(m-4)}{1}\left(\dfrac{-5}{m-4}\right) = \dfrac{(m-4)}{1}\left(3 - \dfrac{5}{m-4}\right)$ Multiply both sides of the equation by the LCD.

$\dfrac{(m-4)}{1}\left(\dfrac{-5}{m-4}\right) = \dfrac{(m-4)}{1}(3) - \dfrac{(m-4)}{1}\left(\dfrac{5}{m-4}\right)$ Use the distributive property.

$\dfrac{\cancel{(m-4)}}{1}\left(\dfrac{-5}{\cancel{m-4}}\right) = 3m - 12 - \dfrac{\cancel{(m-4)}}{1}\left(\dfrac{5}{\cancel{m-4}}\right)$ Divide out like factors.

$-5 = 3m - 12 - 5$ Solve the remaining equation.

$-5 = 3m - 17$

$12 = 3m$

$4 = m$

The excluded value is $m = 4$. Let's examine why this solution does not check.

$$\dfrac{-5}{(4) - 4} \stackrel{?}{=} 3 - \dfrac{5}{(4) - 4}$$

$$\dfrac{-5}{0} \stackrel{?}{=} 3 - \dfrac{5}{0}$$ Division by 0 is undefined.

Therefore, this equation has no solution.

c. To find the excluded values, set the denominators equal to 0. Setting $a - 4 = 0$ and solving for a yields the only excluded value of $a = 4$. The LCD is $a - 4$, so multiply both sides of the equation by $a - 4$ and solve the resulting equation.

$\dfrac{15}{a-4} + 5 = a - 1$ The LCD is $a - 4$. Write $a - 4 = \dfrac{a-4}{1}$.

$\dfrac{(a-4)}{1}\left(\dfrac{15}{a-4} + 5\right) = \dfrac{(a-4)}{1}(a-1)$ Multiply both sides of the equation by the LCD.

$\dfrac{(a-4)}{1}\left(\dfrac{15}{a-4}\right) + \dfrac{(a-4)}{1} \cdot 5 = \dfrac{(a-4)}{1}(a-1)$ Use the distributive property on the left side.

$\dfrac{\cancel{(a-4)}}{1}\left(\dfrac{15}{\cancel{a-4}}\right) + (a-4)5 = (a-4)(a-1)$ Divide out like factors.

$15 + 5a - 20 = a^2 - a - 4a + 4$ Simplify both sides. Use FOIL on the right side.

$5a - 5 = a^2 - 5a + 4$ This is a quadratic equation. Set equal to 0.

$\underline{-5a + 5} \quad \underline{-5a + 5}$

$0 = a^2 - 10a + 9$ Factor.

$0 = (a - 1)(a - 9)$ Set each factor equal to 0, and solve for a.

$a - 1 = 0 \quad\quad a - 9 = 0$

$a = 1 \quad\quad\quad a = 9$

Neither of these values is an excluded value. Check these solutions by substituting them into the original equation.

$$\frac{15}{a-4} + 5 = a - 1$$

$a = 1$

$$\frac{15}{(1)-4} + 5 \stackrel{?}{=} (1) - 1$$

$$\frac{15}{-3} + 5 \stackrel{?}{=} 0$$

$$-5 + 5 = 0$$

$a = 9$

$$\frac{15}{(9)-4} + 5 \stackrel{?}{=} (9) - 1$$

$$\frac{15}{5} + 5 \stackrel{?}{=} 8$$

$$3 + 5 = 8$$

The solutions check and are not excluded values. Therefore, $a = 1$ and $a = 9$ are solutions.

PRACTICE PROBLEM FOR EXAMPLE 2

Solve the following rational equations. Check the solutions.

a. $3 + \dfrac{16}{t} = 8 - \dfrac{4}{t}$ **b.** $\dfrac{-2}{x+3} = 3 - \dfrac{2}{x+3}$ **c.** $\dfrac{3}{h+2} = 1 + \dfrac{2}{h+2}$

Example 3 — Solving rational equations with unlike denominators

Solve the following rational equations. Check the solutions.

a. $\dfrac{7}{2} + \dfrac{3}{4n} = 5$ **b.** $\dfrac{5}{b+3} = \dfrac{4}{b-2}$

SOLUTION

a. The only excluded value is $n = 0$. The LCD is $4n$, so multiply *both sides* of the equation by $4n$ to simplify the equation.

$$\frac{7}{2} + \frac{3}{4n} = 5$$

$$4n\left(\frac{7}{2} + \frac{3}{4n}\right) = 5(4n) \quad \text{Multiply both sides of the equation by the LCD.}$$

$$4n\left(\frac{7}{2}\right) + 4n\left(\frac{3}{4n}\right) = 5(4n) \quad \text{Use the distributive property on the left side.}$$

$${}^2\cancel{4n}\left(\frac{7}{\cancel{2}}\right) + \cancel{4n}\left(\frac{3}{\cancel{4n}}\right) = 5(4n) \quad \text{Reduce.}$$

$$14n + 3 = 20n \quad \text{Solve the remaining equation.}$$

$$\underline{-14n \quad\quad -14n}$$

$$3 = 6n$$

$$\frac{3}{6} = \frac{6n}{6}$$

$$\frac{1}{2} = n$$

> **Skill Connection**
>
> **Writing the LCD with a denominator of 1**
>
> In Example 2, we rewrote the LCD with a denominator of 1. This was done to make it clear that when we are multiplying both sides by the LCD, we multiply the numerators by the LCD. We now drop this extra step. If you get confused about what to multiply the LCD by, then continue to include a denominator of 1.

636 CHAPTER 7 Rational Expressions and Equations

This solution is not an excluded value, so check it using the original equation.

$$\frac{7}{2} + \frac{3}{4\left(\frac{1}{2}\right)} \stackrel{?}{=} 5$$

$$\frac{7}{2} + \frac{3}{2} \stackrel{?}{=} 5$$

$$5 = 5$$

The solution checks, so $n = \frac{1}{2}$ is the correct solution.

b. Setting the denominators equal to 0 yields the excluded values $b = -3$ and $b = 2$. The LCD is $(b + 3)(b - 2)$, so start by multiplying both sides by the LCD.

$$\frac{5}{b + 3} = \frac{4}{b - 2} \quad \text{Multiply both sides of the equation by the LCD.}$$

$$(b + 3)(b - 2)\left(\frac{5}{b + 3}\right) = \left(\frac{4}{b - 2}\right)(b + 3)(b - 2)$$

$$\cancel{(b + 3)}(b - 2)\left(\frac{5}{\cancel{b + 3}}\right) = \left(\frac{4}{\cancel{b - 2}}\right)(b + 3)\cancel{(b - 2)} \quad \text{Reduce.}$$

$$5(b - 2) = 4(b + 3)$$

$$5b - 10 = 4b + 12 \quad \text{Solve the equation.}$$

$$\underline{-4b \qquad -4b}$$

$$b - 10 = 12$$

$$\underline{+10 \ +10}$$

$$b = 22$$

This solution is not an excluded value, so check this solution using the original equation.

$$\frac{5}{(22) + 3} \stackrel{?}{=} \frac{4}{(22) - 2}$$

$$\frac{5}{25} \stackrel{?}{=} \frac{4}{20}$$

$$\frac{1}{5} = \frac{1}{5}$$

The solution checks, so $b = 22$ is the correct solution.

PRACTICE PROBLEM FOR EXAMPLE 3
Solve the following rational equations. Check the solutions.

a. $\dfrac{7}{n + 6} = \dfrac{3}{n + 2}$ **b.** $\dfrac{5}{2} - \dfrac{12}{x} = \dfrac{1}{2}$

As we have seen, some rational equations have more than one solution. In that case, both solutions are checked in the original problem.

Example 4 Solving rational equations that have more than one possible solution

Solve the following rational equations. Check the solutions in the original problems.

a. $1 + \dfrac{20}{(a + 2)(a - 10)} = \dfrac{3}{a + 2}$ **b.** $\dfrac{6}{a + 1} + \dfrac{a}{a - 4} = \dfrac{-30}{a^2 - 3a - 4}$

SECTION 7.4 Solving Rational Equations **637**

SOLUTION $\boxed{3-\frac{5}{x}=\frac{2}{x+1}}$

a. Setting the denominators equal to 0 yields the excluded values $a = -2$ and $a = 10$. Multiply both sides of the equation by the LCD, which is $(a+2)(a-10)$.

$$1 + \frac{20}{(a+2)(a-10)} = \frac{3}{a+2} \quad \text{Multiply both sides by the LCD.}$$

$$(a+2)(a-10)\left(1 + \frac{20}{(a+2)(a-10)}\right) = (a+2)(a-10)\left(\frac{3}{a+2}\right) \quad \text{Use the distributive property.}$$

$$(a+2)(a-10)(1) + \cancel{(a+2)(a-10)} \cdot \left(\frac{20}{\cancel{(a+2)(a-10)}}\right) = 3(a-10)$$

$$a^2 - 8a - 20 + 20 = 3a - 30 \quad \text{Simplify.}$$
$$a^2 - 8a = 3a - 30 \quad \text{This is a quadratic equation.}$$
$$a^2 - 8a - 3a + 30 = 0 \quad \text{Set one side equal to 0.}$$
$$a^2 - 11a + 30 = 0 \quad \text{Factor and solve.}$$
$$(a-5)(a-6) = 0$$
$$a = 5 \quad \text{or} \quad a = 6$$

Neither solution is an excluded value, so check both in the original equation.

$a = 5$

$$1 + \frac{20}{(5+2)(5-10)} \stackrel{?}{=} \frac{3}{5+2}$$
$$1 + \frac{20}{(7)(-5)} \stackrel{?}{=} \frac{3}{7}$$
$$1 + \frac{20}{-35} \stackrel{?}{=} \frac{3}{7}$$
$$1 - \frac{4}{7} \stackrel{?}{=} \frac{3}{7}$$
$$\frac{3}{7} = \frac{3}{7}$$

$a = 6$

$$1 + \frac{20}{(6+2)(6-10)} \stackrel{?}{=} \frac{3}{6+2}$$
$$1 + \frac{20}{(8)(-4)} \stackrel{?}{=} \frac{3}{8}$$
$$1 + \frac{20}{-32} \stackrel{?}{=} \frac{3}{8}$$
$$1 - \frac{5}{8} \stackrel{?}{=} \frac{3}{8}$$
$$\frac{3}{8} = \frac{3}{8}$$

Both solutions check, so $a = 5$ and $a = 6$ are solutions.

b. Start by factoring to find the LCD and any excluded values.

$$\frac{6}{a+1} + \frac{a}{a-4} = \frac{-30}{a^2 - 3a - 4}$$

$$\frac{6}{a+1} + \frac{a}{a-4} = \frac{-30}{(a+1)(a-4)}$$

The excluded values are $a = -1$ and $a = 4$, and the LCD is $(a+1)(a-4)$.

$$\frac{6}{a+1} + \frac{a}{a-4} = \frac{-30}{(a+1)(a-4)}$$

$$(a+1)(a-4)\left(\frac{6}{a+1} + \frac{a}{a-4}\right) = \left(\frac{-30}{(a+1)(a-4)}\right)(a+1)(a-4)$$

$$(a+1)(a-4)\left(\frac{6}{a+1}\right) + (a+1)(a-4)\left(\frac{a}{a-4}\right) = \left(\frac{-30}{(a+1)(a-4)}\right)(a+1)(a-4)$$

$$\cancel{(a+1)}(a-4)\left(\frac{6}{\cancel{a+1}}\right) + (a+1)\cancel{(a-4)}\left(\frac{a}{\cancel{a-4}}\right) = \left(\frac{-30}{\cancel{(a+1)(a-4)}}\right)\cancel{(a+1)(a-4)}$$

$$6(a-4) + a(a+1) = -30$$
$$6a - 24 + a^2 + a = -30$$

638 CHAPTER 7 Rational Expressions and Equations

$$a^2 + 7a - 24 = -30$$
$$\underline{+30 \quad +30}$$
$$a^2 + 7a + 6 = 0 \quad \text{This is a quadratic equation.}$$
$$(a + 1)(a + 6) = 0 \quad \text{Solve by factoring.}$$
$$a + 1 = 0 \quad \text{or} \quad a + 6 = 0$$
$$a = -1 \quad \text{or} \quad a = -6$$

Because $a = -1$ is an excluded value, it is not a solution.
Check $a = -6$.

$$\frac{6}{(-6) + 1} + \frac{(-6)}{(-6) - 4} \stackrel{?}{=} \frac{-30}{((-6) + 1)((-6) - 4)}$$

$$\frac{6}{-5} + \frac{-6}{-10} \stackrel{?}{=} \frac{-30}{(-5)(-10)}$$

$$-\frac{6}{5} + \frac{3}{5} \stackrel{?}{=} -\frac{30}{50}$$

$$-\frac{3}{5} = -\frac{3}{5}$$

Since $a = -6$ checks, it is the only solution.

PRACTICE PROBLEM FOR EXAMPLE 4
Solve the following rational equation. Check the solutions.

a. $\dfrac{x}{x + 4} + \dfrac{2}{x - 4} = \dfrac{16}{(x + 4)(x - 4)}$ **b.** $\dfrac{h}{h - 6} = \dfrac{12}{h^2 - 10h + 24}$

■ Setting Up and Solving Shared Work Problems

Working together is a great way to get a job done quickly. Use the following Concept Investigation to consider different ways of calculating the time it takes for two people working together to complete a task.

■ CONCEPT INVESTIGATION
How long will it take to do the job?

Suppose two people are cleaning an office building. The cleaning supervisor can clean the building in 4 hours, and a newer employee can clean the building in 6 hours. Which of the following calculations make sense when you try to calculate how long it will take these two people, working together, to clean the office?

1. Should we add the two times together to determine how long it will take the two of them working together to clean the building? Why or why not?

2. Should we subtract the two times to determine how long it will take the two of them working together to clean the building? Will this work if both people can do the job in 4 hours? Why or why not?

3. Should we average the two times to determine how long it will take the two of them working together to clean the building? Why or why not?

4. Another option is to consider how much work each employee gets done in an hour (their rate of work). The supervisor gets the job done in 4 hours, so he gets

$\frac{1}{4}$ of the job done in one hour. The new employee gets the job done in 6 hours, so he gets $\frac{1}{6}$ of the job done in one hour. Fill in the following table to see how much they get done together.

Hours Worked	Work Done by Supervisor	Work Done by New Employee	Work Done Together
1	$1 \cdot \frac{1}{4} = \frac{1}{4}$ of the job done	$1 \cdot \frac{1}{6} = \frac{1}{6}$ of the job done	$\frac{1}{4} + \frac{1}{6} = \frac{3}{12} + \frac{2}{12} = \frac{5}{12}$ of the job done
2			
3			
4			

5. According to the table, about how many hours will the job take if both employees are working together?

Examining the rate at which each person carries out the job will give us the best estimate of the time it takes for both people working together to get the job done. The work completed is the work rate multiplied by the time to do the job, so we have the following:

$$\text{work completed} = \text{work rate} \cdot \text{time}$$

and

$$\text{work rate} = \frac{1}{\text{time to complete job}}$$

Adding the work rates of two people yields the rate of both of them working together. The resulting rational equation can be used to solve shared work problems.

What's That Mean?

Work Rate

If a person takes 4 hours to do one job, the work rate is $\frac{1}{4}$, which means that the person gets $\frac{1}{4}$ of the job done per hour. The *work rate* is the part of the job done per unit time.

Steps to Solving Shared Work Problems

1. Find the time for each person to do the job and the person's work rate.

Person 1	Person 2
a = time for person 1 to do the job	b = time for person 2 to do the job
$\frac{1}{a}$ = the work rate for person 1	$\frac{1}{b}$ = the work rate for person 2

2. Write an expression to represent the work rate of both people working together.

   t = time for both people working together to do the job

   $\frac{1}{t}$ = the work rate for both people working together

3. Make an equation by adding the rates of each person and setting the sum equal to the rate of both people working together.

$$\frac{1}{a} + \frac{1}{b} = \frac{1}{t}$$

4. Solve the remaining equation.
5. Check all solutions.

640 CHAPTER 7 Rational Expressions and Equations

In Section 7.3, we looked at the sum of the expressions $\frac{1}{a} + \frac{1}{b}$ in the second Concept Investigation. In Example 8 of Section 7.3, we learned that $\frac{1}{a} + \frac{1}{b}$ can be used to find the rate of work of two people working together. We now build on that idea to find the time it takes two people working together to complete a job.

Example 5 Cleaning an office together

A cleaning service has been hired to clean an office building. The cleaning supervisor can clean the building in 4 hours, and a new employee can clean the building in 6 hours. Find how long it will take the two employees to clean the office building together.

SOLUTION
First find the rates at which each employee works. Then add the rates together to find their combined rate.

Supervisor	Working Together	New Employee
4 hours = time it takes to do the job	t hours = time it takes to do the job together	6 hours = time it takes to do the job
$\frac{1}{4}$ = work rate	$\frac{1}{t}$ = work rate together	$\frac{1}{6}$ = work rate

Adding the rates, we get

$$\frac{1}{4} + \frac{1}{6} = \frac{1}{t}$$

Solve this equation to find t, the time it takes both people working together to do the job.

$$\frac{1}{4} + \frac{1}{6} = \frac{1}{t}$$

$$24t\left(\frac{1}{4} + \frac{1}{6}\right) = \frac{1}{t} \cdot 24t \qquad \text{Multiply both sides of the equation by the LCD.}$$

$$24t \cdot \frac{1}{4} + 24t \cdot \frac{1}{6} = \frac{1}{t} \cdot 24t \qquad \text{Use the distributive property.}$$

$$\overset{6}{24t} \cdot \frac{1}{\cancel{4}} + \overset{4}{24t} \cdot \frac{1}{\cancel{6}} = \frac{1}{\cancel{t}} \cdot 24\cancel{t} \qquad \text{Reduce.}$$

$$6t + 4t = 24 \qquad \text{Solve the remaining equation.}$$

$$10t = 24$$

$$\frac{10t}{10} = \frac{24}{10}$$

$$t = \frac{12}{5} = 2.4$$

Check this answer by finding how much work each person does in 2.4 hours. Using the formula

$$\text{work completed} = \text{work rate} \cdot \text{time}$$

and substituting in

$$\text{time} = 2.4$$

we get

$$2.4 \cdot \frac{1}{4} = 0.6 \quad \text{Work done by the supervisor}$$

$$2.4 \cdot \frac{1}{6} = 0.4 \quad \text{Work done by the new employee}$$

$$0.6 + 0.4 = 1.0 = 1 \quad \text{Total work done together}$$

Working together, the two employees will get the office cleaned in about 2.4 hours.

PRACTICE PROBLEM FOR EXAMPLE 5

Two painters have been hired to paint a house. The lead painter can paint the house in 12 hours. The apprentice painter can paint the house in 18 hours. Find how long it will take the two painters working together to paint the house.

Although Example 5 discussed two people working together, the same process applies if two machines or other mechanisms are working together.

Example 6 Filling a pool

A pool contractor is filling a recently completed swimming pool. The house's hose can fill the pool in 12 hours. If the contractor uses a water truck, the truck can fill the pool in 4 hours. How long will it take to fill up the pool using both the hose and a water truck?

SOLUTION

First calculate the rates for the hose and water truck.

$$\frac{1}{12} = \text{rate at which hose fills the pool}$$

$$\frac{1}{4} = \text{rate at which a water truck fills the pool}$$

$$t = \text{time in hours for both to fill the pool}$$

$$\frac{1}{t} = \text{rate at which both will fill the pool}$$

$$\frac{1}{12} + \frac{1}{4} = \frac{1}{t}$$

Solve this equation to find the time for both to fill the pool.

$$\frac{1}{12} + \frac{1}{4} = \frac{1}{t}$$

$$12t\left(\frac{1}{12} + \frac{1}{4}\right) = \frac{1}{t} \cdot 12t$$

$$12t \cdot \frac{1}{12} + 12t \cdot \frac{1}{4} = \frac{1}{t} \cdot 12t$$

$$\cancel{12}t \cdot \frac{1}{\cancel{12}} + {}^3\cancel{12}t \cdot \frac{1}{\cancel{4}} = \frac{1}{\cancel{t}} \cdot 12\cancel{t}$$

$$t + 3t = 12$$

$$4t = 12$$

$$t = 3$$

Check this answer by finding how much work the hose and water truck do in 3 hours.

$$3 \cdot \frac{1}{12} = \frac{3}{12} = \frac{1}{4} \quad \text{Work done by the hose}$$

$$3 \cdot \frac{1}{4} = \frac{3}{4} \quad \text{Work done by the water truck}$$

$$\frac{1}{4} + \frac{3}{4} = \frac{4}{4} = 1 \quad \text{Total work done together}$$

Working together, the water truck and hose can fill the pool in 3 hours.

7.4 Vocabulary Exercises

1. The work rate is 1 divided by _____ it takes to do the job working alone.

2. When checking the solution(s) to a rational equation, we remove any that are _____ values.

3. To clear the fractions in a rational equation multiply both sides by the _____.

7.4 Exercises

For Exercises 1 through 18, solve each rational equation. Check each solution in the original problem.

1. $\dfrac{10}{x} + 3 = 7 - \dfrac{2}{x}$

2. $\dfrac{27}{m} + 6 = 12 - \dfrac{3}{m}$

3. $4 + \dfrac{16}{n} = 20 - \dfrac{112}{n}$

4. $11 + \dfrac{6}{t} = 15 - \dfrac{3}{t}$

5. $\dfrac{3}{x} + 3 = 5 - \dfrac{1}{x}$

6. $5 - \dfrac{6}{x} = 2 + \dfrac{9}{x}$

7. $2 - \dfrac{7}{3x} = 1 - \dfrac{4}{3x}$

8. $\dfrac{5}{2x} - 1 = 2 - \dfrac{7}{2x}$

9. $-1 + \dfrac{8}{5x} = -\dfrac{7}{5x} + 2$

10. $1 - \dfrac{25}{4x} = -3 + \dfrac{7}{4x}$

11. $\dfrac{4}{x+3} = 2 + \dfrac{10}{x+3}$

12. $\dfrac{5}{c-6} - 4 = \dfrac{13}{c-6}$

13. $\dfrac{10}{r+2} = 9 - \dfrac{8}{r+2}$

14. $\dfrac{18}{y-3} = 8 - \dfrac{6}{y+3}$

15. $\dfrac{11}{n+8} + 4 = n + 2$

16. $\dfrac{6}{a-4} + 3 = a - 6$

17. $\dfrac{-2}{k+7} - 8 = k - 4$

18. $\dfrac{-3}{h+5} + 2 = h + 3$

For Exercises 19 through 26, use the information given in each exercise to answer the questions.

19. The average cost per person who attends an awards banquet can be estimated by the equation

$$C = \dfrac{2500}{p}$$

where C is the cost per person in dollars when p people attend the awards banquet.

a. Find the cost per person if 100 people attend the banquet.

b. How many people must attend for the cost per person to be $6.25?

20. A conference committee is purchasing tote bags for promotional items. They find that the cost per bag can be found by using the equation

$$c = \dfrac{200 + 3b}{b}$$

where c is the cost per bag in dollars when b bags are purchased.

a. Find the cost per bag when 100 bags are purchased.

b. How many bags would the committee need to purchase for the cost per bag to be $3.50?

21. Some employees of a local company want to honor a fellow employee who is a new father by throwing a baby shower with gifts. The baby gifts will cost $150. The cost per person depends on the number of people who agree to help pay. If n employees agree to pay equal shares, the cost per person in dollars can be found by using the equation

$$c = \frac{150}{n}$$

where c is the cost per person in dollars.

a. Find the cost per person if 10 people agree to help pay for the gift.

b. How many people must agree to help pay so that the cost per person will be $5.00?

22. The lump-sum prize for the state lottery is $4 million. Employees at a local hospital decide to start a pool for tickets. If p people pitch in for tickets and split the prize money evenly, their potential winnings per person can be found by using the equation

$$w = \frac{4000000}{p}$$

where w is each person's winnings in dollars.

a. Find the winnings per person if 40 people pool their ticket money.

b. If the group wants the minimum winnings per person to be $125,000, how many people should they allow to pitch in?

23. The cost per person for a college Mathletics team to travel to the state championships can be found by using the equation

$$c = \frac{500 + 25p}{p}$$

where c is the cost per person in dollars when p people go on the trip.

a. What is the cost per person if eight people go on the trip?

b. If they want the cost per person to be $75, how many people should they recruit to go on the trip?

24. The Lansing Cross Country team is having T-shirts made to raise funds for their awards banquet. The cost per shirt can be found by using the equation

$$c = \frac{125 + 8t}{t}$$

where c is the cost per T-shirt in dollars when t T-shirts are made.

a. What is the per T-shirt cost if 100 T-shirts are made?

b. If they want the per T-shirt cost to be $9.00, how many T-shirts should they make?

25. Debbie's Custom Window Coverings can make up to 10 large shutters for arched windows a week. The company's cost to produce each large shutter can be found by using the equation

$$c = \frac{1200 + 250n}{n}$$

where c is the cost per large shutter in dollars when n large shutters are made in a week.

a. What does each large shutter cost if only four large shutters are made in a week?

b. How many shutters have to be made a week to make the cost per shutter for a week $400?

26. The Lakeshore Hotel and Spa charges $600 for a room rental fee plus $40 per person for a banquet. The cost per person can be found by using the equation

$$c = \frac{600 + 40n}{n}$$

where c is the cost per person in dollars when n people attend the banquet.

a. What is the cost per person if 30 people attend the banquet?

b. A group wants to keep the cost per person at $48. How many people must attend the event?

For Exercises 27 through 38, solve each rational equation. Provide reasons for each step. Check the solution(s) in the original problem.

27. $\dfrac{7}{3} + \dfrac{8}{6r} = 2$

28. $\dfrac{7}{8} + \dfrac{3}{4k} = 1$

29. $\dfrac{3}{p+7} = \dfrac{-6}{p-5}$

30. $\dfrac{-12}{t+9} = \dfrac{5}{t-8}$

31. $\dfrac{9}{m-2} = \dfrac{9}{m-2} - 4$

32. $\dfrac{4}{n+3} = \dfrac{4}{n+3} - 12$

33. $\dfrac{3}{x+2} = \dfrac{x^2 - 4x}{(x+2)(x-4)}$

34. $\dfrac{6}{a+3} = \dfrac{a^2 + 13a}{(a+3)(a-2)}$

35. $\dfrac{6}{a-5} + \dfrac{8}{a+3} = \dfrac{a^2 + 2}{(a-5)(a+3)}$

36. $\dfrac{4}{x+2} + \dfrac{3}{x+6} = \dfrac{x^2}{(x+2)(x+6)}$

37. $\dfrac{8}{h-4} + \dfrac{h}{h-6} = \dfrac{12}{h^2-10h+24}$

38. $\dfrac{2}{p-3} + \dfrac{p}{p+5} = \dfrac{16}{p^2+2p-15}$

For Exercises 39 through 62, solve each rational equation. Check each solution in the original problem.

39. $\dfrac{2}{6g} = 3 + \dfrac{3}{10}$

40. $\dfrac{4}{15w} - 5 = \dfrac{3}{2w}$

41. $6 - \dfrac{3}{x} = 7 + \dfrac{9}{x}$

42. $5 + \dfrac{27}{x} = 2 + \dfrac{16}{x}$

43. $\dfrac{3}{n} - 3 = 2 + \dfrac{3}{n}$

44. $4 - \dfrac{8}{n} = -\dfrac{8}{n} + 6$

45. $\dfrac{10}{x+5} = \dfrac{4}{x-4}$

46. $\dfrac{4}{m-3} = \dfrac{8}{m-2}$

47. $\dfrac{8}{t-3} = \dfrac{9}{t+2}$

48. $\dfrac{3}{t+5} = \dfrac{-2}{2t+2}$

49. $\dfrac{-3}{y+5} = 6 - \dfrac{3}{y+5}$

50. $1 - \dfrac{7}{x-1} = \dfrac{-7}{x-1}$

51. $\dfrac{4}{n+8} = \dfrac{n+3}{(n+8)(n-3)}$

52. $\dfrac{6}{a-4} = \dfrac{a-10}{(a-4)(a+5)}$

53. $\dfrac{-8}{k+7} = \dfrac{k-5}{(k+7)(k+4)}$

54. $\dfrac{-3}{h+5} = \dfrac{h+9}{(h+5)(h+1)}$

55. $\dfrac{1}{m-5} + \dfrac{7}{m+12} = \dfrac{m+5}{(m-5)(m+12)}$

56. $\dfrac{4}{t-7} + \dfrac{6}{t+8} = \dfrac{t-1}{(t-7)(t+8)}$

57. $\dfrac{2}{b+1} + \dfrac{3}{b+4} = \dfrac{b^2+6b-1}{(b+1)(b+4)}$

58. $\dfrac{7}{z+5} + \dfrac{2}{z-2} = \dfrac{z^2+10}{(z+5)(z-2)}$

59. $\dfrac{4}{x+6} = \dfrac{28}{x^2+8x+12}$

60. $\dfrac{6}{d-4} = \dfrac{12}{d^2-9d+20}$

61. $\dfrac{n}{n+2} + \dfrac{5}{n-3} = \dfrac{9}{n^2-n-6}$

62. $\dfrac{6}{y-8} + \dfrac{y}{y-5} = \dfrac{-6}{y^2-13y+40}$

For Exercises 63 through 68, find the work rate for each situation.

63. Brittany can shred one file box of documents in 30 minutes. What is her work rate?

64. Justin can paint a room in 3 hours. What is his work rate?

65. A pump can drain a pool in 12 hours. What is the pump's work rate?

66. Francisco can balance his business's books in 6 hours. What is his work rate?

67. Kasim can complete his lab assignment in 2 hours. What is his work rate?

68. Tendai can complete her lab assignment in 1.5 hours. What is her work rate?

For Exercises 69 through 74, solve each work problem.

69. Sherri can scan a file box of medical records in 100 minutes. Jen can scan the same file box of medical records in 90 minutes. How long will it take them to scan the file box of medical records if they work together? Assume that Sherri and Jen have their own scanners to use.

70. Julio can stock an aisle at the grocery store in 2 hours. Janet can stock the same aisle in 3 hours. How long will it take them to stock the aisle if they work together?

© Jonathan Feinstein/Shutterstock.com

71. Martin can mow the lawn area of a park in 4 hours using a riding mower. Bill can mow the same lawn using a push mower in 16 hours. How long will it take them to mow the lawn if they work together?

72. Svetlana can paint an order of stacking dolls in 6 hours. Luba can paint the same order in 10 hours. How long will it take them to paint the order if they work together?

73. North Coast Church volunteers can copy the weekend bulletins using their new photocopier in about 8 hours. The Church's older photocopier can copy the bulletins in 10 hours. How long would it take to copy the bulletins using both machines?

74. Jim is draining a swimming pool. If he uses a pump, it will take 6 hours. Jim has a second, smaller pump that will drain the pool in 10 hours. Find how long it will take to drain the pool if Jim uses both pumps.

For Exercises 75 through 86,
a. determine whether you are given an expression or an equation.
b. solve the equation or perform the operation and simplify the expression.

75. $\dfrac{x+5}{x-6} \cdot \dfrac{(x+3)(x-6)}{(x+5)(x+1)}$

76. $\dfrac{x+2}{x-5} = \dfrac{8}{x-5}$

77. $\dfrac{5x}{x+3} + \dfrac{2x+3}{x+3} = \dfrac{5}{x+3}$

78. $\dfrac{5x}{x+3} + \dfrac{2x+3}{x+3}$

79. $\dfrac{4}{x+8} - \dfrac{5}{x+1}$

80. $\dfrac{4}{x+2} = \dfrac{2}{x-3}$

81. $\dfrac{3x+12}{x+4} \div \dfrac{3}{5x}$

82. $\dfrac{5x+10}{7x-21} \cdot \dfrac{4x-12}{x+3}$

83. $\dfrac{6}{x+1} + \dfrac{4}{x+2} = \dfrac{36}{x^2+3x+2}$

84. $\dfrac{4}{x+2} + \dfrac{5}{x-9}$

85. $\dfrac{(x+2)(x+4)}{(x+5)(x-2)} \div \dfrac{(x+4)(x-2)}{(x+5)(x+1)}$

86. $\dfrac{4}{x+2} - \dfrac{6}{x-5} = \dfrac{-44}{(x+2)(x-5)}$

For Exercises 87 through 90, graph each equation. Clearly label and scale the axes.

87. $y = -0.4x + 6$

88. $y = -0.3x - 4$

89. $-\dfrac{1}{3}x + 2y = 4$

90. $\dfrac{5}{7}x - 2y = 8$

91. Find the equation of the line that passes through the points (2, 7) and (8, 11).

92. Find the equation of the line that passes through the points $(-5, 3)$ and $(4, -4)$.

For Exercises 93 through 100, solve each system of equations. Provide reasons for each step. Check the solution in both original equations.

93. $2x + 5y = 10$
$3x - 4y = 26.5$

94. $x + 3y = 15$
$2x - 6y = -12$

95. $y = 7x - 8$
$-5x + y = 4$

96. $x = 0.2y + 7$
$-5x + y = 15$

97. The perimeter of a rectangle is 40 inches. The length is 2 more than twice the width. Find the length and width of the rectangle.

98. The perimeter of a rectangle is 38 inches. The length is 4 more than the width. Find the length and width of the rectangle.

99. Simplify: $(4x^7y^3)^2(5x^2y^2)^4$.

100. Simplify: $(5x^2 + 7x - 2) - (4x - 9)$.

7.5 Proportions, Similar Triangles, and Variation

LEARNING OBJECTIVES

- Solve problems involving ratio, rates, and proportions.
- Solve problems involving similar triangles.
- Solve variation problems.

■ Ratios, Rates, and Proportions

In everyday life, we compare many quantities. Two of the terms mathematicians use to compare quantities are *ratios* and *rates*. **Ratios** and **rates** are comparisons of two quantities, often expressed as fractions.

> **DEFINITIONS**
>
> **Ratio** A ratio is a comparison of two quantities by division. The units may be the same or different.
>
> **Rate** A rate is a type of ratio that compares two quantities, often with different units.
>
> Ratios and rates may be written in one of the following forms:
>
> $$4:5 \quad 4 \text{ to } 5 \quad \frac{4}{5}$$

The ratio of 7 inches to 1 foot can be written as $\frac{7 \text{ inch}}{1 \text{ foot}}$. Changing 1 foot to 12 inches, we write the ratio as

$$\frac{7 \text{ inch}}{1 \text{ foot}} = \frac{7 \text{ inch}}{12 \text{ inch}} = \frac{7}{12}$$

There are six boys and eight girls in the class. The boy/girl ratio is 6 to 8, or $\frac{6 \text{ boys}}{8 \text{ girls}}$.

There are about 163 tablets sold to 208 laptops sold. The tablets to laptops ratio is 163 to 208, or $\frac{163 \text{ tablets}}{208 \text{ laptops}}$.

Rates appeared in Chapter 3, when we discussed slope as a rate of change. We also saw rates in Sections 7.1 and 7.2, especially in the context of unit conversions. Rates are used in everyday language. A speed of 65 mph means 65 miles per hour, or $\frac{65 \text{ miles}}{1 \text{ hour}}$. Remember that *per* means to divide.

Ratios or rates can be reduced once they are expressed in fraction form.

> **What's That Mean?**
>
> **5 to 9**
>
> When we read a ratio or rate in words, the numerator always comes first. Therefore, the ratio of 5 to 9 may be written as 5:9, or as 5 to 9, or in fraction form as $\frac{5}{9}$.

Example 1 Writing ratios in fraction form

Write the following as ratios in simplified fraction form.

a. The ratio of 6 to 8   **b.** 2 feet to 4 yards

SOLUTION

a. The 6 comes first in the wording, so it is in the numerator.

$$\frac{6}{8} = \frac{3}{4}$$

b. To simplify this ratio, convert units so both quantities have the same units. Convert 4 yards to feet, and then write the fraction.

$$\frac{4 \text{ yd}}{1} \cdot \frac{3 \text{ ft}}{1 \text{ yd}} = 12 \text{ ft}$$

2 feet to 4 yards = 2 feet to 12 feet

$$\frac{2 \text{ ft}}{4 \text{ yd}} = \frac{2 \text{ ft}}{12 \text{ ft}} = \frac{2}{12} = \frac{1}{6}$$

PRACTICE PROBLEM FOR EXAMPLE 1

Write the following as ratios in simplified fraction form.

a. The ratio of 9 to 30   **b.** 20 seconds to 3 minutes

> **What's That Mean?**
>
> **Ratios, Rates, and Rationals**
>
> A ratio is a quotient that compares two quantities. A rate is a special type of ratio in which the quantities often have different units. The word *rational* has *ratio* embedded in it, as rational numbers are ratios of integers and rational expressions are ratios of polynomials.

Example 2 Interpreting statements as rates

Interpret the following statements as rates.

a. 45 miles in 2 hours

b. $500 for 8 ounces of saffron (Saffron, a spice that is harvested from the stamens of the saffron crocus, is the world's most expensive spice.)

SOLUTION

a. Since 45 miles comes first, it is the numerator.

$$\frac{45 \text{ mi}}{2 \text{ hr}} = \frac{45}{2} \text{ mph} = 22.5 \text{ mph}$$

b. The $500 will be in the numerator.

$$\frac{\$500}{8 \text{ oz}} = \frac{\$125}{2 \text{ oz}} = \$62.50 \text{ per ounce}$$

PRACTICE PROBLEM FOR EXAMPLE 2

Interpret the following statements as rates.

a. Weight gain of 6 pounds in 1 month

b. $22.00 for 8 ounces of Aleppo pepper (Aleppo pepper is a type of pepper grown in Syria and usually sold crushed into flakes.)

Writing prices as rates is one of the best ways to compare prices. When you are at the grocery store, comparing prices of different sizes should be done by looking at the **unit rate** for the item. The unit rate will be the price per unit, such as dollars per pound, cents per ounce, or dollars per foot. To find a unit rate, divide the numerator by the denominator to get a denominator of 1.

> **DEFINITION**
>
> **Unit Rate**   A rate that has a denominator of 1 is called a unit rate.

Example 3 Comparing unit rates

Find the unit rates (round to the nearest cent). Determine which item is the better buy.

a. The 16-ounce jar of Jif Extra Crunchy peanut butter for $2.60
The 40-ounce jar of Jif Extra Crunchy peanut butter for $5.48

b. The 75-ounce bottle of Tide laundry detergent for $14.35
The 150-ounce jug of Tide laundry detergent for $21.49

SOLUTION

a. First find the unit rates. Then compare the unit rates to determine which item is the better buy.

$$\frac{\$2.60}{16 \text{ oz}} = \$0.1625 \text{ per ounce} \approx \$0.16 \text{ per ounce} \quad \text{Divide to find unit rate.}$$

$$\frac{\$5.48}{40} = \$0.137 \text{ per ounce} \approx \$0.14 \text{ per ounce} \quad \text{Divide to find unit rate.}$$

The 40-ounce jar of Jif Extra Crunchy peanut butter is the better buy.

b. First find the unit rates. Then compare the unit rates.

$$\frac{\$14.35}{75 \text{ oz}} \approx \$0.1913 \text{ per ounce} \approx \$0.19 \text{ per ounce}$$

$$\frac{\$21.49}{150 \text{ oz}} \approx \$0.1433 \text{ per ounce} \approx \$0.14 \text{ per ounce}$$

The 150-ounce carton of Tide laundry detergent is the better buy.

PRACTICE PROBLEM FOR EXAMPLE 3

Find the unit rates (round to the nearest cent). Determine which item is the better buy.

a. The 18-ounce box of General Mills Cheerios for $3.69

b. The 21-ounce box of General Mills Cheerios for $3.99

As with other useful applications of mathematics, we want to find out whether there are helpful relationships between rates and ratios. A **proportion** is a statement that two ratios or rates are equal. Proportions can be used to solve many types of applications.

DEFINITION

Proportion A statement that two ratios or rates are equal is called a proportion.

$$\frac{a}{b} = \frac{c}{d}$$

$$\frac{3}{5} = \frac{12}{20} \qquad \frac{x}{8} = \frac{7}{34}$$

Notice that a proportion has a special form, which is a fraction equal to a fraction, or $\frac{a}{b} = \frac{c}{d}$. This results in a rational equation that can be solved for a missing value. This means that $\frac{x}{5} = \frac{1}{2}$ is a proportion, but $\frac{x}{5} = \frac{1}{2} + x$ is not, as there are two terms on the right side.

SECTION 7.5 Proportions, Similar Triangles, and Variation 649

Proportions are usually solved in one of two ways. The first method uses the technique of solving rational equations from Section 7.4. Multiply both sides by the LCD to clear the fractions and solve the resulting equation. The second method works only for proportions and involves cross-multiplication.

   Cross-multiply.

$ad = bc$ Solve this equation for the variable.

Skill Connection

Cross-multiplication

Cross-multiplication works only when solving a proportion problem. In other words, it works only when the problem has the following form:

a fraction equals a fraction

Example 4 Solving a proportion in two different ways

Solve the proportion: $\dfrac{9}{x-3} = \dfrac{3}{4}$.

SOLUTION

The first method to solve a proportion was introduced in Section 7.4. Since this is a rational equation, multiply both sides by the LCD and solve for the variable.

$\dfrac{9}{x-3} = \dfrac{3}{4}$ LCD is $4(x-3)$.

$4(x-3) \cdot \dfrac{9}{x-3} = 4(x-3) \cdot \dfrac{3}{4}$ Multiply both sides by the LCD. Reduce.

$4 \cdot 9 = (x-3) \cdot 3$ Use the distributive property on the right side.

$36 = 3x - 9$ Solve the remaining equation.

$+9 +9$

$45 = 3x$

$\dfrac{45}{3} = \dfrac{3x}{3}$

$15 = x$

The second method uses cross-multiplication.

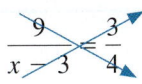   Cross-multiply.

$9 \cdot 4 = 3(x-3)$ Solve for x.

$36 = 3x - 9$

$+9 +9$

$45 = 3x$

$\dfrac{45}{3} = \dfrac{3x}{3}$

$15 = x$

Checking $x = 15$ in the original equation yields

$\dfrac{9}{x-3} = \dfrac{3}{4}$ Substitute in $x = 15$.

$\dfrac{9}{15-3} \stackrel{?}{=} \dfrac{3}{4}$ Simplify the left side.

$\dfrac{9}{12} \stackrel{?}{=} \dfrac{3}{4}$ Reduce the left side.

$\dfrac{3}{4} = \dfrac{3}{4}$ The solution checks.

Connecting the Concepts

Why does cross-multiplication work?

Cross-multiplication is a shortcut procedure to multiply both sides by a common denominator. In Example 4, we saw that cross-multiplication gave us

$9 \cdot 4 = 3(x-3)$

Multiplying both sides by the common denominator $4(x-3)$ yields

$\dfrac{9}{x-3} = \dfrac{3}{4}$

$4(x-3) \cdot \dfrac{9}{x-3} = 4(x-3) \cdot \dfrac{3}{4}$

$4 \cdot 9 = (x-3) \cdot 3$

$4 \cdot 9 = 3 \cdot (x-3)$

Notice that the last line of both techniques yields the same equation.

Example 5 — Finding the cost to fill a sand volleyball court

A local park is installing a sand volleyball court. The park district has been quoted $141 for 3 cubic yards of sand. If the court will need 200 cubic yards of sand to fill the volleyball court, how much will the sand cost?

SOLUTION
Let d be the total cost in dollars for the sand. Set up a proportion to find the cost.

$$\frac{3 \text{ yd}^3}{\$141} = \frac{200 \text{ yd}^3}{\$d}$$

$$141d \cdot \frac{3}{141} = \frac{200}{d} \cdot 141d \qquad \text{Multiply both sides by the LCD.}$$

$$\cancel{141}d \cdot \frac{3}{\cancel{141}} = \frac{200}{\cancel{d}} \cdot 141\cancel{d} \qquad \text{Divide out common factors.}$$

$$3d = 28200 \qquad \text{Solve the resulting equation.}$$

$$\frac{3d}{3} = \frac{28200}{3}$$

$$d = 9400$$

Check this answer by confirming that the ratios are equal.

$$\frac{3}{141} = \frac{1}{47} \qquad \frac{200}{9400} = \frac{1}{47}$$

The sand for the volleyball court will cost $9400.

PRACTICE PROBLEM FOR EXAMPLE 5
Two pounds of bananas cost $1.16. What will 7 pounds of bananas cost?

Example 6 — Solving proportions

Solve each proportion for the variable. Check each solution in the original problem.

a. $\dfrac{y}{6} = \dfrac{3}{36}$ **b.** $\dfrac{a+2}{a-4} = \dfrac{5}{7}$

SOLUTION

a. To solve $\dfrac{y}{6} = \dfrac{3}{36}$, find the LCD, which is 36. Multiplying both sides of the proportion by the LCD yields

$$36 \cdot \frac{y}{6} = 36 \cdot \frac{3}{36} \qquad \text{Multiply both sides by the LCD.}$$

$$6 \cdot \cancel{6} \cdot \frac{y}{\cancel{6}} = \cancel{36} \cdot \frac{3}{\cancel{36}} \qquad \text{Divide out common factors.}$$

$$6y = 3 \qquad \text{Solve the equation.}$$

$$\frac{6y}{6} = \frac{3}{6}$$

$$y = \frac{1}{2}$$

Checking the solution yields

$$\frac{y}{6} = \frac{3}{36}$$ Substitute in $y = \frac{1}{2}$.

$$\frac{\left(\frac{1}{2}\right)}{6} \stackrel{?}{=} \frac{3}{36}$$ Simplify both sides independently.

$$\frac{1}{2} \cdot \frac{1}{6} \stackrel{?}{=} \frac{3}{36}$$ On the left side, invert $\frac{6}{1}$ and multiply.

$$\frac{1}{12} = \frac{1}{12}$$ The answer checks.

b. To solve $\dfrac{a+2}{a-4} = \dfrac{5}{7}$, we find the LCD which is $7(a-4)$.

Note, on the left side, the terms in the numerator and denominator. To multiply the entire left side by the LCD, put parentheses around the numerator and denominator on the left.

$$\frac{(a+2)}{(a-4)} = \frac{5}{7}$$ Put parentheses around the expressions with multiple terms.

Multiplying both sides of the proportion by the LCD yields

$$7(a-4) \cdot \frac{(a+2)}{(a-4)} = 7(a-4) \cdot \frac{5}{7}$$ Multiply both sides by the LCD, which is $7(a-4)$.

$$7\cancel{(a-4)} \cdot \frac{(a+2)}{\cancel{(a-4)}} = \cancel{7}(a-4) \cdot \frac{5}{\cancel{7}}$$ Divide out common factors.

$$7(a+2) = (a-4)5$$ Use the distributive property.

$$7a + 14 = 5a - 20$$ Solve the remaining equation.

$$\underline{-14 -14}$$

$$7a = 5a - 34$$

$$\underline{-5a -5a}$$

$$2a = -34$$

$$\frac{2a}{2} = \frac{-34}{2}$$

$$a = -17$$

Checking the solution yields

$$\frac{a+2}{a-4} = \frac{5}{7}$$ Substitute in $a = -17$.

$$\frac{-17+2}{-17-4} \stackrel{?}{=} \frac{5}{7}$$ Simplify both sides independently.

$$\frac{-15}{-21} \stackrel{?}{=} \frac{5}{7}$$

$$\frac{5}{7} = \frac{5}{7}$$ The answer checks.

PRACTICE PROBLEM FOR EXAMPLE 6

Solve each proportion for the variable. Check the solution in the original problem.

a. $\dfrac{6}{7} = \dfrac{x+4}{x+6}$ **b.** $\dfrac{2}{3} = \dfrac{4}{z+2}$

Another way to solve unit conversion problems is by using proportions.

Example 7 — Using proportions to convert units

Use a proportion to convert units as indicated. Solve. Round to the hundredths place if necessary.

a. 120 kilograms to pounds   **b.** 1520 feet to meters

SOLUTION

a. Let x be the number of pounds in 120 kg. The conversion from kilograms to pounds is about 1 kg ≈ 2.2 lb. Set up the proportion as follows:

$$\frac{120 \text{ kg}}{x \text{ lb}} \approx \frac{1 \text{ kg}}{2.2 \text{ lb}}$$

Make sure that the units are consistent: kilograms in the numerator and pounds in the denominator on both sides of the equal sign. It is all right to have the units of pounds in the numerator and kilograms in the denominator on both sides of the equal sign. Just be consistent on both sides.

$$\frac{120 \text{ kg}}{x \text{ lb}} \approx \frac{1 \text{ kg}}{2.2 \text{ lb}} \qquad \text{Cross-multiply.}$$

$$x \approx 120(2.2) \qquad \text{Solve for } x.$$

$$x \approx 264$$

Therefore, 120 kilograms is equal to about 264 pounds.

b. Let x be the number of meters in 1520 ft. The conversion from feet to meters is 1 ft ≈ 0.3048 m. Set up the proportion as follows:

$$\frac{1520 \text{ ft}}{x \text{ m}} \approx \frac{1 \text{ ft}}{0.3048 \text{ m}}$$

$$\frac{1520 \text{ ft}}{x \text{ m}} \approx \frac{1 \text{ ft}}{0.3048 \text{ m}} \qquad \text{Cross-multiply.}$$

$$x \approx 1520(0.3048) \qquad \text{Solve for } x.$$

$$x \approx 463.30 \qquad \text{Round to the hundredths place.}$$

Therefore, 1520 feet is equal to about 463.30 meters.

PRACTICE PROBLEM FOR EXAMPLE 7

Use a proportion to convert units as indicated. Solve. Round to the hundredths place if necessary.

a. 59 inches to centimeters   **b.** 59 grams to ounces

Proportions can also be applied to the idea of scaling an object. When a photographer wants to change the size of a photograph, she will scale the image to make it either larger or smaller. An architect or designer might make a scale model of a building or car. Maps and blueprints are scale drawings that help us to follow directions.

Example 8 — Scaling a photograph

Mark has taken a photograph and trimmed it to the subject he wants. Now he wants to resize the photograph by scaling it to a larger size before printing. The original photo is

8 inches tall and 12 inches wide. If Mark wants the printed photo to be 10 inches tall, how wide will the printed photo be?

SOLUTION

Because the original size of the photo is given, set up a ratio for the height and width. Then set up a proportion for the desired height and width for the printed photo.

Let w be the width of the printed photo in inches.

$$\frac{\text{ratio of}}{\text{original photo}} = \frac{\text{ratio of}}{\text{printed photo}}$$

$$\frac{\text{height}}{\text{width}} = \frac{\text{height}}{\text{width}}$$

$$\frac{8}{12} = \frac{10}{w}$$

$$12w \cdot \frac{8}{12} = \frac{10}{w} \cdot 12w \qquad \text{Multiply both sides of the equation by the LCD, which is } 12w.$$

$$8w = 120$$

$$\frac{8w}{8} = \frac{120}{8}$$

$$w = 15$$

The printed photo will be 15 inches wide.

PRACTICE PROBLEM FOR EXAMPLE 8

A blueprint is being made at a $\frac{1}{4}$-inch scale. On the blueprint scale, 1 inch is 4 feet in the building. If the height of a real door opening is 7 feet, how big will it be when shown on the blueprint?

■ Similar Triangles

In geometry, two triangles are said to be **similar triangles** if their corresponding angles are equal. Two similar triangles are scaled-up or scaled-down versions of the same triangle. It has been shown that the ratios of **corresponding sides** of similar triangles are proportional. Scaling is the same concept that we used in Example 8 on scaling a photograph.

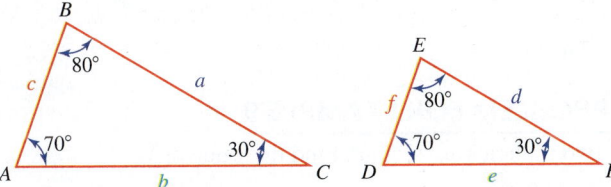

For the similar triangles above, the following are true proportions

$$\frac{a}{d} = \frac{b}{e} \qquad \frac{a}{d} = \frac{c}{f} \qquad \frac{b}{e} = \frac{c}{f}$$

In similar triangles, the ratios of corresponding sides are equal. Similar triangles can be used to help find the lengths of missing sides of triangles or other objects.

> **What's That Mean?**
>
> **Corresponding parts**
>
> The corresponding angles in similar triangles are the angles that have the same measure. The corresponding sides in similar triangles are the sides that are in between angles of the same measure. In the triangles to the left, side c corresponds to side f because they are both in between the 70° and 80° angles. The corresponding sides are proportional.

654 CHAPTER 7 Rational Expressions and Equations

> **DEFINITION**
>
> **Similar Triangles** Two triangles with equal corresponding angles are called similar triangles. The ratios of corresponding sides of similar triangles are proportional.

Example 9 — Finding the height of a tree

A tree casts a 10-foot shadow at the same time that a 6-foot tall man casts a 2-foot shadow. Find the height of the tree.

SOLUTION
Let h be the height of the tree in feet. Draw similar triangles, using the shadows of the tree and man. Set up a proportion comparing the ratio of the shadows to the ratio of the heights.

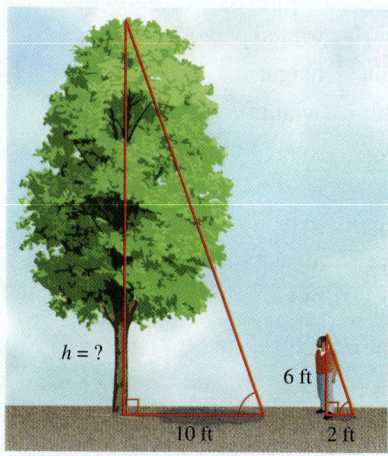

$$\frac{\text{shadow of man}}{\text{shadow of tree}} = \frac{\text{height of man}}{\text{height of tree}}$$

$$\frac{2}{10} = \frac{6}{h}$$

Solve this proportion for the height of the tree.

$$\frac{2}{10} \times \frac{6}{h}$$

$2 \cdot h = 10 \cdot 6$ *Cross-multiply.*

$2h = 60$ *Solve the remaining equation.*

$h = 30$

Check this result by confirming that the ratios are the same.

$$\frac{2}{10} = \frac{1}{5} \qquad \frac{6}{30} = \frac{1}{5}$$

The tree is 30 feet tall.

PRACTICE PROBLEM FOR EXAMPLE 9

The following two triangles are similar. Find the value of x.

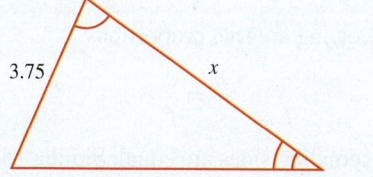
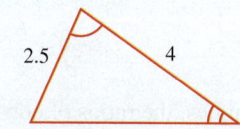

SECTION 7.5 Proportions, Similar Triangles, and Variation 655

■ Variation

In Chapter 3, we studied linear equations in two variables. We now revisit that topic, concentrating on one type of linear equation in two variables where the *y*-intercept is (0, 0). Recall that the slope-intercept form of a line is $y = mx + b$, where *m* is the slope and *b* is the *y*-intercept. Setting $b = 0$ yields $y = mx + 0$ or $y = mx$.

Some fields that use mathematics, such as science and engineering, have renamed this equation, $y = mx$ and slope. The equation is often called a **direct variation equation,** and the slope *m* is renamed the **constant of proportionality,** *k*. Direct variation appears in problems such as how the force needed to stretch a spring depends on the distance the spring is stretched. Another example of direct variation is that the amount of calories burned during exercising depends on how much time a person exercises.

> **DEFINITION**
>
> **Direct Variation** The variable *y* varies directly with *x* if $y = kx$. The value *k* is called the constant of proportionality.

■ CONCEPT INVESTIGATION
What does it mean to vary directly?

Suppose a long-haul trucker drives at an average speed of 65 mph for *t* hours. Using the distance formula, $d = r \cdot t$, the distance, *d*, he travels in miles is given by the equation $d = 65t$. Fill out the following table for the direct variation equation $d = 65t$.

t	$d = 65t$
1	
2	
3	
4	
8	

a. Are the values of *t* increasing or decreasing (reading from top to bottom)?

b. As *t* increases, does the value of *d* increase or decrease?

c. Do the values of *t* and *d* increase together, do they decrease together, or does one increase while the other decreases?

As we see in this Concept Investigation, in a direct variation situation (with positive *k*), both variables increase and decrease in tandem.

Example 10 Paycheck varies with the number of hours worked

Before taxes, Isabella's weekly paycheck (*W*) varies directly with the number of hours (*h*) she worked that week.

a. If Isabella works 32 hours in a week and earns $368.00, find an equation to represent this relationship.

b. What is Isabella's paycheck if she works 25 hours in a week?

Connecting the Concepts

How are direct variation and proportions related?

The direct variation equation is $y = kx$. If we divide both sides by x, we get the equation

$$\frac{y}{x} = \frac{kx}{x} \quad \text{Solve for } k.$$

$$\frac{y}{x} = k \quad \text{Simplify.}$$

$$\frac{y}{x} = \frac{k}{1} \quad \text{Rewrite right side.}$$

This last equation, $\frac{y}{x} = \frac{k}{1}$, shows that a direct variation problem may be thought of as a proportion. That is why we often study these two topics together.

SOLUTION

a. Let $h =$ number of hours Isabella works in a week and $W =$ weekly paycheck amount in dollars (before taxes). Since this is a direct variation relationship, we have that $W = kh$. To find k, substitute in $h = 32$ and $W = 368$ and solve for k.

$$W = kh \quad \text{Substitute in } h = 32 \text{ and } W = 368.$$

$$368 = k(32) \quad \text{Solve for } k.$$

$$\frac{368}{32} = \frac{k(32)}{32}$$

$$11.50 = k$$

Therefore, $W = 11.50h$.

b. Isabella's weekly paycheck for 25 hours of work is found by substituting $h = 25$ into $W = 11.50h$ and simplifying.

$$W = 11.50h \quad \text{Substitute in } h = 25.$$

$$W = 11.50(25)$$

$$W = 287.50$$

Isabella will earn $287.50 for 25 hours of work.

PRACTICE PROBLEM FOR EXAMPLE 10

Before taxes, Hamid's weekly paycheck (W) varies directly with the number of hours (h) he worked that week.

a. If Hamid works 28 hours in a week and earns $343.00, find an equation to represent this relationship.

b. What is the amount of Hamid's weekly paycheck if he works 35 hours?

Example 11 Direct variation

The variable y varies directly with x. If $y = 4$ when $x = 8$, find y when $x = 10$.

SOLUTION

Because y varies directly with x, the equation is

$$y = kx$$

The value of k is unknown. Use the first piece of information to find k: $y = 4$ when $x = 8$. Substitute $y = 4$ and $x = 8$ into the equation $y = kx$ and solve for k.

$$y = kx \quad \text{Substitute in } y = 4 \text{ and } x = 8.$$

$$4 = k(8) \quad \text{Solve the equation for } k.$$

$$\frac{4}{8} = \frac{k(8)}{8}$$

$$\frac{1}{2} = k$$

Therefore, $y = \frac{1}{2}x$.

We now find y when $x = 10$.

$$y = \frac{1}{2}x \quad \text{Substitute in } x = 10.$$

$$y = \frac{1}{2}(10)$$

$$y = 5$$

PRACTICE PROBLEM FOR EXAMPLE 11

The variable y varies directly with x. If $y = 12$ when $x = 4$, find y when $x = -2$.

CONCEPT INVESTIGATION

What does it mean when something varies inversely?

Parvati, a college student, needs to drive 100 miles to attend a cousin's wedding. From the $d = r \cdot t$ formula, substitute in $d = 100$ to get $100 = r \cdot t$. To solve for t, divide both sides by r to get the following equation:

$$\frac{100}{r} = \frac{r \cdot t}{r} \qquad \text{To solve for } t, \text{ divide both sides by } r.$$

$$\frac{100}{r} = t \qquad \text{or} \qquad t = \frac{100}{r}$$

Fill in the following table with the times it will take Parvati to drive 100 miles if she drives at an average speed of r miles per hour.

r	$t = \dfrac{100}{r}$
25	
45	
55	
65	
75	

a. Are the values of r increasing or decreasing (reading from top to bottom)?
b. As r increases, does the value of t increase or decrease?
c. Do the values of r and t increase together, do they decrease together, or does one increase while the other decreases?

Inverse variation occurs when the two variables have the relationship $y = \dfrac{k}{x}$. For positive values of k, if one variable increases (its value goes up), the other will decrease (move down). This up–down relationship is the *inverse* described in the term *inverse variation*. An example of quantities that vary inversely is the speed of a car being driven varies inversely with the time spent driving.

DEFINITION

Inverse Variation The variable y varies inversely with x if $y = \dfrac{k}{x}$. k is called the *constant of proportionality*.

Example 12 Speed and time, an inverse relationship

The speed (s) in miles per hour that Rafael drives his car is inversely proportional to the time (t) spent driving.

a. If Rafael drives 65 miles per hour for 3 hours, find an equation to represent this relationship.
b. What is Rafael's speed if he drives the same distance in 4 hours?

658 CHAPTER 7 Rational Expressions and Equations

SOLUTION

a. The equation that represents this situation is $s = \dfrac{k}{t}$. Substituting in $s = 65$ and $t = 3$ yields

$$s = \frac{k}{t} \quad \text{Substitute in } s = 65 \text{ and } t = 3.$$

$$65 = \frac{k}{3} \quad \text{Solve for } k.$$

$$3 \cdot 65 = 3 \cdot \frac{k}{3}$$

$$195 = k$$

Therefore, $s = \dfrac{195}{t}$.

b. If Rafael drives 4 hours, substitute in $t = 4$ and find s as follows.

$$s = \frac{195}{t} \quad \text{Substitute in } t = 4.$$

$$s = \frac{195}{4}$$

$$s = 48.75 \text{ mph}$$

If Rafael drives at 48.75 miles per hour, he will drive the same distance in 4 hours.

PRACTICE PROBLEM FOR EXAMPLE 12

The speed (s) in miles per hour that Mariam drives her car is inversely proportional to the time (t) spent driving.

a. If Mariam drives 70 miles per hour for 3 hours, find an equation to represent this relationship.

b. What is Mariam's speed if she drives the same distance in 2.5 hours?

Example 13 | Inverse variation

The variable y varies inversely with x. If $y = 2$ when $x = 5$, find y when $x = 18$.

SOLUTION
We are told that y varies inversely with x. This means that the equation is of the form

$$y = \frac{k}{x}$$

Use the first piece of information to find the value of k, namely, $y = 2$ when $x = 5$. Substitute $y = 2$ and $x = 5$ into the equation $y = \dfrac{k}{x}$ and solve for k.

$$y = \frac{k}{x} \quad \text{Substitute in } y = 2 \text{ and } x = 5.$$

$$2 = \frac{k}{5} \quad \text{Solve the equation for } k.$$

$$5 \cdot 2 = 5 \cdot \frac{k}{5}$$

$$10 = k$$

Therefore, $y = \dfrac{10}{x}$.

We now find y when $x = 18$.

$$y = \dfrac{10}{x} \quad \text{Substitute in } x = 18.$$

$$y = \dfrac{10}{18}$$

$$y = \dfrac{5}{9}$$

PRACTICE PROBLEM FOR EXAMPLE 13

The variable y varies inversely with x. If $y = 2$ when $x = 5$, find y when $x = 18$.

7.5 Vocabulary Exercises

1. A ratio is a comparison of two quantities with the _____ units.

2. A rate is a comparison of two quantities that may have _____ units.

3. A(n) _____ is a statement that two rates (or ratios) are equal.

4. Two triangles that have equal angles are called _____ triangles.

5. The distance we travel and the time it takes to travel that distance vary _____. When the distance goes up, the time goes _____.

6. The time it takes to travel a distance and the rate at which we travel vary _____. When the rate goes up, the time goes _____.

7.5 Exercises

For Exercises 1 through 10, write the following as ratios in simplified fraction form. Remember that units must be the same before simplifying.

1. The ratio of 4 to 15
2. The ratio of 5 to 17
3. The ratio of 6 to 20
4. The ratio of 9 to 42
5. 5 feet to 4 yards
6. 4 feet to 3 yards
7. 45 minutes to 2 hours
8. 25 minutes to 1 hour
9. 150 meters to 2 kilometers
10. 300 centimeters to 4 meters

For Exercises 11 through 18, interpret the statements as rates.

11. $35 for 9 ounces of chocolate
12. $4 for 3 pounds of oranges
13. $2400 for 2 ounces of gold
14. $140 for 8 ounces of silver
15. 15 miles in 2 hours
16. 26 miles in 3 hours
17. 40 feet in 30 minutes
18. 75 feet in 8 minutes

For Exercises 19 through 24, find the unit rates (round to the nearest cent). Determine which item is the better buy.

19. 20 Duracell AA batteries for $15.79
    16 Duracell AA batteries for $14.79

20. A 15.2-fluid-ounce bottle of Minute Maid orange juice for $1.47
    A 10-fluid-ounce bottle of Minute Maid orange juice for $2.50

21. A 10-ounce bag of Lay's Classic potato chips for $2.50
    A 15.25-ounce bag of Lay's Classic potato chips for $3.48

22. A 17.5-ounce pack of Mission flour tortillas for $1.88
    A 35-ounce pack of Mission flour tortillas for $3.74

23. A 16-ounce bottle of Herdez salsa casera for $2.03
    A 24-ounce bottle of Herdez salsa casera for $2.87

24. A 45-ounce jar of Ragu sauce for $3.14
    A 24-ounce jar of Ragu sauce for $1.77

For Exercises 25 through 36, use the information given in the exercises to answer the questions. Watch the units and make sure your answers have the correct units of measure.

25. Most dollhouses have a 1/12th scale. This means that the ratio of the toy to the actual item is $\frac{1}{12}$. If a real couch is 40 inches tall, how tall will the toy couch be?

26. In a 1/12th scale dollhouse, how tall will a 78-inch doorway be?

27. A blueprint is drawn to a $\frac{1}{4}$-inch scale, so 1 inch on the blueprint equals 4 feet in the building. If the length of a real wall is 12 feet, what will its length be on the blueprint?

28. How long will a 6-foot window be when drawn on the blueprint described in Exercise 27?

29. A football stadium is drawn on a $\frac{1}{8}$-inch scale blueprint (1 inch on the blueprint is 8 feet on the stadium). The blueprint shows a 2.5-inch window for a concession stand. How long will the window be in the stadium?

30. If a section of bleachers is 15 inches when drawn on the blueprint described in Exercise 29, how long will it be in the stadium?

31. Vinyl fencing costs $79 for a 6-foot section. How much would it cost to buy 306 feet of fencing?

32. A new vinyl fencing that looks like a stone wall costs $149 for a 6-foot section. How much would it cost to buy 192 feet of fencing?

33. 78 square feet of roofing shingles costs $62.40. How much do 3900 square feet of shingles cost?

34. Enough paint to cover 150 square feet costs $25. How much will paint cost to cover 320 square feet?

35. A small order of french fries from McDonald's has the nutritional facts listed on the label. A medium order of french fries at McDonald's has 370 calories.

    Source: Nutritional data and images courtesy of www. NutritionData.com

    a. Find the amount of total fat in a medium order of french fries.

    b. Find the amount of saturated fat in a medium order of french fries.

    Nutrition Facts
    Serving Size 1 small serving 71g (71 g)

    Amount Per Serving
    Calories 224   Calories from Fat 103

    % Daily Value*
    Total Fat 11g   18%
      Saturated Fat 1g   7%
      Trans Fat 0g
    Cholesterol 0mg   0%
    Sodium 161mg   7%
    Total Carbohydrate 28g   9%
      Dietary Fiber 3g   12%
      Sugars 0g
    Protein 3g

    Vitamin A   0% • Vitamin C   9%
    Calcium   1% • Iron   3%
    *Percent Daily Values are based on a 2,000 calorie diet. Your daily values may be higher or lower depending on your calorie needs.

    ©www.NutritionData.com

36. The nutritional values for a fun-size Snickers Bar are given on the label. A king-size Snickers Bar has 537 calories.

    Source: Nutritional data and images courtesy of www. NutritionData.com

    a. Find the total amount of fat in a king-size Snickers Bar.

    b. Find the amount of cholesterol in a king-size Snickers Bar.

    Nutrition Facts
    Serving Size 1 bar, fun size 15g (15 g)

    Amount Per Serving
    Calories 71   Calories from Fat 32

    % Daily Value*
    Total Fat 4g   6%
      Saturated Fat 1g   7%
      Trans Fat 0g
    Cholesterol 2mg   1%
    Sodium 37mg   2%
    Total Carbohydrate 9g   3%
      Dietary Fiber 0g   1%
      Sugars 8g
    Protein 1g

    Vitamin A   1% • Vitamin C   0%
    Calcium   1% • Iron   1%
    *Percent Daily Values are based on a 2,000 calorie diet. Your daily values may be higher or lower depending on your calorie needs.

    ©www.NutritionData.com

SECTION 7.5 Proportions, Similar Triangles, and Variation **661**

For Exercises 37 through 48, solve each proportion. Check each solution in the original problem.

37. $\dfrac{3}{5} = \dfrac{9}{x}$

38. $\dfrac{6}{x} = \dfrac{5}{20}$

39. $\dfrac{-2}{3} = \dfrac{-8}{b}$

40. $\dfrac{-b}{3} = \dfrac{-8}{6}$

41. $\dfrac{y+6}{3} = \dfrac{5}{3}$

42. $\dfrac{y-5}{-2} = \dfrac{14}{4}$

43. $\dfrac{-4}{3} = \dfrac{6x-8}{6}$

44. $\dfrac{-4}{10} = \dfrac{x-7}{5}$

45. $\dfrac{y+2}{6} = \dfrac{y-2}{3}$

46. $\dfrac{y-5}{2} = \dfrac{6y+14}{4}$

47. $\dfrac{x-3}{x+4} = \dfrac{8}{4}$

48. $\dfrac{6x+1}{4x-14} = \dfrac{1}{2}$

For Exercises 49 through 56, convert the units using a proportion. Solve the proportions. Round to the nearest hundredths place if necessary. See Example 7 for help if needed.

49. Convert 25 miles to kilometers.

50. Convert 120 kilometers to miles.

51. Convert 165 pounds to kilograms.

52. Convert 94 kilograms to pounds.

53. Convert 45 inches to centimeters.

54. Convert 34 centimeters to inches.

55. Convert 60 acres to square feet.

56. Convert 80,000 square feet to acres.

57. The 8-by-10 photo is to be blown up to a larger size. The photo will now be 10 inches high.

8 × 10   10 × ?

Source: Photos by Mark Clark, January 2008.

Find the measurement of the width of the enlarged photo.

58. The 11-by-14 photo is to be reduced to a smaller size. The photo will now be $10\dfrac{1}{2}$ inches high.

Find the measurement of the width of the reduced photo.

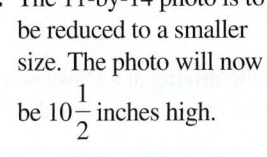

11 × 14   ? × $10\dfrac{1}{2}$

Source: Photos by Mark Clark, August 2005.

For Exercises 59 through 62, the pairs of triangles are similar. Find the value of each unknown side x.

59.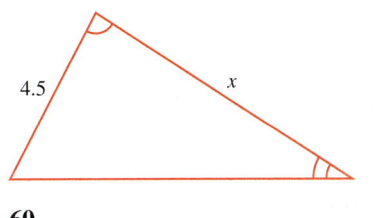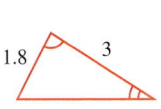

60.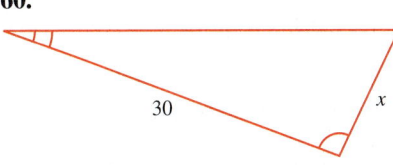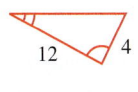

61.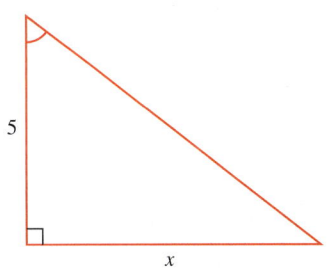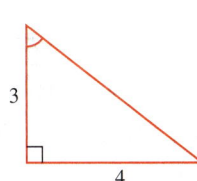

62.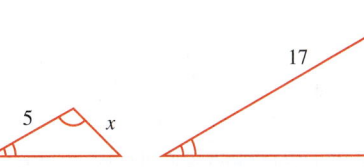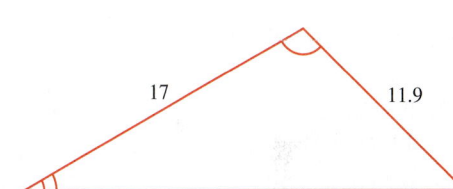

63. The Victory Obelisk was erected in Victory Park in Moscow in 1995 to commemorate the 1418 days and nights of the Great Patriotic War (World War II). If the obelisk casts a shadow 709 meters long at the same time that a 1.6-meter woman casts an 8-meter shadow, find the height of the obelisk.

64. The Washington Monument was erected to commemorate the first president of the United States, George Washington. If the monument casts a shadow 1388.75 feet long at the same time that a 5.5-foot-tall person casts a 13.75-foot shadow, find the height of the memorial.

For Exercises 65 through 68, use the following recipe to find the needed amount of each ingredient.

Fesenjan (Persian-style chicken with walnut, onion, and pomegranate sauce) (Serves 5)

- 1/4 cup unsalted butter
- 2 1/2-pound chicken, cut into serving pieces
- 2 onions, sliced thin
- 1 teaspoon ground cinnamon
- 2 cups coarsely ground toasted walnuts
- 1 pomegranate (about 8 to 10 ounces), halved and squeezed gently to yield enough seeds and juice to measure about 2/3 cup
- 1/2 cup tomato sauce
- 1 1/2 cups chicken broth
- 1 tablespoon plus 1 teaspoon fresh lemon juice
- 1/4 teaspoon salt
- 1/4 teaspoon pepper
- 1 tablespoon unsulfured molasses

65. Find the amount of unsalted butter needed to make 16 servings.

66. Find the amount of tomato sauce needed to make 16 servings.

67. Find the amount of chicken broth needed to make 12 servings.

68. Find the amount of chicken needed to make 12 servings.

For Exercises 69 through 72, use the following recipe to find the needed amount of each ingredient.

Rice Pudding (Serves 4)

- 4 cups milk
- 3/4 cup pure maple syrup
- 1/2 cup long grain white rice
- 1/2 teaspoon ground cinnamon
- 1/2 cup golden raisins
- 1 1/2 teaspoons vanilla
- 1/4 teaspoon salt

69. Find the amount of milk needed to make 16 servings.

70. Find the amount of rice needed to make 16 servings.

71. Find the amount of maple syrup needed to make 12 servings.

72. Find the amount of vanilla needed to make 12 servings.

For Exercises 73 through 88, solve each direct variation problem.

73. The variable y varies directly with x. Find an equation to represent this relationship if $x = 9$ when $y = 30$. Use the equation to find y when $x = 15$.

74. The variable y varies directly with x. Find an equation to represent this relationship if $x = 20$ when $y = 27$. Use the equation to find y when $x = 12$.

75. The variable d varies directly with t. Find an equation to represent this relationship if $t = 7$ when $d = 21$. Use the equation to find d when $t = 10$.

76. The variable d varies directly with t. Find an equation to represent this relationship if $t = 4$ when $d = 36$. Use the equation to find d when $t = 7$.

77. The variable y varies directly with x. Find an equation to represent this relationship if $x = 8$ when $y = 5$. Use the equation to find y when $x = 17$.

78. The variable y varies directly with x. Find an equation to represent this relationship if $x = 2$ when $y = 11$. Use the equation to find y when $x = 4.5$.

79. The number of calories burned (C) while mountain biking varies directly with the number of hours (h) spent biking.
    a. Find an equation to express this relationship if the number of calories burned during 2 hours of mountain biking is 1152 (for a 150-pound person).
    b. Use this equation to find how many calories are burned during 3.5 hours of mountain biking.

80. The number of calories burned (C) while playing full-court basketball varies directly with the number of hours (h) spent playing.
    a. Find an equation to express this relationship if the number of calories burned during 1 hour of playing full-court basketball is 871 (for a 175-pound person).
    b. Use this equation to find how many calories are burned during 2.5 hours of playing full-court basketball.

81. The distance traveled (D) while driving at a constant rate varies directly with the number of hours (t) spent driving.
    a. Find an equation to express this relationship if the distance traveled is 130 miles in 2 hours.
    b. Use this equation to find how many miles are traveled in 5.5 hours.

82. The distance traveled (D) while driving at a constant rate varies directly with the number of hours (t) spent driving.
    a. Find an equation to express this relationship if the distance traveled is 110 miles in 2.5 hours.
    b. Use this equation to find out how many miles are traveled in 8.2 hours.

83. At a community college, the amount of tuition (T) owed by a student varies directly with the number of credits (c) the student enrolls in.

   a. Find the equation to express this relationship if the tuition owed is $368 for 8 credits.

   b. How much tuition will a student owe who enrolls in 15 credits?

84. At a community college, the amount of tuition (T) owed by a student varies directly with the number of credits (c) the student enrolls in.

   a. Find the equation to express this relationship if the tuition owed is $360 for 9 credits.

   b. How much tuition will a student who enrolls in 14 credits owe?

85. Hooke's law says that the force to stretch a spring varies directly with the distance the spring is stretched.

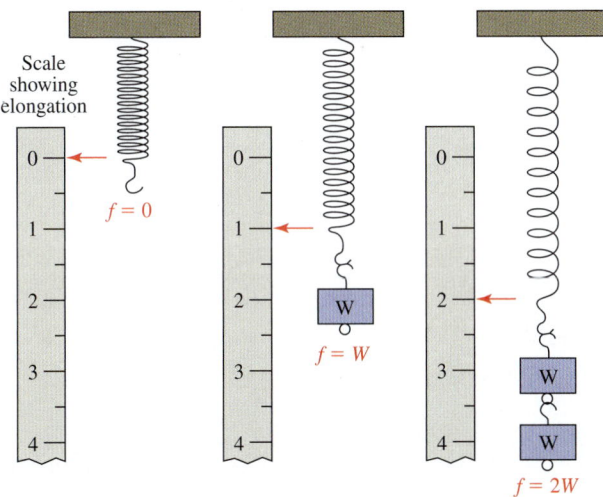

   In symbols, $F = kx$, where F is the force and x is the distance.

   a. Find the equation to express this relationship if a force of 6 pounds stretches a spring 5 inches.

   b. How many inches will a force of 10 pounds stretch the spring?

86. Hooke's law says that the force to stretch a spring varies directly with the distance the spring is stretched. In symbols, $F = kx$, where F is the force and x is the distance.

   a. Find an equation to express this relationship if a force of 9 pounds stretches a spring 7 inches.

   b. How many inches will a force of 15 pounds stretch the spring?

87. The amount of a weekly paycheck (before taxes) (W) varies directly with the amount of time (t) spent working.

   a. Find an equation to express this relationship if the amount of the paycheck is $320 for 20 hours of work.

   b. Use this equation to find the amount of the weekly paycheck for 25 hours of work.

88. The amount of a weekly paycheck (before taxes) (W) varies directly with the amount of time (t) spent working.

   a. Find an equation to express this relationship if the amount of the paycheck is $180.00 for 16 hours of work.

   b. Use this equation to find the amount of the weekly paycheck for 22 hours of work.

For Exercises 89 through 100, solve each inverse variation problem.

89. The variable y varies inversely with x. Find an equation to represent this relationship if $x = 2.5$ when $y = 6$. Use the equation to find y when $x = 5.5$.

90. The variable y varies inversely with x. Find an equation to represent this relationship if $x = 4$ when $y = 3$. Use the equation to find y when $x = 7$.

91. The variable P varies inversely with n. Find an equation to represent this relationship if $n = 5$ when $P = 200$. Use the equation to find P when $n = 50$.

92. The variable P varies inversely with n. Find an equation to represent this relationship if $n = \frac{1}{2}$ when $P = 8$. Use the equation to find P when $n = 17$.

93. The variable C varies inversely with h. Find an equation to represent this relationship if $h = 6$ when $C = 30$. Use the equation to find C when $h = 3$.

94. The variable P varies inversely with c. Find an equation to represent this relationship if $c = 200$ when $P = 1200$. Use the equation to find P when $c = 125$.

95. In a balloon, the pressure of the air inside is inversely proportional to the volume of the balloon. The equation that describes this relationship is $P = \dfrac{k}{V}$.

   a. If $P = 15$ pounds per square inch (psi) when $V = 2$ in^3, find k.

   b. Find the pressure P when $V = 5$ in^3.

96. In a ball, the pressure of the air inside is inversely proportional to the volume of the balloon. The equation that describes this relationship is $P = \dfrac{k}{V}$.

   a. If $P = 0.5$ pound per square inch (psi) when $V = 15$ in^3, find k.

   b. Find the pressure P when $V = 25$ in^3.

97. For a given distance, the speed, *s*, in miles per hour of a car being driven is inversely proportional to the time, *t*, it has been driven.

    a. Find an equation to express this relationship if the speed is 55 miles per hour for 5 hours of driving.

    b. Use this equation to find the speed for 3.5 hours of driving.

98. For a given distance, the speed, *s*, in miles per hour of a car being driven is inversely proportional to the time, *t*, it has been driven.

    a. Find an equation to express this relationship if the speed is 45 miles per hour for 2 hours of driving.

    b. Use this equation to find the speed for 1.5 hours of driving.

99. The illumination, *I*, of a light varies inversely with the square of the distance *d* from the light. The equation that describes this relationship is $I = \dfrac{k}{d^2}$.

    a. Find the equation to express this relationship for a light that has an illumination of 36 foot-candles at a distance of 5 feet from the light.

    b. Use this equation to find the illumination at a distance of 10 feet from the light.

100. The illumination, *I*, of a light varies inversely with the square of the distance *d* from the light. The equation that describes this relationship is $I = \dfrac{k}{d^2}$.

    a. Find the equation to express this relationship for a light that has an illumination of 15.4 foot-candles at a distance of 10 feet from the light.

    b. Use this equation to find the illumination at a distance of 3 feet from the light.

Chapter Summary

Section 7.1 — The Basics of Rational Expressions and Equations

- An expression of the form $\dfrac{P}{Q}$, where P and Q are polynomials and $Q \neq 0$ is called a **rational expression**.
- An equation that contains one or more rational expressions is called a **rational equation**.
- **Excluded values** are any values that make a rational expression undefined.
- Use the following steps to **find excluded values for a rational expression**.
  1. Set the denominator equal to zero, and solve for the variable.
  2. Check the solution(s) in the original expression to verify the expression is undefined.
- Use the following steps to **simplifying** rational expressions:
  1. Factor the numerator and denominator (if needed).
  2. Divide out any common factors in the numerator and denominator.

Example 1 Simplify $\dfrac{x^2 + 3x - 28}{x^2 - x - 12}$.

SOLUTION Start by factoring both the numerator and denominator then divide out any common factors.

$$\dfrac{x^2 + 3x - 28}{x^2 - x - 12} = \dfrac{(x+7)(x-4)}{(x+3)(x-4)} \quad \text{Factor.}$$

$$= \dfrac{(x+7)\cancel{(x-4)}}{(x+3)\cancel{(x-4)}} \quad \text{Divide out common factors.}$$

$$= \dfrac{x+7}{x+3}$$

Section 7.2 — Multiplication and Division of Rational Expressions

- **Steps to Multiplying Rational Expressions**
  1. Factor the numerator and denominator of each fraction (if needed).
  2. Divide out any common factors.
  3. Rewrite as a single fraction.
- **Steps to Divide Rational Expressions**
  1. Multiply the first fraction by the reciprocal of the second fraction.
  2. Factor the numerator and denominator of each fraction (if needed).
  3. Divide out any common factors.
  4. Rewrite as a single fraction.

Example 2 Multiply $\dfrac{x^2 + 8x - 9}{x^2 - 4x + 3} \cdot \dfrac{x - 3}{x + 7}$.

SOLUTION

$\dfrac{x^2 + 8x - 9}{x^2 - 4x + 3} \cdot \dfrac{x - 3}{x + 7} = \dfrac{(x + 9)(x - 1)}{(x - 3)(x - 1)} \cdot \dfrac{(x - 3)}{(x + 7)}$ Factor.

$= \dfrac{(x + 9)\cancel{(x - 1)}}{\cancel{(x - 3)}\cancel{(x - 1)}} \cdot \dfrac{\cancel{(x - 3)}}{(x + 7)}$ Reduce.

$= \dfrac{x + 9}{x + 7}$

Example 3 Divide $\dfrac{(x + 5)(x - 4)}{(x + 2)(x + 7)} \div \dfrac{x^2 - 16}{(x + 2)(x + 3)}$.

SOLUTION $\dfrac{(x + 5)(x - 4)}{(x + 2)(x + 7)} \div \dfrac{x^2 - 16}{(x + 2)(x + 3)} = \dfrac{(x + 5)(x - 4)}{(x + 2)(x + 7)} \cdot \dfrac{(x + 2)(x + 3)}{x^2 - 16}$ Multiply by the reciprocal.

$= \dfrac{(x + 5)(x - 4)}{(x + 2)(x + 7)} \cdot \dfrac{(x + 2)(x + 3)}{(x + 4)(x - 4)}$ Factor.

$= \dfrac{(x + 5)\cancel{(x - 4)}}{\cancel{(x + 2)}(x + 7)} \cdot \dfrac{\cancel{(x + 2)}(x + 3)}{(x + 4)\cancel{(x - 4)}}$ Reduce.

$= \dfrac{(x + 5)(x + 3)}{(x + 7)(x + 4)}$

Section 7.3 Addition and Subtraction of Rational Expressions

- **Steps to Adding or Subtracting Rational Expressions with a Common Denominator**
  1. Check that all fractions have common denominators.
  2. Add or subtract the numerators. Write the sum or difference with the common denominator.
  3. Factor the numerator and reduce (if possible).

  When subtracting rational expressions, be sure to use the distributive property to multiply the entire numerator of the second fraction by -1 when needed.

- **Steps to Finding the Least Common Denominator for Rational Expressions**
  1. Factor the denominators of each fraction (if needed). If the denominators involve numerical factors, find the prime factorization.
  2. The LCD includes each distinct factor. Include the highest power of each factor in the LCD.
  3. The LCD is the product of the factors from step 2. Leave the LCD in factored form.

- **Steps to Writing Rational Expressions with the LCD**
  1. Find the least common denominator.
  2. Compare the denominator with the LCD to determine which factors are missing.
  3. Multiply the numerator and denominator of the fraction by the missing factors.

- **Steps to Adding or Subtracting Rational Expressions with Unlike Denominators**
  1. Find the LCD.
  2. Rewrite all fractions with the LCD.
  3. Add or subtract the numerators. Write the sum or difference with the common denominator.
  4. Factor the numerator and simplify (if possible).

- **Steps to Simplify Complex Fractions**
    1. Find the LCD for the entire expression. Include in the LCD all factors in the denominators of both the numerator and the denominator.
    2. Multiply the numerator and denominator by the LCD.
    3. Factor and reduce (if possible).

Example 4 Add or subtract the following rational expressions.

a. $\dfrac{3x + 5}{(x + 2)(x + 4)} - \dfrac{x - 3}{(x + 2)(x + 4)}$ b. $\dfrac{a}{a^2 - 9} + \dfrac{a + 2}{a^2 + 10a + 21}$

SOLUTION a. These rational expressions have the same denominator, so subtract the numerators.

$$\dfrac{3x + 5}{(x + 2)(x + 4)} - \dfrac{x - 3}{(x + 2)(x + 4)} = \dfrac{3x + 5 - (x - 3)}{(x + 2)(x + 4)} \quad \text{Subtract the numerators.}$$

$$= \dfrac{3x + 5 - x + 3}{(x + 2)(x + 4)} \quad \text{Distribute the } -1 \text{ through the parentheses.}$$

$$= \dfrac{2x + 8}{(x + 2)(x + 4)}$$

$$= \dfrac{2(x + 4)}{(x + 2)(x + 4)} \quad \text{Factor the numerator.}$$

$$= \dfrac{2\cancel{(x + 4)}}{(x + 2)\cancel{(x + 4)}} \quad \text{Reduce.}$$

$$= \dfrac{2}{x + 2}$$

b. First factor the denominators to determine the LCD.

$$\dfrac{a}{a^2 - 9} + \dfrac{a + 2}{a^2 + 10a + 21} = \dfrac{a}{(a + 3)(a - 3)} + \dfrac{a + 2}{(a + 3)(a + 7)} \quad \text{Factor the denominators.}$$

The LCD is $(a + 3)(a - 3)(a + 7)$. Rewrite each fraction using the LCD, and add the numerators.

$$\dfrac{a}{a^2 - 9} + \dfrac{a + 2}{a^2 + 10a + 21}$$

$$= \dfrac{a}{(a + 3)(a - 3)} + \dfrac{a + 2}{(a + 3)(a + 7)}$$

$$= \dfrac{(a + 7)}{(a + 7)} \cdot \dfrac{a}{(a + 3)(a - 3)} + \dfrac{a + 2}{(a + 3)(a + 7)} \cdot \dfrac{(a - 3)}{(a - 3)} \quad \text{Rewrite each fraction over the LCD.}$$

$$= \dfrac{a(a + 7)}{(a + 3)(a - 3)(a + 7)} + \dfrac{(a + 2)(a - 3)}{(a + 3)(a - 3)(a + 7)}$$

$$= \dfrac{a^2 + 7a}{(a + 3)(a - 3)(a + 7)} + \dfrac{a^2 - a - 6}{(a + 3)(a - 3)(a + 7)} \quad \text{Add the numerators.}$$

$$= \dfrac{a^2 + 7a + a^2 - a - 6}{(a + 3)(a - 3)(a + 7)} \quad \text{Simplify.}$$

$$= \dfrac{2a^2 + 6a - 6}{(a + 3)(a - 3)(a + 7)}$$

$$= \dfrac{2(a^2 + 3a - 3)}{(a + 3)(a - 3)(a + 7)}$$

668 CHAPTER 7 Rational Expressions and Equations

Section 7.4 Solving Rational Equations

- **Steps to Solving Rational Equations**
  1. Find any excluded values by setting the denominators equal to zero.
  2. Find the LCD of all the rational expressions.
  3. Multiply both sides of the equation by the LCD and simplify.
  4. Solve the remaining equation.
  5. Check all solutions. Remove any that are excluded values.

- **Steps to Solving Shared Work Problems**
  1. Find the time for each person to do the job and their work rate.

Person 1	Person 2
a = time for person 1 to do the job	b = time for person 2 to do the job
$\dfrac{1}{a}$ = the work rate for person 1	$\dfrac{1}{b}$ = the work rate for person 2

  2. Write an expression to represent the work rate of both people doing the job together.

  t = time for both people working together to do the job

  $\dfrac{1}{t}$ = work rate for both people working together

  3. Write an equation by adding the rates of each person and setting the sum equal to the rate of both people working together.

  $$\frac{1}{a} + \frac{1}{b} = \frac{1}{t}$$

  4. Solve the remaining equation.
  5. Check all solutions

Example 5 Solve the following rational equations.

a. $\dfrac{4}{n+3} = 7 + \dfrac{11}{n+3}$

b. $\dfrac{5}{h+2} + \dfrac{h+3}{h-4} = \dfrac{8h+10}{h^2-2h-8}$

SOLUTION a. The LCD is $n+3$, and the excluded value is $n = -3$. Multiply both sides of the equation by the LCD, and solve.

$$\frac{4}{n+3} = 7 + \frac{11}{n+3}$$

$$(n+3) \cdot \frac{4}{n+3} = \left(7 + \frac{11}{n+3}\right) \cdot (n+3) \qquad \text{Multiply both sides of the equation by the LCD.}$$

$$(n+3) \cdot \frac{4}{n+3} = 7(n+3) + \frac{11}{n+3} \cdot (n+3) \qquad \text{Distribute. Then divide out common factors.}$$

$$\cancel{(n+3)} \cdot \frac{4}{\cancel{n+3}} = 7(n+3) + \frac{11}{\cancel{n+3}} \cdot \cancel{(n+3)}$$

$$4 = 7n + 21 + 11$$

$$4 = 7n + 32$$
$$\underline{-32 \qquad\quad -32}$$
$$-28 = 7n$$
$$\frac{-28}{7} = \frac{7n}{7}$$
$$-4 = n$$

Because $n = -4$ is not an excluded value, check $n = -4$ in the original equation.

$$\frac{4}{(-4) + 3} \stackrel{?}{=} 7 + \frac{11}{(-4) + 3}$$
$$\frac{4}{-1} \stackrel{?}{=} 7 + \frac{11}{-1}$$
$$-4 \stackrel{?}{=} 7 - 11$$
$$-4 = -4$$

The answer checks, so $n = -4$ is a solution.

b. Factor the denominators to find the LCD.

$$\frac{5}{h + 2} + \frac{h + 3}{h - 4} = \frac{8h + 10}{h^2 - 2h - 8}$$

$$\frac{5}{h + 2} + \frac{h + 3}{h - 4} = \frac{8h + 10}{(h + 2)(h - 4)}$$

The LCD is $(h + 2)(h - 4)$, and the excluded values are $h = -2$ and $h = 4$. Multiply both sides of the equation by the LCD and solve.

$$\frac{5}{h + 2} + \frac{h + 3}{h - 4} = \frac{8h + 10}{(h + 2)(h - 4)}$$

$$(h + 2)(h - 4) \cdot \left(\frac{5}{h + 2} + \frac{h + 3}{h - 4}\right) = \left(\frac{8h + 10}{(h + 2)(h - 4)}\right) \cdot (h + 2)(h - 4)$$

$$(h + 2)(h - 4) \cdot \frac{5}{h + 2} + (h + 2)(h - 4) \cdot \frac{h + 3}{h - 4} = \left(\frac{8h + 10}{(h + 2)(h - 4)}\right) \cdot (h + 2)(h - 4)$$

$$\cancel{(h + 2)}(h - 4) \cdot \frac{5}{\cancel{(h + 2)}} + (h + 2)\cancel{(h - 4)} \cdot \frac{h + 3}{\cancel{(h - 4)}} = \left(\frac{8h + 10}{\cancel{(h + 2)(h - 4)}}\right) \cdot \cancel{(h + 2)(h - 4)}$$

$$5(h - 4) + (h + 2)(h + 3) = 8h + 10$$
$$5h - 20 + h^2 + 5h + 6 = 8h + 10$$
$$h^2 + 10h - 14 = 8h + 10$$
$$\underline{\qquad\quad -8h \qquad\quad -8h}$$
$$h^2 + 2h - 14 = 10$$
$$\underline{\qquad\qquad -10 \; -10}$$
$$h^2 + 2h - 24 = 0$$
$$(h + 6)(h - 4) = 0$$
$$h + 6 = 0 \quad \text{or} \quad h - 4 = 0$$
$$h = -6 \quad \text{or} \quad h = 4$$

Because $h = 4$ is an excluded value, it is not a solution. Check $h = -6$.

$$\frac{5}{(-6) + 2} + \frac{(-6) + 3}{(-6) - 4} \stackrel{?}{=} \frac{8(-6) + 10}{(-6)^2 - 2(-6) - 8}$$

$$\frac{5}{-4} + \frac{-3}{-10} \stackrel{?}{=} \frac{-48 + 10}{36 + 12 - 8}$$

$$\frac{-5}{4} + \frac{3}{10} \stackrel{?}{=} \frac{-38}{40}$$

$$\frac{-25}{20} + \frac{6}{20} \stackrel{?}{=} \frac{-19}{20}$$

$$\frac{-19}{20} = \frac{-19}{20}$$

Because $h = -6$ checks, it is an answer to the equation.

Example 6

Mary can water the plants at a local nursery in 8 hours. Alexa can water the plants in 5 hours. How long will it take Mary and Alexa working together to water the plants?

SOLUTION

8 hours = time it takes Mary to water the plants

$\frac{1}{8}$ = work rate for Mary

5 hours = time it takes Alexa to water the plants

$\frac{1}{5}$ = work rate for Alexa

t = time it takes for them working together to water the plants

$\frac{1}{t}$ = work rate when they are working together

Adding the rates together, we can write an equation to solve for t.

$\frac{1}{8} + \frac{1}{5} = \frac{1}{t}$	Add the individual work rates to get the combined work rate.
$40t\left(\frac{1}{8} + \frac{1}{5}\right) = 40t\left(\frac{1}{t}\right)$	Multiply by the LCD, which is $8 \cdot 5 \cdot t = 40t$.
$40t \cdot \frac{1}{8} + 40t \cdot \frac{1}{5} = 40t \cdot \frac{1}{t}$	Distribute $40t$ before you reduce.
$\overset{5}{\cancel{40t}} \cdot \frac{1}{\cancel{8}} + \overset{8}{\cancel{40t}} \cdot \frac{1}{\cancel{5}} = 40\cancel{t} \cdot \frac{1}{\cancel{t}}$	Reduce.
$5t + 8t = 40$	Solve the remaining equation.
$13t = 40$	
$t = \frac{40}{13} \approx 3.077$	

It will take a little over 3 hours for Mary and Alexa to water the plants when working together.

CHAPTER 7 Summary 671

Section 7.5 Proportions, Similar Triangles, and Variation

- A **ratio** is a comparison of two quantities that have the same units.
- A **rate** is a comparison of two quantities that often have different units.
- A **unit** rate is a rate with a denominator equal to 1.
- Unit rates can be used to compare prices of different quantities of the same product.
- A statement that two ratios or rates are equal is called a **proportion**.
- Two triangles with equal corresponding angles are called **similar triangles**. The ratios of corresponding sides of similar triangles are proportional.
- **Direct variation** between two variables x and y is described by the equation $y = kx$.

  Inverse variation between two variables x and y is described by the equation $y = \dfrac{k}{x}$.
  k is called the **constant of proportionality**.

Example 7

Find the unit rates and compare to find the product that is cheaper.
A 2-gallon can of Kilz 2 latex paint for $32.54
A 5-gallon can of Kilz 2 latex paint for $75.00

SOLUTION Find the unit rates by dividing the price by the number of gallons.

$$\frac{\$32.54}{2 \text{ gallons}} = \$16.27 \text{ per gallon}$$

$$\frac{\$75.00}{5 \text{ gallons}} = \$15.00 \text{ per gallon}$$

The 5-gallon container is cheaper.

Example 8

The following triangles are similar. Find the length of the missing side.

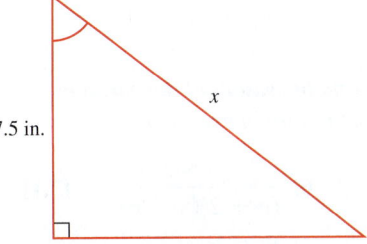
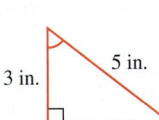

Because the triangles are similar, set up a proportion using the corresponding sides.

$$\frac{3}{7.5} = \frac{5}{x}$$

$$7.5x \cdot \frac{3}{7.5} = \frac{5}{x} \cdot 7.5x \qquad \text{Multiply both sides of the equation by the LCD.}$$

$$7.5x \cdot \frac{3}{7.5} = \frac{5}{x} \cdot 7.5x \qquad \text{Reduce.}$$

$$3x = 37.5$$

$$\frac{3x}{3} = \frac{37.5}{3}$$

$$x = 12.5$$

The missing side has a length of 12.5 inches.

Chapter Review Exercises

1. Evaluate: $\dfrac{m + 7}{m^2 + 3m - 8}$ for $m = 2$. [7.1]

2. Evaluate: $\dfrac{d - 5}{d^2 - 6d + 5}$ for $d = 1$. [7.1]

3. The cost per person for a band to travel to a competition can be found by using the equation
$$c = \dfrac{1100 + 50m}{m}$$
where c is the cost per person in dollars when m members of the band go to the competition. What is the cost per person when 50 members go to the competition? [7.1]

4. The average amount of money spent by the state of Washington per resident can be estimated by using the equation
$$s = \dfrac{3053.7t + 26545.6}{0.107t + 6.636}$$
where s is the spending per person by Washington in dollars t years since 2010. Use the equation to find spending per person by Washington in 2020. [7.1]
Source: Equation derived from data found at www.fiscal.wa.gov

For Exercises 5 through 8, find the values of the variable that must be excluded for each rational expression.

5. $\dfrac{84}{x + 7}$

6. $\dfrac{65}{(n + 2)(n - 6)}$ [7.1]

7. $\dfrac{b + 2}{b^2 + 4b + 3}$

8. $\dfrac{t^2 + 2t - 3}{t^2 - 25}$ [7.1]

For Exercises 9 through 14, simplify each rational expression.

9. $\dfrac{-8x^3 y}{64x^4 y}$

10. $\dfrac{1 - x}{x^2 - 1}$ [7.1]

11. $\dfrac{(y + 2)(y - 8)}{(y + 3)(y + 2)}$

12. $\dfrac{g - 8}{g^2 - 64}$ [7.1]

13. $\dfrac{5 - h}{(h - 5)(h + 3)}$

14. $\dfrac{4n - 24}{n^2 - 8n + 12}$ [7.1]

15. A 130-pound rower will burn approximately 708 calories per hour when competing in a crewing competition. How many calories will the rower burn during a 6-minute event? (*Hint:* First change the units to calories per minute.) [7.2]
Source: Information found at www.nutristrategy.com

16. A runner is going at a rate of 10 miles per hour. What is the runner's speed in feet per minute? [7.2]

For Exercises 17 through 26, multiply or divide the rational expressions.

17. $\dfrac{3h^2}{25hk^4} \cdot \dfrac{-5k^3}{6h}$

18. $\dfrac{-7xy^3}{6y^2} \cdot \dfrac{4xy}{14y^2}$ [7.2]

19. $\dfrac{-10n^2}{9m} \div \dfrac{35n}{3m^2}$

20. $\dfrac{2p^5}{-7q} \div \dfrac{16p^7}{63q^2}$ [7.2]

21. $\dfrac{(x - 4)}{(x + 3)} \cdot \dfrac{(x + 2)}{(x - 4)}$

22. $\dfrac{(d - 7)}{(d + 6)} \cdot \dfrac{(d + 9)}{d^2 + 2d - 63}$ [7.2]

23. $\dfrac{(k + 5)}{(k + 2)} \div \dfrac{(k + 3)(k + 5)}{(k - 2)(k + 2)}$ [7.2]

24. $\dfrac{(p + 8)}{(p + 3)} \div \dfrac{p^2 + 10p + 16}{p^2 - 9}$ [7.2]

25. $\dfrac{z^2 + 4z - 5}{z^2 - 3z + 2} \div \dfrac{z^2 - 8z + 15}{z^2 - 4}$ [7.2]

26. $\dfrac{g^2 - 7g + 10}{g^2 + 3g - 10} \div \dfrac{g^2 - 25}{g^2 + 10g + 25}$ [7.2]

For Exercises 27 through 34, add or subtract the rational expressions.

27. $\dfrac{5x}{x + 2} + \dfrac{2x + 7}{x + 2}$ [7.3]

28. $\dfrac{6n}{5n + 1} + \dfrac{3 + n}{5n + 1}$ [7.3]

29. $\dfrac{2x + 5}{(x - 1)(x + 4)} + \dfrac{x - 8}{(x - 1)(x + 4)}$ [7.3]

30. $\dfrac{4h - 9}{(h - 5)(h - 4)} + \dfrac{h - 11}{(h - 5)(h - 4)}$ [7.3]

31. $\dfrac{8x - 9}{x + 6} - \dfrac{2x - 4}{x + 6}$ [7.3]

32. $\dfrac{5y + 1}{y - 3} - \dfrac{3y - 4}{y - 3}$ [7.3]

CHAPTER 7 Review Exercises **673**

33. $\dfrac{8n+10}{(n+3)(n-4)} - \dfrac{2n-8}{(n+3)(n-4)}$ [7.3]

34. $\dfrac{3x-2}{(x+3)(x+5)} - \dfrac{x-7}{(x+3)(x+5)}$ [7.3]

For Exercises 35 through 42, find the LCD for each of the following sets of rational expressions. Rewrite each fraction in terms of the LCD.

35. $\dfrac{4}{3a^2b}$, $\dfrac{11a}{18ab^3c}$ [7.3]

36. $\dfrac{-9}{8xy^3}$, $\dfrac{7}{4x^3y^2}$ [7.3]

37. $\dfrac{7}{2v+8}$, $\dfrac{11}{v+4}$ [7.3]

38. $\dfrac{-3}{3y-9}$, $\dfrac{11}{y-3}$ [7.3]

39. $\dfrac{5}{(p+2)(p-3)}$, $\dfrac{3p}{(p+2)(p-8)}$ [7.3]

40. $\dfrac{-1}{(3x-1)(x+4)}$, $\dfrac{3x}{(3x-1)(x-2)}$ [7.3]

41. $\dfrac{z-5}{z^2+9z+14}$, $\dfrac{4}{z^2-49}$ [7.3]

42. $\dfrac{x-4}{x^2+14x+45}$, $\dfrac{-3}{x^2-81}$ [7.3]

For Exercises 43 through 50, add or subtract the rational expressions.

43. $\dfrac{7c}{10a^3b} - \dfrac{4}{25ab^2}$ [7.3]

44. $\dfrac{5x}{15xy} + \dfrac{10}{5xy^2}$ [7.3]

45. $\dfrac{5}{y+2} + \dfrac{8}{y-7}$ [7.3]

46. $\dfrac{4}{t-3} - \dfrac{5}{t+6}$ [7.3]

47. $\dfrac{k}{k+1} - \dfrac{1}{k^2+3k+2}$ [7.3]

48. $\dfrac{w+3}{w+7} + \dfrac{7w+5}{w^2+3w-28}$ [7.3]

49. $\dfrac{x+1}{x^2-3x-10} + \dfrac{x-3}{x-5}$ [7.3]

50. $\dfrac{y-1}{y-3} - \dfrac{10y-12}{y^2+3y-18}$ [7.3]

51. Use the equation from Chapter Review Exercise 3 to find how many members of the band must attend for the cost per person to be $100. [7.4]

52. Use the equation in Chapter Review Exercise 4 to find what year the state of Washington spent $6800 per resident. [7.4]

For Exercises 53 through 62, solve the following rational equations.

53. $5 + \dfrac{14}{h} = \dfrac{-6}{h}$ [7.4]

54. $\dfrac{4}{f+2} = 5 - \dfrac{1}{f+2}$ [7.4]

55. $\dfrac{2}{15} + 4 = \dfrac{5}{3r}$ [7.4]

56. $\dfrac{2}{3x} - 6 = \dfrac{1}{2}$ [7.4]

57. $\dfrac{7}{w-3} = \dfrac{2}{w+1}$ [7.4]

58. $\dfrac{-5}{b-4} = \dfrac{5}{b+3}$ [7.4]

59. $\dfrac{2}{n+5} = \dfrac{10}{(n-3)(n+5)}$ [7.4]

60. $\dfrac{4}{x+1} = \dfrac{8}{(x+1)(x-2)}$ [7.4]

61. $\dfrac{4}{a-2} + \dfrac{5}{a+1} = \dfrac{a^2+4a}{(a-2)(a+1)}$ [7.4]

62. $\dfrac{d}{d-3} + \dfrac{2}{d+4} = \dfrac{21}{d^2+d-12}$ [7.4]

63. Simplify the complex fraction: $\dfrac{6+\dfrac{2}{x}}{8-\dfrac{3}{x}}$. [7.3]

64. Simplify the complex fraction: $\dfrac{\dfrac{5}{x}+4}{2-\dfrac{3}{x}}$. [7.3]

65. Simplify the complex fraction: $\dfrac{9-\dfrac{3}{n}}{9-\dfrac{1}{n^2}}$. [7.3]

66. Simplify the complex fraction: $\dfrac{1-\dfrac{16}{n^2}}{1+\dfrac{4}{n}}$. [7.3]

67. Amin can typeset a chapter of a book in 12 hours. Danya can typeset a chapter in 10 hours. Find how long it will take them to typeset a chapter if they work together. [7.4]

68. Mike can paint the outside of a house in 18 hours. His assistant can paint the house in 27 hours. Find how long it will take them to paint the house if they work together. [7.4]

69. Write 7 inches to 1 foot as a ratio in fraction form. [7.5]

70. Write the ratio 3 to 8 in fraction form. [7.5]

71. Write 12 lb for $9 as a rate. [7.5]

674 CHAPTER 7 Rational Expressions and Equations

72. Write 42 miles in 2 hours as a rate. [7.5]

73. Find the unit rates (rounded to 4 decimal places) and determine which product is the better buy.

946 milliliters of Concrobium Mold Control for $8.98

3784 milliliters of Concrobium Mold Control for $36.61 [7.5]

74. Find the unit rates (rounded to the nearest cent) and determine which product is the better buy.

18 extra large grade AA eggs for $3.48

24 extra large grade AA eggs for $6.98 [7.5]

For Exercises 75 and 76, use the following information to answer the questions.

A 12-fluid-ounce Pepsi has the nutritional facts listed on the label. A 32-fluid-ounce serving has 400 calories.
Source: Nutritional data and images courtesy of www.NutritionData.com

75. Find the amount of sodium in a 32-fluid-ounce serving. [7.5]

76. Find the amount of sugars in a 32-fluid-ounce serving. [7.5]

For Exercises 77 through 80, solve each proportion.

77. $\dfrac{x-1}{4} = \dfrac{x-2}{6}$

78. $\dfrac{x+3}{8} = \dfrac{x-5}{4}$ [7.5]

79. $\dfrac{x+1}{2x-14} = \dfrac{9}{2}$

80. $\dfrac{2x+3}{3x+18} = \dfrac{-7}{3}$ [7.5]

81. The following triangles are similar. Find the length of the missing side. [7.5]

82. Each side of the Great Pyramid of Giza goes up 6.35 meters for every 5 meters across. Use the following similar triangles to estimate the original height of the pyramid.
Source: www.discoveringegypt.com [7.5]

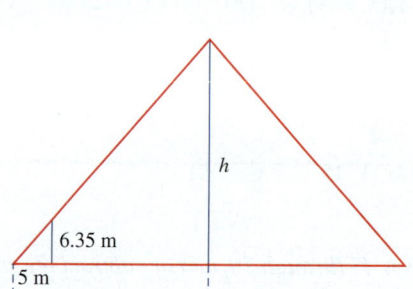

83. The variable y varies directly with x. $y = 12$ when $x = 36$. Find the equation that describes this relationship. Find y when $x = 10$. [7.5]

84. The variable y varies directly with x. $y = 9$ when $x = \dfrac{1}{2}$. Find the equation that describes this relationship. Find y when $x = 6$. [7.5]

85. The variable y varies inversely with x. $y = -\dfrac{1}{3}$ when $x = 12$. Find the equation that describes this relationship. Find y when $x = -2$. [7.5]

86. The variable y varies inversely with x. $y = -\dfrac{1}{2}$ when $x = 14$. Find the equation that describes this relationship. Find y when $x = 21$. [7.5]

Chapter Test

This chapter test should take approximately one hour to complete. Read each question carefully. Show all of your work.

For Exercises 1 and 2, simplify each rational expression.

1. Simplify: $\dfrac{(b+2)(b-9)}{(b-4)(b-9)}$.

2. Simplify: $\dfrac{c^2 - 49}{c^2 + 9c + 14}$.

3. Find the values of x that must be excluded for the rational expression
$$\dfrac{x+4}{x+7}$$

4. Find the values of x that must be excluded for the rational expression
$$\dfrac{x+4}{x^2 - 5x - 24}$$

5. A high-powered model rocket is traveling at 150 mph. What is the rocket's speed in feet per second?

6. Multiply: $\dfrac{(d-3)}{(d+8)} \cdot \dfrac{(d+2)(d+8)}{(d-3)(d+7)}$.

7. Multiply: $\dfrac{(p+1)}{(p-6)} \cdot \dfrac{(p+3)}{p^2 + 5p + 4}$.

8. Divide: $\dfrac{(k-11)}{(k+8)} \div \dfrac{2k-22}{k^2 + 5k - 24}$.

9. Find the LCD of the following set of rational expressions:
$$\dfrac{2}{x+7}, \quad \dfrac{4x}{(x+3)(x+7)}$$

10. Add: $\dfrac{5x}{x+3} + \dfrac{7}{x+3}$.

11. Add: $\dfrac{2}{n-7} + \dfrac{4}{n-8}$.

12. Add: $\dfrac{a}{a-2} + \dfrac{4}{a^2 - 6a + 8}$.

13. Subtract: $\dfrac{4}{g-3} - \dfrac{3}{g+2}$.

14. Subtract: $\dfrac{4}{(t-3)(t+1)} - \dfrac{2}{(t-3)}$.

15. Simplify the complex fraction: $\dfrac{\dfrac{-2}{x+6}}{\dfrac{4}{x^2} + 1}$.

16. Martin is installing a mow curb in his yard. The average cost per foot of mow curb can be calculated using the equation
$$c = \dfrac{150 + 20m}{m}$$
where c is the average cost per foot in dollars when m feet of mow curb are installed.

   a. What is the average cost per foot when 10 feet of mow curb are installed?

   b. How many feet of mow curb must be installed for the average cost to be $25 per foot?

17. Solve: $\dfrac{5}{m} + 3 = \dfrac{17}{m}$.

18. Solve: $\dfrac{3}{r+5} = 2 + \dfrac{1}{r+5}$.

19. Solve: $\dfrac{4}{z+2} = \dfrac{3}{z-8}$.

20. Solve: $\dfrac{h}{h+3} + \dfrac{2}{h+5} = \dfrac{-4}{h^2 + 8h + 15}$.

21. A water truck can be filled by using one pump in 25 minutes. The same truck can be filled from a larger pump in 10 minutes. How long would it take to fill the truck if both pumps were used?

22. Find the unit rates (round to the nearest cent) and determine which is cheaper.

   a. 5 feet of lights for $1.35
   b. 8 feet of lights for $2.24

23. The following triangles are similar. Find the length of the missing side.

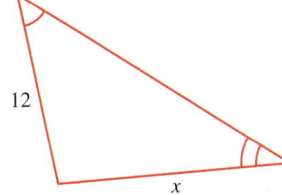 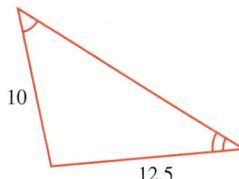

24. It takes 40 seconds to download a 4-megabyte song. How long does it take to download a 5.2-megabyte song?

25. The variable y varies inversely with x. $y = 2$ when $x = 5$. Find the equation that describes this relationship. Find y when $x = 40$.

Chapter Projects

■ Golden Rectangles

Written Project
One or more people

Golden rectangles have been used by artists and architects dating back to the ancient Greeks. Golden rectangles are thought to be very pleasing to the human eye. A rectangle is called a *golden rectangle* if the ratio of the length (l) to the width (w) satisfies the following proportion:

$$\frac{l}{w} = \frac{l + w}{l}$$

It has been shown that for a golden rectangle, the ratio of the length to the width is equal to

$$\frac{l}{w} \approx 1.618$$

The ratio of the length to the width is known as the **golden ratio**. Therefore, the golden ratio is approximately equal to 1.618.

1. Using a tape measure, measure your credit card very carefully. Substitute the length and width into the proportion above. Does it form an approximate golden rectangle?

2. Consider a standard-sized index card, which has measurements 3″ by 5″. Does an index card form an approximate golden rectangle?

3. Some artists have incorporated golden rectangles into their paintings. In his painting *The Sacrament of the Last Supper*, Salvador Dali sized the canvas to the measurements of 105 inches by 65.6 inches. Show that this canvas forms an approximate golden rectangle.

4. The architects of some famous buildings seem to have brought golden rectangles into their plans. Two buildings that are thought to incorporate golden rectangles into their design are the Parthenon, an ancient Greek temple (built in 438 B.C.E. in Athens, Greece) and the Mosque of Uqba, also known as the Great Mosque of Kairouan (built in 670 C.E. in Tunisia). Research either of these two buildings and write three paragraphs describing how golden rectangles were used in the construction.

Golden Ratios and Fibonacci Numbers

Written Project

One or more people

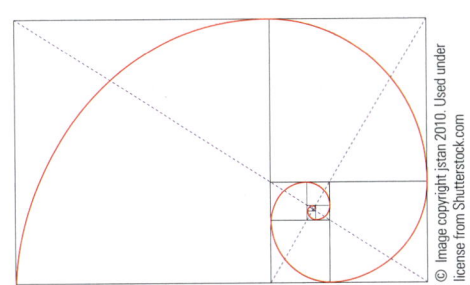

Fibonacci numbers are a sequence of numbers in which a number is equal to the sum of the two previous numbers. The Fibonacci sequence starts with the two numbers 1 and 1.

Fibonacci sequence: 1, 1, ….

Adding 1 and 1 gives the next Fibonacci number, which is 2.

Fibonacci sequence: 1, 1, 2 ….

Adding 1 and 2 gives the next Fibonacci number, which is 3.

Fibonacci sequence: 1, 1, 2, 3….

Adding 2 and 3 gives the next Fibonacci number, which is 5.

Fibonacci sequence: 1, 1, 2, 3, 5, ….

1. Using the pattern above, find the next seven Fibonacci numbers.

2. The ratio of Fibonacci numbers divided by the previous Fibonacci numbers has an interesting pattern. Complete the following table to see the pattern.

3.

Fibonacci Numbers	Ratio of $\dfrac{\text{Fibonacci Number}}{\text{Previous Fibonacci Number}}$	Decimal Approximation of the Ratio
1	$\dfrac{1}{1}$	1
1	$\dfrac{2}{1}$	2
2	$\dfrac{3}{2}$	1.5
3		
5		
8		
13		
21		
34		
55		
89		
144		

4. Write down the golden ratio (see the first project).

5. What is the relationship between the golden ratio and the ratios of the Fibonacci numbers?

Computing BMI (Body Mass Index)

Written Project
One or more people

Suppose a nurse weighs a patient in pounds (lb) and measures the patient's height in inches (in.). The BMI of that patient is defined as

$$\text{BMI} = \frac{\text{mass(lb)} \cdot 703}{(\text{height(in.)})^2}$$

1. Find the BMI for the patients below by filling in the chart.

Patient Number	Weight (lb)	Height (in.)	BMI
1	135	66	
2	234	72	
3	110	62	
4	175	69	
5	265	70	
6	128	65	

2. BMI is used to classify obesity as follows:

   a. BMI < 19.5 means that the person is underweight.
   b. 19.5 ≤ BMI ≤ 25 means that the person is of normal weight.
   c. BMI > 25 means that the person is overweight.
   d. BMI ≥ 30 means that the person is obese.

   Classify each patient above as underweight, normal weight, overweight, or obese.

Write a Review of a Section for Presentation

Research Project
One or more people

What you will need
- This presentation may be done in class or as an online video.
- If it is done as a video, post it on a website where it may be easily viewed by the class.
- You may want to use homework or other review problems from the book.
- Make it creative and fun.

Create a 5-minute review presentation of one section from Chapter 7. The format of the presentation can be a poster presentation, a blackboard presentation from notes, an online video, or a game format (e.g., math jeopardy). The presentation should include the following:

- Any important formulas in the chapter
- Important skills in the chapter, backed up by examples
- Explanation of terms used (e.g., what is *modeling*?)
- Common mistakes and how to avoid them

Radical Expressions and Equations

8

8.1 From Squaring a Number to Roots and Radicals

8.2 Basic Operations with Radical Expressions

8.3 Multiplying and Dividing Radical Expressions

8.4 Solving Radical Equations

The distance, in miles, to the horizon that can be seen by a person on the surface depends on the height, in feet, of the person above the surface of the Earth. A 6-foot-tall person standing on the surface of the Earth can see about 3 miles to the horizon. A person standing on top of Denali (Mt. McKinley) in Alaska, which is 20,310 feet tall, can see approximately 174.5 miles to the horizon. Computing the distance to the horizon involves the use of square roots. In this chapter, we begin to study square roots and other radicals.

8.1 From Squaring a Number to Roots and Radicals

LEARNING OBJECTIVES
- Calculate square roots.
- Evaluate radical expressions.
- Evaluate radical equations.
- Simplify radical expressions.
- Calculate cube roots.

■ Finding Square Roots

Throughout this book, the operation of squaring a number has been used. For example, $5^2 = 25$. To "undo" the operation of squaring a number, we use an operation called a square root. The symbol for the square root is the **radical symbol**, which looks like this: $\sqrt{\ }$. Therefore, to "undo" $5^2 = 25$, we write $\sqrt{25} = 5$. This is read as "the square root of 25 is 5." In the expression $\sqrt{25}$, we call the number underneath the radical symbol, in this case 25, the **radicand**.

$$\underset{\text{radicand}}{\sqrt{25}}\ \text{— radical}$$

We call 25 a **perfect square** because it is 5^2 and it has a whole number square root ($\sqrt{25} = 5$). The following Concept Investigation lists the first seven perfect squares and their square roots.

CONCEPT INVESTIGATION
Do squares and square roots undo each other?

Fill in the three columns headed "Perfect Square," "Perfect Square" (repeated), and "Square Root." The first two rows have been done to get you started.

Number	Perfect Square		Perfect Square	Square Root
1	$1^2 = 1$		1	$\sqrt{1} = 1$
2	$2^2 = 4$		4	$\sqrt{4} = 2$
3				
4				
5				
6				
7				

What is the relationship between the numbers in the first column on the left and your final answer in the column headed "Square Root"?

Since both $5^2 = 25$ and $(-5)^2 = 25$, this means that $\sqrt{25} = -5$ and $\sqrt{25} = 5$. Having two possible values for the expression $\sqrt{25}$ causes confusion in writing the answer. Which answer do we write: $\sqrt{25} = -5$ or $\sqrt{25} = 5$? Therefore, to avoid confusion, the agreement is to give only the positive answer. The answer to the expression $\sqrt{25}$ is 5. This is called the **principal square root**. If a negative answer is desired, we make it very obvious by putting a negative sign in front of the radical: $-\sqrt{25} = -5$.

> **DEFINITION**
>
> **Principal Square Root** The principal square root of b is a, that is, $\sqrt{b} = a$, if $a^2 = b$ and a is positive or 0 ($a \geq 0$).

What's That Mean?

Square Root and Principal Square Root

When asked to find the square root of a number, we find the principal square root. Therefore, $\sqrt{16} = 4$. The principal square root includes only the nonnegative value. If we want the negative answer, we put a negative sign in front of the radical. So $-\sqrt{16} = -4$.

Example 1 Evaluating principal square roots

Evaluate each square root.

a. $\sqrt{16}$ b. $\sqrt{81}$ c. $\sqrt{\dfrac{4}{9}}$

SOLUTION

a. $\sqrt{16} = 4$ because $4^2 = 16$.
b. $\sqrt{81} = 9$ because $9^2 = 81$.
c. $\sqrt{\dfrac{4}{9}} = \dfrac{2}{3}$ because $\left(\dfrac{2}{3}\right)^2 = \dfrac{2}{3} \cdot \dfrac{2}{3} = \dfrac{4}{9}$

PRACTICE PROBLEM FOR EXAMPLE 1

Evaluate each square root.

a. $\sqrt{100}$ b. $\sqrt{\dfrac{25}{9}}$ c. $\sqrt{0}$

An expression such as $\sqrt{-25}$ does not make sense in the real number system. If $\sqrt{-25} = a$, then $a^2 = -25$. Looking closely at the equation $a^2 = -25$, notice that the left side, a^2, is a positive number (or zero) because whenever a real number is squared, the result is positive (or zero). The right side of $a^2 = -25$ is -25, a negative number. This equation is stating that a positive number (or zero) equals a negative number. Since that situation is not possible in the real number system, the answer is **not a real number** or **not real**.

What's That Mean?

Not Real

In Chapter R, we learned what the real number system is. Some numbers, such as $\sqrt{-25}$, do not exist in the real number system. Therefore, we say that a value such as $\sqrt{-25}$ is "not real."

Example 2 Negative signs and square roots

Evaluate each square root, if it exists.

a. $-\sqrt{36}$ b. $\sqrt{-49}$ c. $-\sqrt{0}$

SOLUTION

a. The negative sign in front of the radical means that the coefficient is -1. Therefore:

$$-\sqrt{36} = -1 \cdot \sqrt{36} = -1 \cdot 6 = -6$$

It can also be said that $-\sqrt{36}$ means to find the *opposite* of $\sqrt{36}$, which is -6.

b. Remember that a negative number is not allowed underneath a square root in the real number system.

$\sqrt{-49}$ is not a real number.

c. $-\sqrt{0} = -0 = 0$

682 CHAPTER 8 Radical Expressions and Equations

PRACTICE PROBLEM FOR EXAMPLE 2
Evaluate each square root, if it exists.

a. $\sqrt{-25}$ b. $-\sqrt{81}$

Connecting the Concepts

What are irrational numbers?

Irrational numbers have infinite, nonrepeating decimal expansions. Numbers such as π and $\sqrt{7}$ are said to be irrational. See Section R.5 for more review of irrational numbers.

Looking at the table of the square roots of perfect squares, we notice that a lot of numbers are missing.

For example, we see that we can find the values of $\sqrt{1}$ and $\sqrt{4}$, but what about $\sqrt{2}$ and $\sqrt{3}$? Only perfect squares have rational numbers for their square roots. When numbers that are not perfect squares are placed under the radical, their square roots are **irrational numbers**. In these cases, we can use a calculator to come up with a decimal approximation.

Square Root
$\sqrt{1} = 1$
$\sqrt{4} = 2$
$\sqrt{9} = 3$
$\sqrt{16} = 4$
$\sqrt{25} = 5$
$\sqrt{36} = 6$
$\sqrt{49} = 7$

Calculator Details

On some scientific calculators, the square root operation is above the 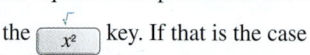 key. If that is the case on your calculator, then you have to use the `2nd` or Shift key to access the square root key.

Example 3 Approximating square roots using a calculator

Use a calculator to approximate the following square roots. Round the answer to two decimal places.

a. $\sqrt{3}$ b. $\sqrt{45}$ c. $\sqrt{-7}$

SOLUTION

a. Using a calculator, $\sqrt{3} \approx 1.732050808$. Rounding to two decimal places yields

$$\sqrt{3} \approx 1.732050808$$
$$\sqrt{3} \approx 1.73$$

Recall from Chapter R that the $\approx$ symbol means "approximately equal to" and lets the reader know that the number has been rounded.

b. Using the calculator, we find that $\sqrt{45} \approx 6.708203932$. Rounding to two decimal places yields

$$\sqrt{45} \approx 6.708203932$$
$$\sqrt{45} \approx 6.71$$

Skill Connection

Rounding

To round a number to two decimal places, underline the second decimal place. Look at the number immediately to the right of the second decimal place. If that number is 0, 1, 2, 3, or 4, we round down, which means that we do not change the value of the second decimal place. If the number to the right is 5, 6, 7, 8, or 9, then we add 1 to the second decimal place (we round up).

c. Input $\sqrt{-7}$ on your calculator. Your calculator may read "error" or "err: non real answer." Remember a negative number is not allowed under a square root in the real number system. The answer is: not a real number.

PRACTICE PROBLEM FOR EXAMPLE 3
Use a calculator to approximate the following square roots. Round the answer to two decimal places.

a. $\sqrt{7}$ b. $\sqrt{65}$

It is important to understand the difference between $\sqrt{3}$ and its decimal approximation 1.73. The expression $\sqrt{3}$ is the **exact** answer, while 1.73 is an **approximation** to the exact answer.

■ Evaluating Radical Expressions

Expressions that have a variable under the radical are called **radical expressions**. For instance, $\sqrt{2x - 1}$ is an example of a radical expression.

SECTION 8.1 From Squaring a Number to Roots and Radicals   683

> **DEFINITION**
> **Radical Expression** A radical expression is an expression in which a variable expression is under the radical.

To evaluate a radical expression, substitute in the given value of the variable(s) and evaluate using the order-of-operations agreement.

Example 4 Evaluating radical expressions

Evaluate the radical expression for the given value of the variable.

a. $\sqrt{2x - 1}$ for $x = 13$ **b.** $\sqrt{x^2 - x}$ for $x = -2$

c. $\sqrt{x^2 - x}$ for $x = \dfrac{1}{2}$

SOLUTION

a. To evaluate $\sqrt{2x - 1}$ for $x = 13$, substitute $x = 13$ into the expression for x and evaluate using the order-of-operations agreement.

$\sqrt{2x - 1}$ Substitute in $x = 13$.
$\sqrt{2(13) - 1}$ Simplify, using the order-of-operations agreement.
$= \sqrt{26 - 1}$
$= \sqrt{25}$
$= 5$

b. To evaluate $\sqrt{x^2 - x}$ for $x = -2$, substitute $x = -2$ into the expression for x in both terms where it appears and evaluate using the order-of-operations agreement.

$\sqrt{x^2 - x}$ Substitute in $x = -2$.
$\sqrt{(-2)^2 - (-2)}$ Simplify, using the order-of-operations agreement.
$= \sqrt{4 + 2}$
$= \sqrt{6}$

> **Connecting the Concepts**
>
> **What are excluded values?**
>
> In Section 7.1, we learned that some values of the variable cannot be substituted into *rational* expressions because they make the denominator 0. There are also excluded values for *radical* expressions, namely, those values that result in negative numbers under the square root.

c. To evaluate $\sqrt{x^2 - 2x}$ for $x = \dfrac{1}{2}$, substitute $x = \dfrac{1}{2}$ into the expression for x in both terms where it appears and evaluate using the order-of-operations agreement.

$\sqrt{x^2 - x}$ Substitute in $x = \dfrac{1}{2}$.

$\sqrt{\left(\dfrac{1}{2}\right)^2 - \left(\dfrac{1}{2}\right)}$ Simplify, using the order-of-operations agreement.

$= \sqrt{\dfrac{1}{4} - \dfrac{1}{2}}$ The LCD is 4.

$= \sqrt{\dfrac{1}{4} - \dfrac{2}{4}}$

$= \sqrt{\dfrac{-1}{4}}$ This is not a real number.

PRACTICE PROBLEM FOR EXAMPLE 4

Evaluate the radical expression for the given value of the variable.

a. $\sqrt{x^2 + 5}$ for $x = 2$ **b.** $\sqrt{3x + 5}$ for $x = -2$

Evaluating Radical Equations

Square roots often show up in applications and equations. A **radical equation** is an equation in which the variable is in the radicand (under the radical).

> **DEFINITION**
> **Radical Equation** A radical equation is an equation in which a variable expression is under the radical.

For instance, $\sqrt{x} - 6 = 2$ is a radical equation. We now learn how to *evaluate* radical equations. Solving radical equations is covered in Section 8.4.

Example 5 — Using a formula to model the height of a falling object

A science class tests how long it takes a ball to fall to the ground when dropped from various heights. The class determines that the time, T, in seconds that it takes the ball to fall to the ground after being dropped from a height of h feet is given by the equation

$$T = 0.243\sqrt{h}$$

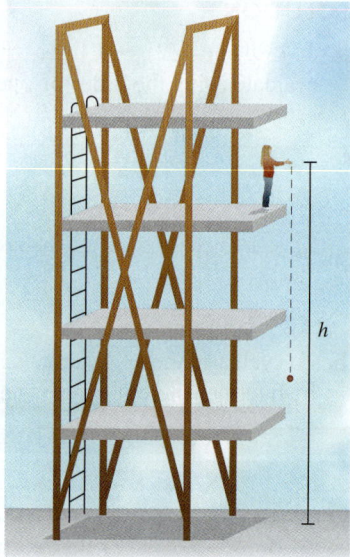

a. What units does h represent? What units does T represent?

b. How long will it take the ball to fall to the ground when the ball is dropped from 15 feet? Round the answer to the hundredths place if necessary.

c. How long will it take the ball to fall to the ground when the ball is dropped from 50 feet? Round the answer to the hundredths place if necessary.

SOLUTION

a. T, in seconds, is the time it takes the ball to fall to the ground. The variable h, in feet, is the initial height the ball is dropped from.

b. When the ball is dropped from 15 feet, substitute in $h = 15$ to find time T.

$T = 0.243\sqrt{h}$ Substitute in $h = 15$.
$T = 0.243\sqrt{15}$
$T \approx 0.9411349531\ldots$ Round only at the very end.
$T \approx 0.94$ Round to the hundredths place.

It will take the object about 0.94 second to fall to the ground.

c. When the ball is dropped from 50 feet, substitute in $h = 50$ to find time T.

$T = 0.243\sqrt{h}$ Substitute in $h = 50$.
$T = 0.243\sqrt{50}$
$T \approx 1.718269478\ldots$
$T \approx 1.72$ Round to the hundredths place.

It will take the ball about 1.72 seconds to fall to the ground.

Calculator Details

Entering Radical Expressions

Evaluating an expression such as

$$0.243\sqrt{15}$$

on your calculator depends on the programming in your calculator. On most calculators, enter the expression from left to right,

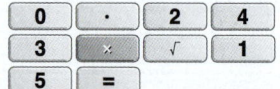

or

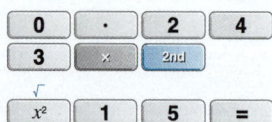

The answer should approximately equal $0.243\sqrt{15} \approx 0.9411$. If this does not work, you might have an older calculator. Older calculators require that you enter these expressions "backward." On these models, $0.243\sqrt{15}$ is entered as

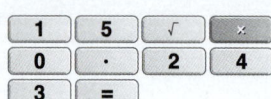

The answer should be the same.

SECTION 8.1 From Squaring a Number to Roots and Radicals 685

PRACTICE PROBLEM FOR EXAMPLE 5

Police use a formula involving square roots when investigating traffic accidents. By measuring the length of the skid mark left by the car, the police can determine the speed of the car at the time of the accident (speed measured in miles per hour). The speed of the car in miles per hour, s, is related to the length of the skid mark in feet, d, by the equation

$$s = \sqrt{30d}$$

Assume that the car is being driven on a dry asphalt road.

a. What units does s represent? What units does d represent?

b. What was the speed of the car when the length of the skid mark was 175 feet?

c. What was the speed of the car when the length of the skid mark was 115 feet?

The distance formula gives the straight-line distance between two points (x_1, y_1) and (x_2, y_2).

> **THE DISTANCE FORMULA**
>
> **The distance, d, of the line segment between two points (x_1, y_1) and (x_2, y_2) is**
>
> $$d = \sqrt{(x_2 - x_1)^2 + (y_2 - y_1)^2}$$
>
> Here the subscript 1 on the coordinates of the first point, namely, (x_1, y_1), indicates that it is point 1, and the subscript 2 on the coordinates of the second point, namely, (x_2, y_2), indicates that it is point 2.

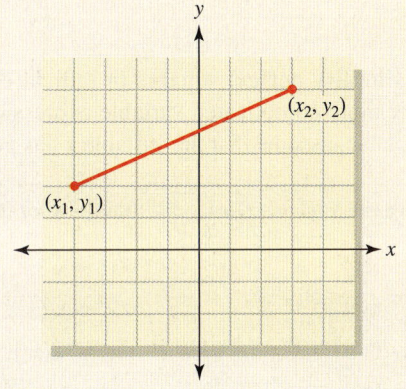

What's That Mean?

Subscripts

A *subscript* is a small number printed below the variable, such as x or y in the distance formula. The subscript distinguishes which point the variable value came from. Subscripts have no mathematical meaning; they do not affect the computation. Subscripts are unlike superscripts, such as 3^5, which indicate an exponent.

Example 6 The distance between two points

Find the distance between the two points. Round the answer to two decimal places if necessary.

a. $(1, 0)$ and $(4, 4)$ **b.** $(3, -2)$ and $(-6, -5)$

SOLUTION

a. Set (x_1, y_1) to be the first point, so $(x_1, y_1) = (1, 0)$. Set (x_2, y_2) to be the second point, so $(x_2, y_2) = (4, 4)$. Substitute these values into the distance formula and simplify.

$d = \sqrt{(x_2 - x_1)^2 + (y_2 - y_1)^2}$ Substitute in $(x_1, y_1) = (1, 0)$ and $(x_2, y_2) = (4, 4)$.
$d = \sqrt{(4 - 1)^2 + (4 - 0)^2}$ Simplify under the radical.
$d = \sqrt{3^2 + 4^2}$
$d = \sqrt{9 + 16} = \sqrt{25}$ Evaluate the square root.
$d = 5$

b. Set (x_1, y_1) to be the first point, so $(x_1, y_1) = (3, -2)$. Set (x_2, y_2) to be the second point, so $(x_2, y_2) = (-6, -5)$. Substitute these values into the distance formula and simplify.

$d = \sqrt{(x_2 - x_1)^2 + (y_2 - y_1)^2}$ Substitute in $(x_1, y_1) = (3, -2)$ and $(x_2, y_2) = (-6, -5)$.
$d = \sqrt{(-6 - 3)^2 + (-5 - (-2))^2}$ Simplify under the radical.
$d = \sqrt{(-9)^2 + (-3)^2}$
$d = \sqrt{81 + 9} = \sqrt{90}$ Evaluate the square root.
$d \approx 9.486832981$ Use a calculator to approximate $\sqrt{90}$.
$d \approx 9.49$ Round to two decimal places.

PRACTICE PROBLEM FOR EXAMPLE 6

Find the distance between the two points. Round the answer to two decimal places if necessary.

a. $(1, 2)$ and $(9, 8)$ **b.** $(-2, 5)$ and $(-1, -8)$

■ Simplifying Radical Expressions That Contain Variables

Now we learn how to simplify perfect squares in radical expressions. In the next section, we will simplify more complicated variable expressions that occur under the radical. In the following table, assume that the variable x has positive values.

Square the Variable (Do)	Square Root (Undo)
$(x)^2 = x^2$	$\sqrt{x^2} = x$
$(x^2)^2 = x^4$	$\sqrt{x^4} = \sqrt{(x^2)^2} = x^2$
$(x^3)^2 = x^6$	$\sqrt{x^6} = \sqrt{(x^3)^2} = x^3$
$(x^4)^2 = x^8$	$\sqrt{x^8} = \sqrt{(x^4)^2} = x^4$

Keep in mind that to simplify a variable expression under a radical, first rewrite the variable expression as a perfect square. Then the square root will undo the operation of squaring, yielding the answer.

In this chapter, we assume that variables under a square root take only nonnegative values.

Example 7 Simplifying radical expressions

Simplify each radical expression.

a. $\sqrt{x^{10}}$ **b.** $\sqrt{25x^4}$ **c.** $\sqrt{\dfrac{x^6}{9}}$ **d.** $\sqrt{x^2 y^2}$

Skill Connection

Exponent rules

In Section 5.1, we learned the power rule for exponents. Recall that $(x^m)^n = x^{mn}$. To simplify a power to a power, multiply the exponents. Therefore, to simplify $(x^3)^2$, multiply the exponents to get $(x^3)^2 = x^6$.

We also use the powers of products and quotients rules for exponents. Recall that

$(xy)^m = x^m y^m$ and $\left(\dfrac{x}{y}\right)^m = \dfrac{x^m}{y^m}$

SOLUTION

a. $x^{10} = (x^5)^2$ *Rewrite radicand as a perfect square.*
$\sqrt{x^{10}} = \sqrt{(x^5)^2} = x^5$ *Simplify the radical.*

b. $25x^4 = (5x^2)^2$ *Rewrite radicand as a perfect square.*
$\sqrt{25x^4} = \sqrt{(5x^2)^2} = 5x^2$ *Simplify the radical.*

c. $\dfrac{x^6}{9} = \left(\dfrac{x^3}{3}\right)^2$ *Rewrite radicand as a perfect square.*
$\sqrt{\dfrac{x^6}{9}} = \sqrt{\left(\dfrac{x^3}{3}\right)^2} = \dfrac{x^3}{3}$ *Simplify the radical.*

d. $x^2y^2 = (xy)^2$ *Rewrite radicand as a perfect square.*
$\sqrt{x^2y^2} = \sqrt{(xy)^2} = xy$ *Simplify the radical.*

PRACTICE PROBLEM FOR EXAMPLE 7

Simplify each radical expression.

a. $\sqrt{x^{14}}$ **b.** $\sqrt{49x^6}$ **c.** $\sqrt{\dfrac{x^2}{16}}$ **d.** $\sqrt{a^4b^4}$

■ Finding Cube Roots

Square roots undo squares. To undo a cube (a quantity raised to the third power), we use a **cube root**. To distinguish a square root from a cube root, the number 3 is placed on the upper left of the radical sign. The number 3 is called the **index** of the root and indicates that we are undoing a cube.

$$\underset{\text{index}}{}\sqrt[3]{8}\underset{\text{radicand}}{}$$

Just as the square root of a perfect square was a rational number, the cube root of a perfect cube is a rational number. The following tables list the first four perfect cubes and their cube roots.

Number	Perfect Cube
1	$1^3 = 1$
2	$2^3 = 8$
3	$3^3 = 27$
4	$4^3 = 64$

Perfect Cube	Cube Root
1	$\sqrt[3]{1} = 1$
8	$\sqrt[3]{8} = 2$
27	$\sqrt[3]{27} = 3$
64	$\sqrt[3]{64} = 4$

Cube roots differ from square roots. The radicand (the number underneath the radical sign) can be negative. To see why this makes sense, remember that $(-2)^3 = -8$. Cubing a negative number results in a negative number. Therefore, we have that $\sqrt[3]{-8} = -2$. It makes sense to have negative numbers under cube roots.

> **DEFINITION**
> **Cube Root** The cube root of b is a (i.e., $\sqrt[3]{b} = a$) if $a^3 = b$.

■ Connecting the Concepts

What is the index for a square root?

Since a square root undoes a square, the index on a square root should be a 2. Therefore, $\sqrt{x} = \sqrt[2]{x}$. However, since square roots are used so often, it has become common practice to drop the 2 in the index.

Example 8 — Cube roots and their signs

Evaluate each expression.

a. $\sqrt[3]{-27}$ b. $-\sqrt[3]{125}$ c. $-\sqrt[3]{-1}$ d. $\sqrt[3]{343}$

SOLUTION

Calculator Details

Some calculators have a cube root key. Some have a "general" root key. Consult your calculator manual, or ask your instructor for help on how to use it.

a. $-27 = (-3)^3$ *Rewrite radicand as a perfect cube.*
$\sqrt[3]{-27} = \sqrt[3]{(-3)^3} = -3$ *Simplify the radical.*

It is okay to have a negative number under a cube root.

b. $-\sqrt[3]{125}$ means to find the opposite of the cube root of 125. Since this is a value that is not commonly known, let's use guess and check to find the answer. Using a calculator, we make the following list:

$$1^3 = 1$$
$$2^3 = 8$$
$$3^3 = 27$$
$$4^3 = 64$$
$$5^3 = 125$$

Therefore, $-\sqrt[3]{125} = -1 \cdot \sqrt[3]{(5)^3} = -1 \cdot 5 = -5$.

c. $-\sqrt[3]{-1}$ is asking us to find the opposite of the cube root of -1. First find $\sqrt[3]{-1}$. The expression $\sqrt[3]{-1} = -1$ because $(-1)^3 = -1$. Therefore, the opposite of the cube root of -1 is found as $-\sqrt[3]{-1} = -(-1) = 1$.

d. To find the value of $\sqrt[3]{343}$, continue the list started in part b.

$$4^3 = 64$$
$$5^3 = 125$$
$$6^3 = 216$$
$$7^3 = 343$$

Therefore, $\sqrt[3]{343} = \sqrt[3]{(7)^3} = 7$.

PRACTICE PROBLEM FOR EXAMPLE 8

Evaluate each expression.

a. $\sqrt[3]{-125}$ b. $\sqrt[3]{216}$ c. $-\sqrt[3]{-8}$

8.1 Vocabulary Exercises

1. The symbol $\sqrt{\ }$ is called a(n) _____.
2. Square roots of negative numbers are _____.
3. The input under the radical symbol is called the _____.
4. To _____ a radical expression means to substitute in the given value of the variable.
5. A(n) _____ undoes a cube (an exponent of 3).

8.1 Exercises

For Exercises 1 through 18, find the square root of each number without using a calculator. If the square root of the number does not exist, write "Not real" as the answer.

1. $\sqrt{25}$
2. $\sqrt{16}$
3. $\sqrt{64}$
4. $\sqrt{144}$
5. $\sqrt{\dfrac{16}{9}}$
6. $\sqrt{\dfrac{25}{36}}$
7. $\sqrt{\dfrac{100}{49}}$
8. $\sqrt{\dfrac{36}{81}}$
9. $\sqrt{\dfrac{1}{16}}$
10. $\sqrt{\dfrac{1}{81}}$
11. $-\sqrt{49}$
12. $-\sqrt{4}$
13. $-\sqrt{\dfrac{25}{64}}$
14. $-\sqrt{\dfrac{4}{81}}$
15. $\sqrt{-9}$
16. $\sqrt{-25}$
17. $\sqrt{\dfrac{-1}{4}}$
18. $\sqrt{\dfrac{-4}{25}}$

For Exercises 19 through 26, use your calculator to approximate each square root. Round the answer to two decimal places (the hundredths place). Make sure you use the notation ≈ to indicate that you rounded your answer. If the square root of the number does not exist, write "Not real" as the answer.

19. $\sqrt{15}$
20. $\sqrt{35}$
21. $\sqrt{51}$
22. $\sqrt{65}$
23. $\sqrt{7}$
24. $\sqrt{29}$
25. $\sqrt{-15}$
26. $\sqrt{-4}$

For Exercises 27 through 38, evaluate each radical expression for the given value of the variable. If the square root does not exist, write "Not real" as the answer.

27. $\sqrt{4x}$ for $x = 9$
28. $\sqrt{3x}$ for $x = 12$
29. $\sqrt{x + 6}$ for $x = -5$
30. $\sqrt{x^2 + 1}$ for $x = 0$
31. $\sqrt{x + 5}$ for $x = -1$
32. $\sqrt{x^2 + 8}$ for $x = -1$
33. $\sqrt{a^2 - 2a}$ for $a = 1$
34. $\sqrt{a^2 + 3a}$ for $a = -2$
35. $\sqrt{h^2 - 36}$ for $h = 2$
36. $\sqrt{h^2 - 16}$ for $h = -2$
37. $\sqrt{1 - 4a}$ for $a = -3$
38. $\sqrt{3 - 2a}$ for $a = -1$

For Exercises 39 through 50, answer the questions for each application problem.

39. A science class determines that a ball dropped from a height of *h* feet will take *t* seconds to fall, which is modeled by the equation $t = \sqrt{\dfrac{h}{16}}$.

    a. How long will it take the ball to fall when dropped from 50 feet? Round the answer to two decimal places if necessary.

    b. How long will it take the ball to fall when dropped from 80 feet? Round the answer to two decimal places if necessary.

40. A science class determines that a rock dropped from a height of *h* feet taking *t* seconds to fall is modeled by the equation $t = \sqrt{\dfrac{h}{16}}$.

    a. How long will it take the rock to fall when dropped from 60 feet? Round the answer to two decimal places if necessary.

    b. How long will it take the rock to fall when dropped from 100 feet? Round the answer to two decimal places if necessary.

41. Police use a formula involving square roots when investigating traffic accidents. By measuring the length of the skid mark left by the car, the police can determine the speed of the car at the time of the accident (speed measured in miles per hour). The equation that gives the speed of the car in miles per hour, *s*, related to the length of the skid mark in feet, *d*, is

$$s = \sqrt{13.32d}$$

The coefficient of *d* in the equation changes according to the road surface the car is being driven on. The road can be asphalt, dirt, or concrete. It can also be wet or dry.

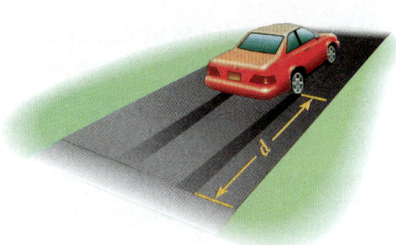

a. What units does *s* represent? What units does *d* represent?

b. What was the speed of the car when the length of the skid mark was 200 feet?

c. What was the speed of the car when the length of the skid mark was 145 feet?

42. Police use a formula involving square roots when investigating traffic accidents. By measuring the length of the skid mark left by the car, the police can determine the speed of the car at the time of the accident (speed measured in miles per hour). The equation that gives the speed of the car in miles per hour, *s*, related to the length of the skid mark in feet, *d*, is

$$s = \sqrt{22.5d}$$

The coefficient of *d* in the equation changes according to the road surface the car is being driven on. The road can be asphalt, dirt, or concrete. It can also be wet or dry.

a. What units does *s* represent? What units does *d* represent?

b. What was the speed of the car when the length of the skid mark was 175 feet?

c. What was the speed of the car when the length of the skid mark was 115 feet?

43. The time *t* in seconds that it takes an object to fall when dropped from a height of *s* feet is given by the equation $t = \dfrac{\sqrt{s}}{4}$. How much time does it take a ball to fall when dropped from 8 feet?

44. The time *t* in seconds that it takes an object to fall when dropped from a height of *s* feet is given by the equation $t = \dfrac{\sqrt{s}}{4}$. How much time does it take a ball to fall 21 feet?

45. The profit *P* in dollars that Eamond earns from selling *n* leather wristbands is given by $P = \sqrt{4n + 25}$. How much profit will he earn from selling 50 wristbands?

46. The profit *P* in dollars that Mary earns from selling *n* bouquets of flowers is given by $P = \sqrt{3n + 18}$. How much profit will she earn from selling 17 bouquets?

47. The length of a side of a cube, *x*, and the volume of the cube, *V*, are related by the equation $x = \sqrt[3]{V}$. If the cube has volume $V = 2.7 \text{ in}^3$, find the length of the side *x* of the cube.

48. Suppose a cylinder has height 6 cm. The radius of the bottom of a cylinder, *r*, and the volume of the cylinder, *V*, are related by the equation $r = \sqrt{\dfrac{V}{6\pi}}$. Find the radius *r* of the cylinder if the cylinder has volume $V = 54\pi \text{ cm}^3$.

49. The distance in miles, *d*, that can be seen on the surface of the earth is given by $d = 1.5\sqrt{h}$, where *h* is the height in feet of a person above the surface of the earth. How far can a person see if the person is 36 feet above the surface of the earth?

50. The distance in miles, *d*, that can be seen on the surface of the earth is given by $d = 1.5\sqrt{h}$, where *h* is the height in feet of a person above the surface of the earth. How far can a person see if the person is 49 feet above the surface of the earth?

For Exercises 51 through 58, find the distance between the two points. Round the answer to two decimal places if necessary.

51. $(3, 4)$ and $(10, 8)$

52. $(5, 9)$ and $(16, 12)$

53. $(-4, 7)$ and $(-6, -3)$

54. $(-5, -3)$ and $(-1, -2)$

55. $(0, 6)$ and $(-5, 0)$

56. $(-12, 0)$ and $(0, -1)$

57. $(11, 2)$ and $(-6, -1)$

58. $(-4, -7)$ and $(8, 9)$

For Exercises 59 through 76, simplify each radical expression. Assume that the variables have positive values.

59. $\sqrt{x^2}$
60. $\sqrt{x^{12}}$
61. $\sqrt{a^6}$
62. $\sqrt{a^4}$
63. $\sqrt{16x^2}$
64. $\sqrt{81x^{10}}$
65. $-\sqrt{x^{20}}$
66. $-\sqrt{x^8}$
67. $-\sqrt{a^2}$
68. $-\sqrt{a^4}$
69. $-\sqrt{4y^6}$
70. $-\sqrt{36y^{10}}$
71. $\sqrt{25x^{22}}$
72. $\sqrt{81x^{18}}$
73. $\sqrt{\dfrac{x^4}{49}}$
74. $\sqrt{\dfrac{x^2}{100}}$
75. $\sqrt{\dfrac{y^{10}}{36}}$
76. $\sqrt{\dfrac{y^6}{144}}$

For Exercises 77 through 88, find each cube root without a calculator.

77. $\sqrt[3]{8}$
78. $\sqrt[3]{27}$
79. $\sqrt[3]{1}$
80. $\sqrt[3]{0}$
81. $\sqrt[3]{-8}$
82. $\sqrt[3]{-64}$
83. $\sqrt[3]{-125}$
84. $\sqrt[3]{-27}$
85. $-\sqrt[3]{-8}$
86. $-\sqrt[3]{-27}$
87. $-\sqrt[3]{-1}$
88. $-\sqrt[3]{-64}$

For Exercises 89 through 112, simplify each expression. Assume that the variables have positive values. If the square root does not exist, write "Not real" as the answer.

89. $\sqrt[3]{64}$
90. $\sqrt[3]{729}$
91. $\sqrt{100}$
92. $\sqrt{49}$
93. $\sqrt{\dfrac{1}{49}}$
94. $\sqrt{\dfrac{1}{64}}$
95. $\sqrt{y^8}$
96. $\sqrt{x^{12}}$
97. $\sqrt{a^2}$
98. $\sqrt{b^4}$
99. $-\sqrt[3]{64}$
100. $-\sqrt[3]{8}$
101. $-\sqrt{25}$
102. $-\sqrt{16}$
103. $\sqrt[3]{-8}$
104. $\sqrt[3]{-125}$
105. $\sqrt{-4}$
106. $\sqrt{-36}$
107. $\sqrt{a^2 b^8}$
108. $\sqrt{x^4 y^8}$
109. $\sqrt{49 z^2}$
110. $\sqrt{25 y^{10}}$
111. $\sqrt{\dfrac{x^4}{81}}$
112. $\sqrt{\dfrac{a^6}{4}}$

For Exercises 113 through 116, answer as indicated.

113. Graph the line. Clearly label and scale the axes.
$$5x + 4y = -16$$

114. Graph the line. Clearly label and scale the axes.
$$-7x + 2y = -6$$

115. Solve the system. State which method you used. Check the solution.
$$x + y = -3$$
$$4x - y = 18$$

116. Solve the system. State which method you used. Check the solution.
$$6x + 5y = -30$$
$$-7x + 5y = 35$$

For Exercises 117 through 120, use the model $A = 8.09t + 189.9$, where A is the admissions in millions at the top 25 theme parks worldwide t years since 2010.

Source: Model based on data from TEA/AECOM Theme Index and Museum Index.

117. Find the number of admissions at the top 25 theme parks worldwide in 2017.

118. Find when the number of admissions at the top 25 theme parks worldwide is predicted to be 270.8 million.

119. Find and interpret the slope of this model.

120. Find and interpret the y-intercept of this model.

For Exercises 121 through 124, simplify each expression.

121. $\dfrac{-9p^4 q^2}{27 pq^5}$
122. $\dfrac{18 r^3 t}{-42 rt^3}$
123. $(3p + 4q)^2$
124. $(r - 5t)^2$

For Exercises 125 through 128, answer as indicated. Round to the hundredths place if necessary.

125. Find the slope of the line passing through the points $(3, -6)$ and $(-5, 2)$.

126. Find the distance between the points $(3, -6)$ and $(-5, 2)$.

127. Find the distance between the points $(3, -6)$ and $(7, 10)$.

128. Find the slope of the line passing through the points $(3, -6)$ and $(7, 10)$.

8.2 Basic Operations with Radical Expressions

LEARNING OBJECTIVES
- Simplify more complicated radical expressions.
- Add radical expressions.
- Subtract radical expressions.

Simplifying More Complicated Radical Expressions

In Section 8.1, we learned how to simplify radicals where the radicand was a perfect square. For example, we saw that $\sqrt{36} = 6$ because $6^2 = 36$. Likewise we have $\sqrt{x^4} = x^2$ because $(x^2)^2 = x^4$. However, we have not learned how to simplify a radical when the radicand is *not* a perfect square. To simplify a radical, we must first study an important property of radicals.

CONCEPT INVESTIGATION
How do we multiply radicals?

1. Fill out the following table. The first row has been done for you to get you started.

Column 1	Column 2	Are columns 1 and 2 the same?
$\sqrt{25 \cdot 4} = \sqrt{100} = 10$	$\sqrt{25} \cdot \sqrt{4} = 5 \cdot 2 = 10$	Yes
$\sqrt{9 \cdot 49} = \sqrt{} = $	$\sqrt{9} \cdot \sqrt{49} = \cdot = $	
$\sqrt{4 \cdot 36} = \sqrt{} = $	$\sqrt{4} \cdot \sqrt{36} = \cdot = $	
$\sqrt{25 \cdot 100} = \sqrt{} = $	$\sqrt{25} \cdot \sqrt{100} = \cdot = $	

2. Look at the patterns from part 1. In column 1, you first simplified under the square root and then evaluated the radical. Is this the same as breaking the problem up into two radicals, evaluating the radicals, and multiplying the results as in column 2?

From this Concept Investigation, notice that $\sqrt{25 \cdot 4} = \sqrt{25} \cdot \sqrt{4}$. What we have observed in this Concept Investigation is called the **product property of radicals**.

> **THE PRODUCT PROPERTY OF RADICALS**
> If a and b are positive or zero, then
> $$\sqrt{a \cdot b} = \sqrt{a} \cdot \sqrt{b}$$

We use the product property of radicals to simplify radical expressions where the radicand is not a perfect square. To simplify a radical expression, first rewrite the radicand using perfect square factors. We then use the product property of radicals and simplify the radicals that contain perfect squares. Therefore, to simplify $\sqrt{8}$, rewrite with a perfect square factor and use the product property of radicals.

Skill Connection

Perfect squares

The first eight perfect squares are as follows:

Number	Perfect Square
1	1
2	4
3	9
4	16
5	25
6	36
7	49
8	64

$$\sqrt{8}$$
$$= \sqrt{4 \cdot 2} \qquad \text{Rewrite with a perfect square factor.}$$
$$= \sqrt{4} \cdot \sqrt{2} \qquad \text{Use the product property of radicals.}$$
$$= 2 \cdot \sqrt{2} \qquad \text{Simplify the perfect square radical.}$$
$$= 2\sqrt{2}$$

> **Steps to Simplify an Expression under a Square Root**
> 1. Factor the radicand using perfect square factors. (*Hint*: Factor using the largest perfect square factor.)
> 2. Rewrite the radical expression using the product property of radicals.
> 3. Simplify the radicals that contain perfect squares.
>
> *Note:* Make sure there are no perfect square factors in the remaining radicand. If a perfect square factor of the radicand remains, repeat steps 1 through 3.

Example 1 Simplify radical expressions

Simplify each radical expression.

a. $\sqrt{27}$ **b.** $\sqrt{48}$

SOLUTION

a. Step 1 Factor the radicand using perfect square factors.

The factorizations of 27 are $27 \cdot 1$ and $9 \cdot 3$. Because 9 is a perfect square, use this factorization.

$$\sqrt{27}$$
$$= \sqrt{9 \cdot 3} \qquad \text{Rewrite 27 as 9} \cdot \text{3.}$$

Step 2 Rewrite the radical expression using the product property of radicals.

$$= \sqrt{9} \cdot \sqrt{3} \qquad \text{Use the product property of radicals.}$$

Step 3 Simplify the radicals that contain perfect squares.

$$= 3\sqrt{3} \qquad \text{Simplify } \sqrt{9} = 3.$$

This is simplified, since there are no more perfect square factors in the radicand.

b. There are many factorizations of 48. Let's look at this problem in two ways. They are equally valid; one just involves an extra step in simplifying. Suppose the factorization of 48 that comes to mind is $48 = 4 \cdot 12$. Because 4 is a perfect square factor of 48, this is a valid place to start. Applying the steps above yields

$$\sqrt{48}$$
$$= \sqrt{4 \cdot 12} \qquad \text{Rewrite 48 as 4} \cdot \text{12.}$$
$$= \sqrt{4} \cdot \sqrt{12} \qquad \text{Use the product property of radicals.}$$
$$= 2\sqrt{12} \qquad \text{Simplify } \sqrt{4} = 2.$$

However, this answer is not completely simplified. Because $\sqrt{12}$ can be simplified further, factor once again and simplify one last time.

$$\sqrt{48}$$
$$= 2\sqrt{12} \qquad \text{Rewrite 12 as 4} \cdot \text{3.}$$
$$= 2\sqrt{4 \cdot 3}$$
$$= 2\sqrt{4} \cdot \sqrt{3} \qquad \text{Use the product property of radicals.}$$
$$= 2 \cdot 2 \cdot \sqrt{3} \qquad \text{Simplify } \sqrt{4} = 2.$$
$$= 4\sqrt{3} \qquad \text{Multiply 2} \cdot \text{2.}$$

There is a shorter way to approach this problem. The largest perfect square factor of 48 is 16 because $48 = 16 \cdot 3$. Using this factorization, we can simplify $\sqrt{48}$ using fewer steps.

$$\sqrt{48}$$
$$= \sqrt{16 \cdot 3} \qquad \text{Rewrite 48 as } 16 \cdot 3.$$
$$= \sqrt{16} \cdot \sqrt{3} \qquad \text{Use the product property of radicals.}$$
$$= 4\sqrt{3} \qquad \text{Simplify } \sqrt{16} = 4.$$

Using the largest perfect square factor will make simplifying a quicker process.

PRACTICE PROBLEM FOR EXAMPLE 1

Simplify each radical expression.

a. $\sqrt{75}$ **b.** $\sqrt{72}$ **c.** $\sqrt{30}$

Another technique that can be used to simplify radical expressions involves factoring the radicand into prime factors. This technique is explored in the next example.

Example 2 Using prime factorization to simplify a radical

Factor each radicand into its prime factors. Then simplify the radical expression.

a. $\sqrt{24}$ **b.** $\sqrt{2520}$

SOLUTION

a. First find the prime factorization of 24. This can be done by using either a horizontal format or a factor tree.

Horizontal format: $\qquad 24 = 4 \cdot 6 = 2 \cdot 2 \cdot 2 \cdot 3 = 2^3 \cdot 3$

Factor tree:

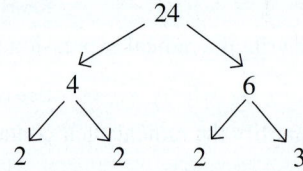

> **Connecting the Concepts**
>
> **What is another way to simplify radicals?**
>
> To simplify $\sqrt{24}$, we can look for repeated factors in the radicand.
>
> $\sqrt{24} = \sqrt{2 \cdot 2 \cdot 2 \cdot 3}$
> $\qquad = 2\sqrt{2 \cdot 3}$
> $\qquad = 2\sqrt{6}$
>
> This is because
>
> $\sqrt{2 \cdot 2} = \sqrt{4} = 2$

This problem involves a square root. Look for exponents that are multiples of 2 in the prime factorization.

$$24 = 4 \cdot 6 = 2 \cdot 2 \cdot 2 \cdot 3 = 2^3 \cdot 3 = 2^2 \cdot 2 \cdot 3$$

Rewriting the radicand and simplifying yields

$\sqrt{24}$ $\qquad$ Write the prime factorization of 24.
$= \sqrt{2^3 \cdot 3}$ $\qquad$ Look for exponents that are multiples of 2.
$= \sqrt{2^2 \cdot 2 \cdot 3}$ $\qquad$ Rewrite using the product property of radicals.
$= \sqrt{2^2} \cdot \sqrt{2 \cdot 3}$ $\qquad$ Simplify each radical.
$= 2\sqrt{6}$

b. First find the prime factorization of 2520. Another way to find the prime factorization is by repeated division. Recall from Chapter R that even numbers are divisible by 2, and that a number ending in 0 or 5 is divisible by 5.

The factorization is $2^3 \cdot 3^2 \cdot 5 \cdot 7$. To find the square root, first look for exponents that are multiples of 2 in the factorization.

$$2^3 \cdot 3^2 \cdot 5 \cdot 7 = 2 \cdot 2^2 \cdot 3^2 \cdot 5 \cdot 7$$

$2 \overline{)2520}$
$2 \overline{)1260}$
$2 \overline{)630}$
$5 \overline{)315}$
$3 \overline{)63}$
$3 \overline{)21}$
$\phantom{3 \overline{)}}7$

Rewriting the radicand and simplifying yields

$$\sqrt{2520}$$
$$= \sqrt{2 \cdot 2^2 \cdot 3^2 \cdot 5 \cdot 7}$$ *Write the prime factorization of 2520.*
$$= \sqrt{2^2} \cdot \sqrt{3^2} \cdot \sqrt{2 \cdot 5 \cdot 7}$$ *Rewrite, using the product property of radicals.*
$$= 2 \cdot 3\sqrt{70}$$ *Simplify each radical.*
$$= 6\sqrt{70}$$

> **What's That Mean?**
>
> **Simplify versus Approximate**
>
> *Simplify* means to find the exact answer in the form that is the least complicated.
>
> *Approximate* means to find a decimal approximation that is close to the exact answer but is not the exact answer.

To simplify a square root means to rewrite the radicand in a form that has no perfect square factors in it. Simplifying does *not* mean using a calculator to come up with a decimal estimation. The process of using the calculator to estimate the value is called *approximating* the radical.

Example 3   Simplifying versus approximating radical expressions

Simplify or approximate each radical expression, as indicated.

a. Simplify $\sqrt{20}$.

b. Approximate $\sqrt{20}$. Round to two decimal places.

SOLUTION

a. *Simplify* means to use the product property of radicals to factor out the perfect square factors of the radicand. Because the largest perfect square factor of 20 is 4, rewrite the radicand as $20 = 4 \cdot 5$.

$$\sqrt{20}$$
$$= \sqrt{4 \cdot 5}$$ *Rewrite 20 as 4 · 5.*
$$= \sqrt{4} \cdot \sqrt{5}$$ *Use the product property of radicals.*
$$= 2\sqrt{5}$$ *Simplify $\sqrt{4} = 2$.*

b. The instructions say to *approximate*, so use a calculator to come up with a numerical approximation.

$$\sqrt{20} \approx 4.472135955$$
$$\sqrt{20} \approx 4.47$$ *Round to two decimal places.*

PRACTICE PROBLEM FOR EXAMPLE 3

Simplify or approximate each radical expression, as indicated.

a. Approximate $\sqrt{45}$. Round to two decimal places.

b. Simplify $\sqrt{45}$.

When simplifying radicals with variables in the radicand, we also look for perfect square factors. However, with the radical part, the perfect square factors are the *even* powers of the variable.

696 CHAPTER 8 Radical Expressions and Equations

Perfect Square	Square Root of a Perfect Square
$(x)^2 = x^2$	$\sqrt{x^2} = x$
$(x^2)^2 = x^4$	$\sqrt{x^4} = x^2$
$(x^3)^2 = x^6$	$\sqrt{x^6} = x^3$
$(x^4)^2 = x^8$	$\sqrt{x^8} = x^4$

Skill Connection

Product rule

Recall from Section 5.1 that $x^m \cdot x^n = x^{m+n}$. In this section, we will use this rule in reverse. Therefore, $x^{m+n} = x^m \cdot x^n$.

What do you notice about this pattern? If a variable has an even power, its square root will simplify nicely to a form in which there are no radicals. However, to simplify a variable expression under a radical that involves an odd power on the variable, first factor out the perfect square factor (the even power). We do this because even powers of the variable are perfect squares, and they can be simplified to a form without radicals.

For example, to simplify $\sqrt{x^3}$, first rewrite the radicand with an even power (perfect square factor).

$$x^3 = x^2 \cdot x^1 \quad \text{Use the product rule for exponents: } 3 = 2 + 1.$$
$$x^3 = x^2 \cdot x \quad \text{Substitute } x \text{ for } x^1.$$

Using this factorization for x^3, namely, that $x^3 = x^2 \cdot x$, simplify the radical as follows.

$$\sqrt{x^3}$$
$$= \sqrt{x^2 \cdot x^1} \quad \text{Rewrite with an even power on } x.$$
$$= \sqrt{x^2} \cdot \sqrt{x} \quad \text{Use the product property of radicals.}$$
$$= x\sqrt{x} \quad \text{Simplify.}$$

To simplify an expression such as $\sqrt{16x^3}$, use several applications of the product property for radicals.

$$\sqrt{16x^3}$$
$$= \sqrt{16 \cdot x^2 \cdot x^1} \quad \text{Rewrite with an even power on } x.$$
$$= \sqrt{16} \cdot \sqrt{x^2} \cdot \sqrt{x} \quad \text{Use the product property of radicals.}$$
$$= 4x\sqrt{x} \quad \text{Simplify.}$$

Example 4 Simplifying radical expressions with variables

Simplify each radical expression.

a. $\sqrt{x^9}$ **b.** $\sqrt{18x^6}$ **c.** $\sqrt{700x^7}$

Simplifying Radicals

Factor the radicand into factors raised to powers that are a multiple of the index of the radical. Simplify the radicals. Multiply remaining radicands and multiply anything factored out.

$$\sqrt{20x^2y^7}$$
$$= \sqrt{2^2x^2y^6}\sqrt{5y}$$
$$= 2xy^3\sqrt{5y}$$

SOLUTION

a. $\sqrt{x^9}$
$$= \sqrt{x^8 \cdot x} \quad \text{Rewrite } x^9 = x^8 \cdot x^1 = x^8 \cdot x.$$
$$= \sqrt{x^8} \cdot \sqrt{x} \quad \text{Use the product property of radicals.}$$
$$= x^4\sqrt{x}$$

b. $\sqrt{18x^6}$
$$= \sqrt{9 \cdot 2 \cdot x^6} \quad \text{Rewrite the radicand with perfect square factors.}$$
$$= \sqrt{9} \cdot \sqrt{2} \cdot \sqrt{x^6} \quad \text{Use the product property of radicals.}$$
$$= 3\sqrt{2} \cdot x^3 \quad \text{Simplify.}$$
$$= 3x^3\sqrt{2} \quad \text{We usually put the radical part last.}$$

c. $\sqrt{700x^7}$
$= \sqrt{100 \cdot 7 \cdot x^6 \cdot x}$ Rewrite the radicand with perfect square factors.
$= \sqrt{100 \cdot x^6 \cdot 7 \cdot x}$ Use the commutative property of multiplication.
$= \sqrt{100}\sqrt{x^6}\sqrt{7x}$ Use the product property of radicals.
$= 10x^3\sqrt{7x}$ Simplify the perfect squares.

PRACTICE PROBLEM FOR EXAMPLE 4

Simplify each radical expression.

a. $\sqrt{y^{15}}$ **b.** $\sqrt{12a^2}$ **c.** $\sqrt{45x^{11}}$

To simplify, first factor out the perfect squares from the radical and simplify. Leave the other factors under the radical.

We can use the same ideas to simplify a radicand that has more than one variable.

Example 5 Simplifying radical expressions with two or more variables

Simplify each radical expression.

a. $\sqrt{x^2 y^4}$ **b.** $\sqrt{16x^5 y^2}$ **c.** $\sqrt{18x^3 y^5 z^2}$

SOLUTION

a. $\sqrt{x^2 y^4}$ Both x and y have even powers, so no rewriting is needed.
$= \sqrt{x^2} \cdot \sqrt{y^4}$ Use the product property of radicals.
$= xy^2$

b. $\sqrt{16x^5 y^2}$
$= \sqrt{16 \cdot x^4 \cdot x \cdot y^2}$ Rewrite the radicand with perfect square factors.
$= \sqrt{16} \cdot \sqrt{x^4} \cdot \sqrt{x} \cdot \sqrt{y^2}$ Use the product property of radicals.
$= 4x^2 \cdot \sqrt{x} \cdot y$ Simplify.
$= 4x^2 y \sqrt{x}$ Put the radical part last.

c. $\sqrt{18x^3 y^5 z^2}$
$= \sqrt{9 \cdot 2 \cdot x^2 \cdot x \cdot y^4 \cdot y \cdot z^2}$ Rewrite with perfect square factors.
$= \sqrt{9 \cdot x^2 \cdot y^4 \cdot z^2 \cdot 2 \cdot x \cdot y}$ Use the commutative property of multiplication.
$= \sqrt{9} \cdot \sqrt{x^2} \cdot \sqrt{y^4} \cdot \sqrt{z^2} \cdot \sqrt{2 \cdot x \cdot y}$ Use the product property of radicals.
$= 3xy^2 z \sqrt{2xy}$ Simplify the perfect squares.

PRACTICE PROBLEM FOR EXAMPLE 5

Simplify each radical expression.

a. $\sqrt{x^6 y^8}$ **b.** $\sqrt{12a^2 b^3}$ **c.** $\sqrt{45x^{11} y^2 z^3}$

■ Adding and Subtracting Radical Expressions

In Chapter 1, we learned how to add variable expressions by combining like terms. The basis for doing so is the distributive property.

$2x^2 + 5x^2$ Expression to be simplified.
$= (2 + 5)x^2$ Use the distributive property.
$= 7x^2$ Simplified form.

■ Connecting the Concepts

Why do we simplify radicals?

One reason we simplify a radical is to be able to look for like terms. Another reason we simplify radicals involves a close relationship between radicals and exponents. This relationship will be studied in intermediate algebra. Just as we simplify exponential expressions, we simplify radical expressions.

The shortcut used to add variable expressions is to add like terms by adding their coefficients rather than writing out the distributive property each time. Recall that terms are like when they have the same variables raised to the same powers. Adding coefficients yields $2x^2 + 5x^2 = 7x^2$.

Likewise to add or subtract terms involving radicals, we combine the coefficients of like terms, which now involve *like radicals*. Like radicals have the same variable, index on the radical, and radical part.

$2\sqrt{6} + 5\sqrt{6}$ Expression to be simplified.
$= (2 + 5)\sqrt{6}$ Use the distributive property.
$= 7\sqrt{6}$ Simplified form.

To add or subtract radicals, add or subtract terms with like radicals by adding or subtracting their coefficients.

$2\sqrt{6} + 5\sqrt{6}$ These are like radicals.
$= 7\sqrt{6}$ Add the coefficients.

> **Steps to Add or Subtract General Radical Expressions**
> 1. Simplify each radical by factoring out the perfect square factors.
> 2. Add or subtract terms with like radicals by combining their coefficients.

Connecting the Concepts

What are like terms and like radicals?

Like terms have the same variables raised to the same powers. Now that variables appear under the radical sign, like terms have to have the same radical part as well as the same variable part outside of the radical. This means that the index on the radical is the same, and the radicands are the same.

Like radicals:

$3x\sqrt{2x}$ and $-x\sqrt{2x}$

Not like radicals:

$5x\sqrt{6}$ and $3x\sqrt{6x}$

Example 6   Basic addition or subtraction of radicals

Simplify each radical expression by adding or subtracting like terms.

a. $3\sqrt{2} + \sqrt{2}$ **b.** $-5\sqrt{a} + 16\sqrt{a}$ **c.** $4\sqrt{x} + 2\sqrt{y}$

SOLUTION

a. Step 1 Simplify each radical by factoring out the perfect square factors.

All radicals are simplified.

Step 2 Add or subtract terms with like radicals by combining their coefficients.

$3\sqrt{2} + \sqrt{2}$ If there is no coefficient on a term, the coefficient is 1.
$= 3\sqrt{2} + 1\sqrt{2}$
$= 4\sqrt{2}$ Add like terms.

b. $-5\sqrt{a} + 16\sqrt{a}$ These are like terms, as they have the same radical part.
$= 11\sqrt{a}$ Add the coefficients.

c. The radical parts in these two terms are $\sqrt{x}$ and $\sqrt{y}$. These are not like radicals. This expression cannot be simplified further.

$4\sqrt{x} + 2\sqrt{y}$ This is the simplified form.

PRACTICE PROBLEM FOR EXAMPLE 6

Simplify each radical expression by adding or subtracting like terms.

a. $4\sqrt{21} - \sqrt{21}$ **b.** $-6\sqrt{x} - 7\sqrt{x}$ **c.** $-3\sqrt{b} + 3\sqrt{a}$

Example 7 Adding or subtracting radicals

Simplify each radical expression by adding or subtracting like terms.

a. $3\sqrt{5} + 8 - 7\sqrt{5}$ **b.** $9\sqrt{x} + 7\sqrt{2} - 3\sqrt{x} + 18\sqrt{2}$

c. $4x - 3x\sqrt{7} + 5x - 8x\sqrt{7}$

SOLUTION

a. $3\sqrt{5} + 8 - 7\sqrt{5}$

$= (3\sqrt{5}) + 8 (-7\sqrt{5})$ Like terms are in red.

$= -4\sqrt{5} + 8$ Combine the coefficients of the like terms.

b. $9\sqrt{x} + 7\sqrt{2} - 3\sqrt{x} + 18\sqrt{2}$ First identify like terms.

$= (9\sqrt{x}) + (7\sqrt{2}) - (3\sqrt{x}) + (18\sqrt{2})$ The variable and radical parts must be the same.

$= 6\sqrt{x} + 25\sqrt{2}$ Combine coefficients of like terms.

c. $4x - 3x\sqrt{7} + 5x - 8x\sqrt{7}$ Identify like terms.

$= (4x) - (3x\sqrt{7}) + (5x) - (8x\sqrt{7})$ Combine coefficients of like terms.

$= 9x - 11x\sqrt{7}$ These are not like terms.

The last part of this expression, $9x - 11x\sqrt{7}$, does not have any like terms. The variable parts are the same, but only the second term has $\sqrt{7}$. Since these terms do not have the same radical part, they are not like terms and cannot be combined.

PRACTICE PROBLEM FOR EXAMPLE 7

Simplify each radical expression by adding or subtracting like terms.

a. $-9 + 4\sqrt{11} + 16 - 3\sqrt{11}$ **b.** $3\sqrt{a} - 7\sqrt{5} - 5\sqrt{a} + 3\sqrt{5}$

c. $16y\sqrt{x} - 9xy + 3y\sqrt{x} + xy$

An expression sometimes has to be simplified first before like terms can be combined. For example, to simplify the expression $(5x)^2 + 2x^2$, first use the power of a product exponent rule to simplify the first term. Then add like terms.

$$(5x)^2 + 2x^2$$
$$= 5^2 x^2 + 2x^2$$
$$= 25x^2 + 2x^2$$
$$= 27x^2$$

> **Skill Connection**
>
> **Power rule for exponents**
>
> Recall from Section 5.1 the power rule for exponents.
>
> $$(x^m)^n = x^{m \cdot n}$$

With radical expressions, we also simplify first before combining like terms. Radical expressions must be first simplified to determine whether they have like terms when adding or subtracting them. If there is a perfect square factor in the radicand, first simplify using the product property of radicals. Then combine like terms.

To simplify the expression $\sqrt{36x^3} + 5x\sqrt{x}$, first simplify the radicals.

$\sqrt{36x^3} + 5x\sqrt{x}$ The first radicand can be simplified.

$= \sqrt{36 \cdot x^2 \cdot x} + 5x\sqrt{x}$ Rewrite, using perfect squares.

$= 6x\sqrt{x} + 5x\sqrt{x}$ Use the product property of radicals.

$= 11x\sqrt{x}$ Combine like terms.

Example 8 Simplifying to combine like terms

Simplify each radical expression by first simplifying the radical part. Then add or subtract like terms.

a. $\sqrt{32} + 5\sqrt{2}$

b. $\sqrt{27} - 5\sqrt{12}$

SOLUTION

a. $\sqrt{32} + 5\sqrt{2}$
$= \sqrt{16 \cdot 2} + 5\sqrt{2}$ Rewrite the first radical with a perfect square factor.
$= \sqrt{16} \cdot \sqrt{2} + 5\sqrt{2}$ Use the product property of radicals.
$= 4\sqrt{2} + 5\sqrt{2}$ Simplify $\sqrt{16} = 4$.
$= 9\sqrt{2}$ Add like terms.

b. $\sqrt{27} - 5\sqrt{12}$ First simplify both radicals.
$= \sqrt{9 \cdot 3} - 5 \cdot \sqrt{4 \cdot 3}$ Rewrite the radicals with perfect square factors.
$= \sqrt{9} \cdot \sqrt{3} - 5 \cdot \sqrt{4} \cdot \sqrt{3}$ Use the product property of radicals.
$= 3\sqrt{3} - 5 \cdot 2 \cdot \sqrt{3}$ Simplify the square roots of the perfect square factors.
$= 3\sqrt{3} - 10\sqrt{3}$ These are like terms.
$= -7\sqrt{3}$

PRACTICE PROBLEM FOR EXAMPLE 8

Simplify each radical expression by first simplifying the radical part. Then add or subtract like terms.

a. $\sqrt{28} + 6\sqrt{7}$

b. $-2\sqrt{20} + \sqrt{45}$

When variables are involved in the expression, the procedure is the same. Simplify each term first and then combine like terms.

Example 9 Combining general radical expressions

Simplify each radical expression by first simplifying the radical part. Then add or subtract like terms.

a. $k\sqrt{45} - k\sqrt{5}$

b. $\sqrt{32x^2} + x\sqrt{2}$

c. $\sqrt{25a^3} + 3a\sqrt{a} - 9 + \sqrt{64}$

SOLUTION

a. $k\sqrt{45} - k\sqrt{5}$ The first radical can be simplified.
$= k\sqrt{9 \cdot 5} - k\sqrt{5}$ Factor 45 as 9 · 5.
$= 3k\sqrt{5} - k\sqrt{5}$ Simplify the radical using the product property of radicals.
$= 2k\sqrt{5}$ Combine like terms.

b. $\sqrt{32x^2} + x\sqrt{2}$ The first radical can be simplified.
$= \sqrt{16 \cdot 2 \cdot x^2} + x\sqrt{2}$ Factor 32 as 16 · 2.
$= 4x\sqrt{2} + x\sqrt{2}$ Simplify the radical using the product property of radicals.
$= 5x\sqrt{2}$ Combine like terms.

c. $\sqrt{25a^3} + 3a\sqrt{a} - 9 + \sqrt{64}$ *The first and last radicals can be simplified.*
$= \sqrt{25 \cdot a^2 \cdot a} + 3a\sqrt{a} - 9 + 8$ *Rewrite the first radical in terms of perfect squares.*
$= 5a\sqrt{a} + 3a\sqrt{a} - 1$ *Combine like terms.*
$= 8a\sqrt{a} - 1$

PRACTICE PROBLEM FOR EXAMPLE 9

Simplify each radical expression by first simplifying the radical part. Then add or subtract like terms.

a. $-y\sqrt{48} + y\sqrt{12}$
b. $\sqrt{8b^2} + b\sqrt{50}$
c. $\sqrt{64x^5} + \sqrt{81} - 7\sqrt{x^5} - \sqrt{36}$

8.2 Vocabulary Exercises

1. When simplifying radicals, the _____ property is used to factor out perfect squares under the radicand.

2. Like terms means the terms have the same variable part, including the same _____ part.

3. Writing $\sqrt{5} \approx 2.24$ is a(n) _____ of the exact value.

8.2 Exercises

For Exercises 1 through 18, simplify each radical expression.

1. $\sqrt{18}$
2. $\sqrt{32}$
3. $\sqrt{125}$
4. $\sqrt{150}$
5. $\sqrt{54}$
6. $\sqrt{63}$
7. $\sqrt{80}$
8. $\sqrt{112}$
9. $\sqrt{98}$
10. $\sqrt{300}$
11. $\sqrt{75}$
12. $\sqrt{40}$
13. $\sqrt{34}$
14. $\sqrt{42}$
15. $\sqrt{148}$
16. $\sqrt{135}$
17. $\sqrt{288}$
18. $\sqrt{338}$

For Exercises 19 through 26, simplify each radical expression and approximate to two decimal places.

19. $\sqrt{50}$
   a. Simplify.
   b. Approximate to two decimal places.

20. $\sqrt{18}$
   a. Simplify.
   b. Approximate to two decimal places.

21. $\sqrt{12}$
   a. Simplify.
   b. Approximate to two decimal places.

22. $\sqrt{40}$
   a. Simplify.
   b. Approximate to two decimal places.

23. $\sqrt{28}$
   a. Simplify.
   b. Approximate to two decimal places.

24. $\sqrt{175}$
   a. Simplify.
   b. Approximate to two decimal places.

25. $\sqrt{54}$
   a. Simplify.
   b. Approximate to two decimal places.
26. $\sqrt{90}$
   a. Simplify.
   b. Approximate to two decimal places.

For Exercises 27 through 36, simplify or approximate each radical expression as directed. Approximate to two decimal places as needed.

27. Approximate: $\sqrt{32}$
28. Simplify: $\sqrt{44}$
29. Simplify: $\sqrt{45}$
30. Approximate: $\sqrt{13}$
31. Approximate: $\sqrt{37}$
32. Simplify: $\sqrt{75}$
33. Simplify: $\sqrt{72}$
34. Approximate: $\sqrt{30}$
35. Approximate: $\sqrt{45}$
36. Simplify: $\sqrt{40}$

For Exercises 37 through 60, simplify each radical expression. Assume that all variables under radicals represent positive values.

37. $\sqrt{b^9}$
38. $\sqrt{b^3}$
39. $\sqrt{a^4}$
40. $\sqrt{x^{12}}$
41. $\sqrt{x^{15}}$
42. $\sqrt{x^{11}}$
43. $\sqrt{y^2}$
44. $\sqrt{r^6}$
45. $\sqrt{4x^3}$
46. $\sqrt{9y^3}$
47. $\sqrt{16x^6}$
48. $\sqrt{25y^6}$
49. $\sqrt{64a^3}$
50. $\sqrt{25a^5}$
51. $\sqrt{49a^4}$
52. $\sqrt{100a^{10}}$
53. $\sqrt{45a^4}$
54. $\sqrt{50a^2}$
55. $\sqrt{60y^6}$
56. $\sqrt{200y^{12}}$
57. $\sqrt{40x^3}$
58. $\sqrt{48x^5}$
59. $\sqrt{75b^{11}}$
60. $\sqrt{288b^{15}}$

For Exercises 61 through 80, simplify each radical expression by adding or subtracting like terms. Assume that all variables under radicals represent positive values.

61. $-3\sqrt{5} + \sqrt{5}$
62. $\sqrt{2} - 3\sqrt{2}$
63. $11\sqrt{3} + 7\sqrt{3} - 6\sqrt{3}$
64. $-5\sqrt{7} + \sqrt{7} - 22\sqrt{7}$
65. $2\sqrt{3} + 3\sqrt{2}$
66. $-5\sqrt{6} - 6\sqrt{5}$
67. $5\sqrt{3} + 3\sqrt{5}$
68. $8\sqrt{2} + \sqrt{7}$
69. $2\sqrt{x} - 7\sqrt{2}$
70. $2\sqrt{5} + 7\sqrt{a}$
71. $8\sqrt{a} - 9\sqrt{a}$
72. $-6\sqrt{x} + \sqrt{x}$
73. $2\sqrt{a} + 2\sqrt{b}$
74. $4\sqrt{x} - 3\sqrt{y}$
75. $3x\sqrt{2} + 2y\sqrt{5}$
76. $-2a\sqrt{3} - 4b\sqrt{11}$
77. $4a\sqrt{b} + 5b - a\sqrt{b} + b$
78. $-3x + 2\sqrt{xy} - 5\sqrt{xy} + x$
79. $-15 + 3x\sqrt{3} - \sqrt{3} - 17 - 2x\sqrt{3}$
80. $-6y\sqrt{7} - 9 + 3y\sqrt{7} - 8$

For Exercises 81 through 86, find the perimeter of the geometric shape. Simplify.

81.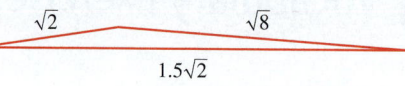

82.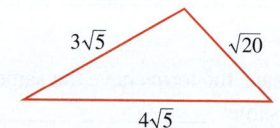

83.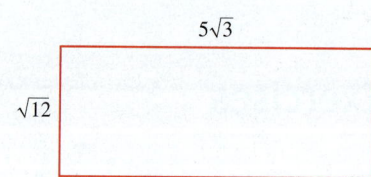

84.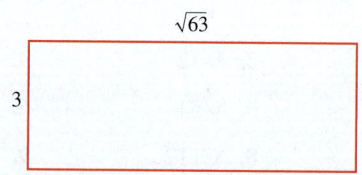

85.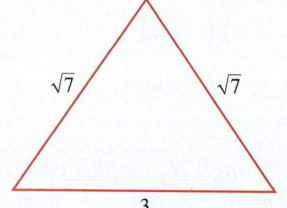

86.

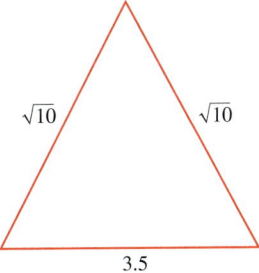

For Exercises 87 through 104, simplify each radical expression by first simplifying all radicals and then adding or subtracting like terms. Assume that all variables under radicals represent positive values.

87. $7\sqrt{2} + \sqrt{8}$

88. $6\sqrt{5} - \sqrt{20}$

89. $4\sqrt{20} - 3\sqrt{18} + 5\sqrt{32}$

90. $7\sqrt{24} - \sqrt{48} + 3\sqrt{54}$

91. $\sqrt{49x^2} - 5x$

   a. Simplify all radicals in each term.
   b. Add or subtract like terms.

92. $9y + \sqrt{16y^2}$

   a. Simplify all radicals in each term.
   b. Add or subtract like terms.

93. $-4a^2 - \sqrt{16a^4}$

   a. Simplify all radicals in each term.
   b. Add or subtract like terms.

94. $-\sqrt{25x^6} + 8x^3$

   a. Simplify all radicals in each term.
   b. Add or subtract like terms.

95. $\sqrt{16x^3} - x\sqrt{25x}$

   a. Simplify all radicals in each term.
   b. Add or subtract like terms.

96. $y^2\sqrt{100y} - \sqrt{25y^5}$

   a. Simplify all radicals in each term.
   b. Add or subtract like terms.

97. $-\sqrt{16} + \sqrt{4x^3} + \sqrt{25x^3}$

   a. Simplify all radicals in each term.
   b. Add or subtract like terms.

98. $\sqrt{81a^5} + \sqrt{36a^5} - \sqrt{9}$

   a. Simplify all radicals in each term.
   b. Add or subtract like terms.

99. $-x\sqrt{32} + x\sqrt{50}$

   a. Simplify all radicals in each term.
   b. Add or subtract like terms.

100. $y^2\sqrt{72} + y^2\sqrt{98}$

   a. Simplify all radicals in each term.
   b. Add or subtract like terms.

101. $\sqrt{25} + \sqrt{12x^3} - 8 + \sqrt{27x^3}$

   a. Simplify all radicals in each term.
   b. Add or subtract like terms.

102. $\sqrt{18a^5} - \sqrt{64b^4} + 9b^2 - a^2\sqrt{2a}$

   a. Simplify all radicals in each term.
   b. Add or subtract like terms.

103. $\sqrt{64y^3} - \sqrt{50x^2} - \sqrt{25y^3} + \sqrt{18x^2}$

   a. Simplify all radicals in each term.
   b. Add or subtract like terms.

104. $\sqrt{100a^5} - \sqrt{45b^4} - \sqrt{36a^5} + \sqrt{80b^4}$

   a. Simplify all radicals in each term.
   b. Add or subtract like terms.

For Exercises 105 through 132, simplify the given expressions. Assume that all variables under radicals represent positive values. If the square root does not exist, write "Not real" as the answer.

105. $-3\sqrt{7} + 9\sqrt{7}$

106. $5\sqrt{13} - 13\sqrt{13}$

107. $-3\sqrt{5} + 5\sqrt{3}$

108. $11\sqrt{2} + 2\sqrt{11}$

109. $\sqrt{98}$

110. $\sqrt{128}$

111. $\sqrt{x^{20}}$

112. $\sqrt{y^{16}}$

113. $\sqrt{27a^5}$

114. $\sqrt{32b^7}$

115. $\sqrt{36y^2} + 7y$

116. $6x - \sqrt{49x^2}$

117. $\sqrt{100b^4} + 5b^2 - 6b$

118. $25x^2 + \sqrt{49x^4} - 6x$

119. $\sqrt{a^9}$

120. $\sqrt{k^7}$

121. $\sqrt{8b^3}$

122. $\sqrt{27b^5}$

123. $\sqrt{50x^7}$

124. $\sqrt{98x^5}$

125. $6a^3 + \sqrt{81a^6}$

126. $\sqrt{16y^2} + y$

127. $\sqrt{-8}$

128. $\sqrt{-20}$

129. $\sqrt[3]{-8}$

130. $\sqrt[3]{-27}$

131. Approximate $\sqrt{35}$ to two decimal places.

132. Approximate $\sqrt{26}$ to two decimal places.

8.3 Multiplying and Dividing Radical Expressions

LEARNING OBJECTIVES
- Multiply radical expressions.
- Divide radical expressions.
- Rationalize the denominator.

■ Multiplying Radical Expressions

In Section 8.2, we learned the product property for radicals and used it to simplify when adding and subtracting radical expressions. We will now use that rule in the reverse order to multiply and divide radical expressions. Since the product rule states that $\sqrt{a \cdot b} = \sqrt{a} \cdot \sqrt{b}$, it is also true that $\sqrt{a} \cdot \sqrt{b} = \sqrt{a \cdot b}$. This means that when multiplying radicals, rewrite the product as a single radical, multiply the two radicands, and then simplify.

> **Steps to Multiply Radical Expressions**
> 1. Use the product property of radicals to rewrite as a single radical expression; $\sqrt{a} \cdot \sqrt{b} = \sqrt{a \cdot b}$. Here both a and b are values that are positive or zero.
> 2. Simplify the radicand by factoring into perfect square factors. Factor out the perfect squares.

Example 1 Multiplying radical expressions

Multiply the radicals and then simplify.

a. $\sqrt{3} \cdot \sqrt{27}$ b. $\sqrt{8} \cdot \sqrt{3}$ c. $\sqrt{5} \cdot \sqrt{10}$

SOLUTION

a. **Step 1** Use the product property of radicals to rewrite as a single radical expression.

$\sqrt{3} \cdot \sqrt{27}$
$= \sqrt{3 \cdot 27}$ *Multiply, using the product property of radicals.*

Step 2 Simplify the radicand by factoring into perfect square factors. Factor out the perfect squares.

$= \sqrt{81} = 9$ *Simplify the radical.*

b. $\sqrt{8} \cdot \sqrt{3}$
$= \sqrt{8 \cdot 3}$ *Multiply, using the product property of radicals.*
$= \sqrt{24}$ *Rewrite with a perfect square factor.*
$= \sqrt{4 \cdot 6}$ *Simplify.*
$= 2\sqrt{6}$

c. $\sqrt{5} \cdot \sqrt{10}$
$= \sqrt{5 \cdot 10}$ *Use the product property of radicals.*
$= \sqrt{50}$ *Rewrite with a perfect square factor.*
$= \sqrt{25 \cdot 2}$ *Simplify.*
$= 5\sqrt{2}$

PRACTICE PROBLEM FOR EXAMPLE 1

Multiply the radicals, and then simplify.

a. $\sqrt{12} \cdot \sqrt{3}$ b. $\sqrt{5} \cdot \sqrt{20}$ c. $\sqrt{6} \cdot \sqrt{2}$ d. $\sqrt{8} \cdot \sqrt{10}$

When there are variables involved, follow the same procedure to multiply and simplify. This means to first rewrite as a single radical and then multiply the radicands. Next rewrite the single radicand in terms of any perfect square factors. Finally, simplify the perfect square parts of the radical.

Example 2 Multiplying radicals involving variables

Multiply the radicals and then simplify.

a. $\sqrt{2x} \cdot \sqrt{18x}$
b. $\sqrt{5x^2} \cdot \sqrt{10x^3}$

SOLUTION

a. $\sqrt{2x} \cdot \sqrt{18x}$
$= \sqrt{2x \cdot 18x}$ Use the product property of radicals.
$= \sqrt{36x^2}$ Rewrite with perfect square factors.
$= \sqrt{36} \cdot \sqrt{x^2}$ Simplify.
$= 6x$

b. $\sqrt{5x^2} \cdot \sqrt{10x^3}$
$= \sqrt{5x^2 \cdot 10x^3}$
$= \sqrt{50x^5}$ Multiply the radicands.
$= \sqrt{25 \cdot 2 \cdot x^4 \cdot x}$ Simplify under the radical.
$= 5x^2\sqrt{2x}$

PRACTICE PROBLEM FOR EXAMPLE 2

Multiply the radicals and then simplify.

a. $\sqrt{8x} \cdot \sqrt{8x^3}$ b. $\sqrt{3a^2} \cdot \sqrt{21a^5}$

CONCEPT INVESTIGATION
What pattern do you notice?

1. Multiply each of the following radicals and then simplify. The first one has been done to help you get started.

   a. $\sqrt{7} \cdot \sqrt{7} = ?$ $\sqrt{7} \cdot \sqrt{7} = \sqrt{7 \cdot 7} = \sqrt{49} = 7$
   b. $\sqrt{x} \cdot \sqrt{x} = ?$
   c. $\sqrt{5a} \cdot \sqrt{5a} = ?$
   d. $\sqrt{x^3} \cdot \sqrt{x^3} = ?$

2. Use your calculator to complete the following table. The first row has been done to get you started.

Expression	Simplified Expression	Input	Evaluate the Expression Using a Calculator	Evaluate the Simplified Expressions Using a Calculator
$\sqrt{y} \cdot \sqrt{y}$	$\sqrt{y^2}$	$y = 3$	$\sqrt{3} \cdot \sqrt{3} = 3$	$\sqrt{3^2} = 3$
$\sqrt{x} \cdot \sqrt{x}$		$x = 5$		
$\sqrt{2y} \cdot \sqrt{2y}$		$y = 7$		
$\sqrt{6x} \cdot \sqrt{6x}$		$x = 8$		

3. What pattern do you notice? When you multiply a radical by itself, what is the result?

We saw that $\sqrt{7} \cdot \sqrt{7} = \sqrt{7 \cdot 7} = \sqrt{49} = 7$. The original problem can also be rewritten by using exponents: $\sqrt{7} \cdot \sqrt{7} = (\sqrt{7})^2$. These two expressions must both equal 7, so $\sqrt{7} \cdot \sqrt{7} = (\sqrt{7})^2 = 7$. In words, when we square a square root, we are left with the original number or expression under the radical.

> **SQUARING A SQUARE ROOT**
> If a is positive or zero, then
> $$\sqrt{a} \cdot \sqrt{a} = \sqrt{a \cdot a} = a$$
> or
> $$\sqrt{a} \cdot \sqrt{a} = (\sqrt{a})^2 = a$$

Example 3 Radicals multiplied by themselves

Multiply the radicals and then simplify.

a. $\sqrt{11v} \cdot \sqrt{11v}$ **b.** $\sqrt{16} \cdot \sqrt{16}$ **c.** $(\sqrt{3x})^2$ **d.** $(4\sqrt{x})^2$

SOLUTION
a. $\sqrt{11v} \cdot \sqrt{11v} = 11v$
b. $\sqrt{16} \cdot \sqrt{16} = 16$
c. $(\sqrt{3x})^2 = \sqrt{3x} \cdot \sqrt{3x} = 3x$
d. $(4\sqrt{x})^2$
  $= 4\sqrt{x} \cdot 4\sqrt{x}$ To square something means to multiply it by itself.
  $= 4 \cdot 4 \cdot \sqrt{x} \cdot \sqrt{x}$ Use the commutative property of multiplication.
  $= 16\sqrt{x^2}$ Multiply the numbers and the radicals.
  $= 16x$

PRACTICE PROBLEM FOR EXAMPLE 3
Multiply the radicals, and then simplify.

a. $\sqrt{5y} \cdot \sqrt{5y}$ **b.** $(\sqrt{15a})^2$ **c.** $(6\sqrt{b})^2$

In Example 3d, we multiplied $(4\sqrt{x})^2$. Another way to approach this problem is to apply exponent rules.

$$(4\sqrt{x})^2$$
$$= (4 \cdot \sqrt{x})^2 \qquad \text{Use the power of a product exponent rule.}$$
$$= 4^2 \cdot (\sqrt{x})^2 \qquad \text{Square each factor.}$$
$$= 16x$$

We can multiply expressions with different coefficients. To multiply $3\sqrt{x} \cdot 5\sqrt{x}$, use the commutative property as follows.

$$3\sqrt{x} \cdot 5\sqrt{x}$$
$$= 3 \cdot 5 \cdot \sqrt{x} \cdot \sqrt{x} \qquad \text{Use the commutative property of multiplication.}$$
$$= 15\sqrt{x^2} \qquad \text{Use the product property of radicals.}$$
$$= 15x \qquad \text{Simplify.}$$

Example 4 Multiplying radicals with different coefficients

Multiply each expression, and simplify.

a. $2\sqrt{x^3} \cdot (-4\sqrt{x})$

b. $(-6\sqrt{8a})(-5\sqrt{3a})$

SOLUTION

a. To multiply this expression, we make use of the commutative property of multiplication.

$$2\sqrt{x^3} \cdot (-4\sqrt{x})$$
$$= 2 \cdot (-4) \cdot \sqrt{x^3} \cdot \sqrt{x} \qquad \text{Use the commutative property of multiplication.}$$
$$= -8\sqrt{x^3 \cdot x} \qquad \text{Recall that } x^3 \cdot x^1 = x^{3+1} = x^4.$$
$$= -8\sqrt{x^4} \qquad \text{Recall that } \sqrt{x^4} = \sqrt{(x^2)^2} = x^2.$$
$$= -8x^2$$

b. Using the commutative property of multiplication yields

$$(-6\sqrt{8a})(-5\sqrt{3a})$$
$$= (-6)(-5) \cdot \sqrt{8a} \cdot \sqrt{3a} \qquad \text{Use the commutative property of multiplication.}$$
$$= 30\sqrt{8a \cdot 3a} \qquad \text{Use the product property of radicals.}$$
$$= 30\sqrt{24a^2}$$
$$= 30\sqrt{4 \cdot 6 \cdot a^2} \qquad \text{Simplify the radical using perfect square factors.}$$
$$= 30 \cdot 2 \cdot a\sqrt{6}$$
$$= 60a\sqrt{6}$$

PRACTICE PROBLEM FOR EXAMPLE 4

Multiply each expression, and simplify.

a. $-3\sqrt{y} \cdot (-5\sqrt{y})$ **b.** $(2\sqrt{6b})(-9\sqrt{3b^3})$

The distributive property shows up when multiplying radical expressions that involve multiplying a single term by two or more terms. To multiply $\sqrt{7}(\sqrt{2} + \sqrt{7})$, we use the distributive property.

> **Skill Connection**
>
> **The distributive property**
> Recall from Section 1.3 that we use the distributive property to multiply a single term by an expression that contains two or more terms.
> Therefore,
>
> $$2(3x^2 - 5x + 7)$$
> $$= 2 \cdot 3x^2 - 2 \cdot 5x + 2 \cdot 7$$
> $$= 6x^2 - 10x + 14$$

708 CHAPTER 8 Radical Expressions and Equations

$$\sqrt{7}(\sqrt{2} + \sqrt{7}) = \sqrt{7} \cdot \sqrt{2} + \sqrt{7} \cdot \sqrt{7}$$ Use the distributive property.
$$= \sqrt{14} + \sqrt{49}$$ Use the product property of radicals.
$$= \sqrt{14} + 7$$ Simplify each term.
$$= 7 + \sqrt{14}$$

Example 5   Multiplying radicals by using the distributive property

Multiply each expression, and simplify.

a. $\sqrt{5}(\sqrt{x} - \sqrt{5})$ **b.** $\sqrt{2}(\sqrt{8} + \sqrt{6})$ **c.** $3(2\sqrt{a} - 4)$

SOLUTION

a.
$$\sqrt{5}(\sqrt{x} - \sqrt{5})$$
$$= \sqrt{5} \cdot \sqrt{x} - \sqrt{5} \cdot \sqrt{5}$$ Use the distributive property.
$$= \sqrt{5x} - 5$$ Recall that $\sqrt{5} \cdot \sqrt{5} = 5$.

b.
$$\sqrt{2}(\sqrt{8} + \sqrt{6})$$
$$= \sqrt{2} \cdot \sqrt{8} + \sqrt{2} \cdot \sqrt{6}$$ Use the distributive property.
$$= \sqrt{16} + \sqrt{12}$$
$$= 4 + \sqrt{4 \cdot 3}$$ Simplify both terms.
$$= 4 + 2\sqrt{3}$$

c.
$$3(2\sqrt{a} - 4) = 3 \cdot 2\sqrt{a} - 3 \cdot 4$$
$$= 6\sqrt{a} - 12$$

PRACTICE PROBLEM FOR EXAMPLE 5

Multiply each expression, and simplify.

a. $\sqrt{10}(\sqrt{y} + \sqrt{10})$ **b.** $\sqrt{2}(\sqrt{10} - \sqrt{32})$ **c.** $-2(4\sqrt{x} - 1)$

In Chapter 5, we used FOIL to multiply two expressions that each had two terms. In multiplying two radical expressions when each has two terms, we will still use the FOIL method.

Example 6   Multiplying radicals by using FOIL

Multiply each expression and simplify.

a. $(\sqrt{5} - 3)(\sqrt{5} + \sqrt{8})$ **b.** $(\sqrt{x} - 3)(2\sqrt{x} + 7)$ **c.** $(3 + \sqrt{2})^2$

SOLUTION

a. $(\sqrt{5} - 3)(\sqrt{5} + \sqrt{8})$
$$= \sqrt{5} \cdot \sqrt{5} + \sqrt{5} \cdot \sqrt{8} - 3 \cdot \sqrt{5} - 3 \cdot \sqrt{8}$$ Use FOIL to multiply.
$$= \sqrt{25} + \sqrt{40} - 3\sqrt{5} - 3\sqrt{8}$$ Multiply within each term.
$$= 5 + \sqrt{4 \cdot 10} - 3\sqrt{5} - 3\sqrt{4 \cdot 2}$$ Simplify radicals within each term.
$$= 5 + 2\sqrt{10} - 3\sqrt{5} - 3 \cdot 2\sqrt{2}$$
$$= 5 + 2\sqrt{10} - 3\sqrt{5} - 6\sqrt{2}$$ There are no like terms to combine.

b. $(\sqrt{x} - 3)(2\sqrt{x} + 7)$
$= \sqrt{x} \cdot 2\sqrt{x} + \sqrt{x} \cdot 7 - 3 \cdot 2\sqrt{x} - 3 \cdot 7$ Use FOIL to multiply the expressions.
$= 2\sqrt{x \cdot x} + 7\sqrt{x} - 6\sqrt{x} - 21$ Multiply each term.
$= 2x + 7\sqrt{x} - 6\sqrt{x} - 21$ Simplify each term.
$= 2x + 1\sqrt{x} - 21$ Combine like terms.
$= 2x + \sqrt{x} - 21$ Recall that $1\sqrt{x} = \sqrt{x}$.

c. $(3 + \sqrt{2})^2$
$= (3 + \sqrt{2})(3 + \sqrt{2})$ To square means to multiply the expression by itself.
$= 3 \cdot 3 + 3 \cdot \sqrt{2} + 3\sqrt{2} + \sqrt{2} \cdot \sqrt{2}$ Use FOIL to multiply the expressions.
$= 9 + 3\sqrt{2} + 3\sqrt{2} + 2$ Multiply each term.
$= 11 + 6\sqrt{2}$ Add like terms.

> **Skill Connection**
>
> **Using FOIL**
> Recall from Section 5.4 that we use the distributive property to multiply an expression with two terms by an expression with two terms. We also learned FOIL, a handy acronym. FOIL stands for First, Outer, Inner, Last.
> Therefore,
>
>
>
> $(3x + 5)(2x + 1)$
> $= 3x \cdot 2x + 3x \cdot 1 + 5 \cdot 2x + 5 \cdot 1$
> $= 6x^2 + 3x + 10x + 5$
> $= 6x^2 + 13x + 5$

PRACTICE PROBLEM FOR EXAMPLE 6

Multiply each expression and simplify.

a. $(\sqrt{3} + 9)(\sqrt{5} - 2)$ **b.** $(\sqrt{7} + 4)(\sqrt{7} - \sqrt{10})$
c. $(\sqrt{y} + 6)(\sqrt{y} - 5)$ **d.** $(x - \sqrt{5})^2$

Recall that when simplifying expressions such as $(5x)^2$ and $(5 + x)^2$, we paid careful attention to which operation was used in the parentheses. This operation determines the multiplication rule to be used.

To simplify $(5x)^2$, use the power of a product rule for exponents.

$$(5x)^2 = (5 \cdot x)^2 = 5^2 x^2 = 25x^2$$

To simplify $(5 + x)^2$, use the distributive property (FOIL).

$$(5 + x)^2 = (5 + x)(5 + x) = 25 + 5x + 5x + x^2 = 25 + 10x + x^2$$

We will pay the same careful attention when there is a radical inside the parentheses.

Example 7   Which multiplication rule should be used?

Multiply each expression, and simplify.

a. $(7\sqrt{x})^2$ **b.** $(7 + \sqrt{x})^2$

SOLUTION

a. This is the square of a *product*.

$$(7\sqrt{x})^2 = (7\sqrt{x})(7\sqrt{x}) = 49\sqrt{x^2} = 49x$$

This problem can also be multiplied out by using the powers of products rule for exponents.

$$(7\sqrt{x})^2 = 7^2(\sqrt{x})^2 = 49x$$

b. This is the square of a *binomial*. Use the distributive property (FOIL).

$(7 + \sqrt{x})^2 = (7 + \sqrt{x})(7 + \sqrt{x})$
$= 49 + 7\sqrt{x} + 7\sqrt{x} + \sqrt{x} \cdot \sqrt{x}$ Use FOIL to multiply the expressions.
$= 49 + 14\sqrt{x} + x$

PRACTICE PROBLEM FOR EXAMPLE 7

Multiply each expression, and simplify.

a. $(3\sqrt{a})^2$ **b.** $(3 + \sqrt{a})^2$

An interesting pattern shows up when we multiply two special kinds of radical expressions. Let's take a look.

Example 8 Multiplying radical expressions with two terms

Multiply each expression and simplify.

a. $(2 + \sqrt{3})(2 - \sqrt{3})$ **b.** $(5 + \sqrt{a})(5 - \sqrt{a})$

SOLUTION

a. To multiply two radical expressions that both involve two terms, use the FOIL method as above.

$(2 + \sqrt{3})(2 - \sqrt{3})$
$= 2 \cdot 2 - 2 \cdot \sqrt{3} + \sqrt{3} \cdot 2 - \sqrt{3} \cdot \sqrt{3}$ Use FOIL to multiply.
$= 4 - 2\sqrt{3} + 2\sqrt{3} - 3$ Use the commutative property: $\sqrt{3} \cdot 2 = 2\sqrt{3}$.
$= 4 - 2\sqrt{3} + 2\sqrt{3} - 3$ Add like terms.
$= 4 - 3$ Add like terms.
$= 1$

b. To multiply two radical expressions when both involve two terms uses the FOIL method as above.

$(5 + \sqrt{a})(5 - \sqrt{a})$
$= 5 \cdot 5 - 5 \cdot \sqrt{a} + \sqrt{a} \cdot 5 - \sqrt{a} \cdot \sqrt{a}$ Use FOIL to multiply.
$= 25 - 5\sqrt{a} + 5\sqrt{a} - a$ Use the commutative property: $\sqrt{a} \cdot 5 = 5\sqrt{a}$.
$= 25 - 5\sqrt{a} + 5\sqrt{a} - a$ Add like terms.
$= 25 - a$

PRACTICE PROBLEM FOR EXAMPLE 8

Multiply each expression and simplify.

a. $(\sqrt{5} + 3)(\sqrt{5} - 3)$

b. $(\sqrt{x} + 2)(\sqrt{x} - 2)$

Radical expressions that involve two terms and differ only by the sign between the first and second terms are called **conjugates** of each other. The expressions $2 + \sqrt{3}$ and $2 - \sqrt{3}$ are conjugates. In Example 7, we saw that when a radical expression is multiplied by its conjugate, we are left with the squares of the first and last terms only, with no radicals remaining.

> **DEFINITION**
>
> The **conjugate** of $a + b$ is $a - b$.
>
> The conjugate of $2 + \sqrt{3}$ is $2 - \sqrt{3}$.

SECTION 8.3 Multiplying and Dividing Radical Expressions

Example 9 — Multiplying an expression by its conjugate

Consider the radical expression $5 + \sqrt{7}$.

a. Find the conjugate of $5 + \sqrt{7}$.

b. Multiply the expression by its conjugate.

c. Are there any radicals remaining in the product?

SOLUTION

a. The conjugate of $5 + \sqrt{7}$ is $5 - \sqrt{7}$.

b. Multiplying $5 + \sqrt{7}$ by its conjugate yields

$$(5 + \sqrt{7})(5 - \sqrt{7})$$
$$= 5 \cdot 5 - 5\sqrt{7} + 5\sqrt{7} - \sqrt{7} \cdot \sqrt{7}$$
$$= 25 - 7$$
$$= 18$$

c. The product is 18. No radicals remain.

PRACTICE PROBLEM FOR EXAMPLE 9

Consider the radical expression $2 - \sqrt{5}$.

a. Find the conjugate of $2 - \sqrt{5}$.

b. Multiply the expression by its conjugate.

c. Are there any radicals remaining in the product?

Examples 8 and 9 demonstrate that when we multiply a radical expression with two terms by its conjugate, there are no radicals left in the product. We make use of this idea later in this section.

■ Dividing Radical Expressions

■ CONCEPT INVESTIGATION
How do we divide radicals?

1. Fill out the following table. The first row has been done to get you started.

Column 1	Column 2	Are columns 1 and 2 the same?
$\sqrt{\dfrac{36}{4}} = \sqrt{9} = 3$	$\dfrac{\sqrt{36}}{\sqrt{4}} = \dfrac{6}{2} = 3$	Yes
$\sqrt{\dfrac{16}{4}} = \sqrt{} = $	$\dfrac{\sqrt{16}}{\sqrt{4}} = \dfrac{}{} = $	
$\sqrt{\dfrac{100}{25}} = \sqrt{} = $	$\dfrac{\sqrt{100}}{\sqrt{25}} = \dfrac{}{} = $	

2. Look at the patterns from part 1. In column 1, you first divided the values under the radical and then evaluated the square root. In column 2, the square roots of the numerator and denominator were found separately, and then those results were divided. Were the results the same?

712 CHAPTER 8 Radical Expressions and Equations

In this Concept Investigation, we noticed that $\sqrt{\frac{36}{4}} = \frac{\sqrt{36}}{\sqrt{4}}$. What we have observed in this Concept Investigation is called the **quotient property of radicals**.

> **THE QUOTIENT PROPERTY OF RADICALS**
> If $a \geq 0$ and $b > 0$, then
> $$\sqrt{\frac{a}{b}} = \frac{\sqrt{a}}{\sqrt{b}}$$
>
> *Note:* $b \neq 0$ because division by 0 is undefined.

Example 10 Dividing radical expressions

Divide each radical, and then simplify.

a. $\dfrac{\sqrt{18}}{\sqrt{2}}$ b. $\dfrac{\sqrt{75}}{\sqrt{3}}$

c. $\dfrac{\sqrt{2}}{\sqrt{32}}$ d. $\dfrac{\sqrt{32x^3}}{\sqrt{2x}}$

SOLUTION

a. $\dfrac{\sqrt{18}}{\sqrt{2}}$

$= \sqrt{\dfrac{18}{2}}$ Rewrite, using the quotient property of radicals.

$= \sqrt{9}$ Simplify.

$= 3$

b. $\dfrac{\sqrt{75}}{\sqrt{3}}$

$= \sqrt{\dfrac{75}{3}}$ Rewrite, using the quotient property of radicals.

$= \sqrt{25}$ Simplify.

$= 5$

c. The quotient property of radicals can be used in reverse order. The quotient property of radicals is $\sqrt{\dfrac{a}{b}} = \dfrac{\sqrt{a}}{\sqrt{b}}$. Rewriting from right to left yields $\dfrac{\sqrt{a}}{\sqrt{b}} = \sqrt{\dfrac{a}{b}}$.

$\dfrac{\sqrt{2}}{\sqrt{32}}$

$= \sqrt{\dfrac{2}{32}}$ Rewrite, using the quotient property of radicals.

$= \sqrt{\dfrac{1}{16}}$ Simplify.

$= \dfrac{\sqrt{1}}{\sqrt{16}}$ Use the quotient property in reverse order.

$= \dfrac{1}{4}$

d. $\dfrac{\sqrt{32x^3}}{\sqrt{2x}}$ Use the quotient property of radicals to rewrite as a single radical.

$= \sqrt{\dfrac{32x^3}{2x}}$ Simplify inside the radical.

$= \sqrt{16x^2}$ Take the square root of each factor.

$= 4x$ Use the product property of radicals.

PRACTICE PROBLEM FOR EXAMPLE 10

Divide each radical, and then simplify.

a. $\dfrac{\sqrt{50}}{\sqrt{2}}$ **b.** $\dfrac{\sqrt{300}}{\sqrt{3}}$ **c.** $\dfrac{\sqrt{3}}{\sqrt{27}}$ **d.** $\dfrac{\sqrt{27x^5}}{\sqrt{3x}}$

■ Rationalizing the Denominator

To simplify the expression $\dfrac{\sqrt{2}}{\sqrt{10}}$, use the quotient property of radicals as follows.

$$\dfrac{\sqrt{2}}{\sqrt{10}} = \sqrt{\dfrac{2}{10}} = \sqrt{\dfrac{1}{5}} = \dfrac{\sqrt{1}}{\sqrt{5}} = \dfrac{1}{\sqrt{5}}$$

Notice that in the final answer, $\dfrac{1}{\sqrt{5}}$, there is a radical in the denominator.

Mathematics instructors have traditionally not liked expressions such as $\dfrac{1}{\sqrt{5}}$ with a radical remaining in the denominator. When we think of fractions using the "part-to-whole" description, it makes sense why. For example, using the part-to-whole description, think of the fraction $\dfrac{1}{5}$ as telling us we are considering one part out of a whole that has been divided into five equal-sized parts.

Now consider the fraction $\dfrac{1}{\sqrt{5}}$. The part-to-whole description tells us to divide an object into $\sqrt{5}$ equal-sized parts! Even if we approximate the radical as $\sqrt{5} \approx 2.24$, how do you divide 1 whole into about 2.24 equal sized parts? This is difficult for some of us to visualize.

This is one reason students of mathematics do not like fractions with radicals in the denominator. To rewrite a fraction such as $\dfrac{1}{\sqrt{5}}$ without a radical in the denominator, we use a process called **rationalizing the denominator**. To see how this process works, we rationalize the denominator of $\dfrac{1}{\sqrt{5}}$.

$\dfrac{1}{\sqrt{5}}$

$= \dfrac{1}{\sqrt{5}} \cdot \dfrac{\sqrt{5}}{\sqrt{5}}$ Multiply the numerator and denominator by the denominator.

$= \dfrac{\sqrt{5}}{\sqrt{25}}$ Simplify the numerator and denominator.

$= \dfrac{\sqrt{5}}{5}$ The radical is now absent from the denominator.

We can now divide the circle into five parts and shade in $\sqrt{5} \approx 2.24$ parts. This would look approximately like the circle graph.

> **What's That Mean?**
>
> **Rationalizing**
>
> In our daily language, to *rationalize* means to cause something to seem reasonable. If you are rationalizing your actions, you are trying to make them seem reasonable. In mathematics, to *rationalize* a denominator means to get rid of its radical parts and rewrite as a rational number.

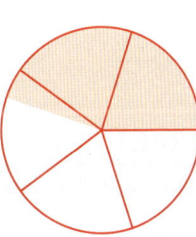

714 CHAPTER 8 Radical Expressions and Equations

> **Steps to Rationalize the Denominator of a Single Radical**
> 1. Multiply both the numerator and the denominator by the radical part of the denominator.
> 2. Simplify the numerator and denominator separately.
>
> *Note:* When the simplification is done, there should be no radicals in the denominator.

Rationalize Denominators with a Single Term

Simplify the numerator and denominator. Multiply the numerator and denominator by the radical factor(s) that allow(s) the radical in the denominator to simplify completely.

$$\frac{7}{\sqrt{2xy^3}}$$

$$= \frac{7}{\sqrt{2xy^3}} \cdot \frac{\sqrt{2xy}}{\sqrt{2xy}}$$

$$= \frac{7\sqrt{2xy}}{\sqrt{2^2 x^2 y^4}}$$

$$= \frac{7\sqrt{2xy}}{2xy^2}$$

Example 11 Rationalizing the denominator of a single radical

Rationalize the denominator of each fraction and simplify.

a. $\dfrac{2}{\sqrt{3}}$ b. $\dfrac{3\sqrt{5}}{\sqrt{6}}$ c. $\dfrac{7}{3\sqrt{x}}$

SOLUTION

a. **Step 1** Multiply both the numerator and denominator by the radical part of the denominator.

$$\frac{2}{\sqrt{3}}$$

$$= \frac{2}{\sqrt{3}} \cdot \frac{\sqrt{3}}{\sqrt{3}} \quad \text{Multiply the numerator and denominator by the denominator.}$$

Step 2 Simplify the numerator and denominator separately.

$$= \frac{2\sqrt{3}}{\sqrt{9}} \quad \text{Simplify.}$$

$$= \frac{2\sqrt{3}}{3}$$

b. $\dfrac{3\sqrt{5}}{\sqrt{6}}$

$$= \frac{3\sqrt{5}}{\sqrt{6}} \cdot \frac{\sqrt{6}}{\sqrt{6}} \quad \text{Multiply the numerator and denominator by the denominator.}$$

$$= \frac{3\sqrt{30}}{\sqrt{36}}$$

$$= \frac{3\sqrt{30}}{6} \quad \text{Simplify.}$$

$$= \frac{\sqrt{30}}{2} \quad \text{Divide out the common factor of 3.}$$

c. $\dfrac{7}{3\sqrt{x}}$

$$= \frac{7}{3\sqrt{x}} \cdot \frac{\sqrt{x}}{\sqrt{x}} \quad \text{Multiply the numerator and denominator by the radical part of the denominator.}$$

$$= \frac{7\sqrt{x}}{3\sqrt{x^2}}$$

$$= \frac{7\sqrt{x}}{3x} \quad \text{Simplify.}$$

PRACTICE PROBLEM FOR EXAMPLE 11

Rationalize the denominator of each fraction and simplify.

a. $\dfrac{5}{\sqrt{7}}$ b. $\dfrac{4\sqrt{3}}{\sqrt{8}}$ c. $\dfrac{-2a}{5\sqrt{b}}$

If the denominator of a fraction has *two terms*, one of which includes a radical, the process of rationalizing involves a new step. Multiplying a denominator by its conjugate gets rid of radicals in the denominator. The goal of rationalizing the denominator is to have no radicals in the denominator, so we multiply the numerator and denominator by the conjugate of the denominator.

> **Steps to Rationalize the Denominator with Two Terms**
> 1. Multiply both the numerator and denominator by the conjugate of the denominator. (*Hint*: Be sure to use parentheses.)
> 2. Multiply and simplify the numerator and denominator separately.
>
> *Note:* When the simplification is done, there should be no radicals in the denominator.

Example 12 Rationalizing a denominator with two terms

Rationalize the denominator of each fraction and simplify.

a. $\dfrac{2}{4 + \sqrt{3}}$ b. $\dfrac{-5}{\sqrt{2} + \sqrt{3}}$

SOLUTION

a. **Step 1** Multiply both the numerator and denominator by the conjugate of the denominator.

To rationalize the denominator of $\dfrac{2}{4 + \sqrt{3}}$, first find the conjugate of the denominator. To do that, switch the sign between the two terms. The conjugate of $4 + \sqrt{3}$ is $4 - \sqrt{3}$. Multiplying the numerator and denominator of the original fraction by the conjugate, $4 - \sqrt{3}$, yields

$\dfrac{2}{4 + \sqrt{3}}$ *Multiply numerator and denominator by the conjugate of the denominator.*

$= \dfrac{2}{(4 + \sqrt{3})} \cdot \dfrac{(4 - \sqrt{3})}{(4 - \sqrt{3})}$

Step 2 Multiply and simplify the numerator and denominator separately.

$= \dfrac{2(4 - \sqrt{3})}{(4 + \sqrt{3})(4 - \sqrt{3})}$ *Use the distributive property in the numerator and FOIL in the denominator.*

$= \dfrac{8 - 2\sqrt{3}}{16 - 4\sqrt{3} + 4\sqrt{3} - \sqrt{3} \cdot \sqrt{3}}$ *Combine like terms in the denominator.*

$= \dfrac{8 - 2\sqrt{3}}{16 - 3}$ *Combine like terms in the denominator again.*

$= \dfrac{8 - 2\sqrt{3}}{13}$

b. To rationalize the denominator of $\dfrac{-5}{\sqrt{2} + \sqrt{3}}$, multiply both the numerator and the denominator by the conjugate of the denominator. The conjugate of $\sqrt{2} + \sqrt{3}$ is $\sqrt{2} - \sqrt{3}$.

Rationalize Denominators with Two Terms

Multiply the numerator and denominator by the conjugate of the denominator. Simplify completely.

$\dfrac{2}{5 + \sqrt{2x}}$

$= \dfrac{2}{(5 + \sqrt{2x})} \cdot \dfrac{(5 - \sqrt{2x})}{(5 - \sqrt{2x})}$

$= \dfrac{10 - 2\sqrt{2x}}{25 - 5\sqrt{2x} + 5\sqrt{2x} - 2x}$

$= \dfrac{10 - 2\sqrt{2x}}{25 - 2x}$

$$\frac{-5}{\sqrt{2} + \sqrt{3}}$$

$$= \frac{-5}{(\sqrt{2} + \sqrt{3})} \cdot \frac{(\sqrt{2} - \sqrt{3})}{(\sqrt{2} - \sqrt{3})}$$ Multiply the numerator and denominator by the conjugate of the denominator.

$$= \frac{-5(\sqrt{2} - \sqrt{3})}{(\sqrt{2} + \sqrt{3})(\sqrt{2} - \sqrt{3})}$$ Use the distributive property in the numerator and FOIL in the denominator.

$$= \frac{-5\sqrt{2} + 5\sqrt{3}}{\sqrt{4} - \sqrt{6} + \sqrt{6} - \sqrt{9}}$$ Combine like terms in the denominator.

$$= \frac{-5\sqrt{2} + 5\sqrt{3}}{2 - 3}$$ Combine like terms in the denominator again.

$$= \frac{-5\sqrt{2} + 5\sqrt{3}}{-1}$$ Dividing by -1 is equivalent to multiplying the numerator by -1.

$$= -1(-5\sqrt{2} + 5\sqrt{3})$$
$$= 5\sqrt{2} - 5\sqrt{3}$$

PRACTICE PROBLEM FOR EXAMPLE 12

Rationalize the denominator of each fraction, and simplify.

a. $\dfrac{7}{1 - \sqrt{5}}$ b. $\dfrac{-2}{\sqrt{5} + \sqrt{2}}$

8.3 Vocabulary Exercises

1. To multiply radical expressions, use the _____ rule for radicals.
2. The _____ of $a + b$ is $a - b$.
3. To rationalize a denominator with two terms, multiply the numerator and denominator by the _____ of the denominator.
4. To divide radicals, use the _____ property of radicals.

8.3 Exercises

For Exercises 1 through 12, multiply the radicals, using the product property of radicals, and simplify.

1. $\sqrt{3} \cdot \sqrt{12}$
2. $\sqrt{2} \cdot \sqrt{50}$
3. $\sqrt{18} \cdot \sqrt{2}$
4. $\sqrt{32} \cdot \sqrt{2}$
5. $\sqrt{2} \cdot \sqrt{22}$
6. $\sqrt{5} \cdot \sqrt{10}$
7. $\sqrt{8} \cdot \sqrt{10}$
8. $\sqrt{6} \cdot \sqrt{8}$
9. $\sqrt{18} \cdot \sqrt{8}$
10. $\sqrt{3} \cdot \sqrt{20}$
11. $\sqrt{3} \cdot \sqrt{50}$
12. $\sqrt{6} \cdot \sqrt{18}$

For Exercises 13 through 36, multiply the radicals, using the product property of radicals, and simplify.

13. $\sqrt{3x} \cdot \sqrt{6x}$
14. $\sqrt{5x} \cdot \sqrt{10x}$
15. $\sqrt{5a} \cdot \sqrt{8a^3}$
16. $\sqrt{3y^3} \cdot \sqrt{15y}$
17. $\sqrt{3x^2} \cdot \sqrt{12x^2}$
18. $\sqrt{7a^5} \cdot \sqrt{7a}$
19. $\sqrt{50x^2} \cdot \sqrt{2x^3}$
20. $\sqrt{18a^2} \cdot \sqrt{2a^3}$
21. $\sqrt{3a^2} \cdot \sqrt{20a^7}$
22. $\sqrt{14a^2} \cdot \sqrt{2a^5}$
23. $\sqrt{8x^{11}} \cdot \sqrt{10x^6}$
24. $\sqrt{27y^{11}} \cdot \sqrt{3y^8}$

25. $\sqrt{17} \cdot \sqrt{17}$
26. $\sqrt{21} \cdot \sqrt{21}$
27. $\sqrt{6x} \cdot \sqrt{6x}$
28. $\sqrt{18z} \cdot \sqrt{18z}$
29. $\sqrt{xyz} \cdot \sqrt{xyz}$
30. $\sqrt{15y^5z} \cdot \sqrt{15y^5z}$
31. $3\sqrt{y} \cdot 7\sqrt{y}$
32. $-5\sqrt{z} \cdot 9\sqrt{z}$
33. $-7\sqrt{a} \cdot (-2\sqrt{a^5})$
34. $4\sqrt{x^3} \cdot (-8\sqrt{x^5})$
35. $(\sqrt{12a^2})(7\sqrt{2a})$
36. $(-8\sqrt{5x^5})(6\sqrt{10x^2})$

For Exercises 37 through 40, find the area of each triangle. Simplify the radical expression. Recall that the area of a triangle is $A = \frac{1}{2} \cdot b \cdot h$.

37.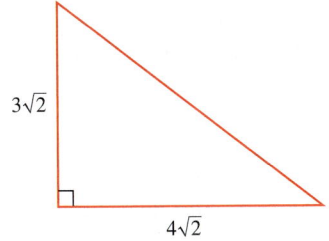

38.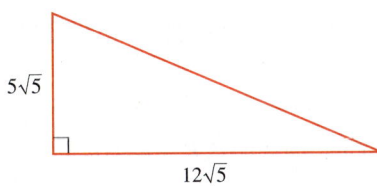

39.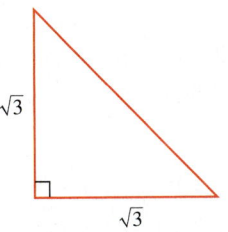

40.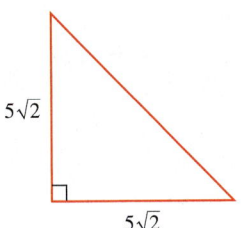

For Exercises 41 through 50, multiply the radical expressions, using the distributive property, and simplify.

41. $\sqrt{2}(\sqrt{2} - 5)$
42. $\sqrt{3}(6 + \sqrt{3})$
43. $\sqrt{3}(\sqrt{6} - \sqrt{5})$
44. $\sqrt{5}(\sqrt{8} - \sqrt{5})$
45. $\sqrt{2}(\sqrt{x} - 2)$
46. $\sqrt{7}(\sqrt{7} - \sqrt{y})$
47. $-2(3\sqrt{8} - \sqrt{5})$
48. $-7(-\sqrt{2} + 3\sqrt{6})$
49. $\sqrt{x}(\sqrt{x} + \sqrt{11})$
50. $\sqrt{a}(\sqrt{15} - \sqrt{a})$

For Exercises 51 through 68, multiply the radical expressions and simplify.

51. $(2 + \sqrt{6})(3 - \sqrt{3})$
52. $(5 - \sqrt{8})(2 - \sqrt{5})$
53. $(9 - \sqrt{7})(3 + \sqrt{7})$
54. $(2 - \sqrt{11})(4 + \sqrt{11})$
55. $(\sqrt{x} + \sqrt{3})(\sqrt{x} - \sqrt{5})$
56. $(\sqrt{y} - \sqrt{2})(\sqrt{y} + \sqrt{7})$
57. $(1 - \sqrt{2})(1 + \sqrt{2})$
58. $(2 - \sqrt{6})(2 + \sqrt{6})$
59. $(3 - \sqrt{x})(3 + \sqrt{x})$
60. $(x - \sqrt{5})(x + \sqrt{5})$
61. $(4 + \sqrt{3})^2$
62. $(5 + \sqrt{7})^2$
63. $(a + \sqrt{11})^2$
64. $(3 + \sqrt{b})^2$
65. $(2 - \sqrt{y})^2$
66. $(7 - \sqrt{x})^2$
67. $(x - \sqrt{5})^2$
68. $(y - \sqrt{3})^2$

For Exercises 69 through 78, divide the radical expressions and simplify.

69. $\dfrac{\sqrt{125}}{\sqrt{5}}$
70. $\dfrac{\sqrt{28}}{\sqrt{7}}$
71. $\dfrac{\sqrt{72}}{\sqrt{2}}$
72. $\dfrac{\sqrt{500}}{\sqrt{5}}$
73. $\dfrac{\sqrt{45x^3}}{\sqrt{5x}}$
74. $\dfrac{\sqrt{50y^5}}{\sqrt{2y}}$
75. $\dfrac{\sqrt{3}}{\sqrt{27}}$
76. $\dfrac{\sqrt{2}}{\sqrt{128}}$
77. $\dfrac{\sqrt{3x^3}}{\sqrt{48x}}$
78. $\dfrac{\sqrt{2a^5}}{\sqrt{200a^3}}$

CHAPTER 8 Radical Expressions and Equations

For Exercises 79 through 94, rationalize the denominator of each radical expression and simplify.

79. $\dfrac{4}{\sqrt{3}}$

80. $\dfrac{5}{\sqrt{7}}$

81. $\dfrac{-15}{\sqrt{3}}$

82. $\dfrac{-8}{\sqrt{2}}$

83. $\dfrac{\sqrt{5}}{3\sqrt{x}}$

84. $\dfrac{\sqrt{6}}{7\sqrt{y}}$

85. $\dfrac{\sqrt{2}}{5\sqrt{a}}$

86. $\dfrac{\sqrt{5}}{\sqrt{7}}$

87. $\dfrac{2}{2 + \sqrt{3}}$

88. $\dfrac{3}{4 - \sqrt{2}}$

89. $\dfrac{7}{\sqrt{7} + \sqrt{3}}$

90. $\dfrac{4}{\sqrt{5} - \sqrt{2}}$

91. $\dfrac{4}{\sqrt{x} + 3}$

92. $\dfrac{4}{5 - \sqrt{b}}$

93. $\dfrac{2}{\sqrt{7} + 10}$

94. $\dfrac{4}{\sqrt{11} + 6}$

For Exercises 95 through 98, use the given information to find the missing side of each rectangle.

95. The area of the rectangle is $6\sqrt{3}$. The width of the rectangle is $\sqrt{2}$. Find the length.

96. The area of the rectangle is $5\sqrt{15}$. The width of the rectangle is $\sqrt{3}$. Find the length.

97. The area of the rectangle is $12\sqrt{7}$. The length of the rectangle is $3\sqrt{7}$. Find the width.

98. The area of the rectangle is $9\sqrt{10}$. The length of the rectangle is $3\sqrt{5}$. Find the width.

For Exercises 99 through 114, simplify each expression.

99. $\sqrt{3} \cdot \sqrt{18}$

100. $\sqrt{3} \cdot \sqrt{20}$

101. $-4\sqrt{x^2} \cdot 3\sqrt{x^2}$

102. $(-2\sqrt{x}) \cdot (-3\sqrt{x})$

103. $(2\sqrt{5y^3})(3\sqrt{5y^2})$

104. $(-4\sqrt{3x^3})(5\sqrt{5x^4})$

105. $\sqrt{5}(\sqrt{7} + \sqrt{5})$

106. $\sqrt{6}(\sqrt{8} + \sqrt{2})$

107. $(2 - \sqrt{3})(3 + \sqrt{6})$

108. $(\sqrt{2} + 1)(\sqrt{10} - 4)$

109. $\sqrt{15x^2} \cdot \sqrt{3x^3}$

110. $\sqrt{7y^5} \cdot \sqrt{2y}$

111. $\dfrac{\sqrt{20}}{\sqrt{5}}$

112. $\dfrac{\sqrt{180}}{\sqrt{5}}$

113. $\dfrac{\sqrt{2a^5}}{\sqrt{50a^3}}$

114. $\dfrac{\sqrt{5y^3}}{\sqrt{45y}}$

115. Rationalize the denominator of $\dfrac{-6}{\sqrt{3}}$ and simplify.

116. Rationalize the denominator of $\dfrac{5}{\sqrt{10}}$ and simplify.

117. Rationalize the denominator of $\dfrac{-2}{5 + \sqrt{3}}$ and simplify.

118. Rationalize the denominator of $\dfrac{6}{\sqrt{3} - 1}$ and simplify.

For Exercises 119 through 130, answer as indicated.

119. Write the equation of the line passing through the points $(7, 3)$ and $(-14, 9)$.

120. Write the equation of the line passing through the points $(3, 1)$ and $(-6, -14)$.

121. Solve for the variable c: $P = 2a + 6b - 2c$.

122. Solve for the variable d: $P = a + b + c + 3d$.

123. Convert 0.5 acre to square feet. There are 43,560 square feet in 1 acre.

124. Convert 11,000 square feet to acres. There are 43,560 square feet in 1 acre.

125. Simplify: $(2t^2)^2$.

126. Simplify: $(2 + t^2)^2$.

127. Simplify: $(-3 + n^2)^2$.

128. Simplify: $(-3n^2)^2$.

129. Solve the inequality: $\dfrac{-5t + 6}{2} \leq 8$.

130. Solve the inequality: $\dfrac{3t - 2}{3} \geq -6$.

8.4 Solving Radical Equations

LEARNING OBJECTIVES
- Check solutions to radical equations.
- Solve radical equations.
- Solve applications problems involving radicals.

■ Checking Solutions to Radical Equations

In Section 8.1, radical equations were defined as equations that have a variable under the radical. An example of a radical equation is $\sqrt{x} + 8 = 16$. A *solution* to a radical equation is a value for the variable that, when substituted into the equation, makes the equation true (both sides equal).

Example 1 Is the value a solution?

Check the given value of each variable to see whether it is a solution of the radical equation.

a. $\sqrt{x} + 1 = 5$ for $x = 16$
b. $\sqrt{x^2 - 9} = 4$ for $x = -5$
c. $\sqrt{x - 3} + 7 = 6$ for $x = 4$

SOLUTION

a. Substituting $x = 16$ into $\sqrt{x} + 1 = 5$ yields

$\sqrt{x} + 1 = 5$ Substitute $x = 16$ into the original equation.
$\sqrt{16} + 1 \stackrel{?}{=} 5$ Simplify.
$4 + 1 = 5$ This is a true statement, so $x = 16$ is a solution.

b. Substituting $x = -5$ into $\sqrt{x^2 - 9} = 4$ yields

$\sqrt{x^2 - 9} = 4$ Substitute $x = -5$ into the original equation.
$\sqrt{(-5)^2 - 9} \stackrel{?}{=} 4$ Simplify.
$\sqrt{25 - 9} \stackrel{?}{=} 4$
$\sqrt{16} = 4$ This is a true statement, so $x = -5$ is a solution.

c. Substituting $x = 4$ into $\sqrt{x - 3} + 7 = 6$ yields

$\sqrt{x - 3} + 7 = 6$ Substitute $x = 4$ into the original equation.
$\sqrt{4 - 3} + 7 \stackrel{?}{=} 6$ Simplify.
$\sqrt{1} + 7 \stackrel{?}{=} 6$
$1 + 7 \neq 6$ $x = 4$ is not a solution.

PRACTICE PROBLEM FOR EXAMPLE 1

Check the given value of each variable to see whether it is a solution of the equation.

a. $\sqrt{a} - 2 = 1$ for $a = 9$
b. $\sqrt{3 - y} = 5$ for $y = 28$

Solving Radical Equations

We now learn how to solve radical equations. An example of a radical equation is $\sqrt{x} = 4$. First examine the equation to verify it is a radical equation. A radical equation has a variable under the radical sign in the equation. To solve a radical equation, the operation of the square root must be undone. In Section 8.1, we saw that squares and square roots undo each other. Applying that idea to the problem above yields

$\sqrt{x} = 4$ The original equation

$(\sqrt{x})^2 = 4^2$ Square both sides. The left side means $(\sqrt{x})^2 = \sqrt{x} \cdot \sqrt{x} = \sqrt{x^2} = x$.

$x = 16$

Check this answer in the original problem.

$\sqrt{x} = 4$ The original equation

$\sqrt{16} \stackrel{?}{=} 4$ Substitute in $x = 16$.

$4 = 4$ This is a true statement. The answer checks.

Here we are using the squaring property of equality to solve an equation.

> **SQUARING PROPERTY OF EQUALITY**
>
> If $a = b$ then $a^2 = b^2$.
>
> *Note:* This property states that it is valid to square both sides of an equation.

Making use of the squaring property of equality leads to the following steps to solve radical equations.

> **Steps to Solving Radical Equations**
> 1. Isolate a radical on one side of the equal sign.
> 2. Use the squaring property of equality to square both sides of the equation. If a radical remains, repeat steps 1 and 2.
> 3. Solve for the variable.
> 4. Check the answer in the original equation.

Example 2   How far is it to the horizon?

The distance that a person can see to the horizon depends on the person's height above the ground. The distance can be found by using the formula

$$d = \sqrt{1.5h}$$

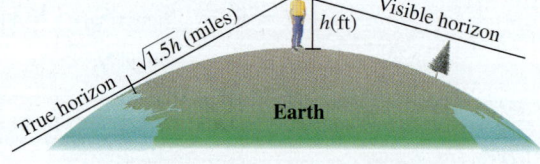

where d is the distance in miles when the person's line of sight is h feet above the ground.

a. Find how far a person can see out to the horizon if the person is 6 feet above the ground.

b. How high off the ground must a person be to see 30 miles to the horizon?

SOLUTION

a. When a person is 6 feet above the ground, substitute in $h = 6$ and evaluate the equation for d.

$$d = \sqrt{1.5h} \quad \text{Substitute in } h = 6.$$
$$d = \sqrt{1.5(6)}$$
$$d = \sqrt{9}$$
$$d = 3$$

The person can see 3 miles out to the horizon.

b. When the person sees 30 miles to the horizon, it means that $d = 30$. Substitute this into the equation, and solve for h.

$$d = \sqrt{1.5h} \quad \text{Substitute in } d = 30.$$
$$30 = \sqrt{1.5h}$$

Step 1 Isolate a radical on one side of the equal sign.

The radical is already isolated.

$$30 = \sqrt{1.5h}$$

Step 2 Use the squaring property of equality to square both sides of the equation.

$$30^2 = (\sqrt{1.5h})^2 \quad \text{Square both sides to undo the square root.}$$
$$900 = 1.5h$$

Step 3 Solve for the variable.

$$\frac{900}{1.5} = \frac{1.5h}{1.5} \quad \text{Solve for } h.$$
$$600 = h$$

The person's line of sight must be 600 feet above the ground.

Step 4 Check the answer in the original problem.

Substituting $h = 600$ into the original equation and evaluating yields

$$30 = \sqrt{1.5h}$$
$$30 \stackrel{?}{=} \sqrt{1.5(600)} \quad \text{Substitute in } h = 600.$$
$$30 \stackrel{?}{=} \sqrt{900} \quad \text{Evaluate the right side.}$$
$$30 = 30 \quad \text{The answer checks.}$$

PRACTICE PROBLEM FOR EXAMPLE 2

The distance that a person can see to the horizon depends on the person's height above the ground. The distance can be found by using the formula

$$d = \sqrt{1.5h}$$

where d is the distance in miles when the person's line of sight is h feet above the ground.

a. Find how far a person can see out to the horizon if the person is 5 feet above the ground.

b. How high off the ground must a person be to see 18 miles to the horizon?

722 CHAPTER 8 Radical Expressions and Equations

Radical Equations

Raise Both Sides to the Reciprocal Exponent

Isolate a radical on one side of the equation and then square both sides of the equation. This may need to be done more than once to clear multiple radicals.

$$\sqrt{x+2} - 7 = 2$$
$$\sqrt{x+2} = 9$$
$$(\sqrt{x+2})^2 = 9^2$$
$$x + 2 = 81$$

Example 3 Solving radical equations

Solve each radical equation. Check the answer(s) in the original problem.

a. $\sqrt{x} = 7$ **b.** $2\sqrt{a} - 3 = 13$ **c.** $\sqrt{2x-1} = 5$

SOLUTION

a.
$\sqrt{x} = 7$ The original equation
$(\sqrt{x})^2 = 7^2$ Square both sides.
$x = 49$

Check: $x = 49$
$\sqrt{49} \stackrel{?}{=} 7$ Substitute $x = 49$ into the original equation.
$7 = 7$ This is a true statement. The answer checks.

b.
$2\sqrt{a} - 3 = 13$ Isolate the radical.
$\underline{ +3 \quad +3}$
$2\sqrt{a} = 16$ To completely isolate the radical, divide both sides by 2.
$\dfrac{2\sqrt{a}}{2} = \dfrac{16}{2}$
$\sqrt{a} = 8$
$(\sqrt{a})^2 = (8)^2$ Square both sides.
$a = 64$

Check: $a = 64$
$2\sqrt{64} - 3 \stackrel{?}{=} 13$ Substitute $a = 64$ into the original problem.
$2 \cdot 8 - 3 \stackrel{?}{=} 13$ Simplify, using the order-of-operations agreement.
$16 - 3 = 13$ This is a true statement. The answer checks.

c.
$\sqrt{2x-1} = 5$ The radical is isolated.
$(\sqrt{2x-1})^2 = 5^2$ Square both sides.
$2x - 1 = 25$
$\underline{ +1 \quad +1}$ Solve for x.
$2x = 26$
$\dfrac{2x}{2} = \dfrac{26}{2}$
$x = 13$

Check: $x = 13$
$\sqrt{2(13) - 1} \stackrel{?}{=} 5$ Substitute $x = 13$.
$\sqrt{26 - 1} \stackrel{?}{=} 5$
$\sqrt{25} \stackrel{?}{=} 5$
$5 = 5$ Both sides are equal. The answer checks.

PRACTICE PROBLEM FOR EXAMPLE 3

Solve each radical equation. Check the answer(s) in the original problem.

a. $\sqrt{3x} = 9$ **b.** $\sqrt{y} - 3 = 10$
c. $2\sqrt{x} + 4 = 12$ **d.** $\sqrt{3x+1} = 4$

Suppose we use the same technique to solve the radical equation $\sqrt{x} = -6$. Squaring both sides and solving for x yields

$$\sqrt{x} = -6 \quad \text{The original equation}$$
$$(\sqrt{x})^2 = (-6)^2 \quad \text{Square both sides.}$$
$$x = 36 \quad \text{Remember that } (-6)^2 = (-6)(-6) = 36.$$

Checking the answer $x = 36$ yields

$$\sqrt{x} = -6 \quad \text{The original equation}$$
$$\sqrt{36} \stackrel{?}{=} -6 \quad \text{Substitute in } x = 36.$$
$$6 \neq -6 \quad \text{The answer does not check.}$$

Why does the answer not check in the original equation $\sqrt{x} = -6$? The left side of this equation does not have a sign, so it has a positive value: $\sqrt{x} = +\sqrt{x}$. The right side of this equation is a negative number, -6. The original equation is asking, "When does a positive number equal a negative number: $+\sqrt{x} = -6$?" The answer is never.

Not all solutions to radical equations check. In fact, sometimes when solving a radical equation, an extra or **extraneous** solution is introduced. This is what happened when we solved the equation $\sqrt{x} = -6$. The solution $x = 36$ is an extraneous solution that did not check in the original problem. For this reason, always check the solution(s) to radical equations in the original problem.

■ **What's That Mean?**

Extraneous (Extra) Solution

An *extraneous solution* is a value that is introduced during the solving process and does not check in the original problem. It is *not* a solution to the problem.

Example 4 Does the solution check?

Solve each radical equation. Check the answer(s) in the original problem.

a. $\sqrt{y - 1} - 4 = -8$ **b.** $\sqrt{3x - 5} = \sqrt{x + 5}$

SOLUTION

a.
$$\sqrt{y - 1} - 4 = -8 \quad \text{Add 4 to both sides to isolate the radical.}$$
$$\underline{\phantom{\sqrt{y-1}} +4 \quad +4}$$
$$\sqrt{y - 1} = -4$$
$$(\sqrt{y - 1})^2 = (-4)^2 \quad \text{Square both sides.}$$
$$y - 1 = 16$$
$$\underline{+1 \quad +1}$$
$$y = 17$$

Check: $y = 17$

$$\sqrt{17 - 1} - 4 \stackrel{?}{=} -8 \quad \text{Substitute } y = 17 \text{ into the original problem to check.}$$
$$\sqrt{16} - 4 \stackrel{?}{=} -8$$
$$4 - 4 \stackrel{?}{=} -8$$
$$0 = -8 \quad \text{This is not a true statement. The answer does not check.}$$

Therefore, the equation $\sqrt{y - 1} - 4 = -8$ has **no solution**.

b.
$$\sqrt{3x - 5} = \sqrt{x + 5} \quad \text{Each radical is isolated on one side of the equal sign.}$$
$$(\sqrt{3x - 5})^2 = (\sqrt{x + 5})^2 \quad \text{Square both sides.}$$
$$3x - 5 = x + 5 \quad \text{Solve the resulting equation for x.}$$
$$\underline{-x \qquad -x}$$
$$2x - 5 = 5$$
$$\underline{+5 \quad +5}$$
$$2x = 10$$
$$\frac{2x}{2} = \frac{10}{2}$$
$$x = 5$$

■ **Connecting the Concepts**

When do we have to solve that radical equation?

When solving a radical equation that contains a square root, if the radical is isolated on one side of the equal sign and there is a negative number on the other side of the equal sign, the equation has no solution. It does not need to be solved.

For example, in the equation $\sqrt{x} = -6$, the left side $\sqrt{x}$ is positive, and the right side -6 is negative. Since a positive value cannot be equal to a negative value, this equation has no solution.

When solving, if we do not notice that a radical equals a negative number and complete the solving process, we will end with an extraneous solution.

$$\sqrt{3(5) - 5} \stackrel{?}{=} \sqrt{(5) + 5}$$
$$\sqrt{15 - 5} \stackrel{?}{=} \sqrt{10}$$
$$\sqrt{10} = \sqrt{10}$$

Check: $x = 5$
Substitute $x = 5$ into the original problem.

The answer checks, as both sides are equal.

The solution is $x = 5$.

PRACTICE PROBLEM FOR EXAMPLE 4

Solve each radical equation. Check the answer(s) in the original problem.

a. $2 + \sqrt{x - 5} = -10$ **b.** $\sqrt{4x + 1} = \sqrt{5x - 7}$

■ Solving Applications Involving Radical Equations

In Section 8.1, we looked at some applications that involved radicals. We will revisit some of those applications, but this time we will *solve* the radical equation rather than just evaluating it.

Example 5 A falling object

A science class tests how long it takes a ball to fall to the ground when dropped from various heights. The class members determine that the time, T, in seconds that it takes the ball to fall when dropped from a height of h feet is given by the equation

$$T = 0.243\sqrt{h}$$

Approximate to two decimal places if needed.

a. If the ball falls for 2.5 seconds, from what height was it originally dropped?

b. If the ball falls for 10 seconds, from what height was it originally dropped?

SOLUTION

a. We are given that the time $T = 2.5$ seconds and are trying to determine the height h. Substituting $T = 2.5$ into the equation and solving for h yields

$$T = 0.243\sqrt{h}$$
$$2.5 = 0.243\sqrt{h}$$ 　Substitute in $T = 2.5$.
$$\frac{2.5}{0.243} = \frac{0.243\sqrt{h}}{0.243}$$ 　Isolate the radical part.
$$\frac{2.5}{0.243} = \sqrt{h}$$
$$\left(\frac{2.5}{0.243}\right)^2 = (\sqrt{h})^2$$ 　Square both sides.
$$h = \left(\frac{2.5}{0.243}\right)^2 \approx 105.84$$ 　Approximate to two decimal places.

Therefore, the ball should originally be at 105.84 feet.

b. The object fell for 10 seconds. Substituting $T = 10$ into the equation and solving for h, we get

$$T = 0.243\sqrt{h}$$
$$10 = 0.243\sqrt{h} \quad \text{Substitute in } T = 10.$$
$$\frac{10}{0.243} = \frac{0.243\sqrt{h}}{0.243} \quad \text{Isolate the radical part.}$$
$$\frac{10}{0.243} = \sqrt{h}$$
$$\left(\frac{10}{0.243}\right)^2 = (\sqrt{h})^2 \quad \text{Square both sides.}$$
$$h = \left(\frac{10}{0.243}\right)^2 \approx 1693.51 \quad \text{Approximate to two decimal places.}$$

The height must be about 1693.51 feet. This is not an experiment that the average science class is be able to do.

PRACTICE PROBLEM FOR EXAMPLE 5

Police officers use a formula involving square roots when investigating traffic accidents. By measuring the length of the skid mark left by the car, the police can determine the speed of the car at the time of the accident (speed measured in miles per hour). The equation that relates the speed of the car in miles per hour, s, to the length of the skid mark in feet, d, is

$$s = \sqrt{30d}$$

Assume that the car is being driven on a dry asphalt road. Approximate to two decimal places if needed.

a. If a car was traveling at 55 miles per hour at the time of the accident, how long a skid mark would the police expect to find at the scene of the accident?

b. If a car was traveling at 75 miles per hour at the time of the accident, how long a skid mark would the police expect to find at the scene of the accident?

■ Solving More Radical Equations

In some problems, after squaring the radical, we may have a quadratic equation to solve, or a radical may be left in the equation. If a quadratic remains, solve by factoring, as was done in Chapter 6. If a radical remains, then isolate the radical and square both sides of the equation again.

Example 6 Solving radical equations with quadratics or multiple radicals

Solve each radical equation. Check the answer(s) in the original problem.

a. $\sqrt{x^2 - x - 8} = 2$ **b.** $\sqrt{x + 20} = \sqrt{x} + 2$

SOLUTION

a.
$$\sqrt{x^2 - x - 8} = 2 \quad \text{The radical is isolated.}$$
$$(\sqrt{x^2 - x - 8})^2 = 2^2 \quad \text{Square both sides.}$$
$$x^2 - x - 8 = 4 \quad \text{This is a quadratic equation because of the } x^2\text{-term.}$$
$$\underline{ -4 \;\; -4}$$
$$x^2 - x - 12 = 0 \quad \text{Set the quadratic equation equal to 0 and factor.}$$

726 CHAPTER 8 Radical Expressions and Equations

$$(x-4)(x+3) = 0$$

$x - 4 = 0 \quad\quad x + 3 = 0$ *Solve each resulting equation for x.*

$x = 4 \quad\quad\quad\quad x = -3$

Check: $x = 4$ Check: $x = -3$ *Remember to check both answers.*

$\sqrt{(4)^2 - (4) - 8} \stackrel{?}{=} 2 \quad\quad \sqrt{(-3)^2 - (-3) - 8} \stackrel{?}{=} 2$

$\sqrt{16 - (4) - 8} \stackrel{?}{=} 2 \quad\quad \sqrt{9 + 3 - 8} \stackrel{?}{=} 2$

$\sqrt{4} \stackrel{?}{=} 2 \quad\quad\quad\quad \sqrt{4} \stackrel{?}{=} 2$

$2 = 2 \quad\quad\quad\quad\quad 2 = 2$ *Both answers check.*

> **Skill Connection**
>
> **FOIL**
>
> Recall from Section 5.4 that we use FOIL to multiply two expressions together that each have two terms. FOIL is used when we square an expression with two terms.
>
> $(x - 6)^2 = (x - 6)(x - 6)$
> $ = x^2 - 6x - 6x + 36$
> $ = x^2 - 12x + 36$

b. There are two radicals in this example. The right side has two terms being added, one of which is a radical, namely, $\sqrt{x}$ and 2. When squaring both sides, use FOIL on the right side.

$\sqrt{x + 20} = \sqrt{x} + 2$

$(\sqrt{x + 20})^2 = (\sqrt{x} + 2)^2$ *Square both sides.*

$x + 20 = (\sqrt{x} + 2)(\sqrt{x} + 2)$ *To multiply the right side, use FOIL.*

$x + 20 = \sqrt{x} \cdot \sqrt{x} + 2\sqrt{x} + 2\sqrt{x} + 4$

$x + 20 = x + 4\sqrt{x} + 4$ *Combine like terms on the right side.*

There is still a radical remaining in this equation. Once again, isolate the radical part, and square both sides.

$x + 20 = x + 4\sqrt{x} + 4$

$\underline{-x \quad -4 \quad -x \quad\quad\quad -4}$

$16 = 4\sqrt{x}$ *Isolate the radical part that remains.*

$\dfrac{16}{4} = \dfrac{4\sqrt{x}}{4}$ *Divide both sides by 4 to isolate the radical completely.*

$4 = \sqrt{x}$

$4^2 = (\sqrt{x})^2$ *Square both sides to undo the square root.*

$16 = x$

Check: $x = 16$

$\sqrt{x + 20} = \sqrt{x} + 2$

$\sqrt{16 + 20} \stackrel{?}{=} \sqrt{16} + 2$

$\sqrt{36} \stackrel{?}{=} 4 + 2$

$6 = 6$ *The answer checks.*

PRACTICE PROBLEM FOR EXAMPLE 6

Solve each radical equation. Check the answer(s) in the original problem.

a. $\sqrt{x^2 + x - 21} = 3$ **b.** $\sqrt{x - 27} = \sqrt{x} - 3$

8.4 Vocabulary Exercises

1. The _____ of equality says if $a = b$ then, $a^2 = b^2$.

2. _____ solutions are those potential solutions that do not check in a radical equation.

3. Equations containing a square root are called _____ equations.

4. After _____ the radical we _____ both sides of the equation to undo the square root.

8.4 Exercises

For Exercises 1 through 12, determine whether each given value of the variable is a solution of the radical equation.

1. $\sqrt{x} - 5 = -1$ for $x = 16$
2. $3 - \sqrt{x} = -7$ for $x = 100$
3. $\sqrt{x} = -9$ for $x = 81$
4. $\sqrt{x} = -1$ for $x = 1$
5. $\sqrt{x-1} + 9 = 4$ for $x = 26$
6. $\sqrt{x+2} - 5 = -1$ for $x = 34$
7. $\sqrt{x^2 + 11} = 6$ for $x = 5$
8. $\sqrt{x^2 - 7} = 3$ for $x = 4$
9. $\sqrt{x^2 - 1} = -1$ for $x = 0$
10. $\sqrt{x^2 + 4} = -4$ for $x = 0$
11. $\sqrt{x^2 + x - 20} + 9 = \sqrt{x} + 7$ for $x = 4$
12. $\sqrt{x^2 - 3x + 2} - 3 = \sqrt{x} - 4$ for $x = 1$

For Exercises 13 through 18, determine whether the work done by the students that are shown is correct or not. If it is not correct, then find the student's error.

13. Determine whether $x = -9$ is a solution of $\sqrt{x} = -3$.

   Maria
   $\sqrt{x} = -3$ for $x = -9$
   $\sqrt{-9} \stackrel{?}{=} -3$
   $-3 = -3$

14. Determine whether $x = -25$ is a solution of $\sqrt{x} = -5$.

   Robert
   $\sqrt{x} = -5$ for $x = -25$
   $\sqrt{-25} \stackrel{?}{=} -5$
   $-5 = -5$

15. Determine whether $x = 49$ is a solution of $-\sqrt{x} = -7$.

   Marilyn
   $-\sqrt{x} = -7$ for $x = 49$
   $-\sqrt{49} \stackrel{?}{=} -7$
   $-7 = -7$

16. Determine whether $x = 36$ is a solution of $-\sqrt{x} = -6$.

   Truong
   $-\sqrt{x} = -6$ for $x = 36$
   $-\sqrt{36} \stackrel{?}{=} -6$
   $-6 = -6$

17. Determine whether $x = -1$ is a solution of $-\sqrt{x} = 1$.

   Jaydy
   $-\sqrt{x} = 1$ for $x = -1$
   $-\sqrt{-1} \stackrel{?}{=} 1$
   $1 = 1$

18. Determine whether $x = -64$ is a solution of $-\sqrt{x} = 8$.

   Gabriela
   $-\sqrt{x} = 8$ for $x = -64$
   $-\sqrt{-64} \stackrel{?}{=} 8$
   $8 = 8$

For Exercises 19 through 30, solve each radical equation and check the answer.

19. $\sqrt{a} = 4$
20. $\sqrt{a} = 9$
21. $\sqrt{y} = -12$
22. $\sqrt{y} = -3$
23. $\sqrt{x-1} = 6$
24. $\sqrt{x+4} = 2$
25. $\sqrt{-y+4} = 2$
26. $\sqrt{5-y} = 4$
27. $\sqrt{a} - 6 = 9$
28. $\sqrt{a} + 11 = 15$
29. $2\sqrt{x} + 3 = 11$
30. $-4 + 5\sqrt{x} = 21$

For Exercises 31 through 38, solve each radical equation. Provide reasons for each step. Check the answer.

31. $\sqrt{x} + 10 = 6$
32. $3\sqrt{x} + 35 = 20$
33. $2\sqrt{a} + 19 = 15$
34. $3\sqrt{x} + 7 = 4$
35. $\sqrt{3y+1} = 5$
36. $\sqrt{4-5y} = 7$
37. $\sqrt{-2a+7} = 9$
38. $\sqrt{15-5a} = 10$

For Exercises 39 through 46, solve each application problem. Round to two decimal places if necessary.

39. The time t in seconds it takes a ball to fall when dropped from a height of h meters is given by the equation $t = \sqrt{\dfrac{h}{4.5}}$.

    a. If the ball falls for 3.4 seconds, from what height was it originally dropped?

    b. If the ball falls for 9 seconds, from what height was it originally dropped?

40. The time t in seconds it takes a ball to fall when dropped from a height of h meters is given by the equation $t = 0.47\sqrt{h}$.

    a. If the ball falls for 4 seconds, from what height was it originally dropped?

    b. If the ball falls for 20 seconds, from what height was it originally dropped?

41. Police investigating traffic accidents use the fact that the speed, in miles per hour, of a car traveling on an asphalt road can be determined by the length of the skid mark left by the car after sudden braking. The speed is given by the formula $s = \sqrt{30d}$, where s is the speed of the car in miles per hour and d is the length in feet of the skid mark.

    a. If the car was traveling at 45 miles per hour, what was the length of the skid mark?

    b. If the car was traveling at 37 miles per hour, what was the length of the skid mark?

42. Police investigating traffic accidents use the fact that the speed, in miles per hour, of a car traveling on an asphalt road can be determined by the length of the skid mark left by the car after sudden braking. The speed is given by the formula $s = \sqrt{30d}$, where s is the speed of the car in miles per hour and d is the length in feet of the skid mark.

    a. If the car was traveling at 52 miles per hour, what was the length of the skid mark?

    b. If the car was traveling at 29 miles per hour, what was the length of the skid mark?

43. Andrew makes beaded wristbands to sell at an upcoming music festival. The profit P he makes when he sells x wristbands is given by $P = 4\sqrt{x-2}$. How many wristbands does he have to sell to make $20.00 profit?

44. Shrimati makes earrings to sell at a bead shop. The profit P she makes when selling x earrings is given by $P = 5\sqrt{x+1}$. How many earrings does she have to sell to make $40.00 profit?

45. A farmer wants to purchase an irrigation system for her small organic farm. A sales representative for an irrigation supply company tells her that the system will water a circular region with an area and radius given by the equation $r = 0.56\sqrt{A}$, where r is the radius in feet and A is the area in square feet. If she would like the system to water a circular area with radius 200 feet, what area does the irrigation system cover?

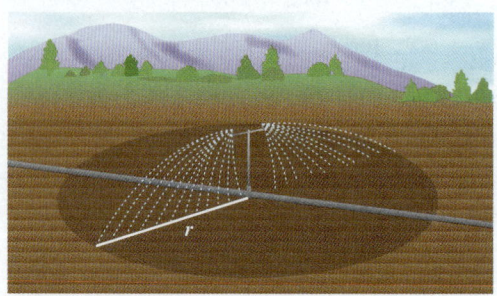

46. A rancher wants to purchase an irrigation system for his horse farm. A sales representative for an irrigation supply company tells him that the system will water a circular region with an area and radius given by the equation $r = 0.56\sqrt{A}$, where r is the radius in feet and A is the area in square feet. If the rancher would like the system to water a circular area with radius 225 feet, what area does the irrigation system cover?

For Exercises 47 through 60, solve each radical equation and check the answer.

47. $\sqrt{x+2} = 4$

48. $\sqrt{x-6} = 6$

49. $\sqrt{x+5} = -1$

50. $\sqrt{x-4} = -2$

51. $\sqrt{z-2} = \sqrt{2z-5}$

52. $\sqrt{-z+2} = \sqrt{-3z+1}$

53. $\sqrt{y-2} + 2 = 0$

54. $\sqrt{y-4} + 4 = 0$

55. $\sqrt{b+4} - 4 = 6$

56. $\sqrt{b-3} + 6 = 10$

57. $2 - \sqrt{a+14} = 1$

58. $-\sqrt{3-a} + 5 = 3$

59. $\sqrt{x-2} + 4 = 3$

60. $6 + \sqrt{x+9} = 4$

For Exercises 61 through 72, find the mistakes in the student work shown. Find the correct answers.

61. Solve the radical equation $2\sqrt{x} = 4$.

> **Mackenzie**
> $2\sqrt{x} = 4$
> $(2\sqrt{x})^2 = 4^2$
> $2x = 16$
> $\dfrac{2x}{2} = \dfrac{16}{2}$
> $x = 8$

62. Solve the radical equation $3\sqrt{x} = 6$.

> **Jazmine**
> $3\sqrt{x} = 6$
> $(3\sqrt{x})^2 = 6^2$
> $3x = 36$
> $\dfrac{3x}{3} = \dfrac{36}{3}$
> $x = 12$

63. Solve the radical equation $\sqrt{x} + 1 = 5$.

> **Andrew**
> $\sqrt{x} + 1 = 5$
> $(\sqrt{x} + 1)^2 = 5^2$
> $x + 1 = 25$
> $\underline{-1 \quad -1}$
> $x = 24$

64. Solve the radical equation $\sqrt{x} + 2 = 8$.

> **Keegan**
> $\sqrt{x} + 2 = 8$
> $(\sqrt{x} + 2)^2 = 8^2$
> $x + 2 = 64$
> $\underline{-2 \quad -2}$
> $x = 62$

65. Solve the radical equation $\sqrt{x} = -6$.

> **Victor**
> $\sqrt{x} = -6$
> $(\sqrt{x})^2 = (-6)^2$
> $x = 36$

66. Solve the radical equation $\sqrt{x} = -1$.

> **Matthew**
> $\sqrt{x} = -1$
> $(\sqrt{x})^2 = (-1)^2$
> $x = 1$

67. Solve the radical equation $\sqrt{-x} = 9$.

> **Margaret**
> $\sqrt{-x} = 9$
> $(\sqrt{-x})^2 = 9^2$
> $x = 81$

68. Solve the radical equation $\sqrt{-x} = 5$.

> **Yevgenia**
> $\sqrt{-x} = 5$
> $(\sqrt{-x})^2 = 5^2$
> $x = 25$

69. Find the distance between the points $(3, 6)$ and $(-5, 2)$.

> **Rio**
> $\dfrac{2 - 6}{-5 - 3} = \dfrac{-4}{-8} = \dfrac{1}{2}$

70. Find the distance between the points $(-1, 5)$ and $(8, -7)$.

> **Ernesto**
> $\dfrac{-7 - 5}{8 - (-1)} = \dfrac{-12}{9} = -\dfrac{4}{3}$

71. Find the distance between the points $(3, 2)$ and $(9, 4)$.

> **Ethan**
> $\sqrt{(2-3)^2 + (4-9)^2} =$
> $\sqrt{(-1)^2 + (-5)^2} =$
> $\sqrt{1 + 25} = \sqrt{26}$

72. Find the distance between the points $(4, 7)$ and $(-3, 5)$.

> **Parker**
> $\sqrt{(7-4)^2 + (5-(-3))^2} =$
> $\sqrt{(3)^2 + (8)^2} =$
> $\sqrt{9 + 64} = \sqrt{73}$

For Exercises 73 through 84, solve each radical equation and check the answer.

73. $\sqrt{x^2 - x + 7} = 3$
74. $\sqrt{x^2 + x - 4} = 4$
75. $\sqrt{2x^2 - x + 1} = 2$
76. $\sqrt{3x^2 + 4x + 21} = 5$
77. $\sqrt{x + 45} = \sqrt{x} + 5$
78. $\sqrt{x + 9} = \sqrt{x} + 3$
79. $\sqrt{y + 3} = \sqrt{y} - 3$
80. $\sqrt{y - 4} = \sqrt{y} + 2$
81. $-1 = \sqrt{a^2 - 7a + 12}$
82. $\sqrt{a^2 + 12a + 35} = -4$
83. $\sqrt{x + 32} = 4 + \sqrt{x}$
84. $\sqrt{x + 33} = 3 + \sqrt{x}$

For Exercises 85 through 92, determine whether each given equation is linear, quadratic, rational, or radical. Do not solve the equations.

85. $\sqrt{6x + 1} = 3$
86. $\sqrt{1 - x^2} = 4$
87. $\dfrac{5}{x - 4} = \dfrac{3}{2x + 5}$
88. $\dfrac{-5}{x^2 - 4} = \dfrac{2}{x + 2}$
89. $-6(2y + 5) - 3 = -2y + 4$
90. $-6y(-4 + 3y) = 8y - 7$
91. $3x + 6 = -2x^2$
92. $1 - 5x = -6x + 9$

For Exercises 93 through 100,

a. Determine whether each problem is an expression or an equation.

b. Solve the equations and simplify the expressions.

93. $\sqrt{x + 6} = 5$
94. $\sqrt{18x^2y^4}$
95. $x^2 + 8x + 12 = 0$
96. $\dfrac{10}{x + 2} = \dfrac{12}{x + 3}$
97. $(n + 5)(3n - 7) + 8n^2 - 4n$
98. $5(h + 4) - 3 = 77$
99. $\sqrt{x - 4} + 1 = \sqrt{x + 3}$
100. $\dfrac{x + 5}{x^2 + 3x + 2} \cdot \dfrac{x + 2}{x^2 + 6x + 5}$

Chapter Summary

Section 8.1 From Squaring a Number to Roots and Radicals

- **Square roots** are used to "undo" a square. We say that the **principal** square root of b is a ($\sqrt{b} = a$) if $a^2 = b$ and $a \geq 0$.
- A **radical symbol** is $\sqrt{}$. The value under the radical symbol is called the **radicand**.
- **Perfect squares** are values that have rational square roots.
- **Square roots** of negative numbers are not real numbers.
- **Square roots** of values that are not perfect squares may be approximated by using a calculator.
- A **radical expression** is an expression in which a variable expression is under the radical.
- A **radical equation** is an equation where a variable expression is under the radical.
- Square roots show up in formulas like the **distance formula.** The distance formula gives the distance between two points (x_1, y_1) and (x_2, y_2). The distance d between these two points is given by $d = \sqrt{(x_2 - x_1)^2 + (y_2 - y_1)^2}$.
- **Cube roots** "undo" cubes or third powers. The cube root of b is a ($\sqrt[3]{b} = a$) if $a^3 = b$.
- **Perfect cubes** are values that have rational cube roots.
- To **evaluate a radical expression involving square roots,** substitute in the given value of the variable. Simplify the resulting expression.
- To **simplify a radical expression involving square roots,** rewrite the radicand with perfect square factors. Then take the square root.

Example 1 Find each square root or cube root without a calculator.

a. $\sqrt{36x^4}$ b. $\sqrt[3]{-64}$

SOLUTION

a. $\sqrt{36x^4} = 6x^2$ because $(6x^2)^2 = 6^2(x^2)^2 = 36x^4$

b. $\sqrt[3]{-64} = -4$ because $(-4)^3 = -64$

Example 2 Find the distance between the two points $(-3, 6)$ and $(1, -2)$. Round to two decimal places if necessary.

SOLUTION To find the distance between the two points, we use the distance formula. Let $(x_1, y_1) = (-3, 6)$ and $(x_2, y_2) = (1, -2)$. Substituting these values in the distance formula yields

$$d = \sqrt{(x_2 - x_1)^2 + (y_2 - y_1)^2}$$
$$d = \sqrt{(1 - (-3))^2 + (-2 - 6)^2}$$
$$d = \sqrt{(4)^2 + (-8)^2}$$
$$d = \sqrt{16 + 64}$$
$$d = \sqrt{80} \approx 8.94$$

Section 8.2 Basic Operations with Radical Expressions

- **The product property of radicals** states that $\sqrt{a \cdot b} = \sqrt{a} \cdot \sqrt{b}$ when both a and b are positive or zero.
- To **simplify expressions under a square root**, use the following steps:
  1. Factor the radicand using perfect square factors.
  2. Rewrite the radical expression using the product property of radicals.
  3. Simplify the radicals that contain perfect squares.
- Radicals may **be simplified or approximated.** The simplified form is the exact answer, and the approximated form is a rounded answer.
- **To simplify variables under square roots,** first write the radicand with perfect square factors. Then use the product property of radicals to rewrite into the product of two or more radicals. Finally, evaluate any radicals with perfect squares in the radicand.
- **Like terms for radicals** means the terms (after simplification) have the same variable part and the same radical part.
- **To add or subtract square roots,** use the following steps:
  1. Simplify each radical by factoring out the perfect square factors.
  2. Add or subtract terms with like radicals by combining their coefficients.

Example 3 Simplify each radical expression.

a. $\sqrt{150}$ b. $\sqrt{40x^5}$

SOLUTION

a. $\sqrt{150} = \sqrt{25 \cdot 6} = \sqrt{25} \cdot \sqrt{6} = 5\sqrt{6}$

b. $\sqrt{40x^5} = \sqrt{4 \cdot 10 \cdot x^4 \cdot x}$
$= \sqrt{4} \cdot \sqrt{10} \cdot \sqrt{x^4} \cdot \sqrt{x}$
$= 2 \cdot \sqrt{10} \cdot x^2 \cdot \sqrt{x}$
$= 2x^2\sqrt{10x}$

Example 4 Simplify each radical expression by adding or subtracting like terms.

a. $-17 + 3\sqrt{x} + 22 - 5\sqrt{x}$ b. $\sqrt{45x^3} - 8x\sqrt{5x}$

SOLUTION

a. $-17 + 3\sqrt{x} + 22 - 5\sqrt{x} = 5 - 2\sqrt{x}$

b. $\sqrt{45x^3} - 8x\sqrt{5x} = \sqrt{9 \cdot 5 \cdot x^2 \cdot x} - 8x\sqrt{5x}$
$= 3x\sqrt{5x} - 8x\sqrt{5x}$
$= -5x\sqrt{5x}$

Section 8.3 Multiplying and Dividing Radical Expressions

- To **multiply radical expressions,** use the following steps:
  1. Use the product property of radicals to rewrite as a single radical expression; $\sqrt{a} \cdot \sqrt{b} = \sqrt{a \cdot b}$. Here both a and b are values that are positive or zero.
  2. Simplify the radicand by factoring into perfect square factors. Factor out the perfect squares.

CHAPTER 8 Summary   733

- **Squaring undoes a square root.** Therefore, $\sqrt{x} \cdot \sqrt{x} = (\sqrt{x})^2 = x$. Assume that $x \geq 0$.
- **Multiplying radical expressions involving square roots** may involve either the distributive property or FOIL, depending on how many terms are in each expression. Simplify each term by factoring radicands into perfect square factors. Then apply the product property for radicals to factor out the perfect squares, which can be evaluated. Finally, add or subtract any like terms.
- The **quotient property of radicals** is $\sqrt{\dfrac{a}{b}} = \dfrac{\sqrt{a}}{\sqrt{b}}$. Assume that $a \geq 0$ and $b > 0$.
- To **divide radicals involving square roots,** use the quotient property to rewrite as a single radical. Then simplify the radicand.
- The **conjugate** of the radical expression $\sqrt{a} + \sqrt{b}$ is $\sqrt{a} - \sqrt{b}$. Assume that $a \geq 0$ and $b \geq 0$.
- To **rationalize a denominator that contains just one term,** use the following steps:
  1. Multiply both the numerator and the denominator by the radical part of the denominator.
  2. Simplify the numerator and denominator separately. When the simplification is done, there should be no radicals in the denominator.
- To **rationalize a denominator that contains two terms,** use the following steps:
  1. Multiply by the numerator and the denominator by the conjugate of the denominator.
  2. Multiply and simplify the numerator and denominator separately.

Example 5 Multiply or divide as indicated and simplify.

  a. $\sqrt{3x^3} \cdot \sqrt{8x^4}$ b. $\dfrac{\sqrt{108}}{\sqrt{3}}$

SOLUTION   a. $\sqrt{3x^3} \cdot \sqrt{8x^4} = \sqrt{3x^3 \cdot 8x^4} = \sqrt{24x^7}$
$= \sqrt{4 \cdot 6 \cdot x^6 \cdot x}$
$= 2x^3\sqrt{6x}$

  b. $\dfrac{\sqrt{108}}{\sqrt{3}} = \sqrt{\dfrac{108}{3}} = \sqrt{36} = 6$

Example 6 Rationalize the denominator of each radical expression and simplify.

  a. $\dfrac{2}{\sqrt{5}}$ b. $\dfrac{3}{4 - \sqrt{2}}$

SOLUTION   a. $\dfrac{2}{\sqrt{5}} = \dfrac{2}{\sqrt{5}} \cdot \dfrac{\sqrt{5}}{\sqrt{5}} = \dfrac{2\sqrt{5}}{5}$

  b. $\dfrac{3}{4 - \sqrt{2}} = \dfrac{3}{(4 - \sqrt{2})} \cdot \dfrac{(4 + \sqrt{2})}{(4 + \sqrt{2})}$
$= \dfrac{3 \cdot 4 + 3\sqrt{2}}{16 + 4\sqrt{2} - 4\sqrt{2} - 2}$
$= \dfrac{12 + 3\sqrt{2}}{16 - 2} = \dfrac{12 + 3\sqrt{2}}{14}$

Section 8.4 Solving Radical Equations

- **A solution to a radical equation** is a value of the variable that, when substituted into the equation, makes the equation true.
- The **squaring property of equality** says that if $a = b$ then $a^2 = b^2$. This means that it is valid to square both sides of an equation.
- **To solve a radical equation,** use the following steps:
    1. Isolate a radical on one side of the equal sign.
    2. Use the squaring property of equality to square both sides of the equation. If a radical remains, repeat steps 1 and 2.
    3. Solve for the variable.
    4. Check the answer in the original equation.
- **Check all answers to radical equations,** as some answers may not check.

Example 7   Solve and check: $3\sqrt{x} - 7 = 14$.

SOLUTION   First isolate the variable part, which is under the radical.

$3\sqrt{x} - 7 = 14$ Add 7 to both sides to isolate the radical term.
$\phantom{3\sqrt{x}} +7 \ +7$
$3\sqrt{x} = 21$ Divide both sides by 3 to completely isolate the radical.
$\dfrac{3\sqrt{x}}{3} = \dfrac{21}{3}$
$\sqrt{x} = 7$ Square both sides to undo the square root.
$(\sqrt{x})^2 = 7^2$
$x = 49$

Check: $x = 49$
$3\sqrt{49} - 7 \stackrel{?}{=} 14$
$3 \cdot 7 - 7 \stackrel{?}{=} 14$
$14 = 14$ The answer checks.

Example 8   Solve and check: $\sqrt{3y + 2} = \sqrt{5y - 2}$.

SOLUTION   This equation has radicals on both sides that are isolated. Therefore, square both sides to undo the square roots and solve for the variable.

$\sqrt{3y + 2} = \sqrt{5y - 2}$ Square both sides to undo the square roots.
$(\sqrt{3y + 2})^2 = (\sqrt{5y - 2})^2$
$3y + 2 = 5y - 2$ Solve the linear equation for y.
$-3y -3y$
$2 = 2y - 2$
$+2 +2$
$4 = 2y$
$\dfrac{4}{2} = \dfrac{2y}{2}$ Divide both sides by 2 to solve for y.
$2 = y$

Check: $y = 2$
$\sqrt{3(2) + 2} \stackrel{?}{=} \sqrt{5(2) - 2}$ Check the solution by substituting $y = 2$ into the original equation.
$\sqrt{8} = \sqrt{8}$ The answer checks.

Chapter Review Exercises

For Exercises 1 through 4, find the square root of each expression without using a calculator. If the square root does not exist, write "Not real" as the answer.

1. $-\sqrt{36}$
2. $-\sqrt{100}$ [8.1]
3. $\sqrt{-9}$
4. $\sqrt{-25}$ [8.1]

For Exercises 5 through 8, find the cube root of each expression without using a calculator.

5. $\sqrt[3]{27}$
6. $\sqrt[3]{125}$ [8.1]
7. $\sqrt[3]{-64}$
8. $\sqrt[3]{-1}$ [8.1]

For Exercises 9 and 10, answer the questions for each application problem.

9. A science class determines that the time t in seconds it takes for a ball to fall to the ground when dropped from a height of h meters can be modeled by the equation

$$t = \sqrt{\frac{h}{4.9}}$$

   a. How long will it take the ball to fall when dropped from 25 meters? Round the answer to two decimal places if necessary.
   b. How long will it take the ball to fall when dropped from 12 meters? Round the answer to two decimal places if necessary. [8.1]

10. A science class determines that the time t in seconds it takes for a ball to fall to the ground when dropped from a height of h meters can be modeled by the equation

$$t = \sqrt{\frac{h}{4.9}}$$

    a. How long will it take the ball to fall when dropped from 32 meters? Round the answer to two decimal places if necessary.
    b. How long will it take the ball to fall when dropped from 9 meters? Round the answer to two decimal places if necessary. [8.1]

11. Find the distance between the two points $(-3, 1)$ and $(-2, -6)$. Round to two decimal places if necessary. [8.1]

12. Find the distance between the two points $(-5, -2)$ and $(-3, 1)$. Round to two decimal places if necessary. [8.1]

For Exercises 13 through 16, evaluate each radical expression for the given value of the variable. If the square root does not exist, write "Not real" as the answer.

13. $\sqrt{4x - 4}$ for $x = 5$ [8.1]
14. $\sqrt{-2x}$ for $x = -2$ [8.1]
15. $\sqrt{1 - x^2}$ for $x = 3$ [8.1]
16. $\sqrt{x^2 - 6}$ for $x = 4$ [8.1]

For Exercises 17 through 22, simplify each radical expression. Assume that the variables have positive values.

17. $\sqrt{y^{10}}$
18. $\sqrt{y^6}$ [8.2]
19. $\sqrt{25x^4}$
20. $\sqrt{36x^6}$ [8.2]
21. $\sqrt{x^6y^2}$
22. $\sqrt{x^2y^4}$ [8.2]

For Exercises 23 through 30, simplify each radical expression.

23. $\sqrt{75}$
24. $\sqrt{54}$ [8.2]
25. $\sqrt{x^7}$
26. $\sqrt{x^{11}}$ [8.2]
27. $\sqrt{4y^3}$
28. $\sqrt{9y^{15}}$ [8.2]
29. $\sqrt{20x^3}$
30. $\sqrt{24x^9}$ [8.2]

For Exercises 31 through 34, simplify and then approximate each radical expression by rounding to two decimal places.

31. $\sqrt{27}$
32. $\sqrt{45}$ [8.2]
33. $\sqrt{200}$
34. $\sqrt{250}$ [8.2]

For Exercises 35 through 38, add or subtract as indicated. Assume that the variables have positive values.

35. $-\sqrt{3x} + 7\sqrt{3x}$ [8.2]
36. $2\sqrt{6y} + 9\sqrt{6y}$ [8.2]
37. $\sqrt{16x^3} + 7 - 2x\sqrt{x} - 3$ [8.2]
38. $11y - 2\sqrt{y^3} + 6y - y\sqrt{9y}$ [8.2]

For Exercises 39 through 46, multiply the radicals and simplify. Assume that the variables have positive values.

39. $\sqrt{3} \cdot \sqrt{6}$
40. $\sqrt{5} \cdot \sqrt{10}$ [8.3]
41. $\sqrt{2y} \cdot \sqrt{8y}$
42. $\sqrt{5x} \cdot \sqrt{8x}$ [8.3]
43. $\sqrt{3x^3} \cdot \sqrt{8x^2}$
44. $\sqrt{2x^2} \cdot \sqrt{22x^5}$ [8.3]
45. $\sqrt{7a} \cdot \sqrt{7a}$
46. $\sqrt{3b^3} \cdot \sqrt{3b^3}$ [8.3]

For Exercises 47 through 54, multiply the radical expressions and simplify.

47. $\sqrt{5}(\sqrt{10} + \sqrt{5})$
48. $\sqrt{3}(\sqrt{6} + \sqrt{3})$ [8.3]
49. $-2(\sqrt{8} - 3\sqrt{x})$
50. $-7(4\sqrt{a} + \sqrt{7})$ [8.3]
51. $(3 - \sqrt{2})(1 + \sqrt{6})$ [8.3]
52. $(5 + \sqrt{3})(2 - \sqrt{12})$ [8.3]
53. $(4 + \sqrt{6})(4 - \sqrt{6})$ [8.3]
54. $(5 - \sqrt{11})(5 + \sqrt{11})$ [8.3]

For Exercises 55 through 60, find the conjugate of each expression. Assume that the variables have positive values.

55. $3 + \sqrt{5}$
56. $2 + \sqrt{7}$ [8.3]
57. $9 - \sqrt{3}$
58. $6 - \sqrt{11}$ [8.3]
59. $-5 - \sqrt{x}$
60. $-9 - \sqrt{a}$ [8.3]

For Exercises 61 through 64, divide the radical expressions and simplify. Assume that the variables have positive values.

61. $\dfrac{\sqrt{32}}{\sqrt{2}}$
62. $\dfrac{\sqrt{75}}{\sqrt{3}}$ [8.3]
63. $\dfrac{\sqrt{54x^3}}{\sqrt{6x}}$
64. $\dfrac{\sqrt{98x^5}}{\sqrt{2x}}$ [8.3]

For Exercises 65 through 68, rationalize the denominator of each radical expression and simplify.

65. $\dfrac{5}{\sqrt{7}}$
66. $\dfrac{-3}{\sqrt{5}}$ [8.3]
67. $\dfrac{4}{3 - \sqrt{2}}$
68. $\dfrac{-3}{1 + \sqrt{5}}$ [8.3]

For Exercises 69 through 72, determine whether each given value of the variable is an answer of the radical equation.

69. $\sqrt{3x - 3} + 5 = 8$ for $x = 4$ [8.4]
70. $\sqrt{5x - 9} - 3 = 1$ for $x = 5$ [8.4]
71. $\sqrt{5x - 1} + 1 = 2$ for $x = 2$ [8.4]
72. $\sqrt{3x - 10} + 3 = 5$ for $x = 2$ [8.4]

For Exercises 73 through 80, solve each radical equation. Check the answer in the original equation.

73. $\sqrt{1 - a} = 3$ [8.4]
74. $\sqrt{b - 7} = 6$ [8.4]
75. $\sqrt{5 - 2x} = 7$ [8.4]
76. $\sqrt{3y - 8} = 4$ [8.4]
77. $\sqrt{x^2 - 10x + 41} = 4$ [8.4]
78. $\sqrt{x^2 - 14x + 58} = 3$ [8.4]
79. $\sqrt{3a - 5} = -3$ [8.4]
80. $\sqrt{1 - 2a} = -1$ [8.4]

For Exercises 81 through 84, solve each application problem.

81. Nelson's profit P for selling x baskets of blueberries is given by $P = \sqrt{x + 24}$. How many baskets of blueberries does he need to sell to make a profit of $12.00? [8.4]

82. Tisha's profit P for selling x baskets of tomatoes is given by $P = \sqrt{x - 2}$. How many baskets of tomatoes does she need to sell to make a profit of $10.00? [8.4]

83. The time, t, in seconds that a rock falls to the ground if dropped from a height of h feet is given by $t = \dfrac{\sqrt{h}}{4}$. From what height must the rock be dropped if it falls to the ground in 6 seconds? [8.4]

84. The time, t, in seconds that a rock falls to the ground if dropped from a height of h feet is given by $t = \dfrac{\sqrt{h}}{4}$. From what height must the rock be dropped if it falls to the ground in 4.5 seconds? [8.4]

For Exercises 85 through 92, determine whether each equation is linear, quadratic, rational, or radical. Do not solve the equations.

85. $6x + 2 = -x^2$
86. $9 - x = 3x^2$ [8.4]
87. $\sqrt{x^2 - 4x} = 5$ [8.4]
88. $-3 = -6 + \sqrt{3x - 7}$ [8.4]
89. $\dfrac{1}{-6x + 5} = 3 - \dfrac{1}{x}$ [8.4]
90. $4 + \dfrac{3}{x - 1} = 7$ [8.4]
91. $-3x + 2(x - 1) = 0$ [8.4]
92. $12 = 7(3 - 5x) + 1$ [8.4]

Chapter Test

The chapter test should take approximately one hour to complete. Show all of your work. Assume that variable expressions under radicals have positive values.

1. Find the square root without a calculator: $-\sqrt{81}$.
2. Find the cube root without a calculator: $\sqrt[3]{-27}$.
3. Simplify the radical expression: $-\sqrt{36a^8}$.
4. Simplify the radical expression: $\sqrt{a^8 b^2}$.
5. Find the distance between the points $(-3, 5)$ and $(4, -2)$.
6. Evaluate the radical expression for the given value of the variable: $\sqrt{6x+1}$ for $x = 4$.
7. Simplify the radical expression: $\sqrt{\dfrac{x^2}{9}}$.
8. Simplify the radical expression: $\sqrt{60}$.
9. Approximate to two decimal places: $\sqrt{24}$.
10. Simplify the radical expression: $\sqrt{28b^5}$.
11. Simplify the radical expression: $\sqrt{25a^3 b^2}$.
12. Add and subtract as indicated: $7\sqrt{3} - 11 - \sqrt{3} + 8$.
13. Simplify, then add as indicated: $\sqrt{50} + 9\sqrt{2}$.
14. Simplify, then subtract as indicated: $\sqrt{24x^2} - x\sqrt{6}$.
15. Multiply and simplify: $\sqrt{6} \cdot \sqrt{8}$.
16. Multiply and simplify: $\sqrt{5x^3} \cdot \sqrt{10x^2}$.
17. Multiply and simplify: $\sqrt{8}(\sqrt{2} + \sqrt{5})$.
18. Multiply and simplify: $(1 + 3\sqrt{5})(2 - 4\sqrt{5})$.
19. Multiply and simplify: $(5 - \sqrt{2})(5 + \sqrt{2})$.
20. Divide and simplify: $\dfrac{\sqrt{125}}{\sqrt{5}}$.
21. Rationalize the denominator and simplify: $\dfrac{-3}{\sqrt{6}}$.
22. Rationalize the denominator and simplify: $\dfrac{5}{2 + \sqrt{3}}$.
23. Solve and check: $\sqrt{a} - 7 = 15$.
24. Solve and check: $2\sqrt{y} + 6 = 3$.
25. Solve and check: $\sqrt{x^2 + 2x - 6} = 3$.

Chapter Projects

■ Speed of a Tsunami

Tsunamis are giant ocean waves that are most often caused by water being displaced by undersea volcanoes or large earthquakes. Tsunamis travel extremely fast over the deep ocean, at speeds up to 500 miles per hour. As they approach land, tsunamis slow down to 20 to 30 mph.

The formula to calculate the speed of a tsunami is

$$s = \sqrt{g \cdot d}$$

Here s is the speed of the tsumani in feet per second, $g = 32.2 \, \dfrac{\text{ft}}{\text{sec}^2}$ is the acceleration due to gravity, and d is the depth in feet of water it is traveling through.

On December 26, 2004, a huge earthquake of magnitude 9.1 struck the Indian Ocean. This was the third largest earthquake ever recorded, and it lasted 8 to 10 minutes. The quake triggered an enormous tsunami that killed approximately 250,000 people. To get an idea of the speed this wave traveled at, answer the following questions.

1. Suppose the water depth the tsunami traveled over was 1500 meters. Convert this depth to feet.
2. Use the formula to determine the speed of the tsunami at this depth.
3. The units of the speed found in step 2 are _____ per _____.
4. Convert the speed to miles per hour.
5. Research online why the Pacific Rim is subject to tsunamis. Write a few paragraphs explaining why tsunamis occur in that part of the world.

■ Heart Rate of Mammals

Written Project
One or more people

What you will need
- Information regarding another mammal. You can use online resources or other library resources.
- Follow the MLA style guide for all citations. If your school recommends another style guide, use that.

The formula that models the relationship between body size and the resting heart rate of a mammal is

$$H = \frac{241}{\sqrt[4]{W}}$$

Here H is the heart rate in beats per minute, and W is the weight in kilograms.

1. An average-sized Labrador retriever (dog) weighs 75 pounds. Convert this weight to kilograms.
2. Use the formula to estimate the heart rate of a 75-pound dog.
3. A horse weighs 1200 pounds. Convert this weight to kilograms.
4. Use the formula to estimate the heart rate of a 1200-pound horse.
5. Find online the average weight of one more mammal (nonhuman). Use the formula to estimate the heart rate. Cite any sources you used, following the MLA style guide or your school's recommended style guide.
6. Find online the real resting heart rate for the mammal you picked in step 5. How does that compare to the estimate made by using the formula?

■ Write a Review of a Section for Presentation

Research Project
One or more people

What you will need
- This presentation may be done in class or as an online video.
- If it is done as a video, post it on a website where it can be easily viewed by the class.
- You might want to use homework or other review problems from the book.
- Make it creative and fun.

Create a 5-minute review presentation of one section from Chapter 8. The format of the presentation can be a poster presentation, a blackboard presentation from notes, an online video, or a game format (e.g., math jeopardy). The presentation should include the following:

- Any important formulas in the chapter
- Important skills in the chapter, backed up by examples
- Explanation of terms used (e.g., what does *simplifying a radical* mean to do?)
- Common mistakes and how to avoid them

Cumulative Review — Chapters 1-8

1. Write in exponential form:
   $(-3a) \cdot (-3a) \cdot (-3a) \cdot (-3a)$

2. Find the value without a calculator: $\left(-\dfrac{2}{3}\right)^4$.

3. Simplify: $-6 \cdot 4 \div 12 + (-4)^2 - (-2 - 5)$.

4. Evaluate the expression for the given value of the variable: $-5x + 6 - x^2$ for $x = -2$.

5. Simplify: $-2(3x^2 - 6x + 1) - 7(3x + 4)$.

6. Solve for y: $-7x + 3y = 6$.

7. Solve: $9x - 4(x + 2) = 2x - 23$.

8. Solve: $\dfrac{7n}{2} - \dfrac{7}{4} = n + 2$.

9. Gustavo works as a tutor and makes $11 per hour after taxes. He can compute his monthly salary using the equation
   $$S = 11h$$
   where S is the monthly salary in dollars that Gustavo makes if he works h hours that month.
   a. Find Gustavo's monthly salary if he works 48 hours.
   b. Find Gustavo's monthly salary if he works 32 hours.
   c. How many hours does Gustavo have to work a month if he wants to earn $800 a month?

10. Solve the inequality: $-3(x - 5) > 4$.

11. Donna is a coordinator for a boys' and girls' club summer program. She would like to take a group of kids to see a 3-D movie. The tickets cost $17.50 each plus an additional fee of $3.00 per person for the 3-D glasses. If she plans on taking 20 children to see a 3-D movie, how much will it cost?

12. Use the equation $y = |x - 3|$ to create a table of nine or more points and graph them. Connect the points with a smooth curve. Clearly label and scale the axes. Is the graph linear or nonlinear?

13. Find the equation of the line that goes through the two points $(6, 0)$ and $(-3, 6)$. Graph the line, clearly labeling and scaling the axes.

14. Find the x-intercept and the y-intercept of the equation $4x - 5y = 20$. Graph the line using the intercepts. Clearly label and scale the axes.

15. Find the slope and the y-intercept of the line $3x - 5y = 10$.

16. Use the graph to find the equation of the line shown in the graph. Put the answer in $y = mx + b$ form.

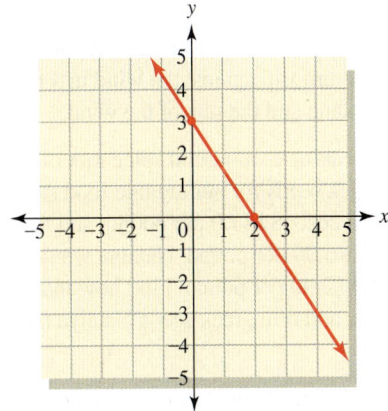

17. Lilja is a medical assistant. She can room five patients an hour. Let the input h represent the number of hours Lilja is working, and let the output P represent the number of patients Lilja can room. Find the linear relationship between h and P.

18. Determine the equation of the line by using point-slope form. The line has slope $m = -\dfrac{1}{5}$ and passes through the point $(-5, 6)$. Put the final answer in $y = mx + b$ form.

19. Find the equation of the line parallel to $y = -4$, passing through the point $(5, 9)$.

20. Find the equation of the line perpendicular to $y = -4x + 7$, passing through the point $(8, -1)$.

21. Simplify, using the order-of-operations agreement. Reduce to simplest form.
    a. $\dfrac{2}{3} \cdot \dfrac{9}{22} + \dfrac{3}{4}$
    b. $6 - \dfrac{3}{10} \div \dfrac{1}{5}$

22. $R = 4.00b$ is an equation that represents the revenue R, in dollars, a youth soccer team takes in for selling b boxes of candy for $4.00 each. Find R when $b = 67$ and explain its meaning in terms of the problem situation.

23. Solve the system, using the substitution method. Check the answer in both original equations.
    $$y = \dfrac{3}{4}x + 5$$
    $$5x + 4y = -12$$

739

24. Solve the system, using the elimination method. Check the answer in both original equations.

$$5x - 3y = 9$$
$$x - y = 5$$

25. A plane travels 360 miles with the wind in 3 hours. It makes the return trip against the wind in 4 hours. Find the rate of the plane and the rate of the wind.

26. Natalya is making a 10-pound spice mix of cumin seeds and fenugreek seeds for her organic food store. The cumin seeds costs $18.00 per pound, and the fenugreek seeds costs $6.00 per pound. Natalya would like to price the mix at $12.00 per pound. Find how many pounds of cumin seeds and fenugreek seeds Natalya has to put in the mix.

27. Solve the system. Check the answer in both original equations.

$$2x - y = 7$$
$$-6x + 3y = -21$$

28. The perimeter of a rectangle is 28 centimeters. The length is 3 centimeters more than the width. Find the length and the width of the rectangle.

29. Aivy has $8000 to invest. She puts part of the money in a CD earning 3.5% simple interest and part in a bond fund earning 5% simple interest. Aivy would like to earn $325.00 a year in interest. How much money should she put in the CD, and how much money should she put in the bond fund?

30. Graph the solution set of the linear inequality $4x - 5y > -20$.

31. Graph the solution set for each system of inequalities. Check the solution set with a sample test point.

$$x \geq -3$$
$$x - 4y < 8$$

32. Simplify.

  a. $(-5x)^2$
  b. $-(5x)^2$
  c. $(3x^3y)(-4xy^4)$

33. Simplify.

  a. $(-3x^4y^3)^3$
  b. $-6a^4 \cdot 3ab^5$
  c. $(-3xy^2)^3(2x^3y)^2$

34. Simplify.

  a. $\dfrac{t^6}{t^2}$

  b. $\dfrac{3b^3}{-12b^2}$

  c. $\dfrac{-15x^3y^2}{-3x^3y}$

35. Simplify: $\left(\dfrac{-15x^4y^2}{50x^3y}\right)^2$.

36. Rewrite the expression using positive exponents.

  a. $-3x^{-2}$

  b. $\dfrac{5}{a^{-4}}$

37. Simplify. Write the final answer using positive exponents only.

  a. $\dfrac{-4a^2b}{2a^3b^3}$ **b.** $(-2x^2y^{-3})^2$ **c.** $\left(\dfrac{5x}{y}\right)^{-3}$

38. Write the number using scientific notation.

  a. 0.000195
  b. 52,890,000,000

39. Convert the number in scientific notation to standard notation.

  a. 3.004×10^{-5}
  b. 9.27×10^8

40. Identify the number of terms and find the degree of the polynomial.

$$8a^3 - a^2 + 3a$$

41. Find the degree of each term and of the polynomial.

$$5a^3b + 3a^5 - 6b + ab^2$$

42. Add: $(-3b^2 + 2b + 7) + (-b^2 + 5b - 1)$.

43. Subtract: $(3x - 6) - (-x + 5)$.

44. Multiply: $5(5y - 8)$.

45. Multiply and simplify.

  a. $(2x + 5)(x - 7)$
  b. $(y - 1)(3y^2 + 2y - 1)$
  c. $(3x - 4)^2$
  d. $3(7a - 2) - 4(-3a + 5)$

46. Divide and simplify: $\dfrac{3x^2 - 6x + 9}{-3x}$.

47. Divide using long division: $(3x^2 - 2x - 5) \div (x + 1)$.

48. Factor.

  a. $6x^3y - 18xy + 24x$
  b. $5xy + 15x + 2y + 6$

49. Factor the polynomial completely:
$6ab + 36a - 4b - 24$.

50. Factor the trinomials.
  a. $x^2 - 5x - 14$
  b. $-n^2 - 3n + 18$
  c. $20y^2 + 11y - 3$

51. Factor the polynomial completely: $16z^3 + 20z^2 - 6z$.

52. Factor the polynomials.
  a. $25 - x^2$
  b. $9y^2 - 1$
  c. $9t^2 - 12t + 4$
  d. $x^2 + 16$
  e. $3y^2 - 12$

53. Determine whether the following equations are linear, quadratic, or neither. *Do not* solve the equations.
  a. $-3(x + 5) - 2x = 9$
  b. $x^5 - 2x = 5$
  c. $(x + 2)(x - 1) = 0$

54. Solve the quadratic equations by factoring.
  a. $3x^2 - 10x - 8 = 0$
  b. $3x(x + 5) = 0$
  c. $n^2 - 81 = 0$

55. Darren purchases an iPhone for $649.00. The sales tax in his area is 6%. What is the total amount he paid for his new iPhone?

56. Sami purchases a condo for $205,000. Two years later, she sells it for $235,000. What is the percent increase?

For Exercises 57 through 59, use the following model. Disney's annual park and resort operating income is modeled by $I = 0.335t + 1.249$, where I is the annual park and resort operating income in billions of dollars t years since 2010.

Source: Model based on data from the Walt Disney Company.

57. Use the model to estimate Disney's annual park and resort operating income in the year 2020.

58. What is the slope of this model and what does it represent in this situation?

59. What is the vertical intercept of this model and what does it represent in this situation?

60. Evaluate the rational expression for the given value of the variable.

$$\frac{2x - 4}{x + 2} \quad \text{for} \quad x = 2$$

61. The parents of a youth soccer team are purchasing a plaque for their championship team. The plaque and the engraving will cost $60.00. The cost per person depends on the number of people who agree to help pay. If *n* parents agree to pay equal shares, the cost per person in dollars can be found using the equation

$$c = \frac{60}{n}$$

where *c* is the cost per person in dollars.
  a. Find the cost per person if 10 parents agree to help pay for the plaque.
  b. If there are 15 parents total on the team, what is the lowest possible cost per parent?

62. For each rational expression, find the value(s) of the variable that must be excluded.
  a. $\dfrac{3x + 1}{2x - 7}$
  b. $\dfrac{5t}{(t + 3)(3t - 4)}$
  c. $\dfrac{3y}{6y^2 + 13y - 5}$

63. Simplify each rational expression.
  a. $\dfrac{-6(2x + 5)}{x(2x + 5)}$
  b. $\dfrac{n - 4}{n^2 - 3n - 4}$
  c. $\dfrac{t - 2}{t^2 - 4}$

64. An experienced office specialist can do a week's filing for his corporation in 6 hours. When he works with a new, inexperienced office specialist, the two office specialists can complete the job in time *T* (in hours).

$$T = \frac{6x}{6 + x}$$

Here *x* is the time it takes the inexperienced office specialist to do the job alone.
  a. How long will it take the two office specialists working together if the inexperienced specialist can complete the job alone in 10 hours?
  b. How long will it take the two office specialists working together if the inexperienced specialist can complete the job alone in 8 hours?

65. Simplify each rational expression.

a. $\dfrac{5n + 7}{2n^2 + 17n + 35}$

b. $\dfrac{4x^2 - 19x - 5}{(x + 7)(x - 5)}$

c. $\dfrac{12 - 4x}{x - 3}$

66. Multiply the rational expressions. Simplify.

a. $\dfrac{-4}{9} \cdot \dfrac{27}{30}$

b. $\dfrac{-5xy}{15x^2} \cdot \dfrac{45x}{12y^2}$

c. $\dfrac{x^2 + x - 12}{x^2 + 3x - 10} \cdot \dfrac{(x - 2)}{(x + 4)}$

d. $\dfrac{y^2 + 5y + 6}{y^2 + y - 20} \cdot \dfrac{y^2 + 4y - 5}{y^2 + 8y + 12}$

67. A 1-pound jar of barbeque sauce is priced at $6.29. Convert this price per pound (lb) to dollars per ounce (oz) so that it can be compared to the price of other containers of barbeque sauce. (Round to two decimal places.)

68. A car is traveling at 35 mph. What is the car's speed in feet per minute?

69. Divide the rational expressions. Simplify.

a. $\dfrac{-3}{7} \div \dfrac{9}{28}$

b. $\dfrac{-16x}{5y^2} \div \dfrac{24x^2}{35y}$

c. $\dfrac{(t + 2)(t - 3)}{(t + 1)(t - 4)} \div \dfrac{(t^2 - 9)}{(t^2 - 1)}$

d. $\dfrac{a^2 + 4a + 3}{a^2 + 2a - 8} \div \dfrac{a^2 - a - 2}{a^2 - 16}$

70. Simplify each complex fraction.

a. $\dfrac{\dfrac{-2x^2}{15y}}{\dfrac{12x^3}{5y^2}}$

b. $\dfrac{\dfrac{b + 2}{2b^2 - 8}}{\dfrac{b + 5}{b - 2}}$

71. Add or subtract the rational expressions. Simplify.

a. $\dfrac{6x}{x + 1} + \dfrac{6}{x + 1}$

b. $\dfrac{6t^2}{t - 5} - \dfrac{30t}{t - 5}$

c. $\dfrac{y + 6}{(y - 2)(y + 3)} - \dfrac{(y - 2)}{(y - 2)(y + 3)}$

d. $\dfrac{x - 7}{(x - 1)(x + 5)} + \dfrac{3x}{(x - 1)(x + 5)}$

72. Find the LCD for each set of fractions.

a. $\dfrac{-4}{5x^2y}$, $\dfrac{1}{10xy}$

b. $\dfrac{-7b}{b - 1}$, $\dfrac{6}{b - 2}$

c. $\dfrac{15}{4h^2 - 12h}$, $\dfrac{-2}{h - 3}$

d. $\dfrac{-2}{a^2 - 8a + 15}$, $\dfrac{5}{a^2 + a - 12}$

73. Find the LCD for each set of fractions. Rewrite each fraction as an equivalent fraction over the LCD.

a. $\dfrac{3}{x - 1}$, $\dfrac{-2x}{x + 4}$

b. $\dfrac{5k}{k^2 - 36}$, $\dfrac{8}{k^2 - 11k + 30}$

74. Add or subtract the rational expressions. Reduce the final answers.

a. $\dfrac{3x}{2xy} + \dfrac{7x}{xy}$

b. $\dfrac{y}{y - 1} - \dfrac{5y - 2}{y^2 - 5y + 4}$

75. Simplify each complex fraction.

a. $\dfrac{\dfrac{10}{x} - 5}{\dfrac{6}{x} - 3}$

b. $\dfrac{\dfrac{3}{x} - \dfrac{2}{y}}{\dfrac{1}{x} + \dfrac{4}{y}}$

76. Solve each rational equation. Check the answer in the original problem.

a. $6 + \dfrac{3}{x} = \dfrac{23}{3} - \dfrac{2}{x}$

b. $\dfrac{4}{y - 3} + 2 = \dfrac{4}{y - 3}$

c. $\dfrac{2}{k - 2} + 4 = k + 3$

77. A local business is throwing a retirement party for an employee. The cost of the gift is $100, and the cost of lunch for each person attending is $10.00. The average cost per person for the retirement party and gift can be estimated by the equation

$$C = \dfrac{100 + 10n}{n}$$

where C is the cost per person in dollars when n people attend the retirement lunch.

a. Find the average cost per person if 25 people attend the retirement lunch.
b. How many people must attend for the average cost per person to be $12?

78. Solve each rational equation. Check the answer in the original problem.

a. $\dfrac{6}{a+1} = \dfrac{-3}{a-2}$

b. $\dfrac{3}{5b} - \dfrac{1}{5} = \dfrac{-1}{b}$

c. $\dfrac{3}{x+4} + \dfrac{4}{x-2} = \dfrac{x^2 - 8}{(x+4)(x-2)}$

79. Kenzo can complete his lab assignment in 3 hours. What is his work rate?

80. The Women's Club volunteers can copy the monthly announcements using their new photocopier in about 4 hours. The Club's older photocopier can copy the announcements in 6 hours. How long would it take to copy the announcements using both machines?

81. Write the ratio of 9 to 45 in simplified fraction form.

82. Interpret 400 miles for 16 gallons as a rate in fraction form.

83. Determine the unit rates (round to the nearest cent). Determine which item is the better buy: 36 ounces of coffee beans for $31.00 or 48 ounces of coffee beans for $49.50.

84. A blueprint is drawn to a $\dfrac{1}{16}$-inch scale, so 1 inch on the blueprint equals 16 feet in the building. If the length of a real wall is 40 feet, what will its length be on the blueprint?

85. Solve each proportion. Check the answer in the original problem.

a. $\dfrac{3}{n} = \dfrac{-15}{2}$

b. $\dfrac{x-2}{2} = \dfrac{13x+10}{14}$

c. $\dfrac{3t-2}{t+5} = \dfrac{25}{14}$

86. The variable d varies directly with t. Find an equation to represent this relationship if $t = 3$ when $d = 150$. Use the equation to find d when $t = 7.5$.

87. The amount of a weekly paycheck (before taxes), W, varies directly with the amount of time, t, spent working.

a. Find an equation to express this relationship if the amount of the paycheck is $231 for 22 hours of work.

b. Use the equation to find the amount of the weekly paycheck for 32 hours of work.

88. The variable y varies inversely with x. Find an equation to represent this relationship if $x = 12$ when $y = -\dfrac{1}{4}$. Use the equation to find y when $x = 2$.

89. The speed, s, in miles per hour of a car being driven is inversely proportional to the time, t, it has been driven.

a. Find an equation to express this relationship if the speed is 50 mph for 4 hours of driving.

b. Use this equation to find the speed for 6 hours of driving.

90. Find the square root of each number. If the square root does not exist, write "Not real" as the answer.

a. $\sqrt{49}$

b. $\sqrt{\dfrac{25}{64}}$

c. $\sqrt{-100}$

91. Use a scientific calculator to approximate each square root to two decimal places. If the square root does not exist, write "Not real" as the answer.

a. $\sqrt{30}$

b. $\sqrt{-15}$

92. Evaluate each radical expression for the given value of the variable.

a. $\sqrt{x-4}$ for $x = 20$

b. $\sqrt{2x+1}$ for $x = 24$

c. $\sqrt{x^2+1}$ for $x = 0$

93. The distance in miles, d, that can be seen on the surface of the earth is given by $d = 1.5\sqrt{h}$, where h is the height in feet of a person above the surface of the earth. How far can a person see if the person is 64 feet above the surface of the earth?

94. Find the distance between the two points. Round to two decimal places if necessary: $(5, -1)$ and $(-2, 6)$.

95. Simplify each radical expression. Assume that the variables have positive values.

a. $\sqrt{x^6}$

b. $\sqrt{25b^4}$

c. $-\sqrt{81y^{12}}$

96. Find each cube root without a calculator.

a. $\sqrt[3]{64}$ b. $\sqrt[3]{-1}$

97. Simplify each radical expression.

a. $\sqrt{45}$ b. $\sqrt{40}$ c. $\sqrt{147}$

98. Simplify: $\sqrt{44}$.

99. Approximate to two decimal places: $\sqrt{44}$.

100. Simplify each radical expression. Assume that all variables under radicals represent positive values.

a. $\sqrt{x^{11}}$ b. $\sqrt{20b^5}$ c. $\sqrt{300c^9}$

101. Simplify each radical expression by adding or subtracting like terms. Assume that all variables under radicals represent positive values.

a. $-4\sqrt{3} + 9\sqrt{3}$
b. $3\sqrt{t} - 5\sqrt{7}$
c. $6\sqrt{x} - 6y - 4\sqrt{x} + 2y$

102. $-\sqrt{64n^2} + 9n$

a. Simplify all radicals in each term.
b. Add or subtract like terms.

103. $\sqrt{36} + \sqrt{16y^5} - \sqrt{49y^5}$

a. Simplify all radicals in each term.
b. Add or subtract like terms.

104. Multiply the radicals using the product property of radicals. Simplify.

a. $\sqrt{27} \cdot \sqrt{3}$
b. $\sqrt{5} \cdot (-\sqrt{15})$
c. $\sqrt{3x^3} \cdot \sqrt{12x}$
d. $\sqrt{2n} \cdot \sqrt{2n}$
e. $\sqrt{7xy^2} \cdot \sqrt{7xy}$

105. Find the area of the triangle. Simplify the radical expression.

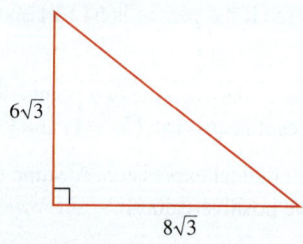

106. Multiply each radical expression, using the distributive property and simplify.

a. $\sqrt{7}(8 - \sqrt{7})$
b. $\sqrt{x}(\sqrt{8} + \sqrt{x})$

107. Multiply the radical expressions and simplify.

a. $(6 - \sqrt{2})(6 + \sqrt{2})$
b. $(3 - \sqrt{5})^2$
c. $(\sqrt{n} + \sqrt{2})(\sqrt{n} - \sqrt{3})$

108. Divide the rational expressions and simplify.

a. $\dfrac{\sqrt{125}}{\sqrt{5}}$ b. $\dfrac{\sqrt{2}}{\sqrt{81}}$ c. $\dfrac{\sqrt{2x^3}}{\sqrt{128x}}$

109. Rationalize the denominator of each radical expression and simplify.

a. $\dfrac{9}{\sqrt{5}}$

b. $\dfrac{\sqrt{3}}{\sqrt{10}}$

c. $\dfrac{2}{3 + \sqrt{7}}$

110. Determine whether the given value of the variable is a solution of the radical equation.

a. $\sqrt{x - 2} + 7 = 11$ for $x = 18$
b. $\sqrt{b^2 - 1} = 1$ for $b = 0$

111. Solve each radical equation. Check the answer.

a. $\sqrt{y} = 7$
b. $\sqrt{n - 3} = 5$
c. $\sqrt{x - 11} = 2$
d. $3\sqrt{x} + 5 = 2$

112. The time t in seconds it takes a ball to fall to the ground when dropped from a height of h meters is given by the equation $t = \sqrt{\dfrac{h}{4.5}}$.

a. If the ball falls for 2.1 seconds, from what height was it originally dropped?
b. If the ball falls for 3.7 seconds, from what height was it originally dropped?

113. Solve each radical expression. Check the answer.

a. $\sqrt{y + 1} = 3$
b. $\sqrt{2x - 1} = \sqrt{2 + x}$
c. $\sqrt{n - 7} - 3 = -2$

114. Determine whether each equation is linear, quadratic, rational, or radical. *Do not* solve the equations.

a. $\dfrac{2}{x - 5} = \dfrac{4}{1 - x}$
b. $-5x + 1 = 3x^2$
c. $\sqrt{2n - 3} = 8$
d. $-2x + 5 = 4x - 3$

Quadratic Equations

9

9.1 Graphing Quadratic Equations

9.2 Solving Quadratic Equations by Using the Square Root Property

9.3 Solving Quadratic Equations by Completing the Square and the Quadratic Formula

9.4 Graphing Quadratic Equations Including Intercepts

9.5 Working with Quadratic Models

9.6 The Basics of Functions

After a ball has been hit, it follows a U-shaped path called a *parabola*. A model rocket, launched from ground level, will also follow a U-shaped path, or a parabolic path. The algebraic formula for a parabola is a quadratic equation in two variables. To determine how high a ball or a rocket goes can be done using algebra. Algebra can also be used to find when the ball or rocket hits the ground.

9.1 Graphing Quadratic Equations

LEARNING OBJECTIVES
- Graph a quadratic equation by plotting points.
- Understand the relationship between the leading coefficient a and the graph.
- Find the vertex.
- Use the axis of symmetry.

■ Graphing Quadratic Equations by Plotting Points

We now turn our attention toward graphing a **quadratic equation in two variables**.

> **DEFINITION**
>
> **Quadratic Equation in Two Variables** A quadratic equation in two variables has the form $y = ax^2 + bx + c$, where $a \neq 0$.

This definition looks similar to the quadratic equations we were solving in Section 6.4, which were in the form $ax^2 + bx + c = 0$. What is different is that there are now two variables, x and y. Remember that a linear equation in two variables can be graphed. The same is true for a quadratic equation in two variables.

Name	Degree	What We Do	Example
Quadratic equation in one variable	2	Solve	$9x^2 + 30x + 25 = 0$
Quadratic equation in two variables	2	Graph, solve	$y = x^2 + 4x + 4$
Quadratic expression in one variable	2	Simplify, evaluate, factor	$x^2 - 2x + 6x - 12$

In Chapter 3, we learned how to graph an equation in two variables by creating a table of values. We use that technique here to graph the quadratic equation in two variables, $y = x^2$. This graph will be our *baseline* graph, which will be used to discuss important components of the graph of a quadratic equation in two variables.

To graph the equation $y = x^2$, create a table of values, plot those points, and connect them with a smooth curve. Because x can take on negative values, zero, and positive values, select x-values that are negative, zero, and positive to generate the table.

■ **What's That Mean?**

Baseline

A *baseline* is a standard that we can make comparisons to. A *baseline* graph is the simplest form of a graph. We can compare all similar graphs of $y = ax^2 + bx + c$ to $y = x^2$.

x	$y = x^2$	(x, y)
-3	$(-3)^2 = 9$	$(-3, 9)$
-2	$(-2)^2 = 4$	$(-2, 4)$
-1	$(-1)^2 = 1$	$(-1, 1)$
0	$(0)^2 = 0$	$(0, 0)$
1	$(1)^2 = 1$	$(1, 1)$
2	$(2)^2 = 4$	$(2, 4)$
3	$(3)^2 = 9$	$(3, 9)$

Plot these points in the (x, y) plane and connect the points with a smooth curve. This curve has a special name. It is called a **parabola.** The graph of a quadratic equation in two variables always has this shape.

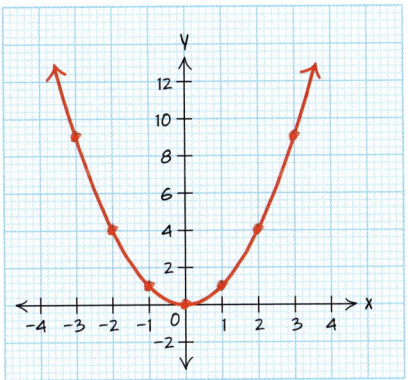

> **DEFINITION**
>
> **Parabola** A parabola is the graph of a quadratic equation in two variables.

To graph a quadratic equation in two variables, create a table of values, plot those points, and connect the points with a smooth curve. Until we learn a bit more about graphing quadratic equations, we will plot seven or more points to make sure we get the shape of the graph correct.

Example 1 Graphing a quadratic equation by plotting points

Graph each quadratic equation in two variables by plotting points. Plot seven or more points. Clearly label and scale the axes.

a. $y = x^2 + 2$ **b.** $y = -x^2$ **c.** $y = x^2 - 2x$

SOLUTION

a. We first create a table of values for this quadratic equation, $y = x^2 + 2$.

x	$y = x^2 + 2$	(x, y)
-3	$(-3)^2 + 2 = 9 + 2 = 11$	$(-3, 11)$
-2	$(-2)^2 + 2 = 4 + 2 = 6$	$(-2, 6)$
-1	$(-1)^2 + 2 = 1 + 2 = 3$	$(-1, 3)$
0	$(0)^2 + 2 = 0 + 2 = 2$	$(0, 2)$
1	$(1)^2 + 2 = 1 + 2 = 3$	$(1, 3)$
2	$(2)^2 + 2 = 4 + 2 = 6$	$(2, 6)$
3	$(3)^2 + 2 = 9 + 2 = 11$	$(3, 11)$

Plotting these points in the (x, y) plane and connecting them with a smooth curve yields the following graph.

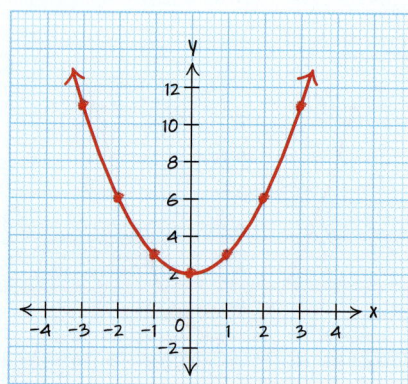

b. Creating a table of values for this quadratic equation, $y = -x^2$, yields the following:

x	$y = -x^2$	(x, y)
-3	$-(-3)^2 = -1 \cdot (-3)^2 = -1 \cdot 9 = -9$	$(-3, -9)$
-2	$-(-2)^2 = -1 \cdot (-2)^2 = -1 \cdot 4 = -4$	$(-2, -4)$
-1	$-(-1)^2 = -1 \cdot (-1)^2 = -1 \cdot 1 = -1$	$(-1, -1)$
0	$-(0)^2 = -1 \cdot (0)^2 = -1 \cdot 0 = 0$	$(0, 0)$
1	$-(1)^2 = -1 \cdot (1)^2 = -1 \cdot 1 = -1$	$(1, -1)$
2	$-(2)^2 = -1 \cdot (2)^2 = -1 \cdot 4 = -4$	$(2, -4)$
3	$-(3)^2 = -1 \cdot (3)^2 = -1 \cdot 9 = -9$	$(3, -9)$

Plotting these points in the (x, y) plane and connecting them with a smooth curve yields the following graph.

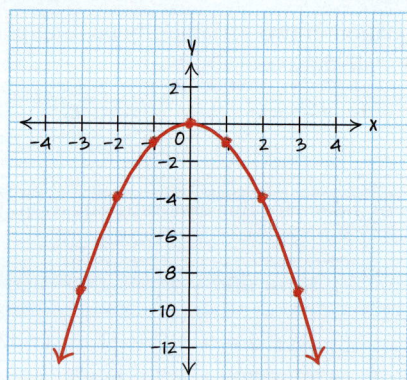

c. Creating a table of values for this quadratic equation, $y = x^2 - 2x$, yields the following:

x	$y = x^2 - 2x$	(x, y)
-3	$(-3)^2 - 2(-3) = 9 + 6 = 15$	$(-3, 15)$
-2	$(-2)^2 - 2(-2) = 4 + 4 = 8$	$(-2, 8)$
-1	$(-1)^2 - 2(-1) = 1 + 2 = 3$	$(-1, 3)$
0	$(0)^2 - 2(0) = 0 - 0 = 0$	$(0, 0)$
1	$(1)^2 - 2(1) = 1 - 2 = -1$	$(1, -1)$
2	$(2)^2 - 2(2) = 4 - 4 = 0$	$(2, 0)$
3	$(3)^2 - 2(3) = 9 - 6 = 3$	$(3, 3)$

Plotting these points in the (x, y) plane and connecting them with a smooth curve yields the following graph.

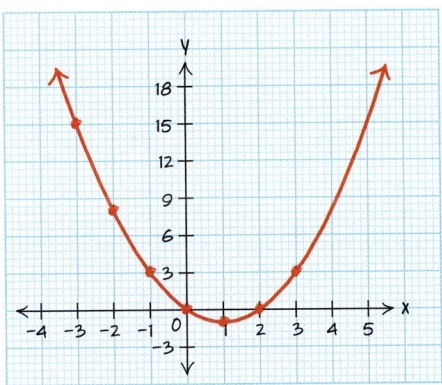

PRACTICE PROBLEM FOR EXAMPLE 1

Graph each quadratic equation in two variables by plotting points. Plot seven or more points. Clearly label and scale the axes.

a. $y = x^2 - 4$ **b.** $y = -x^2 + 1$

The Relationship between the Leading Coefficient *a* and the Graph

In Example 1, notice that some of the parabolas open upward and some open downward. In this Concept Investigation, we will find a way to tell from the equation when a parabola will open upward and when it will open downward.

CONCEPT INVESTIGATION
When does a parabola open upward and when does it open downward?

1. Fill in the following tables of points.

x	$y = x^2$
-3	
-2	
-1	
0	
1	
2	
3	

x	$y = 2x^2$
-3	
-2	
-1	
0	
1	
2	
3	

x	$y = -x^2$
-3	
-2	
-1	
0	
1	
2	
3	

x	$y = -3x^2$
-3	
-2	
-1	
0	
1	
2	
3	

750 CHAPTER 9 Quadratic Equations

2. Sketch graphs using the four tables of points from part 1.

$y = x^2$

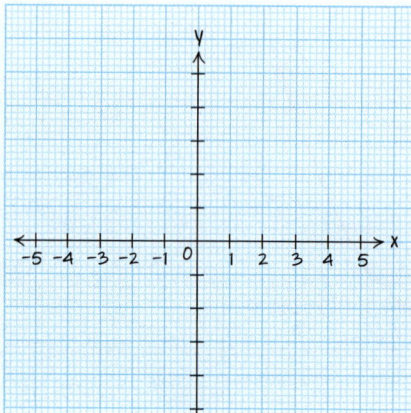

$y = 2x^2$

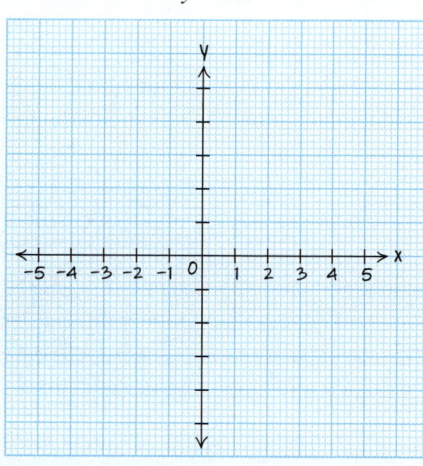

$y = -x^2$

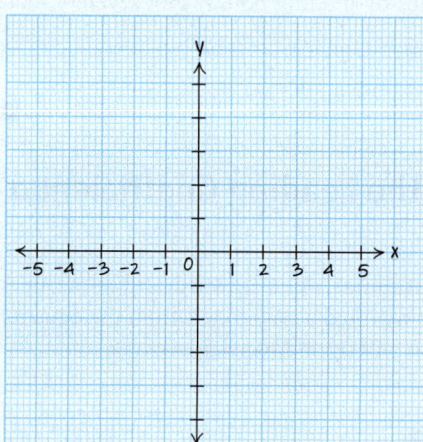

$y = -3x^2$

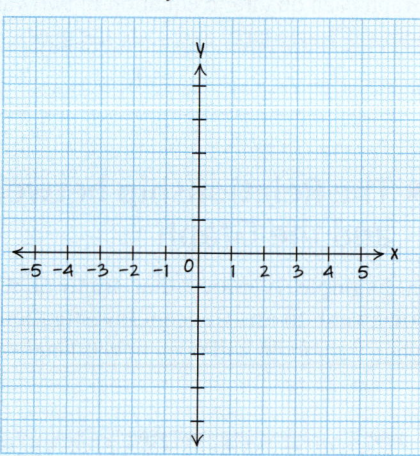

Connecting the Concepts

Why do we say a is positive when $a > 0$?

To interpret $a > 0$, it is useful to think of the number line. The numbers greater than 0 are to the right of 0.

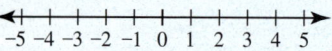

These are the positive numbers. In mathematics, we write $a > 0$, but it is common to say that a is positive.

3. What part of the equations seems to make the parabolas open upward or downward?
_____

From this Concept Investigation, we can see that for a quadratic equation in *standard form* $y = ax^2 + bx + c$ if $a > 0$ (positive), the parabola opens upward. If $a < 0$ (negative), the parabola opens downward. The values of b and c will affect the position of the parabola, but they do not determine whether the parabola opens upward or downward.

Example 2 Determine whether a parabola opens upward or downward

Use the equations to determine whether each parabola will open upward or downward.

a. $y = x^2 + 2x - 6$ **b.** $y = -3x^2 + 5$ **c.** $y = \dfrac{2}{3}x^2 - 2x - 8$

SOLUTION

a. The quadratic $y = x^2 + 2x - 6$ is in standard form $y = ax^2 + bx + c$, and $a = 1$. Since $a > 0$, the parabola opens upward.

b. The quadratic $y = -3x^2 + 5$ is in standard form, and $a = -3$. Since $a < 0$, the parabola opens downward.

c. The quadratic $y = \frac{2}{3}x^2 - 2x - 8$ is in standard form, and $a = \frac{2}{3}$. Since $a > 0$, the parabola opens upward.

PRACTICE PROBLEM FOR EXAMPLE 2

Use the equations to determine whether each parabola will open upward or downward.

a. $y = -7x^2 + 5x - 10$ **b.** $y = 4x^2 + 3$ **c.** $y = x^2 - 7x + 2$

■ Locating the Vertex

The graph of a quadratic equation in two variables is a smooth curve called a parabola that can open upward or downward. Parabolas have several defining characteristics that we want to study in a bit more detail. The first important characteristic of the graph is the **vertex**.

> **DEFINITION**
> **Vertex** The vertex of a parabola is the lowest point on the graph for a parabola that opens upward, and it is the highest point on the graph for a parabola that opens downward.

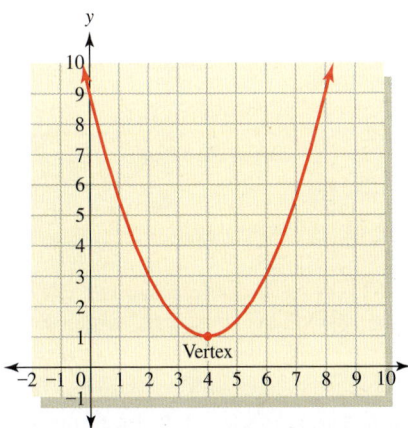 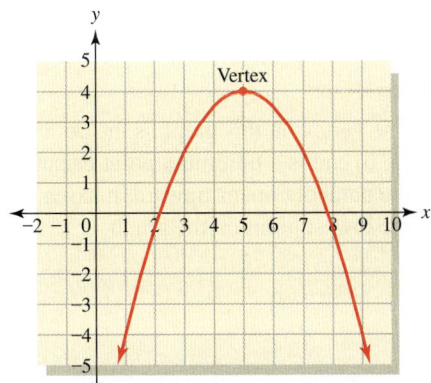

The vertex is often called the **maximum point** or **minimum point** because it is the highest or lowest point on the parabola.

> **DEFINITIONS**
> **Maximum Point** The vertex of a parabola that opens downward is called the maximum point.
> **Minimum Point** The vertex of a parabola that opens upward is called the minimum point.

> **What's That Mean?**
>
> **Vertex**
> The point where a parabola changes direction has many names.
> - Vertex
> - Highest point
> - Maximum point
> - Lowest point
> - Minimum point
>
> All of these names may be used in problems, so keep them in mind.

Example 3 Estimating a vertex from a graph

Estimate the vertex of each of the following parabolas and determine whether it is a maximum or minimum point.

a.

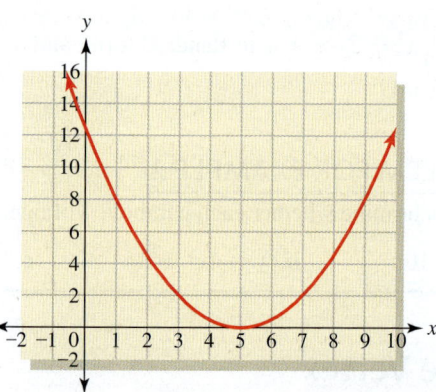

b.

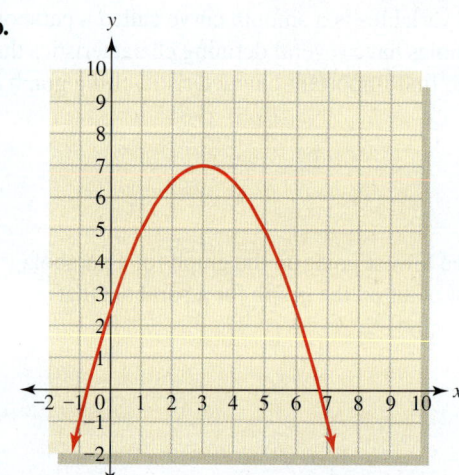

c.

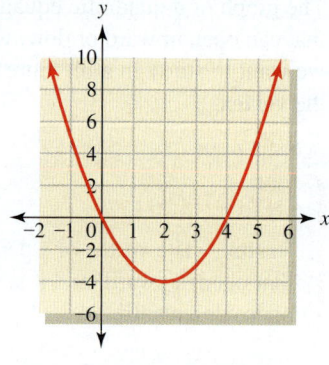

SOLUTION

a. The vertex on this graph is the point (5, 0) and is a minimum point because the parabola opens upward, making it the lowest point on the graph.

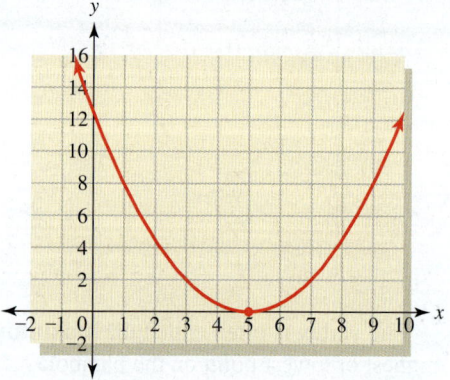

b. The vertex on this graph is the point (3, 7) and is a maximum point because the parabola opens downward, making it the highest point on the graph.

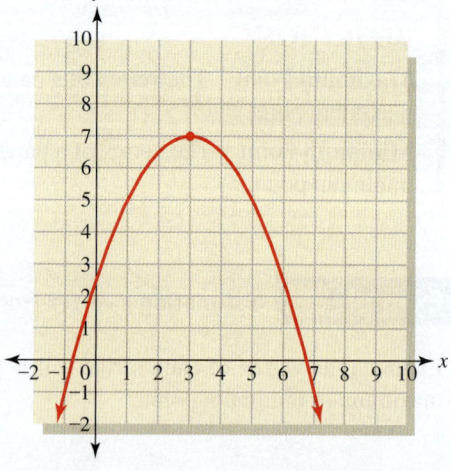

c. The vertex on this graph is the point (2, −4) and is a minimum point because the parabola opens upward, making it the lowest point on the graph.

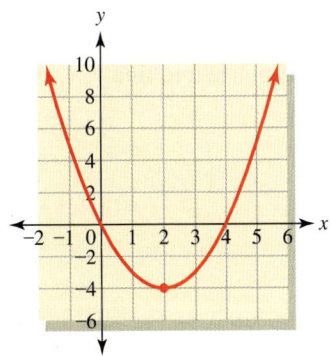

When graphing a quadratic equation, we could try to find the vertex by plotting a lot of points and looking for either the highest or the lowest point on the graph. Doing so might not always work because we may or may not be lucky enough to plot the point that corresponds to the vertex. A more efficient way to find the x-coordinate of the vertex is to use the following formula.

VERTEX FORMULA

The x-coordinate of the vertex of the quadratic equation in two variables, $y = ax^2 + bx + c$, is given by

$$x = \frac{-b}{2a}$$

The y-coordinate is found by substituting the x-coordinate into the equation and evaluating to find y.

Note: The vertex is a point (x, y). It is important to find both the x- and y-coordinates.

Example 4 — Find the vertex of a quadratic equation

Find the vertex of each of the following quadratic equations. Determine whether it is a maximum or a minimum point.

a. $y = 2x^2 - 8x + 7$ b. $y = -x^2 + x - 1$

SOLUTION

a. To find the x-coordinate of the vertex, use the vertex formula $x = \frac{-b}{2a}$. To use this formula, identify the values of a and b in the quadratic equation $y = 2x^2 - 8x + 7$. We have that $a = 2$ and $b = -8$. Substituting these values into the vertex formula yields the following value for the x-coordinate of the vertex.

$$x = \frac{-b}{2a} = \frac{-(-8)}{2(2)} = \frac{8}{4} = 2$$

Now that we have found the x-coordinate of the vertex, find the y-coordinate of the vertex. To find the y-coordinate, substitute $x = 2$ into the quadratic equation and evaluate it to find y.

$$y = 2x^2 - 8x + 7 \quad \text{Substitute in } x = 2.$$
$$y = 2(2)^2 - 8(2) + 7$$
$$y = 2 \cdot 4 - 16 + 7$$
$$y = 8 - 16 + 7$$
$$y = -8 + 7$$
$$y = -1$$

Therefore, the vertex is the point $(2, -1)$. Since $a > 0$, the parabola opens upward, so the vertex is a minimum point.

b. To find the x-coordinate of the vertex, use the vertex formula $x = \dfrac{-b}{2a}$. To use this formula, identify the values of a and b in the quadratic equation $y = -x^2 + x - 1 = -1 \cdot x^2 + 1 \cdot x - 1$. We have that $a = -1$ and $b = 1$. Substituting these values into the vertex formula yields the following value for the x-coordinate of the vertex.

$$x = \frac{-b}{2a} = \frac{-(1)}{2(-1)} = \frac{-1}{-2} = \frac{1}{2}$$

Now that we have found the x-coordinate of the vertex, find the y-coordinate of the vertex. To find the y-coordinate, substitute $x = \dfrac{1}{2}$ into the quadratic equation and evaluate it to find y.

$$y = -x^2 + x - 1 \quad \text{Substitute in } x = \tfrac{1}{2}.$$
$$y = -\left(\frac{1}{2}\right)^2 + \left(\frac{1}{2}\right) - 1$$
$$y = -\frac{1}{4} + \frac{1}{2} - 1 \quad \text{The LCD is 4.}$$
$$y = -\frac{1}{4} + \frac{2}{4} - \frac{4}{4}$$
$$y = -\frac{3}{4}$$

Therefore, the vertex is the point $\left(\dfrac{1}{2}, -\dfrac{3}{4}\right)$. Since $a < 0$, the parabola opens downward, so the vertex is a maximum point.

PRACTICE PROBLEM FOR EXAMPLE 4

Find the vertex of each of the following quadratic equations. Determine whether it is a maximum or a minimum point.

a. $y = -3x^2 + 6x$ **b.** $y = 4x^2 + 5$

When interpreting a vertex in an application problem, be sure to explain both the x- and y-value of the vertex in terms of the problem. Also, indicate whether the vertex represents a maximum or minimum value.

Example 5 Interpreting a vertex

a. The height, h, in feet of a ball t seconds after being hit can be estimated by the equation $h = -16t^2 + 64t + 4$. Find the vertex of this equation and interpret it in terms of time and height.

b. The **unit cost**, C, in dollars for producing m motorcycles in a single production run can be estimated by using the equation $C = 0.05m^2 - 6m + 3000$. Find the vertex of this equation and interpret it in terms of the number of motorcycles produced and unit production costs.

What's That Mean?

Unit Cost

The *unit cost* of producing an item is the average cost of producing one of that particular item. In business, costs include fixed costs (such as rent) and variable costs (such as the price of parts).

SOLUTION

a. This quadratic equation is in standard form, so use the vertex formula to find the input value for the vertex.

$$t = \frac{-b}{2a} = \frac{-(64)}{2(-16)} = 2$$

Find the output value by substituting $t = 2$ into the equation $h = -16t^2 + 64t + 4$.

$$h = -16(2)^2 + 64(2) + 4$$
$$h = -16(4) + 128 + 4$$
$$h = 68$$

Therefore, the vertex is $(t, h) = (2, 68)$. Because $a = -16$ and $a < 0$, the parabola opens downward. The vertex is a maximum point.

Two seconds after being hit, the ball reaches its maximum height of 68 feet.

b. Find the input value of the vertex using the vertex formula.

$$m = \frac{-(-6)}{2(0.05)} = 60$$

Substitute $m = 60$ to find the output value.

$$C = 0.05(60)^2 - 6(60) + 3000$$
$$C = 180 - 360 + 3000$$
$$C = 2820$$

Therefore, the vertex is $(m, C) = (60, 2820)$. Because $a = 0.05$ and $a > 0$, the parabola opens upward. The vertex is a minimum point.

With a single production run, the unit cost reaches its minimum of $2820 when 60 motorcycles are built.

PRACTICE PROBLEM FOR EXAMPLE 5

The profit, P, in dollars from selling t tickets for a local commuter flight can be estimated by using the equation $P = -15t^2 + 600t - 4000$. Find the vertex of this equation and interpret it in terms of the number of tickets sold and the profit from ticket sales.

■ Using the Axis of Symmetry

Another important feature of the graph of a parabola is symmetry. A parabola is symmetric about a vertical line drawn through its vertex. Being symmetric means that each side of the parabola is a mirror image, or a reflection, of the other through this vertical line. This vertical line that goes through the vertex of a parabola has a special name. It is called the **axis of symmetry** of the parabola.

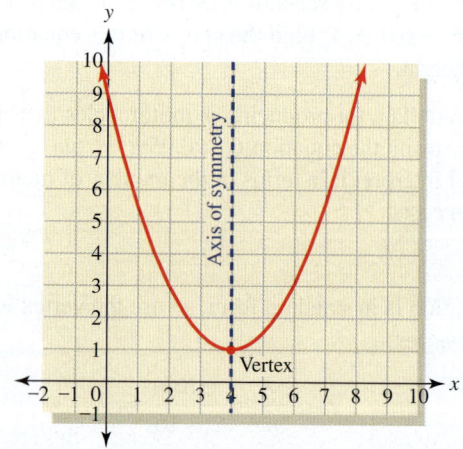

When discussing the axis of symmetry of a parabola, we want to write the equation of that line. From Section 3.1, the equation of a vertical line is of the form $x = c$. Since the axis of symmetry goes through the vertex and is a vertical line, the equation of the axis of symmetry contains the x-coordinate of the vertex.

AXIS OF SYMMETRY FORMULA

$$x = \frac{-b}{2a}$$

Example 6   Finding the equation of the axis of symmetry

For each quadratic equation, determine the equation of the axis of symmetry.

a. $y = -3x^2 + 12x + 11$ **b.** $y = \frac{1}{2}x^2 + x - 2$

SOLUTION

a. To use the axis of symmetry formula, we identify the values of a and b in the quadratic.

$$y = -3x^2 + 12x + 11$$

Therefore, $a = -3$ and $b = 12$. Substituting these values into the formula yields the following equation for the axis of symmetry.

$$x = \frac{-b}{2a} = \frac{-(12)}{2(-3)} = \frac{-12}{-6} = 2$$

The equation of the axis of symmetry is $x = 2$.

b. To use the axis of symmetry formula, we identify the values of a and b in the quadratic.

$$y = \frac{1}{2}x^2 + 1x - 2$$

Therefore, $a = \dfrac{1}{2}$ and $b = 1$. Substituting these values into the formula yields the following equation for the axis of symmetry.

$$x = \dfrac{-b}{2a} = \dfrac{-(1)}{2\left(\dfrac{1}{2}\right)} = -\dfrac{1}{1} = -1$$

The equation of the axis of symmetry is $x = -1$.

PRACTICE PROBLEM FOR EXAMPLE 6

For each quadratic equation, determine the equation of the axis of symmetry.

a. $y = -5x^2 + 9$ **b.** $y = -\dfrac{1}{4}x^2 + 3x + 1$

The axis of symmetry can be used to expedite the graphing process. If we find points on one side of the axis of symmetry, then symmetry can be used to plot points on the other side. If a point is 4 units to the left of the axis of symmetry, there will be a point 4 units to the right with the same y-value. These points are sometimes called *symmetric points*.

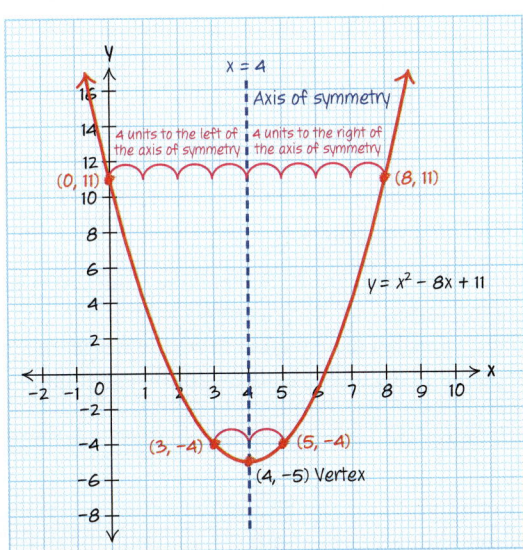

Using this idea, we need only to calculate points on one side of the parabola. We can graph the points on the other side of the parabola using symmetry.

> ### Steps to Graph a Quadratic Equation in Two Variables
> 1. Find the equation of the axis of symmetry.
> 2. Find the coordinates of the vertex.
> 3. Graph a total of seven points. Find the values of additional symmetric points as necessary. There should be at least three points to the left and right of the vertex.
> 4. Connect the points with a smooth curve.

Example 7 Graphing a quadratic using symmetry

Graph each given quadratic equation. First find the equation of the axis of symmetry and the vertex. Generate a table of values. Plot the points in the table and their symmetric points. Clearly label and scale the axes.

a. $y = x^2 - 8x + 7$ **b.** $y = -x^2 + 2x + 24$

SOLUTION

a. Step 1 Find the equation of the axis of symmetry. $y = x^2 - 8x + 7$.

$$x = \frac{-b}{2a}$$
$$x = \frac{-(-8)}{2(1)}$$
$$x = 4$$

Step 2 Find the coordinates of the vertex.

To find the vertex, substitute $x = 4$ into the equation to find y. Remember the x-coordinate of the vertex is the same as the equation of the axis of symmetry.

$y = x^2 - 8x + 7$ Substitute in $x = 4$.
$y = (4)^2 - 8(4) + 7$
$y = 16 - 32 + 7$
$y = -9$

The vertex of the parabola is at the point $(x, y) = (4, -9)$.

Step 3 Graph a total of seven points.

Generate a table of values to graph. Select three values for x near the x-coordinate of the vertex, $x = 4$. Selecting $x = 0$, $x = 1$, and $x = 3$ and substituting these values in the equation to find y, we get the table below. Plotting these points in the table yields the following graph.

x	$y = x^2 - 8x + 7$
0	7
1	0
3	−8
4	−9

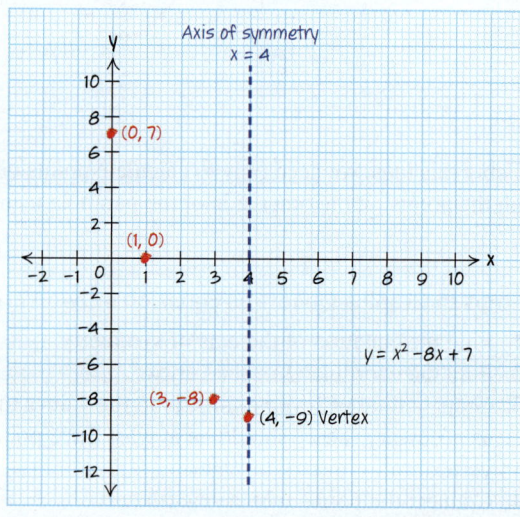

Next find the points that are symmetric to the points already plotted. The point $(0, 7)$ is 4 units away (horizontally) from the axis of symmetry. Count 4 units from the axis of symmetry to the right to find the symmetric pair to $(0, 7)$, which is $(8, 7)$. Count over to the other side of the axis of symmetry and plot each of the symmetric points.

Step 4 Connect the points with a smooth curve.

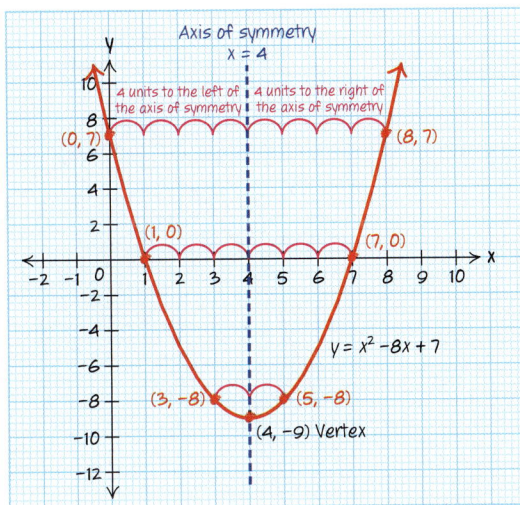

The final graph does not need the axis of symmetry or the extra notation.

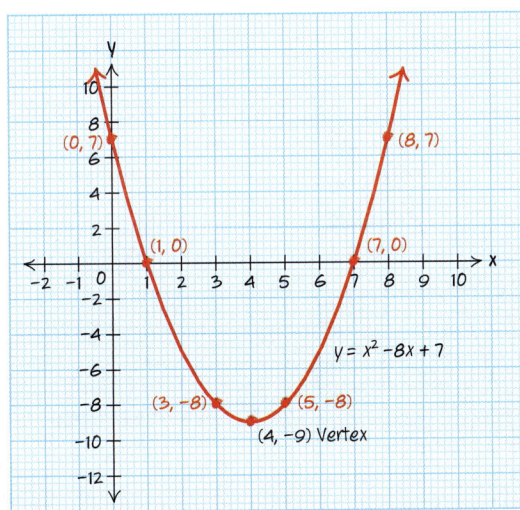

b. Find the equation of the axis of symmetry for the quadratic equation $y = -x^2 + 2x + 24$.

$$x = \frac{-2}{2(-1)}$$

$$x = 1$$

To find the vertex, substitute $x = 1$ into the equation to find y.

$y = -x^2 + 2x + 24$ Substitute in x = 1.
$y = -(1)^2 + 2(1) + 24$
$y = -1 + 2 + 24$
$y = 25$

The vertex of the parabola is at the point $(x, y) = (1, 25)$.

We generate a table of values to graph. Select three values for x near the x-coordinate of the vertex, $x = 1$. Selecting $x = -4$, $x = -2$, and $x = 0$ and substituting these values in the equation to find y, we get the table below. Plotting these points in the table and their symmetric points yields the following graph.

x	$y = -x^2 + 2x + 24$
−4	0
−2	16
0	24
1	25

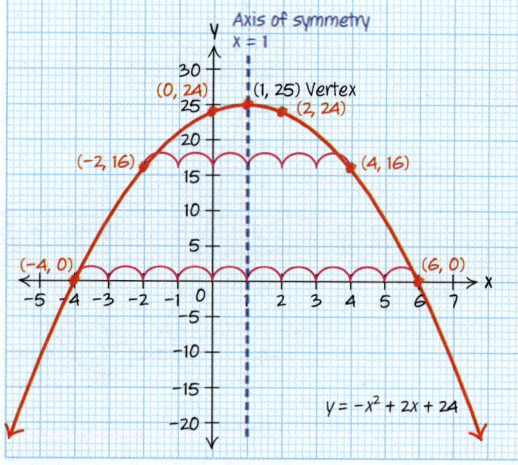

Cleaning up this graph gives us the following parabola.

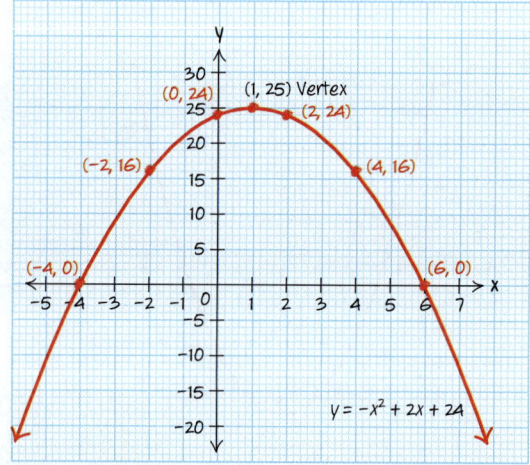

PRACTICE PROBLEM FOR EXAMPLE 7

Graph the quadratic equation. First find the equation of the axis of symmetry and the vertex. Generate a table of values. Plot the points in the table and their symmetric points. Clearly label and scale the axes.

$$y = x^2 + 8x + 12$$

9.1 Vocabulary Exercises

1. A(n) _____ equation in two variables has the form $y = ax^2 + bx + c, a \neq 0$.

2. The graph of a quadratic equation in two variables is called a(n) _____.

3. The _____ of a parabola is either the highest or lowest point on the graph.

4. The vertical line that passes through the vertex is called the _____.

9.1 Exercises

For Exercises 1 through 12, graph the given equations by plotting points. Plot at least seven points. Clearly label and scale the axes.

1. $y = 2x^2$
2. $y = -3x^2$
3. $y = x^2 + 3$
4. $y = x^2 - 7$
5. $y = x^2 - 8x$
6. $y = x^2 - 6x$
7. $y = -x^2 + 10$
8. $y = -x^2 + 6$
9. $y = x^2 - 4x$
10. $y = x^2 + 2x$
11. $y = -x^2 + x$
12. $y = -x^2 - 2x$

For Exercises 13 through 22, use the graphs to estimate each vertex and determine whether it is a maximum or minimum point.

13.
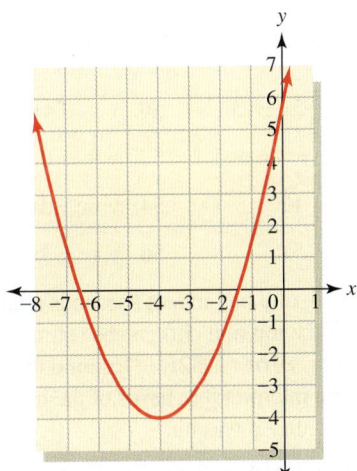

14.
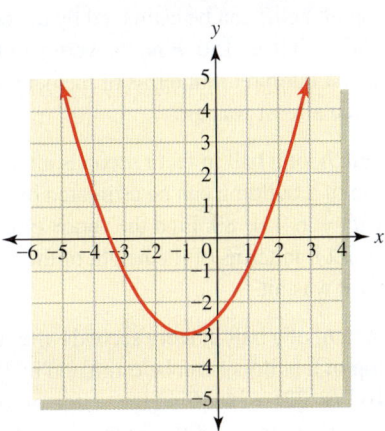

15.
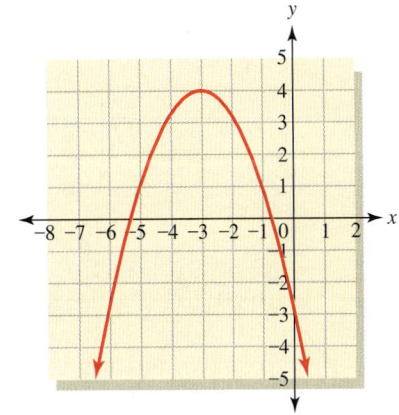

16.
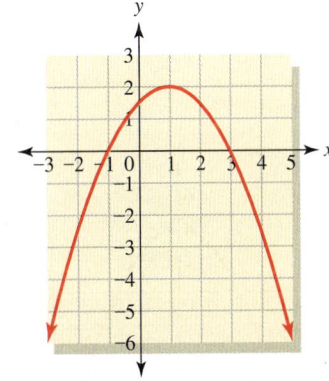

17.
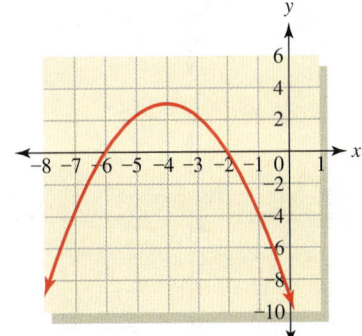

18.
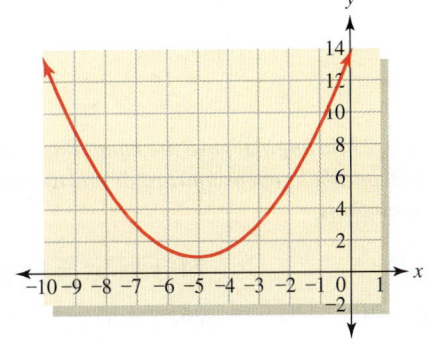

762 CHAPTER 9 Quadratic Equations

19.
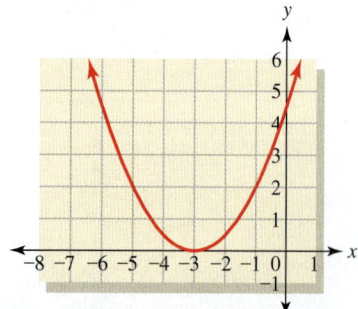

20.
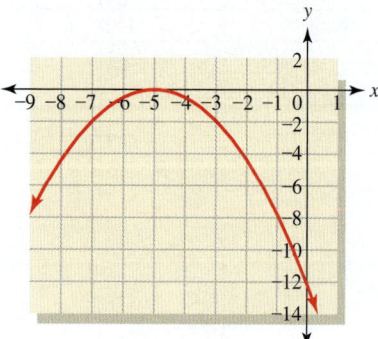

21.
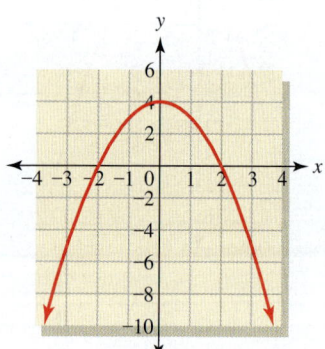

22.
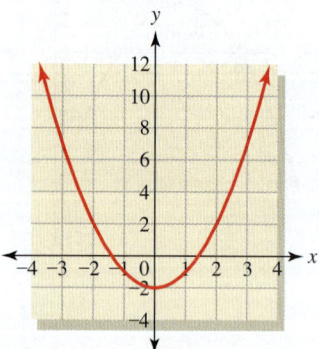

For Exercises 23 through 38, use the equations to determine whether each parabola opens upward or downward. Do not graph the equations.

23. $y = 2x^2 + 5$

24. $y = 5x^2 + 6x - 2$

25. $y = x^2 - 8$

26. $y = x^2 - 6$

27. $y = -4x^2 + 3x - 2$

28. $y = -7x^2 - 5x + 6$

29. $y = \frac{1}{2}x^2$

30. $y = \frac{3}{5}x^2$

31. $y = -\frac{2}{3}x^2 + 7$

32. $y = -\frac{1}{4}x^2 + 3x + 7$

33. $y = 0.5x^2 + 3x - 4$

34. $y = 0.1x^2 + 8x + 2.1$

35. $y = -0.25x^2 - 1$

36. $y = -0.8x^2 + 7.4$

37. $y = 5x + 3 - x^2$

38. $y = 8 + x^2$

For Exercises 39 through 56, find the vertex of each given equation and determine whether it is a maximum or minimum point. Do not graph the parabolas.

39. $y = 2x^2 + 8x - 7$

40. $y = x^2 + 6x - 2$

41. $y = x^2 - 8x + 5$

42. $y = x^2 - 6x - 12$

43. $y = -4x^2 + 32x - 2$

44. $y = -x^2 - 6x + 9$

45. $y = \frac{1}{2}x^2$

46. $y = \frac{3}{5}x^2$

47. $y = \frac{1}{2}x^2 - 3x + 4$

48. $y = \frac{3}{4}x^2 + 6x - 2$

49. $y = 0.5x^2 + 5x - 3$

50. $y = 0.4x^2 - 12x - 3.2$

51. $y = -0.5x^2 + 7x + 1.2$

52. $y = -0.1x^2 + 4x - 4.7$

53. $y = -3x^2 + 12$

54. $y = -2x^2 + 7$

55. $y = -2x^2 + 5x$

56. $y = -3x^2 + 9x$

57. The height, h, of a ball in feet t seconds after being hit from off the top of a cliff can be estimated by the equation $h = -16t^2 + 32t + 75$. Find the vertex of this equation and interpret it in terms of the height of the ball and the time after it was hit.

58. The height, h, of a ball in feet t seconds after being hit off the top of a cliff can be estimated by the equation $h = -16t^2 + 64t + 150$. Find the vertex of this equation and interpret it in terms of the height of the ball and the time after it was hit.

59. The height, h, of a ball in feet t seconds after being hit off the top of a building can be estimated by the equation $h = -16t^2 + 64t + 80$. Find the vertex of this equation and interpret it in terms of the height of the ball and the time after it was hit.

60. The height, h, of a ball in feet t seconds after being hit off the top of a building can be estimated by the equation $h = -16t^2 + 32t + 48$. Find the vertex of this equation and interpret it in terms of the height of the ball and the time after it was hit.

61. An appliance manufacturer's average profit per appliance, P, in dollars when a appliances are sold in a single order can be estimated by the equation

$P = -4a^2 + 80a$. Find the vertex of this equation and interpret it in terms of the number of appliances and the average profit.

62. A company's average cost per item, C, in dollars when n items are produced in a day can be estimated by the equation $C = -6n^2 + 96n$. Find the vertex of this equation and interpret it in terms of the number of items produced and average cost.

63. A plane manufacturer's average profit per plane, P, in millions of dollars when n planes are ordered by an airline at one time can be estimated by the equation $P = -n^2 + 16n$. Find the vertex of this equation and interpret it in terms of the number of planes ordered and the average profit.

64. A customized paint manufacturer's average profit per quart, P, in dollars when q quarts of paint are mixed for one order can be estimated by the equation $P = -7q^2 + 98q$. Find the vertex of this equation and interpret it in terms of the quarts of paint ordered and the average profit.

65. The revenue in dollars from selling art at a flea market can be estimated by the equation $R = -5p^2 + 400p$, where R is the revenue in dollars when p is the price in dollars. Find the vertex of this equation and interpret it in terms of the price and the revenue for that price.

66. A local charity is having a fund-raising dinner and is trying to determine the best price to charge for tickets. The profit brought in by the dinner can be estimated by the equation $p = -10n^2 + 400n + 5000$, where p is the profit, in dollars when the original $50 price is increased by $5 n times. Find the vertex of this equation and interpret it in terms of the profit for the number of $5.00 price increases.

For Exercises 67 through 72, find the axis of symmetry for the given equations.

67. $y = 2x^2 + 12x + 5$
68. $y = x^2 + 8x - 9$
69. $y = -4x^2 + 24x - 7$
70. $y = -3x^2 + 30x + 8$
71. $y = 5x^2 + 10$
72. $y = -7x^2 + 20$

For Exercises 73 through 98, graph each quadratic equation using symmetry. Find the equation of the axis of symmetry and the vertex. Graph at least seven points. Label the vertex. Clearly label and scale the axes. See Example 7 for help if needed.

73. $y = 2x^2 + 5$
74. $y = x^2 - 8$
75. $y = -3x^2 + 12$
76. $y = -2x^2 + 10$
77. $y = x^2 - 8x + 5$
78. $y = x^2 - 6x - 12$
79. $y = x^2 + 4x + 3$
80. $y = x^2 + 10x + 24$
81. $y = -x^2 + 6x + 7$
82. $y = -x^2 + 10x + 5$
83. $y = -x^2 - 4x - 3$
84. $y = -x^2 - 8x + 20$
85. $y = 2x^2 - 9x + 9$
86. $y = 4x^2 - 21x + 5$
87. $y = 2x^2 + 7x + 6$
88. $y = 5x^2 + 16x + 3$
89. $y = -7x^2 - 42x + 6$
90. $y = -3x^2 - x + 2$
91. $y = \frac{1}{2}x^2 - 3x - 4$
92. $y = \frac{1}{2}x^2 - x + 6$
93. $y = -\frac{1}{4}x^2 - 5x + 12$
94. $y = -\frac{3}{4}x^2 - 6x + 2$
95. $y = 0.5x^2 + 5x - 3$
96. $y = 0.4x^2 - 12x - 3.2$
97. $y = -0.5x^2 + 7x + 1.2$
98. $y = -0.1x^2 + 4x - 4.7$

99. Graph the equation for the height of the ball in Exercise 57. Clearly scale and label the axes.

100. Graph the equation for the height of the ball in Exercise 60. Clearly scale and label the axes.

101. Graph the equation for the revenue in Exercise 61. Clearly scale and label the axes.

102. Graph the equation for the profit in Exercise 64. Clearly scale and label the axes.

103. Graph the equation for the revenue in Exercise 65. Clearly scale and label the axes.

104. Graph the equation for the profit in Exercise 66. Clearly scale and label the axes.

For Exercises 105 through 112, identify each equation as linear or quadratic. Do not graph the equation.

105. $x = -5$
106. $x = 3$
107. $y = 2x^2 + 3$
108. $y = -x^2 - 5$
109. $y = 4$
110. $y = -8$
111. $2x + 4x^2 = y$
112. $-3x + x^2 = y$

9.2 Solving Quadratic Equations by Using the Square Root Property

LEARNING OBJECTIVES
- Solve quadratic equations using the square root property.
- Solve applications using the Pythagorean theorem and the distance formula.

Solving Quadratic Equations by Using the Square Root Property

Recall from Section 8.4, we learned how to solve equations involving radicals by squaring both sides to undo the square root (the radical). We now use that idea in a slightly different order. In a quadratic equation, the highest-degree term is x^2. If x^2 is the *only* variable term, then we can apply the idea that squares and square roots undo each other to solve the equation. We will take the square root of both sides of the equation to undo the square on x. The key property that justifies taking the square root of both sides is called the **square root property**.

> **SQUARE ROOT PROPERTY**
>
> If $x^2 = a$, then $x = \sqrt{a}$ or $x = -\sqrt{a}$.
>
> *Note:* The result is often written $x = \pm\sqrt{a}$.

Connecting the Concepts

What is a quadratic equation?

In Section 6.4, a quadratic equation was defined as an equation that can be put in the form $ax^2 + bx + c = 0$, where $a \neq 0$. We call this the **general form** of a quadratic equation.

Let's examine why there are two solutions. Consider the equation $x^2 = 4$. There are two solutions to the equation $x^2 = 4$. Substituting $x = 2$ and $x = -2$ into the original equation yields

$$x = 2 \qquad\qquad x = -2$$
$$(2)^2 \stackrel{?}{=} 4 \qquad (-2)^2 \stackrel{?}{=} 4$$
$$4 = 4 \qquad\qquad 4 = 4$$

When a negative number is squared, the answer is positive.

There are two solutions to the equation $x^2 = 4$: $x = 2$ and $x = -2$. These two solutions can be written as $x = \pm 2$. Solving the quadratic equation $x^2 = 4$ using the square root property yields

$$x^2 = 4 \qquad \text{The only variable term, } x^2\text{, is isolated.}$$
$$x = \pm\sqrt{4} \qquad \text{Use the square root property.}$$
$$x = \pm 2 \qquad \text{This means } x = 2 \text{ or } x = -2.$$

Check: Check both solutions in the original problem.

$$x = 2 \qquad\qquad x = -2$$
$$(2)^2 \stackrel{?}{=} 4 \qquad (-2)^2 \stackrel{?}{=} 4$$
$$4 = 4 \qquad \text{Both answers check.} \qquad 4 = 4$$

This new notation we are using, $x = \pm 2$, means that $x = 2$ or $x = -2$. It is a shorthand way to write both solutions compactly.

Steps to Solving Quadratic Equations Using the Square Root Property

1. Determine whether the quadratic equation is of the form $x^2 = a$.
2. Isolate the squared variable term if needed.
3. Use the square root property. If $x^2 = a$ **and** $a > 0$ **then** $x = \pm\sqrt{a}$.
4. Solve for the variable.
5. Check both the answers in the original equation.

Example 1 The falling rock

The height, h, in feet from the ground, of a falling rock t seconds after being dropped from a 40-foot cliff is given by the equation
$$h = -16t^2 + 40$$

a. How high above the ground is the rock after falling 0.5 second?

b. How long will it take for the rock to fall to 24 feet above the ground? Round the answer to two decimal places if necessary.

c. How long will it take for the rock to hit the ground? Round the answer to two decimal places if necessary.

SOLUTION

a. The phrase "after falling 0.5 second" means to substitute $t = 0.5$ into the equation and find h.
$$h = -16(0.5)^2 + 40$$
$$h = 36$$

The rock is 36 feet above the ground.

b. We are told that the rock is 24 feet above the ground. This means that $h = 24$. The question asks, "How long will it take the rock to fall 24 feet above the ground?" When we are asked "how long," it means to find the time t. Therefore, substitute $h = 24$ into the equation and solve for t.

$$h = -16t^2 + 40 \quad \text{Substitute in } h = 24.$$
$$24 = -16t^2 + 40$$

Step 1 Determine whether the quadratic equation is of the form $x^2 = a$.

This quadratic equation has only the variable term t^2, so it is of the correct form.

Step 2 Isolate the squared variable term if needed.

$$24 = -16t^2 + 40 \quad \text{Isolate the } t^2 \text{ term by subtracting 40 from both sides.}$$
$$\underline{-40 \qquad\qquad -40}$$
$$-16 = -16t^2$$
$$\frac{-16}{-16} = \frac{-16t^2}{-16} \quad \text{Isolate the } t^2 \text{ term by dividing both sides by } -16.$$
$$1 = t^2$$

Step 3 Use the square root property: If $x^2 = a$ **and** $a > 0$ **then** $x = \pm\sqrt{a}$.
$$t = \pm\sqrt{1}$$

Step 4 Solve for the variable.
$$t = \pm 1$$

In this application, the answer $t = -1$ does not make sense because it represents time before the rock was dropped. The answer $t = -1$ is extraneous. The answer is $t = 1$ second.

Step 5 Check both answers in the original equation.

$$24 \stackrel{?}{=} -16(1)^2 + 40 \quad \text{Substitute in } t = 1.$$
$$24 \stackrel{?}{=} -16 + 40 \quad \text{Simplify the right side.}$$
$$24 = 24 \quad \text{The answer checks.}$$

It will take 1 second for the rock to fall to 24 feet above the ground.

c. When the rock hits the ground, it is at no height above the ground. Therefore, $h = 0$. The question asks, "How long will it take?" When we are asked "How long," it means to find time, which is represented by the variable t. We substitute $h = 0$ in the equation and solve for t. As this equation has only a t^2 variable term, we can solve it by using the square root property.

$$h = -16t^2 + 40 \quad \text{Substitute } h = 0.$$
$$0 = -16t^2 + 40 \quad \text{Isolate the } t^2\text{-term by subtracting 40 from both sides.}$$
$$\underline{-40 \qquad\qquad -40}$$
$$-40 = -16t^2$$
$$\frac{-40}{-16} = \frac{-16t^2}{-16} \quad \text{Isolate } t^2 \text{ by dividing both sides by } -16.$$
$$\frac{5}{2} = t^2$$
$$t = \pm\sqrt{\frac{5}{2}} \quad \text{Use the square root property.}$$
$$t = \pm\sqrt{\frac{5}{2}} \approx \pm 1.58$$

Once again, the negative answer for time does not make sense.

The rock will hit the ground in $t \approx 1.58$ seconds.

PRACTICE PROBLEM FOR EXAMPLE 1

The height in meters from the ground of a ball t seconds after being dropped from a 35-meter cliff is given by the equation $h = -4.9t^2 + 35$.

a. How high above the ground is the ball after 1.5 seconds?

b. How long will it take for the ball to fall to 15 meters above the ground? Round the answer to two decimal places if necessary.

c. How long will it take for the ball to hit the ground? Round the answer to two decimal places if necessary.

> **Quadratic Equations**
>
> **Square Root Property**
> Use when there is a squared term but no first degree term. Isolate the squared term and use a plus or minus symbol to indicate both answers.
> $$x^2 = 25$$
> $$x = \pm\sqrt{25}$$
> $$x = \pm 5$$
>
> **Factoring**
> Use when the quadratic equation has small coefficients that factor easily. Set the quadratic equal to zero, factor and then use the zero factor property to write two or more simpler equations.
> $$x^2 + 7x + 10 = 0$$
> $$(x + 5)(x + 2) = 0$$
> $$x + 5 = 0 \quad \text{or} \quad x + 2 = 0$$

Example 2 Solving a quadratic equation by using the square root property

Solve each quadratic equation, using the square root property. Give both the exact answers and approximate answers rounded to two decimal places. Check the answer(s) in the original equation.

a. $a^2 = 49$ b. $y^2 - 144 = 0$ c. $2x^2 = 100$

SOLUTION

a. Solving $a^2 = 49$ using the steps above yields

$$a^2 = 49 \quad \text{The } a^2\text{-term is isolated.}$$
$$a = \pm\sqrt{49} \quad \text{Don't forget the } \pm \text{ on the right side.}$$
$$a = \pm 7 \quad \text{The two solutions are } a = 7 \text{ or } a = -7.$$

SECTION 9.2 Solving Quadratic Equations by Using the Square Root Property

Check:

$a = 7$	$a = -7$
$(7)^2 \stackrel{?}{=} 49$	$(-7)^2 \stackrel{?}{=} 49$
$49 = 49$ Both answers check.	$49 = 49$

b. Solving $y^2 - 144 = 0$ using the steps above yields

$y^2 - 144 = 0$ First isolate the y^2-term.
$\underline{+144 \quad +144}$
$y^2 = 144$ Apply the square root property.
$y = \pm\sqrt{144}$ Don't forget the $\pm$ on the right side.
$y = \pm 12$

Check:

$y = 12$	$y = -12$
$(12)^2 - 144 \stackrel{?}{=} 0$	$(-12)^2 - 144 \stackrel{?}{=} 0$
$144 - 144 = 0$ Both answers check.	$144 - 144 = 0$

c. Solving $2x^2 = 100$ using the steps above yields

$\dfrac{2x^2}{2} = \dfrac{100}{2}$ Divide both sides by 2 to isolate x^2.
$x^2 = 50$ The x^2-term is isolated.
$x = \pm\sqrt{50}$ Don't forget the $\pm$ on the right side.
$x = \pm\sqrt{25 \cdot 2}$ Simplify the radical.
$x = \pm 5\sqrt{2}$ These are the exact answers.
$x \approx \pm 7.07$ These are the approximate answers.

Check:

$x = 5\sqrt{2}$	$x = -5\sqrt{2}$
$(5\sqrt{2})^2 \stackrel{?}{=} 50$	$(-5\sqrt{2})^2 \stackrel{?}{=} 50$
$5^2(\sqrt{2})^2 \stackrel{?}{=} 50$	$(-5)^2(\sqrt{2})^2 \stackrel{?}{=} 50$
$25 \cdot 2 = 50$ Both answers check.	$25 \cdot 2 = 50$

■ Connecting the Concepts

Can we factor instead?

In Example 2 part b we solve the equation

$y^2 - 144 = 0$

using the square root property. This equation also could have been solved using factoring.

$y^2 - 144 = 0$
$(y - 12)(y + 12) = 0$
$y - 12 = 0 \quad \text{or} \quad y + 12 = 0$
$y = 12 \quad \text{or} \quad y = -12$

When solving quadratic equations often there is more than one tool that can be used to solve. The square root property is most easily used when there is only one variable term and it is squared.

PRACTICE PROBLEM FOR EXAMPLE 2

Solve each quadratic equation, using the square root property. Give both the exact answers and approximate answers rounded to two decimal places. Check the answer(s) in the original equation.

a. $b^2 = 81$ **b.** $x^2 - 36 = 0$ **c.** $5m^2 = 100$

We can solve quadratic equations by taking square roots when there is a squared variable expression such as $(ax - h)^2$. After isolating the squared variable expression, we can solve for the variable using the square root property.

Example 3 Solving quadratic equations by using the square root property

Solve each quadratic equation, using the square root property. Give both the exact answers and approximate answers rounded to two decimal places. Check the answer(s) in the original equation.

a. $(x - 1)^2 = 25$ **b.** $(3y + 4)^2 - 3 = 4$

768 CHAPTER 9 Quadratic Equations

SOLUTION

a. To solve $(x - 1)^2 = 25$, first notice that the squared variable term, $(x - 1)^2$, has been isolated on the left side of the equal sign. Solve the equation by using the square root property.

$(x - 1)^2 = 25$ The factored form with the variable has been isolated.
$(x - 1) = \pm\sqrt{25}$ Use the square root property.
$(x - 1) = \pm 5$ Don't forget the $\pm$ on the right side.

There are two equations to solve.

$x - 1 = 5$ $x - 1 = -5$
$\underline{+1 \quad +1}$ $\underline{+1 \quad\quad +1}$
$x = 6$ $x = -4$

The two answers are $x = 6$ or $x = -4$. Checking both answers yields

$(x - 1)^2 = 25$ Check in the original equation.

Check:

$x = 6$ $x = -4$
$(6 - 1)^2 \stackrel{?}{=} 25$ $(-4 - 1)^2 \stackrel{?}{=} 25$
$5^2 \stackrel{?}{=} 25$ $(-5)^2 \stackrel{?}{=} 25$
$25 = 25$ Both answers check. $25 = 25$

b. First isolate the squared variable term.

$(3y + 4)^2 - 3 = 4$
$\underline{\quad\quad\quad\quad\quad +3 \;\; +3}$
$(3y + 4)^2 = 7$

Solve the equation by using the square root property.

$(3y + 4)^2 = 7$ The factored form with the variable has been isolated.
$(3y + 4) = \pm\sqrt{7}$ Use the square root property. Don't forget the $\pm$ on the right side.

There are two equations to solve.

$3y + 4 = \sqrt{7}$ $3y + 4 = -\sqrt{7}$
$\underline{\;\; -4 \quad -4}$ $\underline{\;\; -4 \quad\quad -4}$
$3y = -4 + \sqrt{7}$ $3y = -4 - \sqrt{7}$
$\dfrac{3y}{3} = \dfrac{-4 + \sqrt{7}}{3}$ $\dfrac{3y}{3} = \dfrac{-4 - \sqrt{7}}{3}$
$y = \dfrac{-4 + \sqrt{7}}{3}$ $y = \dfrac{-4 - \sqrt{7}}{3}$ These are the exact answers.
$y \approx -0.45$ $y \approx -2.22$ These are the approximate answers.

Check in the original equation: $(3y + 4)^2 = 7$

$y \approx -2.22$ $y \approx -0.45$
$(3(-2.22) + 4)^2 \stackrel{?}{=} 7$ $(3(-0.45) + 4)^2 \stackrel{?}{=} 7$
$(-2.66)^2 \stackrel{?}{=} 7$ $(2.65)^2 \stackrel{?}{=} 7$
$7.0756 \approx 7$ Both answers check. $7.0225 \approx 7$

Both answers approximately check because we checked using the approximate answer.

Skill Connection

Adding an integer to a radical

In Example 3b, we write $-4 + \sqrt{7}$. Addition is commutative, so we could also write $\sqrt{7} - 4$. However, sometimes people have trouble distinguishing what belongs under the radical. The expression $\sqrt{7} - 4$ is sometimes taken incorrectly to mean $\sqrt{7-4}$. For this reason, we will put the integer first and the radical part second when solving quadratic equations.

SECTION 9.2 Solving Quadratic Equations by Using the Square Root Property   769

PRACTICE PROBLEM FOR EXAMPLE 3

Solve each quadratic equation using the square root property. Give both the exact answers and approximate answers rounded to two decimal places. Check the answer(s) in the original equation.

a. $(y + 2)^2 = 16$ **b.** $(2x - 1)^2 = 9$ **c.** $(5a - 2)^2 = 11$

Applying the square root property works only on equations that can be put in the form $x^2 = a$. To demonstrate why this is true, consider solving the problem $x^2 - 4x - 21 = 0$. In Section 6.4, we learned how to solve equations of this form by factoring. Therefore, the solution to $x^2 - 4x - 21 = 0$ is

$$x^2 - 4x - 21 = 0$$
$$(x - 7)(x + 3) = 0$$
$$x - 7 = 0 \qquad x + 3 = 0$$
$$x = 7 \qquad x = -3$$

However, solving this same equation by isolating x^2 and taking square roots yields

$$x^2 - 4x - 21 = 0$$
$$x^2 = 4x + 21$$
$$x = \pm\sqrt{4x + 21}$$

Notice that x is still on both sides of the equal sign. Therefore, this equation has not been solved for x. By using this technique, x cannot be isolated. The first-degree variable term, x, is getting in the way of our being able to solve this equation by using the square root property.

■ Using the Pythagorean Theorem

In geometry, a **right triangle** is a triangle with one 90° angle. The two other angles have to be less than 90° each. The longest side of a right triangle is the side opposite the 90° angle and is called the **hypotenuse**. The other two shorter sides are called the **legs**.

There is an important theorem that relates the three sides of a right triangle. It is called the Pythagorean theorem.

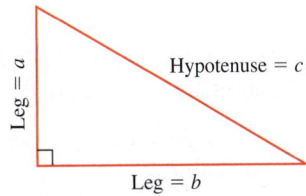

> **What's That Mean?**
>
> **Hypotenuse**
>
> In a triangle, the longest side of the triangle is always opposite the largest angle. In a right triangle, the largest angle measures 90°. The side opposite the 90° angle is called the **hypotenuse**. The word *hypotenuse* comes from the Greek words *hypo*, meaning "under" and *teinein*, meaning "to stretch." In the Pythagorean theorem, we will always label the hypotenuse as c.

> **What's That Mean?**
>
> **Theorem**
>
> According to Wolfram's MathWorld, a *theorem* is a "statement that can be demonstrated to be true by accepted mathematical operations and arguments. The process of showing a theorem to be true is called a *proof*."
>
> **Note:** Stephen Wolfram is a physicist and mathematician. He was the developer of the mathematical software Mathematica as well as the search engine Wolfram | Alpha. Wolfram's MathWorld is a website dedicated to mathematics.

> **THE PYTHAGOREAN THEOREM**
>
> In a right triangle with hypotenuse of length c and sides of lengths a and b,
>
> $$a^2 + b^2 = c^2$$

Example 4 The Pythagorean theorem

Use the Pythagorean theorem to find the unknown side in each right triangle. Round answers to two decimal places if necessary.

a.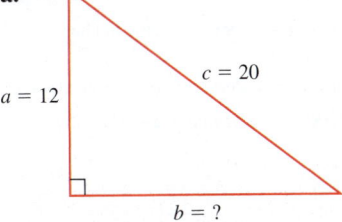

b.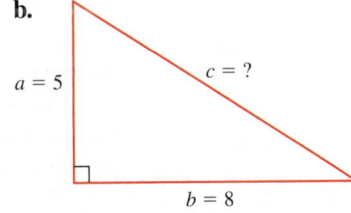

770 CHAPTER 9 Quadratic Equations

c.

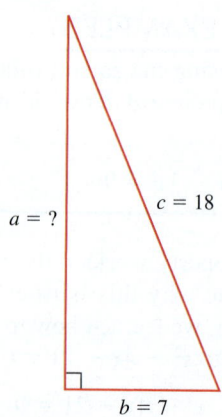

SOLUTION

We use the positive answer only in a problem involving the Pythagorean theorem, as we are solving for the length of a side of a triangle.

a. In this triangle, the missing side is b. Substituting $a = 12$ and $c = 20$ into the Pythagorean theorem and solving for b, we get

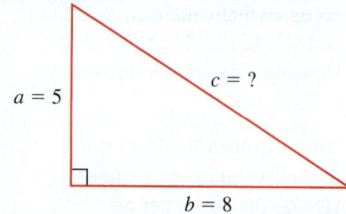

$a^2 + b^2 = c^2$	
$(12)^2 + b^2 = (20)^2$	Substitute the given values into the Pythagorean theorem.
$144 + b^2 = 400$	
$\underline{-144 \qquad -144}$	Isolate the b^2-term.
$b^2 = 256$	
$b = \pm\sqrt{256}$	Solve for b by using the square root property.
$b = \pm 16$	
$b = 16$	Because b represents the length of a side of a triangle, use the positive value.

b. The side opposite the right angle is missing. This is the hypotenuse, or side c. Substituting $a = 5$ and $b = 8$ into the Pythagorean theorem and solving for c, we get

$a^2 + b^2 = c^2$	
$(5)^2 + (8)^2 = c^2$	Substitute the given values into the Pythagorean theorem.
$25 + 64 = c^2$	Simplify the left side.
$89 = c^2$	
$c = \pm\sqrt{89}$	Solve for c by using the square root property. This is the exact answer.
$c = \sqrt{89}$	Because c is the length of a side of a triangle, use the positive value only.
$c \approx 9.43$	This is the approximate answer, rounded to two decimal places.

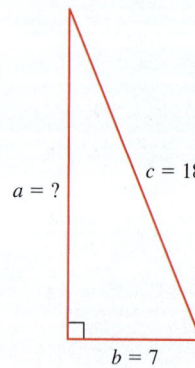

c. In this triangle, the missing side is a. Substituting $b = 7$ and $c = 18$ into the Pythagorean theorem and solving for a, we get

$a^2 + b^2 = c^2$	
$a^2 + (7)^2 = (18)^2$	Substitute the given values into the Pythagorean theorem.
$a^2 + 49 = 324$	Subtract 49 from both sides to isolate a^2.
$\underline{\qquad -49 \quad -49}$	
$a^2 = 275$	
$a = \pm\sqrt{275}$	Solve for a by using the square root property. This is the exact answer.
$a \approx \pm 16.58$	This is the approximate answer, rounded to two decimal places.
$a \approx 16.58$	Because a is the length of a side of a triangle, use the positive value only.

SECTION 9.2 Solving Quadratic Equations by Using the Square Root Property 771

PRACTICE PROBLEM FOR EXAMPLE 4

Use the Pythagorean theorem to find the unknown side in each right triangle. Round answers to two decimal places if necessary.

a.
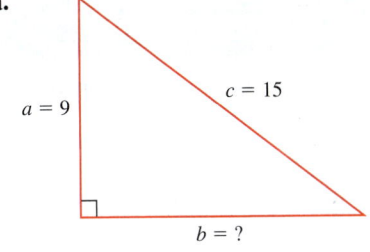

b.
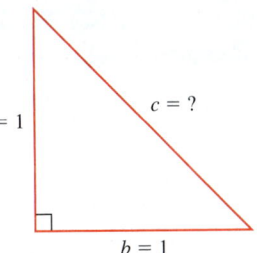

c.
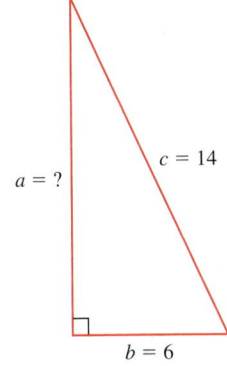

Example 5 — Right triangles in real-life applications

A 25-foot pole is anchored to the ground with a guy wire. The wire is fastened to the ground 10 feet away from the pole. How long is the wire? Round the answer to two decimal places.

SOLUTION

The pole, the wire, and the distance from the pole to where the wire is fastened to the ground form a right triangle. Labeling the height of the pole as 25 feet and the distance from the base of the pole to where the wire is fastened as 10 feet, we see the unknown in the right triangle is the hypotenuse. Call the hypotenuse c. Substituting these values into the Pythagorean theorem and solving for c, we get

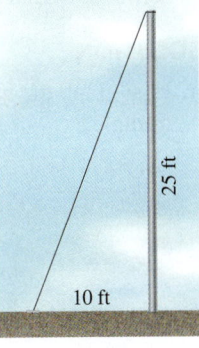

$a^2 + b^2 = c^2$
$25^2 + 10^2 = c^2$ Substitute in the values for a and b.
$625 + 100 = c^2$
$725 = c^2$
$\pm\sqrt{725} = c$ Use the square root property.
$c \approx 26.93$ Take the positive square root and round to two decimal places.

The guy wire is approximately 26.93 feet long.

PRACTICE PROBLEM FOR EXAMPLE 5

A 50-inch TV is one on which the diagonal measurement of the screen is 50 inches. A particular 50-inch TV has a width of 44.17 inches. What is the length of the TV?

772 CHAPTER 9 Quadratic Equations

In the following Concept Investigation, we explore the relationship between the distance formula and the Pythagorean theorem.

CONCEPT INVESTIGATION
Where does the distance formula come from?

1. To find the distance between the two points (2, 4) and (5, 6), use the following steps.
   a. Graph the two points on graph paper.
   b. Connect the points with a straight line. Just like computing slope, start with one point and draw a right triangle to the other point.
   c. Find the lengths of both legs of the right triangle.
   d. Find the length of the hypotenuse, using the Pythagorean theorem.

2. Use the distance formula to find the distance between the two points (2, 4) and (5, 6).
$$d = \sqrt{(x_2 - x_1)^2 + (y_2 - y_1)^2}$$

3. How are the calculations from parts 1 and 2 similar?

This Concept Investigation illustrates that the distance between two points can be found by using either the Pythagorean theorem or the distance formula.

Example 6 Finding the distance between two points

a. Find the distance between the two points (5, 2) and (3, 1) using the Pythagorean theorem. Round to two decimal places if necessary.

b. Find the distance between the two points $(-3, 5)$ and $(2, -1)$ using the distance formula. Round to two decimal places if necessary.

SOLUTION

a. First graph the given points (5, 2) and (3, 1). Connect the two points with a right triangle.

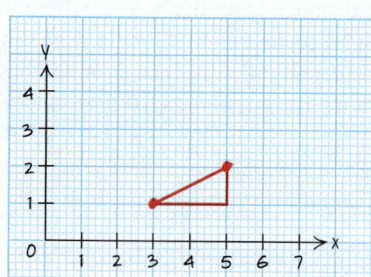

The lengths of the two legs are 2 and 1. Using the Pythagorean theorem, we have that

$a^2 + b^2 = c^2$
$2^2 + 1^2 = c^2$ Substitute in the values for *a* and *b*.
$4 + 1 = c^2$
$5 = c^2$
$\sqrt{c^2} = \sqrt{5}$ Use the square root property.
$c \approx 2.24$ Take the positive square root and round to two decimal places.

The distance between the two points is the same as the length of the hypotenuse, which is approximately 2.24.

b. Set (x_1, y_1) to be the first point, so $(x_1, y_1) = (-3, 5)$. Set (x_2, y_2) to be the second point, so $(x_2, y_2) = (2, -1)$. Substitute these values into the distance formula and simplify.

$d = \sqrt{(x_2 - x_1)^2 + (y_2 - y_1)^2}$ Substitute in $(x_1, y_1) = (-3, 5)$ and $(x_2, y_2) = (2, -1)$.
$d = \sqrt{(2 - (-3))^2 + (-1 - 5)^2}$ Simplify under the radical.
$d = \sqrt{5^2 + (-6)^2}$
$d = \sqrt{25 + 36} = \sqrt{61}$ Evaluate the square root.
$d \approx 7.81$

PRACTICE PROBLEM FOR EXAMPLE 6

a. Find the distance between the two points $(-2, 4)$ and $(-1, 6)$, using the Pythagorean theorem. Round to two decimal places if necessary.

b. Find the distance between the two points $(7, -2)$ and $(-5, -3)$, using the distance formula. Round to two decimal places if necessary.

9.2 Vocabulary Exercises

1. The _____ theorem only applies to right triangles.

2. In a right triangle, the longest side is called the _____.

3. A right triangle has a(n) _____ degree angle.

4. To solve an equation of the form $x^2 = a$, use the _____ property.

9.2 Exercises

For Exercises 1 through 8, warm up by determining whether each given equation is linear, quadratic, rational, or radical. Do not solve the equations.

1. $\dfrac{-6x + 1}{x - 1} = 3$
2. $-\dfrac{1}{2} = \dfrac{x + 2}{3x + 5}$
3. $-6x + 5 = 3(x - 1) + 5$
4. $-3 + 4(x - 1) = 7x + 8$
5. $\sqrt{x^2 - 3x + 4} = 1$
6. $\sqrt{6x + 2} = \sqrt{x + 4}$
7. $-x^2 + 3x = 8$
8. $x^2 - 2x + 64 = 1$

For Exercises 9 through 16, solve each quadratic equation by using the square root property. Check the answer(s) in the original problem.

9. $x^2 = 64$
10. $x^2 = 144$
11. $a^2 = 169$
12. $b^2 = 100$
13. $a^2 - 25 = 0$
14. $b^2 - 16 = 0$
15. $x^2 - 49 = 0$
16. $y^2 - 121 = 0$

For Exercises 17 through 24, solve each quadratic equation by using the square root property. Provide reasons for each step. Check the answer(s) in the original problem.

17. $4r^2 = 100$
18. $6x^2 = 54$
19. $3t^2 = 75$
20. $2y^2 = 72$
21. $2x^2 - 32 = 0$
22. $5b^2 - 125 = 0$
23. $8b^2 - 128 = 0$
24. $3x^2 - 48 = 0$

For Exercises 25 through 32, solve each application problem.

25. A ball is dropped from a building that is 60 feet tall. The height of the ball, h, in feet after t seconds is given by the equation

$$h = -16t^2 + 60$$

a. How long will it take for the ball to be at a height of 20 feet? Round the answer to two decimal places if necessary.

b. How long will it take for the ball to hit the ground? Round the answer to two decimal places if necessary.

774 CHAPTER 9 Quadratic Equations

26. A ball is dropped from a bridge that is 80 feet above a river. The height of the ball, h, in feet after t seconds is given by the equation
$$h = -16t^2 + 80$$
   a. How long will it take for the ball to be at a height of 25 feet above the river? Round the answer to two decimal places if necessary.
   b. How long will it take for the ball to hit the surface of the water? Round the answer to two decimal places if necessary.

27. A rock is dropped from a 25-meter cliff. The height of the rock, h, in meters after t seconds is given by the equation
$$h = -4.9t^2 + 25$$
   a. How long will it take for the rock to be at a height of 10 meters? Round the answer to two decimal places if necessary.
   b. How long will it take for the rock to hit the ground? Round the answer to two decimal places if necessary.

28. A rock is dropped from a 45-meter cliff. The height of the rock, h, in meters after t seconds is given by the equation
$$h = -4.9t^2 + 45$$
   a. How long will it take for the rock to be at a height of 5 meters? Round the answer to two decimal places if necessary.
   b. How long will it take for the rock to hit the ground? Round the answer to two decimal places if necessary.

29. A circular wheat field with area 3,000,000 ft² is irrigated by a center pivot irrigation system that will irrigate a circular region whose area and radius are given by the equation $A = \pi r^2$. What is the radius of this circular region? Round the answer to two decimal places if necessary.

30. A circular potato field with area 500,000 ft² is irrigated by a center pivot irrigation system that will irrigate a circular region whose area and radius are given by the equation $A = \pi r^2$. What is the radius of this circular region? Round the answer to two decimal places if necessary.

31. A square room has area 225 ft². What is the length of each side of the room?

32. A square room has area 256 ft². What is the length of each side of the room?

For Exercises 33 through 52, solve each quadratic equation by using the square root property. Give both the exact answers and the approximate answers rounded to 2 decimal places. Check the answer(s) in the original problem.

33. $(x + 1)^2 = 9$
34. $(x - 4)^2 = 16$
35. $(y - 2)^2 = 81$
36. $(y + 6)^2 = 100$
37. $(x - 1)^2 - 4 = 0$
38. $(x - 3)^2 - 49 = 0$
39. $(x + 1)^2 = -16$
40. $(x - 5)^2 = -1$
41. $(2a - 1)^2 - 25 = 0$
42. $(3b - 2)^2 - 49 = 0$
43. $(6x + 2)^2 - 64 = 0$
44. $(3x - 4)^2 - 9 = 0$
45. $(y - 2)^2 = 5$
46. $(y + 7)^2 = 21$
47. $(a - 3)^2 = -5$
48. $(a + 7)^2 = -12$
49. $(5x - 3)^2 = 12$
50. $(5a - 7)^2 = 24$
51. $(5x + 1)^2 = 48$
52. $(2x - 3)^2 = 45$

For Exercises 53 through 60, solve each triangle for the unknown side using the Pythagorean theorem. If needed, round any answers to two decimal places.

53.

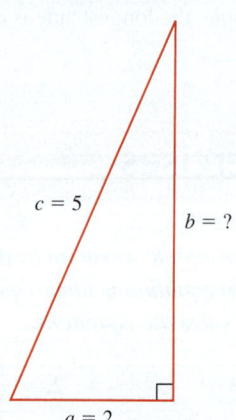

54.

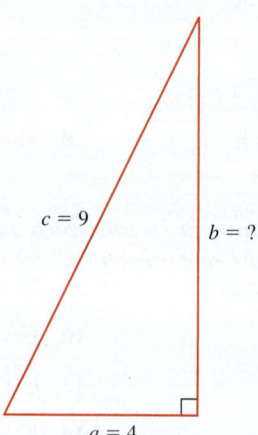

SECTION 9.2 Solving Quadratic Equations by Using the Square Root Property   775

55.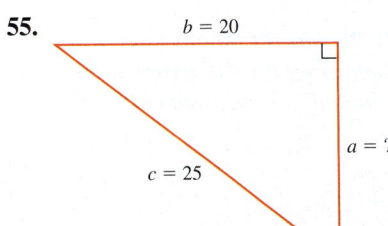

56.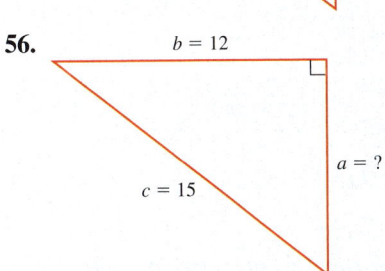

57.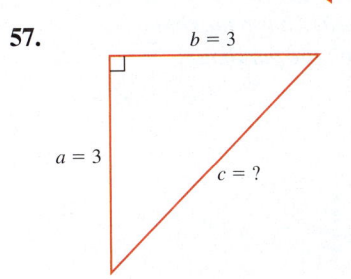

58.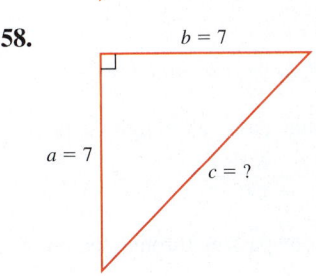

59.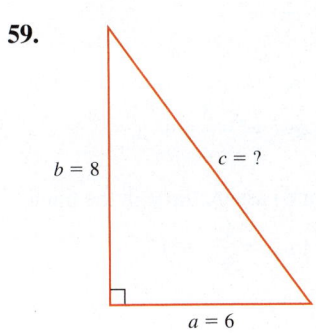

60.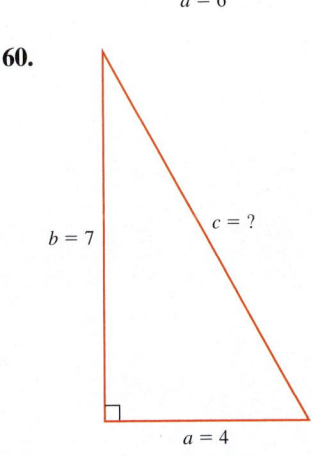

For Exercises 61 through 64, use the diagram to solve each application problem. Round any answers to two decimal places if necessary.

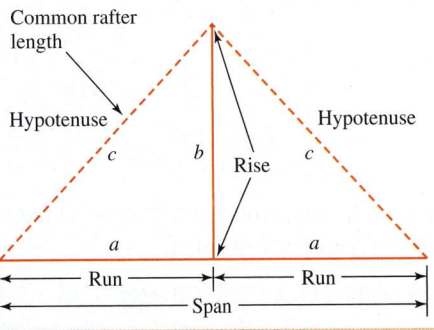

61. A roof has a 30-foot span and has a pitch of $\frac{10}{12}$. The pitch is the $\frac{\text{rise}}{\text{run}}$. Find the following:

   a. The value of b, which is the height of the roof. (*Hint*: Use the ideas you learned about proportions in Section 7.5.)

   b. The hypotenuse of the roof, which is also known as the common rafter length.

62. A roof has a 380-foot span and has a pitch of $\frac{8}{12}$. The pitch is the $\frac{\text{rise}}{\text{run}}$. Find the following:

   a. The value of b, which is the height of the roof. (*Hint*: Use the ideas you learned about proportions in Section 7.5.)

   b. The hypotenuse of the roof, which is also known as the common rafter length.

63. A roof has a 34-foot span and has a pitch of $\frac{9}{12}$. The pitch is the $\frac{\text{rise}}{\text{run}}$. Find the following:

   a. The value of b, which is the height of the roof. (*Hint*: Use the ideas you learned about proportions in Section 7.5.)

   b. The hypotenuse of the roof, which is also known as the common rafter length.

64. A roof has a 38-foot span and has a pitch of $\frac{10}{12}$. The pitch is the $\frac{\text{rise}}{\text{run}}$. Find the following:

   a. The value of b, which is the height of the roof. (*Hint*: Use the ideas you learned about proportions in Section 7.5.)

   b. The hypotenuse of the roof, which is also known as the common rafter length.

For Exercises 65 through 68, use the diagram of the foundation of a room, to solve each application problem. Round any answers to two decimal places if necessary.

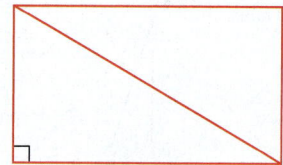

65. The foundation for a room is 16 feet by 12 feet. How long is the diagonal of the room?

66. The foundation for a room is 20 feet by 15 feet. How long is the diagonal of the room?

67. The foundation for a new warehouse is 300 feet by 400 feet. How long is the diagonal of the warehouse? What is the area of the warehouse?

68. The foundation for a new warehouse is 250 feet by 450 feet. How long is the diagonal of the warehouse? What is the area of the warehouse?

For Exercises 69 through 82, solve each quadratic equation. Give both the exact answers and approximate answers rounded to 2 decimal places. Check the answer(s) in the original problem.

69. $(x - 3)^2 = 36$
70. $(5x + 4)^2 = 64$
71. $7x^2 - 63 = 0$
72. $5x^2 - 80 = 0$
73. $3y^2 = 243$
74. $2y^2 = 50$
75. $(x + 7)^2 = 75$
76. $(x - 5)^2 = 98$
77. $(x + 4)^2 - 16 = 0$
78. $(x - 7)^2 - 25 = 0$
79. $(2x - 5)^2 - 81 = 0$
80. $(1 - 3x)^2 - 100 = 0$
81. $r^2 = 144$
82. $p^2 = 100$

For Exercises 83 through 90, solve each quadratic equation either by factoring or by using the square root property. Provide reasons for each step. Give both the exact answers and approximate answers rounded to 2 decimal places. (Hint: See Section 6.4 to review how to solve a quadratic equation by factoring.)

83. $x^2 - 4x - 5 = 0$
84. $x^2 + 5x - 14 = 0$
85. $(x + 3)^2 - 16 = 0$
86. $(x - 6)^2 - 1 = 0$
87. $3x^2 - 10x - 8 = 0$
88. $2x^2 - 7x + 5 = 0$
89. $3x^2 - 48 = 0$
90. $5x^2 - 80 = 0$

For Exercises 91 through 96, find the distance between each set of two points using either the Pythagorean theorem or the distance formula. Round to two decimal places if necessary.

91. (5, 1) and (3, 7)
92. (2, 6) and (4, 8)
93. (−7, 3) and (2, −8)
94. (4, −3) and (−1, 9)
95. (−4, 0) and (−6, −1)
96. (−5, −4) and (0, −1)

For Exercises 97 through 100, find the vertex of each parabola and determine whether it is a maximum or minimum point. Do not graph the parabolas.

97. $y = -x^2 + 4x$
98. $y = 2x^2 - 9x$
99. $y = 0.5x^2 - 6.5$
100. $y = -0.2x^2 + 12$

For Exercises 101 through 104, answer as indicated.

101. Write the equation of the line passing through the point $(-5, 4)$ and parallel to $y = -\frac{2}{5}x + 7$.

102. Write the equation of the line passing through the point $(0, -3)$ and parallel to $y = \frac{1}{3}x - 4$.

103. Write the equation of the line passing through the point $(-6, 1)$ and perpendicular to $y = -\frac{3}{2}x + 5$.

104. Write the equation of the line passing through the point $(8, -3)$ and perpendicular to $y = \frac{4}{5}x - 17$.

9.3 Solving Quadratic Equations by Completing the Square and by the Quadratic Formula

LEARNING OBJECTIVES
- Solve quadratic equations by completing the square.
- Solve quadratic equations using the quadratic formula.

Solving Quadratic Equations by Completing the Square

So far, we have learned two methods to solve quadratic equations. The methods are solving by factoring (Section 6.4) and solving by the square root property (Section 9.2). Both of these methods have limitations. To solve by factoring, the given quadratic equation must be factorable over the rational numbers. In Chapter 6, we discussed that some quadratics cannot be factored over the rationals (such quadratics were called *prime*). We saw that some quadratics are hard to factor. To solve by the square root property requires that the quadratic equation be in the form $(ax - h)^2 = c$. Not all quadratic equations are in that form.

We now learn two general methods to solve quadratic equations. The first of these two methods is called solving by *completing the square*.

In Section 9.2, we learned how to solve an equation such as $(x + 2)^2 = 25$ by using the square root property.

$$(x + 2)^2 = 25$$
$$\sqrt{(x + 2)^2} = \pm\sqrt{25} \quad \text{Use the square root property.}$$
$$x + 2 = \pm 5 \quad \text{Don't forget the } \pm \text{ on the right side.}$$

Solve both equations.

$$x + 2 = 5 \qquad x + 2 = -5$$
$$x = 3 \qquad x = -7$$

The left side of this equation is the factored form of a perfect square trinomial. If a quadratic equation could be rewritten as a perfect square trinomial on one side and a constant on the other side, then it could be solved by using the square root property. That is the idea behind completing the square.

> **Skill Connection**
>
> **Factoring perfect square trinomials**
>
> In Section 6.3, we learned how to factor perfect square trinomials. A trinomial of the form $a^2 + 2ab + b^2$ or $a^2 - 2ab + b^2$ is said to be a perfect square trinomial. To factor a perfect square trinomial, we use the formula
>
> $$a^2 + 2ab + b^2 = (a + b)^2$$
>
> or
>
> $$a^2 - 2ab + b^2 = (a - b)^2.$$

CONCEPT INVESTIGATION

Can that quadratic equation be rewritten as a perfect square trinomial?

1. Solve the quadratic equations by using the following steps. The first row has been done for you.

Equation	Add	Factor Left Side as Perfect Square Trinomial	Solve by Using the Square Root Property
$x^2 + 6x = 7$	Add $3^2 = 9$ to both sides. $x^2 + 6x + 9 = 7 + 9$ $x^2 + 6x + 9 = 16$	$(x + 3)^2 = 16$	$(x + 3)^2 = 16$ $\sqrt{(x + 3)^2} = \pm\sqrt{16}$ $x + 3 = \pm 4$ $x + 3 = 4 \quad x + 3 = -4$ $x = 1 \qquad x = -7$
$x^2 + 10x = 11$	Add $5^2 = 25$ to both sides.		
$x^2 - 8x = 9$	Add $(-4)^2 = 16$ to both sides.		

778 CHAPTER 9 Quadratic Equations

2. Look carefully at the coefficient of *x* in the first column and the number that was squared and added in the second column (both in blue). What is the relationship between these two values?

_____

These quadratic equations were turned into perfect square trinomials when half of the coefficient of *x* was squared and added to both sides. This process is called *completing the square*, and it is summarized in the following steps.

> **Steps to Solve a Quadratic Equation $ax^2 + bx + c = 0$ by Completing the Square**
>
> 1. Isolate the variable terms on one side of the equation.
> 2. If the coefficient of x^2 is not 1 ($a \neq 1$), then divide both sides by *a*.
> 3. Find $\frac{1}{2}$ the coefficient of *x*, square it, and add the result to both sides of the equation.
> 4. Factor the quadratic on the left side. It is now a perfect square trinomial.
> 5. Solve for the variable using the square root property.
> 6. Check the answers in the original problem.

Quadratic Equations

Square Root Property

Use when there is a squared term but no first degree term. Isolate the squared term and use a plus or minus symbol to indicate both answers.

$$x^2 = 25$$
$$x = \pm\sqrt{25}$$
$$x = \pm 5$$

Completing the Square

Use if the vertex form is required. Finish solving with the square root property.

$$x^2 + 6x + 4 = 0$$
$$x^2 + 6x = -4$$
$$x^2 + 6x + 9 = -4 + 9$$
$$x^2 + 6x + 9 = 5$$
$$(x + 3)^2 = 5$$

Factoring

Use when the quadratic equation has small coefficients that factor easily. Set the quadratic equal to zero, factor and then use the zero factor property to write two or more simpler equations.

$$x^2 + 7x + 10 = 0$$
$$(x + 5)(x + 2) = 0$$
$$x + 5 = 0 \quad or \quad x + 2 = 0$$

Example 1 Solving a quadratic equation by completing the square

Solve each quadratic equation by completing the square. Give both the exact answers and approximate answers rounded to two decimal places. Check the answer(s) in the original equation.

a. $x^2 + 8x = 20$ **b.** $x^2 - 2x = 4$

SOLUTION

a. Solving $x^2 + 8x = 20$ using the steps above yields the following:

Step 1 Isolate the variable terms on the left side of the equation.

This step does not need to be done here, as the variable terms are isolated on the left side.

$$x^2 + 8x = 20$$

Step 2 If the coefficient of x^2 is not 1 ($a \neq 1$), then divide both sides by *a*.

The coefficient of x^2 is 1($a = 1$). This step is not required in this problem.

Step 3 Find $\frac{1}{2}$ the coefficient of *x*, square it, and add the result to both sides of the equation.

The coefficient of *x* is 8. Finding $\frac{1}{2}$ the coefficient of *x* yields $\frac{1}{2}(8) = 4$. Add $(4)^2 = 16$ to both sides.

$$x^2 + 8x = 20 \qquad \text{Add } (4)^2 \text{ to both sides.}$$
$$x^2 + 8x + (4)^2 = 20 + (4)^2$$
$$x^2 + 8x + 16 = 20 + 16$$
$$x^2 + 8x + 16 = 36 \qquad \text{Simplify both sides.}$$

Step 4 Factor the quadratic on the left side of the equation.

$$x^2 + 8x + 16 = 36 \qquad \text{Factor the left side.}$$
$$(x + 4)^2 = 36 \qquad \text{Here } a = x \text{ and } b = 4.$$

SECTION 9.3 Solving Quadratic Equations by Completing the Square and by the Quadratic Formula   779

Step 5 Solve for the variable using the square root property.

$$(x + 4)^2 = 36$$
$$x + 4 = \pm\sqrt{36}$$
$$x + 4 = \pm 6$$

$$x + 4 = 6 \qquad x + 4 = -6$$
$$\underline{-4 \ -4} \qquad \underline{-4 \ \ -4}$$
$$x = 2 \qquad x = -10$$

Step 6 Check the answers in the original problem.

$$x = 2 \qquad\qquad\qquad x = -10$$
$$(2)^2 + 8(2) \stackrel{?}{=} 20 \qquad (-10)^2 + 8(-10) \stackrel{?}{=} 20$$
$$4 + 16 \stackrel{?}{=} 20 \qquad\qquad 100 - 80 \stackrel{?}{=} 20$$
$$20 = 20 \qquad \text{Both answers check.} \qquad 20 = 20$$

b. Solving $x^2 - 2x = 4$ using the steps above yields

Step 1 Isolate the variable terms on the left side of the equation. This step does not need to be done here, as the variable terms are both on the left side.

$$x^2 - 2x = 4$$

Step 2 The coefficient of x^2 is 1 ($a = 1$). This step is not required in this problem.

Step 3 The coefficient of x is -2. Finding $\frac{1}{2}$ the coefficient of x yields $\frac{1}{2}(-2) = -1$. Add $(-1)^2 = 1$ to both sides.

$$x^2 - 2x = 4 \qquad\qquad \text{Add } (-1)^2 \text{ to both sides.}$$
$$x^2 - 2x + (-1)^2 = 4 + (-1)^2$$
$$x^2 - 2x + 1 = 4 + 1$$
$$x^2 - 2x + 1 = 5 \qquad\qquad \text{Simplify both sides.}$$

Step 4 Now factor the quadratic on the left side of the equation.

$$x^2 - 2x + 1 = 5 \qquad \text{Factor the left side.}$$
$$(x - 1)^2 = 5 \qquad \text{Here } a = x \text{ and } b = 1.$$

Step 5 Solve for the variable using the square root property.

$$(x - 1)^2 = 5$$
$$x - 1 = \pm\sqrt{5}$$

$$x - 1 = \sqrt{5} \qquad x - 1 = -\sqrt{5}$$
$$\underline{+1 \ \ +1} \qquad \underline{+1 \ \ \ +1}$$
$$x = 1 + \sqrt{5} \qquad x = 1 - \sqrt{5}$$

Step 6 To check the answers, find decimal approximations to the exact answer and substitute those values into the original equation.

$$x = 1 + \sqrt{5} \approx 3.24$$
$$x = 1 - \sqrt{5} \approx -1.24$$

Substituting these values into the original equation yields

$$x \approx 3.24 \qquad\qquad\qquad x \approx -1.24$$
$$(3.24)^2 - 2(3.24) \stackrel{?}{=} 4 \qquad (-1.24)^2 - 2(-1.24) \stackrel{?}{=} 4$$
$$10.4976 - 6.48 \stackrel{?}{=} 4 \qquad\qquad 1.5376 + 2.48 \stackrel{?}{=} 4$$
$$4.0176 \approx 4 \qquad \text{Both answers check.} \qquad 4.0176 \approx 4$$

Owing to roundng, we will not get exactly 4.

Calculator Details

Rounding

To be as accurate as possible, always enter the whole expression into your calculator. Round only the final answer shown on the calculator.

To approximate $1 + \sqrt{5}$ correctly on a scientific calculator, enter this entire expression in order from left to right.

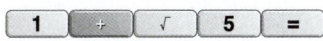

This will yield 3.236067977..., which, rounded to two decimal places, gives 3.24.

To approximate $\dfrac{1 + \sqrt{5}}{2}$ on your calculator, use parentheses around the numerator.

This yields 1.618033989..., which, rounded to two decimal places, gives 1.62.

Calculators vary in how to enter expressions. Consult your manual or instructor for help.

PRACTICE PROBLEM FOR EXAMPLE 1

Solve each quadratic equation by completing the square. Give both the exact answers and approximate answers rounded to two decimal places. Check the answer(s) in the original equation.

a. $x^2 - 4x + 1 = 0$ **b.** $x^2 + 8x + 11 = 0$

When solving a quadratic by completing the square, the lead coefficient of x^2 must be 1. When it is not 1 in the original problem, divide each term by the lead coefficient and then complete the square.

Example 2 — Solving a quadratic equation by completing the square when $a \neq 1$

Solve the quadratic equation by completing the square. Give both the exact answers and the approximate answers rounded to two decimal places. Check the answer(s) in the original equation.

$$2x^2 + 8x - 6 = 0$$

SOLUTION

The equation $2x^2 + 8x - 6 = 0$ is in general form. We can solve it by completing the square.

Step 1 Isolate the variable terms on the left side of the equation.

$$2x^2 + 8x - 6 = 0$$
$$\underline{\ +6\ \ +6} \quad \text{Add 6 to both sides to isolate x-terms on the left.}$$
$$2x^2 + 8x = 6$$

Step 2 The coefficient of x^2 is 2 ($a = 2$). Divide each term on both sides by 2 to get the coefficient of x^2 to be 1.

$$2x^2 + 8x = 6 \quad \text{Divide each term on both sides by 2.}$$
$$\frac{2x^2}{2} + \frac{8x}{2} = \frac{6}{2} \quad \text{Simplify each term.}$$
$$x^2 + 4x = 3 \quad \text{The coefficient of } x^2 \text{ is now 1.}$$

Step 3 The coefficient of x is 4. Finding $\frac{1}{2}$ the coefficient of x yields $\frac{1}{2}(4) = 2$. Add $(2)^2 = 4$ to both sides.

$$x^2 + 4x = 3 \quad \text{Add } (2)^2 \text{ to both sides.}$$
$$x^2 + 4x + (2)^2 = 3 + (2)^2$$
$$x^2 + 4x + 4 = 3 + 4$$
$$x^2 + 4x + 4 = 7 \quad \text{Simplify both sides.}$$

Step 4 Now we factor the quadratic on the left side of the equation.

$$x^2 + 4x + 4 = 7 \quad \text{Factor the left side.}$$
$$(x + 2)^2 = 7$$

Step 5 Solve for the variable, using the square root property.

$$(x + 2)^2 = 7$$
$$x + 2 = \pm\sqrt{7}$$

$$\begin{array}{ll} x + 2 = \sqrt{7} & x + 2 = -\sqrt{7} \\ \underline{\ -2\ \ -2} & \underline{\ -2\ \ -2} \\ x = -2 + \sqrt{7} & x = -2 - \sqrt{7} \end{array}$$

Step 6 To check the answers, we find decimal approximations to the exact answers and substitute those values into the original equation.

$$x = -2 + \sqrt{7} \approx 0.65$$
$$x = -2 - \sqrt{7} \approx -4.65$$

Substituting these values into the original equation yields

$x \approx 0.65$

$2(0.65)^2 + 8(0.65) - 6 \stackrel{?}{=} 0$ *Owing to rounding, we do not get exactly 0.*

$0.845 + 5.2 - 6 \stackrel{?}{=} 0$

$0.045 \approx 0$ *Both answers check.*

$x \approx -4.65$

$2(-4.65)^2 + 8(-4.65) - 6 \stackrel{?}{=} 0$

$43.245 - 37.2 - 6 \stackrel{?}{=} 0$

$0.045 \approx 0$

PRACTICE PROBLEM FOR EXAMPLE 2

Solve the quadratic equation by completing the square. Give both the exact answers and approximate answers rounded to two decimal places. Check the answer(s) in the original equation.

$$3x^2 - 30x + 69 = 0$$

■ Solving Quadratic Equations by Using the Quadratic Formula

There is a formula that will solve *any* quadratic equation. It is called the **quadratic formula**. If a quadratic equation is in the form $ax^2 + bx + c = 0$, in which one side of the equation must be equal to 0, the quadratic formula may be used. The quadratic formula may be proved by using completing the square.

> **THE QUADRATIC FORMULA**
> The solutions to $ax^2 + bx + c = 0$, $a \neq 0$, are given by
> $$x = \frac{-b \pm \sqrt{b^2 - 4ac}}{2a}$$

To solve a quadratic equation using the quadratic formula, we use the following steps.

> **Steps to Use the Quadratic Formula**
> 1. If needed, set the right side of the equation equal to 0. (The equation must be in the form $ax^2 + bx + c = 0$, $a \neq 0$.)
> 2. Identify the values of a, b, and c.
> 3. Substitute the values of a, b, and c into the quadratic formula
> $$x = \frac{-b \pm \sqrt{b^2 - 4ac}}{2a}$$
> 4. Find the value(s) of x by simplifying the expression on the right side using the order-of-operations agreement.
>   a. Find the *opposite* of b. (That is what $-b$ indicates.)
>   b. Calculate $b^2 - 4ac$ to simplify the radicand.
>   c. Find the square root of the simplified radicand.
>   d. Calculate $2a$ to simplify the denominator.
> 5. Rewrite as two equations. Simplify the right side to find the value(s) of x.
> 6. Check the answer(s) in the original problem.

What's That Mean?

Quadratic

In this section, we have mentioned quadratic expressions, quadratic equations, and now the quadratic formula. To keep all of these straight, remember that the "quadratic" part refers to a polynomial of degree 2; that is, the highest-degree term is x^2.

A quadratic expression is of the form

$$ax^2 + bx + c$$

Note that it has no equal sign.

A quadratic equation can be written in the form

$$ax^2 + bx + c = 0$$

The quadratic formula is

$$x = \frac{-b \pm \sqrt{b^2 - 4ac}}{2a}$$

which finds the solutions to a quadratic equation.

782 CHAPTER 9 Quadratic Equations

Quadratic Equations

Square Root Property

Use when there is a squared term but no first degree term. Isolate the squared term and use a plus or minus symbol to indicate both answers.

$$x^2 = 25$$
$$x = \pm\sqrt{25}$$
$$x = \pm 5$$

Completing the Square

Use if the vertex form is required. Finish solving with the square root property.

$$x^2 + 6x + 4 = 0$$
$$x^2 + 6x = -4$$
$$x^2 + 6x + 9 = -4 + 9$$
$$x^2 + 6x + 9 = 5$$
$$(x + 3)^2 = 5$$

Factoring

Use when the quadratic equation has small coefficients that factor easily. Set the quadratic equal to zero, factor and then use the zero factor property to write two or more simpler equations.

$$x^2 + 7x + 10 = 0$$
$$(x + 5)(x + 2) = 0$$
$$x + 5 = 0 \quad or \quad x + 2 = 0$$

Quadratic Formula

Use when the quadratic equation has fractions, decimals, or large numbers. Set the quadratic equal to zero.

$$x = \frac{-b \pm \sqrt{b^2 - 4ac}}{2a}$$

Example 3 — Using the quadratic formula

Solve each quadratic equation, using the quadratic formula. Give both the exact answers and the approximate answers rounded to two decimal places. Check the answers in the original equation.

a. $3x^2 + 2x - 1 = 0$ **b.** $4x^2 - 6x = -1$

SOLUTION

a. Solving $3x^2 + 2x - 1 = 0$ using the steps above yields the following:

Step 1 If needed, set the right side of the equation equal to 0.

The equation is a quadratic in general form and is set equal to 0.

Step 2 Identify the values of a, b, and c.

To identify the values of a, b, and c, we compare the given problem $3x^2 + 2x - 1 = 0$ to the general form $ax^2 + bx + c = 0$ and equate coefficients.

$$ax^2 + bx + c = 0$$
$$3x^2 + 2x - 1 = 0 \quad \text{Include the sign with the coefficient.}$$
$$a = 3 \quad b = 2 \quad c = -1$$

Therefore, we have that $a = 3$, $b = 2$, and $c = -1$.

Step 3 Substitute the values of $a = 3$, $b = 2$, and $c = -1$ in the quadratic formula.

$$x = \frac{-b \pm \sqrt{b^2 - 4ac}}{2a} \quad \text{Substitute in } a = 3, b = 2, c = -1.$$

$$x = \frac{-(2) \pm \sqrt{(2)^2 - 4(3)(-1)}}{2(3)}$$

Step 4 Find the value(s) of x by simplifying the expression on the right side using the order-of-operations agreement.

a. The opposite of b is $-b = -(2) = -2$.

b. Calculate $b^2 - 4ac$ to simplify the radicand.

$$b^2 - 4ac = (2)^2 - 4(3)(-1)$$
$$= 4 + 12$$
$$= 16$$

Substituting these results in the quadratic formula yields

$$x = \frac{-(2) \pm \sqrt{(2)^2 - 4(3)(-1)}}{2(3)}$$

$$x = \frac{-2 \pm \sqrt{16}}{2(3)}$$

c. Find the square root of the simplified radicand.

$$\sqrt{b^2 - 4ac} = \sqrt{16} = 4$$

Therefore,

$$x = \frac{-2 \pm \sqrt{16}}{2(3)}$$

$$x = \frac{-2 \pm 4}{2(3)}$$

d. Calculate $2a$ to simplify the denominator.

$$2a = 2(3) = 6$$

SECTION 9.3 Solving Quadratic Equations by Completing the Square and by the Quadratic Formula

So we now have that
$$x = \frac{-2 \pm 4}{2(3)}$$
$$x = \frac{-2 \pm 4}{6}$$

Step 5 Rewrite as two equations. Simplify the right side to find the value(s) of x.

$$x = \frac{-2 \pm 4}{6}$$

$$x = \frac{-2 + 4}{6} \qquad x = \frac{-2 - 4}{6}$$

$$x = \frac{2}{6} = \frac{1}{3} \qquad x = \frac{-6}{6} = -1$$

Step 6 Check the answer(s) in the original problem.

The original problem was $3x^2 + 2x - 1 = 0$. Checking the two answers yields:

$$x = \frac{1}{3} \qquad\qquad x = -1$$

$$3\left(\frac{1}{3}\right)^2 + 2\left(\frac{1}{3}\right) - 1 \stackrel{?}{=} 0 \qquad 3(-1)^2 + 2(-1) - 1 \stackrel{?}{=} 0$$

$$3\left(\frac{1}{9}\right) + \frac{2}{3} - 1 \stackrel{?}{=} 0 \qquad 3 - 2 - 1 \stackrel{?}{=} 0$$

$$\frac{1}{3} + \frac{2}{3} - 1 \stackrel{?}{=} 0 \qquad\qquad 0 = 0$$

$$0 = 0 \quad \text{Both answers check.}$$

b. Solving $4x^2 - 6x = -1$ using the steps above yields the following:

Step 1 The right side of this equation is not set equal to 0. Add 1 to both sides to put the right side equal to 0.

$$4x^2 - 6x = -1$$
$$\underline{\; +1 \quad +1}$$
$$4x^2 - 6x + 1 = 0$$

Step 2 To identify the values of a, b, and c, we compare the given problem $4x^2 - 6x + 1 = 0$ to the general form $ax^2 + bx + c = 0$ and equate coefficients.

$$ax^2 + bx + c = 0$$
$$4x^2 - 6x + 1 = 0 \quad \text{Include the sign with the coefficient.}$$
$$a = 4 \quad b = -6 \quad c = 1$$

Therefore, we have that $a = 4$, $b = -6$, and $c = 1$.

Step 3 We are ready to substitute the values of $a = 4$, $b = -6$, and $c = 1$ in the quadratic formula.

$$x = \frac{-b \pm \sqrt{b^2 - 4ac}}{2a} \quad \text{Substitute in } a = 4, b = -6, c = 1.$$

$$x = \frac{-(-6) \pm \sqrt{(-6)^2 - 4(4)(1)}}{2(4)}$$

784 CHAPTER 9 Quadratic Equations

Step 4 To find the values of x, we simplify the right side of this formula.

a. The opposite of b is $-b = -(-6) = 6$.
$$x = \frac{6 \pm \sqrt{(-6)^2 - 4(4)(1)}}{2(4)}$$

b. Calculate $b^2 - 4ac$ to simplify the radicand.
$$b^2 - 4ac = (-6)^2 - 4(4)(1)$$
$$= 36 - 16$$
$$= 20$$

Substitute these values in the quadratic formula.
$$x = \frac{6 \pm \sqrt{(-6)^2 - 4(4)(1)}}{2(4)}$$
$$x = \frac{6 \pm \sqrt{20}}{2(4)}$$

c. Find the square root of the simplified radicand.
$$\sqrt{b^2 - 4ac} = \sqrt{20} = \sqrt{4 \cdot 5} = 2\sqrt{5}$$

Substituting $2\sqrt{5}$ in the quadratic formula yields
$$x = \frac{6 \pm \sqrt{20}}{2(4)}$$
$$x = \frac{6 \pm 2\sqrt{5}}{2(4)}$$

d. Calculate $2a$ to simplify the denominator.
$$2a = 2(4) = 8$$

Substituting this result, 8, in the quadratic formula yields
$$x = \frac{6 \pm 2\sqrt{5}}{2(4)}$$
$$x = \frac{6 \pm 2\sqrt{5}}{8}$$

Step 5 Rewrite as two equations. Simplify the right side to find the value(s) of x. Keep in mind that $-b$ means the "opposite of b," which means that the sign must be changed on b.

$$x = \frac{6 \pm 2\sqrt{5}}{8}$$

$$x = \frac{6 + 2\sqrt{5}}{8} \approx 1.31 \qquad x = \frac{6 - 2\sqrt{5}}{8} \approx 0.19$$

Step 6 Check the answer(s) in the original problem. The original problem was $4x^2 - 6x = -1$. Checking the two answers yields

$x \approx 1.31$ $\qquad\qquad\qquad\qquad\qquad\qquad$ $x \approx 0.19$
$4(1.31)^2 - 6(1.31) \stackrel{?}{=} -1$ $\qquad\qquad\quad$ $4(0.19)^2 - 6(0.19) \stackrel{?}{=} -1$
$-0.9956 \approx -1$ $\quad$ Both answers check. $\quad$ $-0.9956 \approx -1$

PRACTICE PROBLEM FOR EXAMPLE 3

Solve each quadratic equation, using the quadratic formula. Give both the exact answers and the approximate answers rounded to two decimal places. Check the answers in the original equation.

a. $5x^2 - 6x + 1 = 0$ $\qquad\qquad$ **b.** $-2x^2 + 6x = 3$

Calculator Details

Rounding

For the sake of accuracy, always enter the entire expression into your calculator. Round only the final answer shown on the calculator.

To approximate $\dfrac{6 + 2\sqrt{5}}{8}$ on your calculator, use parentheses around the numerator.

This yields 1.309016994....
Rounding it to two decimal places gives 1.31.

SECTION 9.3 Solving Quadratic Equations by Completing the Square and by the Quadratic Formula

Example 4: Using the quadratic formula to determine whether there are real solutions

Solve the quadratic equation, using the quadratic formula. Give both the exact answers and the approximate answers rounded to two decimal places. Check the answers in the original equation.

$$7x^2 - x + 6 = 0$$

SOLUTION

Solving $7x^2 - x + 6 = 0$ using the steps above yields the following:

Step 1 The right side of this equation is set to 0, so we do not need to rearrange any terms.

Step 2 To identify the values of a, b, and c, we compare the given problem, $7x^2 - x + 6 = 0$, to the general form $ax^2 + bx + c = 0$ and equate coefficients.

$$ax^2 + bx + c = 0$$
$$7x^2 - x + 6 = 0 \quad \text{Remember that } -x = -1 \cdot x.$$
$$a = 7 \quad b = -1 \quad c = 6$$

Therefore, we have that $a = 7$, $b = -1$, and $c = 6$.

Step 3 We are ready to substitute the values of $a = 7$, $b = -1$, and $c = 6$ into the quadratic formula.

$$x = \frac{-b \pm \sqrt{b^2 - 4ac}}{2a} \quad \text{Substitute in } a = 7, b = -1, c = 6.$$

$$x = \frac{-(-1) \pm \sqrt{(-1)^2 - 4(7)(6)}}{2(7)}$$

Step 4 To find the values of x, we simplify the right side of this formula.

a. The opposite of b is $-b = -(-1) = 1$.

$$x = \frac{1 \pm \sqrt{(-1)^2 - 4(7)(6)}}{2(7)}$$

b. Calculate $b^2 - 4ac$ to simplify the radicand.

$$b^2 - 4ac = (-1)^2 - 4(7)(6)$$
$$= 1 - 168$$
$$= -167$$

Substituting the result into the quadratic formula yields

$$x = \frac{1 \pm \sqrt{(-1)^2 - 4(7)(6)}}{2(7)}$$

$$x = \frac{1 \pm \sqrt{-167}}{2(7)}$$

Find the square root of the simplified radicand.

$$\sqrt{b^2 - 4ac} = \sqrt{-167}$$

This is *not* a real number. Remember that we cannot have a negative value under a square root. There are no real solutions to this quadratic equation.
Therefore, the quadratic equation $7x^2 - x + 6 = 0$ has *no real solutions*.

PRACTICE PROBLEM FOR EXAMPLE 4

Solve the quadratic equation, using the quadratic formula. Give both the exact answers and the approximate answers rounded to two decimal places. Check the answers in the original equation.

$$4x^2 + 5x + 2 = 0$$

Quadratic equations show up in many types of application problems. Since the quadratic formula leads to the solution(s) to a quadratic equation, it can be used to solve for the variable in an application problem.

Example 5 — When does the rocket hit the ground?

A local model rocket club launches a rocket that follows the path given by the equation $h = -16t^2 + 180t + 1$. Here t is the time in seconds after the launch, and h is the height of the rocket in feet t seconds after launch. The students in the club want to know when the rocket will hit the ground.

a. The phrase "when the rocket will hit the ground" contains the word *when*. Does *when* indicate that we are trying to determine the height, h, of the rocket or the time, t?

b. When the rocket hits the ground, what is the value of the height h? Substitute this value into the equation $h = -16t^2 + 180t + 1$.

c. What kind of equation do we have in part b? Is it linear, quadratic, rational, or radical?

d. Solve the equation from part b.

e. If there is more than one answer, which answer makes sense for this problem?

SOLUTION

a. *When* indicates that we are trying to determine the time t when the rocket hits the ground.

b. *When the rocket hits the ground, its height above the ground is 0.*
Therefore, $h = 0$. Substituting $h = 0$ into the equation $h = -16t^2 + 180t + 1$ yields

$$0 = -16t^2 + 180t + 1$$

or

$$-16t^2 + 180t + 1 = 0$$

c. This is a quadratic equation in the general form $ax^2 + bx + c = 0$.

d. To solve the equation $-16t^2 + 180t + 1 = 0$, we can use the quadratic formula. Comparing $-16t^2 + 180t + 1 = 0$ to the general equation $ax^2 + bx + c = 0$, we see that $a = -16$, $b = 180$, and $c = 1$. Substituting these values into the quadratic formula and simplifying, we have that

$$-16t^2 + 180t + 1 = 0$$

$$t = \frac{-(180) \pm \sqrt{(180)^2 - 4(-16)(1)}}{2(-16)}$$

$$t = \frac{-(180) \pm \sqrt{32{,}464}}{-32}$$

$$t = \frac{-(180) + \sqrt{32464}}{-32} \qquad t = \frac{-(180) - \sqrt{32464}}{-32}$$

$$t \approx \frac{-180 + 180.17769}{-32} \qquad t \approx \frac{-180 - 180.17769}{-32}$$

$$t \approx \frac{0.17769}{-32} \qquad t \approx \frac{-360.17769}{-32}$$

$$t \approx -0.005553 \qquad t \approx 11.25555$$

$$t \approx -0.006 \qquad t \approx 11.26$$

SECTION 9.3 Solving Quadratic Equations by Completing the Square and by the Quadratic Formula 787

e. The first answer, $t \approx -0.006$ second, does not make sense in this problem because there is no "negative" time before the rocket launches.

The second answer, $t \approx 11.26$ seconds, is the approximate time when the rocket hits the ground.

PRACTICE PROBLEM FOR EXAMPLE 5

A local model rocket club launches a rocket from a tower 40 feet off the ground that follows the path given by the equation $h = -16t^2 + 140t + 40$. Here t is the time in seconds after the launch, and h is the height of the rocket in feet t seconds after launch. The students in the club want to know when the rocket will hit the ground.

a. The phrase "when the rocket will hit the ground" contains the word *when*. Does *when* indicate that we are trying to determine the height, h, of the rocket or the time, t?

b. When the rocket hits the ground, what is the value of the height h? Substitute this value in the equation $h = -16t^2 + 140t + 40$.

c. What kind of equation do we have in part b? Is it linear, quadratic, rational, or radical?

d. Solve the equation from part b.

e. If there is more than one answer, which answer makes sense for this problem?

9.3 Vocabulary Exercises

1. $-2x^2 + 4x$ is an example of a quadratic _____.

2. $-2x^2 + 4x = 0$ is an example of a quadratic _____.

3. $y = -2x^2 + 4x$ is an example of a quadratic equation in _____ variables.

4. To solve a quadratic equation, we can use the quadratic _____.

9.3 Exercises

For Exercises 1 through 8, identify the values of a, b, and c in the general form of each quadratic equation, $ax^2 + bx + c = 0$. Do not solve the equations. (Hint: The equation must be set equal to zero before you find a, b, or c.)

1. $x^2 - 5x + 21 = 0$
2. $x^2 - 3x + 17 = 0$
3. $-x^2 + 12x - 16 = 0$
4. $-3x^2 - 7x + 5 = 0$
5. $t^2 - 6t = -12$
6. $t^2 - t = 17$
7. $-4m^2 + 10 = 3m$
8. $7b + 2 = -6b^2$

For Exercises 9 through 14, identify and compute the quantities listed. The coefficient of x is b.

9. For the quadratic equation $x^2 + 8x + 15 = 0$, find

   a. $\frac{1}{2}b$ **b.** $\left(\frac{1}{2}b\right)^2$

10. For the quadratic equation $x^2 + 4x + 4 = 0$, find

    a. $\frac{1}{2}b$ **b.** $\left(\frac{1}{2}b\right)^2$

11. For the quadratic equation $x^2 - 14x + 21 = 0$, find

    a. $\frac{1}{2}b$ **b.** $\left(\frac{1}{2}b\right)^2$

12. For the quadratic equation $x^2 - 12x + 18 = 0$, find

    a. $\frac{1}{2}b$ **b.** $\left(\frac{1}{2}b\right)^2$

13. For the quadratic equation $x^2 + 5x - 6 = 0$, find

    a. $\frac{1}{2}b$ **b.** $\left(\frac{1}{2}b\right)^2$

14. For the quadratic equation $x^2 - 7x - 8 = 0$, find

 a. $\frac{1}{2}b$ **b.** $\left(\frac{1}{2}b\right)^2$

For Exercises 15 through 34, solve each quadratic equation by completing the square. Give both the exact answer(s) and the approximate answer(s) rounded to two decimal places. Check the answer(s) in the original equation.

15. $x^2 + 10x + 21 = 0$ **16.** $x^2 - 2x - 42 = 0$

17. $x^2 - 8x + 14 = 0$ **18.** $x^2 - 4x - 3 = 0$

19. $y^2 - 22y + 111 = 0$ **20.** $y^2 + 12y + 33 = 0$

21. $a^2 - 16a = -61$ **22.** $b^2 + 18b = -75$

23. $x^2 - 2x = 63$ **24.** $a^2 - 7a = 60$

25. $2x^2 + 12x + 8 = 0$ **26.** $2x^2 + 28x + 86 = 0$

27. $3y^2 - 54y + 192 = 0$ **28.** $3y^2 - 48y + 177 = 0$

29. $5t^2 - 10t - 70 = 0$ **30.** $5t^2 + 110t + 590 = 0$

31. $2n^2 - 8n = 30$ **32.** $2n^2 + 4n = 28$

33. $-x^2 + 12x = 34$ **34.** $-x^2 + 10x = 18$

For Exercises 35 through 38, the students were asked to solve by completing the square. Find the error in each student's work. Then correctly solve each quadratic equation by completing the square.

35. $x^2 + 2x = 3$

> **Austin**
> $x^2 + 2x = 3$
> $x^2 + 2x + (1)^2 = 3$
> $(x + 1)^2 = 3$
> $x + 1 = \pm\sqrt{3}$
> $x = -1 \pm \sqrt{3}$

36. $x^2 - 6x = 2$

> **Aynzley**
> $x^2 - 6x = 2$
> $x^2 - 6x + 9 = 2$
> $(x - 3)^2 = 2$
> $x - 3 = \pm\sqrt{2}$
> $x = 3 \pm \sqrt{2}$

37. $x^2 - 4x = 12$

> **Diego**
> $x^2 - 4x = 12$
> $x^2 - 4x + 4 = 12 + 4$
> $(x - 2)^2 = 16$
> $x - 2 = 4$
> $x = 6$

38. $x^2 + 10x = 11$

> **Jeffrey**
> $x^2 + 10x = 11$
> $x^2 + 10x + 25 = 11 + 25$
> $(x + 5)^2 = 36$
> $x + 5 = 6$
> $x = 1$

For Exercises 39 through 48, solve each quadratic equation by using the quadratic formula. Give both the exact answer(s) and the approximate answer(s) rounded to two decimal places. Check the answer(s) in the original equation.

39. $a^2 + 2a - 7 = 0$ **40.** $a^2 + 6a + 2 = 0$

41. $2x^2 - 14x + 23 = 0$ **42.** $2x^2 + 10x + 11 = 0$

43. $9x^2 - 6x - 4 = 0$ **44.** $9x^2 + 12x - 2 = 0$

45. $15a^2 - 22a = 5$ **46.** $28a^2 - 17a = 3$

47. $25y^2 + 60y = 29$ **48.** $49y^2 - 112y = -59$

For Exercises 49 through 52, the students were asked to solve by using the quadratic formula. Find the error in each student's work. Then correctly solve each quadratic equation by using the quadratic formula.

49. $2x + x^2 + 1 = 0$

> **Liliana**
> $2x + x^2 + 1 = 0$
> $a = 2 \quad b = 1 \quad c = 1$
> $x = \dfrac{-(1) \pm \sqrt{(1)^2 - 4(2)(1)}}{2(2)}$
> $x = \dfrac{-(1) \pm \sqrt{-7}}{4}$
> No real solution.

SECTION 9.3 Solving Quadratic Equations by Completing the Square and by the Quadratic Formula **789**

50. $-6x + 9 + x^2 = 0$

> **Andres**
>
> $-6x + 9 + x^2 = 0$
> $a = -6 \quad b = 9 \quad c = 1$
> $x = \dfrac{-(9) \pm \sqrt{(9)^2 - 4(-6)(1)}}{2(-6)}$
> $x = \dfrac{-9 \pm \sqrt{105}}{-12}$

51. $3x^2 - 5x - 12 = 0$

> **Cassandra**
>
> $3x^2 - 5x - 12 = 0$
> $a = 3 \quad b = -5 \quad c = -12$
> $x = \dfrac{-5 \pm \sqrt{(-5)^2 - 4(3)(-12)}}{2(3)}$
> $x = \dfrac{-5 \pm \sqrt{169}}{6}$
> $x = \dfrac{-5 \pm 13}{6}$
> $x = \dfrac{4}{3}, x = -3$

52. $2x^2 - x - 10 = 0$

> **William**
>
> $2x^2 - x - 10 = 0$
> $a = 2 \quad b = -1 \quad c = -10$
> $x = \dfrac{-1 \pm \sqrt{(-1)^2 - 4(2)(-10)}}{2(2)}$
> $x = \dfrac{-1 \pm \sqrt{81}}{4}$
> $x = \dfrac{-1 \pm 9}{4}$
> $x = -\dfrac{5}{2}, x = 2$

For Exercises 53 through 60, solve each application problem. If necessary, round the answer(s) to two decimal places.

53. A landscape photographer sells 11 × 14 prints of his work. The profit, P, he makes from selling x copies of his photographs is given by $P = -2x^2 + 250x$. What is the least number of copies of his photographs that he must sell to make a profit of $5000.00?

   a. Since we are interested in a profit of $5000.00, substitute in the equation $P = 5000$.

   b. Solve the quadratic equation from part a.

   c. What is the least number of photographs he must sell to make a profit of $5000.00?

54. A jewelry designer sells bead necklaces in a local farmer's market. The profit, P, that she makes from selling x necklaces is given by $P = -x^2 + 45x$. What is the least number of her necklaces she must sell to make a profit of $500.00?

   a. Since we are interested in a profit of $500.00, substitute $P = 500$ in the equation.

   b. Solve the quadratic equation from part a.

   c. What is the least number of her necklaces she must sell to make a profit of $500.00?

55. A freelance web designer estimates the weekly revenue, R, that she makes from working x hours designing websites is given by $R = -2x^2 + 80x$. What is the least number of hours she must work to make a weekly revenue of $500.00?

56. A barber estimates that the weekly revenue, R, he makes from giving x haircuts is given by $R = -x^2 + 40x$. What is the least number of haircuts he must give to make a weekly revenue of $400.00?

57. A local model rocket club launches a rocket that follows the path given by the equation $h = -16t^2 + 140t + 5$. Here t is the time in seconds after the launch and h is the height of the rocket in feet t seconds after launch. The students in the club want to know when the rocket will hit the ground.

   a. The phrase "when the rocket will hit the ground" contains the word *when*. Does *when* indicate that we are trying to determine the height, h, of the rocket or the time, t?

   b. When the rocket hits the ground, what is the value of the height, h? Substitute this value in the equation $h = -16t^2 + 140t + 5$.

   c. What kind of equation do we have in part b? Is it linear, quadratic, rational, or radical?

   d. Solve the equation from part b.

   e. If there is more than one answer, which answer makes sense for this problem?

58. A local model rocket club launches a rocket that follows the path given by the equation $h = -16t^2 + 120t + 2$.

Here t is the time in seconds after the launch, and h is the height of the rocket in feet t seconds after launch. The students in the club want to know when the rocket will hit the ground.

a. The phrase "when the rocket will hit the ground" contains the word *when*. Does *when* indicate that we are trying to determine the height, h, of the rocket or the time, t?

b. When the rocket hits the ground, what is the value of the height, h? Substitute this value in the equation $h = -16t^2 + 120t + 2$.

c. What kind of equation do we have in part b? Is it linear, quadratic, rational, or radical?

d. Solve the equation from part b.

e. If there is more than one answer, which answer makes sense for this problem?

59. A rock is dropped from a 60-meter cliff. The height of the rock, h, in meters after t seconds is given by the equation

$$h = -4.9t^2 + 60$$

a. How long will it take for the rock to be at a height of 52 meters? Round the answer to two decimal places if necessary.

b. How long will it take for the rock to hit the ground? Round the answer to two decimal places if necessary.

60. A rock is dropped from a 36-meter cliff. The height of the rock, h, in feet after t seconds is given by the equation

$$h = -4.9t^2 + 36$$

a. How long will it take for the rock to be at a height of 24 meters? Round the answer to two decimal places if necessary.

b. How long will it take for the rock to hit the ground? Round the answer to two decimal places if necessary.

For Exercises 61 through 72, solve each quadratic equation using any of the methods learned so far: (a) factoring, (b) solving using the square root property, (c) completing the square, and (d) the quadratic formula. Give both the exact answer(s) and approximate answer(s) rounded to two decimal places.

Three hints:

- *Does the problem have only one variable term with an x^2 or $(ax - h)^2$? Consider solving using the square root property.*
- *Does the problem have three terms and the coefficient of x^2 is 1? Consider solving by factoring.*
- *Does the problem have three terms and cannot be factored easily or the coefficient of x^2 is not 1? Solve by completing the square or the quadratic formula.*

61. $4x^2 - 28 = 0$
62. $6x^2 - 30 = 0$
63. $y^2 + 10y + 24 = 0$
64. $y^2 - 14y + 45 = 0$
65. $4x^2 + 12x + 3 = 0$
66. $3x^2 + 2x - 6 = 0$
67. $a^2 + 2a - 63 = 0$
68. $a^2 - 4a - 32 = 0$
69. $15p^2 - 375 = 0$
70. $5p^2 - 405 = 0$
71. $8x^2 + 12x - 1 = 0$
72. $3x^2 + 14x + 13 = 0$

For Exercises 73 through 76, find the x- and y-intercepts of each equation.

73. $y = -\dfrac{8}{3}x + 4$
74. $y = \dfrac{3}{5}x - 1$
75. $7x - 3y = -21$
76. $-2x + 5y = -15$

For Exercises 77 through 82, use the model $S = 1.72t + 13.96$, where S is the annual sales for Zara in billions of dollars t years since 2010. Zara is one of the world's leading fast fashion retailers.

Source: Model based on data from Zara annual reports.

77. Find when the annual sales for Zara were $18 billion.

78. Find when the annual sales for Zara are predicted to be $30 billion.

79. Estimate the annual sales for Zara in 2017.

80. Find and interpret the slope of this model.

81. Find and interpret the horizontal intercept of this model.

82. Find and interpret the vertical intercept of this model.

For Exercises 83 through 92, identify each equation as linear, quadratic, radical, rational, or other. Do not solve the equations.

83. $\dfrac{t+3}{5} = -\dfrac{t}{2}$

84. $\dfrac{5}{n} = -\dfrac{8}{n+1}$

85. $4 - 2x + x^2 = 6$

86. $-5 + 2x^2 = 4$

87. $|-3x + 5| = 9$

88. $2 = 3 + |x + 3|$

89. $-4(3n + 2) - 1 = 5n - 7$

90. $8 - 16t = 4 - 2(5t + 6)$

91. $\sqrt{2x + 1} = 4$

92. $8 = \sqrt{-x + 3}$

9.4 Graphing Quadratic Equations Including Intercepts

LEARNING OBJECTIVES
- Find the intercepts.
- Sketch the graph (pulling it all together).

Finding Intercepts

Two other key features of the graph of a quadratic equation in two variables are the x- and y-intercepts. In Chapter 3, we learned how to find the intercepts of lines from the graph and the equation. We will review finding intercepts from the graph for a quadratic equation in two variables. Then we will learn how to find the intercepts from the equation.

Example 1 Finding the intercepts from the graph

Find the x- and y-intercepts from the graph for each quadratic equation in two variables.

a. $y = x^2$

b. $y = x^2 - 4$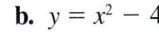

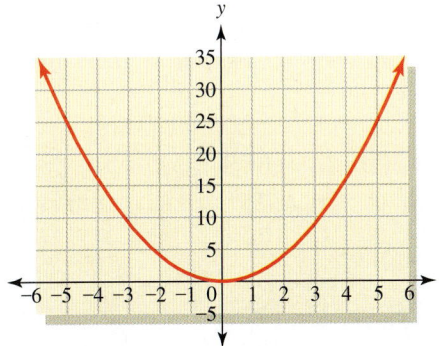

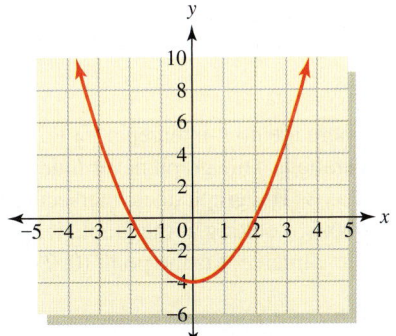

c. $y = x^2 - 4x + 7$

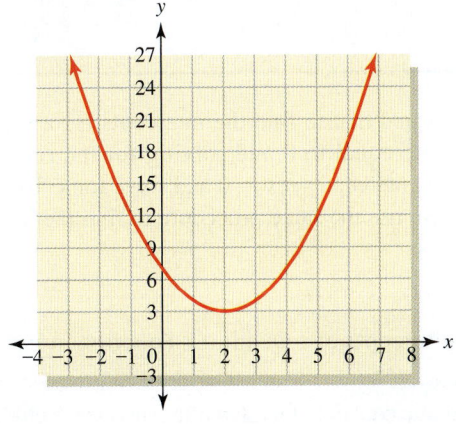

792 CHAPTER 9 Quadratic Equations

SOLUTION

a. To find the *x*-intercept(s) of this graph, look for the point(s) where the graph intersects the *x*-axis. Looking at the graph, we see that the *x*-intercept is the point $(0, 0)$. The *y*-intercept is the point where the graph intersects the *y*-axis. Therefore, the *y*-intercept is the point $(0, 0)$.

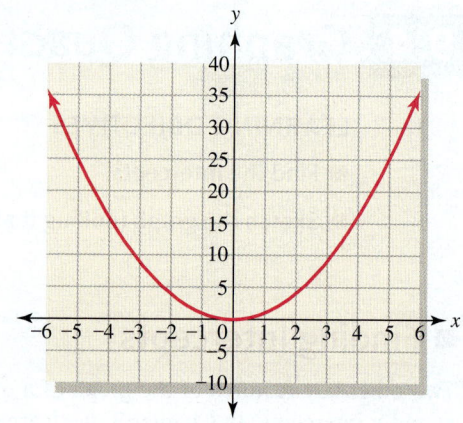

b. There are two *x*-intercepts for this graph because the parabola hits the *x*-axis in two points. The points are $(2, 0)$ and $(-2, 0)$. The *y*-intercept of this graph is the point $(0, -4)$.

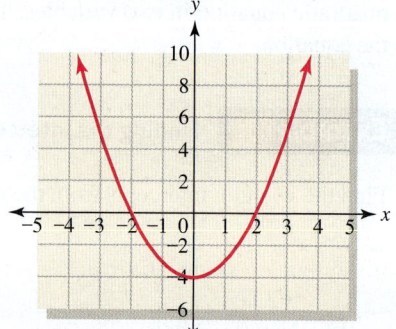

c. There are no *x*-intercepts for this parabola. This graph never touches the *x*-axis. The graph does intersect the *y*-axis. Looking at the graph, it appears the *y*-intercept is the point $(0, 7)$.

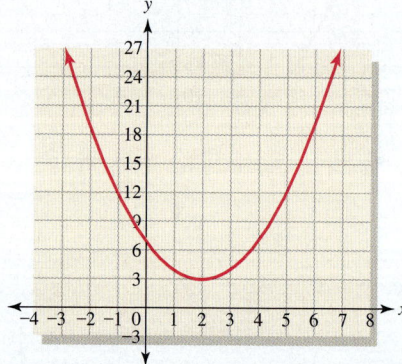

In Chapter 3, we learned that to find the intercepts for a linear equation involved setting the other variable equal to 0 and solving for the remaining variable. More specifically, to find the *x*-intercept, set $y = 0$ and solve for *x*. To find the *y*-intercept, set $x = 0$ and solve for *y*. We use the same procedure to find the intercepts for a quadratic equation in two variables.

> **Steps to Find the Intercepts of a Quadratic Equation in Two Variables**
>
> 1. To find the *x*-intercept, set $y = 0$ and solve for *x*. Write each intercept as a point $(a, 0)$. For a quadratic equation, this step often involves factoring, the square root property, completing the square, or the quadratic formula to solve for *x*.
>
>   *Note:* There may be zero, one, or two *x*-intercepts for a quadratic equation in two variables.
>
> 2. To find the *y*-intercept, set $x = 0$ and solve for *y*. Write this intercept as the point $(0, b)$.

SECTION 9.4 Graphing Quadratic Equations Including Intercepts

Example 2 — Finding the intercepts from the equation

Find the x- and y-intercepts for each quadratic equation. Write the intercepts as points (x, y).

a. $y = x^2 - 2x - 15$ **b.** $y = x^2 - 4x + 4$ **c.** $y = x^2 + 2x + 3$

SOLUTION

a. Step 1 To find the x-intercept, set $y = 0$ and solve for x.

$y = x^2 - 2x - 15$ Set $y = 0$ and solve for x.
$0 = x^2 - 2x - 15$ Factor the right side.
$0 = (x - 5)(x + 3)$ Set each factor to 0 and solve for x.
$x - 5 = 0 \quad x + 3 = 0$
$x = 5 \quad\quad x = -3$

$(5, 0)$ and $(-3, 0)$ are the x-intercepts.

Step 2 To find the y-intercept, set $x = 0$ and solve for y.

$y = x^2 - 2x - 15$ Set $x = 0$ and solve for y.
$y = 0^2 - 2(0) - 15$ Evaluate the right side.
$y = -15$

$(0, -15)$ is the y-intercept.

b. To find the x-intercept, set $y = 0$ and solve for x.

$y = x^2 - 4x + 4$ Set $y = 0$ and solve for x.
$0 = x^2 - 4x + 4$ Factor the right side.
$0 = (x - 2)(x - 2)$ Set each factor to 0 and solve for x.
$x - 2 = 0 \quad x - 2 = 0$
$x = 2 \quad\quad x = 2$

$(2, 0)$ is the x-intercept.

To find the y-intercept, set $x = 0$ and solve for y.

$y = x^2 - 4x + 4$ Set $x = 0$ and solve for y.
$y = 0^2 - 4(0) + 4$ Evaluate the right side.
$y = 4$

$(0, 4)$ is the y-intercept.

c. To find the x-intercept, set $y = 0$ and solve for x.

$y = x^2 + 2x + 3$ Set $y = 0$ and solve for x.
$a = 1 \quad b = 2 \quad c = 3$

$x = \dfrac{-(2) \pm \sqrt{(2)^2 - 4(1)(3)}}{2(1)}$ Substitute the value of a, b, and c into the quadratic formula.

$x = \dfrac{-2 \pm \sqrt{-8}}{2}$ The radicand is negative so there are no real answers.

There are no x-intercepts.
To find the y-intercept, we set $x = 0$ and solve for y.

$y = x^2 + 2x + 3$ Set $x = 0$ and solve for y.
$y = 0^2 + 2(0) + 3$ Evaluate the right side.
$y = 3$

$(0, 3)$ is the y-intercept.

794 CHAPTER 9 Quadratic Equations

PRACTICE PROBLEM FOR EXAMPLE 2

Find the x- and y-intercepts for each quadratic equation. Write the intercepts as points, (x, y).

a. $y = 2x^2 + 5x - 12$ **b.** $y = x^2 - 36$

■ Putting It All Together to Sketch a Graph

We can now sketch a more accurate graph of a quadratic equation in two variables by identifying the vertex, axis of symmetry, and intercepts.

> **Steps to Graph a Quadratic Equation in Two Variables with Intercepts**
> 1. Find the equation of the axis of symmetry.
> 2. Find the coordinates of the vertex.
> 3. Find the x- and y-intercepts.
> 4. Graph a total of seven points. Find the values of additional symmetric points as necessary. There should be at least three points to the left and right of the vertex.
> 5. Connect the points with a smooth curve.

Example 3 Graphing a quadratic equation in two variables

Graph each quadratic equation. Find the equation of the axis of symmetry, the vertex, and the x- and y-intercepts. Graph at least seven points. Clearly label and scale the axes.

a. $y = x^2 - 6x + 5$ **b.** $y = x^2 - 9$

SOLUTION

a. Step 1 Find the equation of the axis of symmetry.

$$x = \frac{-b}{2a} = \frac{-(-6)}{2(1)} = \frac{6}{2}$$

$$x = 3$$

Step 2 Find the coordinates of the vertex.

Recall that the x-coordinate of the vertex is on the axis of symmetry. So $x = 3$. To find the y-coordinate of the vertex, substitute $x = 3$ into the original equation.

$$y = (3)^2 - 6(3) + 5 = 9 - 18 + 5 = -9 + 5 = -4$$

Therefore, the vertex is at the point $(3, -4)$.

Step 3 Find the x- and y-intercepts.

To find the x-intercept, substitute $y = 0$ into the original equation and solve for x.

$0 = x^2 - 6x + 5$ Factor the right side.
$0 = (x - 5)(x - 1)$ Set each factor equal to 0 and solve for x.
$x - 5 = 0 \quad x - 1 = 0$
$x = 5 \quad\quad x = 1$

The x-intercepts are $(5, 0)$ and $(1, 0)$.

To find the y-intercept, substitute $x = 0$ into the original equation and solve for y.

$y = (0)^2 - 6(0) + 5$ Substitute in x = 0.
$y = 5$

The y-intercept is $(0, 5)$.

Step 4 Graph a total of seven points. Find the values of additional symmetric points as necessary. There should be at least three points to the left and right of the vertex.

Let's organize this information in a table and then determine how many more points to find. Put the vertex in the middle of the table to see whether we have three points to the left and right of the vertex.

x	y	(x, y)	Label
0	5	(0, 5)	y-intercept
1	0	(1, 0)	x-intercept
3	-4	$(3, -4)$	Vertex
5	0	(5, 0)	x-intercept

Some missing x-values on either side of the vertex are $x = 2$, $x = 4$, and $x = 6$. Let's substitute those values in the equation and determine the corresponding y-values.

x	y	(x, y)	Label
0	5	(0, 5)	y-intercept
1	0	(1, 0)	x-intercept
2	$y = 2^2 - 6(2) + 5 = -3$	$(2, -3)$	**Additional symmetric point**
3	-4	$(3, -4)$	Vertex
4	$y = 4^2 - 6(4) + 5 = -3$	$(4, -3)$	**Additional symmetric point**
5	0	(5, 0)	x-intercept
6	$y = 6^2 - 6(6) + 5 = 5$	$(6, 5)$	**Symmetric point to y-intercept**

Step 5 Connect the points with a smooth curve.

Now that we have seven points, we can graph the quadratic.

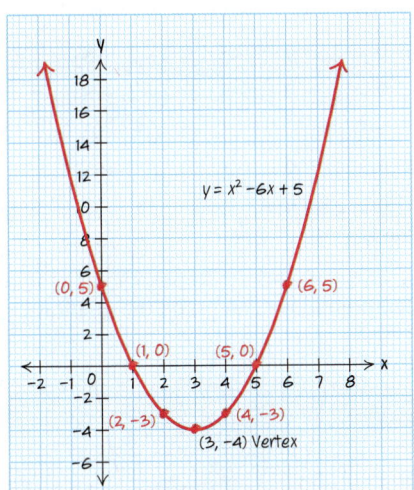

b. Find the equation of the axis of symmetry for the quadratic equation $y = x^2 - 9$.

$$x = \frac{-b}{2a} = \frac{-(0)}{2(1)} = \frac{0}{2}$$

$$x = 0$$

To find the vertex, substitute $x = 0$ into the original equation to find y.

$$y = (0)^2 - 9 = 0 - 9 = -9$$

Therefore, the vertex is at the point $(0, -9)$.

To find the x-intercept, substitute $y = 0$ into the original equation and solve for x.

$0 = x^2 - 9$ Factor the right side as a difference of squares.
$0 = (x - 3)(x + 3)$ Set each factor equal to 0 and solve for x.
$x - 3 = 0 \quad\quad x + 3 = 0$
$\quad x = 3 \quad\quad\quad x = -3$

The x-intercepts are $(3, 0)$ and $(-3, 0)$.

To find the y-intercept, substitute $x = 0$ into the original equation and solve for y.

$y = (0)^2 - 9$ Substitute in $x = 0$.
$y = -9$

The y-intercept is $(0, -9)$.

Let's organize this information in a table and then determine how many more points to find. Put the vertex in the middle of the table to see if we have three points to the left and right of the vertex.

x	y	(x, y)	Label
-3	0	$(-3, 0)$	x-intercept
0	-9	$(0, -9)$	Vertex and y-intercept
3	0	$(3, 0)$	x-intercept

Some missing x-values on either side of the vertex are $x = -2$, $x = -1$, $x = 1$, and $x = 2$. Let's substitute those values into the equation and determine the corresponding y-values.

x	y	(x, y)	Label
-3	0	$(-3, 0)$	x-intercept
-2	$y = (-2)^2 - 9 = 4 - 9 = -5$	$(-2, -5)$	**Additional point**
-1	$y = (-1)^2 - 9 = 1 - 9 = -8$	$(-1, -8)$	**Additional point**
0	-9	$(0, -9)$	Vertex and y-intercept
1	$y = (1)^2 - 9 = 1 - 9 = -8$	$(1, -8)$	**Additional point**
2	$y = (2)^2 - 9 = 4 - 9 = -5$	$(2, -5)$	**Additional point**
3	0	$(3, 0)$	x-intercept

Now that we have seven points, we can graph the quadratic.

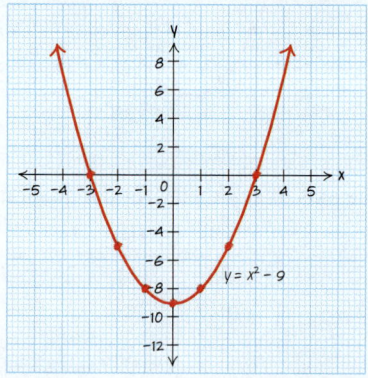

PRACTICE PROBLEM FOR EXAMPLE 3

Graph each quadratic equation. Find the equation of the axis of symmetry, the vertex, and the x- and y-intercepts. Graph at least seven points. Clearly label and scale the axes.

a. $y = x^2 + 2x - 15$ **b.** $y = x^2 - 4x$

9.4 Vocabulary Exercises

1. To find the y-intercept, substitute _____ in the equation and solve for y.

2. To find the x-intercept, substitute _____ in the equation and solve for x.

3. A quadratic equation in two variables may have _____ x-intercepts.

4. On the graph of a parabola, _____ are symmetric about the axis of symmetry.

9.4 Exercises

For Exercises 1 through 8, use the given graphs to estimate the x- and y-intercepts.

1.
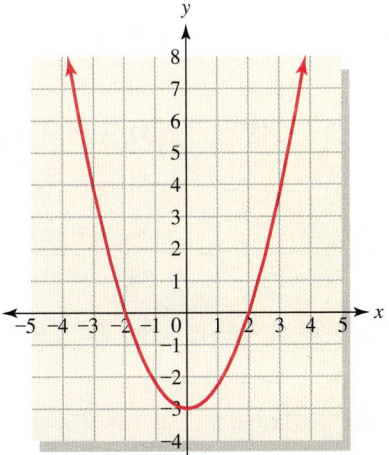

2.
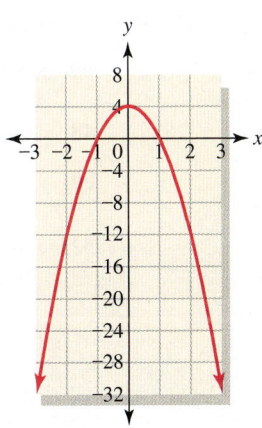

3.
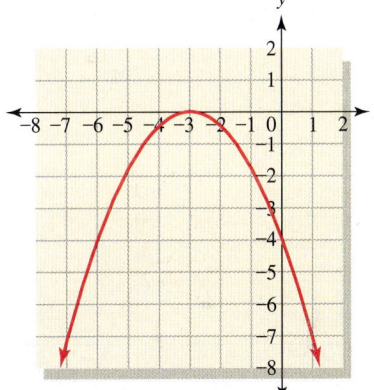

4.
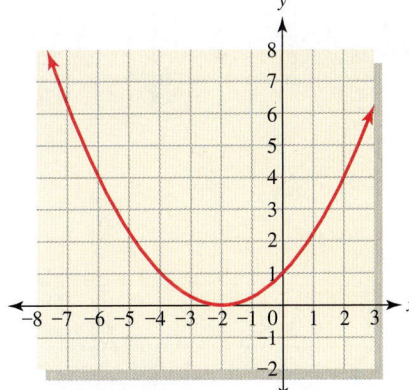

5.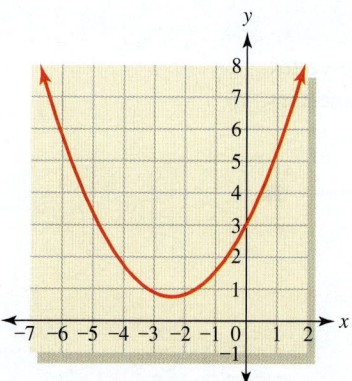

6.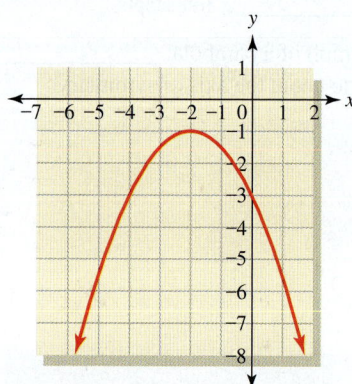

7.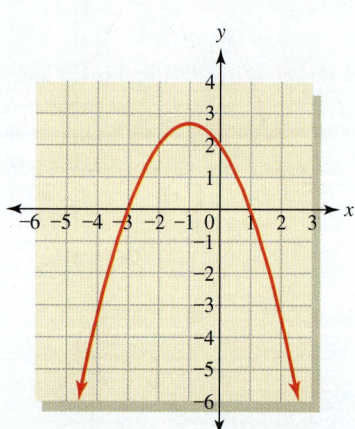

8.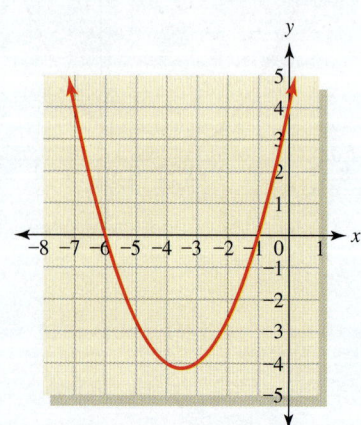

For Exercises 9 through 22, find the x- and y-intercepts of each equation. Write the intercepts as points (x, y). Do not graph the equation.

9. $y = x^2 + 7x - 8$
10. $y = x^2 + 2x - 8$
11. $y = x^2 + 11x + 18$
12. $y = x^2 + 11x + 24$
13. $y = x^2 - 3x - 28$
14. $y = x^2 - 6x - 55$
15. $y = 4x^2 + 12x + 5$
16. $y = 4x^2 + 12x - 7$
17. $y = 14x^2 + 41x + 15$
18. $y = 24x^2 + 25x + 6$
19. $y = \frac{1}{2}x^2$
20. $y = \frac{3}{5}x^2$
21. $y = 30x^2 - 62x + 28$
22. $y = 36x^2 - 77x + 40$

For Exercises 23 through 52, graph the following equations using symmetry. Find the equation of the axis of symmetry, the vertex, and the x- and y-intercepts. Graph at least seven points. Label the vertex and the x- and y-intercepts. Clearly scale and label the axes.

23. $y = x^2 - 7x + 12$
24. $y = x^2 - 14x + 48$
25. $y = -x^2 - 7x + 8$
26. $y = -x^2 - 2x + 8$
27. $y = x^2 - 9$
28. $y = x^2 - 16$
29. $y = -4x^2 + 25$
30. $y = -x^2 + 4$
31. $y = x^2 + 11x + 18$
32. $y = x^2 + 11x + 24$
33. $y = x^2 - 4x$
34. $y = x^2 - x$
35. $y = -x^2 + 2x$
36. $y = -x^2 + 6x$
37. $y = x^2 - 3x - 28$
38. $y = x^2 - 6x - 55$
39. $y = 4x^2 + 12x + 5$
40. $y = 4x^2 + 12x - 7$
41. $y = -14x^2 - 41x - 15$
42. $y = -24x^2 - 25x - 6$
43. $y = \frac{1}{2}x^2$
44. $y = \frac{3}{5}x^2$
45. $y = -\frac{1}{4}x^2$
46. $y = -\frac{1}{5}x^2$
47. $y = 30x^2 - 62x + 28$
48. $y = 36x^2 - 77x + 40$
49. $y = x^2 + 7$
50. $y = x^2 + 4$
51. $y = -2x^2 - 4$
52. $y = -\frac{1}{2}x^2 - 6$

For Exercises 53 through 60, identify each equation as linear or quadratic. Then find the x- and y-intercepts. Do not graph the equations.

53. $y = 2x^2 + 7x - 4$
54. $y = 3x^2 - 5x - 2$
55. $-2x + 5y = 6$
56. $3x - 7y = 12$
57. $y = \frac{2}{3}x + 8$
58. $y = \frac{2}{7}x + 6$
59. $y = -x^2 + 3x$
60. $y = 7x^2 - 14x$

For Exercises 61 through 66, use the model $P = 0.018t^2 + 0.355t + 9.915$, where P represents the number of people, in millions, in U.S. suburban areas living in poverty t years since 2000. The model was derived from data collected in the 100 largest U.S. metro areas.

Source: Model derived from data from Scott Allard's Places in Need: The Changing Geography of Poverty, 2017.

61. Is this a linear or quadratic model?
62. Find the vertex. Interpret it in the context of the problem.
63. Estimate P in the year 2015. *Hint:* Recall that t is the number of years since 2000.
64. Estimate P in the year 2020. *Hint:* Recall that t is the number of years since 2000.
65. Find the vertical intercept of the model. Interpret it in the context of the problem.
66. Find the horizontal intercept(s) of the model. Interpret the horizontal intercept(s) in the context of the problem.

For Exercises 67 through 72, convert units as indicated. Round to two decimal places if necessary.

67. Convert 12 gallons of gasoline to liters. There are approximately 3.79 liters in 1 gallon.
68. Convert 20 liters of gasoline to gallons. There are approximately 3.79 liters in 1 gallon.
69. Convert 98 kilograms to pounds. There are approximately 2.2 pounds in 1 kilogram.
70. Convert 205 pounds to kilograms. There are approximately 2.2 pounds in 1 kilogram.
71. Convert 96 hours to days.
72. Convert 12.8 days to hours.

For Exercises 73 through 78, simplify the expression. Reduce the answer to simplest form.

73. $\frac{2}{5} + \frac{1}{3} + 4$
74. $\frac{8}{9} + 3 + \frac{2}{3}$
75. $1\frac{5}{8} \div \frac{3}{32}$
76. $\frac{3}{4} \div 1\frac{1}{6}$
77. $\frac{7}{2} \cdot \frac{3}{14} - \frac{7}{8}$
78. $\frac{3}{5} + 3 \cdot \frac{5}{6}$

For Exercises 79 through 82, solve each system using the indicated method. Check answer(s) in the original system. Round to the tenths place if necessary.

79. Solve the system using the substitution method:

$$y = 1.8x + 4.5$$
$$1.8x + y = -4.5.$$

80. Solve the system using the substitution method:

$$x + 2y = 7$$
$$x - 2y = -7.$$

81. Solve the system using the elimination method:

$$2x + y = 7$$
$$3x - 2y = 14.$$

82. Solve the system using the elimination method:

$$x + y = 2$$
$$7x + 3y = -6.$$

For Exercises 83 through 86, use the following information: The equation to convert between degrees Fahrenheit (F) and degrees Celsius (C) is $C = \frac{5}{9}(F - 32)$. Round to the tenths place if necessary.

83. Convert 75°F to degrees Celsius.
84. Convert 100°F to degrees Celsius.
85. Convert 32°C to degrees Fahrenheit.
86. Convert 17°C to degrees Fahrenheit.

For Exercises 87 through 90, answer as indicated. Round to the hundredths place, if necessary.

87. What is 30% of 125?
88. What is 25% of 26?
89. 15 is what percent of 90?
90. 8 is what percent of 36?

9.5 Working with Quadratic Models

LEARNING OBJECTIVES
- Determine whether or not a scatterplot is parabolic.
- Use quadratic models to make estimates.
- Determine when model breakdown happens.

In this section, we work with quadratic models. Finding a quadratic model is beyond the scope of this course. Our focus will be on working with given models of quadratic data. We will use the model to make estimates and determine model breakdown.

■ Determining Whether or Not the Graph of a Data Set Is Shaped Like a Parabola

In Section 3.6, we looked at data sets that followed a straight-line pattern when graphed. Such displays of data are said to be linear. The equation of the line, or the model for the data, can then be found. Using that equation, we made predictions about values that did not show up in the data, such as future and past values.

We now look at data sets that, when graphed, follow a *parabolic* pattern. In this chapter, we have graphed quadratic equations and have seen that the graphs were curves called *parabolas*. Data sets that, when graphed, have the shape of a parabola can be modeled by quadratic equations in two variables.

Example 1   Does the data's scatterplot look parabolic?

The data in the following table are the revenue (in billions of dollars) of Gap Inc.

a. Graph the data on a scatterplot.

b. Does the scatterplot follow a linear pattern? Do the points lie on a more or less straight line?

c. Does the scatterplot look parabolic? This question asks whether the points seem to have the shape of a parabola.

Year	Revenue
2013	15.65
2014	16.15
2015	16.44
2016	15.8
2017	15.52

Source: www.marketwatch.com

SOLUTION

a. Let R represent the revenue for Gap Inc. in billions of dollars t years since 2010. Graphing these data on a scatterplot yields the following graph.

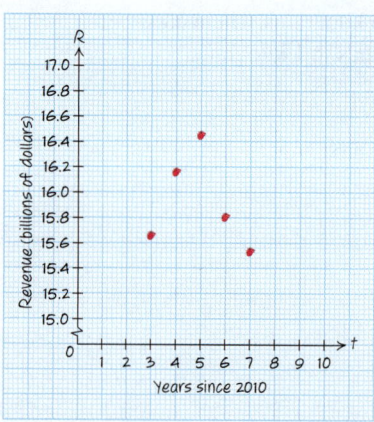

b. These points do not follow a linear pattern. The last two data points turn around and start going down.

c. These points do follow a somewhat parabolic pattern. A parabola will better fit these points.

PRACTICE PROBLEM FOR EXAMPLE 1

The data in the following table are the average daily high temperatures in Atlanta, Georgia.

a. Adjust the data. In the month column, replace the name of the month by the number that represents the month in the year. For example, April is the fourth month in the year, so replace April with 4.

b. Graph the data on a scatterplot.

c. Do the points follow a linear pattern? That is, check whether the points lie more or less on a straight line.

d. Do the points follow a parabolic pattern? That is, check whether the points seem to have the shape of a parabola.

Month	Temperature (degrees Fahrenheit)
April	73
May	80
June	87
July	89
August	88
September	82

Source: www.intellicast.com

Example 2   What pattern do the data trace?

For each graph, determine whether the data follow a linear pattern, a parabolic pattern, or neither.

a.

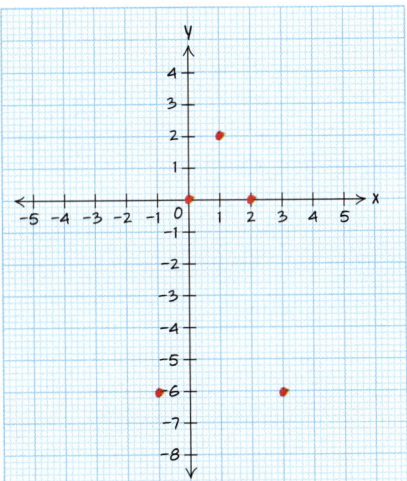

b.

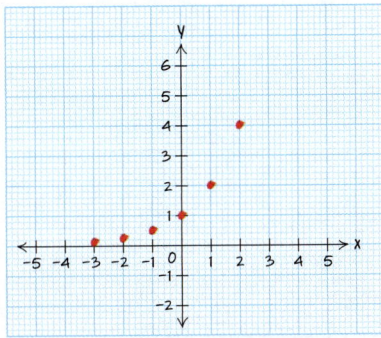

c.

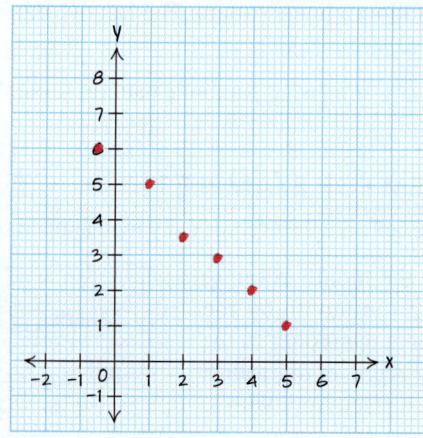

SOLUTION

a. These data follow a parabolic shape. The data are in a quadratic pattern.

b. These data follow neither a parabolic shape nor a linear pattern. The answer is neither.

c. These data follow a linear pattern.

PRACTICE PROBLEM FOR EXAMPLE 2

For each graph, determine whether the data follow a linear pattern, a parabolic pattern, or neither.

a.

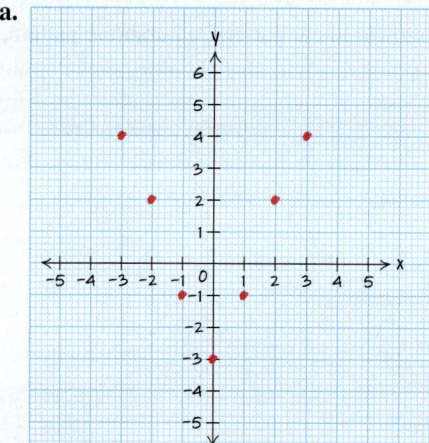

b.

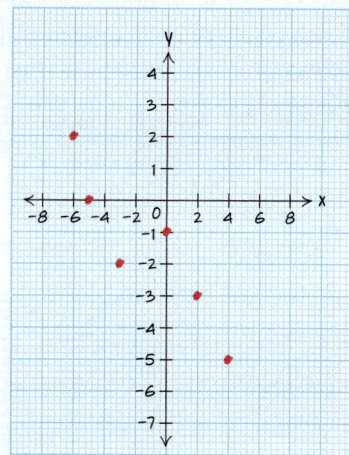

c.

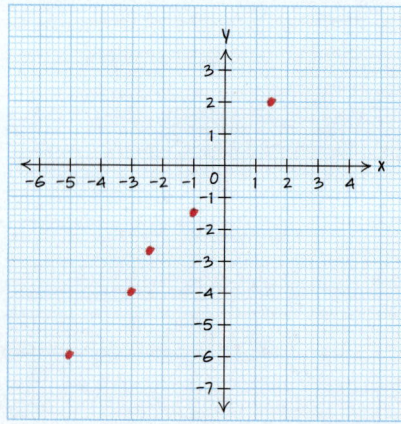

Now that we recognize data that follow a quadratic pattern, we will work with some quadratic models that will be given to us.

Using Quadratic Models to Make Estimates

When data follow a parabolic pattern, the model for the data can be used to make estimates that are not listed in the data table. A **quadratic model** is the equation of the quadratic equation in two variables that fits the data. We will now learn how to make predictions and estimates when given the quadratic model for a data set.

Skill Connection

A quadratic equation in two variables

A quadratic equation in two variables has the form $y = ax^2 + bx + c$. In Sections 9.1 and 9.4, we discovered that the value of a controls whether the parabola opens up or down. If $a > 0$, or a is positive, then the parabola opens up. If $a < 0$, or a is negative, then the parabola opens down. The value of c equals the value of the y-intercept.

We can use these connections between the quadratic equation and their graphs to compare the given model to the pattern in the graphed data.

Example 3 Predicting temperatures using a model

The data in the accompanying table are the average daily high temperatures, in degrees Fahrenheit, in Minneapolis, Minnesota.

a. Define variables, and adjust the data. In the month column, replace the name of the month by the number that represents the month in the year. For example, April is the fourth month in the year, so replace April by 4.

b. Graph the data on a scatterplot. Are the data in a quadratic pattern?

Month	Temperature (degrees Fahrenheit)
April	57
May	70
June	79
July	83
August	80
September	71

Source: www.intellicast.com

c. A quadratic model for the data is $T = -2.82m^2 + 39.65m - 56.96$. The input m represents the month in the year, and the output T represents the average daily high temperature in degrees Fahrenheit. Use this model to predict the average daily high temperature in Minneapolis, Minnesota, in March. Compare the prediction from the model to the real average daily high temperature of 41°F. (*Hint*: First convert March to its corresponding month number.)

d. Where is the vertex of this model $T = -2.82m^2 + 39.65m - 56.96$? Interpret the meaning of the vertex in terms of average daily high temperatures in Minneapolis, Minnesota.

SOLUTION

a. Replacing each month by the number that represents it yields the following table.

Month	Temperature (degrees Fahrenheit)
4	57
5	70
6	79
7	83
8	80
9	71

Let m represent the numerical value of each month and let T represent the temperature in degrees Fahrenheit.

b. Graphing the data on a scatterplot yields the following graph.

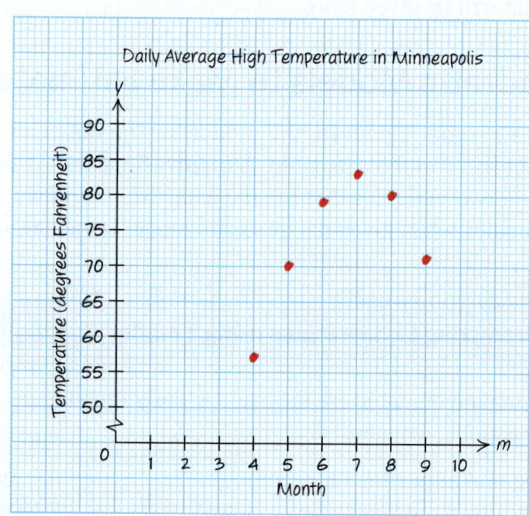

The data follow a parabolic shape, so the data set can be modeled by a quadratic.

c. To use the model to predict the average daily temperature in March, first convert March to the input value of 3. Substituting $m = 3$ into the model yields

$T = -2.82m^2 + 39.65m - 56.96$ Substitute in $m = 3$.

$T = -2.82(3)^2 + 39.65(3) - 56.96$ Evaluate

$T = 36.61$

The model is predicting the monthly high temperature in March as $T = 36.61°F$. This prediction is a little lower than the real average monthly high temperature of $41°F$.

d. To find the vertex, use the vertex formula $x = \dfrac{-b}{2a}$ to first find the x-coordinate. Here the x-coordinate is represented by m. The model is $T = -2.82m^2 + 39.65m - 56.96$, so we have that

$y = ax^2 + bx + c$

$T = -2.82m^2 + 39.65m - 56.96$

$a = -2.82 \quad b = 39.65 \quad c = -56.96$

Substituting $a = -2.82$ and $b = 39.65$ into the vertex formula yields

$$m = \dfrac{-(39.65)}{2(-2.82)}$$

$$m \approx 7.03$$

To find the output value, T, substitute $m \approx 7.03$ in the original equation and evaluate.

$T = -2.82m^2 + 39.65m - 56.96$ Substitute in $m = 7.03$.

$T = -2.82(7.03)^2 + 39.65(7.03) - 56.96$

$T \approx 82.41$

The vertex is the point $(m, T) \approx (7.03, 82.41)$.

In July (which corresponds to $m \approx 7.03$), the average daily high temperature is about $82.41°F$.

> **Skill Connection**
>
> **The vertex formula**
>
> A quadratic equation in two variables has the form $y = ax^2 + bx + c$. In Section 9.1, the vertex formula for the x-coordinate is given as $x = -\dfrac{b}{2a}$.
>
> To find the y-coordinate, substitute the x-coordinate of the vertex into the equation $y = ax^2 + bx + c$ and evaluate.

PRACTICE PROBLEM FOR EXAMPLE 3

A quadratic model for the daily high temperature data for Atlanta, Georgia, is $T = -1.73m^2 + 24.55m + 1.85$, where m is the number corresponding to the month of the year and T is the average daily high temperature in degrees Fahrenheit. The data for this situation were given in the Practice Problem for Example 1.

a. Use this model to predict the average daily high temperature in Atlanta, Georgia, in November. Compare the prediction from the model to the real average daily high temperature of 63.0°F. (*Hint:* First convert November to its corresponding month number.)

b. Where is the vertex of this model $T = -1.73m^2 + 24.55m + 1.85$? Interpret the meaning of the vertex in terms of the average daily high temperatures in Atlanta, Georgia.

In Example 3, we looked at a quadratic model for the average daily high temperatures in Minneapolis, Minnesota. Graphing the model on the following scatterplot shows that the graph of the model (the parabola itself) fits the data fairly well but is a little bit off. This is similar to a linear model, which goes through some of the points from the data set. It will fit the rest of the points as closely as possible.

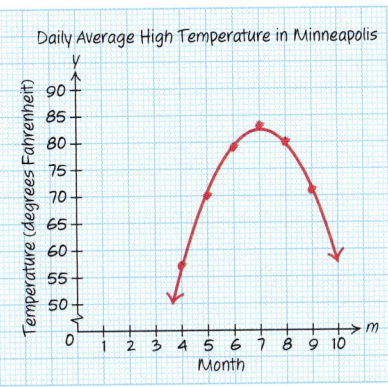

CONCEPT INVESTIGATION

Which model fits better?

1. Does the following data set follow a linear pattern or a quadratic pattern?

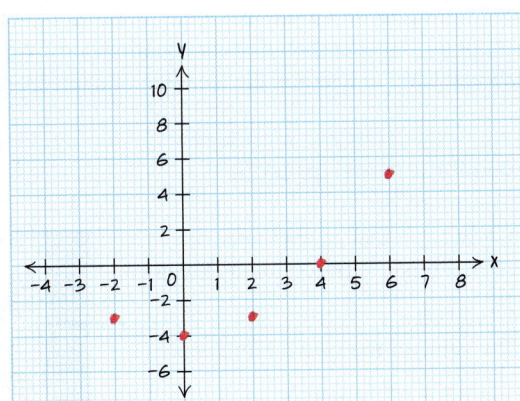

2. Which of the two models below, graphed on the data, fits the data better?

Model 1 $y = x - 4$

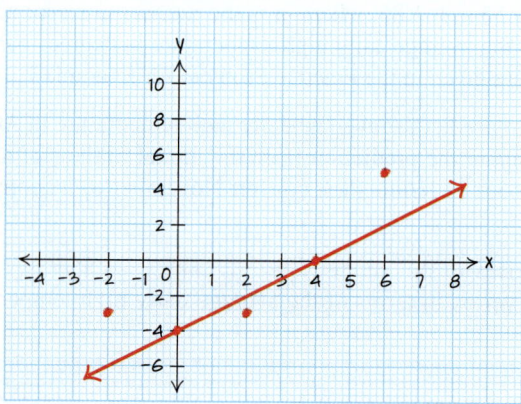

Model 2 $y = 0.25x^2 - 4$

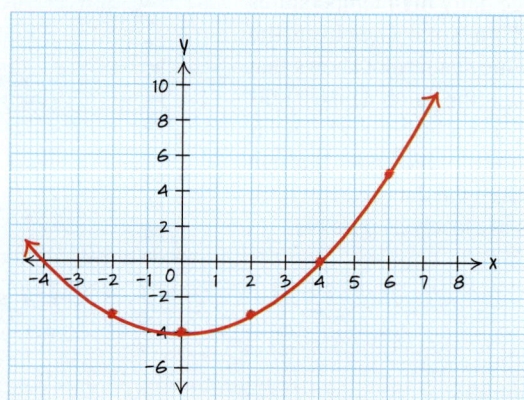

3. Is the model that you selected in part 2 a linear equation or a quadratic equation? Remember that the *model* is the equation.

From the Concept Investigation, we see that a quadratic equation models data that follow a parabolic pattern much better than a linear equation does. We now look at some more quadratic models and learn to interpret the meaning of intercepts and the vertex.

Example 4 A quadratic model for profit

The average profit in dollars, A, that an appliance manufacturer makes when building n refrigerators is given by $A = n(100 - 2n)$.

a. Find and explain the value of the vertical intercept.

b. Find and explain the value of the vertex.

c. Find the number of refrigerators the manufacturer must make to earn an average of $500 profit per refrigerator.

SOLUTION

a. To find the value of the vertical intercept, let $n = 0$ and solve for A.

$A = n(100 - 2n)$ Let n = 0.
$A = 0(100 - 2(0))$
$A = 0$

This means that when 0 refrigerators are made, the average profit is 0.

b. To find the vertex, use the vertex formula $x = \dfrac{-b}{2a}$ to first find the x-coordinate. Here the x-coordinate is represented by n. The model is $A = n(100 - 2n)$, which is not in standard form. Multiplying out the model first to put it in standard form yields

$A = n(100 - 2n)$
$A = 100n - 2n^2$
$A = -2n^2 + 100n$

> **Skill Connection**
>
> **Standard form**
>
> A quadratic equation in two variables is said to be in standard form when it is in the form $y = ax^2 + bx + c$.

We have that

$y = ax^2 + bx + c$
$A = -2n^2 + 100n$
$a = -2 \quad b = 100 \quad c = 0$

Substituting $a = -2$ and $b = 100$ into the vertex formula yields

$n = \dfrac{-(100)}{2(-2)}$

$n = \dfrac{-100}{-4}$

$n = 25$

To find the output value, A, we substitute $n = 25$ into the original equation and evaluate.

$A = -2n^2 + 100n$ Substitute in n = 25.
$A = -2(25)^2 + 100(25)$
$A = 1250$

The vertex is the point $(n, A) = (25, 1250)$.

This means that when 25 refrigerators are produced, the average profit reaches a maximum of $1250.

c. The question asks us to first "find the number of refrigerators the manufacturer must make." The number of refrigerators is represented by the variable n, so we must find n (it is the unknown value here). The second part of the sentence says "to earn an average of $500 profit per refrigerator." The variable A represents the average profit, so let $A = 500$. Substituting $A = 500$ into the equation yields $500 = n(100 - 2n)$.

Multiplying the right side of this equation out and solving for n, we get

$500 = n(100 - 2n)$
$500 = 100n - 2n^2$ This is a quadratic equation because of the n^2-term.
$500 = -2n^2 + 100n$ Set the equation equal to 0.
$\underline{+2n^2 \quad\quad +2n^2}$
$2n^2 + 500 = 100n$
$\underline{-100n \quad\quad\quad -100n}$
$2n^2 - 100n + 500 = 0$

Solving this using the quadratic formula yields

$$2n^2 - 100n + 500 = 0$$

$$n = \frac{-(-100) \pm \sqrt{(-100)^2 - 4(2)(500)}}{2(2)}$$

$$n = \frac{100 \pm \sqrt{6000}}{4}$$

$$n = \frac{100 + \sqrt{6000}}{4} \qquad n = \frac{100 - \sqrt{6000}}{4}$$

$$n \approx \frac{100 + 77.46}{4} \qquad n \approx \frac{100 - 77.46}{4}$$

$$n \approx 44.365 \qquad n \approx 5.635$$

Since *n* represents the number of refrigerators built, it does not make sense to produce 44.365 refrigerators. Therefore, round each answer to the nearest whole number.

To make an average profit of $500 per refrigerator, the manufacturer must product either (approximately) 44 refrigerators or 6 refrigerators.

PRACTICE PROBLEM FOR EXAMPLE 4

The revenue in dollars, *R*, that a clothing shop makes when selling *n* pairs of premium jeans is given by

$$R = n(180 - 6n)$$

a. Find and explain the value of the vertical intercept.

b. Find and explain the value of the vertex.

c. Find the number of pairs of jeans the shop must sell to earn revenue of $1000.

■ Determining when Model Breakdown Happens

Quadratic models can break down, just like linear models. The breakdown in the model occurs when the model outputs values that no longer make sense in the situation we are examining in the problem. In the next example, we will examine model breakdown.

Example 5   Canadian car sales

The new motor vehicle sales data for Canada during the years 2012–2016 are given in the following table.

Year	Number of New Vehicles Sold in Canada (millions)
2012	1.717
2013	1.776
2014	1.890
2015	1.940
2016	1.984

Source: www.statcan.gc.ca

Letting t be years since 2010 yields the following table.

Time	Number of New Vehicles Sold in Canada (millions)
2	1.717
3	1.776
4	1.890
5	1.940
6	1.984

A quadratic model for the data is $N = -0.0067t^2 + 0.1235t + 1.4882$, where t represents year since 2010 and N represents the number of new vehicles sold in Canada in millions. The graph of the data and model are as follows.

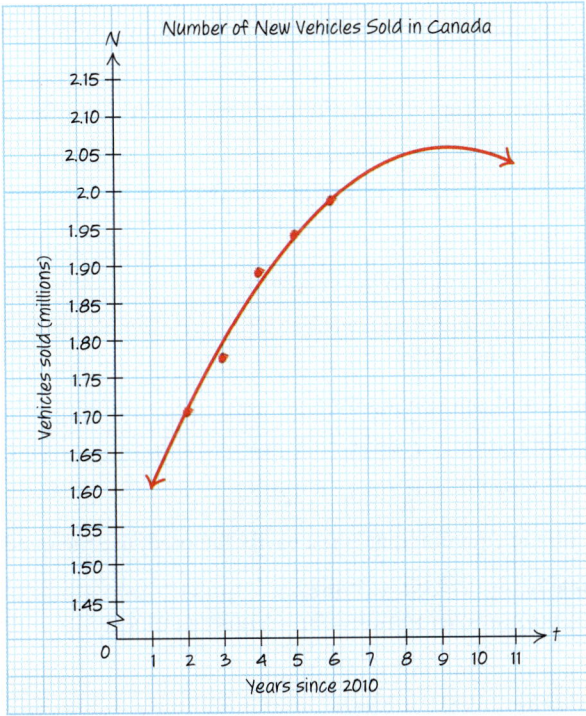

a. Find the vertex for the model $N = -0.0067t^2 + 0.1235t + 1.4882$.

b. Is the vertex value of the model a maximum or minimum?

c. Find the vertical intercept for the model. Does it make sense in regard to Canadian car sales? Does model breakdown happen? Explain.

d. Find the horizontal intercepts for the model. Do they make sense in regard to Canadian car sales? Does model breakdown happen? Explain.

SOLUTION

a. To find the vertex, use the vertex formula $x = \dfrac{-b}{2a}$ to first find the x-coordinate. The model is $N = -0.0067t^2 + 0.1235t + 1.4882$, so we have that

$$y = ax^2 + bx + c$$
$$N = -0.0067t^2 + 0.1235t + 1.4882$$
$$a = -0.0067 \qquad b = 0.1235 \qquad c = 1.4882$$

Substituting $a = -0.0067$ and $b = 0.1235$ into the vertex formula yields

$$t = \frac{-(0.1235)}{2(-0.0067)}$$

$$t \approx 9.2164$$

To find the output value, N, substitute $t \approx 9.2164$ into the original equation and evaluate.

$N = -0.0067t^2 + 0.1235t + 1.4882$ Substitute in $t = 9.2164$.

$N = -0.0067(9.2164)^2 + 0.1235(9.2164) + 1.4882$

$N \approx 2.0573$

The vertex is the point $(t, N) \approx (9.2164, 2.0573)$.

This means the model is estimating sales in 2019 as 2.0573 million new vehicles in Canada.

b. The graph of the parabola given by the model opens down, so the vertex is a maximum value.

c. To find the vertical intercept, we substitute in $t = 0$ and evaluate the model to find N.

$N = -0.0067t^2 + 0.1235t + 1.4882$ Substitute in $t = 0$.

$N = -0.0067(0)^2 + 0.1235(0) + 1.4882$

$N = 1.4882$

Recall that $t = 0$ corresponds to the year 2010.

In the year 2010, Canadian new vehicles sales were estimated to be 1.4882 million. This answer seems reasonable.

d. Substituting in $N = 0$ and solving for t, we get

$N = -0.0067t^2 + 0.1235t + 1.4882$ Substitute in $N = 0$.

$0 = -0.0067t^2 + 0.1235t + 1.4882$

To solve a quadratic equation that is not easily factored, use the quadratic formula.

$0 = -0.0067t^2 + 0.1235t + 1.4882$

$$t = \frac{-(0.1235) \pm \sqrt{(0.1235)^2 - 4(-0.0067)(1.4882)}}{2(-0.0067)}$$

$$t = \frac{-(0.1235) \pm \sqrt{0.05513601}}{-0.0134}$$

$$t = \frac{-(0.1235) + \sqrt{0.05513601}}{-0.0134} \qquad t = \frac{-(0.1235) - \sqrt{0.05513601}}{-0.0134}$$

$$t \approx 26.74 \qquad\qquad\qquad\qquad\quad t \approx -8.31$$

As t represents years since 2010, the two solutions translate into approximately the years 2036 and 2001.

The model estimates the number of new motor vehicle sales in Canada in 2001 and 2036 is 0.

It does not make sense to predict no new car sales in Canada; model breakdown has occurred.

What's That Mean?

Horizontal and Vertical Intercepts

In Sections 3.3 and 9.4, we learned to find intercepts. To find the y-intercept, substitute in $x = 0$ and solve for y. The y-axis is the vertical axis, so y-intercepts are also known as vertical intercepts. When the output value is no longer called y, we still substitute in an input value of 0 and solve for the output value. Likewise, the x-intercepts can be called the horizontal intercepts. To find them, substitute in 0 for the output value, and solve for the input value.

SECTION 9.5 Working with Quadratic Models

PRACTICE PROBLEM FOR EXAMPLE 5

The median home price data for Las Vegas, Nevada, during the years 2005–2009 are given in the following table.

Year	Median Home Price (thousands of dollars)
2005	258
2006	289
2007	279
2008	229
2009	159

Source: www.zillow.com

A quadratic model for the data is $M = -15t^2 + 60180t - 60360251$, where t = year and M = median home price in Las Vegas in thousands of dollars. A graph of the data and model are as follows.

> **What's That Mean?**
>
> **Units in Thousands of Dollars**
>
> When we say that the median home price in Las Vegas in the year 2005 was 258 thousands, we mean that the price was $258,000. To convert units of thousands to units of dollars, just multiply 258 by 1000.

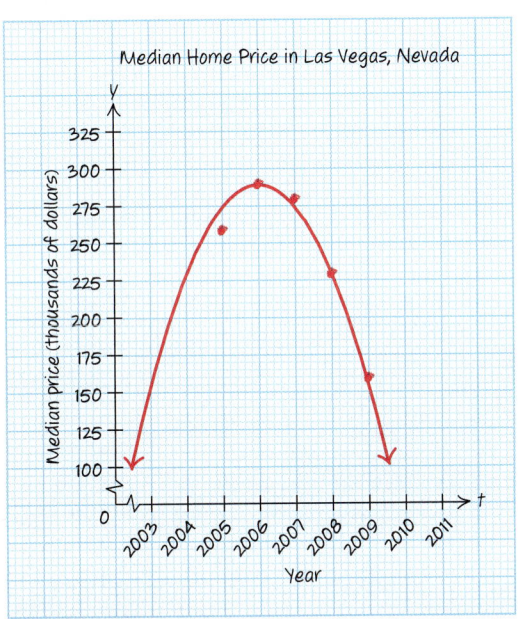

a. Find the vertex for the model $M = -15t^2 + 60180t - 60360251$.

b. Compare the vertex coordinates determined from the model with the real data in the table above.

c. Is the vertex value of the model a maximum or minimum?

d. Find the vertical intercept for the model. Does it make sense in the context of Las Vegas median home prices? Does model breakdown happen? Explain.

e. Find the horizontal intercepts for the model. Do they make sense in the context of Las Vegas median home prices? Does model breakdown happen? Explain.

Quadratic models may experience model breakdown more quickly than linear models do. We saw this in Example 5 when finding the x-intercepts. The data for Example 5 covered the years 2012 through 2016. The x-intercepts were approximately the years 2001 and 2036, and model breakdown had already occurred. Quadratic models may break down more quickly because the trends that are given in a quadratic model, whether increasing or decreasing, occur more rapidly.

9.5 Vocabulary Exercises

1. A scatterplot with a(n) _____ pattern can be modeled using a quadratic equation in two variables.

2. The _____ of the quadratic equation will give the maximum or minimum value of the model.

3. When outputs do not make sense in the context of the problem, _____ has occurred.

9.5 Exercises

For Exercises 1 through 12, decide whether each graphed data set follows a linear pattern, a parabolic pattern, or neither.

1.
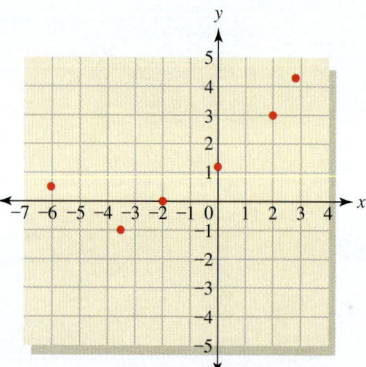

2.
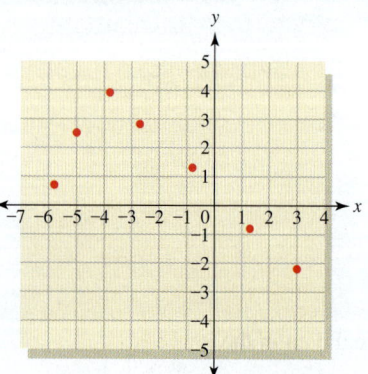

3.
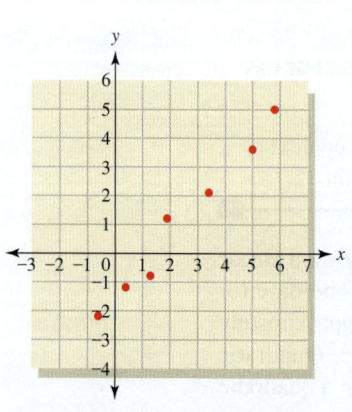

4.
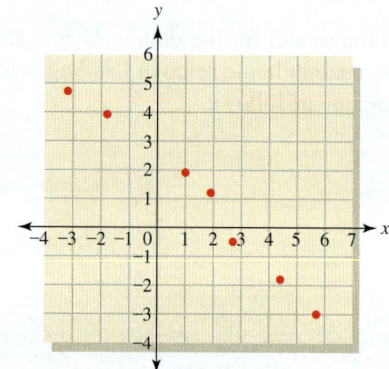

5.
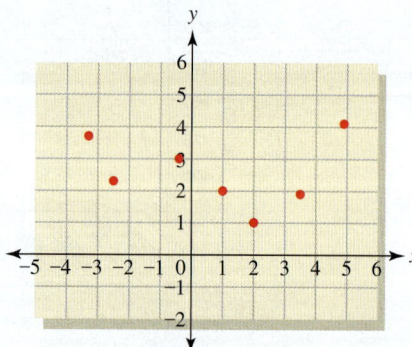

6.
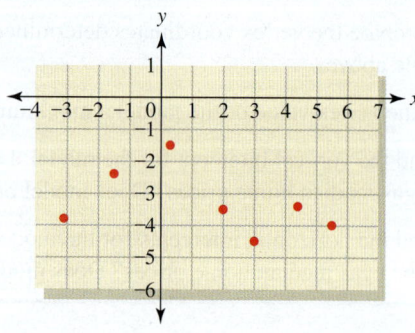

SECTION 9.5 Working with Quadratic Models 813

7.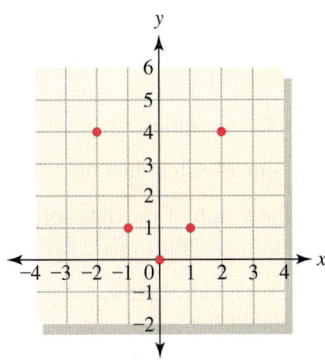

8.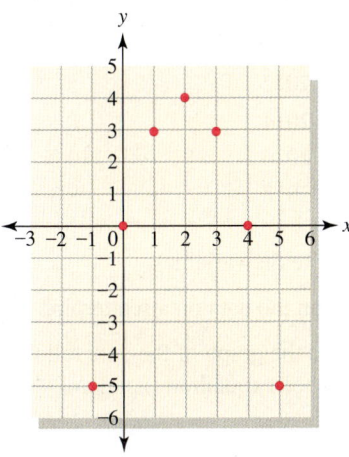

9.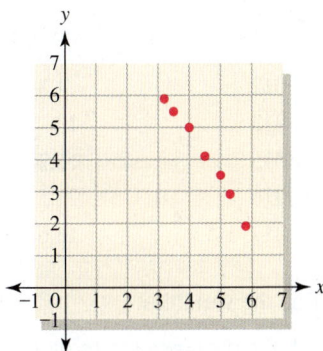

10.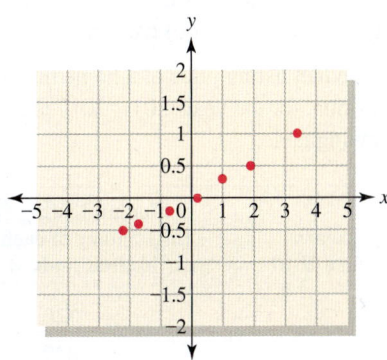

11.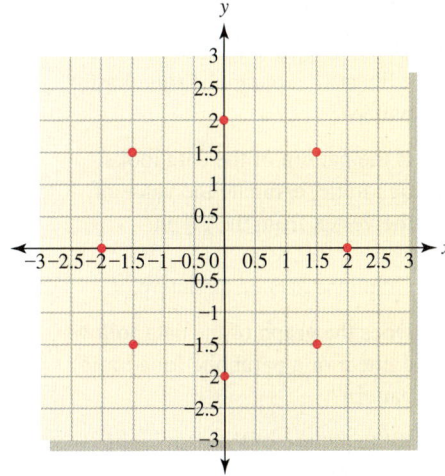

12.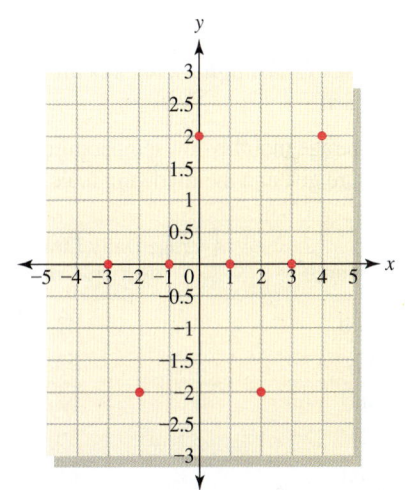

For Exercises 13 through 16, graph each data set on a scatterplot and answer the following questions.

13. a. Graph the data on a scatterplot.

b. Does the graph of the data follow a linear or a parabolic (quadratic) pattern?

c. If the pattern of the data looks quadratic, estimate the value of the vertex from the graph.

x	y
−3	21
−1	5
0	0
1	−3
2	−4
4	0

14. a. Graph the data on a scatterplot.

b. Does the graph of the data follow a linear or a parabolic (quadratic) pattern?

c. If the pattern of the data looks quadratic, estimate the value of the vertex from the graph.

x	y
−1.25	−9
0	4
0.25	5
1.25	−2
2	−14
3	−41

15.
a. Graph the data on a scatterplot.
b. Does the graph of the data follow a linear or a parabolic (quadratic) pattern?
c. If the pattern of the data looks quadratic, estimate the value of the vertex from the graph.

x	y
−2	19
0	7
2	−5
4	−17
6	−29

16.
a. Graph the data on a scatterplot.
b. Does the graph of the data follow a linear or a parabolic (quadratic) pattern?
c. If the pattern of the data looks quadratic, estimate the value of the vertex from the graph.

x	y
−15	−19
−10	−15
−5	−11
0	−7
5	−3

For Exercises 17 through 24, answer the questions for each given data set.

17. The average monthly low temperatures for Cincinnati, Ohio, are given in the following table.

Month	Average Daily Low Temperature (degrees Fahrenheit)
April	43
May	53
June	62
July	66
August	65
September	57

Source: www.intellicast.com

a. Adjust the data. Replace each month's name by the number that represents its location in the year (May = 5).
b. Define the variables.
c. Graph the adjusted data on a scatterplot.
d. Does the graph of the data follow a linear or a parabolic (quadratic) pattern?

18. The average monthly low temperatures for Portland, Maine, are given in the following table.

Month	Average Daily Low Temperature (degrees Fahrenheit)
April	35
May	44
June	53
July	59
August	57
September	49

Source: www.intellicast.com

a. Adjust the data. Replace each month's name by the number that represents its location in the year (August = 8).
b. Define the variables.
c. Graph the adjusted data on a scatterplot.
d. Does the graph of the data follow a linear or a parabolic (quadratic) pattern?

19. The profit for a video game company is given in the following table.

Year	Profit (millions of dollars)
2010	2.4
2012	5
2014	8.1
2016	10.2
2018	8.3

a. Define the variables.
b. Graph the data on a scatterplot.
c. Does the graph of the data follow a linear or a parabolic (quadratic) pattern?

20. The cost of running a network installation and maintenance company is given in the following table.

Year	Cost (millions of dollars)
2012	4.8
2013	10
2015	16.2
2016	20.5
2017	16.6

a. Define the variables.
b. Graph the data on a scatterplot.
c. Does the graph of the data follow a linear or a parabolic (quadratic) pattern?

21. The median listing price of a home in Santa Barbara, California, for the years 2011–2017 is given in the following table.

Years since 2010	Median Listing Price (in January of each listed year in thousands of dollars)
3	740
4	877
5	976
6	1010
7	1060

Source: www.zillow.com

a. Define the variables.

b. Graph the data on a scatterplot.

c. Does the graph of the data follow a linear or a parabolic (quadratic) pattern?

22. The median listing price of a home in Raleigh, North Carolina, for the years 2010–2016 is given in the following table.

Years since 2010	Median Listing Price (in January of each listed year in thousands of dollars)
0	144
1	139
2	136
3	137
4	140
5	142
6	148

Source: www.zillow.com

a. Define the variables.

b. Graph the data on a scatterplot.

c. Does the graph of the data follow a linear or a parabolic (quadratic) pattern?

23. The revenue for Home Depot stores is given in the following table.

Year	Revenue (billions of dollars)
2012	74.8
2013	74.75
2014	78.81
2015	83.18
2016	88.53
2017	94.6

Source: Home Depot's annual reports.

a. Define the variables.

b. Graph the data on a scatterplot.

c. Does the display of data follow a linear or a parabolic (quadratic) pattern?

24. The revenue for Lowe's stores is given in the following table.

Year	Revenue (billions of dollars)
2011	50.21
2012	50.52
2013	50.52
2014	53.24
2015	56.22
2016	59.07
2017	65.02

Source: Lowes' annual reports.

a. Define the variables.

b. Graph the data on a scatterplot.

c. Does the display of data follow a linear or a parabolic pattern?

For Exercises 25 through 30, use each given quadratic model to answer the questions.

25. A quadratic model for the monthly average low temperatures in Cincinnati, Ohio, is given by $T = -2.32m^2 + 33.32m - 54.07$. Here T is the monthly average low temperature in Cincinnati, Ohio, and m is the month (January = 1). See Exercise 17 for the data.

   a. Use the model to predict the average monthly low temperature in Cincinnati, Ohio, in the month of February. Compare this predicted value to the real reported value of 25°F.

   b. Find the vertex of the model. Explain the vertex in terms of the weather in Cincinnati, Ohio.

26. A quadratic model for the monthly average low temperatures in Portland, Maine, is given by $T = -2.30m^2 + 33.23m - 62.46$. Here T is the monthly average low temperature in Portland, Maine, and m is the month (January = 1). See Exercise 18 for the data.

   a. Use the model to predict the average monthly low temperature in Portland, Maine, in the month of November. Compare this predicted value to the real reported value of 30.0°F.

   b. Find the vertex of the model. Explain the vertex in terms of the weather in Portland, Maine.

27. A quadratic model for the median listing price of homes in Santa Barbara, California, is given by $P = -17x^2 + 248x + 153.5$. Here P is the median listing price in thousands of dollars x years since 2010. See Exercise 21 for the data.

   a. Use the model to predict the median listing price of homes in Santa Barbara, California, in the year 2020.

   b. Find the vertex of the model. Explain the vertex as it relates to the median listing price of homes in Santa Barbara, California.

28. A quadratic model for the median listing price of homes in Raleigh, North Carolina, is given by $P = 1.00x^2 - 5.21x + 143.5$. Here P is the median listing price of homes in thousands of dollars x years since 2000. See Exercise 22 for the data.

   a. Use the model to predict the median listing price of homes in Raleigh, North Carolina, in the year 2020.

   b. Find the vertex of the model. Explain the vertex in terms of the median listing price of homes in Raleigh, North Carolina.

29. A quadratic model for the revenue, in billions of dollars, for the Home Depot Corporation is given by $R = 0.64t^2 - 1.61t + 74.91$. Here R is the revenue in billions of dollars t years since 2010.

   a. Use the model to estimate the revenue of Home Depot in 2018.

   b. Find the vertex of the model. Explain the vertex in terms of the revenue of Home Depot.

30. A quadratic model for the revenue, in millions of dollars, for Lowe's Corporation is given by $R = 0.51t^2 - 1.69t + 51.51$. Here R is the revenue in billions of dollars t years since 2010.

   a. Use the model to estimate the revenue of Lowe's in 2018.

   b. Find the vertex of the model. Explain the vertex in terms of the revenue of Lowe's.

For Exercises 31 through 34, use each given model to answer the following questions.

31. The average profit in dollars, A, that a local jewelry shop makes when creating n necklaces is given by $A = n(140 - 7n)$.

   a. Find and explain the value of the vertical intercept.

   b. Find and explain the value of the vertex.

   c. Find the number of necklaces the shop must make to earn an average of $525.00 profit per necklace.

32. The average profit in dollars, A, that a local jewelry shop makes when creating n earrings is given by $A = n(56 - 4n)$.

   a. Find and explain the value of the vertical intercept.

   b. Find and explain the value of the vertex.

   c. Find the number of earrings the shop must make to earn an average profit of $160.00 per earring.

33. The revenue from selling n cameras can be modeled by $R = n(210 - 7n)$, where R is the revenue in dollars when n cameras are sold.

   a. Find and explain the value of the vertex.

   b. Find the number of cameras the shop must sell to have a revenue of $1400.00.

34. The revenue from selling n tablets can be modeled by $R = n(816 - 24n)$, where R is the revenue in dollars, when n tablets are sold.

   a. Find and explain the value of the vertex.

   b. Find the number of tablets the shop must sell to have a revenue of $5400.00.

For Exercises 35 and 36, use each given model to answer the following questions. Be sure to think about the context of the problem to determine whether the solutions make sense. If the solutions do not, answer that model breakdown has occurred.

35. A quadratic model for the median listing price of homes in Santa Barbara, California, is given by $P = -17x^2 + 248x + 153.5$. Here P is the median listing price in thousands of dollars x years since 2010. See Exercise 21 for the data.

   a. Use the model to predict the median listing price of a home in Santa Barbara in the year 2025. Is this answer reasonable? Explain.

   b. Use the model to estimate the median listing price of a home in Santa Barbara in the year 2000. Is this answer reasonable? Explain.

36. A quadratic model for the median listing price of a home in Raleigh, North Carolina, is given by $P = 1.00x^2 - 5.21x + 143.5$. Here P is the median listing price in thousands of dollars x years since 2010. See Exercise 22 for the data.

   a. Use the model to estimate the median listing price of a home in Raleigh in the year 2000. Is this answer reasonable? Explain.

   b. Use the model to predict the median listing price of a home in Raleigh in the year 2025. Is this answer reasonable? Explain.

9.6 The Basics of Functions

LEARNING OBJECTIVES

- Recognize a relation.
- Recognize a function.
- Find the domain.
- Find the range.
- Use the vertical line test.
- Use function notation.
- Evaluate a function.

■ Relations

So far in this text, we have studied several types of equations such as linear equations and quadratic equations. These equations gave us a *relationship* between the input value x and the output value y. For example, $y = -2x + 1$ has the solutions shown in the table.

Written as ordered pairs, these solutions are

$$(-2, 5), (-1, 3), (0, 1), (1, -1), (2, -3)$$

A list of ordered pairs like this has a special name. It is called a **relation**.

x	y
-2	5
-1	3
0	1
1	-1
2	-3

> **DEFINITIONS**
>
> **Relation** A **relation** is a set of ordered pairs (x, y).
>
> **Domain** The set of the x-components (inputs) is called the **domain** of the relation.
>
> **Range** The set of the y-components (outputs) is called the **range** of the relation.

The ordered pairs above, $(-2, 5), (-1, 3), (0, 1), (1, -1),$ and $(2, -3)$, have the domain consisting of the set of inputs $-2, -1, 0, 1,$ and 2. The range consists of the set of outputs $5, 3, 1, -1,$ and -3.

Relations can be any relationship between an input variable and an output variable, not just from linear equations. The relationship between a person and his or her height is a relation. There are many examples of relations in life.

Example 1 Finding parts of a relation

Name of Child	Grade in School
Reuben	8
Damian	2
Brianna	5
Lupe	4
Caleb	7

a. In a sentence, what is the relationship between the two columns in the table?
b. What is the domain of the relation?
c. What is the range of the relation?

SOLUTION

a. The table gives the relationship between the name of a child and the child's grade in school.
b. The domain is the set of names: Reuben, Damian, Brianna, Lupe, and Caleb.
c. The range is the set of grades: 8, 2, 5, 4, and 7.

PRACTICE PROBLEM FOR EXAMPLE 1

Time (years)	Simple Interest Earned on a $1000 Deposit, Interest Rate = 4%
0	0
1	40
2	80
3	120
4	160

a. What is the relationship between the two columns in the table?
b. What is the domain of the relation?
c. What is the range of the relation?

Example 2 Finding the domain and range for a relation

The following table lists the average monthly electricity bill for several states for the year 2015.

State	Average Monthly Electricity Bill
New Hampshire	$114.90
Illinois	$89.91
South Carolina	$144.04
Idaho	$95.01
Oregon	$96.24

This table of data can also be expressed as a set of ordered pairs. The pairs represent (state, electricity bill). So the relation can be written as

{(New Hampshire, 144.90), (Illinois, 89.91), (South Carolina, 144.04), (Idaho, 95.01), (Oregon, 96.24)}

a. What is the domain of the relation?
b. What is the range of the relation?
c. Does every input (value in the domain) result in exactly one output (value in the range)?

SOLUTION

a. The domain consists of the set of states: New Hampshire, Illinois, South Carolina, Idaho, and Oregon.
b. The range consists of the set of average monthly electric bills (in dollars): 114.90, 89.91, 144.04, 95.01, and 96.24.
c. Yes, every input results in exactly one output.

Functions

We now are going to learn about a very special kind of relation. This type of relation has exactly one output for each input. Relations with this association are called **functions.**

> **DEFINITIONS**
>
> **Function** A function is a relation in which each input (*x*) is associated with only one output (*y*).
> **Domain of a Function** The set of all inputs is called the domain of the function.
> **Range of a Function** The set of all outputs is called the range of the function.

What's That Mean?

Domain and Range

To help remember that the *domain* represents the set of inputs (*x*) and the *range* represents the set of outputs (*y*), use alphabetical order. In the alphabet, *x* comes before *y*. Likewise, *d* (domain) comes before *r* (range). Pair *x* with domain (first alphabetically) and *y* with range (second alphabetically).

Example 3 Identifying functions

Determine whether or not each of the following relations is a function. Explain your reasoning.

a. The output is the number of gold medals won by a country (the input) during the 2016 Summer Olympics held in Rio de Janeiro, Brazil.
b. The output is the cost of an iPhone during a given year (the input).
c. The output is the height of all students of a given age (the input) at your college.
d. The output is the total amount of taxes you paid in a given year (the input).

SOLUTION

a. Each country won a fixed number of gold medals, so this is a function. For example, the host country, Brazil, won seven gold medals.
b. In the year 2017, iPhones sold for various prices. Some of those prices were $649, $769, and $869. Since there is more than one output for a given input (the year 2017), this is NOT a function.
c. Twenty-year-old students (an input value) come in a wide range of heights, such as 5′6″, 5′3″, and 6′2″. For a given input, there is more than one output, so this is NOT a function.
d. For each given year (the input), you pay only a fixed total amount of taxes. Since each input results in exactly one output, this is a function.

PRACTICE PROBLEM FOR EXAMPLE 3

Determine whether each of the following relations is a function or not. Explain your reasoning.

a. The output is the price of a gallon of gas at noon each day in Salt Lake City, Utah, and the input is the day of the week.
b. The input is the number of college credits a student is enrolled in, and the output is the tuition owed to the college.
c. The output is your weight in pounds, and the input is the number of days since you started a diet.

820 CHAPTER 9 Quadratic Equations

■ Vertical Line Test

There is a visual way to look at whether a relation is a function. In the next Concept Investigation, we explore this.

CONCEPT INVESTIGATION
What do graphs of functions have in common?

1. Look at the graphs below and answer the following questions.

   Graph Is a Function   **Graph Is NOT a Function**

   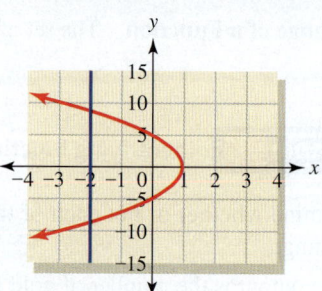

   How many times does a vertical line intersect the graph of a function?   How many times does a vertical line intersect the graph of a relation that is not a function?

2. Look at the graphs below and answer the following questions.

   Graph Is a Function   **Graph Is NOT a Function**

   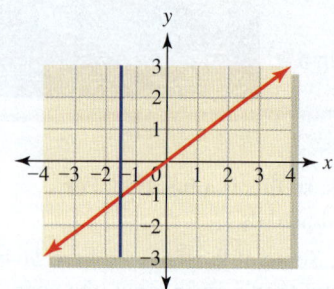 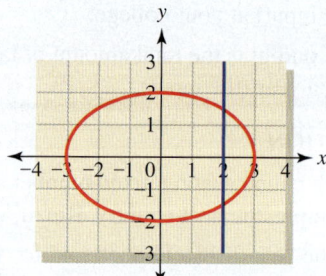

   How many times does a vertical line intersect the graph of a function?   How many times does a vertical line intersect the graph of a relation that is not a function?

3. Using the graph below, find the values of x and y for both points.

   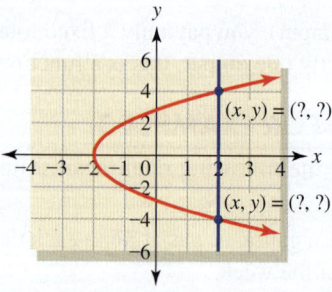

4. The definition of a function says that each input x is associated with only one output y. Is that true for the graph in part 3 when $x = 2$? Does the graph in part 3 represent a function or not?

5. How many times will a vertical line intersect the graph in part 3?

The preceding Concept Investigation illustrates that a vertical line will intersect a function in at most one point (0 or 1 point). There is a graphical way to check whether a given equation is a function. It is called the **vertical line test.**

> **DEFINITION**
>
> **The Vertical Line Test for a Function** If any vertical line intersects the graph of an equation in at most one point, then the equation is that of a function.

If we look at the graph of the equation $y = -2x + 1$ and draw vertical lines through the graph, we see that each vertical line intersects the graph in only one point.

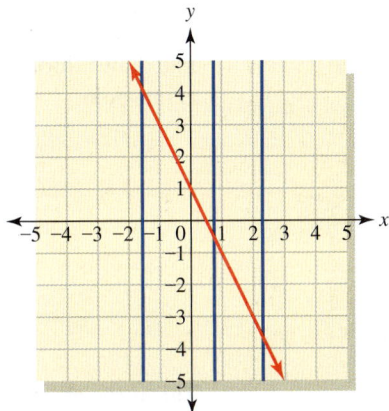

Example 4 Using the vertical line test

Use the vertical line test to determine whether the graph represents a function.

a.
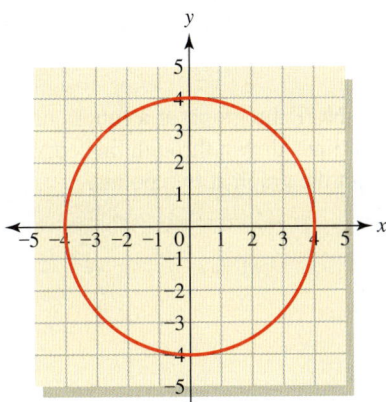

b.
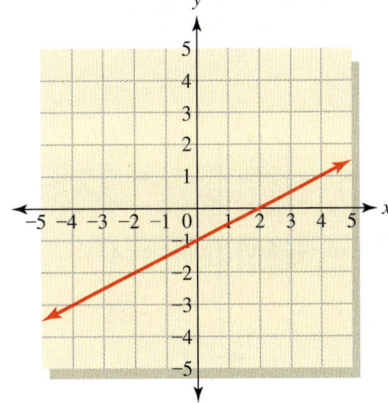

c.
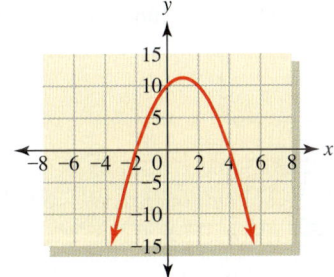

d.
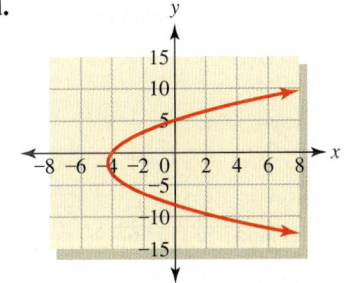

822 CHAPTER 9 Quadratic Equations

SOLUTION

a. 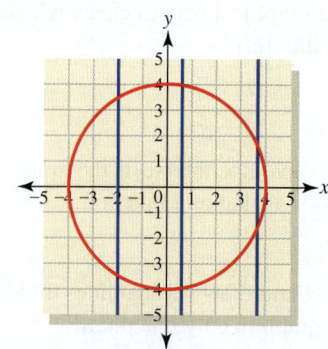 This is NOT the graph of a function. Most vertical lines drawn through this curve will intersect the graph at two points.

b. 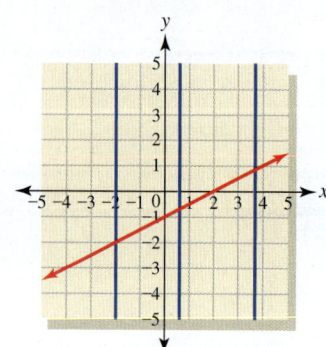 This is the graph of a function. Any vertical line drawn through this line will intersect it at one point.

c. 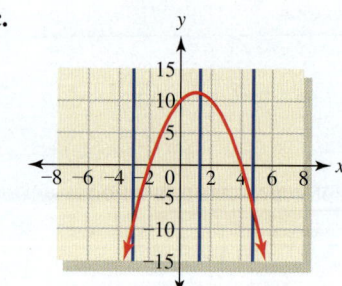 This is the graph of a function. Any vertical line drawn through this curve will intersect it at, at most, one point.

d. 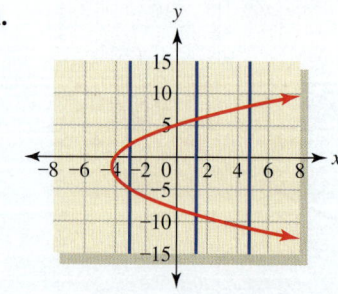 This is NOT the graph of a function. Most vertical lines drawn through this curve will intersect the graph at two points.

PRACTICE PROBLEM FOR EXAMPLE 4

Use the vertical line test to determine whether the graph represents a function.

a.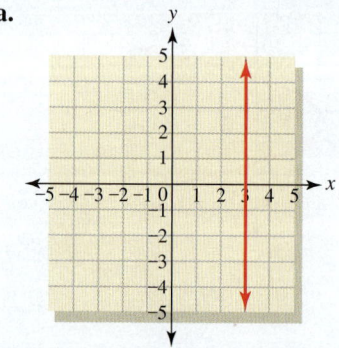

b.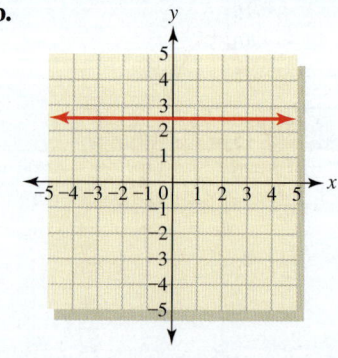

c.
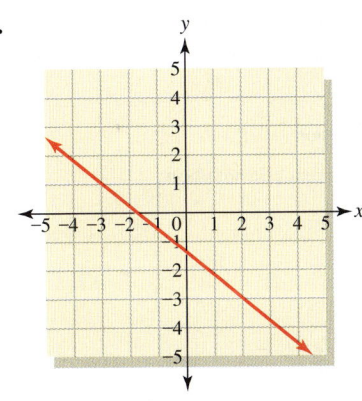

d.
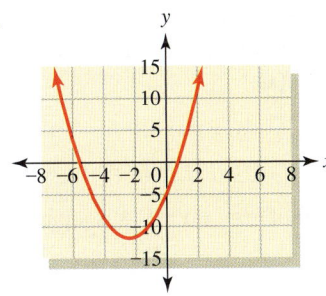

Function Notation

Once we have determined that a relation is a function, we would like to be able to convey this information without writing "this relation is a function" repeatedly. Mathematicians have come up with a special notation to convey that *this relation is a function*. This is called **function notation.**

> **DEFINITION**
>
> **Function Notation** If a relation between the variables x and y is a function, we write $y = f(x)$ to denote that y is a function of x.

In general, to rewrite a function using function notation, simply replace the variable y with $f(x)$.

Example 5 Using function notation

Determine if each relation is a function. Then rewrite each function using function notation.

a. $y = -2x + 1$ **b.** $y = -x^2 + 4$

SOLUTION

a. The graph of $y = -2x + 1$ is a line.

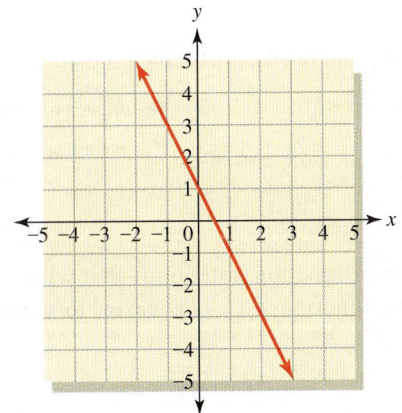

What's That Mean?

Function Notation

The *function notation* $y = f(x)$ means that y is a function of x. $f(x)$ is read, "f of x." f is not a variable, and this does NOT mean to multiply f by x. It is just a shorthand way to say y is a function of x.

This graph passes the vertical line test. Therefore, this relation is a function. Each input is associated with exactly one output. To rewrite the equation in function notation, replace y by $f(x)$.

$$y = -2x + 1 \quad \text{The original relation}$$
$$f(x) = -2x + 1 \quad \text{Rewritten using function notation}$$

b. The graph of $y = -x^2 + 4$ is a parabola.

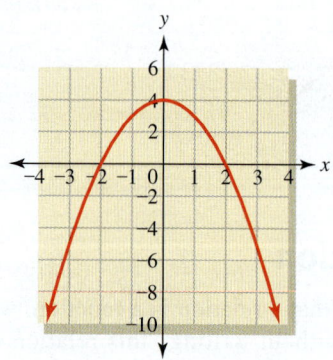

■ **Connecting the Concepts**

When do we use function notation?

In algebra, we tend to use the "$y =$" notation for graphing. When an equation is a function and we are not graphing, we tend to use the function notation, $f(x)$.

This parabola passes the vertical line test, so this relation is a function. Each input is associated with exactly one output. To rewrite the relation in function notation, replace y by $f(x)$.

$$y = -x^2 + 4 \quad \text{The original relation}$$
$$f(x) = -x^2 + 4 \quad \text{Rewritten using function notation}$$

PRACTICE PROBLEM FOR EXAMPLE 5

Determine if each relation is a function. Then rewrite each function using function notation.

a. $y = x^2 - 16$ **b.** $y = -3$

■ Evaluating Functions

Function notation is very useful when it comes to evaluating functions. Suppose we want to evaluate $y = -2x + 1$ when $x = 2$.

$$y = -2x + 1 \quad \text{Substitute in } x = 2.$$
$$y = -2(2) + 1$$
$$y = -4 + 1$$
$$y = -3$$

Using function notation, we can rewrite this same problem as

$$f(x) = -2x + 1 \quad \text{The original function}$$
$$f(2) = -2(2) + 1 \quad f(2) \text{ means to evaluate for } x = 2.$$
$$f(2) = -4 + 1$$
$$f(2) = -3$$

One of the advantages of function notation is that when we write $f(2)$, it means to substitute $x = 2$ in the function $f(x)$. It is easier to write "$f(2)$" than "evaluate $y = -2x + 1$ when $x = 2$."

Example 6 Evaluating functions

Evaluate each of the functions for the given value of x.
a. Find $f(1)$ where $f(x) = 3x - 5$.
b. Find $f(-2)$ where $f(x) = x^2$.
c. Find $f(5)$ where $f(x) = 6$.

SOLUTION

a. Substituting $x = 1$ into the function $f(x)$ yields

$$f(x) = 3x - 5 \quad \text{Substitute in } x = 1.$$
$$f(1) = 3(1) - 5 \quad \text{Simplify the right side.}$$
$$f(1) = -2$$

Be very careful when evaluating $f(1)$. Only substitute $x = 1$ in the right side of the equation and simplify only the right side of the equation. The left side is kept as $f(1)$. The notation $f(1)$ tells us the equation represents a function and that we are substituting in $x = 1$.

b. Substituting $x = -2$ in the function $f(x)$ yields

$$f(x) = x^2 \quad \text{Substitute in } x = -2.$$
$$f(-2) = (-2)^2 \quad \text{Simplify the right side.}$$
$$f(-2) = 4$$

c. Substituting $x = 5$ in the function $f(x)$ yields

$$f(x) = 6 \quad \text{Substitute in } x = 5.$$
$$f(5) = 6 \quad \text{The output of this function is always 6.}$$

PRACTICE PROBLEM FOR EXAMPLE 6

Evaluate each of the functions for the given value of x.

a. Find $f(0)$ where $f(x) = \frac{3}{4}x + 2$.

b. Find $f(-1)$ where $f(x) = x^3$.

c. Find $f(-2)$ where $f(x) = 3$.

Function notation can be used to represent functions that come from problem situations. For example, the weekly pay for an employee who works h hours a week and gets paid $12.50 per hour can be represented by the function

$$P(h) = 12.50h$$

where $P(h)$ represents the weekly pay in dollars when an employee works h hours a week. The function notation can then be used to ask questions or write solutions in a simple manner. With this function, $P(20)$ asks what the weekly pay is when the employee works 20 hours per week. Because the 20 is inside the parentheses, it represents an h-value, and thus, $h = 20$ hours per week. The function can then be evaluated to find the weekly pay.

$$P(20) = 12.50(20) \quad \text{Substitute in } h = 20.$$
$$P(20) = 250$$

Therefore, if an employee works 20 hours a week, that person will earn $250.

826 CHAPTER 9 Quadratic Equations

Example 7 — Using functions in applications

a. Let $P(h) = 11h$ be the weekly pay in dollars when an employee works h hours a week. Find $P(30)$ and explain its meaning.

b. Let $C(t) = 4t + 150$ be the cost in dollars to print a company logo on t T-shirts. Find t such that $C(t) = 350$ and explain its meaning.

c. Let $R(p) = -1.5p^2 + 180p + 300$ be the weekly revenue in dollars from selling vitamin supplements at a price of p dollars. Find $R(50)$, and explain its meaning.

SOLUTION

a. The 30 is inside the parentheses, so it is an h-value.

$$P(30) = 11(30) \quad \text{Substitute in } h = 30.$$
$$P(30) = 330$$

If an employee works 30 hours a week, he or she will earn $330.

b. The 350 is what the function equals, so we will set the function equal to 350 and solve for t.

$$C(t) = 4t + 150$$
$$350 = 4t + 150 \quad \text{Set the function equal to 350.}$$
$$\underline{-150 \qquad -150} \quad \text{Solve for } t.$$
$$200 = 4t$$
$$\frac{200}{4} = \frac{4t}{4}$$
$$50 = t$$

It costs $350 to print a company logo on 50 T-shirts.

c. The 50 is inside the parentheses, so it is a p-value.

$$R(50) = -1.5(50)^2 + 180(50) + 300 \quad \text{Substitute in } p = 50.$$
$$R(50) = 5550$$

When the price is $50, the weekly revenue from selling vitamin supplements is about $5550.

PRACTICE PROBLEM FOR EXAMPLE 7

a. Let $W(d) = 190 - d$ be Philip's weight in pounds after d days on a diet. Find $W(10)$ and explain its meaning.

b. Let $C(p) = 2.5p + 2.50$ be the cost in dollars to make p large pizzas. Find p such that $C(p) = 315$ and explain its meaning.

c. Let $h(t) = -16t^2 + 30t + 3.5$ be the height of a ball thrown straight up in the air t seconds after being thrown. Find $h(1.25)$, and explain its meaning.

9.6 Vocabulary Exercises

1. A relation is a set of _____.
2. The _____ of a relation is the set of all inputs.
3. The _____ of a relation is the set of all outputs.
4. A(n) _____ is a relation where each input is associated with exactly one output.
5. The _____ test is used to determine if a graph is that of a function.

9.6 Exercises

For Exercises 1 through 10, state the domain and the range for each relation. Then determine whether every input results in exactly one output.

1.

Name	Month of Birth
Eduardo	July
Sophie	March
Ana	October
Ashanti	January
Taye	May

2.

Name	Annual Salary
Ahmad	$49,999
Jasmyn	$35,000
Eli	$32,000
Jade	$45,999
Trent	$42,000

3.

Dog Breed	Average Weight (pounds)
German Shepherd	85
Labrador Retriever	75
Cocker Spaniel	25
Chihuahua	4
Papillon	7

4.

Country	Population
China	1,371,220,000
India	1,311,050,530
United States	321,418,820
South Africa	55,011,980
Mexico	127,017,000

Source: data.worldbank.org

5. $\{(4, 5), (6, 9), (-2, 3), (9, 1), (0, 0)\}$
6. $\{(2, 3), (-1, -1), (5, -7), (3, 3), (0, 15)\}$
7. $\{(5, 5), (6, 6), (-2, 4), (4, -2), (5, 0)\}$
8. $\{(-3, -3), (-4, -4), (-3, -4), (-4, -3)\}$
9. $\{(4, 1), (4, 2), (4, 3), (4, 4)\}$
10. $\{(-2, 1), (-2, 2), (-2, 3), (2, 4)\}$

For Exercises 11 through 20, determine if each relation is a function.

11. The input is the date (day/month/year), and the output is the price of a barrel of oil.
12. The input is the number of units a student is enrolled in, and the output is the tuition for that semester.
13. The input is the date, and the output is the cost of gasoline on that date in your city.
14. The input is the semester, and the output is the cost of your math book for that semester.
15. For a single year, the input is the country, and the output is the average income (per capita) of people in that country.
16. The input is the day, and the output is the closing stock price of a particular company on that day.
17. The input is the year, and the output is the cost of a ticket to see a National Football League (NFL) game that year.
18. The input is the day, and the output is the total number of red cars in the parking lot at the local high school.
19. The input is any number (x), and the output is twice that number.
20. The input is any number (x), and the output is four times that number plus one.

For Exercises 21 through 30, determine whether each graph represents a function.

21.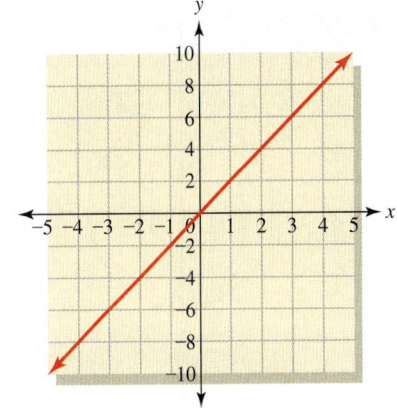

828 CHAPTER 9 Quadratic Equations

22.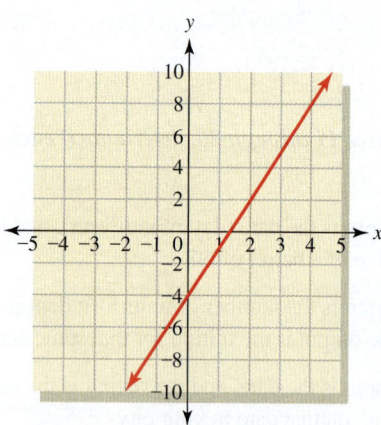

23.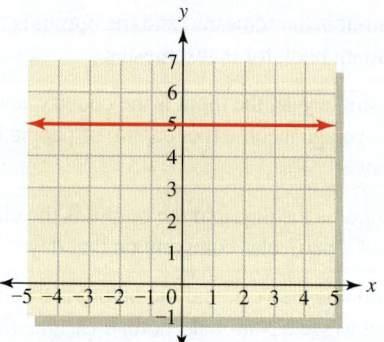

24.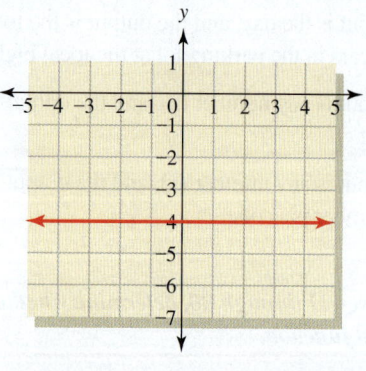

25.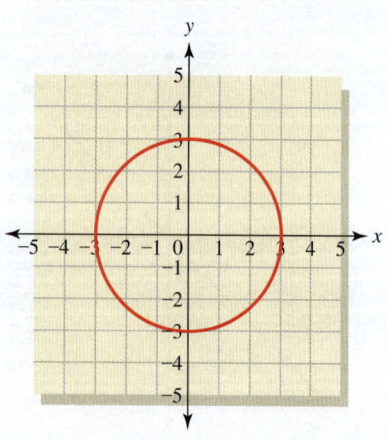

26.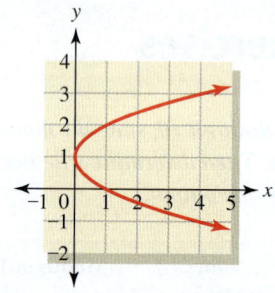

27.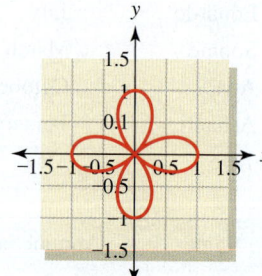

28.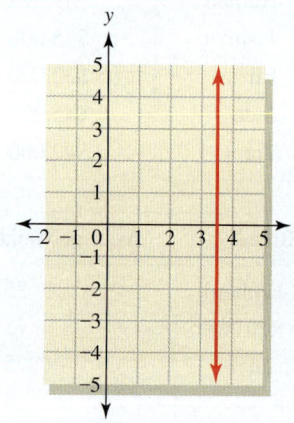

29.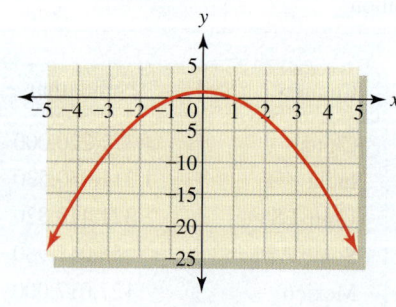

30.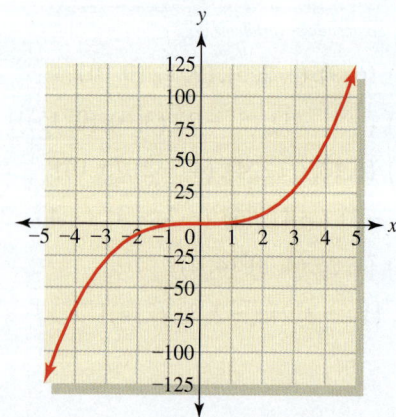

For Exercises 31 through 40, explain why each of the following equation represents a function. Then rewrite each function using function notation.

31. $y = 2x - 9$
32. $y = -5x + 7$
33. $y = 4$
34. $y = -8$
35. $y = x^2 - 8$
36. $y = -x^2 + 2$
37. $y = \frac{2}{3}x - 1$
38. $y = \frac{2}{5}x + \frac{1}{3}$
39. $y = x^3$
40. $y = x^4$

For Exercises 41 through 50, evaluate each function for the given value.

41. $f(x) = 5x + 1$; find $f(2)$
42. $f(x) = -2x + 4$; find $f(5)$
43. $f(x) = -4x + 7$; find $f(-6)$
44. $f(x) = -9x + 10$; find $f(-3)$
45. $f(x) = 16$; find $f(-1)$
46. $f(x) = -2$; find $f(0)$
47. $f(x) = x^2$; find $f(3)$
48. $f(x) = x^2 + 5$; find $f(2)$
49. $f(x) = x^2 - 3$; find $f(-4)$
50. $f(x) = -x^2$; find $f(-1)$

For Exercises 51 through 54, evaluate each function for the given value of the variable and explain its meaning in terms of the problem situation.

51. $R(t) = 9.50t$ is a function that represents the revenue a movie theater takes in for selling t tickets at $9.50 each. Find $R(125)$ and explain its meaning in terms of the problem situation.

52. $R(t) = 36t$ is a function that represents the revenue a baseball team takes in for selling t tickets at $36.00 each. Find $R(22000)$, and explain its meaning in terms of the problem situation.

53. $A(u) = 50.00 - 2.50u$ is a function that represents the amount left on a $50.00 prepaid bridge toll card, where for each use u of the bridge, it costs $2.50. Find $A(10)$ and explain its meaning in terms of the problem situation.

54. $A(m) = 100 - 0.07m$ is a function that represents the amount left on a $100.00 prepaid cell phone, where each minute used costs $0.07. Find $A(25)$ and explain its meaning in terms of the problem situation.

55. Let $C(p) = 1.50p + 2.75$ be the cost in dollars to make p pies. Find p such that $C(p) = 77.75$ and explain its meaning in terms of the problem situation.

56. Let $C(n) = 14.50n + 250$ be the cost in dollars for a company's Christmas party, where n represents the number of people attending. Find n such that $C(n) = 627$ and explain its meaning in terms of the problem situation.

Chapter Summary

Section 9.1 Graphing Quadratic Equations

- A **quadratic equation in two variables** has the form $y = ax^2 + bx + c$, where $a \neq 0$.
- A **parabola** is the shape of the graph of a quadratic equation in two variables.
- For a quadratic equation in the form $y = ax^2 + bx + c$, if $a > 0$ (positive), the parabola opens upward, and if $a < 0$ (negative), the parabola opens downward.
- The **vertex** of a parabola is the lowest point on the graph for a parabola that opens upward, and it is the highest point on the graph for a parabola that opens downward.

    The vertex of a parabola that opens upward is called the **minimum point**.

    The vertex of a parabola that opens downward is called the **maximum point**.

- The ***x*-coordinate of the vertex** of the quadratic equation in two variables, $y = ax^2 + bx + c$, is given by $x = \dfrac{-b}{2a}$. The *y*-coordinate is found by substituting the *x*-coordinate into the equation and simplifying.
- To interpret the meaning of the vertex in an application, be sure to explain what both coordinates of the vertex mean. Indicate if the vertex is a maximum or minimum point.
- The vertical line that passes through the vertex of a parbola is called the **axis of symmetry**. The axis of symmetry formula is $x = \dfrac{-b}{2a}$.
- To graph a **quadratic equation in two variables**, use these steps:
    1. Find the equation of the axis of symmetry.
    2. Find the coordinates of the vertex.
    3. Graph a total of seven points. Find the values of additional symmetric points as necessary. There should be at least three points to the left and right of the vertex.
    4. Connect the points with a smooth curve.

Example 1 The height, *h*, of a ball in feet *t* seconds after being hit can be estimated by the equation $h = -16t^2 + 48t + 3$. Find the vertex of this equation and interpret it in terms of height and time.

SOLUTION Using the vertex formula, we have that

$$t = \frac{-(48)}{2(-16)} = \frac{-48}{-32} = 1.5 \quad \text{Use the vertex formula to find the input value.}$$

$$h = -16(1.5)^2 + 48(1.5) + 3 \quad \text{Substitute 1.5 for } t \text{ to find } h.$$

$$h = 39$$

The vertex is (1.5, 39).

This means that 1.5 seconds after being hit, the ball reaches its maximum height of 39 feet.

CHAPTER 9 Summary

Example 2 Graph the quadratic equation $y = -x^2 + 6x + 16$ by finding the equation of the axis of symmetry and plotting the given points and their symmetric points.

x	$y = -x^2 + 6x + 16$
-3	-11
0	16
1	21
3	25

SOLUTION First find the axis of symmetry and then sketch the graph by plotting the given points and their symmetric points.

$$x = \frac{-(6)}{2(-1)}$$

$$x = 3$$

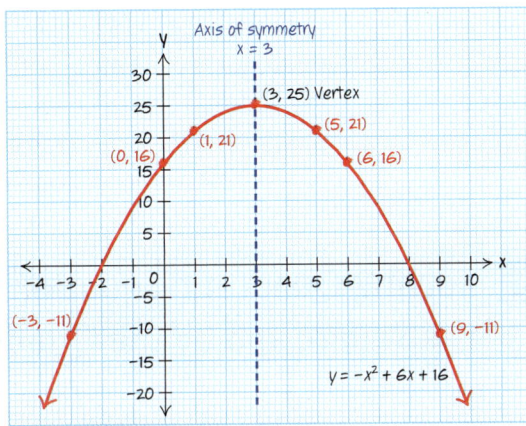

Section 9.2 Solving Quadratic Equations by Using the Square Root Property

- **The square root property** states that if $x^2 = a$, then $x = \pm\sqrt{a}$ for $a \geq 0$.
- **To solve a quadratic equation that can be put in the form** $x^2 = a$, use these steps:
  1. Determine whether the quadratic equation is of the form $x^2 = a$.
  2. Isolate the squared variable term if needed.
  3. Use the square root property. If $x^2 = a$ and $a > 0$ then $x = \pm\sqrt{a}$.
  4. Solve for the variable.
  5. Check both the answers in the original equation.
- **To solve a quadratic equation involving factors** such as $(ax - h)^2 = c$, apply the square root property, and solve for the variable.
- The **Pythagorean theorem** states that $a^2 + b^2 = c^2$, where a and b are the legs of a right triangle, and c is the hypotenuse (the longest side opposite the right angle).

Example 3 Solve the equation $(x + 3)^2 = 49$, using the square root property. Check the answer.

SOLUTION

$(x + 3)^2 = 49$ *Use the square root property to undo the square.*
$x + 3 = \pm\sqrt{49}$ *Don't forget the $\pm$ on the right side.*
$x + 3 = \pm 7$

$x + 3 = 7$ $x + 3 = -7$ *Separate into two equations and solve for x.*
$\underline{-3 \quad -3}$ $\underline{-3 \quad -3}$
$x = 4$ $x = -10$

Check the answers in the original equation:

$x = 4$ $x = -10$

$(4 + 3)^2 \stackrel{?}{=} 49$ $(-10 + 3)^2 \stackrel{?}{=} 49$

$7^2 = 49$ Both answers check. $(-7)^2 = 49$

Example 4

Solve the right triangle for the unknown side by using the Pythagorean theorem. If necessary, round to two decimal places.

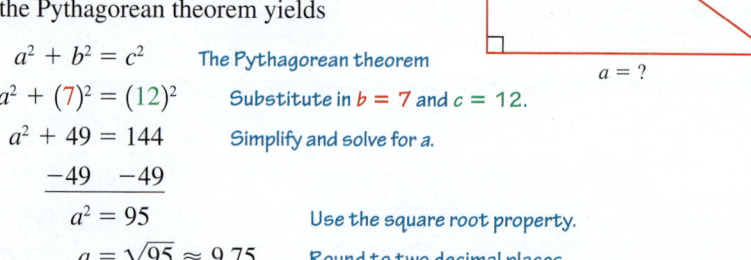

SOLUTION Substituting in the given values of $b = 7$ and $c = 12$ into the Pythagorean theorem yields

$a^2 + b^2 = c^2$ The Pythagorean theorem

$a^2 + (7)^2 = (12)^2$ Substitute in $b = 7$ and $c = 12$.

$a^2 + 49 = 144$ Simplify and solve for a.

$ \underline{-49 -49}$

$a^2 = 95$ Use the square root property.

$a = \sqrt{95} \approx 9.75$ Round to two decimal places.

Section 9.3   Solving Quadratic Equations by Completing the Square and by the Quadratic Formula

- To solve a quadratic equation of the form $ax^2 + bx + c = 0$ by **completing the square**, use the following steps:
    1. Isolate the variable terms on one side of the equation.
    2. If the coefficient of x^2 is not 1 ($a \neq 1$), then divide both sides by a.
    3. Find $\dfrac{1}{2}$ the coefficient of x, square it, and add the result to both sides of the equation.
    4. Factor the quadratic on the left side. It is now a perfect square trinomial.
    5. Solve for the variable using the square root property.
    6. Check the answers in the original problem.

- To find the solutions of a quadratic equation $ax^2 + bx + c = 0, a \neq 0$ using the **quadratic formula,** use the following steps:
    1. If needed, set the right side of the equation equal to 0. (The equation must be in the form $ax^2 + bx + c = 0, a \neq 0$.)
    2. Identify the values of a, b, and c.
    3. Substitute the values of a, b, and c into the quadratic formula
    $$x = \frac{-b \pm \sqrt{b^2 - 4ac}}{2a}$$
    4. Find the value(s) of x by simplifying the expression on the right side using the order-of-operations agreement.
        a. Find the *opposite* of b. (That is what $-b$ indicates.)
        b. Calculate $b^2 - 4ac$ to simplify the radicand.
        c. Find the square root of the simplified radicand.
        d. Calculate $2a$ to simplify the denominator.
    5. Rewrite as two equations. Simplify the right side to find the value(s) of x.
    6. Check the answer(s) in the original problem.

CHAPTER 9 Summary

Example 5 Solve the quadratic equation by completing the square. Give both the exact answer and the approximate answer rounded to two decimal places. Check the answer(s) in the original equation.

$$x^2 + 6x - 14 = 0$$

SOLUTION

$x^2 + 6x - 14 = 0$ Isolate the x-terms on the left.

$x^2 + 6x = 14$ Find $\frac{1}{2}b = \frac{1}{2}(6) = 3$.

$x^2 + 6x + 9 = 14 + 9$ Square 3 and add it to both sides.

$(x + 3)^2 = 23$

$x + 3 = \pm\sqrt{23}$ Separate into two equations and solve for x.

$x + 3 = \sqrt{23}$ $x + 3 = -\sqrt{23}$

$x = -3 + \sqrt{23}$ $x = -3 - \sqrt{23}$

Check: Use decimal approximations to check.

$x = -3 + \sqrt{23} \approx 1.80$ $x = -3 - \sqrt{23} \approx -7.80$

$(1.80)^2 + 6(1.80) - 14 \stackrel{?}{=} 0$ $(-7.80)^2 + 6(-7.80) - 14 \stackrel{?}{=} 0$

$0.04 \approx 0$ $0.04 \approx 0$

Example 6 Solve the quadratic equation by using the quadratic formula. Give both the exact answer and the approximate answer rounded to two decimal places. Check the answer(s) in the original equation.

$$2x^2 + x - 7 = 0$$

SOLUTION

$2x^2 + x - 7 = 0$ $a = 2$, $b = 1$, and $c = -7$.

$x = \dfrac{-b \pm \sqrt{b^2 - 4ac}}{2a}$ Substitute a, b, and c in the quadratic formula.

$x = \dfrac{-(1) \pm \sqrt{(1)^2 - 4(2)(-7)}}{2(2)}$

$x = \dfrac{-1 \pm \sqrt{57}}{4}$ Break into two equations and solve for x.

$x = \dfrac{-1 + \sqrt{57}}{4}$ $x = \dfrac{-1 - \sqrt{57}}{4}$

Check: Use decimal approximations to check.

$x = \dfrac{-1 + \sqrt{57}}{4} \approx 1.64$ $x = \dfrac{-1 - \sqrt{57}}{4} \approx -2.14$

$2(1.64)^2 + (1.64) - 7 \stackrel{?}{=} 0$ $2(-2.14)^2 + (-2.14) - 7 \stackrel{?}{=} 0$

$0.0192 \approx 0$ $0.0192 \approx 0$

Section 9.4   Graphing Quadratic Equations Including Intercepts

- To **find the intercepts of a quadratic equation in two variables**, use the following steps:
    1. To find the x-intercept, set $y = 0$ and solve for x. Write the intercept(s) as a point $(a, 0)$.
    2. To find the y-intercept, set $x = 0$ and solve for y. Write this intercept as the point $(0, b)$.
- There may be zero, one, or two x-intercepts for a quadratic equation in two variables.

834 CHAPTER 9 Quadratic Equations

- To **graph a quadratic equation in two variables,** use the following steps:
  1. Find the equation of the axis of symmetry.
  2. Find the coordinates of the vertex.
  3. Find the x- and y-intercepts.
  4. Graph a total of seven points. Find the values of additional symmetric points as necessary. There should be at least three points to the left of the vertex and three to the right of the vertex.
  5. Connect the points with a smooth curve.

Example 7 Graph the quadratic equation $y = 3x^2 + 12x - 15$. Find the equation of the axis of symmetry, the vertex, and the x- and y-intercepts. Graph at least seven points. Clearly label and scale the axes.

SOLUTION Find the equation for the axis of symmetry and the vertex.

$$x = \frac{-(12)}{2(3)}$$

$$x = -2 \qquad \text{The axis of symmetry}$$

To find the vertex, substitute $x = -2$ into the equation to find y.

$$y = 3(-2)^2 + 12(-2) - 15$$

$$y = -27$$

The vertex is the point $(-2, -27)$.

To find the x-intercepts, substitute $y = 0$ into the original equation and solve for x. To find the y-intercept, substitute $x = 0$ into the original equation and solve for y.

x-intercept	y-intercept
$0 = 3x^2 + 12x - 15$	$y = 3(0)^2 + 12(0) - 15$
$\dfrac{0}{3} = \dfrac{3x^2 + 12x - 15}{3}$	$y = -15$
$0 = x^2 + 4x - 5$	$(0, -15)$
$0 = (x + 5)(x - 1)$	
$x + 5 = 0 \quad$ or $\quad x - 1 = 0$	
$x = -5 \quad$ or $\quad x = 1$	
$(-5, 0) \quad$ and $\quad (1, 0)$	

Use a table to organize this information and find any needed additional symmetric points.

x	y	(x, y)	Label
-5	0	$(-5, 0)$	x-intercept
-4	-15	$(-4, -15)$	**Symmetric point to the y-intercept**
-3	-24	$(-3, -24)$	Additional symmetric point
-2	-27	$(-2, -27)$	Vertex
-1	-24	$(-1, -24)$	**Additional symmetric point**
0	-15	$(0, -15)$	y-intercept
1	0	$(1, 0)$	x-intercept

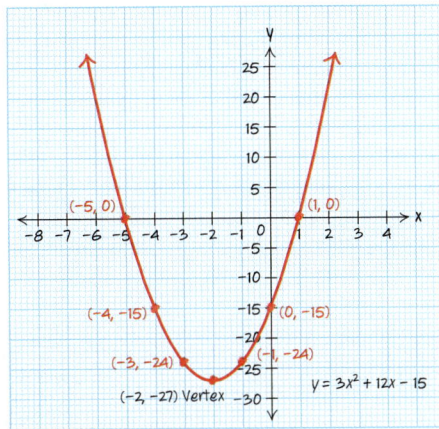

Section 9.5 Working with Quadratic Models

- A scatterplot in which the data follow **a parabolic pattern can be modeled by a quadratic equation**. The parabola may open up or down.
- A **quadratic model** is the quadratic equation in two variables that best fits the data.
- **Estimates** or **predictions** can be made using a quadratic model.
- The vertex of a quadratic model gives a **maximum or minimum value,** depending on whether the graph opens down or up.
- A quadratic model can have **horizontal** (or x) **intercepts** and **vertical** (or y) **intercepts.**
- **Model breakdown** occurs when the model's output values do not make sense in the context of the problem.

Example 8 The average daily high temperature for Chicago, Illinois, is given in the following table.

Month	Average Daily High Temperature (degrees Fahrenheit)
March	47
April	59
May	70
June	80
July	84
August	83

Source: www.intellicast.com

a. Adjust the data by representing each month by its order in the year (January = 1).

b. Draw a scatterplot of the data. Does the display of data look linear or quadratic?

SOLUTION a. The adjusted table is below.

Month	Average Daily High Temperature (degrees Fahrenheit)
3	47
4	59
5	70
6	80
7	84
8	83

Source: www.intellicast.com

b.

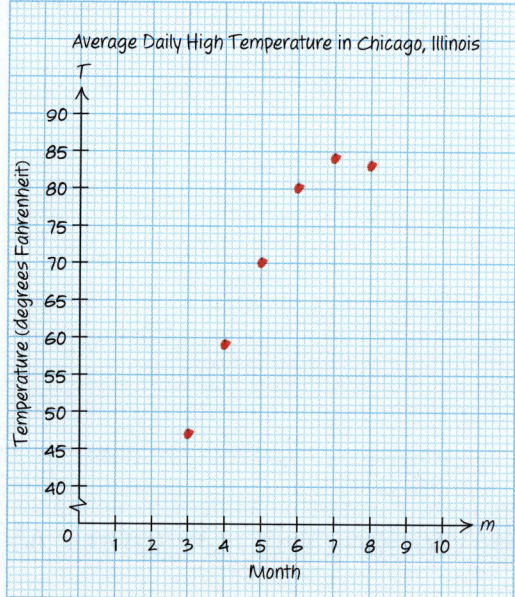

The data follows a parabolic pattern, so the data set can be modeled by a quadratic equation in two variables.

Example 9 A quadratic model for the data given in Example 8 is $T = -1.66m^2 + 25.8m - 16.5$. Here the variable m represents the month in the year, and T represents the average daily high temperature. Use this model to answer the following questions.

a. Use the model to predict the average daily high temperature in Chicago in October. Compare this prediction to the reported value of 64°F.

b. Find the vertex of the model. Explain the meaning of the vertex in terms of average daily high temperatures in Chicago.

c. Does the vertex give a maximum or a minimum value?

SOLUTION a. The month of October corresponds to $m = 10$. Substituting this value in the model and simplifying, we get

$T = -1.66m^2 + 25.8m - 16.5$ Substitute in $m = 10$.
$T = -1.66(10)^2 + 25.8(10) - 16.5$
$T = 75.5$

The model is predicting a daily high temperature of 75.5°F, which is slightly above the reported high of 64°F.

b. To find the vertex, use the vertex formula $x = \dfrac{-b}{2a}$ to first find the x-coordinate. Here the x-coordinate is represented by m. The model is $T = -1.66m^2 + 25.8m - 16.5$, so we have that

$y = ax^2 + bx + c$
$T = -1.66m^2 + 25.8m - 16.5$
$a = -1.66 \qquad b = 25.8 \qquad c = -16.5$

Substituting $a = -1.66$ and $b = 25.8$ into the vertex formula yields

$$m = \dfrac{-(25.8)}{2(-1.66)}$$

$$m \approx 7.8$$

To find the output value, T, we substitute $m \approx 7.8$ into the original equation and evaluate.

$$T = -1.66m^2 + 25.8m - 16.5 \quad \text{Substitute in } m = 7.8.$$
$$T \approx -1.66(7.8)^2 + 25.8(7.8) - 16.5$$
$$T \approx 83.7$$

The vertex is the point $(m, T) \approx (7.8, 83.7)$.

The model is predicting the high temperature in late July to be 83.7°F.

c. The vertex gives a maximum value. The scatterplot is opening down, so the vertex corresponds to the highest point on the graph.

Section 9.6 The Basics of Functions

- A **relation** is a set of ordered pairs (x, y).
- The **domain** of a relation is the set of all inputs of the relation.
- The **range** of a relation is the set of all outputs of the relation.
- A **function** is a special kind of relation in which each input x is associated with exactly one output y.
- **The vertical line test** states that if every vertical line hits the graph of an equation in at most one point, then the equation is that of a function.
- **Function notation** is used when a relation between the variables x and y is a function. To denote this, write $y = f(x)$ to indicate that y is a function of x.
- To **evaluate a function in function notation** $y = f(x)$, substitute the given value of x into the right side for x, and simplify the right side only.

Example 10

Explain why the following equation represents a function. Then rewrite the function using function notation.

$$y = -3x + 5$$

SOLUTION This equation represents a function because when it is graphed, it passes the vertical line test. See the graph.

Rewriting in function notation yields $f(x) = -3x + 5$.

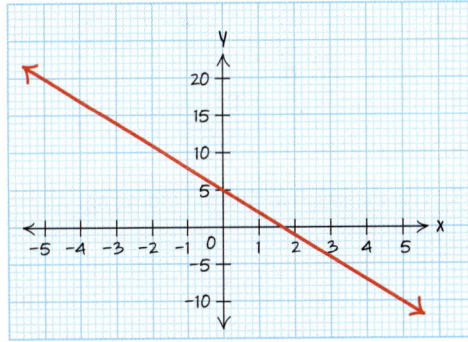

Example 11 Evaluate the function for the given value of the input. $f(x) = -4x + 9$ for $x = -3$.

SOLUTION

$$f(x) = -4x + 9 \quad \text{Substitute in } x = -3.$$
$$f(-3) = -4(-3) + 9 \quad \text{Simplify the right side.}$$
$$f(-3) = 12 + 9$$
$$f(-3) = 21$$

Chapter Review Exercises

1. The profit, P, in thousands of dollars that the Bean Company makes from selling c hundred pounds of coffee beans can be estimated by the equation $P = -c^2 + 8c - 9$. [9.1]
   a. Estimate the profit for this coffee shop if it sells 200 pounds of coffee beans.
   b. Find the number of pounds of coffee beans sold to generate a profit of $6 thousand for this coffee shop.
   c. Find the vertex of this equation and interpret it in terms of coffee beans and profit.

2. The height, h, above the ground in feet of a ball t seconds after being dropped from a building can be estimated by the equation $h = -16t^2 + 100$. [9.1]
   a. Find the vertex of this equation and interpret it in terms of time and the height of the ball.
   b. Find the time when the ball will be 36 feet above the ground.

8. Use the graph to find the following. [9.1]

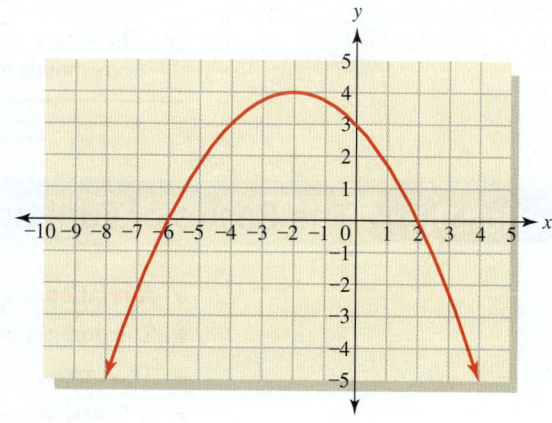

   a. Estimate the vertex.
   b. Estimate the y-intercept.
   c. Estimate the x-intercepts.

For Exercises 3 through 6, find the vertex and x- and y-intercepts of each equation. Determine whether the vertex is a maximum or minimum point. Do not graph.

3. $y = 3x^2 - 24x + 21$ [9.1]
4. $y = x^2 - 4x - 45$ [9.1]
5. $y = -x^2 - 7x - 12$ [9.1]
6. $y = -x^2 + 18x - 72$ [9.1]

7. Use the graph to find the following. [9.1]

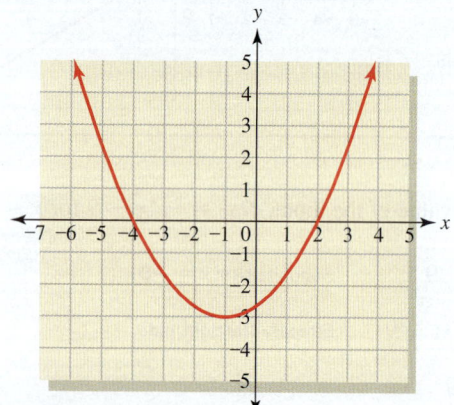

   a. Estimate the vertex.
   b. Estimate the y-intercept.
   c. Estimate the x-intercepts.

For Exercises 9 through 14, solve each quadratic equation by using the square root property. Check the answer(s) in the original equation.

9. $3x^2 = 75$ [9.2]
10. $2x^2 = 72$ [9.2]
11. $(5x + 3)^2 = 36$ [9.2]
12. $(7 - 2x)^2 = 1$ [9.2]
13. $-x^2 = 16$ [9.2]
14. $-2x^2 = 50$ [9.2]

For Exercises 15 and 16, solve each right triangle for the unknown side using the Pythagorean theorem. If necessary, round any answers to two decimal places.

15. [9.2]

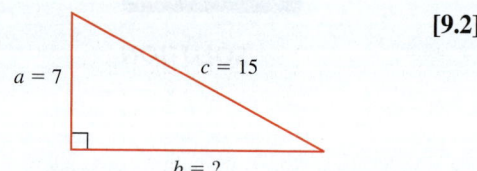

16. [9.2]

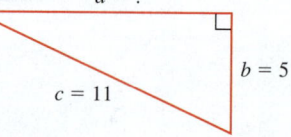

For Exercises 17 through 20, solve each quadratic equation by completing the square. Give both the exact answer(s) and the approximate answer(s) rounded to two decimal places. Check the answer(s) in the original equation.

17. $x^2 - 4x - 11 = 0$ [9.3]
18. $x^2 - 10x + 8 = 0$ [9.3]
19. $3y^2 - 12y + 9 = 0$ [9.3]
20. $2y^2 - 16y + 8 = 0$ [9.3]

For Exercises 21 through 26, solve each quadratic equation by using the quadratic formula. Give both the exact answer(s) and the approximate answer(s) rounded to two decimal places. Check the answer(s) in the original equation.

21. $x^2 + 6x + 2 = 0$ [9.3]
22. $x^2 - 10x + 15 = 0$ [9.3]
23. $8y^2 - 4y + 1 = 0$ [9.3]
24. $4y^2 + 12y + 3 = 0$ [9.3]
25. $36x^2 + 36x + 7 = 0$ [9.3]
26. $x^2 - 8x + 10 = 0$ [9.3]

For Exercises 27 and 28, solve each application problem.

27. A photographer sells 16 × 20 inch prints of her work. The profit, P, that she makes from selling x copies of her photographs is given by $P = -2x^2 + 225x$. How many copies of her photographs must she sell to make a profit of $4500.00? [9.3]
    a. Since we are interested in a profit of $4500.00, substitute into the equation $P = 4500$.
    b. Solve the quadratic equation from part a.
    c. How many photographs must she sell to make a profit of $4500.00?

28. A rock is dropped from an 18-meter cliff. The height of the rock, h, in meters after t seconds is given by the equation [9.3]
$$h = -4.9t^2 + 18$$
    a. How long will it take for the rock to be at a height of 13 meters? Round the answer to two decimal places if necessary.
    b. How long will it take for the rock to hit the ground? Round the answer to two decimal places if necessary.

For Exercises 29 through 36, graph the following equations using symmetry. Find the vertex, axis of symmetry, x- and y-intercepts. Label, on the graph, the vertex and the x- and y-intercepts. Clearly label and scale the axes. Graph at least seven points.

29. $y = x^2 - 25$ [9.4]
30. $y = -x^2 + 64$ [9.4]
31. $y = x^2 + 4x$ [9.4]
32. $y = x^2 + 6x$ [9.4]
33. $y = 3x^2 - 24x + 21$ [9.4]
34. $y = x^2 - 4x - 45$ [9.4]
35. $y = -x^2 - 7x - 12$ [9.4]
36. $y = -x^2 + 18x - 72$ [9.4]

For Exercises 37 through 42, decide whether the data as graphed on the scatterplot follow a linear pattern, a quadratic pattern, or neither.

37. [9.5]

38. [9.5]

39. [9.5]

840 CHAPTER 9 Quadratic Equations

40. [9.5]

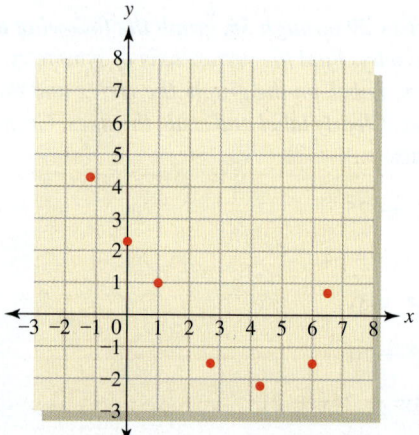

41. [9.5]

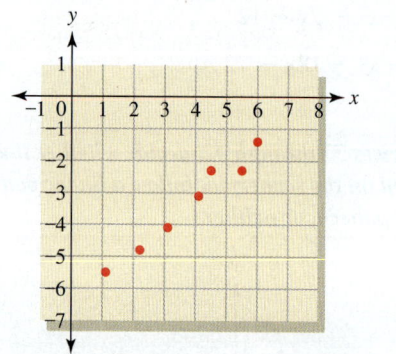

42. [9.5]

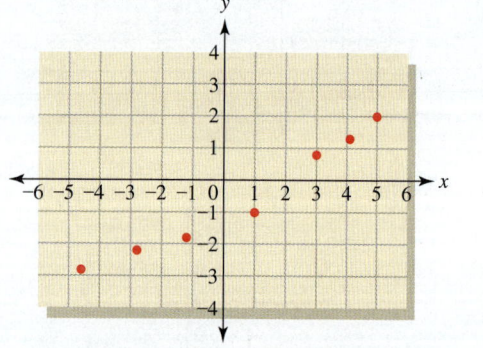

For Exercises 43 and 44, graph the given data on a scatterplot and answer the questions.

43. The average daily high temperature for Rapid City, South Dakota, is given in the following table. [9.5]

Month	Average Daily High Temperature (degrees Fahrenheit)
March	45
April	55
May	65
June	75
July	83
August	82

Source: www.intellicast.com

a. Adjust the data by representing each month by its order in the year (January = 1).
b. Graph the data on a scatterplot. Does the display of data look linear or quadratic?

44. The median price of a home in Detroit, Michigan, for the years 2005–2009 is given in the following table. [9.5]

Years since 2000	Median Price (dollars)
5	100,600
6	108,900
7	107,300
8	104,500
9	82,700

Source: www.zillow.com

a. Graph the data on a scatterplot.
b. Does the display of data look linear or quadratic?

For Exercises 45 through 48, use the given quadratic model to answer the questions.

45. A quadratic model for the daily high temperature in degrees Fahrenheit for Rapid City, South Dakota, is given by $T = -1.125m^2 + 20.35m - 7.09$, where T is the temperature in degrees Fahrenheit and m is the month. [9.5]

a. Find the vertex of the model. Explain the meaning of the vertex in the context of the average daily high temperature in Rapid City, South Dakota.
b. Use the model to estimate the average high temperature in Rapid City in December. Is this answer reasonable? Compare it to the reported value of 37°F.
c. Use the model to estimate the average high temperature in Rapid City in February. Is this answer reasonable? Compare it to the reported value of 38°F.

46. A quadratic model for the median price of a home in Detroit, Michigan, is given by $P = -4385.7t^2 + 57380t - 77188.6$, where P is the median price of a home in Detroit t years since 2000 (time). [9.5]

a. Find the vertex of the model. Explain the meaning of the vertex in terms of median home prices in Detroit, Michigan.
b. Use the model to estimate the median home price in Detroit in the year 2010. Is this answer reasonable? Explain.
c. Use the model to estimate the median home price in Detroit in the year 2012. Is this answer reasonable? Explain.

47. The average profit in dollars, A, that a local frame shop makes when framing n pictures is given by $A = n(36 - 3n)$. [9.5]
   a. Find and explain the value of the vertical intercept.
   b. Find and explain the value of the vertex.
   c. Find the number of pictures the shop must frame to make an average profit of $60 per frame.

48. The average profit in dollars, A, that a local nursery makes when selling apple trees is given by $A = n(32 - 4n)$. [9.5]
   a. Find and explain the value of the vertical intercept.
   b. Find and explain the value of the vertex.
   c. Find the number of apple trees the nursery must sell to make $50.00 per tree.

For Exercises 49 through 52, state the domain and range for each relation. Then determine whether every input results in exactly one output.

49. [9.6]

Name	Hourly Wage
Colin	$18.50
Lily	$9.25
Isaiah	$11.75
Gabrielle	$18.50
Seth	$14.50

50. [9.6]

Amount of Time Traveled (hours)	Distance Traveled (miles)
1	50
2	100
3	150
4	200
5	250

51. $\{(0, 2), (3, -1), (5, 4), (3, -3), (9, 10)\}$ [9.6]

52. $\{(10, 0), (-4, -3), (5, -3), (10, 1), (1, 10)\}$ [9.6]

For Exercises 53 and 54, determine whether or not the relation is a function.

53. The input is the age of a person, and the output is their weight. [9.6]

54. The input is the age of a child, and the output is their grade in school. [9.6]

For Exercises 55 through 58, determine whether or not each graph represents a function.

55. [9.6]

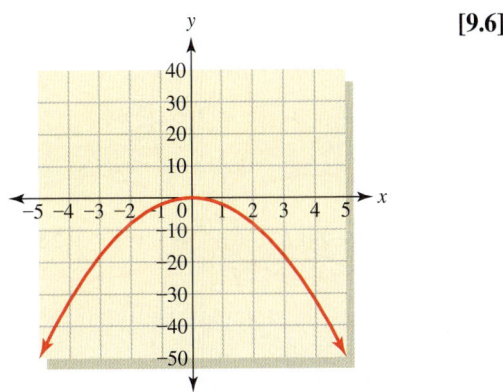

56. [9.6]

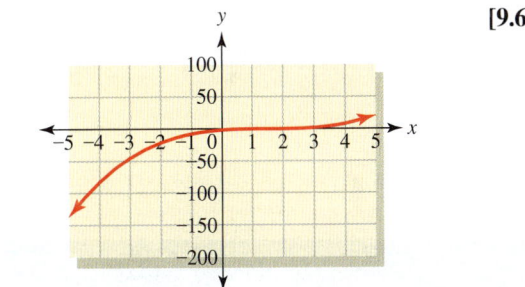

57. [9.6]

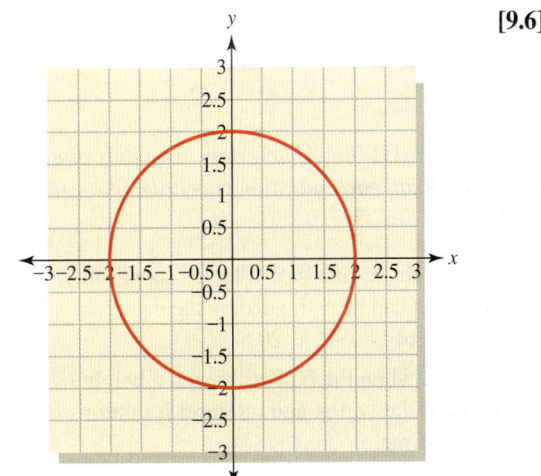

842 CHAPTER 9 Quadratic Equations

58. 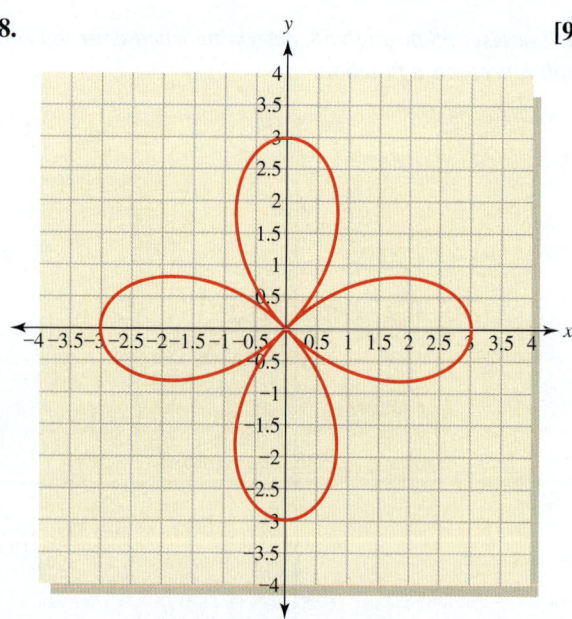 [9.6]

For Exercises 103 and 104, explain why each equation represents a function. Then rewrite each function using function notation.

59. $y = -5x + 3$ [9.6]

60. $y = 7$ [9.6]

For Exercises 105 and 106, evaluate the function for the given value of the input x.

61. $f(x) = -3x + 6$; for $x = -4$ [9.6]

62. $f(x) = \dfrac{2}{3}x - 12$; for $x = -6$ [9.6]

Chapter Test

This chapter test should take approximately one hour to complete. Read each question carefully. Show all of your work.

1. The height, h, of a ball in feet t seconds after being hit can be estimated by the equation $h = -16t^2 + 64t + 20$.
   a. Find when the ball will reach a height of 68 feet.
   b. Find the vertex and interpret it in terms of height and time.

2. A company's revenue, R, in thousands of dollars t years since 2010 can be estimated by the equation $R = 2t^2 + 16t + 60$. Find the vertex of this equation, and interpret it in terms of the year and the revenue for that year.

3. Solve and check: $(3x - 11)^2 = 81$.

4. Solve for the unknown side in the right triangle using the Pythagorean theorem.

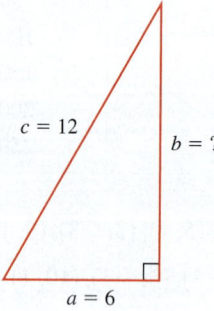

5. Solve by completing the square. Give both the exact answer(s) and the approximate answer(s) rounded to two decimal places. Check the answer(s) in the original equation $a^2 - 18a = -17$.

CHAPTER 9 Test 843

6. Solve by using the quadratic formula. Give both the exact answer(s) and the approximate answer(s) rounded to two decimal places. Check the answer(s) in the original equation $9x^2 + 24x + 11 = 0$.

7. A local model rocket club launches a rocket that follows the path given by the equation $h = -16t^2 + 150t + 4$. Here t is the time in seconds after the launch, and h is the height of the rocket in feet t seconds after launch. The students in the club want to know when the rocket will hit the ground.

   a. The phrase "when the rocket will hit the ground" contains the word *when*. Does *when* indicate that we are trying to determine the height, h, of the rocket or the time, t?
   b. When the rocket hits the ground, what is the value of the height h? Substitute this value into the equation $h = -16t^2 + 150t + 4$.
   c. What kind of equation do we have in part b? Is it linear, quadratic, rational, or radical?
   d. Solve the equation from part b.
   e. If there is more than one answer, which answer makes sense for this problem?

For Exercises 8 and 9, find the vertex and x- and y-intercepts of each equation. Determine whether the vertex is a maximum or minimum point. Do not graph.

8. $y = x^2 + 2x - 24$ 9. $y = -5x^2 + 20x - 15$

10. Use the graph to find the following.

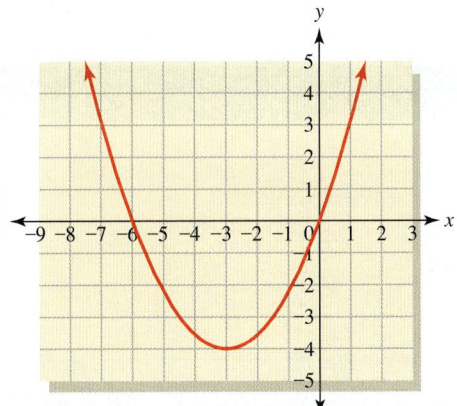

   a. Estimate the vertex.
   b. Estimate the y-intercept.
   c. Estimate the x-intercepts.

For Exercises 11 and 12, graph the following equations using symmetry. Label the vertex and the x- and y-intercepts. Clearly label and scale the graph. Graph at least seven points.

11. $y = x^2 + 2x - 24$ 12. $y = -2x^2 + 12x + 32$

13. Label each scatterplot of a data set as linear, quadratic, or neither.

   a.
   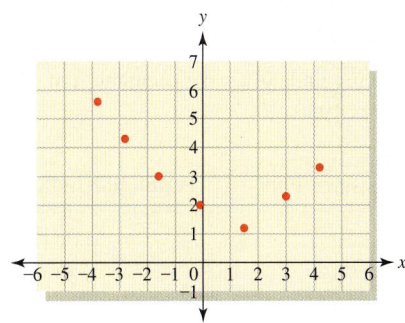

   b.
   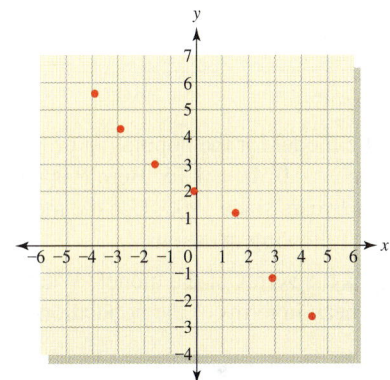

   c.
   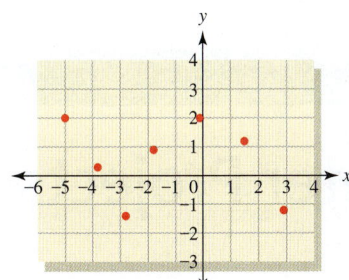

14. Draw a scatterplot for the data in the table and answer the following questions.

   a. Does the graph of the data look linear or quadratic?
   b. If the graph of the data looks quadratic, estimate the vertex from the graph.

x	y
−1	−9
0	0
1	3
2	0
3	−9

15. The following table gives the height, h, in feet above the ground of a falling rock dropped from an 85-foot cliff t seconds after it was dropped.

t	h
0	85
0.5	81
1	69
1.5	49
2	21

   a. Graph the data on a scatterplot.
   b. Does the graph of the data look linear or quadratic?
   c. If the graph of the data looks quadratic, estimate the vertex from the scatterplot.

16. The following table gives the population of Mexico City for the years 1950–2007.

Years since 1900	Population (millions)
50	3.05
70	6.87
80	8.83
105	14.01
107	19

   Source: msn.encarta.com

   A quadratic model for the population of Mexico City is $P = 0.002t^2 - 0.071t + 1.75$. Here P is the population in millions t years since 1900. Use the model to answer each of the following questions.

   a. Where is the vertex of the model? What does the vertex mean in terms of the population of Mexico City?
   b. What is the P-intercept of this model? Explain the P-intercept in terms of the population of Mexico City. Does it make sense in this situation?
   c. For which year(s) does the model predict the population of Mexico City will reach 10 million people? Do both of these answers make sense in this situation?

17. A quadratic model for the height, h, in feet above the ground of a rock dropped from a 50-foot cliff x seconds after it was dropped is given by $h = -16x^2 + 50$. Use the model to answer the following questions.

   a. Find the vertex of the model. Explain the meaning of the vertex in terms of the situation.
   b. Find the x-intercepts of the model. Do they both make sense in terms of this situation? Explain.

18. Explain why the equation $y = 6x - 1$ is a function. Rewrite using function notation.

19. Evaluate $f(x) = 3x^2 + 2x$ for $x = -2$.

20. State the domain and range for the relation.

State	State Sales Tax Rate (as of 2017)
Florida	6%
Texas	6.25%
Michigan	6%
Washington	6.5%
New Mexico	5.125%

Chapter Projects

■ How Do I Create a Quadratic Model?

Written Project
One or more people

What you will need
- A Texas Instruments 83 or 84 graphing calculator.

The standard form of a quadratic equation in two variables is $y = ax^2 + bx + c$. There is another form of a quadratic equation in two variables. It is called the **vertex form** and looks like $y = a(x - h)^2 + k$. The two variables are x and y. The vertex of the parabola is at (h, k). The value of a has the same meaning in both standard form and vertex form. When a is positive, the parabola opens up. When a is negative, the parabola opens down. As we learned in Section 9.1, the value of a also controls the width of the parabola.

To find a quadratic model for the following data set, use the given steps.

The data in the table are the average daily high temperature, in degrees Fahrenheit, in Minneapolis, Minnesota.

Month	Temperature (degrees Fahrenheit)
April	57
May	70
June	79
July	83
August	80
September	71

Source: www.intellicast.com

1. Adjust the data. Let April = 4, May = 5, and so on.
2. Graph the adjusted data on a scatterplot.
3. Pick the highest point on the graph if the parabola opens down or the lowest point on the graph if the parabola opens up as the vertex. Set the coordinates of the vertex equal to (h, k).
4. Substitute the values of h and k into the equation $y = a(x - h)^2 + k$.
5. Substitute a point that is not near the vertex in for (x, y). Solve for a. (*Hint:* Use the point that represents the temperature during the month of May or September.)
6. Write the quadratic model, $y = a(x - h)^2 + k$, with the values of a, h, and k substituted in.
7. Graph the model on the scatterplot. Does it fit the data well?
8. Use the model to predict the temperature in October. Look up the average daily high temperature in October in Minneapolis. Does your answer seem reasonable?
9. Use the model to predict the temperature in January. Look up the average daily high temperature in January in Minneapolis. Does your answer seem reasonable?

Proving the Quadratic Formula

Written Project

One or more people

To solve $ax^2 + bx + c = 0$, $a \neq 0$, complete the square.

1. Divide each term on both sides of the equation by a to get the coefficient of x^2 to be 1. Since $a \neq 0$, it is permissible to divide by a.
2. Isolate the x^2- and x-terms on the left side of the equation.
3. Take $\frac{1}{2}$ of the coefficient of x, square it, and add it to both sides of the equation.
4. The left side of the equation is a perfect square trinomial. Rewrite the left side of the equation in factored form.
5. Rewrite the right side of the equation as a single fraction. (*Hint:* The least common denominator is $4a^2$.)
6. Apply the square root property. Don't forget the $\pm$ on the right side.
7. Use the quotient property of radicals to simplify the right side.
8. Solve for x by subtracting $\frac{b}{2a}$ from both sides.
9. Rewrite the right side as a single fraction. (*Hint:* The least common denominator is $2a$.)
10. You have just derived the quadratic formula: $x = \dfrac{-b \pm \sqrt{b^2 - 4ac}}{2a}$.

How Big Should That Horse Corral Be?

Written Project
One or more people

Jenna wants to build a rectangular corral for her horse. She is planning on placing the corral next to an existing wall. This means she will need to provide only three sides of fencing for the corral.

1. Draw a picture of the problem situation. Label the length of the rectangular corral y, and label the width x.
2. Jenna has 78 feet of fencing to use. Write an equation for the perimeter of the corral, using x for the width and y for the length.
3. Jenna wants to maximize the area enclosed for her horse. Write an expression for the area of the corral in terms of x and y.
4. Solve the equation from step 2 for y. Substitute y in the expression for area in step 3. The area expression should now have only one variable, namely, x.
5. Let A be the area of the corral in square feet. Set the area expression from step 4 equal to A. Multiply out the expression.
6. What kind of an equation is A: linear or quadratic? If A is quadratic, does the corresponding graph of the parabola open up or down?
7. Where does the maximum value for a quadratic equation occur? Find the x-value where the maximum area occurs.
8. What is the maximum area?
9. What is the value of y, the length?

Write a Review of a Section for Presentation

Research Project
One or more people

What you will need
- This presentation may be done in class or as an online video.
- If it is done as a video, post it on a website where it can be easily viewed by the class.
- You might want to use homework or other review problems from the book.
- Make it creative and fun.

Create a 5-minute review presentation of one section from Chapter 9. The format of the presentation can be a poster presentation, a blackboard presentation from notes, an online video, or a game format (e.g., math jeopardy). The presentation should include the following:

- Any important formulas in the chapter.
- Important skills in the chapter, backed up by examples.
- Explanation of terms used (e.g., what is *modeling*?).
- Common mistakes and how to avoid them.

Answers to Practice Problems

APPENDIX A

CHAPTER R
Section R.1

PP for Example 1

a. Natural numbers (if there are no children enrolled, there would not be a kindergarten)

b. Integers

c. Whole numbers

PP for Example 2

The scale is 3.

PP for Example 3

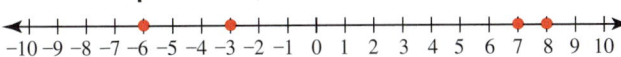

PP for Example 4

a.	−5	>	−8
b.	3	>	−3
c.	−2	<	0

PP for Example 5

a. 10   b. 16

PP for Example 6

$-2, |0|, |-4|, |6|, 8$

PP for Example 7

a. The opposite of 8; $-(8) = -8$

b. The opposite of -1; $-(-1) = 1$

c. The opposite of 0; $-(0) = 0$

PP for Example 8

a. -34 b. 14

PP for Example 9

a. -7 b. 17

PP for Example 10

a. 14,494

The elevation of Mount Whitney is 14,494 ft.

b. -282

The elevation of Death Valley is -282 ft.

c. 14,776

The difference in elevation between the highest and lowest points in the state of California is 14,776 ft.

PP for Example 11

7

PP for Example 12

a. -8 b. 15

PP for Example 13

a. -5 b. 2   c. -9

PP for Example 14

a. $(4) + ((-2) \cdot 6) + (8)$

b. $(3 \cdot 7) + ((-10) \div 2 \cdot 3) + ((-5) \cdot 4) + (1)$

c. $(-65 \div 5 \cdot 4)$

PP for Example 15

a. $24 \div 2 \cdot 3 = 36$

b. $9 + 5 \cdot 6 - 8 \div 4 = 37$

c. $6 \div 3 + 4 \cdot 5 - 2 \cdot 2 + 24 \div 8 = 21$

Section R.2

PP for Example 1

a. $32 = 2 \cdot 2 \cdot 2 \cdot 2 \cdot 2$ b. $10 = 2 \cdot 5$

c. $100 = 2 \cdot 2 \cdot 5 \cdot 5$

PP for Example 2

a. 2   b. 32

* PP = Practice Problem

A-2 APPENDIX A Answers to Practice Problems

PP for Example 3
a. $\dfrac{1}{9}$ b. $\dfrac{7}{10}$ c. $\dfrac{2}{3}$

PP for Example 4
a. $\dfrac{6}{11} = \dfrac{18}{33}$ b. $\dfrac{3}{8} = \dfrac{21}{56}$

PP for Example 5
a. The scale is $\dfrac{1}{4}$.

b.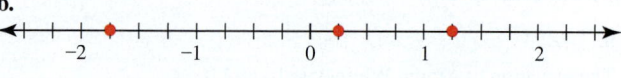

PP for Example 6
a. $\dfrac{5}{6} + 2\dfrac{4}{6} = 3\dfrac{1}{2}$ b. $\dfrac{7}{15} - \dfrac{2}{15} = \dfrac{5}{15} = \dfrac{1}{3}$

PP for Example 7
$\dfrac{1}{2}$ gallon

Henry buys a total of $\dfrac{1}{2}$ gallon.

PP for Example 9
$\dfrac{57}{8}$ feet $= 7\dfrac{1}{8}$ feet

Don had $7\dfrac{1}{8}$ feet remaining.

PP for Example 11
a. $\dfrac{2}{3} \cdot \dfrac{12}{17} = \dfrac{8}{17}$ b. $\dfrac{5}{27} \div \dfrac{20}{9} = \dfrac{1}{12}$

PP for Example 12
a. $\left(\dfrac{1}{3} \div \dfrac{5}{6}\right) + \left(\dfrac{5}{2}\right) = \dfrac{29}{10}$ b. $\boxed{7} + \left(\left(-\dfrac{2}{5}\right) \cdot \dfrac{15}{4}\right) + \boxed{\dfrac{1}{3}} = \dfrac{35}{6}$

Section R.3

PP for Example 1
a. The digit in the tenths place is 3.

b. The digit in the ten-thousandths place is 5.

PP for Example 2
a. $2.46 = 2\dfrac{46}{100} = 2\dfrac{23}{50}$ b. $1.025 = 1\dfrac{25}{1000} = 1\dfrac{1}{40}$

c. $10.9 = 10\dfrac{9}{10}$

PP for Example 3
a. $\dfrac{2}{5} = 0.4$ b. $\dfrac{3}{4} = 0.75$ c. $\dfrac{1}{9} = 0.\overline{1}$

PP for Example 4
a. 0.25

b.

PP for Example 6
a. 10.99 b. 7.16

c. $7.32

Laurence received $7.32 in change.

PP for Example 7
a. $4.21 \cdot 2.3 = 9.683$

b. $3 \cdot 12.99 = \$38.97$

If you purchased three shirts for $12.99 each, you spent a total of $38.97.

PP for Example 8
a. $6.24 \div 2.6 = 2.4$ b. $30.65 \div 1.6 \approx 19.16$

PP for Example 9
a. $\boxed{3.2} + \boxed{4.1 \cdot 2.5} = 13.45$ b. $\boxed{6 \div 1.5} + \boxed{(-3.1)} + \boxed{4} = 4.9$

Section R.4

PP for Example 1
a. $55\% = \dfrac{55}{100} = \dfrac{11}{20}$ b. $\dfrac{84}{100} = 84\%$

c. $\dfrac{1}{5} = \dfrac{20}{100} = 20\%$

PP for Example 2
a. $8.25\% = 0.0825$ b. $0.36\% = 0.0036$

c. $0.22 = 22\%$ d. $110 = 11000\%$

PP for Example 4
a. $0.08 \cdot 95 = 7.6$ b. $0.30 \cdot 50 = 15$

c. $0.045 \cdot 1600 = 72$

PP for Example 5
The percent increase for the condo is $0.2 = 20\%$.

PP for Example 6
a. The markup on this cell phone is $130.

b. The retail price for this cell phone is $330.

PP for Example 7
a. A new car, purchased for $33,275.00, will decrease in value $6322.25 in the first year.

b. The new car is worth $26,952.75 at the end of the first year.

Section R.5

PP for Example 1
a. Integers or rational numbers b. Integers

c. Integers or rational numbers

PP for Example 2

	Natural Num	Whole Num	Integer	Rational Num	Irrational Num	Real Num
8.3				✓		✓
$-\frac{4}{3}$				✓		✓
56	✓	✓	✓	✓		✓
π					✓	✓
-15			✓	✓		✓

PP for Example 3

a. $0.255 = \frac{255}{1000} = \frac{51}{200}$. Rational b. $\frac{1}{9} = 0.\overline{1}$. Rational

PP for Example 4

a. $\sqrt{7} \approx 2.6$ b. $\frac{2}{7} \approx 0.29$

PP for Example 5

a. The unit price for this bologna is $0.16 per ounce.

b. The unit price for this jar of peanut butter is $0.32 per ounce.

CHAPTER 1

Section 1.1

PP for Example 1

a. $16^5 = 16 \cdot 16 \cdot 16 \cdot 16 \cdot 16$ b. $29 \cdot 29 \cdot 29 \cdot 29 = 29^4$

c. $(-2)^5 = -32$ d. $5^6 = 15{,}625$

PP for Example 2

a. $(-4)^2 = 16$ b. $(-5)^3 = -125$

c. $38^0 = 1$ d. $-7^2 = -49$

PP for Example 3

a. $1.02 \times 10^6 = 1{,}020{,}000$

b. $4.51 \times 10^9 = 4{,}510{,}000{,}000$

PP for Example 4

a. $\boxed{3^4} + \boxed{\left(-\frac{10}{2}\right)} + \boxed{5 \cdot 4}$ b. $\boxed{9^4} + \boxed{(-16) \cdot 3} + \boxed{9} + \boxed{(-6)}$

PP for Example 5

a. $6 + 3 \cdot 4 - 2 \div 2 = 17$

b. $8 \div 2 + 6 \cdot 3 - 3 \cdot 3 + 28 \div 4 = 20$

c. $5^2 - 16 + 9 = 18$

PP for Example 6

a. $16 \div (-8) \cdot 3 - (5 \cdot 3) + 12 \cdot 5 = 39$

b. $(-3 - 4)^2 = 49$

c. $|9 - 2 \cdot 3| + (-6) \div (-18 \div 3) = 4$

d. $\frac{(-2)^2 + 4 \cdot 3}{7 - 2 \cdot 3} - [10 \div 2] = 11$

PP for Example 7

a. Associative property of addition

b. Commutative property of multiplication

c. Associative property of multiplication

d. Commutative property of addition

PP for Example 9

a. $5(16 - 7) = 5 \cdot 16 - 5 \cdot 7 = 80 - 35 = 45$

b. $5(16 - 7) = 5 \cdot 9 = 45$

c. Yes

PP for Example 10

a. $7(2 + 8) - 25 = 45$ b. $-(17 - 8) + 5 = -4$

Section 1.2

PP for Example 1

a. $50 \cdot 4.5 = 225$ miles b. $50 \cdot 7.25 = 362.5$ miles

PP for Example 3

a. The cost of printing 5 pages is $0.50, the cost of printing 10 pages is $1.00, and the cost of printing 25 pages is $2.50.

b. A home air conditioner uses 14,000 watt-hours when on for 4 hours; 28,000 watt-hours when on for 8 hours; and 84,000 watt-hours when on for 24 hours.

PP for Example 4

a. $-3(7) + 12 = -9$

b. $-3(-2)^2 + 6(-2) + 1 = -23$

c. $|-(5) + 6| + 2(5)^2 - 1 = 50$

PP for Example 5

a. $500 \text{ ft} \cdot \frac{1 \text{ m}}{3.28 \text{ ft}} \approx 152.44 \text{ m}$

b. $42 \text{ oz} \cdot \frac{28.35 \text{ g}}{1 \text{ oz}} \approx 1190.7 \text{ g}$

c. $156 \text{ lb} \cdot \frac{1 \text{ kg}}{2.2 \text{ lb}} \approx 70.91 \text{ kg}$

APPENDIX A Answers to Practice Problems

PP for Example 6
a. Let t represent the time of day and let T represent the temperature in degrees Fahrenheit.

b. Let c represent the number of credits enrolled in and let T represent the tuition paid in dollars.

PP for Example 7
a. $5n$ b. $4n + 9$ c. $\dfrac{6}{n}$ d. $2n - 16$

PP for Example 8
a. $x - 29$

b. Let x represent the number of tickets sold. $9x$

c. $150 + 3x$

PP for Example 9
a. Let m represent the number of miles.

m miles $\cdot \dfrac{1.609 \text{ kilometers}}{1 \text{ mile}} \approx 1.609m$ kilometers

b. Let t represent the number of teaspoons.

t teaspoons $\cdot \dfrac{1 \text{ tablespoon}}{3 \text{ teaspoons}} = \dfrac{t}{3}$ tablespoons

PP for Example 10
a. 15 ft $\cdot$ 18 ft = 270 ft^2 The area of the living room is 270 ft^2.

b. ft^2, or square feet

c. $270 \cdot 2.38 = \$642.60$ The total cost of the flooring for the living room is \$642.60.

PP for Example 12
a. Let n represent the number of people who attend the reception. $425 + 22n$

b. To stay within their budget of \$3500.00, they can invite 139 people.

Section 1.3

PP for Example 1
a. There are four terms. $-5x^2 + \boxed{7c} + 4d^2 + \boxed{8c}$

b. There are three terms. There are no like terms. $45x^2 + 7x - 9$

c. There are four terms. $\boxed{23g^2k} + \boxed{45g^2k^2} + \boxed{(-16g^2k)} + \boxed{57g^2k^2}$

PP for Example 2
a. $8p + 5p = 13p$

b. $2w + 7 - 9w + 3 = -7w + 10$

c. $3x^2y + 4xy - 7xy + 15x^2y = 18x^2y - 3xy$

d. $8z^2 + 5x - 2z - 20$

PP for Example 3
a. $-3d + 9c = 9c - 3d$

b. $16 - 2x + 3x^2 = 3x^2 - 2x + 16$

c. $7 - 4b^2 - 2a + 3a^2 = 3a^2 - 2a - 4b^2 + 7$

PP for Example 4
a. $125h$ b. $23h$

c. $125h + 23h = 148h$ d. $2b$ e. No

PP for Example 6
a. Multiplication property; $-9(4m^3) = -36m^3$

b. Distributive property; $7(-4x + 6y) = -28x + 42y$

c. Distributive property; $-\dfrac{4}{5}(25c^2 + 5c - 10) = -20c^2 - 4c + 8$

d. Distributive property; $0.4(3x - 6) = 1.2x - 2.4$

PP for Example 7
a. $5(4x - 9) + 8x + 20 = 28x - 25$

b. $1.5(2w + 7) + 3.5(4w - 6) = 17w - 10.5$

PP for Example 8
a. $(4x^2 + 5x) - (x^2 - 3x) = 3x^2 + 8x$

b. $4(8t + 9u) - 3(6t - 5u) = 14t + 51u$

Section 1.4

PP for Example 1
a. The maximum number of science and engineering bachelor's degrees was 589,330. They were awarded in 2012.

b. The minimum number of science and engineering bachelor's degrees was 496,168. They were awarded in 2008.

PP for Example 2
a. In 2010, approximately 4% of New Yorkers worked in manufacturing.

b. In 1960, the highest percentage of New Yorkers worked in manufacturing.

PP for Example 4

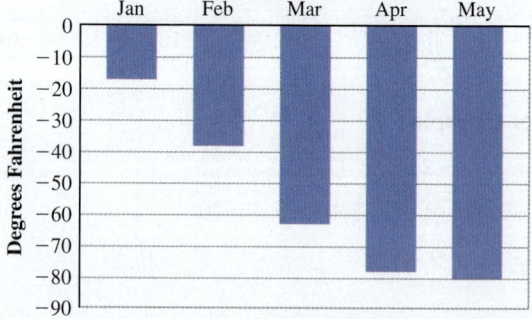

Average High Temperature at Vostok Research Station in Antarctica

PP for Example 5

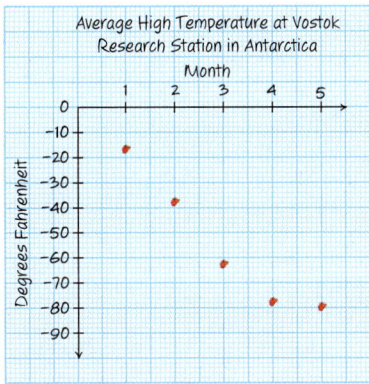

PP for Example 6

Point	Ordered Pair
A	(−3, 5)
B	(−5, −1)
C	(3, −4)
D	(5, 4)
E	(4, 0)
F	(0, 2)
G	(−4, 0)
H	(0, 0)

PP for Example 7

The x-axis has scale 10, and the y-axis has scale 5.

PP for Example 8

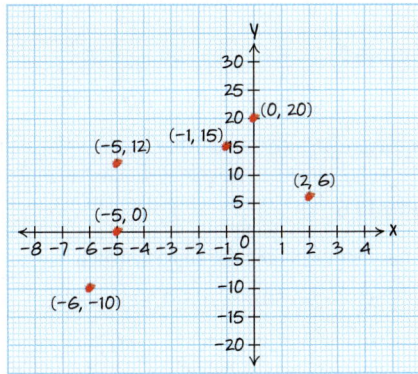

PP for Example 10

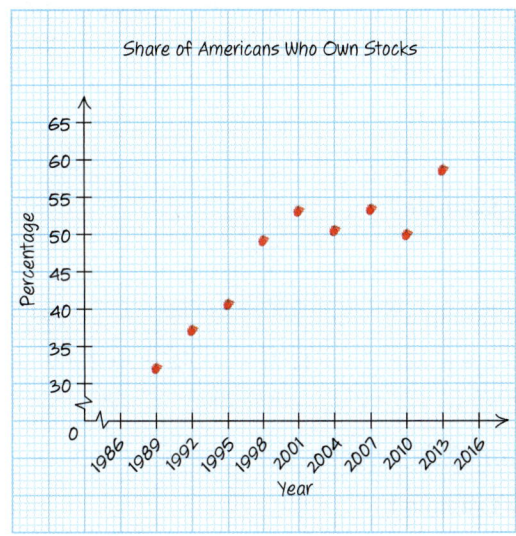

CHAPTER 2

Section 2.1

PP for Example 1

a. Equation **b.** Expression

PP for Example 2

a. $P = 25n$ Substitute in $P = 4000$ and $n = 120$.

$4000 \stackrel{?}{=} 25(120)$

$4000 \neq 3000$

Because the final results are not equal, $n = 120$ and $P = 4000$ do not give a solution to the equation $P = 25n$. These values state that if 120 batteries are sold, then the profit is $4000. These values do not make the equation true, so they do not make sense in this situation.

b. This is a solution to the equation.

c. This is not a solution of the equation.

PP for Example 3

a. If Jennifer takes 11 general education credits this semester, she should take 4 elective credits.

b. If Jennifer takes 2.5 credits of electives this semester, she should take 12.5 credits of general education.

PP for Example 4

a. $r = 15$ **b.** $p = 32.6$

PP for Example 5

$$P = R - C$$

a. $P = 42{,}500$. If the company's monthly revenue is $65,000 and its monthly cost is $22,500, its monthly profit is $42,500.

b. $R = 210,000$. If the company's monthly profit is $135,000 and its monthly cost is $75,000, its monthly revenue is $210,000.

PP for Example 6

$$x = \frac{1}{6}$$

PP for Example 7

a. $m = 13$ **b.** $3 = c$

PP for Example 8

a. $W = e + f$ for e
$e = W - f$

b. $M + N - P = 200$ for N
$N = 200 - M + P$

Section 2.2

PP for Example 1

$$P = 6s$$

a. $P = 120$. Ann gets paid $120 for 20 subscriptions.

b. $40 = s$. For Ann to make $240, she must sell 40 subscriptions.

PP for Example 2

a. $d = 13$ **b.** $m = 6.5$ **c.** $n = -8.42$

PP for Example 3

a. $n = -\frac{3}{2}$ **b.** $x = -12$

PP for Example 4

a. $M = 61145.8$. In the year 2018 ($t = 18$), there were 61,145.8 million passenger miles of travel on transit in the United States.

b. In the year 2014 ($t \approx 14.6$), there were about 58,000 million passenger miles of travel on transit in the United States.

PP for Example 5

a. $b = 6$ **b.** $75 = x$ **c.** $x = 17.5$

PP for Example 6

a. Half a number minus 13 is equal to 7.

$$\frac{1}{2}n - 13 = 7$$
$$n = 40$$

b. The product of 3 and a number plus 8 is 23.

$$3n + 8 = 23$$
$$n = 5$$

PP for Example 7

Let t represent time in years and let V represent the value of the backhoe in dollars. $V = 61000 - 5000t$.

When the backhoe is 5 years old ($t = 5$), it will be worth $36,000.

PP for Example 8

a. $\frac{V}{lh} = w$ **b.** $\frac{2A}{(b_1 + b_2)} = h$

PP for Example 9

a. $y = \frac{1}{3}x - 3$ **b.** $-\frac{y}{8} + \frac{3}{8} = x$

Section 2.3

PP for Example 1

a. $x = -20$ **b.** $x = \frac{39}{19}$ **c.** $-29 = k$

PP for Example 2

a. $x = -24$ **b.** $c = \frac{-45}{2} = -22.5$

PP for Example 3

a. No solution   **b.** $x = 0$ **c.** All real numbers

PP for Example 4

a. $\frac{2}{3}$ of a number plus 8 equals -2 times the number plus 40.

$$\frac{2}{3}x + 8 = -2x + 40$$
$$x = 12$$

b. A number plus twice the difference between the number and 10 is equal to 5 times the number.

$$n + 2(n - 10) = 5n$$
$$n = -10$$

PP for Example 5

$$x + (x + 25) + (x + 20) = 180$$

The angles measure 45°, 65°, and 70°.

PP for Example 6

Let x be the number of people the Clarks invite.

$$5x + 2(0.50)x + 1.5(0.20)x = 500$$
$$x \approx 79$$

If the Clarks spend their entire budget, they can invite 79 people.

Section 2.4

PP for Example 2

a. $x < -4$ **b.** $x \geq -13$

PP for Example 4

a. $3 + 5c \leq 58$

b. $c \leq 11$. Marcella can take at most 12 Zumba classes, the first class for $3 and 11 classes for $5 each, to stay within her budget of $58.00.

PP for Example 5

a. $x \geq -7$

b. $x > -8$

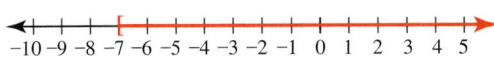

PP for Example 6

a. $(-\infty, 7]$ $x \leq 7$

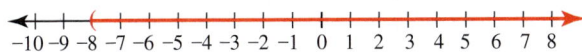

b. $(-6, \infty)$ $x > -6$

PP for Example 7

a. $40 \leq h \leq 47$ **b.** $0 \leq n \leq 2265$

PP for Example 8

a. $2 < x < 5$, $(2, 5)$

b. $-16 \leq x < 0$

CHAPTER 3
Section 3.1
PP for Example 1

r = riders	$C = 50r + 4000$
0	4000
50	6500
100	9000
150	11,500
200	14,000
225	15,250
250	16,500

PP for Example 2

x	$y = -3x + 6$	(x, y)
-2	12	$(-2, 12)$
-1	9	$(-1, 9)$
0	6	$(0, 6)$
1	3	$(1, 3)$
2	0	$(2, 0)$

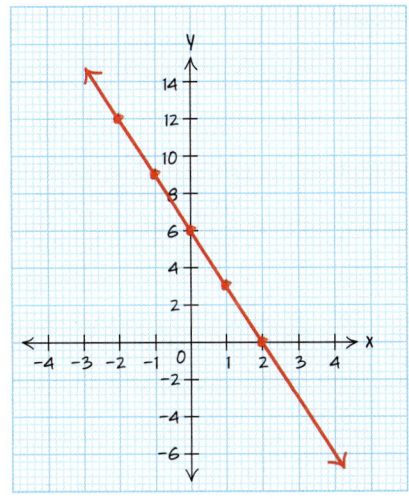

PP for Example 3

x	$y = 3x - 6$	(x, y)
-2	-12	$(-2, -12)$
-1	-9	$(-1, -9)$
0	-6	$(0, -6)$
1	-3	$(1, -3)$
2	0	$(2, 0)$
3	3	$(3, 3)$

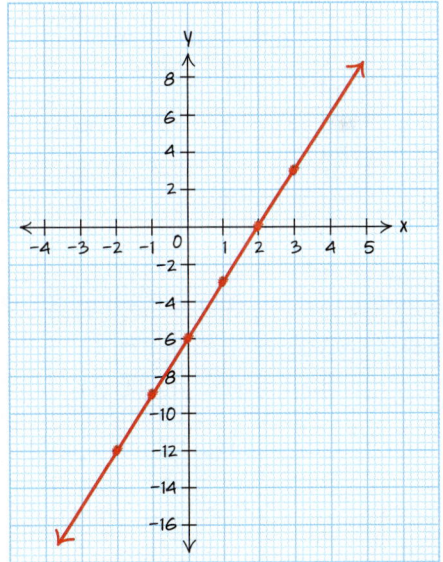

PP for Example 4

a.

n	W = 210 − 5n	(x, y)
0	210	(0, 210)
1	205	(1, 205)
2	200	(2, 200)
4	190	(4, 190)
6	180	(6, 180)
8	170	(8, 170)

b.

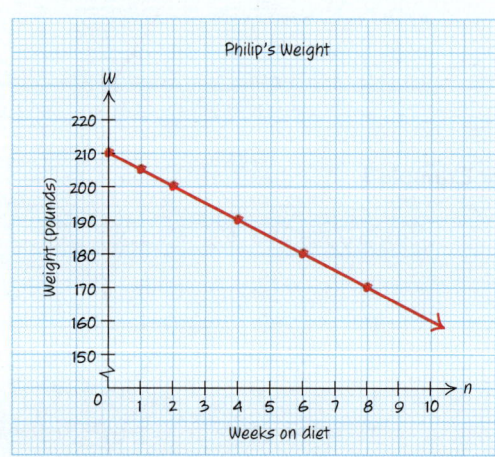

PP for Example 5

a.

x	y = x² − 5	(x, y)
−4	11	(−4, 11)
−3	4	(−3, 4)
−2	−1	(−2, −1)
−1	−4	(−1, −4)
0	−5	(0, −5)
1	−4	(1, −4)
2	−1	(2, −1)
3	4	(3, 4)
4	11	(4, 11)

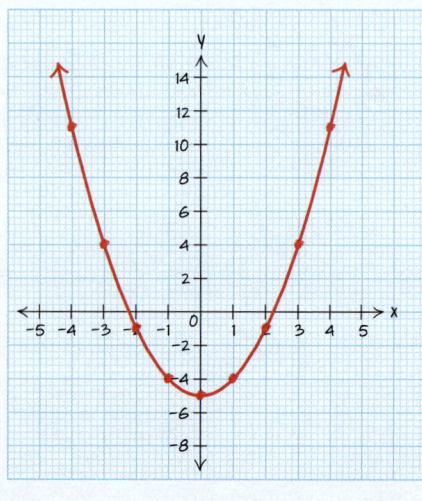

b.

x	y = |x + 4|	(x, y)
−7	3	(−7, 3)
−6	2	(−6, 2)
−5	1	(−5, 1)
−4	0	(−4, 0)
−3	1	(−3, 1)
−2	2	(−2, 2)
−1	3	(−1, 3)
0	4	(0, 4)
1	5	(1, 5)

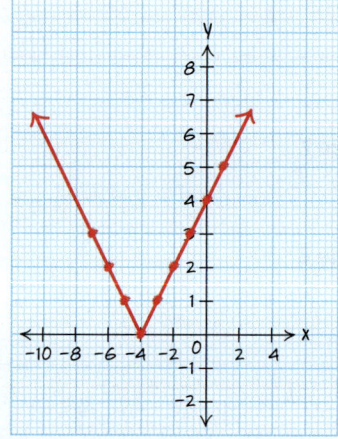

Section 3.2

PP for Example 1

a. Austria has a higher share of renewable energy in the electricity mix, which we can see because the line for Austria is higher than the line for Denmark.

b. The line for Denmark is steeper, so it is showing a faster rate of growth.

PP for Example 2

$$\text{slope} = \frac{\text{rise}}{\text{run}} = \frac{2}{4} = \frac{1}{2}$$

PP for Example 3

$$\text{slope} = \frac{\text{rise}}{\text{run}} = -\frac{1}{3}$$

PP for Example 4

$$\text{slope} = \frac{9 - (-5)}{7 - 2} = \frac{14}{5}$$

PP for Example 5

a. $\text{slope} = \dfrac{23.5 - 24}{2 - 0} = \dfrac{-0.5}{2} = -\dfrac{1}{4}$

b. $\text{slope} = \dfrac{0 - (-3)}{7 - 7} = \dfrac{3}{0}$

The slope is undefined.

PP for Example 6

$$\text{slope} = \frac{0 - 20}{4 - 0} = -\frac{20}{4} = -\frac{5}{1}$$

The slope means that 5000 fewer tickets are available every hour.

Section 3.3

PP for Example 1

a. The x-intercept is $(-1.5, 0)$, and the y-intercept is $(0, 3)$.

b. The x-intercepts are $(-2.5, 0)$, $(0, 0)$, and $(5, 0)$. The y-intercept is $(0, 0)$.

PP for Example 2

a. The vertical intercept is about the point $(0, 1)$. It means that in the year 2010, the income for Starbucks was about $1 billion.

b. The vertical intercept is the point $(0, 400)$. It means at the start of summer, there is 400 feet of water in the town's reservoir. The horizontal intercept is the point $(15, 0)$. It means that after 15 weeks of summer, there is no water left in the town's reservoir.

PP for Example 3

a. The x-intercept is the point $(3, 0)$.

The y-intercept is the point $(0, 21)$.

b. The x-intercept is the point $(14, 0)$.

The y-intercept is the point $(0, -8)$.

PP for Example 4

a. The D-intercept is the point $(0, 60)$. It means that after 0 hours of walking, Deb and Christine have 60 miles to walk.

b. The h-intercept is the point $(20, 0)$. It means that after 20 hours of walking, Deb and Christine will have 0 miles to walk.

c. Yes

PP for Example 5

a. The slope is $m = -\frac{2}{3}$. The x-intercept is the point $(6, 0)$. The y-intercept is the point $(0, 4)$.

b. The slope is $m = \frac{1}{4}$. The x-intercept is the point $\left(\frac{8}{5}, 0\right)$. The y-intercept is the point $\left(0, -\frac{2}{5}\right)$.

PP for Example 6

a. When $f = 0$, $T = 2$. This means that it takes 2 hours to produce 0 bike frames. The 2 hours may be the production setup time.

b. The slope is $m = 0.5 = \frac{0.5 \text{ hour}}{1 \text{ bike frame}}$. This means that every half hour, a bike frame is produced. This could also be interpreted as $m = \frac{1}{2} = \frac{1 \text{ hour}}{2 \text{ bike frames}}$, which means that each hour, 2 bike frames are produced.

PP for Example 7

a. $y = 4x + \frac{5}{2}$

$m = 4$

The y-intercept is $\left(0, \frac{5}{2}\right)$.

b. $y = \frac{2}{3}x + 12$

$m = \frac{2}{3}$

The y-intercept is $(0, 12)$.

c. $y = -\frac{4}{3}x + \frac{1}{3}$

$m = -\frac{4}{3}$

The y-intercept is $\left(0, \frac{1}{3}\right)$.

Section 3.4

PP for Example 1

a. $y = 3x - 4$

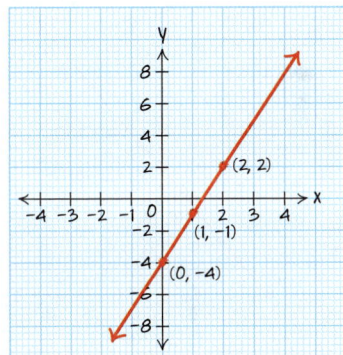

b. $y = -\frac{3}{4}x - 3$

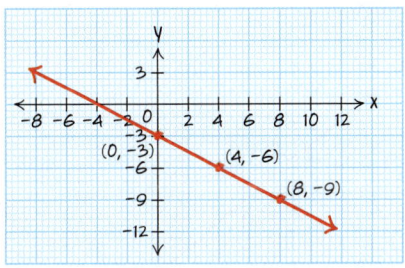

A-10 APPENDIX A Answers to Practice Problems

PP for Example 2

$$y - 8 = \frac{2}{3}(x - 6)$$

$$y = \frac{2}{3}x + 4$$

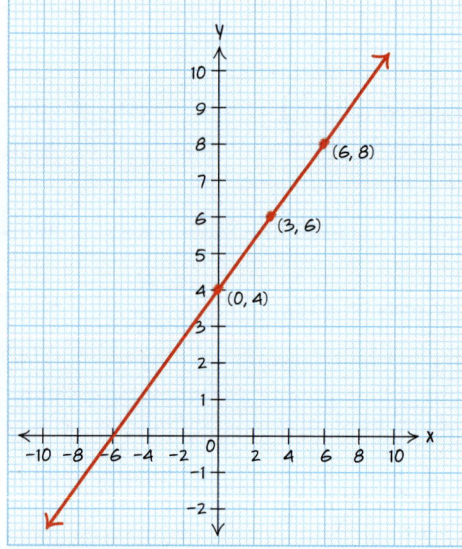

PP for Example 3

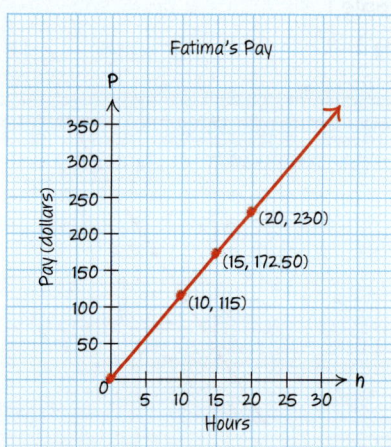

PP for Example 4

a.

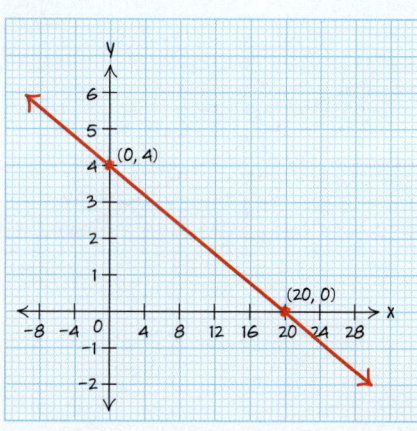

b.

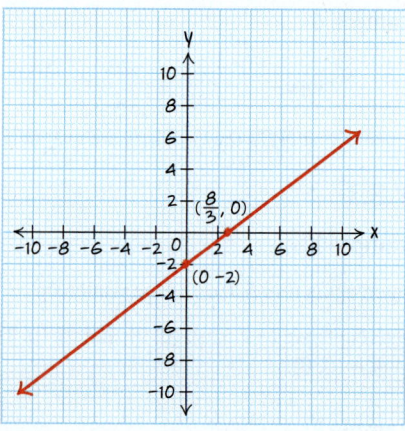

PP for Example 5

a.

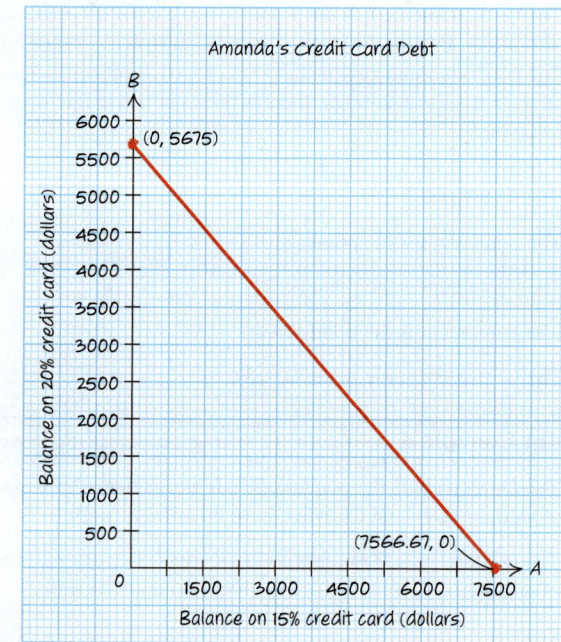

b. If Amanda owes $4500.00 on the credit card that charges 15% interest, then she owes $2300.00 on the credit card that charges 20% interest.

PP for Example 7

a. Linear; $y = 4x - 7$ b. Not a linear equation

PP for Example 8

a. The two lines are parallel.   b. The two lines are not parallel.

PP for Example 9

a. The two lines are not perpendicular.

b. The two lines are perpendicular.

Section 3.5

PP for Example 1

$$y = 2x - 4$$

APPENDIX A Answers to Practice Problems A-11

PP for Example 2

a. $y = -\dfrac{7}{4}x + \dfrac{19}{4}$

b. $y = \dfrac{3}{4}x - \dfrac{7}{2}$

PP for Example 3

$$d = \dfrac{1}{24}h$$

PP for Example 5

a. $d = \dfrac{1}{5}s$ or $d = \dfrac{s}{5}$

b. $N = -w + 195$ or $N = 195 - w$

PP for Example 6

a. The point-slope form is $y - 5 = -\dfrac{1}{2}(x - (-2))$.

The slope-intercept form is $y = -\dfrac{1}{2}x + 4$.

b. The point-slope form is $y - (-1) = \dfrac{1}{6}(x - 6)$.

The slope-intercept form is $y = \dfrac{1}{6}x - 2$.

PP for Example 7

$$y = -\dfrac{1}{2}x - 2$$

PP for Example 8

$$y = -\dfrac{7}{3}x + 13$$

Section 3.6

PP for Example 2

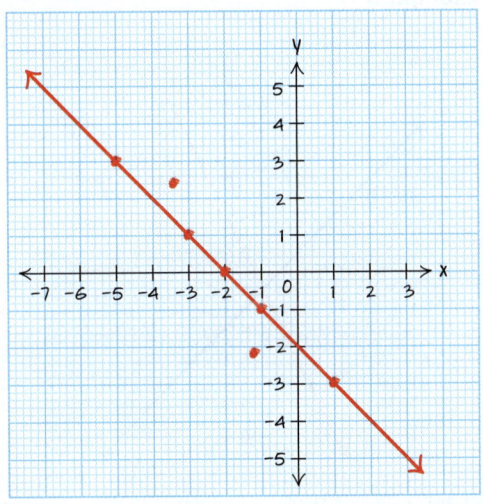

PP for Example 3

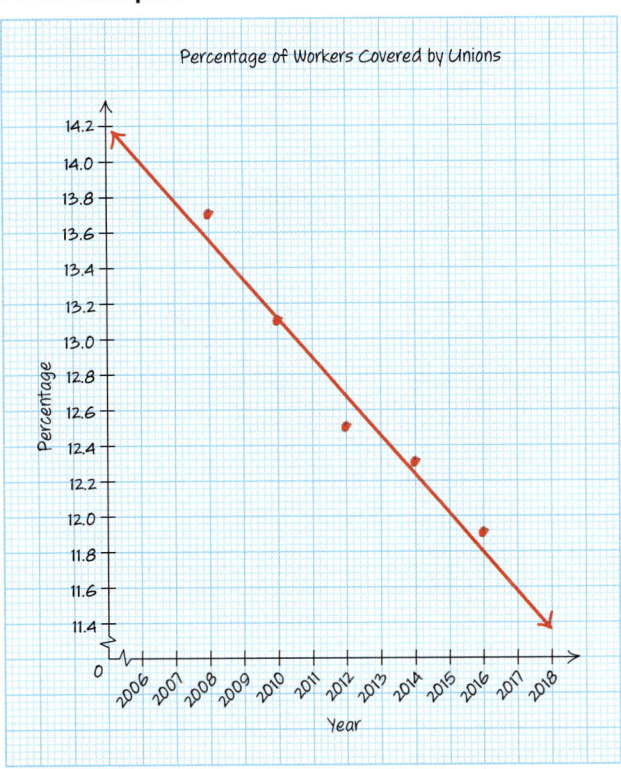

PP for Example 4

a. An equation of a best-fit line is $y = 0.4x - 7$.

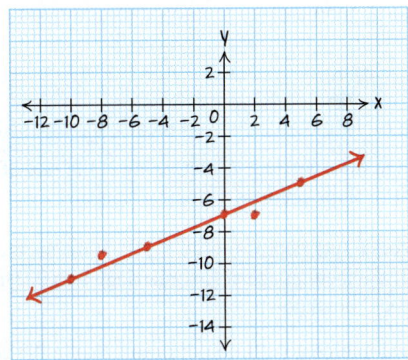

b. An equation of a best-fit line is $y = -\dfrac{1}{3}x + 1$.

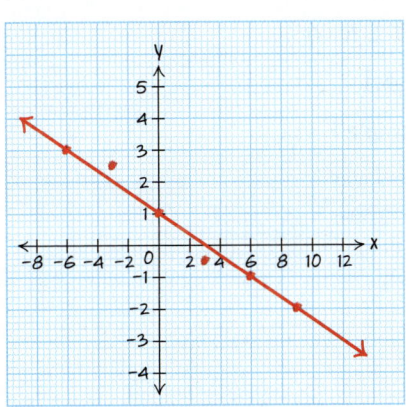

A-12 APPENDIX A Answers to Practice Problems

PP for Example 5

a. Let t represent years since 1950 (the input value) and let M represent the median age at first marriage for women (the output value). This means that the ordered pairs will have the form (t, M).

b.
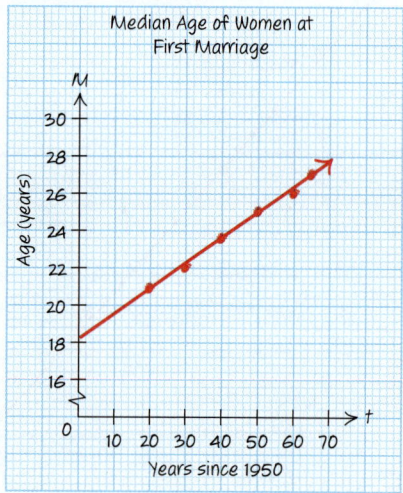

c. The equation of the model is $M = 0.14t + 18$.

PP for Example 7

a. The model predicts that the approximate age at first marriage for women in 1985 was 22.9 years.

b. The model forecasts that the approximate age at first marriage for women in 2020 will be 27.8 years.

PP for Example 9

a. The slope of the model is $\dfrac{0.14}{1}$. It means that the median age of women at first marriage is increasing at the rate of 0.14 year per year. The slope predicts a positive trend.

b. The horizontal intercept is $(-128.57, 0)$. It means that in the year 1821, the median age at first marriage was 0. This is a case of model breakdown.

c. The vertical intercept is $(0, 18)$. It means that in the year 1950, the median age at first marriage for women was 18 years. This is possibly a case of model breakdown.

CHAPTER 4

Section 4.1

PP for Example 1

a. $S = 55 + 2.30m$
$I = 50 + 2.50m$

b.

m = number of miles over 15 miles	S = cost of shipping for Ship Quick ($)
0	55.00
5	66.50
10	78.00
15	89.50
20	101.00
25	112.50
30	124.00

m = number of miles over 15 miles	I = cost of shipping for In Town Shipping ($)
0	50.00
5	62.50
10	75.00
15	87.50
20	100.00
25	112.50
30	125.00

c. When $m = 25$ miles over 15 miles, or 40 miles, the shipping costs are the same.

d. She should use In Town Shipping.

PP for Example 2

a. Yes   b. Yes   c. No

PP for Example 3

a. About 2009; the debt was approximately $0.8 trillion.

b. About 2005; the debt was approximately $0.4 trillion.

PP for Example 4

The solution is $(2, -3)$.

PP for Example 5

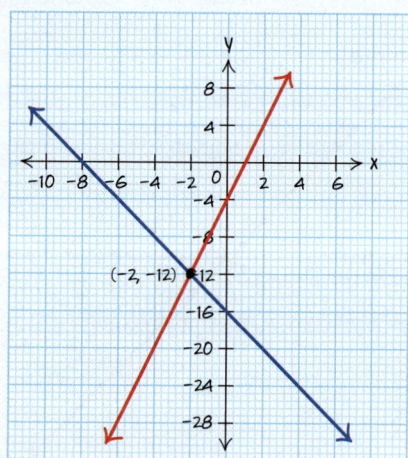

The solution is $(-2, -12)$.

PP for Example 7

a. The equation for Help-U-Stor is $C = 160m + 10$. The equation for Local Storage Solutions is $C = 150m + 30$.

b.
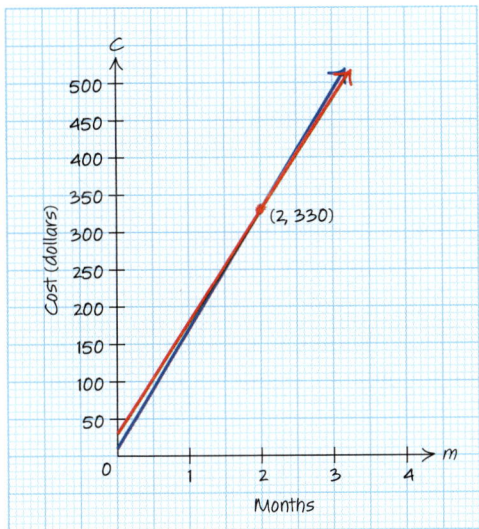

c. The solution is the point $(m, C) = (2, 330)$.

d. You should use Local Storage Solutions.

Section 4.2

PP for Example 1

a. The solution is the point $(x, y) = (-11, -9)$.

b. The solution is the point $(x, y) = \left(13, \dfrac{9}{2}\right)$.

PP for Example 2

The solution is the point $(m, C) = (2, 330)$. Two months of storage costs $330 at both Help-U-Stor and Local Storage Solutions.

PP for Example 3

a. The lines are dependent. There are infinitely many solutions to the system.

b. There are no solutions to the system.

PP for Example 4

a. The speed of the plane with a tailwind is 505 mph.

b. The speed of the ship traveling upstream is 11 mph.

PP for Example 5

The rate of the boat is $\dfrac{35}{12} \approx 2.92$ mph, and the rate of the current is $\dfrac{5}{12} \approx 0.42$ mph.

PP for Example 6

The two integers are -24 and -78.

PP for Example 7

a. The value of the walnuts is $112.35. The value of the hazelnuts is $169.80.

b. The unit price of the mix is $8.06.

PP for Example 8

They should include approximately 31.52 pounds of hazelnuts and approximately 18.48 pounds of almonds.

Section 4.3

PP for Example 1

The solution is the point $(x, y) = (4, -1)$.

PP for Example 2

a. The solution is the point $(x, y) = (-5, 3)$.

b. There is no solution to this system.

PP for Example 3

The solution is the point $(x, y) = (4, 5)$.

PP for Example 4

Each food coloring kit costs $9.00, and each cake decorating kit costs $24.00.

PP for Example 5

The simple interest earned is $6000.00.

PP for Example 6

Russ should invest $35,833.33 in the stock fund and $9166.67 in the bond fund.

PP for Example 8

The farmer should use 166.67 gallons of the 4% milk and 333.33 gallons of the 1% milk in the mixture.

PP for Example 9

The measure of the first angle is 18°, and the measure of the second angle is 72°.

Section 4.4

PP for Example 2

a. The point is in the solution set.

b. The point is in the solution set.

c. The point is not in the solution set.

d. The point is not in the solution set.

PP for Example 4

a.
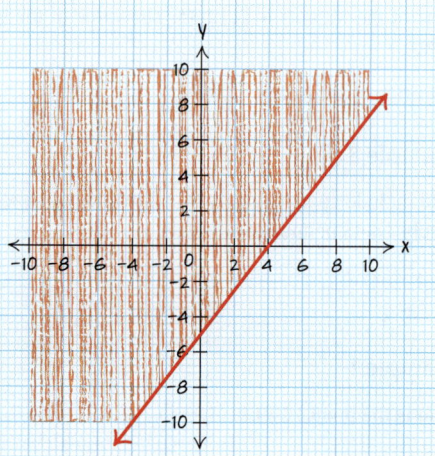

b.
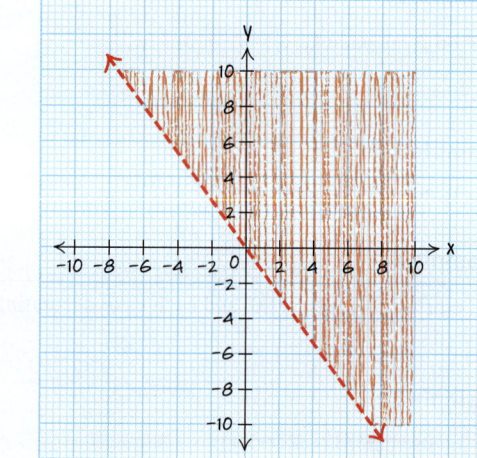

PP for Example 6

Let f represent the amount in pounds of the first type of candy Kelli purchases and let s represent the amount in pounds of the second type of candy she purchases. The inequality is $2.25f + 4.25s \leq 20$.

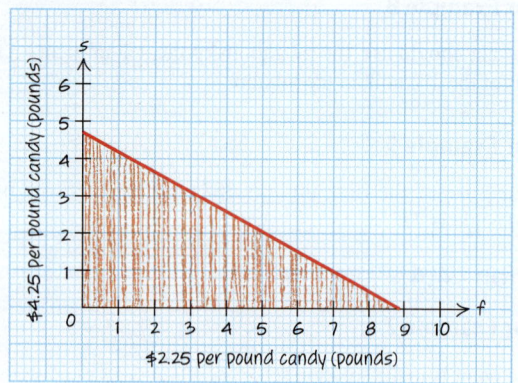

PP for Example 7

a.
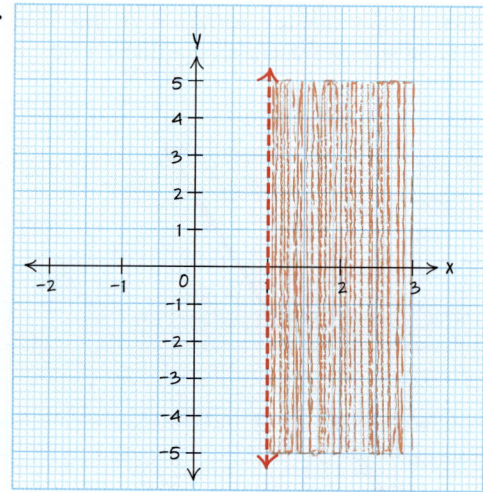

b.
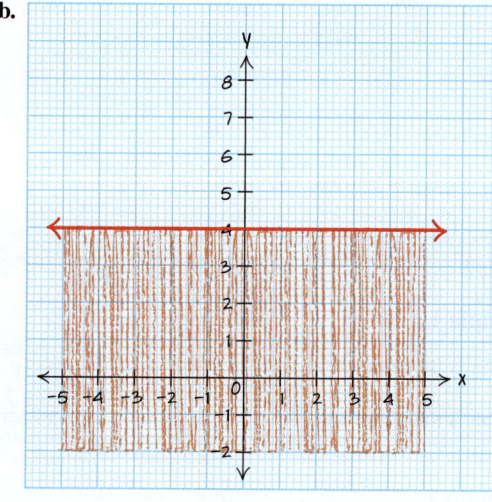

Section 4.5

PP for Example 2

Let b represent the number of bathroom cabinets made per week and let k represent the number of kitchen cabinets made per week.

a. The two inequalities are
$$b + k \geq 60$$
$$3b + 8k \leq 500$$

b.
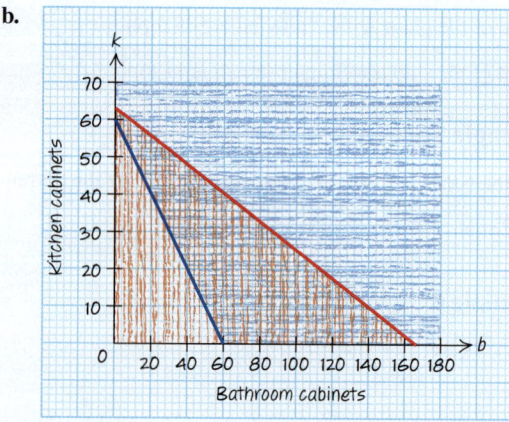

c. A test point is $(b, k) = (62, 5)$.
$$62 + 5 = 67 \geq 60$$
$$3(62) + 8(5) = 226 \leq 500$$

PP for Example 3

a.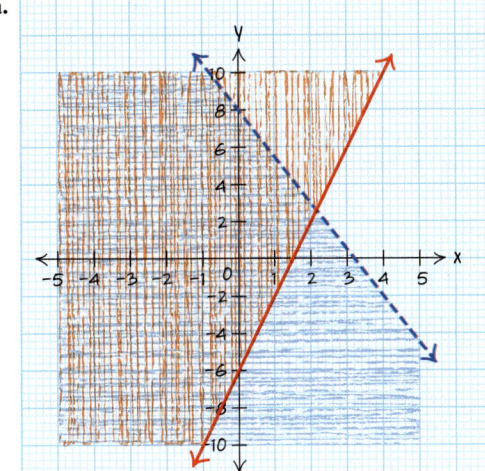

b.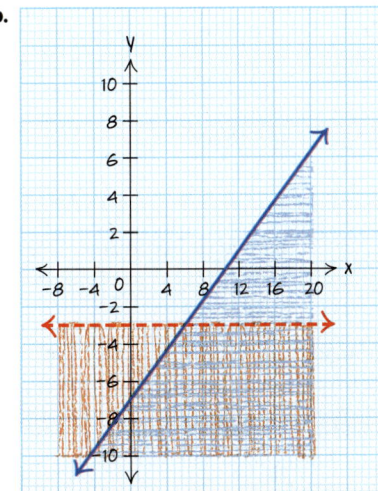

PP for Example 4

a. Let N represent the number of Nutter Bars sold, and let P represent the number of Peanut Bars sold. The system is
$$N + P \leq 1000$$
$$1.50N + 2.50P \geq 2000$$

b.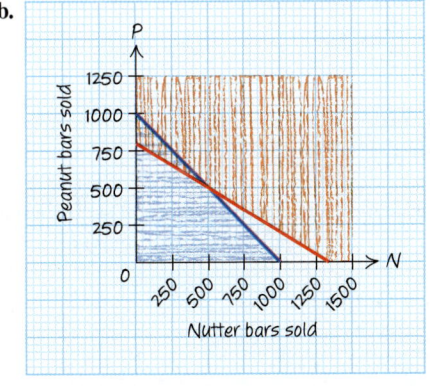

c. A test point is $(N, P) = (1, 900)$.
$$1 + 900 = 901 \leq 1000$$
$$1.5(1) + 2.5(900) = 2251.50 \geq 2000$$

CHAPTER 5

Section 5.1

PP for Example 1

a. $2^5 = 32$
b. $(-2)^4 = 16$
c. $-2^4 = -16$
d. $(-a)^3 = -a^3$

PP for Example 2

a. $x^5 x^6 = x^{11}$
b. $t^3 t = t^4$
c. $7^4 7^5 = 7^9 = 40{,}353{,}607$

PP for Example 3

a. $(5x^6 y^9)(6x^5 y^2) = 30 x^{11} y^{11}$
b. $(-9a^4 b^3)(4a^2 b^6) = -36 a^6 b^9$

PP for Example 4

a. $\dfrac{x^9}{x^2} = x^7$
b. $\dfrac{4^9}{4^3} = 4^6 = 4096$

PP for Example 5

a. $\dfrac{a^3 a^5 b^7}{a^2 a^4 b^2} = a^2 b^5$
b. $\dfrac{x^5 y^6}{x^2 y^4} \cdot \dfrac{x^8 y^4}{x y^3} = x^{10} y^3$

PP for Example 6

a. $(c^4)^6 = c^{24}$
b. $(8^3)^2 = 8^6 = 262{,}144$

PP for Example 7

a. $(-5a)^4 = 625 a^4$
b. $(4ac^4)^3 = 64 a^3 c^{12}$
c. $\left(\dfrac{a^2}{b^5}\right)^2 = \dfrac{a^4}{b^{10}}$

PP for Example 8

a. $\left(-\dfrac{3m^7 n^5}{m^2 n^3}\right)^4 = 81 m^{20} n^8$
b. $\dfrac{(5x^3)^2}{(2x^2 y^4)^3} = \dfrac{25}{8y^{12}}$
c. $(-2a^5 b)^3 (6a^3 b^5)^2 = -288 a^{21} b^{13}$
d. $(4m)^5 (4m)^8 = (4m)^{13}$

Section 5.2

PP for Example 1

a. $b^{-6} = \dfrac{1}{b^6}$
b. $7h^{-2} = \dfrac{7}{h^2}$
c. $4x^{-3} = \dfrac{4}{x^3}$

PP for Example 2

a. $\dfrac{35 x^4}{14 x^{10}} = \dfrac{5}{2 x^6}$
b. $-12 x^{-4} y^3 = -\dfrac{12 y^3}{x^4}$
c. $\dfrac{5 a^{-5} b^4}{7 a^{-2} b^{-1}} = \dfrac{5 b^5}{7 a^3}$

PP for Example 3

a. $72{,}000{,}000{,}000 = 7.2 \times 10^{10}$; Mark Zuckerberg's Facebook stock was approximately worth $\$7.2 \times 10^{10}$.

b. $0.00000002 = 2 \times 10^{-8}$ second

PP for Example 4

a. $4(1.25 \times 10^8) = 5 \times 10^8$

b. $25(3.4 \times 10^{13}) = 8.5 \times 10^{14}$

c. $\dfrac{7.4 \times 10^8}{1480} = 5 \times 10^5$

PP for Example 5

The average distance of Uranus from the Sun is 2.870824×10^9 km.

PP for Example 6

a. $(8.4 \times 10^6)(7.0 \times 10^8) = 5.88 \times 10^{15}$

b. $(1.5 \times 10^{-9})(4.0 \times 10^{-17}) = 6 \times 10^{-26}$

c. $\dfrac{6.4 \times 10^{15}}{2.0 \times 10^4} = 3.2 \times 10^{11}$

PP for Example 7

The mass of Jupiter is about 317.39 times the mass of the Earth.

Section 5.3

PP for Example 1

a. Binomial. The terms are $5x^2$ and $7x$.

b. Trinomial. The terms are $4a^3b^2$, $-2ab$, and b.

PP for Example 2

a. The degree is 2. $\quad -x^2 + 4$

b. The degree is 3. $\quad 3x^3 + x^2 + x - 4$

c. The degree is 1. $\quad -7x + 5$

PP for Example 3

a. Working from left to right, the degree of the first term is 5. The degree of the second term is 4. The degree of the third term is 2. The degree of the fourth term is 0. The degree of the polynomial is 5.

b. Working from left to right, the degree of the first term is 6. The degree of the second term is 9. The degree of the third term is 4. The degree of the polynomial is 9.

PP for Example 4

a. Linear equation in two variables

b. Linear expression in one variable

c. Nonlinear equation in two variables

PP for Example 5

a. $6x^2 - 6x + 11$ $\qquad$ **b.** $5a^2 + 8ab + 7b^2$

PP for Example 6

a. $9x^2 + 2x + 10$ $\qquad$ **b.** $3y^2 - 3y + 4$

PP for Example 7

a. $9c + 25$

b. It will cost $187 to purchase art projects without the winter projects for the 18 children in her class.

Section 5.4

PP for Example 1

a. $6x(4x - 8) = 24x^2 - 48x$

b. $4a(5a^2 - 6a + 8) = 20a^3 - 24a^2 + 32a$

PP for Example 2

a. $20x^2 - 17x^2 - 53x + 14$

b. $12x^6y^2 + 8x^5y^3 - 20x^4y^4 - 15x^4y - 10x^3y^2 + 25x^2y^3$

PP for Example 3

$$-24x^3 + 4x^2y + 160xy^2$$

PP for Example 4

a. $30x^2 + 8x - 6$ $\qquad$ **b.** $12x^2 - 9xy - 30y^2$

PP for Example 6

a. $(2.16t + 127.98)(2.32t + 309.32) = 5.0112t^2 + 965.045t + 39586.8$

b. Substitute in $t = 10$ to find 49,738.3 million pounds of fresh fruit will be consumed by Americans in the year 2020.

PP for Example 7

$$(2x + 3)(7x - 4) = 14x^2 + 13x - 12$$

PP for Example 8

a. $(3x - 8)^2 = 9x^2 - 48x + 64$

b. $(3x^2 + 5y)^2 = 9x^4 + 30x^2y + 25y^2$

c. $\left(\dfrac{t}{4} - 5\right)^2 = \dfrac{t^2}{16} - \dfrac{5t}{2} + 25$

PP for Example 9

a. This is a product. $\quad (8t)^2 = 64t^2$

b. This is a sum. $\quad (8 + t)^2 = 64 + 16t + t^2$ $\quad$ or $\quad t^2 + 16t + 64$

PP for Example 10
a. $(2x + 1)(2x - 1) = 4x^2 - 1$
b. $(x + 5y)(x - 5y) = x^2 - 25y^2$

Section 5.5
PP for Example 1
a. $\dfrac{30x^2 + 4x}{5x} = 6x + \dfrac{4}{5}$
b. $\dfrac{24m^4 - 36m}{8m} = 3m^3 - \dfrac{9}{2}$

PP for Example 2
a. $\dfrac{40x^2 + 12x - 72}{4x} = 10x + 3 - \dfrac{18}{x}$
b. $\dfrac{60m^4n^3 - 51m^3n^2 + 30m^2n}{3m^2n} = 20m^2n^2 - 17mn + 10$

PP for Example 3
a. $(x^2 + 12x + 32) \div (x + 4) = x + 8$
b. $(20x^2 - 9x - 18) \div (4x + 3) = 5x - 6$

PP for Example 4
The length is $4x + 2$.

PP for Example 5
a. $8x + 5 - \dfrac{4}{x - 3}$
b. $3x^2 - 20x + 33 - \dfrac{62}{x + 3}$

PP for Example 6
a. $4x^2 - 8x + 13 - \dfrac{19}{x + 2}$
b. $x^2 + 3x + 9$

CHAPTER 6
Section 6.1
PP for Example 1
a. The GCF is 5.   b. The GCF is $4y$.   c. The GCF is $3xy$.

PP for Example 2
a. The GCF is $8x^2$.   b. The GCF is $25y^2$.

PP for Example 3
a. The GCF is $3xy$.   b. The GCF is $10pq$.

PP for Example 4
a. $5x - 35 = 5(x - 7)$
b. $2y^2(y - 3 + 6y^2) = 2y^2(6y^2 + y - 3)$
c. $6x^3 - 2x^2 + 2x = 2x(3x^2 - x + 1)$

PP for Example 5
a. $-x^2 - 5 = -1(x^2 + 5)$ b. $-y + 16 = -1(y - 16)$
c. $-16n + 48 = -16(n - 3)$

PP for Example 6
a. $a(2 - b) - 3(2 - b) = (2 - b)(a - 3)$
b. $x(x + 1) - 5(x + 1) = (x + 1)(x - 5)$

PP for Example 7
a. $(x + 7)(2x - 5)$ b. $(2a - b)(3a + b)$

PP for Example 8
a. $16y^2 - 48y = 16y(y - 3)$
b. $3x^2 + 15x - 3x - 15 = 3(x - 1)(x + 5)$

Section 6.2
PP for Example 1
a. $b = -9$ and $c = -12$
b. $b = -3$ and $c = -5$

PP for Example 2
a. $x^2 + 11x + 24 = (x + 3)(x + 8)$
b. $-35 + 2x + x^2 = (x + 7)(x - 5)$
c. $x^2 - 13x + 36 = (x - 4)(x - 9)$

PP for Example 3
a. $a = -3, b = 0, c = -16$ b. $a = -1, b = -4, c = 3$
c. $a = -4, b = 8, c = 16$

PP for Example 4
a. $3x^2 + 23x + 14 = (3x + 2)(x + 7)$
b. $2x^2 + 5x - 42 = (2x - 7)(x + 6)$
c. $3x^2 - 11x + 10 = (3x - 5)(x - 2)$

PP for Example 6
a. $2(3m - n)(2m + n)$ b. $-3(t + 7)(t - 1)$
c. $2(x^2y - x - 5y)$

Section 6.3
PP for Example 2
a. This is a difference of squares.
$p^2 - 100 = (p - 10)(p + 10)$
b. This is a difference of squares.
$9x^2 - 64 = (3x - 8)(3x + 8)$

PP for Example 3

a. This is a perfect square trinomial.

b. This is a perfect square trinomial.

c. This is not a perfect square trinomial.

PP for Example 4

a. $x^2 + 16x + 64 = (x + 8)^2$ b. $9x^2 - 6x + 1 = (3x - 1)^2$

PP for Example 5

a. $5(q - 4)(3p + 2)$ b. $3(x - 12)(x + 12)$

c. $-3a(2a - 1)(a + 4)$

Section 6.4

PP for Example 1

a. Quadratic equation   b. Linear equation

PP for Example 2

a. $x = 6$ b. $y = 0, y = 1$ c. $y = \dfrac{1}{2}$

PP for Example 3

a. $d = -\dfrac{1}{3}, d = -\dfrac{3}{2}$ b. $t = \pm 12$

c. $n = 6, n = -\dfrac{1}{2}$

PP for Example 4

a. JT's Fishing Supplies earns $5000 profit when they sell 400 fishing kits.

b. JT's Fishing Supplies must sell 500 fishing kits to earn $6500 profit.

PP for Example 5

a. $x = \dfrac{1}{4}, x = -\dfrac{2}{5}$ b. $x = 0, x = 5, x = -7$

CHAPTER 7

Section 7.1

PP for Example 1

a. The cost is $40.00 per student when 20 students go on the trip.

b. The cost is $20.00 per student when 60 students go on the trip.

PP for Example 2

a. $\dfrac{12}{-4} = -3$ b. $\dfrac{2}{126} = \dfrac{1}{63}$

c. $\dfrac{52}{0}$. This is undefined.

PP for Example 3

a. The excluded value is $x = -3$.

b. The excluded values are $x = 5, x = -2$.

c. There are no excluded values.

PP for Example 4

a. $\dfrac{a - 4}{a + 6}$ b. $\dfrac{1}{x - 5}$

PP for Example 5

a. $-\dfrac{x + 5}{x + 10}$ b. $\dfrac{5}{y + 4}$

Section 7.2

PP for Example 1

a. $\dfrac{14}{13}$ b. $\dfrac{x - 6}{x - 7}$

PP for Example 2

a. $-\dfrac{a + 8}{(a + 9)(a - 3)}$ b. $\dfrac{(x - 3)(x + 6)}{(x - 9)(x + 3)}$

PP for Example 3

The cereal is priced at $0.19 per ounce.

PP for Example 4

The car's speed is 33.528 meters per second.

PP for Example 5

a. $\dfrac{24 \text{ meters}}{1 \text{ hour}} \cdot \dfrac{1 \text{ mile}}{1609.344 \text{ meters}} = \dfrac{0.01491 \text{ mile}}{1 \text{ hour}}$. The machine can lay asphalt at a rate of 0.01491 mph.

b. $\dfrac{0.01491 \text{ mile}}{1 \text{ hour}} \cdot 8 \text{ hours} = 0.1193 \text{ mile}$. In an 8-hour workday the crew can lay about 0.1193 miles of asphalt.

PP for Example 6

a. $\dfrac{9}{4}$ b. $\dfrac{5}{3y^2}$ c. $\dfrac{y - 3}{y - 7}$

PP for Example 7

a. $\dfrac{(b + 10)(b - 8)}{(b + 11)(b + 6)}$ b. $\dfrac{x + 2}{x + 5}$

PP for Example 8

$\dfrac{x + 6}{2(x + 4)}$

Section 7.3

PP for Example 1

a. $\dfrac{3a + 7}{5b^2}$ b. $\dfrac{4x + 5}{x - 3}$ c. $\dfrac{2}{m + 5}$

PP for Example 2

a. $\dfrac{7x - 5}{3y^2}$ b. 3 c. $\dfrac{4(x + 2)}{(x + 3)(x - 4)}$

PP for Example 3

The LCD is $600m^2n$.

PP for Example 4

a. The LCD is $(k + 1)(k - 12)(k - 3)$.

b. The LCD is $(t - 4)(t - 5)(t + 8)$.

PP for Example 5

a. The LCD is $150abc^3$.

$$\dfrac{24a^2c}{150abc^3} \text{ and } \dfrac{35}{150abc^3}$$

b. The LCD is $(x + 6)(x - 2)$.

$$\dfrac{8x(x - 2)}{(x + 6)(x - 2)} \text{ and } \dfrac{5(x + 6)}{(x - 2)(x + 6)}$$

PP for Example 6

The LCD is $(t + 3)^2(t - 4)$.

$$\dfrac{t - 5}{t^2 + 6t + 9} = \dfrac{(t - 5)(t - 4)}{(t + 3)^2(t - 4)}$$

$$\dfrac{6t}{t^2 - t - 12} = \dfrac{6t(t + 3)}{(t + 3)^2(t - 4)}$$

PP for Example 7

a. $\dfrac{3x^2 + 21x + 6}{x + 7}$ b. $\dfrac{5m^2 - 37m - 28}{(m + 4)(m - 6)}$

c. $\dfrac{a^2 + 8a + 18}{(a + 2)(a + 3)(a + 5)}$

PP for Example 8

a. $\dfrac{b + a}{ab}$

b. $\dfrac{15 + 10}{(10)(15)} = \dfrac{25}{150} = \dfrac{1}{6}$. Working together, Jackson and Noah can create about 1/6 of the budget per hour.

PP for Example 9

a. $-\dfrac{2z + 8}{z - 5}$ b. $\dfrac{n^2 - 6}{(n - 4)(n + 4)(n - 1)}$

PP for Example 11

a. $\dfrac{1}{12}$ b. $\dfrac{x}{x + 5}$

Section 7.4

PP for Example 1

a. It will cost $24.25 per person if 40 students go on the trip.

b. A total of 50 students will need to go on the trip for the cost to be $21.00 per person.

PP for Example 2

a. $t = 4$

b. $x = -3$. It does not check. There is no solution.

c. $h = -1$

PP for Example 3

a. $n = 1$ b. $x = 6$

PP for Example 4

a. The two solutions are $x = 4$, $x = -2$. Because $x = 4$ is an excluded value, the only solution is $x = -2$.

b. The two solutions are $h = 6$, $h = -2$. Because $h = 6$ is an excluded value, the only solution is $h = -2$.

PP for Example 5

Working together, the two painters will take 7.2 hours to paint the house.

Section 7.5

PP for Example 1

a. $\dfrac{9}{30} = \dfrac{3}{10}$ b. $\dfrac{1}{9}$

PP for Example 2

a. $\dfrac{6 \text{ pounds}}{1 \text{ month}}$

b. $\dfrac{\$22.00}{8 \text{ ounces}} = \$2.75$ per ounce

PP for Example 3

a. $\dfrac{\$3.69}{18 \text{ ounces}} \approx \$0.21$ per ounce

b. $\dfrac{\$3.99}{21 \text{ ounces}} = \$0.19$ per ounce

The 21-ounce package of General Mills Cheerios is the better buy.

PP for Example 5

Let $b = $ pounds of bananas

$$\dfrac{2 \text{ pounds}}{\$1.16} = \dfrac{7 \text{ pounds}}{b}$$

$$b = \$4.06$$

7 pounds of bananas will cost $4.06.

PP for Example 6

a. $x = 8$ b. $z = 4$

PP for Example 7

a. 149.86 centimeters   b. 2.08 ounces

APPENDIX A Answers to Practice Problems

PP for Example 8

The 7-foot-tall door opening will be shown as 1.75 inches on the blueprint.

PP for Example 9

$$x = 6$$

PP for Example 10

a. $W = 12.25h$

b. $W = 12.25(35) = \$428.75$. If Hamid works 35 hours in a week, his paycheck will be $428.75.

PP for Example 11

$$y = 3x \quad y = -6 \quad \text{when} \quad x = -2$$

PP for Example 12

a. $s = \dfrac{210}{t}$

b. $s = \dfrac{210}{2.5} = 84$. If it takes Mariam 2.5 hours to drive the same distance, she drove at an average speed of 84 miles per hour.

PP for Example 13

$$y = \dfrac{10}{x} \quad y = \dfrac{5}{9} \quad \text{when} \quad x = 18$$

CHAPTER 8

Section 8.1

PP for Example 1

a. $\sqrt{100} = 10$ b. $\sqrt{\dfrac{25}{9}} = \dfrac{5}{3}$ c. $\sqrt{0} = 0$

PP for Example 2

a. Not real   b. $-\sqrt{81} = -9$

PP for Example 3

a. $\sqrt{7} \approx 2.65$ b. $\sqrt{65} \approx 8.06$

PP for Example 4

a. $\sqrt{(2)^2 + 5} = \sqrt{9} = 3$

b. $\sqrt{3(-2) + 5} = \sqrt{-1}$. Not real.

PP for Example 5

a. The units represented by s are the speed of the car in miles per hour. The units represented by d are the length of the skid mark in feet.

b. When the skid mark was 175 feet the speed of the car was approximately 72.46 mph.

c. When the skid mark was 115 feet the speed of the car was approximately 58.74 mph.

PP for Example 6

a. The distance is 10.

b. The distance is approximately 13.04.

PP for Example 7

a. x^7 b. $7x^3$ c. $\dfrac{x}{4}$ d. a^2b^2

PP for Example 8

a. -5 b. 6 c. 2

Section 8.2

PP for Example 1

a. $5\sqrt{3}$ b. $6\sqrt{2}$ c. $\sqrt{30}$

PP for Example 3

a. $\sqrt{45} \approx 6.71$ b. $\sqrt{45} = 3\sqrt{5}$

PP for Example 4

a. $y^7\sqrt{y}$ b. $2a\sqrt{3}$ c. $3x^5\sqrt{5x}$

PP for Example 5

a. x^3y^4 b. $2ab\sqrt{3b}$ c. $3x^5yz\sqrt{5xz}$

PP for Example 6

a. $3\sqrt{21}$ b. $-13\sqrt{x}$

c. There are no like terms. $-3\sqrt{b} + 3\sqrt{a}$

PP for Example 7

a. $7 + \sqrt{11}$ b. $-2\sqrt{a} - 4\sqrt{5}$ c. $19y\sqrt{x} - 8xy$

PP for Example 8

a. $8\sqrt{7}$ b. $-\sqrt{5}$

PP for Example 9

a. $-2y\sqrt{3}$ b. $7b\sqrt{2}$ c. $3 + x^2\sqrt{x}$

Section 8.3

PP for Example 1

a. 6 b. 10 c. $2\sqrt{3}$ d. $4\sqrt{5}$

PP for Example 2

a. $8x^2$ b. $3a^3\sqrt{7a}$

PP for Example 3

a. $5y$ b. $15a$ c. $36b$

PP for Example 4

a. $15y$ b. $-54b^2\sqrt{2}$

PP for Example 5

a. $\sqrt{10y} + 10$ b. $2\sqrt{5} - 8$ c. $-8\sqrt{x} + 2$

APPENDIX A Answers to Practice Problems A-21

PP for Example 6
a. $\sqrt{15} - 2\sqrt{3} + 9\sqrt{5} - 18$
b. $7 - \sqrt{70} + 4\sqrt{7} - 4\sqrt{10}$
c. $y + \sqrt{y} - 30$
d. $x^2 - 2x\sqrt{5} + 5$

PP for Example 7
a. $9a$ b. $9 + 6\sqrt{a} + a$

PP for Example 8
a. -4 b. $x - 4$

PP for Example 9
a. $2 + \sqrt{5}$ b. $(2 + \sqrt{5})(2 - \sqrt{5}) = -1$ c. No

PP for Example 10
a. 5 b. 10 c. $\dfrac{1}{3}$ d. $3x^2$

PP for Example 11
a. $\dfrac{5\sqrt{7}}{7}$ b. $\sqrt{6}$ c. $\dfrac{-2a\sqrt{b}}{5b}$

PP for Example 12
a. $\dfrac{-7(1 + \sqrt{5})}{4}$ b. $\dfrac{-2(\sqrt{5} - \sqrt{2})}{3}$

Section 8.4
PP for Example 1
a. Yes, it is a solution. b. No, it is not a solution.

PP for Example 2
a. A person 5 feet above the ground can see out approximately 2.74 miles to the horizon.
b. To see 18 miles to the horizon, a person would need to be about 216 feet off the ground.

PP for Example 3
a. $x = 27$ b. $y = 169$
c. $x = 16$ d. $x = 5$

PP for Example 4
a. No solution b. $x = 8$

PP for Example 5
a. A car traveling at 55 miles per hour would leave a skid mark of about 100.83 feet.
b. A car traveling at 75 miles per hour would leave a skid mark of about 187.5 feet.

PP for Example 6
a. $x = 5, x = -6$ b. $x = 36$

CHAPTER 9
Section 9.1
PP for Example 1
a.

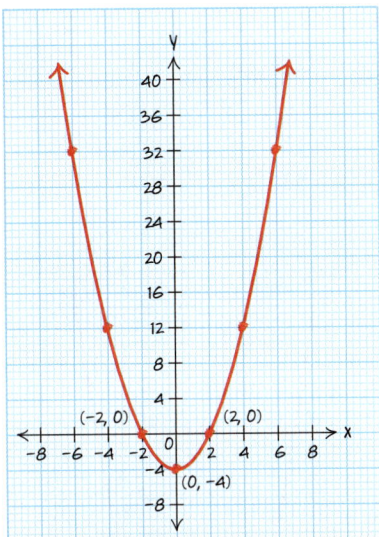

b.

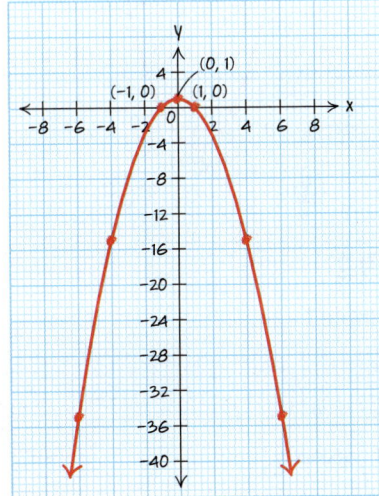

PP for Example 2
a. Because $a = -7$, the parabola opens down.
b. Because $a = 4$, the parabola opens up.
c. Because $a = 1$, the parabola opens up.

PP for Example 4
a. The vertex is the point $(x, y) = (1, 3)$. It is a maximum value.
b. The vertex is the point $(x, y) = (0, 5)$. It is a minimum value.

PP for Example 5
The vertex is the point $(t, P) = (20, 2000)$. A maximum profit of $2000 is earned when 20 tickets are sold.

PP for Example 6

a. $x = 0$ b. $x = 6$

PP for Example 7

x	y
−10	32
−8	12
−6	0
−4	−4
−2	0
0	12
2	32

Axis of symmetry: $x = -4$

Vertex: $(-4, -4)$

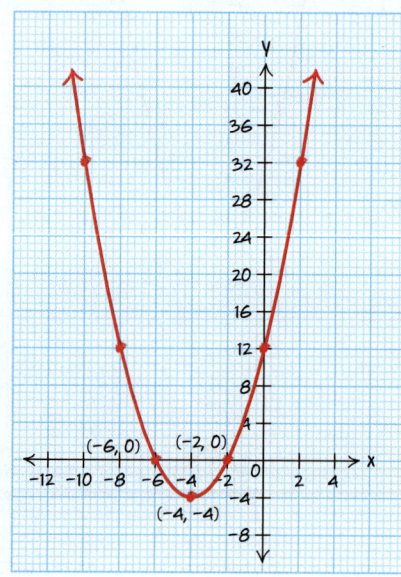

Section 9.2

PP for Example 1

a. 23.975 meters

b. About 2.02 seconds

c. About 2.67 seconds

PP for Example 2

a. $b = \pm 9$

b. $x = \pm 6$

c. $m = \pm\sqrt{20} = \pm 2\sqrt{5} \approx \pm 4.47$

PP for Example 3

a. $y = 2, y = -6$

b. $x = 2, x = -1$

c. $a = \dfrac{2 \pm \sqrt{11}}{5}$

PP for Example 4

a. $b = 12$

b. $c = \sqrt{2}$

c. $a = \sqrt{160} \approx 12.65$

PP for Example 5

The length is about 23.43 inches.

PP for Example 6

a. $d = \sqrt{5} \approx 2.24$

b. $d = \sqrt{145} \approx 12.04$

Section 9.3

PP for Example 1

a. $x = 2 \pm \sqrt{3}$

$x \approx 3.73, 0.27$

b. $x = -4 \pm \sqrt{5}$

$x \approx -1.76, -6.24$

PP for Example 2

$x = 5 \pm \sqrt{2}$

$x \approx 3.59, 6.41$

PP for Example 3

a. $x = 1, \dfrac{1}{5}$

b. $x = \dfrac{-6 \pm 2\sqrt{3}}{-4} = \dfrac{-2(3 \pm \sqrt{3})}{-4} = \dfrac{3 \pm \sqrt{3}}{2}$

$x \approx 0.64, 2.37$

PP for Example 4

This equation has no real solutions.

PP for Example 5

a. The time t

b. When the rocket hits the ground, $h = 0$. The equation becomes $0 = -16t^2 + 140t + 40$.

c. This is a quadratic equation.

d. $t \approx -0.28, 9.03$

e. Since negative time does not make sense for this problem, the solution is $t \approx 9.03$.

Section 9.4

PP for Example 2

a. $\left(\dfrac{3}{2}, 0\right), (-4, 0), (0, -12)$ b. $(6, 0), (-6, 0), (0, -36)$

PP for Example 3

a. Axis of symmetry: $x = -1$
Vertex: $(-1, -16)$
x-intercepts: $(-5, 0)$ and $(3, 0)$
y-intercept: $(0, -15)$

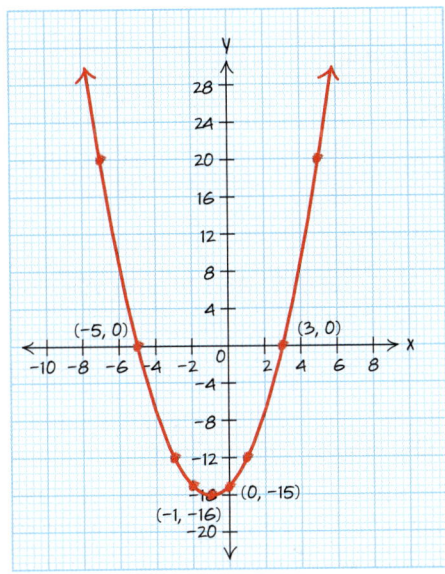

b. Axis of symmetry: $x = 2$
Vertex: $(2, -4)$
x-intercepts: $(0, 0)$ and $(4, 0)$
y-intercept: $(0, 0)$

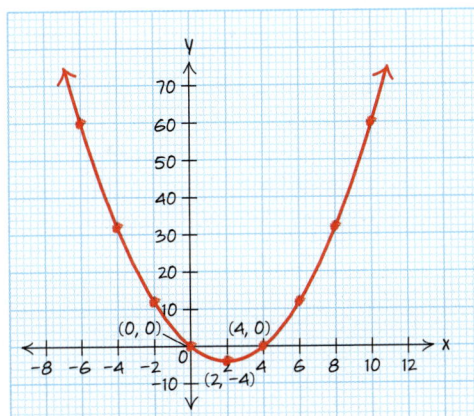

Section 9.5

PP for Example 1

a.

Month	Temperature
4	73
5	80
6	87
7	89
8	88
9	82

b.

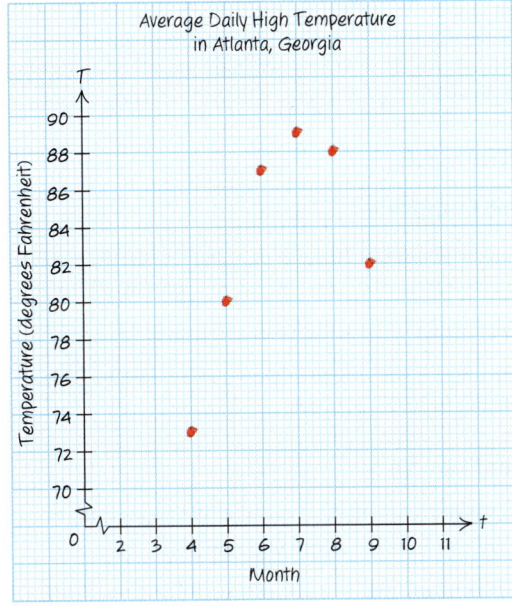

c. No, the points do not follow a linear pattern.

d. Yes, the points do seem to follow a parabolic pattern.

PP for Example 2

a. Parabolic pattern

b. Neither

c. Linear pattern

PP for Example 3

a. The model predicts the average daily high temperature to be 62.57°F. This is very close to the real value of 63.0°F.

b. The vertex is (7.10, 88.95). The model is predicting the average daily high temperature in July to be about 88.95°F.

PP for Example 4

a. The vertical intercept is (0, 0). It means that when the shop sells 0 pairs of jeans, the revenue is $0.00.

b. The vertex is (15, 1350). It means that when the shop sells 15 pairs of jeans, the revenue is $1350.00.

c. To make a revenue of $1000, the shop must sell either (approximately) 7 or 23 pairs of jeans.

PP for Example 5

a. The vertex is at (2006, 289).

b. The numbers are the same.

c. The vertex is a maximum.

d. The vertical intercept is at the point $(0, -60360251)$. This means that in the year 0, the median home price in thousands of dollars in Las Vegas was $-60,360,251$. This is a case of model breakdown.

e. The horizontal intercepts are (2001.6, 0) and (2010.4, 0). This means that the model is predicting that in the years 2001 and 2010, the median home price in thousands of dollars in Las Vegas is $0. This is a case of model breakdown.

Section 9.6

PP for Example 1

a. The relationship is between time and the interest earned on a $1000 deposit into a savings account earning 4% simple interest.

b. The domain is the set of years {0, 1, 2, 3, 4}.

c. The range is the simple interest earned: {$0.00, $40.00, $80.00, $120.00, $160.00}.

PP for Example 3

a. This is not a function. At noon on Wednesday, the price of gas is often different at different gas stations.

b. This is a function.   c. This is a function.

PP for Example 4

a. This is not the graph of a function.

b. This is the graph of a function.

c. This is the graph of a function.

d. This is the graph of a function.

PP for Example 5

a. The graph of $y = x^2 - 16$ passes the vertical line test, so it is a function. In function notation, it is $f(x) = y = x^2 - 16$.

b. The graph of $y = -3$ passes the vertical line test, so it is a function. In function notation, it is $f(x) = -3$.

PP for Example 6

a. $f(0) = \dfrac{3}{4}(0) + 2 = 2$ b. $f(-1) = (-1)^3 = -1$

c. $f(-2) = 3$

PP for Example 7

a. $W(10) = 190 - 10 = 180$. After 10 days on the diet, Philip's weight is 180 pounds.

b. $315 = 2.5p + 2.50$

$p = 125$

It costs $315 to make 125 large pizzas.

c. $h(1.25) = -16(1.25)^2 + 30(1.25) + 3.5 = 16$. 1.25 seconds after being thrown, the ball is about 16 feet off the ground.

Answers to Selected Exercises

APPENDIX B

CHAPTER R

Section R.1

1. Whole number, integer
3. Integer
5. Natural number, whole number, integer
7. Integer
9. Natural number, whole number, integer
11. Natural number, whole number, integer
13. Natural number, whole number, integer
15. Whole number, integer; there could be 0 kittens at the shelter
17. Whole number, integer; the employee could have a week off or otherwise work 0 hours in a week
19. Integer; the temperature could be positive or negative, so this belongs to the set of integers
21. $\{-5, -4, -3, -2, -1, 0, 1\}$
23. $\{1, 2, 3, 4\}$
25. $\{0\}$
27. 4
29. 0.1
31. number line with point at −5
33. number line with point at −2
35. number line with point at −3
37. $-5 > -15$
39. $-3 \leq -3$ or $-3 \geq -3$
41. $-4 < -3$
43. $0 > -7$
45. 3
47. 8
49. 16
51. 7
53. $-\dfrac{3}{2}, -1, \dfrac{1}{3}, 0.5, 2.95, 3$
55. $-5, -2, |-2|, |-5|$
57. $-3.5, \dfrac{2}{3}, |-4|, 4.2, |-5|$
59. The expression translates as "the opposite of five." $-(5) = -5$
61. The expression translates as "the opposite of the absolute value of negative 2." $-|-2| = -2$
63. The expression translates as "the opposite of negative six." $-(-6) = 6$
65. -8
67. 9
69. -7
71. -16
73. 4
75. -1
77. $-26°F$
79. -50 feet
81. 9
83. 8
85. 19
87. -6
89. -20
91. 18
93. 3
95. -6
97. Undefined
99. 0
101. 2 terms: -3 + $5 \cdot 6$
103. 1 term: $-24 \div 3 \cdot 7$
105. 3 terms: $3 + 22 \div 11 + (-16) \div 4$
107. 4 terms: $92 \cdot (-8) + 26 \div 13 + (-1) + 7(-3)$
109. 1 term; -6
111. 2 terms; 6
113. 3 terms; 23
115. 3 terms; -4
117. 4 terms; 30

Section R.2

1. $2 \cdot 2 \cdot 2 \cdot 2$
3. $17 \cdot 2$
5. $7 \cdot 3 \cdot 2$
7. $2 \cdot 2 \cdot 2 \cdot 2 \cdot 2 \cdot 2$
9. $7 \cdot 5 \cdot 2$
11. 3
13. 15
15. 9
17. 12
19. 3
21. $\dfrac{2}{5}$
23. $\dfrac{4}{21}$
25. $\dfrac{1}{3}$
27. $-\dfrac{1}{12}$
29. $-5\dfrac{2}{7}$
31. $\dfrac{10}{15}$
33. $\dfrac{40}{56}$
35. $-\dfrac{10}{32}$
37. $4\dfrac{10}{45}$
39. $\dfrac{15}{3}$
41. number line with point at 2
43. number line with point at −3
45. number line with point at −3
47. number line with point at −2
49. number line with point at −4
51. 2
53. $\dfrac{3}{5}$
55. $\dfrac{3}{5}$
57. 5
59. $3\dfrac{1}{3}$
61. $1\dfrac{1}{2}$ or $\dfrac{3}{2}$
63. $-\dfrac{1}{6}$
65. $-1\dfrac{5}{12}$ or $-\dfrac{17}{12}$
67. $4\dfrac{5}{6}$
69. $-5\dfrac{3}{4}$
71. $3\dfrac{1}{2}$ gallons
73. $\dfrac{3}{4}$ teaspoon
75. $9\dfrac{5}{8}$ feet
77. $-\dfrac{1}{3}$
79. $\dfrac{3}{5}$
81. 9
83. $\dfrac{1}{8}$
85. Does not exist
87. 2
89. $\dfrac{5}{12}$
91. $\dfrac{9}{7}$ or $1\dfrac{2}{7}$
93. 5
95. $-\dfrac{3}{2}$ or $-1\dfrac{1}{2}$
97. $\dfrac{1}{54}$

B-1

99. 2 terms; $\dfrac{-29}{2}$ or $-14\dfrac{1}{2}$ 101. 2 terms; $\dfrac{13}{5}$ or $2\dfrac{3}{5}$

103. 3 terms; $\dfrac{-1}{5}$ 105. 3 terms; $\dfrac{35}{12}$ or $2\dfrac{11}{12}$

107. 1 term; $\dfrac{12}{5}$ or $2\dfrac{2}{5}$

Section R.3

1. 4   3. 4   5. 7   7. 8

9. 0   11. $2\dfrac{9}{10}$ 13. $\dfrac{21}{25}$ 15. $7\dfrac{71}{100}$

17. $1\dfrac{1}{125}$ 19. $1\dfrac{33}{100}$ 21. 0.6   23. 1.25

25. $0.41\overline{6}$ 27. $3.1\overline{6}$ 29. 0.875

31. [number line with point at 0.4]
33. [number line with point at 1.6]
35. [number line with point at −1.2]
37. [number line with point at −0.4]
39. [number line with point at −1.8]

41. 8.3   43. 9.21   45. 210   47. 0.1
49. 1   51. 4.326   53. 6.5176   55. 5.76
57. 7.29   59. 96.5985   61. 5.33   63. 6.749

65. The two books cost Michelle $274.90.
67. Rami received $9.50 in change.
69. 5.418   71. 21.245   73. 0.112
75. I spent a total of $2.97 for the three apps.
77. Sam spent a total of $1.95 for the three candy bars.

79. 820   81. 0.06   83. 8.4   85. 0.037
87. 2 terms; 13.35   89. 2 terms; 19.57
91. 3 terms; 214.3   93. 3 terms; 47.42
95. 1 term; 134.4

97. a. $1.25 + 0.99 + 2(0.59)$ b. Addition and multiplication
    c. 3 terms   d. Multiplication; addition
    e. The total amount of Luc's purchase was $3.42.

99. a. $2(1.59) + 2(3.99)$ b. Addition and multiplication
    c. 2 terms   d. Multiplication; addition
    e. The total of Kym's purchase was $11.16.

Section R.4

1. 0.34   3. 47%   5. 1.7%   7. 36%

9.
Percent	Decimal	Fraction
16%	0.16	$\dfrac{4}{25}$

11.
Percent	Decimal	Fraction
28.5%	0.285	$\dfrac{57}{200}$

13.
Percent	Decimal	Fraction
0.12%	0.0012	$\dfrac{3}{2500}$

15. 10.24   17. 0.56

19. There are 10 Labrador retrievers in this puppy class.
21. There are six diners at the restaurant that have a coupon.
23. One serving of yellow split peas contains 12 g of dietary fiber.
25. One serving of pasta sauce contains 600 mg of sodium.
27. The tax on Brianna's $65 purchase is $3.90.
29. The sales tax on Yerica's tire purchase is $36.48.
31. The percent increase for this house is about 23.1%.
33. The percent decrease for this house is about 5.4%.
35. The markup on this princess cut 0.8 carat diamond necklace is $1949.06.
37. This new car is worth about $18,029.79 at the end of its first year.
39. The discounted price for this shirt is $33.75.

Section R.5

1.
	Whole Number	Integer	Rational Number	Irrational Number	Real Number
1.47			✓		✓
$\dfrac{16}{49}$			✓		✓
-7		✓	✓		✓
$\sqrt{5}$				✓	✓
9	✓	✓	✓		✓

3. Rational number   5. Rational number   7. Both
9. Rational number   11. Both   13. Both

15. $-3, 0, \dfrac{5}{2}$ 17. $\pi, \sqrt{2}$ 19. 16, 0, 2

21. 3, 111   23. 56

25. 0.625; rational; the number $\dfrac{5}{8}$ is rational because it is a fraction, a ratio of integers.

27. $1\dfrac{17}{50}$; rational; the number 1.34 is rational because it is a finite decimal.

29. $0.\overline{09}$; rational; the number $\dfrac{1}{11}$ is rational because it is a fraction, a ratio of integers.

31. $\dfrac{133}{10000}$; rational; the number 0.0133 is rational because it is a finite decimal.

33. $0.\overline{273}$; rational; the number $\dfrac{3}{11}$ is rational because it is a fraction, a ratio of integers.

35. 3.3166; irrational; $\sqrt{11}$ is irrational because it can be expressed as an infinite, nonrepeating decimal and it cannot be expressed as a ratio of integers.

37. 3.141593; irrational; π is irrational because it can be expressed as an infinite, nonrepeating decimal and it cannot be expressed as a ratio of integers.

39. This 40-ounce container of mixed nuts costs $0.67 per ounce.

41. This 16-ounce package of angel hair pasta costs $0.09 per ounce.

43. This 24-ounce jar of marinara sauce costs $0.11 per ounce.

45. 2 terms; $\dfrac{3}{5}$ **47.** 3 terms; 11.4

49. 2 terms; $5\dfrac{1}{10}$ or $\dfrac{51}{10}$ or 5.1

51. 2 terms; $4\dfrac{9}{50}$ or $\dfrac{209}{50}$ or 4.18

53. 3 terms; $-1\dfrac{7}{12}$ or $-\dfrac{19}{12}$ or $-1.58\overline{3}$

55. 3 terms; $-\dfrac{153}{28}$ or $-5\dfrac{13}{28}$

Chapter R Review Exercises

1. a. Integer
  b. Whole number, integer
  c. Natural number, whole number, integer

2. a. Natural number, whole number, integer
  b. Integer
  c. Whole number, integer

3. Whole number, integer; there could be 0 students in the chemistry lab.

4. Whole number, integer; there could be 0 trees in the park.

5. {0, 1, 2, 3, 4}   **6.** {−6, −5, −4, −3}   **7.** 2   **8.** 15

9. number line from −10 to 10 with points marked

10. number line from −10 to 10 with points marked

11. $-3 > -5$ **12.** $-9 < -2$ **13.** 9   **14.** 6.7

15. $-2, \dfrac{2}{5}, 3, |-7|$ **16.** $-4, -0.1, \left|-\dfrac{2}{3}\right|, 1.8$

17. The opposite of negative 8; 8

18. The opposite of negative 1; 1

19. a. -67 **b.** -26 **c.** -10 **d.** -5
20. a. -18 **b.** 68   **c.** -14 **d.** -7
21. a. -150 **b.** 8   **c.** -16 **d.** 5
22. a. 30   **b.** -18 **c.** -3 **d.** 8
23. 3 terms   **24.** 3 terms   **25.** 3 terms; 25
26. 3 terms; -7 **27.** $3 \cdot 3 \cdot 3$ **28.** $5 \cdot 3 \cdot 3$
29. 4   **30.** 15   **31.** $\dfrac{5}{11}$
32. $\dfrac{2}{5}$ **33.** $\dfrac{35}{56}$ **34.** $\dfrac{18}{42}$

35. a. Rational number   **b.** Both   **c.** Rational number
36. a. Rational number   **b.** Rational number   **c.** Both
37. a. $\dfrac{2}{3}$ **b.** $\dfrac{5}{12}$ **c.** $-\dfrac{1}{6}$ **d.** $2\dfrac{1}{6}$
38. a. $\dfrac{1}{2}$ **b.** $\dfrac{4}{15}$ **c.** $-\dfrac{3}{20}$ **d.** $\dfrac{7}{10}$
39. a. $\dfrac{1}{6}$ **b.** 4
40. a. $-\dfrac{1}{5}$ **b.** $\dfrac{2}{3}$
41. a. $\dfrac{2}{11}$ **b.** $\dfrac{1}{45}$ **c.** $\dfrac{2}{3}$
42. a. $\dfrac{3}{7}$ **b.** $\dfrac{3}{14}$ **c.** $\dfrac{3}{4}$
43. 2 terms; $-5\dfrac{1}{2}$ or $-\dfrac{11}{2}$ **44.** 2 terms; $\dfrac{1}{3}$
45. Hundredths   **46.** Thousandths
47. $2\dfrac{3}{4}$ **48.** $4\dfrac{17}{25}$ **49.** $0.\overline{4}$ **50.** 0.625
51. a. 10.644   **b.** 0.074
52. a. 21.071   **b.** 1.246
53. a. 4.841   **b.** 2.7
54. a. 34.96   **b.** 9.5
55. 3 terms; 4.1   **56.** 3 terms; -7.65 **57.** 0.02
58. 5.4   **59.** 0.0875   **60.** 0.0545
61. $\dfrac{7}{25}$ **62.** $\dfrac{9}{20}$ **63.** 4.56%
64. 7.75%   **65.** 190.5%   **66.** 201%
67. 80%   **68.** 6%   **69.** 120%
70. $333\dfrac{1}{3}\%$ **71.** 3.7125   **72.** 0.3105
73. 1.1   **74.** 1.632   **75.** $19.98
76. The sales tax on the $130 purchase is $5.85.
77. The percent increase for this house is about 41.5%.
78. The percent decrease for this house is about 11.5%.
79. The ring has a retail price of $875.00.
80. The dress has a retail price of $46.55.
81. $\dfrac{4}{3}, -16.9, 0$ **82.** $-7.5, 2, \dfrac{5}{4}$ **83.** $\sqrt{5}$
84. π **85.** 0.43   **86.** 4.583
87. $-2\dfrac{3}{5}$ or $-\dfrac{13}{5}$ or -2.6 **88.** $12\dfrac{71}{100}$ or $\dfrac{1271}{100}$ or 12.71
89. $3\dfrac{1}{6}$ or $\dfrac{19}{6}$ **90.** $-\dfrac{2}{3}$
91. $\dfrac{19}{50}$ or 0.38   **92.** $1\dfrac{6}{25}$ or $\dfrac{31}{25}$ or 1.24
93. $55\dfrac{3}{5}$ or $\dfrac{278}{5}$ or 55.6   **94.** 39.53
95. $\dfrac{2}{3}$ **96.** $\dfrac{11}{5}$ or $2\dfrac{1}{5}$

Chapter R Test

1.

	Whole Number	Integer	Rational Number	Irrational Number	Real Number
2.25			✓		✓
$\frac{2}{3}$			✓		✓
-6		✓	✓		✓
$\sqrt{2}$				✓	✓
10	✓	✓	✓		✓

2. 5

3. number line with points marked at $-6, -5, -3, 0, 4, 5$ (approximately) on range -10 to 10

4. $-4, \frac{13}{5}, -(-6.8), |-11|$

5. a. $-9 > -11$ b. $\frac{1}{3} > 0.3$

6. a. -3 b. 6

7. $2 \cdot 2 \cdot 3 \cdot 3$

8. a. 4 b. -1.05 c. -4

9. a. -48 b. 9 c. 9.2

10. 4 terms; 7 11. 3 terms; 35 12. 2 terms; -29

13. 3 terms; 127 14. 3 terms; $10\frac{2}{3}$ or $\frac{32}{3}$

15. 3 terms; 4.8 16. $2 \cdot 5 \cdot 5$

17. 18 18. $\frac{6}{7}$ 19. $\frac{18}{63}$

20. 3 21. $\frac{21}{20}$ or $1\frac{1}{20}$

22. a. $-\frac{3}{2}$ or $-1\frac{1}{2}$ b. $-\frac{4}{3}$ or $-1\frac{1}{3}$

23. Thousandths 24. 5.023

25.

Percent	Decimal	Fraction
1.05%	0.0105	$\frac{21}{2000}$

26. 18.54

27. There are 16 grams of sugar in one tablespoon of ketchup.

28. The percent increase for this house is about 9.7%.

CHAPTER 1
Section 1.1

1. 7^5 3. $5 \cdot 5 \cdot 5$ 5. $(-7) \cdot (-7) \cdot (-7)$

7. $(-3)^5$ 9. $\left(\frac{2}{3}\right)^3$ 11. $\frac{5}{8} \cdot \frac{5}{8} \cdot \frac{5}{8}$

13. 8 15. 16 17. -25

19. 1 21. 2187 23. 25

25. -2401 27. -1 29. $\frac{27}{64}$

31. 0.1225 33. -216 35. 1,034,000

37. 93,000,000 39. 4,001,000,000 41. $-54,870$

43. Three terms: $(4 \div 2) + (3 \cdot 27) + (-27)$

45. Two terms: $(8 \cdot 9) + ((-4) \cdot 16^2)$

47. Two terms: $\left(\frac{3}{7}\right) + \left(\left(-\frac{1}{2}\right) \div 8\right)$

49. One term: $((-4)^5 \cdot 2 \div 32)$

51. One term: $((-64) \div 4 \div 2)$

53. 20 55. $-\frac{2}{5}$ 57. 33

59. $1\frac{4}{25}$ or $\frac{29}{25}$ 61. -32 63. -17

65. -35 67. 17 69. 8

71. 3 73. 1728 75. $-4\frac{2}{5}$ or -4.4

77. -14 79. 91

81. Associative property of addition

83. Commutative property of multiplication

85. Distributive property

87. Commutative property of addition

89. Distributive property

91. Associative property of multiplication

93. a. Two walls have an area of 108 ft² each, and the other two walls have an area of 90 ft² each. The total area of all four walls is 396 ft².
   b. The total area of all four walls is 396 ft².
   c. The distributive property

95. a. The first corral has an area of 9600 ft². The second corral has an area of 7800 ft². The third corral has an area of 6000 ft². The three corrals have a total area of 23,400 ft².
   b. The three corrals have a total area of 23,400 ft².
   c. The distributive property

97. a. The wall molding for each 12-ft wall will cost $26.16. The wall molding for each 10-ft wall will cost $21.80. The total costs for the wall molding will be $95.92.
   b. The total costs for the wall molding will be $95.92.
   c. The distributive property

99. The corrals in Exercise 95 also need the fencing replaced. Each linear foot of fencing costs $5.59.
   a. The four 120-ft fences will cost $670.80 each. The two 80-ft fences will cost $447.20 each. The two 65-ft fences will cost $363.35 each. The two 50-ft fences will cost $279.50 each. The total cost for the fences will be $4863.30.
   b. The total cost for the fences will be $4863.30.
   c. The distributive property

101.

Operation	Commutative Property	Associative Property
Addition	x	x
Subtraction		
Multiplication	x	x
Division		

103. $2 \cdot 5 + 2 \cdot 9$ **105.** $16 \cdot 9 - 16 \cdot 5$
107. $-2 \cdot 8 + (-2) \cdot 7$ **109.** $-2 \cdot 7 - (-2) \cdot 16$
111. $\frac{7}{9} \cdot 11 + \frac{7}{9} \cdot 15$ **113.** 120
115. -165 **117.** -42
119. 12 **121.** -6

Section 1.2

1. If Sarah worked 15 hours in a week, her weekly pay would be $168.75.
3. If Thomas sold 53 rolls of wrapping paper, he would earn $265.00 in revenue.
5. If Marcus sells $1045.00 worth of merchandise in a week, his commission would be $73.15.
7. If Vinny runs for 2.5 hours at a pace of 7.5 miles per hour, he will run 18.75 miles.
9. If Judy is currently listing 5 houses, she estimates spending $1250.00 on advertising.
11. If Fusako drives for 6.5 hours at an average speed of 75 miles per hour, she will drive 487.5 miles.
13. -15 **15.** 3 **17.** 14
19. -9 **21.** $\frac{3}{2}$ or $1\frac{1}{2}$ **23.** 916
25. -140 **27.** $-\frac{7}{5}$ or $-1\frac{2}{5}$ **29.** 105
31. 5 **33.** $7\frac{1}{3}$ yd or $7.\overline{3}$ yd   **35.** ≈ 0.2 mi
37. $\approx \$2.06$ **39.** ≈ 17.358 lb   **41.** 56.775 l
43. $0.\overline{50}$ acre   **45.** $601{,}000$ thousand
47. 5.397 billion   **49.** 9000 billion
51. 2560 MB

For Exercises 53 through 97, the letters chosen to represent the variables can vary.

53. Let E = the amount owed in dollars on a monthly electric bill and k = the number of kilowatt-hours of electricity used during that month.
55. Let T = the amount paid yearly (in dollars) in federal taxes and d = income (in dollars) for that year.
57. Let F = the temperature outdoors in degrees Fahrenheit and t = the time of day.
59. Let p = the amount of first-class postage you owe (in dollars) for a package and w = the weight of the package you are shipping (in pounds).
61. Let c = the cost in dollars of attending a community college and u = the number of credits a student enrolls in per semester.

Let x represent an unknown number.

63. $x + 62$ or $62 + x$ **65.** $2x + 9$ or $9 + 2x$
67. $\frac{16}{x}$ **69.** $\frac{1}{3}x - 7$
71. $8x - 12$ **73.** $-5x^2$
75. $16 + \frac{x}{4}$ or $\frac{x}{4} + 16$ **77.** $\frac{6}{x - 2}$
79. $\frac{1}{2}x + 6$ or $6 + \frac{1}{2}x$ **81.** $7 - 2x$

83. Let c = the number of credit hours a part-time resident student is enrolled in for one semester.
$187c + 10c + 11c + 20$
85. Let L = the number of hours the house-painting company charges for labor. $475 + 23L$
87. Let x = the number of ounces a piece of first-class mail weighs in excess of 1 ounce.

$$0.49 + 0.21x$$

or

Let z = the number of ounces a piece of first-class mail weighs. $0.49 + 0.21(z - 1)$

89. Let n = the number of pounds. $\frac{n}{2.2}$ kg
91. Let n = the number of miles. $5280n$ ft
93. Let n = the number of liters. $\frac{n}{3.785}$ gal
95. Let n = the number of acres. $43{,}560n$ ft^2
97. Let n = the number of seconds. $\frac{n}{60}$ min
99. **a.** The area of Bill's living room is 300 ft^2. The units are in square feet.
   b. 300 ft$^2 \approx$ 34 yd^2 rounded up to the nearest full square yard.
   c. The Berber carpet Bill wants will cost $798.66.
   d. The padding will cost Bill $93.50.
   e. The total cost of the carpet and padding will be $892.16.
101. **a.** The total weight of the luggage is 72 lb.
   b. The total weight of the luggage is 74.2 lb.
103. $4.5 - t$

Time to Drive from San Diego to Los Angeles	Time to Drive from Los Angeles to San Diego
1	$4.5 - 1 = 3.5$
2	$4.5 - 2 = 2.5$
2.5	$4.5 - 2.5 = 2$
3	$4.5 - 3 = 1.5$
t	$4.5 - t$

B-6 APPENDIX B Answers to Selected Exercises

105. $2500 - d$

Amount Invested in the 3.5% Account	Amount Invested in the 7% Account
0	$2500 - 0 = 2500$
500	$2500 - 500 = 2000$
100	$2500 - 100 = 2400$
1500	$2500 - 1500 = 1000$
d	$2500 - d$

107. $65 - c$

Pounds of Colombian Coffee	Pounds of Kenyan Coffee
0	$65 - 0 = 65$
10	$65 - 10 = 55$
20	$65 - 20 = 45$
50	$65 - 50 = 15$
c	$65 - c$

For Exercises 107 through 110, the letters chosen to represent the variables can vary.

109. a. $12p$ (when p represents the number of people invited to the party)

b. Sandy can invite approximately 41 people to the party.

111. a. $18a + 15a$ (when a represents the number of people attending the reception)

b. Approximately 51 people can attend the conference reception, and Tricia will stay within the budget.

Section 1.3

1. Three terms; $2x$ and $7x$ are like terms.
3. Four terms; 3 and -5 are like terms; $2h$ and $7h$ are like terms.
5. Four terms; $3x^2$ and $7x^2$ are like terms; $5x$ and $3x$ are like terms.
7. Three terms; no like terms.
9. Four terms; $7xy$ and $-2xy$ are like terms.
11. $20x$ **13.** $-6p$ **15.** $-2x + 4y$
17. $-11n - 8$ **19.** $-x^2 + 11x - 6$ **21.** $3x - 8y$
23. $-8m + 7n + 12$ **25.** $7x^2 - x + 5$
27. $-7x^5 - 3x^2 + 5x + 8$ **29.** $-\frac{1}{2}a^6 + 3a^2 + a - \frac{7}{5}b$
31. a. $4.19k$ **b.** $1.39k$
   c. $4.19k + 1.39k$ **d.** $5.58k$
33. a. $8.79k$ **b.** $1.19k$
   c. $1.39k$ **d.** $11.37k$
35. a. $0.15l$ **b.** $0.08l$ **c.** $0.23l$
37. $8x + 7$ **39.** $4t + 14$
41. $7x^2 + 9x$ **43.** $4k^2 + 5k - 9$
45. $5x - xy + 7y$ **47.** $5a^2b^3 + 8a^2b^2 + 6ab^2$
49. $7s^5t^3 + 4s^3t^5 + 9s^2t$
51. Distributive property; $6x + 14$
53. Multiplication property; $10xy$
55. Distributive property; $-28x^2 - 36x + 24$
57. Multiplication property; $-20a^2bc$
59. Distributive property; $6xy - 14y$
61. Incorrect; $6xy$ **63.** Incorrect; $6k - 12$
65. Incorrect; $21mn$ **67.** Correct
69. Incorrect; $4x + 6$ **71.** $23x + 30$
73. $-10g - 12$ **75.** $-3x + 12$
77. $-2k^2 - 15k + 5$ **79.** $5x + 8xy + 14$
81. $-4x^2 - 3x$ **83.** $26w^2 + 36w - 9$
85. $7r^2 + 3r + 9$ **87.** $8.9t + 15.8$
89. $3h + 43$ **91.** $9x - 7$ **93.** $11a - 16$
95. $5b$ **97.** $2l + 2w$ **99.** $3x + 5y$
101. $6x + 16y - 12$ **103.** $12x + 9$ **105.** $12a$

Section 1.4

1. a. The maximum number of iPads sold in a quarter was 26.04 million in the first quarter of 2014.

b. The minimum number of iPhones sold was 35.2 million in the third quarter of 2014.

3. a. The fastest marathon time is 2:02:57. This happened in the year 2014.

b. Most of the top 10 runners come from Kenya.

5. a. The percentage of obese adults was highest during the year 2014. Approximately 38% of the adult population was obese that year.

b. The percentage of obese adults was lowest during the year 2004. Approximately 32% of the adult population was obese that year.

7. a. New home sales were at their lowest point during the year 2011. There were approximately 300,000 sales made.

b. New home sales were at their highest point during the year 2005. There were approximately 1,300,000 sales made.

9.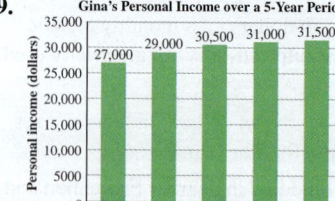

11.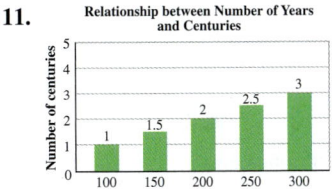

13.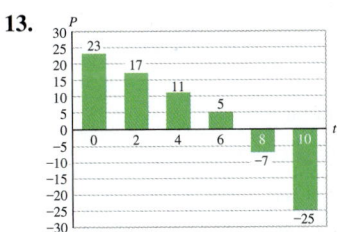

15.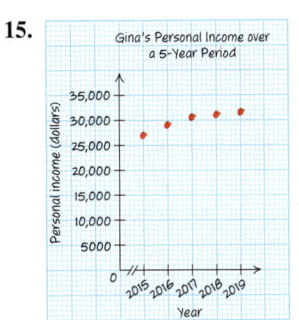

17.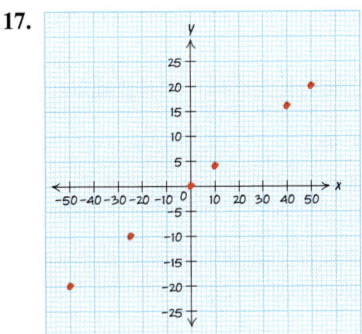

19.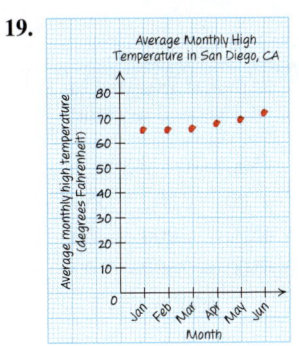

21. A: (5, 4)   B: (−8, 7)   C: (−4, −6)   D: (2, −3)

23. A: (2, 0)   B: (0, 3)   C: (−5, 0)   D: (0, −4)

25. x-axis: The scale is 5.   y-axis: The scale is 1.

27. x-axis: The scale is 0.5.   y-axis: The scale is 0.1.

29. A: (24, 6)   B: (0, 21)   C: (−16, 12)   D: (12, 0)
    E: (−28, −9)

31.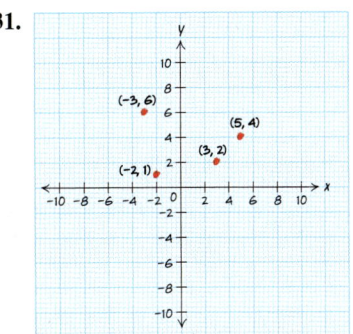

33.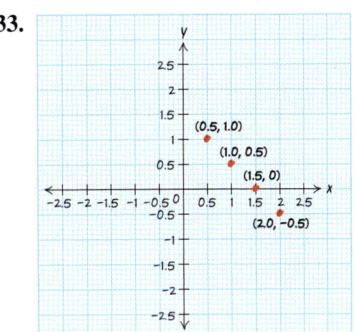

35.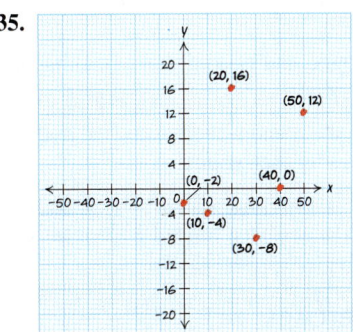

37.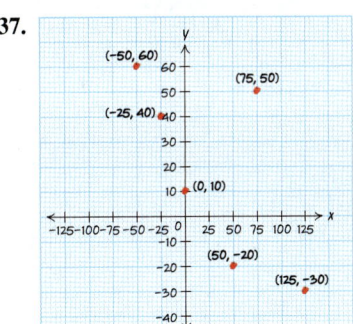

39.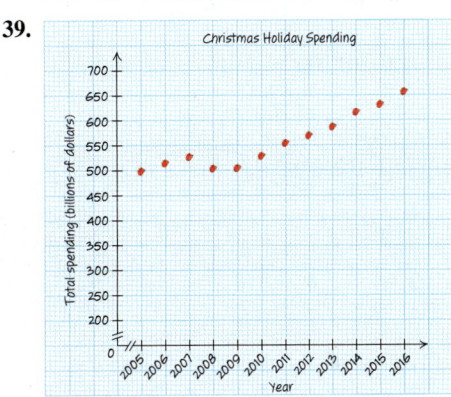

APPENDIX B Answers to Selected Exercises

41.
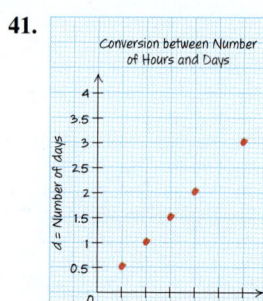

43.
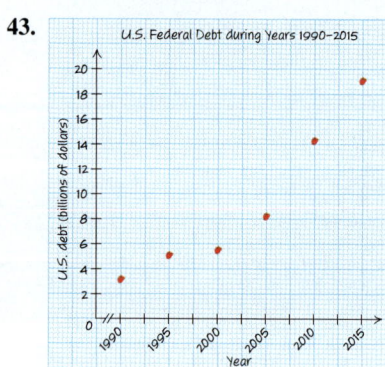

45. The student transposed the x- and y-coordinates, graphing the point (4, 3) instead of (3, 4).

47. The student is confusing the x- and y-coordinates and has transposed the coordinates for each point.

49. The student was inconsistent with the scale of the y-axis, with a scale of 10 for positive values and a scale of 1 for negative values.

Chapter 1 Review Exercises

1. 6^5 **2.** 8^4 **3.** $\left(\dfrac{3}{8}\right)^3$

4. $\left(-\dfrac{4}{5}\right)^5$ **5.** $5 \cdot 5 \cdot 5 \cdot 5 \cdot 5 \cdot 5$

6. $(-2) \cdot (-2) \cdot (-2)$

7. $\left(-\dfrac{2}{5}\right) \cdot \left(-\dfrac{2}{5}\right) \cdot \left(-\dfrac{2}{5}\right) \cdot \left(-\dfrac{2}{5}\right)$

8. $\left(\dfrac{1}{11}\right) \cdot \left(\dfrac{1}{11}\right) \cdot \left(\dfrac{1}{11}\right)$

9. -27 **10.** 36 **11.** $\dfrac{16}{81}$

12. $-\dfrac{8}{27}$ **13.** $207,000$ **14.** $5,670,000$

15. $(-3) + (4 \cdot (-2)) + (24 \div 8) + ((-3^2)) + (-2)$; -19

16. $(27 \div 3) + ((-5^2)) + (-10) + (20) + ((-3) \cdot (-5))$; 9

17. $\left(5 \div \dfrac{15}{7}\right) + \left(\dfrac{2}{3}\right) + (-1)$; 2

18. $(-4) + \left(\left(-\dfrac{1}{5}\right) + \dfrac{3}{2} \cdot \dfrac{4}{15}\right)$; $-3\dfrac{4}{2}$ or $-\dfrac{19}{5}$

19. -6 **20.** -36 **21.** $-3\dfrac{2}{3}$ or $-\dfrac{11}{3}$

22. $-\dfrac{2}{5}$ **23.** 21 **24.** $5\dfrac{1}{2}$ or $\dfrac{11}{2}$

25. -48 **26.** 60 **27.** 3

28. 3 **29.** 63 **30.** 29

31. Commutative property of multiplication

32. Commutative property of addition

33. Distributive property

34. Distributive property

35. Associative property of addition

36. Associative property of multiplication

37. Commutative property of addition

38. Commutative property of multiplication

For Exercises 39–44, the letters chosen to represent the variables can vary.

39. Let D = the amount of money (in dollars) you earn in a month and h = the number of hours you work in a month.

40. Let R = the amount of money (revenue in dollars) a local playhouse earns for a night and t = the number of tickets sold for that night.

Let n = an unknown number.

41. $3n - 6$ **42.** $5n + 8$ **43.** $\dfrac{1}{2}(n - 4)$

44. $\dfrac{n + 12}{16}$ **45.** 22.86 cm **46.** ≈ 34.18 miles

47. $84.0\overline{9}$ kg **48.** 220 lb **49.** ≈ 1.94 oz

50. 141.75 g **51.** $9\dfrac{5}{6}$ **52.** -28

53. -5 **54.** -23 **55.** -25

56. -9 **57.** 8.25 **58.** 29

59. -86 **60.** 12

61. $4500 - d$

Amount Invested in the First Account	Amount Invested in the Second Account
0	$4500 - 0 = 4500$
1000	$4500 - 1000 = 3500$
2000	$4500 - 2000 = 2500$
3500	$4500 - 3500 = 1000$
d	$4500 - d$

62. $20 - d$

First Piece	Second Piece
0	$20 - 0 = 20$
4	$20 - 4 = 16$
8	$20 - 8 = 12$
16	$20 - 16 = 4$
d	$20 - d$

APPENDIX B Answers to Selected Exercises   B-9

63. 800,000 thousand
64. 700,000 million
65.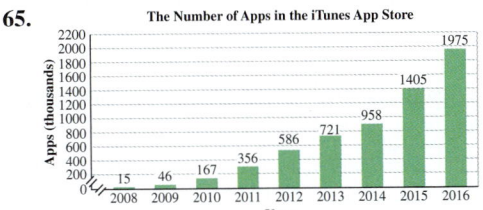
66.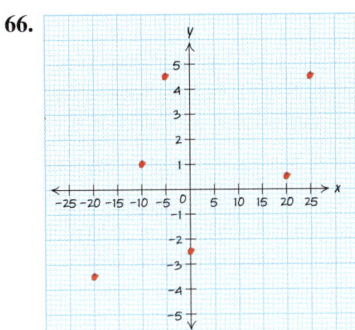
67. Four terms; $-6p$ and $7p$ are like terms; 4 and -9 are like terms.
68. Four terms; $5x$ and $-\dfrac{2}{3}x$ are like terms; 16 and 12 are like terms.
69. Four terms; $-3x^2$ and $-4x^2$ are like terms.
70. Four terms; $-\dfrac{2}{5}y$ and $-6y$ are like terms.
71. a. $3.25s$ b. $1.25s$ c. $3.25s + 1.25s$ d. $4.5s$
72. a. $d - 500$ b. $4(85)$ c. $d - 500 - 4(85)$ d. $d - 840$
73. $4y^3 + 4y^2$ 74. $4a^2b - 13ab$ 75. $-x - 5$
76. $x - 12$ 77. $-6.25a + 2.1b$ 78. $-0.3x + 1.9y$
79. $2x^2 + x + 12$ 80. $5y^2 - 7y + 19$
81. a. Multiplication property b. $-48ab^2$
82. a. Multiplication property b. $-12xy$
83. a. Distributive property b. $-15x + 30y$
84. a. Distributive property b. $6a - 16b$
85. a. Multiplication property b. $13x$
86. a. Distributive property b. $-5y + 7$
87. $-6x + 27$ 88. $x + 7$ 89. $2x - 3$
90. $-9y + 5$ 91. $-6a + 8b$ 92. $7x - 18y$
93. $11x - 9$ 94. $4h - 6k$ 95. 0
96. $0.5x + 8.3y$ 97. $6x + 8y + 6$ 98. $8m + 2w$
99. $(6, 1), (4, 5), (-1, 6), (-2, 0), (-5, -2), (3, -4)$
100. $(5, 0), (6, 3), (0, 6), (-4, 3), (-3, -5), (2, -1)$
101. x-axis: The scale is 30.
     y-axis: The scale is 7.
102. x-axis: The scale is 10.
     y-axis: the scale is 200.
103.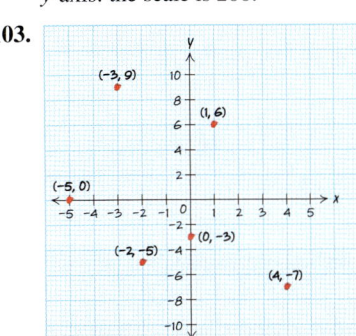
104.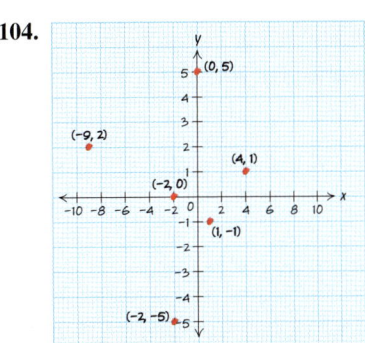
105.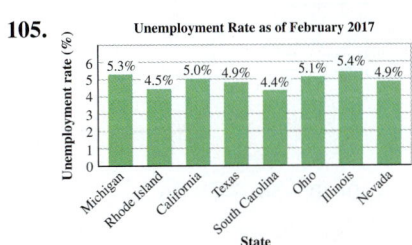
106.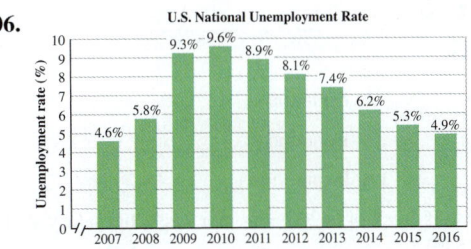

Chapter 1 Test

1. 3^6
2. $\left(-\dfrac{5}{8}\right) \cdot \left(-\dfrac{5}{8}\right) \cdot \left(-\dfrac{5}{8}\right)$
3. -8
4. $42{,}990{,}000$
5. Four terms; 7
6. Three terms; 41
7. Two terms; 5
8. Four terms; 34
9. Associative property of addition
10. Distributive property

11. Commutative property of multiplication
12. Lucas's weekly salary would be $330.00 if he worked 24 hours in a week.
13. Let $L =$ the number of hours charged for labor; $20 + 65L$
14. ≈ 91.74 m
15. 2
16. $6x^2 - 15x + 3$
17. $7h - 3hk + 3k^2 + 22$
18. $3x + 2y$
19. a. The index was at its highest approximately during the third quarter of 2003.
    b. The index was at its lowest approximately during the third quarter of 2006.
    c. The scale on the vertical axis is 5.
    d. The scale on the horizontal axis is 1.
20. x-axis: The scale is 12.
    y-axis: The scale is 5.
21.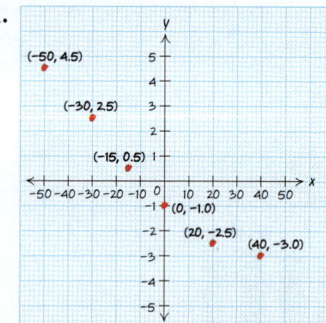
22. $(4, 50), (-4, 50), (-10, 0), (0, -50), (4, -10)$

CHAPTER 2
Section 2.1

1. Expression
3. Equation
5. Expression
7. Equation
9. Equation
11. Expression
13. a. $S = 215$. If Joe works 20 hours in a week, his salary will be $215.00.
    b. Yes, $S = 376.25$ when $h = 35$ is a solution to the equation $S = 10.75h$. These values represent that if Joe works 35 hours in a week, his salary will be $376.25.
    c. No, $S = 260$ when $h = 25$ is not a solution to the equation $S = 10.75h$. These values represent Joe's salary and the hours worked. If Joe works 25 hours in a week, his salary would be $268.75, not $260.00.
    d. Yes, $S = 430$ when $h = 40$ is a solution to the equation $S = 10.75h$. These values represent that if Joe works 40 hours in a week, his salary will be $430.00.
15. a. Yes, $D = 240$ when $t = 4$ is a solution to the equation $D = 60t$.
    b. These values represent that 240 miles is the distance traveled when driving for 4 hours.

17. a. Yes, $C = 37.7$ when $r = 6$ is a solution to the equation $C = 2\pi r$.
    b. These values represent that a circle with a radius of 6 has a circumference of 37.7 (rounded to one decimal place).
19. a. Yes, $F = 4500$ when $L = 300$ is a solution to the equation $F = 15L$.
    b. These values represent that a loan of $300,000.00 would have an origination fee of $4500.00.
21. a. No, $F = 2000$ when $L = 225$ is not a solution to the equation $F = 10L$.
    b. These values represent that a loan of $225,000.00 would have an origination fee of $2000.00, but a loan of that value would have an origination fee of $2250.00.
23. a. Yes, $S = 155.6$ when $f = 2$ is a solution to the equation $S = 77.8f$.
    b. These values represent that it would require 155.6 cubic yards of sand to fill a volleyball court 2 feet deep.
25. Yes, $x = 11$ is a solution to the equation.
27. No, $m = 8$ is not a solution to the equation.
29. Yes, $x = 28$ is a solution to the equation.
31. Yes, $x = 5$ is a solution to the equation.
33. Yes, $x = 1$ is a solution to the equation.
35. No, $t = 3$ is not a solution to the equation.
37. Yes, $x = 5$ and $y = 1$ are solutions to the equation.
39. Yes, $x = -4$ and $y = 0$ are solutions to the equation.
41. Yes, $m = 3$ and $n = 5$ are solutions to the equation.
43. No, $x = 0$ and $y = 5$ are not solutions to the equation.
45. a. An additional 45 lb of other cargo could be stored on the roof of the car.
    b. An additional 85 lb of luggage could be stored on the roof of the car.
47. The company's profit is $8000.
49. The company's profit is $-$$3000.00; therefore, the company is operating at a loss.
51. The company must generate $29,000 in revenue.
53. $x = 25$
55. $x = 24.5$
57. $m = 19$
59. $p = 27$
61. $k = -15$
63. $x = 8$
65. $a = 2.5$
67. $x = 1$
69. $m = \dfrac{5}{3}$
71. $m = \dfrac{1}{9}$
73. $x = 9$
75. $d = 8.25$
77. $x = -3.4$
79. $x = \dfrac{1}{35}$
81. $y = -\dfrac{1}{2}$
83. $R = C + P$
85. $l = 220 - s - c$
87. $A = 90 - B$
89. $M = T - 45$
91. $y = -x - 4$
93. $G = D - 3F + 30$
95. $w = t + 3u$
97. $b = P - a - c$
99. $x = 3y + 5$

101. Equation; $x = 14$

103. Expression; $2x + 7$

105. Expression; $17a^2 + 6a + 10$

107. Equation; $c = 4$

109. Equation; $p = 34$

111. $p + 6$

113. $t = 35$ represents 35 years since 1960. So $1960 + 35 = 1995$. In the year 1995, the population of South America was 321.05 million.

115. $t = 58$ represents 58 years since 1960. So $1960 + 58 = 2018$. In the year 2018, the population of South America was 435.82 million.

117. a. 25
b. 25
c. 271.15 million

119. a. 56
b. 56
c. 425.84 million

Section 2.2

1. $x = -27$
3. $y = 5\frac{1}{2}$
5. $g = 13$
7. $x = 9$
9. $x = 15$
11. $n = 2.5$
13. $c = -7$
15. $t = 70$
17. $x = -15$
19. $n = -\frac{14}{25}$
21. $k = 14$

23. a. $S = 1360$. If Victoria works 80 hours, her salary will be $1360.
b. $h = 120$. Victoria has to work 120 hours to make $2040 a month.
c. $h \approx 41.18$. Victoria has to work approximately 41.18 hours to earn $700.

25. a. $F = 5000$. The loan fees will be $5000 for a loan of $250,000.
b. $L = 200,000$. Pablo can suggest a maximum loan amount of $200,000 to keep the fees at $4000.

27. a. $D = 9000$. There will be 9000 bricks that do not meet the quality standards if 500 million bricks are produced.
b. $D = 734.4$. Approximately 734 bricks will not meet the quality standards of the 40.8 million bricks produced in a day.
c. $b = 1200$. LEGO produced 1200 million (or 1.2 billion) bricks if there are 21,600 bricks that do not meet the company's quality standards.

29. a. $B = 5$. A total of 5 buses would be needed to transport 300 people.
b. $P = 21,000$. Public transportation can handle 21,000 people at a time with 350 city buses.
c. $P = 36,000$. The city can transport 36,000 people at a time with 600 school and city buses.

31. a. $M = 1,314,225.8$. The total amount of mortgage debt in 2018 for single- and multifamily homes in the United States was $1,314,225.8 million.
b. $t = 10$. In the year 2020, the total amount of mortgage debt for multifamily homes in the United States will be $1,464,767.6 million.

33. a. $f = 1452.00$. The fee is $1452.00 if the final price of a tractor is $17,900.
b. $p = 1050$. The final price of the log splitter was $1050 if the total fee charged for selling the item was $104.
c. $p = 13,500$. The final price of the tractor was $13,500 if the total fee charged for selling the item was $1100.

35. a. $F = 3.20$. The fee for a $100 transaction is $3.20.
b. $p = 360$. The total transaction was $360 if the fee charged was $10.74.
c. $p = 1500$. The total transaction was $1500 if the fee charged was $43.80.

37. Isolate the variable term using the subtraction property of equality.

$$-2.5c = -20$$
$$\frac{-2.5c}{-2.5} = \frac{-20}{-2.5}$$
$$c = 8$$

39. $5x + 20 = 60$ Use the subtraction property of equality to isolate the variable term.
$\underline{-20 \ -20}$
$5x = 40$
$\frac{5x}{5} = \frac{40}{5}$ Use the division property of equality to isolate the variable.
$x = 8$

41. $p = 45$
43. $t = -3$
45. $w = 24.8$
47. $m = 6\frac{1}{2}$
49. $h = 31$
51. $x = 25$
53. $n \approx -0.53$
55. $m = 3.4$
57. $z = -3\frac{1}{5}$
59. $x = 12$
61. $x = 40$
63. $x = \frac{9}{10}$

65. a. Let $C =$ the total cost in dollars to ship n number of items.
$$C = 6.99 + 1.99n$$
b. $C = 16.94$. It costs $16.94 to ship five items UPS Ground.
c. $n = 3$. A total of three items can be shipped UPS Ground for $12.96.

67. a. Let $T =$ the total time in minutes it will take the machine shop to produce c cranks.
$$T = 360 + 5c$$
b. $T = 610$. It will take 610 minutes, or 10 hours and 10 minutes, to produce 50 cranks.
c. $c = 163$. The machine shop made 163 cranks if it took 1175 minutes to produce the cranks.

69. Let "a number" = x.
$3x = 45$
$x = 15$

71. Let l = the length of the garden.
$56 = 2l + 2 \cdot 8$
$l = 20$
The length of the garden is 20 feet.

73. Let h = the height of the triangle, using the formula $A = \frac{1}{2}bh$ to find the area of a triangle, where A = area and b = base.
$40 = \frac{1}{2} \cdot 8h$
$h = 10$
The height of the triangle is 10 inches.

75. Let "a number" = x.
$\frac{x}{7} = 13$
$x = 91$

77. Let "a number" = x.
$6 + 7x = 62$
$x = 8$

79. Let "a number" = x.
$\frac{1}{3}x - 8 = 28$
$x = 108$

81. Let "a number" = x.
$2x - 15 = 40$
$x = 27.5$

83. Let "an unknown number" = x.
$8 - x = 19$
$x = -11$

85. Let "a number" = x.
$2x - 7 = 11$
$x = 9$

87. $l = \frac{P - 2w}{2}$ or $l = \frac{P}{2} - w$

89. $h = \frac{V}{\pi r^2}$

91. $b = \frac{2A}{h}$

93. $t = \frac{W + 20}{3}$

95. $B = \frac{P - 2b}{2}$ or $B = \frac{P}{2} - b$

97. $y = \frac{5}{3}x + 5$

99. $y = \frac{x - 9}{5}$ or $y = \frac{1}{5}x - \frac{9}{5}$

101. $x = \frac{y - 23}{3}$

103. $x = \frac{y - 5}{9}$

105. $r = \frac{I}{Pt}$

107. Expression; $5a^2 + 18a - 5$

109. Equation; $x = 7$

111. Equation; $r = 20$

113. Expression; $20m - 3n + 11$

115. Equation; $y = 5.6$

Section 2.3

1. Let "a number" = x.
$4x = x + 33$
$x = 11$

3. Let "a number" = x.
$5x + 8 = 3x - 4$
$x = -6$

5. Let "a number" = x.
$2x = 5(x + 6)$
$x = -10$

7. Let "a number" = x.
$20 - 3x = 4x - 22$
$x = 6$

9. Let "a number" = x.
$\frac{1}{2}x = x + 5$
$x = -10$

11. Let "a number" = x.
$5 + \frac{x}{2} = x - 7$
$x = 24$

13. Let "a number" = x.
$x - 8 = 5x + 4$
$x = -3$

15. Let "a number" = x.
$\frac{1}{4}x - 8 = 2x - 29$
$x = 12$

17. Use the distributive property to simplify each side of the equation.
$2x - 11 = -5x - 18$

Use the addition property of equality to combine the variable terms.

$7x - 11 = -18$
$\underline{+11 \quad +11}$
$7x = -7$

Use the division property of equality to isolate the variable.

19. $3x + 12 = 7x - 28$ Use the subtraction property of equality to move variable terms to one side.
$\underline{-7x \quad\quad -7x}$
$-4x + 12 = -28$
$-4x + 12 = -28$
$\underline{-12 \quad -12}$ Use the subtraction property of equality to isolate the variable term.
$-4x = -40$
$\frac{-4x}{-4} = \frac{-40}{-4}$ Use the division property of equality to solve for the variable.
$x = 10$

21. $t = 8$ 23. $b = -6$ 25. No solution
27. $x = 10$ 29. $x = 6$ 31. All real numbers
33. $h = -8$ 35. $x = 7$ 37. $h = -5$

39. $x = 6$ **41.** No solution **43.** $d = -6$
45. $x = 20$ **47.** $x = 12$ **49.** $y = 1.4$
51. $t = -0.01$
53. $x = 22.5$. The measures of the three angles are 22.5°, 45°, and 112.5°.
55. $x = 4$. The length and width of the yard measure 13 ft and 20 ft.
57. $x = 7\frac{7}{12}$. The length of each side of the yard measures $20\frac{3}{4}$ ft.
59. $x = 11$. The length and width of the lawn measure 6 ft and 44 ft.
61. $x = 36$. The measures of the three angles are 36°, 72°, and 72°.
63. All real numbers **65.** No solution
67. $r = 0$ **69.** All real numbers
71. $g = 25$ **73.** $x = 20$
75. No solution
77. $x = 55$. The measures of the three angles of the triangle are 75°, 55°, and 50°.
79. $x = 63$. The measures of the three angles of the triangle are 63°, 68°, and 49°.
81. $x = 100$. The measures of the four angles of the quadrilateral are 100°, 75°, 65°, and 120°.
83. $p = 80$. The lengths of the sides are 77 ft, 100 ft, 80 ft, and 85 ft.
85. $s = 60$. The lengths of the sides are 60 ft, 120 ft, 55 ft, and 110 ft.
87. Carlotta can supply 32 students for the bike tour.
89. a. Let S = the total amount in dollars needed for supplies.
$S = 23.76 + 1.22k + 2.01k + 0.05k$ or $S = 23.76 + 3.28k$
b. $S = \$269.76$. If 75 kids came to build the project, the cost would be \$269.76.
c. $k = 120.00$. The manager can buy supplies for 120 projects.
91. For a book that costs the bookstore \$120, the sales price will be \$156 after the 30% markup.
93. A pair of headphones that originally cost \$20 will have a discount of \$7 and a final sales price of \$13 after the 35% discount.
95. This car had an 18% decrease in value during that 1 year.
97. This house had about a 12.3% increase in value.
99. Expression; $6a^2 - 4a + 17ab + 3$
101. Equation; $x = -7.\overline{3}$
103. Equation; no solution
105. Expression; $-2x + 19$
107. Equation; $x = 15\frac{3}{7}$

Section 2.4

1. $x \geq 7$ **3.** $d < 7.5$ **5.** $x \leq -192$
7. $a > 61$ **9.** $g \leq 15$ **11.** $m > -25$

13. $x \geq -8$ **15.** $x < 2$ **17.** $x \leq -2$
19. $k < 11$ **21.** $x < -7$
23. a. $11h \geq 275$
b. $h \geq 25$. Kati needs to work at least 25 hours this week so she can pay her rent.
c. $11h \geq 400$
$h \geq 36.\overline{36}$. Kati must work at least 36.36 hours to earn at least \$400 a week.
25. a. $3s + 25 \leq 5000$
b. $s \leq 1658.33$; Kevin can refinish no more than 1658 square feet of floor to stay within his budget of \$5000.
c. $3s + 25 \leq 4000$; $s \leq 1325$: Kevin can refinish no more than 1325 square feet of floor to stay within his budget of \$4000.
27. a. $19 + 20h \leq 119$
b. $h \leq 5$; Eddie can rent the flat-bed truck for up to 6.5 hours and keep his rental costs to at most \$119.
c. $19 + 20h \leq 89$; $h \leq 3.5$; Eddie can rent the flat-bed truck for up to 4.5 hours and keep his rental costs to at most \$89. You should round h to 3, since you will be charged for an entire hour if you keep it the 0.5 hour extra.
29. a. $75 + 3m \leq 115$
b. $m \leq 13\frac{1}{3}$. Alicia can have her car towed up to 23 miles in order to spend no more than \$115 for the tow.
31. $t \geq 7.0$. The population of Arkansas is projected to be 3 million or more in the year 2017 and after.
33. $t \geq 8.3$. The population of Florida is projected to be 21 million or more after the year 2008.
35. $x < 4$ **37.** $n > -9$ **39.** $t \leq 0$
41. $2x < 6$
$x < 3$
43. $2x + 1 \geq 5$
$x \geq 2$
45. $x > 5$
$(5, \infty)$
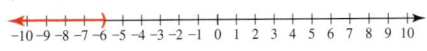
47. $m < -6$
$(-\infty, -6)$
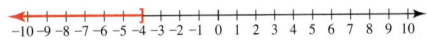
49. $x \leq -4$
$(-\infty, -4]$
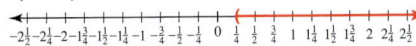
51. $x > \frac{1}{4}$
$\left(\frac{1}{4}, \infty\right)$

B-14 APPENDIX B Answers to Selected Exercises

53. $a \geq 12$
    $[12, \infty)$
    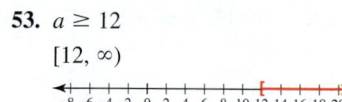

55. $k \leq 3$
    $(-\infty, 3]$

57. $z \leq 12$
    $(-\infty, 12]$
    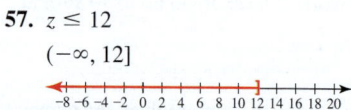

59. $c \geq -2.5$
    $[-2.5, \infty)$
    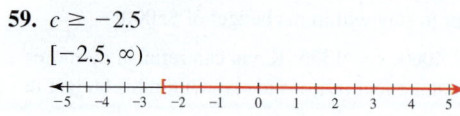

61. $w \leq 7.6$
    $(-\infty, 7.6]$
    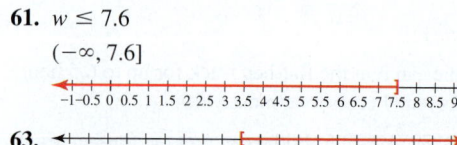

63.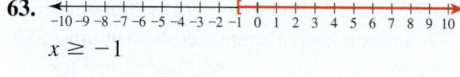
    $x \geq -1$

65.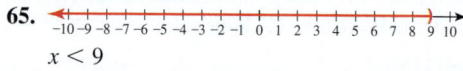
    $x < 9$

67.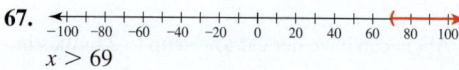
    $x > 69$

69.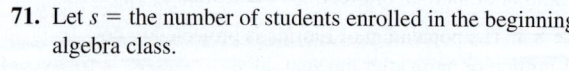
    $x \leq -16$

71. Let $s =$ the number of students enrolled in the beginning algebra class.
    $16 \leq s \leq 38$

73. Let $s =$ the number of students that can ride on the bus.
    $0 \leq s \leq 71$

75. Let $p =$ the number of people the camp can accommodate.
    $200 \leq p \leq 450$

77. Let $h =$ the height requirements in inches for children to ride on some attractions with an adult.
    $42 \leq h \leq 52$

79. Let $r =$ the average adult's resting heart rate in beats per minute.
    $66 \leq r \leq 100$

81. $3 < x < 4$
    $(3, 4)$

83. $-6 < y < -2$
    $(-6, -2)$
    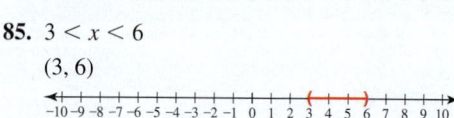

85. $3 < x < 6$
    $(3, 6)$
    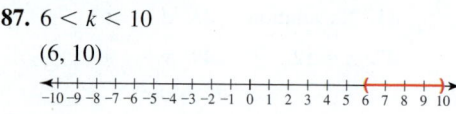

87. $6 < k < 10$
    $(6, 10)$
    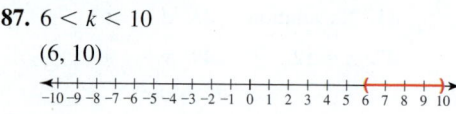

89. $6 \leq x \leq 10$
    $[6, 10]$
    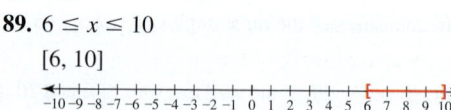

91. $-9 \leq m \leq -3$
    $[-9, -3]$
    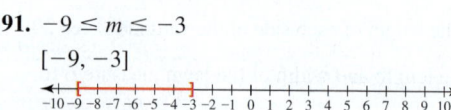

93. $-20 \leq x \leq 4$
    $[-20, 4]$
    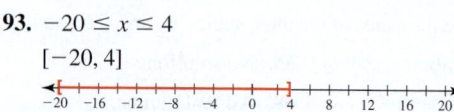

95. $8 < x < 60$
    $(8, 60)$
    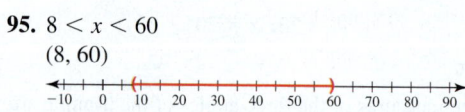

97. $-\dfrac{28}{3} < x < -\dfrac{4}{3}$
    $\left(-\dfrac{28}{3}, -\dfrac{4}{3}\right)$
    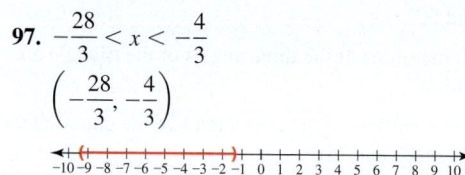

99. $2.75 \leq x \leq 5.25$
    $[2.75, 5.25]$
    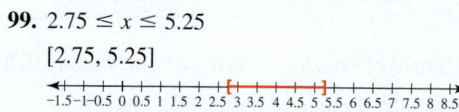

101. $-5 \leq v < -1$
     $[-5, -1)$
     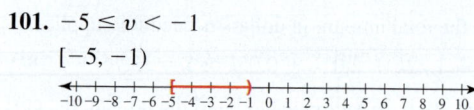

103. $\dfrac{3}{5} < a \leq \dfrac{33}{5}$
     $\left(\dfrac{3}{5}, \dfrac{33}{5}\right)$
     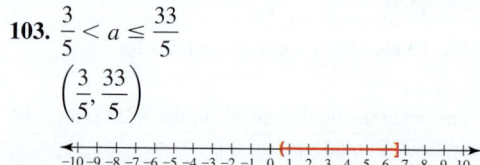

105. Inequality
     $x > 17$
     $(17, \infty)$

107. Expression
     $-8x + 7$

109. Equation
     $h = -14$

111. Inequality
     $a \leq 6$
     $(-\infty, 6]$

113. Equation
     $x = 17$

115. Inequality
     $-5.5 < a < 2.5$
     $(-5.5, 2.5)$

117. Expression
     $\dfrac{10}{9}b + 3$

119. Inequality
     $-14 \leq g < -8$
     $[-14, -8)$

Chapter 2 Review Exercises

1. **a.** Yes, $D = 262.5$ when $t = 3.5$ is a solution to the equation $D = 75t$.
   b. These values represent that 262.5 miles is the distance traveled when driving for 3.5 hours.
2. **a.** Yes, $F = 6900$ when $L = 600$ is a solution to the equation $F = 11.5L$.
   b. These values represent that a loan of $600,000 would have an origination fee of $6900.
3. No, the values $x = -2$ and $y = 3$ are not solutions to the equation $-2x + 4y = -8$.
4. Yes, the values $x = -3$ and $y = 14$ are solutions to the equation $y = -3x + 5$.
5. $P = 23,000$. The company has a profit of $23,000.
6. $P = 39,000$. The company has a profit of $39,000.
7. $x = -6$
8. $p = 7$
9. $a = -17$
10. $b = -23$
11. $k = 2$
12. $x = \dfrac{6}{5}$
13. $y = 14.4$
14. $h = 1.02$
15. $q = -2p + 9$
16. $x = y + 2$
17. $y = \dfrac{1}{3}x - 2$
18. $p = \dfrac{5}{3}q - 10$
19. $y = 12$
20. $x = -16$
21. $a = -7$
22. $x = 8$
23. $b = 11$
24. $y = -7$
25. $x = -24$
26. $n = 12$
27. $t \approx -1.35$
28. $k = 33$
29. **a.** $J \approx 4.13$. To transport 520 people, 5 airplanes would be needed.
   b. $P = 1890$. If the airline carrier has 15 planes, they can transport 1890 people at a time.
30. **a.** $S = 1960$. Devora's monthly salary will be $1960 if she works 140 hours.
   b. $h = 150$. Devora has to work 150 hours a month to make $2100 a month.
   c. $h = 95$. Devora has to work 95 hours a month to earn $1330 a month.
31. $x = -9$
32. $y = 10$
33. $a = 7$
34. $k = -3$
35. $x = -20$
36. $x = 18$
37. $x = -16$
38. $y = -24$
39. Let "a number" $= x$.
    $9x = 76$
    $x = 8\dfrac{4}{9}$
40. Let "a number" $= x$.
    $12 = -2x$
    $x = -6$
41. Let "a number" $= x$.
    $2x + 16 = -44$
    $x = -30$
42. Let "a number" $= x$.
    $\dfrac{1}{2}x + 4 = -18$
    $x = -44$
43. Let "a number" $= x$.
    $\dfrac{x}{4} = 0.25$
    $x = 1$
44. Let "a number" $= x$.
    $-\dfrac{x}{3} = 0.75$
    $x = -2.25$
45. Let "a number" $= x$.
    $8 - 3x = 17$
    $x = -3$
46. Let "a number" $= x$.
    $2x - 9 = 7$
    $x = 8$
47. $m = \dfrac{D}{ca}$
48. $h = \dfrac{V}{lw}$
49. $d = \dfrac{C}{\pi}$
50. $h = \dfrac{2A}{b}$
51. $y = 3x + 4$
52. $y = 2x - 6$
53. Let "a number" $= x$.
    $2x + 1 = 3x - 4$
    $x = 5$
54. Let "a number" $= x$.
    $\dfrac{1}{2}x + 10 = x - 4$
    $x = 28$
55. $x = 4$
56. $y = -3$
57. $a = 5\dfrac{3}{7}$
58. $x = -10$
59. No solution
60. No solution
61. All real numbers
62. All real numbers
63. $x = 62$. The measures of the three angles are 62°, 68°, and 50°.
64. $x = 34$. The measures of the three angles are 102°, 54°, and 24°.
65. $x > 12$
    $(12, \infty)$

66. $x \leq 7$
    $(-\infty, 7]$
    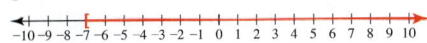
67. $x \geq -7$
    $[-7, \infty)$
    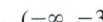
68. $y \leq -3$
    $(-\infty, -3]$
    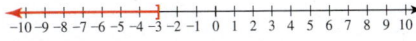

B-16 APPENDIX B Answers to Selected Exercises

69. $y \leq \dfrac{17}{5}$

$\left(-\infty, \dfrac{17}{5}\right]$

70. $x > -\dfrac{2}{3}$

$\left(-\dfrac{2}{3}, \infty\right)$

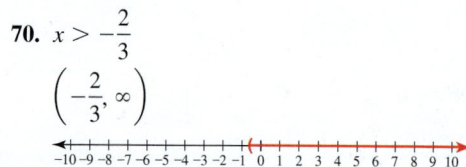

71. $y \geq 20$

$[20, \infty)$

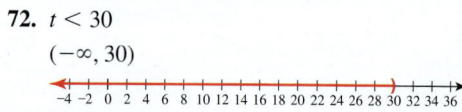

72. $t < 30$

$(-\infty, 30)$

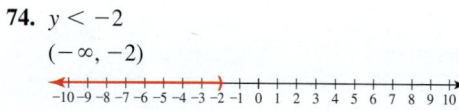

73. $x > -12$

$(-12, \infty)$

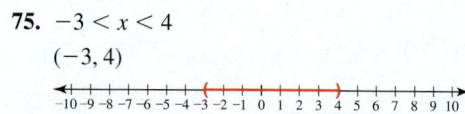

74. $y < -2$

$(-\infty, -2)$

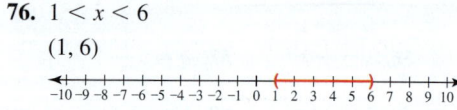

75. $-3 < x < 4$

$(-3, 4)$

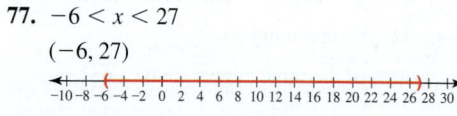

76. $1 < x < 6$

$(1, 6)$

77. $-6 < x < 27$

$(-6, 27)$

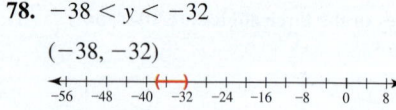

78. $-38 < y < -32$

$(-38, -32)$

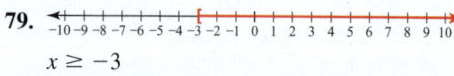

79.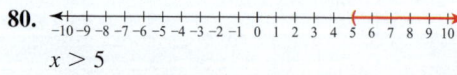
$x \geq -3$

80. $x > 5$

81. $x < 6$

82. $x \leq -12$

83. $x < -6$ 84. $y \geq 0$

85. $n > 30$. Kiano must sell more than 30 stained-glass windows to make a profit.

86. $d \leq 3$. Jessie can rent a car for 3 days or less if she wants to spend only $119.12 or less.

87. **a.** $135 + 40h \leq 235$

   b. $h \leq 2.5$. To stay within her budget, Karyn can hire the bartender for no more than 2.5 hours after the 3 initial hours.

88. **a.** $50 + 1.5m \leq 65$

   b. $m \leq 10$. Trey can have the TV delivered and stay within his budget for the delivery charge if he lives 10 miles or less from the store.

Chapter 2 Test

1. No, $b = -2$ is not a solution of $5b + 8 = 2$.
2. Yes, $t = 4.2$ and $D = 294$ are solutions to the equation $D = 70t$.
3. $x = -12$ 4. $c = P - a - b$
5. $P = 162{,}000$. The company has a profit of $162,000.
6. $w = -90$ 7. $p = 36$ 8. $x = -36$
9. All real numbers 10. No solution
11. Let $l =$ the length of the building lot.

    $330 = 2l + 2 \cdot 40$

    $l = 125$. The length of the building lot is 125 feet.
12. $x = \dfrac{y - b}{m}$
13. Let "a number" $= x$.

    $4x + 7 = -17$

    $x = -6$
14. $x = 2$ 15. All real numbers
16. $x = 10$. The measures of the three angles are $10°$, $110°$, and $60°$.
17. $x \geq -16$
18. $x < 2$

19. $6 \leq x \leq 11$
20. **a.** $10.25h > 287$

    b. $h > 28$. To earn more than $287.00 a week to meet his expenses, Aiden must work more than 28 hours.

Cumulative Review for Chapters 1 and 2

1. $\left(\dfrac{9}{4}\right)^6$ 2. $(-4) \cdot (-4) \cdot (-4)$
3. -1 4. $304{,}500{,}000$
5. $\boxed{\dfrac{84}{10 - 6} \cdot 3} + \boxed{(-2^2)} - \boxed{(-5)} + \boxed{20 \div 4 \cdot (-3)}$

   49
6. 20
7. Commutative property of addition

APPENDIX B Answers to Selected Exercises B-17

8. Let R = the amount of revenue in dollars the soccer club makes and let c = the number of candy bars sold.

9. Let "a number" = x.
   $8x - 7$

10. 28.956 m

11. -40

12. $50 - n$

Number of Biscuits for Scout	Number of Biscuits for Shadow
0	$50 - 0 = 50$
5	$50 - 5 = 45$
10	$50 - 10 = 40$
15	$50 - 15 = 35$
n	$50 - n$

13. $-12x^2 + 5x + 15xy$

14. $3x + 13$

15. $4b - 8c$

16. $(4, 0), (1, 2), (-3, 4), (-3, -5), (0, -3), (6, -2)$

17.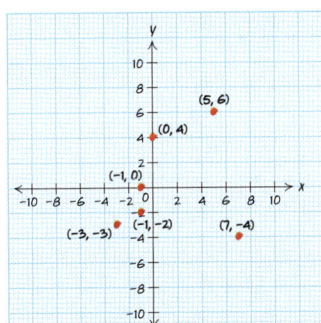

18.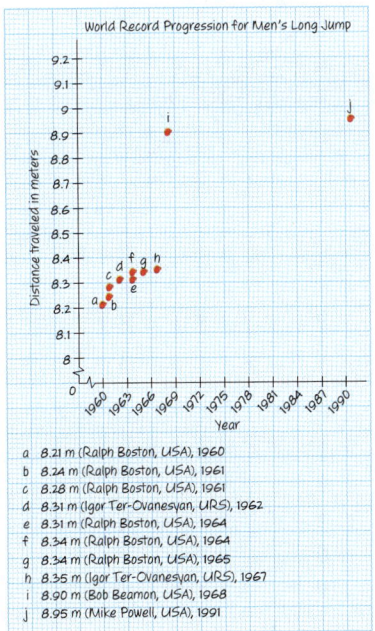

19. Yes, $x = -1.5$ and $y = 0$ are solutions to the equation $4x - 7y = -6$.

20. $y = -5$ 21. $x = \frac{7}{6}$ or $x = 1\frac{1}{6}$

22. $n = 4.057$ 23. $y = \frac{3}{5}x + 6$

24. $x = 3$ 25. $n = -12$

26. $z = -1.59$

27. a. $S = 1920$. If Corinthia works 120 hours in a month, her monthly salary would be $1920.

    b. $h = 140$. Corinthia would need to work 140 hours in a month to make $2240 a month.

    c. $h = 100$. Corinthia would need to work 100 hours in a month if she wants to earn only $1600 a month.

28. $y = 4$ 29. $x = 65$ 30. $x = 4.5$

31. Let "a number" = x.
    $2x + 20 = 6$
    $x = -7$

32. $m = \frac{P}{rg}$ 33. $y = 2x + 10$

34. Let "a number" = x.
    $2x + 8 = 4x - 18$
    $x = 13$

35. $y = -4$ 36. $n = -16$ 37. $w = 2$

38. $x = 55$. The measures of the three angles are 55°, 73°, and 52°.

39. $x > 8$ 40. $r \geq -28$

41. $3 < x < 7$ 42. $t < \frac{112}{3}$ or $t < 37\frac{1}{3}$

43. Let s = the number of students enrolled in the class.
    $15 \leq s \leq 25$

44. Let n = the course number.
    $200 \leq n < 400$

CHAPTER 3

Section 3.1

1.

s = sales (dollars)	P = monthly pay (dollars)
0	300
1000	350
8000	700
30,000	1800
50,000	2800

B-18 APPENDIX B Answers to Selected Exercises

3.

h = hours since opening	C = number of customers for the day
0	2
1	10
3	26
6	50
10	82

5.

n = number of studs purchased	C = cost (dollars)
4	14
6	21
8	28
11	38.50
18	63

7.

h = hours of bowling	C = calories burned
1	259
2	518
3	777
5	1295
6	1554

9.

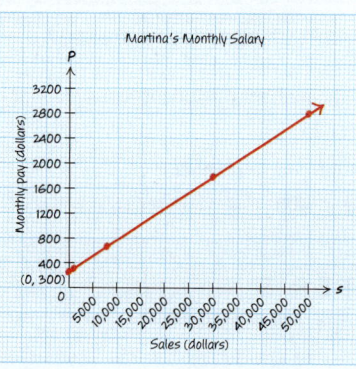

11.

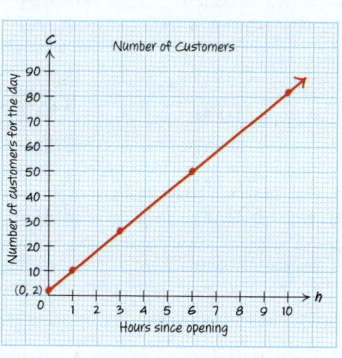

13.

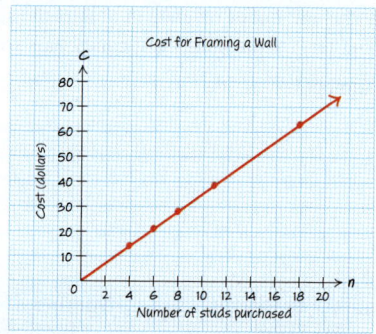

15.

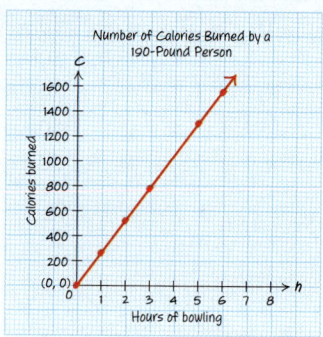

17.

t	$P = -8t + 17$
-4	49
-2	33
0	17
2	1
4	-15

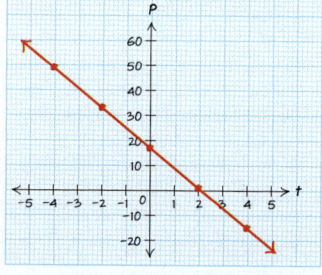

The graph is linear.

19.

x	$y = |x - 2|$
-6	8
-3	5
-1	3
0	2
2	0
3	1
4	2
6	4
10	8

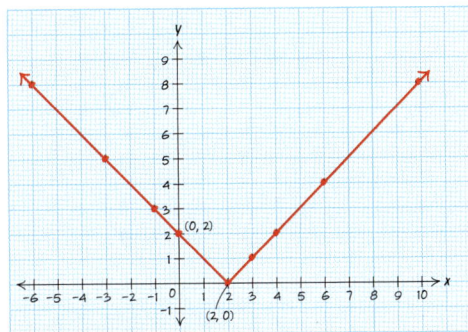

The graph is nonlinear.

21.

s	$R = s^2 + 7$
−4	23
−3	16
−2	11
−1	8
0	7
1	8
2	11
3	16
4	23

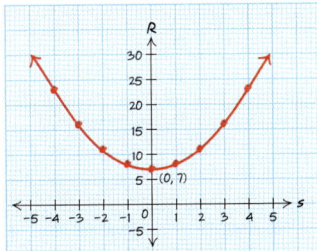

The graph is nonlinear.

23.

x	$y = -2x^2 + 3$
−4	−29
−3	−15
−2	−5
−1	1
0	3
1	1
2	−5
3	−15
4	−29

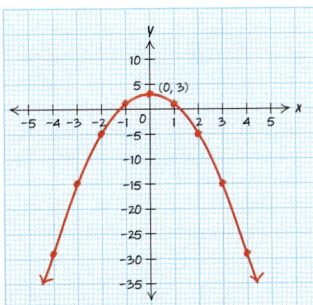

The graph is nonlinear.

25.

x	$y = 2x - 8$
−1	−10
0	−8
2	−4
4	0
7	6

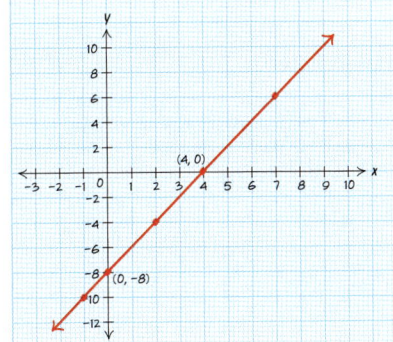

The graph is linear.

27.

x	$y = 0.5x + 3$
−8	−1
−4	1
0	3
4	5
8	7

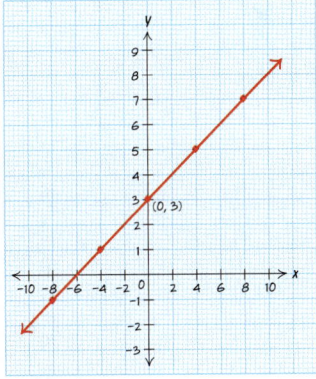

The graph is linear.

29.

x	$y = -2x + 10$
−2	14
0	10
2	6
4	2
7	−4

B-20 APPENDIX B Answers to Selected Exercises

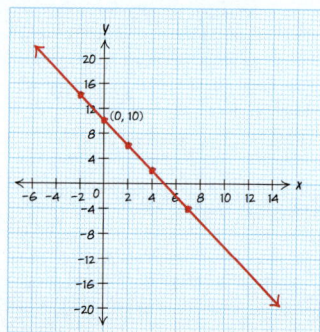

The graph is linear.

31.

x	$y = -0.4x + 2$
-10	6
-5	4
0	2
5	0
10	-2

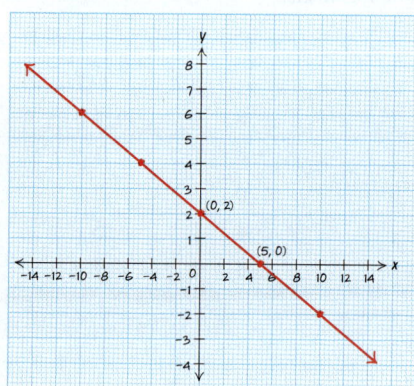

The graph is linear.

33.

x	$y = 3x$
-3	-9
-1	-3
0	0
1	3
3	9

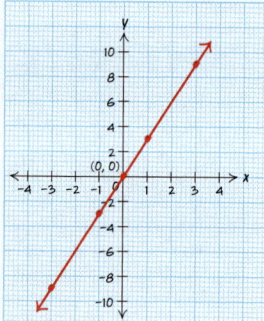

The graph is linear.

35.

x	$y = -3x$
-3	9
-1	3
0	0
1	-3
3	-9

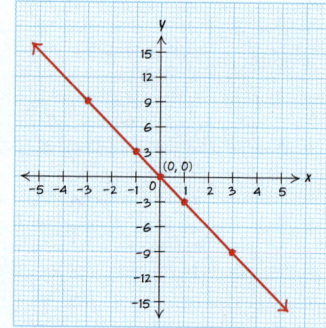

The graph is linear.

37.

x	$y = \frac{1}{2}x + 3$
-10	-2
-6	0
0	3
6	6
10	8

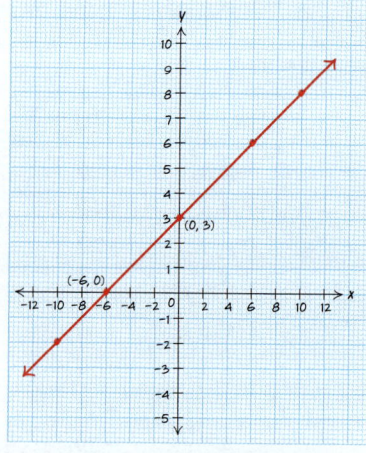

The graph is linear.

39.

x	$y = \frac{2}{3}x - 5$
-6	-9
0	-5
3	-3
6	-1
9	1

APPENDIX B Answers to Selected Exercises   B-21

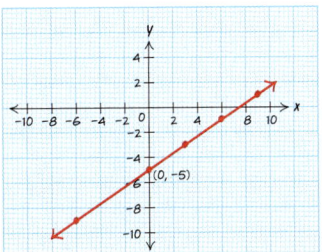

The graph is linear.

41.

x	$y = -\dfrac{1}{5}x - 2$
-10	0
-5	-1
0	-2
5	-3
10	-4

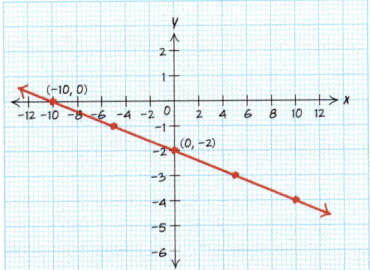

The graph is linear.

43.

d	$T = 5d$
0	0
1	5
2	10
5	25
7	35

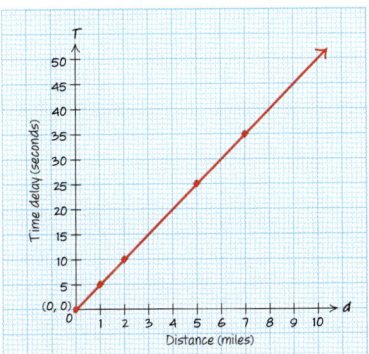

45.

h	$T = 75 - 3h$
0	75
1	72
2	69
4	63
6	57
8	51

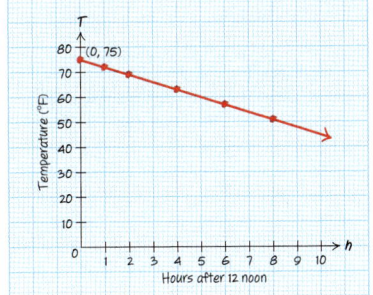

47.

m	$g = 24{,}000 - 100m$
0	$24{,}000$
50	$19{,}000$
100	$14{,}000$
175	6500
240	0

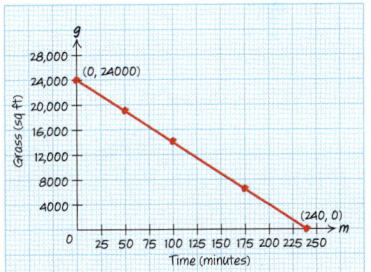

49.

x	$y = x^2 + 1$
-5	26
-3	10
-2	5
-1	2
0	1
1	2
2	5
3	10
5	26

B-22 APPENDIX B Answers to Selected Exercises

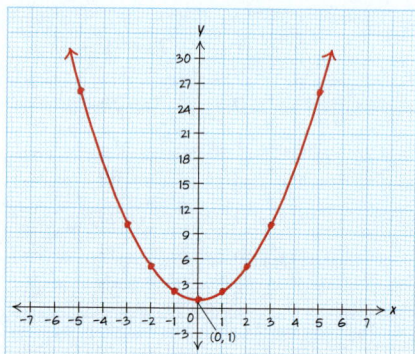

The graph is nonlinear.

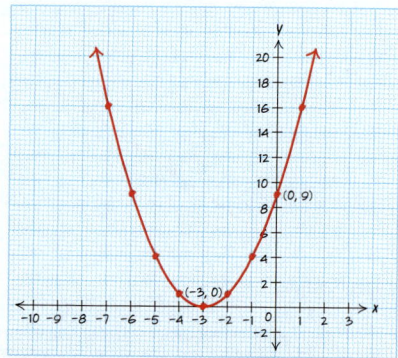

The graph is nonlinear.

51.

x	$y = x^2 - 4$
−5	21
−3	5
−2	0
−1	−3
0	−4
1	−3
2	0
3	5
5	21

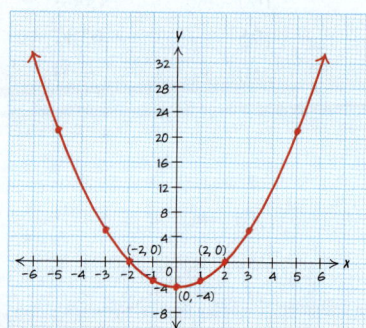

The graph is nonlinear.

55.

x	$y = (x - 2)^2$
−2	16
−1	9
0	4
1	1
2	0
3	1
4	4
5	9
6	16

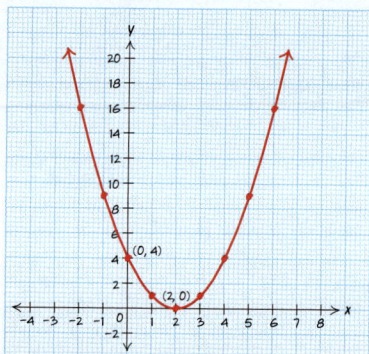

The graph is nonlinear.

53.

x	$y = (x + 3)^2$
−7	16
−6	9
−5	4
−4	1
−3	0
−2	1
−1	4
0	9
1	16

57.

x	$y = |x| + 2$
−8	10
−6	8
−4	6
−2	4
0	2
2	4
4	6
6	8
8	10

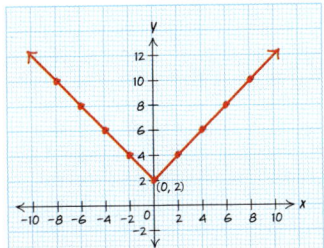

The graph is nonlinear.

59.

x	$y = |x + 2|$
-9	7
-7	5
-5	3
-3	1
-2	0
-1	1
1	3
3	5
5	7

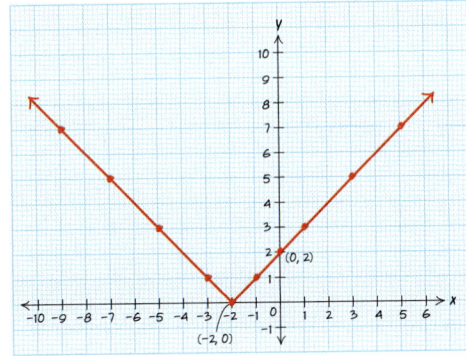

The graph is nonlinear.

61.

x	$y = |x - 3|$
-3	6
-1	4
0	3
2	1
3	0
4	1
6	3
7	4
9	6

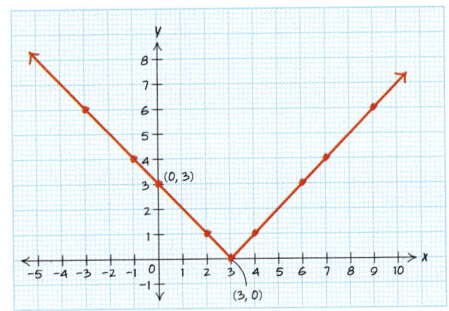

The graph is nonlinear.

63.

x	$y = 6$
-9	6
-5	6
0	6
5	6
9	6

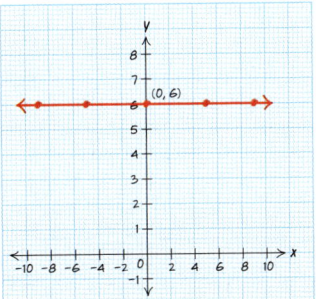

The graph is linear.

65.

$x = 7$	y
7	-9
7	-5
7	0
7	5
7	9

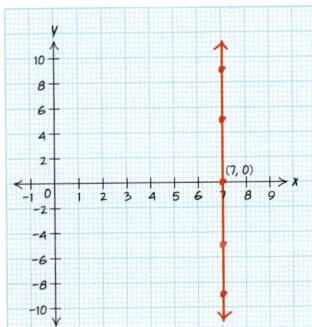

The graph is linear.

B-24 APPENDIX B Answers to Selected Exercises

67.

x	$y = -3.5$
-9	-3.5
-5	-3.5
0	-3.5
5	-3.5
9	-3.5

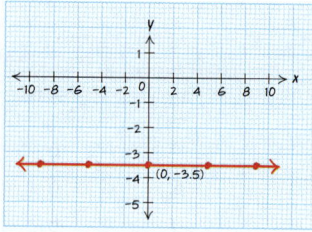

The graph is linear.

69.

$x = -1.5$	y
-1.5	-9
-1.5	-5
-1.5	0
-1.5	5
-1.5	9

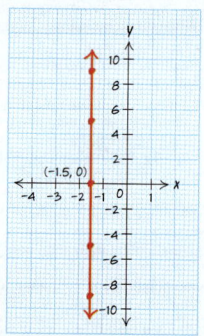

The graph is linear.

71.

$x = 0$	y
0	-9
0	-5
0	0
0	5
0	9

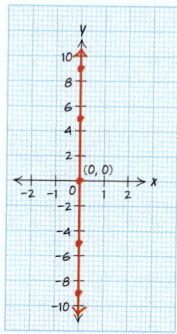

The graph is linear.

73. -19

75. $-\dfrac{4}{15}$

77. About 4.72 in.

79. 83.91 kg

81. $x + 60$

83. $-12n^2 + 6n - 21$

85. $\dfrac{1}{2}t + \dfrac{1}{2} = -1$ Use the multiplication property of equality to multiply both sides by the common denominator to clear the fractions.

$2\left(\dfrac{1}{2}t\right) + 2\left(\dfrac{1}{2}\right) = 2(-1)$

$t + 1 = -2$ Use the subtraction property of equality to isolate the variable.

$\underline{ -1 \quad -1}$

$t = -3$

87. $k = 13$

89. $x = 2$

91. $y = \dfrac{3}{5}x - 3$

93. Inequality: $x \geq -3$. Interval: $[-3, \infty)$. Number line:

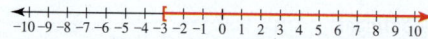

Section 3.2

1. **a.** New York City, New York, has the larger population than Los Angeles, California, during these years. We can conclude this from the graph because the line for New York City is above the line for Los Angeles.

   b. During the given time period, the population of New York City is growing faster than the population of Los Angeles. We can conclude this from the graph because the line for New York City is rising at a steeper rate than the line for Los Angeles.

3. Archery burns more calories per hour than billiards does. We can conclude this from the graph because the line for archery is steeper than the line for billiards.

5. **a.** Polyester has the higher production worldwide. We can conclude this from the graph because the line for polyester is above the line for cotton.

b. The worldwide production of cotton is growing more slowly than that of polyester. We can conclude this from the graph because the line for cotton is rising at a slower rate than the line for polyester.

7. a. The Rock-On Live Tour had more tickets available at the start of sales. We can conclude this from the graph because the line for the Rock-On Live Tour at the start of sales is above the line for the Classics Rule Tour.

b. The Rock-On Live Tour sold out first. We can conclude this from the graph because the line reaches zero on the graph before the line for the Classics Rule Tour.

c. The available tickets for the Rock-On Live Tour decreased faster than the available tickets for the Classics Rule Tour. We can conclude this from the graph because the line for Rock-On Live Tour is decreasing at a steeper rate than the line for Classics Rule Tour.

9. Slope = 2
11. Slope = $\frac{3}{2}$
13. Slope = 0
15. Slope = $\frac{1}{4}$
17. Slope is undefined.
19. Slope = $-\frac{1}{2}$
21. Slope = -3
23. Slope = 2
25. Slope = $\frac{1}{4}$
27. Slope = 3
29. Slope = 0
31. Slope = $\frac{1}{2}$
33. Slope is undefined.
35. Slope = -4

37. The blue line has a slope that is closer to zero than that of the red line. The blue line is closer to horizontal than the red line.

39. The blue line has a slope that is closer to being undefined than that of the red line. The blue line is closer to vertical than the red line.

41. Slope = 3
43. Slope = $\frac{3}{2}$
45. Slope is undefined.
47. Slope = $\frac{1}{3}$
49. Slope = -4
51. Slope = 0
53. Slope = $-\frac{1}{2}$

55. a. Slope = 0.075 million people per year

b. The population of New York City, New York, is increasing at a rate of approximately 0.075 million, or 75,000, people per year.

57. a. Slope = $9\frac{2}{3}$ thousand people per year

b. The population of Oklahoma City, Oklahoma, is increasing at a rate of approximately $9\frac{2}{3}$ thousand, or 9667, people per year.

59. a. Slope ≈ 200 calories burned per hour

b. A 130-pound person burns approximately 200 calories per hour of archery.

61. a. Slope ≈ 400 calories burned per hour

b. A 130-pound person burns approximately 400 calories per hour of competitive badminton.

63. a. Slope = 0.22 hundred million metric tons per year

b. The worldwide production of cotton is increasing by about 22 million metric tons per year.

65. a. Slope = 0.37 percentage points per year

b. The percentage of full-time employees in the computer and mathematical sciences fields is increasing by about 0.37 percentage points per year.

67. a. Slope = $-178\frac{4}{7}$ tickets per hour

b. The number of available tickets for the Rock-On Live Tour declined by approximately 179 tickets per hour.

69. a. Slope ≈ $-91\frac{41}{49}$ dollars per day

b. The amount that Deb still needed to raise for breast cancer declined by approximately $92 each day after she began fund-raising.

71. The blue line has the greater slope. The blue line is steeper than the red line.

73. The red line has the greater slope. The red line is steeper than the blue line.

75. The line is rising from left to right; therefore, it is increasing.

77. The line is falling from left to right; therefore, it is decreasing.

79. The line is horizontal; therefore, it is neither increasing nor decreasing.

81. When graphed, the line is falling from left to right; therefore, it is decreasing. The slope is negative.

83. The equation has no x-value, only a y-value. For any value of x, the y-value would remain the same; therefore, the line is horizontal and is neither increasing nor decreasing.

85. When graphed, the line is rising from left to right; therefore, it is increasing. The slope is positive.

87. The output value is increasing.
89. The output value is decreasing.
91. The output value is decreasing.
93. The output value is increasing.

95. a. The red line appears to be decreasing more rapidly.

b. Red line slope = -1

Blue line slope = $-\frac{1}{2}$

c. The line that is decreasing more rapidly has a slope that is less (has a greater negative value) than the other line.

Section 3.3

1. x-intercept is the point (2, 0).

y-intercept is the point (0, -1).

3. x-intercept is the point (3, 0).

y-intercept is the point (0, 2).

5. x-intercept is the point $(2, 0)$.
y-intercept is the point $(0, 10)$.

7. x-intercepts are the points $(0, 0)$ and $(-6, 0)$.
y-intercept is the point $(0, 0)$.

9. x-intercepts are the points $(-4, 0), (-2, 0)$, and $(4, 0)$.
y-intercept is the point $(0, -50)$.

11. a. The vertical intercept is the point $(0, 8.2)$. The population of New York City, New York, was about 8.2 million in the year 2010.

b. The vertical intercept is the point $(0, 3.8)$. The population of Los Angeles, California, was about 3.8 million in the year 2010.

13. The intercept for these lines is the point $(0, 0)$, meaning that zero calories are burned during zero hours of exercise.

15. a. The vertical intercept is the point $(0, 26.25)$. The worldwide production of cotton was about 2625 million metric tons in the year 2015.

b. The vertical intercept is the point $(0, 46)$. The worldwide production of polyester was about 4600 million metric tons in the year 2015.

17. a. The vertical intercept for the Rock-On Live Tour line is the point $(0, 2500)$, meaning that there were 2500 tickets available at the start of ticket sales. The horizontal intercept is the point $(14, 0)$, meaning that 14 hours after going on sale, the tickets were sold out.

b. The vertical intercept for the Classics Rule Tour line is the point $(0, 2000)$, meaning that there were 2000 tickets available at the start of ticket sales. The horizontal intercept is the point $(18, 0)$, meaning that 18 hours after going on sale, the tickets were sold out.

19. x-intercept is the point $(-4, 0)$.
y-intercept is the point $(0, 20)$.

21. x-intercept is the point $(9, 0)$.
y-intercept is the point $(0, -36)$.

23. x-intercept is the point $(-6, 0)$.
y-intercept is the point $(0, 3)$.

25. x-intercept is the point $(-9, 0)$.
y-intercept is the point $(0, 6)$.

27. x-intercept is the point $(-15, 0)$.
y-intercept is the point $(0, 3.75)$.

29. x-intercept is the point $(0, 0)$.
y-intercept is the point $(0, 0)$.

31. x-intercept is the point $(15, 0)$.
y-intercept is the point $(0, 3)$.

33. x-intercept is the point $\left(2\frac{1}{2}, 0\right)$.
y-intercept is the point $(0, -5)$.

35. x-intercept is the point $\left(7\frac{1}{2}, 0\right)$.
y-intercept is the point $(0, 10)$.

37. There is no x-intercept.
y-intercept is the point $(0, 6)$.

39. x-intercept is the point $(-5, 0)$.
There is no y-intercept.

41. a. The P-intercept is the point $(0, 5058)$. In 2010, there were about 5058 thousand, or 5,058,000, American Indians and Alaska Natives alone or in combination in the United States.

b. The t-intercept is approximately $(-66.2, 0)$. In about 1944, there were no American Indians and Alaska Natives alone or in combination in the United States.

c. The t-intercept does not make sense because there were more than 0 American Indians and Alaska Natives alone or in combination in the United States in 1944. The P-intercept makes sense in this situation.

43. a. The P-intercept is the point $(0, 59)$. In 2015, 59% of college enrollees were white.

b. The t-intercept is approximately $(168.57, 0)$. In about 2184, 0% of college enrollees will be white.

c. The t-intercept does not make sense because that is too far in the future and having no white college students seems unreasonable. The P-intercept seems to make sense in this situation.

45. a. The C-intercept is the point $(0, 25)$, meaning that there were 25 cars sold during week 0 of the year.

b. The w-intercept is the point $\left(12\frac{1}{2}, 0\right)$, meaning that no cars were sold during week 13, of the year.

c. The C-intercept could mean that the last week of the previous year the car lot sold 25 cars. The w-intercept could make sense if there were no cars sold during a given week.

47. a. The D-intercept is the point $(0, 12{,}500)$, meaning that José had $12,500 in credit card debt before he began to pay his cards off.

b. The m-intercept is the point $(50, 0)$, meaning that José had no credit card debt 50 months after starting to pay his cards off.

c. The D-intercept makes sense because it is reasonable to have $12,500 in credit card debt. The m-intercept makes sense in this situation, as it is reasonable that it would take José 50 months to pay off the credit cards and have no debt.

49. Slope $= -3$
x-intercept is the point $(9, 0)$.
y-intercept is the point $(0, 27)$.

51. Slope $= 8$
x-intercept is the point $(2, 0)$.
y-intercept is the point $(0, -16)$.

53. Slope $= \dfrac{1}{2}$
x-intercept is the point $(-12, 0)$.
y-intercept is the point $(0, 6)$.

55. Slope $= -\dfrac{1}{3}$

x-intercept is the point (15, 0).

y-intercept is the point (0, 5).

57. Slope $= \dfrac{2}{3}$

x-intercept is the point (6, 0).

y-intercept is the point (0, −4).

59. Slope = 1.5

x-intercept is the point (2, 0).

y-intercept is the point (0, −3).

61. Slope = 0

There is no x-intercept.

y-intercept is the point (0, −5).

63. Slope is undefined.

x-intercept is the point (6, 0).

There is no y-intercept.

65. The slope is 38, which represents that there is 38 thousand, or 38,000, more American Indians and Alaska Natives in the United States for each year.

67. The slope is 0.05, which represents that the price for navel oranges increases by about $0.05 each year.

69. The slope is −2, which represents that during each successive week of the year, Big Jim's Car Lot sells 2 fewer cars than the week before.

71. **a.** The R-intercept is the point (0, 18.2). In 2010, Nike, Inc. earned a total revenue of about $18.2 billion.

b. The slope is 2.4. The total revenue for Nike, Inc. is increasing by about $2.4 billion per year.

73. Slope $= -3$

y-intercept is the point (0, 18).

75. Slope $= -\dfrac{3}{2}$

y-intercept is the point (0, 3).

77. Slope $= \dfrac{2}{7}$

y-intercept is the point (0, −3).

79. Slope $= \dfrac{1}{2}$

y-intercept is the point $\left(0, \dfrac{7}{2}\right)$.

81. Slope = 2

y-intercept is the point (0, −13).

83. Slope $= \dfrac{1}{2}$

y-intercept is the point (0, 12).

85. Slope = 0

y-intercept is the point (0, 7).

87. $h = \dfrac{2A}{b_1 + b_2}$

89. $y = \dfrac{3}{5}x - 3$

91. About 281.64 km

93. About 1559.22 g

Section 3.4

1. $y = 2x + 3$ Graph A

   $y = -2x + 3$ Graph C

   $y = 2x - 3$ Graph B

3. $y = \dfrac{1}{2}x + 2$ Graph C

   $y = \dfrac{1}{2}x - 3$ Graph A

   $y = \dfrac{1}{2}x + 5$ Graph B

5.

7.

9.

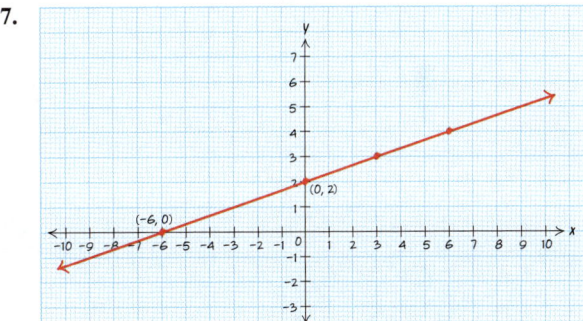

B-28 APPENDIX B Answers to Selected Exercises

11.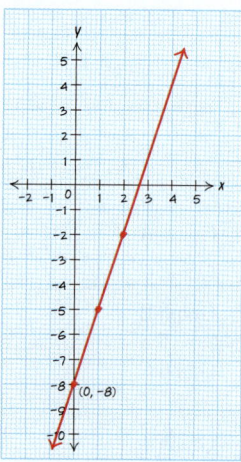

13.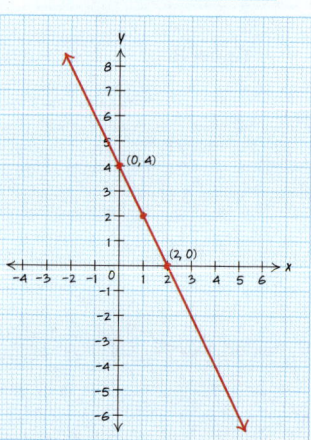

15.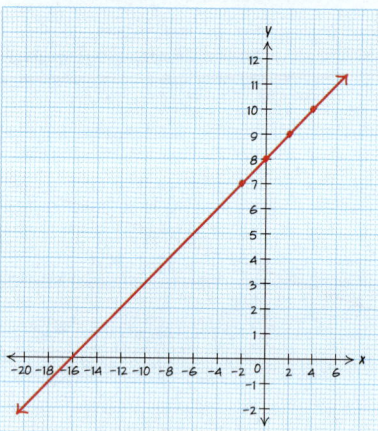

17.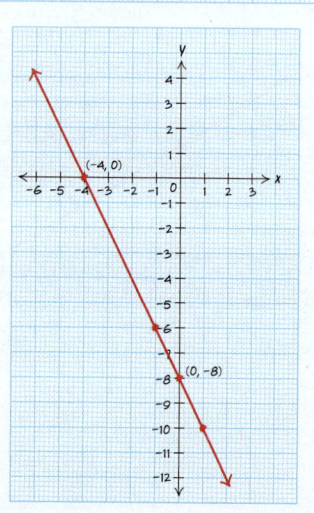

19.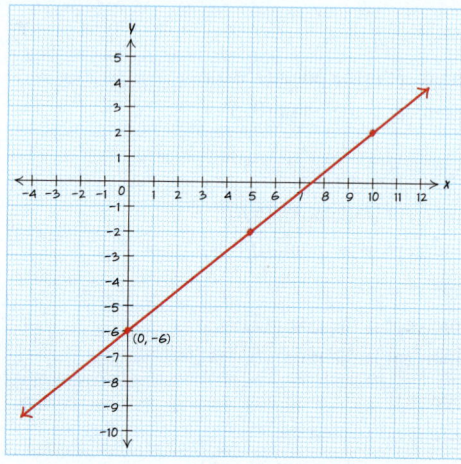

21.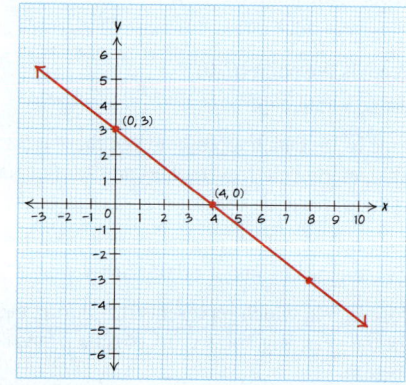

23. Jill graphed a slope of $-\dfrac{1}{2}$ instead of the slope of $\dfrac{1}{2}$.

25. Kwong graphed a slope of 2 instead of the slope of -2.

27. Marybeth graphed a slope of $\dfrac{1}{2}$ instead of the slope of 2.

29. José graphed a y-intercept of $(0, -1)$ instead of the correct y-intercept of $(0, 2)$.

31.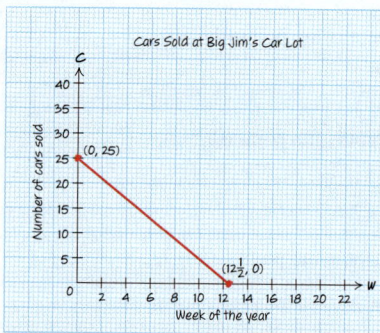

33.

35.

37.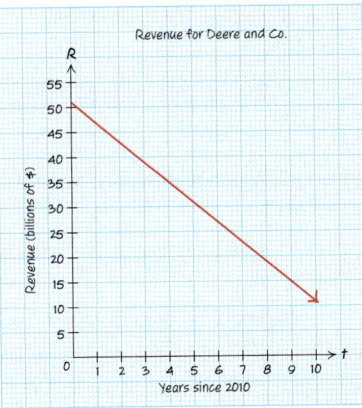

39.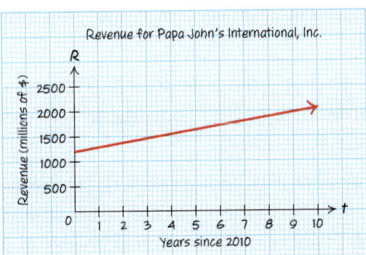

41.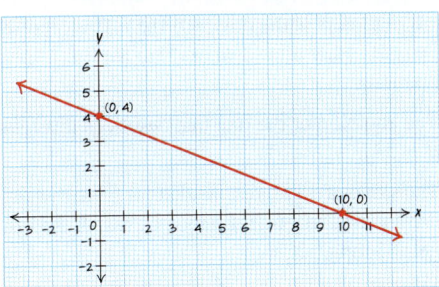

43.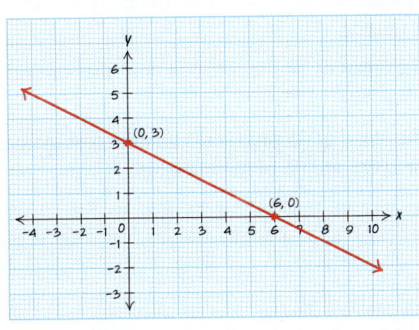

45.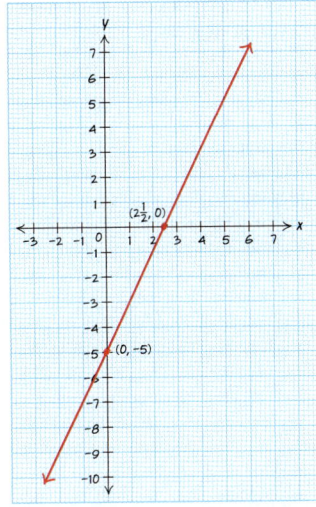

47.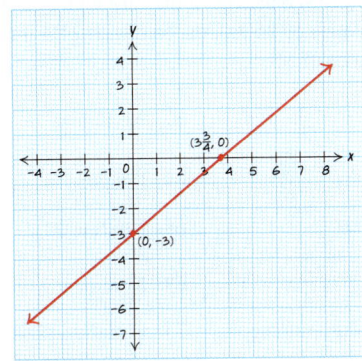

49.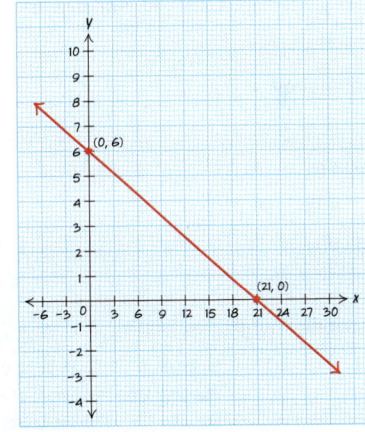

51. a.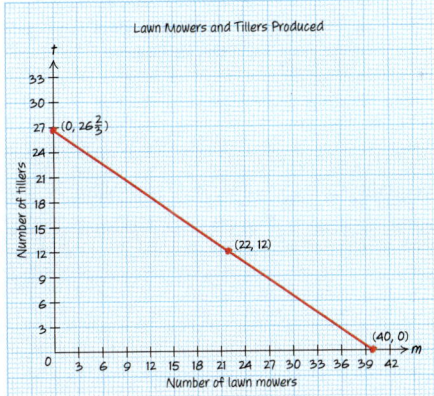

b. The factory can produce 22 lawn mowers if the workers make 12 tillers that week.

53. a.

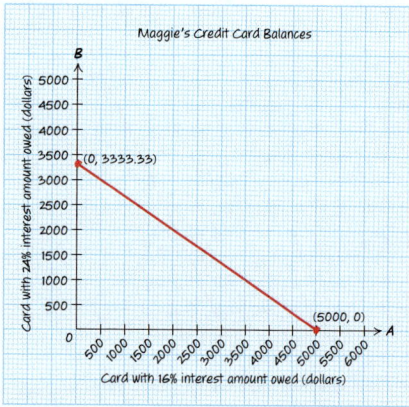

b. The A-intercept represents one possible distribution of credit card balances: if the card with 16% interest had a balance of $5000, the card with 24% interest had a balance of $0.

55. Linear; $y = \frac{3}{2}x + 7$ **57.** Nonlinear

59. Nonlinear   **61.** Linear; $y = \frac{1}{2}x - 7\frac{1}{4}$

63. Nonlinear   **65.** Nonlinear

67. Linear; $y = \frac{3}{4}x + 4\frac{1}{4}$ **69.** Nonlinear

71. Slope and y-intercept   **73.** x- and y-intercepts

75. Table of points

77. Slope and y-intercept, or x- and y-intercepts

79. x- and y-intercepts

81. Table of points

83. Parallel

85. Perpendicular

87. Neither parallel nor perpendicular

89. Perpendicular

91. Parallel

93. Neither parallel nor perpendicular

95. $-7t + 11 = -10$ Subtract 11 from both sides using the subtraction property of equality to isolate the variable term.
$\underline{-11 \quad -11}$
$-7t = -21$ Divide both sides by -7 using the division property of equality to isolate the variable.
$\frac{-7t}{-7} = \frac{-21}{-7}$
$t = 3$

97. $x = 1$

99. Inequality: $x \geq 5$. Interval: $[5, \infty)$. Number line:

101. 97.9

Section 3.5

1. $y = \frac{1}{6}x + 1$ **3.** $y = -\frac{5}{2}x$ **5.** $y = 4$

7. $y = \frac{110}{3}x - 86\frac{2}{3}$ **9.** $y = -\frac{5}{36}x + 4\frac{11}{12}$

11. The equation of the line that describes this relationship is $c = \frac{1}{100}y$.

13. The equation of the line that describes this relationship is $F = -2t + 77$.

15. The equation of the line that describes this relationship is $W = -2t + 165$.

17. The slope is $\frac{1}{100}$, which represents that there is 1 century per 100 years.

19. The slope is -2, which represents that the temperature drops by 2 degrees per hour after 1:00 P.M.

21. The slope is -2, which represents that Irina is losing 2 pounds per week.

23. The y- and c-intercepts are both the point $(0, 0)$. This represents that for 0 years, there are 0 centuries.

25. The t-intercept is the point $(38.5, 0)$. This represents that 38.5 hours after noon the temperature would be 0 degrees Fahrenheit. This value does not make sense in the given situation.

The F-intercept is the point $(0, 77)$. This represents that when $t = 0$, presumably 12:00 noon, the temperature would be 77 degrees Fahrenheit.

27. The t-intercept is the point $(82.5, 0)$. This represents that 82.5 weeks after starting her diet, Irina will weigh 0 pounds. This value does not make sense in the given situation.

The W-intercept is the point $(0, 165)$. This represents that at the start of her diet (0 weeks into her diet), Irina weighs 165 pounds.

29. a. The equation of the line that describes this relationship is $d = 53g$.

b. The slope of 53 represents that for 53 highway miles driven, the Prius uses 1 gallon of gas, or the Prius gets 53 mpg.

c. The g- and d-intercepts are both the point $(0, 0)$. This represents that for 0 miles driven, there are 0 gallons of gas used.

31. a. The equation of the line that describes this relationship is $c = 45m + 125$.

b. The slope of 45 represents that the cost increases by $45 for every month of membership.

APPENDIX B Answers to Selected Exercises   B-31

c. The m-intercept is the point $\left(-2\frac{7}{9}, 0\right)$. This represents that $2\frac{7}{9}$ months before beginning a membership, the cost would be $0. This value does not make sense in the problem situation.

The c-intercept is the point $(0, 125)$. This represents that for 0 months of membership, the cost is $125.

33. $y = x - 9$ 35. $y = \frac{2}{3}x + 1$ 37. $y = -4x - 2$

39. $y = 5$ 41. $y = 3x - 15$ 43. $y = -2x - 7$

45. $y = 6x - 9$ 47. $y = \frac{2}{3}x$ 49. $y = 5$

51. $y = 3x + 2$ 53. $y = -3x + 4$ 55. $y = \frac{1}{2}x + 1$

57. $y = -6$ 59. $2x + 3y = 21$ 61. $y = \frac{3}{4}x - 6$

63. $y = -6x - 63$ 65. $-7x + y = 9$ 67. $x = 5$

69. $x = 0$ 71. $y = 6x + 42$ 73. $y = \frac{1}{5}x + 4$

75. $y = -\frac{8}{3}x - 28$ 77. $y = -\frac{4}{5}x - 3$ 79. $y = 6$

81. $y = 5$ 83. $x = 1$ 85. $x = 5$

87. 19   89. $\frac{23}{2}$ 91. $5x - 9$

93. $\frac{n^2}{4}$

95. $4 - 2(3 - t) = 5(2t - 1) + 7$ Simplify each side using the distributive property.
     $4 - 6 + 2t = 10t - 5 + 7$
     $-2 + 2t = 10t + 2$ Simplify each side by combining like terms.

     $\underline{-2t \quad -2t}$ Subtract $2t$ from both sides using the subtraction property of equality to combine the variable terms.
     $-2 = 8t + 2$

     $\underline{-2 \quad\quad -2}$ Subtract 2 from both sides using the subtraction property of equality to isolate the variable term.
     $-4 = 8t$

     $\frac{-4}{8} = \frac{8t}{8}$ Divide both sides by 8 using the division property of equality to isolate the variable.

     $-\frac{1}{2} = t$

97. $x = 9$

99.

Percent	Decimal	Fraction
12.5%	0.125	$\frac{125}{1000} = \frac{1}{8}$

101. Inequality: $-2 < x < 4$. Interval: $(-2, 4)$. Number line:

Section 3.6

1.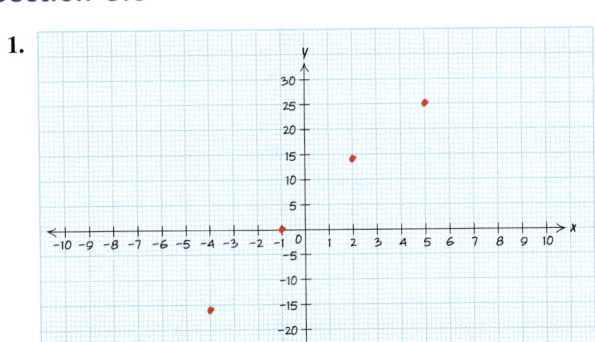

Yes, the data are approximately linear.

3.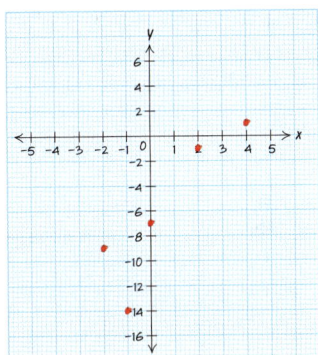

No, the data are not linear.

5.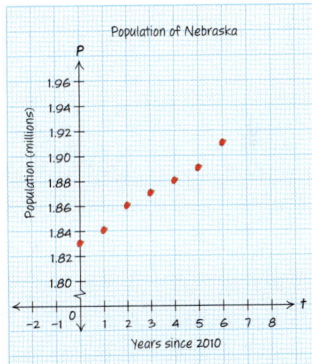

Yes, the data are approximately linear.

7.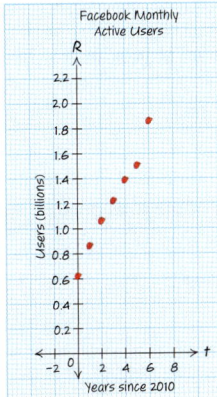

Yes, the data are approximately linear.

9. Graph 1   11. Graph 1

B-32 APPENDIX B Answers to Selected Exercises

For Exercises 13–30 answers may vary slightly.

13. a, b.

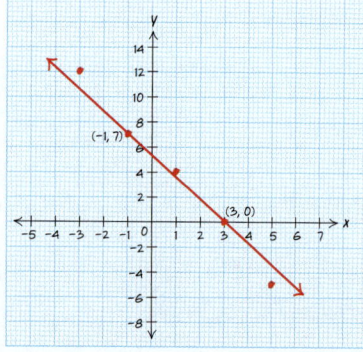

Yes, the data are approximately linear.

c. $y = -\dfrac{7}{4}x + \dfrac{21}{4}$

15. a, b.

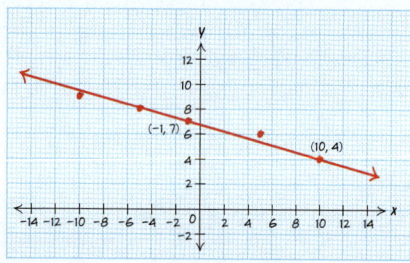

Yes, the data are approximately linear.

c. $y = -\dfrac{3}{11}x + \dfrac{74}{11}$

17. a, b.

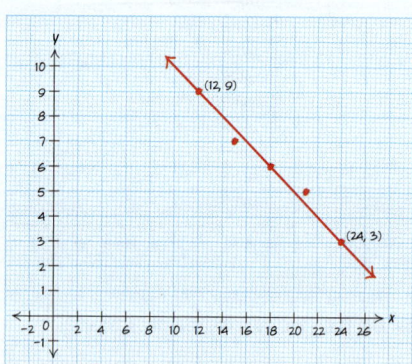

Yes, the data are approximately linear.

c. $y = -\dfrac{1}{2}x + 15$

19. a. Let P = the number of immigrants in the United States as a percentage of the total population, and let t = years since 1950.

b, c.

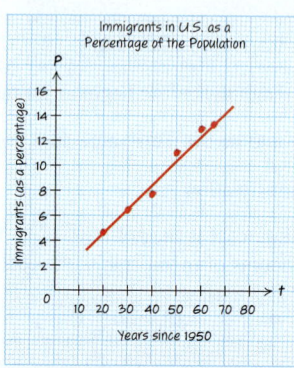

Yes, the data are approximately linear.

d. $P = 0.191t + 0.878$

21. a. Let A = the age of the person in years, and let R = the target heart rate (in beats per minute) during exercise.

b, c.

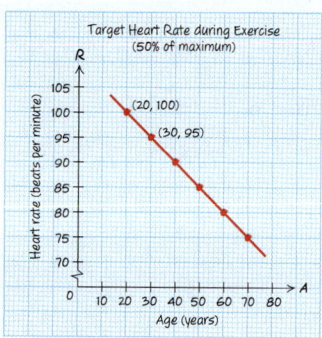

Yes, the data are linear.

d. $R = -\dfrac{1}{2}A + 110$

23. a. $P = 9.473$. According to the model, the estimated percentage of the U.S. population who were immigrants in the year 1995 was 9.47%.

b. $P = 14.248$. According to the model, the predicted percentage of the U.S. population who are immigrants in the year 2020 will be 14.25%.

c. Slope = 0.191. The percentage of the U.S. population that are immigrants is increasing by about 0.191 percentage points per year.

25. a. $R = 97.5$. According to the model, the estimated target heart rate of a 25-year-old is 97.5 beats per minute.

b. $R = 70$. According to the model, the estimated target heart rate of an 80-year-old is 70 beats per minute.

c. Slope = $-\dfrac{1}{2}$. A person's target heart rate decreases by $\dfrac{1}{2}$ beats per minute per year of age.

27. a.

Month	Unemployment Rate
1 (= January)	5.7
3 (= March)	5.6
5 (= May)	5.5
7 (= July)	5.4
9 (= September)	5.3

b. Let t = the month and let R = the unemployment rate in California for the year 2016.

c.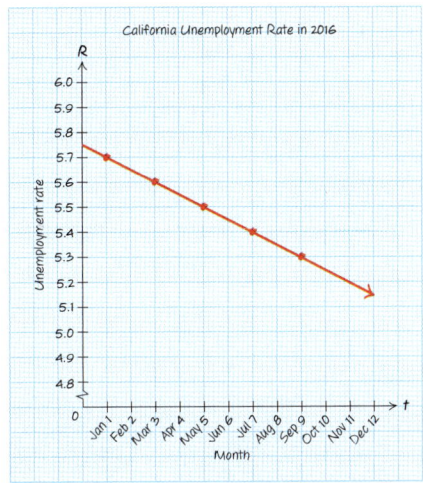

d. $R = -0.1t + 5.8$

e. $R = 4.6$. According to the model, the predicted unemployment rate in December 2016 was 4.6%.

29. a. Let t = years since 2010 and let R = annual revenues for JetBlue Airways Corporation in millions of dollars.

b. $R = 459t + 4064$

c. Slope = 459. The annual revenues for JetBlue Airways Corporation is increasing by about $459 million per year.

d. $R = 7736$. In 2018, JetBlue Airways Corporation had annual revenues of about $7736 million.

31. a. $P = 2.04$. In 2018, about 2.04% of twelfth grade students smoked.

b. $P = -0.4$. In 2020, about −0.4% of twelfth grade students will smoke. This is model breakdown, since there cannot be a negative percentage of students who smoke.

33. a. $P = 13.75$. In 2010, about 13.75% of adults age 20 and over in the United States had high cholesterol.

b. $P = 8.5$. In 2020, about 8.5% of adults age 20 and over in the United States will have high cholesterol.

35. a. $M = -804$. In 2000, there were about −804 billion mobile payment transactions in the United States. This is model breakdown, since there cannot be a negative number of transactions.

b. $M = 214.06$. In 2020, there were about 214.06 billion mobile payment transactions in the United States.

37. a. Slope = −1.22. The percentage of twelfth grade students who smoke is decreasing by about 1.22 percentage points per year.

b. Horizontal intercept ≈ (9.67, 0). In about 2020, no twelfth grade students will smoke. This is model breakdown, since this is probably very unlikely.

c. Vertical intercept = (0, 11.8). In 2010, about 11.8% of twelfth grade students smoked.

39. a. Slope = −0.525. The percentage of adults age 20 and over in the United States that have high cholesterol is decreasing by about 0.525 percentage point per year.

b. Horizontal intercept ≈ (26.19, 0). In about 2036, no adults in the United States will have high cholesterol. This is model breakdown, since this is probably very unlikely.

c. Vertical intercept = (0, 13.75). In 2010, about 13.75% of adults age 20 and over in the United States had high cholesterol.

41. a. Slope = 50.903. The number of mobile payment transactions in the United States is increasing by about 50.903 billion per year.

b. Horizontal intercept ≈ (0.79, 0). In about 2016, there were no mobile payment transactions in the United States. This is model breakdown, since these transactions began before 2016.

c. Vertical intercept = (0, −40.455). In 2015, there were about −40.455 billion mobile payment transactions in the United States. This is model breakdown, since we cannot have a negative number of transactions.

Chapter 3 Review Exercises

1.

h = hours of walking	C = calories burned
1	87
2	174
3	261
4	348
6	522

2.

n = number of plants purchased	C = cost (dollars)
2	10.50
4	21.00
8	42.00
14	73.50
20	105.00

3.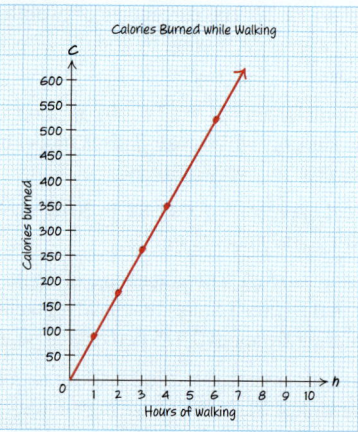

4.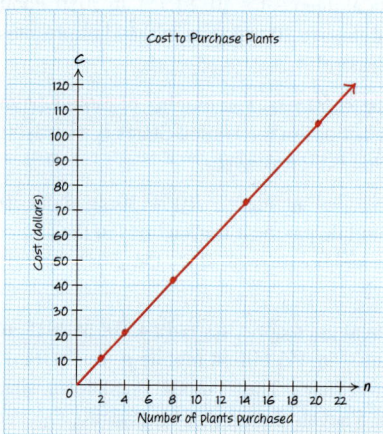

5.

x	y = −3x + 8
−1	11
0	8
1	5
3	−1
5	−7

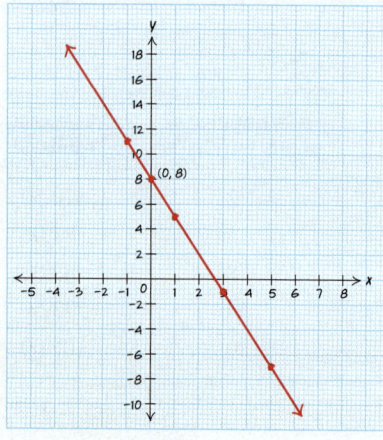

6.

x	y = −5x − 3
−2	7
−1	2
0	−3
1	−8
2	−13

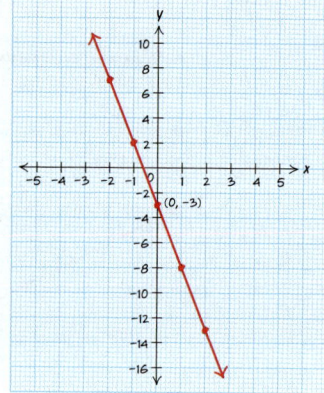

7.

x	$y = \frac{3}{4}x + 1$
−8	−5
−4	−2
0	1
4	4
8	7

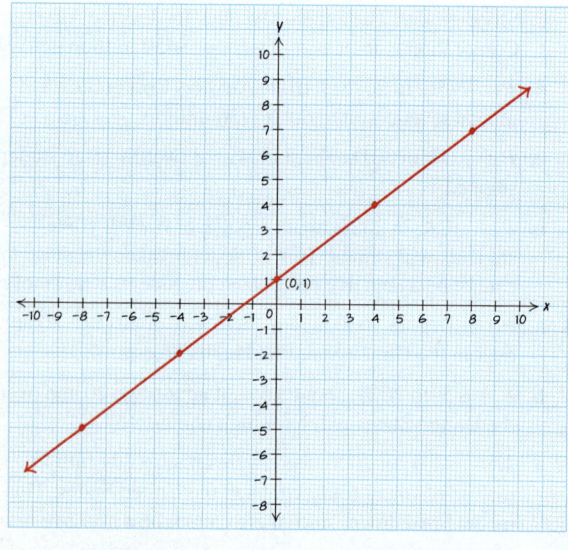

APPENDIX B Answers to Selected Exercises   B-35

8.

x	$y = -\dfrac{1}{5}x - 6$
-30	0
-10	-4
0	-6
10	-8
20	-10

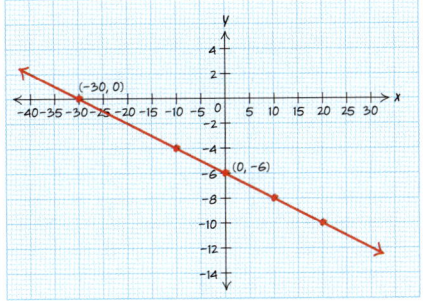

9.

x	$y = 5$
-9	5
-5	5
0	5
5	5
9	5

10.

x	$y = -4$
-9	-4
-5	-4
0	-4
5	-4
9	-4

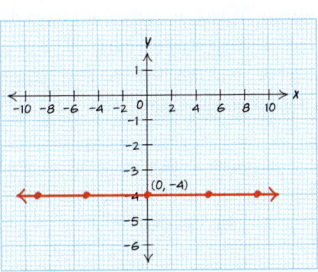

11.

$x = -3$	y
-3	-9
-3	-5
-3	0
-3	5
-3	9

12.

$x = 5$	y
5	-9
5	-5
5	0
5	5
5	9

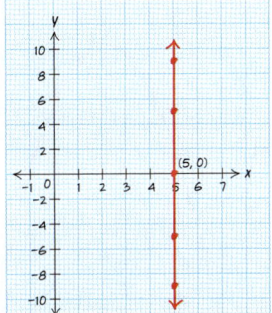

13.

x	$y = x^2 + 2$
-4	18
-3	11
-2	6
-1	3
0	2
1	3
2	6
3	11
4	18

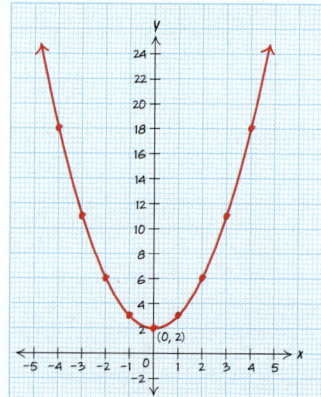

14.

x	$y = -x^2 + 5$
-4	-11
-3	-4
-2	1
-1	4
0	5
1	4
2	1
3	-4
4	-11

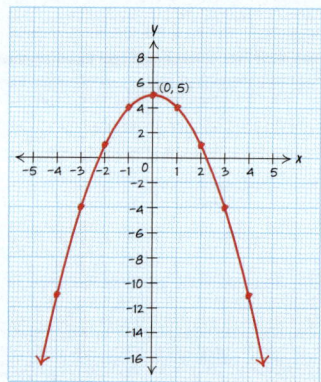

15.

x	$y = |9 - 3x|$
-1	12
0	9
1	6
2	3
3	0
4	3
5	6
6	9
7	12

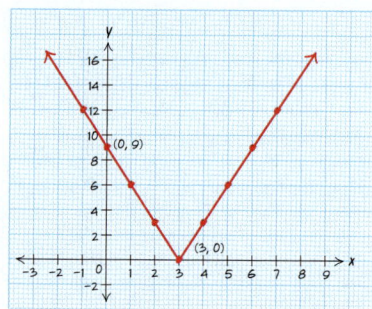

16.

x	$y = |x + 2|$
-5	3
-4	2
-3	1
-2	0
-1	1
0	2
1	3
2	4
3	5

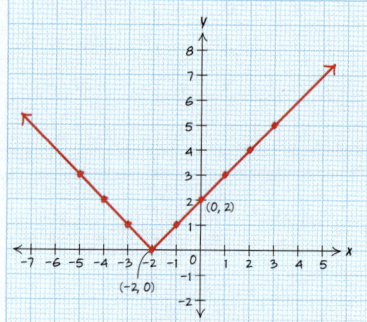

17. a. Wichita, Kansas, has a larger population than Bakersfield, California, during these years. We can conclude this from the graph because the line for Wichita is above the line for Bakersfield.

   b. Bakersfield, California, is growing faster than Wichita, Kansas, during these years. We can conclude this from the graph because the line for Bakersfield is rising at a steeper rate than the line for Wichita.

18. a. San Antonio, Texas, has a larger population than St. Paul, Minnesota, during these years. We can conclude this from the graph because the line for San Antonio is above the line for St. Paul.

   b. San Antonio, Texas, is growing faster than St. Paul, Minnesota, during these years. We can conclude this from the graph because the line for San Antonio is rising at a steeper rate than the line for St. Paul.

19. Slope $= \dfrac{1}{2}$ 20. Slope $= -\dfrac{1}{4}$ 21. Slope $= -\dfrac{3}{4}$

22. Slope $= \dfrac{3}{8}$ 23. Slope $= \dfrac{2}{5}$ 24. Slope $= \dfrac{1}{3}$

25. The slope for Bakersfield, California, is 5 thousand people per year. This means that the population of Bakersfield is increasing at a rate of approximately 5000 people per year.

26. The slope for San Antonio, Texas, is $26\frac{1}{3}$ thousand, or 26,333, people per year. This means that the population of San Antonio is increasing at a rate of approximately 26,333 people per year.

27. x-intercept is the point $(-3, 0)$.
    y-intercept is the point $(0, -1)$.

28. x-intercepts are the points $(-2, 0), (0, 0), (3, 0),$ and $(4, 0)$.
    y-intercept is the point $(0, 0)$.

29. a. The vertical intercept for the Bakersfield line is $(0, 349)$. This means that the population of Bakersfield was approximately 349 thousand, or 349,000, people in the year 2010.
    b. The vertical intercept for the Wichita line is $(0, 383)$. This means that the population of Wichita was approximately 383 thousand, or 383,000, people in the year 2010.

30. a. The vertical intercept is $(0, 2000)$. This means that the balance of the credit card debt is \$2000 at the beginning of the year.
    b. The horizontal intercept is $(16, 0)$. This means that the balance of the credit card debt was paid down to zero 16 weeks since the beginning of the year.

31. x-intercept is the point $(4, 0)$.
    y-intercept is the point $(0, 3)$.

32. x-intercept is the point $(3, 0)$.
    y-intercept is the point $(0, -6)$.

33. a. The T-intercept is $(0, 1)$. This means that it would take 1 hour to paint 0 dolls. In this problem situation, it means that in addition to the time it takes to paint each doll, there is an additional hour of work to do.
    b. The d-intercept is $\left(-\frac{2}{3}, 0\right)$. This means that it would take 0 hours to paint $-\frac{2}{3}$ dolls. This does not make sense in terms of the problem situation.

34. a. The C-intercept is $(0, 10)$. This means that there is a base cost of \$10 to produce the dolls regardless of how many dolls are painted.
    b. The d-intercept is $\left(-\frac{2}{5}, 0\right)$. This means that it would cost nothing to paint $-\frac{2}{5}$ dolls. This does not make sense in terms of the problem situation.

35. x-intercept is the point $(3, 0)$.
    y-intercept is the point $(0, 9)$.
    Slope $= -3$

36. x-intercept is the point $(-6, 0)$.
    y-intercept is the point $(0, 12)$.
    Slope $= 2$

37. x-intercept is the point $(7.077, 0)$.
    y-intercept is the point $(0, -7.112)$.
    Slope $= 1.005$

38. x-intercept is the point $(39, 0)$.
    y-intercept is the point $(0, 9.75)$.
    Slope $= -0.25$

39. x-intercept is the point $(-4, 0)$.
    y-intercept is the point $(0, -2)$.
    Slope $= -\frac{1}{2}$

40. x-intercept is the point $(-12, 0)$.
    y-intercept is the point $(0, -4)$.
    Slope $= -\frac{1}{3}$

41. x-intercept is the point $(-5, 0)$.
    y-intercept is the point $(0, 4)$.
    Slope $= \frac{4}{5}$

42. x-intercept is the point $(-10, 0)$.
    y-intercept is the point $(0, 2)$.
    Slope $= \frac{1}{5}$

43. a. The slope is 4.7 thousand, or 4700, people per year. This means that the population of Alaska is increasing at a rate of approximately 4700 people per year.
    b. The P-intercept is $(0, 714)$. This means that the population of Alaska was approximately 714 thousand, or 714,000 people in the year 2010.

44. a. The slope is 2.5. This means that the cost increases by \$2.50 for every set of clamps produced.
    b. The C-intercept is $(0, 200)$. This means that there is a base cost of \$200 regardless of how many sets of clamps are produced.

45.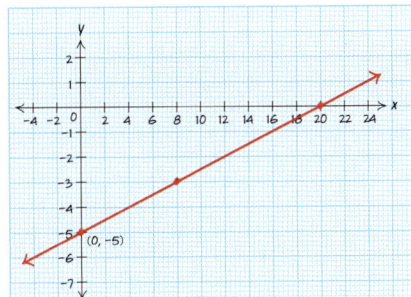

46.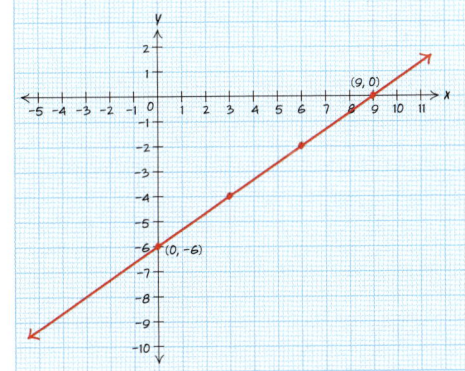

B-38 APPENDIX B Answers to Selected Exercises

47.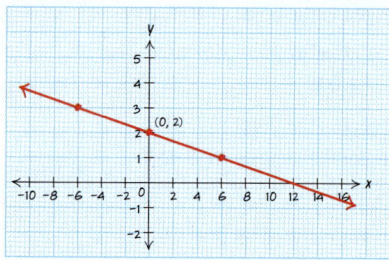

48.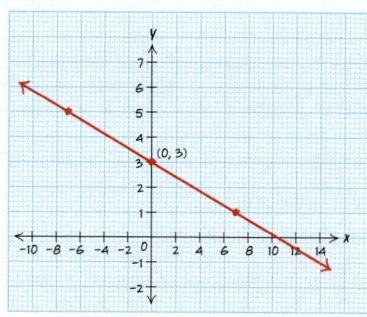

49.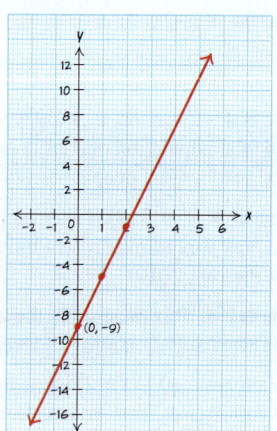

50.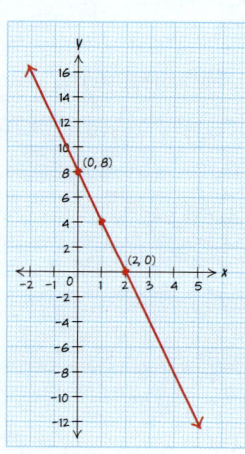

51.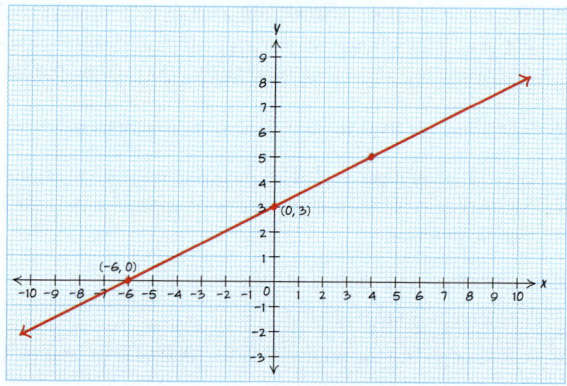

52.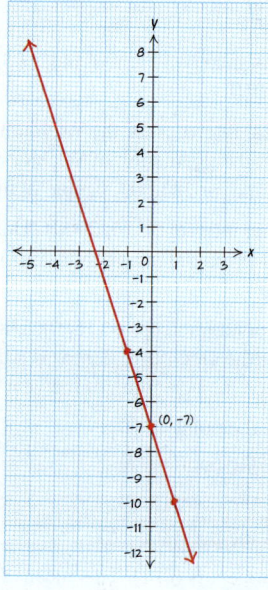

53.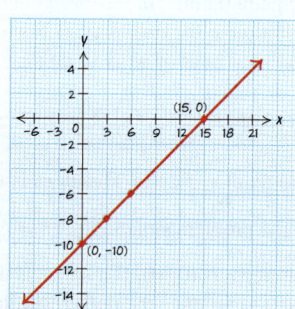

54.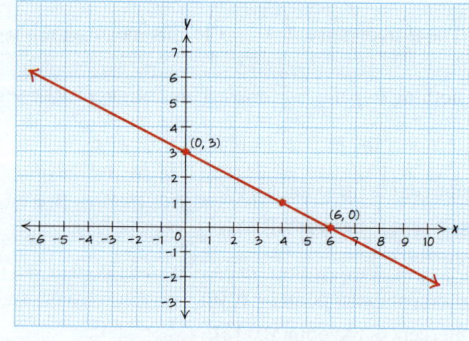

55.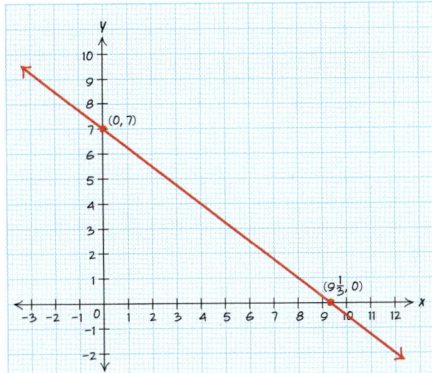

56.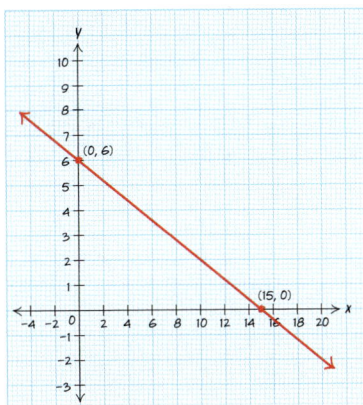

57.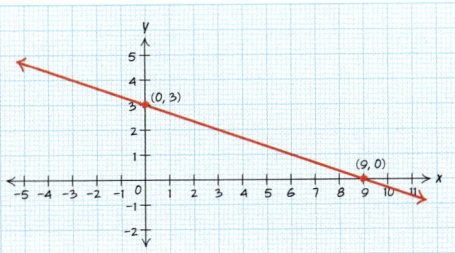

58.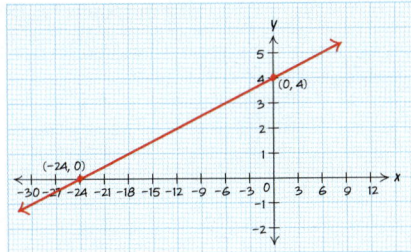

59. $y = -4x + 3$

60. $y = \frac{2}{3}x - 6$

61. Let $C =$ the number of calories burned and let $t =$ the time in hours spent running. The equation of the line that describes this relationship is $C = 852t$.

62. Let $C =$ the number of calories burned and let $t =$ the time in hours spent golfing using a handcart. The equation of the line that describes this relationship is $C = 264t$.

63. The slope of 852 means that a 160- to 170-pound person will burn 852 calories per hour while running at a pace of 8 mph.

64. The slope of 264 means that a 160- to 170-pound person will burn 264 calories per hour while golfing using a handcart.

65. $y = -x + 8$ **66.** $y = -\frac{2}{3}x + 5$ **67.** $y = \frac{1}{10}x + 5$

68. $y = \frac{1}{2}x - 7$ **69.** $y = -11$ **70.** $y = 12$

71. $x = 1$ **72.** $x = -4$ **73.** $y = \frac{1}{2}x + 6$

74. $y = -\frac{1}{3}x + 2$ **75.** $y = -\frac{3}{4}x + 2$ **76.** $y = -\frac{1}{6}x + 3$

77. Parallel **78.** Parallel **79.** Perpendicular

80. Perpendicular

81. $y = \frac{1}{8}x + 4$ **82.** $y = \frac{5}{2}x - 26$ **83.** $y = \frac{3}{4}x - 14$

84. $y = 2x - 1$ **85.** $5x + 6y = 42$ **86.** $4x + 5y = -45$

87. $y = \frac{6}{5}x - 6$ **88.** $y = \frac{1}{2}x + 10$ **89.** $2x + y = -7$

90. $7x + y = 32$ **91.** $y = \frac{2}{3}x - 8$ **92.** $y = -\frac{5}{7}x - 3$

93.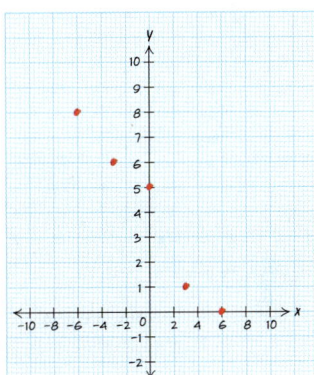

Yes, the data show a linear pattern.

94.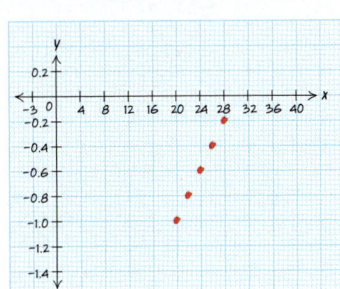

Yes, the data show a linear pattern.

B-40 APPENDIX B Answers to Selected Exercises

95.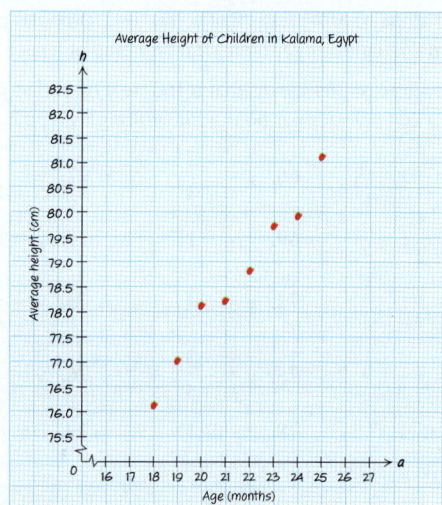

Yes, the data show a linear pattern.

96.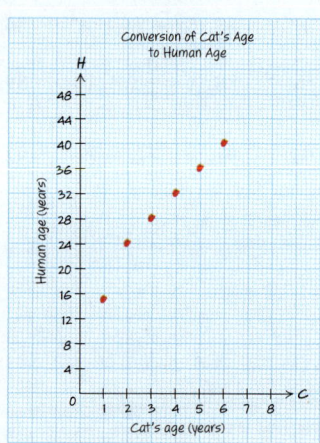

Yes, the data show a linear pattern.

97. Graph 2   **98.** Graph 2

99.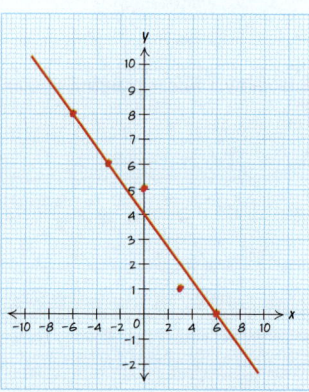

100.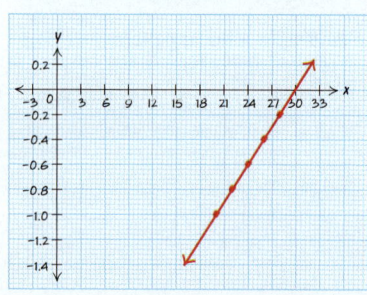

101.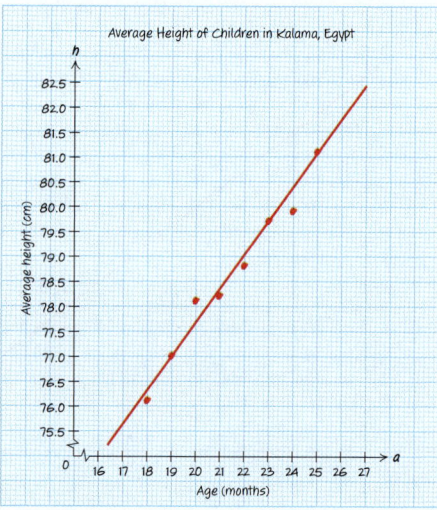

102.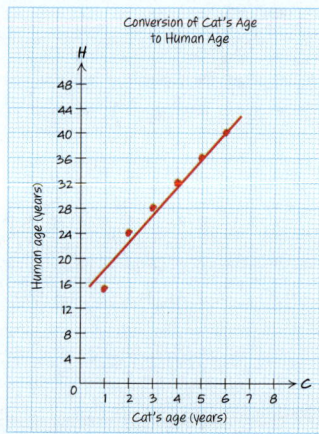

103. $y = -\dfrac{2}{3}x + 4$ **104.** $y = \dfrac{1}{10}x - 3$

105. Let a = the age in months and let h = the average height in centimeters. $h = 0.675a + 64.175$

106. Let C = the cat's age in years and let H = the human age in years. $H = 4.5C + 13.5$ using the points $(1, 18)$ and $(5, 36)$

107. a. Let h = the horse's age in years and let a = the human age in years.

b.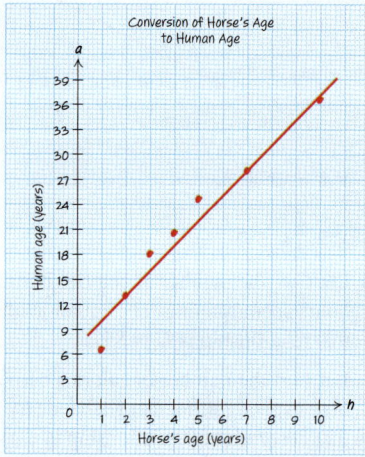

c. $a = 3h + 7$

APPENDIX B Answers to Selected Exercises   B-41

d. $a = 82$. According to the model, a 25-year-old horse is 82 years old in human years.

e. $h = -1$. According to the model, a 4-year-old human correlates to an age of -1 for a horse. This answer does not make sense, which means that this is an example of model breakdown.

108. a. Let $t =$ years since 2010 and let $p =$ the population of Montana in thousands.

b.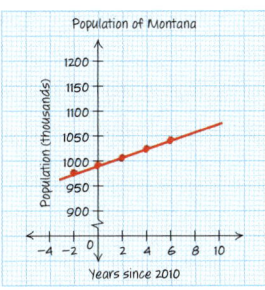

c. $p = 8t + 991$

d. $p = 1071$. According to the model, the population of Montana in the year 2020 is estimated to be 1071 thousand, or 1,071,000, people. This answer seems reasonable.

e. $p = -209$. According to the model, the population of Montana in the year 1860 is estimated to be -209 thousand people. This answer is not reasonable, which means that it is an example of model breakdown.

Chapter 3 Test

1.

x	$y = -5x + 2$
-2	12
-1	7
0	2
1	-3
2	-8

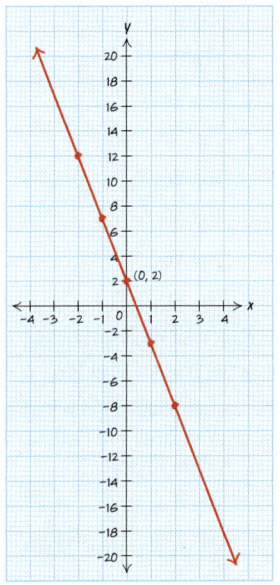

2.

x	$y = x^2 - 1$
-4	15
-3	8
-2	3
-1	0
0	-1
1	0
2	3
3	8
4	15

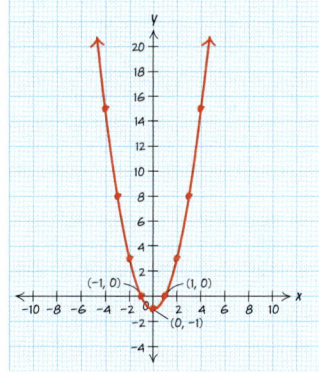

3. Slope $= 1$

4. The slope is 18,744.75. This means that during the years 1999–2003, there were approximately 18,745 more children diagnosed with autism each year.

5. x-intercept is the point $(3, 0)$.

   y-intercept is the point $(0, -2)$.

6. a. The vertical intercept is $(0, 967)$. In 2010, 967 trillion Btus of wind energy were consumed.

   b. The vertical intercept is $(0, 30)$. In 2010, 30 trillion Btus of solar energy were consumed.

7. Slope $= \dfrac{3}{5}$

   x-intercept is the point $\left(\dfrac{2}{3}, 0\right)$.

   y-intercept is the point $\left(0, -\dfrac{2}{5}\right)$.

8.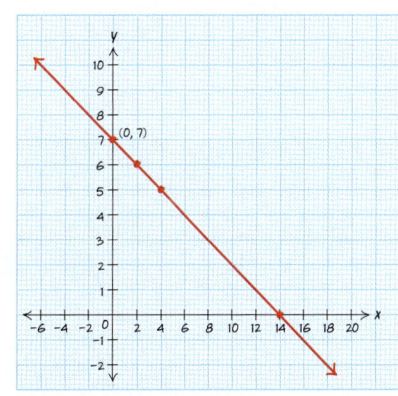

B-42 APPENDIX B Answers to Selected Exercises

9.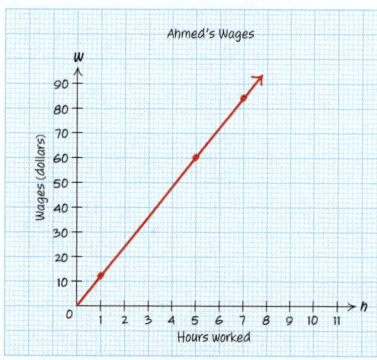

10.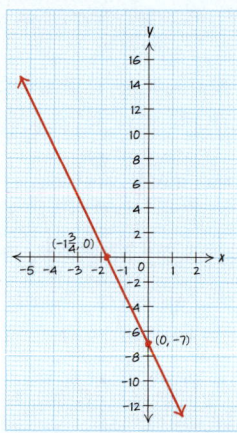

11. a. Nonlinear
  b. Linear
  $y = 3x + 14$

12. $y = 5x - 7$ **13.** $y = \dfrac{1}{4}x + 2\dfrac{1}{2}$

14. $W = -5t + 235$

15. The slope is -5, which represents that Paul is losing weight at a rate of 5 pounds per week.

16. $y = \dfrac{3}{4}x - 5$ **17.** Perpendicular

18. $y = \dfrac{2}{5}x - 1$

19. a. Let $d =$ the dog's age in years, and let $h =$ the human age in years.
  b.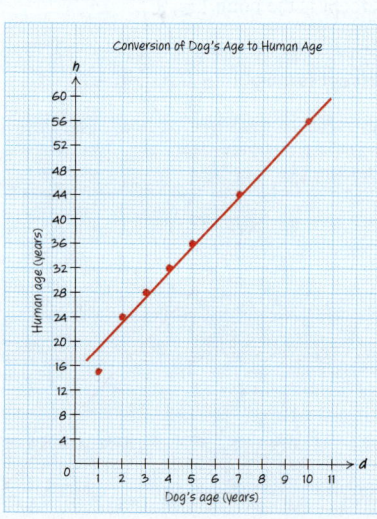

  c. See scatterplot in part b.
  d. $h = 4.17d + 14.83$ using the points $(1, 19)$ and $(7, 44)$
  e. $h = 48.19$. According to the model, the age of an 8-year-old dog approximately corresponds to a human age of 48 years. This answer seems reasonable.
  f. $d \approx -2.36$. According to the model, the age of a 5-year-old human approximately corresponds to a dog's age of -2 years. This answer is not reasonable because you cannot have a negative age. This is a case of model breakdown.

20. a. Let $t =$ years since 2010 and let $p =$ the population of West Virginia in millions.
  b.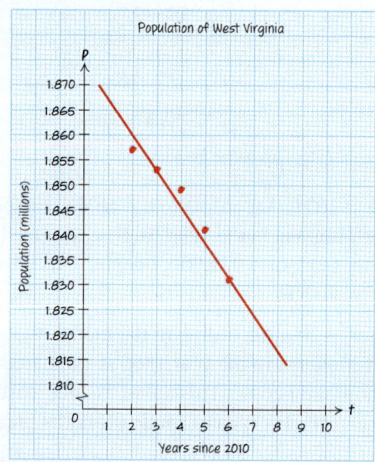

  Yes, the scatterplot looks linear.
  c. See scatterplot in part b.
  d. $p = -0.0073t + 1.875$
  e. $p = 1.802$. According to the model, the population of West Virginia in the year 2020 will be about 1.802 million. This answer seems reasonable.

CHAPTER 4
Section 4.1

1. $(-5, 6)$ **3.** $(2, 6)$
5. $(0, 3)$ **7.** $(9, 1)$
9. a. No **b.** Yes
11. a. No **b.** No
13. a. Yes **b.** Yes
15. a. No **b.** No
17. a. No **b.** Yes
19. a. Yes **b.** Yes

21. The CHD mortality rate for Serbia and Japan are both about 4% when the serum cholesterol level is about 125 mg/dL.

23. a. Let $h =$ hours the music studio is rented for, $M =$ cost to rent Music by You studio in dollars, and $B =$ cost to rent Bling Music Studio in dollars.

$$M = 100 + 60h$$
$$B = 150 + 50h$$

b.

Hours	Cost for Music by You
1	160
2	220
3	280
4	340
5	400
6	460
7	520

Hours	Cost for Bling Music Studio
1	200
2	250
3	300
4	350
5	400
6	450
7	500

c. Both studios charge $400 to rent their studio for 5 hours.

d. If you need a studio for 7 hours, you should rent it from Bling Music Studio, since their per hour price is lower and that results in a lower total cost.

25. a. Let m = number of months;

let C = cost in dollars.

Fitness Plus: $C = 15m + 100$

Rock's Gym: $C = 20m + 80$

b.

Fitness Plus

m	C
0	100
1	115
2	130
3	145
4	160
5	175
6	190

Rock's Gym

m	C
0	80
1	100
2	120
3	140
4	160
5	180
6	200

c. The two clubs have the same cost of $160 after four months.

d. Fitness Plus is the cheaper club to join for 6 months.

27. $(1, 1)$ 29. $(100, -4)$ 31. $(-3, 2)$ 33. No solution

35. a. Yes   b. No

37. a. No   b. Yes

39.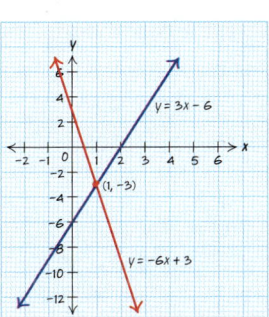

$(1, -3)$, consistent

41.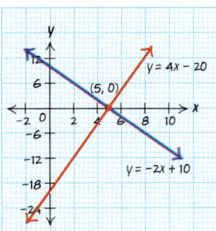

$(5, 0)$, consistent

43.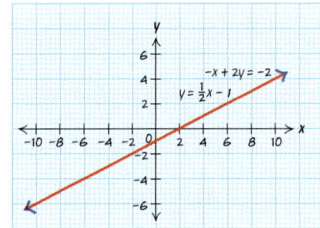

Infinitely many solutions, consistent

45.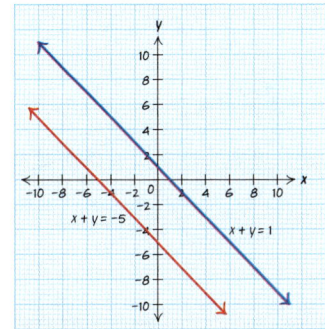

No solution, inconsistent

47.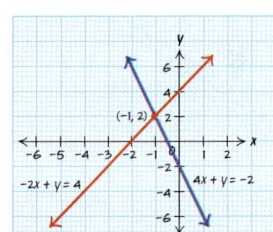

$(-1, 2)$, consistent

49.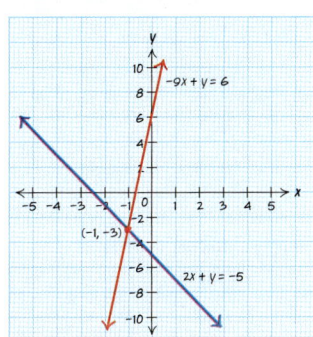

$(-1, -3)$, consistent

51.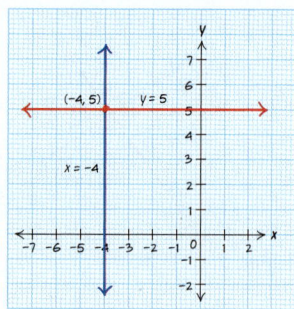
(−4, 5), consistent

53.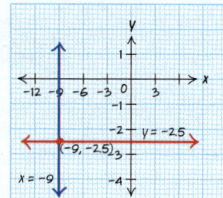
(−9, −2.5), consistent

55.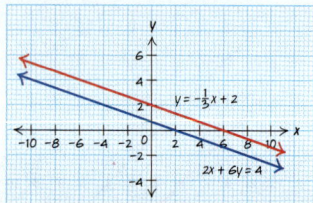
No solution, inconsistent

57.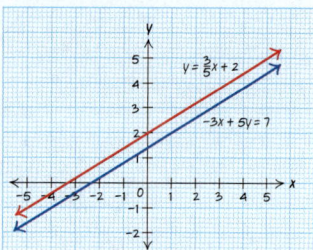
No solution, inconsistent

59.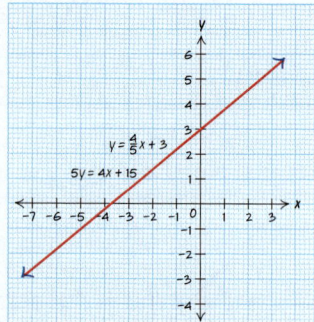
Infinitely many solutions, consistent

61.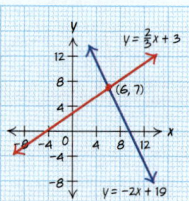
(6, 7), consistent

63. Jason's mobile auto repair breaks even after 12 hours with a cost and revenue of $900.

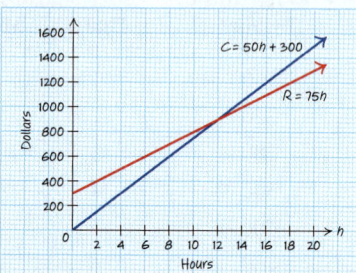

65. Hummus R Us breaks even after selling 115 products for a total cost and revenue of $690.

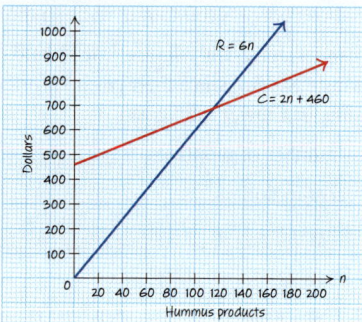

67. a. Let h = number of hours; let C = cost in dollars.
$$C = 60h + 10$$
b. $C = 55h + 15$
c.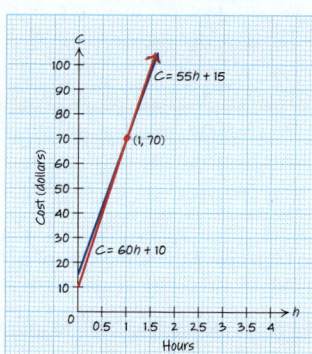

(1, 70) The cost to hire Mike or Josh is the same after 1 hour. They both cost $70.

69. The oven costs $800, and the microwave costs $400.00.
71. The length is 9 cm, and the width is 3 cm.
73. a. Let p = number of people; let C = cost in dollars;
$$C = 5.75p + 255$$
b. Let R = revenue; $R = 10.00p$.
c.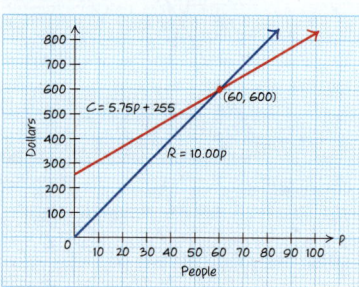

(60, 600)

d. The track team has to sell 60 tickets for the fund-raiser to break even.

75. Parallel 77. Neither
79. Perpendicular 81. Parallel
83. Parallel 85. One solution
87. No solutions 89. Infinitely many solutions
91. $y = \dfrac{5}{3}x + 5$ 93. $x = 6y + 14$ 95. $r = \dfrac{c}{2\pi}$
97. About 7.40 ml 99. About 0.10 l

Section 4.2

1. $y = -3x - 4$ 3. $x = -4y - 3$
5. $y = \dfrac{1}{2}x - 3$ 7. $x + 3y = 7$
   $x = -3y + 7$
9. $-a + 9b = 6$ 11. $y = -2x + 1$
   $a = 9b - 6$
13.
$x = 3y - 1$
$2x + y = 12$
$2(3y - 1) + y = 12$ Substitute $3y - 1$ for x in the second equation.
$6y - 2 + y = 12$ Distribute the 2 on the left side to simplify.
$7y - 2 = 12$ Combine like terms to simplify.
$\underline{+2 \quad +2}$ Use addition property of equality to isolate the variable term.
$7y = 14$
$\dfrac{7y}{7} = \dfrac{14}{7}$ Use the division property of equality to isolate the variable.
$y = 2$
$x = 3(2) - 1$ Find the value for x by substituting in the value for y.
$x = 5$
$(5, 2)$

15. $(-9, 2)$ 17. $\left(\dfrac{15}{8}, \dfrac{25}{4}\right)$
19. $(2, -7)$ 21. $(-2, -2)$
23. a. Let x = the first number; let y = the second number.
    b. $x + y = 65$
       $x = 12y$
    c. $x = 60, y = 5$
25. The numbers are 5 and 20.
27. The numbers are 2 and 10.
29. The numbers are 2 and 9.
31. The numbers are 4 and 8.
33. a. Let w = weekly sales; let P = pay in dollars.
    b. $P = 100 + 0.05w$
       $P = 120 + 0.04w$
    c. (2000, 200). If Ethan makes $2000 in weekly sales, he will earn $200 in either job.

35. a. Let f = number of hours of the first job; let s = number of hours of the second job.
    b. $f + s = 20$
       $10.25f + 11.50s = 220$
    c. $f = 8, s = 12$
       Marcus should work 8 hours at the first job and 12 hours at the second job.
37. $(1, 1)$ 39. $(-2, -3)$
41. $(2, 1)$ 43. $\left(\dfrac{11}{5}, \dfrac{2}{5}\right)$
45. $\left(\dfrac{7}{4}, \dfrac{3}{8}\right)$ 47. $(1, 1)$
49. $(-2, 3)$ 51. No solution
53. Infinitely many solutions.
55. No solution 57. $(-2.1, 1.3)$ 59. $(-10, 0)$
61. Infinitely many solutions.
63. $\left(\dfrac{37}{6}, \dfrac{-34}{3}\right)$
65. When flying into a headwind, this plane is traveling at a rate of 105 mph.
67. When flying with a tailwind, this plane is traveling at a rate of 105 mph.
69. When traveling downstream, this boat is going at a rate of 8.5 mph.
71. When traveling upstream, this boat is going at a rate of 7 mph.
73. a. 5.5 pounds of Oolong tea has a value of $318.73.
    b. t pounds of Oolong tea has a value of $57.95t$.
75. a. 26 pounds of French Roast coffee has a value of $440.70.
    b. c pounds of French Roast coffee has a value of $16.95c$.
77. The coffee's unit price is $15.95 per pound.
79. a. Let s = rate of the boat in miles per hour; let c = rate of the current in miles per hour.
    b. $2.5(s + c) = 50$
       $5(s - c) = 50$
    c. The rate of the boat in still water is 15 mph, and the rate of the current is 5 mph.
81. a. Let p = rate of the plane in miles per hour; let w = rate of the wind in miles per hour.
    b. $3.4(p - w) = 1275$
       $3(p + w) = 1275$
    c. The rate of the plane in still air is 400 mph, and the rate of the wind is 25 mph.
83. a. Let c = amount of Ceylon tea in pounds; let b = amount of black currant tea in pounds.
    b. $c + b = 50$
       $23c + 27b = 1250$
    c. Julie should mix 25 lb of each tea to get the 50-pound mix she wants.

85. a. Let p = amount of papaya in pounds; let g = amount of ginger in pounds.
   b. $p + g = 25$
   $3.80p + 9.50g = 180.50$
   c. Graciela should mix 10 lb of papaya with 15 lb of ginger to get the 25-pound mix she wants.

87. Rita's Auto Detail will break even when 28 cars are detailed with revenue and costs of $4200.

89. Mobile Style will break even when about 269 accessories are sold with revenue and costs of about $6588.50.

91. a. Infinitely many solutions
   b.
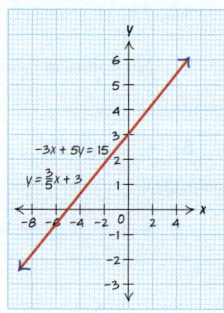
   c. Yes
   d. No
   e. Points on the line

93. a. No solutions
   b.
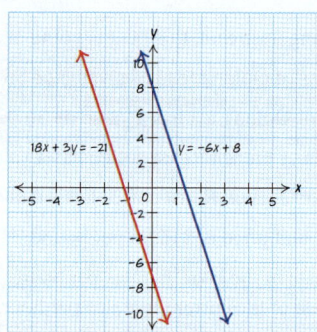
   c. No
   d. No
   e. A solution is a point on both lines at the same time. These lines do not intersect.

95. $2x - 4x - 8 = 6$ should have been written
$$2x - 4x + 8 = 6$$

97. $-5(-4y + 11) + 2y = -1$ should have been written
$$-5\left(-\dfrac{4}{3}y + \dfrac{11}{3}\right) + 2y = -1$$

99. Inconsistent   **101.** Inconsistent   **103.** Consistent

Section 4.3

1. $2x - y = 16$ Add the two equations so that the y variables
   $3x + y = 9$ will eliminate.
   $5x = 25$ Solve for x using the division property of equality.
   $x = 5$

$3(5) + y = 9$ Substitute 5 for x and solve for y.
$15 + y = 9$
$y = -6$
$(5, -6)$

3. $(-9, -2)$ **5.** $(-7, -8)$
7. $(0, 4)$ **9.** $(1, -3)$
11. Infinitely many solutions
13. $(-2, 4)$ **15.** $(-2, -7)$ **17.** $(4, -6)$
19. No solution
21. Infinitely many solutions
23. No solution
25. a. Let f = the first integer; let s = the second integer.
   b. $f + s = 95$
   $4f - s = 15$
   c. $f = 22, s = 73$
27. a. Let t = cost of the T-shirt in dollars; let w = cost of the water bottle in dollars.
   b. $8t + 6w = 186$
   $10t + 8w = 237$
   c. $t = 16.5, w = 9$
29. a. Let m = price of the milk in dollars; let b = price of a bread loaf in dollars.
   b. $3m + 4b = 23.40$
   $2m + 3b = 16.35$
   c. $m = \$4.80, b = \$2.25$
31. No solution   **33.** $(4, 7)$
35. $(-3, 4)$ **37.** $(-5, -1)$
39. Infinitely many solutions
41. Infinitely many solutions
43. $(2, 9)$ **45.** No solution
47. Infinitely many solutions
49. No solution
51. Elimination because $4y - 4y = 0$
53. Substitution because one equation is already solved for y
55. Substitution because one equation is already solved for y
57. Elimination because the coefficients of y are opposites
59. Tomas will earn $85 interest over 4 years with an investment of $500 at 4.25%.
61. Steven will earn $1111.80 interest over 3 years with an investment of $8720 at 4.25%.
63. There is 0.85 liter of sodium in the 10 liters of saline.
65. There is 0.64 liter of sucrose in the 8 liters of solution.
67. A 25° angle has a complementary angle of 65°.
69. A 110° angle has a supplementary angle of 70°.
71. $10000 - n$ **73.** $3800 - n$
75. a. Let p = rate of the plane in miles per hour; let w = rate of the wind in miles per hour.

b. $3(p + w) = 375$
   $5(p - w) = 375$
c. The rate of the plane in still air is 100 mph, and the rate of the wind is 25 mph.

77. a. Let b = rate of the boat in miles per hour; let c = rate of the current in miles per hour.
    b. $3(b + c) = 60$
       $6(b - c) = 60$
    c. The rate of the boat in still water is 15 mph, and the rate of the current is 5 mph.

79. a. Let w = amount of whole-wheat flour in pounds; let f = amount of white flour in pounds.
    b. $w + f = 50$
       $4.25w + 3.75f = 200$
    c. The baker should use 25 lb each of whole-wheat and white flour in the mix.

81. a. Let f = number of hours of first job; let s = number of hours of second job.
    b. $f + s = 25$
       $10.25f + 11.50s = 275$
    c. Renata should work 10 hours at her first job and 15 hours at her second job each week to earn the $275 she needs.

83. Cate should invest $1500 in the CD and $2500 in the bond fund to earn the $185 in interest per year.

85. Michael owes $1000 on the MasterCard and $2500 on the store credit card.

87. Greg should mix 5 liters of 15% solution with 15 liters of 20% solution to get the 20 liters of 18.75% hydrochloric acid he needs.

89. Christine should mix 40 liters of 30% solution with 60 liters of 50% solution to get the 100 liters of 42% antiseptic alcohol solution she needs.

91. One angle is 29°, and the other angle is 61°.

93. One angle is 37°, and the other angle is 143°.

95. The coefficients of x are not opposites, so they do not eliminate when added.

97. $-2(x + 5y) = -2$ should have been written $-2(x + 5y) = 4$.

Section 4.4

1. a. Yes   b. Yes
3. a. No   b. No
5. a. No   b. Yes
7. a. No   b. Yes
9. a. Yes   b. No
11. a. Dashed b. No
13. a. Solid   b. Yes
15. a. Solid   b. Yes
17. a. Dashed b. No
19. a. $(0, 0)$
    b. $0 < -3$, No

c.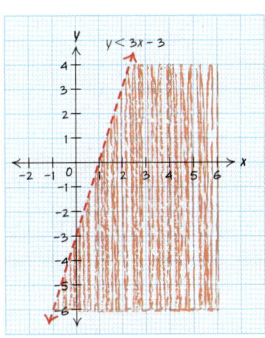

21. a. $(0, 0)$
    b. $0 \geq 12$, No
    c.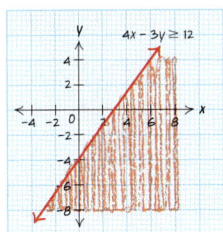

23. a. $(1, 1)$
    b. $-2 < 3$, Yes
    c.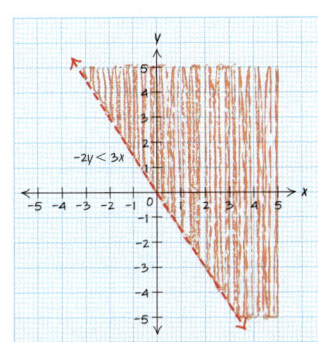

25. a. $(0, 0)$
    b. $0 > 2$, No
    c.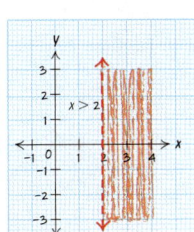

27. a. $(0, 0)$
    b. $0 \geq -2$, Yes
    c.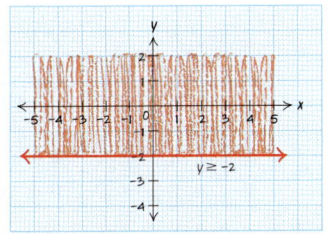

B-48 APPENDIX B Answers to Selected Exercises

29.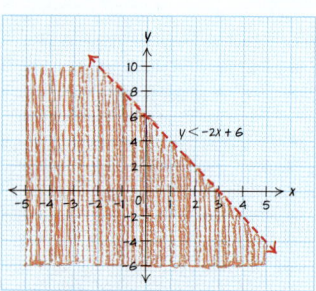

31.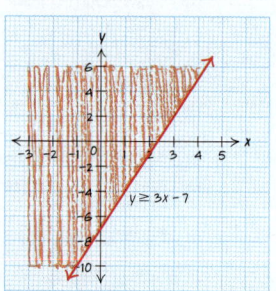

33.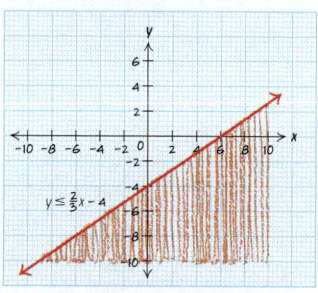

35.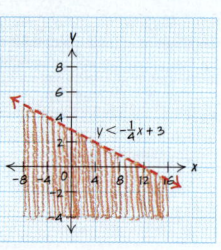

37.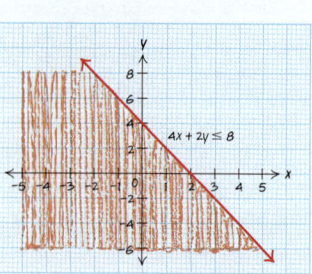

39.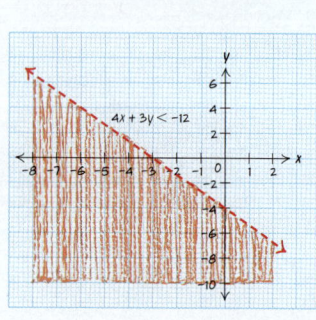

41.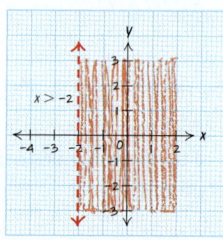

43.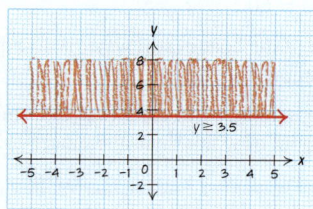

45.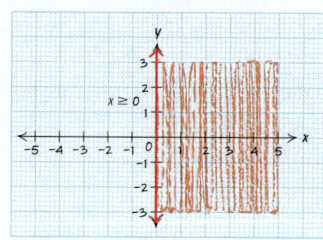

47. **a.** Let $m =$ number of oatmeal cookies; let $c =$ number of chocolate chip cookies.
   b. $4.00m + 5.50c \leq 25.00$
   c.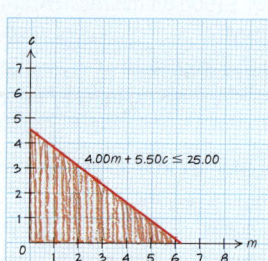

49. **a.** Let $f =$ number of hours of the first job; let $s =$ number of hours of the second job.
   b. $12f + 10s > 360$
   c.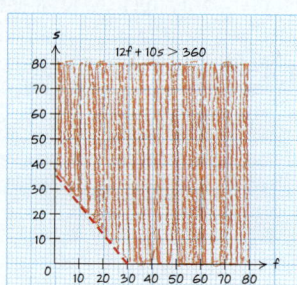

51. **a.** Let $f =$ amount in first account in dollars; let $s =$ amount in second account in dollars.
   b. $0.05f + 0.06s > 500$

c.
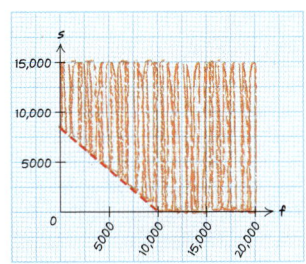

53. a. Let w = width in ft; let l = length in ft.
    b. $2w + 2l \leq 200$
    c.
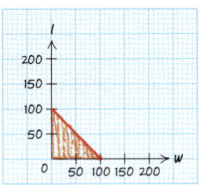

55. (1) Graph B, (2) Graph A
57. (1) Graph A, (2) Graph B
59. (1) Graph A, (2) Graph B
61. a. $y = -1$ b. $\geq$ or $\leq$ c. $y \geq -1$
63. a. $y = 3x$ b. $>$ or $<$ c. $y > 3x$
65. a. $y = -2x$ b. $\geq$ or $\leq$ c. $y \geq -2x$
67. a. $x = 2$ b. $>$ or $<$ c. $x < 2$
69. Inequality   71. Equality
73. Inequality   75. Inequality
77. Slope, $m = \dfrac{3}{8}$; y-intercept, $(0, -3)$
79. $y = -\dfrac{2}{5}x + 1$ 81. 5
83. 14 students are male.
85. 788.6 km   87. $x = -4$
89. $5(x + 12) = 10$; $x = -10$

Section 4.5

1. a. No   b. Yes
3. a. No   b. No
5. a. Yes   b. No
7. a. No   b. Yes
9. a. Yes   b. No
11. a. Yes b. No
13. a. Yes b. Yes
15.
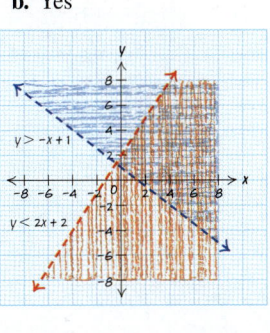

17.
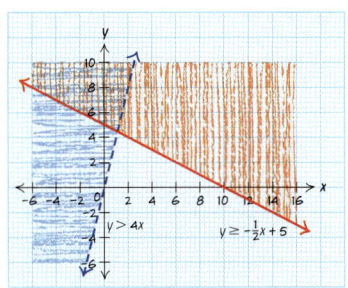

19.
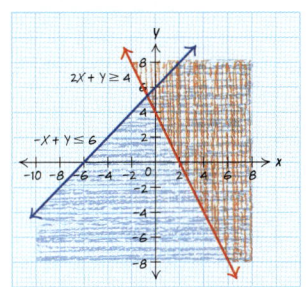

21.
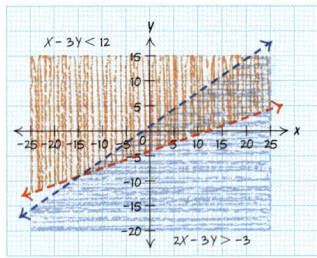

23.
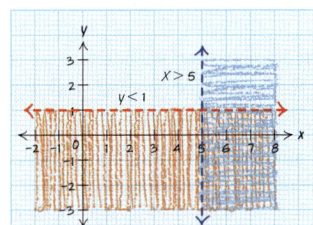

25.
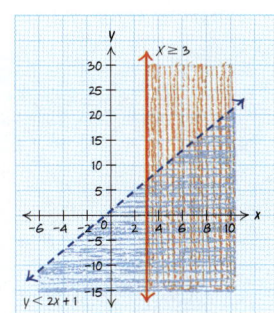

27.
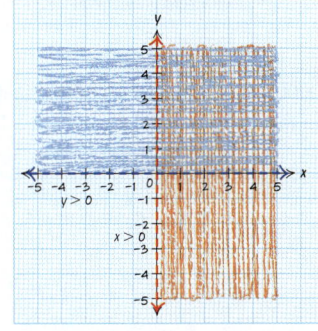

29.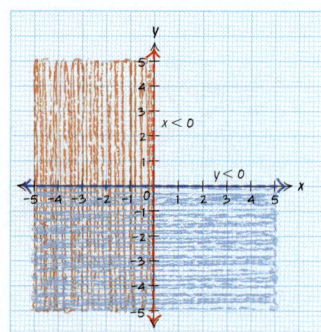

31. a. Let f = number of hours of first job; let s = number of hours of second job.

$$11f + 11.50s > 175$$
$$f + s \leq 20$$

b.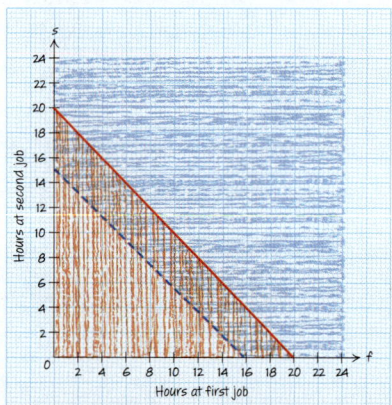

c. (9, 10). If Samuel works 9 hours at the first job and 10 hours at the second job, he will work less than 20 hours and earn $214.

33. a. Let f = amount in the first fund in dollars; let s = amount in the second fund in dollars.

$$f + s \leq 5000$$
$$0.08f + 0.10s > 100$$

b.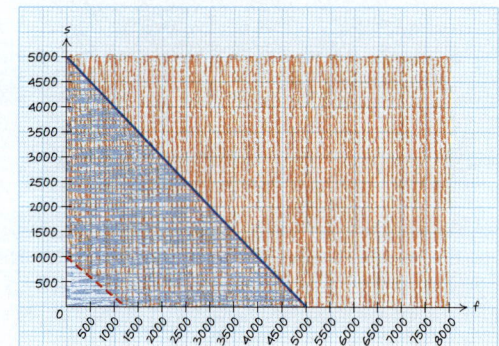

c. (1000, 3000). If Darren invests $1000 in the first fund and $3000 in the second fund, he will earn $380 in interest.

35. a. Let c = number of hours producing cars; let t = number of hours producing trucks.

$$c + t \leq 80$$
$$c + t \geq 40$$

b.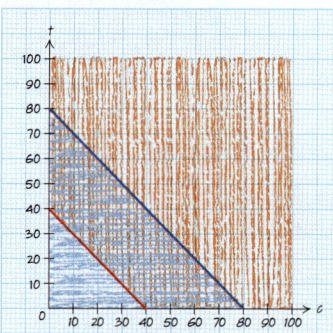

c. (50, 20). 50 hours are scheduled for cars, and 20 hours are scheduled for trucks.

37. a. Let s = number of student tickets; let g = number of general tickets.

$$s + g \leq 55000$$
$$s \leq 10000$$

b.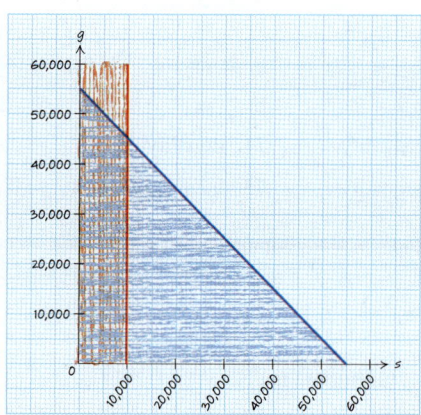

c. (5000, 20,000). They would meet their requirements if they sold 5000 student tickets and 20,000 general tickets.

39. inequality **41.** intersection

43. dashed **45.** $(-3, 13)$

47. $\left(2, -\dfrac{9}{2}\right)$ **49.** Yes

51. Yes

53.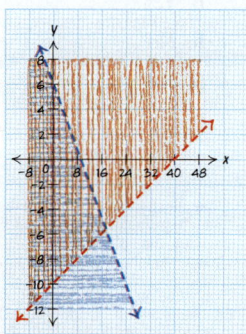

55.

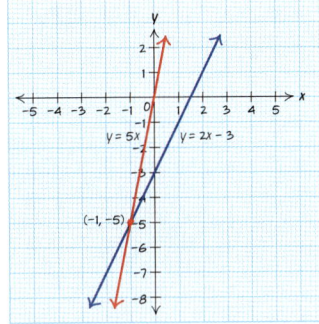

Chapter 4 Review Exercises

1. $(2, -4)$
2. $(2, 3)$
3. a. No   b. Yes
4. a. Yes   b. No
5.

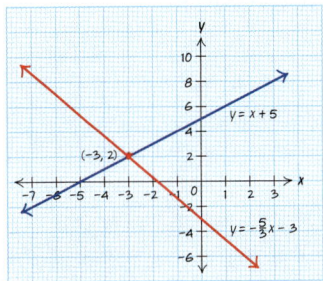

$(-1, -5)$

6.

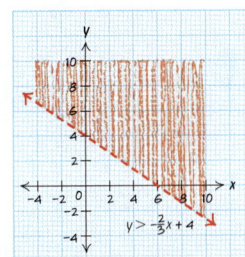

$(-3, 2)$

7. $(3, 10)$ 8. $(-15, 26)$ 9. $(0, 1)$
10. Infinitely many solutions
11. No solution   12. No solution
13. The cyclist's rate is 20 mph, and the wind's rate is 4 mph.
14. The cyclist's rate is 12 mph, and the wind's rate is 3 mph.
15. The plane's rate is 100 mph, and the wind's rate is 20 mph.
16. The boat's rate is 10.5 mph, and the current's rate is 3.5 mph.
17. The first number is 15, and the second number is 7.
18. The first number is 9, and the second number is -11.
19. The first number is 39, and the second number is 13.
20. The first number is 24, and the second number is 4.
21. $(1, -3)$ 22. $(-3, 6)$ 23. $(2, 5)$
24. $(-3, -1)$
25. No solution
26. No solution
27. Infinitely many solutions
28. Infinitely many solutions
29. Elimination because $-2x + 2x = 0$
30. Substitution because it is easy to solve for x in $x - 4y = 9$
31. The jerky treats cost $8.00, and the leash costs $16.50.
32. The candy costs $14.25, and the metronome costs $25.00.
33. Lance should invest $1500 in bonds and $500 in stocks.
34. Ming-Yue should invest $3300 in bonds and $1700 in stocks.
35. One angle is 150°, and the other angle is 30°.
36. One angle is 145°, and the other angle is 35°.
37. One angle is 81°, and the other angle is 9°.
38. One angle is 60°, and the other angle is 30°.
39. Tarin should use 4 ounces of the 5% solution and 16 ounces of the 30% solution.
40. Shelby should use 6.67 gallons of the 6% juice and 13.33 gallons of the 12% juice.
41. Inna should use 6 pounds of peanuts and 4 pounds of raisins.
42. Lorraine should use $13\frac{1}{3}$ pounds of French Roast and $6\frac{2}{3}$ pounds of Guatemalan coffee.
43. a. No   b. Yes
44. a. Yes   b. Yes
45.

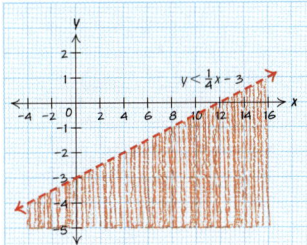

46.

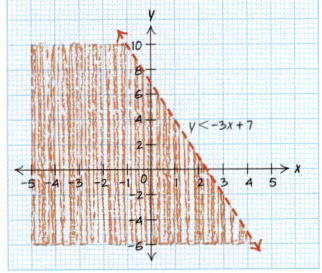

47.

B-52 APPENDIX B Answers to Selected Exercises

48.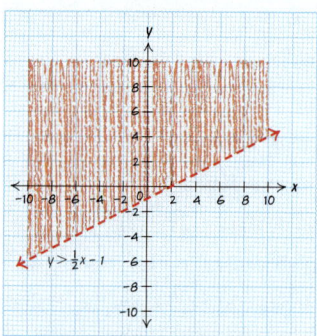

49.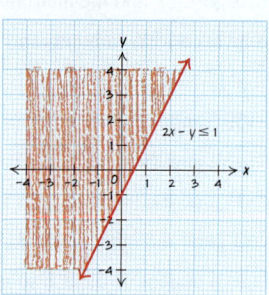

50.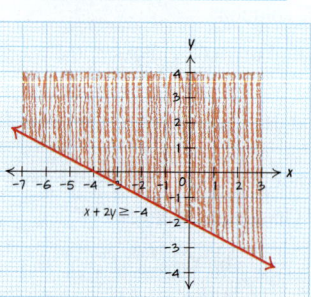

51.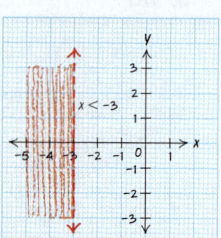

52.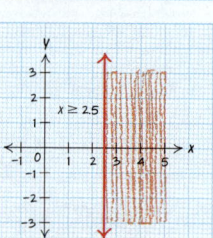

53.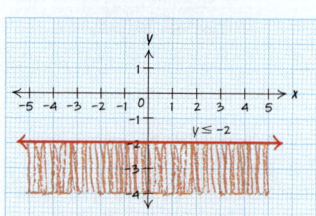

54.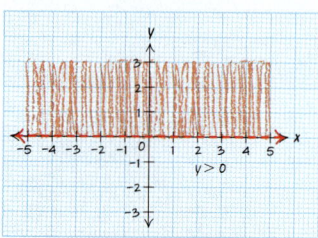

55. a. Let f = number of hours of the first job; let s = number of hours of the second job.
   b. $10.75f + 11.25s > 500$
   c.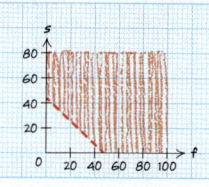

56. a. Let f = number of hours of the first job; let s = number of hours of the second job.
   b. $11f + 11.5s > 450$
   c.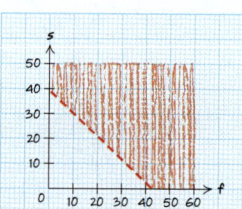

57. (1) Graph B, (2) Graph A
58. (1) Graph A, (2) Graph B
59. a. No   **b.** Yes
60. a. No   **b.** Yes
61.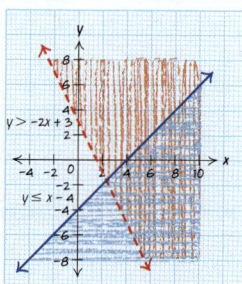

62.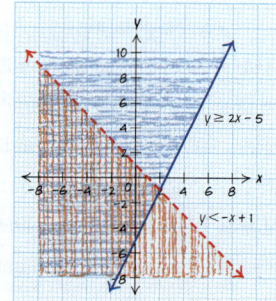

APPENDIX B Answers to Selected Exercises   **B-53**

63. a. Let $f =$ the amount in the first fund in dollars; let $s =$ the amount in the second fund in dollars.

$$f + s \leq 6000$$
$$0.05f + 0.07s > 200$$

b.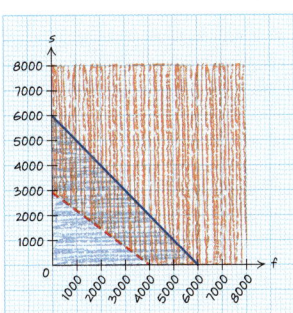

c. Test: (4200, 1500)

$4200 + 1500 = 5700 \leq 6000$

$0.05(4200) + 0.07(1500) = 315 > 200$

If Darius puts $4200 in the first fund and $1500 in the second fund, he will earn $315 in interest.

64. a. Let $f =$ amount in the first fund in dollars; let $s =$ amount in the second fund in dollars.

$$f + s \leq 7500$$
$$0.07f + 0.08s > 110$$

b.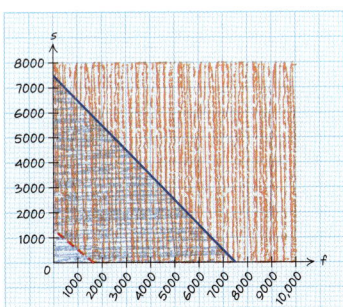

Test: (1000, 6000)

$1000 + 6000 = 7000 \leq 7500$

$0.07(1000) + 0.08(6000) = 550 > 110$

If Leila invests $1000 in the first fund and $6000 in the second fund, she will earn $550 in interest.

Chapter 4 Test

1.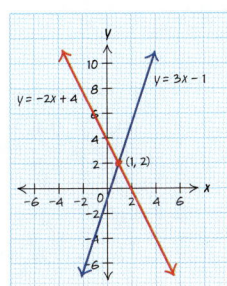

Consistent, (1, 2)

2. $y = \dfrac{7}{3}x + 7$ **3.** $(-2, -3)$ **4.** $(5, 2)$

5. The monitor costs $33, and the water bottle costs $12.

6. The boat's rate is 10 mph, and the current's rate is 2 mph.

7. Laurent should invest $800 in the bond fund and $1200 in the stock fund.

8. No solution

9. Infinitely many solutions

10.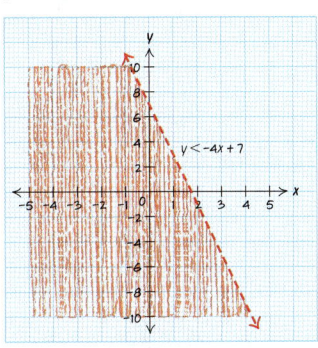

11.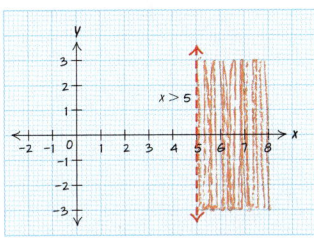

12.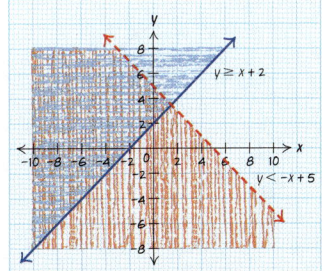

Cumulative Review for Chapters 1–4

1. $\left(-\dfrac{1}{5}\right)^4$ **2.** 16 **3.** 6,103,000

4. 93

5. Let $c =$ the amount of the commission in dollars; let $n =$ the amount of sales in dollars.

6. $3(9 + x)$ **7.** 39.37 feet **8.** -11

9. $3x + 25$ **10.** $-5x$

11.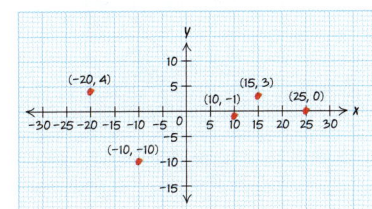

12. Yes **13.** $x = -24$ **14.** $y = -3x - 9$

15. $x = 6$

16. a. $1430 **b.** 140 hours

B-54　APPENDIX B　Answers to Selected Exercises

17. $x = 2.5$

18. $8 + 3n = 2n - (-6), n = -2$

19. No solution

20. $x < -7$

21. a.

x	$y = -\dfrac{3}{5}x - 7$
-5	-4
0	-7
5	-10
15	-16
20	-19

b.

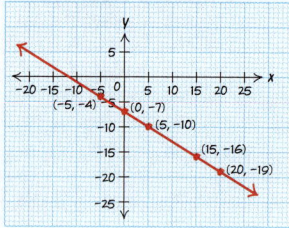

c. Linear

22.

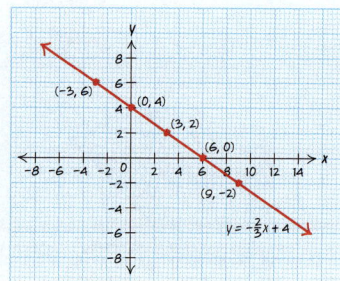

Linear

23.

d	C
1	4000
2	8000
3	12,000
4	16,000

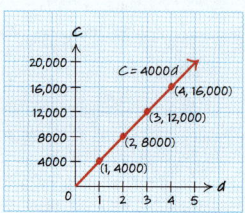

24.

x	y
-4	14
-3	7
-2	2
-1	-1
0	-2
1	-1
2	2
3	7
4	14

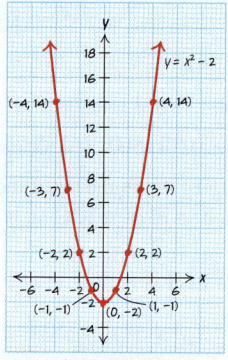

Nonlinear

25.

x	y
-2	-5
-1	-5
0	-5
1	-5
2	-5
3	-5

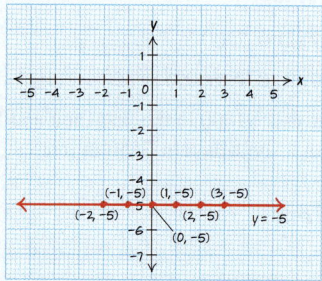

Linear

26.

x	y
6	0
6	1
6	2
6	3
6	4

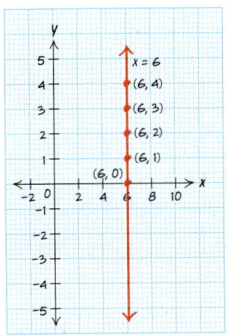

Linear

27. $m = \dfrac{2}{3}$ **28.** $m = -\dfrac{12}{5}$ **29.** $m = -\dfrac{2}{5}$

30. $m = 0$

31. The red line because it rises faster when the x-values are increasing than the blue line

32. Decreasing because the y-values get smaller as the x-values get bigger

33. Increasing because $m > 0$

34. $(-3, 0)$ and $(0, -2)$

35. **a.** In 2010, Minnesota had a population of 5.31 million.
b. In 2010, Washington had a population of 6.74 million.
c. $m = 0.034$. Minnesota's population increases by about 34,000 people a year.
d. $m = 0.092$. Washington's population increases by about 92,000 people a year.

36. y-intercept: $(0, 15)$; x-intercept: $(25, 0)$

37. $m = \dfrac{1}{2}$; y-intercept: $\left(0, -\dfrac{3}{2}\right)$; x-intercept: $(3, 0)$

38.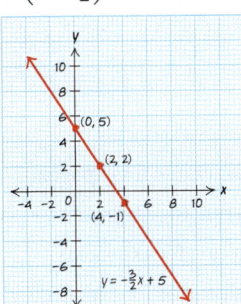

39.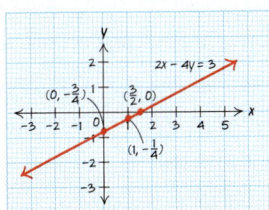

40.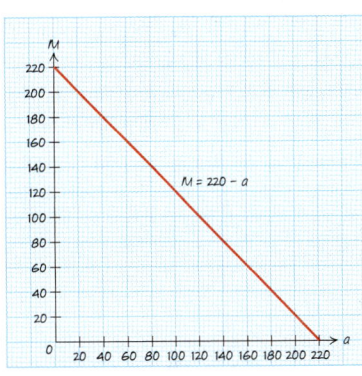

41.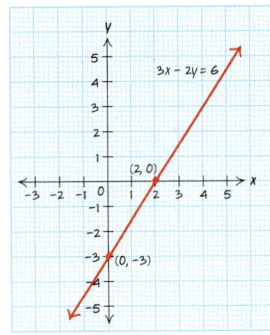

42. **a.** Linear because the exponents on the variables are all 1
b. Not linear because the exponent on the variable is not 1
c. Not linear because there is an absolute value that involves the variable
d. Linear because the exponents on the variables are all 1

43. Parallel   **44.** Perpendicular

45. $y = \dfrac{7}{2}x - 2$ **46.** $W = -\dfrac{25}{6}t + 155$

47. $y = -\dfrac{2}{5}x + 9$ **48.** $y = \dfrac{2}{3}x + 4$ **49.** $y = -5$

50. $x = 3$ **51.** $y = -5x - 47$ **52.** $y = -\dfrac{1}{3}x - 2$

53. This new car had a 19.2% decrease in value over the first year.

54. This house had a 131.8% increase in value.

55. The markup on this bat is $24 for a total sales price of $84.

56. **a.** Let $t =$ years since 2010, and let $M =$ worldwide makeup sales in billions of dollars.

b.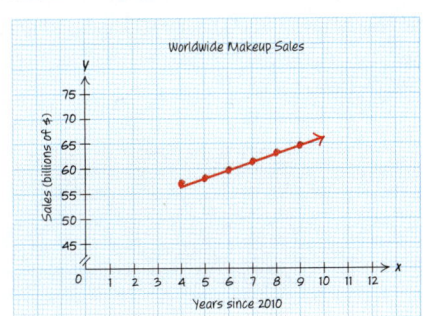

c. $M = 1.63t + 49.93$

d. In 2025, worldwide makeup sales will reach an estimated $74.38 billion.

B-56 APPENDIX B Answers to Selected Exercises

  e. Slope = 1.63. Worldwide makeup sales are increasing by about $1.63 billion per year.
57. $T = 380$ when $c = 30$. It takes 380 minutes, 6 hours and 20 minutes, to decorate 30 Christmas ornaments.
58. $R = 884$ when $t = 68$. If they sell 68 tickets, their revenue is $884.
59. $(-1, 6)$
60. a. No   b. Yes
61. $(0, 1)$
62. a. No   b. Yes
63.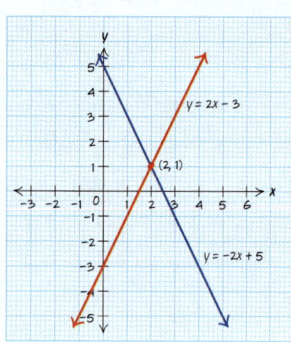
64.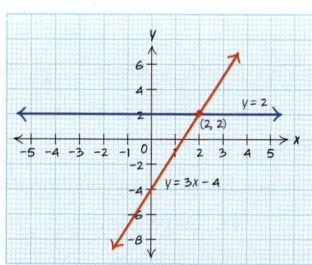
65. $(-4, 6)$ 66. $(-5, -13)$
67. No solution
68. The bottles cost $6, and the packages cost $25.
69. One number is 13, and the other number is 19.
70. The plane's rate is 125 mph, and the wind's rate is 25 mph.
71. Francis needs to use 3 pounds of turmeric and 2 pounds of ground coriander.
72. $(4, -5)$ 73. $(-1, -4)$
74. Infinitely many solutions
75. The scrapbooks cost $18.00, and the stickers cost $12.50.
76. One angle is 30°, and the other angle is 150°.
77. Eric should invest $8000 in the CD and $4000 in the bond fund.
78. Yes
79.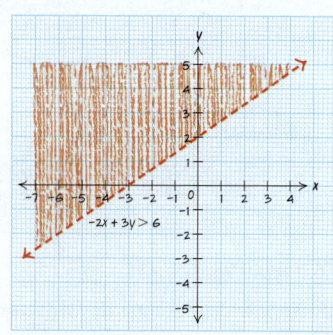

80.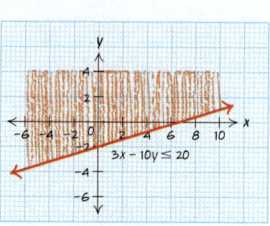

81. a. Let $r =$ the number of spring rolls; let $g =$ the number of gyozas.
    b. $3.99r + 4.25g \leq 20$
    c.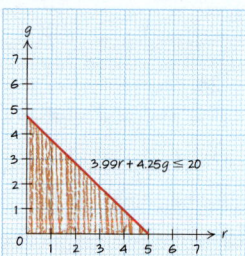

82. Yes
83.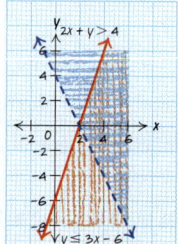

CHAPTER 5

Section 5.1

1. 4   3. -8
5. -25 7. -1
9. x^9; product rule for exponents
11. 5,764,801; product rule for exponents
13. $15w^{12}$; product rule for exponents
15. $40x^6y^5$; product rule for exponents
17. $-15m^{10}n^7$; product rule for exponents
19. $-18x^6y^4$; power rule and product rule for exponents
21. $1679616x^8$; power rule and product rule for exponents
23. $117649b^6$; power rule and product rule for exponents
25. $-1024m^5$; power rule and product rule for exponents
27. x^7; quotient rule for exponents
29. $5k^{14}$; quotient rule for exponents
31. x^{10}; power rule for exponents
33. $a^{10}b^{30}$; powers of products rule for exponents
35. Elizabeth simplified the expression incorrectly. The student multiplied exponents instead of adding.
37. Elba simplified the expression incorrectly. The student subtracted coefficients instead of dividing.

39. Luis simplified the expression incorrectly. The student multiplied the coefficient by the exponent instead of multiplying the coefficient by itself.

41. Yomaira simplified the expression incorrectly. The student did not apply the exponent to the negative sign.

43. Andrew simplified the expression incorrectly. The student did not apply the exponents before dividing, or the student ignored the exponents altogether.

45. x^3y^3; product and quotient rules for exponents

47. m^4n; product and quotient rules for exponents

49. $\dfrac{15x^4}{7}$; product and quotient rules for exponents

51. $\dfrac{x^2}{5}$; quotient rule for exponents

53. b^3; product and quotient rules for exponents

55. $\dfrac{a^3b^3}{3}$; product and quotient rules for exponents

57. $9c^4d^{10}$; power rule for exponents and powers of products

59. $25a^{14}b^{10}$; power rule for exponents and powers of products

61. $36x^6y^6$; power rule for exponents and powers of products and quotients

63. $\dfrac{169x^4y^8}{9}$; power rule for exponents and powers of products and quotients

65. $45x^5y^{11}$; power rule for exponents and powers of products

67. $432m^{11}n^{21}$; power rule for exponents and powers of products

69. $648a^{11}b^9$; power rule for exponents and powers of products

71. $\dfrac{16x^{10}}{25}$; power rule for exponents and powers of products and quotients

73. $2x^4y^3$; power rule for exponents and powers of products and quotients

75. Elias mistakenly applied the power to the negative sign.

77. Beverly multiplied the bases before applying the exponent and included the exponent incorrectly. In the order of operations, exponents come before multiplication.

79. Alexis applied the exponent over addition, but exponents can be applied only over operations of multiplication or division.

81. Victoria divided the bases before applying the exponent and included the exponent incorrectly. In the order of operations, exponents come before division.

83. a. Yes, the expression $(x + 3)^2(x + 3)^5$ can be simplified to $(x + 3)^7$ by using the product rule for exponents. In this case, $(x + 3)$ is the base, and the exponents are added.

  b. No, the exponent 7 in the expression $(x + 3)^7$ cannot be applied to x and 3 to get $x^7 + 3^7$. This would be applying the power rule to a sum, but the power rule applies only to products and quotients.

85. No, $x^2 + x^3$ cannot be simplified. These are not like terms because they do not have the same exponent and therefore cannot be combined using addition.

87. $A = 49x^2$ 89. $V = 125x^3$
91. $A = 4\pi x^2$ 93. $A = 6x^2$

Section 5.2

1. $\dfrac{1}{x^3}$ 3. $\dfrac{1}{9}$ 5. $\dfrac{1}{32}$

7. $\dfrac{4}{d^6}$ 9. $-\dfrac{5}{a^4}$ 11. $\dfrac{5x^2y^3}{2}$

13. $\dfrac{x^4}{5}$ 15. $\dfrac{1}{2a^3}$ 17. $\dfrac{1}{9x}$

19. $\dfrac{y^3}{x^2}$

21. Julia treated the negative sign as though it were a negative exponent. The expression is already in simplest form.

$$-\dfrac{2x}{y}$$

23. Frida applied the negative exponent to the coefficient, but it applies only to the variable.

$$\dfrac{42}{a}$$

25. Christopher applied the negative exponent only to the variable, but it applies to the whole term.

$$\dfrac{1}{25x^2}$$

27. Madalyn applied a positive exponent to the coefficient and applied the negative exponent only to the variable, but the negative exponent applies to the whole term.

$$\dfrac{1}{9x^2}$$

29. $\dfrac{1}{x^4}$ 31. $\dfrac{1}{m^3n}$ 33. $\dfrac{5y^5}{3x^6}$

35. $\dfrac{4}{x^2y^8}$ 37. $\dfrac{12n^8}{5m^5}$ 39. $-\dfrac{5x}{2y^7}$

41. $\dfrac{8}{x^6}$ 43. $\dfrac{25}{x^6}$ 45. $-\dfrac{54}{a^5b^3}$

47. $\dfrac{4y^8}{9x^4}$ 49. $\dfrac{b^6}{a^2}$

51. Joseph subtracted 5 for the exponent in the denominator rather than subtracting -5.

$$x^8$$

53. Mackenzie did not retain the negative sign for the coefficient and incorrectly moved the coefficient in the denominator to the numerator.

$$-\dfrac{1}{8x^7}$$

55. Tristan did not apply the power of products rule correctly, missing the first variable.

$$\dfrac{1}{x^6y^2}$$

57. 623,000,000

59. The mass of the earth: 5,980,000,000,000,000,000,000,000 kg.

61. 0.000000003745

63. An electron's mass: 0.000000000000000000000000000167 kg.

65. 6.2×10^{10} **67.** 4.3×10^{-6}

69. The 2017 first-quarter revenue for Alphabet, Inc.: 2.475×10^{10}.

71. The 2017 first-quarter U.S. GDP: 1.90276×10^{10}.

73. The average width of a human hair: 7×10^{-5} meter.

75. The width of pits where data on compact discs are stored: 5×10^{-7} meter.

77. 8.4×10^{11} **79.** 7.5×10^{-12}

81. 2.52×10^{14} **83.** 3.6×10^{-6}

85. 5.0×10^{7} **87.** 2.0×10^{14}

89. 1.5×10^{25} **91.** 3.6×10^{-22}

93. 6.0×10^{27} **95.** 2.4×10^{-17}

97. -1.35×10^{12} **99.** -1.8×10^{24}

101. The mass of Saturn is approximately 5.681×10^{26} kg.

103. The mass of Mercury is approximately 3.289×10^{23} kg.

105. The average distance from Pluto to the Sun is approximately 5.9092×10^{9} km.

107. The average distance from Mars to the Sun is approximately 2.244×10^{8} km.

109. The diameter of Mars is approximately 6.4×10^{3} km.

111. The electric charge of 200 protons is 3.204×10^{-17} coulomb.

113. The mass of 500 electrons is 8.35×10^{-25} kg.

115. The mass of 3 million electrons is 5.01×10^{-21} kg.

117. 8.4×10^{24} **119.** 3.75×10^{5}

121. 3.78×10^{34} **123.** 2.13×10^{-14}

125. 2.0×10^{4} **127.** 9.0×10^{-14}

129. 4.8×10^{19}

131. The mass of Venus is 0.81 times as large as the mass of the Earth.

133. The diameter of Jupiter is 11.156 times larger than the diameter of the Earth.

135. Each person's share of the U.S. national debt in May 2017 was approximately $60,736.20.

Section 5.3

1. Terms: x^2, $2x$, and 5
This is a trinomial.

3. Terms: $4x$ and 8
This is a binomial.

5. Term: $8x^3y$
This is a monomial.

7. Terms: $7m^2$ and -9
This is a binomial.

9. Terms: $14x^3y$, $7x^2y^2$, and $-9xy$
This is a trinomial.

11. Terms: $2m^7n^3$ and $-4m^2n^5$
This is a binomial.

13. Terms: $-45mn^2$ and 56
This is a binomial.

15. Two terms
Descending order: $\frac{3}{2}x - 12$
Degree is 1.

17. Three terms
Descending order: $-6x^2 + 3x - 9$
Degree is 2.

19. One term
Descending order: -45
Degree is 0.

21. Four terms
Descending order: $-y^3 + 9y^2 + 4y + 7$
Degree is 3.

23. Three terms
Descending order: $-3b^8 + 2b^7 + 9b$
Degree is 8.

25. Degree of the term $-6xy^2$ is 3.
Degree of the polynomial is 3.

27. Degree of the term $3a$ is 1.
Degree of the term $2ab$ is 2.
Degree of the term $-6b^2$ is 2.
Degree of the polynomial is 2.

29. Degree of the term 1 is 0.
Degree of the term $-x^3y^3$ is 6.
Degree of the polynomial is 6.

31. Degree of the term $-24x^2yz^3$ is 6.
Degree of the term $6z$ is 1.
Degree of the term $-7y$ is 1.
Degree of the term 5 is 0.
Degree of the polynomial is 6.

33. Degree of the term -29 is 0.
Degree of the polynomial is 0.

35. $x^2 + 2x + 5$ is an expression that has a second-degree term, so it is not linear. There is no equal sign, so it is an expression, not an equation.

37. $4x + 8$ is a polynomial in 1 variable with degree 1, so it is a linear expression in one variable. There is no equal sign, so it is an expression, not an equation.

39. $y = 4x + 3$ is an equation in two variables that has degree 1, so it is a linear equation in two variables. It is an equation because there is an equal sign.

41. $5x - (2x + 8) = 4x$ is a linear equation in one variable that has degree 1. It is an equation because there is an equal sign.

43. $4xy + 7x - 9xy$ is a polynomial in two variables that has degree-2 terms. It is not linear because it is not of degree 1. There is no equal sign, so it is an expression, not an equation.

45. $y = 2x^2 + 3x - 20$ is an equation in two variables that has a second-degree term, so it is not linear. It is an equation because there is an equal sign.

47. $2x + 7y = 68$ is an equation in two variables that has degree 1, so it is a linear equation in two variables. It is an equation because there is an equal sign.

49. $7x + 14$ **51.** $5t^2 + 8t - 5$
53. $-9m^2 + 13m - 9$ **55.** $2x + 6$
57. $13n - 9$ **59.** $2h^2 - 2h + 4$

61. Reina treated this as a product, adding exponents rather than a sum, and should have added coefficients. Correctly simplified, it is $3x$.

63. Fari did not subtract the entire quantity of $(4x + 5)$, subtracting only the first term and not the second term. Correctly simplified, it is $-x - 7$.

65. Anya did not subtract the entire quantity of $(2x^2 - 5x)$, subtracting only the first term and not the second term. (Subtracting a negative changes it to adding the term.) Correctly simplified, it is $-x^2 + 9x - 8$.

67. a. The total cost for a class to see the life of a pumpkin demo and visit the petting zoo is $3.5k + 15$.

b. The cost for 33 students to see the life of a pumpkin demo and visit the petting zoo is \$130.50.

69. a. The total cost for a class to see the life of a pumpkin demo, take a tractor ride, and take home a pumpkin is $7.25k + 10$.

b. The cost for 60 students to see the life of a pumpkin demo, take a tractor ride, and take home a pumpkin is \$445.

71. a. The total cost for the Grizzlies to purchase all the gear is $26m + 95$.

b. The cost for 15 members to purchase all the gear is \$485.

73. a. The total cost for the Grizzlies to purchase T-shirts and go on the Dos Picos campout is $37.5m + 155$.

b. The cost for 12 members to purchase T-shirts and go on the Dos Picos campout is \$605.

75. a. The total cost for a 5-day trip is $150m + 350$.

b. The cost for 30 ski club members to go on the 5-day trip is \$4850.

77. $3x^2 + 2x + 20$ **79.** $P = 12x + 26$
81. $7x^2 + 26x$ **83.** $5a^2 - 4ab - 2b$
85. $-10t^2 + 6t - 20$ **87.** $-12a + 6b$
89. $-6m^2 + 13mn - 4n$ **91.** $45a^2 + 9a - 24$
93. $-x^2 + 15x - 3$ **95.** $2m^3 + 3m^2 + 20m - 6$

97.

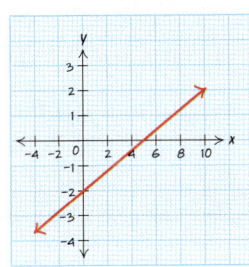

99. $(5, -1)$

101. The percentage share of 25- to 29-year-old employees in the United States with a bachelor's degree or higher in 1995 was 30.25%.

103. Slope: $m = \dfrac{0.457}{1}$. The percentage share of 25- to 29-year-old employees in the United States with a bachelor's degree or higher is increasing at the rate of 0.457 percentage points per year.

Section 5.4

1. $10x + 35$ **3.** $14x^2 + 12x$
5. $-12x - 6$ **7.** $12x^2 + 18x - 30$
9. $-7x^2 - 28x + 21$ **11.** $15x^3 - 9x^2 + 12x$

13. The number of screws needed to hold down a lid after an accident can be represented by the expression $6P$.

15. $x^3 + 7x^2 + 7x - 15$ **17.** $5x^3 - 13x^2 - 2x + 16$
19. $8x^3 + 16x^2 - 36x + 21$ **21.** $16x^3 - 36x^2 + 6x + 9$
23. $2x^4 + 3x^3 + 2x^2 + 9x - 12$ **25.** $x^2 + 8x + 15$
27. $x^2 + 4x - 12$ **29.** $a^2 + 7a + 10$
31. $2m^2 + 3mn + n^2$ **33.** $5x^3 - 4x^2 + 15x - 12$

35. a. The total amount of high-fructose corn syrup consumed by people in the United States is represented by the expression $-2.714t^2 - 250.452t + 14859.7$ million pounds.

b. The total amount of high-fructose corn syrup consumed by Americans in 2020 will be approximately 12,083.8 million pounds, or 12,083,800,000 pounds.

37. a. The average income of a family in the United States in 2018 is estimated at \$71,490.

b. The number of families in the United States in 2018 is estimated at 83,020,000.

c. The total income in billions of dollars of all families in the United States can be represented by the expression $0.5124t^2 + 123.1118t + 4917.4118$.

d. The total income of all families in the United States in 2020 is estimated at \$6,199,769,800,000.

39. $A = 15x^2 - 4x - 32$ **41.** $x^2 + 12x + 36$
43. $x^2 - 6x + 9$ **45.** $9x^2 + 30x + 25$
47. $25x^2 - 30x + 9$ **49.** $4a^2 + 28ab + 49b^2$
51. $9x^4 + 30x^2 + 25$ **53.** $16d^6 - 40cd^3 + 25c^2$
55. $x^2 - 25$ **57.** $4a^2 - 25$
59. $\dfrac{x^2}{4} - 1$ **61.** $\dfrac{16}{25}x^2 - \dfrac{24}{5}x + 9$
63. $9a^2 - 4b^2$ **65.** $x^4 - 25$
67. $4a^{10} - 9b^2$ **69.** Product $4x^2$

71. Sum $y^2 + 12y + 36$ **73.** Sum $x^2 + 2xy + y^2$

75. Product $\dfrac{1}{4}n^2$ **77.** Product $\dfrac{1}{25}a^2$

79. Tristan left out the exponent when multiplying the first terms, which should be x^2.
$$x^2 + x - 12$$

81. Dolores left out the middle step when squaring a binomial, adding twice the product of the two terms.
$$x^2 + 8x + 16$$

83. Tyler did not square the coefficient of the first term and left out the middle step when squaring a binomial, adding twice the product of the two terms.
$$\frac{4}{9}a^2 - \frac{4}{3}a + 1$$

85. $5x^3 + 50x^2 + 105x$ **87.** $24x^3 - 10x^2 - 56x$
89. $7a^3 - 28a$ **91.** $7h^3 + 70h^2 + 175h$
93. $200x^4 - 320x^3 + 128x^2$ **95.** $x^2 - 2x - 16$
97. $3x^2 - 12$ **99.** $x^2 - 3x + 7$
101. $14x + 7$ **103.** $7x^2 + 6x - 33$
105. $-22t^2 - t - 15$

Section 5.5

1. Monomial
3. Monomial
5. Polynomial of more than one term
7. Polynomial of more than one term
9. Polynomial of more than one term
11. $6x^2 + 8x - 4$ **13.** $5m^2 - 4m + 12$
15. $\frac{1}{2}x^4 + \frac{7}{4}x^2 - \frac{15}{4}x + \frac{3}{2}$ **17.** $3h^3 + \frac{12}{7}h - \frac{2}{7}$
19. $5x + 6 - \frac{1}{x}$ **21.** $\frac{12}{7}d^3 - \frac{3}{2}d + \frac{1}{d}$
23. $\frac{4x}{y} - 2 + \frac{3y}{x}$ **25.** $3m^3n - \frac{10}{3}m + \frac{5}{2n}$
27. Correct; $3(x - 2) = 3x - 6$
29. Correct; $(x + 3)(x - 2) = x^2 + x - 6$
31. Correct; $(x^2 + x + 1)(x - 1) = x^3 - 1$
33. $x + 9$ **35.** $x - 7$
37. $m - 4$ **39.** $2x + 5$
41. $5d - 7$ **43.** Length $= x + 7$
45. Length $= 7a - 8$
47. $x + 9 + \frac{7}{x - 4}$ **49.** $b - 7 - \frac{8}{b - 3}$
51. $x + 4 - \frac{11}{2x + 5}$ **53.** $3a - 2 + \frac{5}{4a + 7}$
55. $3x - 7 - \frac{6}{5x - 2}$ **57.** $x^2 + 5x + 7$
59. $2a^2 + 4a - 5$ **61.** $5x^2 + 15x + 47 + \frac{153}{x - 3}$
63. $a^2 - 2a + 4$ **65.** $4x^2 + 6x + 9$

67. $-\frac{17}{9}$ **69.** 18
71. 1.024 kilobites per second
73. 525,000 gigabytes.
75. $-\frac{x^3}{2y}$ **77.** $108m^7n^5$
79. $t = -\frac{2}{3}$ **81.** $n = \frac{4}{5}$
83. 324.75% **85.** $y = \frac{5}{3}x + 12$

87.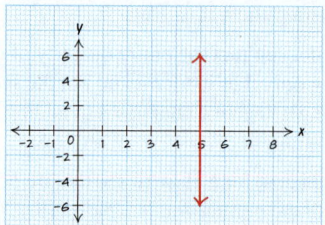

89.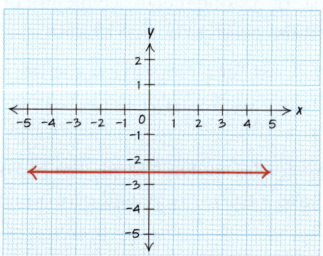

Chapter 5 Review Exercises

1. $28x^8$ **2.** $-24y^5$
3. $24x^9y^5$ **4.** $-28x^3y^{10}$
5. $-\frac{xy^2}{4}$ **6.** $-\frac{x^6}{2}$
7. $c^{11}d^4$ **8.** $\frac{1}{2ab^2}$
9. $125a^6b^{15}$ **10.** $4x^4y^6$
11. $-\frac{a^3b^9}{8}$ **12.** $\frac{10000m^{12}n^8}{2401}$
13. $-3087x^{12}y^{17}$ **14.** $22500x^{18}y^8$
15. $-\frac{5y^4}{2x^5}$ **16.** $\frac{a^3}{3b^2}$
17. $-\frac{9}{x^4y^{13}}$ **18.** $-\frac{12m^4}{n^{10}}$
19. $\frac{c^6}{d^8}$ **20.** y^4z^{13}
21. 5.26×10^{10} **22.** 6.512×10^7
23. The average distance from Neptune to the Sun is about 4.55×10^9 km.
24. The width of one helix of the DNA molecule is 2.0×10^{-9} meter.

25. 0.000006024
26. 43,000,000
27. The width of a flu virus is 0.0000003 meter.
28. The approximate width of a carbon nanotube shaft of one of the first nanomotors created is 0.0000000075 meter.
29. 1.4×10^{20}
30. -1.71×10^{-6}
31. 5.0×10^{-21}
32. 3.1×10^{16}
33. 1.5×10^{19}
34. 1.44×10^{-16}
35. Three terms
    Degree is 2.
36. Two terms
    Degree is 1.
37. Degree of the term $-5a^3b$ is 4.
    Degree of the term $6a^5$ is 5.
    Degree of the term $-7b^2$ is 2.
    Degree of the term 8 is 0.
    Degree of the poynomial is 5.
38. Degree of the term $20a^2b^3c$ is 6.
    Degree of the term $18ab^2c^2$ is 5.
    Degree of the term $-9bc^8$ is 9.
    Degree of the polynomial is 9.
39. $7x - 16y = -23$ is linear because it has degree 1.
40. $y = 3x - 8$ is linear because it has degree 1.
41. $y = 3x^2 + 10$ is not linear because it has a second-degree term.
42. $y = 25 - x^2$ is not linear because it has a second-degree term.
43. $P = 11x + 19$
44. $P = 6x + 8$
45. There are $17 - 0.1w$ more students that attend the math class than the English class.
46. a. The cost for one student to go on the trip for 5 days is $850.
    b. The cost for n students to go on the trip for 5 days is $850n$.
47. $8a^2 - 4a - 7$
48. $10x^2 - 5x + 12$
49. $3x^2 - 8x + 10$
50. $-7n^2 - 3n - 5$
51. $A = 36x^2 - 43x - 35$
52. The cost for gasoline used by Vista Unified School District during week w of the year can be represented by the expression $0.2w^2 + 16.75w + 322.5$.
53. $35x^2 + 40x$
54. $-56x^3 - 16x^2 + 32x$
55. $6x^2 + 3x - 30$
56. $8x^2 - 26x + 21$
57. $9x^2 - 49$
58. $16a^2 - 25b^2$
59. $x^3 - x^2 + x + 14$
60. $x^3 + x^2 - 23x + 12$
61. $9x^2 + 42x + 49$
62. $36x^2 - 108x + 81$
63. $15m^3 + 20m^2 - 160m$
64. $-30a^3 - 2a^2 + 80a$
65. Length = $2x - 7$
66. Length = $7x - 3$
67. $10x + 3 - \dfrac{4}{x}$
68. $5ab^2 - 6b + \dfrac{3}{a}$
69. $x + 7$
70. $3a + 7$
71. $m - 5 + \dfrac{5}{m - 3}$
72. $4p^2 - 10p + 25$

Chapter 5 Test

1. $20x^{10}y^7$
2. m^2n
3. $\dfrac{8b^9}{a^{12}}$
4. $\dfrac{10y^7}{7x^8}$
5. The diameter of an average red blood cell is 7.0×10^{-6} meter.
6. The average distance from the Sun to Pluto is roughly 5,900,000,000 km.
7. 3.65×10^{10}
8. 1.35×10^{12}
9. 1.3×10^{-14}
10. $5x^2 + 5x - 4$
11. $3a^2 - 13a + 13$
12. $14x^3 + 21x^2 - 49x$
13. $4x^2 - 25$
14. $12x^3 - 10x^2 + 7x - 20$
15. $36x^2 - 84x + 49$
16. $2m + 3 - \dfrac{4}{m}$
17. $2d - 3$
18. $3x^2 + x - 2 - \dfrac{14}{x + 2}$
19. $A = 40x^2 - 17x - 12$
20. $P = 26x - 2$

CHAPTER 6

Section 6.1

1. 3
3. 8
5. $3a$
7. $27y^2$
9. xy
11. 3
13. $5a$
15. $-3x$
17. $-2ab$
19. $3x$
21. $5(2x + 11)$
23. $2(3x + 1)$
25. $-5y(3 + 4y)$
27. $16a^2(3a^2 - 1)$
29. $-6(x + y)$
31. $3p(4p^5 - 2p^2 + 1)$
33. $16a^2b^2(3a^2 - b^2)$
35. $(x + 4)(2 + y)$
37. $(x - 2)(9x + 5)$
39. $(x + 8)(2x - 1)$
41. $(y + 5)(3 - 2y)$
43. $(5y + 7)(-3x + 2)$
45. Commutative property of multiplication
47. Commutative property of addition
49. Distributive property; $-6(x + 2)$
51. $-9(x - 3)$
53. $-(y + 2)$
55. $-(-5 + x)$ or $-(x - 5)$
57. $-(x^2 - 7)$
59. $(x + 5)(x + 4)$
61. $(x + 7)(x + 11)$
63. $(x - 2)(x + 6)$
65. $(x + 8)(x - 3)$
67. $(xy + 8)(x - 3)$
69. $(n - 1)(2m - 5)$
71. $(3b - 2)(a - 3)$
73. $(2x - 1)(y + 3)$
75. $(a - 3b)(2a + b)$
77. Javier: no, Joey: yes
79. Abi: yes, Noah: no
81. Kyla: yes, Lucia: no
83. $6b^2(ab - 2a^2 + 3)$
85. $-3x(2x - 1)$
87. $-17(p - 2q)$
89. $27x(3x^2 - 1)$
91. $2(p + 3)(p + 2)$

B-62 APPENDIX B Answers to Selected Exercises

93. $6(n - 2)(n - 1)$
95. Didn't factor; multiplied
97. Forgot $+1$ in the parentheses: $2x(3x - 4y + 1)$
99. Didn't multiply; factored
101. Karlee should mix 7.5 liters of the 20% hydrochloric acid solution with 2.5 liters of the 12% hydrochloric acid solution to get the 10 liters of 18% hydrochloric acid solution she needs.
103. Christopher must make at least $3333.33 worth of sales per week to earn at least $200 per week to pay his bills.

Section 6.2

1. $a = 1, b = -6, c = 5$
3. $a = 1, b = -6, c = -1$
5. $a = 1, b = 0, c = -7$
7. $a = 1, b = -2, c = 0$
9. $a = 2, b = 6, c = -8$
11. $a = -1, b = -6, c = 7$
13. $a = -6, b = -3, c = 5$
15. $a = 1, b = -2, c = 16$
17. $(x + 1)(x + 5)$
19. $(x + 6)(x - 3)$
21. $(x - 8)(x + 5)$
23. $(x - 8)(x - 7)$
25. Prime
27. $(2x + 3)(x + 2)$
29. Prime
31. $(7x - 1)(x + 6)$
33. $(3x - 1)(x - 6)$
35. $(x - 3)(4x - 5)$
37. $-(x - 3)(x + 2)$
39. Prime
41. $4y(x - 2y)$
43. $2b(3a^2 - 6a + 1)$
45. $3(x + 4)(x - 6)$
47. $5(3a + b)(a - b)$
49. $3(4x - 1)(x + 2)$
51. $t(t - 6)(t + 5)$
53. $2a(a - 4)(a - 5)$
55. $4c(2c + 5)(3c - 8)$
57. $y(x - 9)(x - 7)$
59. $(5m + 9n)(2m - 3n)$
61. $x > 24$

   $(24, \infty)$

   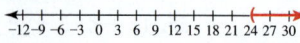

63. $x \leq \dfrac{17}{3}$

   $\left(-\infty, \dfrac{17}{3}\right]$

   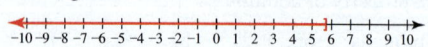

65.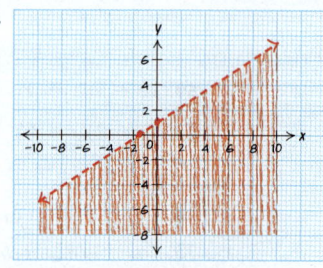

67.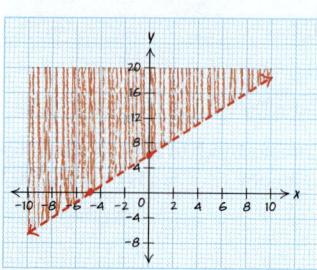

69. $y = -\dfrac{3}{2}x + 10$

71. $2x + 4y = 10$
    $-3x + 8y = 34$

    $-2(2x + 4y) = (10)(-2)$ — Multiply the first equation by -2 to make the coefficients of y to be opposites.
    $-3x + 8y = 34$

    $-4x - 8y = -20$ — Add the two equations to eliminate y.
    $-3x + 8y = 34$
    $-7x = 14$ — Divide both side of the equation to isolate x.
    $x = -2$

    $-3(-2) + 8y = 34$ — Substitute $x = -2$ into one of the equations to solve for y.
    $6 + 8y = 34$
    $8y = 28$
    $y = 3.5$

    $(-2, 3.5)$ — Write the solution as a coordinate pair.

73. No solution. This is an inconsistent system.
75. $25x^4y^{14}$
77. $\dfrac{9x^{14}}{y^6}$

Section 6.3

1. 2
3. 1
5. 4
7. 3
9. Yes
11. Yes
13. No
15. Yes
17. No
19. $(x + 12)(x - 12)$
21. $(1 + y)(1 - y)$
23. Prime
25. $(10x + 7)(10x - 7)$
27. Prime
29. $(9 - a)(9 + a)$
31. $(4x - 7)(4x + 7)$
33. $(3t - 2)(3t + 2)$
35. Yes
37. Yes
39. No
41. No
43. Yes
45. $(x + 1)^2$
47. $(y - 4)^2$
49. $(3p - 1)^2$
51. $(t + 3)(t + 2)$
53. $(2x + 5)(3x - 4)$
55. $(2x + 5y)^2$
57. $(3x + 4y)^2$
59. $-(t - 3)^2$

APPENDIX B Answers to Selected Exercises B-63

61. Prime
63. $3(x + 5)(x - 5)$
65. $17x(y - 2x)$
67. $5(x + 5)(x - 5)$
69. $2(2p - 1)(r + 3)$
71. $3(2x + 1)(x - 5)$
73. $-2(3a + 1)^2$
75. $5(x^2 + 4)$. The polynomial $x^2 + 4$ is prime.
77. $3(t + 1)^2$
79. $5x(y + 4)^2$
81. $5(x - 7)(x + 3)$
83. $-2(2x - 3)(5x - 4)$
85. $a(3a - 5)(a + 4)$
87. $\left(\dfrac{x}{2} + 5\right)\left(\dfrac{x}{2} - 5\right)$
89. $\left(x + \dfrac{5}{4}\right)\left(x - \dfrac{5}{4}\right)$

91.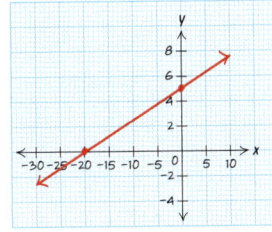

93.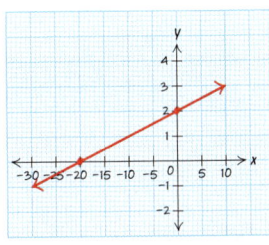

95. $-6x^2 + x + 40$
97. $x + 8$
99. $7x - 5$

47. **a.** $P = 256$. If the company sells four appliances in an order, the average profit per appliance will be $256.
   b. To make an average profit of $300 per appliance, the company needs to sell either 5 or 15 appliances in a single order.
49. **a.** $P = 39$. If the company sells three planes to an airline, they will earn an average profit of $39 million per plane.
   b. An order of 6 or 10 planes will result in an average profit per plane of $60 million.
51. Width = 2 feet, length = 12 feet
53. Width = 9 cm, length = 16 cm
55. Base = 10 inches, height = 6 inches
57. The ball will hit the ground after 3 seconds.
59. The ball at a height of 128 ft at 1 second and 3 seconds.
61. $x = -1, x = -3$
63. $a = \dfrac{3}{2}, a = 3$
65. $x = -2, x = 7$
67. $n = -5, n = -4$
69. $x = -4, x = 3$
71. $y = -8, y = 2$
73. $t = -5, t = 5$
75. $g = -10, g = 10$
77. $x = \dfrac{20}{3}, x = -3$
79. $t = 0, t = -3, t = 3$
81. $x = 0, x = -3, x = -4$
83. $t = 0, t = 3, t = 5$
85. $(x + 8)^2$
87. $n = \dfrac{5}{2}, n = -\dfrac{5}{2}$
89. $(a + 7)(a + 1)$
91. $x = 0, x = -3, x = 2$
93. $m = -\dfrac{1}{2}, m = \dfrac{3}{2}$
95. $(4b + 5)(10b - 3)$
97. $m = 5, m = -5$
99. Prime

Section 6.4

1. Linear
3. Neither
5. Quadratic
7. Quadratic
9. Linear
11. Linear
13. Quadratic
15. $x = -3$
17. $n = \dfrac{5}{2}$
19. $a = 0, a = 8$
21. $y = 0, y = -\dfrac{7}{2}$
23. $t = 0, t = -\dfrac{5}{2}$
25. $x = -5, x = 6$
27. $p = 0, p = -\dfrac{3}{2}, p = 9$
29. $x = -2, x = -3$
31. $a = -3, a = -4$
33. $x = -3, x = 1$
35. $t = 3, t = 4$
37. $x = -2, x = -1$
39. $x = \dfrac{1}{2}, x = -5$
41. $x = -\dfrac{2}{3}, x = 4$
43. $x = -\dfrac{5}{2}, x = -\dfrac{4}{3}$
45. $a = -\dfrac{2}{3}, a = -\dfrac{3}{8}$

Chapter 6 Review Exercises

1. $2x^2(6x^3 + 5x - 3)$
2. $3b^2(a^3 + 2a - 3)$
3. $(3a + 5)(2a - 7)$
4. $(x + 3)(2y - 5)$
5. $(4x + 3)(x + 2)$
6. $(2t + 7)(t - 6)$
7. $(5m - 7)(m + 2n)$
8. $(2x - 1)(3y + 5)$
9. $(a + 4)(a + 6)$
10. $(m + 8)^2$
11. $(x - 4)(x - 5)$
12. $(h - 9)(h - 4)$
13. $(x - 7)(x + 3)$
14. $(y - 9)(y + 2)$
15. $(2x + 3)(x + 5)$
16. $(3a + 1)(a + 6)$
17. $(5x - 2)(x + 3)$
18. $(3b - 4)(b + 7)$
19. $(4x + 5)(x - 8)$
20. $(3y - 4)^2$
21. $-(5b - 2)(2b - 3)$
22. $-(2t - 5)(t - 5)$
23. $g(g - 3)(g - 5)$
24. $(3x - 4y)(2x + 5y)$
25. $(n - 9)(n + 9)$
26. $(t + 8)(t - 8)$
27. $(3a + 10)(3a - 10)$
28. $(2x - 7)(2x + 7)$
29. Prime
30. Prime
31. $(3x - 7)(x - 9)$
32. $(4y - 5)(x^2 + 2)$
33. $3c(2a + 1)(b - 4)$
34. $2z(x - 6)(2y + 1)$

B-64 APPENDIX B Answers to Selected Exercises

35. $x = -7, \dfrac{5}{3}$
36. $x = 3, \dfrac{9}{2}$
37. $y = 0, \dfrac{3}{2}$
38. $x = 0, \dfrac{1}{3}$
39. $x = 4, -7$
40. $x = 9, -6$
41. $x = -7, \dfrac{5}{2}$
42. $y = -4, \dfrac{1}{3}$
43. $m = 4, -8$
44. $n = -7, 5$
45. $a = 9, -9$
46. $x = 7, -7$
47. $y = 2, -2$
48. $a = 3, -3$
49. $t = 0, 5, -6$
50. $x = 0, -\dfrac{3}{2}, 5$
51. $x = 4, -9$
52. $x = \dfrac{1}{3}, -1$

53. The length is 9 cm, and the width is 4 cm.
54. **a.** If RC Toys, Inc. sells 8000 toy cars, they will earn a profit of $144 thousand.
   b. To earn $120 thousand profit, RC Toys, Inc. needs to sell either 5000 or 12,000 toy cars.

11. **a.** 422.5 miles **b.** 7 hours
12. $x = -\dfrac{32}{11}$
13. $x > 10$
14. **a.**

x	$y = \dfrac{2}{3}x - 4$
-6	-8
-3	-6
0	-4
3	-2
6	0

b.

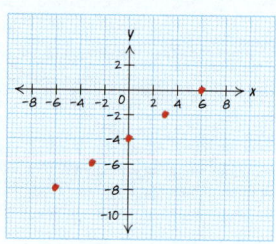

c. Linear

15.

m	$C = 124m$
0	0
1	124
2	248
3	372
4	496

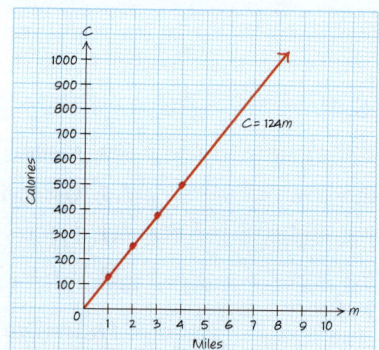

16.

x	$y = x^2 + 3$
-4	19
-3	12
-2	7
-1	4
0	3
1	4
2	7
3	12
4	19

Chapter Test

1. $(x + 6)(x - 4)$
2. $(a - 6)(a - 8)$
3. $3(2x + 5)(x - 4)$
4. Prime
5. Prime
6. $(x + 10)(x - 10)$
7. $(5m + 1)(5m - 1)$
8. $(3x - 1)(2x + 3)$
9. $(2y + 5)^2$
10. $t(4t - 7)(t + 3)$
11. $2x(x - 4)(x + 1)$
12. $(5a + 2)(b - 2)$
13. $(2m - 3)(n + 8)$
14. $t = -3, -10$
15. $w = 2, -2$
16. $x = 0, 6, -8$
17. $x = 5, -\dfrac{4}{3}$
18. $x = 4, \dfrac{8}{5}$
19. **a.** If XY-Wireless sells 5000 Bluetooth earpieces, they will have a profit of $190,000.
   b. To have $160 thousand profit, XY-Wireless must sell 4000 Bluetooth earpieces.
20. The length of the rectangle is 9 meters, and the width is 7 meters.

Cumulative Review for Chapters 1–6

1. $(2x)^5$
2. $-\dfrac{1}{8}$
3. -18
4. 11
5. $-15a + b$
6. $4x + 6$
7. $y = \dfrac{6}{5}x - 7$
8. $y = -20$
9. $b = P - a - c$
10. $n = \dfrac{2}{3}$

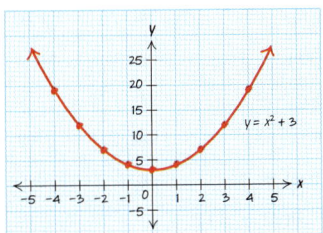

Nonlinear

17. $m = -2$

18. x-intercept: $(8, 0)$, y-intercept: $(0, -3)$

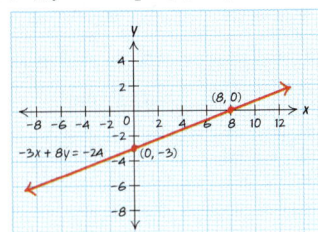

19. $m = -\dfrac{2}{7}$, $(0, 2)$, $(7, 0)$

20.

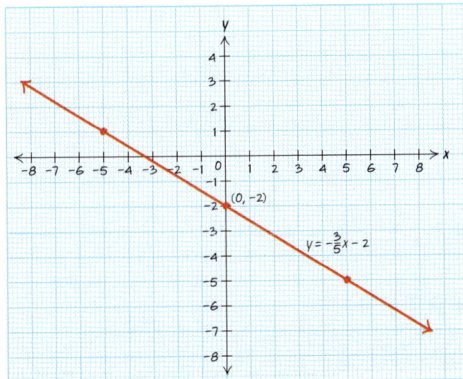

21.

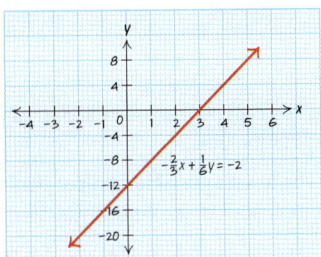

22. Parallel 23. $y = \dfrac{2}{3}x + 3$ 24. $C = 600h$

25. $y = -\dfrac{5}{2}x + 6$ 26. $y = 4$ 27. $y = -\dfrac{7}{2}x + 3$

28. $y = \dfrac{1}{2}x - 3$ 29. $y = -\dfrac{5}{3}x + 2$

30. The markup rate for these pet toys is 87.5%.

31. This computer declined in value by about 72.7% in the first year.

32. The discount on the dress is $36, and the sales price is $84.

33. The tax on the $8000 used car is $600.

34. $(-5, 3)$

35. a. No   b. Yes

36. $(1, -4)$

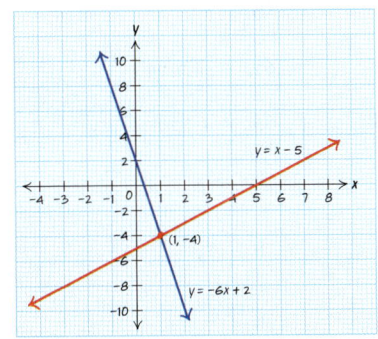

37. No solution   38. $(-7, 16)$

39. Each T-shirt costs $9, and a basketball costs $16.

40. The boat's rate is 9 mph, and the current rate is 1 mph.

41. Emma should put 25 lb of papaya and 15 lb of pineapple in the mix to make 40 lb of $9.49 per pound mix.

42. $(6, -5)$ 43. No solution

44. The first angle is 35°, and the second angle is 55°.

45. Seamus should invest $2000 in the CD and $8000 in the bonds.

46. No

47.

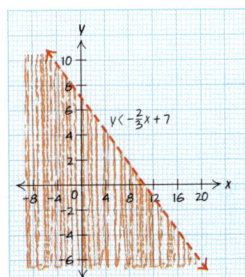

48. Yes

49.

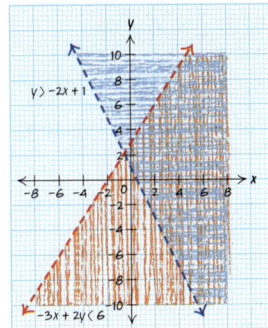

50. 9   51. -9

52. a. x^8 b. $12a^5$ c. $15a^5b^4$

53. a. n^6 b. $35t^5$ c. $-27x^3y^7$

54. a. $27x^{12}$ b. $-8x^6$ c. $4x^6y^8$

55. a. $25y^8$ b. $-243y^{10}$ c. $64a^{15}b^6$

56. a. $-63x^5$ b. $-72x^2y^{11}$

57. a. $-64n^{19}$ b. $-25x^9y^{22}$
58. a. x^6 b. $-\frac{1}{2}a^4$ c. x^2y^2
59. a. y^2 b. $-5y$ c. a^2b^4
60. $\frac{9}{100}x^2y^2$
61. $-343n^9m^3$
62. $A = 16x^2$
63. $A = 9\pi n^2$
64. a. $\frac{5}{t^2}$ b. $-2x^3$
65. a. $\frac{6}{n^5}$ b. $-7y^4$
66. a. $\frac{x^2}{y}$ b. $-\frac{y}{3x^4}$ c. $\frac{9}{n^6}$ d. $\frac{y^2}{x^4}$
67. a. $\frac{t^4}{r}$ b. $-\frac{2m^2}{3n}$ c. $-\frac{64}{m^6}$ d. $\frac{y^{12}}{x^3}$
68. a. 8.951×10^{11} b. 4.13×10^{-4}
69. a. 6.487×10^8 b. 6.92×10^{-6}
70. a. $67{,}500{,}000$ b. 0.0000358
71. a. $401{,}000{,}000$ b. 0.000843
72. 1×10^9
73. 0.0000000025
74. a. 2.88×10^9 b. 1.7×10^{12} c. 8.5×10^{-8}
    d. 2.666×10^{22} e. 2.0×10^8
75. Four terms, third-degree
76. Two terms, second-degree
77. $4xy^2$: degree = 3
    $-xy$: degree = 2
    $2y^4$: degree = 4
    $9x$: degree = 1
    Degree of polynomial is 4.
78. $8a^5b$: degree = 6
    $3a^3$: degree = 3
    $-6b^5$: degree = 5
    5: degree = 0
    Degree of polynomial is 6.
79. $-x^2 - 4x - 3$
80. $2y^2 + 6y - 12$
81. $-6x^2 + 5x - 16$
82. $-7a^2 - 2a + 40$
83. $-10n + 18$
84. $6n - 3$
85. a. $x^2 + 3x - 28$ b. $6x^2 - 13x - 5$
    c. $x^3 - x^2 - 2x + 8$ d. $n^2 + 8n + 16$
    e. $4x^2 - 1$
86. a. $y^2 + 6y - 27$ b. $-8x^2 - 10x + 7$
    c. $x^3 + 3x^2 - 6x + 2$ d. $n^2 - 6n + 9$
    e. $9x^2 - 25$
87. $-7x - 12$
88. $-10y + 22$
89. $-2n + 4 - \frac{5}{n}$
90. $-y + 2 - \frac{4}{y}$
91. $2x - 1$
92. $3x + 1$
93. a. $-3ab(4a - b + 10)$ b. $(n + 3)(2n - 1)$
94. a. $16x^2y^2(xy + 3)$ b. $(m - 4)(-3m + 2)$
95. $(3a + 1)(b - 4)$
96. $(2x - 5)(y + 6)$
97. $3(x + 1)(x - 5)$
98. $2(y - 7)(y - 2)$
99. a. $(x + 6)(x - 5)$ b. $-(x + 3)(x - 2)$
    c. $(2x + 3)(2x - 1)$
100. a. $(x - 9)(x + 4)$ b. $-(x - 5)(x + 4)$
     c. Prime
101. a. $2n(n - 5)(n + 2)$ b. $5(x + 3y)(2x + y)$
102. a. $3t(t - 9)(t - 6)$ b. $3(x - 7y)(2x + y)$
103. a. $(t - 7)(t + 7)$ b. $(3 - 2y)(3 + 2y)$
     c. $(2x + 5)^2$ d. Prime
     e. $5(x + 3)(x - 3)$
104. a. $(y + 6)(y - 6)$ b. $(1 - 5n)(1 + 5n)$
     c. $(3y - 2)^2$ d. Prime
     e. $3(y + 4)(y - 4)$
105. a. Quadratic b. Neither c. Linear
106. a. Linear b. Quadratic c. Neither
107. a. $x = -4, -8$ b. $x = \frac{3}{2}, -5$
     c. $x = 7, -5$ d. $x = 7, -7$
108. a. $x = -3, -7$ b. $x = -\frac{1}{4}, 6$
     c. $x = 3, -2$ d. $n = 5, -5$

CHAPTER 7

Section 7.1

1. 2
3. $\frac{25}{34}$
5. $\frac{1}{10}$
7. Undefined
9. 0
11. a. $c = 25$. The cost per person is \$25 if 10 people agree to help pay for the gift.
    b. $c = 5$. The lowest possible cost per person is \$5 if there are 50 employees in the company.
13. a. $c = 75$. The cost per person is \$75 if 8 people go on the trip.
    b. $c = 41$. The lowest possible cost per person is \$41 if the trip is limited to 25 people.
15. a. $c = 200$. The cost of each surfboard is \$200 if 8 surfboards are made in a week.
    b. $c = 180$. The lowest cost per surfboard for a week is \$180 if 10 surfboards can be made in a week.
17. $s = 2117.92$. In 2020, Texas will spend approximately \$2117.92 per person on public assistance.
19. $x = 0$ is excluded.
21. $x = 3$ is excluded.

23. $a = -\frac{4}{3}$ is excluded.

25. No excluded values for this expression

27. $x = 3, -2$ are excluded.

29. $h = -4, 8$ are excluded.

31. $r = 3, 2$ are excluded.

33. $n = -2$ is excluded.

35. $b = -2$ is excluded.

37. $\frac{4}{x}$

39. $7n$

41. $\frac{1}{a+5}$

43. $\frac{4x}{x-3}$

45. $\frac{h+6}{h+4}$

47. $\frac{1}{r+3}$

49. a. $T = 2.4$. The two landscapers working together can complete the job in 2.4 hours, or 2 hours and 24 minutes, if it takes the assistant 6 hours to complete the job alone.

b. $T \approx 2.32$. The two landscapers working together can complete the job in 2.32 hours, or 2 hours and 19 minutes, if it takes the assistant 5.5 hours to complete the job alone.

51. a. $T = \frac{40}{13} \approx 3.08$. The two staff members working together can complete the job in 3.08 hours, or 3 hours and 5 minutes, if it takes the office specialist 8 hours to complete the job alone.

b. $T = \frac{65}{23} \approx 2.83$. The two staff members working together can complete the job in 2.83 hours, or 2 hours and 50 minutes, if it takes the assistant 5.5 hours to complete the job alone.

53. a. $y = 52.5$. Jack's speed on his return trip is 52.5 mph if he drives at 70 mph to the city.

b. $y = 66$. Jack's speed on his return trip is 66 mph if he drives at 55 mph to the city.

55. a. $y = \frac{4500}{11} \approx 409.09$. The pilot's speed on her return trip is 409.09 mph if she flies at 500 mph to the city.

b. $y = \frac{3825}{8} \approx 478.13$. The pilot's speed on her return trip is 478.13 mph if she flies at 425 mph to the city.

57. The student divided out common terms, but only common factors can be divided. Correctly simplified: $\frac{x+2}{x+4}$.

59. The student divided out common terms, but only common factors can be divided. Correctly simplified: $\frac{x}{x+7}$.

61. The student divided out common terms, but only common factors can be divided. Correctly simplified: $\frac{(x+2)(x-2)}{(x-2)} = x+2$

63. $\frac{1}{n+2}$

65. $\frac{2}{b+5}$

67. $\frac{x+15}{(x+2)(x+7)}$

69. $\frac{1}{a-3}$

71. $\frac{r-9}{r-7}$

73. $\frac{y+12}{y-11}$

75. $-(x-1)$

77. $-(y-4)$

79. $-(3x^2 + 2x - 5)$

81. $-(2x^2 - 3x + 8)$

83. $-(8x^2 - 4)$

85. -3

87. -1

89. $\frac{4}{3}$

91. $-\frac{w+7}{w+3}$

93. $\frac{1}{a+4}$

95. $-\frac{1}{x-5}$

97. $\frac{m-1}{3}$

99. $-\frac{3}{t-6}$

101. $\frac{8}{15}$

103. $-\frac{6}{5}$

105. $\frac{5}{3}$

107. $-\frac{44}{45}$

Section 7.2

1. $\frac{32}{15}$

3. $-\frac{18}{115}$

5. $\frac{3}{10y}$

7. $-\frac{2n}{5}$

9. $\frac{x+5}{x+2}$

11. $\frac{(a+7)(a+3)}{(a+2)(a-5)}$

13. $\frac{x+2}{x-4}$

15. $\frac{m-5}{m+9}$

17. $\frac{h+3}{h+2}$

19. $\frac{1}{(b+3)(b-6)}$

21. $\frac{k-5}{k+3}$

23. $\frac{h+4}{h-3}$

25. $\frac{x+6}{x+1}$

27. $\frac{t-4}{t-6}$

29. $\frac{1}{4x+8}$ or $\frac{1}{4(x+2)}$

31. $-\frac{4}{(x-1)(x+2)}$

33. $\frac{3(x-7)}{x(x+4)}$

35. The child's drug dosage is 13.6 mg/lb.

37. The adult's drug dosage is 20.5 mg/lb.

39. The price for the 5-pound bag of rice is about $0.06 per ounce.

41. The price for the 1-pound jar of peanut butter is $0.24 per ounce.

43. The car's speed is 96.6 kilometers per hour.

45. The car's speed is 4400 feet per minute.

47. The car's speed is 22.4 meters per second.

49. a. The dosage is 9 mg per hour.

b. The patient will receive 216 mg of dopamine in a 24-hour period.

51. a. The cyclist's rate is 24.85 miles per hour.

b. The cyclist can ride 149.1 miles during a 6-hour ride.

53. a. The Solartaxi travels at about 55.9 mph.

b. The Solartaxi travels about 279.5 miles in 5 hours.

55. $\frac{18}{25}$

57. $\frac{85}{12}$

59. $\frac{4}{11x^2}$

61. $-\dfrac{5m^2n}{3}$
63. $\dfrac{x+4}{x-3}$
65. $\dfrac{(a+3)^2}{(a+4)(a+1)}$
67. $\dfrac{x-7}{x+5}$
69. $\dfrac{n-5}{n+3}$
71. $\dfrac{(h+9)^2}{(h+7)(h+2)}$
73. $\dfrac{(b+7)^2}{(b+5)(b+4)}$
75. $\dfrac{(k+9)(k+2)}{(k-7)(k+6)}$
77. $-\dfrac{2}{xy}$
79. 5
81. $\dfrac{1}{5x-15}$ or $\dfrac{1}{5(x-3)}$
83. $-\dfrac{15m^2n^2}{2}$
85. $\dfrac{x+7}{2x-16}$ or $\dfrac{x+7}{2(x-8)}$
87. $-\dfrac{7x+21}{3}$ or $-\dfrac{7(x-3)}{3}$
89. $-\dfrac{12b}{7a}$
91. $\dfrac{z+2}{z+5}$
93. 30
95. $a+2$
97. $\dfrac{(a+3)^2}{(a-3)(a+8)}$
99. $\dfrac{(b+5)(b-4)}{(b+4)(b-5)}$
101. $\dfrac{2}{5}$
103. $\dfrac{4}{15}$
105. $\dfrac{23}{42}$
107. $-\dfrac{1}{6}$

49. $\dfrac{26}{45}$
51. $\dfrac{48x+5z^2}{6xyz^2}$
53. $-\dfrac{3m^2+4n^3}{9m^3n^2}$
55. $\dfrac{4h^2-21h-27}{(h-9)(h-6)}$
57. 4
59. $\dfrac{3b-7}{b-7}$
61. $\dfrac{17}{4(a-2)}$
63. $\dfrac{x^2+4x+3}{(x+5)(x+4)}$
65. $\dfrac{m^2+8m+15}{(m+10)(m+2)}$
67. $\dfrac{3x^2+x}{(x+3)(x-3)(x+2)}$
69. $\dfrac{8m+55}{(m+10)(m+5)(m-2)}$
71. a. $\dfrac{a+b}{ab}$
   b. The two brothers working together can mow $\dfrac{5}{6}$ of the lawn in one hour.
73. a. $\dfrac{m+n}{mn}$
   b. The two employees working together can file $\dfrac{9}{20}$ of the fall applications in one day.
75. $\dfrac{11x+21}{(x+3)(x+1)}$
77. $\dfrac{32x}{(x+3)(x+2)}$
79. $\dfrac{n+5}{2(n+3)}$
81. $\dfrac{8h^2-26h-42}{(h-7)(h-4)}$
83. 7
85. $\dfrac{(c+1)(5c+3)}{c-1}$
87. $\dfrac{3}{2}$
89. $\dfrac{7b}{3(a-2)}$
91. $\dfrac{7}{5}$
93. $-\dfrac{4a+3}{7a+5}$
95. $\dfrac{4x^2-x}{3x^2-2}$
97. 1
99. $\dfrac{5x+4y}{-7x+6y}$
101. a. The average cost for food is $90.91 if 20 players go on the trip.
   b. The average cost for transportation and lodging is $190.91 if 20 players go on the trip.
   c. $\dfrac{230n+1600}{n+2}$
   d. The average cost for all expenses is $272.22 if 25 players go on the trip.

Section 7.3

1. $\dfrac{10}{7c^2}$
3. $\dfrac{3}{a^2}$
5. $\dfrac{2x+3}{x+5}$
7. $\dfrac{4h-3}{h-9}$
9. 5
11. $2a$
13. $\dfrac{1}{h+4}$
15. $\dfrac{1}{z-3}$
17. $\dfrac{1}{x+2}$
19. $\dfrac{2a+4}{(2a+3)(a-4)}$
21. 150
23. $7xy^3z^2$
25. $(h-9)(h-6)$
27. $2a-6$ or $2(a-3)$
29. $(h+4)(h+7)(h+2)$
31. $(z+2)(z-3)(z-5)$
33. $(m+10)(m-2)(m+5)$
35. $\dfrac{9}{168}$ $\dfrac{32}{168}$
37. $\dfrac{63n^3}{180m^3n^2}$ $\dfrac{50m^2}{180m^3n^2}$
39. $\dfrac{10x+20}{(x+4)(x+2)}$ $\dfrac{5x^2+20x}{(x+4)(x+2)}$
41. $\dfrac{5}{7(a-2)}$ $\dfrac{63}{7(a-2)}$
43. $\dfrac{z^2-4z-32}{(z+5)(z-8)(z-2)}$ $\dfrac{z^2+6z+5}{(z+5)(z-8)(z-2)}$
45. $\dfrac{2x^2+4x}{(x+3)(x-3)(x+2)}$ $\dfrac{x^2-3x}{(x+3)(x-3)(x+2)}$
47. $\dfrac{15m-15}{(m+8)(m-2)(m-1)}$ $\dfrac{3m-6}{(m+8)(m-2)(m-1)}$

Section 7.4

1. $x=3$
3. $n=8$
5. $x=2$
7. $x=1$
9. $x=1$
11. $x=-6$
13. $r=0$
15. $n=3,-9$
17. $k=-5,-6$
19. a. The cost per person is $25 if 100 people attend the banquet.
   b. There must be 400 people attending the banquet for the cost per person to be $6.25.
21. a. The cost per person is $15 if 10 people agree to help pay for the gift.
   b. There must be 30 people who agree to help pay for the gift for the cost per person to be $5.

23. **a.** The cost per person is $87.50 if eight people go on the trip.
   b. If they want the cost per person to be $75, they should recruit 10 people to go on the trip.

25. **a.** The cost for each large shutter is $550 if four large shutters are made in a week.
   b. For the cost per shutter to be $400, there must be eight large shutters made in a week.

27. $\dfrac{7}{3} + \dfrac{8}{6r} = 2$ Multiply both sides by the LCD to clear the fractions.

   $6r\left(\dfrac{7}{3}\right) + 6r\left(\dfrac{8}{6r}\right) = 6r(2)$

   $14r + 8 = 12r$ Use the subtraction property of equality to combine variable terms.
   $\underline{-12r \quad\quad -12r}$

   $2r + 8 = 0$ Use the subtraction property of equality to isolate the variable term.
   $2r = -8$

   $r = -4$ Use the division property of equality to isolate r.

29. $p = -3$
31. No solution
33. $x = 3$
35. $a = 2, 12$
37. $h = -10$
39. $g = \dfrac{10}{99}$
41. $x = -12$
43. No solution
45. $x = 10$
47. $t = 43$
49. No solution
51. $n = 5$
53. $k = -3$
55. $m = 4$
57. $b = 3$
59. $x = 5$
61. $n = -1$

63. Brittany's work rate is 1 per 30 minutes or 2 per hour.

65. The pump's work rate is $\dfrac{1}{12}$ per hour.

67. Kasim's work rate is $\dfrac{1}{2}$ per hour.

69. Sherri and Jen can scan the file box of medical records in $\dfrac{15}{19}$ of an hour, or about 47 minutes, if they work together.

71. Martin and Bill can mow the lawn in $3\dfrac{1}{5}$ hours, or 3 hours and 12 minutes, if they work together.

73. Using both copy machines, the bulletins can be copied in $4\dfrac{4}{9}$ hours, or about 4 hours and 27 minutes.

75. **a.** Expression   **b.** $\dfrac{x+3}{x+1}$

77. **a.** Equation   **b.** $x = \dfrac{2}{7}$

79. **a.** Expression   **b.** $-\dfrac{x+36}{(x+8)(x+1)}$

81. **a.** Expression   **b.** $5x$

83. **a.** Equation   **b.** $x = 2$

85. **a.** Expression   **b.** $\dfrac{x^2 + 3x + 2}{(x-2)^2}$

87.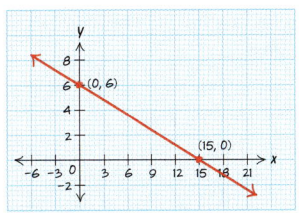

89.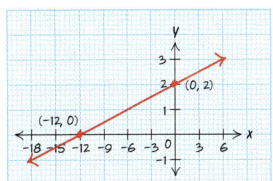

91. $y = \dfrac{2}{3}x + \dfrac{17}{3}$ 93. $(7.5, -1)$ 95. $(6, 34)$

97. The rectangle is 6 inches wide and 14 inches long.

99. $10000x^{22}y^{14}$

Section 7.5

1. $\dfrac{4}{15}$ 3. $\dfrac{3}{10}$ 5. $\dfrac{5}{12}$

7. $\dfrac{3}{8}$ 9. $\dfrac{3}{40}$

11. $3.89 per ounce
13. $1200 per ounce
15. 7.5 miles per hour
17. $1\dfrac{1}{3}$ feet per minute

19. 20 Duracell AA batteries for $15.79 cost $0.79 for each battery.

   16 Duracell AA batteries for $14.79 cost $0.92 for each battery.

   20 Duracell AA batteries for $15.79 are the better buy.

21. An 10-oz bag of Lay's Classic potato chips for $2.50 costs $0.25 per ounce.

   A 15.25-oz bag of Lay's Classic potato chips for $3.48 costs $0.23 per ounce.

   An 15.25-oz bag of Lay's Classic potato chips for $3.48 is the better buy.

23. A 16-oz bottle of Herdez salsa casera for $2.03 costs $0.13 per ounce.

   A 24-oz bottle of Herdez salsa casera for $2.87 costs $0.12 per ounce.

   A 24-oz bottle of Herdez salsa casera for $2.87 is the better buy.

25. The toy couch will be $3\dfrac{1}{3}$ inches tall.

27. The length of a 12-foot wall will be 3 inches on the blueprint.

29. The length of the window in the stadium will be 20 feet.

31. The cost of 306 feet of vinyl fencing is $4029.

33. The cost of 3900 square feet of shingles is $3120.

B-70 APPENDIX B Answers to Selected Exercises

35. **a.** The total fat in a medium order of french fries is about 18.17 grams.

  b. The amount of saturated fat in a medium order of french fries is about 1.65 grams.

37. $x = 15$ 39. $b = 12$ 41. $y = -1$
43. $x = 0$ 45. $y = 6$ 47. $x = -11$
49. 25 miles is approximately equal to 40.23 km.
51. 165 pounds is equal to 75 kilograms.
53. 45 inches is equal to 114.3 centimeters.
55. 60 acres is equal to 2,613,600 square feet.
57. The width of the enlarged photo will be 12.5 inches.
59. $x = 7.5$ 61. $x = 6\frac{2}{3}$
63. The Victory Obelisk is 141.8 meters tall.
65. The amount of unsalted butter needed to make 16 servings is $\frac{4}{5}$ cup.
67. The amount of chicken broth needed to make 12 servings is $3\frac{3}{5}$ cups.
69. The amount of milk needed to make 16 servings is 16 cups.
71. The amount of maple syrup needed to make 12 servings is $2\frac{1}{4}$ cups.
73. $y = \frac{10}{3}x$; $y = 50$ when $x = 15$
75. $d = 3t$; $d = 30$ when $t = 10$
77. $y = \frac{5}{8}x$; $y = 10.625$ when $x = 17$
79. **a.** $C = 576h$
  b. $C = 2016$. A 150-pound person burns 2016 calories during 3.5 hours of mountain biking.
81. **a.** $D = 65t$
  b. $D = 357.5$. 357.5 miles are traveled in 5.5 hours.
83. **a.** $T = 46c$
  b. $T = 690$. A student who enrolls in 15 credits will owe $690 for tuition.
85. **a.** $F = \frac{6}{5}x$
  b. $x = 8\frac{1}{3}$. A force of 10 pounds will stretch the spring $8\frac{1}{3}$ inches.
87. **a.** $W = 16t$
  b. $W = 400$. The amount of the paycheck would be $400.00 for 25 hours of work.
89. $y = \frac{15}{x}$; $y \approx 2.73$ when $x = 5.5$
91. $P = \frac{1000}{n}$; $P = 20$ when $n = 50$

93. $C = \frac{180}{h}$; $C = 60$ when $h = 3$
95. **a.** $k = 30$
  b. $P = 6$. The inside air pressure of the balloon is 6 psi when the volume of the balloon is 5 in^3.
97. **a.** $s = \frac{275}{t}$
  b. $s \approx 78.57$. The speed is approximately 78.57 mph for 3.5 hours of driving.
99. **a.** $I = \frac{900}{d^2}$
  b. $I = 9$. The illumination is 9 foot-candles at a distance of 10 feet from the light.

Chapter 7 Review Exercises

1. $\frac{9}{2}$
2. Undefined
3. The cost per person is $72 when 50 members go to the competition.
4. The average spending per person by the state of Washington in 2020 will be $7407.55.
5. $x = -7$ is excluded.
6. $n = -2, 6$ are excluded.
7. $b = -3, -1$ are excluded.
8. $t = \pm 5$ is excluded.
9. $-\frac{1}{8x}$ 10. $-\frac{1}{x+1}$ 11. $\frac{y-8}{y+3}$
12. $\frac{1}{g+8}$ 13. $-\frac{1}{h+3}$ 14. $\frac{4}{n-2}$
15. A 130-pound rower will burn approximately 70.8 calories during a 6-minute event.
16. The runner's speed is 880 feet per minute.
17. $-\frac{1}{10k}$ 18. $-\frac{x^2}{3}$ 19. $-\frac{2mn}{21}$
20. $-\frac{9q}{8p^2}$ 21. $\frac{x+2}{x+3}$ 22. $\frac{1}{d+6}$
23. $\frac{k-2}{k+3}$ 24. $\frac{p-3}{p+2}$ 25. $\frac{z^2+7z+10}{(z-5)(z-3)}$
26. 1 27. $\frac{7x+7}{x+2}$ 28. $\frac{7n+3}{5n+1}$
29. $\frac{3}{x+4}$ 30. $\frac{5}{h-5}$ 31. $\frac{6x-5}{x+6}$
32. $\frac{2y+5}{y-3}$ 33. $\frac{6}{n-4}$ 34. $\frac{2x+5}{(x+3)(x+5)}$
35. $\frac{24b^2c}{18a^2b^3c}$, $\frac{11a^2}{18a^2b^3c}$ 36. $-\frac{9x^2}{8x^3y^3}$, $\frac{14y}{8x^3y^3}$
37. $\frac{7}{2(v+4)}$, $\frac{22}{2(v+4)}$ 38. $-\frac{1}{y-3}$, $\frac{11}{y-3}$

39. $\dfrac{5p-40}{(p+2)(p-3)(p-8)}$ $\dfrac{3p^2-9p}{(p+2)(p-3)(p-8)}$

40. $-\dfrac{x-2}{(3x-1)(x+4)(x-2)}$ $\dfrac{3x^2+12x}{(3x-1)(x+4)(x-2)}$

41. $\dfrac{z^2-12z+35}{(z+7)(z+2)(z-7)}$ $\dfrac{4z+8}{(z+7)(z+2)(z-7)}$

42. $\dfrac{x^2-13x+36}{(x+9)(x-9)(x+5)}$ $-\dfrac{3x+15}{(x+9)(x-9)(x+5)}$

43. $-\dfrac{8a^2-35bc}{50a^3b^2}$

44. $\dfrac{xy+6}{3xy^2}$

45. $\dfrac{13y-19}{(y+2)(y-7)}$

46. $-\dfrac{t-39}{(t+6)(t-3)}$

47. $\dfrac{k^2+2k-1}{(k+2)(k+1)}$

48. $\dfrac{w-1}{w-4}$

49. $\dfrac{x^2-5}{(x-5)(x+2)}$

50. $\dfrac{y-2}{y+6}$

51. $m=22$. There must be 22 members of the band who travel to the competition in order for the cost per person to be \$100.

52. $t \approx 8$. In the year 2018, the state of Washington spent \$4241 per resident.

53. $h=-4$ 54. $f=-1$ 55. $r=\dfrac{25}{62}$

56. $x=\dfrac{4}{39}$ 57. $w=-\dfrac{13}{5}$ 58. $b=\dfrac{1}{2}$

59. $n=8$ 60. $x=4$ 61. $a=3$

62. $d=-9$ 63. $\dfrac{6x+2}{8x-3}$ 64. $\dfrac{4x+5}{2x-3}$

65. $\dfrac{3n}{3n+1}$ 66. $\dfrac{n-4}{n}$

67. $t=5\dfrac{5}{11}$. Working together, Amin and Danya can typeset a chapter in $5\dfrac{5}{11}$ hours, or about 5 hours and 27 minutes.

68. $t=10\dfrac{4}{5}$. Working together, Mike and his assistant can paint the house in $10\dfrac{4}{5}$ hours, or 10 hours and 48 minutes.

69. $\dfrac{7}{12}$ 70. $\dfrac{3}{8}$

71. \$0.75 per pound

72. 21 miles per hour

73. 946 milliliters of Concrobium Mold Control for \$8.98 costs \$0.0095 per milliliter.

    3784 milliliters of Concrobium Mold Control for \$36.61 costs \$0.0097 per milliliter.

    946 milliliters of Concrobium Mold Control for \$8.98 is the better buy.

74. 18 extra large grade AA eggs for \$3.48 cost \$0.19 per egg.

    24 extra large grade AA eggs for \$6.98 cost \$0.29 per egg.

    The package of 18 extra-large grade AA eggs for \$3.48 is the better buy.

75. There are approximately 93 milligrams of sodium in a 32 fl. oz. serving of Pepsi.

76. There are approximately 109 grams of sugar in a 32 fl. oz. serving of Pepsi.

77. $x=-1$ 78. $x=13$ 79. $x=8$
80. $x=-5$ 81. $x=30$

82. $h \approx 146.41$. The original height of the Great Pyramid of Giza is approximately 146.41 meters.

83. $y=\dfrac{1}{3}x$; $y=\dfrac{10}{3}$ when $x=10$

84. $y=18x$; $y=108$ when $x=6$

85. $y=-\dfrac{4}{x}$; $y=2$ when $x=-2$

86. $y=-\dfrac{7}{x}$; $y=-\dfrac{1}{3}$ when $x=21$

Chapter 7 Test

1. $\dfrac{b+2}{b-4}$ 2. $\dfrac{c-7}{c+2}$

3. $x=-7$ is excluded. 4. $x=8, -3$ are excluded.

5. The rocket is traveling at 220 feet per second.

6. $\dfrac{d+2}{d+7}$ 7. $\dfrac{p+3}{(p-6)(p+4)}$ 8. $\dfrac{k-3}{2}$

9. $(x+3)(x+7)$ 10. $\dfrac{5x+7}{x+3}$ 11. $\dfrac{6n-44}{(n-7)(n-8)}$

12. $\dfrac{a-2}{a-4}$ 13. $\dfrac{g+17}{(g-3)(g+2)}$

14. $-\dfrac{2t-2}{(t-3)(t+1)}$ 15. $-\dfrac{2x^2}{(x+6)(x^2+4)}$

16. a. $c=35$. The average cost per foot is \$35 when 10 feet of mow curb are installed.

    b. $m=30$. There must be 30 feet of mow curb installed for the average cost to be \$25 per foot.

17. $m=4$ 18. $r=-4$

19. $z=38$ 20. $h=-2$

21. It would take $7\dfrac{1}{7}$ minutes, or about 7 minutes and 9 seconds, to fill the truck if both pumps were used.

22. a. 5 feet of lights for \$1.35 cost \$0.27 per foot.

    b. 8 feet of lights for \$2.24 cost \$0.28 per foot. 5 feet of lights for \$1.35 is cheaper.

23. $x=15$

24. It takes 52 seconds to download a 5.2-megabyte song.

25. $y=\dfrac{10}{x}$; $y=\dfrac{1}{4}$ when $x=40$

CHAPTER 8

Section 8.1

1. 5
3. 8
5. $\frac{4}{3}$
7. $\frac{10}{7}$
9. $\frac{1}{4}$
11. -7
13. $-\frac{5}{8}$
15. Not real
17. Not real
19. $\sqrt{15} \approx 3.87$
21. $\sqrt{51} \approx 7.14$
23. $\sqrt{7} \approx 2.65$
25. Not real
27. 6
29. 1
31. 2
33. Not real
35. Not real
37. $\sqrt{13}$
39. a. $t \approx 1.77$. It will take the ball approximately 1.77 seconds to fall 50 feet.
    b. $t \approx 2.24$. It will take the ball approximately 2.24 seconds to fall 80 feet.
41. a. The variable s represents the speed of the car in miles per hour, and d represents the length in feet of the skid mark left by the car.
    b. $s \approx 51.61$. The speed of the car was approximately 51.61 mph when the length of the skid mark was 200 feet.
    c. $s \approx 43.95$. The speed of the car was approximately 43.95 mph when the length of the skid mark was 145 feet.
43. a. $t \approx 0.71$. It takes the ball approximately 0.71 second to fall 8 feet.
45. $P = 15$. Eamond will earn $15 in profit from selling 50 wristbands.
47. $x \approx 1.39$. The length of the side of the cube is about 1.39 inches.
49. $d = 9$. A person can see a distance of 9 miles if the person is 36 feet above the surface of the earth.
51. $d \approx 8.06$
53. $d \approx 10.20$
55. $d \approx 7.81$
57. $d \approx 17.26$
59. x
61. a^3
63. $4x$
65. $-x^{10}$
67. $-a$
69. $-2y^3$
71. $5x^{11}$
73. $\frac{x^2}{7}$
75. $\frac{y^5}{6}$
77. 2
79. 1
81. -2
83. -5
85. 2
87. 1
89. 4
91. 10
93. $\frac{1}{7}$
95. y^4
97. a
99. -4
101. -5
103. -2
105. Not real
107. ab^4
109. $7z$
111. $\frac{x^2}{9}$

113.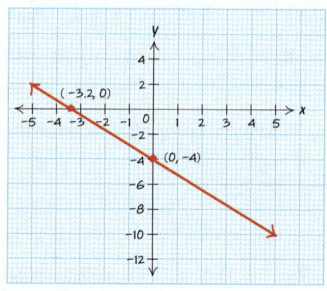

115. $(3, -6)$
117. In the year 2017, there were 247.53 million admissions to the top 25 theme parks worldwide.
119. $m = \frac{8.09}{1}$. The number of admissions at the top 25 theme parks worldwide is increasing at the rate of 8.09 million per year.
121. $-\frac{p^3}{3q^3}$
123. $9p^2 + 24pq + 16q^2$
125. $m = -1$
127. $\sqrt{272} \approx 16.49$

Section 8.2

1. $3\sqrt{2}$
3. $5\sqrt{5}$
5. $3\sqrt{6}$
7. $4\sqrt{5}$
9. $7\sqrt{2}$
11. $5\sqrt{3}$
13. $\sqrt{34}$
15. $2\sqrt{37}$
17. $12\sqrt{2}$
19. a. $5\sqrt{2}$ b. $\sqrt{50} \approx 7.07$
21. a. $2\sqrt{3}$ b. $\sqrt{12} \approx 3.46$
23. a. $2\sqrt{7}$ b. $\sqrt{28} \approx 5.29$
25. a. $3\sqrt{6}$ b. $\sqrt{54} \approx 7.35$
27. $\sqrt{32} \approx 5.66$
29. $3\sqrt{5}$
31. $\sqrt{37} \approx 6.08$
33. $6\sqrt{2}$
35. $\sqrt{45} \approx 6.71$
37. $b^4\sqrt{b}$
39. a^2
41. $x^7\sqrt{x}$
43. y
45. $2x\sqrt{x}$
47. $4x^3$
49. $8a\sqrt{a}$
51. $7a^2$
53. $3a^2\sqrt{5}$
55. $2y^3\sqrt{15}$
57. $2x\sqrt{10x}$
59. $5b^5\sqrt{3b}$
61. $-2\sqrt{5}$
63. $12\sqrt{3}$
65. $2\sqrt{3} + 3\sqrt{2}$
67. $5\sqrt{3} + 3\sqrt{5}$
69. $2\sqrt{x} - 7\sqrt{2}$
71. $-\sqrt{a}$
73. $2\sqrt{a} + 2\sqrt{b}$
75. $3x\sqrt{2} + 2y\sqrt{5}$
77. $6b + 3a\sqrt{b}$
79. $-32 + x\sqrt{3} - \sqrt{3}$
81. $4.5\sqrt{2}$
83. $14\sqrt{3}$
85. $3 + 2\sqrt{7}$
87. $9\sqrt{2}$
89. $8\sqrt{5} + 11\sqrt{2}$
91. a. $7x - 5x$ b. $2x$
93. a. $-4a^2 - 4a^2$ b. $-8a^2$
95. a. $4x\sqrt{x} - 5x\sqrt{x}$ b. $-x\sqrt{x}$
97. a. $-4 + 2x\sqrt{x} + 5x\sqrt{x}$ b. $-4 + 7x\sqrt{x}$
99. a. $-4x\sqrt{2} + 5x\sqrt{2}$ b. $x\sqrt{2}$
101. a. $5 + 2x\sqrt{3x} - 8 + 3x\sqrt{3x}$ b. $-3 + 5x\sqrt{3x}$

103. a. $8y\sqrt{y} - 5x\sqrt{2} - 5y\sqrt{y} + 3x\sqrt{2}$
 b. $3y\sqrt{y} - 2x\sqrt{2}$
105. $6\sqrt{7}$
107. $-3\sqrt{5} + 5\sqrt{3}$
109. $7\sqrt{2}$
111. x^{10}
113. $3a^2\sqrt{3a}$
115. $13y$
117. $15b^2 - 6b$
119. $a^4\sqrt{a}$
121. $2b\sqrt{2b}$
123. $5x^3\sqrt{2x}$
125. $15a^3$
127. Not real
129. -2
131. $\sqrt{35} \approx 5.92$

Section 8.3

1. 6
3. 6
5. $2\sqrt{11}$
7. $4\sqrt{5}$
9. 12
11. $5\sqrt{6}$
13. $3x\sqrt{2}$
15. $2a^2\sqrt{10}$
17. $6x^2$
19. $10x^2\sqrt{x}$
21. $2a^4\sqrt{15a}$
23. $4x^8\sqrt{5x}$
25. 17
27. $6x$
29. xyz
31. $21y$
33. $14a^3$
35. $14a\sqrt{6a}$
37. $A = 12$
39. $A = \dfrac{3}{2}$
41. $2 - 5\sqrt{2}$
43. $3\sqrt{2} - \sqrt{15}$
45. $\sqrt{2x} - 2\sqrt{2}$
47. $-12\sqrt{2} + 2\sqrt{5}$
49. $x + \sqrt{11x}$
51. $6 - 2\sqrt{3} + 3\sqrt{6} - 3\sqrt{2}$
53. $20 + 6\sqrt{7}$
55. $x - \sqrt{5x} + \sqrt{3x} - \sqrt{15}$
57. -1
59. $9 - x$
61. $19 + 8\sqrt{3}$
63. $a^2 + 2a\sqrt{11} + 11$
65. $4 - 4\sqrt{y} + y$
67. $x^2 - 2x\sqrt{5} + 5$
69. 5
71. 6
73. $3x$
75. $\dfrac{1}{3}$
77. $\dfrac{x}{4}$
79. $\dfrac{4\sqrt{3}}{3}$
81. $-5\sqrt{3}$
83. $\dfrac{\sqrt{5x}}{3x}$
85. $\dfrac{\sqrt{2a}}{5a}$
87. $4 - 2\sqrt{3}$
89. $\dfrac{7\sqrt{7} - 7\sqrt{3}}{4}$
91. $\dfrac{-12 + 4\sqrt{x}}{x - 9}$
93. $\dfrac{-2\sqrt{7} + 20}{93}$
95. The length of the rectangle is $3\sqrt{6}$.
97. The width of the rectangle is 4.
99. $3\sqrt{6}$
101. $-12x^2$
103. $30y^2\sqrt{y}$
105. $5 + \sqrt{35}$
107. $6 + 2\sqrt{6} - 3\sqrt{3} - 3\sqrt{2}$
109. $3x^2\sqrt{5x}$
111. 2
113. $\dfrac{a}{5}$
115. $-2\sqrt{3}$
117. $\dfrac{-5 + \sqrt{3}}{11}$
119. $y = -\dfrac{2}{7}x + 5$
121. $c = \dfrac{2a + 6b - P}{2}$
123. $21{,}780 \text{ ft}^2$
125. $4t^4$
127. $9 - 6n^2 + n^4$
129. $t \geq -2$

Section 8.4

1. $x = 16$ is a solution.
3. $x = 81$ is not a solution.
5. $x = 26$ is not a solution.
7. $x = 5$ is a solution.
9. $x = 0$ is not a solution.
11. $x = 4$ is a solution.
13. Maria's work is not correct. She concludes that $\sqrt{-9} = -3$, but this is not correct because a negative number under the radical sign ($\sqrt{-9}$) is not a real number and cannot be equal to a real number (-3).
15. Marilyn's work is correct.
17. Jaydy's work is not correct. She concludes that $-\sqrt{-1} = 1$, but this is not correct because a negative number under the radical sign ($\sqrt{-1}$) is not a real number and cannot be equal to a real number (1).
19. $a = 16$
21. No solution
23. $x = 37$
25. $y = 0$
27. $a = 225$
29. $x = 16$
31. $\sqrt{x} + 10 = 6$ Isolate the radical using the subtraction
    $\phantom{\sqrt{x} + } -10 \ -10$ property of equality.
    $\sqrt{x} = -4$ A square root cannot equal a negative number.
    No solution
33. No solution
35. $y = 8$
37. $a = -37$
39. a. $h = 52.02$. The ball was dropped from a height of 52.02 meters.
 b. $h = 364.5$. The ball was dropped from a height of 364.5 meters.
41. a. $d = 67.5$. The length of the skid mark was 67.5 feet.
 b. $d \approx 45.63$. The length of the skid mark was approximately 45.63 feet.
43. $x = 27$. Andrew must sell 27 wristbands to make a profit of $20.
45. $A \approx 127551.02$. The irrigation system covers an area of approximately 127,551.02 square feet.
47. $x = 14$
49. No solution
51. $z = 3$
53. No solution
55. $b = 96$
57. $a = -13$
59. No solution
61. Mackenzie's work is not correct. She did not isolate the radical before squaring. She then simplified $(2\sqrt{x})^2$ incorrectly (correctly simplified, this would be $4x$, not $2x$).
63. Andrew's work is not correct. He did not isolate the radical before squaring. He then simplified $(\sqrt{x} + 1)^2$ incorrectly (correctly simplified, this would be $x + 2\sqrt{x} + 1$, not $x + 1$). The equation would be easier to solve correctly if the radical were isolated first.
65. Victor's work is not correct. If he had checked the solution, this would have revealed that $x = 36$ is not a solution to the equation. There is no solution for the equation $\sqrt{x} = -6$.
67. Margaret's work is incorrect. She did not square the left side correctly, since $(\sqrt{-x})^2 = -x$. The correct answer is $x = -81$.
69. Rio did not use the distance formula correctly. The differences between the x- and y-coordinates squared should be added under the radical sign. The distance formula does not use a quotient.

71. Ethan found the differences between the x and y of the ordered pairs ($y_1 - x_1$ and $y_2 - x_2$) rather than correctly finding the difference between the x-coordinates and y-coordinates ($x_2 - x_1$ and $y_2 - y_1$).

73. $x = 2, x = -1$ 75. $x = \dfrac{3}{2}, x = -1$ 77. $x = 4$

79. No solution 81. No solution 83. $x = 4$

85. Radical 87. Rational 89. Linear

91. Quadratic 93. a. Equation b. $x = 19$

95. a. Equation b. $x = -6, x = -2$

97. a. Expression b. $11n^2 + 4n - 35$

99. a. Equation b. $x = 13$

Chapter 8 Review Exercises

1. -6 2. -10 3. Not real
4. Not real 5. 3 6. 5
7. -4 8. -1
9. a. $t \approx 2.26$. The ball will fall 25 meters in approximately 2.26 seconds.
   b. $t \approx 1.56$. The ball will fall 12 meters in approximately 1.56 seconds.
10. a. $t \approx 2.56$. The ball will fall 32 meters in approximately 2.56 seconds.
    b. $t \approx 1.36$. The ball will fall 9 meters in approximately 1.36 seconds.
11. $d \approx 7.07$ 12. $d \approx 3.61$ 13. 4
14. 2 15. Not real 16. $\sqrt{10}$
17. y^5 18. y^3 19. $5x^2$
20. $6x^3$ 21. x^3y 22. xy^2
23. $5\sqrt{3}$ 24. $3\sqrt{6}$ 25. $x^3\sqrt{x}$
26. $x^5\sqrt{x}$ 27. $2y\sqrt{y}$ 28. $3y^7\sqrt{y}$
29. $2x\sqrt{5x}$ 30. $2x^4\sqrt{6x}$ 31. $3\sqrt{3} \approx 5.20$
32. $3\sqrt{5} \approx 6.71$ 33. $10\sqrt{2} \approx 14.14$ 34. $5\sqrt{10} \approx 15.81$
35. $6\sqrt{3x}$ 36. $11\sqrt{6y}$ 37. $4 + 2x\sqrt{x}$
38. $17y - 5y\sqrt{y}$ 39. $3\sqrt{2}$ 40. $5\sqrt{2}$
41. $4y$ 42. $2x\sqrt{10}$ 43. $2x^2\sqrt{6x}$
44. $2x^3\sqrt{11x}$ 45. $7a$ 46. $3b^3$
47. $5 + 5\sqrt{2}$ 48. $3 + 3\sqrt{2}$
49. $-4\sqrt{2} + 6\sqrt{x}$ 50. $-28\sqrt{a} - 7\sqrt{7}$
51. $3 + 3\sqrt{6} - \sqrt{2} - 2\sqrt{3}$ 52. $4 - 8\sqrt{3}$
53. 10 54. 14 55. $3 - \sqrt{5}$
56. $2 - \sqrt{7}$ 57. $9 + \sqrt{3}$ 58. $6 + \sqrt{11}$
59. $-5 + \sqrt{x}$ 60. $-9 + \sqrt{a}$ 61. 4
62. 5 63. $3x$ 64. $7x^2$
65. $\dfrac{5\sqrt{7}}{7}$ 66. $\dfrac{-3\sqrt{5}}{5}$ 67. $\dfrac{12 + 4\sqrt{2}}{7}$
68. $\dfrac{3 - 3\sqrt{5}}{4}$

69. Yes, $x = 4$ is a solution to the radical equation.
70. Yes, $x = 5$ is a solution to the radical equation.
71. No, $x = 2$ is not a solution to the radical equation.
72. No, $x = 2$ is not a solution to the radical equation.
73. $a = -8$ 74. $b = 43$ 75. $x = -22$
76. $y = 8$ 77. $x = 5$ 78. $x = 7$
79. No solution 80. No solution
81. $x = 120$. Nelson needs to sell 120 baskets of blueberries to make a profit of $12.00.
82. $x = 102$. Tisha needs to sell 102 baskets of tomatoes to make a profit of $10.00.
83. $h = 576$. The rock falls 576 feet in 6 seconds.
84. $h = 324$. The rock falls 324 feet in 4.5 seconds.
85. Quadratic 86. Quadratic 87. Radical
88. Radical 89. Rational 90. Rational
91. Linear 92. Linear

Chapter 8 Test

1. -9 2. -3 3. $-6a^4$
4. a^4b 5. $d \approx 9.90$ 6. 5
7. $\dfrac{x}{3}$ 8. $2\sqrt{15}$ 9. 4.90
10. $2b^2\sqrt{7b}$ 11. $5ab\sqrt{a}$ 12. $-3 + 6\sqrt{3}$
13. $14\sqrt{2}$ 14. $x\sqrt{6}$ 15. $4\sqrt{3}$
16. $5x^2\sqrt{2x}$ 17. $4 + 2\sqrt{10}$ 18. $-58 + 2\sqrt{5}$
19. 23 20. 5 21. $-\dfrac{\sqrt{6}}{2}$
22. $10 - 5\sqrt{3}$ 23. $a = 484$ 24. No solution
25. $x = -5, x = 3$
26. $x = \dfrac{20}{3}, x = \dfrac{2}{3}$

Cumulative Review for Chapters 1–8

1. $(-3a)^4$ 2. $\dfrac{16}{81}$ 3. 21
4. 12 5. $-6x^2 - 9x - 30$ 6. $y = \dfrac{7}{3}x + 2$
7. $x = -\dfrac{25}{3}$ 8. $n = \dfrac{3}{2}$
9. a. $S = 528$. Gustavo's monthly salary is $528 if he works 48 hours.
   b. $S = 352$. Gustavo's monthly salary is $352 if he works 32 hours.
   c. $h \approx 72.73$. Gustavo must work approximately 72.73 hours if he wants to earn $800.
10. $x < \dfrac{11}{3}$
11. It will cost $410 for Donna to take 20 children to see the 3-D movie.

12.

x	$y = |x - 3|$
-5	8
-3	6
-1	4
1	2
3	0
5	2
6	3
8	5
9	6

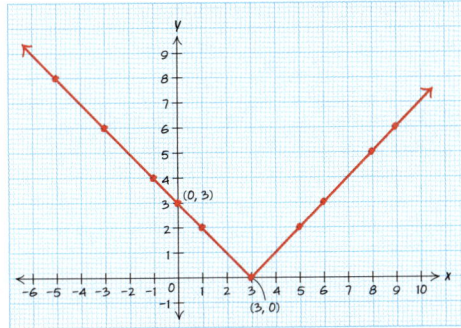

The graph is nonlinear.

13. $y = -\dfrac{2}{3}x + 4$

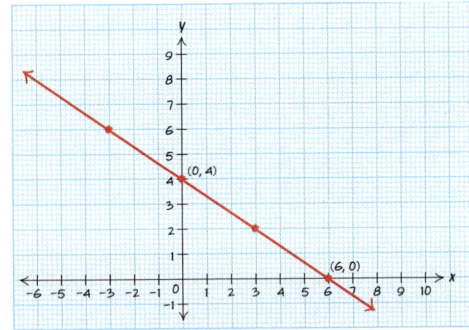

14. x-intercept: $(5, 0)$, y-intercept: $(0, -4)$

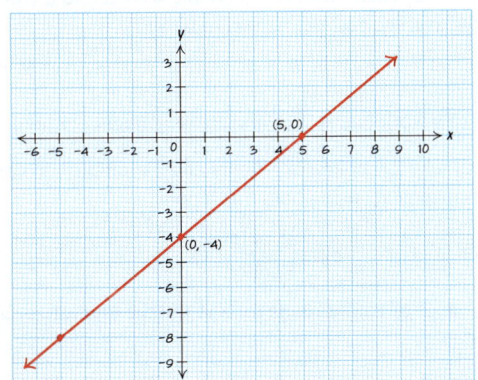

15. Slope: $\dfrac{3}{5}$, y-intercept: $(0, -2)$

16. $y = -\dfrac{3}{2}x + 3$

17. $P = 5h$

18. $y = -\dfrac{1}{5}x + 5$

19. $y = 9$

20. $y = \dfrac{1}{4}x - 3$

21. a. $\dfrac{45}{44}$ b. $\dfrac{9}{2}$

22. $R = 268.00$. This means that the youth soccer team will take in $268.00 in revenue for selling 67 boxes of candy.

23. $x = -4$ $y = 2$

24. $x = -3$ $y = -8$

25. $s = 105$ $w = 15$

The rate of the plane is 105 mph, and the rate of the wind is 15 mph.

26. Natalya needs to use 5 pounds of cumin seeds and 5 pounds of fenugreek seeds to make a 10-pound spice mix priced at $12.00 per pound.

27. Infinitely many solutions

28. The length of the rectangle is 8.5 centimeters, and the width is 5.5 centimeters.

29. Aivy should invest $5000 in the CD and $3000 in the bond fund to earn $325 a year in interest.

30.

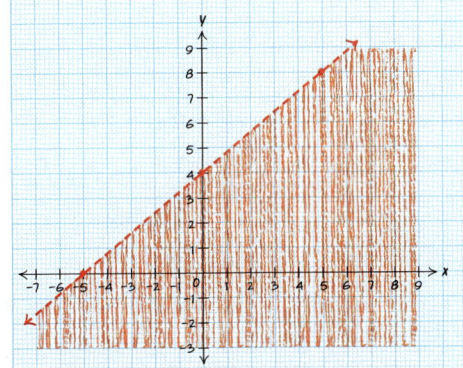

31.

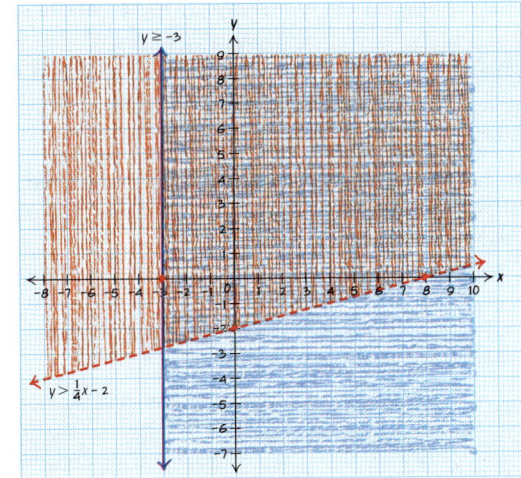

32. a. $25x^2$ b. $-25x^2$ c. $-12x^4y^5$

33. a. $-27x^{12}y^9$ b. $-18a^5b^5$ c. $-108x^9y^8$

34. a. t^4 b. $-\dfrac{b}{4}$ c. $5y$

35. $\dfrac{9x^2y^2}{100}$ **36. a.** $-\dfrac{3}{x^2}$ **b.** $5a^4$

37. a. $-\dfrac{2}{ab^2}$ **b.** $\dfrac{4x^4}{y^6}$ **c.** $\dfrac{y^3}{125x^3}$

38. a. 1.95×10^{-4} **b.** 5.289×10^{10}

39. a. 0.00003004 **b.** $927{,}000{,}000$

40. Three terms; degree is 3.

41. Degree of the term $5a^3b$ is 4.
Degree of the term $3a^5$ is 5. Degree of the term $-6b$ is 1.
Degree of the term ab^2 is 3. Degree of the polynomial is 5.

42. $-4b^2 + 7b + 6$ **43.** $4x - 11$ **44.** $25y - 40$

45. a. $2x^2 - 9x - 35$ **b.** $3y^3 - y^2 - 3y + 1$
c. $9x^2 - 24x + 16$ **d.** $33a - 26$

46. $-x + 2 - \dfrac{3}{x}$ **47.** $3x - 5$

48. a. $6x(x^2y - 3y + 4)$ **b.** $(5x + 2)(y + 3)$

49. $2(3a - 2)(b + 6)$

50. a. $(x - 7)(x + 2)$ **b.** $-(n - 3)(n + 6)$
c. $(5y - 1)(4y + 3)$

51. $2z(4z - 1)(2z + 3)$

52. a. $-(x + 5)(x - 5)$ or $(5 - x)(5 + x)$
b. $(3y + 1)(3y - 1)$ **c.** $(3t - 2)^2$
d. Prime **e.** $3(y + 2)(y - 2)$

53. a. Linear **b.** Neither **c.** Quadratic

54. a. $x = -\dfrac{2}{3}, x = 4$ **b.** $x = 0, x = -5$ **c.** $n = \pm 9$

55. Darren's new iPhone cost a total of $687.94

56. Sami's condo increased in value about 14.6% during the two years she owned it.

57. In the year 2020, the Disney Company's annual park and resort operating income is estimated to be $4.599 billion.

58. The slope is $m = 0.335 = \dfrac{0.335}{1}$. Disney's annual park and resort operating income is increasing at the rate of $0.335 billion per year.

59. The vertical intercept is the point $(0, 1.249)$. In the year 2010 ($t = 0$), the Disney Company's annual park and resort operating income was $1.249 billion.

60. 0

61. a. $c = 6$. The cost is $6 per person if 10 parents agree to help pay for the plaque.
b. $c = 4$. The lowest possible cost per parent is $4.

62. a. $x = \dfrac{7}{2}$ is excluded. **b.** $t = -3, t = \dfrac{4}{3}$ are excluded.
c. $y = -\dfrac{5}{2}, y = \dfrac{1}{3}$ are excluded.

63. a. $-\dfrac{6}{x}$ **b.** $\dfrac{1}{n + 1}$ **c.** $\dfrac{1}{t + 2}$

64. a. $T = 3.75$. The two office specialists working together can complete the job in 3.75 hours, or 3 hours and 45 minutes.
b. $T \approx 3.43$. The two office specialists working together can complete the job in about 3.43 hours, or 3 hours and 26 minutes.

65. a. $\dfrac{5n + 7}{2n^2 + 17n + 35}$ simplified **b.** $\dfrac{4x + 1}{x + 7}$ **c.** -4

66. a. $-\dfrac{2}{5}$ **b.** $-\dfrac{5}{4y}$
c. $\dfrac{x - 3}{x + 5}$ **d.** $\dfrac{(y + 3)(y - 1)}{(y - 4)(y + 6)}$

67. The price for a 1-pound jar of barbeque sauce is $0.39 per ounce.

68. The car's speed is 3080 feet per minute.

69. a. $-\dfrac{4}{3}$ **b.** $-\dfrac{14}{3xy}$
c. $\dfrac{(t + 2)(t - 1)}{(t - 4)(t + 3)}$ **d.** $\dfrac{(a + 3)(a - 4)}{(a - 2)^2}$

70. a. $-\dfrac{y}{18x}$ **b.** $\dfrac{1}{2b + 10}$ or $\dfrac{1}{2(b + 5)}$

71. a. 6 **b.** $6t$
c. $\dfrac{8}{(y - 2)(y + 3)}$ **d.** $\dfrac{4x - 7}{(x - 1)(x + 5)}$

72. a. $10x^2y$ **b.** $(b - 1)(b - 2)$
c. $4h(h - 3)$ **d.** $(a - 5)(a - 3)(a + 4)$

73. a. $\dfrac{3x + 12}{(x - 1)(x + 4)}, \dfrac{-2x^2 + 2x}{(x - 1)(x + 4)}$
b. $\dfrac{5k^2 - 25k}{(k + 6)(k - 6)(k - 5)}, \dfrac{8k + 48}{(k + 6)(k - 6)(k - 5)}$

74. a. $\dfrac{17}{2y}$ **b.** $\dfrac{y^2 - 9y + 2}{(y - 4)(y - 1)}$

75. a. $\dfrac{5}{3}$ **b.** $\dfrac{-2x + 3y}{4x + y}$

76. a. $x = 3$ **b.** No solution **c.** $k = 0, k = 3$

77. a. $C = 14$. The average cost per person is $14 if 25 people attend the retirement lunch.
b. $n = 50$. There must be 50 people who attend the lunch for the average cost per person to be $12.

78. a. $a = 1$ **b.** $b = 8$ **c.** $x = 9, x = -2$

79. Kenzo's work rate is $\dfrac{1}{3}$ of his lab assignment per hour.

80. It will take 2.4 hours, or 2 hours and 24 minutes, to copy the announcements using both machines.

81. $\dfrac{1}{5}$ **82.** $\dfrac{25 \text{ miles}}{1 \text{ gallon}}$

83. 36 ounces of coffee beans for $31.00; the unit rate is $0.86 per ounce.
48 ounces of coffee beans for $49.50; the unit rate is $1.03 per ounce.
The better buy is 36 ounces of coffee beans for $31.00.

84. The length of the wall on the blueprint will be 2.5 inches.

85. a. $n = -\dfrac{2}{5}$ b. $x = -4$ c. $t = 9$

86. $d = 50t$; $d = 375$ when $t = 7.5$

87. a. $W = 10.5t$ b. $W = 336$

    The amount of weekly paycheck is $336 for 32 hours of work.

88. $y = -\dfrac{3}{x}$; $y = -\dfrac{3}{2}$ when $x = 2$

89. a. $s = \dfrac{200}{t}$ b. $s = 33\dfrac{1}{3}$

    The speed is $33\dfrac{1}{3}$ mph, or approximately 33.33 mph, for 6 hours of driving.

90. a. 7 b. $\dfrac{5}{8}$ c. Not real

91. a. $\sqrt{30} \approx 5.48$ b. Not real

92. a. 4 b. 7 c. 1

93. $d = 12$. A person can see a distance of 12 miles if the person is 64 feet above the surface of the earth.

94. $d \approx 9.90$

95. a. x^3 b. $5b^2$ c. $-9y^6$

96. a. 4 b. -1

97. a. $3\sqrt{5}$ b. $2\sqrt{10}$ c. $7\sqrt{3}$

98. $2\sqrt{11}$ 99. $\sqrt{44} \approx 6.63$

100. a. $x^5\sqrt{x}$ b. $2b^2\sqrt{5b}$ c. $10c^4\sqrt{3c}$

101. a. $5\sqrt{3}$ b. $3\sqrt{t} - 5\sqrt{7}$ c. $-4y + 2\sqrt{x}$

102. a. $-8n + 9n$ b. n

103. a. $6 + 4y^2\sqrt{y} - 7y^2\sqrt{y}$ b. $6 - 3y^2\sqrt{y}$

104. a. 9 b. $-5\sqrt{3}$ c. $6x^2$ d. $2n$ e. $7xy\sqrt{y}$

105. $A = 72$

106. a. $-7 + 8\sqrt{7}$ b. $x + 2\sqrt{2x}$

107. a. 34 b. $14 - 6\sqrt{5}$
     c. $n - \sqrt{3n} + \sqrt{2n} - \sqrt{6}$

108. a. 5 b. $\dfrac{\sqrt{2}}{9}$ c. $\dfrac{x}{8}$

109. a. $\dfrac{9\sqrt{5}}{5}$ b. $\dfrac{\sqrt{30}}{10}$ c. $3 - \sqrt{7}$

110. a. Yes, $x = 18$ is a solution.
     b. No, $b = 0$ is not a solution.

111. a. $y = 49$ b. $n = 28$ c. $x = 169$
     d. No solution

112. a. $h \approx 19.85$. If the ball falls for 2.1 seconds, it was dropped from a height of approximately 19.85 meters.
     b. $h \approx 61.61$. If the ball falls for 3.7 seconds, it was dropped from a height of approximately 61.61 meters.

113. a. $y = 8$ b. $x = 3$ c. $n = 8$

114. a. Rational b. Quadratic c. Radical d. Linear

CHAPTER 9

Section 9.1

1.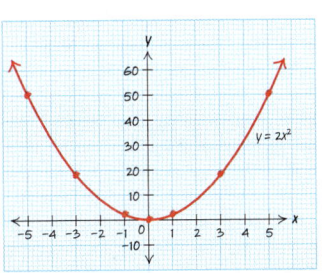

3.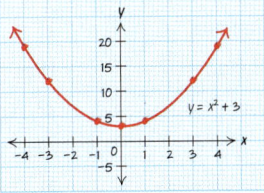

5.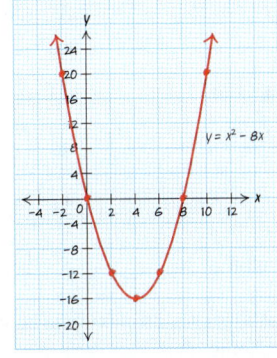

7.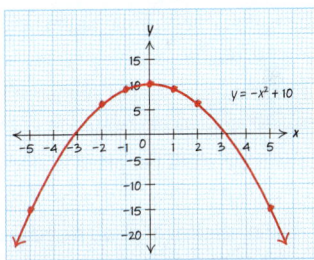

9.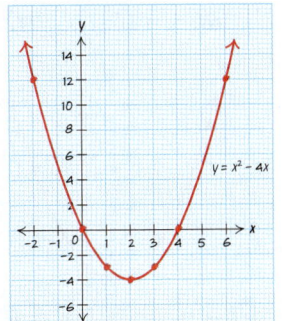

11.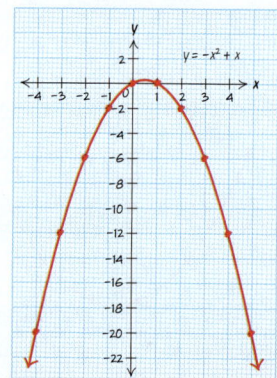

13. $(-4, -4)$, minimum
15. $(-3, 4)$, maximum
17. $(-4, 3)$, maximum
19. $(-3, 0)$, minimum
21. $(0, 4)$, maximum
23. Upward
25. Upward
27. Downward
29. Upward
31. Downward
33. Upward
35. Downward
37. Downward
39. $(-2, -15)$, minimum
41. $(4, -11)$, minimum
43. $(4, 62)$, maximum
45. $(0, 0)$, minimum
47. $\left(3, -\dfrac{1}{2}\right)$, minimum
49. $(-5, -15.5)$, minimum
51. $(7, 25.7)$, maximum
53. $(0, 12)$, maximum
55. $\left(\dfrac{5}{4}, 3\dfrac{1}{8}\right)$, maximum

57. $(1, 91)$. The ball reaches its maximum height of 91 feet 1 second after being hit.
59. $(2, 144)$. The ball reaches its maximum height of 144 feet 2 seconds after being hit.
61. $(10, 400)$. The average profit per appliance is at a maximum of $400 when 10 appliances are sold in one order.
63. $(8, 64)$. The average profit per plane is at a maximum amount of $64 million when 8 planes are ordered by an airline.
65. $(40, 8000)$. The revenue will reach its maximum value of $8000 when the price is $40.

67. $x = -3$
69. $x = 3$
71. $x = 0$

73.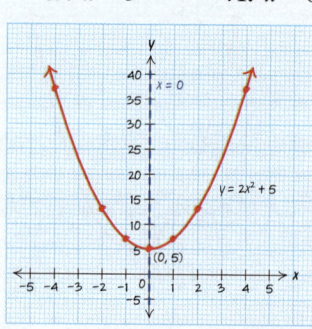

75.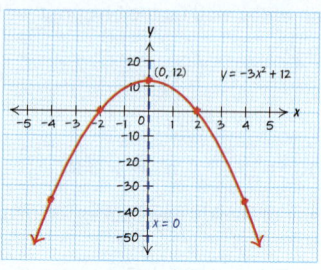

77.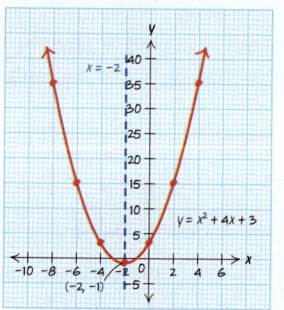

79.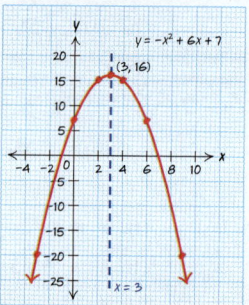

81.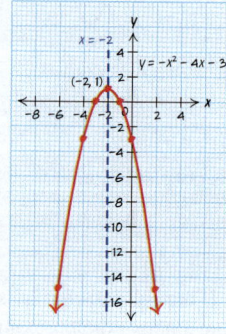

83.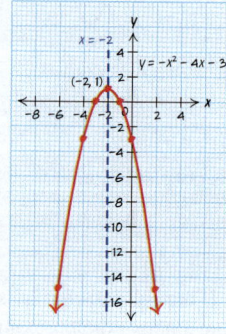

85.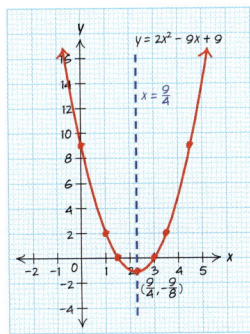

87.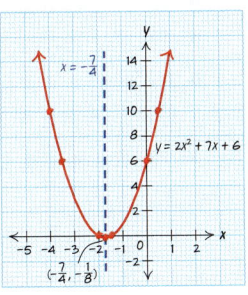

89.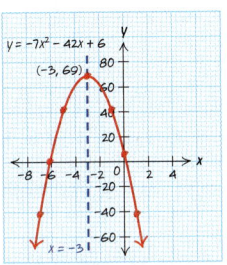

91.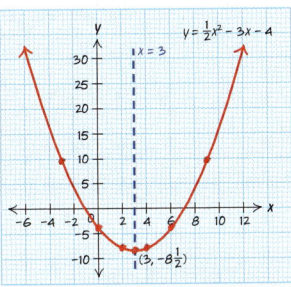

93.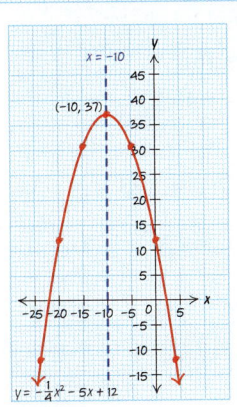

95.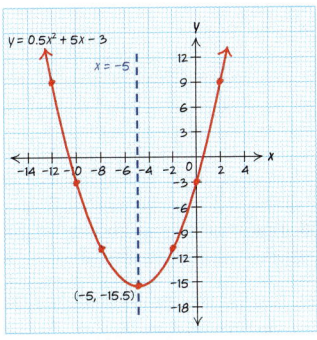

97.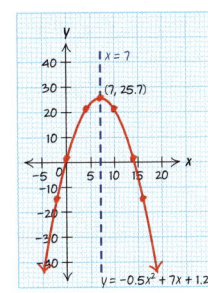

99.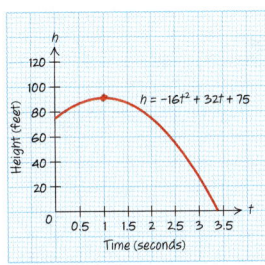

101.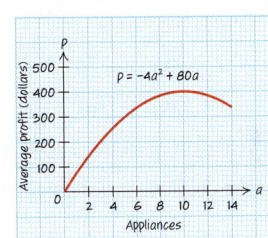

103.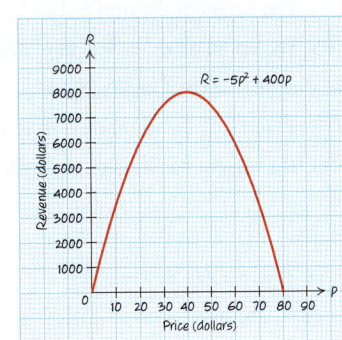

105. The equation is linear.
107. The equation is quadratic.
109. The equation is linear.
111. The equation is quadratic.

Section 9.2

1. Rational
3. Linear
5. Radical
7. Quadratic
9. $x = \pm 8$
11. $a = \pm 13$
13. $a = \pm 5$
15. $x = \pm 7$
17. $r = \pm 5$
19. $t = \pm 5$
21. $x = \pm 4$
23. $b = \pm 4$
25. a. $t \approx 1.58$. It will take approximately 1.58 seconds for the ball to fall to a height of 20 feet.
    b. $t \approx 1.94$. It will take approximately 1.94 seconds for the ball to hit the ground.
27. a. $t \approx 1.75$. It will take approximately 1.75 seconds for the rock to fall to a height of 10 meters.
    b. $t \approx 2.26$. It will take approximately 2.26 seconds for the rock to hit the ground.
29. $r \approx 977.21$. The radius of the circular region is approximately 977.21 feet.
31. $s = 15$. The length of each side of the room is 15 feet.
33. $x = 2, x = -4$
35. $y = 11, y = -7$
37. $x = 3, x = -1$
39. No real solution
41. $a = 3, a = -2$
43. $x = 1, x = -\dfrac{5}{3}$
45. $y = 2 \pm \sqrt{5}; y \approx 4.24, y \approx -0.24$
47. No real solution
49. $x = \dfrac{3 \pm 2\sqrt{3}}{5}; x \approx 1.29, x \approx -0.09$
51. $x = \dfrac{-1 \pm 4\sqrt{3}}{5}; x \approx 1.19, x \approx -1.59$
53. $b \approx 4.58$
55. $a = 15$
57. $c \approx 4.24$
59. $c = 10$
61. a. $b = 12.5$. The height of the roof is 12.5 feet.
    b. $c \approx 19.53$. The common rafter length is approximately 19.53 feet.
63. a. $b = 12.75$. The height of the roof is 12.75 feet.
    b. $c = 21.25$. The common rafter length is 21.25 feet.
65. $c = 20$. The diagonal of the room is 20 feet.
67. $c = 500$. The diagonal of the warehouse is 500 feet.
    $A = 120{,}000$. The area of the warehouse is 120,000 square feet.
69. $x = 9, x = -3$
71. $x = \pm 3$
73. $y = \pm 9$
75. $x = -7 \pm 5\sqrt{3}; x \approx 1.66, x \approx -15.66$
77. $x = 0, x = -8$
79. $x = 7, x = -2$
81. $r = \pm 12$
83. $x = 5, x = -1$
85. $x = 1, x = -7$
87. $x = -\dfrac{2}{3}, x = 4$
89. $x = \pm 4$
91. $d \approx 6.32$
93. $d \approx 14.21$
95. $d \approx 2.24$
97. The vertex is located at $(2, 4)$ and is a maximum point.
99. The vertex is located at $(0, -6.5)$ and is a minimum point.
101. $y = -\dfrac{2}{5}x + 2$
103. $y = \dfrac{2}{3}x + 5$

Section 9.3

1. $a = 1$ $b = -5$ $c = 21$
3. $a = -1$ $b = 12$ $c = -16$
5. $a = 1$ $b = -6$ $c = 12$
7. $a = -4$ $b = -3$ $c = 10$
9. a. 4   b. 16
11. a. -7 b. 49
13. a. 2.5   b. 6.25
15. $x = -3, x = -7$
17. $x = 4 \pm \sqrt{2}; x \approx 5.41, x \approx 2.59$
19. $y = 11 \pm \sqrt{10}; y \approx 14.16, y \approx 7.84$
21. $a = 8 \pm \sqrt{3}; a \approx 6.27, a \approx 9.73$
23. $x = 9, x = -7$
25. $x = -3 \pm \sqrt{5}; x \approx -0.76, x \approx -5.24$
27. $y = 9 \pm \sqrt{17}; y \approx 13.12, y \approx 4.88$
29. $t = 1 \pm \sqrt{15}; t \approx 4.87, t \approx -2.87$
31. $n = 2 \pm \sqrt{19}; n \approx 6.36, n \approx -2.36$
33. $x = 6 \pm \sqrt{2}; x \approx 7.41, x \approx 4.59$
35. Austin did not add $\left(\dfrac{1}{2}b\right)^2$ to the right side of the equation.

$$x^2 + 2x = 3$$
$$x^2 + 2x + (1)^2 = 3 + (1)^2$$
$$(x + 1)^2 = 4$$
$$x + 1 = \pm 2$$
$$x = 1, -3$$

37. Diego did not include both positive and negative possibilities as solutions to $\sqrt{16}$.

$$x^2 - 4x = 12$$
$$x^2 - 4x + 4 = 12 + 4$$
$$(x - 2)^2 = 16$$
$$x - 2 = \pm 4$$
$$x = 6, -2$$

39. $a = -1 \pm 2\sqrt{2}; a \approx 1.83, a \approx -3.83$
41. $x = \dfrac{7 \pm \sqrt{3}}{2}; x \approx 4.37, x \approx 2.63$
43. $x = \dfrac{1 \pm \sqrt{5}}{3}; x \approx 1.08, x \approx -0.41$
45. $a = \dfrac{5}{3}, a = -\dfrac{1}{5}$
47. $y = -\dfrac{6 \pm \sqrt{65}}{5}; y \approx 0.41, y \approx -2.81$

49. Liliana did not put the equation in the form $ax^2 + bx + c = 0$ and improperly identified the values for a, b, and c.

$$x^2 + 2x + 1 = 0$$
$$a = 1 \quad b = 2 \quad c = 1$$
$$x = \frac{-(2) \pm \sqrt{(2)^2 - 4(1)(1)}}{2(1)}$$
$$x = \frac{-(2) \pm \sqrt{0}}{2}$$
$$x = -1$$

51. Cassandra did not take the opposite of b for the first step in using the formula. The opposite of -5 is 5.

$$3x^2 - 5x - 12 = 0$$
$$a = 3 \quad b = -5 \quad c = -12$$
$$x = \frac{-(-5) \pm \sqrt{(-5)^2 - 4(3)(-12)}}{2(3)}$$
$$x = \frac{5 \pm \sqrt{169}}{6}$$
$$x = \frac{5 \pm 13}{6}$$
$$x = 3, x = -\frac{4}{3}$$

53. **a.** $5000 = -2x^2 + 250x$
   b. $x = 25, x = 100$
   c. The photographer must sell at least 25 copies of his photographs to make a profit of $5000.00.

55. $x \approx 7.75, x \approx 32.25$

   The web designer must work at least 7.75 hours to make a revenue of $500.00.

57. **a.** *When* is a reference to time t.
   b. The value of the height h is 0 when the rocket hits the ground.
   $$0 = -16t^2 + 160t + 5$$
   c. The equation in part b is quadratic.
   d. $t \approx -0.04, t \approx 8.79$
   e. The negative solution $t \approx -0.04$ does not make sense for this problem, but $t \approx 8.79$ does make sense. The rocket will hit the ground approximately 8.79 seconds after launch.

59. **a.** $t \approx 1.28$. The rock will fall to a height of 52 meters in approximately 1.28 seconds.
   b. $t \approx 3.50$. The rock will hit the ground in approximately 3.50 seconds.

61. $x = \pm\sqrt{7}; x \approx \pm 2.65$
63. $y = -6, y = -4$
65. $x = \frac{-3 \pm \sqrt{6}}{2}; x \approx -0.28, x \approx -2.72$
67. $a = -9, a = 7$ 69. $p = \pm 5$
71. $x = \frac{-3 \pm \sqrt{11}}{4}; x \approx -0.08, x \approx -1.58$

73. The x-intercept is $\left(\frac{3}{2}, 0\right)$, and the y-intercept is $(0, 4)$.
75. The x-intercept is $(-3, 0)$, and the y-intercept is $(0, 7)$.
77. $t \approx 2.35$. Zara's annual sales were $18 billion in 2012.
79. In 2017, Zara's annual sales were estimated to be $26 billion.
81. Late in 2001, Zara's annual sales were $0 billion. This is model breakdown.
83. This is a linear equation.
85. This is a quadratic equation.
87. This equation is "other."
89. This is a linear equation.
91. This is a radical equation.

Section 9.4

1. $(2, 0), (-2, 0), (0, -3)$ 3. $(-3, 0), (0, -4)$
5. $(0, 3)$, no x-intercept(s)   7. $(1, 0), (-3, 0), (0, 2)$
9. $(-8, 0), (1, 0), (0, -8)$ 11. $(-2, 0), (-9, 0), (0, 18)$
13. $(7, 0), (-4, 0), (0, -28)$ 15. $\left(-\frac{1}{2}, 0\right), \left(-\frac{5}{2}, 0\right), (0, 5)$
17. $\left(-\frac{3}{7}, 0\right), \left(-\frac{5}{2}, 0\right), (0, 15)$
19. $(0, 0)$ 21. $\left(\frac{2}{3}, 0\right), \left(\frac{7}{5}, 0\right), (0, 28)$

23.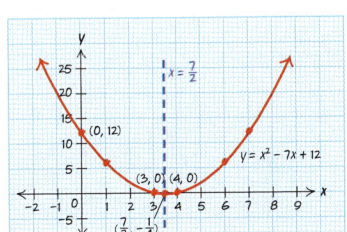

25.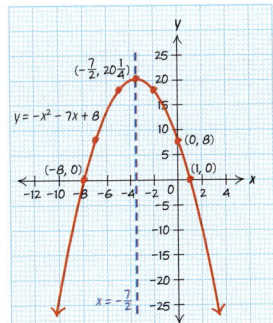

27.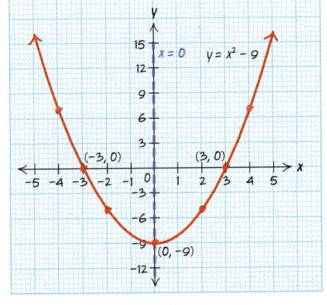

B-82 APPENDIX B Answers to Selected Exercises

29.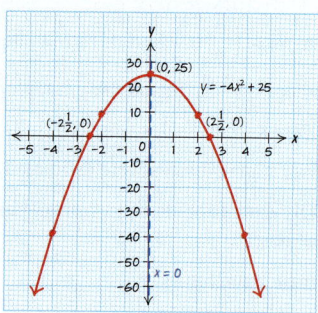

31.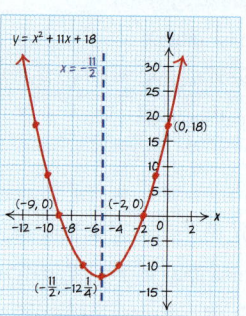

33.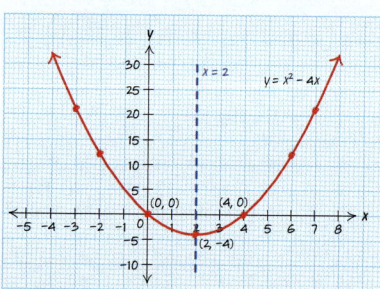

35.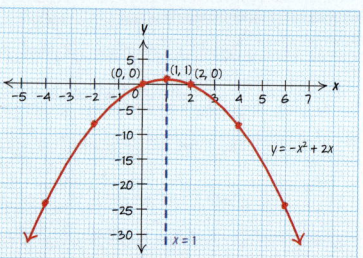

37.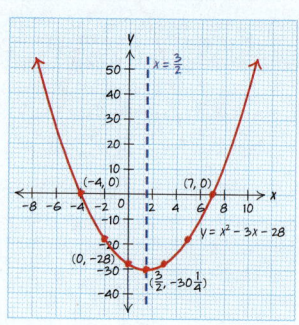

39.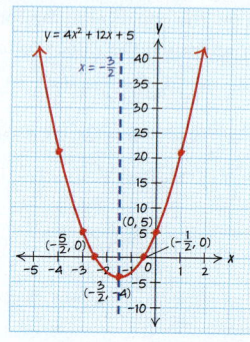

41.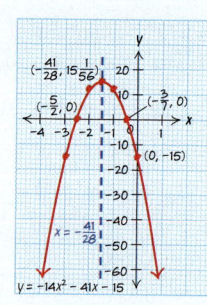

43.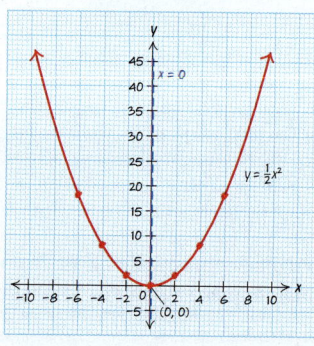

45.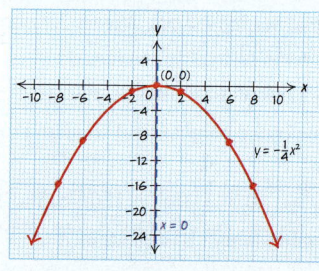

47.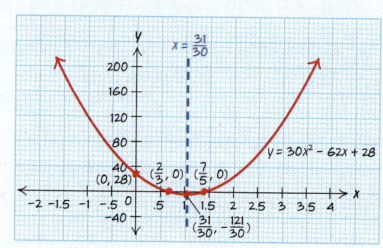

49.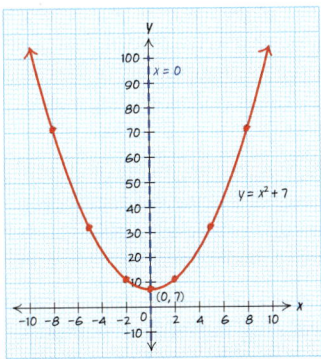

51.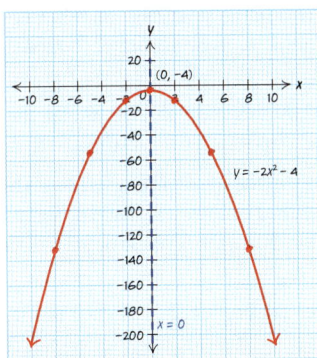

53. This is a quadratic equation. The x-intercepts are $\left(\frac{1}{2}, 0\right)$ and $(-4, 0)$. The y-intercept is $(0, -4)$.

55. This is a linear equation. The x-intercept is $(-3, 0)$. The y-intercept is $\left(0, \frac{6}{5}\right)$.

57. This is a linear equation. The x-intercept is $(-12, 0)$. The y-intercept is $(0, 8)$.

59. This is a quadratic equation. The x-intercepts are $(0, 0)$ and $(3, 0)$. The y-intercept is $(0, 0)$.

61. This model is quadratic.

63. The model estimates that there were 19.29 million people living in poverty in the year 2015.

65. The vertical intercept is $(0, 9.915)$. The model estimates that in the year 2000, there were 9.915 million people living in poverty in U.S. suburban areas.

67. 45.48 liters

69. 215.6 pounds

71. 4 days

73. $\frac{71}{15}$ **75.** $\frac{52}{3}$ **77.** $-\frac{1}{8}$

79. $(-2.5, 0)$ **81.** $(4, -1)$ **83.** 23.9°C

85. 89.6°F   **87.** 37.5   **89.** 16.67%

Section 9.5

1. Parabolic pattern   **3.** Linear pattern
5. Neither   **7.** Parabolic pattern
9. Linear pattern   **11.** Neither

13. a.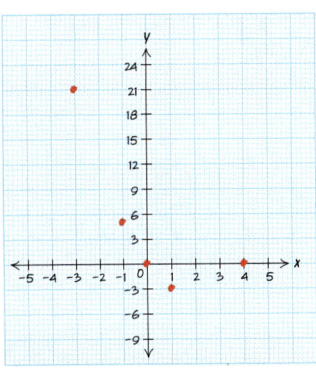

b. The data follow a parabolic pattern.

c. Estimated value of the vertex is $(2, -4)$.

15. a.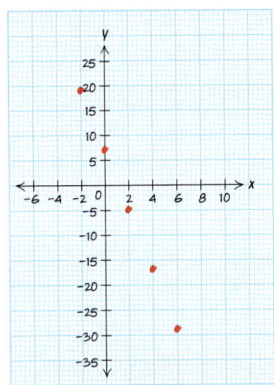

b. The data follow a linear pattern.   **c.** N/A

17. a.

Month	Average Daily Low Temperature (degrees Fahrenheit)
4	43
5	53
6	62
7	66
8	65
9	57

Source: www.intellicast.com

b. Let $m =$ the month, and let $T =$ the average daily low temperature.

c.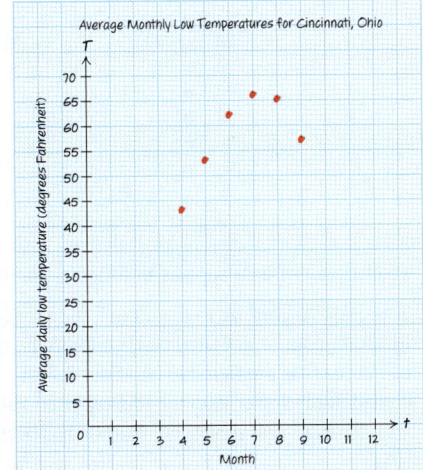

d. The data follow a parabolic pattern.

19. a. Let t = years since 2010, and let p = the profit in millions of dollars.

b.
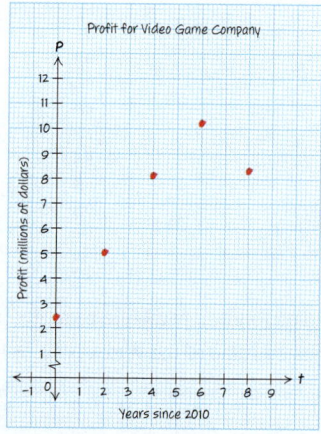

c. The data follow a parabolic pattern.

21. a. Let t = the years since 2010, and let p = the median listing price in thousands of dollars for a home in Santa Barbara, California.

b.
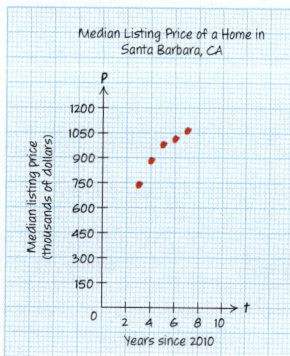

c. The data follow a parabolic pattern.

23. a. Let t = years since 2010, and let R = the revenue in billions of dollars.

b.
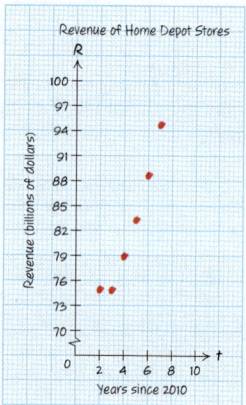

c. The display of data appears to follow a parabolic pattern.

25. a. $T = 3.29$. According to the model, the average monthly low temperature in Cincinnati, Ohio, in the month of February is predicted to be about 3.29°F. This is considerably lower than the real reported value of 25°F, which means that model breakdown has occurred.

b. The vertex of the model is approximately (7.18, 65.57). This means that in the seventh month of the year (July), the average low temperature in Cincinnati, Ohio, is about 65.57°F. The average low temperatures reach a high at this time of year.

27. a. $P = 933.5$. According to the model, the predicted listing price of a home in Santa Barbara, California, in the year 2020 will be $933,500. This is unlikely to be accurate, so model breakdown has occurred.

b. The vertex of the model is approximately (7.29, 1058). This means that the median home price in Santa Barbara, California, reached a maximum of approximately $1,058,000 in the year 2017 and then began to decrease. This is unlikely to be accurate, so model breakdown has occured.

29. a. $R = 102.99$. According to the model, the estimated revenue of Home Depot in 2018 is estimated to be $102.99 billion.

b. The vertex of the model is (1.26, 73.9). This means that Home Depot reached its lowest revenue of about $73.9 billion in 2011. After 2011, the revenue began to increase again.

31. a. The vertical intercept is (0, 0). This means that if 0 necklaces are produced, there is 0 profit.

b. The vertex is (10, 700). This means that the jewelry shop will make the maximum amount of average profit of $700 when they create 10 necklaces.

c. $n = 15, n = 5$. The jewelry shop must create either 15 necklaces or 5 necklaces to make an average profit of $525 per necklace.

33. a. The vertex is (15, 1575). This means that the shop will earn the maximum amount of revenue of $1575 when 15 cameras are sold.

b. $n = 20, n = 10$. The shop must sell either 20 cameras or 10 cameras to earn a revenue of $1400.

35. a. $P = 48.5$. The model predicts that the median listing price of a home in Santa Barbara, California, in the year 2025 will be about $48,500. Model breakdown has likely occured.

b. $P = -4027$. The model estimates that the median listing price of a home in Santa Barbara, California, in the year 2000 was about $-$$4,027,000. Model breakdown has occurred.

Section 9.6

1. The domain is the set of names: Eduardo, Sophie, Ana, Ashanti, and Taye.

The range is the set of months of births: July, March, October, January, and May.

Every input results in exactly one output.

3. The domain is the set of dog breeds: German shepherd, Labrador retriever, cocker spaniel, chihuahua, and papillon.

The range is the set of average weights in pounds: 85, 75, 25, 4, and 7.

Every input results in exactly one output.

5. The domain is the set of numbers: 4, 6, −2, 9, and 0.
   The range is the set of numbers: 5, 9, 3, 1, and 0.
   Every input results in exactly one output.

7. The domain is the set of numbers: 5, 6, −2, and 4.
   The range is the set of numbers: 5, 6, 4, −2, and 0.
   No, each input does not result in exactly one output.

9. The domain is the number 4.
   The range is the set of numbers: 1, 2, 3, and 4.
   No, each input does not result in exactly one output.

11. This relation is a function.
13. This relation is not a function.
15. This relation is a function.
17. This relation is not a function.
19. This relation is a function.
21. The graph represents a function.
23. The graph represents a function.
25. The graph does not represent a function.
27. The graph does not represent a function.
29. The graph represents a function.

31. The graph of $y = 2x - 9$ passes the vertical line test. For each input x, there is only one output y.
    $$f(x) = 2x - 9$$

33. The graph of $y = 4$ passes the vertical line test. For each input x, there is only one output y.
    $$f(x) = 4$$

35. The graph of $y = x^2 - 8$ passes the vertical line test. For each input x, there is only one output y.
    $$f(x) = x^2 - 8$$

37. The graph of $y = \frac{2}{3}x - 1$ passes the vertical line test. For each input x, there is only one output y.
    $$f(x) = \frac{2}{3}x - 1$$

39. The graph of $y = x^3$ passes the vertical line test. For each input x, there is only one output y.
    $$f(x) = x^3$$

41. $f(2) = 11$
43. $f(-6) = 31$
45. $f(-1) = 16$
47. $f(3) = 9$
49. $f(-4) = 13$
51. $R(125) = 1187.50$
    If a movie theater sells 125 tickets, they will take in $1187.50 in revenue.

53. $A(10) = 25.00$
    The bridge toll card has been used 10 times, so there is a remaining balance of $25.00 on the toll card for use of the bridge.

55. $C(50) = 77.75$
    If the cost to make the pies is $77.75, then 50 pies could be made.

Chapter 9 Review Exercises

1. **a.** If the coffee shop sells 200 pounds of coffee beans, its profit will be $3000.
   b. To earn $6000 profit, the shop must sell either 300 or 500 pounds of coffee beans.
   c. (4, 7). If the coffee shop sells 400 pounds of beans, it will earn a maximum profit of $7000.

2. **a.** (0, 100). The ball is at its maximum height of 100 feet just before it is dropped.
   b. $t = 2$ seconds

3. (7, 0), (1, 0), (0, 21); minimum at (4, −27)
4. (9, 0), (−5, 0), (0, −45); minimum at (2, −49)
5. (−3, 0), (−4, 0), (0, −12); maximum at $\left(-\frac{7}{2}, \frac{1}{4}\right)$
6. (6, 0), (12, 0), (0, −72); maximum at (9, 9)
7. **a.** (−1, −3)   **b.** (0, −2.7)
   c. (2, 0), (−4, 0)
8. **a.** (−2, 4)   **b.** (0, 3)
   c. (2, 0), (−6, 0)
9. $x = \pm 5$
10. $x = \pm 6$
11. $x = \frac{3}{5}, x = -\frac{9}{5}$
12. $x = 3, x = 4$
13. No real solution   14. No real solution
15. $b \approx 13.27$ 16. $a \approx 9.80$
17. $x = 2 \pm \sqrt{15}; x \approx 5.87, x \approx -1.87$
18. $x = 5 \pm \sqrt{17}; x \approx 9.12, x \approx 0.88$
19. $y = 3, y = 1$ 20. $y = 4 \pm 2\sqrt{3}; y \approx 7.46, y \approx 0.54$
21. $x = -3 \pm \sqrt{7}; x \approx -0.35, x \approx -5.65$
22. $x = 5 \pm \sqrt{10}; x \approx 8.16, x \approx 1.84$
23. No real solution
24. $y = \frac{-3 \pm \sqrt{6}}{2}; y \approx -0.28, y \approx -2.72$
25. $x = \frac{-3 \pm \sqrt{2}}{6}; x \approx -0.26, x \approx -0.74$
26. $x = 4 \pm \sqrt{6}; x \approx 6.45, x \approx 1.55$
27. **a.** $4500 = -2x^2 + 225x$
    b. $x \approx 86.48, x \approx 26.02$
    c. The photographer must sell at least 27 prints to make a profit of $4500.00.
28. **a.** $t \approx 1.01$. The rock will fall to a height of 13 meters in approximately 1.01 seconds.
    b. $t \approx 1.92$. The rock will hit the ground in approximately 1.92 seconds.

29.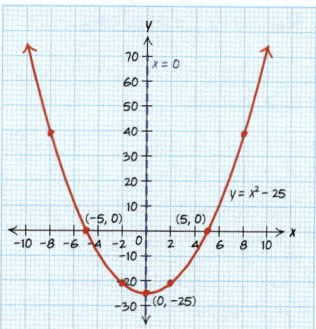

30.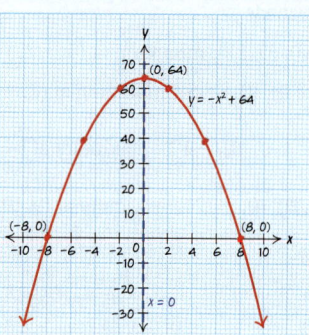

31.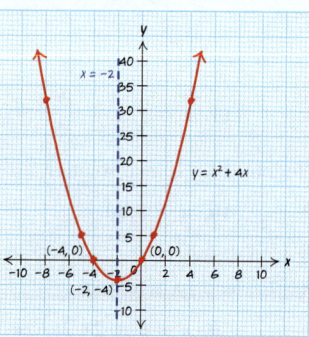

32.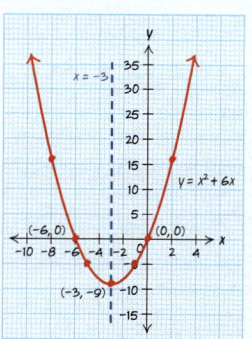

33.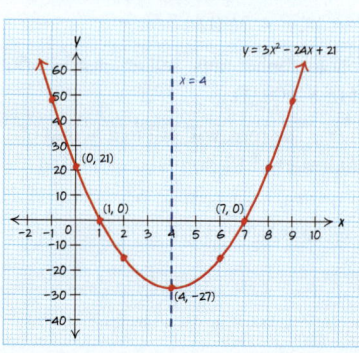

34.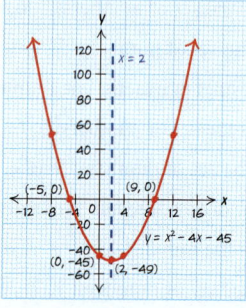

35.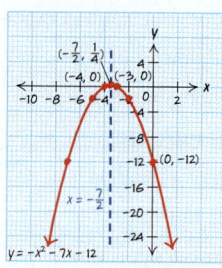

36.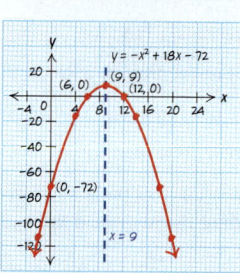

37. Neither **38.** Neither **39.** Quadratic
40. Quadratic **41.** Linear **42.** Linear

43. a.

Month	Average Daily High Temperature (degrees Fahrenheit)
3	45
4	55
5	65
6	75
7	83
8	82

Source: www.weather.com

b.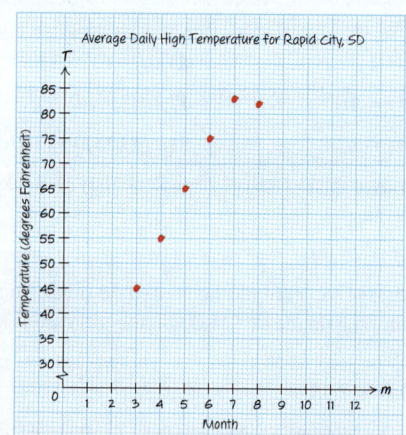

The display of data follows a parabolic pattern.

44. a.

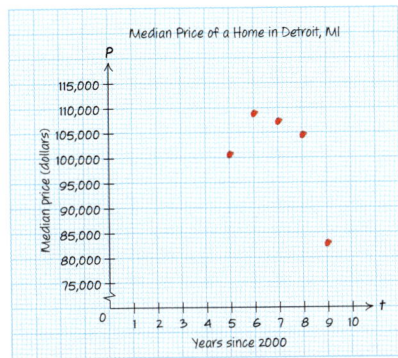

b. The display of data follows a parabolic pattern.

45. a. The vertex of the model is approximately (9, 85). This means that the average daily high temperature in Rapid City, South Dakota, will reach a high of approximately 85°F in the ninth month of the year (September), then the daily average high temperature will begin to decrease.

b. $T = 75.11$. According to the model, the estimated average high temperature in Rapid City, South Dakota, in December would be about 75°F. This answer does not seem reasonable for that city and is considerably different from the reported value of 37°F. This is an example of model breakdown.

c. $T = 29.11$. According to the model, the estimated average high temperature in Rapid City, South Dakota, in February would be about 29°F. This answer seems reasonable and is not considerably different from the reported value of 38°F.

46. a. The vertex of the model is (6.5, 110,493). This means that the median price of a home in Detroit, Michigan, will reach a high of approximately $110,493 in the year 2007, then the median home prices will begin to decrease.

b. $P = 58,041.4$. According to the model, the median price of a home in Detroit, Michigan, in the year 2010 would be about $58,041. This answer is possible, but the real median price in the year 2010 was even lower. This is an example of model breakdown.

c. $P = -20,169.4$. According to the model, the median price of a home in Detroit, Michigan, in the year 2012 would be about −$20,169. This answer is not reasonable because a home price cannot be a negative value. This is an example of model breakdown.

47. a. The vertical intercept is (0, 0). This means that if 0 pictures are framed, there is $0 profit.

b. The vertex is (6, 108). This means that the frame shop will make the maximum amount of average profit of $108 when they frame 6 pictures.

c. $n = 2, n = 10$. The shop must frame either 2 pictures or 10 pictures to make an average profit of $60 per frame.

48. a. The vertical intercept is (0, 0). This means that if 0 apple trees are sold, there is $0 profit.

b. The vertex is (4, 64). This means that the nursery will make the maximum amount of average profit of $64 when they sell four trees.

c. $n \approx 5.87, n \approx 2.13$. The nursery must sell either six trees or two trees to make an average profit of about $50 per tree.

49. The domain is the set of names: Colin, Lily, Isaiah, Gabrielle, and Seth.

The range is the set of hourly wages: $18.50, $9.25, $11.75, $18.50, and $14.50.

Every input results in exactly one output.

50. The domain is the set of amounts of time traveled in hours: 1, 2, 3, 4, and 5.

The range is the set of distances traveled in miles: 50, 100, 150, 200, and 250.

Every input results in exactly one output.

51. The domain is the set of numbers: 0, 3, 5, and 9.

The range is the set of numbers: 2, −1, 4, −3, and 10.

No, each input does not result in exactly one output.

52. The domain is the set of numbers: 10, −4, 5, and 1.

The range is the set of numbers: 0, −3, 1, and 10.

No, each input does not result in exactly one output.

53. The relation is a function assuming one given moment in time.

54. The relation is not a function when taking into account the full year that a child is a certain age.

55. The graph represents a function.

56. The graph represents a function.

57. The graph does not represent a function.

58. The graph does not represent a function.

59. The equation represents a function because for each input x there is exactly one output y.

$$f(x) = -5x + 3$$

60. The equation represents a function because for each input x there is exactly one output y.

$$f(x) = 7$$

61. $f(-4) = 18$ **62.** $f(-6) = -16$

Chapter 9 Test

1. a. 1 and 3 seconds

b. (2, 84). The ball will reach its maximum height of 84 feet after 2 seconds.

2. (−4, 28). The revenue was at a minimum of $28 thousand in 2006.

3. $x = \dfrac{20}{3}, x = \dfrac{2}{3}$

4. $b \approx 10.39$

5. $a = 17, a = 1$

6. $x = \dfrac{-4 \pm \sqrt{5}}{3}; x \approx -0.59, x \approx -2.08$

7. a. When indicates a reference to time t.

b. When the rocket hits the ground, the value of the height h is 0.

$$0 = -16t^2 + 150t + 4$$

B-88 APPENDIX B Answers to Selected Exercises

c. The equation in part b is quadratic.

d. $t \approx 9.40, t \approx -0.03$

e. Only the positive solution makes sense for this problem. The rocket will hit the ground approximately 9.40 seconds after it is launched.

8. $(-6, 0), (4, 0), (0, -24)$; minimum at $(-1, -25)$

9. $(1, 0), (3, 0), (0, -15)$; maximum at $(2, 5)$

10. a. $(-3, -4)$ b. $(0, 0)$ c. $(-6, 0), (0, 0)$

11.
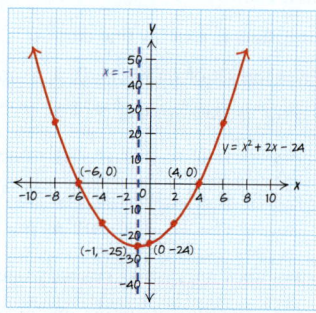

12.
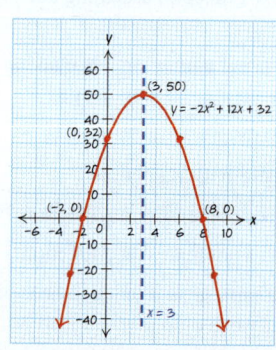

13. a. Quadratic   b. Linear   c. Neither

14.
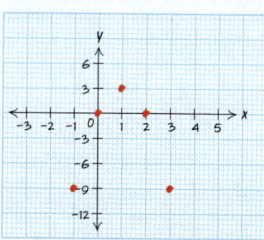

a. The graph of the data follows a parabolic pattern.

b. The estimated vertex from the graph is (1, 3).

15. a.
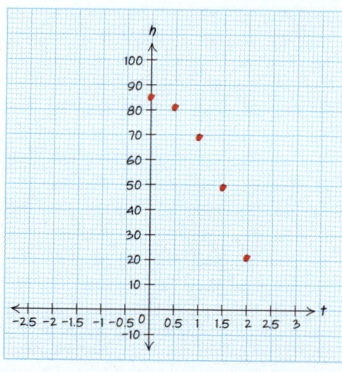

b. The graph of the data follows a parabolic pattern.

c. The estimated vertex from the graph is (0, 85).

16. a. The vertex is (17.75, 1.12). This means that in the year 1918, Mexico City reached its minimum population of 1.14 million, then the population began to rise.

b. The vertical intercept is (0, 1.75). This means that in the year 1900, the population of Mexico City would have been 1.75 million. This may be model breakdown if the actual population in 1900 was not more than the population in 1918.

c. $t \approx 84, t \approx -49$. The model predicts that the population of Mexico City would reach 10 million people in the years 1984 and 1851. 1984 may make sense in this situation. 1851 is too far in the past for that large a population.

17. a. The vertex is (0, 50). This means that the rock is at its maximum height of 50 feet after 0 seconds from the time it is dropped, or before it has been dropped.

b. The x-intercepts are approximately (1.77, 0) and (−1.77, 0). The intercept (1.77, 0) means that the rock will be 0 feet above the ground 1.77 seconds after being dropped. This makes sense in terms of this situation. The intercept (−1.77, 0) means that the rock would be 0 feet above the ground 1.77 seconds before being dropped, and this does not make sense in terms of this situation.

18. The equation is a function because each input has exactly one output. The equation also graphs as a line and passes the vertical line test. $f(x) = 6x - 1$

19. $f(-2) = 3(-2)^2 + 2(-2) = 8$

20. Domain: Florida, Texas, Michigan, Washington, New Mexico; range: 5.125%, 6%, 6.25%, 6.5%

Applications Index

Agriculture/farming
  horse farming/boarding, 86, 87, 105–106
  irrigation system, 774

Banking/finance/investing. *See also* **Salary/pay**
  credit, 140, 280, 281, 288, 296, 297, 298
  currency conversion, 92
  depreciation, 48–49, 50, 181, 201, 460, 585
  federal debt, 141
  investing and interest, 107, 134, 149, 186, 187, 406–408, 413, 414, 430, 439–441, 443, 453, 455, 456, 462, 545, 585, 740, 827
  loans/mortgages, 168, 184, 185, 224
  mobile payments, 338, 339
  personal income, 137
  taxes, 105, 119, 630, 819, 844

Business/industry. *See also* **Buy/sell/rent; Salary/pay**
  accounting/bookkeeping/budgeting, 108, 216, 287–288, 599, 623, 644
  advertising, 103, 104
  airplane manufacturing, 573, 763
  Amazon, 186
  apparel and accessories, 380, 397, 398, 443, 520, 585, 808
  Apple, 316–317, 320–321, 486
  appliance manufacturing, 573, 762–763, 806–808
  bicycles, 113–114, 276, 297
  bookstore, 201
  bottled water, 377
  cabinet maker, 437
  candy, 280, 296
  capital equipment, 186
  cars and car repair, 280, 296, 379, 397, 808–810
  casino, 570–571
  charities, 380, 412, 763
  cleaning, 638–639, 640–641
  coffee shop, 107
  customers, 50, 241
  Deere and Company, 297
  dog trainer, 453
  electronics, 581
  Facebook, 333, 338, 339, 481
  fiber production, 259, 263, 279
  fishing supplies, 571
  food/groceries, 379, 392–394, 396–397, 398, 414, 453, 610, 736, 742, 838
  golf club manufacturer, 167–168
  Google, 486

  holiday spending, 140
  Home Depot, 815, 816
  hours worked, 18
  iPhones, iPads, iTunes, 135, 149
  JetBlue Airways, 337
  jewelry business, 215, 789, 816
  landscaping, 201, 297
  LEGO Group, 184–185
  Lowe's, 815, 816
  machine shop, 187
  makeup/cosmetics, 461
  markups/discounts, 48, 50, 67, 201, 460
  McDonalds, 297
  music industry, 187, 297, 376–377
  networks, 814
  Nike, 281
  office work, 599, 644, 741
  oil prices, 827
  paint, 573–574, 763
  Papa John's International, 297
  personal trainer, 456
  pet supplies, 585
  photographer, 789, 839
  piano teacher, 453
  pie maker, 829
  profit/revenue/costs, 163, 169, 224, 226, 573
  real estate, 186, 200
  remote workers, 259, 263, 279
  screws, 507
  shipping, 100–101, 135, 362
  stained glass, 226
  stocking, 644
  surfboards, 598
  toys, 200–201, 307–308, 351, 443, 580, 645, 660
  trailer hitches, 113
  trucking, 3
  Twitter, 296
  typesetting, 673
  unit prices, 56–57, 59
  video games, 814
  Volkswagen income, 125–126, 126–127
  web designer, 789
  window shutters, 643
  Zara, 790

Buy/sell/rent
  apparel/accessories, 43, 64, 112–113, 176, 690, 728
  appliances, 380
  art, 763
  Boy Scout model rockets, 119
  cameras, 816
  car repair, 105

  electronics/phones, 43, 52, 816, 829
  flowers, 690
  food/groceries, 38, 39, 42, 43, 79, 426, 430, 501–502, 585, 647, 648, 650, 660, 674, 740, 743, 826
  framing, 841
  gas, 26, 43
  gifts, 597, 642, 643
  Girl Scout cookies, 3
  home utilities, 52, 58
  Internet services, 377
  iPhone, 741, 819
  irrigation systems, 728
  postage, 105, 106
  sales tax, 46, 47, 50, 67, 585
  storage, 374, 386
  towing, 215
  vehicle rental, 180, 181, 215, 226, 361–362, 363, 378, 385

Entertainment/sports/recreation/hobbies. *See also* **Transportation/travel/vacations**
  attendance/seating, 52, 58, 212
  biking, 103, 242, 243, 452
  bowling, 228
  boxing, 228, 272–273
  camping, 216
  exercise, 52, 259, 263, 279, 336
  golf, 228
  gym, 314, 377
  hiking, 496–497
  Little League baseball, 631
  long jump, 229
  lottery prizes, 597, 643
  Olympics, 819
  painting, 630
  parties/events, 102, 108, 197–198, 226, 416–417, 424–425, 462, 493, 496, 598, 642, 643, 742, 743, 829
  photography, 652–653, 661
  rafting, 233–234
  rock climbing, 585
  rowing, 672
  running, 103, 135, 136, 352, 584
  sewing/trimming, 26
  soccer, 631, 739, 741
  summer camp, 462
  theme parks, 216, 691, 741
  ticket sales/prices, 98, 105, 106, 148, 259, 263, 269–270, 279, 284–285, 380, 443, 461, 739, 755, 827, 829
  TV size, 771
  uniforms/equipment/supplies, 412, 585
  volleyball, 168, 650

I-1

walking, 273, 349
woodworking, 201

Health/nutrition/healthcare
autism, 355
BMI, 678
calories, 458, 610, 660, 662
cholesterol, 338, 339, 376
diabetes, 338, 339
drugs/dosages, 609, 610
flu tests, 122–123, 125
fund raising, 259–260, 263, 279
heart, 216, 457, 460, 604–605, 738
height/weight/dieting, 136, 212, 238, 265, 308, 313, 353, 355, 460, 647, 819, 826, 841
IV fluids, 105
medical assistant, 739
nutrient contents, 50, 68

Home/community
carpentry, 599
carpeting/rugs, 82–83, 106, 215, 275–276, 574
church work, 645
cooking/recipes, 662
corral, 846
electricity usage, 91, 105
fencing, 243, 660
flooring, 52, 215
food/cooking, 33, 58, 412
gardening/landscaping, 241, 243, 349, 412–413, 599, 629–630, 645, 670, 675, 841
home improvement, 243
home values/prices, 47, 50, 67, 68, 201, 460, 741, 811, 814–815, 816, 830
house painting, 28, 33, 86, 105, 106, 641, 644, 673
lumber, 107, 149, 241
monuments, 661
pets, 229, 827
plumbing, 152, 380
poles, 771
roof, 775
room size, 574, 774, 776
security system, 545
sod installation, 52
swimming pool, 641–642, 644, 645

Population/social statistics/ economics/ military
animals, 3, 17, 50, 241, 353, 354, 356
consumption, 502, 503, 507
crime, 133–134
debt, 364–365, 480, 481, 487
elevation/depth of places, 10, 11, 19, 52, 58, 357
energy/electricity, 246, 256, 355, 818
first names, 827
gross domestic product (GDP), 486

household statistics, 326–327, 507–508
housing market, 136–137, 141
immigration, 124, 335, 336
income per capita, 827
Internet use, 336
labor statistics, 124, 152
life expectancy, 329–331
marriages, 324–326, 328, 332
population groups, 280
population of places, 17, 170, 215, 245–246, 255–256, 258–259, 263, 268, 269–270, 278, 279, 333, 336, 342, 346–348, 349, 351, 354, 356, 459–460, 827, 844
postage, 333
poverty, 799
precipitation, 132, 304
public assistance, 598
road building, 605
severe weather, 136
temperature, 3, 4, 10, 18, 19, 58, 105, 126, 127, 138, 265, 801, 803–804, 805, 814, 815, 835–837, 840, 844–845
transit systems, 176
trees, 65
unemployment, 151, 336–337
unions, 321, 337–338
water usage, 243

Salary/pay
annual salary, 827
budget, 90, 150, 169, 205–206, 208, 214, 227
hourly pay, 172, 355, 841
monthly pay, 148, 184, 225, 230, 233, 458, 739
salespeople, 103, 173, 241, 373–374, 379, 395, 461, 545
weekly pay, 89, 103, 105, 152, 156, 232, 234–235, 237, 265, 285, 395, 396, 414, 425–426, 430, 434, 435–436, 442–443, 454, 655–656, 663, 743, 789, 825, 826

Schools/education
activities, 496, 527
applications, 630
assignments, 644
books/supplies, 118–119, 149, 197, 405–406, 462, 494, 598
college degrees, 123, 497
costs/fees, 42, 71, 90–91, 97, 101–102, 105, 137, 180–181, 314, 663, 827
courses, 216, 230
credits, 161–162, 819
enrollment, 216, 230, 527
fund raising, 103, 380, 441, 643
school buses, 155, 185, 216
smoking by students, 338–339
student statistics, 4, 17, 46, 47, 58, 65, 817–818, 819, 841

Science/technology
air/gas pressure, 212, 663
astronomical units, 483, 486–487
atoms, 483b, 486
blueprints, 653, 660, 743
cells, 481, 528
compact discs, 486
computers, 212
DNA, 526
downloading, 675
Earth, 486–487
electric charge, 482–483, 483b, 486, 487
electrons, 483b, 486, 487
falling objects/projectiles, 684, 689, 690, 724–725, 728, 735, 736, 744, 745, 755, 762, 765–766, 773, 774, 790, 826, 830, 838, 839, 842, 844
golden rectangles, 676
hair, 486
height/altitude, 574, 579
Hooke's law, 663
horizon, 679, 690, 720–721, 743
illumination, 664
Ishango Bone, 69
lightning, 231, 243, 308
light years, 153, 463
mass, 484b, 486
mixtures/solutions, 108, 409–410, 413, 414, 415, 453, 545
nanotubes, 481, 526
planets, 484, 487, 526, 528
protons, 482–483, 483b, 486, 487
rockets, 786–787, 789–790, 843
scaling, 589
spacecraft, 480, 481
temperature, 243, 313
tide, 105
tsunamis, 737–738
units conversions, 93–94, 98, 100–101, 104–105, 137, 140, 281, 305–307, 312–313, 381, 520, 603, 604, 652, 799
viruses, 526

Transportation/travel/vacations
baggage/backpacks, 106–107, 160–161, 168–169
gas costs/prices/mileage, 105, 137, 313, 507, 527, 819, 827
IRS mileage rate, 97
shipping costs, 185, 187, 226
taxi fare, 430
time/distance/speed, 89, 104, 107, 184, 224, 246–247, 247–249, 265, 389–391, 396, 397, 414, 452, 456, 462, 584, 585, 599, 604, 610, 655, 657–658, 662, 664, 740, 742, 743, 841
traffic accidents, 685, 689–690, 725, 728
trip costs, 69–70, 200, 590–591, 597, 632, 633, 643, 672, 755, 829
vehicle capacity, 185, 211, 212, 224

Index

Note: Page numbers with "b" refer to marginal boxes.

A

Absolute value, 6–7, 11–12
   in addition and subtraction of integers, 9–11
   definition of, 6
Absolute value bars (||), 77b
AC method of factoring, 550–552, 551b, 561
Adding vertically, 402b
Addition, 15, 178b
   addition property of equality, 159–160, 160b, 172b, 400, 401b
      inequalities and, 202–203
   addition property of inequalities, 204, 205b
   associative property of addition, 79–80
   commutative property of addition, 111b, 179b
      in adding rational expressions, 622–623
      definition of, 79, 80
   with decimals, 38–39
   definition of, 76
   distributive property in, 115b
   in evaluating expressions, 110–114, 110b
   exponents and, 465b. *See also* Product rule for exponents
   of fractions, 25–28, 612b, 619b
   with integers, 8–9
   of polynomials, 491–492
   of radical expressions, 697–701
      integers and, 768b
   of rational expressions, 612–625
      common denominators and, 612–615
      unlike denominators and, 619–625, 619b
   sum as result of, 9b
   translation of, into expressions, 95
Addition method. *See* Elimination method
Addition property of equality, 159–160, 160b, 172b, 400, 401b
   inequalities and, 202–203
Addition property of inequalities, 204, 205b
Angles, supplementary and complementary, 410b
Applications problems
   addition or subtraction of polynomials and, 493–494
   algebraic expressions for, 95
   constants and variables in, 88–89, 94, 156
   decimals in answers to, 172b
   equations and, 156, 164–165, 178–181, 304–308
      graphing cost equations, 284–285
      radical equations, 724–725
      systems of linear equations, 373–374, 389–394, 405–410
   functions and, 825–826

   inequalities and, 208, 212
      systems of linear inequalities, 439–441
   intercepts and, 268–270, 272, 275–276
   prime numbers in, 20
   shared work problems, 638–642, 639b
   slope and, 275–276
   vertex in, 754–755
Approximate answers, 55–56
Approximately equal to ($\approx$), 5, 5b, 682
Area
   distributive property and, 82–83
   of rectangle, 82b
   units for, 93b
Arithmetic, 2, 14
Associative property of addition, 79–80
Associative property of multiplication, 80
"at least," 205b
"at most," 205b
Atoms, 483b
Axes of graphs. *See also* Intercepts; Scale
   variables on, 234b
Axis of symmetry, in graphing quadratic equations, 755–760, 794–797
Axis of symmetry formula, 756

B

Bar graphs, 122–126
   steps in creation of, 125
Base
   vs. coefficient, 477, 477b
   definition of, 72
"baseline," 746b
Baseline graph, 746, 746b
Basic operations, 14, 76
Binomials
   definition of, 488
   multiplication of polynomials by, 499. *See also* FOIL
Bonds, 408b
Brackets ([ ]), 92b
   inequalities and, 209, 209b, 210
      compound inequalities, 213, 214
Break in graph, 132–134

C

Calculator
   cube roots, 688b
   division by zero, 593b
   exponents, 73b, 74, 464b
   order of operations, 77b
   parentheses, 81b, 114b

   pi, 56b
   radical expressions, 684b
   rounding, 779b, 784b
   square roots, 56b, 682b
   squaring a number, 73b
Capacity, units for, 93b
Cartesian coordinate system, 127b
Center of a circle, 52b
Certificate of deposit (CD), 440b
Checking solutions
   of inequalities, 204, 205
   of linear equations, 157–158, 157b, 160b, 161, 161b, 177
      systems of linear equations, 362–364
   of radical equations, 719, 723–724
Circle, 52b
Circumference of a circle, 52b
Coefficients, 190b
   1 as, 110b
   vs. base, 477, 477b
   definition of, 89, 109
   fractional coefficients, 174–175, 175b
   lead coefficient, 489b, 568b
      in completing the square, 780–781
      graphing quadratic equations and, 749–751, 750b
   in scientific notation, 480
   sign of, 110b
"commission," 373b
Common denominator, 26–27. *See also* Least common denominator
   addition and subtraction of fractions with, 25–28, 612b
   addition and subtraction of rational expressions with, 612–615
   in solving equations with fractions, 191–192
"commutative," 79b
Commutative property of addition, 111b, 179b
   in adding rational expressions, 622–623
   definition of, 79, 80
Commutative property of multiplication, 179b
   definition of, 79, 80
   exponents and, 466
   in multiplying polynomials, 498, 500b
   in multiplying radical expressions, 707
Complementary angles, 410b
Completing the square, 766b, 777–781
Complex fractions, 607–608
   simplifying rational expressions with, 625–628, 626b
Composite numbers, 20

I-3

Compound inequalities, 211–214
  definition of, 211
Compound interest, 406b
Concept Investigation
  addition and subtraction of fractions, 612–613
  commutative properties, 80–81
  difference of squares, 557
  direct variation, 655
  distance between a number and zero, 6
  distance formula, 772
  distributive property, 81–84
  dividing radicals, 711
  equations that are identities or have no solution, 192–194
  equivalent expressions, 622
  exponents
    division and, 466–467
    exponential expressions on calculator, 73
    multiplication and, 464–465
    negative exponents, 476–477
    zero as, 73
  eyeball best-fit line, 317–318
  factored form of polynomials, 535–536
  factoring completely, 542
  functions, 820
  inequalities
    linear inequalities in two variables, 418, 420–421
    properties of equality and, 202–203
    systems of linear inequalities, 434
  inverse variation, 657
  models, 805–806
    model breakdown, 329–330
  multiplication
    in evaluating expressions, 114–115
    of integers, 12–13
    of radicals, 692, 705–706
  numbers, 3
  parabolas, 749–750
  parallel lines, 290–291
  pi, 52
  quadratic equations as perfect square trinomials, 777–778
  scale, 24
  simplifying fractions and rational expressions, 593–594
  slope, 257
    y-intercept, slope, and equation, 273–274
  squares and square roots, 680
  systems of linear equations, solutions to, 368–369
  time, distance, speed, 247–248, 248–249
  zero-product property, 565–567
Conjugates
  definition of, 710
  in multiplying radical expressions, 710–711
  in rationalizing the denominator, 715–716
Consistent systems of linear equations, 369–373
  definition of, 369
  substitution method and, 386–388
Constant of proportionality, 655, 657
Constants, 88–89, 89b, 109b, 248b

definition of, 88
  multiplication of terms by, 114–115
Conventional form, 111–112
  definition of, 111b
Conversion factors, 92b
Converting units. See Unit conversions
Coordinates, 128–130, 128b, 234
  of intercepts, 266
Corresponding angles, 653b
Corresponding parts of triangles, 653b
Corresponding sides, 653, 653b
Cost, 163b
  graphing cost equations, 284–285
  unit cost, 755, 755b
Counting numbers. See Natural numbers
Credits, 90b
Cross-multiplication, 649, 649b
Cube roots
  on calculator, 688b
  definition of, 687
  finding, 687–688
Curly braces ({ }), 2b, 77b
Curve. See Smooth curve

D

Data. See also Data tables; Models
  in bar graphs, 124–127
  in input–output tables, 99, 232–240
  in rectangular coordinate system, 127, 132–134
Data tables, 122–123, 133–134
Decimals. See also Finite decimals; Infinite decimals
  in answers, 189b
  to applications problems, 172b
  fractions and, 35–37, 51, 54–55, 285b
  operations with, 38–41
  percents and, 45–46, 407b
  place value in, 34–35
  rounding of, 37–38, 682b
  for slope, 302b
Degree of polynomials and terms, 488–491
  definition of, 489
Denominator, 22b, 93b. See also Common denominator; Least common denominator
  definition of, 22
  as divisor, 511b
  of equivalent fractions, 23–24
  of rational expressions, 616–617
    unlike denominators, addition and subtraction with, 619–625
  rationalizing the denominator, 713–716
  of whole number, 28b
  zero as, 592b
Dependent variables, 99, 99b, 232, 234b. See also Input–output tables; Output variables
Descartes, René, 127b
Descending order of polynomials, 488, 515–516, 516b
Diameter of a circle, 52b
Difference, 14, 76

"difference," 493b
Difference of squares, 503b, 505–506
  definition of, 556
  factoring of, 556–558, 557b, 561
Digits, 480b
Direct variation, 655–657
  definition of, 655
  proportions and, 656b
Direct variation equation, 655
Distance formula, 389–391, 685–686, 772–773
Distributive property, 81–84, 115–116, 115b, 117, 707b. See also FOIL
  in adding radical expressions, 697–698
  definition of, 82, 115
  in multiplying
    polynomials, 498–506, 498b, 535
    radical expressions, 707–708
  vs. order-of-operations agreement, 83
  in simplifying complex fractions, 626
  in solving rational equations, 633–634, 635–640
  in subtracting
    polynomials, 492–493, 493b
    rational expressions, 614–615, 619–621
Dividend, 14b, 511b
Divisibility rules, 553b
Division, 15, 178b. See also Powers of products and quotients rule for exponents; Quotient rule for exponents
  by 0, 14
  commutative property and, 81, 179b
  with decimals, 40–41
  definition of, 76
  divisibility rules, 553b
  division property of equality, 171–172, 172b
    definition of, 171
    inequalities and, 203
    solving equations with, 173–175
  division property of inequalities, 204, 205b
  of fractions, 29–30
  with integers, 12–14
  long division, 511
    of decimals, 40–41
    dividend and divisor in, 511b
    of polynomials by polynomials, 511–518, 514b
  of polynomials, 509–520
    by monomials, 509–511
    by polynomials, using long division, 511–518, 514b
  of radical expressions, 711–716
  of rational expressions, 605–608
  with scientific notation, 481–484
  symbols for, 14b, 29b
  translation of, into expressions, 96
Division line, 23b. See also Fraction bar
Division property of equality, 171–172, 172b
  definition of, 171
  inequalities and, 203
  solving equations with, 173–175
Division property of inequalities, 204, 205b
Division sign (÷), 14b

Index I-5

Divisor, 14b, 511b
Domain, 817–818
  definition of, 817, 819, 819b

E

Elimination method, with systems of linear equations, 400–415, 400b, 410b
  with applications problems, 405–410
  vs. substitution method, 410–411, 411b
Empty set ({ } or ∅), 2b
End points, 204, 204b
  in interval notation, 209
Equality, properties of
  addition property of equality, 159–160, 160b, 172b, 400, 401b
    inequalities and, 202–203
  division property of equality, 171–172, 172b
    definition of, 171
    inequalities and, 203
    solving equations with, 173–175
  inequalities and, 202–203
  multiplication property of equality, 171–172, 172b
    definition of, 171
    inequalities and, 203
  subtraction property of equality, 159–160, 160b, 172b
    inequalities and, 203
  undoing operations and, 173b
Equal to (=), 5
  translations of, 178
  variable with, 179b
Equations. See also Formulas; Inequalities; Linear equations in one variable; Linear equations in two variables; Literal equations; Quadratic equations; Rational equations; Solving equations
  from applications, 178–181, 179b, 180b, 181b
  definition of, 156
  vs. expressions, 156–157, 156b
  intercepts and, 271–273
  nonlinear equations, 238–239
Equivalent fractions, 22, 23–24
Estimating, 361b
  from linear models, 328, 328b
  from quadratic models, 803–811
    model breakdown and, 808–811
"evaluate," 15b, 159b
Evaluating expressions, 78–79, 90–92, 159b
  with calculator, 114–115, 114b
  distributive property used in, 83–84
  functions, 824–826
  radical expressions, 682–683, 683b
  rational expressions, 590–593, 591b
    excluded values in, 592–593
Exact answers, 55
Excluded values
  definition of, 592
  of radical expressions, 683b
  of rational expressions, 592–593, 633b
    in solving rational equations, 633–638
Expanded form, 72

Exponential form, 72
Exponentiation, 75, 75b, 76
Exponents, 72–74. See also Degree of polynomials and terms; Scientific notation
  1 as exponent, 465, 465b
  on calculator, 73b, 74, 464b
  definition of, 72, 76
  negative numbers as, 73–74, 476–479, 479b
  as powers, 75b
  radicals and, 697b
  rules for, 464–475, 505b, 686b
    power rule for exponents, 469, 469b, 699, 699b
    powers of products and quotients rule for exponents, 470–472, 470b, 707
    product rule for exponents, 464–466, 465b, 498b, 537b
    quotient rule for exponents, 466–468, 467b, 478–479
  zero as exponent, 73–74
Expressions, 8, 8b. See also Evaluating expressions; Radical expressions; Rational expressions; Simplifying expressions
  vs. equations, 156–157, 156b
  exponential expressions, 74–75, 464, 478, 478b
  from input–output tables, 99–102
  as sentences, 8, 95–99
  terms in, 14–16, 76–77
  translation into, 95–98
Extraneous solutions, 723, 723b
Extrapolation, 328b
Eyeball best-fit line, 323b
  equations of, 321–327
  finding, 317–321, 317b

F

Factored form of polynomials, 535–536
Factoring, 20, 20b, 532–583. See also Factors; Polynomials, factoring of; Prime factorization; Trinomials, factoring of
  AC method of, 550–552, 551b, 561
  of difference of squares, 556–558, 557b, 561
  factoring by grouping, 540–541, 540b, 561
  factoring by inspection, 547
  factoring completely, 541–543, 553–554, 553b, 561–562
  of integers, 532
  of integers or monomials, 532–533
  quadratic equations solved by, 567–572, 568b, 766b, 767b
Factoring by grouping, 540–541, 540b, 561
Factoring by inspection, 547
Factoring completely, 541–543, 553–554, 553b, 561–562
Factors, 20b. See also Factoring; Greatest common factor
  definition of, 14, 76, 594b
Factor tree, 21
Finite decimals, 35b

FOIL, 500–501, 500b, 546, 709b, 726b
  in multiplying radical expressions, 708–710
Forecasting, 328b
Formulas, 165–166
  axis of symmetry formula, 756
  definition of, 166
  distance formula, 389–391, 685–686, 772–773
  simple interest formula, 406
  for slope, 252, 253b
  solutions (mixtures) formula, 408
  value formula, 392–394
  vertex formula, 753–755
Fraction bar (—), 77b
  order-of-operations agreement and, 14b, 22, 23b
Fractions, 22b. See also Denominator; Numerator; Rationalizing the denominator; Reciprocals
  in answers, 189b
  complex fractions, 607–608
    simplifying rational expressions with, 625–628, 626b
  decimals and, 35–37, 51, 54–55, 285b
  dividend and divisor of, 511b
  equivalent fractions, 23–24, 618b
  improper fractions, 22b, 25b, 26b
  negative sign in, 25b
  on number lines, 24–25
  operations with, 25–31
    addition and subtraction, 25–28, 612b, 619b
    multiplication and division, 28–30
    order of operations, 30–31
  percents and, 44–45, 46
  proportions and, 648–653
  ratios and rates as, 646–647, 646b
  simplifying, by reduction to lowest terms, 22–23, 23b, 182b, 466b, 593–594, 594b, 595b
  solving equations with, 164, 191–192
    fractional coefficients, 174–175, 174b
  undefined fractions, 592b
  unity fractions
    as conversion factors, 92–94, 92b
    definition of, 92, 603b
Function notation, 823–824, 823b, 824b
Functions, 819–829
  evaluation of, 824–826
  function notation and, 823–824, 823b, 824b
  vertical line test for, 820–823

G

Gauss, Karl, 400b
Gaussian elimination method. See Elimination method
GCF. See Greatest common factor
General form of a line, 285b
  definition of, 285
  elimination method and, 404–405
  graphing from, 285–289
Geometry, 196

Graphing. *See also* Graphs; Scale
   with graph paper, 236b
   linear equations in two variables
      from general form, 285–289
      by plotting points, 234–238
      from slope-intercept form, 282–285, 282b
      systems of linear equations, 364–368
   linear inequalities in two variables, 417–427, 419b
      systems of linear inequalities, 434–441, 438b
      vertical and horizontal inequalities, 426–427
   on number line
      decimals, 37
      fractions, 25
      integers, 5
   ordered pairs, 131–134
   by plotting points, 238b
      linear equations in two variables, 234–238
      nonlinear equations, 238–239
      quadratic equations, 746–749, 746b, 794–797
   of quadratic equations, 746–763, 791–799, 800–803
      axis of symmetry and, 755–760, 794–797
      intercepts and, 791–799
      lead coefficient a and, 749–751, 750b
      by plotting points, 746–749, 746b, 794–797
      vertex and, 751–755, 751b, 794–797
   with smooth curve, 235, 235b
   of vertical and horizontal lines, 239–240
Graph paper, 236b
Graphs. *See also* Axes of graphs; Bar graphs; Graphing; Intercepts; Parabola; Rectangular coordinate system; Rise and run; Scale; Scatterplots; Slope
   bar graphs, 122–126
      steps in creation of, 125
   baseline graph, 746, 746b
   interpretation of, 245–246, 245b
   linear graphs
      vs. nonlinear graphs, 238
      slope of, 249, 299–303, 299b
Greater than (>), 5, 202b, 420–421
Greater than or equal to (≥), 5, 202b, 205b, 420–421
Greatest common factor (GCF), 21–22, 532–543
   definition of, 533, 596b
   of integers or monomials, 532–533
   of polynomials, 533–543
      in factoring by grouping, 540–541, 540b
      in factoring completely, 541–543
      factoring out of, 535–539, 536b, 537b, 561
   in reduction of fractions to lowest terms, 23–24
Grouping symbols, 77–78, 77b
   fraction bar as, 591b

H

Hindu–Arabic number system, 34
Hippasus, 53
Horizontal axis, 123. *See also* Horizontal intercepts
   in rectangular coordinate system, 127–134
Horizontal intercepts (x-intercepts), 267b. *See also* Intercepts *(for general discussions)*
   definition of, 266
Horizontal lines
   definition of, 239, 239b
   graphing of, 239–240
      for inequalities, 426–427, 426b, 438b
   slope of, 255
Hypotenuse, 769, 769b

I

Identities, 193
Improper fractions, 22b, 25b, 26b
Inconsistent systems of linear equations, 369–373
   definition of, 369
   substitution method and, 386–388
Independent variables, 99, 99b, 232, 234b. *See also* Input–output tables; Input variables
Index of a root, 687, 687b
Inequalities, 5, 202–217. *See also* Linear inequalities in two variables; Solving inequalities
   compound inequalities, 211–214
   end point of, 204, 204b
   interval notation and number lines with, 209–211, 209b
   properties of, 202–204, 205b
   symbols for, 202b, 206b
   reversal of, 204–207
Infinite decimals, 35b, 36b, 51, 53
Infinite nonrepeating decimals, 35b
   irrational numbers as, 53
Infinite repeating decimals, 35b, 36b, 51
Infinity symbol (∞), 209, 209b, 210
Information. *See* Data
Input–output tables, 99–102, 232b
   equations from, 181b
   ordered pairs and data represented by, 232–240
   for graphing equations by plotting points, 234–240
Input variables, 234, 234b. *See also* Domain; Independent variables; Input–output tables
   in systems of linear equations, 360
Integers
   definition of, 3
   factoring of, 532
   as fractions, 51
   operations with, 8–17
      addition, 8–9, 768b
      division, 12–14
      multiplication, 12–14
      order of operations, 14–17
      subtraction, 9–11

systems of linear equations and, 391–392
Intercepts, 266–273, 267b, 810b. *See also* General form of a line; Horizontal intercepts; Slope-intercept form of a line; Vertical intercepts
   applications problems and, 268–270, 272, 275–276
   equations and, 271–273, 299–303
   quadratic equations, 791–799
Interpolation, 328b
"intersection," 436b
Interval notation, 209–210, 209b
Inverse variation, 657–659
   definition of, 657
Irrational numbers, 52–53
   definition of, 53
   square roots as, 682, 682b
Ishango Bone, 1, 69
Isolating variables, 158–159, 159b, 165
   equality properties and, 172b
   in multiple-step equations, 175–178, 175b
   in systems of linear equations, 383b
   written form for, 167b

L

LCD. *See* Least common denominator
Lead coefficient, 489b, 568b
   in completing the square, 780–781
   graphing quadratic equations and, 749–751, 750b
Least common denominator, 164b. *See also* Least common multiple
   definition of, 27
   of rational expressions, 615–615, 616b, 617b, 618b, 619b
   in solving rational equations, 631–638, 635b
Least common multiple, 27, 403b
Length, units for, 93b
Less than (<), 5, 202b
Less than or equal to (≤), 5, 202b, 205b
Like radicals, 698–701, 698b
Like terms
   addition of, 110–114, 110b
      with polynomials, 491–492, 493b
   definition of, 109
   in evaluating expressions, 109–114
   exponents and, 465b
   like radicals as, 698–701, 698b
   in simplifying expressions, 109–114
   subtraction of, 110–114, 110b
"linear," 490–491
Linear equations in one variable, 162–165, 490–491, 564–565
   definition of, 162, 564b
   multiple-step equations, 175–178
Linear equations in two variables, 289–290, 490–491. *See also* General form of a line; Slope-intercept form of a line; Systems of linear equations
   definition of, 289
   direct variation and, 655
   finding, 299–315

from applications, 304–308
of parallel lines and perpendicular lines, 310–311
with point-slope form, 308–309
with slope-intercept form, 299–303
graphing of
from general form, 285–289
by plotting points, 234–238, 234b–236b, 238b
from slope-intercept form, 282–285, 282b
systems of linear equations, 364–368
for linear models, 321–327
Linear inequalities in two variables, 416–444. *See also* Systems of linear inequalities
definition of, 416
graphically solving, 417–427
vertical and horizontal inequalities, 426–427
solution sets of, 417–424, 423b
for horizontal lines and vertical lines, 426–427, 426b
strict inequalities, 423b
truth of, 421b
Linear models, 316–339. *See also* Eyeball best-fit line
definition of, 321, 321b
equations found for, 321–327, 322b
estimating from, 328, 328b
model breakdown and, 329–332, 329b
Lines, 235b. *See also* Horizontal lines; Linear equations in one variable; Linear equations in two variables; Linear models; Parallel lines; Perpendicular lines; Vertical lines
forms of, 274b. *See also* General form of a line; Point-slope form of a line; Slope-intercept form of a line
graphing of, 422b
increasing or decreasing lines, 252b
slope of. *See also* Slope-intercept form of a line
calculation of, 249–255, 253b, 256b
interpretation of, 255–258
Literal equations, 165–167, 181–183
definition of, 166
solutions to, 276b
Long division, 511
of decimals, 40–41
dividend and divisor in, 511b
of polynomials by polynomials, 511–518, 514b

M

Markdown, 48
Markup, 47–48
Mass, 484b
MathWorld, 769b
Maximum point, 751–753, 751b
definition of, 751
Metric prefixes, 479
Minimum point, 751–753, 751b
definition of, 751
Mixed numbers, 25b, 26b

Model breakdown, 329–332, 329b, 808–811
Models. *See* Linear models; Model breakdown; Quadratic models
Money market account, 440b
Monomials
definition of, 488
division of polynomials by, 509–511
multiplication of polynomials by, 498
Multiple, 27b
least common multiple, 27, 403b
Multiple-step equations, 175–178, 175b
Multiplication, 15, 178b
associative property of multiplication, 80
commutative property of multiplication, 179b
definition of, 79, 80
exponents and, 466
in multiplying polynomials, 498, 500b
in multiplying radical expressions, 707
with decimals, 39–40
definition of, 76
in evaluating expressions, 114–116
exponents and, 470b
power rule for exponents, 469, 469b, 699, 699b
powers of products and quotients rule for exponents, 470–472, 470b, 707
product rule for exponents, 464–466, 465b, 498b, 537b
of fractions, 29–30
with integers, 12–14
multiplication property, 115–116
definition of, 115
multiplication property of equality, 171–172, 172b
definition of, 171
inequalities and, 203
multiplication property of inequalities, 204, 205b
of polynomials, 498–509, 498b, 500b, 535
of radical expressions, 704–711
of radicals, using product property of radicals, 692–697
of rational expressions, 601–605
unit conversions and, 603–605, 603b, 605b
with scientific notation, 481–484
symbols for, 29b, 171b
translation of, into expressions, 96
of variable and constant, 89b
Multiplication property, 115–116
definition of, 115
Multiplication property of equality, 171–172, 172b
definition of, 171
inequalities and, 203
Multiplication property of inequalities, 204, 205b

N

Natural numbers. *See also* Composite numbers; Prime numbers
definition of, 2
exponents as, 72–73

Negative numbers, 3
in bar graphs, 125–126
as cube roots, 687–688
distributive property and, 117
as exponents, 73–74, 476–479
with exponents, 471b, 479b
in factoring polynomials, 537–538
in inequalities, 204, 205, 205b
on number line, 206b
square roots and, 206b, 680–681, 683b
Negative sign (−), 8, 8b. *See also* Negative numbers
exponents and, 471b
in factoring, 596b
in fractions, 25b
"nonlinear," 490–491
Nonlinear equations, graphing of, 238–239
Nonnegative, 6, 6b
Notation, 5b
function notation, 823–824, 823b, 824b
scientific notation, 74–75, 479–484, 480b, 482b
Not equal to (≠), 5
"not real," 681b
Number lines, 4–5
for addition and subtraction with integers, 8–12
fractions on, 24–25
graphing on
decimals, 37
fractions, 25
integers, 5
inequalities and, 209–211, 209b
negative numbers on, 206b
Numbers. *See* Absolute values; Decimals; Digits; Fractions; Integers; Negative numbers; Number lines; Number systems
Number systems, 2–4. *See also* Irrational numbers; Natural numbers; Rational numbers; Real number system; Whole numbers
Hindu–Arabic number system, 34
Numerator, 22b
definition of, 22
as dividend, 511b

O

"of," 46b
Operations, 178b. *See also* Addition; Division; Multiplication; Order of operations; Subtraction
basic operations, 14, 76
with exponents, 470b
with fractions, 25–31
with integers, 8–17
"of" and, 46b
undoing operations, 160, 173b
Opposite of a number, 8, 401b
Opposite reciprocals, 292–293, 292b
Ordered pairs, 127–131. *See also* Relations
definition of, 127
in input–output tables, 232–240
Order in numbers, 5–6, 7

Order of operations, 75–79. *See also* Order-of-operations agreement
  on calculator, 77b
  with decimals, 41
  with fractions, 30–31
  with integers, 14–17
Order-of-operations agreement, 76–79, 78b
  vs. distributive property, 83
  with fraction bar, 591b
  in simplifying radical expressions, 683
Origin, 127
Output variables, 234, 234b. *See also* Dependent variables; Input–output tables; Range
  in systems of linear equations, 360

P

Parabola, 745. *See also* Vertex of a parabola
  data set and, 800–803
  definition of, 747
  in function notation, 824
Parallel lines, 290–292, 291b
  definition of, 291
  equations found from, 310
  inconsistent systems of linear equations, 370, 371
  slope of, 371, 371b
Parentheses, 77b, 92b
  on calculator, 81b, 114b
  inequalities and, 209, 209b, 210
    compound inequalities, 213, 214
Pattern symbol (. . .), 2b
PEMDAS, 78b, 468, 468b
"per," 646
Percentages, 256b
Percents, 44–49, 256b
  definition of, 44
  fractions and decimals and, 44–46, 407b
  increase and decrease in, 47–49
  problem solving with, 46–47
Percent sign (%), 44
Perfect cubes, 687
Perfect squares, 556, 556b, 680, 682, 692b. *See also* Difference of squares
  multiplying radical expressions and, 704–705
  simplifying radical expressions and, 686–687, 692–694, 695–696
Perfect square trinomials, 503–505, 503b, 777–778
  definition of, 558
  factoring of, 558–561, 560b, 777b
Perpendicular lines, 292–293
  definition of, 292
  equations found from, 310–311
Pi (π), 52–53, 55, 56b
Place value, 34–35
Point-slope form of a line
  definition of, 308
  equations found from, 308–309

Polynomials. *See also* Binomials; Difference of squares; Monomials; Quadratic equations; Rational expressions; Trinomials
  addition of, 491–492
  definition of, 488
  degree of, 489–491
  division of, 509–520
    by monomials, 509–511
    by polynomials, using long division, 511–518, 514b
  factored form of, 535–536
  factoring of, 533–574
    difference of squares, 556–558
    factoring completely, 541–543, 553–554, 553b, 561–562
    by factoring out the greatest common factor, 535–538, 536b, 537b
    by grouping, 540–541, 540b, 561
    perfect square trinomials, 558–561, 560b
    quadratic equations solved by, 564–572
    summary of factoring tools, 561–562
  greatest common factor of, 533–534
  multiplication of, 498–509, 498b, 500b, 535
    special products, 503–506, 503b, 505b
  prime polynomials, 552, 552b
  subtraction of, 492–493
  terminology of, 488–491, 489b
Power, 75
Power rule for exponents, 469, 469b, 699, 699b
Powers of 10, 482b. *See also* Scientific notation
Powers of products and quotients rule for exponents, 470–472, 470b
  in multiplying radical expressions, 707
Prime factorization, 20–22, 532–534, 616b
  least common denominator and, 164b, 616
  in simplifying radical expressions, 694–695, 694b
Prime numbers, 20. *See also* Prime factorization
Prime polynomials, 552, 552b
Principal square root, 681, 681b, 682
  definition of, 681
Product. *See also* Factoring; Zero-product property
  definition of, 14, 76
  of variable and constant, 89b
Product property of radicals
  in multiplying radical expressions, 704–708
  in simplifying radical expressions, 692–697
Product rule for exponents, 464–466, 465b, 498b, 537b
Profit, 163b
Properties, 79b. *See also specific properties—e.g.,* Distributive property, Zero-product property
Proportions, 648–653
  definition of, 648
  direct variation and, 656b
  similar triangles and, 653–654, 653b
Pythagorean theorem, 769–773

Q

Quadrants, 426b
"quadratic," 781b
Quadratic equations, 745–846. *See also under* Solving equations
  definition of, 564
  graphing of, 746–763, 791–799, 800–803
    axis of symmetry and, 755–760, 794–797
    intercepts and, 791–799
    lead coefficient a and, 749–751, 750b
    by plotting points, 746–749, 746b, 794–797
    vertex and, 751–755, 751b, 794–797
  vs. linear equations, 564–565
  as models, 800–816
    in estimating, 803–808
    model breakdown and, 808–811
  in solving radical equations, 725–726
  standard form of, 564, 750, 764b, 807b
Quadratic equations in two variables, 746, 803b. *See also* Quadratic equations
Quadratic formula, 766b, 781–787, 781b
Quadratic models, 800–816
  in estimating, 803–808
  model breakdown and, 808–811
Quotient, 14b, 179
  definition of, 14, 14b, 76
Quotient property of radicals, 712–713
Quotient rule for exponents, 466–468, 467b, 478–479

R

Radical equations. *See also under* Solving equations
  definition of, 684
  evaluation of, 684–686
Radical expressions. *See also* Radical equations
  addition and subtraction of, 697–701
    integers and, 768b
  on calculator, 684b
  definition of, 683
  division of, 711–716
  evaluation of, 682–683, 683b
  multiplication of, 704–711
  simplification of, 683, 692–697, 694b, 696b, 697b
    vs. approximation, 695
    with product property of radicals, 692–697
    with variables, 686–687, 695–697
Radical symbol ($\sqrt{\ }$), 680
Radicand, 680
  of cube roots, 687–688
Radius of a circle, 52b
Range, 817–818
  definition of, 817, 819, 819b
Rate of change, 246–249, 250b. *See also* Slope
  definition of, 247
Rates, 646–648, 646b, 647b. *See also* Proportions; Rate of change

definition of, 246b
"rational," 590b, 647b
Rational equations. *See also under* Solving equations
   definition of, 590
   proportions as, 648–649
Rational expressions, 590–631. *See also* Rational equations
   addition and subtraction of, 612–625
      common denominators and, 612–615
      unlike denominators and, 619–625, 619b
   definition of, 590
   division of, 605–608
      complex fractions and, 607–608, 625–628, 626b
   evaluation of, 590–593, 591b
      excluded values in, 592–593
   least common denominator of, 615–619, 617b, 618b, 619b
   multiplication of, 601–603, 602b
   simplification of, 593–597, 594b, 595b
      with complex fractions, 625–628, 626b
   unit conversions and, 603–605, 603b
"rationalize," 713b
Rationalizing the denominator, 713–716
Rational numbers, 51–52
   definition of, 51
Ratios, 51, 51b, 646–648, 646b, 647b. *See also* Proportions; Rates
   definition of, 646
Real numbers, 53–55
   identities and, 193, 193b
   properties of, 79–84
   solutions to quadratic equations and, 785
Real number system, 53–55. *See also* Real numbers
   definition of, 53
   square roots and, 681
Reciprocals, 28–29, 477b
   definition of, 28
   negative exponents and, 477, 477b
   opposite reciprocals, 292–293, 292b
Rectangle, area of, 82b
Rectangular coordinate system, 127–134
   for graphing equations by plotting points, 127–134, 234–240
   quadrants in, 426b
Reduction to lowest terms, 22–23, 23b, 595b
Relations, 817–818. *See also* Functions
   definition of, 817
Revenue, 163b
"rewrite as a single fraction," 602b
Rise and run, 249–255, 253b
   in graphing cost equations, 284–285
Risk, 407, 407b
Roots. *See* Cube roots; Radical equations; Radical expressions; Square roots
Rounding, 56
   on calculator, 779b, 784b
   of decimals, 37–38, 682b
   of money amounts, 408b
Rounding down, 38
Rounding up, 38
Rules for exponents. *See* Exponents, rules for
Run. *See* Rise and run

S

Scale, 4, 4b, 24–25, 24b, 130–134, 235b
   adjustment of, 237b, 374b
   in graphing cost equations, 285
Scaling, 652–653
   similar triangles and, 653
Scatterplots, 126–127
   linear models and, 316–321
   quadratic models and, 800–803
Scientific calculator. *See* Calculator
Scientific notation, 74–75, 479–484, 480b, 482b
Sentences. *See* Translating sentences
Sets, 2, 2b
"setting two equations equal," 385b
Shared work problems, 638–642, 639b
Signed numbers, 8b. *See also* Opposite of a number
   addition and subtraction with, 9
   multiplication and division with, 13–14
Similar triangles, 653–654, 653b
Simple interest, 406, 406b
Simplest form, 23b
Simplifying, 16b, 159b. *See also* Simplifying expressions
   fractions, 22–23, 23b, 182b, 466b, 468b, 593–594, 594b, 595b
Simplifying expressions, 16–17, 109–122
   complicated expressions, 116–117
   exponential expressions, 464, 478, 478b. *See also* Exponents, rules for
   like terms in, 109–114
   multiplication property and distributive property and, 114–116, 117
   radical expressions, 683, 694b, 696b, 697b
      perfect squares and, 686–687, 692–694, 695–696
      prime factorization in, 694–695, 694b
      with product property of radicals, 692–697
      with variables, 686–687, 695–697, 700–701
   rational expressions, 593–597, 594b, 595b
      with complex fractions, 625–628, 626b
Slope, 250b. *See also* Point-slope form of a line; Slope-intercept form of a line
   applications problems and, 275–276, 304–308
   calculation of, 249–255, 253b, 256b
   definition of, 249
   finding
      from applications, 304–308
      from graphs, 299–303, 299b
   formula for, 252, 253b
   of horizontal and vertical lines, 255
   interpretation of, 255–258
   linear models and, 322b
   of parallel lines, 290–293, 371, 371b
   percentages and, 256b
   of perpendicular lines, 290–293
   units and, 255–257
Slope-intercept form of a line, 273–277, 274b
   equations found from, 299–303
   general form converted to, 288–289
   graphing from, 282–285, 282b
Smooth curve, 235, 235b
Solutions (mixtures), 408–410
Solutions to equations, 157–158. *See also* Checking solutions; Solving equations
   definition of, 157
   extraneous solutions, 723, 723b
   with fractions or decimals, 189b
   systems of linear equations, 360–364
      consistent systems and inconsistent systems, 368–374, 386–388, 387b
   writing answers, 167b
Solving equations, 158b, 159b. *See also* Solutions to equations; Solving inequalities
   with addition and subtraction properties of equality, 158–165
   after translating sentences into equations, 195–198
   with division property of equality, 173–175
   equations that are identities or have no solution, 192–194
   equations with fractions, 164, 191–192
      fractional coefficients, 174–175, 174b
   equations with variables on both sides, 189–190
   literal equations, 165–167, 181–183
   multiple-step equations, 175–178, 180b
   quadratic equations, 764–790, 766b
      by completing the square, 766b, 777–781
      by factoring, 567–572, 568b, 766b, 767b
      with Pythagorean theorem, 769–773
      with quadratic formula, 766b, 781–787, 781b
      with square root property, 764–769, 766b, 768b, 777–781
      with zero-product property, 566–567
   radical equations, 719–730, 723b
      applications involving, 724–725
      checking solutions, 719, 723–724
      with quadratics or multiple radicals, 725–726
      with squaring property of equality, 719–724
   rational equations, 631–642, 633b, 635b
      proportions, 649–653, 649b
      for shared work problems, 638–642, 639b
   systems of linear equations
      with elimination method, 400–415
      graphically solving, 364–368
      with substitution method, 382–399
Solving inequalities, 204–208, 206b, 209b
   linear inequalities in two variables, 417–427
   vertical and horizontal inequalities, 426–427, 426b
   systems of linear inequalities, 433–441, 438b

Speed, 246–248
   reasonable solutions with, 390b
Square brackets ([ ]), 77b
Square root property, 764–769, 766b, 768b, 777–781
Square roots, 55, 56b. *See also* Perfect squares; Product property of radicals; Square root property
   finding, 680–682, 681b, 682b
   index for, 687b
   irrational numbers as, 682, 682b
   negative numbers and, 681, 681b
   squaring of, 706
Square root symbol ($\sqrt{\ }$), 77b
Squaring
   on calculator, 73
   of square roots, 706
Stocks, 408b
Straight lines. *See* Lines
Strict inequalities, 423b
Subscript, 685b
Substitution method with systems of linear equations, 382–399, 401b
   with applications problems, 389–394
   vs. elimination method, 410–411, 411b
   inconsistent systems and consistent systems, 386–388
   setting the two equations equal, 385b
Subtraction, 15, 178b
   commutative property and, 80–81, 96, 179b
   with decimals, 38–39
   definition of, 76
   distributive property in, 115b
   in evaluating expressions, 110–114, 110b
   exponents and, 465b. *See also* Quotient rule for exponents
   of fractions, 25–28, 612b, 619b
   with integers, 9–11
   of polynomials, 492–493, 493b
   of radical expressions, 697–701
   of rational expressions, 612–625
      common denominators and, 612–615
      unlike denominators and, 619–625, 619b
   subtraction property of equality, 159–160, 160b, 172b
      inequalities and, 203
   subtraction property of inequalities, 204, 205b
   translation of, into expressions, 95–96, 95b
Subtraction property of equality, 159–160, 160b, 172b
   inequalities and, 203
Subtraction property of inequalities, 204, 205b
Subtraction symbol (−), 15b. *See also* Negative sign
Sum, 9b
   definition of, 14, 76
Supplementary angles, 410b
Symbols. *See also* Brackets; Equal to; Fraction bar; Grouping symbols; Negative sign; Parentheses
   absolute value (| |), 6
   approximately equal to (≈), 5, 5b, 682
   for checking solutions, 157b

curly braces ({ }), 2b, 77b
   for division, 14b, 29b
   empty set (or Ø), 2b
   greater than (>), 5, 202b, 420–421
   greater than or equal to (≥), 5, 202b, 205b, 420–421
   for inequalities, 5, 202b, 204–207, 206b
   for infinite repeating decimals, 36b
   infinity symbol (∞), 209, 209b, 210
   less than (<), 5, 202b
   less than or equal to (≤), 5, 202b, 205b
   for multiplication, 29b, 171b
   not equal to ($\neq$), 5
   pattern (. . .), 2b
   percent sign (%), 44
   radical symbol ($\sqrt{\ }$), 680
   subtraction symbol (−), 15b
Symmetric points, 757
"system," 360b
Systems of linear equations, 359–415
   applications problems and, 373–374, 389–394
   consistent systems and inconsistent systems, 368–374
   definition of, 360
   solutions to, 360–364
      with elimination method, 400–415
      graphical solutions, 364–368
      with substitution method, 382–399
Systems of linear inequalities, 433–441
   application problems and, 439–441
   definition of, 433
   graphing of, 434–441, 438b

T

Tables. *See* Data tables; Input–output tables
Terms, 109b. *See also* Constants; Like terms; Variables
   definition of, 14, 15, 76, 109, 594b
   degree of, 489–491
Theorem, 769b
"total," 493b
Translating sentences
   into equations, 178–181, 179b, 195–198
   into expressions, 8, 95–99
Triangles, similar, 653–654, 653b
Trinomials. *See also* Perfect square trinomials
   definition of, 488
   factoring of, 546–552, 547b, 550b, 551b
      of form $ax^2 + bx + c$, 549–552
      of form $x^2 + bx + c$, 546–549
      perfect square trinomials, 558–561, 560b, 777b
Truncating numbers, 55b

U

"undefined," 14
Unit conversions, 92–94, 98–99, 603–605, 603b, 605b
   of billions of dollars, 317b
   of thousands of dollars, 811b

Unit cost, 755, 755b
Unit price, 56–57, 811b
Unit rates, 647–648
   definition of, 647
Units (course), 90b
Units (measure). *See also* Unit conversions; Unity fractions
   abbreviations for, 93b
   definition of, 92b
   slope and, 255–257
   in thousands, 255b
Unity fractions
   as conversion factors, 92–94, 92b
   definition of, 92, 603b

V

Value formula, 392–394
Variables, 88–89, 89b, 109b. *See also* Dependent variables; Elimination method; Equations; Independent variables; Inequalities; Input variables; Isolating variables; Output variables
   in applications problems, 88–89, 94, 156
   on axes of graphs, 234b
   defining variables, 94–95, 97–99
   with equal to, 179b
   in equations with variables on both sides, 189–190
   evaluating expressions with, 91–92
   in inequalities, 207–208
      in compound inequalities, 212
   M vs. m, 325b
   in radical expressions
      multiplication of, 705–707
      simplification of, 686–687, 695–697, 700–701
   in systems of linear equations, 383b
Variation. *See* Direct variation; Inverse variation
Vertex formula, 753–755, 804, 804b, 807
Vertex of a parabola, 751–755, 751b, 794–797
   definition of, 751
Vertical axis, 123. *See also* Vertical intercepts
   in rectangular coordinate system, 127–134
Vertical intercepts (*y*-intercepts), 267b. *See also* Intercepts (for general discussions); Slope-intercept form of a line
   definition of, 266
Vertical lines
   definition of, 239, 239b
   graphing of, 239–240
      for inequalities, 426–427, 426b
   slope of, 255
Vertical line test, 820–823
   definition of, 821

W

Weight, units for, 93b
Whole numbers

definition of, 2
denominator of, 28b
Wolfram, Stephen, 769b
Work rate, 639–640, 639b

x-axis. *See* Horizontal axis
x-intercepts. *See* Horizontal intercepts

y-axis. *See* Vertical axis
y-intercepts. *See* Slope-intercept form of a line; Vertical intercepts

Zero (0). *See also* Intercepts
   division by, 14, 592b
      on calculator, 593b

as exponent, 73–74
multiplication and division properties of equality and, 171–172
in ordered pairs, 131b
as solution to equation, 194
Zero-product property, 565–567, 572b
   definition of, 565
   in solving quadratic equations, 567–572, 568b

Equation Solving Tools

Linear Equations
   Addition and Subtraction Property of Equality, 160
   Multiplication and Division Property of Equality, 172
Systems of Equations
   Substitution Method, 383
   Elimination Method, 401
Linear Inequalities
   Addition and Subtraction Property of Inequalities, 205
   Multiplication and Division Property of Inequalities, 205
Quadratic Equations
   Square Root Property, 766
   Completing the Square, 778
   Factoring, 568
   Quadratic Formula, 782
Rational Equations, 633
Radical Equations, 722

Expression Simplifying Tools

Add Like Terms, 110
Distributive Property, 115, 498
Rules for Exponents
   Product Rule, 465
   Quotient Rule, 467
   Power Rule, 469
   Powers of Products and Quotients, 470
   Negative Exponents, 477
   Zero as an Exponent, 73
Simplifying Rationals, 594
Rewriting Rationals with a Common Denominator, 617
Complex Fractions, 626
Simplifying Radicals, 696
Rationalize Denominators with a Single Term, 714
Rationalize Denominators with Two Terms, 715

Factoring Tools

Factoring Polynomials with Four Terms
   Factor by Grouping, 540
Factoring Polynomials with Three Terms
   GCF First, 536
   Coefficient of x^2 is 1, 547
   Coefficient of x^2 is not 1, 550
   Perfect Square Trinomial, 560
Factoring Polynomials with Two Terms
   Difference of Squares, 557

Unit Conversions

Capacity

U.S. system
8 fluid ounces = 1 cup
2 cups = 1 pint
2 pints = 1 quart
4 quarts = 1 gallon

Metric system
1 liter ≈ 1.06 quarts
1 liter ≈ 0.26 gallons
3.79 liters ≈ 1 gallon
0.95 liters ≈ 1 quart
29.57 milliliters ≈ 1 fluid ounce
1,000 milliliters = 1 liter

Metric to U.S. system
1 liter ≈ 1.06 quarts
1 liter ≈ 0.26 gallons
3.79 liters ≈ 1 gallon
0.95 liters ≈ 1 quart
29.57 milliliters ≈ 1 fluid ounce
1,000 milliliters = 1 liter

Length

U.S. system
12 in = 1 ft
3 ft = 1 yd
5,280 ft = 1 mile

Metric system
1 cm = 10 mm
100 cm = 1 m
1 km = 1000 m

Metric to U.S. system
1 meter ≈ 1.09 yards
1 meter ≈ 3.28 feet
1 kilometer ≈ 0.62 miles
2.54 centimeters = 1 inch
0.30 meters ≈ 1 foot
1.61 kilometers ≈ 1 mile
1 AU = 1.496 x 10^8 km

Area

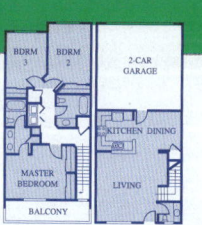

9 ft² = 1 yd²
43,560 ft² = 1 acre

Temperature

$$F = \frac{9}{5}C + 32$$

$$C = \frac{5}{9}(F - 32)$$

Weight

1 kilogram ≈ 2.20 pounds
1 gram ≈ 0.04 ounces
0.45 kilograms ≈ 1 pound
28.35 grams ≈ 1 ounce
1 ton = 2000 pounds

Geometric Formulas

Perimeter, Area, and Angle Formulas

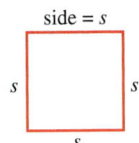

Square
$P = 4s$
$A = s^2$

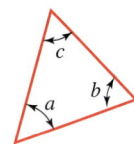

Triangle
$a + b + c = 180°$

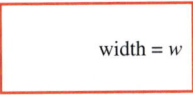

Rectangle
$P = 2l + 2w$
$A = lw$

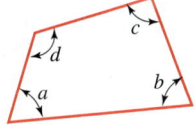

Quadrilateral
$a + b + c + d = 360°$

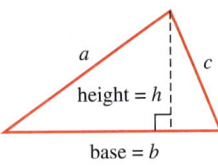

Triangle
$P = a + b + c$
$A = \frac{1}{2}bh$

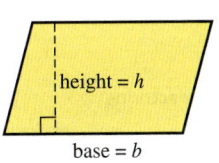

Parallelogram
Area: $A = bh$

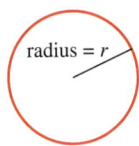

Circle
$C = 2\pi r$ or $C = \pi D$
where $\pi \approx 3.14$,
$A = \pi r^2$

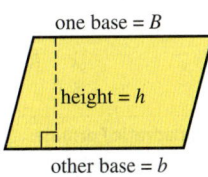

Trapezoid
Area: $A = \frac{1}{2}(b + B)h$

Volume Formulas

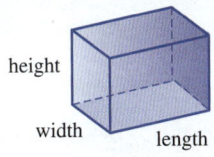

Rectangular Solid
$V = lwh$
$SA = 2lh + 2wh + 2lw$

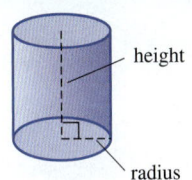

Circular Cylinder
$V = \pi r^2 h$
$SA = 2\pi rh + 2\pi r^2$

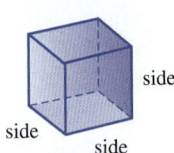

Cube
$V = s^3$
$SA = 6s^2$

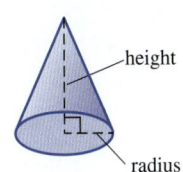

Cone
$V = \frac{1}{3}\pi r^2 h$
$SA = \pi r\sqrt{r^2 + h^2} + \pi r^2$

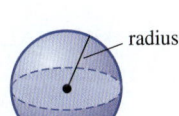

Sphere
$V = \frac{4}{3}\pi r^3$
$SA = 4\pi r^2$

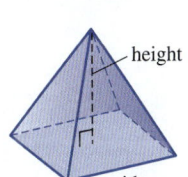

Square-Based Pyramid
$V = \frac{1}{3}s^2 h$

$SA = B + \frac{1}{2}pl$

Where B = area of base,
p = perimeter of base,
h = height, s = side, l = slant height

REF-3

Equation Solving Toolbox

Linear Equations

Addition and Subtraction Property of Equality
Use to isolate a variable term.
$$3x + 7 = 20$$
$$\underline{-7 \quad -7}$$
$$3x = 13$$

Multiplication and Division Property of Equality
Use to isolate a variable.
$$5x = 70$$
$$\frac{5x}{5} = \frac{70}{5}$$
$$x = 14$$

Systems of Equations

Substitution Method
Use when a variable is isolated or can be easily isolated.
$$y = 6x + 8$$
$$2x + 4y = 30$$
$$2x + 4(6x + 8) = 30$$

Elimination Method
Use when the equations are in general form.
$$5x - 7y = 42$$
$$2x + 4y = 30$$
$$4(5x - 7y) = 4(42)$$
$$7(2x + 4y) = 7(30)$$
$$20x - 28y = 168$$
$$\underline{14x + 28y = 210}$$
$$34x \qquad = 378$$

Linear Inequalties

Addition and Subtraction Properties of Inequalities
Use to isolate a variable term.
$$4x + 5 < 21$$
$$\underline{-5 \quad -5}$$
$$4x < 16$$

Multiplication and Division Properties of Inequalities
Use to isolate a variable. When multiplying or dividing by a negative number, reverse the inequality symbol.

$$-3x < 21 \qquad 5x < 45$$
$$\frac{-3x}{-3} > \frac{21}{-3} \qquad \frac{5x}{5} < \frac{45}{5}$$
$$x > -7 \qquad x < 9$$

Quadratic Equations

Square Root Property
Use when there is a squared term but no first degree term. Isolate the squared term and use a plus or minus symbol to indicate both answers.
$$x^2 = 25$$
$$x = \pm\sqrt{25}$$
$$x = \pm 5$$

Completing the Square
Use if the vertex form is required. Finish solving with the square root property.
$$x^2 + 6x + 4 = 0$$
$$x^2 + 6x = -4$$
$$x^2 + 6x + 9 = -4 + 9$$
$$x^2 + 6x + 9 = 5$$
$$(x + 3)^2 = 5$$

Factoring
Use when the quadratic equation has small coefficients that factor easily. Set the quadratic equal to zero, factor and then use the zero factor property to write two or more simpler equations.
$$x^2 + 7x + 10 = 0$$
$$(x + 5)(x + 2) = 0$$
$$x + 5 = 0 \quad or \quad x + 2 = 0$$

Quadratic Formula
Use when the quadratic equation has fractions, decimals, or large numbers. Set the quadratic equal to zero.
$$x = \frac{-b \pm \sqrt{b^2 - 4ac}}{2a}$$

Rational Equations

Multiply by the LCD
Multiply both sides of the equation by the least common denominator. This will clear all fractions from the equation.
$$\frac{3}{x + 1} = \frac{2}{x}$$
$$(x + 1)(x) \cdot \frac{3}{x + 1} = \frac{2}{x} \cdot (x + 1)(x)$$
$$3x = 2x + 2$$

Radical Equations

Raise Both Sides to the Reciprocal Exponent
Isolate a radical on one side of the equation and then square both sides of the equation. This may need to be done more than once to clear multiple radicals.
$$\sqrt{x + 2} - 7 = 2$$
$$\sqrt{x + 2} = 9$$
$$(\sqrt{x + 2})^2 = 9^2$$
$$x + 2 = 81$$

REF-4

Expression Simplifying Toolbox

Add Like Terms

Add or Subtract like terms that have the same variables with the same exponents.
$$2x^2 + 7x + 9x^2 - 3x = 11x^2 + 4x$$

Distributive Property

Distribute terms across addition or subtraction.

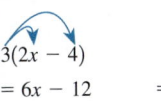
$= 6x - 12$

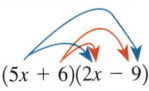
$= 10x^2 - 45x + 12x - 54$
$= 10x^2 - 33x - 54$

Rules for Exponents

Product Rule for Exponents

When multiplying exponential expressions with the same base, add the exponents.
$$x^2 x^5 = x^7$$

Quotient Rule for Exponents

When dividing exponential expressions with the same base, subtract the exponents.
$$\frac{x^5}{x^2} = x^{5-2} = x^3$$

Power Rule for Exponents

When raising an exponential expressions to another power, multiply the exponents.
$$(x^5)^3 = x^{15}$$

Powers of Product and Quotients

When raising an expression to a power, apply the power over multiplication or division.
$$(xy)^3 = x^3 y^3 \qquad \left(\frac{x}{y}\right)^5 = \frac{x^5}{y^5}$$

Negative Exponents

When raising a base to a negative exponent, raise the reciprocal of that base to the absolute value of the exponent.
$$y^{-5} = \frac{1}{y^5}$$

Zero as an Exponent

When any exponential expression with a base other than zero is raised to the power of zero, the expression will equal 1.
$$x^0 = 1 \qquad (x) \neq 0$$

Simplifying Rationals

Factor the numerator and denominator, then divide out any common factors.
$$\frac{5\cancel{(x+2)}}{(x+3)\cancel{(x+2)}}$$
$$= \frac{5}{x+3}$$

Rewriting Rationals with a Common Denominator

Determine the least common denominator, multiply the numerator and denominator by any needed factors. This is an example of multiplying by a version of one.

Rational 1
$$\frac{x+6}{(x+2)(x-5)}$$

Rational 2
$$\frac{5}{(x-5)(x+3)}$$

$$\text{LCD} = (x-5)(x+3)(x+2)$$

Rational 1
$$\frac{(x+3)}{(x+3)} \cdot \frac{x+6}{(x+2)(x-5)}$$
$$= \frac{(x+3)(x+6)}{(x-5)(x+3)(x+2)}$$

Rational 2
$$\frac{5}{(x-5)(x+3)} \cdot \frac{(x+2)}{(x+2)}$$
$$= \frac{5(x+2)}{(x-5)(x+3)(x+2)}$$

Complex Fractions

Determine the least common denominator of all the fractions inside the rational expression. Multiply the numerator and denominator of the rational expression by the LCD. This is an example of multiplying by a version of one.

$$\frac{2 + \frac{3}{x}}{1 - \frac{7}{x^2}}$$

$$\text{LCD} = x^2 = \frac{x^2}{1}$$

$$\frac{\left(2 + \frac{3}{x}\right)}{\left(1 - \frac{7}{x^2}\right)} \cdot \frac{\frac{x^2}{1}}{\frac{x^2}{1}}$$

$$\frac{2x^2 + \frac{3}{x}\frac{x^2}{1}}{x^2 - \frac{7}{x^2}\frac{x^2}{1}}$$

$$\frac{2x^2 + 3x}{x^2 - 7}$$

Simplifying Radicals

Factor the radicand into factors raised to powers that are a multiple of the index of the radical. Simplify the radicals. Multiply remaining radicands and multiply anything factored out.

$$\sqrt{20x^2 y^7}$$
$$= \sqrt{2^2 x^2 y^6} \sqrt{5y}$$
$$= 2xy^3 \sqrt{5y}$$

Rationalize Denominators with a Single Term

Simplify the numerator and denominator. Multiply the numerator and denominator by the radical factor(s) that allow(s) the radical in the denominator to simplify completely.

$$\frac{7}{\sqrt{2xy^3}}$$
$$= \frac{7}{\sqrt{2xy^3}} \cdot \frac{\sqrt{2xy}}{\sqrt{2xy}}$$
$$= \frac{7\sqrt{2xy}}{\sqrt{2^2 x^2 y^4}}$$
$$= \frac{7\sqrt{2xy}}{2xy^2}$$

Rationalize Denominators with Two Terms

Multiply the numerator and denominator by the conjugate of the denominator. Simplify completely.

$$\frac{2}{5 + \sqrt{2x}}$$
$$= \frac{2}{(5 + \sqrt{2x})} \cdot \frac{(5 - \sqrt{2x})}{(5 - \sqrt{2x})}$$
$$= \frac{10 - 2\sqrt{2x}}{25 - 5\sqrt{2x} + 5\sqrt{2x} - 2x}$$
$$= \frac{10 - 2\sqrt{2x}}{25 - 2x}$$

REF-5

Factoring Toolbox

Factoring Polynomials with Four Terms $(x+2)(x-5)$

GCF First
Factor out the GCF first, if there is one.
$$6x^2 + 4x - 8x + 10$$
$$= 2(3x^2 + 2x - 4x + 5)$$

Factor by Grouping
$$12x^2 + 8x - 21x - 14$$
$$= (4x - 7)(3x + 2)$$

Factoring Polynomials with Three Terms $(x+2)(x-5)$

GCF First
Factor out the GCF first, if there is one.
$$6x^2 + 9x - 15 = 3(2x^2 + 3x - 5)$$

Coefficient of x^2 is 1
Trial and error or AC method.
$$x^2 + 7x + 12 = (x + 3)(x + 4)$$

Coefficient of x^2 is not 1
AC method or trial and error.
$$12x^2 + 8x - 15 = (2x + 3)(6x - 5)$$

Perfect Square Trinomial
Use the patterns
$$a^2 + 2ab + b^2 = (a + b)^2$$
$$a^2 - 2ab + b^2 = (a - b)^2$$
$$x^2 + 6x + 9 = (x + 3)^2$$
$$x^2 - 10x + 25 = (x - 5)^2$$

Factoring Polynomials with Two Terms $(x+2)(x-5)$

GCF First
Factor out the GCF first, if there is one.
$$6x^2 + 4x = 2x(3x + 2)$$

Difference of Squares
Use the difference of squares pattern.
$$a^2 - b^2 = (a + b)(a - b)$$
$$x^2 - 25 = (x + 5)(x - 5)$$